AutoCAD® 2002 Companion

Essentials of AutoCAD Plus Solid Modeling

James A. Leach

University of Louisville

Mc Graw Hill

Boston   Burr Ridge, IL   Dubuque, IA   Madison, WI   New York   San Francisco   St. Louis
Bangkok   Bogotá   Caracas   Kuala Lumpur   Lisbon   London   Madrid   Mexico City
Milan   Montreal   New Delhi   Santiago   Seoul   Singapore   Sydney   Taipei   Toronto

The McGraw-Hill Graphics Series

Providing you with the highest quality textbooks that meet your changing needs requires feedback, improvement, and revision. The team of authors and McGraw-Hill are committed to this effort. We invite you to become part of our team by offering your wishes, suggestions, and comments for future editions and new products and texts.

Please mail or fax your comments to: Jim Leach
c/o McGraw-Hill
1333 Burr Ridge Parkway
Burr Ridge, IL 60521
fax 630-789-6946

TITLES IN THE McGRAW-HILL GRAPHICS SERIES INCLUDE:

Graphic Drawing Workbook
Gary Bertoline, 1999

Technical Graphics Communications, 2e
Bertoline, Wiebe, Miller, and Mohler, 1997

Fundamentals of Graphics Communication, 3e
Bertoline, Wiebe, and Mohler, 2001

Engineering Graphics Communication
Bertoline, Wiebe, Miller, and Nasman, 1995

Problems for Engineering Graphics Communications and Technical Graphics Communication (Revised edition)

AutoCAD Instructor Release 12
James A. Leach, 1995

AutoCAD Companion Release 12
James A. Leach, 1995

AutoCAD 13 Instructor
James A. Leach, 1996

AutoCAD 13 Companion
James A. Leach, 1996

AutoCAD 14 Instructor
James A. Leach, 1998

AutoCAD 14 Companion
James A. Leach, 1998

AutoCAD 2000 Instructor
James A. Leach, 2000

AutoCAD 2002 Instructor
James A. Leach, 2002

The Companion for CADKEY 97
John Cherng, 1998

Hands-on CADKEY: A Guide to Versions 5, 6, and 7
Timothy J. Sexton, 1996

Modeling with AutoCAD Designer
Dobek and Ranschaert, 1996

Engineering Design and Visualization Workbook
Dennis Stevenson, 1995

Graphics Interactive CD-ROM
Dennis Lieu, 1997

DEDICATION

To Donna,
for being a good editor,
a great friend,
and a wonderful wife

McGraw-Hill Higher Education
*A Division of The **McGraw-Hill** Companies*

AUTOCAD 2002 COMPANION: ESSENTIALS OF AUTOCAD PLUS SOLID MODELING

Published by McGraw-Hill, a business unit of The McGraw-Hill Companies, Inc., 1221 Avenue of the Americas, New York, NY 10020. Copyright © 2003 by The McGraw-Hill Companies, Inc. All rights reserved. No part of this publication may be reproduced or distributed in any form or by any means, or stored in a database or retrieval system, without the prior written consent of The McGraw-Hill Companies, Inc., including, but not limited to, in any network or other electronic storage or transmission, or broadcast for distance learning.

Some ancillaries, including electronic and print components, may not be available to customers outside the United States.

This book is printed on recycled, acid-free paper containing 10% postconsumer waste.

1 2 3 4 5 6 7 8 9 0 QPD/QPD 0 9 8 7 6 5 4 3 2

ISBN 0–07–252863–X

Publisher: *Elizabeth A. Jones*
Sponsoring editor: *Kelly Lowery*
Executive marketing manager: *John Wannemacher*
Project manager: *Joyce Watters*
Production supervisor: *Sherry L. Kane*
Media project manager: *Sandra M. Schnee*
Senior media technology producer: *Phillip Meek*
Coordinator of freelance design: *Michelle D. Whitaker*
Cover designer: *Jamie E. O'Neal*
Cover image: *© PhotoDisc, U.S. Landmarks and Travel, Image #16290*
Compositor: *Interactive Composition Corporation*
Typeface: *10/13 Times Roman*
Printer: *Quebecor World Dubuque, IA*

Library of Congress Cataloging-in-Publication Data

Leach, James A., 1950–
    AutoCAD 2002 companion : essentials of AutoCAD plus solid modeling / James A. Leach. — 1st ed.
      p. cm.
    ISBN 0–07–252863–X
    1. Computer graphics. 2. AutoCAD. I. Title.

T385 .L382663   2003
620'.0042'02855369—dc21   2002025520
                                                                         CIP

www.mhhe.com

TABLE OF CONTENTS

TABLE OF CONTENTS

PREFACE xv

INTRODUCTION xxi

CHAPTER 1.
GETTING STARTED 1
CONCEPTS 2
    Coordinate Systems 2
    The CAD Database 3
    Angles in AutoCAD 3
    Draw True Size 3
    Plot to Scale 5
STARTING AutoCAD 5
THE AutoCAD DRAWING EDITOR 5
    The *Today* Window and the *Startup*
        Dialog Box 5
    Graphics Area 6
    Command Line 7
    Toolbars 7
    Pull-down Menus 8
    Screen (Side) Menu 8
    Dialog Boxes 9
    Status Bar 10
    Coordinate Display (*Coords*) 10
    Digitizing Tablet Menu 11
COMMAND ENTRY 12
    Methods for Entering Commands 12
    Using "Command Tables" in This Book
        to Locate a Particular Command 12
    Shortcut Menus 13
    Mouse and Digitizing Puck Buttons 14
    Function Keys 14
    Control Key Sequences (Accelerator
        Keys) 15
    Special Key Functions 15
    Drawing Aids 15
    AutoCAD Text Window (F2) 17
    Commands When Multiple Drawings
        Are Open 18
    Model Tab and Layout Tabs 18
COMMAND ENTRY METHODS
    PRACTICE 19
CUSTOMIZING THE AutoCAD
    SCREEN 21
    Toolbars 21
    Command Line 23
    OPTIONS 23
CHAPTER EXERCISES 24

CHAPTER 2.
WORKING WITH FILES 25
AutoCAD DRAWING FILES 26
    Naming Drawing Files 26
    Beginning and Saving an AutoCAD
        Drawing 26
    Accessing File Commands 26
    File Navigation Dialog Box Functions ... 27
    Windows Right-Click Shortcut Menus ... 28
AutoCAD FILE COMMANDS 29
    NEW 29
    OPEN 31
    SAVE 32
    QSAVE 32
    SAVEAS 32
    CLOSE 33
    CLOSEALL 34
    EXIT 34
    AutoCAD Backup Files 34
    AutoCAD Drawing File Management ... 35
CHAPTER EXERCISES 35

CHAPTER 3.
DRAW COMMAND CONCEPTS 37
AutoCAD OBJECTS 38
LOCATING THE DRAW COMMANDS 38
THE FIVE COORDINATE ENTRY
    METHODS 39
DRAWING *LINES* USING THE FIVE
    COORDINATE ENTRY METHODS 40
    LINE 41
    Drawing Horizontal Lines 41
    Drawing Vertical Lines 45
    Drawing Inclined Lines 47
DRAWING *CIRCLES* USING THE FIVE
    COORDINATE ENTRY METHODS 49
    CIRCLE 49
POLAR TRACKING AND POLAR SNAP ... 51
    Polar Tracking 52
    Polar Tracking with Direct Distance
        Entry 53
    Polar Tracking Override 53
    Polar Snap 53
DRAWING *LINES* USING POLAR
    TRACKING AND POLAR SNAP 54
    Using Polar Tracking and Grid Snap ... 54
    Using Polar Tracking and Polar Snap ... 56
    Using Polar Tracking and Direct
        Distance Entry 57
CHAPTER EXERCISES 58

CHAPTER 4.
SELECTION SETS................... 61
MODIFY COMMAND CONCEPTS 62
SELECTION SETS 62
    Accessing Selection Set Options......... 63
    Selection Set Options 63
    Ctrl+Left Button (Object Cycling) 67
    SELECT 68
NOUN/VERB SYNTAX 68
SELECTION SETS PRACTICE 68
    Using *Erase* 69
    ERASE 69
    Using *Move* 70
    MOVE.............................. 70
    Using *Copy*.......................... 71
    COPY 71
    NOUN/VERB Command Syntax
      Practice 72
CHAPTER EXERCISES 73

CHAPTER 5.
HELPFUL COMMANDS 75
CONCEPTS 76
    HELP............................... 76
    ASSIST 79
    OOPS 80
    U 80
    UNDO 80
    REDO 81
    REGEN 82
CHAPTER EXERCISES 82

CHAPTER 6.
BASIC DRAWING SETUP 85
STEPS FOR BASIC DRAWING SETUP...... 86
STARTUP OPTIONS 86
DRAWING SETUP OPTIONS 87
    Start from Scratch..................... 87
    Template 88
    Table of *Start from Scratch* <u>*English*</u>
      Settings (ACAD.DWT) 88
    Table of *Start from Scratch* <u>*Metric*</u>
      Settings (ACADISO.DWT)............ 89
    Wizard.............................. 89
    Quick Setup Wizard 89
    Advanced Setup Wizard 90
SETUP COMMANDS 91
    UNITS.............................. 91
    Keyboard Input of *Units* Values 93
    LIMITS 93
    SNAP 95
    GRID............................... 97
    DSETTINGS......................... 98
    Using *Snap* and Grid 98
INTRODUCTION TO LAYOUTS AND
    PRINTING......................... 98
    Model Tab and Layout Tabs............. 99
    Why Set Up Layouts Before
      Drawing?......................... 101
    Printing and Plotting 101
    Setting Layout Options and Plot
      Options 102
CHAPTER EXERCISES 103

CHAPTER 7.
**OBJECT SNAP AND OBJECT SNAP
    TRACKING** 107
CAD ACCURACY 108
OBJECT SNAP 108
OBJECT SNAP MODES................. 109
    Acquisition Object Snap Modes 113
OSNAP SINGLE POINT SELECTION...... 115
OSNAP RUNNING MODE............... 117
    Running Object Snap 117
    Accessing Running Object Snap........ 117
    Running Object Snap Toggle........... 118
    Object Snap Cycling.................. 118
OBJECT SNAP TRACKING 118
    To Use Object Snap Tracking 120
    Object Snap Tracking with Polar
      Tracking......................... 120
    Object Snap Tracking Settings 121
OSNAP APPLICATIONS................. 122
OSNAP PRACTICE 122
    Single Point Selection Mode 122
    Running Object Snap Mode 124
    Object Snap Tracking 125
CHAPTER EXERCISES 126

CHAPTER 8.
DRAW COMMANDS I 133
CONCEPTS 134
    Draw Commands—Simple and
      Complex 134
    Draw Command Access 134
    Coordinate Entry 135
COMMANDS 135
    LINE 135
    CIRCLE............................ 136
    ARC............................... 138
    Use Arcs or Circles? 142
    POINT 143
    DDPTYPE 143
    PLINE............................. 144
    Pline Arc Segments................... 146
CHAPTER EXERCISES 146

CHAPTER 9.
MODIFY COMMANDS I 151
CONCEPTS 152
COMMANDS 153
    ERASE 153
    MOVE............................. 153
    ROTATE 155
    SCALE 156
    STRETCH.......................... 157
    LENGTHEN 158
    TRIM............................. 160
    EXTEND 161
    Trim and Extend Shift-Select Option 162
    BREAK............................ 162
    COPY 165
    MIRROR........................... 166
    OFFSET 167
    ARRAY............................ 168
    FILLET 171
    CHAMFER......................... 174
CHAPTER EXERCISES 175

CHAPTER 10.
VIEWING COMMANDS 185
CONCEPTS 186
ZOOM and *PAN* with the Mouse Wheel.... 187
    Zoom and *Pan* 187
    Zoom with the Mouse Wheel 187
    Pan with the Mouse Wheel or Third
       Button 188
COMMANDS 188
    ZOOM 188
    PAN.............................. 192
    Scroll Bars 194
    VIEW 194
    VIEWRES.......................... 196
    UCSICON.......................... 197
    VPORTS........................... 197
CHAPTER EXERCISES 200

CHAPTER 11.
LAYERS AND OBJECT PROPERTIES 205
CONCEPTS 206
    Assigning Colors, Linetypes, and
       Lineweights 206
    Object Properties 207
LAYERS AND LAYER PROPERTIES
    CONTROLS........................ 207
    Layer Control Drop-Down List 207
    LAYER 208
    Color, Linetype, Lineweight, and
       Other Properties................. 210

Displaying an Object's Properties and
    Visibility Settings................. 212
    Make Object's Layer Current 213
    LAYERP 213
OBJECT-SPECIFIC PROPERTIES
    CONTROLS........................ 214
    LINETYPE 214
    Linetype Control Drop-Down List...... 214
    LWEIGHT.......................... 215
    Lineweight Control Drop-Down List ... 215
    COLOR............................ 216
    Color Control Drop-Down List 216
CONTROLLING LINETYPE SCALE 217
    LTSCALE 217
    CELTSCALE......................... 218
CHANGING OBJECT PROPERTIES
    RETROACTIVELY 219
    Object Properties Toolbar 219
    PROPERTIES........................ 220
    MATCHPROP 221
CHAPTER EXERCISES 222

CHAPTER 12.
ADVANCED DRAWING SETUP......... 227
CONCEPTS 228
STEPS FOR DRAWING SETUP 228
    1. Set *Units* 228
    2. Set *Limits* 229
    Drawing Scale Factor................. 229
    3. Set *Snap* 230
    4. Set *Grid*........................ 231
    5. Set the *LTSCALE* and *PSLTSCALE*... 231
    6. Create *Layers*; Assign *Linetypes,*
         Lineweights, and *Colors* 231
    7. Create *Text Styles*.................. 231
    8. Create Dimension Styles 232
    9. Activate a *Layout* Tab, Set the Plot
         Device, and Create a Viewport 232
    10. Create a Title Block and Border 232
USING AND CREATING TEMPLATE
    DRAWINGS 232
    Using Template Drawings.............. 232
    Creating Template Drawings 233
    Additional Drawing Setup
       Concepts 234
CHAPTER EXERCISES 234

CHAPTER 13.
LAYOUTS AND VIEWPORTS 237
CONCEPTS 238
    Paper Space and Model Space 238
    Layouts............................. 238

Viewports.............................. 239
Layouts and Viewports Example....... 241
Guidelines for Using Layouts and
    Viewports......................... 244
CREATING AND SETTING UP
    LAYOUTS........................... 245
    LAYOUTWIZARD.................... 245
    LAYOUT.......................... 247
    Inserting Layouts with AutoCAD
        DesignCenter.................. 249
    Setting up the Layout............. 249
    OPTIONS......................... 250
    PAGESETUP....................... 251
USING VIEWPORTS IN PAPER SPACE.... 252
    VPORTS.......................... 252
    PSPACE.......................... 254
    MSPACE.......................... 254
    MODEL........................... 255
    Scaling the Display of Viewport
        Geometry...................... 255
    Locking Viewport Geometry......... 257
    Linetype Scale in Viewports
        —*PSLTSCALE*.................. 257
CHAPTER EXERCISES..................... 258

CHAPTER 14.
PRINTING AND PLOTTING........... 267
CONCEPTS............................. 268
TYPICAL STEPS TO PLOTTING............ 268
USING THE PLOT AND PAGE SETUP
    DIALOG BOXES...................... 269
    PLOT............................ 269
    Plot Device Tab................. 270
    Plot Settings Tab............... 271
    -PLOT........................... 274
    PAGESETUP....................... 275
    Should You Use the *Plot* Dialog Box
        or *Page Setup* Dialog Box?... 275
    PREVIEW......................... 276
PLOTTING TO SCALE.................... 276
    Standard Paper Sizes.............. 277
    Calculating the Drawing Scale
        Factor........................ 277
    Guidelines for Plotting the *Model* Tab
        to Scale...................... 278
    Guidelines for Plotting *Layouts*
        to Scale...................... 278
TABLES OF LIMITS SETTINGS............ 279
    Mechanical Table of *Limits* Settings..... 280
    Architectural Table of *Limits* Settings... 281
    Metric Table of *Limits* Settings for
        Metric Sheet Sizes............ 282

Metric Table of *Limits* Settings for
    Engineering Sheet Sizes........... 283
Civil Table of *Limits* Settings..... 284
Examples for Plotting to Scale....... 285
CHAPTER EXERCISES.................... 286

CHAPTER 15.
DRAW COMMANDS II................ 289
CONCEPTS............................. 290
COMMANDS............................. 290
    XLINE........................... 290
    RAY............................. 293
    POLYGON......................... 293
    RECTANG......................... 294
    DONUT........................... 295
    SPLINE.......................... 296
    ELLIPSE......................... 297
    DIVIDE.......................... 299
    MEASURE......................... 301
    SKETCH.......................... 301
    The *SKPOLY* Variable............. 303
    BOUNDARY........................ 304
    REGION.......................... 305
CHAPTER EXERCISES.................... 306

CHAPTER 16.
MODIFY COMMANDS II.............. 311
CONCEPTS............................. 312
COMMANDS............................. 312
    Considerations for Changing Basic
        Properties of Objects......... 313
    PROPERTIES...................... 314
    DBLCLKEDIT...................... 317
    MATCHPROP....................... 318
    Object Properties Toolbar....... 319
    CHPROP.......................... 320
    CHANGE.......................... 320
    EXPLODE......................... 321
    ALIGN........................... 322
    PEDIT........................... 323
    Vertex Editing.................... 326
    Converting *Lines* and *Arcs* to *Plines*..... 328
    SPLINEDIT....................... 329
    Editing *Spline* Data Points...... 329
    Editing *Spline* Control Points... 331
    Boolean Commands.................. 332
    UNION........................... 332
    SUBTRACT........................ 333
    INTERSECT....................... 333
CHAPTER EXERCISES.................... 334

CHAPTER 17.
INQUIRY COMMANDS 341
CONCEPTS 342
COMMANDS 342
    STATUS 342
    LIST................................ 343
    DBLIST............................. 344
    AREA 344
    DIST 346
    ID 346
    TIME............................... 347
    SETVAR 347
    MASSPROP.......................... 348
CHAPTER EXERCISES 349

CHAPTER 18.
CREATING AND EDITING TEXT 351
CONCEPTS 352
TEXT CREATION COMMANDS........... 353
    DTEXT 354
    MTEXT............................. 357
    Properties Tab...................... 358
    Character Tab 359
    Line Spacing Tab 360
    Mtext Right-Click Shortcut Menu 360
    Text Flow and *Justification*............ 361
    AutoStack Properties Dialog Box 362
    Calculating Text Height for Scaled
       Drawings........................... 363
TEXT STYLES AND FONTS.............. 363
    STYLE.............................. 363
EDITING TEXT 366
    SPELL.............................. 366
    DDEDIT 367
    PROPERTIES....................... 367
    FIND............................... 368
    SCALETEXT 369
    JUSTIFYTEXT 370
    QTEXT 371
    TEXTFILL.......................... 371
CHAPTER EXERCISES 372

CHAPTER 19.
INTERNET TOOLS..................... 375
CONCEPTS 376
    TODAY............................. 377
    Autodesk Point A 379
    BROWSER 380
USING HYPERLINKS.................... 380
    HYPERLINK 381
    SELECTURL 382
    PUBLISHTOWEB................... 382

    ETRANSMIT 385
CHAPTER EXERCISES 387

CHAPTER 20.
BLOCKS AND DesignCenter........... 389
CONCEPTS 390
COMMANDS 392
    BLOCK 392
    Block Color, Linetype, and *Lineweight*
       Settings 393
    INSERT............................ 394
    MINSERT.......................... 395
    EXPLODE 395
    XPLODE............................ 396
    WBLOCK 397
    Redefining *Blocks* 399
    BASE............................... 399
    PURGE 400
    RENAME 401
DesignCenter........................... 401
    ADCENTER........................ 402
    Tree View Options 403
    To Open a Drawing from
       DesignCenter 407
    To *Insert* a *Block* Using DesignCenter ... 407
CHAPTER EXERCISES 408

CHAPTER 21.
GRIP EDITING 413
CONCEPTS 414
GRIPS FEATURES 414
    GRIPS and *DDGRIPS*.............. 414
    Activating Grips on Objects 415
    Warm, Hot, and Cold Grips 415
    Grip Editing Options 417
    Auxiliary Grid....................... 419
    Editing Dimensions 420
GUIDELINES FOR USING GRIPS........ 420
CHAPTER EXERCISES 421

CHAPTER 22.
MULTIVIEW DRAWING 425
CONCEPTS 426
PROJECTION AND ALIGNMENT
   OF VIEWS 426
    Using *ORTHO* and *OSNAP* to Draw
       Projection Lines 426
    Using Polar Tracking to Draw
       Projection Lines 427
    Using *Object Snap Tracking* to Draw
       Projection Lines Aligned with *OSNAP*
       Points............................. 428

Using *Xline* and *Ray* for Construction
  Lines 430
Using *Offset* for Construction of
  Views 430
Realignment of Views Using *Polar
  Snap* and *Polar Tracking* 431
USING CONSTRUCTION LAYERS 431
USING LINETYPES 432
  Drawing Hidden and Center Lines 433
  Managing Linetypes, Lineweights,
    and Colors 434
CREATING FILLETS, ROUNDS, AND
  RUNOUTS 435
GUIDELINES FOR CREATING A TYPICAL
  THREE-VIEW DRAWING 437
CHAPTER EXERCISES 441

CHAPTER 23.
PICTORIAL DRAWINGS 445
CONCEPTS 446
  Types of Pictorial Drawings 446
  Pictorial Drawings Are 2D Drawings ... 447
ISOMETRIC DRAWING IN AutoCAD..... 448
  SNAP 448
  Isometric Ellipses 450
  ELLIPSE 450
  Creating an Isometric Drawing 452
OBLIQUE DRAWING IN AutoCAD....... 454
CHAPTER EXERCISES 456

CHAPTER 24.
SECTION VIEWS 459
CONCEPTS 460
DEFINING HATCH PATTERNS AND
  HATCH BOUNDARIES 460
  Steps for Creating a Section View
    Using the *Bhatch* Command 461
  BHATCH 462
  Boundary Hatch Dialog Box—Quick
    Tab 463
  Selecting the Hatch Area 465
  Advanced Tab 467
  HATCH............................. 469
  Drag and Drop Hatch Patterns 471
EDITING HATCH PATTERNS AND
  BOUNDARIES 471
  HATCHEDIT 471
  Using Grips with Hatch Patterns 472
  DRAWORDER....................... 472
DRAWING CUTTING PLANE LINES 473
CHAPTER EXERCISES 474

CHAPTER 25.
AUXILIARY VIEWS 477
CONCEPTS 478
CONSTRUCTING AN AUXILIARY VIEW . 478
  Setting Up the Principal Views 478
  Using *Snap Rotate* and *ORTHO* 479
  Rotating *SNAP* Back to the Original
    Position 480
  Using *Polar Tracking* 482
  Using the *Offset* Command 483
  OFFSET 484
  Using the *Xline* and *Ray* Commands.... 485
  XLINE.............................. 485
  RAY................................ 485
  Constructing Full Auxiliary Views 486
CHAPTER EXERCISES 487

CHAPTER 26.
DIMENSIONING 491
CONCEPTS 492
DIMENSION DRAWING COMMANDS ... 493
  DIMLINEAR 493
  DIMALIGNED....................... 496
  DIMBASELINE 497
  DIMCONTINUE 498
  DIMDIAMETER 499
  DIMRADIUS 500
  DIMCENTER........................ 501
  DIMANGULAR...................... 502
  LEADER 503
  QLEADER 505
  Annotation Tab 506
  Leader Line & Arrow Tab 507
  Attachment Tab 507
  DIMORDINATE 508
  QDIM 509
EDITING DIMENSIONS 510
  Associative Dimensions 510
  DIMREASSOCIATE 512
  DIMDISASSOCIATE 513
  DIMREGEN......................... 513
  Grip Editing Dimensions 514
  Exploding Associative Dimensions 515
  Dimension Editing Commands 515
  DIMTEDIT.......................... 515
  DIMEDIT........................... 516
  PROPERTIES........................ 517
  Customizing Dimensioning Text 518
DIMENSIONING VARIABLES
  INTRODUCTION 518
CHAPTER EXERCISES 519

CHAPTER 27.
DIMENSION STYLES AND VARIABLES 521
CONCEPTS 522
    Dimension Variables 522
    Dimension Styles 522
    Dimension Style Families 523
    Dimension Style Overrides........... 524
DIMENSION STYLES 525
    DIMSTYLE and *DDIM* 525
    -DIMSTYLE 528
DIMENSION VARIABLES 530
    Changing Dimension Variables Using the Dialog Box Method.............. 530
    Lines and Arrows Tab................. 531
    Text Tab............................ 534
    Fit Tab............................. 537
    Primary Units Tab 541
    Alternate Units Tab 544
    Tolerances Tab...................... 545
    Changing Dimension Variables Using the Command Line Format 547
    DIM...(VARIABLE NAME) 547
    Associative Dimensions 548
MODIFYING EXISTING DIMENSIONS ... 548
    Modifying a Dimension Style.......... 548
    Creating Dimension Style Overrides and Using *Update*................... 548
    UPDATE............................ 549
    DIMOVERRIDE 549
    MATCHPROP 551
    PROPERTIES........................ 551
GUIDELINES FOR DIMENSIONING IN AutoCAD............................ 552
    Strategy 1. Dimensioning a Single Drawing 552
    Strategy 2. Creating Dimension Styles as Part of Template Drawings 554
    Optional Method for Fixed Dimension Text Height in Template Drawings.... 554
DIMENSIONING IN PAPER SPACE LAYOUTS 555
    When to Dimension in Paper Space 556
CHAPTER EXERCISES 557

CHAPTER 28.
3D MODELING BASICS................. 561
CONCEPTS 562
    Wireframe Models 562
    Surface Models 563
    Solid Models 564
3D COORDINATE ENTRY 565
    3D Coordinate Entry Formats 566
COORDINATE SYSTEMS................ 570
    The World Coordinate System (WCS) and WCS Icon...................... 570
    User Coordinate Systems (UCS) and Icons.............................. 571
THE RIGHT-HAND RULE 572
UCS ICON CONTROL 574
    UCSICON.......................... 574
CHAPTER EXERCISES 576

CHAPTER 29.
3D DISPLAY AND VIEWING 577
AutoCAD's 3D VIEWING AND DISPLAY CAPABILITIES 578
    Display Commands 578
    Viewing Commands 578
3D DISPLAY COMMANDS 579
    SHADEMODE....................... 579
    HIDE.............................. 582
3D VIEWING COMMANDS 582
    VIEW 583
    3D Orbit Commands 586
    3DORBIT........................... 586
    3DPAN 589
    3DZOOM.......................... 589
    3DCORBIT......................... 589
    3DDISTANCE 590
    3DCLIP............................ 591
    VPOINT 592
    DDVPOINT 595
    PLAN 596
    ZOOM 596
    VPORTS 596
CHAPTER EXERCISES 598

CHAPTER 30.
USER COORDINATE SYSTEMS......... 603
CONCEPTS 604
UCS COMMANDS...................... 604
    UCS............................... 605
    DDUCSP 610
    DDUCS............................ 611
    UCSMAN.......................... 612
UCS SYSTEM VARIABLES............... 612
    UCSBASE.......................... 613
    UCSVIEW 613
    UCSORTHO....................... 614
    UCSFOLLOW 615
    UCSVP............................ 616
CHAPTER EXERCISES 617

CHAPTER 31.
SOLID MODELING CONSTRUCTION . . 619
CONCEPTS 620
CONSTRUCTIVE SOLID GEOMETRY
    TECHNIQUES 620
       Primitives. 620
       Boolean Operations 621
SOLID PRIMITIVES COMMANDS. 622
    BOX 622
    CONE 624
    CYLINDER. 625
    WEDGE 626
    SPHERE 627
    TORUS 627
    EXTRUDE 628
    REVOLVE 631
COMMANDS FOR MOVING SOLIDS. 632
    MOVE. 632
    ALIGN 633
    ROTATE3D 635
    MIRROR3D 638
    3DARRAY 641
BOOLEAN OPERATION COMMANDS . . . 643
    UNION. 644
    SUBTRACT 645
    INTERSECT. 646
    CHAMFER. 647
    FILLET 649
DESIGN EFFICIENCY. 650
SOLIDS EDITING. 651
    SOLIDEDIT 652
    Face Options 653
    Edge Options 659
    Body Options 660
CHAPTER EXERCISES 663

CHAPTER 32.
ADVANCED SOLIDS FEATURES 673
CONCEPTS 674
SOLID MODELING DISPLAY
    VARIABLES. 674
    ISOLINES. 674
    DISPSILH. 675
    FACETRES. 675
    FACETRATIO 676
ANALYZING SOLID MODELS. 677
    MASSPROP. 677
    INTERFERE. 678
CREATING SECTIONS FROM SOLIDS 679
    SECTION 679
    SLICE 681
CHAPTER EXERCISES 681

APPENDICES. 683

APPENDIX A. AutoCAD 2002 Command
    Alias List Sorted by Command 684

APPENDIX B. BUTTONS AND SPECIAL
    KEYS. 686
    Mouse and Digitzing Puck Buttons. 686
    Function (F) Keys 686
    Control Key Sequences (Accelerator
       Keys) 687
    Special Key Functions 688

INDEX 689

PREFACE

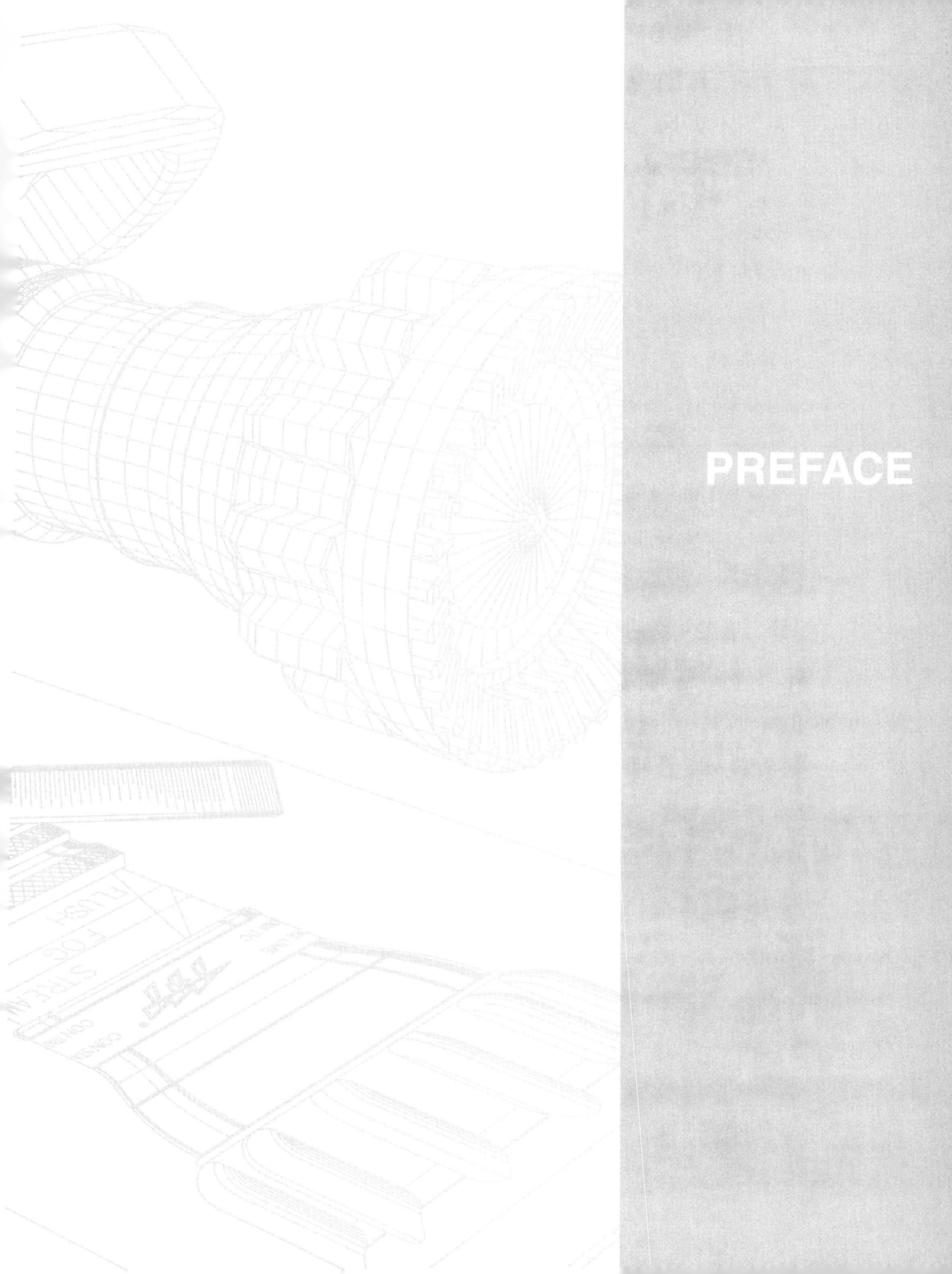

ABOUT THIS BOOK

AutoCAD in One Semester
AutoCAD 2002 Companion is designed to provide you with the material typically covered in a one-semester AutoCAD course. *AutoCAD 2002 Companion* covers the essentials of 2D design and drafting as well as solid modeling.

Graphically Oriented
Because *AutoCAD 2002 Companion* discusses concepts that are graphical by nature, many illustrations (approximately **1000**) are used to communicate the concepts, commands, and applications.

Pedagogical Progression
AutoCAD 2002 Companion begins with small pieces of information explained in a simple form and then builds on that experience to deliver more complex ideas, requiring a synthesis of earlier concepts. The chapter exercises follow the same progression, beginning with a simple tutorial approach and ending with more challenging problems requiring a synthesis of earlier exercises.

Easy Update from AutoCAD Release 14, AutoCAD 2000, and AutoCAD 2000i
AutoCAD 2002 Companion is helpful if you are already an AutoCAD user but are updating from Release 14, 2000, or 2000i. All new commands, concepts, features, and variables are denoted on the edges of the pages by a vertical "2000" bar (denoting an update since Release 14) or a "2002" bar (denoting an update since 2000 or 2000i).

Important "Tips"
Tips, reminders, notes, and cautions are given in the book and denoted by a "TIP!" (light bulb) icon. This feature helps you identify and remember important concepts, commands, procedures, and tricks used by professionals that would otherwise be discovered only after much experience.

Valuable Reference Guide
AutoCAD 2002 Companion is structured to be used as a reference guide to AutoCAD. Every command throughout the book is given with a "command table" listing the possible methods of invoking the command. A complete index gives an alphabetical listing of all AutoCAD commands, command options, system variables, and concepts discussed.

For Students in Diverse Areas
AutoCAD 2002 Companion is written for students in the fields of engineering, architecture, design, construction, manufacturing, and any other field that has a use for AutoCAD. Applications and examples from many fields are given throughout the text. The applications and examples are not intended to have an inclination toward one particular field. Instead, applications to a particular field are used when they best explain an idea or use of a command.

Additional Topics
For instruction in the full range of AutoCAD commands and features, you can purchase *AutoCAD 2002 Instructor* by James A. Leach, McGraw-Hill. *AutoCAD 2002 Instructor* covers all of the topics in this book as well as advanced selection sets, block attributes, external references, object linking and embedding, raster images and other formats, advanced layouts and plotting, wireframe modeling, surface modeling, rendering, creating 2D drawings from 3D models, basic customization, miscellaneous commands and features, menu customization, CAD management, batch plotting, and a variety of additional reference material. *AutoCAD 2002 Instructor* includes 1550 illustrations in 1300 pages.

www.mhhe.com/leach
Please visit our Web page at the above address. Ancillary materials are available for reading or download. Over 400 drawing problems specifically for architectural, mechanical engineering, civil, and electrical applications are available. Solutions for drawing problems and questions can be downloaded by requesting a password on the Web site.

Have Fun
I predict you will have a positive experience learning AutoCAD. Although learning AutoCAD is not a trivial endeavor, you will have fun using this exciting technology. In fact, I predict that more than once in your learning experience you will say to yourself, "Cool!" (or something to that effect).

James A. Leach

ABOUT THE AUTHOR

James A. Leach (B.I.D., M.Ed.) is an associate professor of engineering graphics at the University of Louisville. He began teaching AutoCAD at Auburn University early in 1984 using Version 1.4, the first version of AutoCAD to operate on IBM personal computers. Jim is currently the director of the Authorized Autodesk Training Center (ATC) established at the University of Louisville in 1985, one of the first fifteen centers to be authorized by Autodesk.

In his 26 years of teaching Engineering Graphics and AutoCAD courses, Jim has published numerous journal and magazine articles, drawing workbooks, and textbooks about AutoCAD and engineering graphics instruction. He has designed CAD facilities and written AutoCAD-related course materials for Auburn University, University of Louisville, the ATC at the University of Louisville, and several two-year and community colleges. Jim is the author of eleven AutoCAD textbooks published by Richard D. Irwin and McGraw-Hill.

CONTRIBUTING AUTHORS

Steven H. Baldock is an engineer at a consulting firm in Louisville and operates a CAD consulting firm, Infinity Computer Enterprises (ICE). Steve is an Autodesk Certified Instructor and teaches several courses at the University of Louisville AutoCAD Training Center. He has thirteen years experience using AutoCAD in architectural, civil, and structural design applications. Steve has degrees in engineering, computer science, and mathematics. Steve Baldock prepared material for several sections of *AutoCAD 2002 Companion*, such as text and dimensioning. Steve also created several hundred figures used in *AutoCAD 2002 Companion*.

Michael E. Beall is the owner of Computer Aided Management and Planning in Shelbyville, Kentucky. Michael offers contract services and professional training on AutoCAD as well as CAP products from Sweets Group, a division of McGraw-Hill. He has co-authored *AutoCAD 14 Fundamentals*, and was a contributing author to *Inside AutoCAD 14* from New Riders Publishing. Michael is currently editor for *Inside AutoCAD 2000*. Other efforts include co-authoring *AutoCAD Release 13 for Beginners* and *Inside AutoCAD LT for Windows 95*. Mr. Beall has been presenting CAD training seminars to architects and engineers since 1982 and is currently an Autodesk Certified Instructor (ACI) at the University of Louisville ATC. He was also a presenter for the *Mastering Today's AutoCAD* seminar series from Awareness Learning, Inc., an organization founded by Hugh Bathurst (www.awarenesslearning.com).

Mr. Beall received a Bachelor of Architecture degree from the University of Cincinnati. Michael Beall assisted with several topics in *AutoCAD 2002 Companion*. You can contact Michael at 502.633.3994 or michael.beall@autocadtrainerguy.com.

ACKNOWLEDGMENTS

I want to thank the contributing authors for their assistance in writing *AutoCAD 2002 Companion*. Without their help, this text could not have been as application-specific nor could it have been completed in the short time frame. I especially want to thank Steven H. Baldock for his hard work on earlier editions of this book and for his continual contributions and valuable input on new AutoCAD concepts and related industrial applications.

I am very grateful to Gary Bertoline for his foresight in conceiving the Irwin Graphics Series and for including my efforts in it.

I would like to give thanks to the excellent editorial and production group at McGraw-Hill who gave their talents and support during this project, especially Betsy Jones, Kelley Butcher, and Kelly Lowery.

A special thanks goes to Karen Collins for the layout and design of *AutoCAD 2002 Companion*. She was instrumental in fulfilling my objective of providing the most direct and readable format for conveying concepts.

David Haag deserves credit for preparation of the contents of the "command tables" used throughout this text. David also prepared the solution drawings for the Chapter Exercises that are available on the www.mhhe.com/leach Web site.

I also want to thank all of the readers that have contacted me with comments and suggestions on specific sections of this text. Your comments help me improve this book, assist me in developing new ideas, and keep me abreast of ways AutoCAD and this text are used in industrial and educational settings.

I also acknowledge: my colleague and friend Robert A. Matthews for his support of this project and for doing his job well; Charles Grantham of Contemporary Publishing Company of Raleigh, Inc., for generosity and consultation; and Speed Scientific School Dean's Office for support and encouragement.

Special thanks, once again, to my wife, Donna, for the many hours of copy editing required to produce this and the other texts.

TRADEMARK AND COPYRIGHT ACKNOWLEDGMENTS

The following drawings used for Chapter Exercises appear courtesy of James A. Leach, *Problems in Engineering Graphics Fundamentals, Series A* and *Series B,* ©1984 and 1985 by Contemporary Publishing Company of Raleigh, Inc.: Gasket A, Gasket B, Pulley, Holder, Angle Brace, Saddle, V-Block, Bar Guide, Cam Shaft, Bearing, Cylinder, Support Bracket, Corner Brace, and Adjustable Mount. Reprinted or redrawn by permission of the publisher.

The following are registered trademarks of Autodesk, Inc., in the USA and /or other countries: 3D Studio, 3D Studio MAX, 3D Studio VIZ, Actrix, ADE, ADI, Advanced Modeling Extension, AME, ATC, AutoCAD, AutoCAD Data Extension, AutoCAD Development System, AutoCAD LT, Autodesk, Autodesk (logo), AutoCAD 2000i, Autodesk Authorized Training Center, Autodesk Certified Instructor, AutoCAD DesignCenter, AutoCAD Designer, Autodesk Inventor, Autodesk Point A, Autodesk Premier Authorized Training Center, AutoLISP, AutoShade, AutoSurf, Autodesk Training Center, AutoVision, Heidi, HOOPS, Kinetix, Mechanical Desktop, ObjectARX, *WHIP!, WHIP! (logo)*. The following are trademarks of Autodesk, Inc., in the USA and /or other countries: ACAD, AutoCAD Learning Assistance, AutoCAD SQL Extension, Autodesk Inventor, AutoSnap, AutoTrack, Content Explorer, DesignCenter, DXF, FLI, FLIC, PolarSnap, Visual LISP, Volo. Windows, Notepad, WordPad, Excel, and MS-DOS are registered trademarks of Microsoft Corporation. Corel WordPerfect is a registered trademark of Corel, Inc. Norton Editor is a registered trademark of S. Reifel & Company. City Blueprint, Country Blueprint, EuroRoman, EuroRoman-oblique, PanRoman, SuperFrench, Romantic, Romantic-bold, Sans Serif, Sans Serif-bold, Sans Serif-oblique, Sans Serif-BoldOblique, Technic Technic-light, and Technic-bold are Type 1 fonts, copyright 1992 P. B. Payne.

All other brand names, product names, or trademarks belong to their respective holders.

LEGEND

The following special treatment of characters and fonts in the textual content is intended to assist you in translating the meaning of words or sentences in *AutoCAD 2002 Companion*.

Underline	Emphasis of a word or an idea.
Helvetica font	An AutoCAD prompt appearing on the screen at the Command line or in a text window.
Italic (Upper and Lower)	An AutoCAD command, option, menu, toolbar, or dialog box name.
UPPER CASE	A file name.
UPPER CASE ITALIC	An AutoCAD system variable or a drawing aid (*OSNAP, SNAP, GRID, ORTHO*).

Anything in **Bold** represents user input:

Bold	What you should type or press on the keyboard.
Bold Italic	An AutoCAD command that you should type or menu item that you should select.
BOLD UPPER CASE	A file name that you should type.
BOLD UPPER CASE ITALIC	A system variable that you should type.
PICK	Move the cursor to the indicated position on the screen and press the select button (button #1 or left mouse button).

INTRODUCTION

WHAT IS CAD?

CAD is an acronym for Computer-Aided Design. CAD allows you to accomplish design and drafting activities using a computer. A CAD software package, such as AutoCAD, enables you to create designs and generate drawings to document those designs.

Design is a broad field involving the process of making an idea into a real product or system. The design process requires repeated refinement of an idea or ideas until a solution results—a manufactured product or constructed system. Traditionally, design involves the use of sketches, drawings, renderings, two-dimensional (2D) and three-dimensional (3D) models, prototypes, testing, analysis, and documentation. Drafting is generally known as the production of drawings that are used to document a design for manufacturing or construction or to archive the design.

CAD is a tool that can be used for design and drafting activities. CAD can be used to make "rough" idea drawings, although it is more suited to creating accurate finished drawings and renderings. CAD can be used to create a 2D or 3D computer model of the product or system for further analysis and testing by other computer programs. In addition, CAD can be used to supply manufacturing equipment such as lathes, mills, laser cutters, or rapid prototyping equipment with numerical data to manufacture a product. CAD is also used to create the 2D documentation drawings for communicating and archiving the design.

The tangible result of CAD activity is usually a drawing generated by a plotter or printer but can be a rendering of a model or numerical data for use with another software package or manufacturing device. Regardless of the purpose for using CAD, the resulting drawing or model is stored in a CAD file. The file consists of numeric data in binary form usually saved to a magnetic or optical device such as a diskette, hard disk, tape, or CD.

WHY SHOULD YOU USE CAD?

Although there are other methods used for design and drafting activities, CAD offers the following advantages over other methods in many cases:

1. Accuracy
2. Productivity for repetitive operations
3. Sharing the CAD file with other software programs

Accuracy

Since CAD technology is based on computer technology, it offers great accuracy. When you draw with a CAD system, the graphical elements, such as lines, arcs, and circles, are stored in the CAD file as numeric data. CAD systems store that numeric data with great precision. For example, AutoCAD stores values with fourteen significant digits. The value 1, for example, is stored in scientific notation as the equivalent of 1.0000000000000. This precision provides you with the ability to create designs and drawings that are 100% accurate for almost every case.

Productivity for Repetitive Operations

It may be faster to create a simple "rough" drawing, such as a sketch by hand (pencil and paper), than it would by using a CAD system. However, for larger and more complex drawings, particularly those involving similar shapes or repetitive operations, CAD methods are very efficient. Any kind of shape or operation accomplished with the CAD system can be easily duplicated since it is stored in a CAD file.

In short, it may take some time to set up the first drawing and create some of the initial geometry, but any of the existing geometry or drawing setups can be easily duplicated in the current drawing or for new drawings.

Likewise, making changes to a CAD file (known as editing) is generally much faster than creating the original geometry. Since all the graphical elements in a CAD drawing are stored, only the affected components of the design or drawing need to be altered, and the drawing can be plotted or printed again or converted to other formats.

As CAD and the associated technology advance and software becomes more interconnected, more productive developments are available. For example, it is possible to make a change to a 3D model that automatically causes a related change in the linked 2D engineering drawing. One of the main advantages of these technological advances is productivity.

Sharing the CAD File with Other Software Programs

Of course, CAD is not the only form of industrial activity that is making technological advances. Most industries use computer software to increase capability and productivity. Since software is written using digital information and may be written for the same or similar computer operating systems, it is possible and desirable to make software programs with the ability to share data or even interconnect, possibly appearing simultaneously on one screen.

For example, word processing programs can generate text that can be imported into a drawing file, or a drawing can be created and imported into a text file as an illustration. (This book is a result of that capability.) A drawing created with a CAD system such as AutoCAD can be exported to a finite element analysis program that can read the computer model and compute and analyze stresses. CAD files can be dynamically "linked" to spreadsheets or databases in such a way that changing a value in a spreadsheet or text in a database can automatically make the related change in the drawing, or vice versa.

Another advance in CAD technology is the automatic creation and interconnectivity of a 2D drawing and a 3D model in one CAD file. With this tool, you can design a 3D model and have the 2D drawings automatically generated. The resulting set has bi-directional associativity; that is, a change in either the 2D drawings or the 3D model is automatically updated in the other.

With the introduction of the new Web technologies, designers and related professionals can more easily collaborate by viewing and transferring drawings over the Internet. CAD drawings can contain Internet links to other drawings, text information, or other related Web sites. Multiple CAD users can even share a single CAD session from remote locations over the Internet.

CAD, however, may not be the best tool for every design related activity. For example, CAD may help develop ideas but probably won't replace the idea sketch, at least not with present technology. A 3D CAD model can save much time and expense for some analysis and testing but cannot replace the "feel" of an actual model, at least not until virtual reality technology is developed and refined.

With everything considered, CAD offers many opportunities for increased accuracy, productivity, and interconnectivity. Considering the speed at which this technology is advancing, many more opportunities are rapidly obtainable. However, we need to start with the basics. Beginning by learning to create an AutoCAD drawing is a good start.

WHY USE AutoCAD?

CAD systems are available for a number of computer platforms: laptops, personal computers (PCs), workstations, and mainframes. AutoCAD, offered to the public in late 1982, was one of the first PC-based CAD software products. Since that time, it has grown to be the world leader in market share for <u>all</u> CAD products. Autodesk, the manufacturer of AutoCAD, is the world's leading supplier of PC design software and multimedia tools. At the time of this writing, Autodesk is one of the largest software producers in the world and has over three million customers in more than 150 countries.

Learning AutoCAD offers a number of advantages to you. Since AutoCAD is the most widely used CAD software, using it gives you the highest probability of being able to share CAD files and related data and information with others.

As a student, learning AutoCAD, as opposed to learning another CAD software product, gives you a higher probability of using your skills in industry. Likewise, there are more employers who use AutoCAD than any other single CAD system. In addition, learning AutoCAD as a first CAD system gives you a good foundation for learning other CAD packages because many concepts and commands introduced by AutoCAD are utilized by other systems. In some cases, AutoCAD features become industry standards. The .DXF file format, for example, was introduced by Autodesk and has become an industry standard for CAD file conversion between systems.

As a professional, using AutoCAD gives you the highest possibility that you can share CAD files and related data with your colleagues, vendors, and clients. Compatibility of hardware and software is an important issue in industry. Maintaining compatible hardware and software allows you the highest probability for sharing data and information with others as well as offering you flexibility in experimenting with and utilizing the latest technological advancements. AutoCAD provides you with great compatibility in the CAD domain.

This introduction is not intended as a selling point but to remind you of the importance and potential of the task you are about to undertake. If you are a professional or a student, you have most likely already made up your mind that you want to learn to use AutoCAD as a design or drafting tool. If you have made up your mind, then you can accomplish anything. Let's begin.

1

GETTING STARTED

Chapter Objectives

After completing this chapter you should:

1. understand how the X, Y, Z coordinate system is used to define the location of drawing elements in digital format in a CAD drawing file;

2. understand why you should create drawings full size in the actual units with CAD;

3. be able to start AutoCAD to begin drawing;

4. recognize the areas of the AutoCAD Drawing Editor and know the function of each;

5. be able to use the five methods of entering commands;

6. be able to turn on and off the *SNAP*, *GRID*, *ORTHO*, and *POLAR* drawing aids;

7. know how to customize the AutoCAD for Windows screen to your preferences.

CONCEPTS

Coordinate Systems

Any location in a drawing, such as the endpoint of a line, can be described in X, Y, and Z coordinate values (Cartesian coordinates). If a line is drawn on a sheet of paper, for example, its endpoints can be charted by giving the distance over and up from the lower-left corner of the sheet (Fig. 1-1).

These distances, or values, can be expressed as X and Y coordinates; X is the horizontal distance from the lower-left corner (origin) and Y is the vertical distance from that origin. In a three-dimensional coordinate system, the third dimension, Z, is measured from the origin in a direction perpendicular to the plane defined by X and Y.

Two-dimensional (2D) and three-dimensional (3D) CAD systems use coordinate values to define the location of drawing elements such as lines and circles (called <u>objects</u> in AutoCAD).

In a 2D drawing, a line is defined by the X and Y coordinate values for its two endpoints (Fig. 1-2).

In a 3D drawing, a line can be created and defined by specifying X, Y, and Z coordinate values (Fig. 1-3). Coordinate values are always expressed by the X value first separated by a comma, then Y, then Z.

Figure 1-1

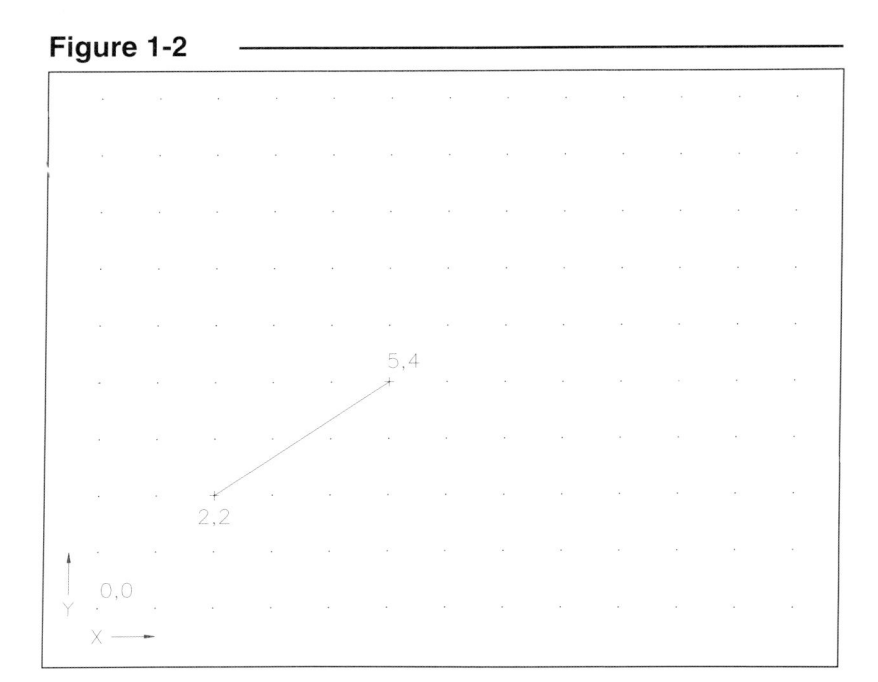

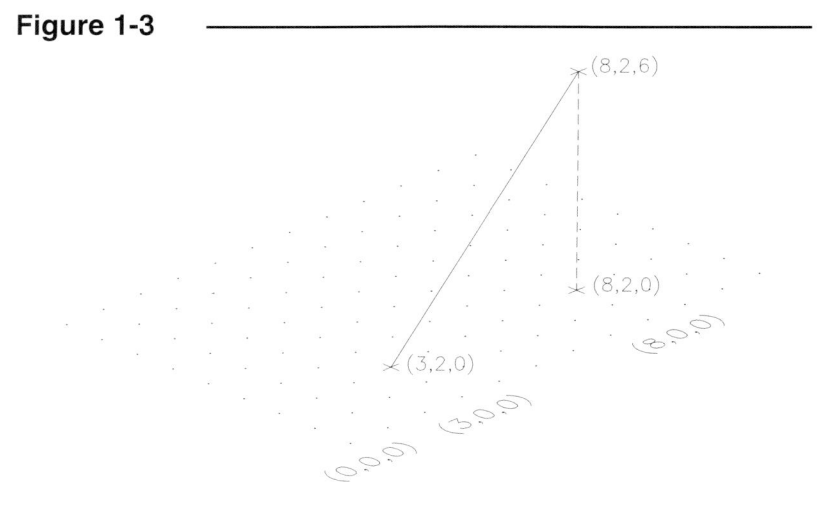

Figure 1-2

Figure 1-3

The CAD Database

A CAD (Computer-Aided Design) file, which is the electronically stored version of the drawing, keeps data in binary digital form. These digits describe coordinate values for all of the endpoints, center points, radii, vertices, etc. for all the objects composing the drawing, along with another code that describes the kinds of objects (line, circle, arc, ellipse, etc.). Figure 1-4 shows part of an AutoCAD DXF (Drawing Interchange Format) file giving numeric data defining lines and other objects. Knowing that a CAD system stores drawings by keeping coordinate data helps you understand the input that is required to create objects and how to translate the meaning of prompts on the screen.

Figure 1-4

```
LINE
  8
0
 62
      8
 10
15.0
 20
8.105789
 30
0.0
 11
15.644291
 21
8.75
 31
0.0
  0
LINE
  8
0
 62
      8
 18
```

Angles in AutoCAD

Figure 1-5

Angles in AutoCAD are measured in a <u>counterclockwise direction</u>. Angle 0 is positioned in a positive X direction, that is, horizontally from left to right. Therefore, 90 degrees is in a positive Y direction, or straight up; 180 degrees is in a negative X direction, or to the left; and 270 degrees is in a negative Y direction, or straight down (Fig. 1-5).

The position and direction of measuring angles in AutoCAD can be changed; however, the defaults listed here are used in most cases.

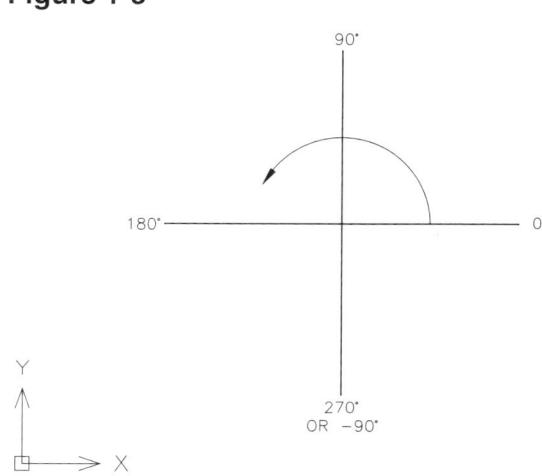

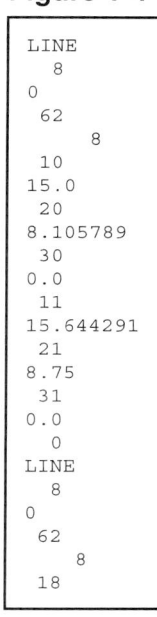

Draw True Size

Figure 1-6

When creating a drawing with pencil and paper tools, you must first determine a scale to use so the drawing will be proportional to the actual object and will fit on the sheet (Fig. 1-6). However, when creating a drawing on a CAD system, there is no fixed size drawing area. The number of drawing units that appear on the screen is variable and is assigned to fit the application.

The CAD drawing is not scaled until it is physically transferred to a fixed size sheet of paper by plotter or printer.

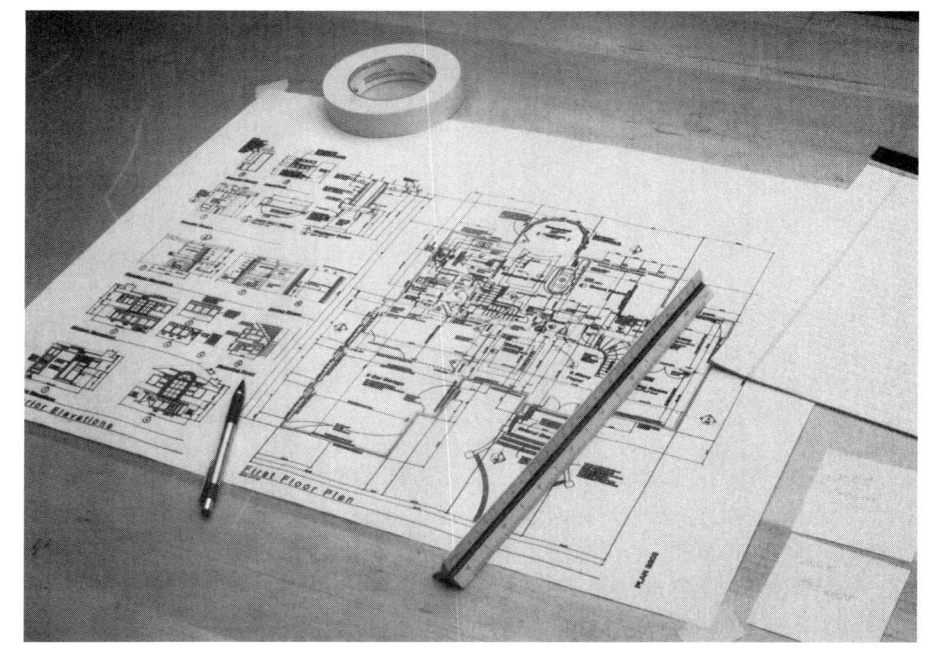

4 Getting Started

TIP: The rule for creating CAD drawings is that the drawing should be created <u>true size</u> using real-world units. The user specifies what units are to be used (architectural, engineering, etc.) and then specifies what size drawing area is needed (in X and Y values) to draw the necessary geometry.

Figure 1-7

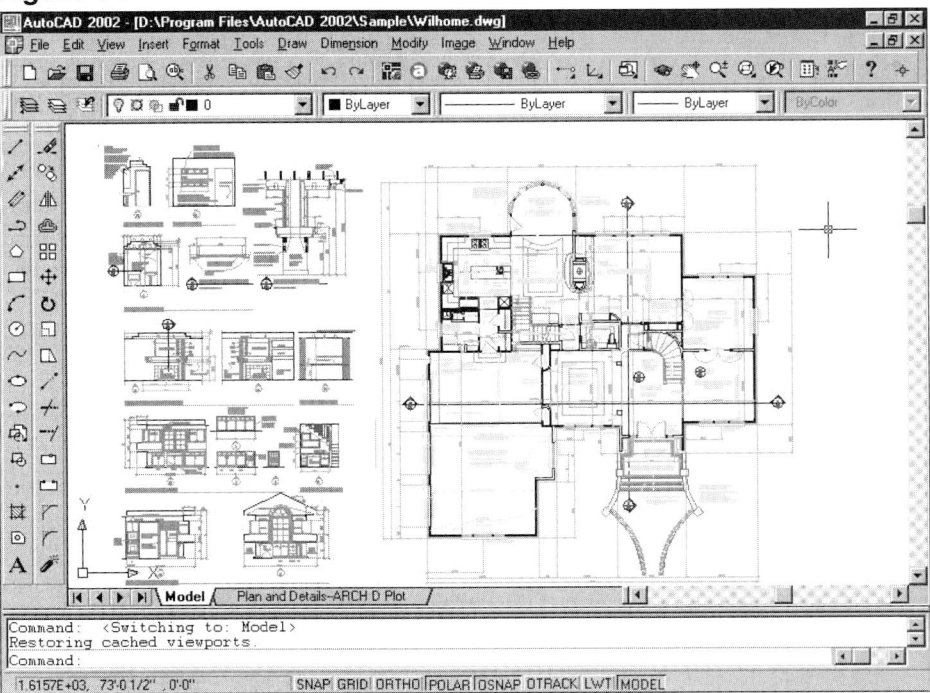

Whatever the specified size of the drawing area, it can be displayed on the screen in its entirety (Fig. 1-7) or as only a portion of the drawing area (Fig. 1-8).

Figure 1-8

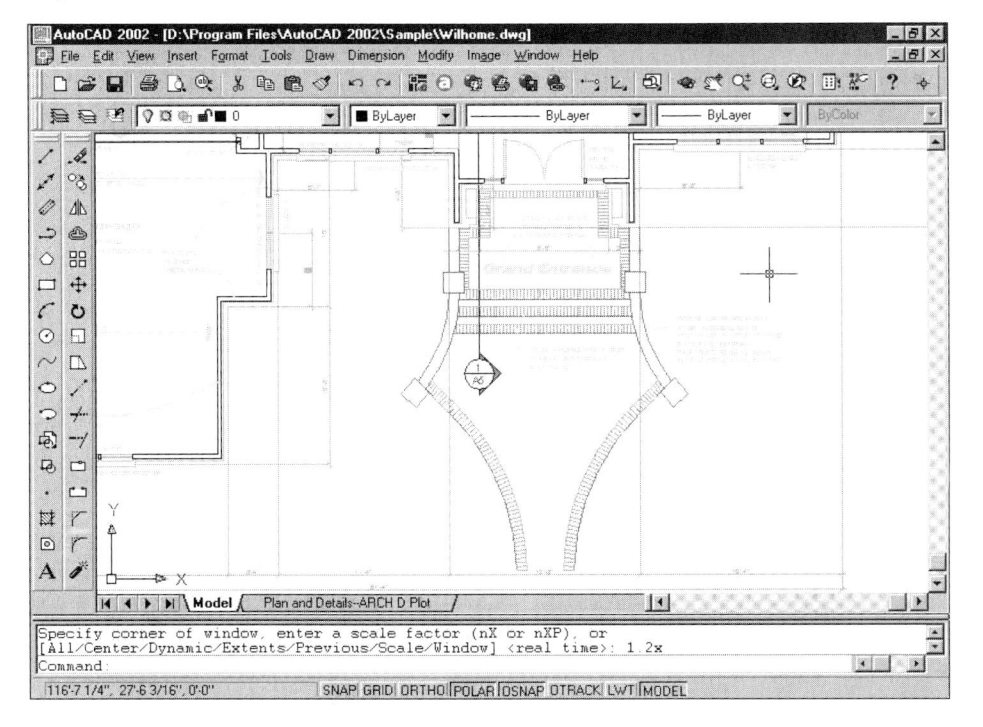

Plot to Scale

As long as a drawing exists as a CAD file or is visible on the screen, it is considered a virtual, full-sized object. Only when the CAD drawing is transferred to paper by a plotter or printer is it converted (usually reduced) to a size that will fit on a sheet. A CAD drawing can be automatically scaled to fit on the sheet regardless of sheet size; however, this action results in a plotted drawing that is not to an accepted scale (not to a regular proportion of the real object). Usually it is desirable to plot a drawing so that the resulting drawing is a proportion of the actual object size. The scale to enter as the plot scale (Fig. 1-9) is simply the proportion of the plotted drawing size to the actual object.

Figure 1-9

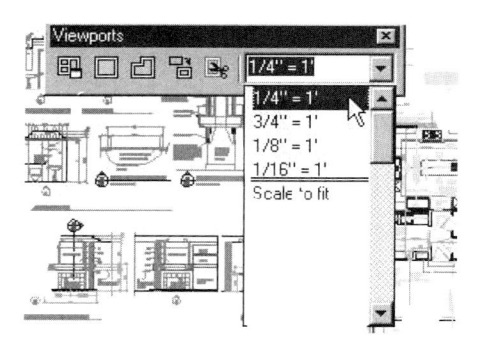

STARTING AutoCAD

Assuming that AutoCAD has been installed and configured properly for your system, you are ready to begin using AutoCAD.

To start AutoCAD for Windows 98, ME, 2000, or NT, locate the "AutoCAD 2002" shortcut icon on the desktop (Fig. 1-10). Double-clicking on the icon (point the arrow and press the left mouse button quickly two times) opens AutoCAD 2002.

Figure 1-10

If you cannot locate a shortcut icon on the desktop, select the "Start" button, highlight "Programs," and search for "AutoCAD 2002" in the menu. From the list that appears, select "AutoCAD 2002."

To exit AutoCAD, either select *Exit* from the *File* pull-down menu or click on the "X" in the very upper-right corner of the AutoCAD screen.

THE AutoCAD DRAWING EDITOR

The *Today* Window and the *Startup* Dialog Box

When you start AutoCAD 2002, you are presented with the *Today* window, the *Startup* dialog box, or a "blank" drawing screen, depending on how your system has been configured. By default, the *Today* window appears.

Figure 1-11

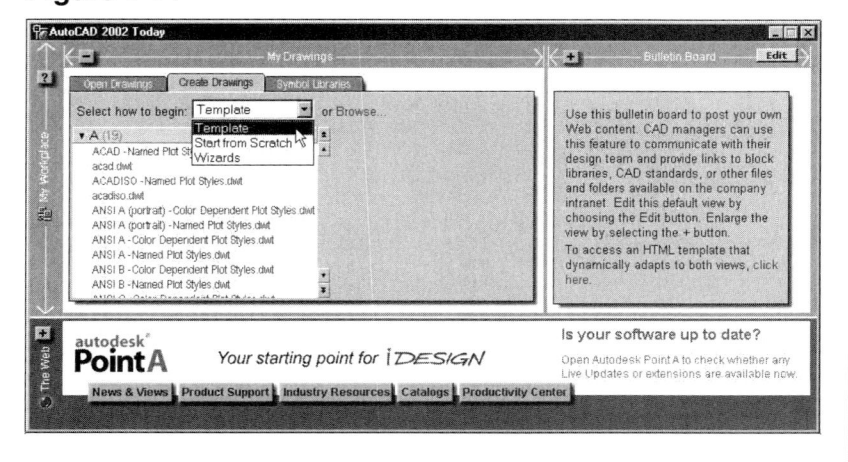

The *Today* Window

The *Today* window (Fig. 1-11) provides several functions. You can begin a new drawing or open an existing drawing. If you begin a new drawing, you can *Start from Scratch*, use a *Template*, or use a *Wizard*. The *Today* window also provides messages on the *Bulletin Board* and gives access to Autodesk's PointA Web site.

6 Getting Started

The *Startup* Dialog Box

This *Startup* dialog box (Fig. 1-12) also allows you to begin a new drawing or open an existing drawing, including providing access to drawing templates and setup *Wizards*.

Figure 1-12

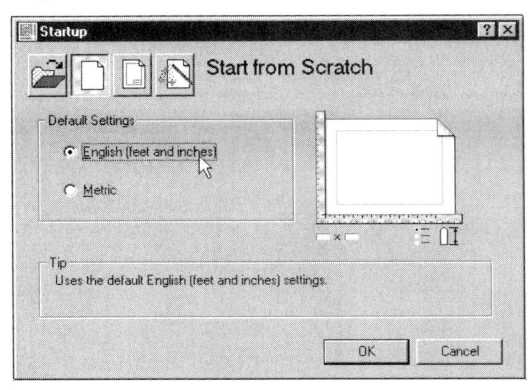

If you want to start AutoCAD with the default English or metric settings using either the *Today* window or the *Startup* dialog box, you can use one of the following options:

1. Select the *Create Drawings* tab and *Start from Scratch* option in the *Today* window or the *Start from Scratch* button in the *Startup* dialog box. Text appears displaying a choice of *English (feet and inches)* or *Metric*. Make your selection, then press *OK*.

2. Select the *Template* option in either the *Today* window or the *Startup* dialog box, and then select the *Acad.dwt* (English) or *Acadiso.dwt* (metric) template drawing.

3. Close the *Today* window or select the *Cancel* button in the *Startup* dialog box.

All three of these methods begin a new drawing using AutoCAD's default settings for use with English or metric units. The English settings (*Acad.dwt*) use a 12" x 9" drawing area (called *Limits*), and the metric settings (*Acadiso.dwt*) use a 420mm x 297mm drawing area. The *MEASUREINIT* system variable setting (0=English, 1=metric) controls which of the two formats is used if you cancel the dialog boxes or if you use the startup *Wizards*.

NOTE: The *Today* window, the *Startup* dialog box, and the *Create New Drawing* dialog box tools are explained in more detail in Chapter 2 (Working With Files), Chapter 6 (Basic Drawing Setup), and Chapter 12 (Advanced Drawing Setup). For the examples and exercises in Chapters 1-5, use the first method above to begin a new drawing.

After specifying your choice in the *Today* window or the *Startup* dialog box, the Drawing Editor appears on the screen and allows you to immediately begin drawing. Figure 1-13 displays the Drawing Editor in AutoCAD 2002.

Figure 1-13

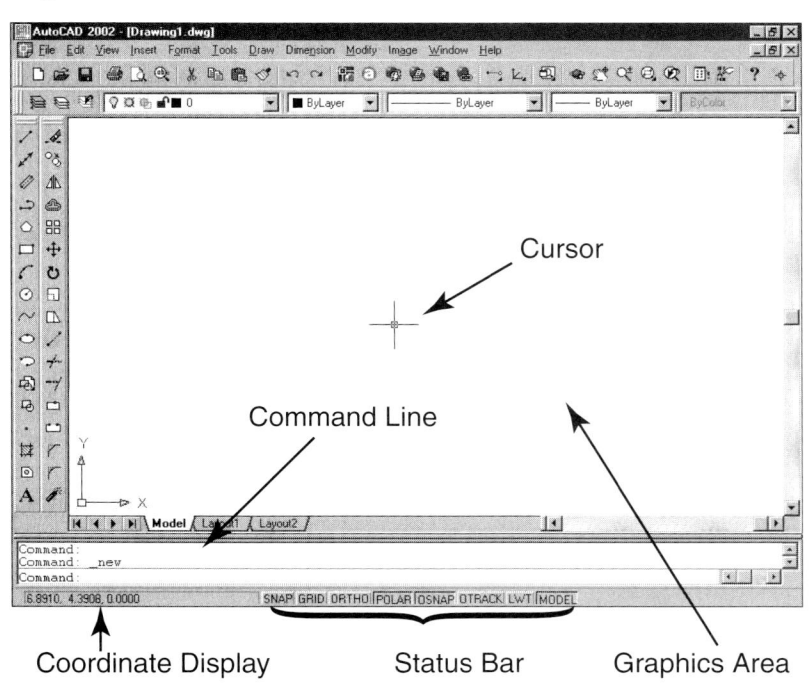

Graphics Area

The large central area of the screen is the Graphics area. It displays the lines, circles, and other objects you draw that will make up the drawing. The cursor is the intersection of the crosshairs (vertical and horizontal lines that follow the mouse or puck movements). The default size of the graphics area for English settings is 12 units (X or horizontal) by 9 units (Y or vertical). This usable drawing area (12 x 9) is called the drawing *Limits* and can be changed to any size to fit the application. As you move the cursor, you will notice the numbers in the Coordinate display change (Fig. 1-13, bottom left).

Command Line

The Command line consists of the three text lines at the bottom of the screen (by default) and is the most important area other than the drawing itself (see Fig. 1-13). Any command that is entered or any prompt that AutoCAD issues appears here. The Command line is always visible and gives the current state of drawing activity. You should develop the habit of glancing at the Command line while you work in AutoCAD. The Command line can be set to display any number of lines and/or moved to another location (see "Customizing the AutoCAD Screen" later in this chapter).

Pressing the F2 key opens a text window displaying the command history. This window is like an expanded Command line because it displays many more than just the three text lines that normally appear at the bottom of the screen. See "AutoCAD Text Window" later in this chapter.

Toolbars

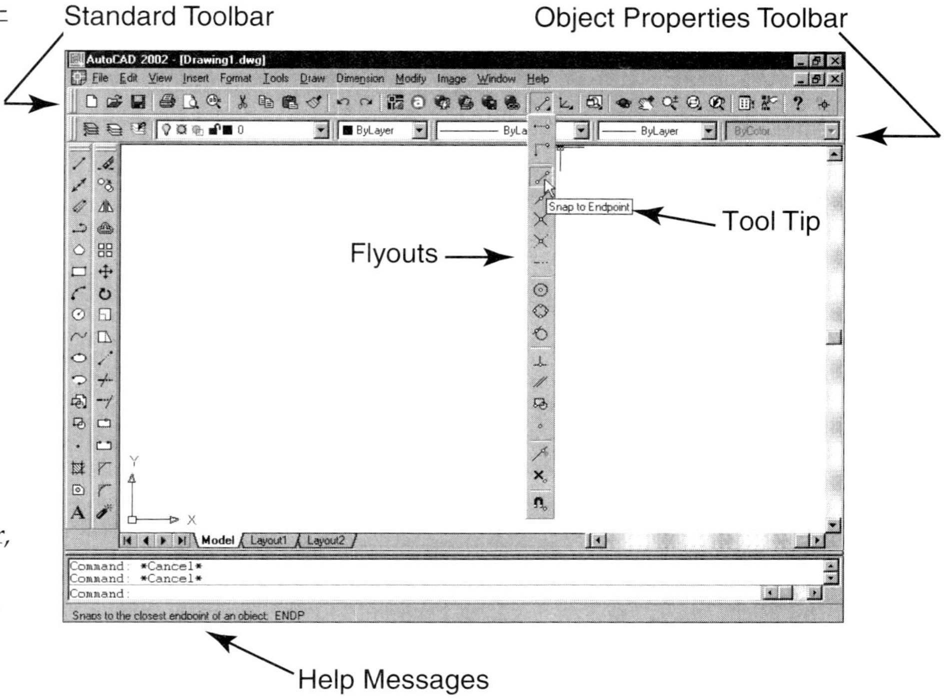

AutoCAD provides a variety of toolbars (Fig. 1-14). Each toolbar contains a number of icon buttons (tools) that can be PICKed to invoke commands for drawing or editing objects (lines, arcs, circles, etc.) or for managing files and other functions. The Standard toolbar is the row of icons nearest the top of the screen. The Standard toolbar contains many standard icons used in other Windows applications (like *New, Open, Save, Print, Cut, Paste,* etc.) and other icons for AutoCAD-specific functions (like *Zoom* and *Pan*). The Object Properties toolbar, located beneath the standard toolbar, is used for managing properties of objects, such as *Layers* and *Linetypes*. The Draw and Modify toolbars also appear (by default) when you first use AutoCAD. As shown in Figure 1-14, the Draw and Modify toolbars (side) and the Standard and Object Properties toolbars (top) are docked. Many other toolbars are available and can be made to float, resize, or dock (see "Customizing the AutoCAD Screen" later in this chapter).

Figure 1-14

If you place the pointer on an any icon and wait momentarily, a Tool Tip and a Help message appear. Tool Tips pop out by the pointer and give the command name (see Figure 1-14). The Help message appears at the bottom of the screen, giving a short description of the function. Flyouts are groups of related icons that pop out in a row or column when one of the group is selected. PICKing any icon that has a small black triangle in its lower-right corner causes the related icons to fly out.

Pull-down Menus

The pull-down menu bar is at the top of the screen just under the title bar (Fig. 1-15). Selecting any of the words in the menu bar activates, or pulls down, the respective menu. Selecting a word appearing with an arrow activates a cascading menu with other options. Selecting a word with ellipsis (...) activates a dialog box (see "Dialog Boxes"). Words in the pull-down menus are not necessarily the same as the formal command names used when typing commands. Menus can be canceled by pressing Escape or PICKing in the graphics area. The pull-down menus do not contain all of the AutoCAD commands and variables but contain the most commonly used ones.

Figure 1-15

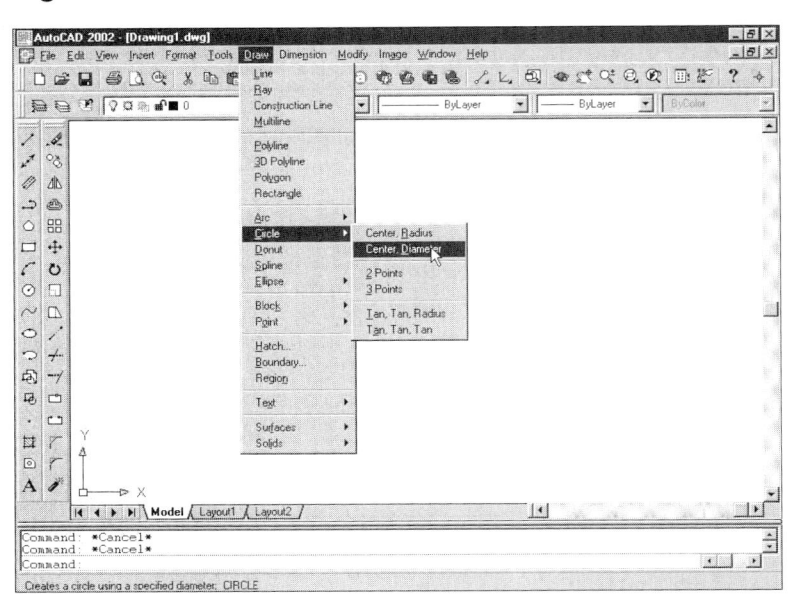

NOTE: Because the words that appear in the pull-down menus are not always the same as the formal commands you would type to invoke a command, AutoCAD can be confusing to learn. The Help message that appears just below the Command line (when a menu is pulled down or the pointer rests on an icon button) can be instrumental in avoiding this confusion. The Help message gives a description of the command followed by colon (:), then the formal command name (see Figures 1-14 and 1-15).

Screen (Side) Menu

The screen menu does not appear by default in AutoCAD, but can be made to appear by selecting *Options...* from the *Tools* pull-down menu. This selection causes the *Options* dialog box to appear (Fig. 1-16). Select the *Display* tab and check the *Display screen menu* option. Keep in mind that this menu is not needed if you prefer to use the icon buttons, pull-down menus, or keyboard to enter commands.

Not all of the AutoCAD commands can be accessed through the menu system located at the right side of the screen (see Figure 1-16). The screen menu has a tree structure.

Figure 1-16

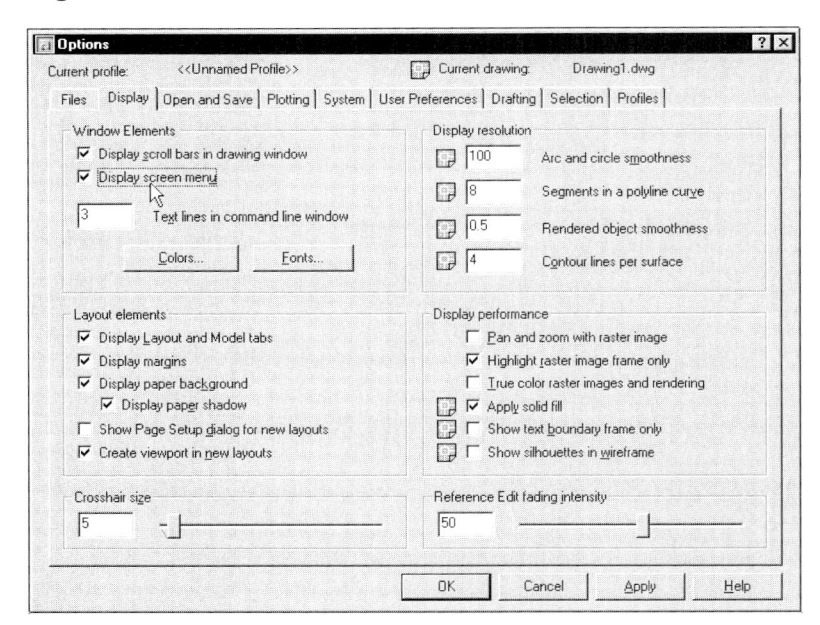

Menus and commands are accessed by branching out from the root or top-level menu (Fig. 1-17). Commands (in uppercase and lowercase) or menus (capital letters only) are selected by moving the cursor to the desired position until the word is highlighted and then pressing the PICK button. The root menu (top level) is accessed from any level of the structure by selecting the word "AutoCAD" at the top of any menu.

The commands in these menus are generally the formal command names that can also be typed at the keyboard. (Some commands names are truncated in the screen menu because only eight characters can be displayed.)

If commands are invoked by typing, pull-downs, or toolbars, the screen menu automatically changes to the current command.

Figure 1-17

Dialog Boxes

Dialog boxes provide an interface for controlling complex commands or a group of related commands. Depending on the command, the dialog boxes allow you to select among multiple options and sometimes give a preview of the effect of selections. The *Layer Properties Manager* dialog box (Fig. 1-18) gives complete control of layer colors, linetypes, and visibility.

Dialog boxes can be invoked by typing a command, selecting an icon button, or PICKing from the menus. For example, typing *Layer*, PICKing the *Layers* button, or selecting *Layer* from the *Format* pull-down menu causes the *Layer Properties Manager* dialog box to appear. In the pull-down menu, all commands that invoke a dialog box end with ellipsis points (...).

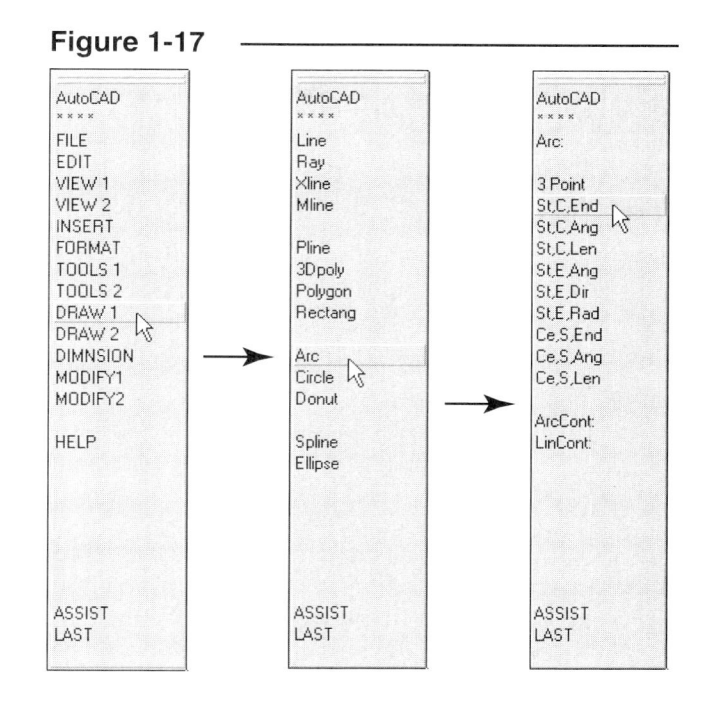

Figure 1-18

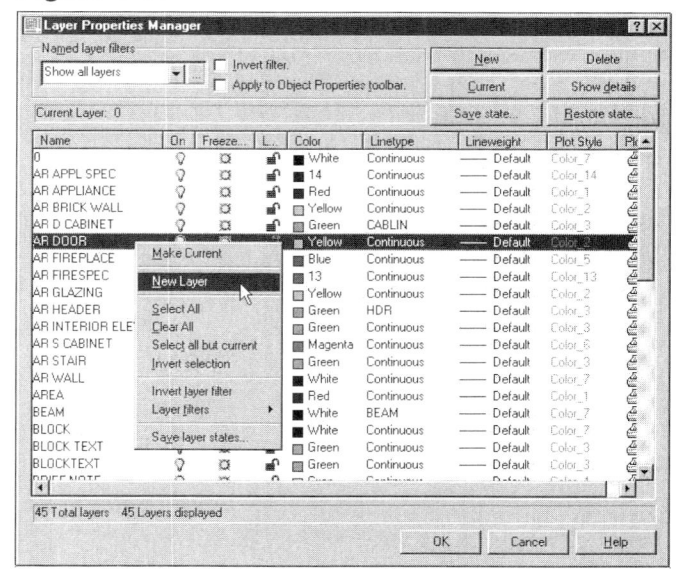

The basic element or smallest component of a dialog box is called a tile. Several types of tiles and the resulting actions of tile selection are listed below.

Button	Resembles a push button and triggers some type of action
Edit box	Allows typing or editing of a single line of text
Image tile	A button that displays a graphical image
List box	A list of text strings from which one or more can be selected
Drop-down list	A text string that drops down to display a list of selections
Radio button	A group of buttons, only one of which can be turned on at a time
Checkbox	A checkbox for turning a feature on or off (displays a check mark when on)

You can change dialog box border colors (for all applications) with the Windows Control Panel. The AutoCAD screen colors can also be customized (see "Customizing the AutoCAD Screen," this chapter). See "File Navigation Dialog Box Functions" in Chapter 2 for more information on dialog boxes that are used to access files.

Status Bar

The Status bar is a set of informative words or symbols that gives the status of the drawing aids. The Status bar appears at the very bottom of the screen (see Figure 1-13). The following drawing aids can be toggled on or off by single-clicking (pressing the left mouse button once) on the desired word or by using Function keys or Ctrl key sequences. The following drawing aids are explained here or in following chapters:

SNAP, GRID, ORTHO, POLAR, OSNAP, OTRACK, LWT, MODEL

Coordinate Display (*Coords*)

The Coordinate display is located in the lower-left corner of the AutoCAD screen (Fig. 1-19). The Coordinate display (*Coords*) displays the current position of the cursor in one of two possible formats explained below. This display can be very helpful when you draw because it can give the X, Y, and Z coordinate position of the cursor or give the cursor's distance and angle from the last point established. The format of *Coords* is controlled by toggling the F6 key, pressing Ctrl+D, or single-clicking on the *Coords* display (numbers) at the bottom left of the screen. *Coords* can also be toggled off.

Cursor tracking
When *Coords* is in this position, the values display the current location of the cursor in absolute (X, Y, and Z) coordinates (see Figure 1-19).

Relative polar display
This display is possible only if a draw or edit command is in use. The values give the distance and angle of the "rubberband" line from the last point established (not shown).

Figure 1-19

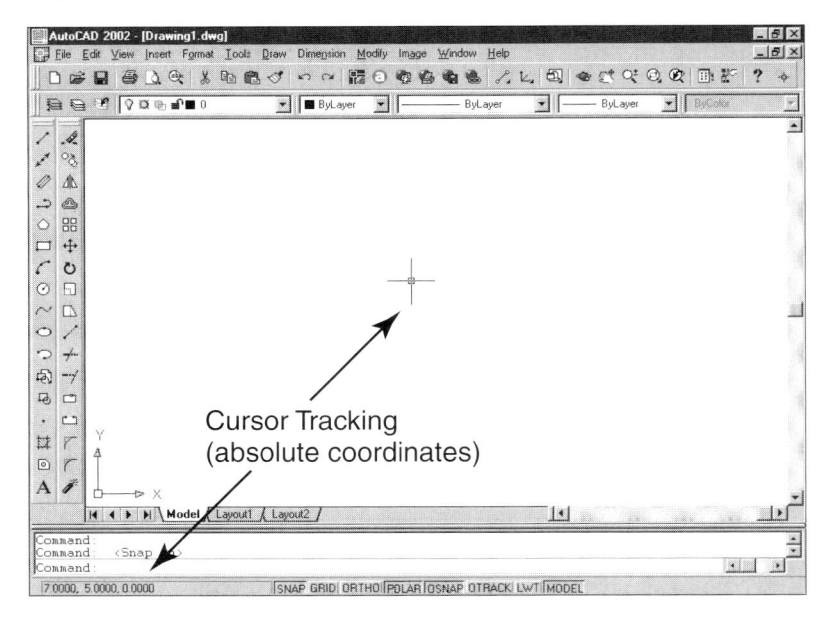

Digitizing Tablet Menu

If you have a digitizing tablet, the AutoCAD commands are available by making the desired selection from the AutoCAD digitizing tablet menu (Fig. 1-20). The open area located slightly to the right of center is called the Screen Pointing area. Locating the digitizing puck there makes the cursor appear on the screen. Locating the puck at any other location allows you to select a command. The icons on the tablet menu are identical to the toolbar icons. Similar to the format of the toolbars, commands are located in groups such as Draw, Edit, View, and Dimension. The tablet menu has the columns numbered along the top and the rows lettered along the left side. The location of each command by column and row is given in the "command tables" in this book. For example, the *Line* command can be found at *10,J*.

Figure 1-20

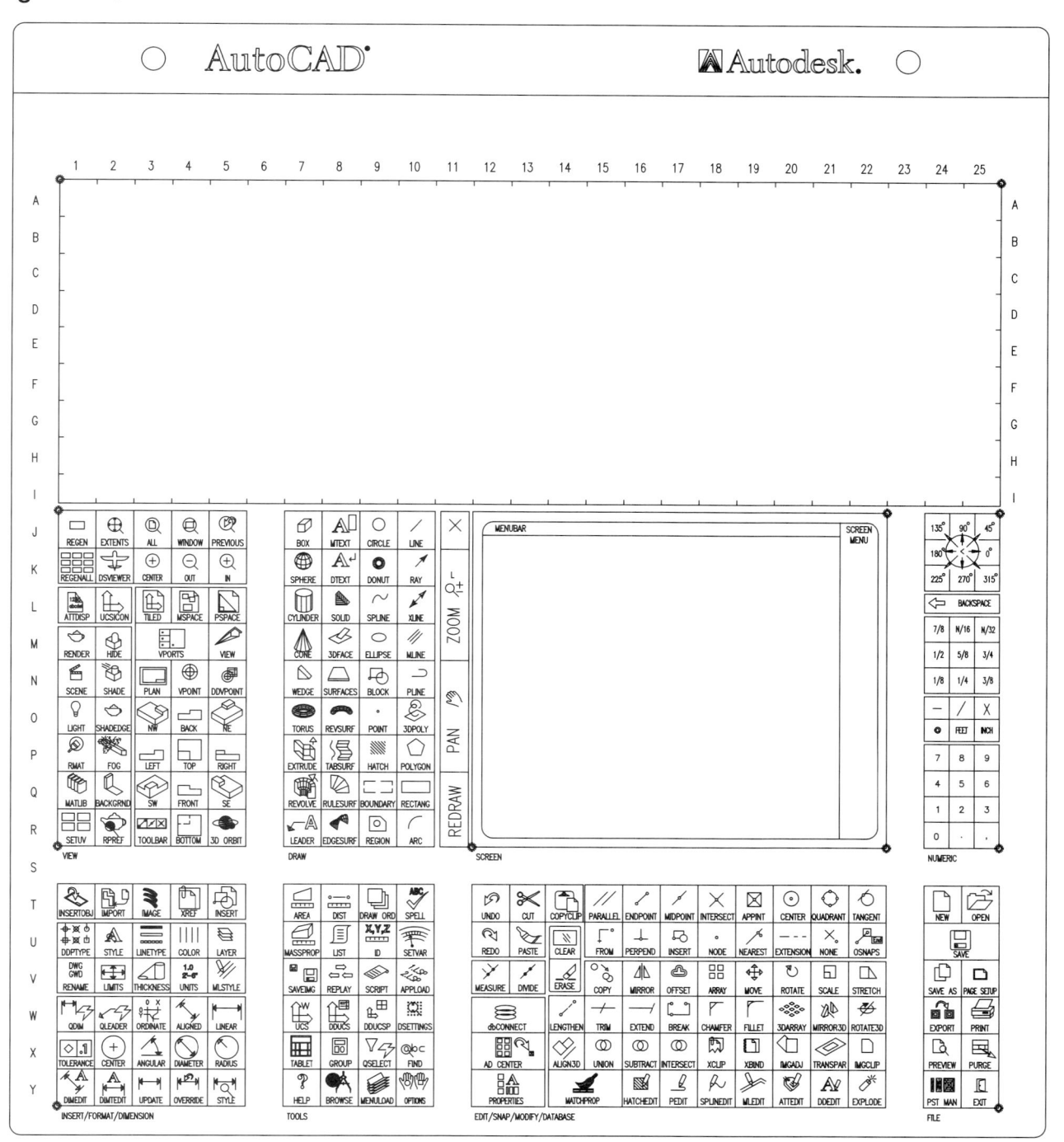

COMMAND ENTRY

Methods for Entering Commands

There are five possible methods for entering commands in AutoCAD depending on your *Options* setting (for the screen menu) and availability of a digitizing tablet. Generally, any one of the five methods can be used to invoke a particular command.

1. **Toolbars** — Select the command or dialog box by PICKing an icon (tool) from a toolbar.
2. **Pull-down menu** — Select the command or dialog box from a pull-down menu.
3. **Screen (side) menu** — Select the command or dialog box from the screen (side) menu.
4. **Keyboard** — Type the command name, command alias, or accelerator (Ctrl) keys at the keyboard.*
5. **Tablet menu** — Select the command from the digitizing tablet menu (if available).

*A command alias is a one- or two-letter shortcut. Accelerator keys use the Ctrl key plus another key and are used for utility operations. The command aliases and accelerator keys are given in the command tables (see below).

All five methods of entering commands accomplish the same goals; however, one method may offer a slightly different option or advantage to another depending on the command used.

All menus, including the digitizing tablet, can be customized by editing the ACAD.MNU file, and command aliases can be changed or added in the ACAD.PGP file so that command entry can be designed to your preference (see AutoCAD 2002 *Help*, Customization Guide).

Using the "Command Tables" in This Book to Locate a Particular Command

Command tables, like the one below, are used throughout this book to show the possible methods for entering a particular command. The table shows the icon used in the toolbars and digitizing tablet, gives the selections to make for the pull-down and screen (side) menus, gives the correct spelling for entering commands and command aliases at the keyboard, and gives the command location (column, row) on the digitizing tablet menu. This example uses the *Copy* command.

COPY

Pull-down Menu	COMMAND (TYPE)	ALIAS (TYPE)	Short-cut	Screen (side) Menu	Tablet Menu
Modify *Copy*	*COPY*	CO or CP	(Edit Mode) Copy Selection	MODIFY1 *Copy*	V,15

Shortcut Menus

AutoCAD makes use of shortcut menus that are activated by pressing the right mouse button (sometimes called right-click menus). Shortcut menus give quick access to command options. There are many shortcut menus to list since they are based on the active command or dialog box. The menus fall into five basic categories listed here.

Default Menu
The default menu appears when you right-click in the drawing area and no command is in progress.

Figure 1-21

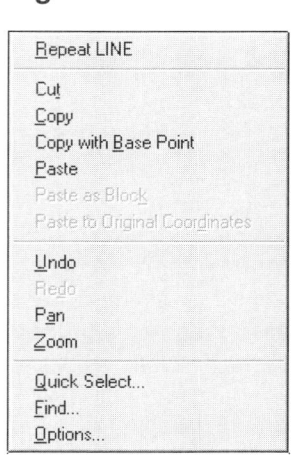

Edit-Mode Menu
This menu appears when you right-click when <u>objects have been selected</u> but no command is in progress.

Figure 1-22

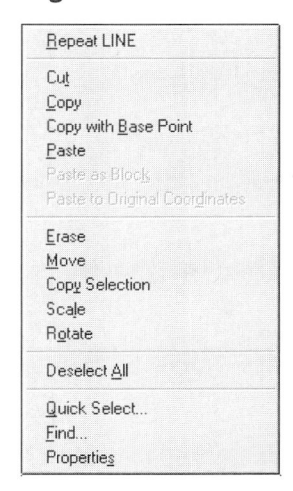

Command-Mode Menu
These menus appear when you right-click <u>when a command is in progress</u>. This menu changes since the options are specific to the command.

Figure 1-23

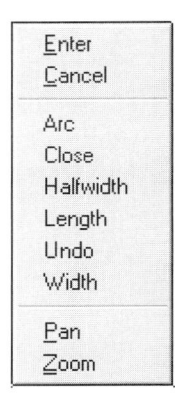

Dialog-Mode Menu
When the pointer is in a dialog box or tab and you right-click, this menu appears. The options on this menu can change based on the current dialog box.

Figure 1-24

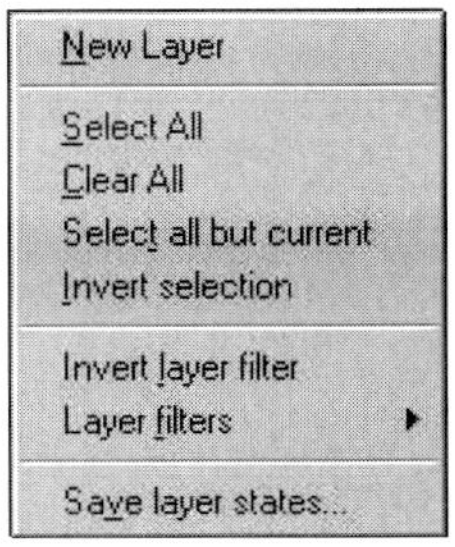

Other Menus
There are other menus that can be invoked. For example, this menu appears if you right-click in the Command line area.

Because there are so many shortcut menus, don't be too concerned about learning these until you have had some experience. The best advice at this time is just remember to experiment by right-clicking often to display the possible options.

Figure 1-25

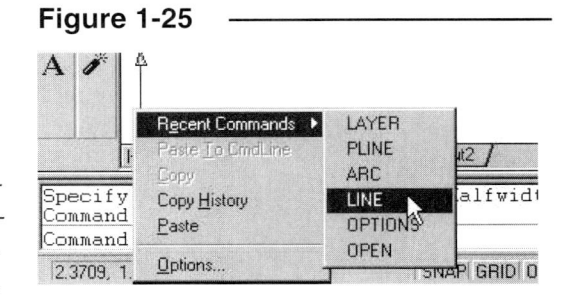

Mouse and Digitizing Puck Buttons
Depending on the type of mouse or digitizing puck used for cursor control, a different number of buttons is available. In any case, the buttons perform the following tasks:

#1 (left button)	PICK	Used to select commands or pick locations on the screen.
#2 (right button)	Enter or shortcut menu	Depending on the status of the drawing or command, this button either performs the same function as the enter key or produces a shortcut menu.
#3 (middle button or wheel)	*Pan*	If you press and drag, you can pan the drawing about on the screen.
	Zoom	If you turn the wheel, you can zoom in and out centered on the location of the cursor.

Function Keys
Several function keys are usable with AutoCAD. They offer a quick method of turning on or off (toggling) drawing aids.

F1	*Help*	Opens a help window providing written explanations on commands and variables.
F2	*Flipscreen*	Activates a text window showing the previous command line activity (command history).
F3	*Osnap Toggle*	If Running Osnaps are set, toggling this tile temporarily turns the Running Osnaps off so that a point can be picked without using Osnaps. If no Running Osnaps are set, F3 produces the *Osnap Settings* dialog box (discussed in Chapter 7).
F4	*Tablet*	Turns the *TABMODE* variable on or off. If *TABMODE* is on, the digitizing tablet can be used to digitize an existing paper drawing into AutoCAD.

F5	*Isoplane*	When using an *Isometric* style SNAP and GRID setting, toggles the cursor (with ORTHO on) to draw on one of three isometric planes.
F6	*Coords*	Toggles the Coordinate Display between cursor tracking mode and off. If used transparently (during a command in operation), displays a polar coordinate format.
F7	*GRID*	Turns the GRID on or off (see "Drawing Aids").
F8	*ORTHO*	Turns ORTHO on or off (see "Drawing Aids").
F9	*SNAP*	Turns SNAP on or off (see "Drawing Aids").
F10	*POLAR*	Turns POLAR on or off (see "Drawing Aids").
F11	*Osnap Tracking*	Turns Object Snap Tracking on or off.

Control Key Sequences (Accelerator Keys)

Several control key sequences (holding down the Ctrl key or Alt key and pressing another key simultaneously) invoke regular AutoCAD commands or produce special functions. See Appendix B.

Special Key Functions

Esc   The Escape key cancels a command, menu, or dialog box or interrupts processing of plotting or hatching.

Space bar   In AutoCAD, the space bar performs the same action as the Enter key. Only when you are entering text into a drawing does the space bar create a space.

Enter   If Enter or Spacebar is pressed when no command is in use (the open Command: prompt is visible), the last command used is invoked again.

Drawing Aids

This section gives a brief introduction to AutoCAD's Drawing Aids. For a full explanation of the related commands and options, see Chapter 6.

SNAP (**F9** or **Status Bar**)
SNAP has two modes in AutoCAD: Grid Snap and Polar Snap. Only one of the two modes can be active at one time. Grid Snap is a function that forces the cursor to "snap" to regular intervals (.5 units is the default English setting), which aids in creating geometry accurate to interval lengths. You can use the *Snap* command or the *Drafting Settings* dialog box to specify any value for the Grid Snap increment. Figure 1-26 displays a Snap setting of .125 (note the values in the coordinate display).

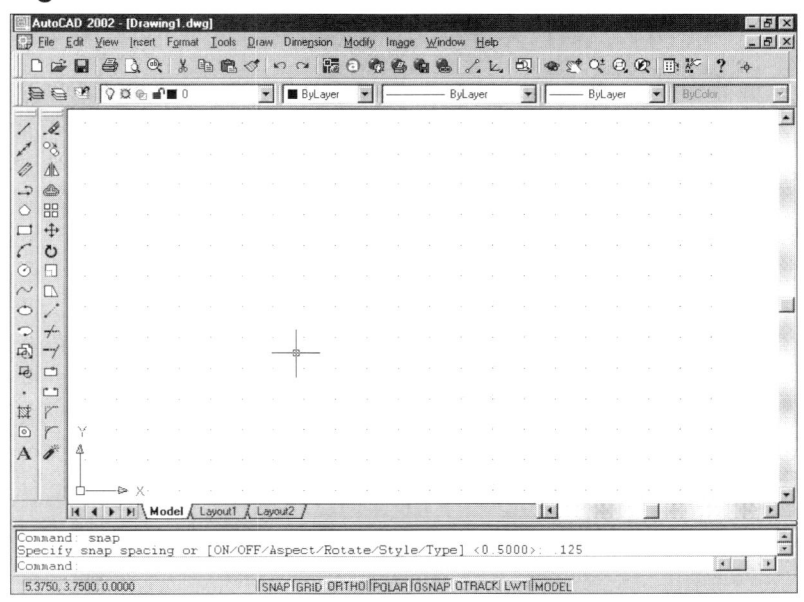

Figure 1-26

16 Getting Started

The other mode of Snap is Polar Snap. Polar Snap forces the cursor to snap to regular intervals along angular lines (Fig. 1-27). Polar Snap is functional only when *POLAR* is also toggled on since it works in conjunction with *POLAR*. The Polar Snap interval uses the Grid Snap setting by default but can be changed to any value using the *Snap* command or the *Drafting Settings* dialog box. Polar Snap is discussed in detail in Chapter 3.

Figure 1-27

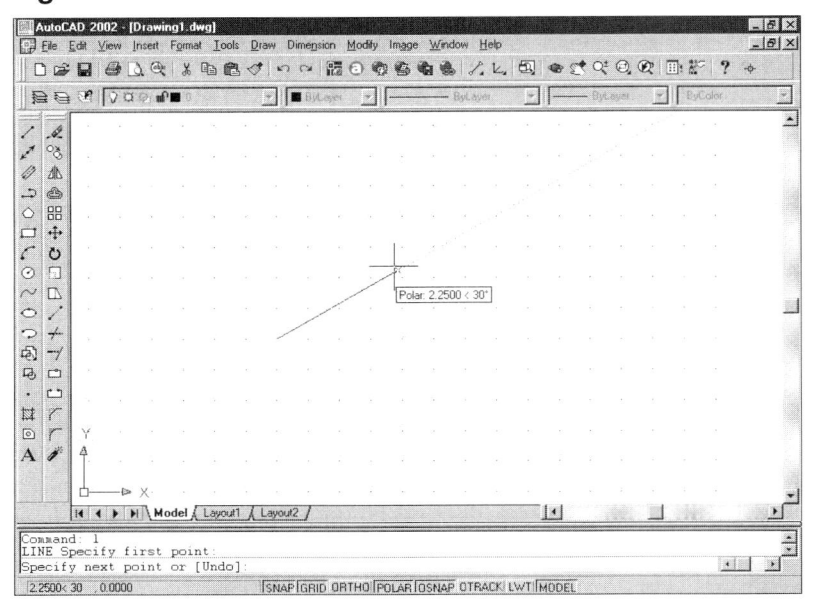

Since you can have only one *SNAP* mode on at a time (Grid Snap or Polar Snap), you can select which of the two is on by right-clicking on the word "*SNAP*" on the Status bar (Fig. 1-28). You can also access the *Drafting Settings* dialog box by selecting *Settings...* from the menu.

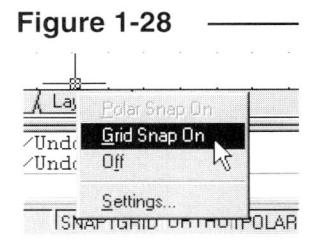

Figure 1-28

GRID (**F7** or **Status Bar**)
A drawing aid called *GRID* can be used to give a visual reference of units of length. The *GRID* default value for English settings is .5 units. The *Grid* command or *Drawing Aids* dialog box allows you to change the interval to any value. The *GRID* is not part of the geometry and is not plotted. Figure 1-26 displays a *GRID* of 1.0. *SNAP* and *GRID* are independent functions—they can be turned on or off independently. However, you can force the *GRID* to have the same interval as *SNAP* by entering a *GRID* value of 0 or you can use a proportion of *SNAP* by entering a *GRID* value followed by an "X".

ORTHO (**F8** or **Status Bar**)
If *ORTHO* is on, lines are forced to an orthogonal alignment (horizontal or vertical) when drawing (Fig. 1-29). *ORTHO* is often helpful since so many drawings are composed mainly of horizontal and vertical lines. *ORTHO* can be turned only on or off.

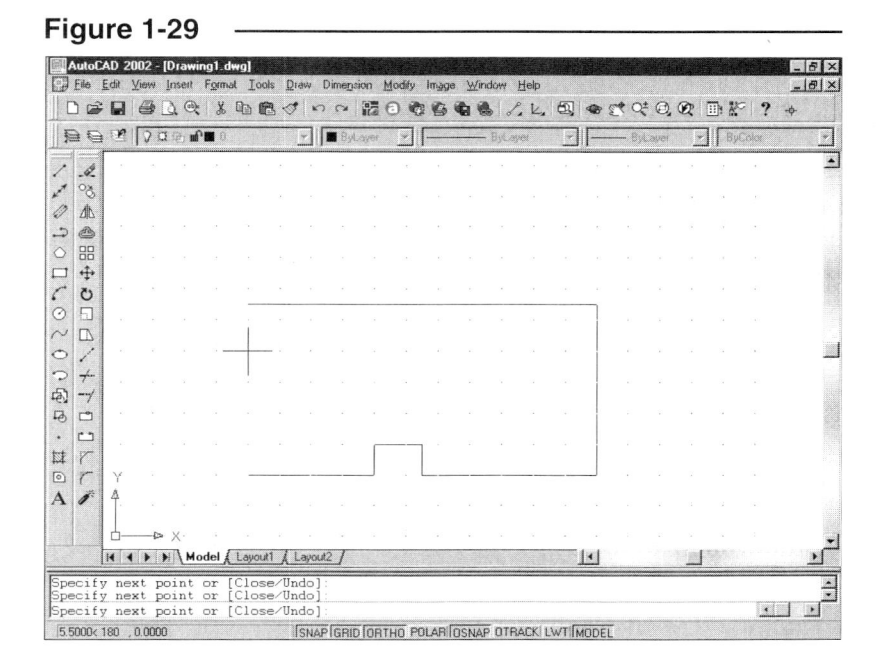

Figure 1-29

AutoCAD Text Window (F2)

Pressing the F2 key activates the *AutoCAD Text Window,* sometimes called the Command History. Here you can see the text activity that occurred at the Command line—kind of an "expanded" Command line. Press F2 again to close the text window.

The *Edit* pull-down menu in this text window provides several options. If you highlight text in the window (Fig. 1-30), you can then *Paste to Cmdline* (Command line), *Copy* it to another program such as a word processor, *Copy History* (entire command history) to another program, or *Paste* text into the window. The *Options* choice invokes the *Options* dialog box (discussed later).

Figure 1-30

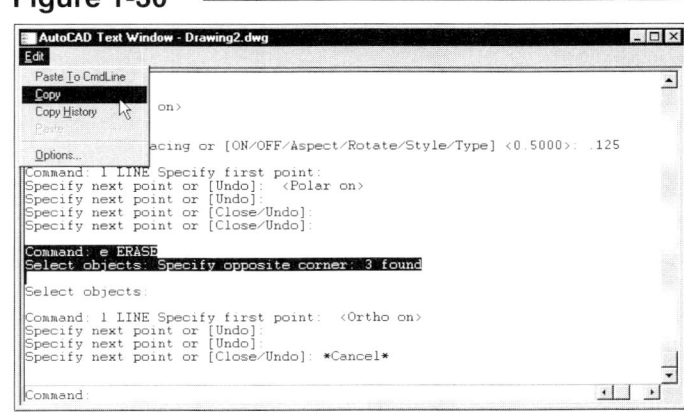

Commands When Multiple Drawings Are Open

In AutoCAD you can have several drawings open at the same time (Fig. 1-31). This feature offers several advantages. Most notably, you can *Copy* and *Paste* from one drawing to another.

You can use the *Syswindows* command to control how you want the drawings displayed in the graphics area. You can *Cascade* or *Tile* them or drag and resize them as you please. If you want each drawing to display as a full screen, use Ctrl+Tab to toggle between drawings.

Figure 1-31

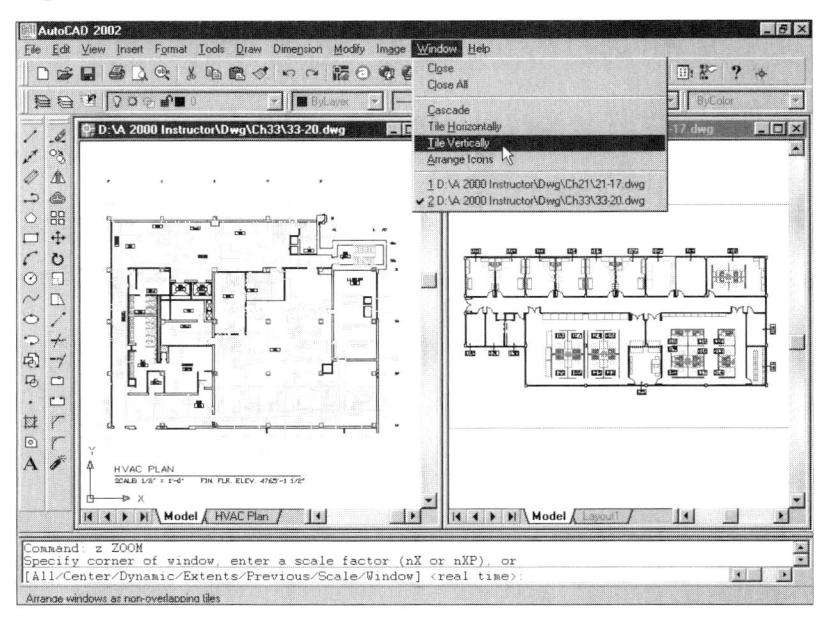

When multiple drawings are open, only one drawing can be active at a time. Any command that is issued (except *Syswindows*) affects only the active drawing. Each drawing has its own command history, so you can go from one drawing to the next and AutoCAD remembers the current command status for each drawing.

Model Tab and *Layout* Tabs

Model Tab

When you create a new drawing, the *Model* tab is the current tab (see Figure 1-32, bottom left). This area is also known as model space. In this area you should create the geometry representing the subject of your drawing, such as a floor plan, a mechanical part, or an electrical schematic. Dimensions are usually created and attached to your objects in model space.

Figure 1-32

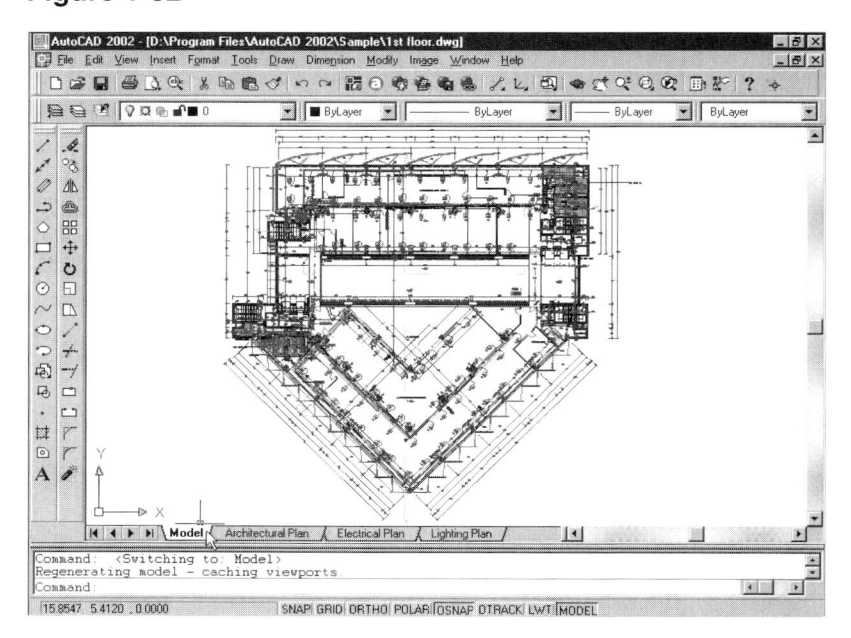

Layout **Tabs**

When you are finished with your drawing, you can plot it directly from the *Model* tab or switch to a *Layout* tab (Fig. 1-33). Layout tabs, sometimes known as paper space, represent sheets of paper that you plot on. You must use several commands to set up the layout to display the geometry and set all the plotting options such as scale, paper size, plot device, and so on. Plotting and layouts are discussed in detail in Chapters 13 and 14.

Figure 1-33

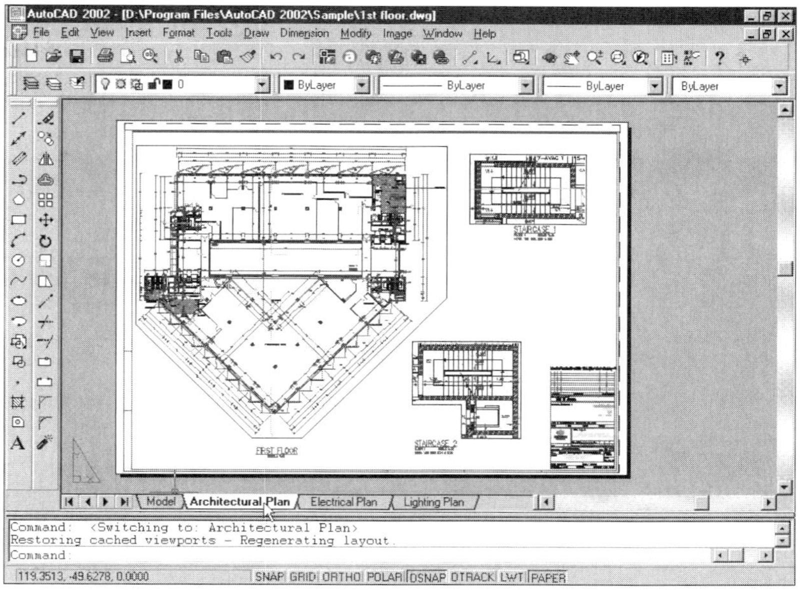

COMMAND ENTRY METHODS PRACTICE

Start AutoCAD. If the *Startup* dialog box or the *Today* window dialog box appears, select *Start from Scratch*, choose *English* as the default setting, and click the *OK* button. Invoke the *Line* command using each of the command entry methods as follows.

1. **Type the command**

STEPS	COMMAND PROMPT	PERFORM ACTION	COMMENTS
1.		press **Escape** if another command is in use	
2.	Command:	type *Line* and press **Enter**	
3.	LINE Specify first point:	**PICK** any point	a "rubberband" line appears
4.	Specify next point or [Undo]:	**PICK** any point	another "rubberband" line appears
5.	Specify next point or [Undo]:	press **Enter**	to complete command

20 Getting Started

2. **Type the command alias**

STEPS	COMMAND PROMPT	PERFORM ACTION	COMMENTS
1.		press **Escape** if another command is in use	
2.	Command:	type *L* and press **Enter**	
3.	LINE Specify first point:	**PICK** any point	a "rubberband" line appears
4.	Specify next point or [Undo]:	**PICK** any point	another "rubberband" line appears
5.	Specify next point or [Undo]:	press **Enter**	to complete command

3. **Pull-down menu**

STEPS	COMMAND PROMPT	PERFORM ACTION	COMMENTS
1.	Command:	select the *Draw* menu from the menu bar on top	
2.	Command:	select *Line*	menu disappears
3.	LINE Specify first point:	**PICK** any point	a "rubberband" line appears
4.	Specify next point or [Undo]:	**PICK** any point	another "rubberband" line appears
5.	Specify next point or [Undo]:	press **Enter**	to complete command

4. **Screen menu** (if activated on your setup)

STEPS	COMMAND PROMPT	PERFORM ACTION	COMMENTS
1.	Command:	select **AutoCAD** from screen menu on the side	only if menu is not at root level
2.	Command:	select *DRAW1* from root screen menu	menu changes to *DRAW 1*
3.	Command:	select *Line*	
4.	LINE Specify first point:	**PICK** any point	a "rubberband" line appears
5.	Specify next point or [Undo]:	**PICK** any point	another "rubberband" line appears
6.	Specify next point or [Undo]:	press **Enter**	to complete command

5. Toolbars

STEPS	COMMAND PROMPT	PERFORM ACTION	COMMENTS
1.	Command:	select the *Line* icon from the Draw toolbar (on the left side of the screen)	the *Line* tool should be located at the top of the toolbar
2.	LINE Specify first point:	**PICK** any point	a "rubberband" line appears
3.	Specify next point or [Undo]:	**PICK** any point	another "rubberband" line appears
4.	Specify next point or [Undo]:	press **Enter**	to complete command

6. **Digitizing tablet menu** (if available)

STEPS	COMMAND PROMPT	PERFORM ACTION	COMMENTS
1.	Command:	select *LINE* on the digitizing tablet menu	located at *10,J*
2.	LINE Specify first point:	**PICK** any point	a "rubberband" line appears
3.	Specify next point or [Undo]:	**PICK** any point	another "rubberband" line appears
4.	Specify next point or [Undo]:	press **Enter**	to complete command

When you are finished practicing, use the *Files* pull-down menu and select *Exit* to exit AutoCAD. You do not have to "Save Changes."

CUSTOMIZING THE AutoCAD SCREEN

Toolbars

Many toolbars are available, each with a group of related commands for specialized functions. For example, when you are ready to dimension a drawing, you can activate the Dimension toolbar for efficiency. Right-clicking on any toolbar displays a list of all toolbars (Fig. 1-34). Selecting any toolbar name makes that toolbar appear on the screen. You can also select *Toolbars...* from the *View* pull-down menu or type the command *Toolbar* to display the *Customize* dialog box. Toolbars can be removed from the screen by clicking once on the "X" symbol in the upper right of a floating toolbar.

Figure 1-34

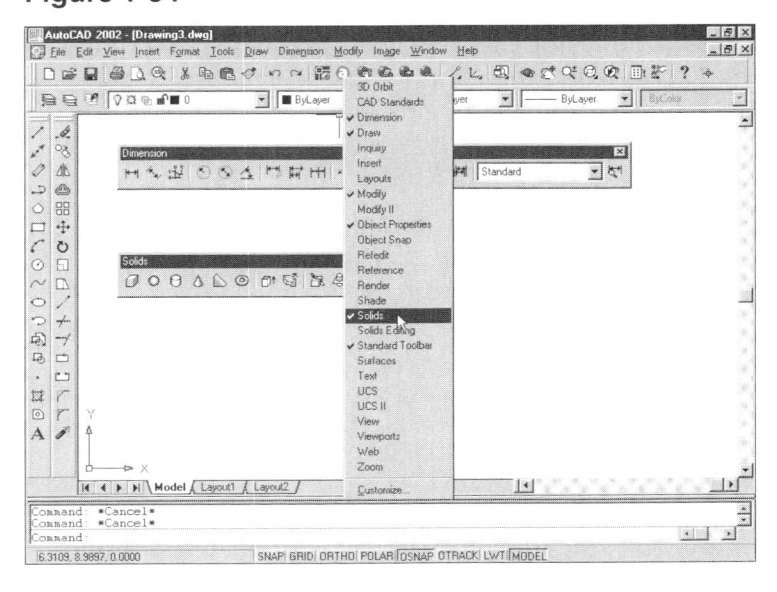

22 Getting Started

By default, the Object Properties and Standard toolbars (top) and the Draw and Modify toolbars (side) are docked, whereas toolbars that are newly activated are floating (see Fig. 1-34). A floating toolbar can be easily moved to any location on the screen if it obstructs an important area of a drawing. Placing the pointer in the title background allows you to move the toolbar by holding down the left button and dragging it to a new location (Fig. 1-35). Floating toolbars can also be resized by placing the pointer on the narrow border until a two-way arrow appears, then dragging left, right, up, or down (Fig 1-36).

Figure 1-35

Figure 1-36

A floating toolbar can be docked against any border (right, left, top, bottom) by dragging it to the desired location (Fig. 1-37). Several toolbars can be stacked in a docked position. By the same method, docked toolbars can be dragged back onto the graphics area. The Object Properties and Standard toolbars can be moved onto the graphics area or docked on another border, although it is wise to keep these toolbars in their standard position. The Object Properties toolbar only displays its full options when located in a horizontal position.

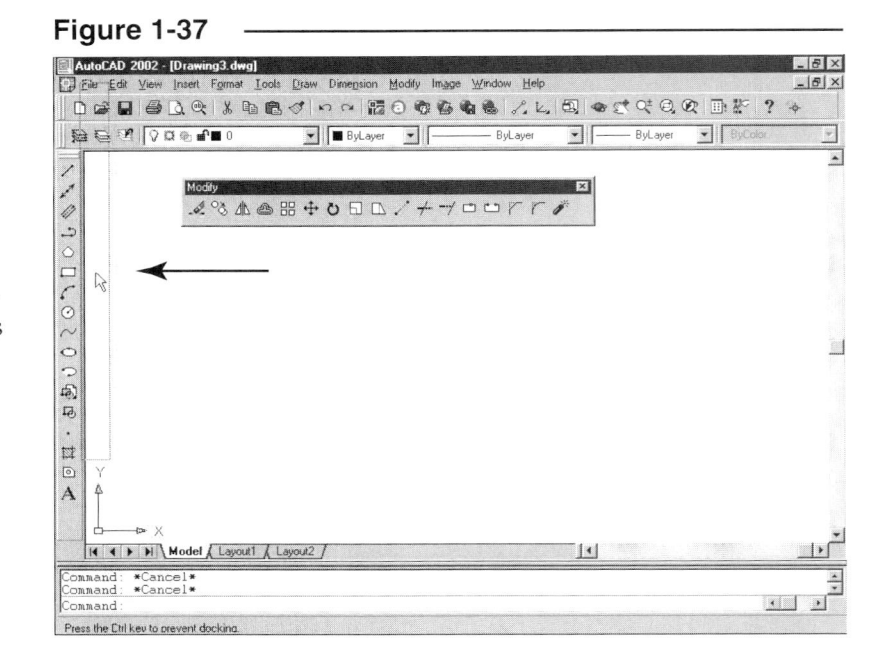

Figure 1-37

Holding down the **Ctrl** key while dragging a toolbar near a border of the drawing window prevents the toolbar from docking. This feature enables you to float a toolbar anywhere on the screen. New toolbars can be created and existing toolbars can be customized. Typically, you would create toolbars to include groups of related commands that you use most frequently or need for special activities. See AutoCAD 2002 *Help*, Customization Guide.

Command Line

The Command line, normally located near the bottom of the screen, can be resized to display more or fewer lines of text. Moving the pointer to the border between the graphics screen and the Command line until two-way arrows appear allows you to slide the border up or down. A <u>minimum of two lines</u> of text is recommended. The Command line text window can also be moved to any location on the screen by pointing to the border, holding down the left button, and dragging to the new position (Fig. 1-38).

Figure 1-38

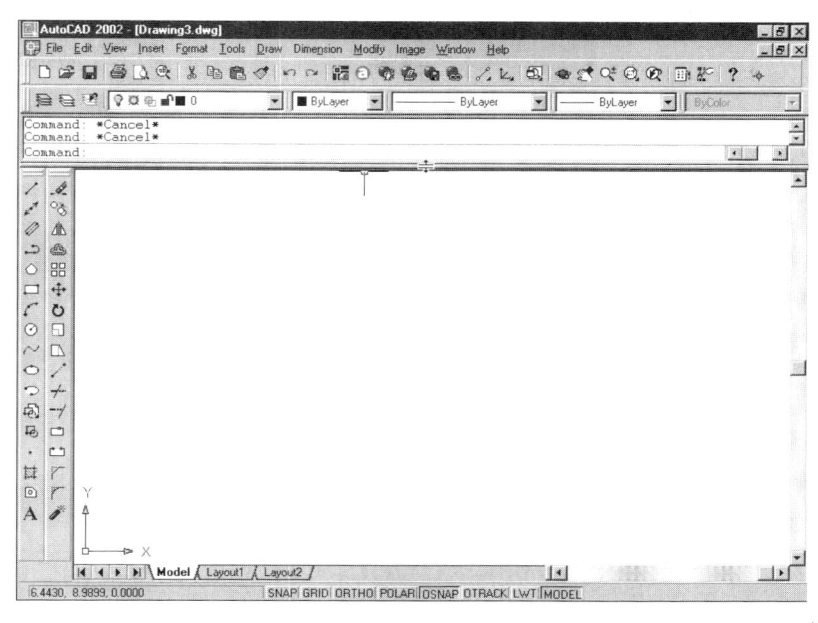

OPTIONS

Fonts, colors, and other features of the AutoCAD drawing editor can be customized to your liking by using the *Options* command and selecting the *Display* tab. Typing *Options* or selecting *Options…* from the *Tools* pull-down menu or default shortcut menu activates the dialog box shown in Figure 1-39. The screen menu can also be activated through this dialog box (second checkbox in the dialog box).

Figure 1-39

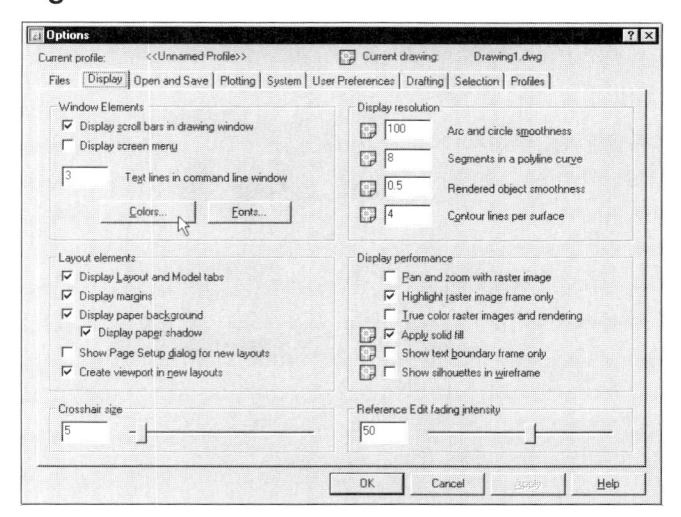

Selecting the *Color…* tile provides a dialog box for customizing the screen colors (Fig. 1-40).

All changes made to the Windows screen by any of the options discussed in this section are automatically saved for the next drawing session. The changes are saved as the current profile (in the system registry). However, if you are working in a school laboratory, it is likely that the computer systems are set up to present the same screen defaults each time you start AutoCAD.

Figure 1-40

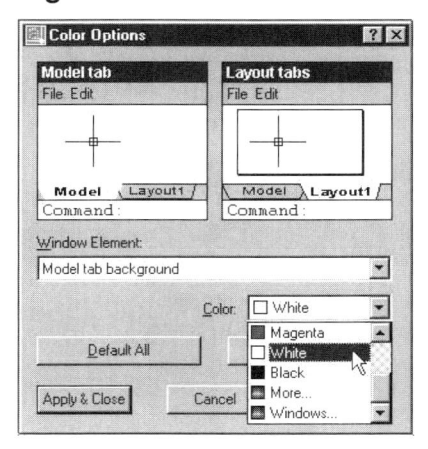

CHAPTER EXERCISES

1. **Starting and Exiting AutoCAD**

   Start AutoCAD by double-clicking the "AutoCAD 2002" shortcut icon or selecting "AutoCAD 2002" from the Programs menu. If the *Startup* dialog box or the *Today* window appears, select *Start from Scratch*, choose *English* as the default setting, and click the *OK* button. Draw a *Line*. Exit AutoCAD by selecting the *Exit* option from the *Files* pull-down menu. Answer *No* to the "Save changes to Drawing1.dwg?" prompt. Repeat these steps until you are confident with the procedure.

2. **Using Drawing Aids**

   Start AutoCAD. Turn on and off each of the following modes:

   SNAP, GRID, ORTHO, POLAR

3. **Understanding Coordinates**

   Begin drawing a *Line* by PICKing a "Specify first point:". Toggle *Coords* to display each of the <u>three</u> formats. PICK several other points at the "Specify next point or [Undo]:" prompt. Pay particular attention to the coordinate values displayed for each point and visualize the relationship between that point and coordinate 0,0 (absolute value) or the last point established (relative polar value). Finish the command by pressing Enter.

4. **Using *Flipscreen***

   Use *Flipscreen* (**F2**) to toggle between the text window and the graphics screen.

5. **Drawing with Drawing Aids**

   Draw four *Lines* using <u>each</u> Drawing Aid: **GRID, SNAP, ORTHO, POLAR**. Toggle on and off each of the drawing aids one at a time for each set of four *Lines*. Next, draw *Lines* using combinations of the Drawing Aids, particularly **GRID + SNAP** and **GRID + SNAP + POLAR**.

2

WORKING WITH FILES

Chapter Objectives

After completing this chapter you should be able to:

1. name drawing files;
2. use file-related dialog boxes;
3. use the Windows right-click shortcut menus in file dialog boxes;
4. create *New* drawings;
5. *Open* and *Close* existing drawings;
6. *Save* drawings to disk;
7. use *SaveAs* to save a drawing under a different name, path, and/or format;
8. practice good file management techniques.

AutoCAD DRAWING FILES

Naming Drawing Files

What is a drawing file? A CAD drawing file is the electronically stored data form of a drawing. The computer's hard disk is the principal magnetic storage device used for saving and restoring CAD drawing files. Diskettes, compact disks (CDs), networks, and the Internet are used to transport files from one computer to another, as in the case of transferring CAD files among clients, consultants, or vendors in industry. The AutoCAD commands used for saving drawings to and restoring drawings from files are explained in this chapter.

An AutoCAD drawing file has a name that you assign and a file extension of ".DWG." An example of an AutoCAD drawing file is:

PART-024.DWG
file name   extension

The file name you assign must be compliant with the Windows file name conventions; that is, it can have a maximum of 256 alphanumeric characters. File names and directory names (folders) can be in UPPER-CASE, Title Case, or lowercase letters. Characters such as _ - $ # ( ) ^ and spaces can be used in names, but other characters such as \ / : * ? < > | are not allowed. AutoCAD automatically appends the extension of .DWG to all AutoCAD-created drawing files.

NOTE: The chapter exercises and other examples in this book generally list file names in UPPERCASE letters for easy recognition. The file names and directory (folder) names on your system may appear as UPPERCASE, Title Case, or lowercase.

Beginning and Saving an AutoCAD Drawing

When you start AutoCAD, the drawing editor appears and allows you to begin drawing even before using any file commands. As you draw, you should develop the habit of saving the drawing periodically (about every 15 or 20 minutes) using *Save*. *Save* stores the drawing in its most current state to disk.

The typical drawing session would involve using *New* to begin a new drawing or using *Open* to open an existing drawing. Alternately, the *Startup* dialog box or *Today* window that appears when starting AutoCAD would be used to begin a new drawing or open an existing one. *Save* would be used periodically, and *Exit* would be used for the final save and to end the session.

Figure 2-1

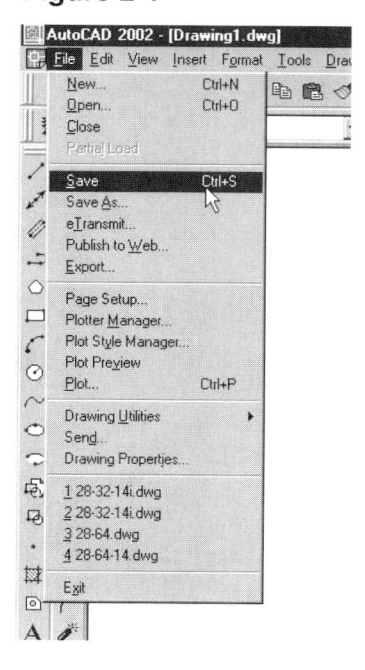

Accessing File Commands

Proper use of the file-related commands covered in this chapter allows you to manage your AutoCAD drawing files in a safe and efficient manner. Although the file-related commands can be invoked by several methods, they are easily accessible via the first pull-down menu option, *File* (Fig. 2-1). Most of the selections from this pull-down menu invoke dialog boxes for selection or specification of file names.

The Standard toolbar at the top of the AutoCAD screen has tools (icon buttons) for *New, Open,* and *Save*. File commands can also be entered at the keyboard by typing the formal command name, the command alias, or using the Ctrl keys sequences.

File Navigation Dialog Box Functions

There are many dialog boxes appearing in AutoCAD that help you manage files. All of these dialog boxes operate in a similar manner. A few guidelines will help you use them. The *Save Drawing As* dialog box for Windows (Fig. 2-2) is used as an example.

- The desired file can be selected by PICKing it, then PICKing *OK*. Double-clicking on the selection accomplishes the same action. File names can also be typed in the *File name:* edit box.
- Every file name has an extension (three letters following the period) called the type. File types can be selected from the *Files of Type:* section of the dialog boxes, or the desired file extension can be entered in the *File name:* edit box.
- The current folder (directory) is listed in the drop-down box near the top of the dialog box. You can select another folder (directory) or drive by using the drop-down list displaying the current path, by selecting the *Back* arrow, or by selecting the *Up One Level* icon to the right of the list. (Rest your pointer on an icon momentarily to make the tool tip appear.)
- Selecting one of the folders displayed on the left side of the file dialog boxes allows you to navigate to the following locations. The files or locations found appear in the list in the central area.
  History: Lists a history of all files you have opened from most to least recent.
  Personal: Lists all files and folders saved in the Windows/Profiles/Personal folder.
  Favorites: Lists all files and folders saved in the Windows/Profiles/Favorites folder.
  PointA, *Buzzsaw*, *RedSpark*: Autodesk services.
  FTP: Allows you to browse FTP locations you may have saved.
  Desktop: Shows files on your desktop.
- You can use the *Search the Web* icon to display the *Browse the Web* dialog box (not shown). The default site is http://www.autodesk.com/.
- Any highlighted file(s) can be deleted using the "X" (*Delete*) button.
- A new folder (subdirectory) can be created within the current folder (directory) by selecting the *New Folder* icon.
- The *View* drop-down list allows you to toggle the listing of files to a *List* or to show *Details*. The *List* option displays only file names (see Figure 2-2), whereas the *Details* option gives file-related information such as file size, file type, and time and date last modified (see Figure 2-3). The detailed list can also be sorted alphabetically by name, by file size, alphabetically by file type, or chronologically by time and date. Do this by clicking the *Name*, *Type*, *Size*, or *Modified* tiles immediately above the list. Double-clicking on one of the tiles reverses the order of the list.

Figure 2-2

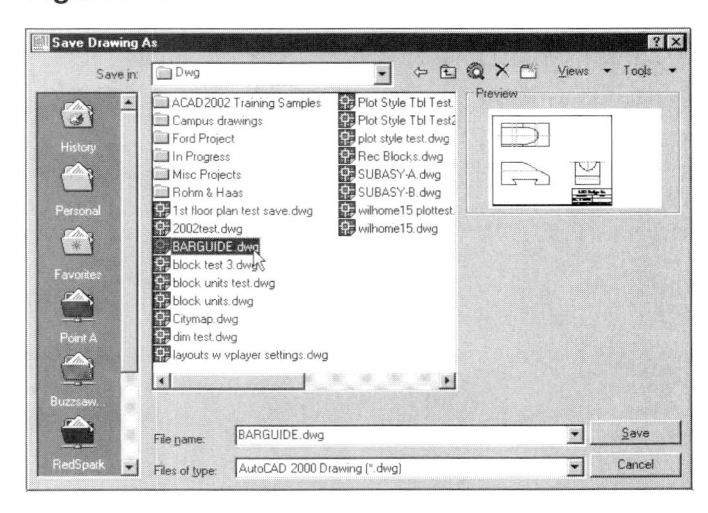

Figure 2-3

Windows Right-Click Shortcut Menus

AutoCAD utilizes the right-click shortcut menus that operate with the Windows operating systems. Activate the shortcut menus by pressing the right mouse button (right-clicking) inside the file list area of a dialog box. There are two menus that give you additional file management capabilities.

Right-Click, No Files Highlighted

When no files are highlighted, right-clicking produces a menu for file list display and file management options (Fig. 2-3). The menu choices are as follows:

View Shows *Large Icons* or *Small Icons* next to the file names, displays a *List* of file names only (see Fig. 2-2), or displays all file *Details* (see Fig. 2-3).
Arrange Icons Sorts the files *by Name, by Type* (file extension), *by Size,* or *by Date*.
Paste If the *Copy* or *Cut* function was previously used (see "Right-Click, File Highlighted," Fig. 2-4), *Paste* can be used to place the copied file into the displayed folder.
Paste Shortcut Use this option to place a *Shortcut* (to open a file) into the current folder.
Undo Rename If the *Rename* function was previously used (see "Right-Click, File Highlighted," Fig. 2-4), this action restores the original file name.
New Creates a new *Folder, Shortcut,* or document.
Properties Displays a dialog box listing properties of the current folder.

Right-Click, File Highlighted

When a file is highlighted, right-clicking produces a menu with options for the selected file (Fig. 2-4). The menu choices are as follows:

Figure 2-4

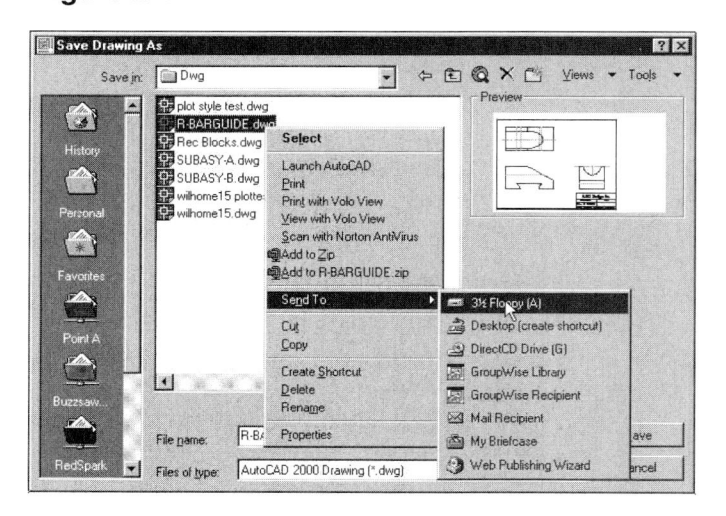

Select Processes the file (like selecting the *OK* button) according to the dialog box function. For example, if the *SaveAs* dialog box is open, *Select* saves the highlighted file; or if the *Select File* dialog box is active, *Select* opens the drawing.
Launch Opens the application (AutoCAD or other program) and loads the selected file. Since AutoCAD can have multiple drawings open, you can select several drawings to open with this feature.
Print Sends the selected file to the configured system printer.
Send To Copies the selected file to the selected device. This is an easy way to copy a file from your hard drive to a diskette in A: drive.
Cut In conjunction with *Paste,* allows you to move a file from one location to another.
Copy In conjunction with *Paste,* allows you to copy the selected file to another location. You can copy the file to the same folder, but Windows renames the file to "Copy of . . .".
Create Shortcut Use this option to create a *Shortcut* (to open the highlighted file) in the current folder. The shortcut can be moved to another location using drag and drop.
Delete Sends the selected file to the Recycle Bin.
Rename Allows you to rename the selected file. Move your cursor to the highlighted file name, click near the letters you want to change, then type or use the backspace, delete, space, or arrow keys.
Properties Displays a dialog box listing properties of the selected file.

AutoCAD FILE COMMANDS

When you start AutoCAD, one of three situations occurs based on how the *Options* are set on your system:

1. The *Today* window appears.
2. The *Startup* dialog box appears.
3. No dialog box or window appears and the session starts with the ACAD.DWT or the ACADISO.DWT template (determined by the *MEASUREINIT* system variable setting, 0 or 1, respectively).

To control which of these options occurs, use the *Options* command to produce the *Options* dialog box, select the *System* tab, and make the desired selection from the *Startup* drop-down list (Fig. 2-5).

Figure 2-5

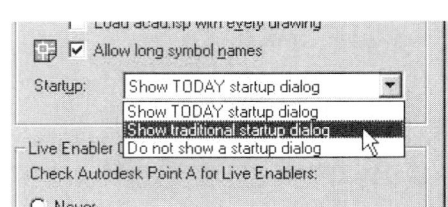

Both the *Startup* dialog box and the *Today* window allow you to create new drawings and open existing drawings. These options are the same as using the *New* and *Open* commands. Therefore, a description of the *New* and *Open* commands explains these functions in the *Startup* dialog box and the *Today* window.

NEW

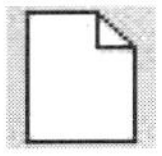

Pull-down Menu	COMMAND (TYPE)	ALIAS (TYPE)	Short-cut	Screen (side) Menu	Tablet Menu
File New...	NEW	...	Ctrl+N	FILE New	T,24

Based on the settings for your system (see previous explanation), the *New* command produces the *Today* window (Fig. 2-6), produces the *Create New Drawing* dialog box (Fig. 2-7, on the next page, identical to the *Startup* dialog box shown in Figure 1-12), or prompts for a template file to use. In both the *Today* window and the *Create New Drawing* dialog box, the options for creating a new drawing are the same: *Start from Scratch*, *Use a Template*, and use a *Wizard*. If your system is set for "Do not show a startup dialog," only the template option is available.

Figure 2-6

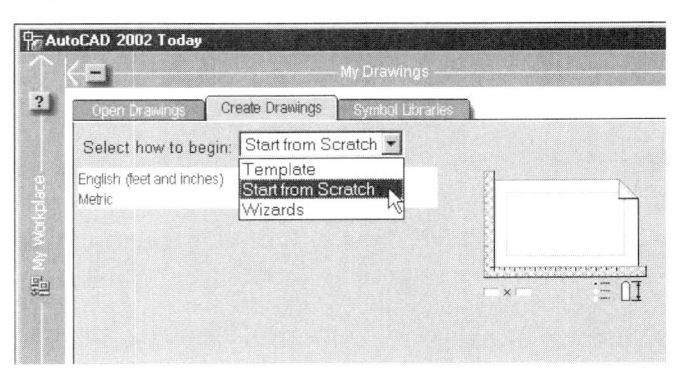

Start from Scratch
In the *Today* window, select the *Create Drawings* tab and the *Start from Scratch* option (see Figure 2-6). In the *Create New Drawing* dialog box, select the *Start from Scratch* option (see Figure 2-7). Next, choose from either the *English (feet and inches)* or *Metric* default settings. Selecting *English* uses the default ACAD.DWT template drawing with a drawing area (called *Limits*) of 12 x 9 units. Choosing *Metric* uses the ACADISO.DWT template drawing, which has *Limits* settings of 420 x 297. For more details on templates, see Chapter 12.

Figure 2-7

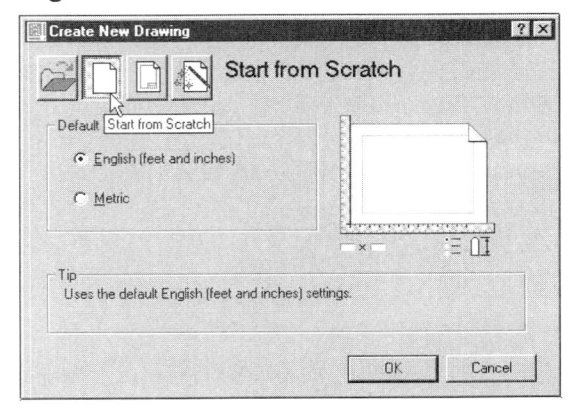

30 Working With Files

Use a Wizard
In the *Today* window, select the *Create Drawings* tab and the *Wizards* option (see Figure 2-6). In the *Create New Drawing* dialog box, select the *Use a Wizard* option (Fig. 2-8). Next, choose between a *Quick Setup Wizard* and an *Advanced Setup Wizard*. The *Quick Setup Wizard* prompts you to select the type of drawing *Units* you want to use and to specify the drawing *Area* (*Limits*). The *Advanced Setup* offers options for units, angular direction and measurement, and area. The *Quick Setup Wizard* and the *Advanced Setup Wizard* are discussed in Chapter 6, Basic Drawing Setup.

Figure 2-8

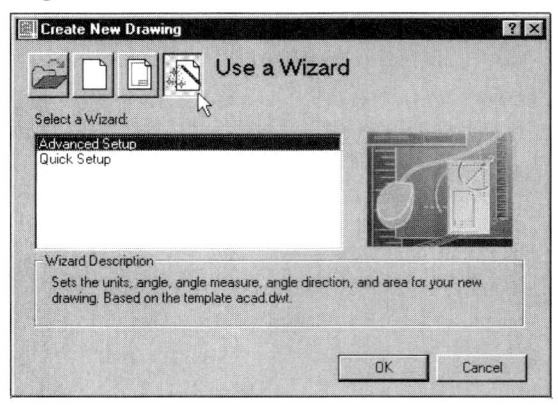

Use a Template
Use this option if you want to create a new drawing based on an existing template drawing. A template drawing is one that may have some of the setup steps performed but contains no geometry (graphical objects). In the *Today* window, select the *Create Drawings* tab and the *Template* option (Fig. 2-9). In the *Create New Drawing* dialog box, select the *Use a Template* option (Fig. 2-10). If your system is set for "Do not show a startup dialog," you are automatically prompted to use either the ACAD.DWT or ACADISO.DWT template, based on your *MEASUREINIT* setting.

Figure 2-9

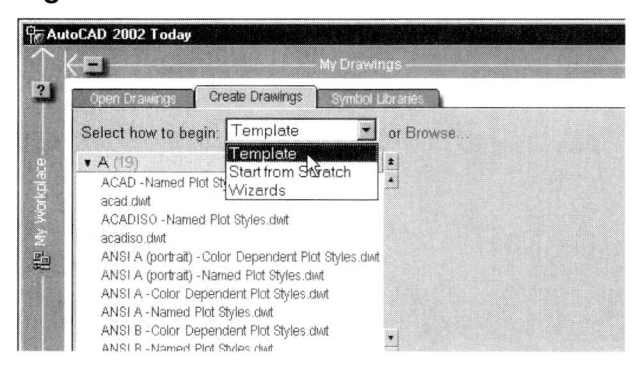

Selecting the ACAD.DWT template begins a new drawing using the English default settings with a drawing area of 12 x 9 units. Selecting the ACADISO.DWT template begins a new drawing using the metric settings with a drawing area of 420 x 279 units.

Figure 2-10

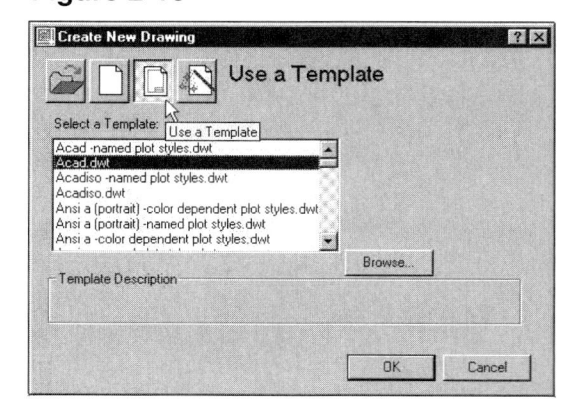

For the purposes of learning AutoCAD starting with the basic principles and commands discussed in this text, it is helpful to use the English settings when beginning a new drawing. Until you read Chapters 6 and 12, you can begin new drawings for completing the exercises and practicing the examples in Chapters 1 through 5 by any of the following methods.

Select *English (feet and inches)* in the *Start from Scratch* option using any method.
Select the ACAD.DWT template drawing using any method.
Set the *MEASURINIT* system variable to 0, select *Do not show a startup dialog* (see Figure 2-5), then use the *New* command.
Set the *MEASURINIT* system variable to 0, then cancel the *Startup* dialog box or *Today* window when AutoCAD starts.

OPEN

Pull-down Menu	COMMAND (TYPE)	ALIAS (TYPE)	Short-cut	Screen (side) Menu	Tablet Menu
File Open...	OPEN	...	Ctrl+O	FILE Open	T,25

Use *Open* to select an existing drawing to be loaded into AutoCAD. Normally you would open an existing drawing (one that is completed or partially completed) so you can continue drawing or to make a print or plot. You can *Open* multiple drawings at one time in AutoCAD 2000 or later versions.

Figure 2-11

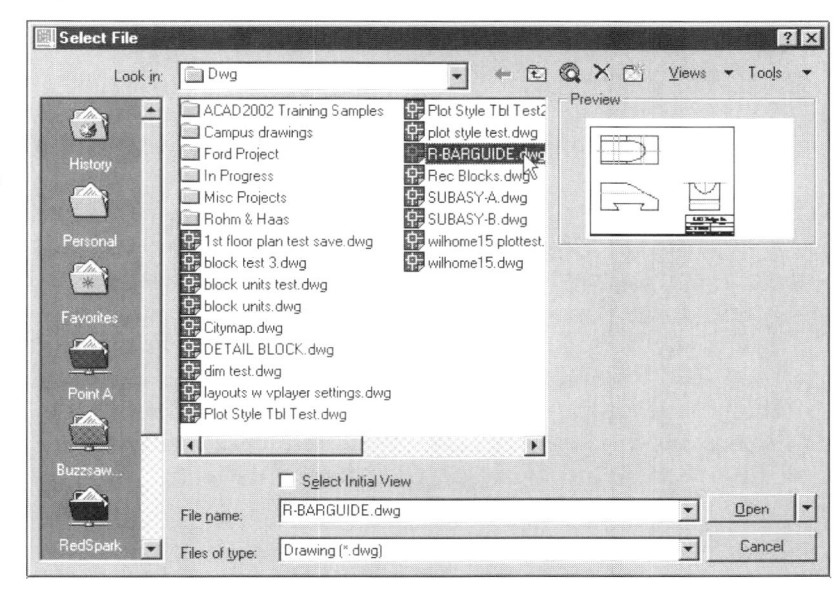

The *Open* command produces the *Select File* dialog box (Fig. 2-11). In this dialog box you can select any drawing from the current directory list. PICKing a drawing name from the list displays a small bitmap image of the drawing in the *Preview* tile. Select the *Open* button or double-click on the file name to open the highlighted drawing. You could instead type the file name (and path) of the desired drawing in the *File name:* edit box and press Enter, but a preview of the typed entry will not appear.

Using the *Open* drop-down list (lower right) you can select from these options:

Open Opens a drawing file and allows you to edit the file.
Open Read-Only Opens a drawing file for viewing, but you cannot edit the file.
Partial Open Allows you to open only a part of a drawing for editing. See *Partialopen*.
Partial Open Read-Only Allows you to open only a part of a drawing for viewing, but you cannot edit the drawing.

Other options in this dialog box are similar to the *Save Drawing As* dialog box described earlier in "File Navigation Dialog Box Functions." For example, to locate a drawing in another folder (directory) or on another drive on your computer or network, select the drop-down list on top of the dialog box next to *Look in:*, or use the *Up one level* button. Remember that you can also display the file details (size, type, modified) by toggling the *Details* option from the *Views* drop-down list. You can open a .DWG (drawing), .DWS (drawing standards), .DXF (drawing interchange format), or .DWT (drawing template) file by selecting from the *Files of type:* drop-down list.

NOTE: It is considered poor practice to *Open* a drawing from a diskette in A: drive. Normally, drawings should be copied to a directory on a fixed (hard) drive, then opened from that directory. Opening a drawing from a diskette and performing saves to a diskette are much slower and less reliable than operating from a hard drive. In addition, some temporary files may be written to the drive or directory from which the file is opened. See other "NOTE"s at the end of the discussion on *Save* and *Saveas*.

32 Working With Files

SAVE

Pull-down Menu	COMMAND (TYPE)	ALIAS (TYPE)	Short-cut	Screen (side) Menu	Tablet Menu
File Save	SAVE	...	Ctrl+S	...	U,24-U,25

The *Save* command is intended to be used periodically during a drawing session (every 15 to 20 minutes is recommended). When *Save* is selected from a menu, the current version of the drawing is saved to disk without interruption of the drawing session. The first time an unnamed drawing is saved, the *Save Drawing As* dialog box (see Figures 2-2 and 2-3) appears, which prompts you for a drawing name. Typically, however, the drawing already has an assigned name, in which case *Save* actually performs a *Qsave* (quick save). A *Qsave* gives no prompts or options, nor does it display a dialog box.

When the file has an assigned name, using *Save* by selecting the command from a menu or icon button automatically performs a quick save (*Qsave*). Therefore, *Qsave* automatically saves the drawing in the same drive and directory from which it was opened or where it was first saved. In contrast, typing *Save* always produces the *Save Drawing As* dialog box, where you can enter a new name and/or path to save the drawing or press Enter to keep the same name and path (see *SaveAs*).

NOTE: If you want to save a drawing directly to a diskette in A:, first *Save* the drawing to the hard drive, then use the *Send To* option in the right-click menu in the *Save*, *Save Drawing As*, or *Select File* dialog box (see "Windows Right-Click Shortcut Menus" and Figure 2-4).

QSAVE

Pull-down Menu	COMMAND (TYPE)	ALIAS (TYPE)	Short-cut	Screen (side) Menu	Tablet Menu
...	QSAVE	...	Ctrl+S	FILE Qsave	...

Qsave (quick save) is normally invoked automatically when *Save* is used (see *Save*) but can also be typed. *Qsave* saves the drawing under the previously assigned file name. No dialog boxes appear nor are any other inputs required. This is the same as using *Save* (from a menu), assuming the drawing name has been assigned. However, if the drawing has not been named when *Qsave* is invoked, the *Save Drawing As* dialog box appears.

SAVEAS

Pull-down Menu	COMMAND (TYPE)	ALIAS (TYPE)	Short-cut	Screen (side) Menu	Tablet Menu
File Save As...	SAVEAS	...	...	FILE Saveas	V,24

The *SaveAs* command can fulfill four functions:

1. save the drawing file under a new name if desired;
2. save the drawing file to a new path (drive and directory location) if desired;
3. in the case of either 1 or 2, assign the new file name and/or path to the current drawing (change the name and path of the current drawing);
4. save the drawing in a format other than the default AutoCAD 2000 format or as a different file type (.DWT, .DWS, or .DXF).

Therefore, assuming a name has previously been assigned, *SaveAs* allows you to save the current drawing under a different name and/or path; but, beware, SaveAs sets the current drawing name and/or path to the last one entered. This dialog box is shown in Figures 2-2 and 2-3.

SaveAs can be a benefit when creating two similar drawings. A typical scenario follows. A design engineer wants to make two similar but slightly different design drawings. During construction of the first drawing, the engineer periodically saves under the name DESIGN1 using *Save*. The first drawing is then completed and *Saved*. Instead of starting a *New* drawing, *SaveAs* is used to save the current drawing under the name DESIGN2. *SaveAs* also resets the current drawing name to DESIGN2. The designer then has two separate but identical drawing files on disk which can be further edited to complete the specialized differences. The engineer continues to work on the current drawing DESIGN2.

NOTE: If you want to save the drawing to a diskette in A: drive, do not use *SaveAs*. Invoking *SaveAs* by any method resets the drawing name and path to whatever is entered in the *Save Drawing As* dialog box, so entering A:NAME would set A: as the current drive. This could cause problems because of the speed and reliability of a diskette as opposed to a hard drive. Instead, you should save the drawing to the hard drive (usually C:), then close the drawing (by using *Close*). Next, use a right-click shortcut menu to copy the drawing file to A:. (See "Windows Right-Click Shortcut Menus" and Figure 2-4.)

You can save an AutoCAD 2002 drawing in several formats other than the default format (*AutoCAD 2000 Drawing *.dwg*). Use the drop-down list at the bottom of the *Save* (or *Save Drawing As*) dialog box (Fig. 2-12) to save the current drawing as an earlier version drawing (.DWG) of AutoCAD or LT, a template file (.DWT), or a .DWS or a .DXF file. In AutoCAD 2002, a drawing can be saved to an earlier release and later *Opened* in AutoCAD 2002 and all new features are retained during the "round trip."

Figure 2-12

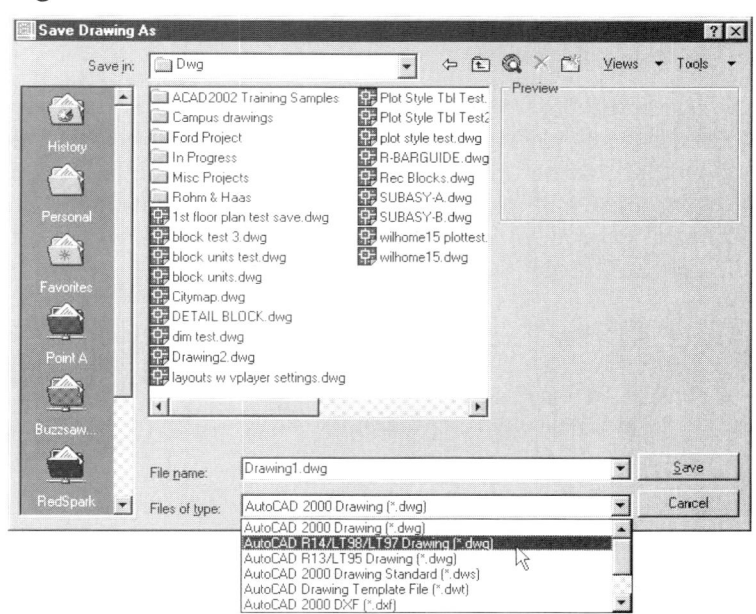

CLOSE

Pull-down Menu	COMMAND (TYPE)	ALIAS (TYPE)	Short-cut	Screen (side) Menu	Tablet Menu
File Close	CLOSE	...	...	...	...

Use the *Close* command to close the current drawing. Because AutoCAD allows you to have several drawings *Open*, *Close* gives you control to close one drawing while leaving others open. If the drawing has been changed but not saved, AutoCAD prompts you to save or discard the changes. In this case, a warning box appears (Fig. 2-13). *Yes* causes AutoCAD to *Save* then close the drawing; *No* closes the drawing without saving; and *Cancel* aborts the close operation so the drawing stays open.

Figure 2-13

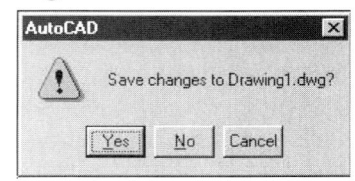

CLOSEALL

Pull-down Menu	COMMAND (TYPE)	ALIAS (TYPE)	Short-cut	Screen (side) Menu	Tablet Menu
Window Closeall	CLOSEALL	...	...	...	...

The *Closeall* command closes all drawings currently open in your AutoCAD session. If any of the drawings have been changed but not saved, you are prompted to save or discard the changes for each drawing.

EXIT

Pull-down Menu	COMMAND (TYPE)	ALIAS (TYPE)	Short-cut	Screen (side) Menu	Tablet Menu
File Exit	EXIT	...	...	...	Y,25

This is the simplest method to use when you want to exit AutoCAD. If any changes have been made to the drawings since the last *Save, Exit* invokes a warning box asking if you want to *Save changes to . . .?* before ending AutoCAD (see Figure 2-14).

Figure 2-14

An alternative to using the *Exit* option is to use the standard Windows methods for exiting an application. The two options are (1) select the **"X"** in the extreme upper-right corner of the AutoCAD window, or (2) select the AutoCAD logo in the extreme upper-left corner of the AutoCAD window. Selecting the logo in the upper-left corner produces a pull-down menu allowing you to *Minimize, Maximize,* etc., or to *Close* the window (Fig. 2-14). Using this option is the same as using *Exit*.

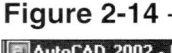

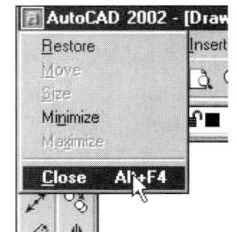

AutoCAD Backup Files

Figure 2-15

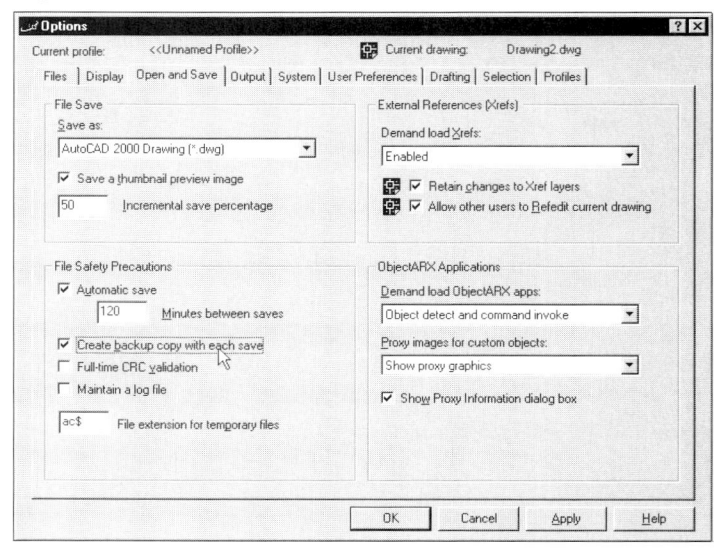

When a drawing is saved, AutoCAD creates a file with a .DWG extension. For example, if you name the drawing PART1, using *Save* creates a file named PART1.DWG. The next time you save, AutoCAD makes a new PART1.DWG and renames the old version to PART1.BAK. One .BAK (backup) file is kept automatically by AutoCAD by default. You can disable the automatic backup function by changing the *ISAVEBAK* system variable to 0 or by accessing the *Open and Save* tab in the *Options* dialog box (Fig. 2-15).

You cannot *Open* a .BAK file. It must be renamed to a .DWG file. Remember that you already have a .DWG file by the same name, so rename the extension and the filename. For example, PART1.BAK could be renamed to PART1OLD.DWG. Use the Windows Explorer or the *Select File* dialog box (use *Open*) with the right-click options to rename the file.

The .BAK files can also be deleted without affecting the .DWG files. The .BAK files accumulate after time, so you may want to periodically delete the unneeded ones to conserve disk space.

AutoCAD Drawing File Management

AutoCAD drawing files should be stored in folders (directories) used exclusively for that purpose. For example, you may use Windows Explorer to create a new folder called "Drawings" or "DWG." This could be a folder in the main C: or D: drive, or could be a subdirectory such as "C:\My Documents\ AutoCAD Drawings." You can create folders for different types of drawings or different projects. For example, you may have a folder named "DWG" with several subdirectories for each project, drawing type, or client, such as "COMPONENTS" and "ASSEMBLIES," or "In Progress" and "Completed," or "TWA," "Ford," and "Dupont." If you are working in an office or school laboratory, a directory structure most likely already has been created for you to use.

It is considered poor practice to create drawings in folders where the AutoCAD system files are kept. Beware, if you install AutoCAD and accept all defaults, when you first use *Save*, the files are saved in the "C:\Program Files\AutoCAD 2002" directory.

Safety and organization are important considerations for storage of AutoCAD drawing files. AutoCAD drawings should be saved to designated folders to prevent accidental deletion of system or other important files. If you work with AutoCAD long enough or if you work in an office or laboratory, most likely many drawings are saved on the computer hard drive or network drives. It is imperative that a logical directory structure be maintained so important drawings can be located easily and saved safely.

CHAPTER EXERCISES

Start AutoCAD. If the *Startup* dialog box or *Today* window appears, select *Start from Scratch*, then use the *English* default settings. NOTE: The chapter exercises in this book list file names in UPPERCASE letters for easy recognition. The file names and directory (folder) names on your system may appear as UPPERCASE, Title Case, or lower case.

1. **Create or determine the name of the folder ("working" directory) for opening and saving AutoCAD files on your computer system**

    If you are working in an office or laboratory, a folder (directory) has most likely been created for saving your files. If you have installed AutoCAD yourself and are learning on your home or office system, you should create a folder for saving AutoCAD drawings. It should have a name like "C:\ACAD\DWG," "C:\My Documents\Dwgs," or "C:\Acad\Drawing Files." (HINT: Use the *SaveAs* command. The name of the folder last used for saving files appears at the top of the *Save Drawing As* dialog box.)

2. *Save* **and name a drawing file**

    Draw 2 vertical *Lines*. Select *Save* from the *File* pull-down, from the Standard toolbar, or by another method. (The *Save Drawing As* dialog box appears since a name has not yet been assigned.) Name the drawing **"CH2 VERTICAL."**

3. **Using** *Qsave*

    Draw 2 more vertical *Lines*. Select *Save* again from the menu or by any other method except typing. (Notice that the *Qsave* command appears at the Command line since the drawing has already been named.) Draw 2 more vertical *Lines* (a total of six lines now). Select *Save* again. Do not *Close*.

4. **Start a *New* drawing**

   Invoke *New* from the *File* pull-down, the Standard toolbar, or by any other method. If the *Startup* dialog box or *Today* window appears, select **Start from Scratch**, then use the *English* default settings. Draw 2 horizontal *Lines*. Use *Save*. Enter **"CH2 HORIZONTAL"** as the name for the drawing. Draw 2 more horizontal *Lines*, but do not *Save*. Continue to exercise 5.

5. ***Close* the current drawing**

   Use *Close* to close CH2 HORIZONTAL. Notice that AutoCAD first forces you to answer *Yes* or *No* to *Save changes to CH2 HORIZONTAL.DWG?* PICK *Yes* to save the changes.

6. **Using *SaveAs***

   The CH2 VERTICAL drawing should now be the current drawing. Draw 2 inclined (angled) *Lines* in the CH2 VERTICAL DRAWING. Invoke *SaveAs* to save the drawing under a new name. Enter **"CH2 INCLINED"** as the new name. Notice the current drawing name displayed in the AutoCAD title bar (at the top of the screen) is reset to the new name. Draw 2 more inclined *Lines* and *Save.*

7. ***Open* an AutoCAD sample drawing**

   Open a drawing named **"COLORWH.DWG"** usually located in the "C:\Programs\AutoCAD 2002\Sample" directory. Use the *Directories* section of the dialog box to change drive and directory if necessary. Do not *Save* the sample drawing after viewing it. Practice the *Open* command by looking at other sample drawings in the Sample directory. Finally, *Close* all the sample drawings. Be careful not to *Save* the drawings.

8. **Check the *SAVETIME* setting**

   At the command prompt, enter *SAVETIME*. The reported value is the interval (in minutes) between automatic saves. Your system may be set to 120 (default setting). If you are using your own system, change the setting to **15** or **20**.

9. **Find and rename a backup file**

   Close all open drawings. Use the *Open* command. When the *Select File* dialog box appears, locate the folder where your drawing files are saved, then enter **"*.BAK"** in the *File name:* edit box (* is a wildcard that means "all files"). All of the .BAK files should appear. Search for a backup file that was created the last time you saved **CH2 VERTICAL.DWG,** named **CH2 VERTICAL.BAK**. **Right-click** on the file name. Select *Rename* from the shortcut menu and rename the file **"CH2 VERT 2.DWG."** Next enter *.DWG in the *File name:* edit box to make all the .DWG files reappear. Highlight **CH2 VERT 2.DWG** and the bitmap image should appear in the preview display. You can now *Open* the file if you wish.

3

DRAW COMMAND CONCEPTS

Chapter Objectives

After completing this chapter you should be able to:

1. use *SNAP*, *GRID*, *ORTHO*, and *POLAR* while drawing *Lines* and *Circles* interactively;

2. create *Lines* by specifying <u>absolute</u> coordinates;

3. create *Lines* by specifying <u>relative rectangular</u> coordinates;

4. create *Lines* by specifying <u>relative polar</u> coordinates;

5. create *Lines* by specifying <u>direct distance entry</u> coordinates;

6. use Polar Tracking and Polar Snap to draw *Lines* interactively.

AutoCAD OBJECTS

The smallest component of a drawing in AutoCAD is called an <u>object</u> (sometimes referred to as an entity). An example of an object is a *Line,* an *Arc,* or a *Circle* (Fig. 3-1). A rectangle created with the *Line* command would contain four objects.

Draw commands <u>create</u> objects. The draw command names are the same as the object names.

Simple objects are *Point, Line, Arc,* and *Circle*.

Complex objects are shapes such as *Polygon, Rectangle, Ellipse, Polyline,* and *Spline* which are created with one command (Fig. 3-2). Even though they appear to have several segments, they are <u>treated</u> by AutoCAD as one object.

Figure 3-1

POINT　　LINE

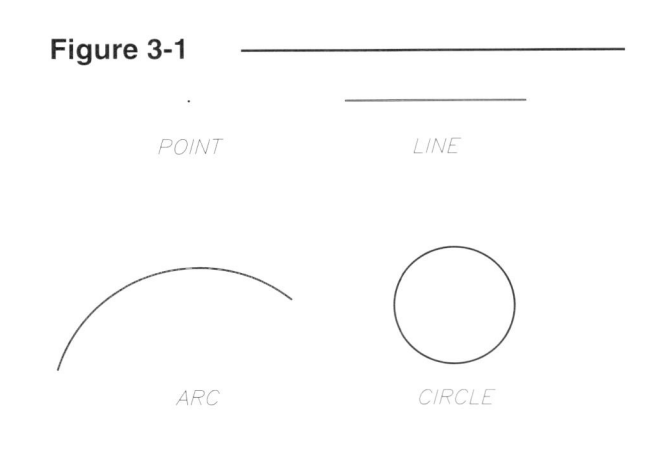

ARC　　CIRCLE

Figure 3-2

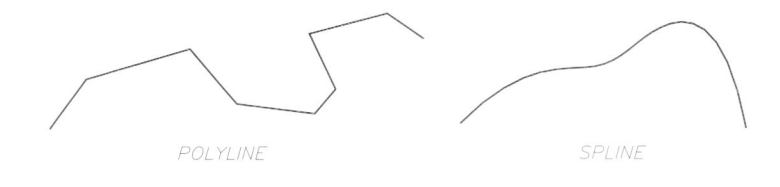

POLYGON　　RECTANGLE　　ELLIPSE

POLYLINE　　SPLINE

It is not always apparent whether a shape is composed of one or more objects. However, if you pick an object with the "pickbox," an object is "highlighted," or shown in a broken line pattern (Fig. 3-3). This highlighting reveals whether the shape is composed of one or several objects.

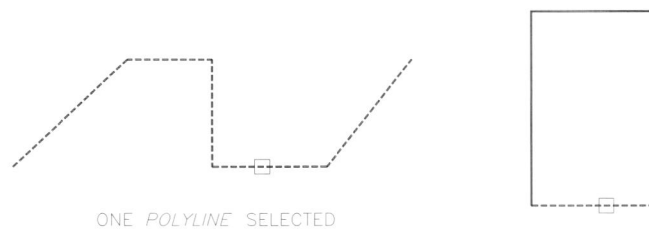

Figure 3-3

ONE *POLYLINE* SELECTED

ONE OF FOUR *LINES* SELECTED

LOCATING THE DRAW COMMANDS

To invoke *Draw* commands, any of the five command entry methods can be used depending on your computer setup.

1. **Toolbars**   Select the icon from the *Draw* toolbar.
2. **Pull-down menu**   Select the command from the *Draw* pull-down menu.
3. **Screen menu**   Select the command from the *Draw* screen menu.
4. **Keyboard**   Type the command name, command alias, or accelerator keys at the keyboard (a command alias is a one- or two-letter shortcut).
5. **Tablet menu**   Select the icon from the digitizing tablet menu (if available).

For example, a draw command can be activated by PICKing its icon button from the *Draw* toolbar (Fig. 3-4).

Figure 3-4

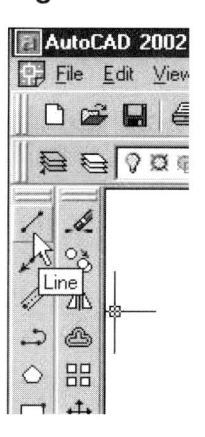

The *Draw* pull-down menu can also be used to select draw commands (Fig. 3-5). Options for a command are found on cascading menus.

Figure 3-5

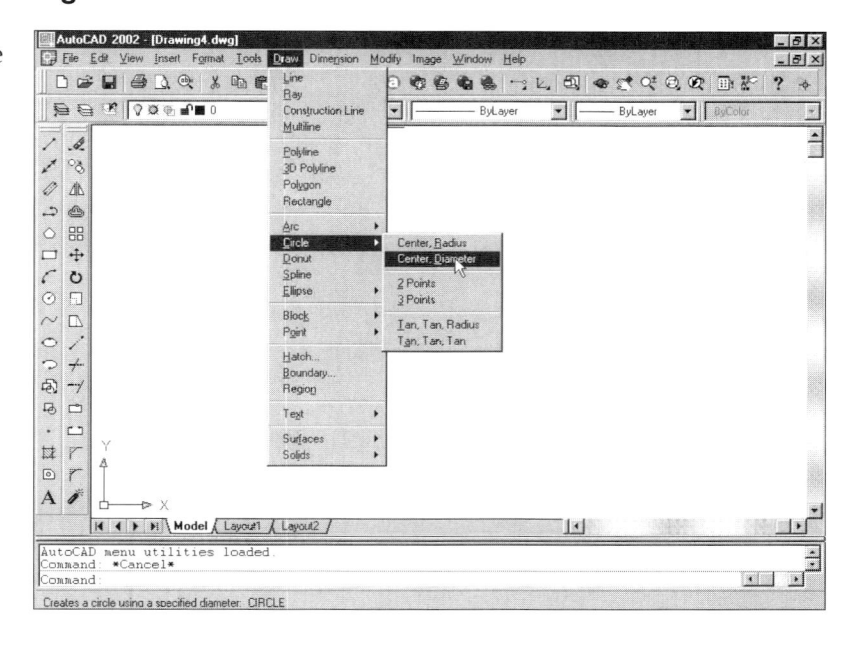

THE FIVE COORDINATE ENTRY METHODS

All drawing commands prompt you to specify points, or locations, in the drawing. For example, the *Line* command prompts you to give the "Specify first point:" and "Specify next point:," expecting you to specify locations for the first and second endpoints of the line. After you specify those points, AutoCAD stores the coordinate values to define the line. A two-dimensional line in AutoCAD is defined and stored in the database as two sets of X, Y, and Z values (with Z values of 0), one for each endpoint.

There are five ways to specify coordinates; that is, there are five ways to tell AutoCAD the location of points when you draw objects.

1. **Interactive method**   PICK   Use the cursor to select points on the screen.
2. **Absolute coordinates**   X,Y   Type explicit X and Y values relative to the origin at 0,0.
3. **Relative rectangular**   @X,Y   Type explicit X and Y values relative to the last point
   coordinates   (@ means "last point").
4. **Relative polar coordinates** @dist<angle   Type a distance value and angle value relative
                                                    to the last point (< means "angle of").
5. **Direct distance entry**   dist,direction   Type a distance value relative to the last point,
                                                    indicate direction with the cursor, then press Enter.

The interactive method of coordinate entry is simple because you PICK the desired locations with your cursor when AutoCAD prompts for a point. AutoCAD 2000 introduced Polar Tracking and Polar Snap. Later in this chapter you will see how these new features greatly enhance your capabilities for using the interactive method.

Absolute coordinates are used when AutoCAD prompts for a point and you know the exact coordinates of the desired location. You simply key in the coordinate values at the keyboard (X and Y values are separated by a comma).

Relative rectangular coordinates are similar to absolute values except the X and Y distances are given in relation to the last point, instead of being relative to the origin. AutoCAD interprets the @ symbol as the "last point." Use this type of coordinate entry when you do not know the exact absolute values, but you know the X and Y distances from the last specified point.

Relative polar coordinate entry is often used when you want to draw a line or specify a point that is at an exact angle with respect to the last point. Interactive coordinate entry is not accurate enough to draw diagonal lines to a specific angle (unless Polar Tracking is used). AutoCAD interprets the @ symbol as last point and the < symbol as an angular designator for the following value. So, "@ 2<45" means "from the last point, 2 units at an angle of 45 degrees."

Direct distance coordinate entry is a combination of relative polar coordinates and interactive specification because a distance value is entered at the keyboard and the angle is specified by the direction of the cursor movement (from the last point). Direct distance entry is useful primarily for orthogonal (horizontal or vertical) operations by toggling ORTHO or POLAR on. For example, assume you wanted to draw a *Line* 7.5 units in a horizontal (positive X) direction from the last point. Using direct distance entry to establish the "Specify next point:" (second endpoint), you would turn on ORTHO or POLAR and move the cursor to the right any distance, then type "7.5" and press Enter.

DRAWING *LINES* USING THE FIVE COORDINATE ENTRY METHODS

To practice using the five coordinate entry methods in this section, you must turn Grid Snap on rather than Polar Snap. Turn Grid Snap on by right-clicking the word *SNAP* on the Status Bar and selecting **Grid Snap On** from the shortcut menu (Fig. 3-6). Also, Polar Tracking, Object Snap, and Object Snap Tracking should be turned off (these are on by default when you install AutoCAD). Turn off Polar Tracking, Object Snap, and Object Snap Tracking by toggling the *POLAR*, *OSNAP*, and *OTRACK* buttons on the Status Bar or using F10, F3, and F11, respectively. The Status Bar indicates the on/off status of these drawing aids such that the button in a recessed (or depressed) position means on and a protruding button means off (see Figure 3-6).

Figure 3-6

Chapter Three 41

Begin a *New* drawing to complete the *Line* exercises. If the *Startup* dialog box or *Today* window appears, select *Start from Scratch,* choose *English* as the default setting, and click the *OK* button. The *Line* command can be activated by any one of the methods shown in the command table below.

LINE

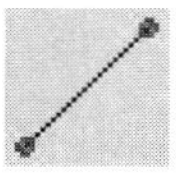

Pull-down Menu	COMMAND (TYPE)	ALIAS (TYPE)	Short-cut	Screen (side) Menu	Tablet Menu
Draw Line	LINE	L	...	DRAW 1 Line	J,10

Drawing Horizontal Lines

1. Draw a horizontal *Line* of 2 units length starting at point 2,2. Use the <u>interactive</u> method. See Figure 3-7.

Steps	Command Prompt	Perform Action	Comments
1.		turn on *GRID* (**F7**)	grid appears
2.		turn on *SNAP* (**F9**)	"SNAP" recessed on Status Bar
3.	Command:	select or type **Line**	use any method
4.	LINE Specify first point:	**PICK** location **2,2**	watch *Coords*
5.	Specify next point:	**PICK** location **4,2**	watch *Coords*
6.	Specify next point:	press **Enter**	completes command

The preceding steps produce a *Line* as shown in Figure 3-7.

Figure 3-7

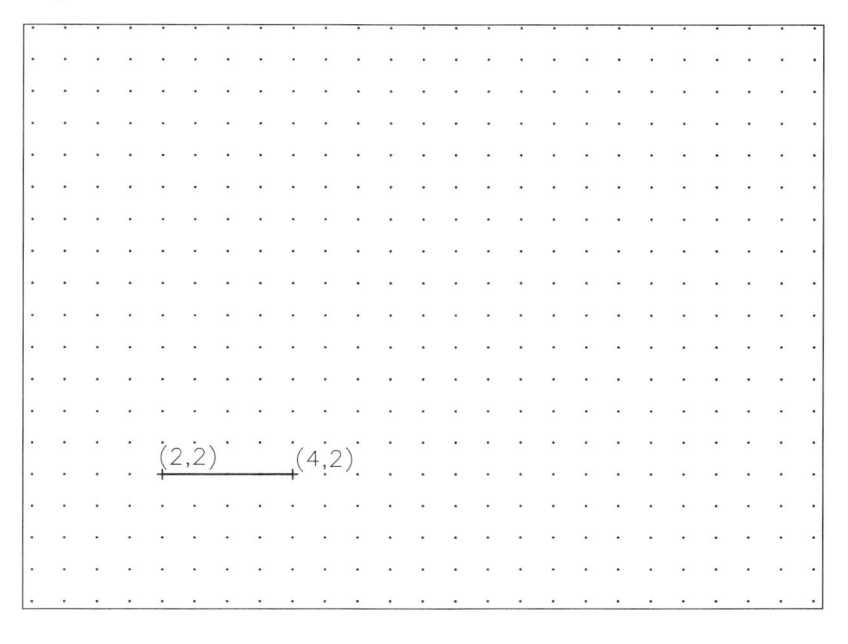

42　Draw Command Concepts

2. Draw a horizontal *Line* of 2 units length starting at point 2,3 using <u>absolute coordinates</u>.

Steps	Command Prompt	Perform Action	Comments
1.	Command:	select *Line*	use any method
2.	LINE Specify first point:	type **2,3** and press **Enter**	establishes the first endpoint
3.	Specify next point:	type **4,3** and press **Enter**	a *Line* should appear
4.	Specify next point:	press **Enter**	completes command

The above procedure produces the new *Line* above the first *Line* as shown (Fig. 3-8).

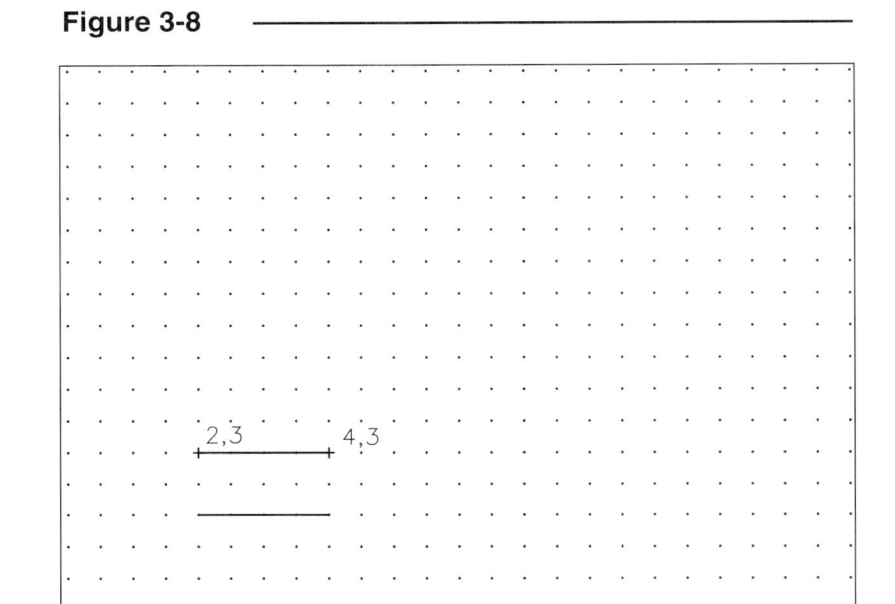

Figure 3-8

3. Draw a horizontal *Line* of 2 units length starting at point 2,4 using <u>relative rectangular coordinates</u>.

Steps	Command Prompt	Perform Action	Comments
1.	Command:	select *Line*	use any method
2.	LINE Specify first point:	type **2,4** and press **Enter**	establishes the first endpoint
3.	Specify next point:	type **@2,0** and press **Enter**	@ means "last point"
4.	Specify next point:	press **Enter**	completes command

The new *Line* appears above the previous two as shown in Figure 3-9.

Figure 3-9

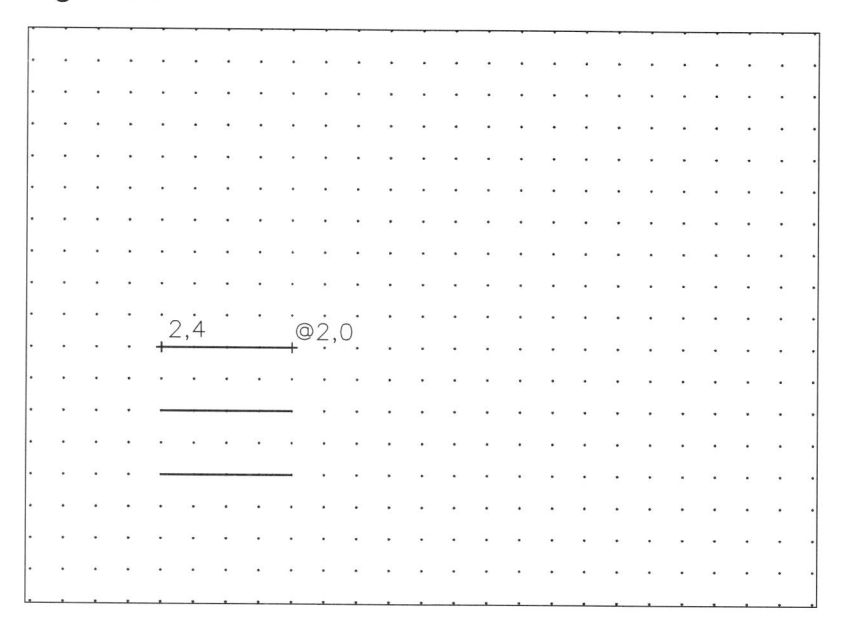

[Figure showing a grid with three horizontal lines; the top line labeled "2,4" at left endpoint and "@2,0" at right endpoint]

4. Draw a horizontal *Line* of 2 units length starting at point 2,5 using <u>relative polar coordinates</u>.

Steps	Command Prompt	Perform Action	Comments
1.	Command:	select *Line*	use any method
2.	LINE Specify first point:	type **2,5** and press **Enter**	establishes the first endpoint
3.	Specify next point:	type **@2<0** and press **Enter**	@ means "last point" < means "angle of"
4.	Specify next point:	press **Enter**	completes command

The new horizontal *Line* appears above the other three. See Figure 3-10.

Figure 3-10

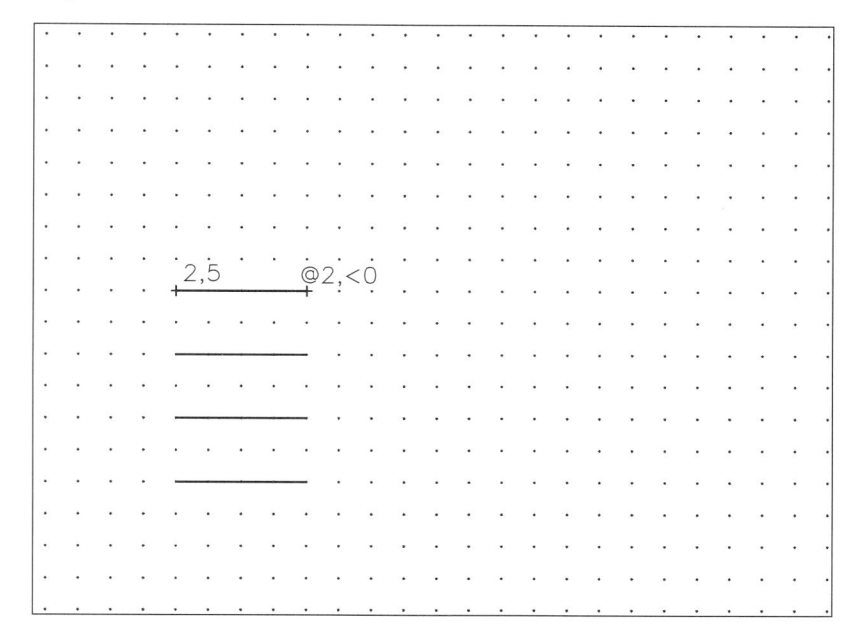

[Figure showing a grid with four horizontal lines; the top line labeled "2,5" at left endpoint and "@2,<0" at right endpoint]

5. Draw a horizontal *Line* of 2 units length starting at point 2,6 using <u>direct distance</u> coordinate entry.

Steps	Command Prompt	Perform Action	Comments
1.	Command:	select *Line*	use any method
2.	LINE Specify first point:	type **2,6** and press **Enter**	establishes the first endpoint
3.	Specify next point:	turn on **ORTHO**, move the cursor to the right, type **2**, then press **Enter**	the cursor movement indicates direction, 2 is the distance
4.	Specify next point:	press **Enter**	completes command

The new line appears above the other four as shown in Figure 3-11.

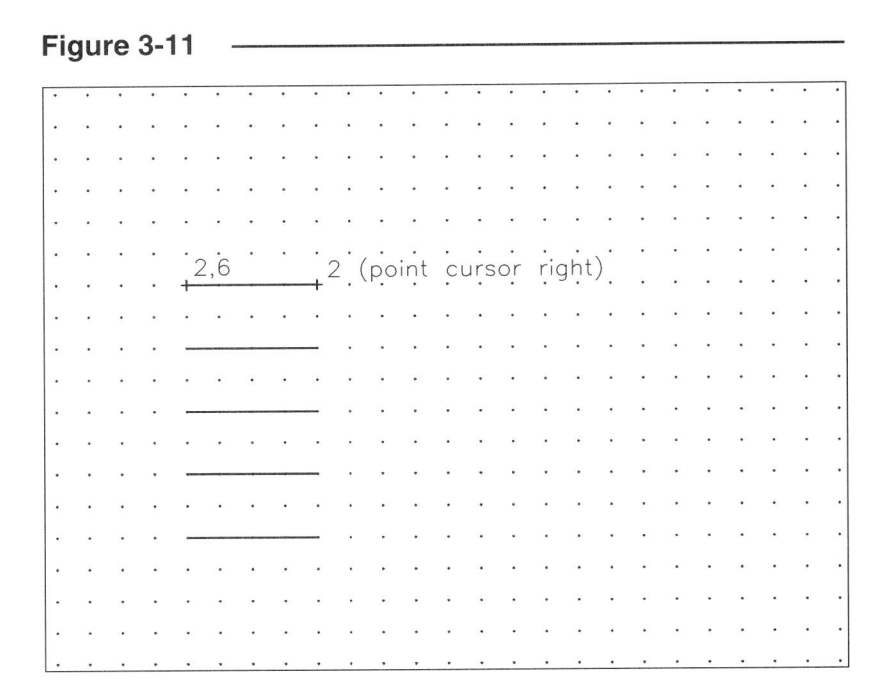

Figure 3-11

One of these methods may be more favorable than another in a particular situation. The interactive method is fast and easy, assuming that OSNAP or POLAR is used or that SNAP and GRID are used and set to appropriate values. SNAP and GRID are used successfully for small drawings where objects have regular interval lengths such as simple mechanical parts; however, SNAP and GRID are not used much for most drawings—those that use a variety of increments or occupy a large area, such as complex mechanical drawings, architectural drawings, or civil engineering drawings. Direct distance coordinate entry is fast, easy, and accurate when drawing horizontal or vertical lines. ORTHO or POLAR should be on for most applications of direct distance entry. Direct distance coordinate entry is preferred for horizontal or vertical drawing and editing in cases when OSNAP cannot be used (no geometry to OSNAP to) and when SNAP and GRID are not appropriate, such as an architectural or civil engineering application. In these applications, objects are relatively large (lines are long) and are not necessarily drawn to regular intervals; therefore, it is convenient and accurate to enter exact lengths as opposed to interactively PICKing points.

Drawing Vertical Lines

Below are listed the steps for drawing vertical lines using each of the five methods of coordinate entry. The following completed problems should look like those in Figure 3-12.

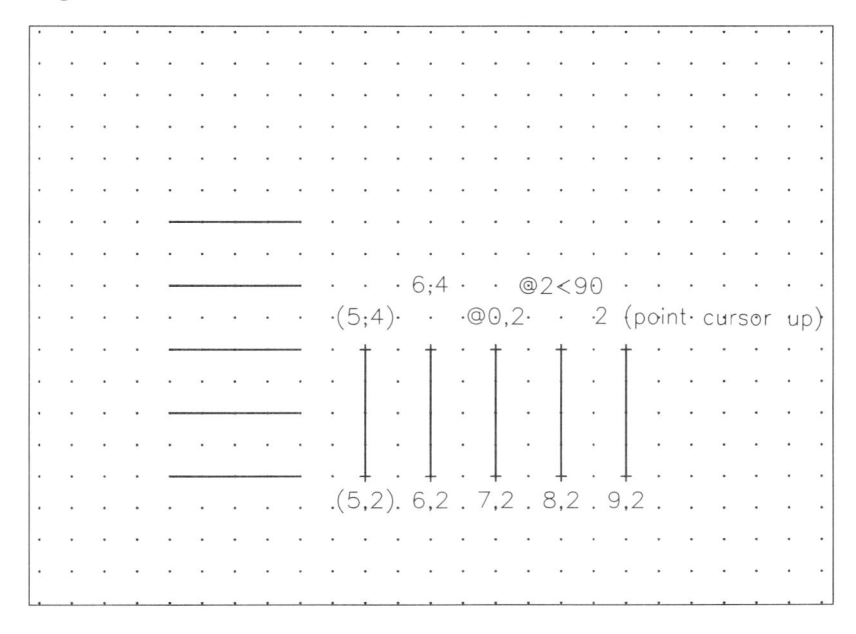

Figure 3-12

```
                              6,4    @2<90
                       (5,4)  @0,2   2 (point cursor up)


                       (5,2)  6,2  7,2  8,2  9,2
```

1. Draw a vertical *Line* of 2 units length starting at point 5,2 using the <u>interactive</u> method.

Steps	Command Prompt	Perform Action	Comments
1.		turn on *GRID* (**F7**)	grid appears
2.		turn on *SNAP* (**F9**)	"SNAP" recessed on Status Bar
3.	Command:	select or type *Line*	use any method
4.	LINE Specify first point:	**PICK** location **5,2**	watch *Coords*
5.	Specify next point:	**PICK** location **5,4**	watch *Coords*
6.	Specify next point:	press **Enter**	completes command

2. Draw a vertical *Line* of 2 units length starting at point 6,2 using <u>absolute coordinates</u>.

Steps	Command Prompt	Perform Action	Comments
1.	Command:	select *Line*	use any method
2.	LINE Specify first point:	type **6,2** and press **Enter**	establishes the first endpoint
3.	Specify next point:	type **6,4** and press **Enter**	a *Line* should appear
4.	Specify next point:	press **Enter**	completes command

46 Draw Command Concepts

3. Draw a vertical *Line* of 2 units length starting at point 7,2 using <u>relative rectangular coordinates</u>.

Steps	Command Prompt	Perform Action	Comments
1.	Command:	select *Line*	use any method
2.	LINE Specify first point:	type **7,2** and press **Enter**	establishes the first endpoint
3.	Specify next point:	type **@0,2** and press **Enter**	@ means "last point"
4.	Specify next point:	press **Enter**	completes command

4. Draw a vertical *Line* of 2 units length starting at point 8,2 using <u>relative polar coordinates</u>.

Steps	Command Prompt	Perform Action	Comments
1.	Command:	select *Line*	use any method
2.	LINE Specify first point:	type **8,2** and press **Enter**	establishes the first endpoint
3.	Specify next point:	type **@2<90** and press **Enter**	@ means "last point," < means "angle of"
4.	Specify next point:	press **Enter**	completes command

5. Draw a vertical *Line* of 2 units length starting at point 9,2 using <u>direct distance</u> coordinate entry.

Steps	Command Prompt	Perform Action	Comments
1.	Command:	select *Line*	use any method
2.	LINE Specify first point:	type **9,2** and press **Enter**	establishes the first endpoint
3.	Specify next point:	turn on **ORTHO**, move the cursor upward, type **2**, then press **Enter**	the cursor movement indicates direction, 2 is the distance
4.	Specify next point:	press **Enter**	completes command

The method used for drawing vertical lines depends on the application and the individual, much the same as it would for drawing horizontal lines. Refer to the discussion after the horizontal lines drawing exercises.

Drawing Inclined Lines

Following are listed the steps in drawing <u>inclined lines</u> using each of the five methods of coordinate entry. The following completed problems should look like those in Figure 3-13.

Figure 3-13

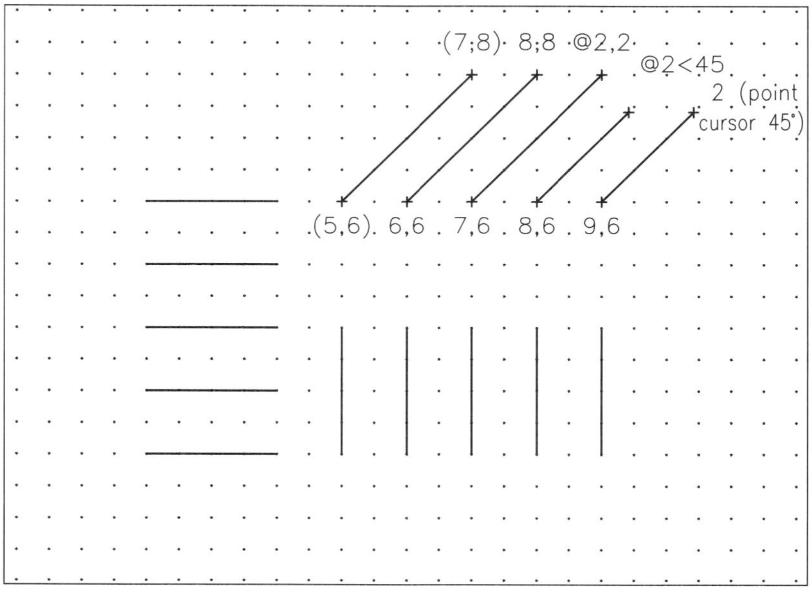

1. Draw an inclined *Line* of from 5,6 to 7,8. Use the <u>interactive</u> method.

Steps	Command Prompt	Perform Action	Comments
1.		turn on *GRID* (**F7**)	grid appears
2.		turn on *SNAP* (**F9**)	"SNAP" recessed on Status Bar
3.		turn off *ORTHO* (**F8**)	in order to draw inclined *Lines*
4.	Command:	select or type **Line**	use any method
5.	LINE Specify first point:	**PICK** location **5,6**	watch *Coords*
6.	Specify next point:	**PICK** location **7,8**	watch *Coords*
7.	Specify next point:	press **Enter**	completes command

2. Draw an inclined *Line* starting at 6,6 and ending at 8,8. Use <u>absolute coordinates</u>.

Steps	Command Prompt	Perform Action	Comments
1.	Command:	select **Line**	use any method
2.	LINE Specify first point:	type **6,6** and press **Enter**	establishes the first endpoint
3.	Specify next point:	type **8,8** and press **Enter**	a *Line* should appear
4.	Specify next point:	press **Enter**	completes command

3. Draw an inclined *Line* starting at 7,6 and ending 2 units over (in a positive X direction) and 2 units up (in a positive Y direction). Use <u>relative rectangular coordinates</u>.

Steps	Command Prompt	Perform Action	Comments
1.	Command:	select *Line*	use any method
2.	LINE Specify first point:	type **7,6** and press **Enter**	establishes the first endpoint
3.	Specify next point:	type **@2,2** and press **Enter**	@ means "last point"
4.	Specify next point:	press **Enter**	completes command

4. Draw an inclined *Line* of 2 units length at a 45 degree angle and starting at 8,6. Use <u>relative polar coordinates</u>.

Steps	Command Prompt	Perform Action	Comments
1.	Command:	select *Line*	use any method
2.	LINE Specify first point:	type **8,6** and press **Enter**	establishes the first endpoint
3.	Specify next point:	type **@2<45** and press **Enter**	@ means "last point" < means "angle of"
4.	To point	press **Enter**	completes command

5. Draw an inclined *Line* of 2 units length at a 45 degree angle starting at point 9,6. Use <u>direct distance</u> coordinate entry.

Steps	Command Prompt	Perform Action	Comments
1.		Turn on *GRID* (**F7**)	grid appears
2.		Turn on *SNAP* (**F9**)	"SNAP" recessed on Status Bar
3.	Command:	select *Line*	use any method
4.	LINE Specify first point:	type **9,6** and press **Enter**	establishes the first endpoint
5.	Specify next point:	turn off **ORTHO**, move the cursor up and to the right (the coordinate display must indicate an angle of 45), type **2**, then press **Enter**	the cursor movement indicates direction, 2 is the distance
6.	Specify next point:	press **Enter**	completes command

Method 4 is suitable for drawing inclined lines when you know the exact length and angle. The other methods are not suitable for drawing angled lines for most cases. Instead, a feature called Polar Tracking should be used for most cases when many lines are to be drawn at regular angles such as 15, 30, or 45. Polar Tracking is described later in this chapter.

DRAWING *CIRCLES* USING THE FIVE COORDINATE ENTRY METHODS

Begin a *New* drawing to complete the *Circle* exercises. If the *Startup* dialog box or *Today* window appears, select *Start from Scratch,* and choose *English* as the default setting. The *Circle* command can be invoked by any of the methods shown in the command table. Make sure *Grid Snap* is on and *POLAR*, *OSNAP*, and *OTRACK* are off for these exercises.

CIRCLE

Pull-down Menu	COMMAND (TYPE)	ALIAS (TYPE)	Short-cut	Screen (side) Menu	Tablet Menu
Draw Circle >	CIRCLE	C	...	DRAW 1 Circle	J,9

Below are listed the steps for drawing *Circles* using the *Center, Radius* method. The circles are to be drawn using each of the four coordinate entry methods and should look like those in Figure 3-14.

Figure 3-14

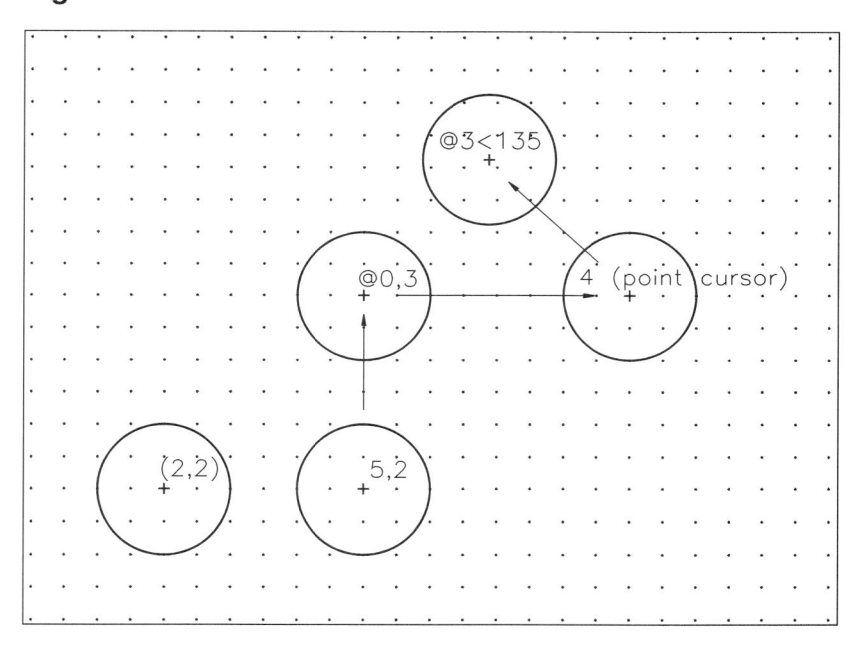

1. Draw a *Circle* of 1 unit radius with the center at point 2,2. Use the <u>interactive</u> method.

Steps	Command Prompt	Perform Action	Comments
1.		turn on *GRID* (**F7**)	grid appears
2.		turn on *SNAP* (**F9**)	"SNAP" recessed on Status Bar
3.		turn on *ORTHO* (**F8**)	"ORTHO" recessed on Status Bar
4.	Command:	select or type **Circle**	use *Center, Radius* method
5.	CIRCLE Specify center point for circle:	**PICK** location **2,2**	watch *Coords*
6.	Specify radius of circle:	move 1 unit and **PICK**	watch *Coords*

50 Draw Command Concepts

2. Draw a circle of 1 unit radius with the center at point 5,2. Use <u>absolute coordinates</u>.

Steps	Command Prompt	Perform Action	Comments
1.	Command:	select *Circle*	use *Center, Radius* method
2.	CIRCLE Specify center point for circle:	type **5,2** and press **Enter**	a *Circle* should appear
3.	Specify radius of circle:	type **1** and press **Enter**	the correct *Circle* appears

3. Draw a circle of 1 unit radius with the center 3 units above the last point (previous *Circle* center). Use <u>relative rectangular coordinates</u>.

Steps	Command Prompt	Perform Action	Comments
1.	Command:	select *Circle*	use *Center, Radius* method
2.	CIRCLE Specify center point for circle:	type **@0,3** and press **Enter**	a *Circle* should appear
3.	Specify radius of circle:	type **1** and press **Enter**	the correct *Circle* appears

(If the new *Circle* is not above the last, type "ID" and enter "5,2." The entered point becomes the last point. Then try again.)

4. Draw a circle of 1 unit radius with the center 4 units to the right of the previous *Circle* center. Use <u>direct distance</u> coordinate entry.

Steps	Command Prompt	Perform Action	Comments
1.		turn on *ORTHO* (**F8**)	"ORTHO" recessed on Status Bar
2.	Command:	select *Circle*	use *Center, Radius* method
3.	CIRCLE Specify center point for circle:	type **4**, move the cursor to the right, and press **Enter**	a *Circle* should appear
4.	Specify radius of circle:	type **1**, then press **Enter**	the correct *Circle* appears

(If the new *Circle* is not directly to the right of the last, type "ID" and enter "5,5." The entered point becomes the last point. Then try again.)

5. Draw a circle of 1 unit radius with the center 3 units at a 135-degree angle above and to the left of the previous *Circle*. Use <u>relative polar coordinates</u>.

Steps	Command Prompt	Perform Action	Comments
1.	Command:	select *Circle*	use *Center, Radius* method
2.	CIRCLE Specify center point for circle:	type **@3<135** and press **Enter**	a *Circle* should appear
3.	Specify radius of circle:	type **1** and press **Enter**	the *Circle* appears

(If the new *Circle* is not correctly positioned with respect to the last, type "ID" and enter "9,5." The entered point becomes the last point. Then try again.)

POLAR TRACKING AND POLAR SNAP

Remember that all draw commands prompt you to specify locations, or coordinates, in the drawing. You can indicate these coordinates interactively (using the cursor), entering values (at the keyboard), or using a combination of both. Features in AutoCAD that provide an easy method for specifying coordinate locations interactively are Polar Tracking and Polar Snap. Another feature, Object Snap Tracking, is discussed in Chapter 7. These features help you draw objects at specific angles and in specific relationships to other objects.

Polar Tracking (*POLAR*) helps the rubberband line snap to angular increments such as 45, 30, or 15 degrees. Polar Snap can be used in conjunction to make the rubberband line snap to incremental lengths such as .5 or 1.0. You can toggle these features on and off with the *SNAP* and *POLAR* buttons on the Status Bar or by toggling F9 and F10, respectively. First, you should specify the settings in the *Drafting Settings* dialog box.

In the previous sections of this chapter you learned and practiced with the five coordinate entry methods listed below.

1. Interactive method
2. Absolute coordinates
3. Relative rectangular coordinates
4. Relative polar coordinates
5. Direct distance entry

Technically, <u>Polar Tracking and Polar Snap would be considered a variation of the Interactive method</u> since the settings are determined and specified beforehand, then points on the screen are PICKed using the cursor.

To practice with Polar Tracking and Polar Snap, you should turn both *OSNAP* and *OTRACK* off (these are on by default when you install AutoCAD). Since all of these new features generate alignment vectors, it can be difficult to determine which of the new features is operating when they are all on.

Polar Tracking

Figure 3-15

Polar Tracking (*POLAR*) simplifies drawing *Lines* or performing other operations such as *Move* or *Copy* at specific angle increments. For example, you can specify an increment angle of 30 degrees. Then when Polar Tracking is on, the rubberband line "snaps" to 30-degree increments when the cursor is in close proximity (within the *Aperture* box) to a specified angle. A dotted "tracking" line is displayed and a "tracking tip" appears at the cursor giving the current distance and angle (Fig. 3-15). In this case, it is simple to draw lines at 0, 30, 60, 90, or 120-degree angles and so on.

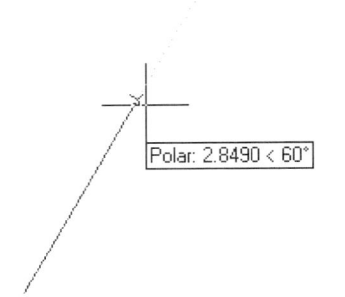

Figure 3-16 displays possible positions for Polar Tracking <u>when a 30-degree angle is specified</u>.

Figure 3-16

Available angle options are 90, 45, 30, 22.5, 18, 15, 10, and 5 degrees, or you can specify any user-defined angle. This is a tremendous aid for drawing angled lines since the only other method for precisely specifying angles is polar coordinate entry.

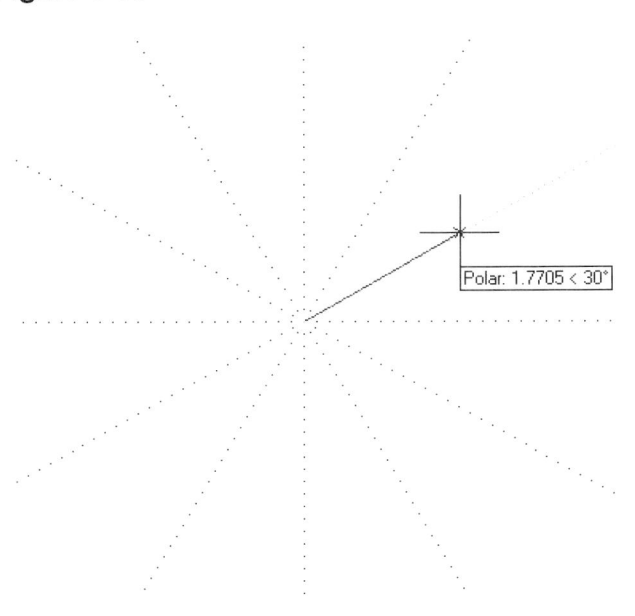

You can access the settings using the *Polar Tracking* tab of the *Drafting Settings* dialog box (Fig. 3-17). Invoke the dialog box by the following methods:

Figure 3-17

1. Enter *Dsettings* or *DS* at the Command line, then select the *Polar Tracking* tab.

2. Right-click on the word *POLAR* on the Status Bar, then choose *Settings...* from the menu.

3. Select *Drafting Settings* from the *Tools* pull-down menu, then select the *Polar Tracking* tab.

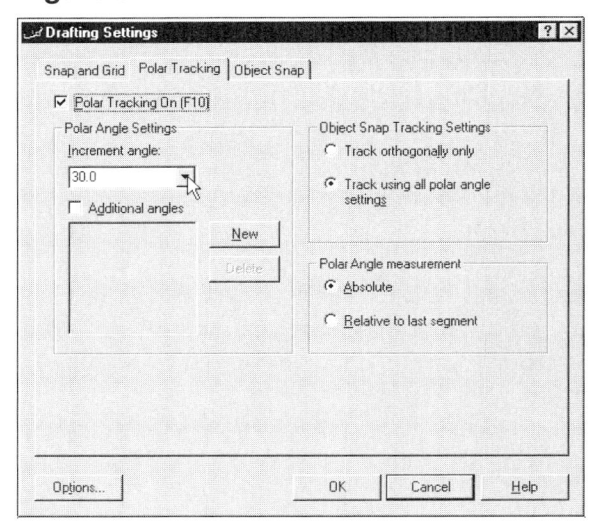

In this dialog box, you can select the *Increment Angle* from the drop-down list (as shown) or specify *Additional Angles*. Use the *New* button to create user-defined angles. Highlighting a user-defined angle and selecting the *Delete* button removes the angle from the list.

The *Object Snap Tracking Settings* are used only with Object Snap (discussed in Chapter 7). When the *Polar Angle Measurement* is set to *Absolute*, the angle reported on the "tracking tip" is an absolute angle (relative to angle 0 of the current coordinate system).

Polar Tracking with Direct Distance Entry

Figure 3-18

With normal Polar Tracking (Polar Snap is off), the line "snaps" to the set angles but can be drawn to any length (distance). This option is particularly useful in conjunction with Direct Distance Entry since you specify the angular increment in the *Polar Tracking* tab of the *Drafting Settings* dialog box, but indicate the distance by entering values at the Command line. In other words, when drawing a *Line*, move the cursor so it "snaps" to the desired angle, then enter the desired distance at the keyboard (Fig. 3-18).

Polar Tracking Override

When Polar Tracking is on, you can override the previously specified angle increment and draw to another specific angle. The new angle is valid only for one point specification. To enter a polar override angle, enter the left angle bracket (<) and an angle value whenever a command asks you to specify a point. The following command prompt sequence shows a 12-degree override entered during a *Line* command.

```
Command: line
Specify first point: PICK
Specify next point or [Undo]: <12
Angle Override: 12
Specify next point or [Undo]: PICK
```

This action forces the line to a 12-degree angle for that segment only.

Figure 3-19

Polar Snap

Polar Tracking makes the rubberband line "snap" to angular increments, whereas Polar Snap makes the rubberband line "snap" to distance increments that you specify. For example, setting the distance increment to 2 allows you to draw lines at intervals of 2, 4, 6, 8, and so on. Therefore, Polar Tracking with Polar Snap allows you to draw at specific angular *and* distance increments (Fig. 3-19). Using Polar Tracking with Polar Snap off, as described in the previous section, allows you to draw at specified angles but not at any specific distance intervals.

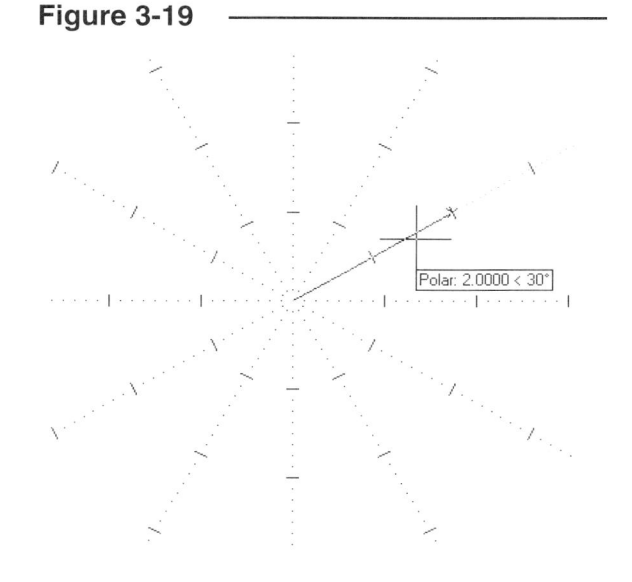

54 Draw Command Concepts

TIP
Only one type of Snap, Polar Snap or Grid Snap, can be used at any one time. Toggle either *Polar Snap* or *Grid Snap* on by setting the radio button in the *Snap and Grid* tab of the *Drafting Settings* dialog box (Fig. 3-20, lower right). Alternately, right-click on the word *SNAP* at the Status Bar and select either *Polar Snap On* or *Grid Snap On* from the shortcut menu (see Figure 3-21). Set the Polar Snap distance increment in the *Polar Distance* edit box (Fig. 3-20, lower left).

Figure 3-20

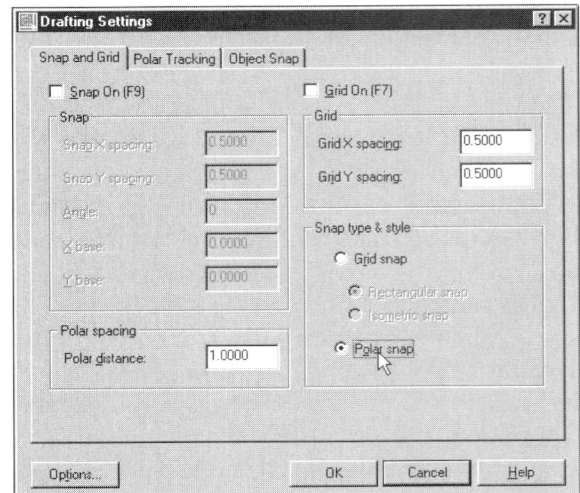

To utilize Polar Snap, both *SNAP* (F9) and *POLAR* (F10) must be turned on, Polar Snap must be on, and a Polar Distance value must be specified in the *Drafting Settings* dialog box. As a check, settings should be as shown in Figure 3-21 (*SNAP* and *POLAR* are recessed), with the addition of some Polar Distance setting in the dialog box. *SNAP* turns on or off whichever of the two snap types is active, Grid Snap or Polar Snap. For example, if Polar Snap is the active snap type, toggling F9 turns only Polar Snap off and on.

Figure 3-21

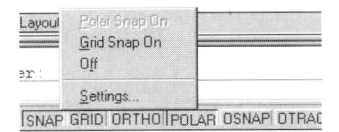

DRAWING *LINES* USING POLAR TRACKING AND POLAR SNAP

For these exercises you will use combinations of Polar Tracking with Polar Snap, Grid Snap and no snap type. Also, Object Snap and Object Snap Tracking should be turned off (these are on by default when you install AutoCAD). Turn off Object Snap and Object Snap Tracking by toggling the *OSNAP* and *OTRACK* buttons on the Status Bar to the protruding position or using F3 and F11 to do so.

Using Polar Tracking and Grid Snap

Draw the shape in Figure 3-22 using Polar Tracking in conjunction with Grid Snap. Follow the steps below.

1. Begin a *New* drawing. Select **Start from Scratch** and the **English** default settings.

2. Toggle on Grid Snap (rather than Polar Snap) by right-clicking on the word **SNAP** on the Status Bar, then selecting **Grid Snap On** from the menu (see Figure 3-6).

Figure 3-22

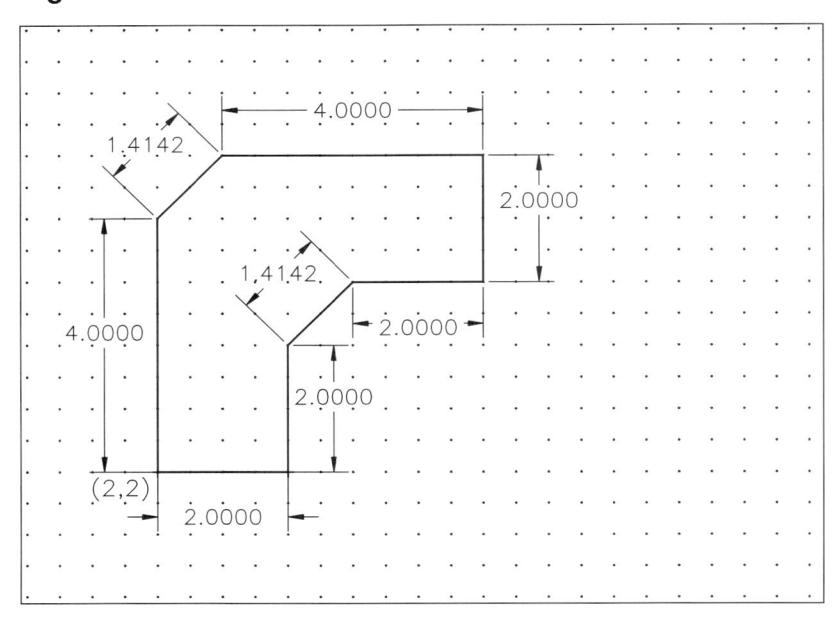

3. Turn on Polar Tracking and make the appropriate settings. Do this by right-clicking on the word *POLAR* on the Status Bar and selecting *Settings* from the menu (Figure 3-23). In the *Polar Tracking* tab of the *Drafting Settings* dialog box that appears, set the *Increment Angle* to **45.0** (see Figure 3-17). Select the *OK* button.

Figure 3-23

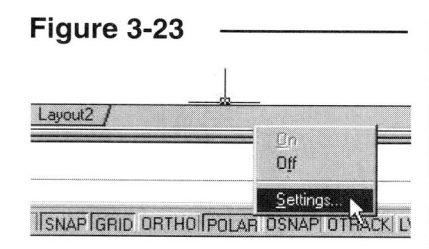

4. Make these other settings by single-clicking the words on the Status Bar or using the Function keys. The Status Bar indicates the on/off status of these drawing aids such that the button in a recessed position means on and a protruding button means off (Fig. 3-24). The command prompt also indicates the on or off status when they are changed.

Figure 3-24

   SNAP (F9) is on
   GRID (F7) is on
   ORTHO (F8) is off
   POLAR (F10) is on
   OSNAP (F3) is off
   OTRACK (F11) is off

   The resulting Status Bar should look like that in Figure 3-24.

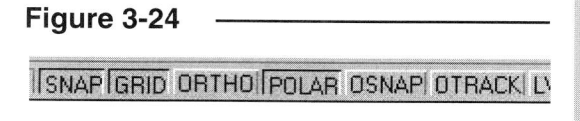

5. Use the *Line* command. Starting at location 2.000, 2.000, 0.000 (watch the *COORDS* display), begin drawing the shape shown in Figure 3-22. The tracking tip should indicate the current length and angle of the lines as you draw (Fig. 3-25).

Figure 3-25

6. Complete the shape. Compare your drawing to that in Figure 3-22. Use *Save* and name the drawing **PTRACK-GRIDSNAP**. *Close* the drawing.

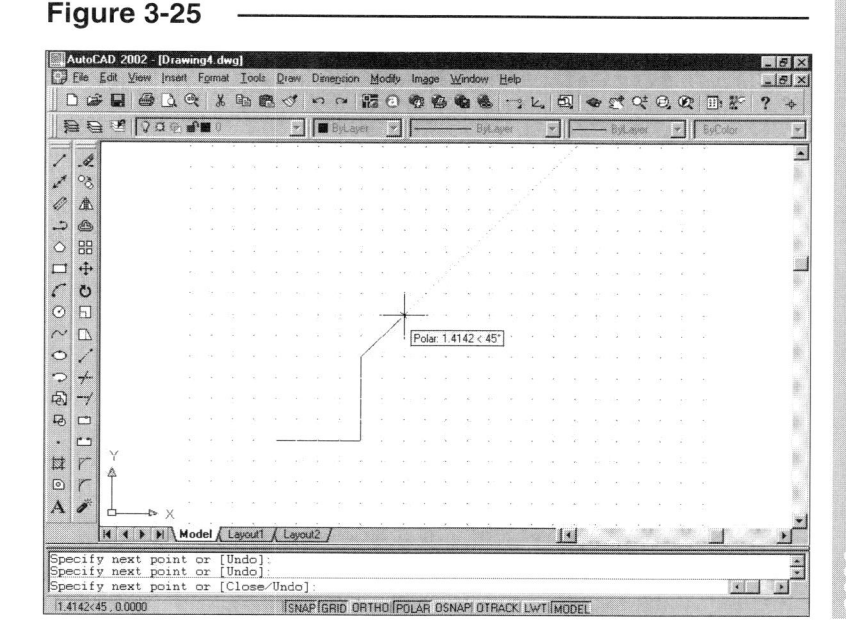

Using Polar Tracking and Polar Snap

Draw the shape in Figure 3-26. Since the lines are regular interval lengths, use Polar Tracking in conjunction with Polar Snap. Follow the steps below.

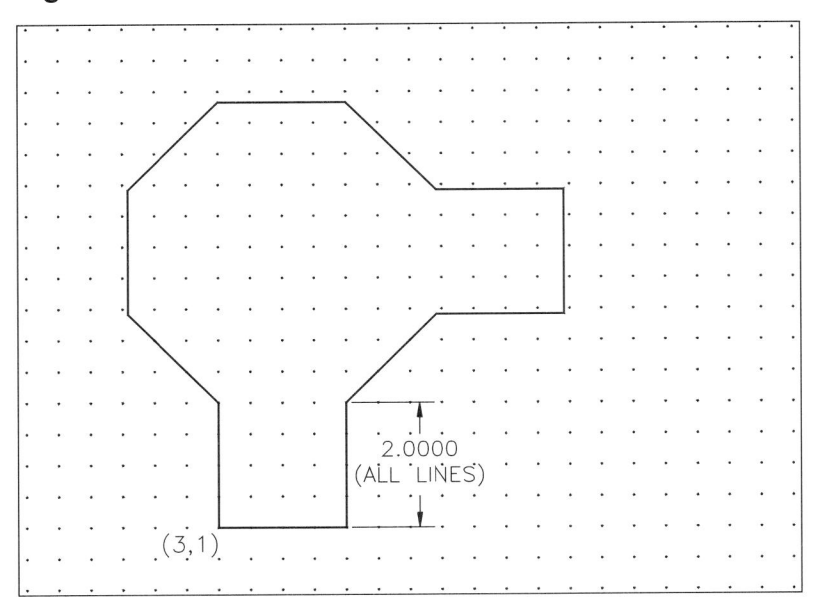

1. Begin a *New* drawing. Select *Start From Scratch* and the *English* default settings.

2. Toggle on Polar Snap (rather than Grid Snap) by right-clicking on the word *SNAP* on the Status Bar, then selecting *Polar Snap On* from the menu.

Figure 3-26

3. Make the appropriate Polar Snap settings. Do this by right-clicking on the word *SNAP* on the Status Bar and selecting *Settings* from the menu. In the *Snap and Grid* tab of the *Drafting Settings* dialog box that appears, set the *Polar Distance* to **1.0000** (see Figure 3-27). Next, access the *Polar Tracking* tab and ensure the *Increment Angle* is set to **45.0**. Select the *OK* button.

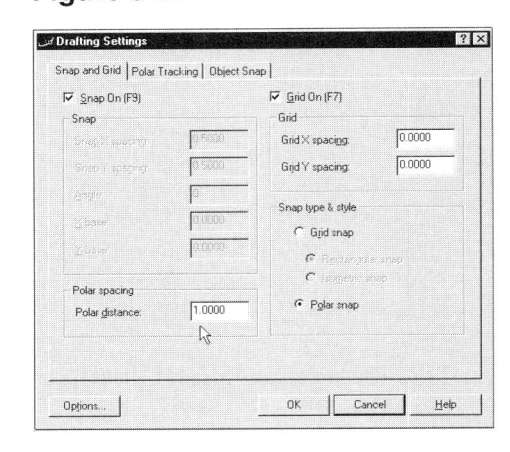

4. Ensure these other settings are correct by clicking the words on the Status Bar or using the Function keys.

   SNAP (F9) is on
   GRID (F7) is on
   ORTHO (F8) is off
   POLAR (F10) is on
   OSNAP (F3) is off
   OTRACK (F11) is off

Figure 3-27

5. Use the *Line* command. Starting at location 3.000, 1.000, 0.000 (enter absolute coordinates of **3,1**), begin drawing the shape shown in Figure 3-26. The tracking tip should indicate the current length and angle of the lines as you draw (Fig. 3-28).

6. Complete the shape. Compare your drawing to that in Figure 3-26. Use *Save* and name the drawing **PTRACK-POLARSNAP**. *Close* the drawing.

Figure 3-28

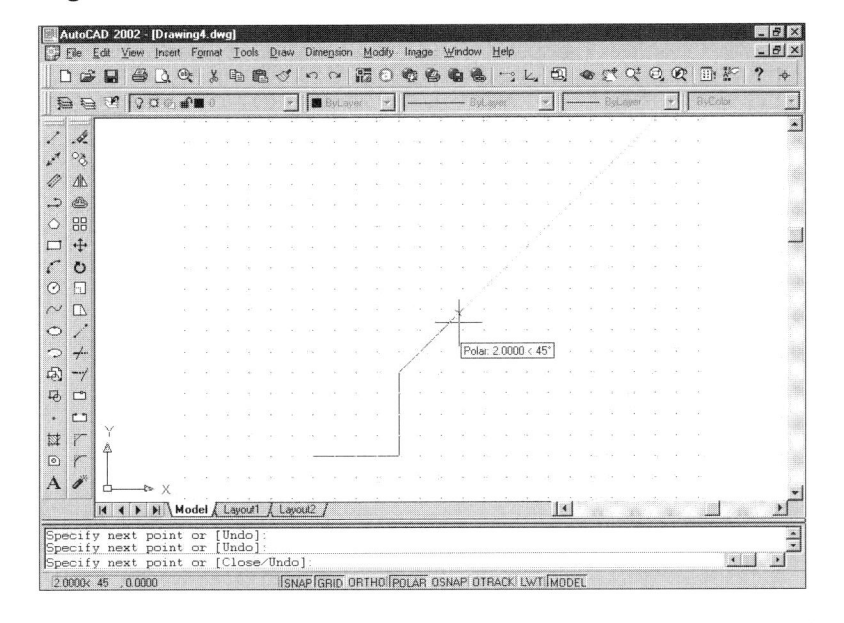

Using Polar Tracking and Direct Distance Entry

Draw the shape in Figure 3-29. Since the lines are at regular angles but irregular lengths, use Polar Tracking in conjunction with Direct Distance Entry. Follow the steps below.

1. Begin a *New* drawing. Select *Start From Scratch* and the *English* default settings.

2. Toggle off *SNAP* on the Status Bar. No Snap type is used since distances will be entered at the keyboard.

3. Turn on Polar Tracking and make the appropriate settings. Do this by right-clicking on the word *POLAR* on the Status Bar and selecting *Settings* from the menu. In the *Polar Tracking* tab of the *Drafting Settings* dialog box, ensure the *Increment Angle* is set to **30.0**, and *Polar Tracking* is *On*. Select the *OK* button.

4. Ensure these other settings are correct by clicking the words on the Status Bar or using the Function keys.

   SNAP (F9) is off
   GRID (F7) is off
   ORTHO (F8) is off
   POLAR (F10) is on
   OSNAP (F3) is off
   OTRACK (F11) is off

5. Use the *Line* command. Starting at location 2.000, 2.000, 0.000 (enter absolute coordinates of **2,2**), begin drawing the shape shown in Figure 3-29. Use Polar Tracking by moving the mouse in the desired direction (Fig. 3-30). When the tracking tip indicates the correct angle for each line, enter the distance for each line (**2.3**) at the keyboard.

6. Complete the shape. Compare your drawing to that in Figure 3-29. Use *Save* and name the drawing **PTRACK-DDE**. *Close* the drawing and *Exit* AutoCAD.

Figure 3-29

Figure 3-30

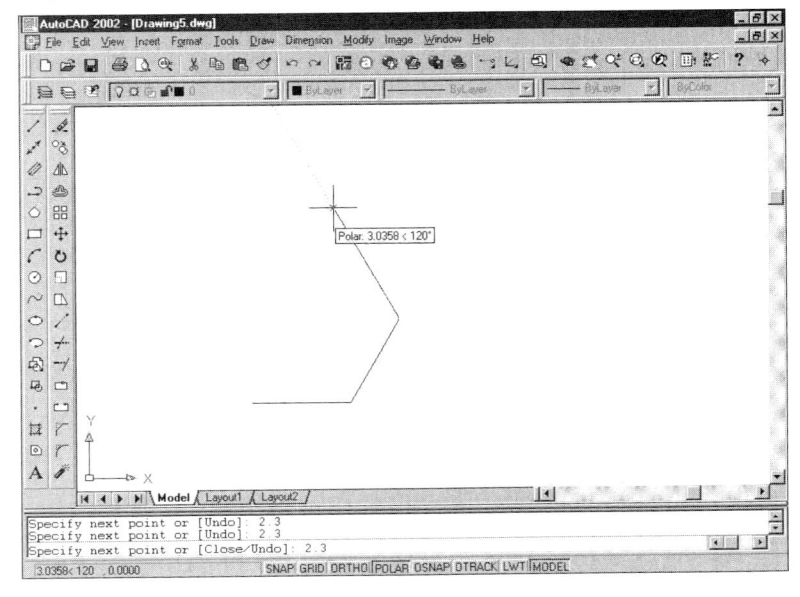

CHAPTER EXERCISES

1. **Start a *New* Drawing**

   Start a *New* drawing. Select the *Start from Scratch* option and select *English* settings. Next, use the *Save* command and save the drawing as "CH3EX1." Remember to *Save* often as you complete the following exercises. The completed exercise should look like Figure 3-31.

   Figure 3-31

2. **Use interactive coordinate entry**

   Draw a square with sides of 2 units length. Locate the lower-left corner of the square at **2,2**. Use the *Line* command with interactive coordinate entry. (HINT: Turn on *SNAP, GRID,* and *ORTHO*.)

3. **Use absolute coordinates**

   Draw another square with 2 unit sides using the *Line* command. Enter absolute coordinates. Begin with the lower-left corner of the square at **5,2**.

4. **Use relative rectangular coordinates**

   Draw a third square (with 2 unit sides) using the *Line* command. Enter relative rectangular coordinates. Locate the lower-left corner at **8,2**.

5. **Use direct distance coordinate entry**

   Draw a fourth square (with 2 unit sides) beginning at the lower-left corner of **2,5**. Complete the square drawing *Lines* using direct distance coordinate entry. Don't forget to turn on *ORTHO* or *POLAR*.

6. **Use relative polar coordinates**

   Draw an equilateral triangle with sides of 2 units. Locate the lower-left corner at **5,5**. Use relative polar coordinates (after establishing the "Specify first point:"). HINT: An equilateral triangle has interior angles of 60 degrees.

7. **Use interactive coordinate entry**

   Draw a *Circle* with a 1 unit radius. Locate the center at **9,6**. Use the interactive method. (Turn on *SNAP* and *GRID*.)

8. **Use relative rectangular, relative polar, or direct distance entry coordinates**

   Draw another *Circle* with a 2 unit diameter. Using any method listed above, locate the center 3 units below the previous *Circle*.

9. **Save your drawing**

   Use *Save*. Compare your results with Figure 3-31. When you are finished, *Close* the drawing.

10. In the next series of steps, you will create the Stamped Plate shown in Figure 3-32. Begin a *New* drawing. Select the *Start from Scratch* option and select *English* settings. Next, use the *Save* command and assign the name **CH3EX2**. Remember to *Save* often as you work.

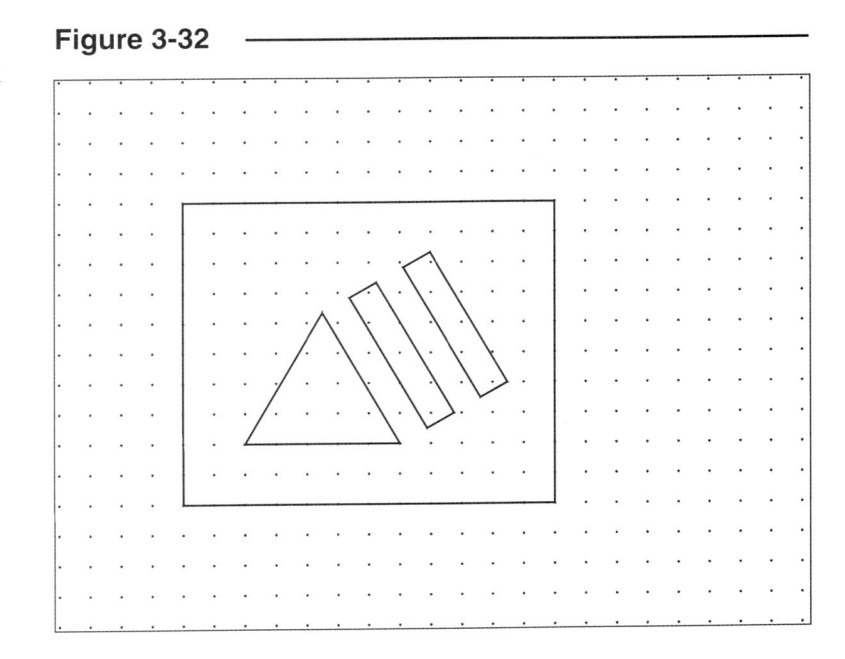

Figure 3-32

11. First, you will create the equilateral triangle as shown in Figure 3-33. All sides are equal and all angles are equal. To create the shape easily, you should use Polar Tracking with Direct Distance Entry. Access the *Polar Tracking* tab of the *Drafting Settings* dialog box and set the *Increment Angle* to **30**. (HINT: Right-click on the word *POLAR* and select *Settings* from the shortcut menu.) On the Status Bar, make sure *POLAR* is on (appears recessed), but not *SNAP*, *ORTHO*, *OSNAP*, or *OTRACK*. *GRID* is optional.

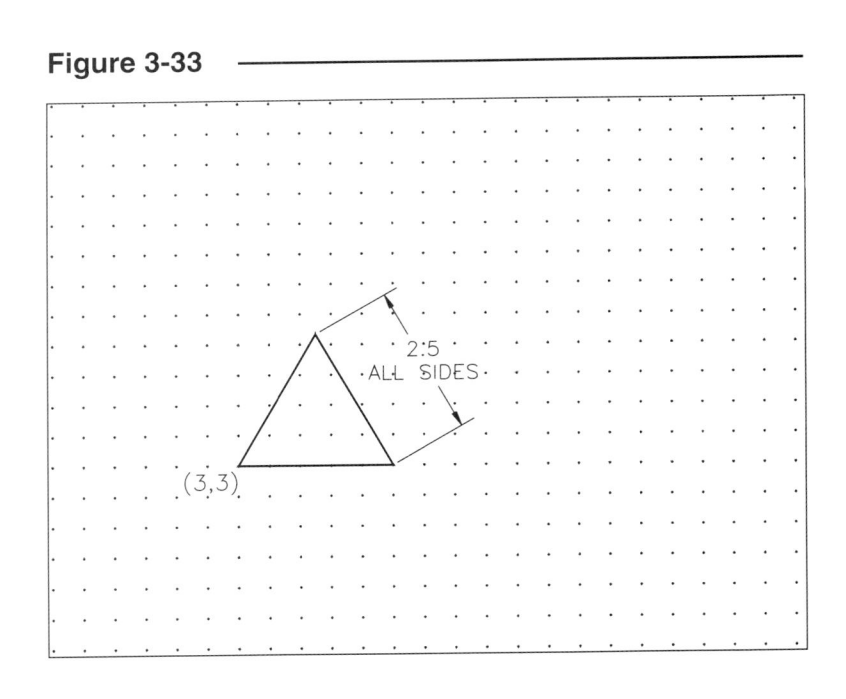

Figure 3-33

12. Use the *Line* command to create the equilateral triangle. Start at position **3,3** by entering absolute coordinates. All sides should be **2.5** units, so enter the distance values at the keyboard as you position the mouse in the desired direction for each line. The drawing at this point should look like that in Figure 3-33 (not including the notation). *Save* the drawing but do not *Close* it.

13. Next, you should create the outside shape (Fig. 3-34). Since the dimensions are all at even unit intervals, use Grid Snap to create the rectangle. (HINT: Right-click on the word *SNAP* and select **Grid Snap On**.) At the Status Bar, make sure that only *SNAP* and *POLAR* are on. Use the *Line* command to create the rectangular shape starting at point **2,2** (enter absolute coordinates).

Figure 3-34

14. The two inside shapes are most easily created using Polar Tracking in combination with Polar Snap. Access the **Grid and Snap** tab of the *Drafting Settings* dialog box. (HINT: Right-click on the word *SNAP* and select **Settings**.) Set the *Snap Type* to **Polar Snap** and set the *Polar Distance* to **.5**. At the Status Bar, make sure that only *SNAP* and *POLAR* are on. Next, use the *Line* command and create the two shapes as shown in Figure 3-35. Specify the starting positions for each shape (**5.93,3.25** and **6.80,3.75**) by entering absolute coordinates.

Figure 3-35

15. When you are finished, **Save** the drawing and **Exit** AutoCAD.

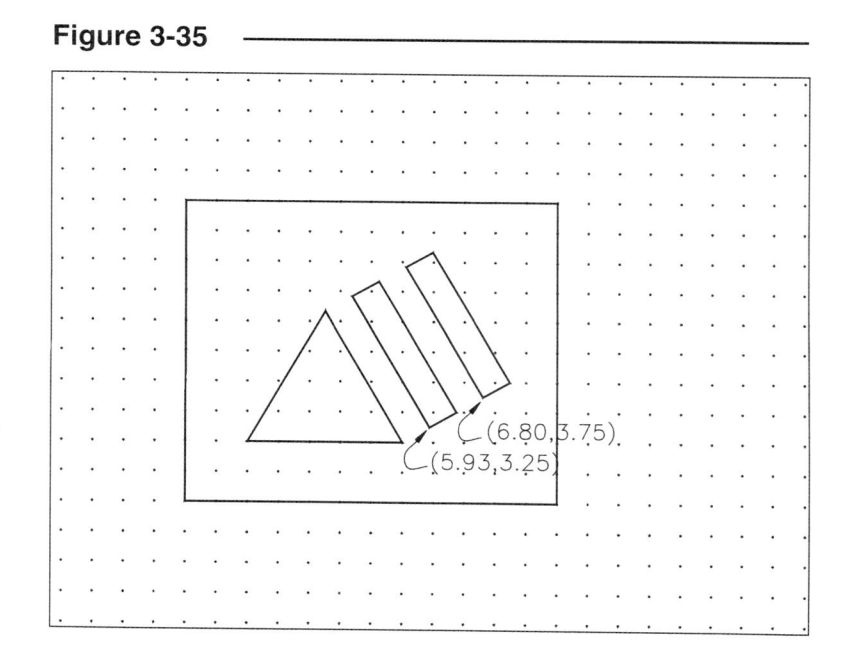

4

SELECTION SETS

Chapter Objectives

After completing this chapter you should:

1. know that *Modify* commands and many other commands require you to select objects;

2. be able to create a selection set using each of the specification methods;

3. be able to *Erase* objects from the drawing;

4. be able to *Move* and *Copy* objects from one location to another;

5. understand Noun/Verb and Verb/Noun order of command syntax.

MODIFY COMMAND CONCEPTS

Draw commands create objects. Modify commands change existing objects or use existing objects to create new ones. Examples are *Copy* an existing *Circle, Move* a *Line,* or *Erase* a *Circle.*

Since all of the Modify commands use or modify existing objects, you must first select the objects that you want to act on. The process of selecting the objects you want to use is called building a selection set. For example, if you want to *Copy, Erase,* or *Move* several objects in the drawing, you must first select the set of objects that you want to act on.

Remember that any of the five command entry methods (depending on your setup) can be used to invoke Modify commands.

1. **Toolbars** — *Modify* or *ModifyII* toolbar
2. **Pull-down menu** — *Modify* pull-down menu
3. **Screen menu** — *MODIFY1* or *MODIFY2* screen menus
4. **Keyboard** — Type the command name or command alias
5. **Tablet menu** — Select the command icon

All of the Modify commands will be discussed in detail later, but for now we will focus on how to build selection sets.

SELECTION SETS

No matter which of the five methods you use to invoke a Modify command, you must specify a selection set during the command operation. There are two ways you can select objects: you can select the set of objects either (1) immediately before you invoke the command or (2) when the command prompts you to select objects. For example, examine the command syntax that would appear at the Command line when the *Erase* command is used (method 2).

    Command: erase
    Select objects:

Figure 4-1

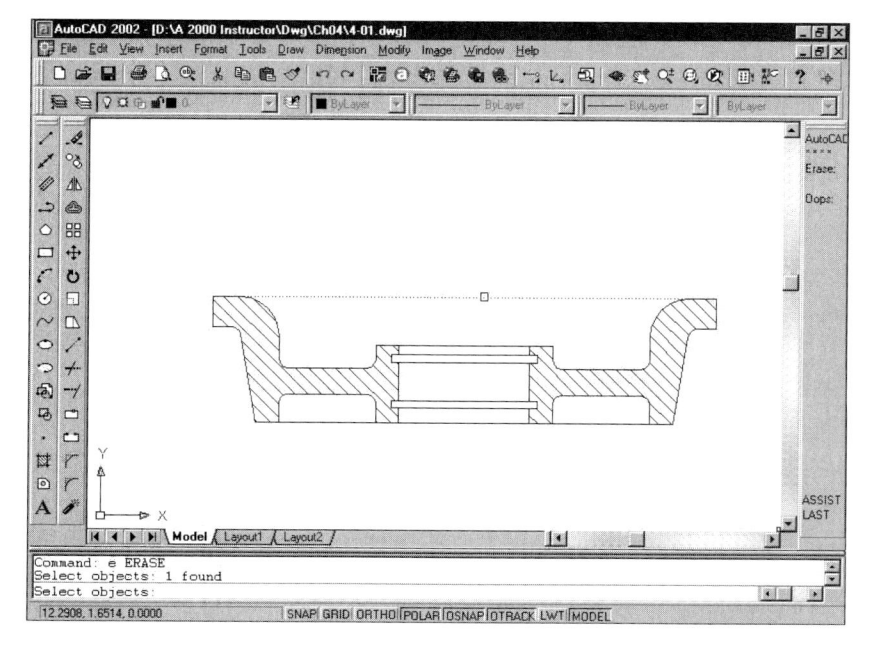

The "Select objects:" prompt is your cue to use any of several methods to PICK the objects to erase. As a matter of fact, every Modify command begins with the same "Select objects:" prompt (unless you selected immediately before invoking the command).

When the "Select objects:" prompt appears, the "crosshairs" cursor disappears and only a small, square pickbox appears at the cursor (Fig. 4-1).

You can PICK objects using only the pickbox or any of several other methods illustrated in this chapter.

Only when you PICK objects at the "Select objects:" prompt can you use all of the selection methods shown here. If you PICK immediately before the command, called Noun/Verb order, only the *AUto* method (pickbox, window, or crossing window) can be used. (See "Noun/Verb Syntax" in this chapter.)

When the objects have been selected, they become highlighted (displayed as a broken line), which serves as a visual indication of the current selection set. Press Enter to indicate that you are finished selecting and are ready to proceed with the command.

Accessing Selection Set Options

When the "Select objects:" prompt appears, you should select objects using the pickbox, window, crossing window, or one of a variety of other methods. Any method can be used independently or in combination to achieve the desired set of objects because object selection is a cumulative process.

The pickbox is the default option, which can automatically be changed to a window or crossing window by PICKing in an open area (PICKing no objects). Since the pickbox, window, and crossing window (sometimes known as the *AUto* option) are the default selection methods, no action is taken on your part to activate these methods. The other methods can be selected from the screen (side) menu or by typing the capitalized letters shown in the option names following.

All the possible selection set options are listed if you enter a question mark symbol (?) at the "Select objects:" prompt.

    Command: *move*
    Select objects: *?*
    Invalid selection
    Expects a point or
    Window/Last/Crossing/BOX/ALL/Fence/WPolygon/CPolygon/Group/Add/Remove/Multiple/
    Previous/Undo/AUto/SIngle
    Select objects:

Use any option by typing the indicated uppercase letters and pressing Enter.

If the screen (side) menu is displayed, you can access many of the object selection options from the *ASSIST* menu. *ASSIST* is displayed at the bottom of all other menus.

In AutoCAD 2002, there are no icon buttons or pull-down menu selections that are easily available for the object selection options. However, a toolbar can be "customized" to display the object selection options (see AutoCAD 2002 *Help*, Customization Guide).

Figure 4-2

Selection Set Options

The options for creating selection sets (PICKing objects) are shown on this and the following pages. Two *Circles* and five *Lines* (as shown in Figure 4-2) are used for every example. In each case, the circles only are selected for editing. If you want to follow along and practice as you read, draw *Circles* and *Lines* in a configuration similar to this. Then use a *Modify* command. (Press Escape to cancel the command after selecting objects.)

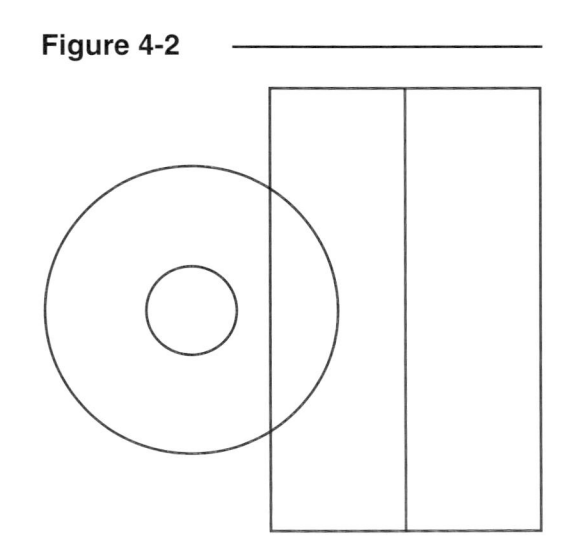

64 Selection Sets

AUto (**pickbox**)
This default option is used for selecting <u>one object</u> at a time. Locate the pickbox so that an object crosses through it and **PICK** (Fig. 4-3). You do not have to type or select anything to use this option.

Figure 4-3

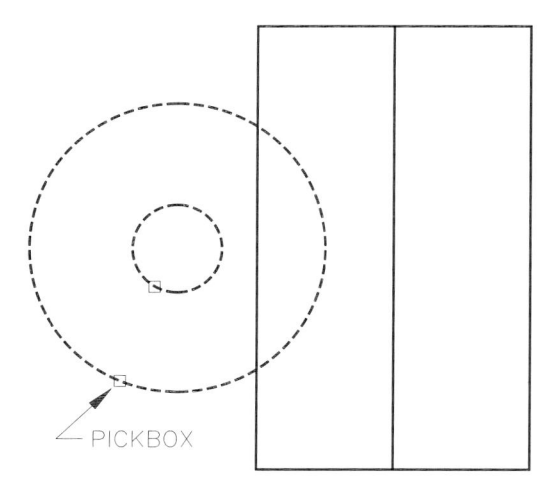

AUto (**window**)
To use this option, you do not have to type or select anything from the *Assist* screen menu. The pickbox must be positioned in an open area so that no objects cross through it; then **PICK** to start a window. If you drag to the <u>right</u>, a *Window* is created (Fig. 4-4). **PICK** the other corner.

Figure 4-4

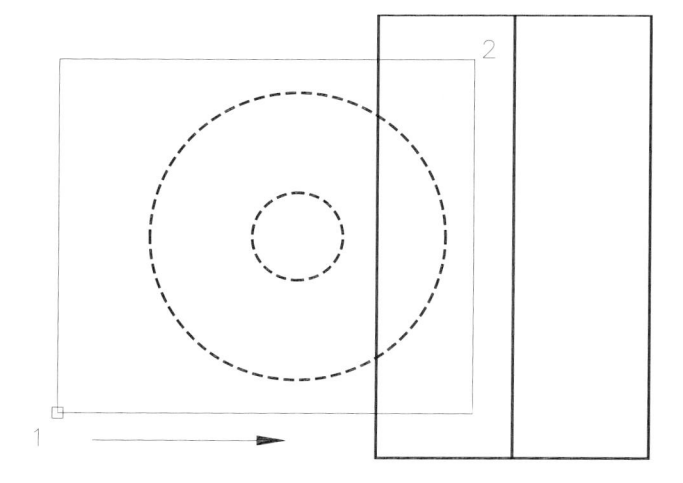

AUto (**crossing window**)
If you drag to the <u>left</u> instead, a *Crossing Window* forms (Fig. 4-5). (See "Window" and "Crossing Window" on the next page.)

Figure 4-5

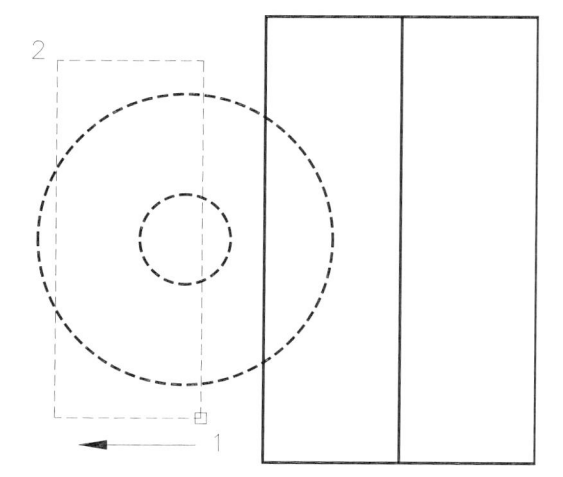

Window
Only objects <u>completely within</u> the *Window* are selected. The *Window* is a solid linetype rectangular box. Select the first and second points (diagonal corners) in any direction as shown in Figure 4-6.

Figure 4-6

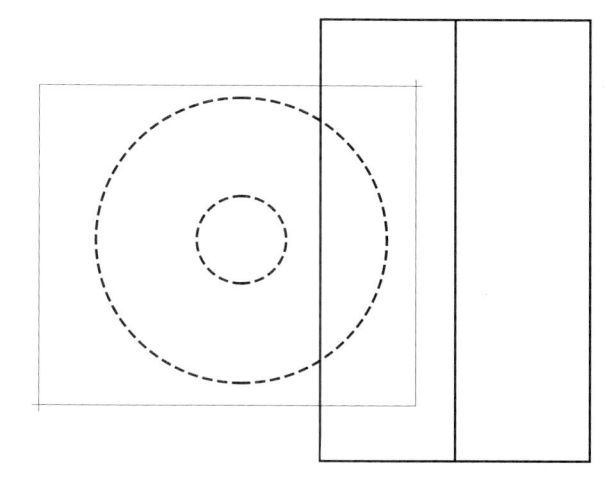

Crossing Window
All objects <u>within and crossing through</u> the window are selected. The *Crossing Window* is displayed as a broken linetype rectangular box (Fig. 4-7). Select two diagonal <u>corners</u>.

Figure 4-7

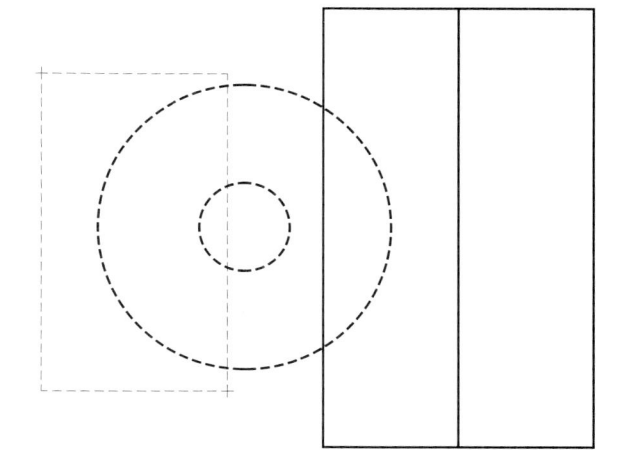

Window Polygon
The *Window Polygon* operates like a *Window,* but the box can be <u>any</u> irregular polygonal shape (Fig. 4-8). You can pick any number of corners to form any shape rather than picking just two corners to form a rectangle as with the *Window* option.

Figure 4-8

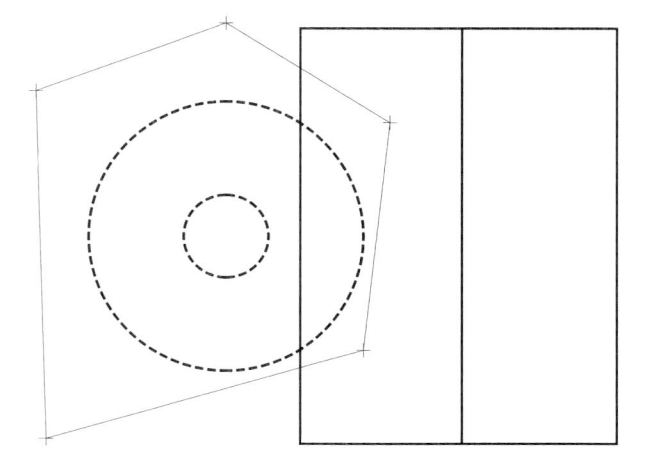

66 Selection Sets

Crossing Polygon
The *Crossing Polygon* operates like a *Crossing Window*, but can have any number of corners like a *Window Polygon* (Fig. 4-9).

Figure 4-9

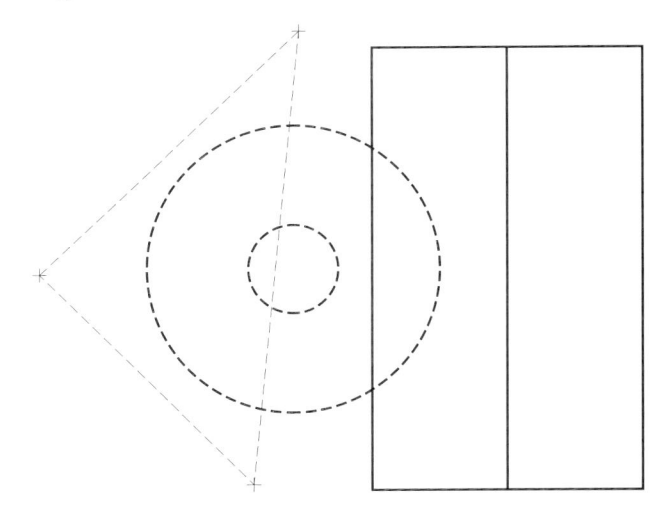

Fence
This option operates like a <u>crossing line</u>. Any objects crossing the *Fence* are selected. The *Fence* can have any number of segments (Fig. 4-10).

Figure 4-10

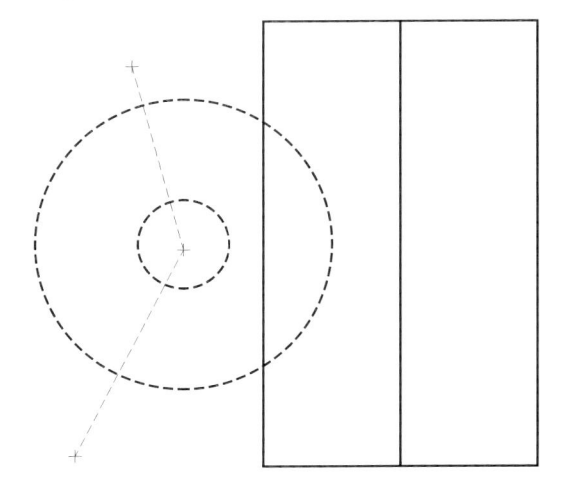

Last
This option automatically finds and selects <u>only</u> the last object <u>created</u>. *Last* does not find the last object modified (with *Move, Stretch*, etc.).

Previous
Previous finds and selects the <u>previous selection set</u>, that is, whatever was selected during the previous command (except after *Erase*). This option allows you to use several editing commands on the same set of objects without having to respecify the set.

ALL
This option selects <u>all objects</u> in the drawing except those on *Frozen* or *Locked* layers (*Layers* are covered in Chapter 11).

BOX
This option is equivalent to the *AUto* window/crossing window option <u>without</u> the pickbox. PICKing diagonal corners from left to right produces a window and PICKing diagonal corners from right to left produces a crossing window (see *AUto*).

Multiple

The *Multiple* option allows selection of objects with the pickbox only; however, the selected objects are not highlighted. Use this method to select very complex objects to save computing time required to change the objects' display to highlighted.

Single

This option allows only a single selection using one of the *AUto* methods (pickbox, window, or crossing window), then automatically continues with the command. Therefore, you can select multiple objects (if the window or crossing window is used), but only one selection is allowed. You do not have to press Enter after the selection is made.

Undo

Use *Undo* to cancel the selection of the object(s) most recently added to the selection set.

Remove

Selecting this option causes AutoCAD to switch to the "Remove objects:" mode. Any selection options used from this time on remove objects from the highlighted set (see Figure 4-11).

Figure 4-11

Add

The *Add* option switches back to the default "Select objects:" mode so additional objects can be added to the selection set.

Shift + Left Button

Holding down the **Shift** key and pressing the left button simultaneously removes objects selected from the highlighted set as shown in Figure 4-11. This method is generally quicker than, but performs the same action as, *Remove*. The advantage here is that the *Add* mode is in effect unless Shift is held down.

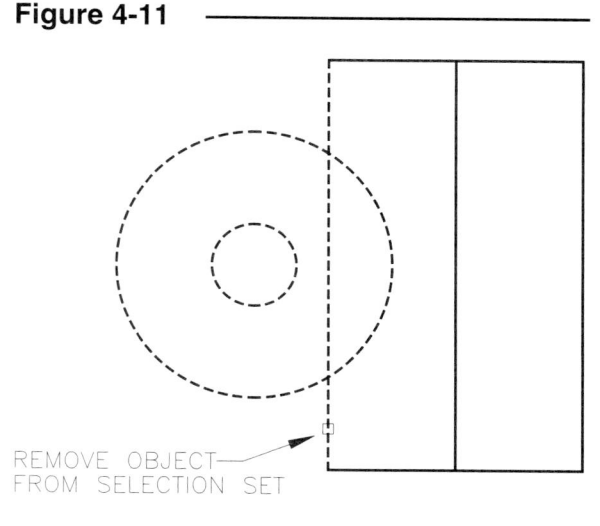

Group

The *Group* option selects groups of objects that were previously specified using the *Group* command. Groups are selection sets to which you can assign a name.

Ctrl + Left Button (Object Cycling)

Holding down the **Ctrl** key and simultaneously pressing the **PICK** (left) button causes AutoCAD to cycle through (highlight one at a time) two or more objects that may cross through the pickbox.

Figure 4-12

For example, if you attempted to PICK an object but other objects were in the pickbox, AutoCAD may not highlight the one you want (Fig. 4-12). In this case, hold down the **Ctrl** key and press the **PICK** button several times (the cursor can be located anywhere on the screen during cycling). All of the objects that passed through the pickbox will be highlighted one at a time. The "<Cycle on>" prompt appears at the Command line. When the object you want is highlighted, press Enter and AutoCAD adds the object to the selection set and returns to the "Select objects:" prompt.

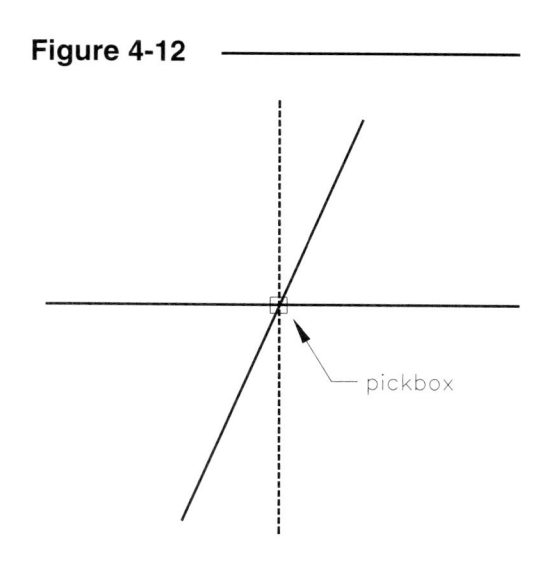

68 Selection Sets

SELECT

Pull-down Menu	COMMAND (TYPE)	ALIAS (TYPE)	Short-cut	Screen (side) Menu	Tablet Menu
...	SELECT	...	...	...	...

The *Select* command can be used to PICK objects to be saved in the selection set buffer for subsequent use with the *Previous* option. Any of the selection methods can be used to PICK the objects.

```
Command: select
Select objects: PICK (use any selection option)
Select objects: Enter (completes the selection process)
Command:
```

The selected objects become unhighlighted when you complete the command by pressing Enter. The objects become highlighted again and are used as the selection set if you use the *Previous* selection option in the next editing command.

NOUN/VERB SYNTAX

An object is the noun and a command is the verb. Noun/Verb syntax order means to pick objects (nouns) first, then use an editing command (verb) second. If you select objects first (at the open Command: prompt) and then immediately choose a Modify command, AutoCAD recognizes the selection set and passes through the "Select objects:" prompt to the next step in the command.

Verb/Noun means to invoke a command and then select objects within the command. For example, if the *Erase* command (verb) is invoked first, AutoCAD then issues the prompt to "Select objects:"; therefore, objects (nouns) are PICKed second. In the previous examples, and with much older versions of AutoCAD, only Verb/Noun syntax order was used.

You can use either order you want (Noun/Verb or Verb/Noun) and AutoCAD automatically understands. If objects are selected first, the selection set is passed to the next editing command used, but if no objects are selected first, the editing command automatically prompts you to "Select objects:".

If you use Noun/Verb order, you are limited to using only the *AUto* options for object selection (pickbox, *Window*, and *Crossing Window*). You can only use the other options (e.g., *Crossing Polygon, Fence, Previous*) if you invoke the desired Modify command first, then select objects when the "Select objects:" prompt appears.

The *PICKFIRST* variable (a very descriptive name) enables Noun/Verb syntax. The default setting is 1 (*On*). If *PICKFIRST* is set to 0 (*Off*), Noun/Verb syntax is disabled and the selection set can be specified only within the editing commands (Verb/Noun).

Setting *PICKFIRST* to 1 provides two options: Noun/Verb and Verb/Noun. You can use either order you want. If objects are selected first, the selection set is passed to the next editing command, but if no objects are selected first, the editing command prompts you to select objects.

SELECTION SETS PRACTICE

NOTE: While learning and practicing with the editing commands, it is suggested that *GRIPS* be turned off. This can be accomplished by typing in **GRIPS** and setting the *GRIPS* variable to a value of **0**. The AutoCAD default has *GRIPS* on (set to 1). *GRIPS* are covered in Chapter 21.

Begin a *New* drawing to complete the selection set practice exercises. If the *Startup* dialog box or *Today* window appears, select *Start from Scratch,* choose *English* as the default setting, and click the *OK* button. The *Erase, Move,* and *Copy* commands can be activated by any one of the methods shown in the command tables that follow.

Using *Erase*

Erase is the simplest editing command. *Erase* removes objects from the drawing. The only action required is the selection of objects to be erased.

ERASE

Pull-down Menu	COMMAND (TYPE)	ALIAS (TYPE)	Short-cut	Screen (side) Menu	Tablet Menu
Modify Erase	ERASE	E	(Edit Mode) Erase	MODIFY1 Erase	V,14

1. Draw several *Lines* and *Circles*. Practice using the object selection options with the *Erase* command. The following sequence uses the pickbox, *Window*, and *Crossing Window*.

STEPS	COMMAND PROMPT	PERFORM ACTION	COMMENTS
1.	Command:	type *E* and press **Spacebar**	*E* is the alias for *Erase*, Spacebar can be used like Enter
2.	ERASE Select objects:	use pickbox to select one or two objects	objects are highlighted
3.	Select objects:	type *W* or use a *Window*, then select more objects	objects are highlighted
4.	Select objects:	type *C* or use a *Crossing Window*, then select objects	objects are highlighted
5.	Select objects:	press **Enter**	objects are erased

2. Draw several more *Lines* and *Circles*. Practice using the *Erase* command with the *AUto Window* and *AUto Crossing Window* options as indicated below.

STEPS	COMMAND PROMPT	PERFORM ACTION	COMMENTS
1.	Command:	select the *Modify* pull-down, then *Erase*	
2.	ERASE Select objects:	use pickbox to **PICK** an open area, drag *Window* to the <u>right</u> to select objects	objects inside *Window* are highlighted
3.	Select objects:	**PICK** an open area, drag *Crossing Window* to the <u>left</u> to Select objects	objects inside and crossing through *Window* are highlighted
4.	Select objects:	press **Enter**	objects are erased

Using *Move*

The *Move* command specifically prompts you to (1) select objects, (2) specify a "base point," a point to move from, and (3) specify a "second point of displacement," a point to move to.

MOVE

Pull-down Menu	COMMAND (TYPE)	ALIAS (TYPE)	Short-cut	Screen (side) Menu	Tablet Menu
Modify Move	MOVE	M	(Edit Mode) Move	MODIFY2 Move	V,19

1. Draw a *Circle* and two *Lines*. Use the *Move* command to practice selecting objects and to move one *Line* and the *Circle* as indicated in the following table.

STEPS	COMMAND PROMPT	PERFORM ACTION	COMMENTS
1.	Command:	type *M* and press **Spacebar**	*M* is the command alias for *Move*
2.	MOVE Select objects:	use pickbox to select one *Line* and the *Circle*	objects are highlighted
3.	Select objects:	press **Spacebar** or **Enter**	
4.	Specify base point or displacement:	**PICK** near the *Circle* center	base point is the handle, or where to move from
5.	Specify second point of displacement:	**PICK** near the other *Line*	second point is where to move to

2. Use *Move* again to move the *Circle* back to its original position. Select the *Circle* with the *Window* option.

STEPS	COMMAND PROMPT	PERFORM ACTION	COMMENTS
1.	Command:	select the *Modify* pull-down, then *Move*	
2.	MOVE Select objects:	type *W* and press **Spacebar**	select only the circle; object is highlighted
3.	Select objects:	press **Spacebar** or **Enter**	
4.	Specify base point or displacement:	**PICK** near the *Circle* center	base point is the handle, or where to move from
5.	Specify second point of displacement:	**PICK** near the original location	second point is where to move to

Using *Copy*

The *Copy* command is similar to *Move* because you are prompted to (1) select objects, (2) specify a "base point," a point to copy from, and (3) specify a "second point of displacement," a point to copy to.

COPY

Pull-down Menu	COMMAND (TYPE)	ALIAS (TYPE)	Short-cut	Screen (side) Menu	Tablet Menu
Modify Copy	COPY	CO or CP	(Edit Mode) Copy Selection	MODIFY1 Copy	V,15

1. Using the *Circle* and 2 *Lines* from the previous exercise, use the *Copy* command to practice selecting objects and to make copies of the objects as indicated in the following table.

STEPS	COMMAND PROMPT	PERFORM ACTION	COMMENTS
1.	Command:	type *CO* and press **Enter** or **Spacebar**	CO is the command alias for *Copy*
2.	COPY Select objects:	type *F* and press **Enter**	F invokes the *Fence* selection option
3.	First fence point:	**PICK** a point near two *Lines*	starts the "fence"
4.	Specify endpoint of line or [Undo]:	**PICK** a second point across the two *Lines*	lines are highlighted
5.	Specify endpoint of line or [Undo]:	Press **Enter**	completes *Fence* option
6.	Select objects:	Press **Enter**	completes object selection
7.	Specify base point or displacement:	**PICK** between the *Lines*	base point is the handle, or where to copy from
8.	Specify second point of displacement:	enter @2<45 and press **Enter**	copies of the lines are created 2 units in distance at 45 degrees from the original location

2. Practice removing objects from the selection set by following the steps given in the table below.

STEPS	COMMAND PROMPT	PERFORM ACTION	COMMENTS
1.	Command:	type **CO** and press **Enter** or **Spacebar**	*CO* is the command alias for *Copy*
2.	COPY Select objects:	**PICK** in an open area near the right of your drawing, drag *Window* to the left to select all objects	all objects within and crossing the *Window* are highlighted
3.	Select objects:	hold down **Shift** and **PICK** all highlighted objects except one *Circle*	holding down Shift while PICKing removes objects from the selection set
4.	Select objects:	press **Enter**	only one circle is highlighted
5.	Specify base point or displacement or [Multiple]:	**PICK** near the circle's center	base point is the handle, or where to copy <u>from</u>
6.	Specify second point of displacement:	turn on **ORTHO**, move the cursor to the right, type **3** and press **Enter**	a copy of the circle is created 3 units to the right of the original circle

Noun/Verb Command Syntax Practice

1. Practice using the *Move* command using Noun/Verb syntax order by following the steps in the table below.

STEPS	COMMAND PROMPT	PERFORM ACTION	COMMENTS
1.	Command:	**PICK** one *Circle*	circle becomes highlighted (grips may appear if not disabled)
2.	Command:	type **M** and press **Enter** or **Spacebar**	*M* is the command alias for *Move*
3.	MOVE 1 found Specify base point or displacement:	**PICK** near the *Circle's* center	command skips the "Select objects:" prompt and proceeds
4.	Specify second point of displacement:	enter **@-3,-3** and press **Enter**	circle is moved -3 X units and -3 Y units from the original location

2. Practice using Noun/Verb syntax order by selecting objects for *Erase*.

STEPS	COMMAND PROMPT	PERFORM ACTION	COMMENTS
1.	Command:	**PICK** one *Line*	line becomes highlighted (grips may appear if not disabled)
2.	Command:	type *E* and press **Enter** or **Spacebar**	*E* is the command alias for *Erase*
3.	ERASE 1 found		command skips the "Select objects:" prompt and erases highlighted line

CHAPTER EXERCISES

Open drawing **CH3EX1** that you created in Chapter 3 Exercises. Turn off *SNAP* (**F9**) to make object selection easier.

1. **Use the pickbox to select objects**

   Invoke the *Erase* command by any method. Select the lower-left square with the pickbox (Fig. 4-13, highlighted). Each *Line* must be selected individually. Press **Enter** to complete *Erase*. Then use the *Oops* command to unerase the square. (Type *Oops*.)

   Figure 4-13

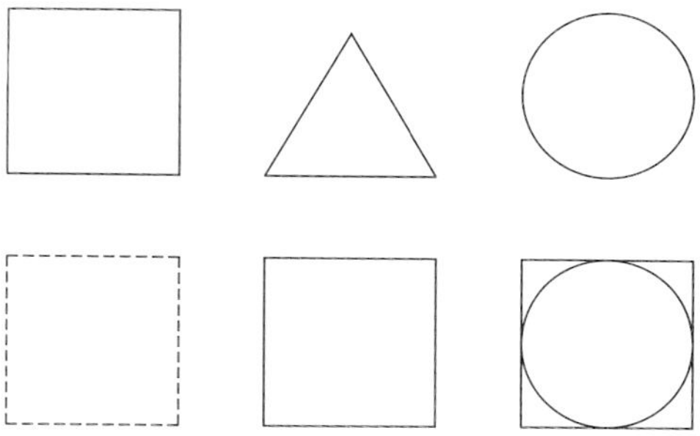

2. **Use the *AUto Window* and *AUto Crossing Window***

   Invoke *Erase*. Select the center square on the bottom row with the *AUto Window* and select the equilateral triangle with the *AUto Crossing Window*. Press **Enter** to complete the *Erase* as shown in Figure 4-14. Use *Oops* to bring back the objects.

   Figure 4-14

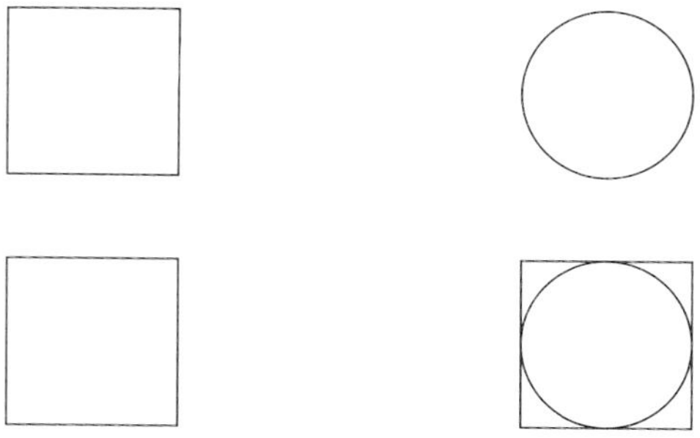

74 Selection Sets

3. **Use the *Fence* selection option**

   Invoke *Erase* again. Use the **Fence** option to select all the vertical *Lines* and the *Circle* from the squares on the bottom row. Complete the *Erase* (see Fig. 4-15). Use *Oops* to unerase.

   Figure 4-15

4. **Use the *ALL* option and deselect**

   Use *Erase*. Select all the objects with *ALL*. Remove the four *Lines* (shown highlighted in Fig. 4-16) from the selection set by pressing **Shift** while PICKing. Complete the *Erase* to leave only the four *Lines*. Finally, use *Oops*.

   Figure 4-16

5. **Use Noun/Verb selection**

   <u>Before</u> invoking *Erase*, use the pickbox or *AUto Window* to select the triangle. (Make sure no other commands are in use.) <u>Then</u> invoke *Erase*. The triangle should disappear. Retrieve the triangle with *Oops*.

6. **Use *Move* with *Wpolygon***

   Invoke the *Move* command by any method. Use the **WP** option (*Window Polygon*) to select only the *Lines* comprising the triangle. Turn on *SNAP* and PICK the lower-left corner as the "Base point:". *Move* the triangle up 1 unit. (See Fig. 4-17.)

   Figure 4-17

7. **Use *Previous* with *Move***

   Invoke *Move* again. At the "Select objects:" prompt, type **P** for *Previous*. The triangle should highlight. Using the same base point, move the triangle back to its original position.

8. *Exit* AutoCAD and do <u>not</u> save changes.

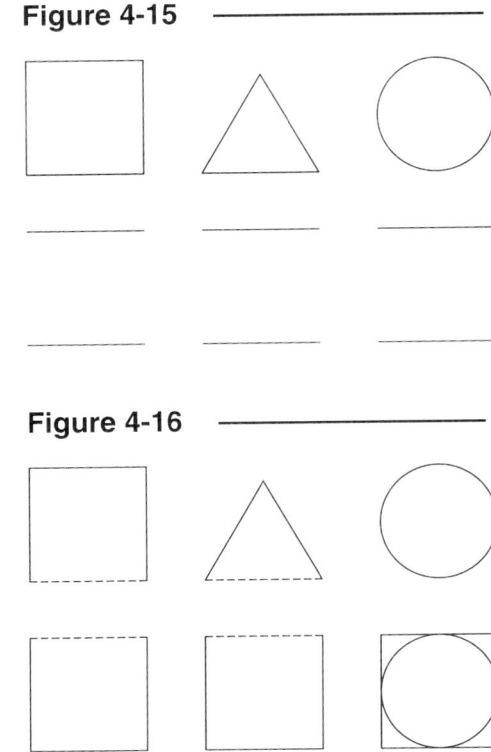

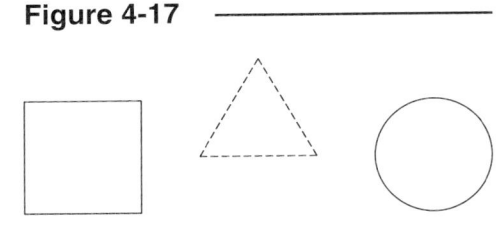

5

HELPFUL COMMANDS

Chapter Objectives

After completing this chapter you should be able to:

1. find *Help* for any command or system variable;

2. use the *Active Assistance* window to provide an active help window based on the *Settings* you prefer;

3. use *Oops* to unerase objects;

4. use *U* to undo one command or use *Undo* to undo multiple commands;

5. *Redo* commands that were undone;

6. regenerate the drawing with *Regen*.

CONCEPTS

There are several commands that do not draw or edit objects in AutoCAD, but are intended to assist you in using AutoCAD. These commands are used by experienced AutoCAD users and are particularly helpful to the beginner. The commands, as a group, are not located in any one menu, but are scattered throughout several menus.

HELP

Pull-down Menu	COMMAND (TYPE)	ALIAS (TYPE)	Short-cut	Screen (side) Menu	Tablet Menu
Help Help	HELP	?	F1	HELP Help	Y,7

Help gives you an explanation for any AutoCAD command or system variable as well as help for using the menus and toolbars. *Help* displays a window that gives a variety of methods for finding the information that you need. There is even help for using *Help*!

Help can be used two ways: (1) entered as a command at the open Command: prompt or (2) used transparently while a command is currently in use.

1. If the *Help* command is entered at an open Command: prompt (when no other commands are in use), the *Help* window appears (Fig. 5-1).

2. When *Help* is used transparently (when a command is in use), it is context sensitive; that is, help on the current command is given automatically. For example, Figure 5-2 displays the window that appears if *Help* is invoked during the *Line* command. (If typing a transparent command, an apostrophe (') symbol is typed as a prefix to the command, such as, '**HELP** or '**?**. If you PICK *Help* from the menus or press **F1**, it is automatically transparent.)

Figure 5-1

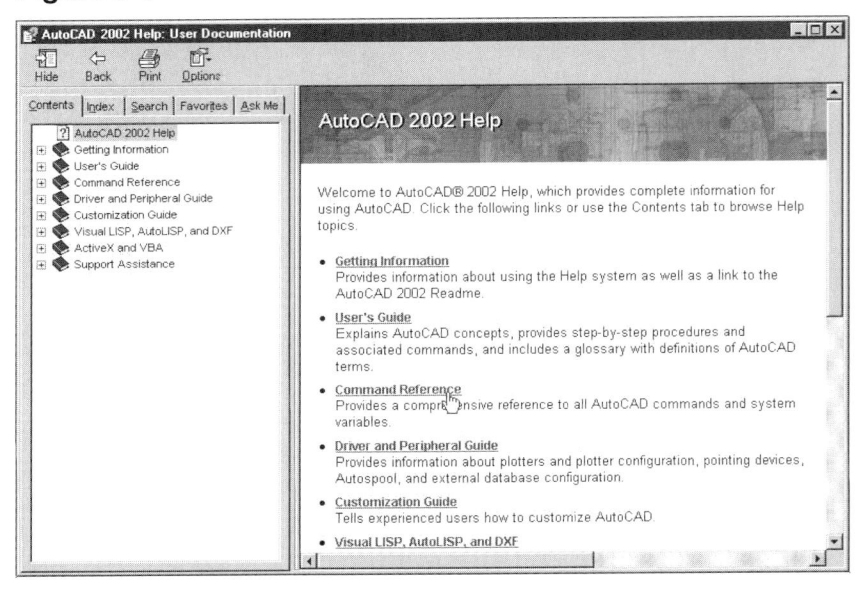

Figure 5-2

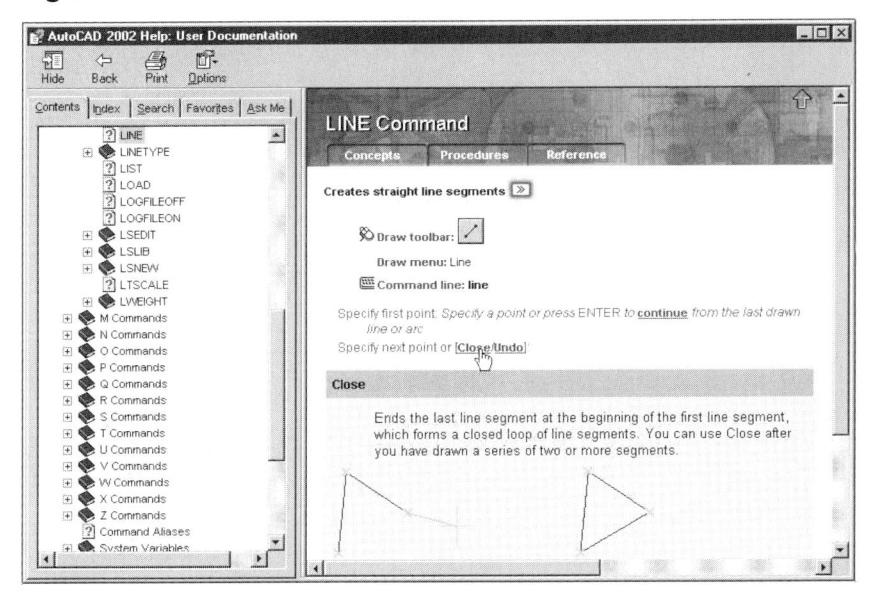

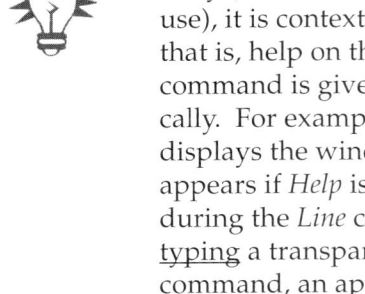

Much of the text that appears in the window can be PICKed to reveal another level of help on that item. This feature, called hypertext, is activated by moving the pointer to a word (usually underlined) or an icon. When the pointer changes to a small hand, click on the item to activate the new information.

The five tabs in the AutoCAD 2002 Help window, *Contents*, *Index*, *Search*, *Favorites*, and *Ask Me*, are described next.

Contents
Several levels of the *Contents* tab are available, offering an overwhelming amount of information (Fig. 5-3). Each main level can be opened to reveal a second (chapter level) or third level of information. The main levels are:

Figure 5-3

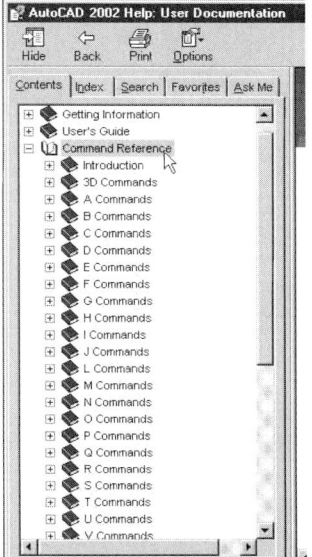

AutoCAD 2002 Help	This section explains the other features listed below.
Getting Information	Get instruction on how to use the Help system and the Active Assistance feature, locate e-learning training sessions available through the Internet, check for interim documentation updates, and view the readme file.
Command Reference	The following informative sections are available:
	Introduction
	Commands
	Command Aliases
	System Variables
	Command Modifiers
	Dimension Variables
	Utilities
	Standard Libraries
	AutoCAD Graphical Objects
	Using Unicode Fonts
User's Guide	Eight "chapters" and a glossary are given, each with several sections explaining concepts and commands.
Driver and Peripheral Guide	Get information on installing, configuring, and optimizing peripheral devices and drivers.
Customization Guide	This section contains information on how to customize linetypes, shapes, fonts, menus, templates, and dialog boxes.
Visual Lisp, AutoLISP, and DXF *ActiveX and VBA*	These two sections connect you to the Autodesk Web site where the most current documentation describing AutoCAD's programming tools is available.
Support Assistance	Get help on support issues and solutions to problems that may occur. Topics range from installation to using normal AutoCAD commands to printing and plotting. The topics are presented in an "Issue" and matching "Solution" format. Some of these options connect you to the Autodesk Web site for the most current information.

78 Helpful Commands

Index
In this section, you can type two or three letters of a word that you want information about (Fig. 5-4). As you type, the list below displays the available contents beginning with the letters that you have entered. The word you type can be a command, system variable, term, or concept used in AutoCAD. Once the word you want information on is found, press Enter or click the *Display* button to display the related information.

Figure 5-4

Search
Search operates like the *Index* function, except you can enter several words describing the topic you want help with (Fig. 5-5). There are three steps to using the *Search* function: (1) type in word or words in the top edit box, (2) select a topic from the list in the center to narrow your search, (3) select an option from below and click *Display*. The related information appears.

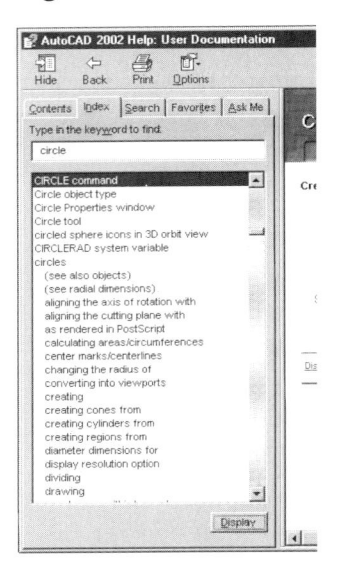

Figure 5-5

Favorites
The *Favorites* section of the AutoCAD 2002 Help window (Fig. 5-6) is used to save sections of Help that you may want to refer to in the future. For example, tables or lists such as the list of command aliases or system variables make good candidates for the *Favorites* section. To save sections to the *Favorites* list: (1) use the *Contents, Index,* or *Search* tabs to locate the information you need, then (2) access the *Favorites* tab and press the *Add* button. To access the information in the *Topics* list, highlight the desired topic and press *Display*.

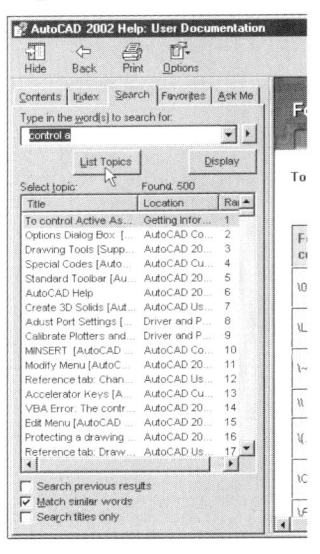

Figure 5-6

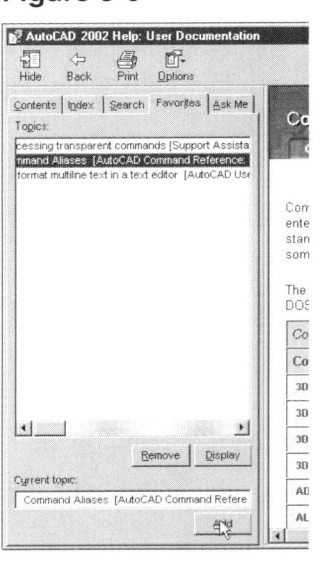

Ask Me
This tool (Fig. 5-7) allows you to access the same information that is available by the other methods, but you can ask for the information in sentence (query) format. In other words, type a question in the top edit box. You can specify which document you want AutoCAD to search in by making a selection from the *List of components to search*. Once the list of matching topics is found, select the desired topic from the list to display the related information.

Figure 5-7

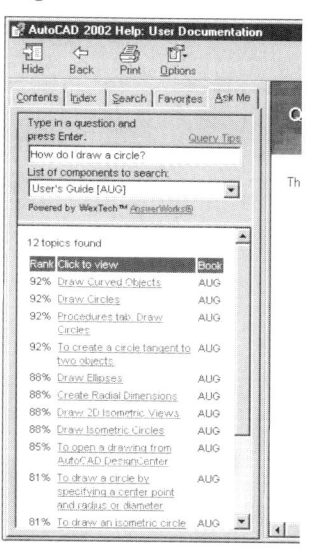

ASSIST

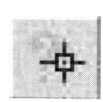

Pull-down Menu	COMMAND (TYPE)	ALIAS (TYPE)	Shortcut	Screen (side) Menu	Tablet Menu
Help Active Assistance	ASSIST	...	...	...	...

An additional form of help is offered in AutoCAD 2002 called *Active Assistance*. Invoke *Active Assistance* by typing *Assist* or *'Assist* (for transparent use) or select it by using any of the methods shown in the command table above. *Active Assistance* is set to appear by default when you first install and start AutoCAD 2002.

Figure 5-8

Active Assistance is an active, rather than passive, help screen on the current command (Fig. 5-8). In other words, *Active Assistance* appears automatically by default when you use a command instead of waiting for you to ask for help. The *Active Assistance* window remains on the screen while you use the command and disappears when you finish the command.

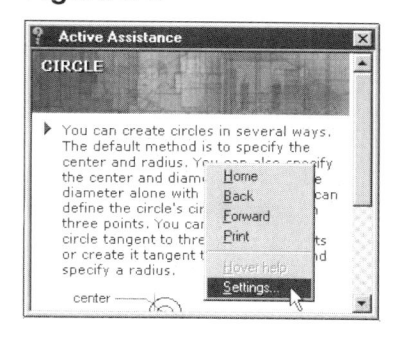

Figure 5-9

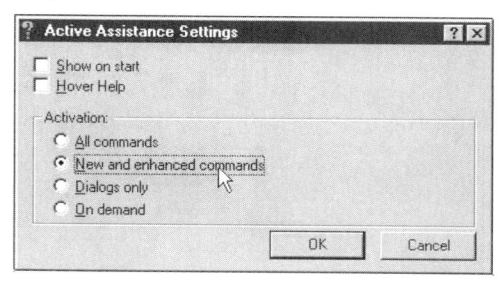

To set your preferences for *Active Assistance*, right-click inside the *Active Assistance* window or take your cursor to the system tray (usually on the lower-right corner of your screen) and right-click on the *Active Assistance* icon. From the menu that appears, select *Settings....* This action causes the *Active Assistance Settings* dialog box to appear (Fig. 5-9). *Active Assistance* can be made to appear when you start AutoCAD and remain on the screen until you close it (*Show on start*), appear with and disappear after each command (*All commands*), appear with only some commands (*New and enhanced commands* or *Dialogs only*), or made to appear only when you use the *Assist* command (*On demand*). For example, if you are an experienced AutoCAD user, you might want to set *Active Assistance* to appear only for *New and enhanced commands*, in which case the dialog box appears only for those features new with AutoCAD 2002.

80 *Helpful Commands*

> 💡 **TIP (2002)** The *Hover Help* option makes the *Active Assistance* window a "smart," dynamic help feature. *Hover Help* causes *Active Assistance* to display information about a specific option in any dialog box when you pass the cursor over that item. As you move the cursor over other options, information about those options is displayed.

OOPS

Pull-down Menu	COMMAND (TYPE)	ALIAS (TYPE)	Short-cut	Screen (side) Menu	Tablet Menu
...	OOPS	...	...	MODIFY1 Erase Oops:	...

The *Oops* command unerases whatever was erased with the <u>last</u> *Erase* command. *Oops* does not have to be used immediately after the *Erase*, but can be used at <u>any time after</u> the *Erase*. *Oops* is typically used after an accidental erase. However, *Erase* could be used intentionally to remove something from the screen temporarily to simplify some other action. For example, you can *Erase* a *Line* to simplify PICKing a group of other objects to *Move* or *Copy*, and then use *Oops* to restore the erased *Line*.

Oops can be used to restore the original set of objects after the *Block* or *Wblock* command is used to combine many objects into one object (explained in Chapter 20).

The *Oops* command is available only from the side menu; there is no icon button or option available in the pull-down menu. *Oops* appears on the side menu only <u>if</u> the *Erase* command is typed or selected. Otherwise, *Oops* must be typed.

U

Pull-down Menu	COMMAND (TYPE)	ALIAS (TYPE)	Short-cut	Screen (side) Menu	Tablet Menu
Edit Undo	U	...	Crtl+Z or (Default Menu) Undo	EDIT Undo	T,12

> 💡 **TIP** The *U* command undoes only the <u>last</u> command. *U* means "undo one command." If used after *Erase*, it unerases whatever was just erased. If used after *Line*, it undoes the group of lines drawn with the last *Line* command. Both *U* and *Undo* do not undo inquiry commands (like *Help*), the *Plot* command, or commands that cause a write-to-disk, such as *Save*.

If you type the letter **U**, select the icon button, or select **Undo** from the pull-down menu or screen (side) menu, only the last command is undone. Typing **Undo** invokes the full *Undo* command. The *Undo* command is explained next.

UNDO

Pull-down Menu	COMMAND (TYPE)	ALIAS (TYPE)	Short-cut	Screen (side) Menu	Tablet Menu
...	UNDO	...	...	ASSIST Undo	...

The full *Undo* command (as opposed to the *U* command) must be typed. This command has the same effect as the *U* command in that it undoes the previous command(s). However, the *Undo* command allows you to <u>undo multiple commands</u> in reverse chronological order.

For example, you can use *Undo* and enter a value of **5** to undo the last 5 commands you used. Entering a value (specifying the number of commands to undo) is the default option of *Undo*. The command syntax is as follows:

    Command: undo
    Enter the number of operations to undo or [Auto/Control/BEgin/End/Mark/Back]:

All of the *Undo* options are listed next.

<number>
Enter a value for the number of commands to *Undo*. This is the default option.

Mark
This option sets a marker at that stage of the drawing. The marker is intended to be used by the *Back* option for future *Undo* commands.

Back
This option causes *Undo* to go back to the last marker encountered. Markers are created by the *Mark* option. If a marker is encountered, it is removed. If no marker is encountered, beware, because *Undo* goes back to the beginning of the session. A warning message appears in this case.

BEgin
This option sets the first designator for a group of commands to be treated as one *Undo*.

End
End sets the second designator for the end of a group.

Auto
If *On*, *Auto* treats each command as one group; for example, several lines drawn with one *Line* command would all be undone with *U*.

Control
This option allows you to disable the *Undo* command or limit it to one undo each time it is used.

REDO

Pull-down Menu	COMMAND (TYPE)	ALIAS (TYPE)	Short-cut	Screen (side) Menu	Tablet Menu
Edit Redo	REDO	...	Crtl+Y or (Default Menu) Redo	EDIT Redo	U,12

The *Redo* command undoes an *Undo*. *Redo* must be used as the next command after the *Undo*. The result of *Redo* is as if *Undo* was never used. Remember, *U* or *Undo* can be used at any time, but *Redo* has no effect unless used immediately after *U* or *Undo*.

REGEN

Pull-down Menu	COMMAND (TYPE)	ALIAS (TYPE)	Short-cut	Screen (side) Menu	Tablet Menu
View Regen	REGEN	RE	...	VIEW 1 Regen	J,1

The *Regen* command reads the database and redisplays the drawing accordingly. A *Regen* is caused by some commands automatically. Occasionally, the *Regen* command is required to update the drawing to display the latest changes made to some system variables. *Regenall* is used to regenerate all viewports when several viewports are being used.

CHAPTER EXERCISES

Start AutoCAD. If the *Startup* dialog box or *Today* window appears, select **Start from Scratch**, choose **English** as the default setting, and click the **OK** button. Complete the following exercises.

1. *Help*

   Use *Help* by any method to find information on the following commands. Use the *Contents* tab and select the *Command Reference* to locate information on each command. (Use *Help* at the open Command: prompt, not during a command in use.) Read the text screen for each command.

   New, Open, Save, SaveAs
   Oops, Undo, U

2. **Context-sensitive** *Help*

   Invoke each of the commands listed below. When you see the first prompt in each command, enter '*Help* or '? (transparently) or select **Help** from the menus or Standard toolbar. Read the explanation for each prompt. Select a hypertext item in each screen (underlined).

   Line, Arc, Circle, Point

3. *Active Assistance*

   If the *Active Assistance* window is not currently visible on your screen, invoke it by using the *Active Assistance* icon button or by selecting it from the *Help* pull-down menu. Right-click inside the *Active Assistance* window to produce the shortcut menu and select the *All commands* option. Next, use the following commands and notice that the *Active Assistance* window appears and disappears with each command.

   Line, Circle, Move, and *Copy*

   Next, access the *Active Assistance* window *Settings* dialog box again and select the *New and enhanced commands* option, then use the commands below (cancel after invoking each command by pressing Esc). The *Active Assistance* window should appear for *Publishtoweb*, but not for *Line*.

   Line and *Publishtoweb*

   Access the *Settings* dialog box again and select both the *Hover Help* and *Dialogs only* options. Use the following commands. Move your pointer over different sections of the dialog boxes that appear and read the text that appears in the *Active Assistance* window. Cancel each command.

Open and *Plot*

Finally, select the *On demand* option in the *Settings* dialog box. Use a few more commands and notice that the *Active Assistance* window does not appear unless you invoke the *Active Assistance* command.

4. *Oops*

    Draw 3 vertical **Lines**. *Erase* one line; then use *Oops* to restore it. Next **Erase** two *Lines*, each with a separate use of the *Erase* command. Use *Oops*. Only the last *Line* is restored. **Erase** the remaining two *Lines*, but select both with a *Window*. Now use *Oops* to restore both *Lines* (since they were *Erased* at the same time).

5. **Delayed** *Oops*

    Oops can be used at any time, not only immediately after the *Erase*. Draw several horizontal **Lines** near the bottom of the screen. Draw a *Circle* on the **Lines**. Then *Erase* the *Circle*. Use *Move*, select the *Lines* with a *Window*, and displace the *Lines* to another location above. Now use *Oops* to make the *Circle* reappear.

6. *U*

    Press the letter **U** (make sure no other commands are in use). The *Circle* should disappear (*U* undoes the last command—*Oops*). Do this repeatedly to *Undo* one command at a time until the *Circle* and *Lines* are in their original position (when you first created them).

7. *Undo*

    Use the **Undo** command and select the **Back** option. Answer **Yes** to the warning message. This action should *Undo* everything.

    Draw a vertical **Line**. Next, draw a square with four **Line** segments (all drawn in the same *Line* command). Finally, draw a second vertical **Line**. *Erase* the first *Line*.

    Now type **Undo** and enter a value of **3**. You should have only one *Line* remaining. *Undo* reversed the following three commands:

Erase	The first vertical *Line* was unerased.
Line	The second vertical *Line* was removed.
Line	The four *Lines* comprising the square were removed.

8. *Redo*

    Invoke **Redo** immediately after the *Undo* (from the previous exercise). The three commands are redone. *Redo* must be used immediately after *Undo*.

9. *Exit* AutoCAD and answer *No* to "Save Changes to...?"

84 **Helpful Commands**

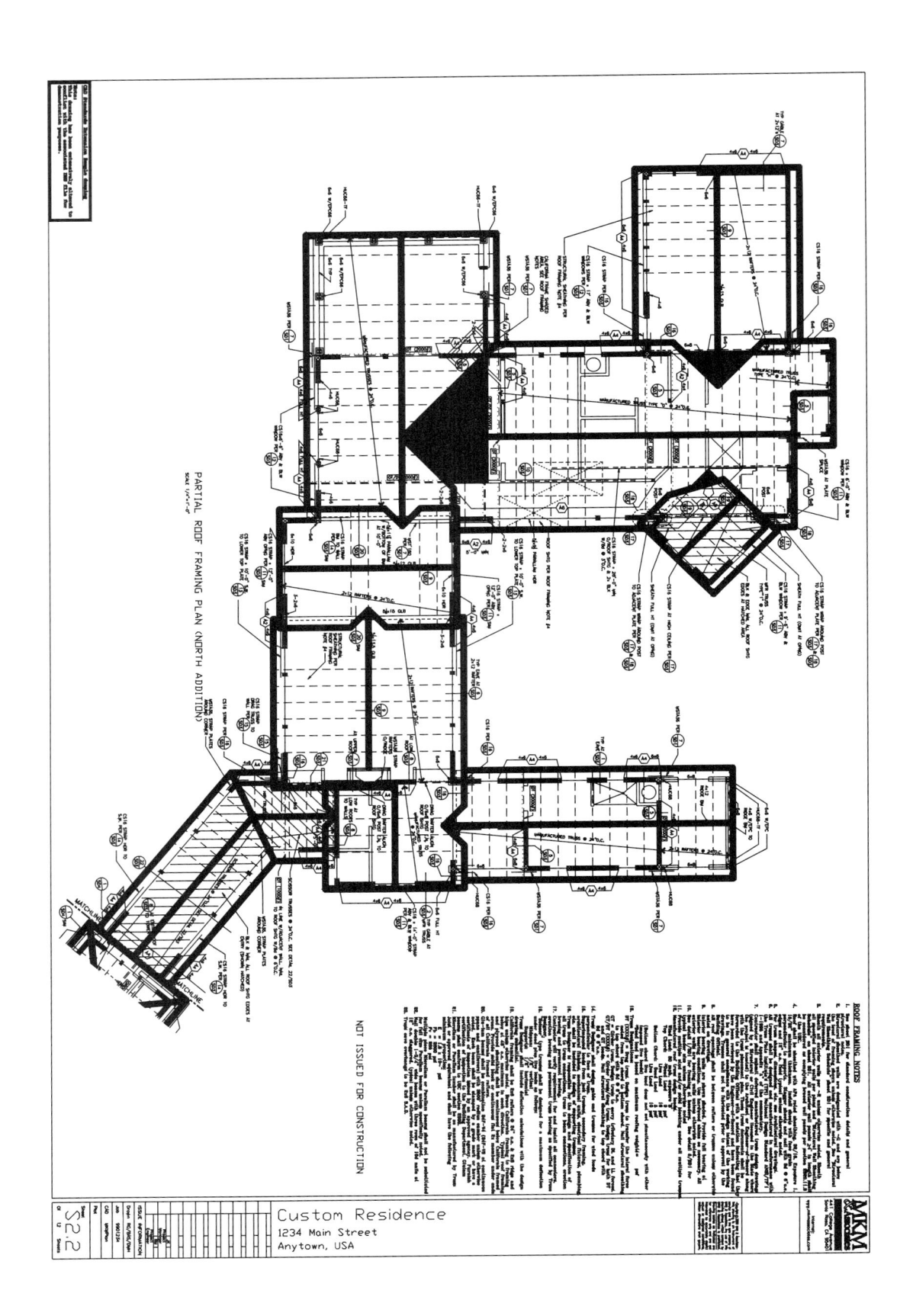

MKNPLAN.DWG, Courtesy of Autodesk, Inc.

6

BASIC DRAWING SETUP

Chapter Objectives

After completing this chapter you should:

1. know the basic steps for setting up a drawing;

2. know how to use the *Start from Scratch* option and the *Quick Setup* and *Advanced Setup* wizards that appear in the *Startup* and *Create New Drawing* dialog boxes;

3. be able to specify the desired *Units*, *Angles* format, and *Precision* for the drawing;

4. be able to specify the drawing *Limits*;

5. know how to specify the *Snap* increment;

6. know how to specify the *Grid* increment;

7. understand the basic function and use of a *Layout*.

STEPS FOR BASIC DRAWING SETUP

Assuming the general configuration (dimensions and proportions) of the geometry to be created is known, the following steps are suggested for setting up a drawing:

1. Determine and set the *Units* that are to be used.
2. Determine and set the drawing *Limits;* then *Zoom All.*
3. Set an appropriate *Snap* type and increment.
4. Set an appropriate *Grid* value to be used.

These additional steps for drawing setup are discussed also in Chapter 12, Advanced Drawing Setup.

5. Change the *LTSCALE* value based on the new *Limits.*
6. Create the desired *Layers* and assign appropriate *linetype* and *color* settings.
7. Create desired *Text Styles* (optional).
8. Create desired *Dimension Styles* (optional).
9. Activate a *Layout* tab, set it up for the plot or print device and paper size, and create a viewport (if not already existing).
10. Create or insert a title block and border in the layout.

When you start AutoCAD for the first time or use the *New* command to begin a new drawing, the *Startup* or *Create New Drawing* dialog box or *Today* window appears. Each of these tools makes available three options for setting up a new drawing. The three options represent three levels of automation/preparation for drawing setup. The *Start from Scratch* and *Use a Wizard* options are described in this chapter. *Use a Template* and creating template drawings are discussed in Chapter 12, Advanced Drawing Setup.

The *Start from Scratch* option requires you to step through each of the individual commands listed above to set up a drawing to your specifications. The *Use a Wizard* option provides two wizards, the *Quick Setup* and the *Advanced Setup* wizard. These two wizards lead you through the first two steps listed above.

STARTUP OPTIONS

When you start AutoCAD 2002 or use the *New* command, you are presented with one of three options, depending on how your system has been configured:

1. the *Today* window;
2. the *Startup* dialog box or *Create New Drawing* dialog box (these two are the same); or
3. a "blank" drawing screen and prompt for the name of a template file to use.

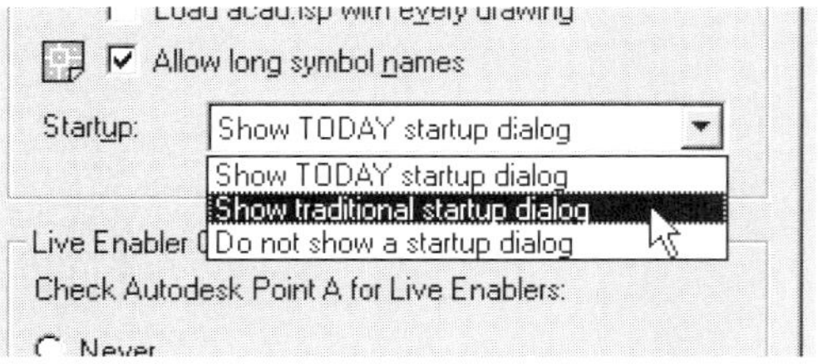

To control which of these options occurs, use the *Options* command to produce the *Options* dialog box, select the *System* tab, and make the desired selection from the *Startup* drop-down list (Fig. 6-1). (See also "The *Today* Window and the *Startup* Dialog Box" in Chapter 1 and "AutoCAD File Commands" in Chapter 2.)

Figure 6-1

DRAWING SETUP OPTIONS

You can use up to three methods to set up a new drawing based on which of the three startup options you have configured. They are:

1. *Start from Scratch*;
2. use a *Wizard*; and
3. use a *Template*.

All three methods, *Start from Scratch*, use a *Wizard*, and use a *Template*, are available in the *Today* window (Fig. 6-2) and the *Startup* or *Create New Drawing* dialog boxes (Fig. 6-3). The *Template* option (in Command line format) is the only method available if your system is configured (in the *Options* dialog box) for *Do not show a startup dialog*, although you can use other commands to accomplish any particular drawing setup.

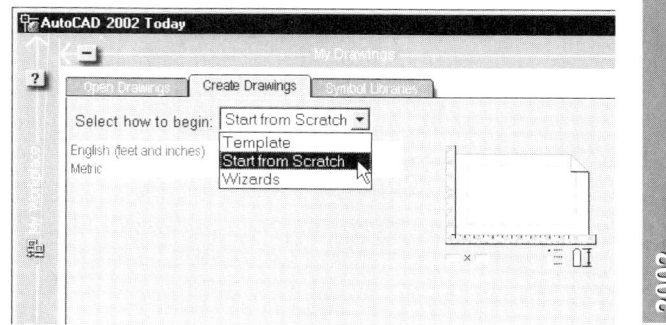

Figure 6-2

Figure 6-3

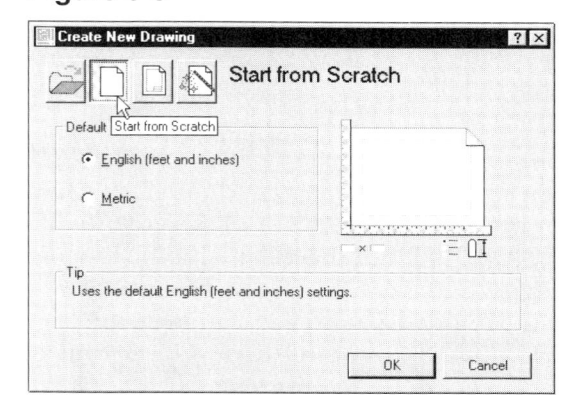

However, no matter which of the three methods you select, you can accomplish essentially the same setup to begin a drawing. Despite all the "bells and whistles" AutoCAD supplies, all of these startup options and drawing setup options generally start the drawing session with one of two setups, either the ACAD.DWT (English) or the ACADISO.DWT (metric) drawing template. In fact, using *Start from Scratch* from any source, or selecting the defaults in either setup *Wizard* from any source, or selecting either of these templates by any method produces the same result—beginning with the ACAD.DWT (English) or the ACADISO.DWT (metric) drawing template! The template used by the *Wizards* is determined by the *MEASUREINIT* system variable setting (0 for ACAD.DWT, 1 for ACADISO.DWT).

Start from Scratch

The *Start from Scratch* option is available in the *Today* window and the *Startup* or *Create New Drawing* dialog box (see Figures 6-2 and 6-3). Use this option to begin a drawing with basic drawing settings, then determine your own system variable settings using the individual setup commands such as *Units*, *Limits*, *Snap* and *Grid*.

English (feet and inches)
Use the *English (feet and inches)* option if you want to begin with the traditional AutoCAD default drawing settings, such as *Limits* settings of 12 x 9. This option actually causes AutoCAD to use the ACAD.DWT template drawing. See the "Table of *Start from Scratch English* Settings (ACAD.DWT)." The same setup can be accomplished by any of the following methods.

Select the *Start from Scratch, English* option using any startup option.
Select the ACAD.DWT by any *Template* method.
Configure your system for *Do not show a startup dialog*, then start AutoCAD (assuming the *MEASUREINIT* system variable is set to 0).
Cancel the *Today* window or *Startup* dialog box that appears when AutoCAD starts (assuming the *MEASUREINIT* system variable is set to 0).

Metric

Use the *Metric* option for setting up a drawing for use with metric units. This option causes AutoCAD to use the ACADISO.DWT template drawing. The drawing has *Limits* settings of 420 x 279, equal to a metric A3 sheet measured in mm. See the "Table of *Start from Scratch Metric* Settings (ACADISO.DWT)." The same setup can be accomplished by any of the following methods.

> Select the *Start from Scratch*, *Metric* option using any startup option.
> Select the ACADISO.DWT by any *Template* method.
> Configure your system for *Do not show a startup dialog*, then start AutoCAD (assuming the MEASUREINIT system variable is set to 1).
> Cancel the *Today* window or *Startup* dialog box that appears when AutoCAD starts (assuming the MEASUREINIT system variable is set to 1).

Template

The *Template* option is available in the *Today* window and the *Startup* or *Create New Drawing* dialog box (see Figures 6-2 and 6-3). If your system is configured for *Do not show a startup dialog*, the template option, in Command line format, is the only option available.

Use the *Template* option if you want to begin a drawing using an existing template drawing (.DWT) as a starting point. A template drawing can have many of the drawing setup steps performed but contains no geometry. Several templates are provided by AutoCAD, including the default English template ACAD.DWT and the default metric template ACADISO.DWT. See the "Table of *Start from Scratch English* Settings (ACAD.DWT)" and the "Table of *Start from Scratch Metric* Settings (ACADISO.DWT)." More information on this option, including creating template drawings and using templates provided by AutoCAD, is given in Chapter 12, Advanced Drawing Setup.

The following tables list the AutoCAD settings for the two template drawings, ACAD.DWT and ACADISO.DWT. Many of these settings, such as *Units*, *Limits*, *Snap*, and *Grid*, are explained in the following sections.

Table of *Start from Scratch English* Settings (ACAD.DWT)

Related Command	Description	System Variable	Default Setting
Units	linear units	LUNITS	2 (decimal)
Limits	drawing area	LIMMAX	12.0000,9.0000
Snap	snap increment	SNAPUNIT	.5000, .5000
Grid	grid increment	GRIDUNIT	.5000, .5000
LTSCALE	linetype scale	LTSCALE	1.0000
DIMSCALE	dimension scale	DIMSCALE	1.0000
Dtext, Mtext	text height	TEXTSIZE	.2000
Hatch	hatch pattern scale	HPSCALE	1.0000

Table of *Start from Scratch Metric* Settings (ACADISO.DWT)

Related Command	Description	System Variable	Default Setting
Units	linear units	LUNITS	2 (decimal)
Limits	drawing area	LIMMAX	420.0000, 297.0000
Snap	snap increment	SNAPUNIT	10.0000, 10.0000
Grid	grid increment	GRIDUNIT	10.0000, 10.0000
LTSCALE	linetype scale	LTSCALE	1.0000
DIMSCALE	dimension scale	DIMSCALE	1.0000
Dtext, Mtext	text height	TEXTSIZE	2.5000
Hatch	hatch pattern scale	HPSCALE	1.0000

The metric drawing setup is intended to be used with ISO linetypes and ISO hatch patterns, which are pre-scaled for these *Limits*, hence the *LTSCALE* and hatch pattern scale of 1. The individual dimensioning variables for arrow size, dimension text size, gaps, and extensions, etc. are changed so the dimensions are drawn correctly with a *DIMSCALE* of 1 with the ISO-25 dimension style.

Wizard

Selecting the *Wizard* option in the *Today* window or in the *Startup* or *Create New Drawing* dialog box (see Figures 6-2 and 6-3) gives a choice of using the *Quick Setup* or *Advanced Setup* wizard. The *Advanced Setup* wizard is an expanded version of the *Quick Setup* wizard.

The wizards use default settings based on either the ACAD.DWT or the ACADISO.DWT template. The template used by the wizards is determined by the *MEASUREINIT* system variable setting for your system (0 = ACAD.DWT, English template, and 1 = ACADISO.DWT, metric template). To change the setting, type MEASUREINIT at the Command prompt.

Quick Setup Wizard

The *Quick Setup* wizard automates only the first two steps listed under "Steps for Basic Drawing Setup" on the chapter's first page. Those functions, simply stated, are:

1. *Units*
2. *Limits*

Choosing the *Quick Setup* wizard invokes the *Quick Setup* dialog box (Fig. 6-4). There are two steps, *Units* and *Area*.

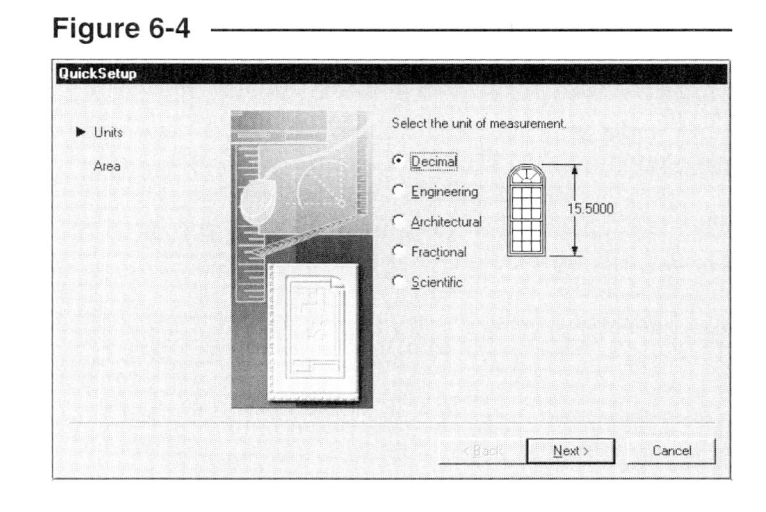

Figure 6-4

90 Basic Drawing Setup

Units
Press the desired radio button to display the units you want to use for the drawing. The options are:

Decimal   Use generic decimal units with a precision 0.0000.
Engineering   Use feet and decimal inches with a precision of 0.0000.
Architectural Use feet and fractional inches with a precision of 1/16 inch.
Fractional   Use generic fractional units with a precision of 1/16 units.
Scientific   Use generic decimal units showing a precision of 0.0000.

Use *Architectural* or *Engineering* units if you want to specify coordinate input using feet values with the apostrophe (') symbol. If you want to set additional parameters for units such as precision or system of angular measurement, use the *Units* or *-Units* command. Keep in mind the setting you select in this step changes only the display of units in the coordinate display area of the Status Bar (*Coords*) and in some dialog boxes, but not necessarily for the dimension text format. Select the *Next* button after specifying *Units*.

Area
Enter two values that constitute the X and Y measurements of the area you want to work in (Fig. 6-5). These values set the *Limits* of the drawing. The first edit box labeled *Width* specifies the X value for *Limits*. The X value is usually the longer of the two measurements for your drawing area and represents the distance across the screen in the X direction or along the long axis of a sheet of paper. The second edit box labeled *Length* specifies the Y value for *Limits* (the Y distance of the drawing area). (Beware: The terms "Width" and "Length" are misleading since "length" is defined as the measurement of something along its greatest dimension. When setting AutoCAD *Limits*, the Y measurement is generally the shorter of the two measurements.) The two values together specify the upper right corner of the drawing *Limits*. When you finish the two steps, you should use *Zoom All* to display the entire *Limits* area in the screen.

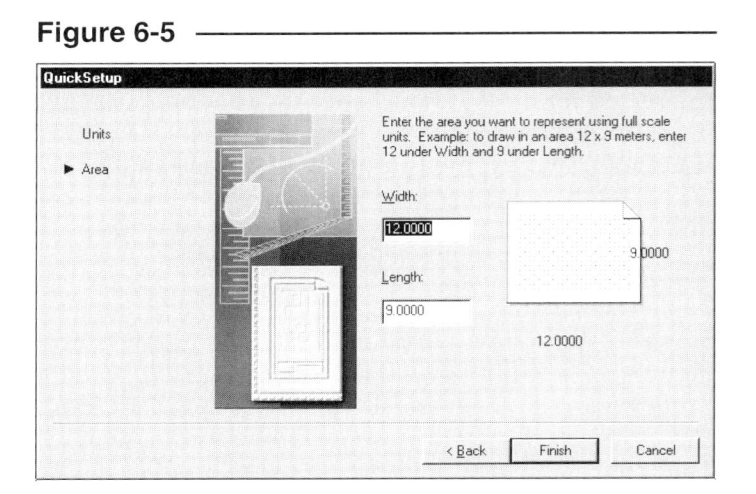

Figure 6-5

Advanced Setup Wizard

The *Advanced Setup* wizard performs the same tasks as the *Quick Setup* wizard with the addition of allowing you to select units precision and other units options (normally available in the *Units* dialog box). Selecting *Advanced Setup* produces a series of dialog boxes (not all shown). There are five steps involved in the series.

The following list indicates the "steps" in the *Advanced Setup* wizard and the related "Steps for Basic Drawing Setup" on the chapter's first page.

1. *Units* and *Precision*   *Units* command
2. *Angle*   *Units* command
3. *Angle Measure*   *Units* command
4. *Angle Direction*   *Units* command
5. *Area*   *Limits* command

Units
You can select the units of measurement for the drawing as well as the unit's *Precision* in this first step (Fig. 6-6). These are the same options available in the *Drawing Units* dialog box (see "*Units,*" this chapter). This is similar to the first step in the *Quick Setup* wizard but with the addition of *Precision*.

Figure 6-6

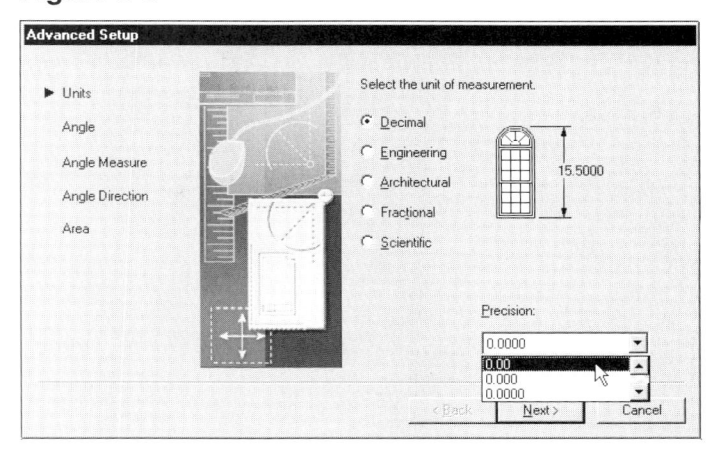

Similar to using the *Units* or *-Units* command, your choices in this and the next three dialog boxes determine the display of units for the coordinate display area of the Status Bar (*Coords*) and in dialog boxes. If you want to use feet units for coordinate input, select *Architectural* or *Engineering*.

Angle
This step (not shown) provides for your input of the desired system of angular measurement. Select the drop-down list to select the angular *Precision*.

Angle Measure
This step sets the direction for angle 0. East (X positive) is the AutoCAD default. This has the same function as selecting the *Angle 0 Direction* in the *Drawing Units* dialog box.

Angle Direction
Select *Clockwise* if you want to change the AutoCAD default setting for measuring angles. This setting (identical to the *Drawing Units* dialog box option) affects the direction of positive and negative angles in commands such as *Rotate*, *Array Polar* and dimension commands that measure angles, but does not affect the direction *Arcs* are drawn (always counterclockwise).

Area
Enter values to define the upper right corner for the *Limits* of the drawing. The *Width* refers to the X *Limits* component and the *Length* refers to the Y component. Generally, the *Width* edit box contains the larger of the two values unless you want to set up a vertically oriented drawing area. If you plan to print or plot to a standard scale, your input for *Area* should be based on the intended plot scale and sheet size. See the "Tables of *Limits* Settings" in Chapter 14 for appropriate values to use.

SETUP COMMANDS

If you want to set up a drawing using individual commands instead of the *Quick Setup* or *Advanced Setup* wizard, use the commands given in this section. The *Quick Setup* and *Advanced Setup* wizards use only the first two commands discussed in this section, *Units* and *Limits*.

UNITS

Pull-down Menu	COMMAND (TYPE)	ALIAS (TYPE)	Short-cut	Screen (side) Menu	Tablet Menu
Format Units...	UNITS or -UNITS	UN or -UN	...	FORMAT Units	V,4

The *Units* command allows you to specify the type and precision of linear and angular units as well as the direction and orientation of angles to be used in the drawing. The current setting of *Units* determines the display of values by the coordinates display (*Coords*) and in some dialog boxes.

You can select *Units ...* from the *Format* pull-down or type *Units* (or command alias *UN*) to invoke the *Drawing Units* dialog box (Fig. 6-7). Type *-Units* (or *-UN*) to produce a text screen (Fig. 6-8).

Figure 6-7

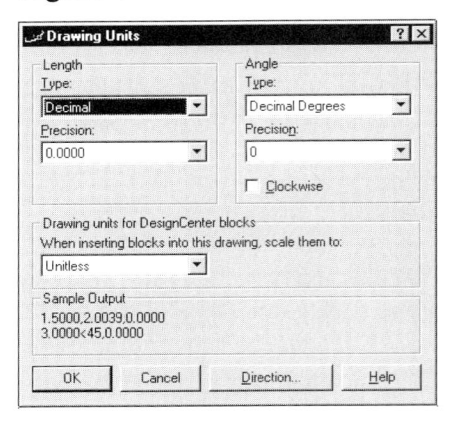

The linear and angular units options are displayed in the dialog box format (Fig. 6-7) and in Command line format (Fig. 6-8). The choices for both linear and angular *Units* are shown in the figures.

Figure 6-8

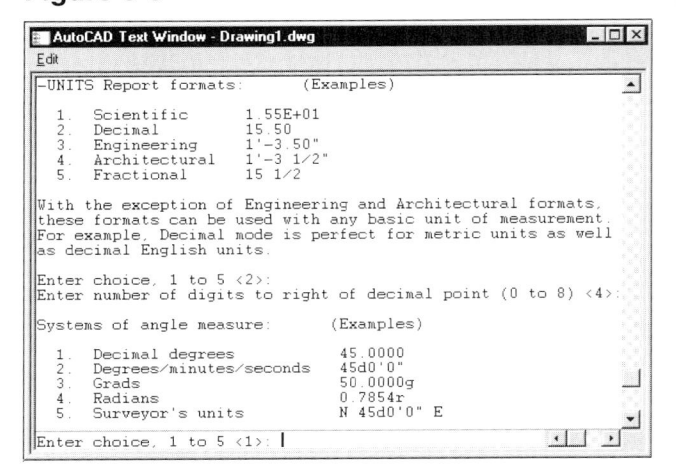

Units	**Format**	
1. Scientific	1.55E + 01	Generic decimal units with an exponent
2. Decimal	15.50	Generic decimal usually used for applications in metric or decimal inches
3. Engineering	1'-3.50"	Explicit feet and decimal inches with notation, one unit equals one inch
4. Architectural	1'-3 1/2"	Explicit feet and fractional inches with notation, one unit equals one inch
5. Fractional	15 1/2	Generic fractional units

Precision

When setting *Units,* you should also set the precision. *Precision* is the number of places to the right of the decimal or the denominator of the smallest fraction to display. The precision is set by making the desired selection from the *Precision* pop-up list in the *Drawing Units* dialog box (Fig. 6-7) or by keying in the desired selection in Command line format.

Precision controls only the display of *COORDS*. The actual precision of the drawing database is always the same in AutoCAD, that is, 14 significant digits.

Angles

You can specify a format other than the default (decimal degrees) for expression of angles. Format options for angular display and examples of each are shown in Figure 6-7 (dialog box format).

The orientation of angle 0 can be changed from the default position (east or X positive) to other options by selecting the *Direction* tile in the *Drawing Units* dialog box. This produces the *Direction Control* dialog box (Fig. 6-9). Alternately, the *Units* command can be typed to select these options in Command line format.

Figure 6-9

The direction of angular measurement can be changed from its default of counterclockwise to clockwise. The direction of angular measurement affects the direction of positive and negative angles in commands such as *Array Polar, Rotate* and dimension commands that <u>measure</u> angular values but does not change the direction *Arcs* are <u>drawn</u>, which is always counterclockwise.

Drawing Units for DesignCenter Blocks
This section specifies the units to use when you drag and drop *Blocks* from AutoCAD DesignCenter into the current drawing. This choice does not affect insertion of *Blocks* using the *Insert* command. There are many choices including *Unitless, Inches, Feet, Millimeters,* and so on. If you intend to insert *Blocks* into the drawing (you are currently setting units for) using drag-and-drop, select a unit from the list that matches the units of the *Blocks* to insert. If you are not sure, select *Unitless*. See Chapter 20, Blocks and DesignCenter, for more information on this subject.

Keyboard Input of *Units* Values

When AutoCAD prompts for a point or a distance, you can respond by entering values at the keyboard. The values can be in <u>any format</u>—integer, decimal, fractional, or scientific—regardless of the format of *Units* selected.

You can <u>type in explicit feet and inch values only if *Architectural* or *Engineering*</u> units have been specified as the drawing units. For this reason, specifying *Units* is the first step in setting up a drawing.

Type in explicit feet or inch values by using the apostrophe (') symbol after values representing feet and the quote (") symbol after values representing inches. If no symbol is used, the values are understood by AutoCAD to be <u>inches</u>.

Feet and inches input <u>cannot</u> contain a blank, so a hyphen (-) must be typed between inches and fractions. For example, with *Architectural* units, key in **6'2-1/2"**, which reads "six feet two and one-half inches." The standard engineering and architectural format for dimensioning, however, places the hyphen between feet and inches (as displayed by the default setting for the *Coords* display).

The *UNITMODE* variable set to **1** changes the display of *Coords* to remind you of the correct format for <u>input</u> of feet and inches (with the hyphen between inches and fractions) rather than displaying the standard format for feet and inch notation (standard format, *UNITMODE* of **0**, is the default setting). If options other than *Architectural* or *Engineering* are used, values are read as generic units.

LIMITS

Pull-down Menu	COMMAND (TYPE)	ALIAS (TYPE)	Short-cut	Screen (side) Menu	Tablet Menu
Format Drawing Limits	LIMITS	...	...	FORMAT Limits	V,2

The *Limits* command allows you to set the size of the drawing area by specifying the lower-left and upper-right corners in X,Y coordinate values.

Command: *limits*
Reset Model space limits
Specify lower left corner or [ON/OFF] <0.0000,0.0000>: **X,Y** or **Enter** (Enter an X,Y value or accept the 0,0 default—normally use 0,0 as lower-left corner.)
Specify upper right corner <12.0000,9.0000>: **X,Y** (Enter new values to change upper-right corner to allow adequate drawing area.)

The default *Limits* values in the ACAD.DWT are 12 and 9; that is, 12 units in the X direction and 9 units in the Y direction (Fig. 6-10). Starting a drawing by the following methods results in *Limits* of 12 x 9:

    Selecting the ACAD.DWT template drawing
    Selecting the *English* defaults in the *Start from Scratch* option

Figure 6-10

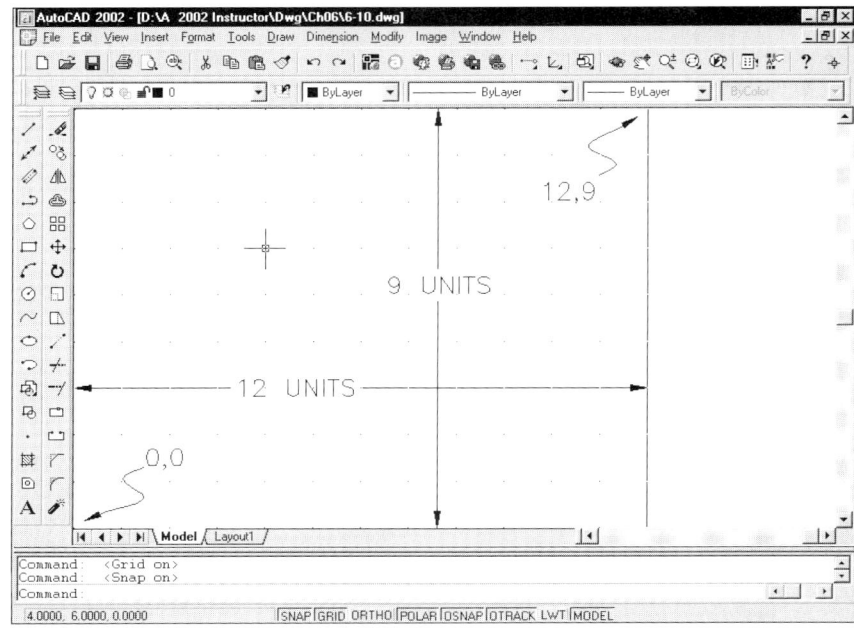

If the *GRID* is turned on, the dots are displayed only over the *Limits*. The AutoCAD screen (default configuration) displays additional area on the right past the *Limits*. The units are generic decimal units that can be used to represent inches, feet, millimeters, miles, or whatever is appropriate for the intended drawing. Typically, however, decimal units are used to represent inches or millimeters. If the default units are used to represent inches, the default drawing size would be 12 x 9 inches.

Remember that when a CAD system is used to create a drawing, the geometry should be drawn <u>full size</u> by specifying dimensions of objects in <u>real-world units</u>. A completed CAD drawing or model is virtually an exact dimensional replica of the actual object. Scaling of the drawing occurs only when plotting or printing the file to an actual fixed-size sheet of paper.

Before beginning to create an AutoCAD drawing, determine the size of the drawing area needed for the intended geometry. After setting *Units,* appropriate *Limits* should be set in order to draw the object or geometry to the <u>real-world size in the actual units</u>. There are no practical maximum or minimum settings for *Limits*.

The X,Y values you enter as *Limits* are understood by AutoCAD as values in the units specified by the *Units* command. For example, if you previously specified *Architectural units,* then the values entered are understood as inches unless the notation for feet (') is given (**240,180** or **20',15'** would define the same coordinate). Remember, you can type in explicit feet and inch values only if *Architectural* or *Engineering* units have been specified as the drawing units.

If you are planning to plot the drawing to scale, *Limits* should be set to a proportion of the <u>sheet size</u> you plan to plot on. For example, setting limits to 22 x 17 (2 times 11 by 8.5) would allow enough room for drawing an object about 20" x 15" and allow plotting at 1/2 size on the 11" x 8.5" sheet. Simply stated, <u>set *Limits* to a proportion of the paper</u>.

Limits also defines the display area for *GRID* as well as the minimum area displayed when a *Zoom All* is used. *Zoom All* forces the full display of the *Limits*. *Zoom All* can be invoked by typing *Z* (command alias) then *A* for the *All* option.

Changing *Limits* does not automatically change the display. As a general rule, you should make a habit of invoking a *Zoom All* immediately following a change in *Limits* to display the area defined by the new limits (Fig. 6-11).

Figure 6-11

When you reduce *Limits* while *Grid* is *ON*, it is apparent that a change in *Limits* does not automatically change the display. In this case, the area covered by the grid is reduced in size as *Limits* are reduced, yet the display remains unchanged.

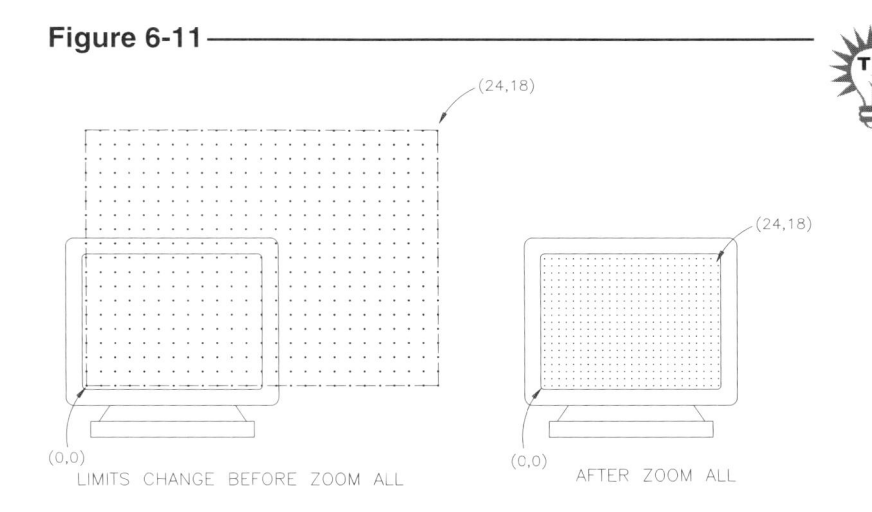

If you are already experimenting with drawing in different *Linetypes*, a change in *Limits* and *Zoom All* affects the display of the hidden and dashed lines. The *LTSCALE* variable controls the spacing of non-continuous lines. The *LTSCALE* is often changed proportionally with changes in *Limits*.

ON/OFF
If the *ON* option of *Limits* is used, limits checking is activated. Limits checking prevents you from drawing objects outside of the limits by issuing an outside-limits error. This is similar to drawing "off the paper." Limits checking is *OFF* by default.

SNAP

Pull-down Menu	COMMAND (TYPE)	ALIAS (TYPE)	Short-cut	Screen (side) Menu	Tablet Menu
Tools Drafting Settings… Snap and Grid	SNAP	SN	F9 or Ctrl+B	TOOLS 2 Grid Snap and Grid	W,10

Snap in AutoCAD has two possible types, *Grid Snap* and *Polar Snap*. When you are setting up the drawing, you should set the desired *Snap type* and increment. You may decide to use both types of Snap, so set both increments initially. You can have only one of the two *Snap types* active at one time. (See Chapter 3, Draw Command Concepts, for further explanation and practice using both Snap types.)

Grid Snap forces the cursor to preset positions on the screen, similar to the *Grid*. *Grid Snap* is like an invisible grid that the cursor "snaps" to. This function can be of assistance if you are drawing *Lines* and other objects to set positions on the drawing, such as to every .5 unit. The default value for *Grid Snap* is .5, but it can be changed to any value.

Polar Snap forces the cursor to move in set intervals from the previously designated point (such as the "first point" selected during the *Line* command) to the next point. *Polar Snap* operates for cursor movement at any previously set *Polar Tracking* angle, whereas *Grid Snap* (since it is rectangular) forces regular intervals only in horizontal or vertical movements. *Polar Tracking* must also be on (*POLAR* or F10) for *Polar Snap* to operate.

Set the desired snap type and increments using the *Snap* command or the *Drafting Settings* dialog box. In the *Drafting Settings* dialog box, select the *Snap and Grid* tab (Fig. 6-12).

Figure 6-12

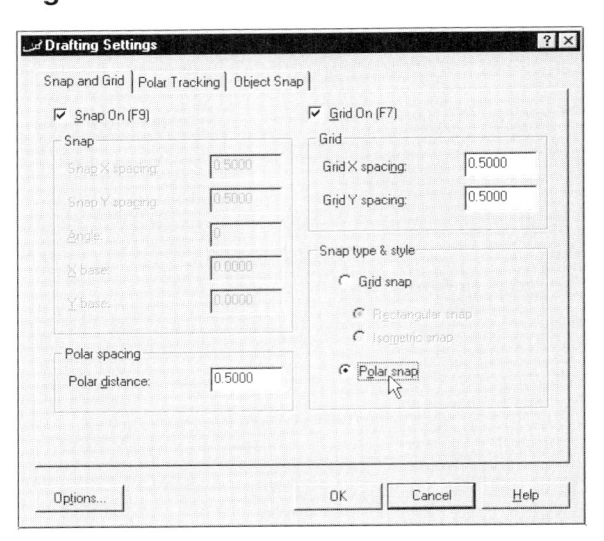

To set the Polar Snap increment, first select *Polar Snap* in the *Snap Type & Style* section (lower right). This action causes the *Polar Spacing* section to be enabled. Next enter the desired *Polar Distance* in the edit box.

To set the Grid Snap increment, first select *Grid Snap* in the *Snap Type & Style* section. This action causes the *Snap* section to be enabled. Next enter the desired *Snap X Spacing* value in the edit box. If you want a non-square snap grid, enter a different value in the *Snap Y Spacing* edit box. You can also rotate the snap grid by entering a value other than 0 in the *Angle* edit box. See the *Rotate* option of *Snap* (Command line version below).

The Command line format of *Snap* is as follows:

    Command: *snap*
    Specify snap spacing or [ON/OFF/Aspect/Rotate/Style/Type] <0.5000>:

Value
Entering a value at the Command line prompt sets the Grid Snap increment only. You must use the *Drafting Settings* dialog box to set the *Polar Snap* distance.

ON/OFF
Selecting *ON* or *OFF* accomplishes the same action as toggling the F9 key or selecting the word *SNAP* on the Status Bar.

Aspect
The *Aspect* option allows specification of unequal X and Y spacing for the snap grid. This is identical to entering unequal *Snap X Spacing* and *Snap Y Spacing* in the dialog box.

Rotate
The *Grid Snap* can be rotated about any point and set to any angle. When Snap has been rotated, the *GRID*, *ORTHO*, and the "crosshairs" automatically follow this alignment. This action facilitates creating objects oriented at the specified angle, for example, creating an auxiliary view of drawing part or a floor plan at an angle. To do this, use the *Rotate* option in Command line format or set the *Angle*, *X Base*, and *Y Base* (point to rotate about) in the *Drafting Settings* dialog box (see Chapter 25, Auxiliary Views).

Style
The *Style* option allows switching between a *Standard* snap pattern (square or rectangular) and an *Isometric* snap pattern. If using the dialog box, toggle *Isometric* snap. When *Snap Style* or *Rotate* (*Angle*) is changed, the *GRID* automatically aligns with it (see Chapter 23, Pictorial Drawings).

Type
This option switches between *Grid Snap* and *Polar Snap*. Remember, you can have only one of the two snap types active at one time.

Once your snap type and increment(s) are set, you can begin drawing using either snap type or no snap at all. While drawing, you can right-click on the word *SNAP* on the Status Bar to invoke the shortcut menu shown in Figure 6-13. Here you can toggle between *Grid Snap* and *Polar Snap* or turn both off (*POLAR* must also be on to use *Polar Snap*). Left-clicking on the word *SNAP* or pressing F9 toggles on or off whichever snap type is current.

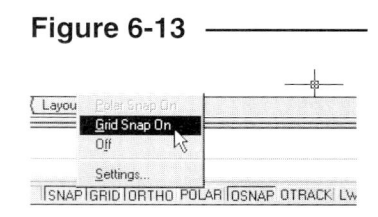

Figure 6-13

The process of setting the *Snap Type*, *Grid Snap Spacing*, *Polar Snap Spacing*, and *Polar Tracking Angle* is usually done during the initial stages of drawing setup, although these settings can be changed at any time. The *Grid Snap Spacing* value is stored in the *SNAPUNIT* system variable and saved in the drawing file. The other settings (*Snap Type*, *Polar Snap Spacing*, and *Polar Tracking Angle*) are saved in the system registry (as the SNAPTYPE, POLARDIST, and POLARANG system variables) so that the settings remain in affect for any drawing until changed.

GRID

Pull-down Menu	COMMAND (TYPE)	ALIAS (TYPE)	Short-cut	Screen (side) Menu	Tablet Menu
Tools Drafting Settings... Snap and Grid	GRID	...	F7 or Ctrl+G	TOOLS 2 Grid Snap and Grid	W,10

GRID is visible on the screen, whereas *Grid Snap* is invisible. *GRID* is only a visible display of some regular interval. *GRID* and *Grid Snap* can be independent of each other. In other words, each can have separate spacing settings and the active state of each (*ON*, *OFF*) can be controlled independently. The *GRID* follows the *SNAP* if *SNAP* is rotated or changed to *Isometric Style*. Although the *GRID* spacing can be different than that of *SNAP*, it can also be forced to follow *SNAP* by using the *Snap* option. The default *GRID* setting is **0.5**.

The *GRID* cannot be plotted. It is not comprised of *Point* objects and therefore is not part of the current drawing. *GRID* is only a visual aid.

Grid can be accessed by Command line format (shown below) or set via the *Drafting Settings* dialog box (Fig. 6-12). The dialog box is invoked by menu selection or by typing *Dsettings* or *DS*. The dialog box allows only *X Spacing* and *Y Spacing* input for *Grid*.

    Command: grid
    Specify grid spacing(X) or [ON/OFF/Snap/Aspect] <0.5000>:

Grid Spacing (X)
If you supply a value for the *Grid spacing*, *GRID* is displayed at that spacing regardless of *SNAP* spacing. If you key in an *X* as a suffix to the value (for example, **2X**), the *GRID* is displayed as that value times the *SNAP* spacing (for example, "2 times" *SNAP*).

ON/OFF
The *ON* and *OFF* options simply make the *GRID* visible or not (like toggling the **F7** key, pressing **Ctrl+G**, or clicking **GRID** on the Status Bar).

Snap
The *Snap* option of the *Grid* command forces the *GRID* spacing to equal that of *SNAP*, even if *SNAP* is subsequently changed.

Aspect

The *Aspect* option of *GRID* allows different X and Y spacing (causing a rectangular rather than a square *GRID*).

DSETTINGS

Pull-down Menu	COMMAND (TYPE)	ALIAS (TYPE)	Short-cut	Screen (side) Menu	Tablet Menu
Tools Drafting Settings...	DSETTINGS	DS	Status Bar (right-click) Settings...	TOOLS 2 Osnap... (Grid or Polar)	W,10

You can access controls to *SNAP* and *GRID* features using the *Dsettings* command. *Dsettings* produces the *Drafting Settings* dialog box described earlier (see Figure 6-12). The three tabs in the dialog box are *Snap and Grid*, *Polar Tracking* and *Object Snap*. Use the *Snap and Grid* tab to control settings for *SNAP* and *GRID* as previously described. The *Polar Tracking* and *Object Snap* tabs are explained in Chapter 7.

Using *Snap* and *Grid*

Using *Snap* and *Grid* for drawing is a personal preference and should be used whenever appropriate for the drawing. Using *Snap* and *Grid* can be beneficial for some drawings, but may not be useful for others.

Generally, *Snap* and *Grid* can be useful in cases where many of the lines to be drawn or other measurements used are at some regular interval, such as 1 mm or 1/2". Typically, small mechanical drawings and some simple architectural drawings may fall into this category. On the other hand, if you anticipate that few of the measurements in the drawing will be at regular interval lengths, *Snap* and *Grid* may be of little value. Drawings such as civil engineering drawings involving site plans would fall into this category.

Also, *Grid* and *Snap* are useful only in cases where the interval is set to a large enough value relative to the screen size (or *Limits*) to facilitate seeing and PICKing points easily. For example, with *Limits* of 12 x 9 it would be relatively easy to see a *Grid* set to .5 and to PICK points at *Snap* intervals of .125. However, in cases where the *Limits* cover a large area and the desired increments are <u>relatively</u> small, *Snap* may not be of much usefulness. For example, an architectural drawing may have *Limits* that represent hundreds of feet, and although all measurements are to be drawn at 1/4" intervals, it would be almost impossible to interactively PICK points at such a small increment relative to the overall drawing size.

INTRODUCTION TO LAYOUTS AND PRINTING

The last several steps listed in the "Steps for Basic Drawing Setup" on this chapter's first page are also listed below.

5. Change the *LTSCALE* value based on the drawing *Limits*.
6. Create the desired *Layers* and assign appropriate *linetype* and *color* settings.
7. Create desired *Text Styles*.
8. Create desired *Dimension Styles*.
9. Activate a *Layout* tab, set it up for the plot or print device and paper size, and create a viewport (if not already created).
10. Create or insert a title block and border in the layout.

Part of the process of setting the *LTSCALE*, creating *Text*, and creating *Dimension Styles*, (steps 5, 7, and 8) is related to the size of the drawing. Steps 9 and 10 prepare the drawing for making a print or plot.

Although steps 9 and 10 are often performed after the drawing is complete and just before making a print or plot, it is wise to consider these steps in the drawing setup process. Because you want hidden line dashes, text, dimensions, hatch patterns, etc. to have the correct size in the finished print or plot, it would be sensible to consider the paper size of the print or plot and the drawing scale before you create text, dimensions, and hatch patterns in the drawing. In this way, you can more accurately set the necessary sizes and system variables before you draw, or as you draw, instead of changing multiple settings upon completion of the drawing.

To put it simply, to determine the size for linetypes (*LTSCALE*), text, and dimensions, use the proportion of the drawing to the paper size. In other words, if the size of the drawing area (*Limits*) is 22 x 17 and the paper size you will print on is 11 x 8.5, then the drawing is 2 times the size of the paper. Therefore, set the *LTSCALE* to 2 and create the text and dimensions twice as large as you want them to appear on the print.

This chapter gives an introduction to layouts and creating paper space viewports. Advanced features of drawing setup, layouts, viewports, and plotting are discussed in Chapter 12, Advanced Drawing Setup, Chapter 13, Layouts and Viewports, and Chapter 14, Printing and Plotting.

Model Tab and *Layout* Tabs

At the bottom of the drawing area you should see one *Model* tab and two *Layout* tabs. If the tabs have been turned off on your system, see "Setting Layout Options and Plotting Options" later in this section.

When you start AutoCAD and begin a drawing, the *Model* tab is active by default. Objects that represent the subject of the drawing (model geometry) are normally drawn in the *Model* tab, also known as "model space." Traditionally, dimensions and notes (text) are also created in model space.

A layout is activated by selecting the *Layout1*, *Layout2*, or other layout tab (Fig. 6-14). A layout represents the sheet of paper that you intend to print or plot on (sometimes called paper space). The dashed line around the "paper sheet" in Figure 6-14 represents the maximum printable area for the configured printer or plotter.

Figure 6-14

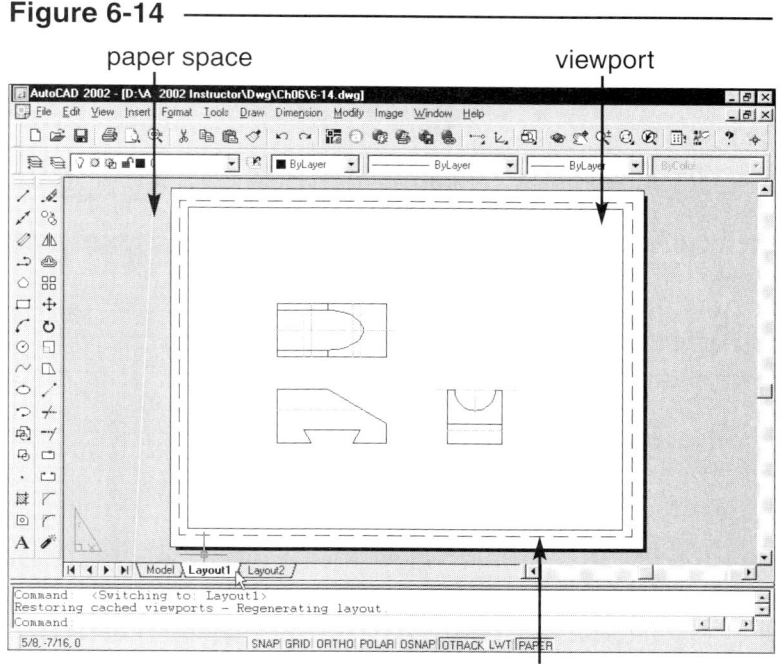

You can have multiple layouts (*Layout1*, *Layout2*, etc.), each representing a different sheet size and/or print or plot device. For example, you may have one layout to print with a laser printer on a 8.5 x 11 inch sheet and another layout set up to plot on an 24 x 18 ("C" size) sheet of paper.

With the default options set when AutoCAD is installed, a viewport is automatically created when you activate a layout. However, you can create layouts using the *Vports* command. A viewport is a "window" that looks into model space. Therefore, you first create the drawing objects in model space (the *Model* tab), then activate a layout and create a viewport to "look into" model space. Typically, only drawing objects such as a title block, border, and some text are created in a layout. Creating layouts and viewports is discussed in detail in Chapter 13, Layouts and Viewports.

Since a layout represents the actual printed sheet, you normally print the layout full size (1:1). However, the view of the drawing objects appearing in the viewport is scaled to achieve the desired print scale. In other words, you can control the scale of the drawing by setting the "viewport scale"—the proportion of the drawing objects that appear in the viewport relative to the paper size, or simply stated, the proportion of model space to paper space. (This is the same idea as the proportion of the drawing area to the paper size, discussed earlier.)

One easy way to set the viewport scale is with the *Viewports* drop-down list (Fig. 6-15). To set the scale of the display of objects in the viewport relative to the paper size, select or enter the desired scale in the drop-down list in the *Viewports* toolbar. Setting the scale of the drawing for printing and plotting using this method and other methods is discussed in detail in Chapter 13, Layouts and Viewports, and Chapter 14, Printing and Plotting.

Figure 6-15

Using a *Layout* tab to set up a drawing for printing or plotting is recommended, although you can also make a print or plot directly from the *Model* tab (print from model space). This method is also discussed in Chapter 14, Printing and Plotting.

Step 10 in the "Steps for Basic Drawing Setup" is to create a titleblock and border for the layout. The titleblock and text that you want to appear only in the print but not in the drawing (in model space) are typically created in the layout (in paper space). Figure 6-16 shows the drawing after a titleblock has been added.

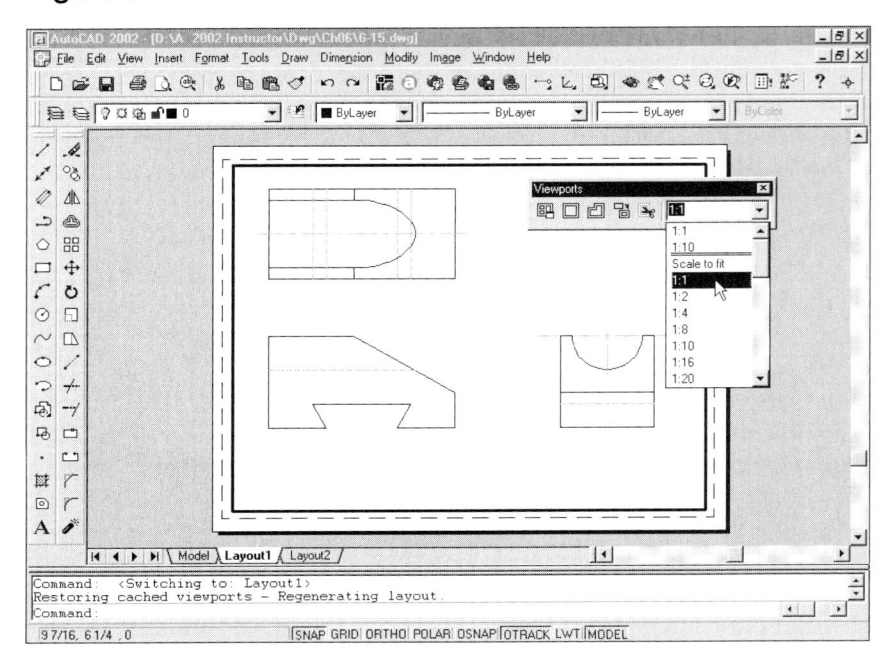

Figure 6-16

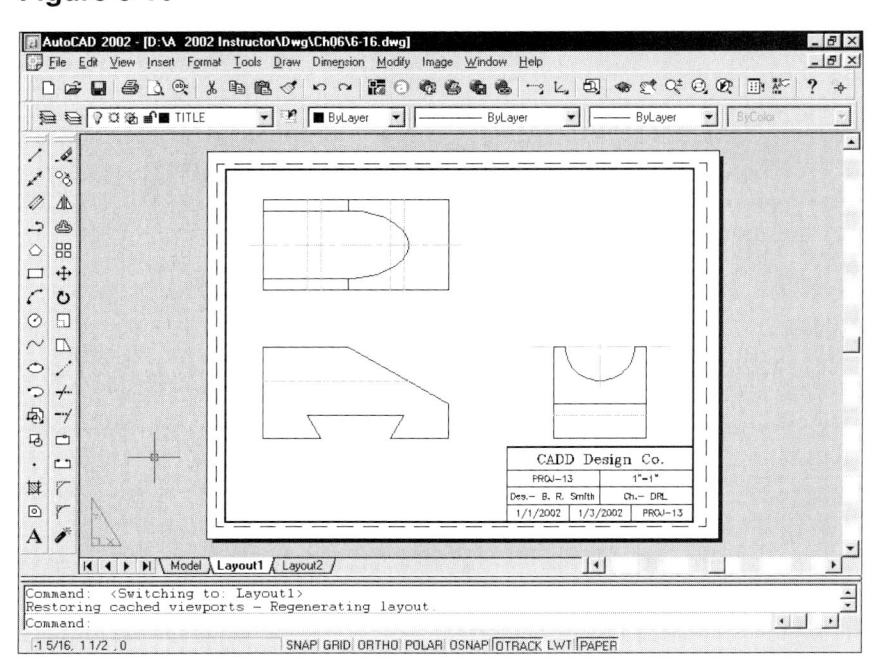

The advantage of creating a titleblock early in the drawing process is knowing how much space the titleblock occupies so you can plan around it. Note that in Figure 6-16 no drawing objects can be created in the lower-right corner because the titleblock occupies that area.

Often creating a titleblock is accomplished by using the *Insert* command to bring a previously created *Block* into the layout. A *Block* in AutoCAD is a group of objects (*Lines*, *Arcs*, *Circles*, *Text*, etc.) that is saved as one object. In this way, you do not have to create the set of objects repeatedly—you need create them only once and save them as a *Block*, then use the *Insert* command to insert the *Block* into your drawing. For example, your school or office may have a standard titleblock saved as a *Block* that you can *Insert*. *Blocks* are covered in Chapter 20, Blocks and DesignCenter.

Why Set Up Layouts Before Drawing?

Knowing the intended drawing scale before completing the drawing helps you set the correct size for linetypes, text, dimensions, hatch patterns, and other size-related drawing objects. Since the text and dimensions must be readable in the final printed drawing (usually 1/8" to 1/4" or 3mm-6mm), you should know the drawing scale to determine how large to create the text and dimensions in the drawing. Although you can change the sizes when you are ready to print or plot, knowing the drawing scale early in the drawing process should save you time.

You may not have to go through the steps to create a layout and viewport to determine the drawing scale, although doing so is a very "visible" method to achieve this. <u>The important element is knowing ahead of time the intended drawing scale.</u> If you know the intended drawing scale, you can calculate the correct sizes for creating text and dimensions and setting the scale for linetypes. This topic is a major theme discussed in Chapters 12, 13, and 14.

Printing and Plotting

The *Plot* command allows you to print a drawing (using a printer) or plot a drawing (using a plotter). The *Plot* command invokes the *Plot* dialog box (Fig. 6-17). You can type the *Plot* command, select the *Plot* icon button from the Standard toolbar, or select *Plot* from the *File* pull-down menu, as well as use other methods (see "*Plot*," Chapter 14).

Figure 6-17

If you want to print the drawing as it appears in model space, invoke *Plot* while the *Model* tab is active. Likewise, if you want to print a layout, invoke *Plot* while the desired *Layout* tab is active.

The *Plot* dialog box allows you to specify several options with respect to printing and plotting, such as selecting the plot device, paper size, and scale for the plot. If you are printing the drawing from the *Model* tab, you would select an appropriate scale (such as *1:1*, *1:2*, or *Scaled to Fit*) in the *Scale* drop-down list so the drawing geometry would fit appropriately on the printed sheet (see Figure 6-17). If you are printing the drawing from a layout, you would select *1:1* in the *Scale* drop-down list. In this case, you want to print the layout (already set to the sheet size for the plot device) at full size. The drawing geometry that appears in the viewport can be scaled by setting the viewport scale (using the *Viewports* drop-down list), as described earlier. Any plot specifications you make with the *Plot* dialog box (or the *Page Setup* dialog box) are saved with each *Layout* tab and the *Model* tab; therefore, you can have several layouts, each saved with a particular print or plot setup (scale, device, etc.).

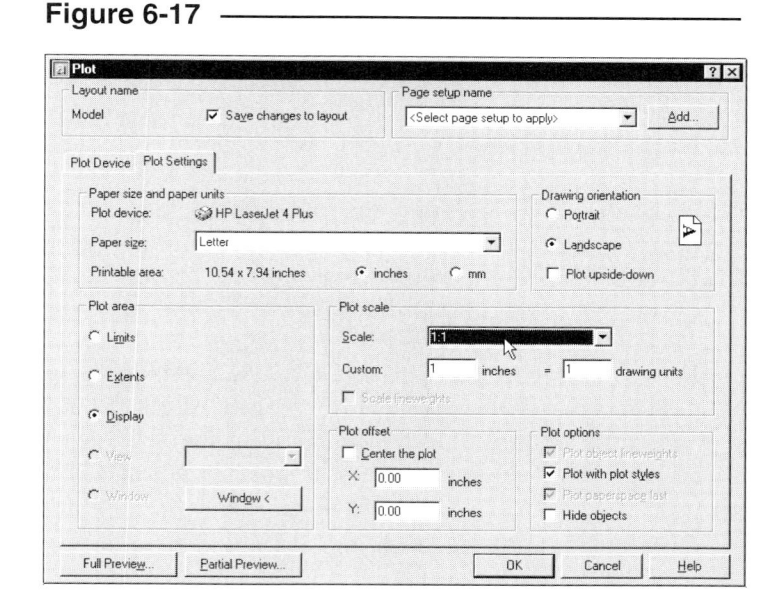

Details of printing and plotting, including all the options of the *Plot* dialog box, printing to scale, and configuring printers and plotters, are discussed in Chapter 14.

Setting Layout Options and Plot Options

In Chapters 12, 13, and 14 you will learn advanced steps in setting up a drawing, how to create layouts and viewports, and how to print and plot drawings. Until you study those chapters, and for printing drawings before that time, you may need to go through a simple process of configuring your system for a plot device and to automatically create a viewport in layouts. However, if you are at a school or office, some settings may have already been prepared for you, so the following steps may not be needed. Activating a *Layout* tab automatically creates a viewport and displays the *Page Setup* dialog box if you use AutoCAD's default options that appear when it is first installed.

As a check, start AutoCAD and draw a *Circle* in model space. When you activate a *Layout* tab for the first time in a drawing, you see either a "blank" sheet with no viewport, or a viewport that already exists, or one that is automatically created by AutoCAD. The viewport allows you to view the circle you created in model space. (If the *Page Setup* dialog box appears, select *Cancel* this time.) If a viewport exists, but no circle appears, double-click <u>inside</u> the viewport and type Z (for *Zoom*), press Enter, then type A (for *All*) and press Enter to make the circle appear.

If your system has not previously been configured for you at your school or office, follow the procedure given in "Configuring a Default Output Device" and "Creating Automatic Viewports." After doing so, you should be able to activate a *Layout* tab and AutoCAD will automatically match the layout size to the sheet size of your output device and automatically create a viewport.

Configuring a Default Output Device
If no default output device has been specified for your system or to check to see if one has already been specified for you, follow the steps below (Fig. 6-18).

1. Invoke the *Options* dialog box by right-clicking in the drawing area and selecting *Options...* from the bottom of the shortcut menu, or selecting *Options...* from the bottom of the *Tools* pull-down menu.

2. In the *Plotting* tab, locate the *Default plot settings for new drawings* cluster near the top-left corner of the dialog box. Select the desired plot device from the list (such as a laser printer), then select the *Use as default output device* button.

3. Select *OK*.

Figure 6-18

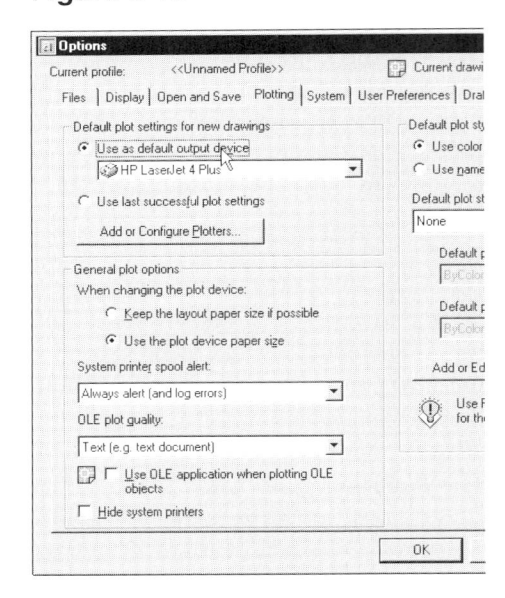

Creating Automatic Viewports
If no viewport exists on your screen, configure your system to automatically create a viewport by following these steps (Fig. 6-19).

1. Invoke the *Options* dialog box again.
2. In the *Display* tab, find the *Layout elements* cluster near the lower-left corner of the dialog box. Select the *Create viewport in new layouts* checkbox from the bottom of the list.
3. Select *OK*.

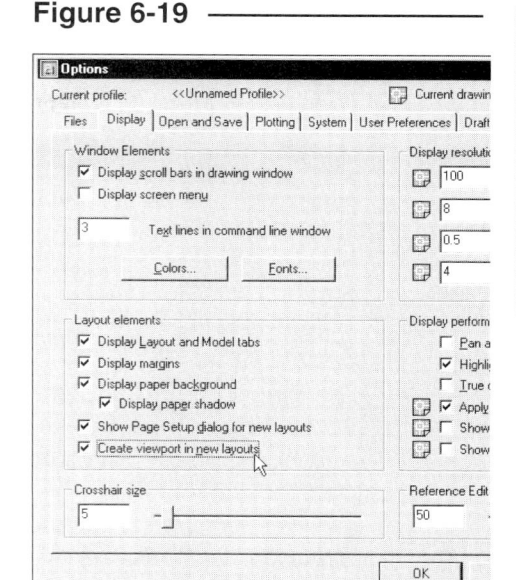

Printing the Drawing
Once an output device has been configured for your system, you are ready to print. For simplicity, you need to perform only three steps to make a print of your drawing, either from the *Model* tab or from a *Layout* tab.

Figure 6-19

1. Invoke the *Plot* dialog box. Set the desired scale in the *Plot Settings* tab, *Scale* drop-down list. If you are printing from a *Layout* tab, select *1:1*. If you are printing from the *Model* tab, select *Scaled to Fit* or other appropriate scale.
2. Select the *Full Preview* button to ensure the drawing will be printed as you expect.
3. Assuming the preview is as you expect, select the *OK* button to produce the print or plot.

This section, "Introduction to Layouts and Printing," will give you enough information to create prints or plots of drawings for practice and for the next several Chapter Exercises. You also have learned some important concepts that will help you understand all aspects of setting up drawings, creating layouts and viewports, and printing and plotting drawings, as discussed in Chapters 12, 13, and 14.

CHAPTER EXERCISES

1. A drawing is to be made to detail a mechanical part. The part is to be manufactured from sheet metal stock; therefore, only one view is needed. The overall dimensions are 16 x 10 inches, accurate to the nearest .125 inch. Complete the steps for drawing setup:

   A. The drawing will be automatically "scaled to fit" the paper (no standard scale).

      1. Begin a *New* drawing. When the *Create New Drawing* dialog box or *Today* window appears, select **Start from Scratch**. Select the **English** default settings.
      2. *Units* should be *Decimal*. Set the *Precision* to **0.000**.
      3. Set *Limits* in order to draw full size. Make the lower-left corner **0,0** and the upper right at **24,18**. This is a 4 x 3 proportion and should allow space for the part and dimensions or notes.
      4. *Zoom All*. (Type **Z** for *Zoom;* then type **A** for *All*.)
      5. Set the *GRID* to **1**.
      6. Set the *Grid Snap* increment to **.125**. Since the *Polar Snap* increment and *Snap Type* are not saved with the drawing (but are saved in the system registry), it is of no use to set these options at this time.
      7. Save this drawing as **CH6EX1A** (to be used again later). (When plotting at a later time, "Scale to Fit" can be specified.)

104 **Basic Drawing Setup**

- B. The drawing will be printed from a layout to scale on engineering A size paper (11" x 8.5").

    1. Begin a *New* drawing. When the *Create New Drawing* dialog box or *Today* window appears, select *Start from Scratch*. Select the *English* default settings.
    2. *Units* should be *Decimal*. Set the *Precision* to **0.000**.
    3. Set *Limits* to a proportion of the paper size, making the lower-left corner **0,0** and the upper-right at **22,17**. This allows space for drawing full size and for dimensions or notes.
    4. *Zoom All*. (Type **Z** for *Zoom*; then type **A** for *All*.)
    5. Set the *GRID* to **1**.
    6. Set the *Grid Snap* increment to **.125**. Since the *Polar Snap* increment and *Snap Type* are not saved with the drawing (but are saved in the system registry), it is of no use to set these options at this time.
    7. Activate a *Layout* tab. Assuming your system is configured for a printer using an 11 x 8.5 sheet and is configured to automatically create a viewport (see "Setting Layout Options and Plot Options" in this chapter), a viewport should appear. If the *Page Setup* dialog box appears, select *OK*.
    8. Double-click inside the viewport, then type **Z** for *Zoom* and press **Enter**, and type **A** for *All* and press **Enter**. Activate the *Viewports* drop-down list by right-clicking on any icon button and selecting *Viewports* from the list of available toolbars that appears. Select *1:2* from the list.
    9. Activate the *Model* tab. *Save* the drawing as **CH6EX1B**.

2. A drawing is to be prepared for a house plan. Set up the drawing for a floor plan that is approximately 50' x 30'. Assume the drawing is to be automatically "Scaled to Fit" the sheet (no standard scale).

    A. Begin a *New* drawing. When the *Create New Drawing* dialog box or *Today* window appears, select *Start from Scratch*. Select the *English* default settings.
    B. Set *Units* to *Architectural*. Set the *Precision* to **0'-0 1/4"**. Each unit equals 1 inch.
    C. Set *Limits* to **0,0** and **80',60'**. Use the apostrophe (') symbol to designate feet. Otherwise, enter **0,0** and **960,720** (size in inch units is: 80x12=960 and 60x12=720).
    D. *Zoom All*. (Type **Z** for *Zoom*; then type **A** for *All*.)
    E. Set *GRID* to **24** (2 feet).
    F. Set the *Grid Snap* increment to **6"**. Since the *Polar Snap* increment and *Snap Type* are not saved with the drawing (but are saved in the system registry), it is of no use to set these options at this time.
    G. *Save* this drawing as **CH6EX2**.

3. A multiview drawing of a mechanical part is to be made. The part is 125mm in width, 30mm in height, and 60mm in depth. The plot is to be made on an A3 metric sheet size (420mm x 297mm). The drawing will use ISO linetypes and ISO hatch patterns, so AutoCAD's *Metric* default settings can be used.

    A. Begin a *New* drawing. When the *Create New Drawing* dialog box or *Today* window appears, select *Start from Scratch*. Select the *Metric* default settings.
    B. *Units* should be *Decimal*. Set the *Precision* to **0.00**.
    C. Calculate the space needed for three views. If *Limits* are set to the sheet size, there should be adequate space for the views. Make sure the lower-left corner is at **0,0** and the upper right is at **420,297**. (Since the *Limits* are set to the sheet size, a plot can be made later at 1:1 scale.)
    D. Set the *Grid Snap* increment to **10**. Make *Grid Snap* current.
    E. *Save* this drawing as **CH6EX3** (to be used again later).

4. Assume you are working in an office that designs many mechanical parts in metric units. However, the office uses a standard laser jet printer for 11" x 8.5" sheets. Since AutoCAD does not have a setup for metric drawings on non-metric sheets, it would help to carry out the steps for drawing setup and save the drawing as a template to be used later.

   A. Begin a *New* drawing. When the *Create New Drawing* dialog box or *Today* window appears, select **Start from Scratch**. Select **English** default settings.
   B. Set the **Units Precision** to **0.00**.
   C. Change the **Limits** to match an 11" x 8.5" sheet. Make the lower-left corner **0,0** and the upper right **279,216** (11 x 8.5 times 25.4, approximately). (Since the *Limits* are set to the sheet size, plots can easily be made at 1:1 scale.)
   D. **Zoom All**. (Type **Z** for *Zoom*, then **A** for *All*).
   E. Change the **Grid Snap** increment to **2**. When you begin the drawing in another exercise, you may want to set the *Polar Snap* increment to 2 and make *Polar Snap* current.
   F. Change **Grid** to **10**.
   G. At the Command prompt, type **LTSCALE**. Change the value to **25**.
   H. Activate a **Layout** tab. Assuming your system is configured for a printer using an 11 x 8.5 sheet and is configured to automatically create a viewport (see "Setting Layout Options and Plot Options" in this chapter), a viewport should appear. (If the *Page Setup* dialog box automatically appears, select **OK** and the viewport should appear.)
   I. Produce the *Page Setup* dialog box by selecting **Page Setup** from the **File** pull-down menu or type **Pagesetup** at the Command prompt. In the *Page Setup* dialog box, activate the **Layout Settings** tab. Select the **mm** button in the *Printable area* section (under the *Paper size* box). Next, select **1:1** in the *Scale* drop-down list near the center of the dialog box. Select the **OK** button. This action (using the *Page Setup* dialog box) saves the print settings for the layout for future use without making a print at this time.
   J. Examining the layout, it appears that the viewport has been reduced in size and exists in the lower-left corner of the sheet. Technically, the sheet size has been increased from 11 x 8.5 units to 279 x 216 units (approximately) and the viewport remained the original size. **Erase** the viewport.
   K. To make a new viewport, type **–Vports** at the Command prompt (don't forget the hyphen) and press **Enter**. Accept the default (**Fit**) option by pressing **Enter**. A viewport appears to fit the printable area.
   L. Double-click inside the viewport, then type **Z** for *Zoom* and press **Enter**, and type **A** for *All* and press **Enter**. Activate the *Viewports* drop-down list by right-clicking on any icon button and selecting **Viewports** from the list of available toolbars that appears. Select **1:1** from the list.
   M. Activate the *Model* tab. *Save* the drawing as **A-METRIC**.

5. Assume you are commissioned by the local parks and recreation department to provide a layout drawing for a major league sized baseball field. Follow these steps to set up the drawing:

   A. Begin a *New* drawing. When the *Create New Drawing* dialog box or *Today* window appears, select **Start from Scratch**, then **English** default units.
   B. Set the **Units** to **Architectural** and the **Precision** to **1/2"**.
   C. Set **Limits** to an area of **512'** x **384'** (make sure you key in the apostrophe to designate feet). *Zoom All*.
   D. Type **DS** to invoke the *Drafting Settings* dialog box. Change the **Grid Snap** to **10'** (don't forget the apostrophe). When you are ready to draw (at a later time), you may want to set the *Polar Snap* increment to 10' and make *Polar Snap* the current *Snap Type*.
   E. Use the **Grid** command and change the value to **20'**. Ensure *SNAP* and *GRID* are on.
   F. *Save* the drawing and assign the name **BALL FIELD CH6**.

106 Basic Drawing Setup

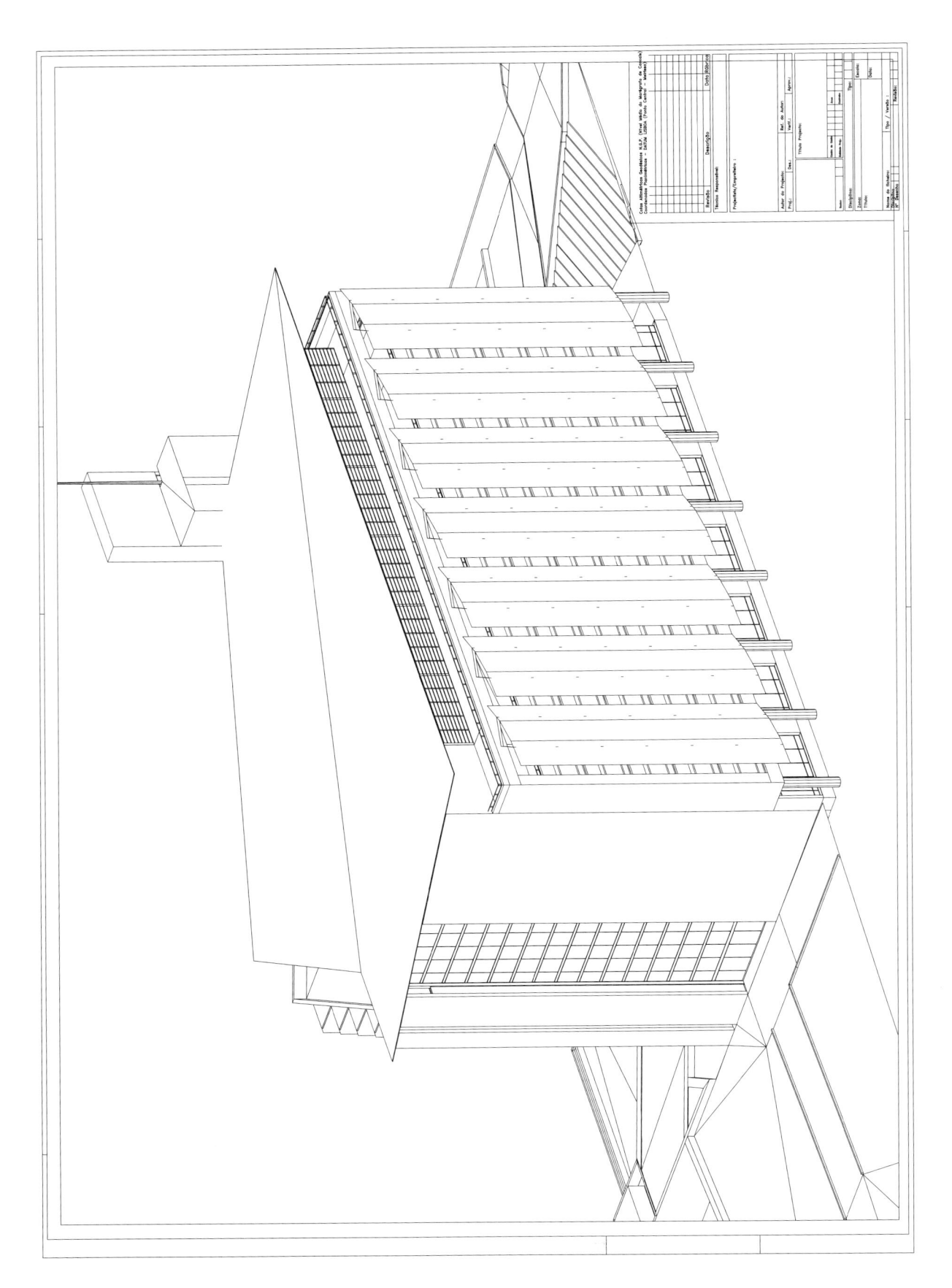

EXPO HEADQUARTERS MODEL.DWG, Courtesy of Autodesk, Inc.

7

OBJECT SNAP AND OBJECT SNAP TRACKING

Chapter Objectives

After completing this chapter you should:

1. understand the importance of accuracy in CAD drawings;

2. know the function of each of the Object Snap (*OSNAP*) modes;

3. be able to recognize the AutoSnap Marker symbols;

4. be able to invoke *OSNAP*s for single point selection;

5. be able to operate Running Object Snap modes;

6. know that you can toggle Running Object Snap off to specify points not at object features;

7. be able to use Object Snap Tracking to create and edit objects that align with existing *OSNAP* points.

CAD ACCURACY

Because CAD databases store drawings as digital information with great precision (fourteen numeric places in AutoCAD), it is possible, practical, and desirable to create drawings that are 100% accurate; that is, a CAD drawing should be created as an exact dimensional replica of the actual object. For example, lines that appear to connect should actually connect by having the exact coordinate values for the matching line endpoints. Only by employing this precision can dimensions placed in a drawing automatically display the exact intended length, or can a CAD database be used to drive CNC (Computer Numerical Control) machine devices such as milling machines or lathes, or can the CAD database be used for rapid prototyping devices such as Stereo Lithography Apparatus. With CAD/CAM technology (Computer-Aided Design/Computer-Aided Manufacturing), the CAD database defines the configuration and accuracy of the finished part. Accuracy is critical. Therefore, in no case should you create CAD drawings with only visual accuracy such as one might do when sketching using the "eyeball method."

OBJECT SNAP

AutoCAD provides a capability called "Object Snap," or *OSNAP* for short, that enables you to "snap" to existing object endpoints, midpoints, centers, intersections, etc. When an *OSNAP* mode (*Endpoint, Midpoint, Center, Intersection,* etc.) is invoked, you can move the cursor near the desired object feature (endpoint, midpoint, etc.) and AutoCAD locates and calculates the coordinate location of the desired object feature. Available Object Snap modes are:

Center
Endpoint
Insert
Intersection
Midpoint
Nearest
Node (Point)
Perpendicular
Parallel
Extension
Temporary Tracking
Quadrant
Tangent
From
Apparent Intersection
    (for 3D use)

Figure 7-1

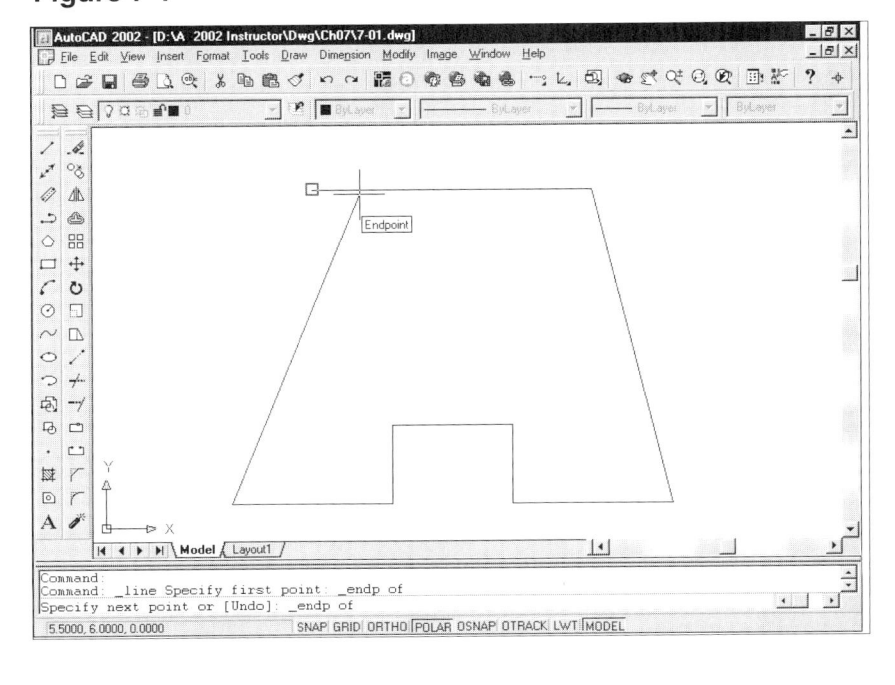

For example, when you want to draw a *Line* and connect its endpoint to an existing *Line*, you can invoke the *Endpoint OSNAP* mode at the "Specify next point or [Undo]:" prompt, then snap exactly to the desired line end by moving the cursor near it and PICKing (Fig. 7-1). *OSNAP*s can be used for any draw or modify operation— whenever AutoCAD prompts for a point (location).

A feature called "AutoSnap" displays a "Snap Marker" indicating the particular object feature (endpoint, midpoint, etc.) when you move the cursor <u>near</u> an object feature. Each *OSNAP* mode (*Endpoint, Midpoint, Center, Intersection,* etc.) has a distinct symbol (AutoSnap Marker) representing the object feature. This innovation allows you to preview and confirm the snap points before you PICK them (Fig. 7-2). The table in Figure 7-27 lists each Object Snap mode, the related AutoSnap Marker, and the icon button used to activate a single Object Snap selection.

Figure 7-2

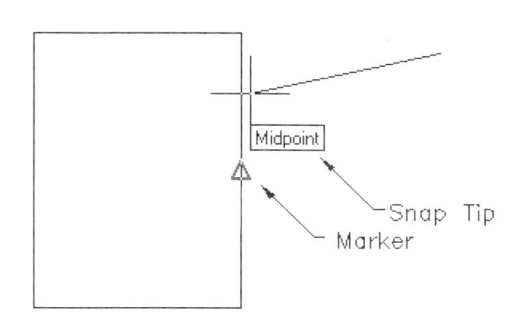

AutoSnap provides two other aids for previewing and confirming *OSNAP* points before you PICK. A "Snap Tip" appears shortly after the Snap Marker appears (hold the cursor still and wait one second). The Snap Tip gives the name of the found *OSNAP* point, such as an *Endpoint, Midpoint, Center,* or *Intersection,* (Fig. 7-2). In addition, a "Magnet" draws the cursor to the snap point if the cursor is within the confines of the Snap Marker. This Magnet feature helps confirm that you have the desired snap point before making the PICK.

A visible target box, or "Aperture," <u>can be displayed</u> at the cursor (invisible by default) whenever an *OSNAP* mode is in effect (Fig. 7-3). The Aperture is a square box larger than the pickbox (default size of 10 pixels square). Technically, this target box (visible or invisible) must be located <u>on an object</u> before a Snap Marker and related Snap Tip appear. The settings for the Aperture, Snap Markers, Snap Tips, and Magnet are controlled in the *Drafting* tab of the *Options* dialog box (discussed later).

Figure 7-3 —

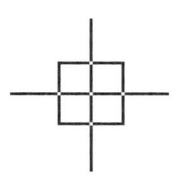

The Object Snap modes are explained in the next section. Each mode, and its relation to the AutoCAD objects, is illustrated.

Object Snaps must be activated in order for you to "snap" to the desired object features. Two methods for activating Object Snaps, Single Point Selection and Running Object Snaps, are discussed in the sections following Object Snap Modes. The *OSNAP* modes (*Endpoint, Midpoint, Center, Intersection,* etc.) operate identically for either method.

OBJECT SNAP MODES

AutoCAD provides the following Object Snap Modes.

Center

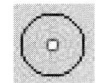

This *OSNAP* option finds the center of a *Circle, Arc,* or *Donut*. You can PICK the *Circle* <u>object</u> or where you think the center is.

Figure 7-4

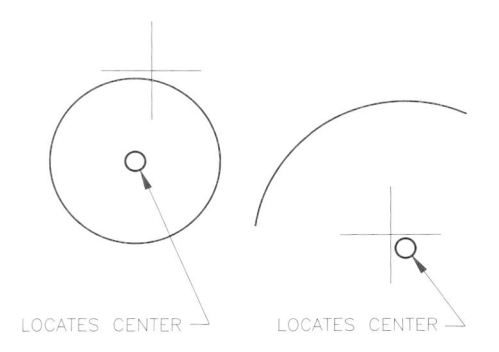

110   Object Snap and Object Snap Tracking

Endpoint

The *Endpoint* option snaps to the endpoint of a *Line*, *Pline*, *Spline*, or *Arc*. PICK the object near the desired end.

Figure 7-5

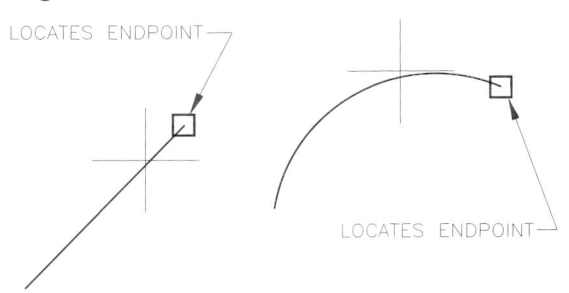

Insert

This option locates the insertion point of *Text* or a *Block*. PICK anywhere on the *Block* or line of *Text*.

Figure 7-6

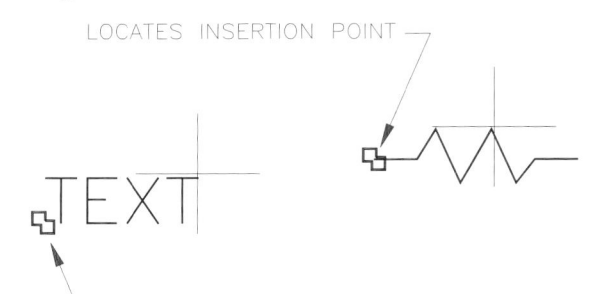

Intersection

Using this option causes AutoCAD to calculate and snap to the intersection of any two objects. You can locate the cursor (Aperture) so that both objects pass near (through) it, or you can PICK each object individually.

Figure 7-7

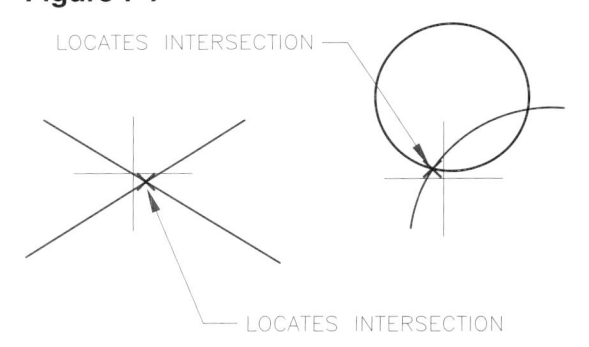

Even if the two objects that you PICK do not physically intersect, you can PICK each one individually with the *Intersection* mode and AutoCAD will find the extended intersection.

Figure 7-8

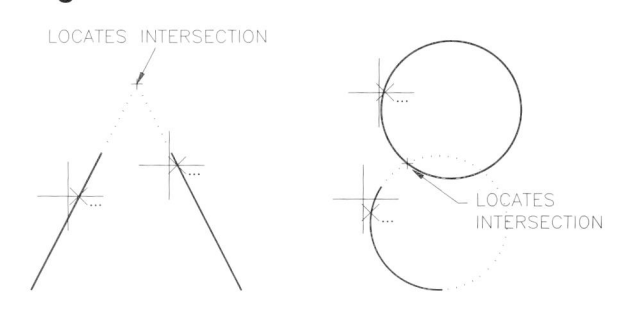

Midpoint

The *Midpoint* option snaps to the point of a *Line* or *Arc* that is halfway between the endpoints. PICK anywhere on the object.

Figure 7-9

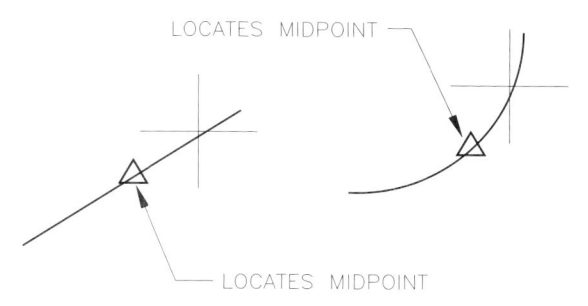

Nearest

The *Nearest* option locates the point on an object nearest to the cursor position. Place the cursor center nearest to the desired location, then PICK.

Nearest cannot be used effectively with *ORTHO* or *POLAR* to draw orthogonal lines because the *Nearest* point takes precedence over *ORTHO* and *POLAR*. However, the *Intersection* mode (in Running Osnap mode only) in combination with *POLAR* does allow you to construct orthogonal lines (or lines at any Polar Tracking angle) that intersect with other objects.

Figure 7-10

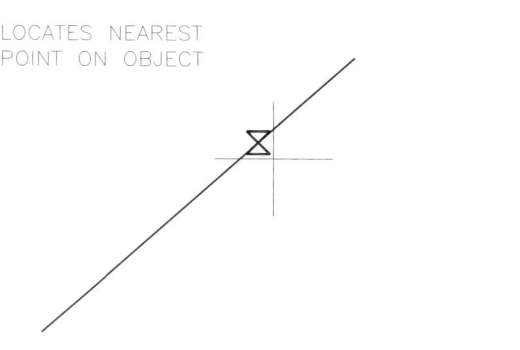

Node

This option snaps to a *Point* object. The *Point* must be within the Aperture (visible or invisible Aperture).

Figure 7-11

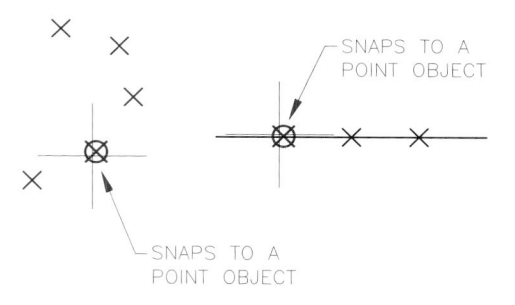

Perpendicular

Use this option to snap perpendicular to the selected object. PICK anywhere on a *Line* or straight *Pline* segment. The *Perpendicular* option is typically used for the second point ("Specify next point:" prompt) of the *Line* command.

Figure 7-12

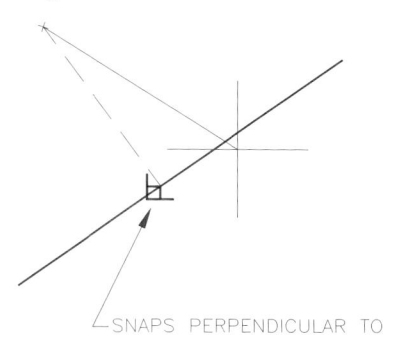

112 Object Snap and Object Snap Tracking

Quadrant

The *Quadrant* option snaps to the 0, 90, 180, or 270 degree quadrant of a *Circle*. PICK <u>nearest</u> to the desired *Quadrant*.

Figure 7-13

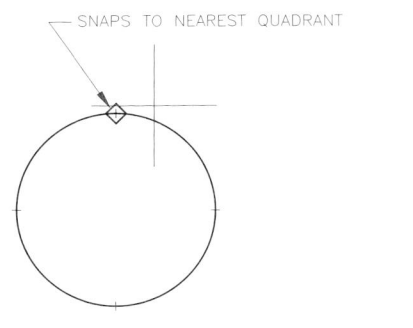

SNAPS TO NEAREST QUADRANT

Tangent

This option calculates and snaps to a tangent point of an *Arc* or *Circle*. PICK the *Arc* or *Circle* as near as possible to the expected *Tangent* point.

Figure 7-14

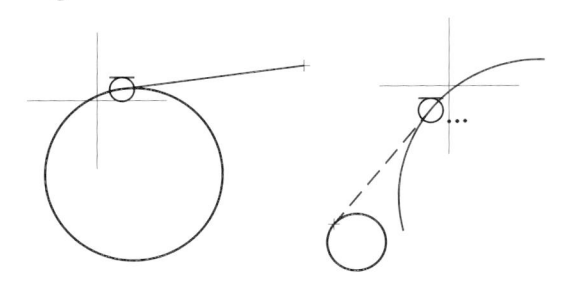

SNAPS TANGENT TO

From

The *From* option is designed to let you snap to a point <u>relative</u> to another point using relative rectangular, relative polar, or direct distance entry coordinates. There are two steps: first select a "Base-point:" (coordinates or another *OSNAP* may be used); then select an "Offset:" (enter relative rectangular, relative polar, or direct distance entry coordinates). The *From* option has effectively been replaced by the newer Object Snap Tracking feature (see "Object Snap Tracking").

Figure 7-15

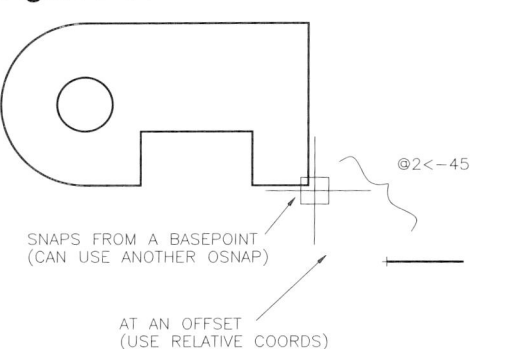

@2<-45

SNAPS FROM A BASEPOINT
(CAN USE ANOTHER OSNAP)

AT AN OFFSET
(USE RELATIVE COORDS)

Apparent intersection

Use this option when you are working with a 3D drawing and want to snap to a point in space where two objects appear to intersect (from your viewpoint) but do not actually physically intersect.

Figure 7-16

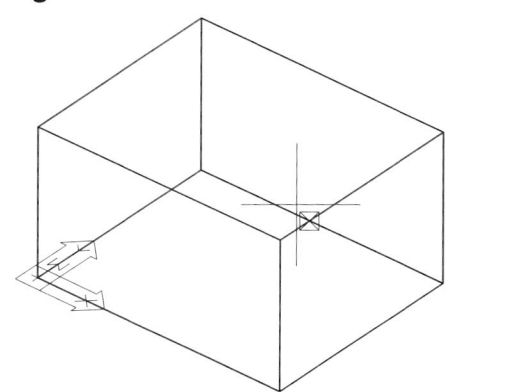

SNAPS TO APPARENT INTERSECTION IN 3D SPACE

Acquisition Object Snap Modes

Three new Object Snap modes were introduced with AutoCAD 2000: *Parallel, Extension,* and *Temporary Tracking*. When you use these Object Snap modes a dotted line appears, called an alignment vector, that indicates a vector along which the selected point will lie. In addition, these modes require an additional step—you must "acquire" (select) a point or object. For example, using the *Parallel* mode, you must "acquire" an object to be parallel to. The process used to "acquire" objects is explained here and is similar to that used for Object Snap Tracking (see "Object Snap Tracking" later in this chapter).

To Acquire an Object

Figure 7-17

To acquire an object to use for an *Extension* or *Parallel* Object Snap mode, move the cursor over the desired object and pause briefly, but do not pick the object. A small plus sign (+) is displayed when AutoCAD acquires the object (Fig. 7-17). A dotted-line "alignment vector" appears as you move the cursor into a parallel or extension position (Fig. 7-18). You can acquire multiple objects to generate multiple vectors.

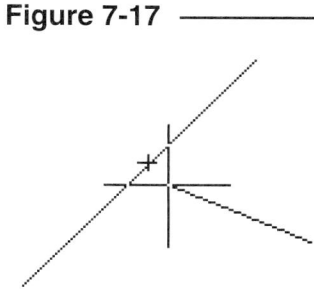

To clear an acquired object (in case you decide not to use that object), move the cursor back over the acquisition marker until the plus sign (+) disappears. Acquired points also clear automatically when another command is issued.

Parallel

Figure 7-18

The *Parallel* Osnap option snaps the rubberband line into a parallel relationship with any acquired object (see "To Acquire an Object," earlier this section).

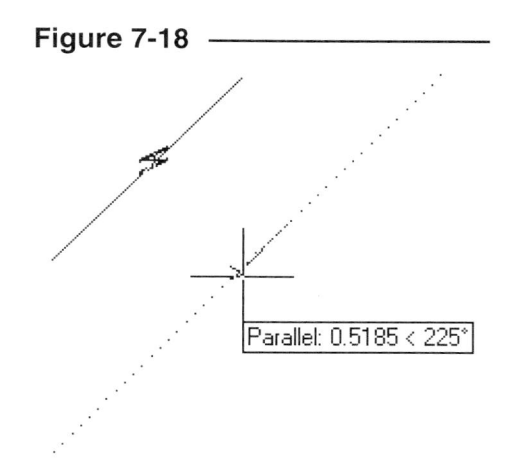

Start a *Line* by picking a start point. When AutoCAD prompts to "Specify next point or [Undo]:," acquire an object to use as the parallel source (a line to draw parallel to). Next, move the cursor to within a reasonably parallel position with the acquired line. The current rubberband line snaps into an exact parallel position (Fig. 7-18). A dotted-line parallel alignment vector appears as well as a tool tip indicating the current *Line* length and angle. The parallel Osnap symbol appears on the source parallel line. Pick to specify the current *Line* length. Keep in mind multiple lines can be acquired, giving you several parallel options. The acquired source objects lose their acquisition markers when each *Line* segment is completed.

Consider the use of *Parallel* Osnap with other commands such as *Move* or *Copy*. Figure 7-19 illustrates using *Move* with a *Circle* in a direction *Parallel* to the acquired *Line*.

Figure 7-19

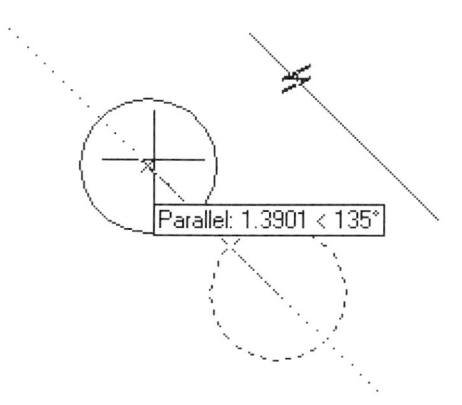

Extension

The *Extension* Osnap option snaps the rubberband line so that it intersects with an extension of any acquired object (see "To Acquire an Object," earlier this section).

For example, assume you use the *Line* command, then the *Extension* Osnap mode. When another existing object is acquired, the current *Line* segment intersects with an extension of the acquired object. Figure 7-20 depicts drawing a *Line* segment (upper left) to an *Extension* of an acquired *Line* (lower right). Notice the acquisition marker (plus symbol) on the acquired object.

Figure 7-20

Consider drawing a *Line* to an *Extension* of other objects, such as shown in Figure 7-21. Here a *Line* is drawn to the *Extension* of an *Arc*.

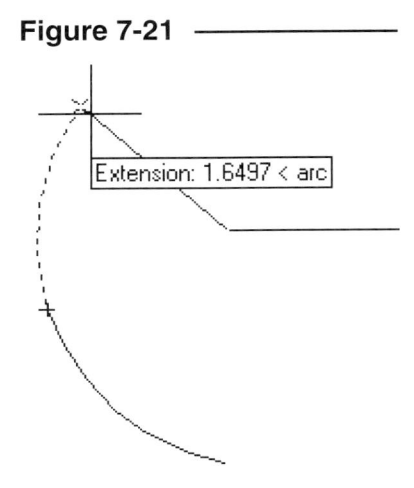

Figure 7-21

If you set *Extension* as a Running Osnap mode, each newly created *Line* segment automatically becomes acquired; therefore, you can draw an extension of the previous segment (at the same angle) easily (Fig. 7-22). (See "Osnap Running Mode" later in this chapter.)

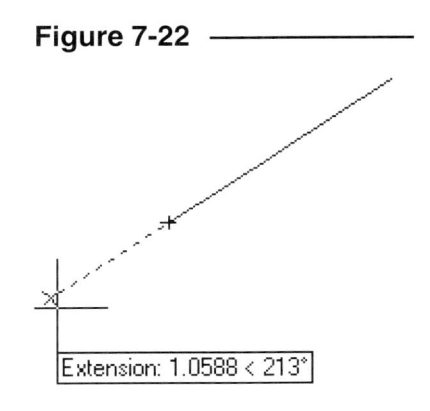

Figure 7-22

Temporary Tracking (**TT**)

Temporary Tracking sets up a temporary Polar Tracking point. This option allows you to "track" in a polar direction from any point you select. Tracking is the process of moving from a selected point in a preset angular direction. The preset angles are those specified in the *Increment Angle* section in the *Polar Tracking* tab of the *Drafting Settings* dialog box (see "Polar Tracking" in Chapter 3, Figure 3-17).

For example, if you use the *Line* command, then use the *Temporary Tracking* button or type **TT** at the "Specify first point:" prompt, you can select a point anywhere and that point becomes the temporary tracking point. The cursor moves in a preset angle from that point (Fig. 7-23). The next point selected along the alignment path becomes the *Line's* first point. POLAR does not have to be on to use TT. Note that the temporary tracking point is indicated by an acquisition marker, so it appears similar to any other acquired point.

Figure 7-23

Temporary Tracking can be used in conjunction with another Osnap mode (use *TT*, then another *Osnap* mode to specify the point). As an example, assume you wanted to begin drawing a *Line* directly above the corner of an existing rectangular shape. To do this, use the *Line* command and at the "Specify first point:" prompt invoke *TT*. Then use *Endpoint* to acquire the desired corner of the rectangle. A temporary tracking alignment vector appears from the corner as indicated in Figure 7-24. To specify an exact distance, enter a value (Direct Distance Entry) or turn on Polar Snap. The resulting point is the point that satisfies the "Specify first point:" prompt for the *Line*.

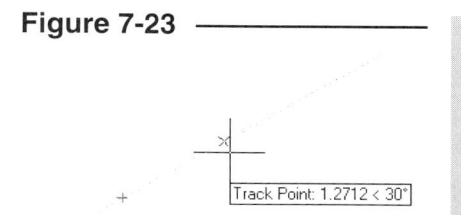

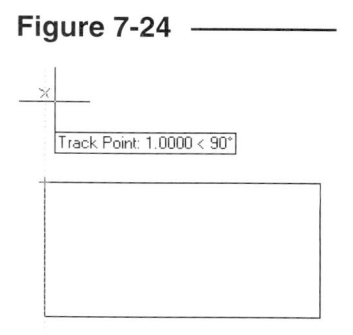

Figure 7-24

Object Snaps must be activated in order for you to snap to the desired object features. Two methods for activating Object Snaps, Single Point Selection and Running Object Snaps, are discussed next. The *OSNAP* modes (*Endpoint, Midpoint, Center, Intersection,* etc.) operate identically for either method.

OSNAP SINGLE POINT SELECTION

OBJECT SNAPS (SINGLE POINT)	Pull-down Menu	COMMAND (TYPE)	ALIAS (TYPE)	Short-cut	Screen (side) Menu	Tablet Menu	Cursor Menu (Shift+button 2)
	...	END, MID, etc. (first three letters)	...	...	****(asterisks)	T,15 - U,22	Endpoint Midpoint, etc.

There are many methods for invoking *OSNAP* modes for single point selection, as shown in the command table. If you prefer to type, enter only the first three letters of the *OSNAP* mode at the "Specify first point: " prompt, "Specify next point or [Undo]:" prompt, or any time AutoCAD prompts for a point.

Figure 7-25

Object Snaps are available from the Standard toolbar (Fig. 7-25). If desired, a separate *Object Snap* toolbar can be activated to float or dock on the screen by right-clicking on any icon button and selecting *Object Snap* from the list of toolbars. The advantage of invoking the *Object Snap* toolbar is that only one PICK is required for an *OSNAP* mode, whereas the Standard toolbar requires two PICKs because the *OSNAP* icons are on a flyout.

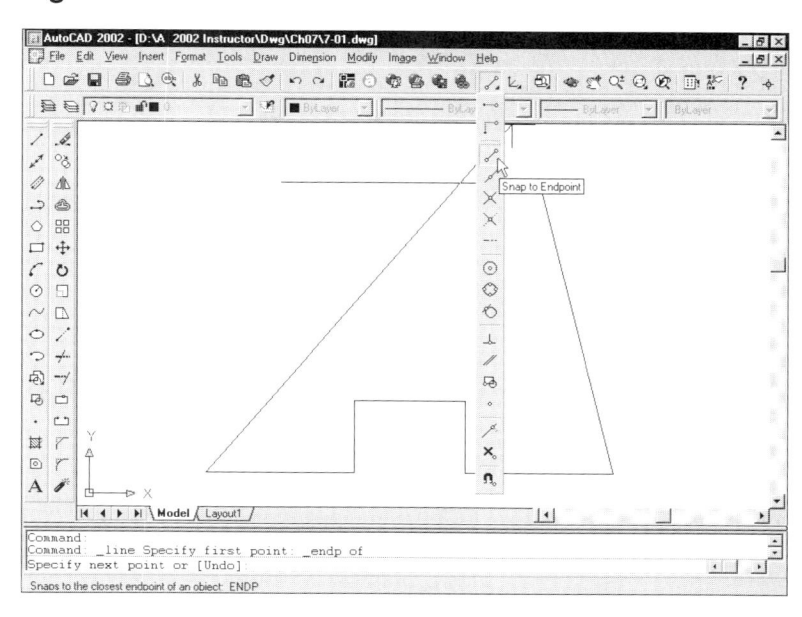

116   Object Snap and Object Snap Tracking

In addition to these options, a special menu called the cursor menu can be used. The cursor menu pops up at the current location of the cursor and replaces the cursor when invoked (Fig. 7-26). This menu is activated by pressing **Shift + #2** (hold down the Shift key while clicking the right mouse button).

(If you have a wheel mouse, you can also change the *MBUTTONPAN* system variable to 0. This action allows you to press the wheel to invoke the *Osnap* cursor menu, but disables your ability to *Pan* by holding down the wheel.)

Figure 7-26

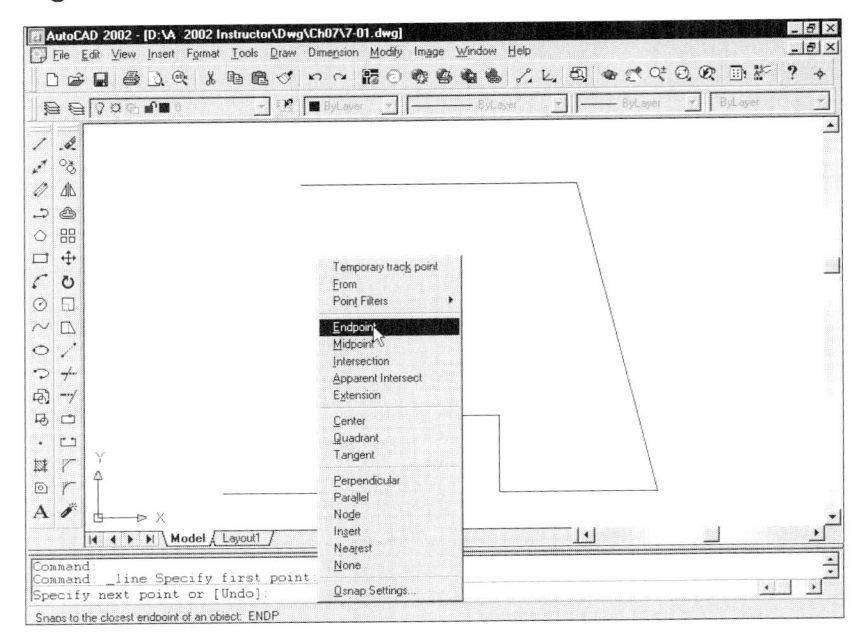

With any of these methods, OSNAP modes are active only for selection of a single point. If you want to OSNAP to another point, you must select an OSNAP mode again for the second point. With this method, the desired OSNAP mode is selected transparently (invoked during another command operation) immediately before selecting a point when prompted. In other words, whenever you are prompted for a point (for example, the "Specify first point:" prompt of the *Line* command), select or type an OSNAP option. Then PICK near the desired object feature (endpoint, center, etc.) with the cursor. AutoCAD snaps to the feature of the object and uses it for the point specification. Using OSNAP in this way allows the OSNAP mode to operate only for that single point selection.

For example, when using OSNAP during the *Line* command, the Command line reads as shown:

    Command: _line Specify first point: endp of (PICK)
    Specify next point or [Undo]: endp of (PICK)
    Specify next point or [Undo]: Enter
    Command:

If you prefer using OSNAP for single point specification, it may be helpful to associate the tools (buttons) with the Marker that appears on the drawing when you use the OSNAP mode. The table in Figure 7-27 displays the OSNAP icons and the respective Markers.

Figure 7-27

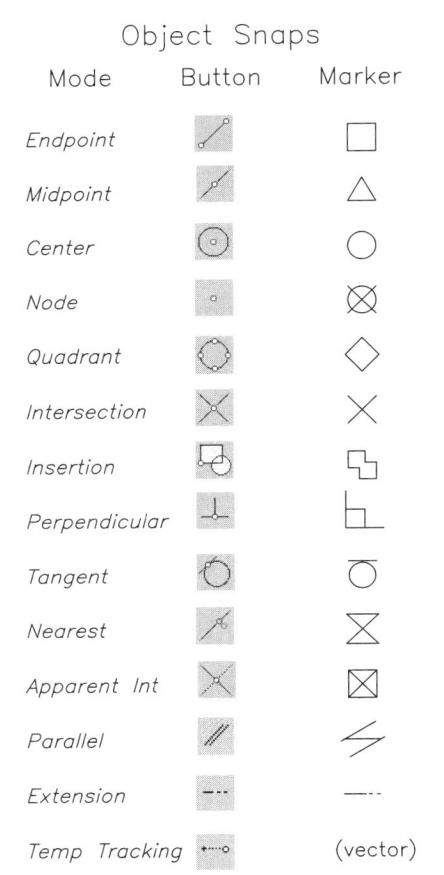

Object Snaps

Mode	Button	Marker
Endpoint		□
Midpoint		△
Center		○
Node		⊗
Quadrant		◇
Intersection		×
Insertion		⌐⌐
Perpendicular		⌐
Tangent		○
Nearest		⋈
Apparent Int		⊠
Parallel		∥
Extension		---
Temp Tracking		(vector)

OSNAP RUNNING MODE

OBJECT SNAPS (RUNNING)

Pull-down Menu	COMMAND (TYPE)	ALIAS (TYPE)	Short-cut	Screen (side) Menu	Tablet Menu	Cursor Menu (Shift+button 2)
Tools Drafting Settings... Object Snap	OSNAP or -OSNAP	OS or -OS	...	TOOLS 2 Osnap...	U,22	Osnap Settings...

Running Object Snap

A more effective method for using Object Snap is called "Running Object Snap" because one or more *OSNAP* modes (*Endpoint, Center, Midpoint,* etc.) can be turned on and kept running indefinitely. This method can obviously be more productive because you do not have to continually invoke an *OSNAP* mode each time you need to use one. For example, suppose you have several *Endpoint*s to connect. It would be most efficient to turn on the running *Endpoint OSNAP* mode and leave it running during the multiple selections. This is faster than continually selecting the *OSNAP* mode each time before you PICK.

You can even have several *OSNAP* modes running at the same time. A common practice is to turn on the *Endpoint, Center, Midpoint* modes simultaneously. In that way, if you move your cursor near a *Circle*, the *Center* Marker appears; if you move the cursor near the end or the middle of a *Line*, the *Endpoint* or the *Midpoint* mode Markers appear.

Accessing Running Object Snap

All features of Running Object Snaps are controlled by the *Drafting Settings* dialog box (Fig. 7-28). This dialog box can be invoked by the following methods (see the previous Command Table):

1. type the *OSNAP* command;
2. type *OS*, the command alias;
3. select *Drafting Settings...* from the *Tools* pull-down menu;
4. select *Osnap Settings...* from the bottom of the cursor menu (Shift + #2 button);
5. select the *Object Snap Settings* icon button from the *Object Snap* toolbar;
6. right-click on the words *OSNAP* or *OTRACK* on the Status Bar, then select *Settings...* from the shortcut menu that appears (see Figure 7-29).

Figure 7-28

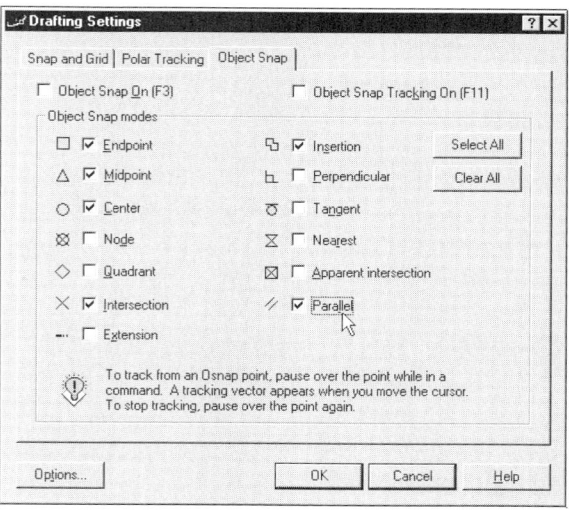

Use the *Object Snap* tab to select the desired Object Snap settings. Try using three or four commonly used modes together, such as *Endpoint, Midpoint, Center,* and *Intersection*. The AutoSnap Markers indicate which one of the modes would be used as you move the cursor near different object features. Using similar modes simultaneously, such as *Center, Quadrant,* and *Tangent,* can sometimes lead to difficulties since it requires the cursor to be placed almost in an exact snap spot, or in some cases it may not find one of the modes (*Quadrant* overrides *Center*). In these cases, the Tab key can be used to cycle through the options (see "Object Snap Cycling").

Figure 7-29

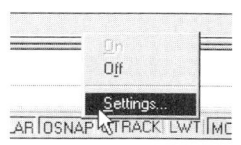

Running Object Snap Toggle

Another feature that makes Running Object Snap effective is Osnap Toggle. If you need to PICK a point without using *OSNAP*, use the Osnap Toggle to temporarily override (turn off) the modes. With Running Osnaps temporarily off, you can PICK any point without AutoCAD forcing your selection to an *Endpoint, Center,* or *Midpoint,* etc. When you toggle Running Object Snap on again, AutoCAD remembers which modes were previously set.

The following methods can be used to toggle Running Osnaps on and off:

1. click the word *OSNAP* that appears on the Status Bar (at the bottom of the screen)
2. press **F3**
3. press **Ctrl+F**

None

The *None OSNAP* option is a Running Osnap <u>override effective for only one PICK</u>. *None* is similar to the Running Object Snap Toggle, except it is effective for one PICK, then Running Osnaps automatically come back on without having to use a toggle. If you have *OSNAP* modes running but want to deactivate them for a <u>single</u> PICK, use *None* in response to "Specify first point:" or other point selection prompt. In other words, using *None* <u>during a draw or edit command</u> overrides any Running Osnaps for that single point selection. *None* can be typed at the command prompt (when prompted for a point) or can be selected from the bottom of the Object Snap toolbar.

Object Snap Cycling

In cases when you have multiple Running Osnaps set, and it is difficult to get the desired AutoSnap Marker to appear, you can use the <u>Tab</u> key to cycle through the possible *OSNAP* modes for the highlighted object. In other words, pressing the Tab key makes AutoCAD highlight the object nearest the cursor, then cycles through the possible snap Markers (for running modes that are set) that affect the object.

For example, when the *Center* and *Quadrant* modes are both set as Running Osnaps, moving the cursor near a *Circle* makes the *Quadrant* Marker appear but not the *Center* Marker. In this case, pressing the Tab key highlights the *Circle,* then cycles through the four *Quadrant* and one *Center* snap candidates.

OBJECT SNAP TRACKING

A feature called Object Snap Tracking helps you draw objects at specific angles and in specific relationships to other objects. Object Snap Tracking works in conjunction with object snaps and displays temporary alignment paths called "tracking vectors" that help you create objects aligned at precise angular positions relative to other objects. You can toggle Object Snap Tracking on and off with the *OTRACK* button on the Status Bar or by toggling F11.

To practice with Object Snap Tracking, try turning the *Extension* and *Parallel* Osnap options off. Since these two new options require point acquisition and generate alignment vectors, it can be difficult to determine which of these features is operating when they are all on.

Object Snap Tracking is similar to Polar Tracking in that it displays and snaps to alignment vectors, but the alignment vectors are generated from <u>other existing objects</u>, not from the current object. These other objects are acquired by Osnapping to them. Once a point (*Endpoint, Midpoint,* etc.) is acquired, alignment vectors generate from them in proximity to the cursor location. This process allows you to construct geometry that has orthogonal or angular relationships to other existing objects.

Using Object Snap Tracking is essentially the same as using the *Temporary Tracking* Osnap option, then using another Osnap mode to acquire the tracking point. (See previous Figure 7-24 and related explanation to refresh your memory.) Object Snap Tracking can be used with either Single Point or Running Osnap mode.

For example, Figure 7-30 displays an alignment vector generated from an acquired *Endpoint*. The current *Line* can then be drawn to a point that is horizontally aligned with the acquired *Endpoint*. Note that in Figure 7-30 and the related figures following, the Osnap Marker (*Endpoint* marker in this case) is anchored to the acquired point. The alignment vector always rotates about, and passes through, the acquired point (at the *Endpoint* marker). The current *Line* being constructed is at the cursor location.

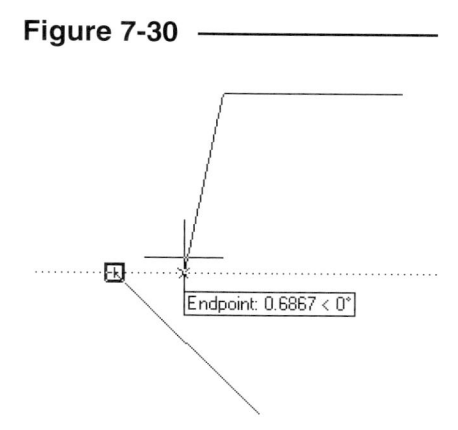

Figure 7-30

Moving the cursor to another location causes a different alignment vector to appear; this one displays vertical alignment with the same *Endpoint*.

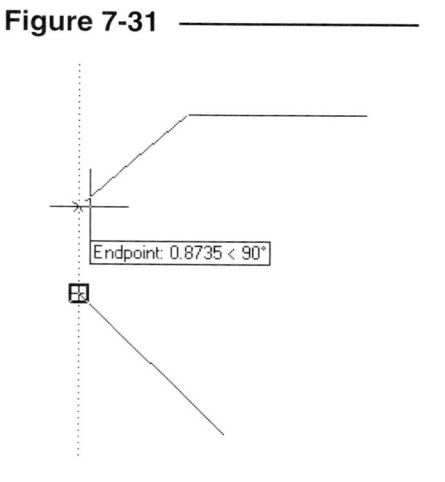

Figure 7-31

In addition, moving the cursor from an acquired point causes an array of alignment vectors to appear based on the current angular increment set in the *Polar Tacking* tab of the *Drafting Settings* dialog box (see Chapter 3, Figure 3-17, and related discussion). In Figure 7-32, alignment vectors are generated from the acquired point in 30-degree increments.

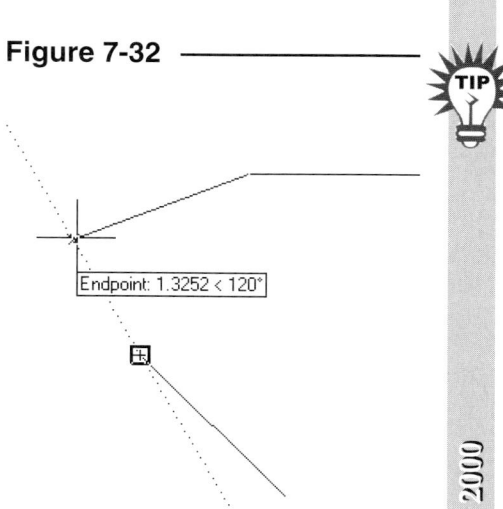

Figure 7-32

Figure 7-33

To acquire a point to use for Object Snap Tracking (when AutoCAD prompts to specify a point, move the cursor over the object point and pause briefly when the Osnap Marker appears, but do not pick the point. A small plus sign (+) is displayed when AutoCAD acquires the point (Fig. 7-33). The alignment vector appears as you move the cursor away from the acquired point. You can acquire multiple points to generate multiple alignment vectors.

In Figure 7-33, the *Endpoint* object snap is on. Start a *Line* by picking its start point, move the cursor over another line's endpoint to acquire it, and then move the cursor along the horizontal, vertical, or polar alignment vector that appears (not shown in Figure 7-33) to locate the endpoint you want for the line you are drawing.

To clear an acquired point (in case you decide not to use an alignment vector from that point), move the cursor back over the point's acquisition marker until the plus sign (+) disappears. Acquired points also clear automatically when another command is issued. You can also toggle the word *OTRACK* on the status bar to clear acquired points.

To Use Object Snap Tracking:

1. Turn on Object Snap and Object Snap Tracking (press F3 and F11 or togle *OSNAP* and *OTRACK* on the Status Bar). You can use *OSNAP* single point selection when prompted to specify a point instead of using Running Osnap.

2. Start a *Draw* or *Modify* command that prompts you to specify a point.

3. Move the cursor over an Object Snap point to temporarily acquire it. Do not PICK the point but only pause over the point briefly to acquire it.

4. Move the cursor away from the acquired point until the desired vertical, horizontal, or polar alignment vector appears, then PICK the desired location for the line along the alignment vector.

Remember, you must set an Object Snap (single or running) before you can track from an object's snap point. Object Snap and Object Snap Tracking must both be turned on to use Object Snap Tracking. Polar tracking can also be turned on, but it is not necessary for Object Snap Tracking to operate.

Object Snap Tracking with Polar Tracking

For some cases you may want to use Object Snap Tracking in conjunction with Polar Tracking. This combination allows you to track from the last point specified (on the current *Line* or other operation) as well as to connect to an alignment vector from an existing object.

Figure 7-34

Figure 7-34 displays both Polar Tracking and Object Snap Tracking in use. The current *Line* (dashed vertical line) is Polar Tracking along a vertical vector from the last point specified (on the horizontal line above). The new *Line's* endpoint falls on the horizontal alignment vector (Object Snap Tracking) acquired from the *Endpoint* of the existing diagonal *Line*. Note that the tool tip displays the Polar Tracking angle ("Polar: <270") and the Object Snap Tracking mode and angle ("Endpoint: <0").

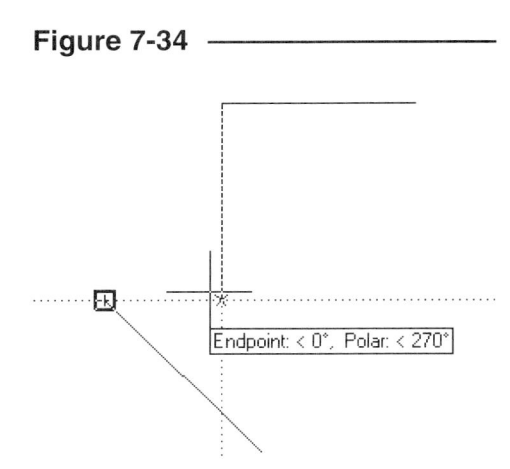

If the cursor is moved from the previous location (in the previous figure), additional Polar Tracking options and Object Snap Tracking options appear. For example in Figure 7-35, the current *Line* endpoint is tracking at 210 degrees (Polar Tracking) and falls on a vertical alignment vector from the *Endpoint* of the diagonal line (Object Snap Tracking).

Figure 7-35

You can accomplish the same capabilities available with the combination of Polar Tracking and Object Snap Tracking by using only Object Snap Tracking and multiple acquired points. For example, Figure 7-36 illustrates the same situation as in Figure 7-34 but only Object Snap tracking is on (Polar Tracking is off). Notice that two *Endpoints* have been acquired to generate the desired vertical and horizontal alignment vectors.

Figure 7-36

Object Snap Tracking Settings

You can set the *Object Snap Tracking Settings* in the *Drafting Settings* dialog box (Fig. 7-37). The only options are to *Track orthogonally only* or to *Track using all polar angle settings*. The Object Snap Tracking vectors are determined by the *Increment angle* and *Additional angles* set for Polar Tracking (left side of dialog box). If you want to track using these angles, select *Track using all polar angle settings*.

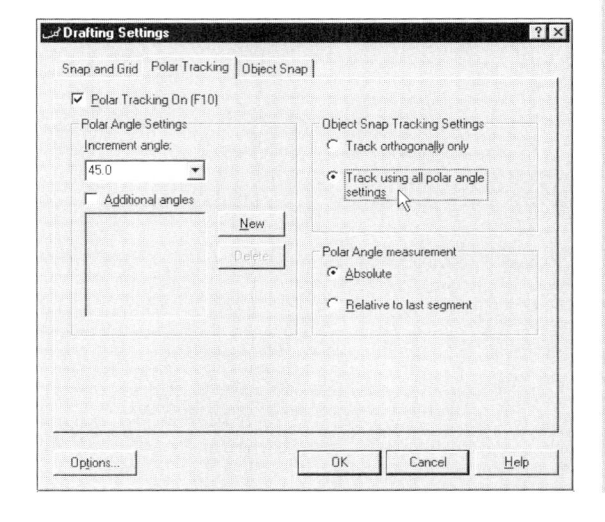

Figure 7-37

OSNAP APPLICATIONS

OSNAP can be used any time AutoCAD prompts you for a point. This means that you can invoke an *OSNAP* mode during any draw or modify command as well as during many other commands. *OSNAP* provides you with the potential to create 100% accurate drawings with AutoCAD. Take advantage of this feature whenever it will improve your drawing precision. Remember, any time you are prompted for a point, use *OSNAP* if it can improve your accuracy.

OSNAP PRACTICE

Single Point Selection Mode

1. Turn off *SNAP*, *POLAR*, *OSNAP*, and *OTRACK*. Draw two (approximately) vertical **Lines**. Follow these steps to draw another *Line* between *Endpoint*s.

STEPS	COMMAND PROMPT	PERFORM ACTION	COMMENTS
1.	Command:	select *Line* by any method	
2.	LINE Specify first point:	type *END* and press **Enter** (or Spacebar)	
3.	endp of	move the cursor near the end of one of the *Lines*	endpoint Marker appears (square box at *Line* end), "Endpoint" Snap Tip may appear
4.		**PICK** while the Marker is visible	rubberband line appears
5.	Specify next point or [Undo]:	type *END* and press **Enter** (or Spacebar)	
6.	endp of	move the cursor near the end of the second *Line*	endpoint Marker appears (square box at *Line* end), "Endpoint" Snap Tip may appear
7.		**PICK** while the Marker is visible	*Line* is created between endpoints
8.	Specify next point or [Undo]:	press **Enter**	completes command

2. Draw two *Circles*. Follow these steps to draw a *Line* between the *Centers*.

STEPS	COMMAND PROMPT	PERFORM ACTION	COMMENTS
1.	Command:	select *Line* by any method	
2.	LINE Specify first point:	invoke the cursor menu (press **Shift+#2** button) and select *Center*	
3.	cen of	move the cursor near a *Circle* object (or where you think the center is)	the AutoSnap Marker appears (small circle at center), "Center" Snap Tip may appear
4.		**PICK** while the Marker is visible	rubberband line appears
5.	Specify next point or [Undo]:	invoke the cursor menu (press **Shift+#2** button) and select *Center*	
6.	cen of	move the cursor near the second *Circle* object (or where you think the center is)	the AutoSnap Marker appears (small circle at center), "Center" Snap Tip may appear
7.		**PICK** the other *Circle* while the Marker is visible	*Line* is created between *Circle* centers
8.	Specify next point or [Undo]:	press **Enter**	completes command

3. *Erase* the *Line* only from the previous exercise. Draw another *Line* anywhere, but <u>not</u> attached to the *Circles*. Follow the steps to *Move* the *Line* endpoint to the *Circle* center.

STEPS	COMMAND PROMPT	PERFORM ACTION	COMMENTS
1.	Command:	select *Move* by any method	
2.	MOVE Select objects:	**PICK** the *Line*	the *Line* becomes highlighted
3.	Select objects:	press **Enter**	completes selection set
4.	Specify base point or displacement:	Select the *Endpoint* icon button from the *Object Snap* toolbar	
5.	endp of	move the cursor near a *Line* endpoint	AutoSnap Marker appears (square box), "Endpoint" Snap Tip may appear
6.		**PICK** while Marker is visible	endpoint becomes the "handle" for *Move*
7.	Specify second point of displacement:	select *Center* from *Object Snap* toolbar	
8.	cen of	move the cursor near a *Circle*	AutoSnap Marker appears (small circle), "Center" Snap Tip may appear
9.		**PICK** the *Circle* object while Marker is visible	selected *Line* is moved to *Circle* center

Running Object Snap Mode

4. Draw several *Lines* and *Circles* at random. To draw several *Lines* to *Endpoints* and *Tangent* to the *Circles*, follow these steps.

STEPS	COMMAND PROMPT	PERFORM ACTION	COMMENTS
1.	Command:	type *OSNAP* or *OS*	*Drafting Settings* dialog box appears
2.		select *Endpoint* and *Tangent* then press *OK*	turns on the Running Osnap modes
3.	Command:	invoke the *Line* command	use any method
4.	LINE Specify first point:	move the cursor near one *Line* endpoint	AutoSnap Marker appears (square box), "Endpoint" Snap Tip may appear
5.		**PICK** while Marker is visible	rubberband line appears, connected to endpoint
6.	Specify next point or [Undo]:	move cursor near endpoint of another *Line*	AutoSnap Marker appears (square box), "Endpoint" Snap Tip may appear
7.		**PICK** while Marker is visible	a *Line* is created between endpoints
8.	Specify next point or [Undo]:	move the cursor near a *Circle* object	AutoSnap Marker appears (small circle with tangent line segment), "Tangent" Snap Tip may appear
9.		**PICK** *Circle* while Marker is visible	a *Line* is created *Tangent* to the *Circle*
10.	Specify next point or [Undo]:	press **Enter**	ends *Line* command
11.	Command:	invoke the *Line* command	use any method
12.	LINE Specify first point:	move cursor near a *Circle*	AutoSnap Marker appears (small circle with tangent line segment), "Deferred Tangent" Snap Tip may appear
13.		**PICK** *Circle* while Marker is visible	rubberband line does NOT appear
14.	Specify next point or [Undo]:	move the cursor near a second *Circle* object	AutoSnap Marker appears (small circle with tangent line segment), "Deferred Tangent" Snap Tip may appear
15.		**PICK** *Circle* while Marker is visible	a *Line* is created tangent to the two *Circles*
16.	Specify next point or [Undo]:	click the word *OSNAP* on the Status Bar, press **F3** or **Ctrl+F**	temporarily toggles Running Osnaps off

(Continued)

17.		**PICK** a point near a *Line* end	*Line* is created, but does not snap to *Line* endpoint
18.	Specify next point or [Undo]:	click the word *OSNAP* on the Status Bar, press **F3** or **Ctrl+F**	toggles Running Osnaps back on
19.		**PICK** near a *Circle* (when Marker is visible	*Line* is created that snaps tangent to *Circle*
20.	Specify next point or [Undo]:	**Enter**	ends *Line* command
21.	Command:	invoke the cursor menu (press **Shift+#2**) and select *Osnap Settings*...	*Osnap Settings* dialog box appears
22.		Press the *Clear all* button, then *OK*	running *OSNAPS* are turned off

Object Snap Tracking

5. Begin a *New* drawing. Create an "L" shape by drawing one vertical and one horizontal *Line*, each 5 units long. To draw several *Lines* that track from *Endpoints* and *Midpoints*, follow these steps.

STEPS	COMMAND PROMPT	PERFORM ACTION	COMMENTS
1.	Command:	type *OSNAP* or *OS*	*Drafting Settings* dialog box appears, *Object Snap* tab
2.		select *Endpoint* and *Midpoint* options	turns on the Running Osnap modes
3.		select *Polar Tracking* tab and set *Increment Angle* to **45**, then select *OK*	sets tracking angle for Object Snap Tracking
4.		toggle <u>on</u> *OSNAP* and *OTRACK* and toggle <u>off</u> *SNAP*, *ORTHO*, and *POLAR*	select the words on the Status Bar (recessed is on and protruding is off
5.	Command:	invoke the *Line* command	
6.	LINE Specify first point:	move the cursor to the right horizontal *Line* endpoint and rest the cursor until the point is aquired	AutoSnap Marker appears (square box) and acquisition marker (plus) appears
7.		**PICK** while Marker is visible	rubberband line appears connected to endpoint and tracking vectors appear at 45 and 90 positions
8.	Specify next point	move the cursor to endpoint of vertical *Line* and rest the cursor until the point is aquired but <u>do</u> not PICK the point	AutoSnap Marker appears (square box) and acquistion marker (plus) appears

(Continued)

9.		move the cursor so two 90-degree alignment vectors appear (forming a square), then **PICK**	*Line* is drawn from first endpoint at 90-degree alignment to both endpoints
10.	Specify next point:	move cursor near top endpoint of vertical *Line* and PICK the point	*Line* is draw from last endpoint at 90-degree alignment to endpoint of vertical *Line* to form a square
11.	Specify next point:	press **Enter**	ends *Line* command
12.	Command:	invoke the ***Line*** command	use any method
13.	LINE Specify next point:	move the cursor to the vertical *Line* endpoint (top) and rest the cursor until the point is acquired	AutoSnap Marker appears (square box) and acquistion marker (plus) appears
14.		**PICK** *Line* while Marker is visible	rubberband line appears connected to endpoint and tracking vectors appear at 45 and 90 positions
15.	Specify next point:	move cursor to midpoint of original horizontal *Line* and rest the cursor until the point is aquired but <u>do not</u> PICK the point	AutoSnap Marker appears (triangle) and acquistion marker (plus) appears
16.		move the cursor upward so vertical alignment vector from horizontal line and 45-degree vector from vertical line appears, then PICK	diagonal *Line* is drawn above midpoint and aligned (45 degrees) to vertical line endpoint (to center of square)
17.	Specify next point:	press **Enter**	ends *Line* command

CHAPTER EXERCISES

1. ***OSNAP* Single Point Selection**

    Figure 7-38

    Open the **CH6EX1A** drawing and begin constructing the sheet metal part. Each unit in the drawing represents one inch.

    A. Create four ***Circles***. All *Circles* have a radius of **1.685**. The *Circles'* centers are located at **5,5**, **5,13**, **19,5**, and **19,13**.

    B. Draw four ***Lines***. The *Lines* (highlighted in Figure 7-38) should be drawn on the outside of the *Circles* by using the ***Quadrant*** *OSNAP* mode as shown for each *Line* endpoint.

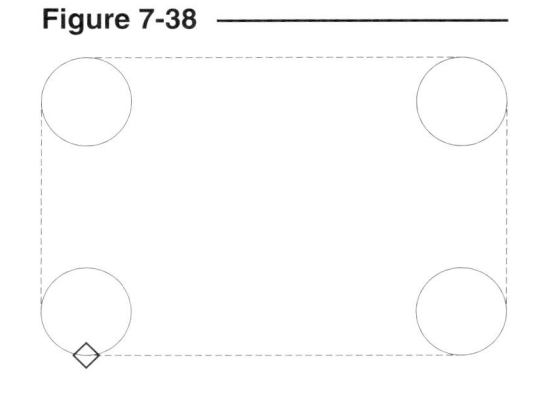

C. Draw two *Lines* from the *Center* of the existing *Circles* to form two diagonals as shown in Figure 7-39.

D. At the *Intersection* of the diagonals create a *Circle* with a **3** unit radius.

Figure 7-39

E. Draw two *Lines*, each from the *Intersection* of the diagonals to the *Midpoint* of the vertical *Lines* on each side. Finally, construct four new *Circles* with a radius of **.25,** each at the *Center* of the existing ones (Fig. 7-40).

F. *SaveAs* **CH7EX1**.

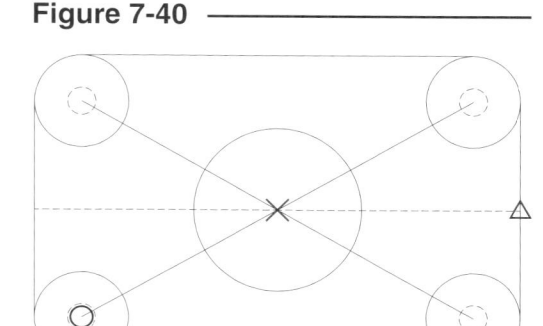

Figure 7-40

2. *OSNAP* **Single Point Selection**

A multiview drawing of a mechanical part is to be constructed using the A-METRIC drawing. All dimensions are in millimeters, so each unit equals one millimeter.

A. *Open* **A-METRIC** from Chapter 6 Exercises. Draw a *Line* from **60,140** to **140,140**. Create two *Circles* with the centers at the *Endpoints* of the *Line,* one *Circle* having a diameter of **60** and the second *Circle* having a diameter of **30**. Draw two *Lines Tangent* to the *Circles* as shown in Figure 7-41. *SaveAs* **PIVOTARM CH7**.

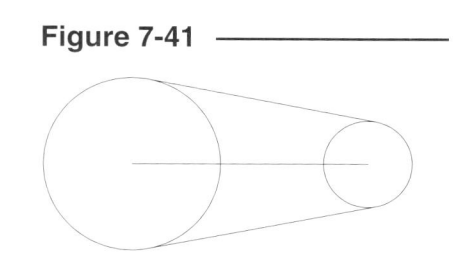

Figure 7-41

B. Draw a vertical *Line* down from the far left *Quadrant* of the *Circle* on the left. Specify relative polar (**@100<270**) or direct distance entry coordinates to make the *Line* 100 units. Draw a horizontal *Line* **125** units from the last *Endpoint* using relative polar or direct distance entry coordinates. Draw another *Line* between that *Endpoint* and the *Quadrant* of the *Circle* on the right. Finally, draw a horizontal *Line* from point **30,70** and *Perpendicular* to the vertical *Line* on the right.

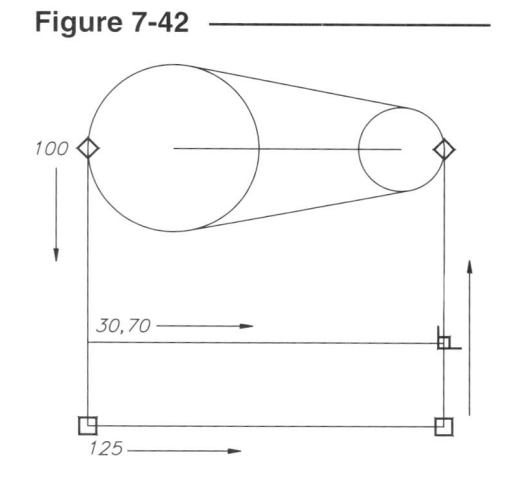

Figure 7-42

C. Draw two vertical *Lines* from the
   Intersections of the horizontal *Line*
   and *Circles* and *Perpendicular* to the
   Line at the bottom. Next, draw two
   Circles concentric to the previous
   two and with diameters of **20** and **10**
   as shown in Figure 7-43.

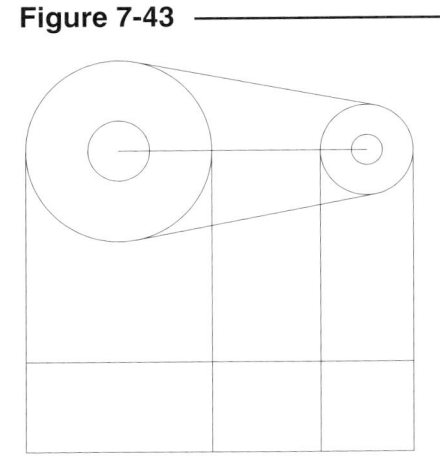

Figure 7-43

D. Draw four more vertical *Lines* as
   shown in Figure 7-44. Each *Line* is
   drawn from the new *Circles'*
   Quadrant and *Perpendicular* to the
   bottom line. Next, draw a miter
   Line from the *Intersection* of the
   corner shown to **@150<45**. *Save* the
   drawing for completion at a later
   time as another chapter exercise.

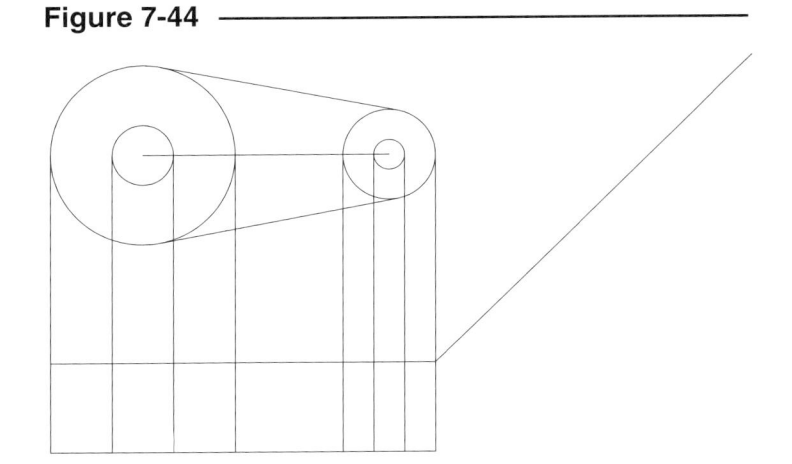

Figure 7-44

3. **Running Osnap**

   Create a cross-sectional view of a door header composed of two 2 x 6
   wooden boards and a piece of 1/2" plywood. (The dimensions of a
   2 x 6 are actually 1-1/2" x 5-3/8".)

   A. Begin a *New* drawing and assign the name **HEADER**. Draw four
      vertical lines as shown in Figure 7-45.

   B. Use the *OSNAP* command or select *Drafting Settings...* from the
      Tools pull-down menu and turn on the *Endpoint* and
      Intersection modes.

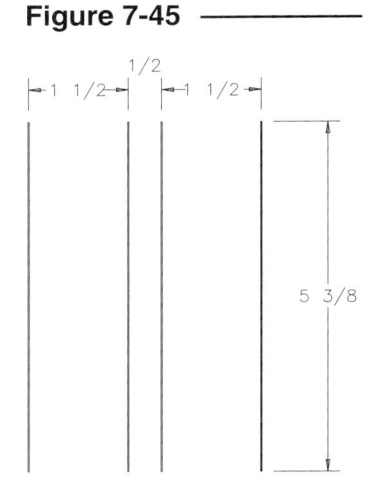

Figure 7-45

C. Draw the remaining lines as shown in Figure 7-46 to complete the header cross-section. *Save* the drawing.

Figure 7-46

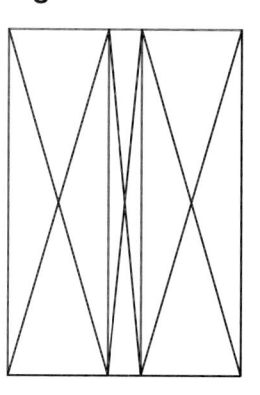

4. **Running Object Snap and *OSNAP* toggle**

   Assume you are commissioned by the local parks and recreation department to provide a layout drawing for a major league sized baseball field. Lay out the location of bases, infield, and outfield as follows.

   A. *Open* the **BALL FIELD CH6.DWG** that you set up in Chapter 6. Make sure *Limits* are set to 512',384', *Snap* is set to 10', and *Grid* is set to 20'. Ensure *SNAP* and *GRID* are on. Use *SaveAs* to save and rename the drawing to **BALL FIELD CH7**.

   B. Begin drawing the baseball diamond by using the *Line* command and using direct distance coordinate entry. **PICK** the "Specify first point:" at **20',20'** (watch *Coords*). Draw the foul line to first base by turning on **ORTHO** or **POLAR**, move the cursor to the right (along the X direction) and enter a value of **90'** (don't forget the apostrophe to indicate feet). At the "Specify next point or [Undo]:" prompt, continue by drawing a vertical *Line* of **90'**. Continue drawing a square with **90'** between bases (Fig. 7-47).

   Figure 7-47

   C. Invoke the *Drafting Settings* dialog box by any method. Turn on the *Endpoint*, *Midpoint*, and *Center* object snaps.

   D. Use the *Circle* command to create a "base" at the lower-right corner of the square. At the "Specify center point for circle or [3P/2P/Ttr]:" prompt, **PICK** the *Line Endpoint* at the lower-right corner of the square. Enter a *Radius* of **1'** (don't forget the apostrophe). Since Running Object Snaps are on, AutoCAD should display the Marker (square box) at each of the *Line Endpoints* as you move the cursor near; therefore, you can easily "snap" the center of the bases (*Circles*) to the corners of the square. Draw *Circles* of the same *Radius* at second base (upper-right corner), and third base (upper-left corner of the square). Create home plate with a *Circle* of a **2'** *Radius* by the same method (see Figure 7-47).

   E. Draw the pitcher's mound by first drawing a *Line* between home plate and second base. **PICK** the "Specify first point:" at home plate (*Center* or *Endpoint*), then at the "Specify next point or [Undo]:" prompt, PICK the *Center* or the *Endpoint* at second base. Construct a *Circle* of **8'** *Radius* at the *Midpoint* of the newly constructed diagonal line to represent the pitcher's mound (see Figure 7-47).

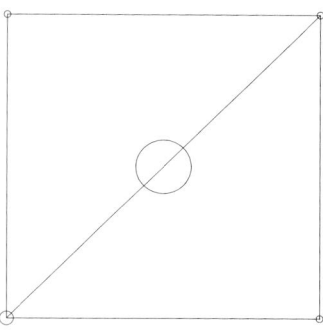

F. *Erase* the diagonal *Line* between home plate and second base. Draw the foul lines from first and third base to the outfield. For the first base foul line, construct a *Line* with the "Specify first point:" at the *Endpoint* of the existing first base line or *Center* of the base. Move the cursor (with *ORTHO* on) to the right (X positive) and enter a value of **240'** (don't forget the apostrophe). Press **Enter** to complete the *Line* command. Draw the third base foul line at the same length (in the positive Y direction) by the same method. (see Fig. 7-48).

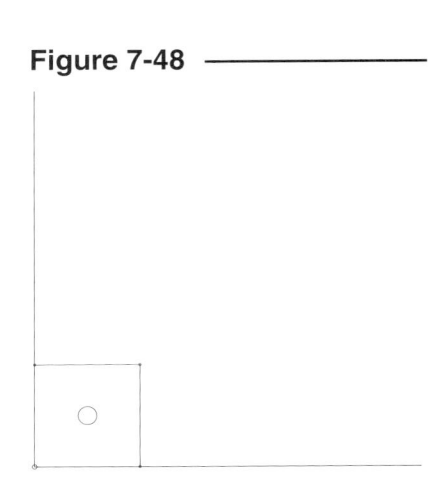

Figure 7-48

G. Draw the home run fence by using the *Arc* command from the *Draw* pull-down menu. Select the **Start, Center, End** method. **PICK** the end of the first base line (*Endpoint* object snap) for the **Start** of the *Arc* (Fig. 7-49, point 1), **PICK** the pitcher's mound (*Center* object snap) for the **Center** of the *Arc* (point 2), and the end of the third base line (*Endpoint* object snap) for the **End** of the *Arc* (point 3). Don't worry about *ORTHO* in this case because *OSNAP* overrides *ORTHO*.

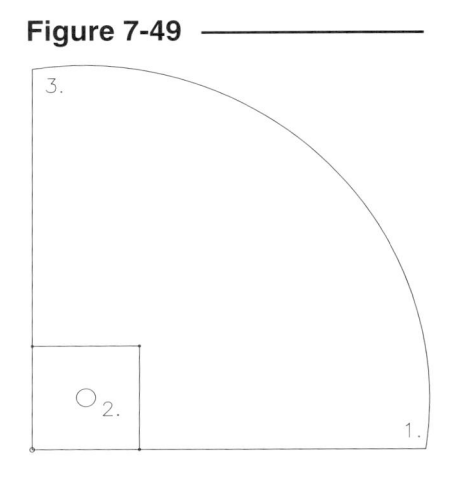

Figure 7-49

H. Now turn off **ORTHO**. Construct an *Arc* to represent the end of the infield. Select the *Arc* **Start, Center, End** method from the *Draw* pull-down menu. For the **Start** point of the *Arc*, toggle Running Osnaps off by pressing **F3, Ctrl+F,** or clicking the word **OSNAP** on the Status Bar and **PICK** location **140', 20', 0'** on the first base line (watch *Coords* and ensure *SNAP* and *GRID* are on). (See Figure 7-50, point 1) Next, toggle Running Osnaps back on, and **PICK** the **Center** of the pitcher's mound as the **Center** of the *Arc* (point 2). Third, toggle Running Osnaps off again and **PICK** the **End** point of the *Arc* on the third base line (point 3).

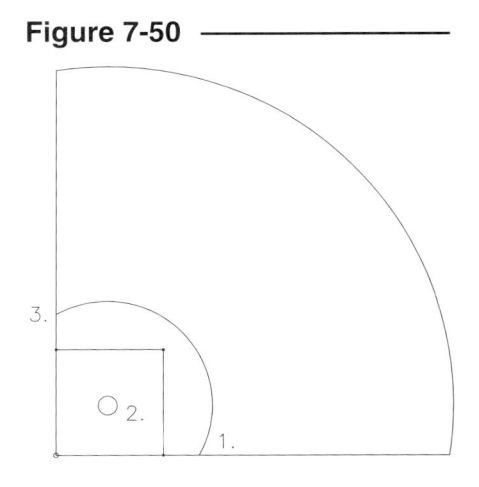

Figure 7-50

I. Lastly, the pitcher's mound should be moved to the correct distance from home plate. Type **M** (the command alias for *Move*) or select *Move* from the *Modify* pull-down menu. When prompted to "Select objects:" **PICK** the *Circle* representing pitcher's mound. At the "Specify base point or displacement:" prompt, toggle Running Osnaps on and **PICK** the **Center** of the mound. At the "Specify second point of displacement:" prompt, enter **@3'2"<225**. This should reposition the pitcher's mound to the regulation distance from home plate (60'-6"). Compare your drawing to Figure 7-50. *Save* the drawing (as BALL FIELD CH7).

J. Activate a *Layout* tab. Assuming your system is configured for a printer using an 11 x 8.5 sheet and is configured to automatically create a viewport (see "Setting Layout Options and Plot Options" in Chapter 6), a viewport should appear. (If the *Page Setup* dialog box automatically appears, select *OK*.) If a viewport appears, proceed to step L. If no viewport appears, complete step K.

K. To make a viewport, type *–Vports* at the Command prompt (don't forget the hyphen) and press **Enter**. Accept the default (*Fit*) option by pressing **Enter**. A viewport appears to fit the printable area.

L. Double-click inside the viewport, then type *Z* for *Zoom* and press **Enter**, and type *A* for *All* and press **Enter**. Activate the *Viewports* drop-down list by right-clicking on any icon button and selecting *Viewports* from the list of available toolbars that appears. Select *1/64"=1'* from the list. Make a print of the drawing on an 11 x 8.5 inch sheet.

M. Activate the *Model* tab. *Save* the drawing as **BALL FIELD CH7**.

5. **Polar Tracking and Object Snap Tracking**

   A. In this exercise, you will create a table base using Polar Tracking and Object Snap Tracking to locate points for construction of holes and other geometry. Begin a *New* drawing and use the **ACAD.DWT** template. *Save* the drawing and assign the name **TABLE-BASE**. Set the drawing limits to **48 x 32**.

   B. Invoke the *Drafting Settings* dialog box. In the *Snap and Grid* tab, set *Polar Distance* to **1.00** and make *Polar Snap* current. In the *Polar Tracking* tab, set the *Increment Angle* to **45**. In the *Object Snap* tab, turn on the *Endpoint*, *Midpoint* and *Center Osnap* options. Select *OK*. On the Status Bar ensure *SNAP, POLAR, OSNAP,* and *OTRACK* are on.

   C. Use a *Line* and create the three line segments representing the first leg as shown in Figure 7-51. Begin at the indicated location. Use Polar Tracking and Polar Snap to assist drawing the diagonal lines segments. (Do not create the dimensions in your drawing.)

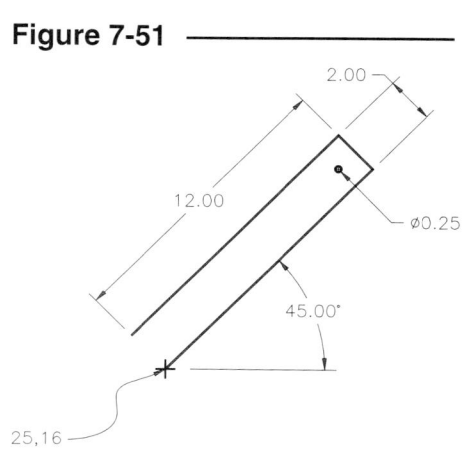

   Figure 7-51

   D. Place the first drill hole at the end of the leg using *Circle* with the *Center, Radius* option. Use Object Snap Tracking to indicate the center for the hole (*Circle*). Track vertically and horizontally from the indicated corners of the leg in Figure 7-52. Use *Endpoint Osnap* to snap to the indicated corners. (See Figure 7-51 for hole dimension.)

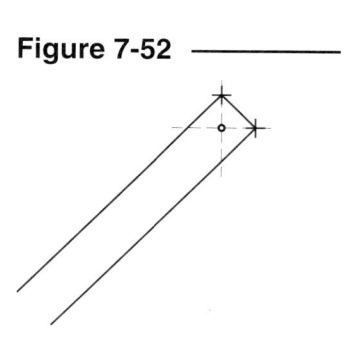

   Figure 7-52

132 Object Snap and Object Snap Tracking

E. Create the other 3 legs in a similar fashion. You can track from the *Endpoints* on the bottom of the existing leg to identify the "first point" of the next *Line* segment as shown in Figure 7-53.

Figure 7-53

F. Use *Line* to create a 24" x 24" square table top on the right side of the drawing as shown in Figure 7-54. Use Object Snap Tracking and Polar Tracking with the *Move* command to move the square's center point to the center point of the 4 legs (HINT: At the "Specify base point or displacement:" prompt, Osnap Track to the square's *Midpoints*. At the "Specify second point of displacement:" prompt, use *Endpoint* Osnaps to locate the legs' center.

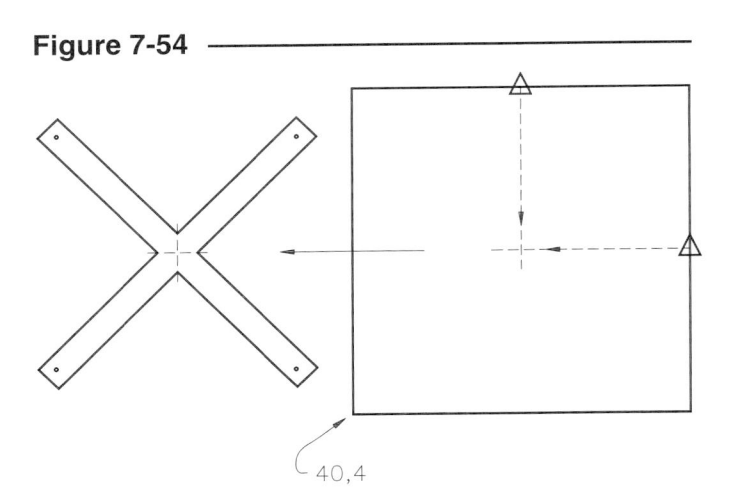

Figure 7-54

40,4

G. Create a smaller square (*Line*) in the center of the table which will act as a support plate for the legs. Track from the *Midpoint* of the legs to create the lines as indicated (highlighted) in Figure 7-55.

Figure 7-55

H. *Save* the drawing. You will finish the table base in Chapter 9.

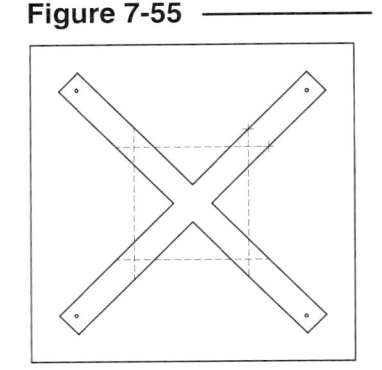

8

DRAW COMMANDS I

Chapter Objectives

After completing this chapter you should:

1. know where to locate and how to invoke the draw commands;
2. be able to draw *Lines*;
3. be able to draw *Circles* by each of the five options;
4. be able to draw *Arcs* by each of the eleven options;
5. be able to create *Point* objects and specify the *Point Style*;
6. be able to create *Plines* with width and combined of line and arc segments.

CONCEPTS

Draw Commands—Simple and Complex

Draw commands create objects. An object is the smallest component of a drawing. The draw commands listed immediately below create simple objects and are discussed in this chapter. Simple objects appear as one entity.

Figure 8-1

Line, Circle, Arc, and *Point*

Other draw commands create more complex shapes. Complex shapes appear to be composed of several components, but each shape is usually one object. An example of an object that is one entity but usually appears as several segments is listed below and is also covered in this chapter:

Pline

Other draw commands discussed in Chapter 15 (listed below) are a combination of simple and complex shapes:

Xline, Ray, Polygon, Rectangle, Donut, Spline, Ellipse, Divide, Mline, Measure, Sketch, Solid, Region, and *Boundary*

Draw Command Access

As a review from Chapter 3, Draw Command Concepts, remember that any of the five methods can be used to access the draw commands: *Draw* toolbar (Fig. 8-1), *Draw* pull-down menu (Fig. 8-2), DRAW1 and DRAW2 screen menus, keyboard entry of the command or alias, and digitizing tablet icons.

Figure 8-2

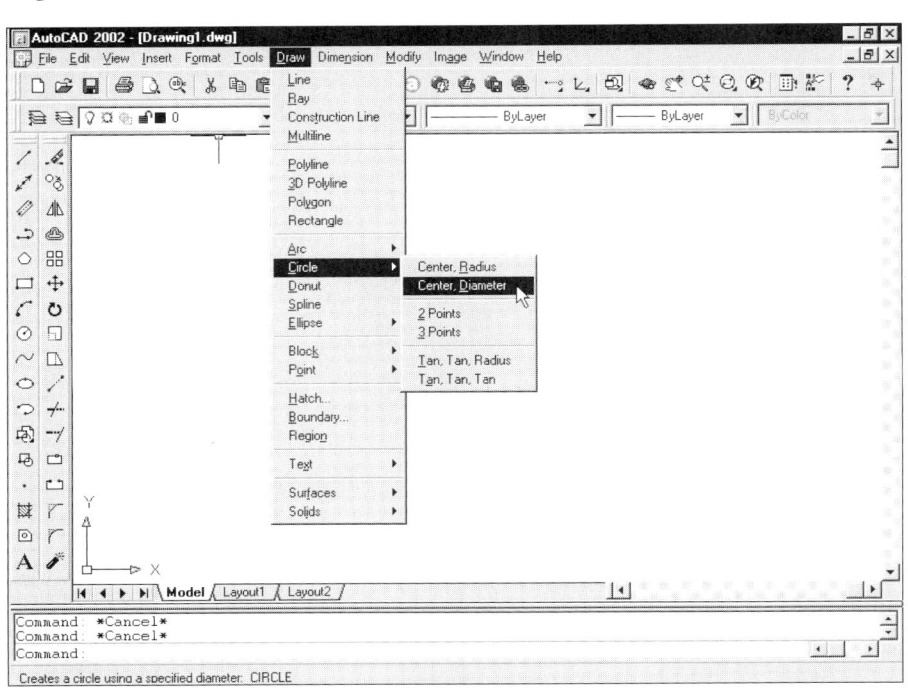

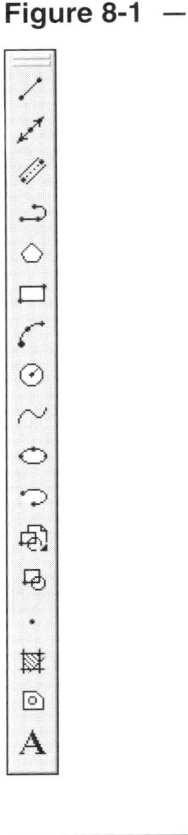

Coordinate Entry

When creating objects with draw commands, AutoCAD always prompts you to indicate points (such as endpoints, centers, radii) to describe the size and location of the objects to be drawn. An example you are familiar with is the *Line* command, where AutoCAD prompts for the "Specify first point:". Indication of these points, called <u>coordinate entry</u> can be accomplished by five formats (for 2D drawings):

1. **Interactive** — PICK points on screen with input device
2. **Absolute coordinates** — X,Y
3. **Relative rectangular coordinates** — @X,Y
4. **Relative polar coordinates** — @distance<angle
5. **Direct distance entry** — dist,direction (Type a distance value relative to the last point, indicate direction with the cursor, then press Enter.)

Any of these methods can be used <u>whenever</u> AutoCAD prompts you to specify points. (For practice with these methods, see Chapter 3, Draw Command Concepts.)

Interactive Entry Tools

Also keep in mind that you can specify points interactively using the following AutoCAD features individually or in combination: Grid Snap, Polar Snap, Ortho, Polar Tracking, Object Snap, and Object Snap Tracking. Use these drawing tools <u>whenever</u> AutoCAD prompts you to select points. (See Chapters 3 and 7 for details on these tools.)

COMMANDS

LINE

Pull-down Menu	COMMAND (TYPE)	ALIAS (TYPE)	Short-cut	Screen (side) Menu	Tablet Menu
Draw Line	LINE	L	...	DRAW 1 Line	J,10

This is the fundamental drawing command. The *Line* command creates straight line segments; each segment is an object. One or several line segments can be drawn with the *Line* command.

    Command: *Line*
    Specify first point: **PICK** or (**coordinates**) (A point can be designated by interactively selecting with the input device or by entering coordinates. Use any of the drawing tools to assist with interactive entry.)
    Specify next point or [Undo]: **PICK** or (**coordinates**) (Again, device input or keyboard input can be used.)
    Specify next point or [Undo]: **PICK** or (**coordinates**)
    Specify next point or [Close/Undo]: **PICK** or (**coordinates**) or *C*
    Specify next point or [Close/Undo]: press **Enter** to finish command
    Command:

Figure 8-3

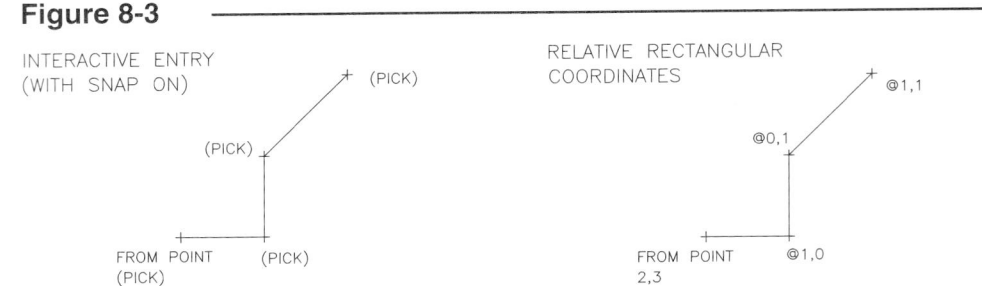

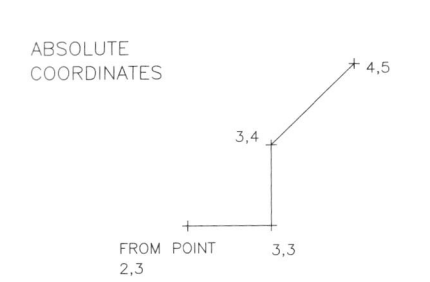

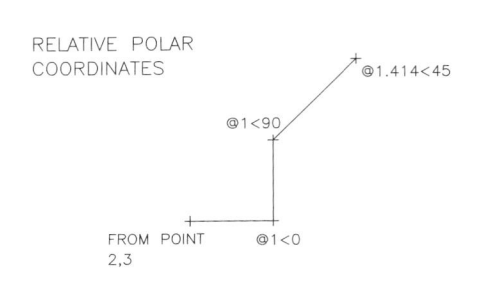

Figure 8-3 shows five examples of creating the same *Line* segments using different methods of coordinate entry.

Refer to Chapter 3, Draw Command Concepts, for examples of drawing vertical, horizontal, and inclined lines using the five formats for coordinate entry.

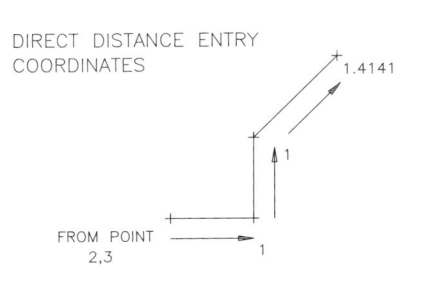

CIRCLE

Pull-down Menu	COMMAND (TYPE)	ALIAS (TYPE)	Short-cut	Screen (side) Menu	Tablet Menu
Draw *Circle >*	CIRCLE	C	...	DRAW 1 *Circle*	J,9

The *Circle* command creates one object. Depending on the option selected, you can provide two or three points to define a *Circle*. As with all commands, the Command line prompt displays the possible options:

Command: *Circle*
Specify center point for circle or [3P/2P/Ttr (tan tan radius)]: **PICK** or (**coordinates**) or (**option**).
(PICKing or entering coordinates designates the center point for the circle. You can enter 3P, 2P, or T for another option.)

As with most commands, the default and other options are displayed on the Command line. The default option always appears first. The other options can be invoked by typing the numbers and/or uppercase letter(s) that appear in brackets [ ].

All of the options for creating *Circles* are available from the *Draw* pull-down menu and from the *DRAW 1* screen (side) menu. The tool (icon button) for only the *Center, Radius* method is included in the *Draw* toolbar. Tools for the other explicit *Circle* and *Arc* options are available, but only if you customize your own toolbar.

The options, or methods, for drawing *Circles* are listed below. Each figure gives several possibilities for each option, with and without *OSNAPs*.

Center, Radius
Specify a center point, then a radius (Fig. 8-4).

The *Radius* (or *Diameter*) can be specified by entering values or by indicating a length interactively (PICK two points to specify a length when prompted). As always, points can be specified by PICKing or entering coordinates. Watch *Coords* for coordinate or distance (polar format) display. Grid Snap, Polar Snap, Polar Tracking, Object Snap, and Object Snap Tracking can be used for interactive point specification.

Figure 8-4

Center, Diameter
Specify the center point, then the diameter (Fig. 8-5).

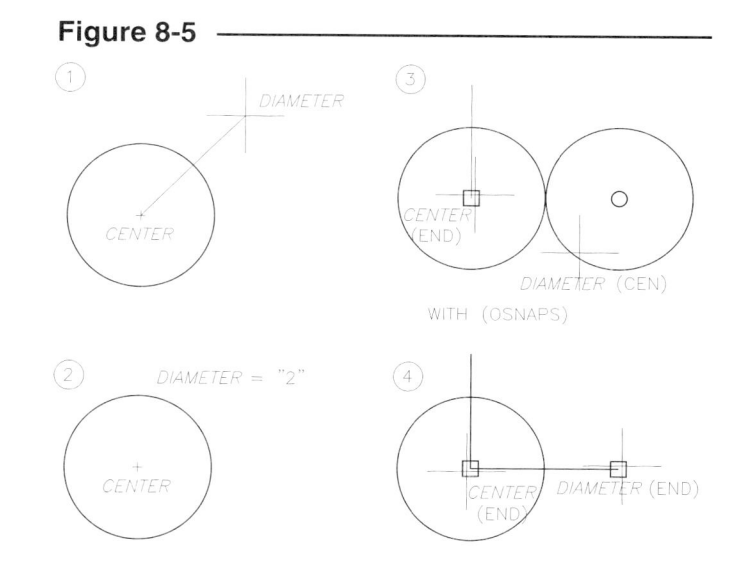

Figure 8-5

2 Points
The two points specify the location and diameter.

The *Tangent OSNAPs* can be used when selecting points with the *2 Point* and *3 Point* options, as shown in Figures 8-6 and 8-7.

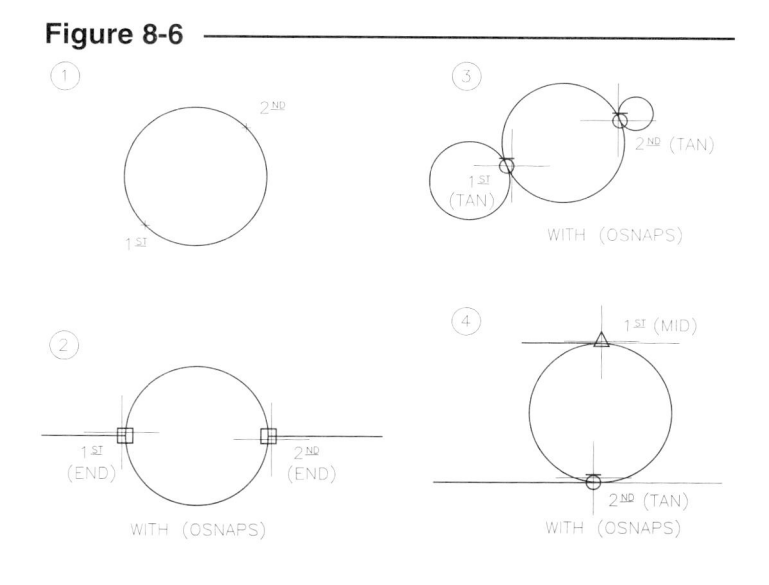

Figure 8-6

3 Points
The *Circle* passes through all three points specified.

Figure 8-7

Tangent, Tangent, Radius
Specify two objects for the *Circle* to be tangent to; then specify the radius.

TIP The *TTR* (Tangent, Tangent, Radius) method is extremely efficient and productive. The *OSNAP Tangent* modes are automatically invoked. This is the only draw command option that automatically calls *OSNAP*s.

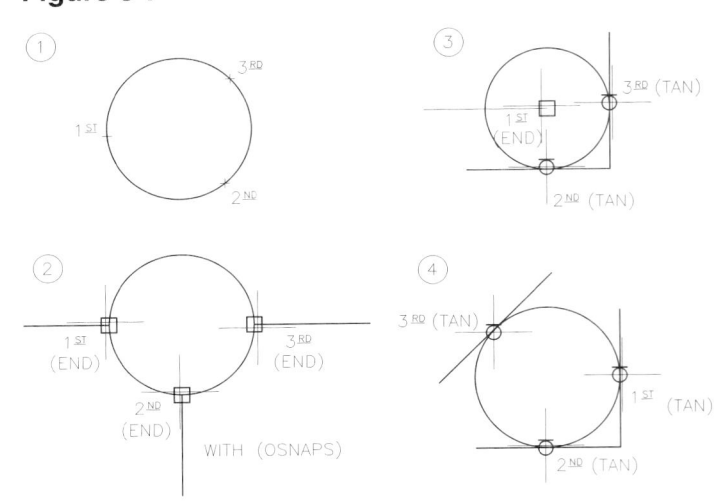

Figure 8-8

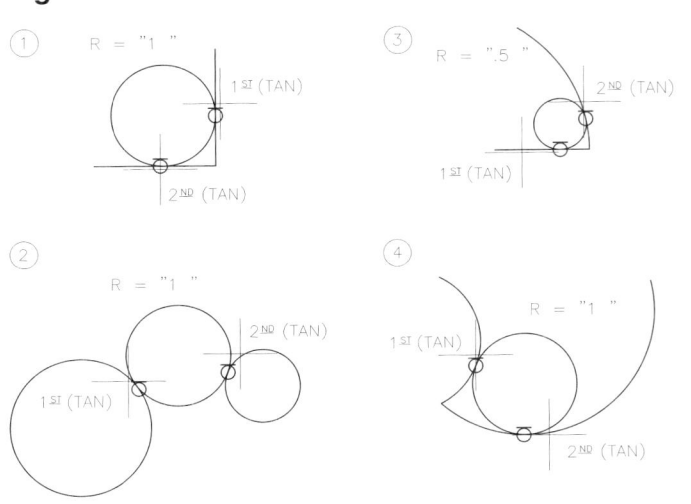

ARC

Pull-down Menu	COMMAND (TYPE)	ALIAS (TYPE)	Short-cut	Screen (side) Menu	Tablet Menu
Draw Arc >	ARC	A	...	DRAW 1 Arc	R,10

An arc is part of a circle; it is a regular curve of less than 360 degrees. The *Arc* command in AutoCAD provides eleven options for creating arcs. An *Arc* is one object. *Arcs* are always drawn by default in a counterclockwise direction. This occurrence forces you to decide in advance which points should be designated as *Start* and *End* points (for options requesting those points). For this reason, it is often easier to create arcs by another method, such as drawing a *Circle* and then using *Trim* or using the *Fillet* command. (See "Use *Arcs* or *Circles*?" at the end of this section on *Arcs*.) The *Arc* command prompt is:

    Command: Arc
    Specify start point of arc or [Center]: PICK or (coordinates) or C (Interactively select or enter coordinates in any format for the start point. Type C to use the Center option.)

The prompts displayed by AutoCAD are different depending on which option is selected. At any time while using the command, you can select from the options listed on the Command line by typing in the capitalized letter(s) for the desired option.

Alternately, to use a particular option of the *Arc* command, you can select from the *Draw* pull-down menu or from the *DRAW 1* screen (side) menu. The tool (icon button) for only the *3 Points* method is included in the *Draw* toolbar. The other tools are available, but only if you customize your own toolbar. These options require coordinate entry of <u>points in a specific order</u>.

3Points
Specify three points through which the *Arc* passes (Fig. 8-9).

Figure 8-9

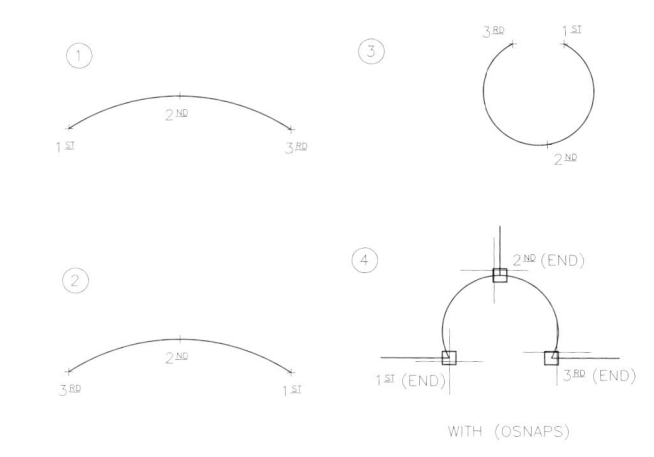

Start, Center, End
The radius is defined by the first two points that you specify (Fig. 8-10).

Figure 8-10

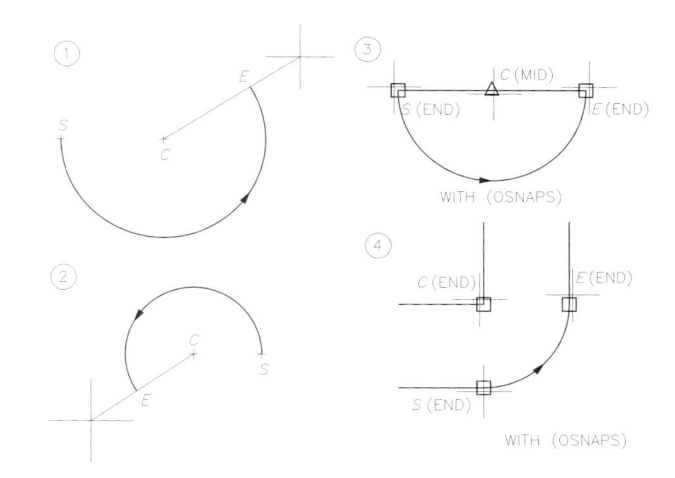

Start, Center, Angle
The angle is the <u>included</u> angle between the sides from the center to the endpoints. A <u>negative</u> angle can be entered to generate an *Arc* in a <u>clockwise</u> direction.

Figure 8-11

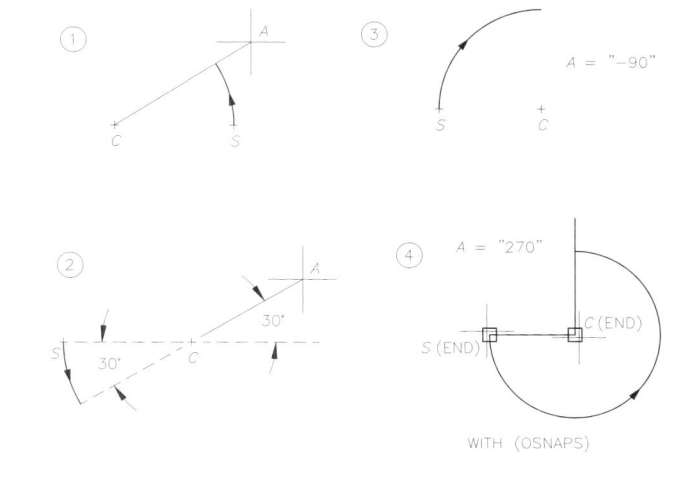

Start, Center, Length

Length means length of chord. The length of chord is between the start and the other point specified. A negative chord length can be entered to generate an *Arc* of 180+ degrees.

Figure 8-12

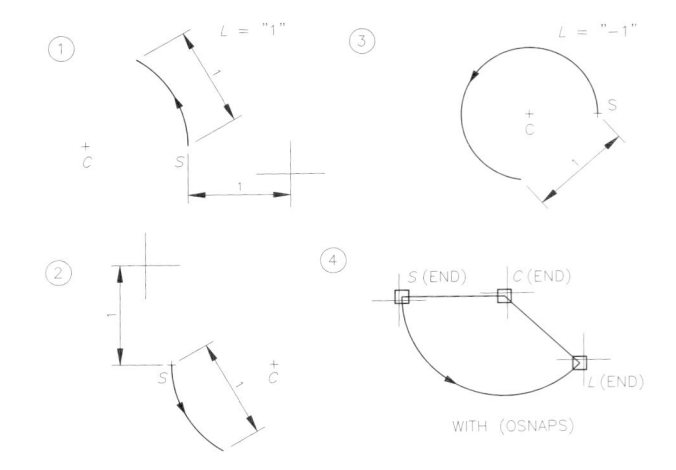

Start, End, Angle

The included angle is between the sides from the center to the endpoints. Negative angles generate clockwise *Arcs*.

Figure 8-13

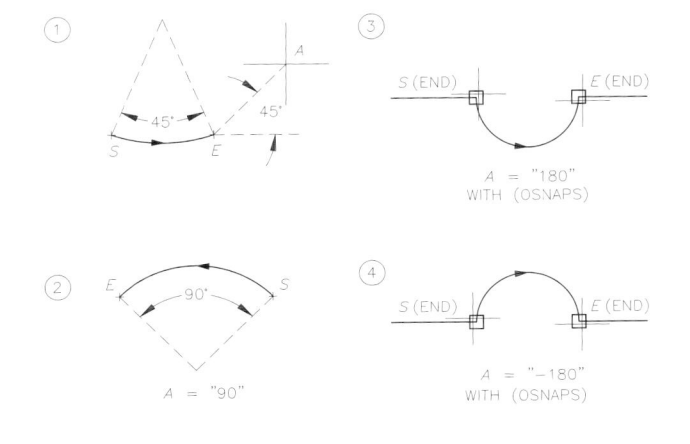

Start, End, Radius

The radius can be PICKed or entered as a value. A negative radius value generates an *Arc* of 180+ degrees.

Figure 8-14

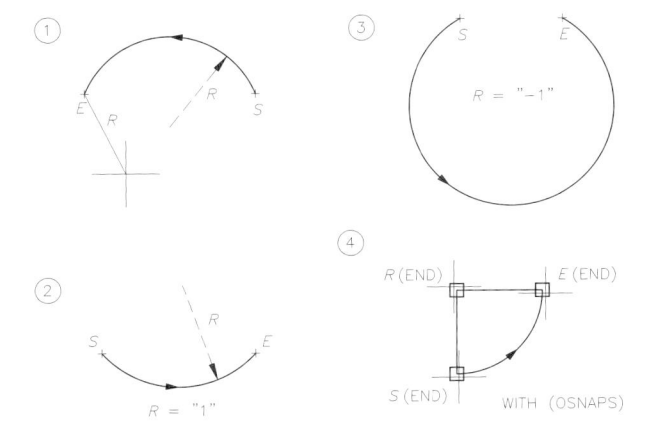

Start, End, Direction
The direction is tangent to the start point.

Figure 8-15

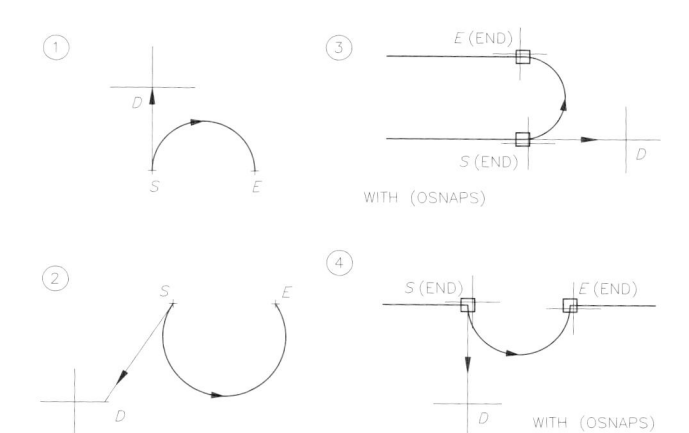

Center, Start, End
This option is like *Start, Center, End* but in a different order.

Figure 8-16

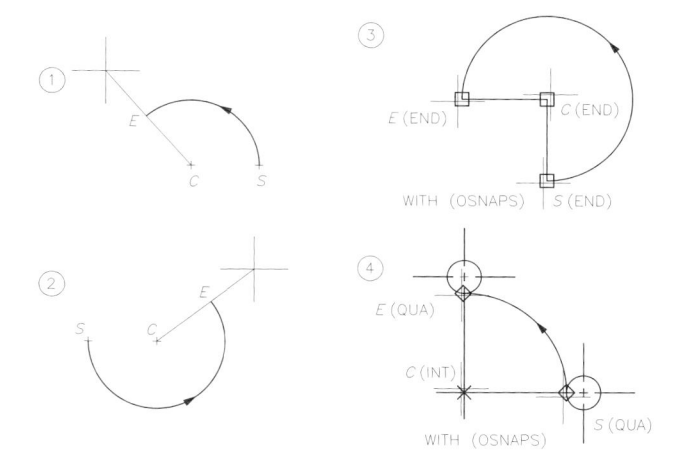

Center, Start, Angle
This option is like *Start, Center, Angle* but in a different order.

Figure 8-17

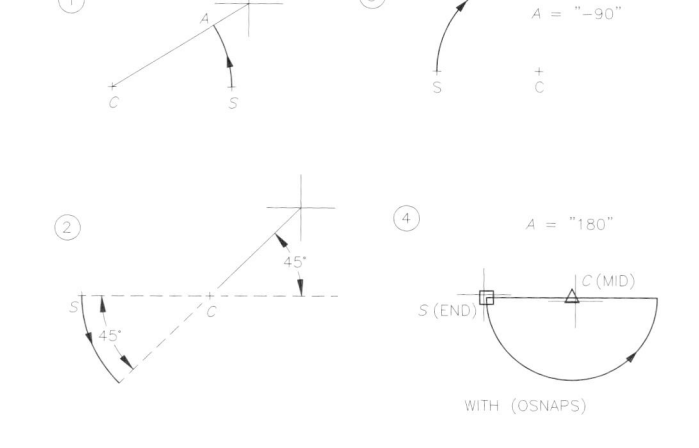

Center, Start, Length
This is similar to the *Start, Center, Length* option but in a different order. *Length* means length of chord.

Figure 8-18

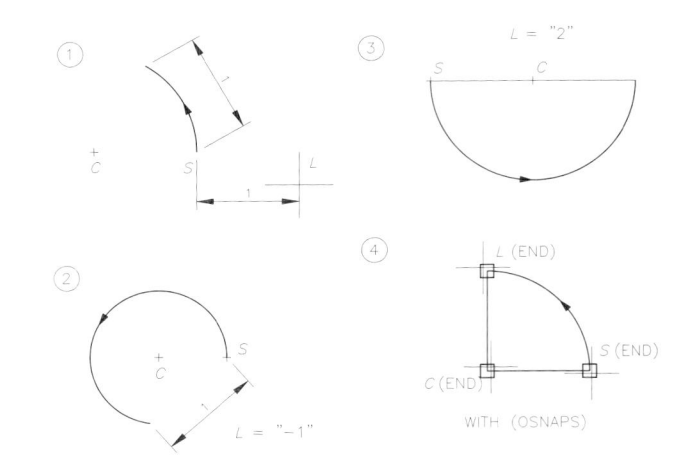

Continue
The new *Arc* continues from and is tangent to the last point. The only other point required is the endpoint of the *Arc*. This method allows drawing *Arcs* tangent to the preceding *Line* or *Arc*.

Figure 8-19

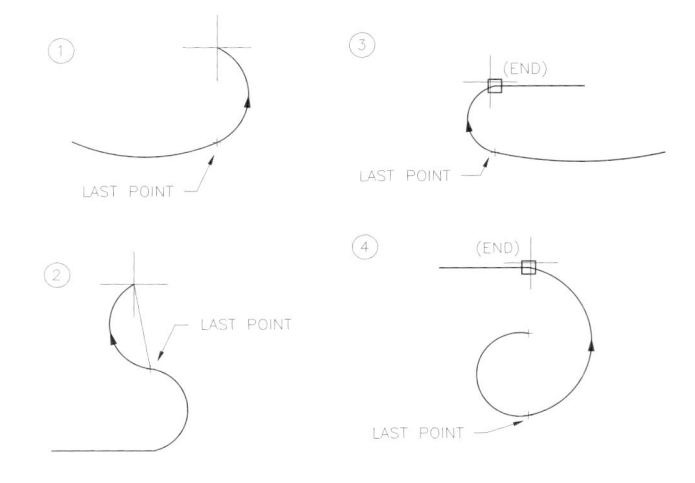

Arcs are always created in a <u>counterclockwise</u> direction. This fact must be taken into consideration when using any method <u>except</u> the *3-Point*, the *Start, End, Direction*, and the *Continue* options. The direction is explicitly specified with *Start, End, Direction* and *Continue* methods, and direction is irrelevant for *3-Point* method.

As usual, points can be specified by PICKing or entering coordinates. Watch *Coords* to display coordinate values or distances. Grid Snap, Polar Snap, Polar Tracking, Object Snap, and Object Snap Tracking can be used when PICKing. The *Endpoint, Intersection, Center, Midpoint,* and *Quadrant OSNAP* options can be used with great effectiveness. The *Tangent OSNAP* option <u>cannot</u> be used effectively with most of the *Arc* options. The *Radius, Direction, Length,* and *Angle* specifications can be given by entering values or by PICKing with or without *OSNAPs*.

Use *Arcs* or *Circles*?

Although there are sufficient options for drawing *Arcs*, <u>usually it is easier to use the *Circle* command</u> followed by *Trim* to achieve the desired arc. Creating a *Circle* is generally an easier operation than using *Arc* because the counterclockwise direction does not have to be considered. The unwanted portion of the circle can be *Trimmed* at the *Intersection* of or *Tangent* to the connecting objects using *OSNAP*. The *Fillet* command can also be used instead of the *Arc* command to add a fillet (arc) between two existing objects (see Chapter 9, Modify Commands I).

Chapter Eight 143

POINT

Pull-down Menu	COMMAND (TYPE)	ALIAS (TYPE)	Short-cut	Screen (side) Menu	Tablet Menu
Draw Point > Single Point or Multiple Point	POINT	PO	...	DRAW 2 Point	O,9

A *Point* is an object that has no dimension; it only has location. A *Point* is specified by giving only one coordinate value or by PICKing a location on the screen.

Figure 8-20 compares *Points* to *Line* and *Circle* objects.

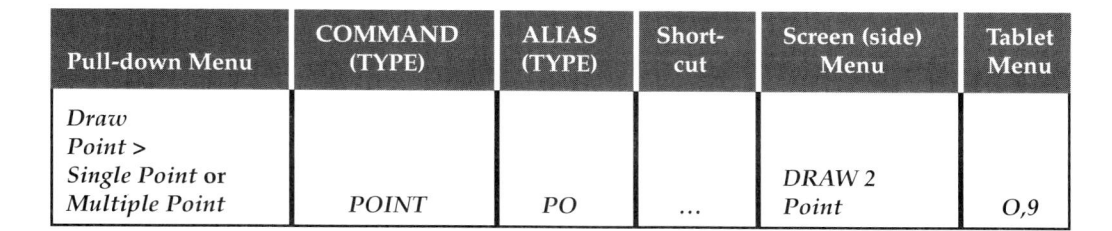

Figure 8-20

    Command: **point**
    Current point modes: PDMODE=0 PDSIZE=0.0000
    Specify a point: **PICK** or (**coordinates**)
    (Select a location for the *Point* object.)

Points are useful in construction of drawings to locate points of reference for subsequent construction or locational verification. The *Node OSNAP* option is used to snap to *Point* objects.

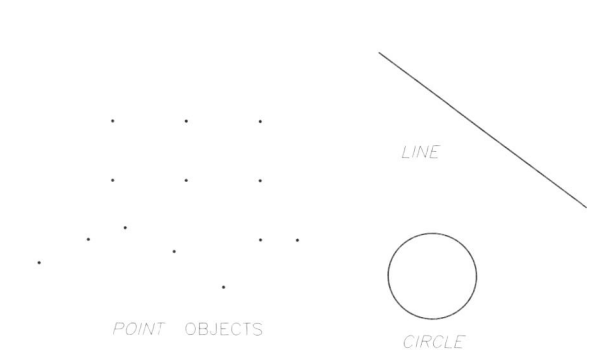

Points are drawing objects and therefore appear in prints and plots. The default "style" for points is a tiny dot. The *Point Style* dialog box can be used to define the format you choose for *Point* objects (the *Point* type [PDMODE] and size [PDSIZE]).

The *Draw* pull-down menu offers the *Single Point* and the *Multiple Point* options. The *Single Point* option creates one *Point,* then returns to the command prompt. This option is the same as using the *Point* command by any other method. Selecting *Multiple Point* continues the *Point* command until you press the Escape key.

DDPTYPE

Pull-down Menu	COMMAND (TYPE)	ALIAS (TYPE)	Short-cut	Screen (side) Menu	Tablet Menu
Format Point Style...	DDPTYPE	...	...	DRAW 2 Point Ddptype:	U,1

The *Point Style* dialog box (Fig. 8-21) is available only through the methods listed in the command table above. This dialog box allows you to define the format for the display of *Point* objects. The selected style is applied immediately to all newly created *Point* objects. The *Point Style* controls the format of *Points* for printing and plotting as well as for the computer display.

Figure 8-21

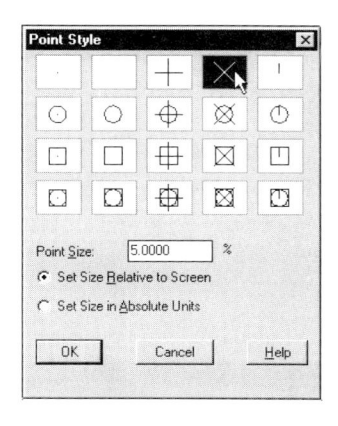

You can set the *Point Size* in *Absolute* units or *Relative to Screen* (default option). The *Relative to Screen* option keeps the *Points* the same size on the display when you *Zoom* in and out, whereas setting *Point Size* in *Absolute Units* gives you control over the size of *Points* for prints and plots. The *Point Size* is stored in the *PDSIZE* system variable. The selected *Point Style* is stored in the *PDMODE* variable.

PLINE

Pull-down Menu	COMMAND (TYPE)	ALIAS (TYPE)	Short-cut	Screen (side) Menu	Tablet Menu
Draw Polyline	PLINE	PL	...	DRAW 1 Pline	N,10

A *Pline* (or *Polyline*) has special features that make this object more versatile than a *Line*. Three features are most noticeable when first using *Pline*s:

1. A *Pline* can have a specified *width*, whereas a *Line* has no width.
2. Several *Pline* segments created with one *Pline* command are treated by AutoCAD as <u>one</u> object, whereas individual line segments created with one use of the *Line* command are individual objects.
3. A *Pline* can contain arc segments.

Figure 8-22 illustrates *Pline* versus *Line* and *Arc* comparisons.

The *Pline* command begins with the same prompt as *Line*; however, <u>after</u> the "start point:" is established, the *Pline* options are accessible.

Figure 8-22

```
Command: Pline
Specify start point: PICK or (coordinates)
Current line-width is 0.0000
Specify next point or [Arc/Close/Halfwidth/Length/Undo/Width]:
```

The options and descriptions follow.

Width
You can use this option to specify starting and ending widths. Width is measured perpendicular to the centerline of the *Pline* segment (Fig. 8-23). *Plines* can be tapered by specifying different starting and ending widths. See NOTE at the end of this section.

Figure 8-23

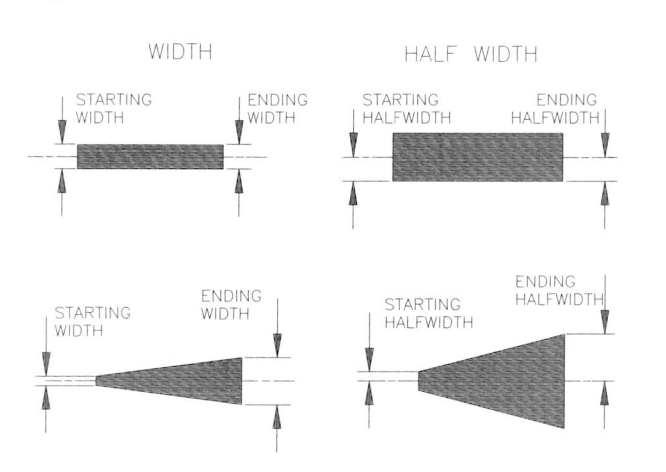

Halfwidth
This option allows specifying half of the *Pline* width. *Plines* can be tapered by specifying different starting and ending widths (Fig. 8-23).

Arc

This option (by default) creates an arc segment in a manner similar to the *Arc Continue* method (Fig. 8-24). Any of several other methods are possible (see "*Pline Arc* Segments").

Figure 8-24

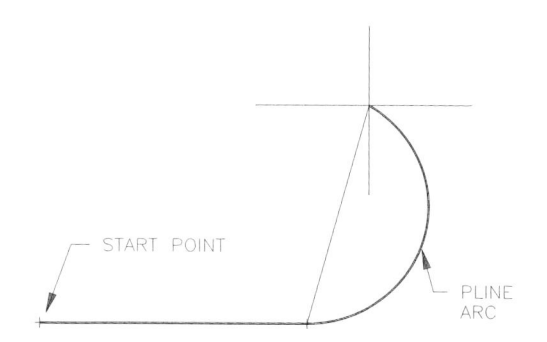

Close

The *Close* option creates the closing segment connecting the first and last points specified with the current *Pline* command as shown in Figure 8-25.

This option can also be used to close a group of connected *Pline* segments into one continuous *Pline*. (A *Pline* closed by PICKing points has a specific start and endpoint.) A *Pline Closed* by this method has special properties if you use *Pedit* for *Pline* editing or if you use the *Fillet* command with the *Pline* option (see *Fillet* in Chapter 9 and *Pedit* in Chapter 16).

Figure 8-25

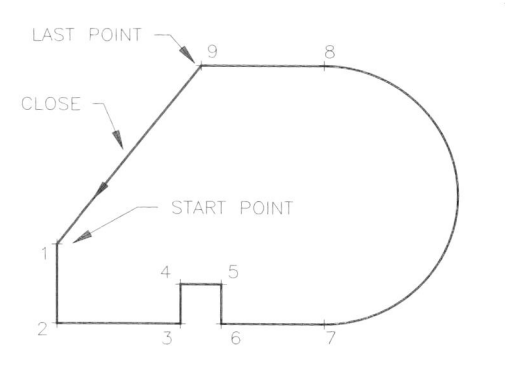

Length

Length draws a *Pline* segment at the same angle as and connected to the previous segment and uses a length that you specify. If the previous segment was an arc, *Length* makes the current segment tangent to the ending direction (Fig. 8-26).

Figure 8-26

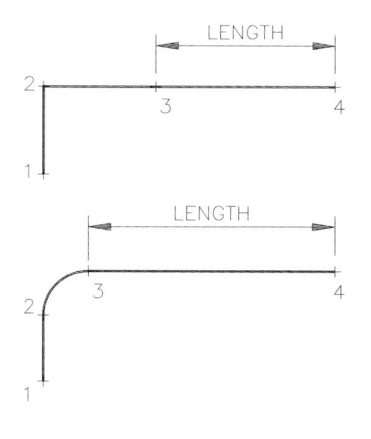

Undo

Use this option to *Undo* the last *Pline* segment. It can be used repeatedly to undo multiple segments.

NOTE: If you change the *Width* of a *Pline*, be sure to respond to <u>both prompts</u> for width ("Specify starting width:" and "Specify ending width:") before you draw the first *Pline* segment. It is easy to hastily PICK the endpoint of the *Pline* segment after specifying the "Specify starting width:" instead of responding with a value (or Enter) for the "Specify ending width:". In this case (if you PICK at the "Specify ending width:" prompt), AutoCAD understands the line length that you interactively specified to be the ending width you want for the next line segment (you can PICK two points in response to the "Specify starting width:" or "Specify ending width:" prompts).

```
Command: Pline
Specify start point: PICK
Current line-width is 0.0000
Specify next point or [Arc/Close/Halfwidth/Length/Undo/Width]: w
Specify starting width <0.0000>: .2
Specify ending width <0.2000>: Enter a value or press Enter—do not PICK the "next point."
```

Pline Arc Segments

When the *Arc* option of *Pline* is selected, the prompt changes to provide the various methods for construction of arcs:

Specify endpoint of arc or [Angle/CEnter/CLose/Direction/Halfwidth/Line/Radius/Second pt/Undo/Width]:

Angle
You can draw an arc segment by specifying the included angle (a negative value indicates a clockwise direction for arc generation).

CEnter
This option allows you to specify a specific center point for the arc segment.

CLose
This option closes the *Pline* group with an arc segment.

Direction
Direction allows you to specify an explicit starting direction rather than using the ending direction of the previous segment as a default.

Line
This switches back to the line options of the *Pline* command.

Radius
You can specify an arc radius using this option.

Second pt
Using this option allows specification of a 3-point arc.

Because a shape created with one *Pline* command is one object, manipulation of the shape is generally easier than with several objects. For some applications, one *Pline* shape can have advantages over shapes composed of several objects (see "*Offset*," Chapter 9). Editing *Plines* is accomplished by using the *Pedit* command. As an alternative, *Plines* can be "broken" back down into individual objects with *Explode*.

Drawing and editing *Plines* can be somewhat involved. As an alternative, you can draw a shape as you would normally with *Line, Circle, Arc, Trim*, etc., and then convert the shape to one *Pline* object using *Pedit*. (See Chapter 16 for details on converting *Lines* and *Arcs* to *Plines*.)

CHAPTER EXERCISES

Create a **New** drawing. When the *Create New Drawing* dialog box or *Today* window appears, select **Start from Scratch** and select **English** settings. Set **Polar Snap On** and set the **Polar Distance** to **.25**. Turn on **SNAP, POLAR, OSNAP,** and **OTRACK**. *Save* the drawing as **CH8EX**. For each of the following problems, **Open CH8EX**, complete one problem, then use *SaveAs* to give the drawing a new name.

1. ***Open* CH8EX**. Create the geometry shown in Figure 8-27. Start the first *Circle* center at point **4,4.5** as shown. Do not copy the dimensions. *SaveAs* **LINK**. (HINT: Locate and draw the two small *Circles* first. Use *Arc, Start, Center, End* or *Center, Start, End* for the rounded ends.)

Figure 8-27

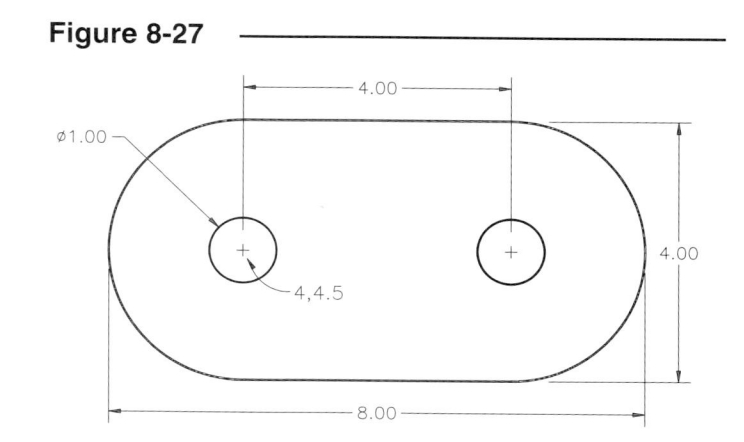

2. ***Open* CH8EX**. Create the geometry as shown in Figure 8-28. Do not copy the dimensions. Assume symmetry about the vertical axis. *SaveAs* **SLOTPLATE CH8**.

Figure 8-28

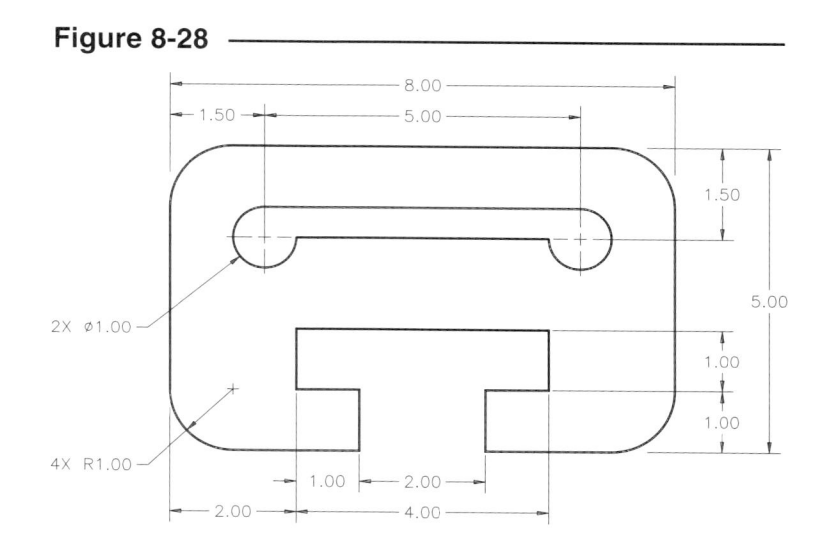

3. ***Open* CH8EX**. Create the shapes shown in Figure 8-29. Do not copy the dimensions. *SaveAs* **CH8EX3**.

    Draw the *Lines* at the bottom first, starting at coordinate **1,3**. Then create *Point* objects at **5,7**, **5.4,7**, **5.8,7**, etc. Change the *Point Style* to an X and *Regen*. Use the NODe OSNAP mode to draw the inclined *Lines*. Create the *Arc* on top with the *Start, End, Direction* option.

Figure 8-29

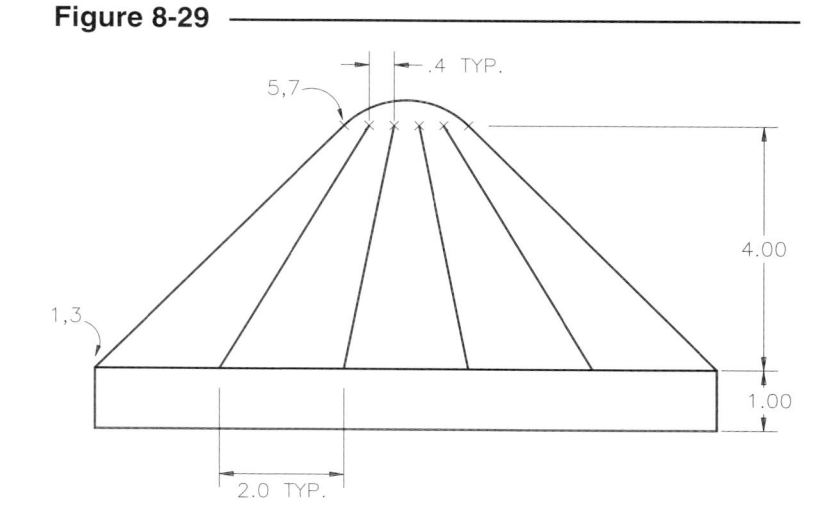

148   Draw Commands I

4. *Open* **CH8EX**. Create the shape shown in Figure 8-30. Draw the two horizontal *Lines* and the vertical *Line* first by specifying the endpoints as given. Then create the *Circle* and *Arcs*. *SaveAs* **CH8EX4**.

   HINT: Use the *Circle 2P* method with *Endpoint* OSNAPs. The two upper *Arcs* can be drawn by the *Start, End, Radius* method.

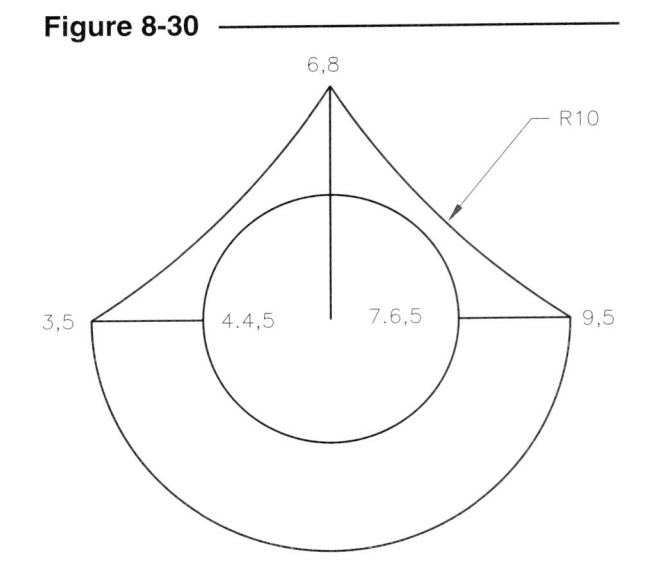

Figure 8-30

5. *Open* **CH8EX**. Draw the shape shown in Figure 8-31. Assume symmetry along a vertical axis. Start by drawing the two horizontal *Lines* at the base.

   Next, construct the side *Arcs* by the *Start, Center, Angle* method (you can specify a negative angle). The small *Arc* can be drawn by the *3P* method. Use OSNAPs when needed (especially for the horizontal *Line* on top and the *Line* along the vertical axis). *SaveAs* **CH8EX5**.

Figure 8-31

6. *Open* **CH8EX**. Complete the geometry in Figure 8-32. Use the coordinates to establish the *Lines*. Draw the *Circles* using the *Tangent, Tangent, Radius* method. *SaveAs* **CH8EX6**.

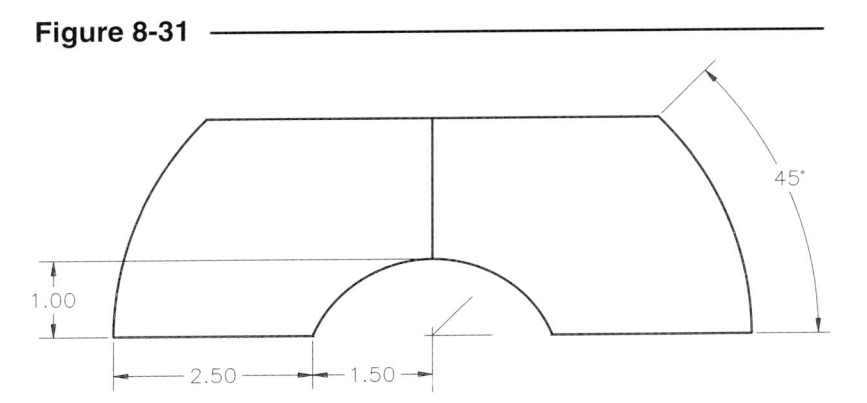

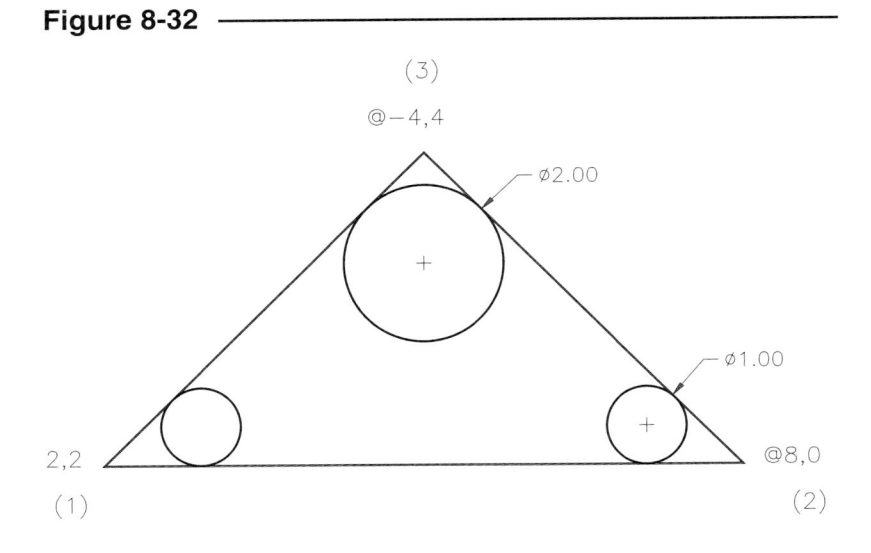

Figure 8-32

7. **Retaining Wall**

Figure 8-33

A contractor plans to stake out the edge of a retaining wall located at the bottom of a hill. The following table lists coordinate values based on a survey at the site. Use the drawing you created in Chapter 6 Exercises named **CH6EX2**. *Save* the drawing as **RET-WALL**.

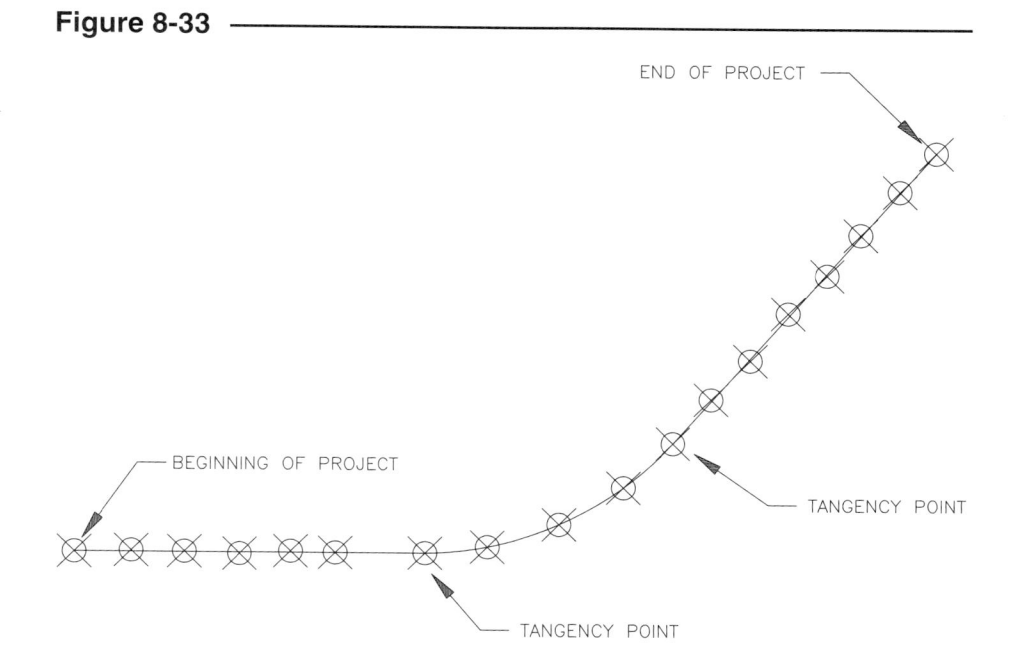

Place *Points* at each coordinate value in order to create a set of data points. Determine the location of one *Arc* and two *Lines* representing the centerline of the retaining wall edge. The centerline of the retaining wall should match the data points as accurately as possible (Fig. 8-33). The data points given in the following table are in inch units.

Pt #	X	Y	
1.	60.0000	108.0000	
2.	81.5118	108.5120	
3.	101.4870	108.2560	
4.	122.2305	107.4880	
5.	141.4375	108.5120	
6.	158.1949	108.0000	
7.	192.0000	108.0000	Recommended tangency point
8.	215.3914	110.5185	
9.	242.3905	119.2603	
10.	266.5006	133.6341	
11.	285.1499	151.0284	Recommended tangency point
12.	299.5069	168.1924	
13.	314.2060	183.4111	
14.	328.3813	201.7784	
15.	343.0811	216.4723	
16.	355.6808	232.2157	
17.	370.3805	249.0087	
18.	384.0000	264.0000	

(Hint: Change the *Point Style* to an easily visible format.)

8. *Pline*

Create the shape shown in Figure 8-34. Draw the outside shape with <u>one continuous</u> *Pline* (with 0.00 width). When finished, *SaveAs* **PLINE1**.

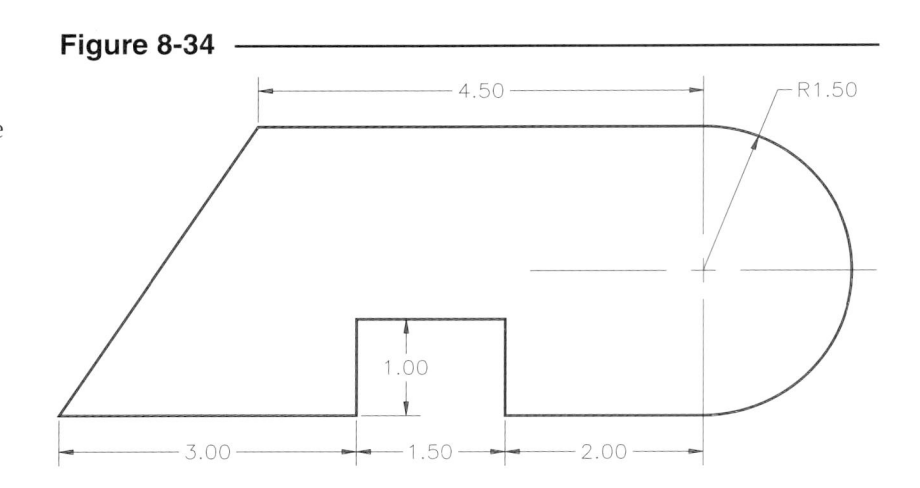

Figure 8-34

9

MODIFY COMMANDS I

Chapter Objectives

After completing this chapter you should:

1. be able to *Erase* objects;

2. be able to *Move* objects from a base point to a second point;

3. know how to *Rotate* objects about a base point;

4. be able to enlarge or reduce objects with *Scale*;

5. be able to *Stretch* objects and change the length of *Lines* and *Arcs* with *Lengthen*;

6. be able to *Trim* away parts of objects at cutting edges and *Extend* objects to boundary edges;

7. know how to use the four *Break* options;

8. be able to *Copy* objects and make *Mirror* images of selected objects;

9. know how to create parallel copies of objects with *Offset*;

10. be able to make *rectangular* and *polar Arrays* of objects;

11. be able to create a *Fillet* and a *Chamfer* between two objects.

152 Modify Commands I

CONCEPTS

Draw commands are used to create new objects. *Modify* commands are used to change existing objects or to use existing objects to create new and similar objects. The *Modify* commands covered first in this chapter (listed below) only <u>change existing objects</u>.

Erase, Move, Rotate, Scale, Stretch, Lengthen, Trim, Extend, and *Break*

The *Modify* commands covered near the end of this chapter (listed below) <u>use existing objects to create new and similar objects</u>. For example, the *Copy* command prompts you to select an object (or set of objects), then creates an identical object (or set).

Copy, Mirror, Offset, Array, Fillet and *Chamfer*

Modify commands can be invoked by any of the five command entry methods: toolbar icons, pull-down menus, screen menus, keyboard entry, and digitizing tablet menu. The *Modify* toolbar is visible and docked to the left side of the screen by default (Fig. 9-1).

Several commands are found in the *Edit* pull-down menu such as *Cut*, *Copy*, and *Paste*. Although these command names appear similar (or the same in the case of *Copy*) to some of those in the *Modify* menu, these AutoCAD command names are actually *Cutclip*, *Copyclip*, and *Pasteclip*. These commands are for OLE operations (cutting and pasting between AutoCAD drawings or other software applications.

Figure 9-1

The *Modify* pull-down menu (Fig. 9-2) contains commands that change existing geometry and that use existing geometry to create new but similar objects. All of the *Modify* commands are contained in this single pull-down menu.

Figure 9-2

TIP Since all *Modify* commands affect or use existing geometry, the first step in using any *Modify* command is to construct a selection set (see Chapter 4). This can be done by one of two methods:

1. Invoking the desired command and then creating the selection set in response to the "Select Objects:" prompt (Verb/Noun syntax order) using any of the select object options;

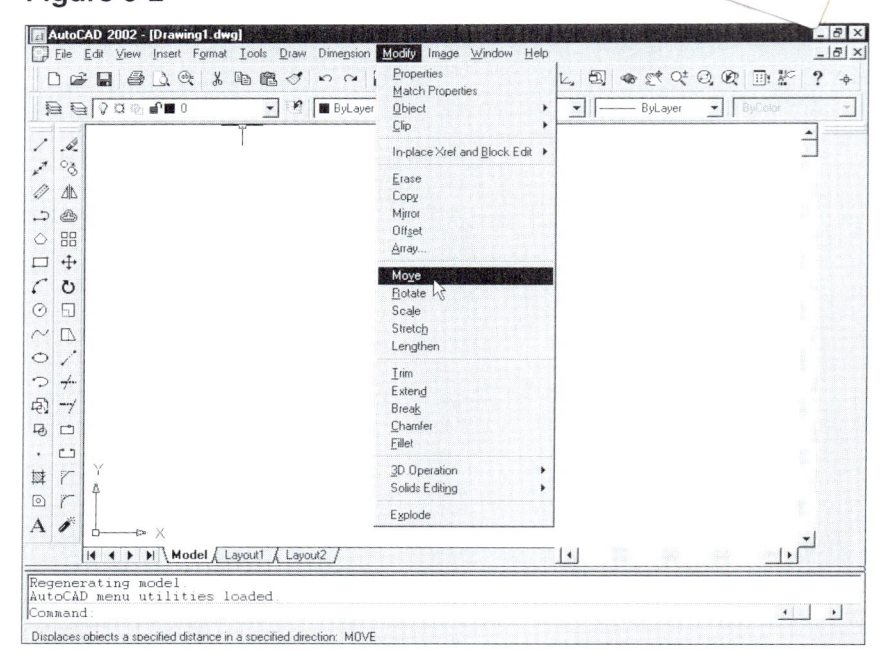

2. Selecting the desired set of objects with the pickbox or *Auto Window* or *Crossing Window* <u>before</u> invoking the edit command (Noun/Verb syntax order).

The first method allows use of any of the selection options (*Last, All, WPolygon, Fence,* etc.), while the latter method allows <u>only</u> the use of the pickbox and *Auto Window* and *Crossing Window*.

COMMANDS

ERASE

Pull-down Menu	COMMAND (TYPE)	ALIAS (TYPE)	Short-cut	Screen (side) Menu	Tablet Menu
Modify Erase	ERASE	E	(Edit Mode) Erase	MODIFY1 Erase	V,14

The *Erase* command deletes the objects you select from the drawing. Any of the object selection methods can be used to highlight the objects to *Erase*. The only other required action is for you to press *Enter* to cause the erase to take effect.

    Command: **erase**
    Select Objects: **PICK** (Use any object selection method.)
    Select Objects: **PICK** (Continue to select desired objects.)
    Select Objects: **Enter** (Confirms the object selection process and causes *Erase* to take effect.)
    Command:

If objects are erased accidentally, *U* can be used immediately following the mistake to undo one step, or *Oops* can be used to bring back into the drawing whatever was *Erased* the last time *Erase* was used (see Chapter 5). If only part of an object should be erased, use *Trim* or *Break*.

MOVE

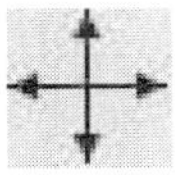

Pull-down Menu	COMMAND (TYPE)	ALIAS (TYPE)	Short-cut	Screen (side) Menu	Tablet Menu
Modify Move	MOVE	M	(Edit Mode) Move	MODIFY2 Move	V,19

Move allows you to relocate one or more objects from the existing position in the drawing to any other position you specify. After selecting the objects to *Move*, you must specify the "base point" and "second point of displacement." You can use any of the five coordinate entry methods to specify these points. Examples are shown in Figure 9-3.

Figure 9-3

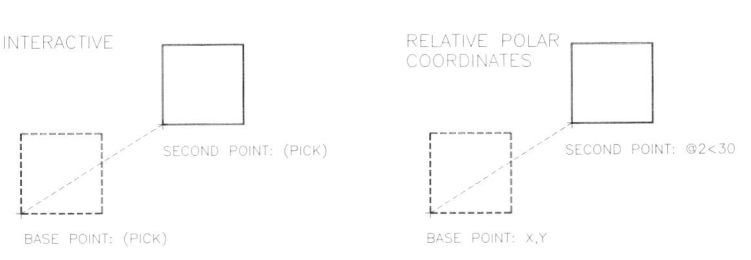

    Command: **move**
    Select Objects: **PICK** (Use any of the object selection methods.)
    Select Objects: **PICK** (Continue to select other desired objects.)
    Select Objects: **Enter** (Press Enter to indicate selection of objects is complete.)
    Specify base point or displacement: **PICK** or (**coordinates**) (This is the point to <u>move from</u>. Select a point to use as a "handle." An *Endpoint* or *Center*, etc., can be used.)
    Specify second point of displacement or <use first point as displacement>: **PICK** or (**coordinates**) (This is the point to <u>move to</u>. *OSNAP*s can also be used here.)
    Command:

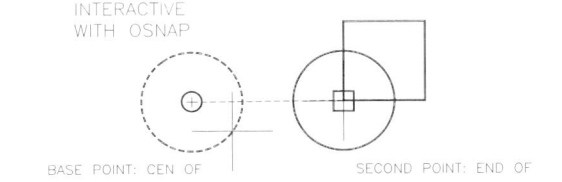

Keep in mind that *OSNAP*s can be used when PICKing any point. It is often helpful to toggle *ORTHO* or *POLAR ON* (**F8**) to force the *Move* in a horizontal or vertical direction.

If you know a specific distance and an angle that the set of objects should be moved, Polar Tracking or relative polar coordinates can be used. In the following sequence, relative polar coordinates are used to move objects 2 units in a 30 degree direction (see Figure 9-3, coordinate entry).

```
Command: move
Select Objects: PICK
Select Objects: PICK
Select Objects: Enter
Specify base point or displacement: X,Y (coordinates)
Specify second point of displacement or <use first point as displacement>: @2<30
Command:
```

Direct distance coordinate entry can be used effectively with *Move*. For example, assume you wanted to move the right side view of a multiview drawing 20 units to the right using direct distance entry (Fig. 9-4). First, invoke the *Move* command and select all of the objects comprising the right side view in response to the "Select Objects:" prompt. The command sequence is as follows:

Figure 9-4

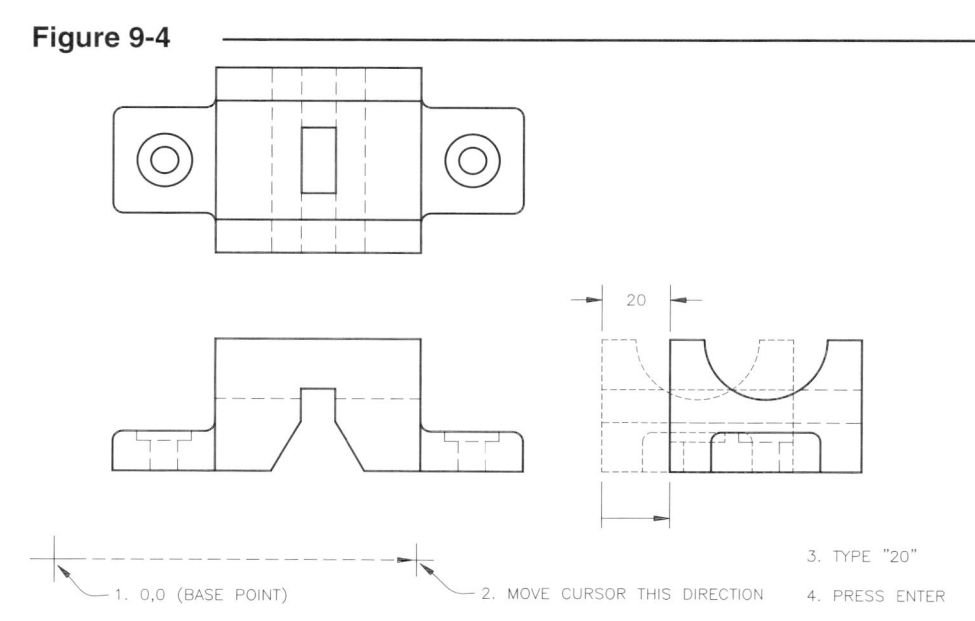

```
Command: move
Select Objects: PICK
Select Objects: Enter
Specify base point or displacement: 0,0 (or any value)
Specify second point of displacement or <use first point as displacement>: 20 (with ORTHO or POLAR
on, move cursor to right), then press Enter
```

In response to the "Specify base point or displacement:" prompt, PICK a point or enter any coordinate pairs or single value. At the "Specify second point of displacement :" prompt, move the cursor to the right any distance (using Polar Tracking or *ORTHO* on), then type "20" and press Enter. The value specifies the distance, and the cursor location from the last point indicates the direction of movement.

In the previous example, note that <u>any value</u> can be entered in response to the "Specify base point or displacement:" prompt. If a single value is entered ("3," for example), AutoCAD recognizes it as direct distance entry. The point designated is 3 units from the last PICK point in the direction specified by wherever the cursor is at the time of entry.

ROTATE

Pull-down Menu	COMMAND (TYPE)	ALIAS (TYPE)	Short-cut	Screen (side) Menu	Tablet Menu
Modify Rotate	ROTATE	RO	(Edit Mode) Rotate	MODIFY2 Rotate	V,20

Selected objects can be rotated to any position with this command. After selecting objects to *Rotate*, you select a "base point" (a point to rotate about) then specify an angle for rotation. AutoCAD rotates the selected objects by the increment specified from the original position.

For example, specifying a value of **45** would *Rotate* the selected objects 45 degrees counterclockwise from their current position; a value of **-45** would *Rotate* the objects 45 degrees in a clockwise direction (Fig. 9-5).

Figure 9-5

Command: *rotate*
Current positive angle in UCS: ANGDIR=counterclockwise ANGBASE=0
Select objects: **PICK**
Select objects: **PICK**
Select objects: **Enter** (Indicates completion of object selection.)
Specify base point: **PICK** or (**coordinates**) (Select a point to rotate about.)
Specify rotation angle or [Reference]: **PICK** or (**coordinates**) (Interactively rotate the set or enter a value for the number of degrees to rotate the object set.)
Command:

The base point is often selected interactively with *OSNAPs*. When specifying an angle for rotation, you can enter an angular value, use Polar Tracking, or turn on *ORTHO* for 90-degree rotation. Note the status of the two related system variables is given: *ANGDIR* (counterclockwise or clockwise rotation) and *ANGBASE* (base angle used for rotation).

The **Reference** option can be used to specify a vector as the original angle before rotation. This vector can be indicated interactively (*OSNAPs* can be used) or entered as an angle using keyboard entry. Angular values that you enter in response to the "New angle:" prompt are understood by AutoCAD as absolute angles for the *Reference* option only.

SCALE

Pull-down Menu	COMMAND (TYPE)	ALIAS (TYPE)	Short-cut	Screen (side) Menu	Tablet Menu
Modify Scale	SCALE	SC	(Edit Mode) Scale	MODIFY2 Scale	V,21

The *Scale* command is used to increase or decrease the size of objects in a drawing. The *Scale* command does not normally have any relation to plotting a drawing to scale.

After selecting objects to *Scale*, AutoCAD prompts you to select a "Base point:", which is the stationary point. You can then scale the size of the selected objects interactively or enter a scale factor (Fig. 9-6).

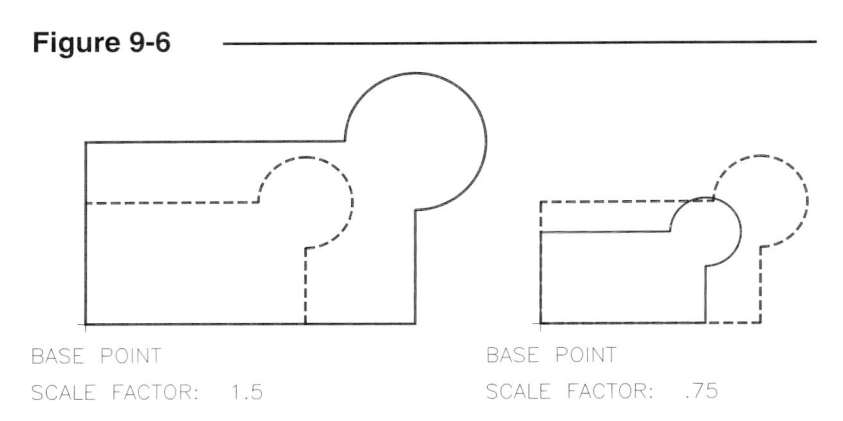

Figure 9-6

BASE POINT
SCALE FACTOR: 1.5

BASE POINT
SCALE FACTOR: .75

Using interactive input, you are presented with a rubberband line connected to the base point. Making the rubberband line longer or shorter than 1 unit increases or decreases the scale of the selected objects by that proportion; for example, pulling the rubberband line to two units length increases the scale by a factor of two.

```
Command: scale
Select objects: PICK or (coordinates) (Select the objects to scale.)
Select objects: Enter (Indicates completion of the object selection.)
Specify base point: PICK or (coordinates) (Select the stationary point.)
Specify scale factor or [Reference]: PICK or (value) or R (Interactively scale the set of objects or
enter a value for the scale factor.)
Command:
```

It may be desirable in some cases to use the **Reference** option to specify a value or two points to use as the reference length. This length can be indicated interactively (*OSNAPs* can be used) or entered as a value. This length is used for the subsequent reference length that the rubberband uses when interactively scaling. For example, if the reference distance is 2, then the "rubberband" line must be stretched to a length greater than 2 to increase the scale of the selected objects.

Scale normally should not be used to change the scale of an entire drawing in order to plot on a specific size sheet. CAD drawings should be created full size in actual units.

STRETCH

Pull-down Menu	COMMAND (TYPE)	ALIAS (TYPE)	Short-cut	Screen (side) Menu	Tablet Menu
Modify Stretch	STRETCH	S	...	MODIFY2 Stretch	V,22

Objects can be made longer or shorter with *Stretch*. The power of this command lies in the ability to *Stretch* groups of objects while retaining the connectivity of the group (Fig. 9-7). When *Stretched*, *Lines* and *Plines* become longer or shorter and *Arcs* change radius to become longer or shorter. *Circles* do not stretch; rather, they move if the center is selected within the Crossing Window.

Objects to *Stretch* are not selected by the typical object selection methods but are indicated by a <u>Crossing Window or Crossing Polygon only</u>. The *Crossing Window* or *Polygon* should be created so the objects to *Stretch* <u>cross</u> through the window. *Stretch* actually moves the object endpoints that are located within the *Crossing Window*.

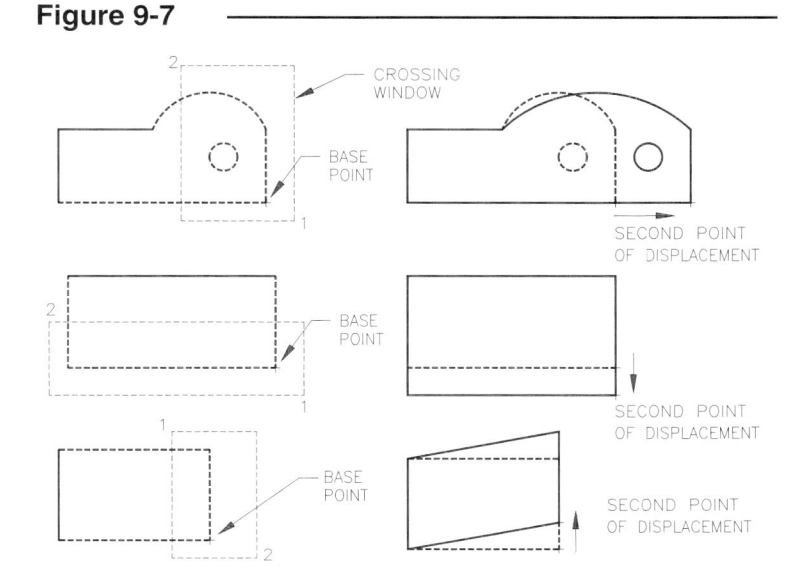

Figure 9-7

Following is the command sequence for *Stretch*.

Command: *stretch*
Select objects to stretch by crossing-window or crossing-polygon...
Select objects: **PICK**
Specify opposite corner: **PICK**
Select objects: **Enter**
Specify base point or displacement: **PICK** or (**coordinates**) (Select a point to stretch from.)
Specify second point of displacement: **PICK** or (**coordinates**) (Select a point to stretch to.)
Command:

Stretch can be used to lengthen one object while shortening another. Application of this ability would be repositioning a door or window on a wall (Fig. 9-8).

In summary, the *Stretch* command <u>stretches objects that cross</u> the selection Window and <u>moves objects that are completely within</u> the selection Window, as shown in Figure 9-8.

Figure 9-8

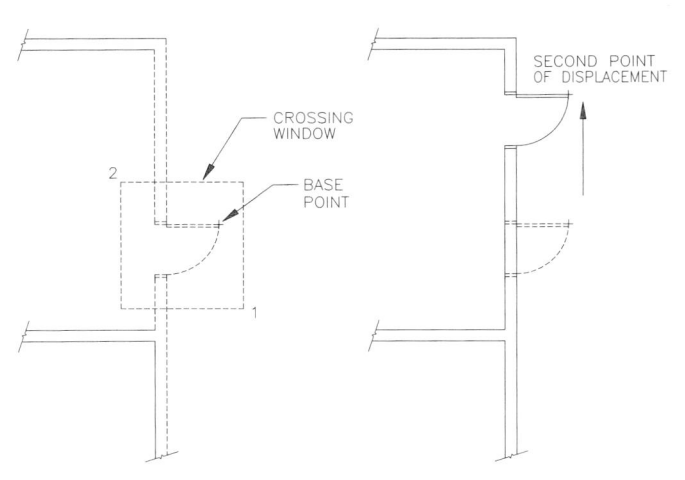

LENGTHEN

Pull-down Menu	COMMAND (TYPE)	ALIAS (TYPE)	Short-cut	Screen (side) Menu	Tablet Menu
Modify Lengthen	LENGTHEN	LEN	...	MODIFY2 Lengthen	W,14

Lengthen changes the length (longer or shorter) of linear objects and arcs. No additional objects are required (as with *Trim* and *Extend*) to make the change in length. Many methods are provided as displayed in the command prompt.

Command: `lengthen`
Select an object or [DElta/Percent/Total/DYnamic]:

Select an object
Selecting an object causes AutoCAD to report the current length of that object. If an *Arc* is selected, the included angle is also given.

DElta
Using this option returns the prompt shown next. You can change the current length of an object (including an arc) by any increment that you specify. Entering a positive value increases the length by that amount, while a negative value decreases the current length. The end of the object that you select changes while the other end retains its current endpoint (Fig. 9-9).

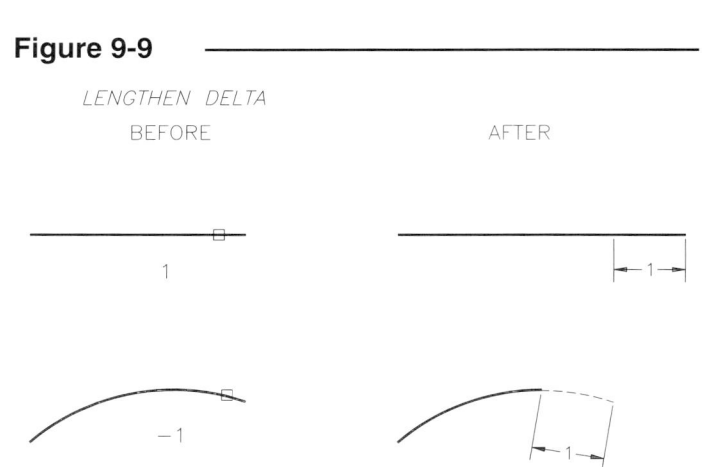

Figure 9-9

LENGTHEN DELTA
BEFORE   AFTER

Enter delta length or [Angle] <0.0000>: (`value`) or `a`

The *Angle* option allows you to change the included angle of an arc (the length along the curvature of the arc can be changed with the *Delta* option). Enter a positive or negative value (degrees) to add or subtract to the current included angle, then select the end of the object to change (Fig. 9-10).

Figure 9-10

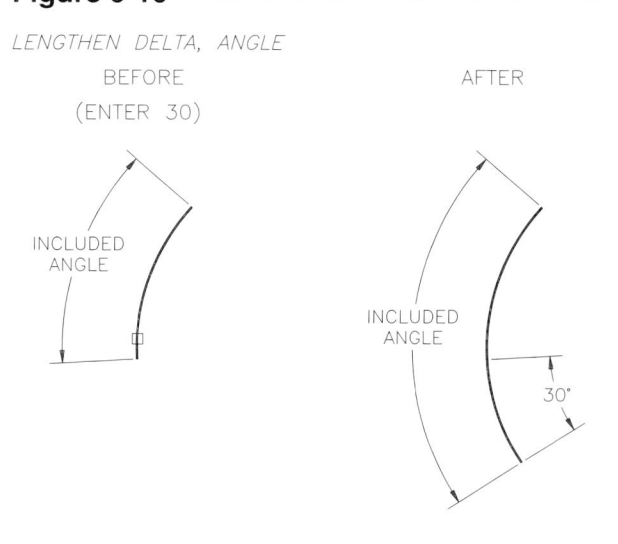

LENGTHEN DELTA, ANGLE
BEFORE
(ENTER 30)

INCLUDED ANGLE

AFTER

INCLUDED ANGLE

30°

Percent

Use this option if you want to change the length by a percentage of the current total length. For arcs, the percentage applied affects the length and the included angle equally, so there is no *Angle* option. A value of greater than 100 increases the current length, and a value of less than 100 decreases the current length. Negative values are not allowed. The end of the object that you select changes.

Figure 9-11

LENGTHEN PERCENT
BEFORE (100%)   AFTER

ENTER 133

133%

ENTER 67

67%

Total

This option lets you specify a value for the new total length. Simply enter the value and select the end of the object to change. The angle option is used to change the total included angle of a selected arc.

Specify total length or [Angle] <1.0000)>: (**value**) or **A**

Figure 9-12

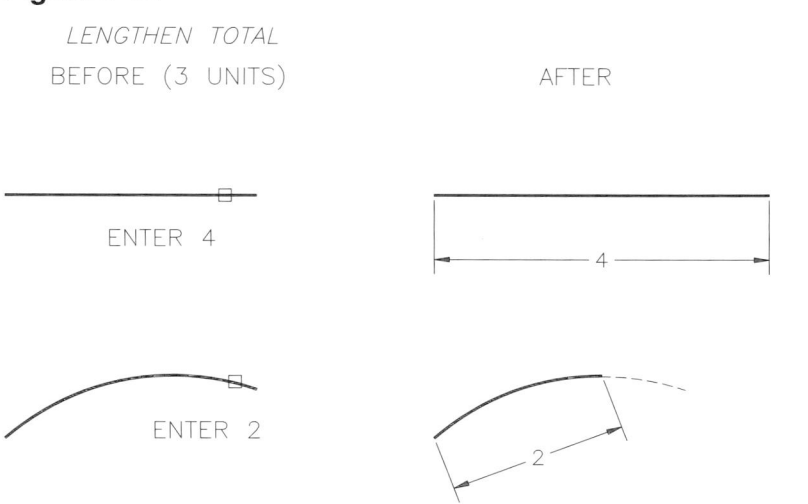

LENGTHEN TOTAL
BEFORE (3 UNITS)   AFTER

ENTER 4

4

ENTER 2

2

DYnamic

This option allows you to change the length of an object by dynamic dragging. Select the end of the object that you want to change. Object Snaps and Polar Tracking can be used.

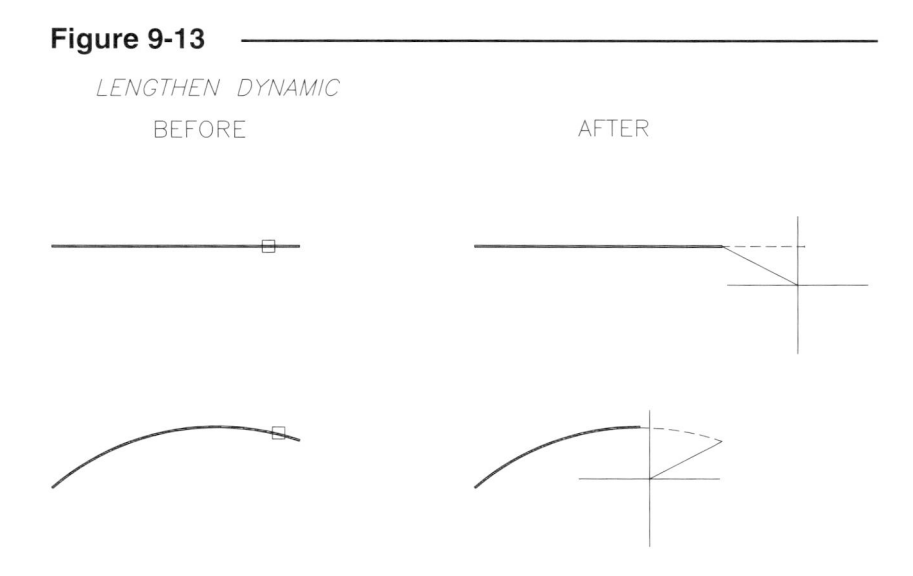

Figure 9-13

LENGTHEN DYNAMIC
BEFORE   AFTER

160   Modify Commands I

TRIM

Pull-down Menu	COMMAND (TYPE)	ALIAS (TYPE)	Short-cut	Screen (side) Menu	Tablet Menu
Modify Trim	TRIM	TR	...	MODIFY2 Trim	W,15

The *Trim* command allows you to trim (shorten) the end of an object back to the intersection of another object (Fig. 9-14). The middle section of an object can also be *Trimmed* between two intersecting objects. There are two steps to this command: first, PICK one or more "cutting edges" (existing objects); then PICK the object or objects to *Trim* (portion to remove). The cutting edges are highlighted after selection. Cutting edges themselves can be trimmed if they intersect other cutting edges, but lose their highlight when trimmed.

Figure 9-14

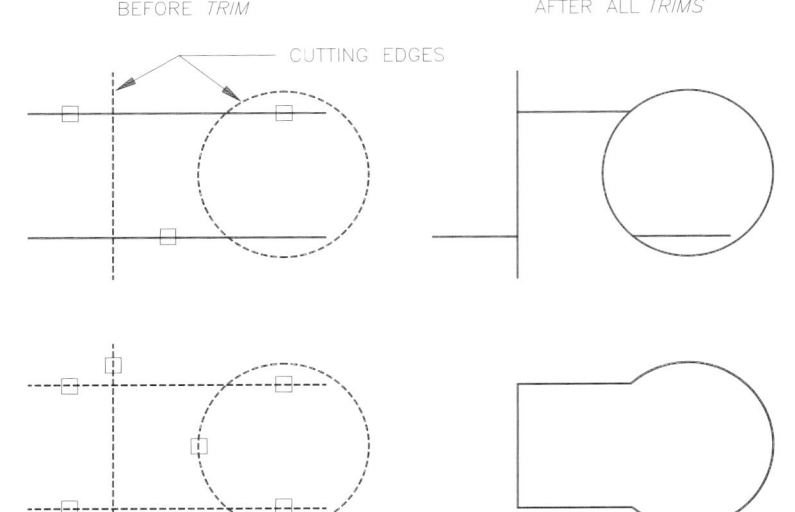

```
Command: Trim
Current settings: Projection=UCS
Edge=None
Select cutting edges ...
Select objects: PICK (Select an
    object to use as a cutting edge.)
Select objects: Enter
Select object to trim or shift-select to extend or [Project/Edge/Undo]: PICK (Select the end of an
    object to trim.)
Select object to trim or shift-select to extend or [Project/Edge/Undo]: PICK
Select object to trim or shift-select to extend or [Project/Edge/Undo]: Enter
Command:
```

Edge

The *Edge* option can be set to *Extend* or *No extend*. In the *Extend* mode, objects that are selected as trimming edges will be <u>imaginarily extended</u> to serve as a cutting edge. In other words, lines used for trimming edges are treated as having infinite length (Fig. 9-15). The *No extend* mode considers only the actual length of the object selected as trimming edges.

Figure 9-15

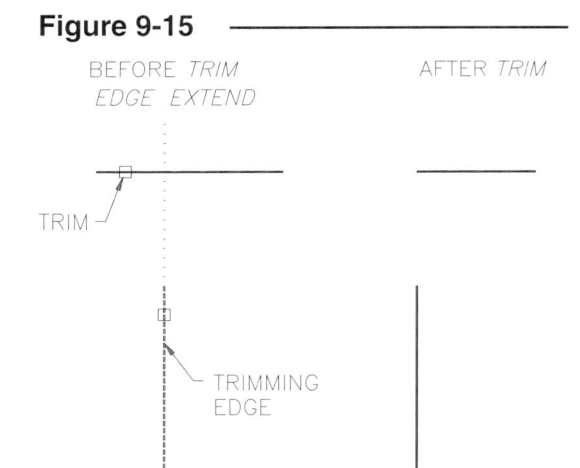

```
Command: Trim
Current settings: Projection=UCS Edge=None
Select cutting edges ...
Select objects: PICK (Select an object to use as a
    cutting edge.)
Select objects: Enter
Select object to trim or shift-select to extend or [Project/Edge/Undo]: e  (Edge)
Enter an implied edge extension mode [Extend/No extend] <No extend>: e  (Extend)
Select object to trim or shift-select to extend or [Project/Edge/Undo]: PICK  (Select an object to trim.)
Select object to trim or shift-select to extend or [Project/Edge/Undo]: Enter
Command:
```

Projection

The *Projection* switch controls how *Trim* and *Extend* operate in 3D space. *Projection* affects the projection of the cutting edge and boundary edge. The three options are useful primarily for wireframe modeling.

Undo

The *Undo* option allows you to undo the last *Trim* in case of an accidental trim.

Shift-Select

See "*Trim* and *Extend* Shift-Select Option."

EXTEND

Pull-down Menu	COMMAND (TYPE)	ALIAS (TYPE)	Short-cut	Screen (side) Menu	Tablet Menu
Modify Extend	EXTEND	EX	...	MODIFY2 Extend	W,16

Extend can be thought of as the opposite of *Trim*. Objects such as *Lines*, *Arcs*, and *Plines* can be *Extended* until intersecting another object called a boundary edge (Fig. 9-16). The command first requires selection of <u>existing</u> objects to serve as boundary edge(s) which become highlighted; then the objects to extend are selected. Objects extend until, and only if, they eventually intersect a boundary edge. An *Extended* object acquires a new endpoint at the boundary edge intersection.

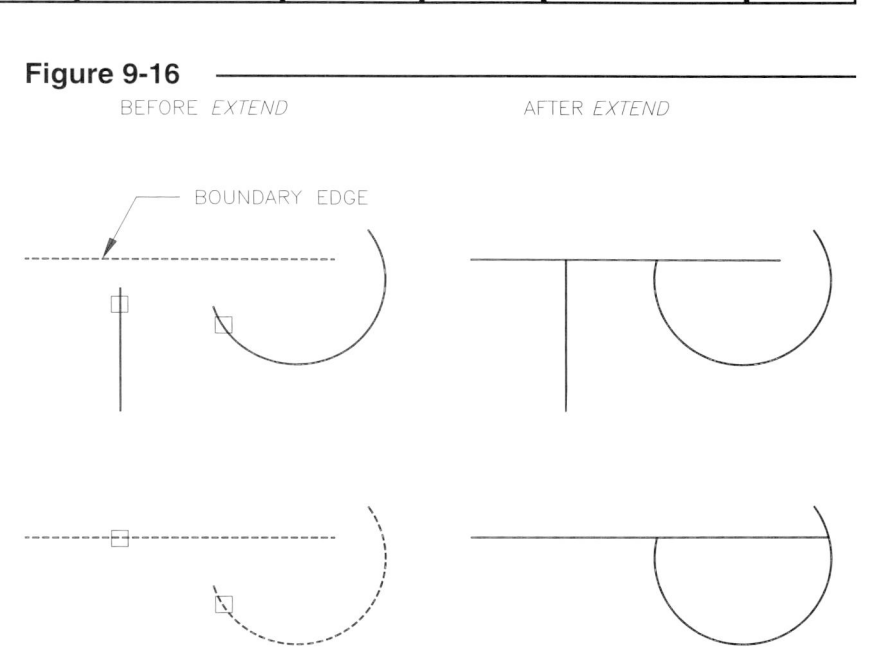

Figure 9-16

```
Command: extend
Current settings: Projection=UCS Edge=None
Select boundary edges ...
Select objects: PICK (Select boundary edge.)
Select objects: Enter
Select object to extend or shift-select to trim or [Project/Edge/Undo]: PICK (Select object to extend.)
Select object to extend or shift-select to trim or [Project/Edge/Undo]: PICK (Select object to extend.)
Select object to extend or shift-select to trim or [Project/Edge/Undo]: Enter
Command:
```

162 *Modify Commands I*

Edge/Projection
The *Edge* and *Projection* switches operate identically to their function with the *Trim* command. Use *Edge* with the *Extend* option if you want a boundary edge object to be imaginarily extended (Fig. 9-17).

Shift-Select
See "*Trim* and *Extend* Shift-Select Option."

Figure 9-17
BEFORE *EXTEND*
EDGE *EXTEND*
AFTER *EXTEND*

EXTEND

BOUNDARY EDGE

Trim and *Extend* **Shift-Select Option**

A new enhancement to the *Trim* and *Extend* commands in AutoCAD 2002 is the shift-select option. With this feature, you can toggle between *Trim* and *Extend* by holding down the Shift key. For example, if you invoked *Trim* and are in the process of trimming lines but decide to then *Extend* a line, you can simply hold down the Shift key to change to the *Extend* command without leaving *Trim*.

    Command: trim
    Current settings: Projection=UCS, Edge=None
    Select cutting edges ...
    Select objects: PICK
    Select objects: Enter
    Select object to trim or shift-select to extend or [Project/Edge/Undo]: PICK (to trim)
    Select object to trim or shift-select to extend or [Project/Edge/Undo]: Shift, then PICK (to extend)

Not only does holding down the Shift key toggle between *Trim* and *Extend*, but objects you selected as *cutting edges* become *boundary edges* and vice versa. Therefore, if you want to use this feature effectively, during the first step of either command (*Trim* or *Extend*), you must anticipate and select edges you might potentially use as both cutting edges and boundary edges.

BREAK

Pull-down Menu	COMMAND (TYPE)	ALIAS (TYPE)	Short-cut	Screen (side) Menu	Tablet Menu
Modify Break	BREAK	BR	...	MODIFY2 Break	W,17

Break allows you to break a space in an object or break the end off an object. You can think of *Break* as a partial erase. If you choose to break a space in an object, the space is created between two points that you specify (Fig. 9-18). In this case, the *Break* creates two objects from one.

Figure 9-18
BEFORE *BREAK*
AFTER *BREAK*

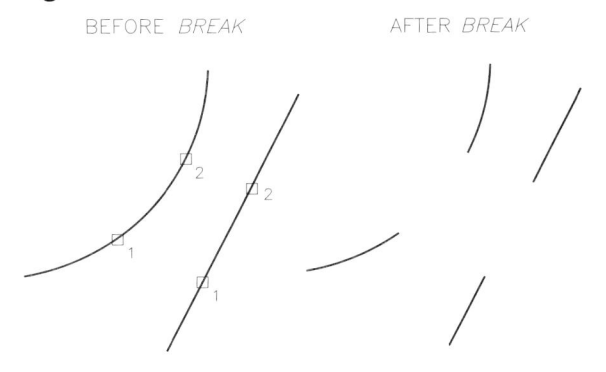

If *Break*ing a circle (Fig. 9-19), the break is in a <u>counterclockwise</u> direction from the first to the second point specified.

Figure 9-19

If you want to *Break* the end off a *Line* or *Arc*, the first point should be specified at the point of the break and the second point should be <u>just off the end</u> of the *Line* or *Arc* (Fig. 9-20).

Figure 9-20

There are four options for *Break*. The *Modify* toolbar includes icons for only the default option (see *Select, Second*) and the *Break at Point* option. All four methods are available only from the screen menu. If you are typing, the desired option is selected by keying the letter *F* (for *First point*) or symbol @ (for "last point"). The four options are explained next in detail.

Select, Second
This method (the default method) has two steps: select the object to break; then select the second point of the break. The first point used to select the object is <u>also</u> the first point of the break (see Figures 9-18, 9-19, 9-20).

    Command: **break**
    Select object: **PICK** (This is the first point of the break.)
    Specify second break point or [First point]: **PICK** (This is the second point of the break.)
    Command:

Select, 2 Points
This method uses the first selection only to indicate the <u>object</u> to *Break*. You then specify the point that is the first point of the *Break*, and the next point specified is the second point of the *Break*. This option can be used with *OSNAP Intersection* to achieve the same results as *Trim*.

Selecting this option from the screen menu automatically sequences through the correct prompts (Fig. 9-21).

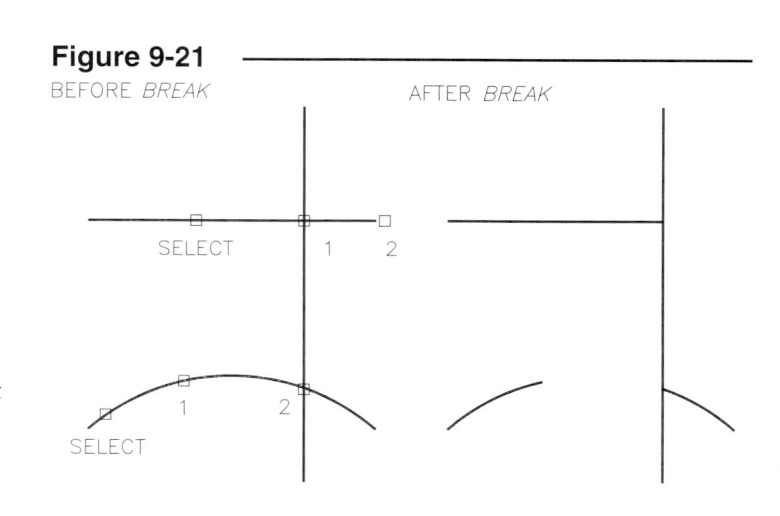

Figure 9-21

If you are typing, the command sequence is as follows:

Command: **break**
Select object: **PICK** (This selects only the object to break.)
Specify second break point or [First point]: **f** (Indicates respecification of the first point.)
Specify first break point: **PICK** (This is the first point of the break.)
Specify second break point: **PICK** (This is the second point of the break.)
Command:

Select Point

This option breaks one object into two separate objects with <u>no space</u> between. You specify only <u>one</u> point with this method. When you select the object, the point indicates the object <u>and</u> the point of the break. When prompted for the second point, the @ symbol (translated as "last point") is specified. This second step is done automatically only if selected from the screen menu. If you are typing the command, the @ symbol (last point) must be typed.

Command: **break**
Select object: **PICK** (This is the first point of the break.)
Specify second break point or [First point]: **@** (Indicates the break to be at the last point.)
Command:

The resulting object should <u>appear</u> as before; however, it has been transformed into <u>two</u> objects with matching endpoints (Fig. 9-22).

Figure 9-22

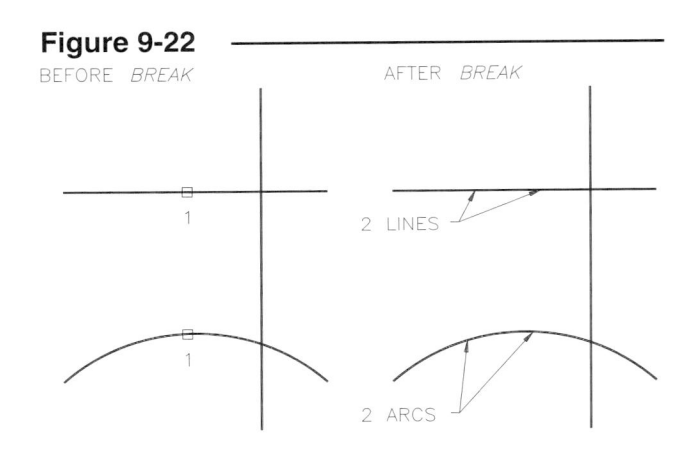

Break at Point

This option creates a break with <u>no space</u>, like the *1 Point* option; however, you can select the object you want to *Break* first and the point of the *Break* next (Fig. 9-23).

Command: **break**
Select object: **PICK** (This selects only the object to break.)
Specify second break point or [First point]: **f** (Indicates respecification of the first point.)
Specify first break point: **PICK** (This is the first point of the break.)
Specify second break point: **@** (The second point of the break is the last point.)

Figure 9-23

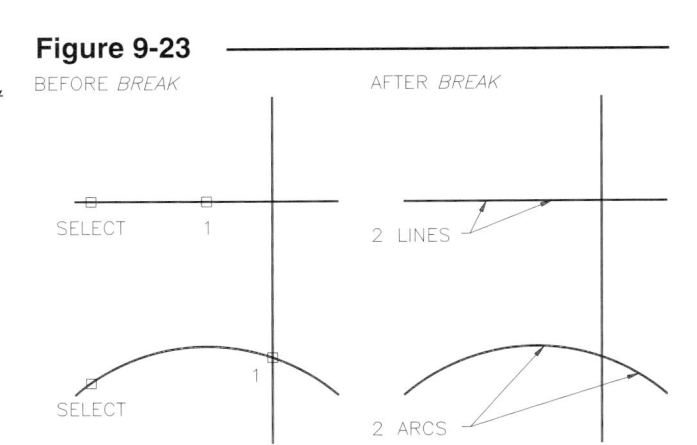

COPY

Pull-down Menu	COMMAND (TYPE)	ALIAS (TYPE)	Short-cut	Screen (side) Menu	Tablet Menu
Modify Copy	COPY	CO or CP	(Edit Mode) Copy Selection	MODIFY1 Copy	V,15

Copy creates a duplicate set of the selected objects and allows placement of those copies. The *Copy* operation is like the *Move* command, except with *Copy* the original set of objects remains in its original location. You specify a "base point:" (point to copy from) and a "specify second point of displacement or <use first point as displacement>:" (point to copy to). See Figure 9-24.

Figure 9-24

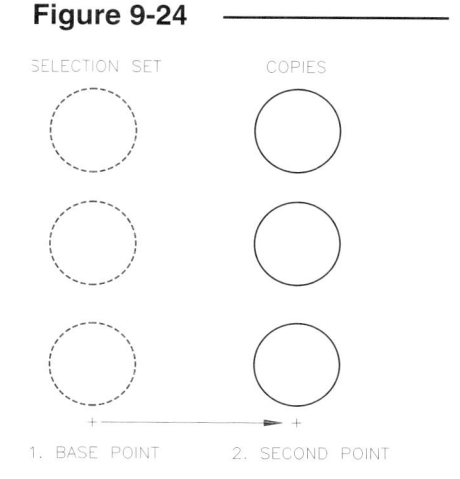

The command syntax for *Copy* is as follows.

   Command: *copy*
   Select Objects: **PICK** (Select objects to be copied.)
   Select Objects: **Enter** (Indicates completion of the selection set.)
   Specify base point or displacement, or [Multiple]: **PICK** or (**coordinates**) (This is the point to copy from. Select a point, usually on the object, to use as a "handle" or reference point. *OSNAP*s can be used. Coordinates in any format can also be entered.)
   Specify second point of displacement or <use first point as displacement>: **PICK** or (**coordinates**) (This is the point to copy to. Select a point. *OSNAP*s can be used. Coordinates in any format can be entered.)
   Command:

In many applications it is desirable to use *OSNAP* options to PICK the "base point:" and the "second point of displacement (Fig. 9-25).

Figure 9-25

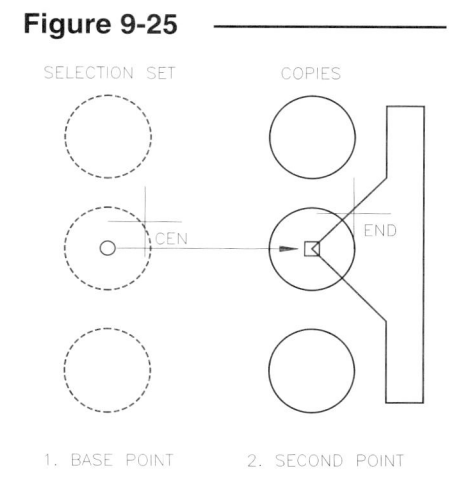

Alternately, you can PICK the "base point:" and use Polar Tracking or enter relative rectangular coordinates, relative polar coordinates, or direct distance entry to specify the "second point of displacement:".

Figure 9-26

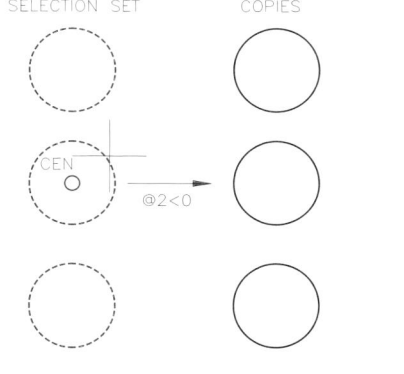

166 Modify Commands I

The *Copy* command has a *Multiple* option. The *Multiple* option allows creating and placing multiple copies of the selection set.

Command: *copy*
Select Objects: **PICK**
Select Objects: **Enter**
Specify base point or displacement, or [Multiple]: *m*
Specify base point: **PICK**
Specify second point of displacement or <use first point as displacement>: **PICK**
Specify second point of displacement or <use first point as displacement>: **PICK**
Specify second point of displacement or <use first point as displacement>: **Enter**
Command:

Figure 9-27

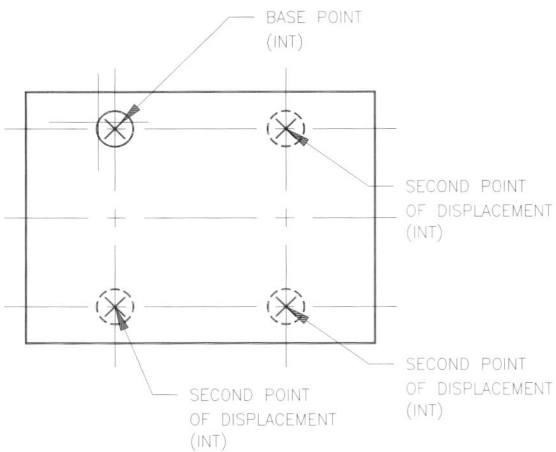

MIRROR

Pull-down Menu	COMMAND (TYPE)	ALIAS (TYPE)	Short-cut	Screen (side) Menu	Tablet Menu
Modify Mirror	MIRROR	MI	...	MODIFY1 Mirror	V,16

This command creates a mirror image of selected existing objects. You can retain or delete the original objects ("old objects"). After selecting objects, you create two points specifying a "rubberband line," or "mirror line," about which to *Mirror*.

The length of the mirror line is unimportant since it represents a vector or axis (Fig. 9-28).

Figure 9-28

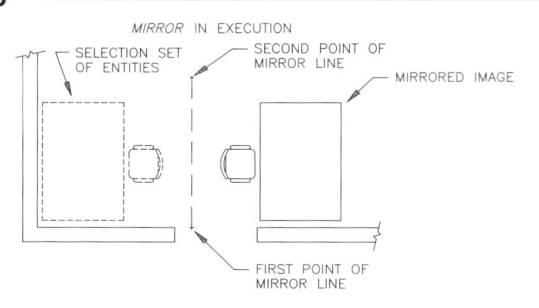

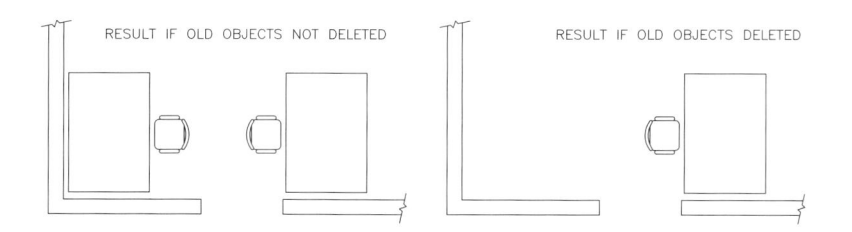

Command: *mirror*
Select Objects: **PICK** (Select object or group of objects to mirror.)
Select Objects: **Enter** (Press Enter to indicate completion of object selection.)
Specify first point of mirror line: **PICK** or (**coordinates**) (Draw first endpoint of line to represent mirror axis by PICKing or entering coordinates.)
Specify second point of mirror line: **PICK** or (**coordinates**) (Select second point of mirror line by PICKing or entering coordinates.)
Delete source objects? <N> **Enter** or **Y** (Press Enter to yield both sets of objects or enter Y to keep only the mirrored set.)
Command:

If you want to *Mirror* only in a vertical or horizontal direction, toggle *ORTHO* or *POLAR On* before selecting the "second point of mirror line."

Mirror can be used to draw the other half of a symmetrical object, thus saving some drawing time (Fig. 9-29).

Figure 9-29

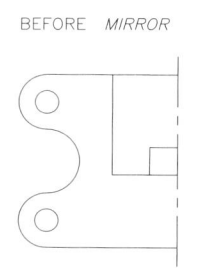

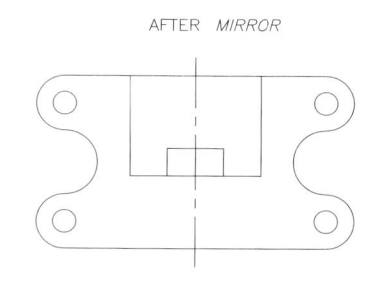

BEFORE MIRROR   AFTER MIRROR

OFFSET

Pull-down Menu	COMMAND (TYPE)	ALIAS (TYPE)	Short-cut	Screen (side) Menu	Tablet Menu
Modify Offset	OFFSET	O	...	MODIFY1 Offset	V,17

Offset creates a <u>parallel copy</u> of selected objects. Selected objects can be *Line*s, *Arc*s, *Circle*s, *Pline*s, or other objects. *Offset* is a very useful command that can increase productivity greatly, particularly with *Pline*s.

Depending on the object selected, the resulting *Offset* is drawn differently (Fig. 9-30). *Offset* creates a parallel copy of a *Line* equal in length and perpendicular to the original. *Arc*s and *Circle*s have a concentric *Offset*. *Offsetting* closed *Pline*s or *Spline*s results in a complete parallel shape.

Figure 9-30

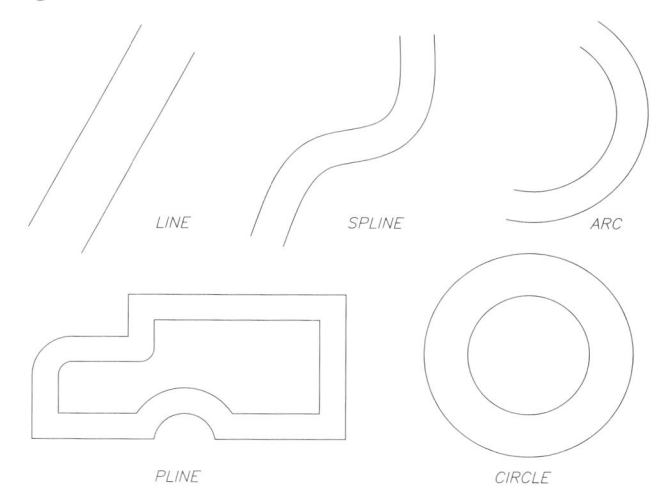

LINE   SPLINE   ARC

PLINE   CIRCLE

Two options are available with *Offset*: (1) *Offset* a specified *distance* and (2) *Offset through* a specified point (Fig. 9-31).

Figure 9-31

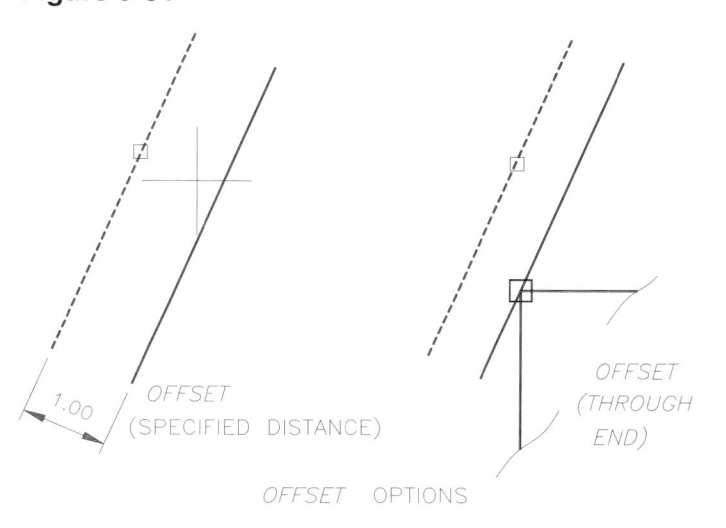

1.00 OFFSET (SPECIFIED DISTANCE)   OFFSET (THROUGH END)

OFFSET OPTIONS

168 Modify Commands I

Distance

The *Distance* option command sequence is as follows:

Command: *offset*
Specify offset distance or [Through] <1.0000>: (**value**) or **PICK** (Indicate distance to offset by entering a value or PICKing two points.)
Select object to offset or <exit>: **PICK** (Only one object can be selected.)
Specify point on side to offset: **PICK** (Select which side of the selected object for the offset to be drawn.)
Select object to offset or <exit>: **PICK** or **Enter** (*Offset* can be used repeatedly to offset at the same distance or press Enter to exit.)
Command:

Through

The command sequence for *Offset through* is as follows:

Command: *offset*
Specify offset distance or [Through] <1.0000>: **t**
Select object to offset or <exit>: **PICK** (Only one object can be selected.)
Specify through point: **PICK** or (**coordinates**) (Select a point for the object to be drawn through. You can use *OSNAPs*, Polar Tracking, and other tools or enter coordinates in any format.)
Select object to offset or <exit>: **PICK** or **Enter** (*Offset* can be used repeatedly by PICKing a new through point or press Enter to exit.)
Command:

TIP Notice the power of using *Offset* with closed *Pline* or *Spline* shapes (Fig. 9-32). Because a *Pline* or a *Spline* is one object, *Offset* creates one complete smaller or larger "parallel" object. Any closed shape composed of *Line*s and *Arc*s can be converted to one *Pline* object (see "*Pedit*," Chapter 16).

Figure 9-32

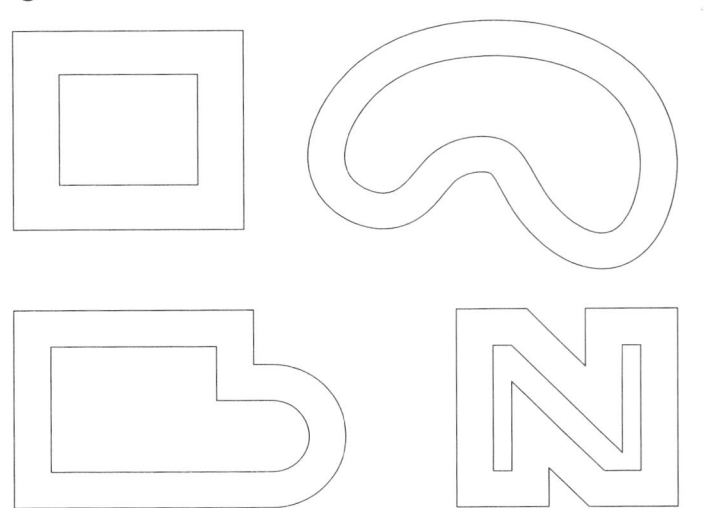

ARRAY

Pull-down Menu	COMMAND (TYPE)	ALIAS (TYPE)	Short-cut	Screen (side) Menu	Tablet Menu
Modify Array	ARRAY	AR	...	MODIFY1 Array	V,18

The *Array* command creates either a *Rectangular* or a *Polar* (circular) pattern of existing objects that you select. The pattern could be created from a single object or from a group of objects. *Array* copies a duplicate set of objects for each "item" in the array.

In AutoCAD 2002, the *Array* command produces a dialog box to enter the desired array parameters and to preview the pattern (Fig. 9-33). The first step is to select the type of array you want (at the top of the box): *Rectangular Array* or *Polar Array*.

Rectangular Array
A rectangular array is a pattern of objects generated into rows and columns. Selecting a *Rectangular Array* produces the central area of the box and preview image as shown in Figure 9-33. Follow these steps to specify the pattern for the *Rectangular Array*.

1. Pick the *Select objects* button in the upper-right corner of the dialog box to select the objects you want to array.
2. Enter values in the *Rows:* and *Columns:* edit boxes to indicate how many rows and columns you want. The image area indicates the number of rows and columns you request, but does not indicate the shape of the objects you selected.
3. Enter values in the *Row offset:* and *Column offset:* edit boxes to indicate the distance between the rows and columns (see Figure 9-34). Enter positive values to create an array to the right (+X) and upward (+Y) from the selected set, or enter negative values to create the array in the –X and –Y directions. You can also use the small select buttons to the far right of these edit boxes to interactively PICK two points for the *Row offset* and *Column offset*. Alternately, select the large button to the immediate right of the *Row offset* and *Column offset* edit boxes to use the "unit cell" method. With this method, you select diagonal corners of a window to specify both directions and both distances for the array (see Figure 9-35).
4. If you want to create an array at an angle (instead of generating the rows vertically and columns horizontally), enter a value in the *Angle of array* edit box or use the select button to PICK two points to specify the angle of the array.
5. Select the *Preview* button to temporarily close the dialog box and return to the drawing where you can view the array and chose to *Accept* or *Modify* the array. Selecting *Modify* returns you to the *Array* dialog box.

Figure 9-33

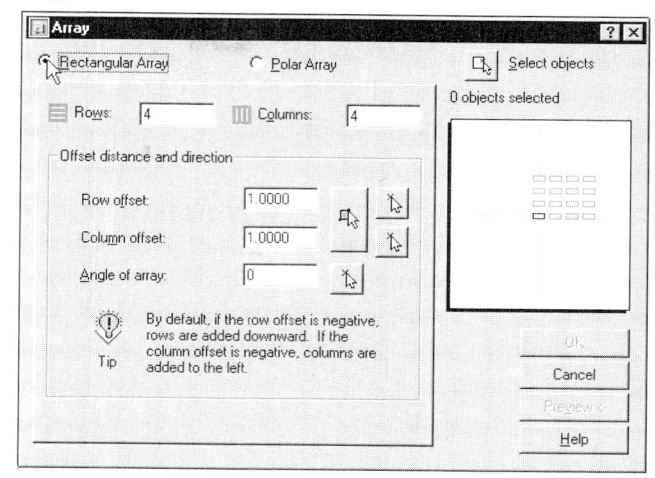

Figure 9-34

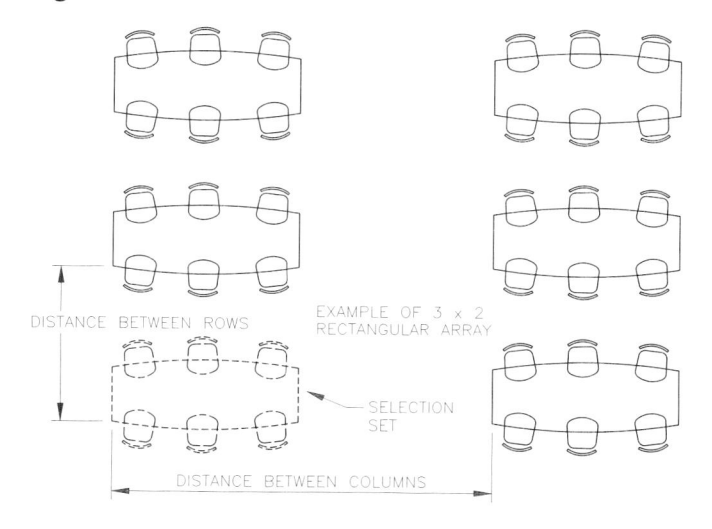

Figure 9-35

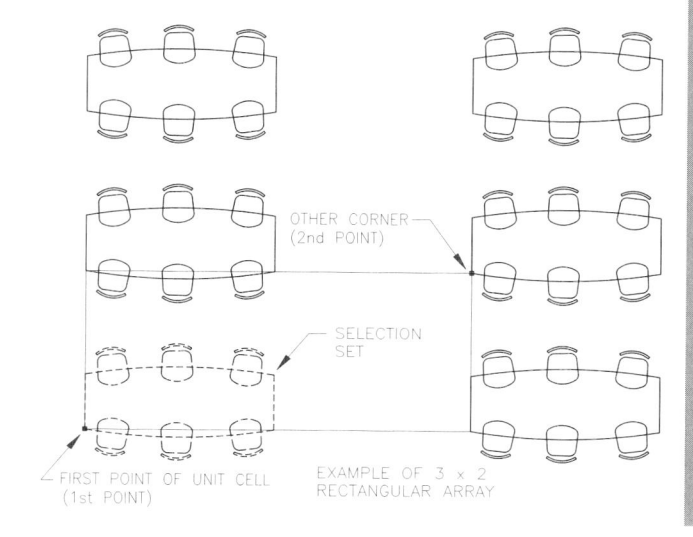

Polar Array

The *Polar Array* option generates a circular pattern of the selection set. Selecting the *Polar Array* button produces the central area of the box and preview image similar to that shown in Figure 9-36. Typical steps are as follows.

Figure 9-36

1. Pick the *Select objects* button in the upper-right corner of the dialog box to select the objects you want to array.
2. Enter values for the *Center point* of the circular pattern in the X and Y edit boxes, or use the select button to PICK a center point interactively.
3. Select the *Method* from the drop-down list. The three possible methods are:
    Total number of items & Angle to fill
    Total number of items & Angle between items
    Angle to fill & Angle between items
    Whichever option you choose enables the two applicable edit boxes below.
4. You must supply two of the three possible parameters following to specify the configuration of the array based on your selection in the *Method* drop-down list.
    Total number of items (the total number <u>includes</u> the selected object set)
    Angle to fill
    Angle between items
    Arrays are generated counterclockwise by default. To produce a clockwise array, enter a negative value in *Angle to fill*; or enter a negative value in *Angle to fill*, then switch to the *Total number of items & Angle between items*. The image area indicates the configuration of the array according to the options you selected and values you specified in the *Method and values* section. The image area does not indicate the shape of the object(s) you selected to array.
5. If you want the set of objects to be rotated but remain in the same orientation, remove the check in the *Rotate items as copied* box.
6. Select the *Preview* button to temporarily close the dialog box and return to the drawing where you can view the array and chose to *Accept* or *Modify* the array. Selecting *Modify* returns you to the *Array* dialog box.

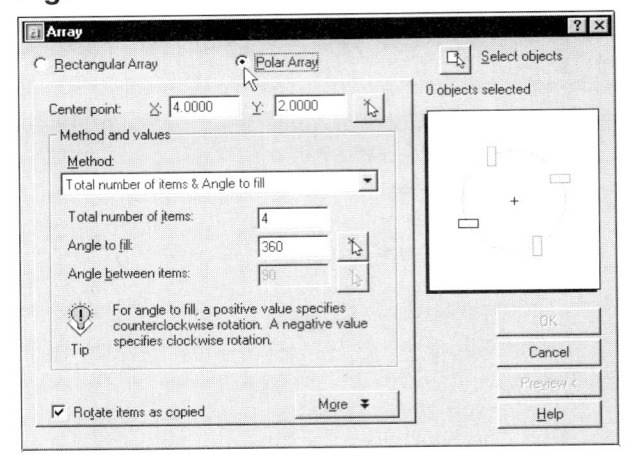

Figure 9-37 illustrates a *Polar Array* created using 8 as the *Total number of items*, 360 as the *Angle to fill*, and selecting *Rotate items as copied*.

Figure 9-37

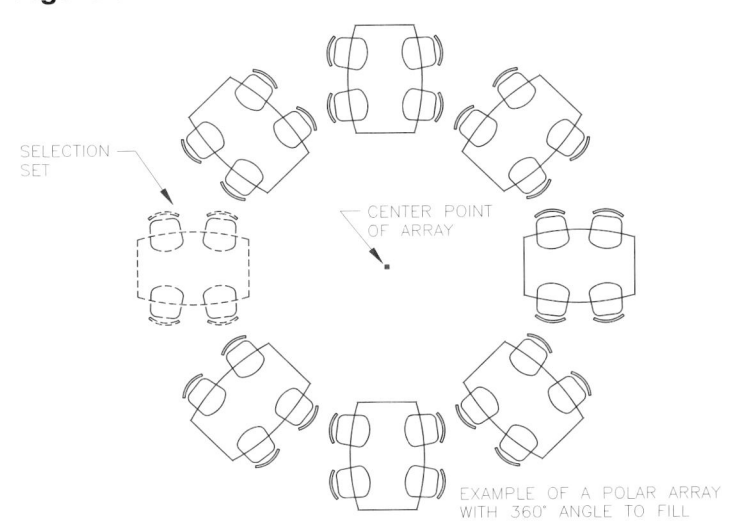

Figure 9-38 illustrates a *Polar Array* created using 5 as the *Total number of items*, 180 as the *Angle to fill*, and selecting *Rotate items as copied*.

Figure 9-38

As an alternative to using the *Array* command to produce the *Array* dialog box, you can type *–Array* to use the Command line version. After selecting objects, enter the letter "r" to produce a *Rectangular Array* as shown in the following sequence.

> Command: *-array*
> Select objects: **PICK**
> Select objects: **Enter**
> Enter the type of array [Rectangular/Polar] <R>: **r**
> Enter the number of rows (—-) <1>: (**value**)
> Enter the number of columns (|||) <1> (**value**)
> Enter the distance between rows or specify unit cell (—-): **PICK** or (**value**)
> Specify opposite corner: **PICK**
> Command:

To create a *Polar Array*, enter the letter "p" at the "Enter type of array" prompt as shown.

> Command: *-array*
> Select objects: **PICK**
> Select objects: **Enter**
> Enter the type of array [Rectangular/Polar] <R>: **p**
> Specify center point of array or [Base]: **PICK** or (**coordinates**)
> Enter the number of items in the array: (**value**)
> Specify the angle to fill (+=ccw, -=cw) <360>: (**value**)
> Rotate arrayed objects? [Yes/No] <Y>: (**option**)
> Command:

FILLET

Pull-down Menu	COMMAND (TYPE)	ALIAS (TYPE)	Short-cut	Screen (side) Menu	Tablet Menu
Modify Fillet	FILLET	F	...	MODIFY2 Fillet	W,19

The *Fillet* command automatically rounds a sharp corner (intersection of two *Lines*, *Arcs*, *Circles*, or *Pline* vertices) with a radius. You specify only the radius and select the objects' ends to be *Filleted*. The objects to fillet do not have to completely intersect but can overlap. You can specify whether or not the objects are automatically extended or trimmed as necessary (Fig. 9-39).

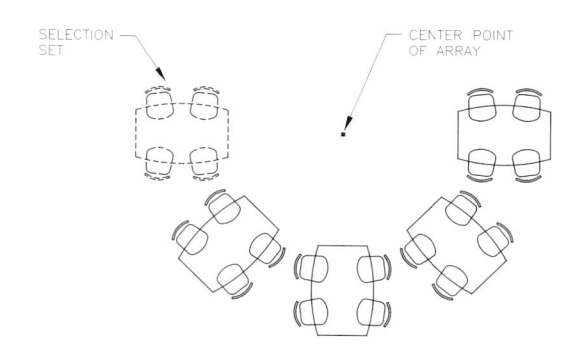

Figure 9-39

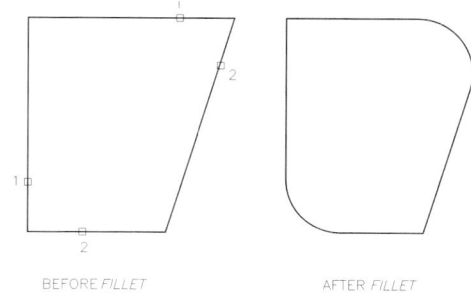

> Command: *fillet*
> Current settings: Mode = TRIM, Radius = 0.5000
> Select first object or [Polyline/Radius/Trim]: **PICK** (Select one *Line*, *Arc*, or *Circle* near the point where the fillet should be created.)
> Select second object: **PICK** (Select second object to fillet near fillet location.)
> Command:

The fillet is created at the corner selected.

Treatment of *Arcs* and *Circles* with *Fillet* is shown in Figure 9-40. Note that the objects to *Fillet* do not have to intersect but can overlap.

Figure 9-40

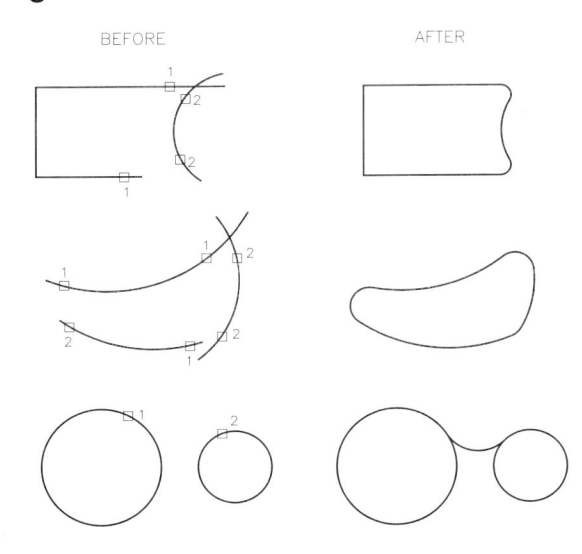

Radius
Use the *Radius* option to specify the desired radius for the fillets. *Fillet* uses the specified radius value for all new fillets until the value is changed.

```
Command: fillet
Current settings: Mode = TRIM, Radius = 0.5000
Select first object or [Polyline/Radius/Trim]: r
Specify fillet radius <0.5000>: .75
Select first object or [Polyline/Radius/Trim]:
```

Since trimming and extending are done automatically by *Fillet* when necessary, using *Fillet* with a *radius* of *0* has particular usefulness. Using *Fillet* with a *0 radius* creates clean, sharp corners even if the original objects overlap or do not intersect (Fig. 9-41).

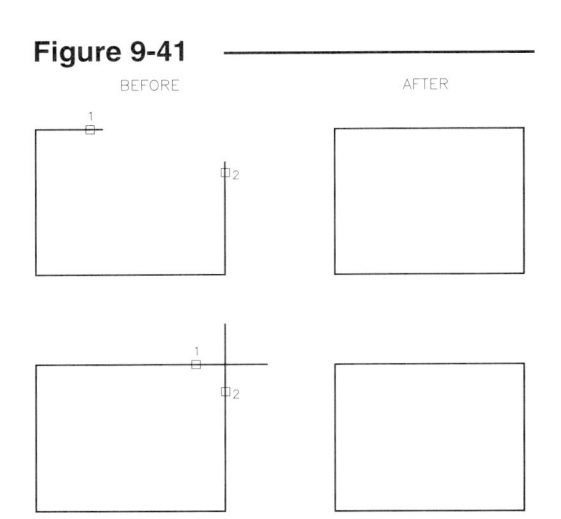

Figure 9-41

If parallel objects are selected, the *Fillet* is automatically created to the correct radius (Fig. 9-42). Therefore, parallel objects can be filleted at any time <u>without specifying a radius value</u>.

Figure 9-42

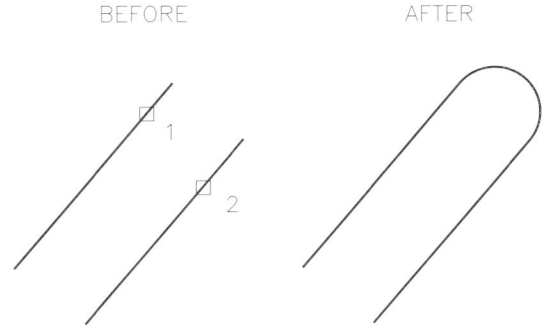

Polyline

Fillets can be created on *Polylines* in the same manner as two *Line* objects. Use *Fillet* as you normally would for *Lines*. However, if you want a fillet equal to the specified radius to be added to each vertex of the *Pline* (except the endpoint vertices), use the *Polyline* option; then select anywhere on the *Pline* (Fig. 9-43).

   Command: *fillet*
   Current settings: Mode = TRIM, Radius = 0.5000
   Select first object or [Polyline/Radius/Trim]: *P*
   (Indicates *Polyline* option.)
   Select 2D polyline: **PICK** (Select the desired polyline.)

Figure 9-43

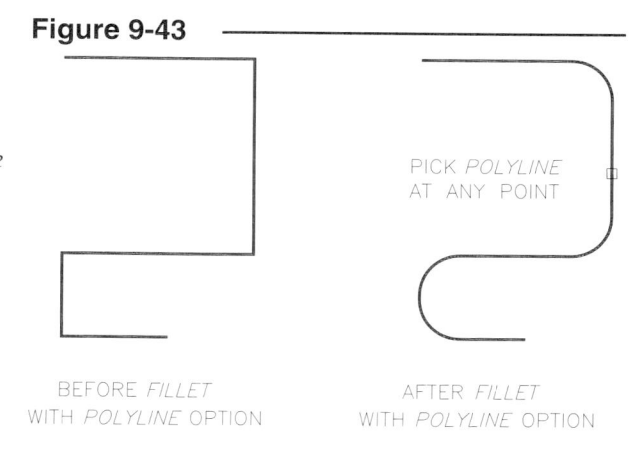

BEFORE *FILLET* WITH *POLYLINE* OPTION

AFTER *FILLET* WITH *POLYLINE* OPTION

Closed Plines created by the *Close* option of *Pedit* react differently with *Fillet* than *Plines* connected by PICKing matching endpoints. Figure 9-44 illustrates the effect of *Fillet Polyline* on a *Closed Pline* and on a connected *Pline*.

Figure 9-44

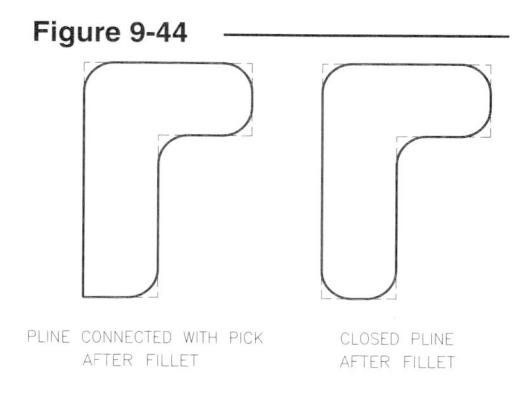

PLINE CONNECTED WITH PICK AFTER FILLET

CLOSED PLINE AFTER FILLET

Trim/Notrim

In the previous figures, the objects are shown having been filleted in the *Trim* mode—that is, with automatically trimmed or extended objects to meet the end of the new fillet radius. The *Notrim* mode creates the fillet without any extending or trimming of the involved objects (Fig. 9-45). Note that the command prompt indicates the current mode (as well as the current radius) when *Fillet* is invoked.

   Command: *fillet*
   Current settings: Mode = TRIM, Radius = 0.5000
   Select first object or [Polyline/Radius/Trim]: *t*
   (Invokes the *Trim/No trim* option.)
   Enter Trim mode option [Trim/No trim] <Trim>: *n* (Sets fillet to *No trim*.)
   Select first object or [Polyline/Radius/Trim]:

Figure 9-45

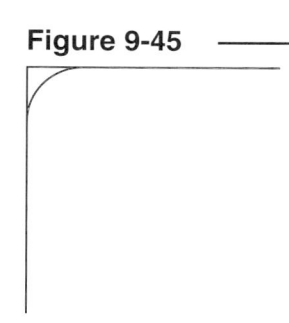

FILLET NOTRIM

The *Notrim* mode is helpful in situations similar to that in Figure 9-46. In this case, if the intersecting *Lines* were trimmed when the first fillet was created, the longer *Line* (highlighted) would have to be redrawn in order to create the second fillet.

Figure 9-46

Changing the *Trim/Notrim* option sets the TRIMMODE system variable to 0 (*Notrim*) or 1 (*Trim*). This variable controls trimming for both *Fillet* and *Chamfer*, so if you set *Fillet* to *Notrim*, *Chamfer* is also set to *Notrim*.

FILLET NOTRIM

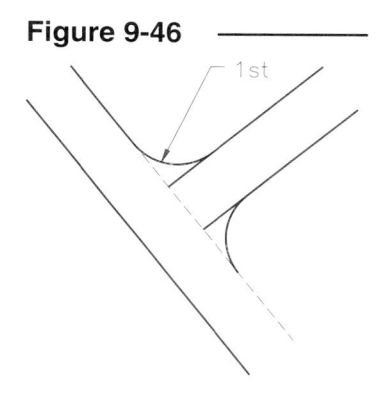

CHAMFER

Pull-down Menu	COMMAND (TYPE)	ALIAS (TYPE)	Short-cut	Screen (side) Menu	Tablet Menu
Modify Chamfer	CHAMFER	CHA	...	MODIFY2 Chamfer	W,18

Chamfering is a manufacturing process used to replace a sharp corner with an angled surface. In AutoCAD, *Chamfer* is commonly used to change the intersection of two *Lines* or *Plines* by adding an angled line. The *Chamfer* command is similar to *Fillet*, but rather than rounding with a radius or "fillet," an angled line is automatically drawn at the distances (from the existing corner) that you specify.

Chamfers can be created by two methods: *Distance* (specify two distances) or *Angle* (specify a distance and an angle). The current method and the previously specified values are displayed at the Command prompt along with the options:

    Command: *chamfer*
    (TRIM mode) Current chamfer Dist1 = 0.5000, Dist2 = 0.5000
    Select first line or [Polyline/Distance/Angle/Trim/Method]:

Method
Use this option to indicate which of the two methods you want to use: *Distance* (specify 2 distances) or *Angle* (specify a distance and an angle).

Distance
The *Distance* option is used to specify the two values applied when the *Distance Method* is used to create the chamfer. The values indicate the distances from the corner (intersection of two lines) to each chamfer endpoint (Fig. 9-47).

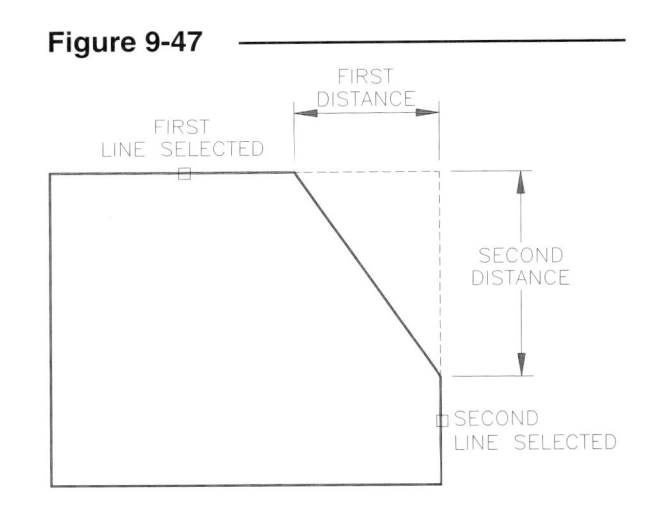

Figure 9-47

    Command: *chamfer*
    (TRIM mode) Current chamfer Dist1 = 0.5000, Dist2 = 0.5000
    Select first line or [Polyline/Distance/Angle/Trim/Method]: **d** (Indicates the *Distance* option.)
    Specify first chamfer distance <0.5000>: (**value**) or **PICK** (Enter a value for the distance from the existing corner to the endpoint of the chamfer on the first line or select two points to interactively specify the distance.)
    Specify second chamfer distance <0.5000>: **Enter**, (**value**) or **PICK** (Press Enter to use the same distance as the first, enter another value, or PICK as before.)
    Select first line or [Polyline/Distance/Angle/Trim/Method]:

Chamfer with distances of 0 reacts like *Fillet* with a radius of 0; that is, overlapping corners can be automatically trimmed and non-intersecting corners can be automatically extended (using the *Trim* mode).

Angle

The *Angle* option allows you to specify the values that are used for the *Angle Method*. The values specify a distance along the first line and an angle from the first line (Fig. 9-48).

Figure 9-48

```
Select first line or [Polyline/Distance/Angle/Trim/Method]: a
Specify chamfer length on the first line <1.0000>: (value)
Specify chamfer angle from the first line <0>: (value)
```

Trim/Notrim

The two lines selected for chamfering do not have to intersect but can overlap or not connect. The *Trim* setting automatically trims or extends the lines selected for the chamfer (Fig. 9-49), while the *Notrim* setting adds the chamfer without changing the length of the selected lines. These two options are the same as those in the *Fillet* command (see "*Fillet*"). Changing these options sets the *TRIMMODE* system variable to 0 (*Notrim*) or 1 (*Trim*). The variable controls trimming for both *Fillet* and *Chamfer*.

Figure 9-49

Polyline

The *Polyline* option of *Chamfer* creates chamfers on all vertices of a *Pline*. All vertices of the *Pline* are chamfered with the supplied distances. The first end of the *Pline* that was drawn takes the first distance. Use *Chamfer* without this option if you want to chamfer only one corner of a *Pline*. This is similar to the same option of *Fillet* (see "*Fillet*").

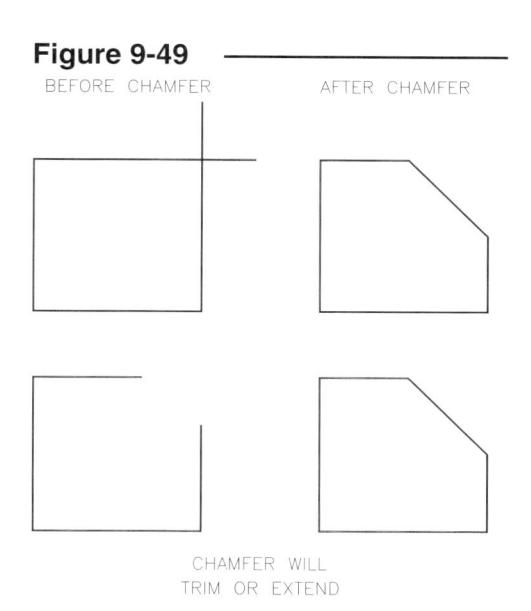

CHAPTER EXERCISES

Figure 9-50

1. *Move*

    Begin a *New* drawing and create the geometry in Figure 9-50A using **Lines** and **Circles**. If desired, set **Polar Distance** to **.25** to make drawing easy and accurate.

    For practice, turn *SNAP OFF* (**F9**). Use the *Move* command to move the *Circles* and *Lines* into the positions shown in Fig. 9-50B. *OSNAP*s are required to *Move* the geometry accurately (since *SNAP* is off). *Save* the drawing as **MOVE1**.

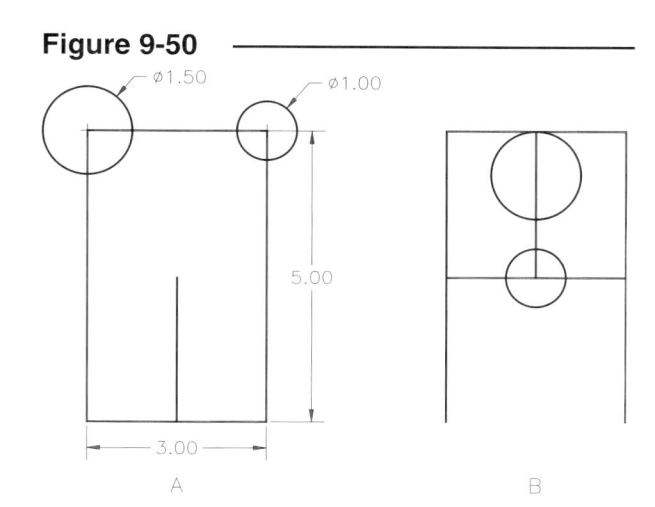

176   Modify Commands I

2. *Rotate*

   Figure 9-51

   Begin a *New* drawing and create the geometry in Figure 9-51A.

   Rotate the shape into position shown in step B. *SaveAs* **ROTATE1**.

   Use the *Reference* option to *Rotate* the box to align with the diagonal *Line* as shown in C. *SaveAs* **ROTATE2**.

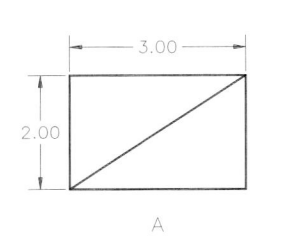

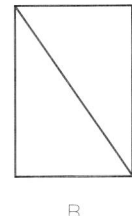

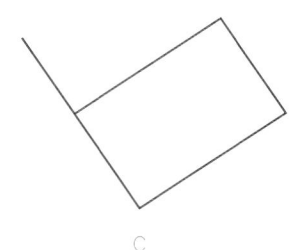

3. *Scale*

   Figure 9-52

   Open **ROTATE1** to again use the shape shown in Figure 9-52B and *SaveAs* **SCALE1**. *Scale* the shape by a factor of 1.5.

   Open **ROTATE2** to again use the shape shown in C. Use the *Reference* option of *Scale* to increase the scale of the three other *Lines* to equal the length of the original diagonal *Line* as shown. (HINT: *OSNAP*s are required to specify the *Reference length* and *New length*.) *SaveAs* **SCALE2**.

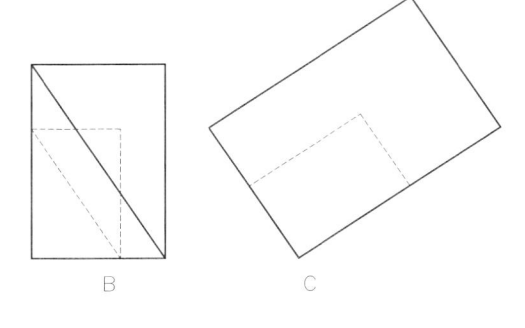

4. *Stretch*

   Figure 9-53

   A design change has been requested. *Open* the **SLOTPLATE CH8** drawing and make the following changes.

   A. The top of the plate (including the slot) must be moved upward. This design change will add 1" to the total height of the Slot Plate. Use *Stretch* to accomplish the change, as shown in Figure 9-53. Draw the *Crossing Window* as shown.

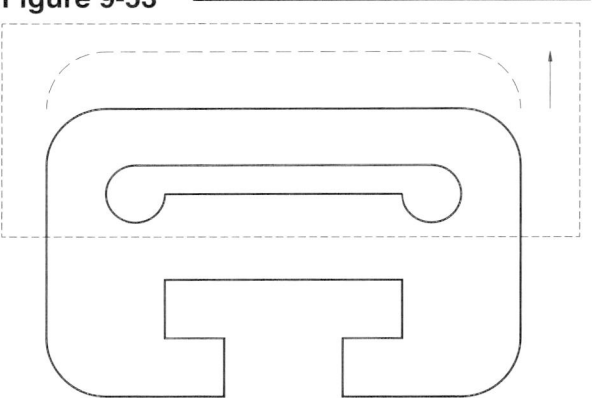

   B. The notch at the bottom of the plate must be adjusted slightly by relocating it .50 units to the right, as shown in Figure 9-54. Draw the *Crossing Window* as shown. Use *SaveAs* to reassign the name to **SLOTPLATE 2**.

   Figure 9-54

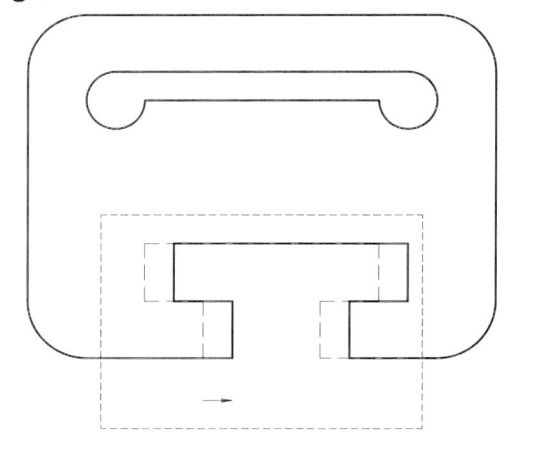

5. *Trim*

   A. Create the shape shown in Figure 9-55A. *SaveAs* **TRIM-EX**.

   B. Use *Trim* to alter the shape as shown in B. *SaveAs* **TRIM1**.

   C. *Open* **TRIM-EX** to create the shapes shown in C and D using *Trim*. *SaveAs* **TRIM2** and **TRIM3**.

6. *Extend*

   Open each of the drawings created as solutions for Figure 9-55 (**TRIM1**, **TRIM2**, and **TRIM3**). Use *Extend* to return each of the drawings to the original form shown in Figure 9-55A. *SaveAs* **EXTEND1**, **EXTEND2**, and **EXTEND3**.

Figure 9-55

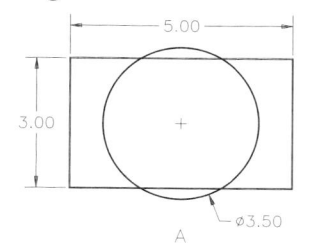

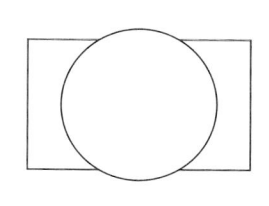

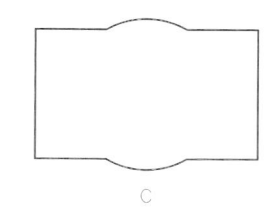

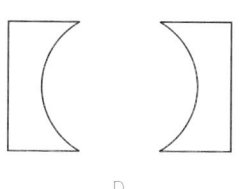

7. *Break*

   A. Create the shape shown in Figure 9-56A. *SaveAs* **BREAK-EX**.

   B. Use *Break* to make the two breaks as shown in B. *SaveAs* **BREAK1**. (HINT: You may have to use *OSNAP*s to create the breaks at the *Intersections* or *Quadrants* as shown.)

   C. Open **BREAK-EX** each time to create the shapes shown in C and D with the *Break* command. *SaveAs* **BREAK2** and **BREAK3**.

Figure 9-56

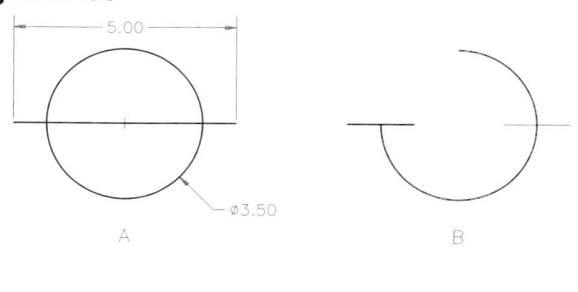

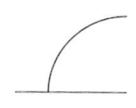

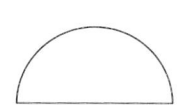

8. **Complete the Table Base**

   A. Open the **TABLE-BASE** drawing you created in the Chapter 7 Exercises. The last step in the previous exercise was the creation of the square at the center of the table base. Now, use *Trim* to edit the legs and the support plate to look like Figure 9-57.

Figure 9-57

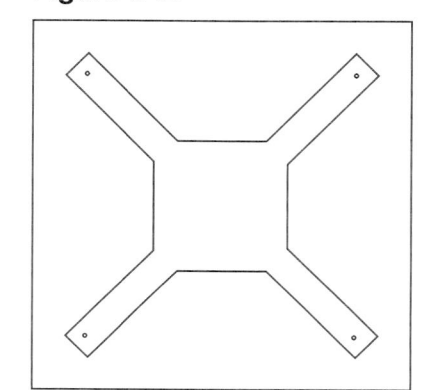

178 Modify Commands I

B. Use *Circle* with the *Center, Radius* option to create a 4" diameter circle to represent the welded support for the center post for this table. Finally, add the other 4 drill holes (.25" diameter) on the legs at the intersection of the vertical and horizontal lines representing the edges of the base. Use **Polar Tracking** or *Intersection OSNAP* for this step. The table base should appear as shown in Figure 9-58. *Save* the drawing.

Figure 9-58

9. *Lengthen*

Five of your friends went to the horse races, each with $5.00 to bet. Construct a simple bar graph (similar to Fig. 9-59) to illustrate how your friends' wealth compared at the beginning of the day.

Modify the graph with *Lengthen* to report the results of their winnings and losses at the end of the day. The reports were as follows: friend 1 made $1.33 while friend 2 lost $2.40; friend 3 reported a 150% increase and friend 4 brought home 75% of the money; friend 5 ended the day with $7.80 and friend 6 came home with $5.60. Use the *Lengthen* command with the appropriate options to change the line lengths accordingly.

Who won the most? Who lost the most? Who was closest to even? Enhance the graph by adding width to the bars and other improvements as you wish (similar to Fig. 9-60). *SaveAs* **FRIENDS GRAPH.**

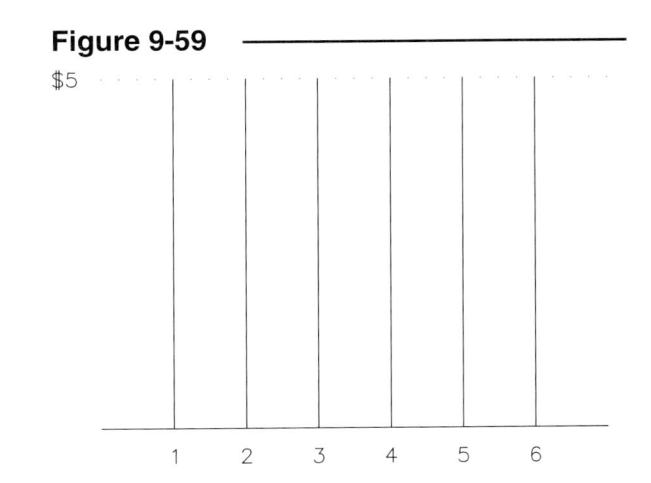

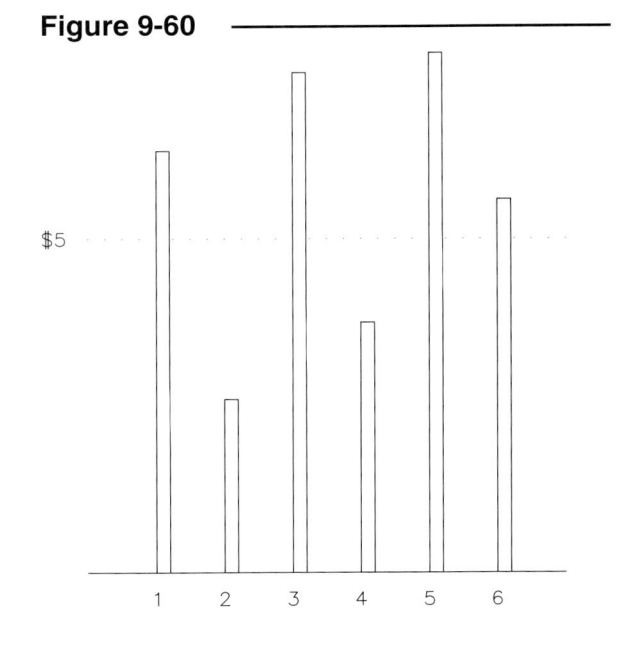

Figure 9-59

Figure 9-60

10. *Trim, Extend*

   A. *Open* the **PIVOTARM CH7** drawing from the Chapter 7 Exercises. Use *Trim* to remove the upper sections of the vertical *Lines* connecting the top and front views. Compare your work to Figure 9-61.

   Figure 9-61

   B. Next, draw a horizontal *Line* in the front view between the *Midpoints* of the vertical *Line* on each end of the view, as shown in Figure 9-62. *Erase* the horizontal *Line* in the top view between the *Circle* centers.

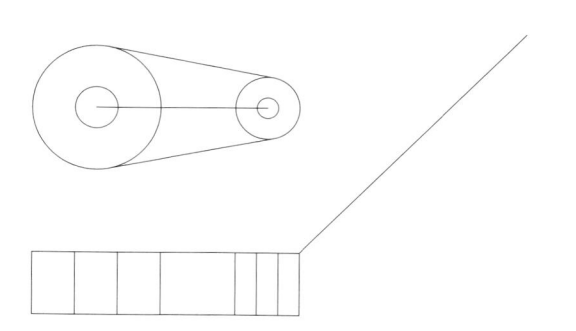

   Figure 9-62

   C. Draw vertical *Lines* from the *Endpoints* of the inclined *Line* (side of the object) in the top view down to the bottom *Line* in the front view, as shown in Figure 9-63. Use the two vertical lines as *Cutting edges* for *Trimming* the *Line* in the middle of the front view, as shown highlighted.

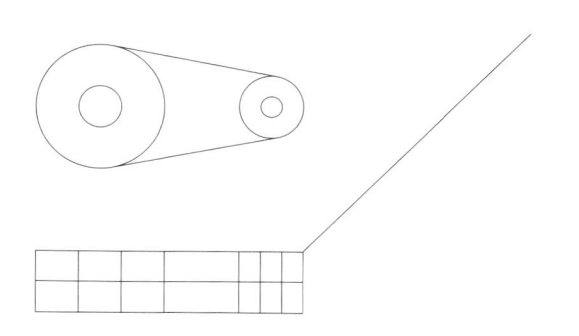

   Figure 9-63

   D. Finally, *Erase* the vertical lines used for *Cutting edges* and then use *Trim* to achieve the object as shown in Figure 9-64. Use *SaveAs* and name the drawing **PIVOTARM CH9**.

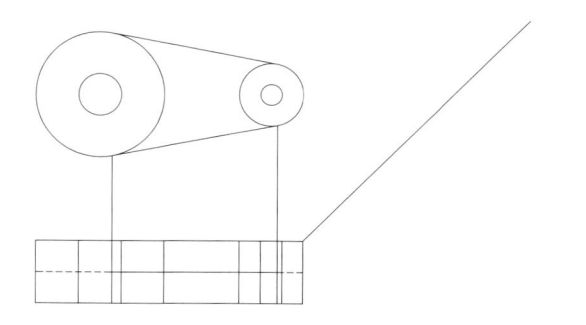

   Figure 9-64

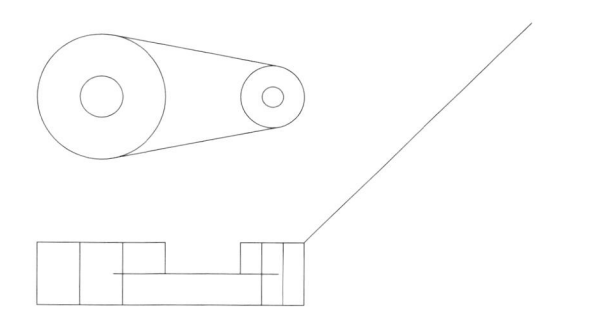

180   Modify Commands I

11. **GASKETA**

    Begin a *New* drawing. Set *Limits* to **0,0** and **8,6**. Set *SNAP* and *GRID* values appropriately. Create the Gasket as shown in Figure 9-65. Construct only the gasket shape, not the dimensions or centerlines. (HINT: Locate and draw the four .5" diameter *Circles*, then create the concentric .5" radius arcs as full *Circles*, then *Trim*. Make use of *OSNAPs* and *Trim* whenever applicable.) *Save* the drawing and assign the name **GASKETA**.

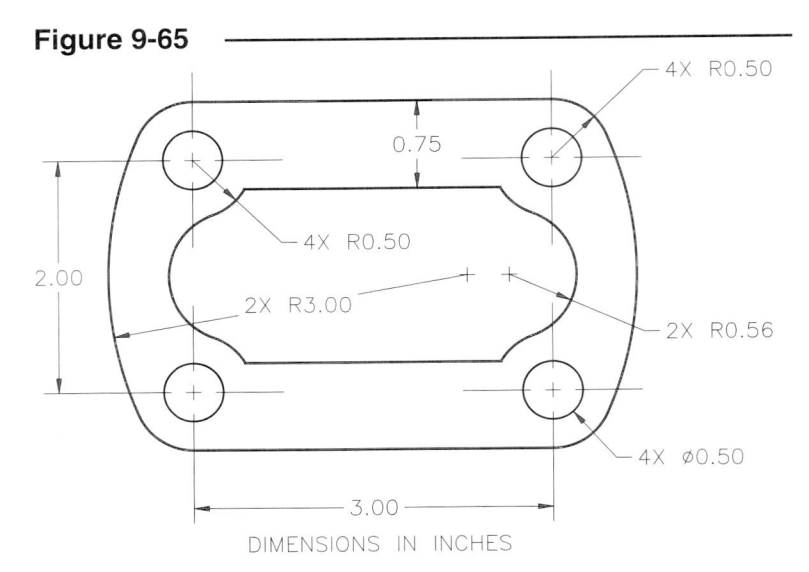

    Figure 9-65

12. **Chemical Process Flow Diagram**

    Begin a *New* drawing and re-create the chemical process flow diagram shown in Figure 9-66. Use *Line, Circle, Arc, Trim, Extend, Scale, Break,* and other commands you feel necessary to complete the diagram. Because this is diagrammatic and will not be used for manufacturing, dimensions are not critical, but try to construct the shapes with proportional accuracy. *Save* the drawing as **FLOWDIAG**.

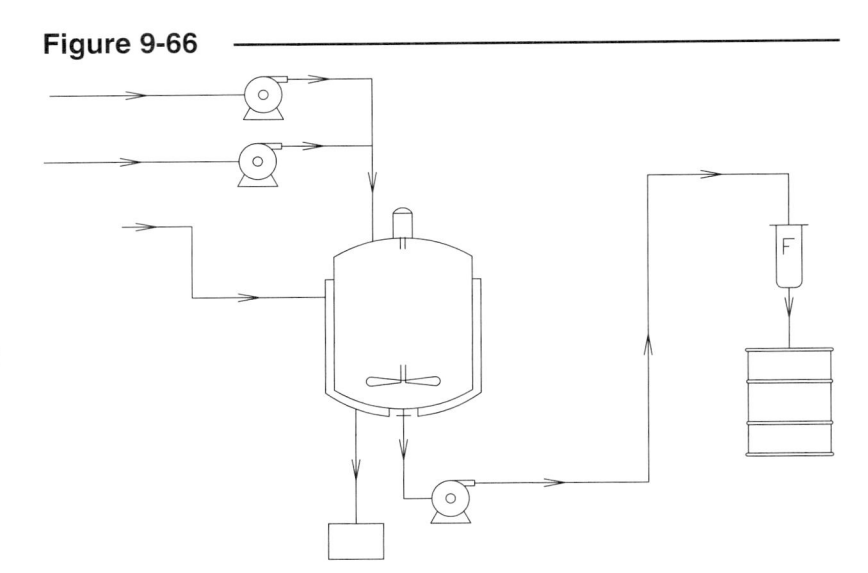

    Figure 9-66

13. *Copy*

    Begin a *New* drawing. Set the *Limits* to **24,18**. *Set Polar Snap* to **.5** and set your desired running *OSNAPs*. Turn On *SNAP, GRID, OSNAP,* and *POLAR*. Create the sheet metal part composed of four *Lines* and one **Circle** as shown in Figure 9-67. The lower-left corner of the part is at coordinate 2,3.

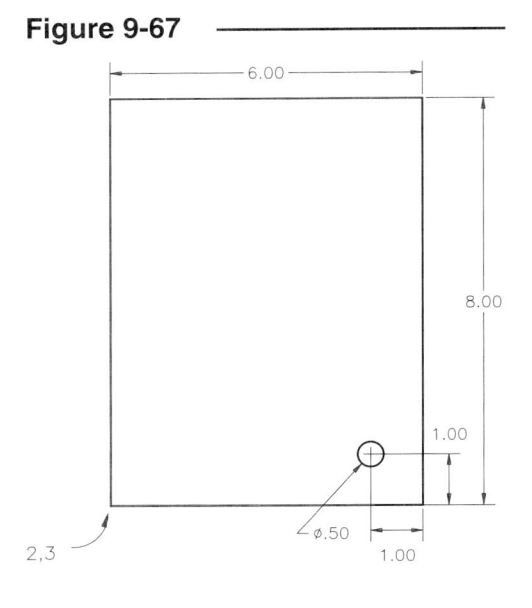

    Figure 9-67

Use *Copy* to create two copies of the rectangle and hole in a side-by-side fashion as shown in Figure 9-68. Allow 2 units between the sheet metal layouts. *SaveAs* **PLATES**.

Figure 9-68

A. Use *Copy* to create the additional 3 holes equally spaced near the other corners of the part as shown in Figure 9-68A. *Save* the drawing.

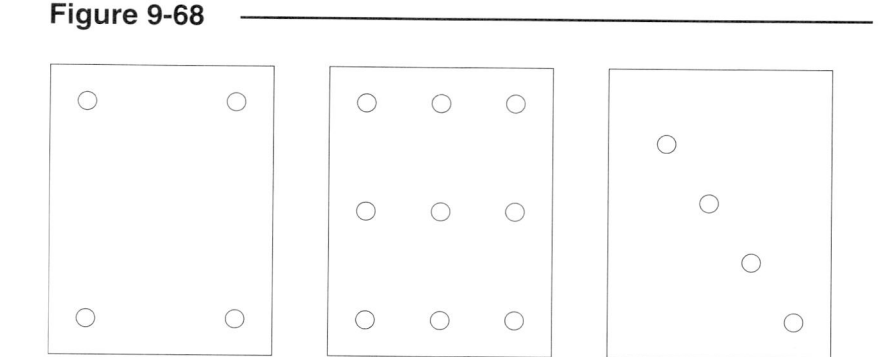

B. Use *Copy Multiple* to create the hole configuration as shown in Figure 9-68B. *Save*.

C. Use *Copy* to create the hole placements as shown in C. Each hole <u>center</u> is **2** units at **125** degrees from the previous one (use relative polar coordinates or set an appropriate *Polar Angle*). *Save* the drawing.

14. *Mirror*

A manufacturing cell is displayed in Figure 9-69. The view is from above, showing a robot <u>centered</u> in a work station. The production line requires 4 cells. Begin by starting a *New* drawing, setting *Units* to *Engineering* and *Limits* to 40' x 30'. It may be helpful to set *Polar Snap* to 6". Draw one cell to the dimensions indicated. Begin at the indicated coordinates of the lower-left corner of the cell. *SaveAs* **MANFCELL**.

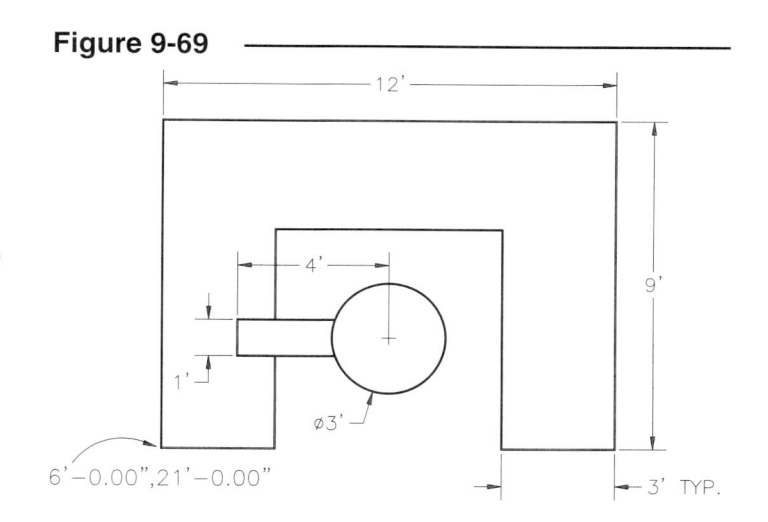

Figure 9-69

Use *Mirror* to create the other three manufacturing cells as shown in Figure 9-70. Ensure that there is sufficient space between the cells as indicated. Draw the two horizontal *Lines* representing the walkway as shown. *Save*.

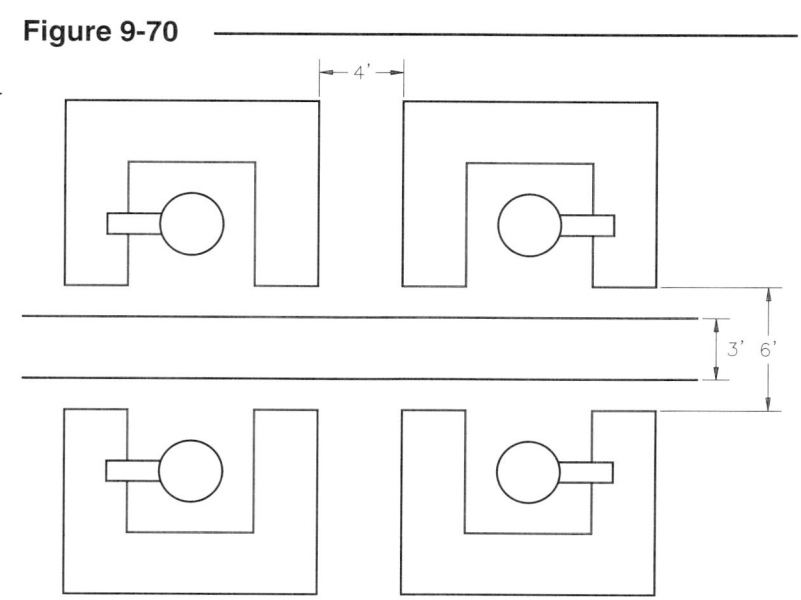

Figure 9-70

182 Modify Commands I

15. *Offset*

    Create the electrical schematic as shown in Figure 9-71. Draw *Lines* and *Plines*, then *Offset* as needed. Use *Point* objects with an appropriate (circular) *Point Style* at each of the connections as shown. Check the *Set Size to Absolute Units* radio button (in the *Point Style* dialog box) and find an appropriate size for your drawing. Because this is a schematic, you can create the symbols by approximating the dimensions. Omit the text. *Save* the drawing as **SCHEMATIC1**. Create a *Plot* and check *Scaled to Fit*.

    Figure 9-71

16. *Array, Polar*

    Begin a *New* drawing. Select *Start from Scratch*, *English* defaults. Create the starting geometry for a Flange Plate as shown in Figure 9-72. *SaveAs* **ARRAY**.

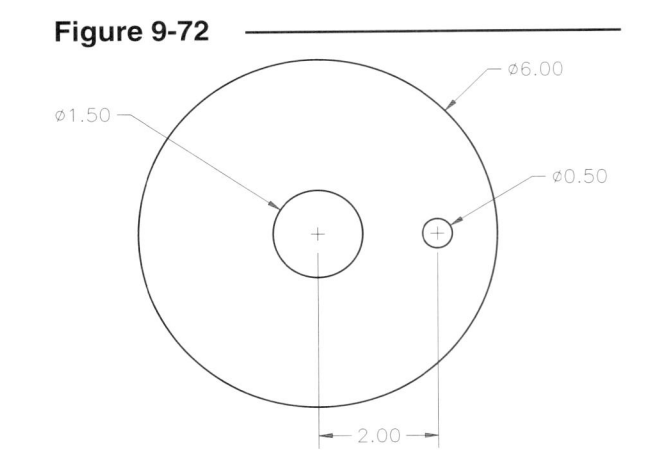

    Figure 9-72

    A. Create the *Polar Array* as shown in Figure 9-73A. *SaveAs* **ARRAY1**.

    B. *Open* **ARRAY**. Create the *Polar Array* as shown in Figure 9-73B. *SaveAs* **ARRAY2**. (HINT: Use a negative angle to generate the *Array* in a clockwise direction.)

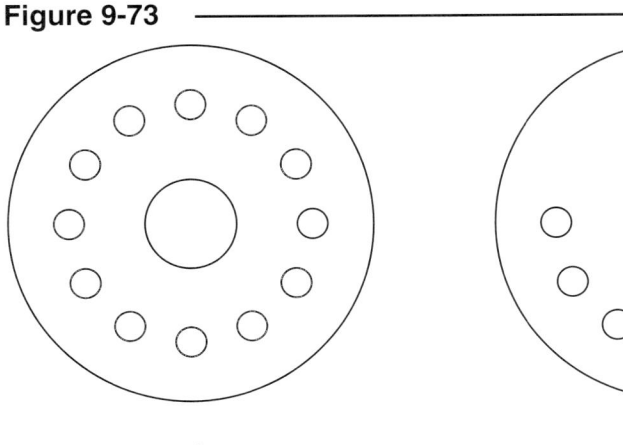

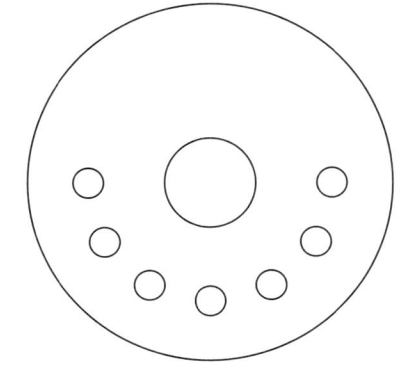

    Figure 9-73

17. *Array, Rectangular*

Begin a *New* drawing. Select *Start from Scratch*, *English* defaults. Use *Save* and assign the name **LIBRARY DESKS**. Create the *Array* of study carrels (desks) for the library as shown in Figure 9-74. The room size is 36' x 27' (set *Units* and *Limits* accordingly). Each carrel is **30" x 42"**. Design your own chair. Draw the first carrel (highlighted) at the indicated coordinates. Create the *Rectangular Array* so that the carrels touch side to side and allow a 6' aisle for walking between carrels (not including chairs).

Figure 9-74

18. *Array, Rectangular*

Create the bolt head with one thread as shown in Figure 9-75A. Create the small line segment at the end (0.40 in length) as a *Pline* with a *Width* of **.02**. *Array* the first thread (both crest and root lines, as indicated in B) to create the schematic thread representation. There is **1** row and **10** columns with **.2** units between each. Add the *Lines* around the outside to complete the fastener. *Erase* the *Pline* at the small end of the fastener (Fig. 9-75B) and replace it with a *Line*. *Save* as **BOLT**.

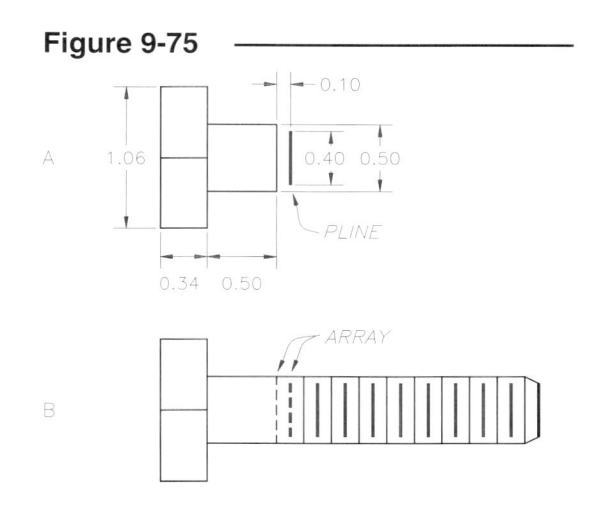

Figure 9-75

19. *Fillet*

Create the "T" Plate shown in Figure 9-76. Use *Fillet* to create all the fillets and rounds as the last step. When finished, *Save* the drawing as **T-PLATE** and make a plot.

Figure 9-76

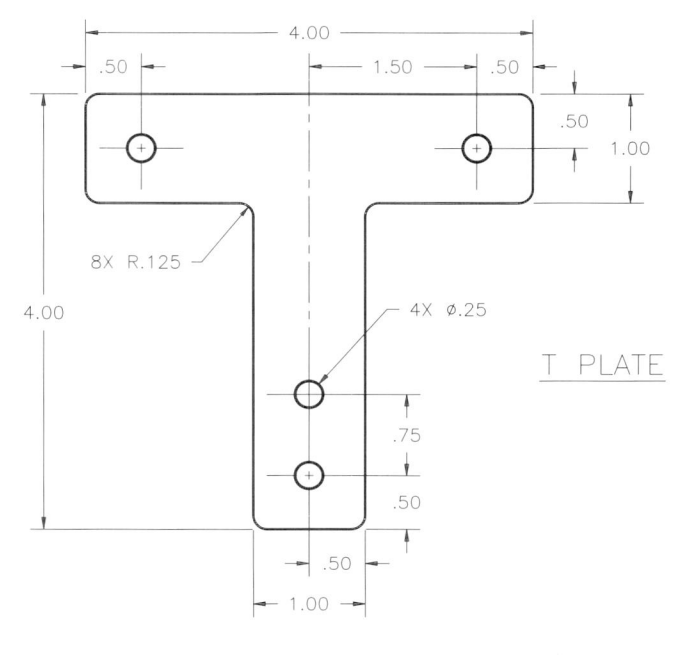

184 Modify Commands I

20. *Copy, Fillet*

    Create the Gasket shown in Figure 9-77. Use the *Start from Scratch, English* settings when you begin. Include center-lines in your drawing. *Save* the drawing as **GASKETB**.

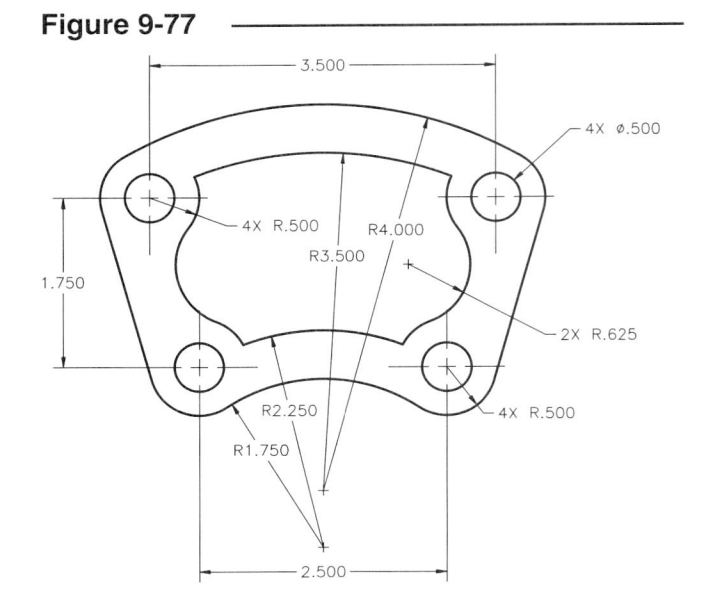

    Figure 9-77

21. *Array*

    Create the perforated plate as shown Figure 9-78. *Save* the drawing as **PERF-PLAT**. (HINT: For the *Rectangular Array* in the center, create 100 holes, then *Erase* four holes, one at each corner for a total of 96. The two circular hole patterns contain a total of 40 holes each.)

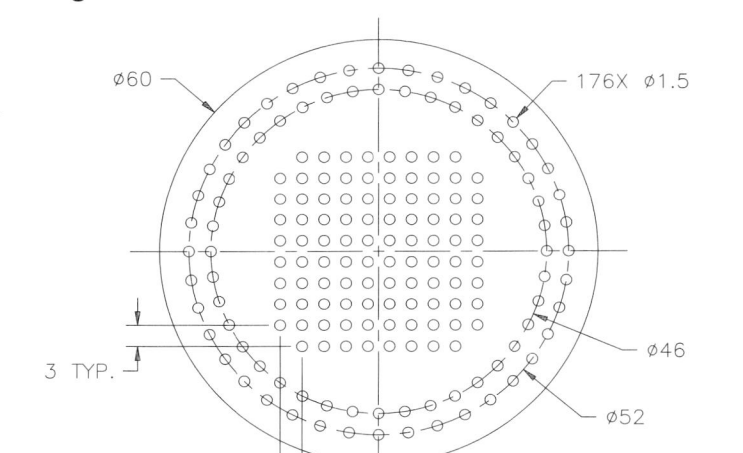

    Figure 9-78

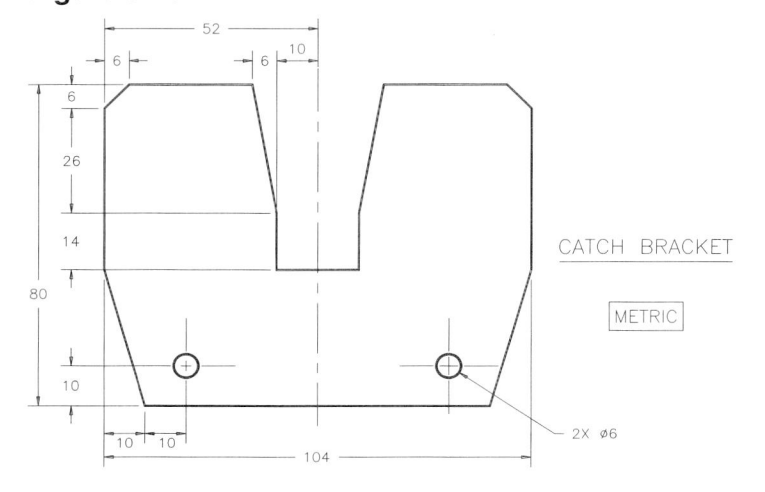

22. *Chamfer*

    Begin a *New* drawing and select *Start from Scratch, English* defaults. Set the *Limits* to **279,216**. Next, type *Zoom* and use the *All* option. Set *Snap* to **2** and *Grid* to **10**. Create the Catch Bracket shown in Figure 9-79. Draw the shape with all vertical and horizontal *Lines*, then use *Chamfer* to create the six chamfers. *Save* the drawing as **CBRACKET** and create a plot.

    Figure 9-79

10

VIEWING COMMANDS

Chapter Objectives

After completing this chapter you should:

1. understand the relationship between the drawing objects and the display of those objects;

2. be able to use all of the *Zoom* options to view areas of a drawing;

3. be able to *Pan* the display about your screen;

4. be able to save and restore *Views*;

5. know how to use *Viewres* to change the display resolution for curved shapes;

6. be able to turn on and off the *UCS Icon*;

7. be able to create, save, and restore model space viewports using the *Vports* command.

CONCEPTS

The accepted CAD practice is to draw full size using actual units. Since the drawing is a virtual dimensional replica of the actual object, a drawing could represent a vast area (several hundred feet or even miles) or a small area (only millimeters). The drawing is created full size with the actual units, but it can be displayed at any size on the screen. Consider also that CAD systems provide for a very high degree of dimensional precision, which permits the generation of drawings with great detail and accuracy.

Combining those two CAD capabilities (great precision and drawings representing various areas), a method is needed to view different and detailed segments of the overall drawing area. In AutoCAD the commands that facilitate viewing different areas of a drawing are *Zoom*, *Pan*, and *View*.

The viewing commands are found in the *View* pull-down menu (Fig. 10-1).

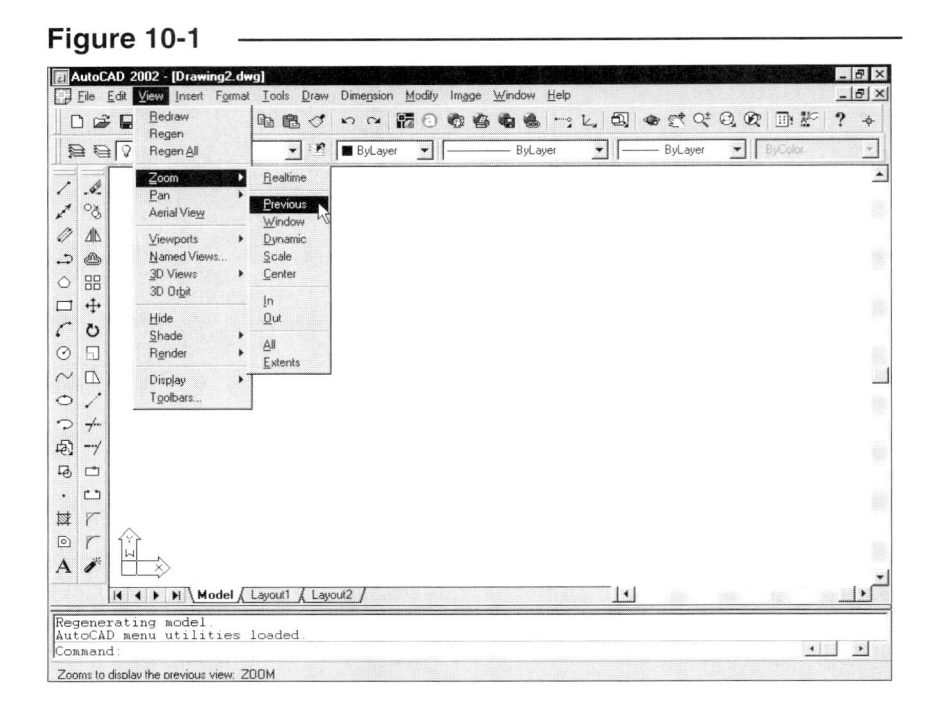

Figure 10-1

The Standard toolbar contains a group of tools (icon buttons) for the viewing commands located near the right end of the toolbar (Fig. 10-2). The *Realtime* options of *Pan* and *Zoom*, and *Zoom Previous* have icons permanently displayed on the toolbar, whereas the other *Zoom* options are located on flyouts.

Figure 10-2

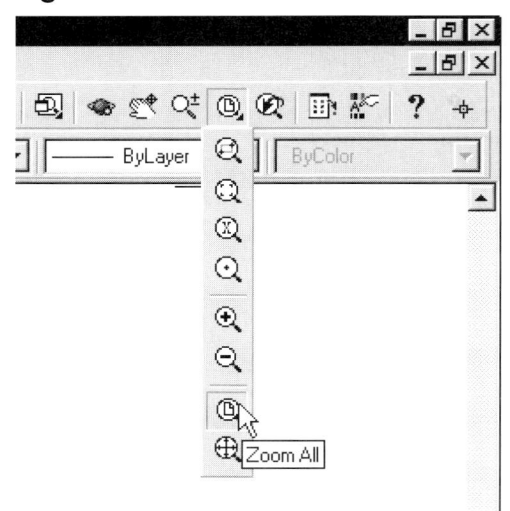

Like other commands, the viewing commands can also be invoked by typing the command or command alias, using the screen menu (*VIEW1*), or using the digitizing tablet menu (if available).

The commands discussed in this chapter are:

Zoom, Pan, View, Viewres, Ucsicon, and *Vports*

The *Realtime* options are the default options for both the *Pan* and *Zoom* commands. *Realtime* options of *Pan* and *Zoom* allow you to <u>interactively</u> change the drawing display by moving the cursor in the drawing area.

ZOOM and PAN with the Mouse Wheel

Zoom and Pan

To zoom in means to magnify a small segment of a drawing and to zoom out means to display a larger area of the drawing. Zooming does not change the size of the drawing objects; zooming changes only display of those objects—the area that is displayed on the screen. All objects in the drawing have the same dimensions before and after zooming. Only your display of the objects changes.

To pan means to move the display area slightly without changing the size of the current view window. Using the pan function, you "drag" the drawing across the screen to display an area outside of the current view in the Drawing Editor.

Although there are many ways to change the area of the drawing you want to view using either the *Zoom* or *Pan* commands, the fastest and easiest method for simple zoom and pan operations is to use the mouse wheel (the small wheel between the two mouse buttons) if you have one. Using the mouse wheel to zoom or pan does not require you to invoke the *Zoom* or *Pan* commands. Additionally, the mouse wheel zoom and pan functions are transparent, meaning that you can use this method while another command is in use. So, if you have a mouse wheel, you can zoom and pan at any time without using any commands.

Figure 10-3

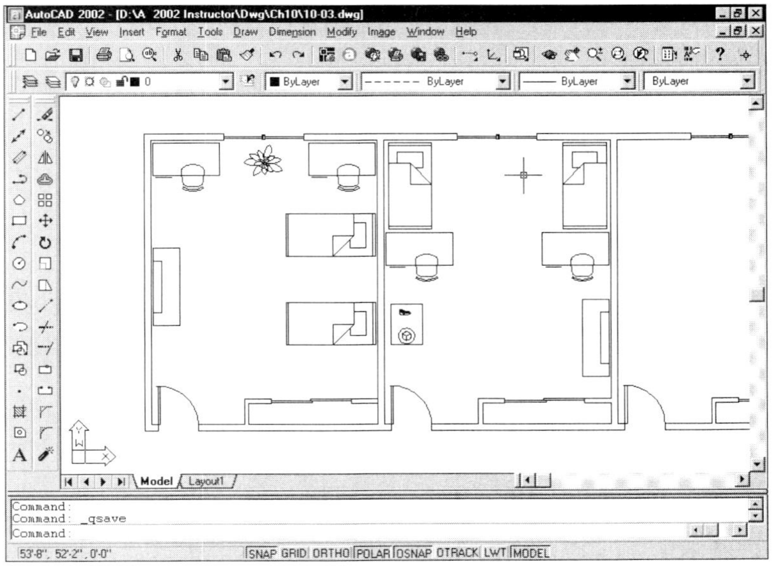

Zoom with the Mouse Wheel

If you have a mouse wheel, you can zoom by simply turning the wheel. Turn the wheel forward to zoom in and turn the wheel backward to zoom out. The current location of the cursor is the center for zooming. In other words, if you want to zoom in to an area, simply locate your cursor on the spot and turn the wheel forward. This type of zooming can be done transparently (during another command operation). You can also pan using the wheel (see "*Pan* with the Mouse Wheel or Third Button").

Figure 10-4

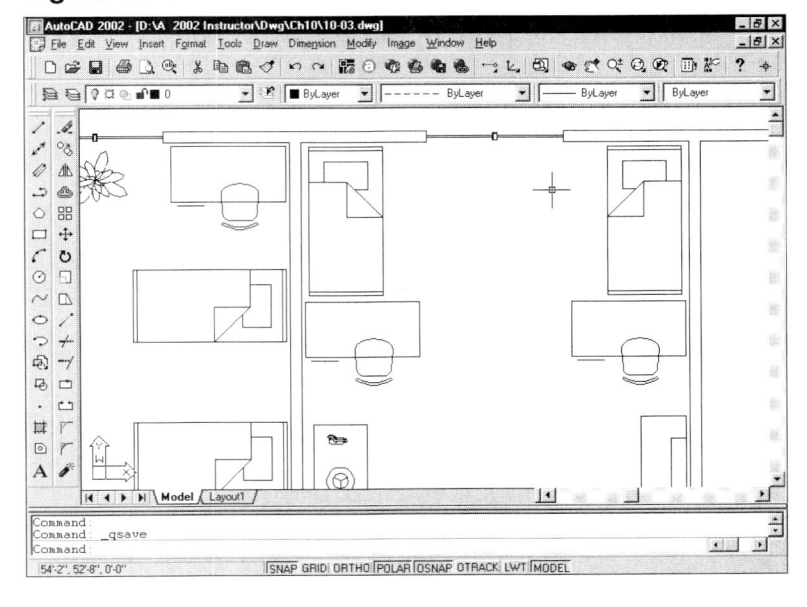

For example, Figure 10-3 displays a floor plan of a dorm room. Notice the location of the cursor in the upper-right corner of the drawing area. By turning the mouse wheel forward, the display changes by zooming in (enlarging) to the area designated by the cursor location. The resulting display (after turning the mouse wheel forward) is shown in Figure 10-4.

188 Viewing Commands

Using the same method, you could zoom out, or change the display from Figure 10-4 to Figure 10-3, by turning the wheel backward. When zooming out, the cursor location also controls the center for zooming.

The amount of zooming in or out that occurs with each turn of the mouse wheel is controlled by the *ZOOMFACTOR* system variable. The default value is 40. Increasing this value causes a greater degree of zooming with each increment of the mouse wheel's forward or backward movement; whereas, decreasing the value results in smaller changes in zooming with the wheel movement. The setting for *ZOOMFACTOR* can be between 3 and 100 and is stored with the individual computer (in the registry).

Pan with the Mouse Wheel or Third Button

If you have a mouse wheel or third mouse button, you can pan by holding down the wheel or button so the "hand" appears (Fig. 10-5), then move the hand cursor in any direction to "drag" the drawing around on the screen. Panning is typically used when you have previously zoomed in to an area but then want to view a different area that is slightly out of the current viewing area (off the screen). Panning with the mouse wheel or third button, similar to zooming with the mouse wheel, does not require you to invoke any commands. In addition, the pan feature is transparent, so you can pan with the wheel or third button while another command is in operation. Using the mouse wheel or third button to pan is essentially the same as using the *Pan* command with the *Realtime* option, except you do not have to invoke the *Pan* command.

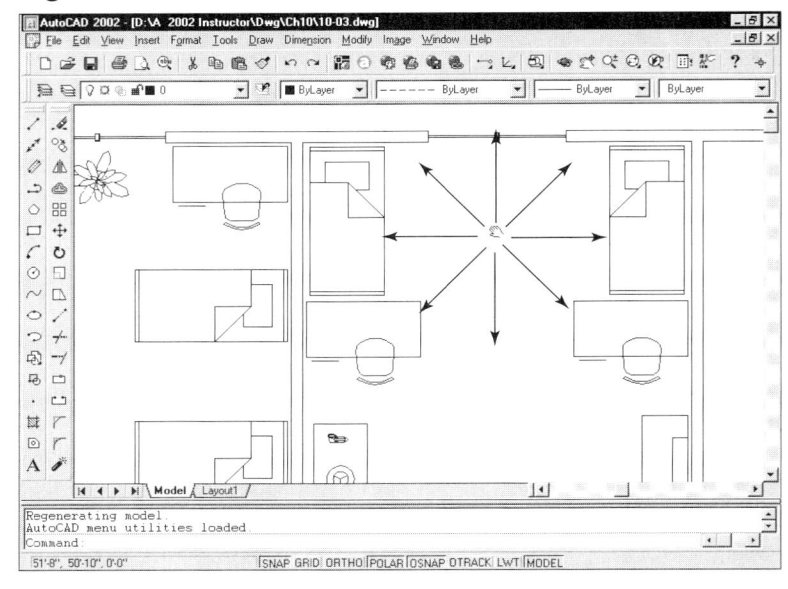

Figure 10-5

Your ability to pan with the wheel or third button is based on the *MBUTTONPAN* system variable setting. If *MBUTTONPAN* is set to 1, then pressing the wheel or third button activates the *Realtime Pan* feature. If *MBUTTONPAN* is set to 0, pressing the wheel or button triggers the action defined in the ACAD.MNU file, normally set to activate the *Osnap* shortcut menu. The *MBUTTONPAN* setting is saved on the individual computer (in the registry).

COMMANDS

ZOOM

Pull-down Menu	COMMAND (TYPE)	ALIAS (TYPE)	Short-cut	Screen (side) Menu	Tablet Menu
View Zoom >	ZOOM	Z	(Default Menu) Zoom	VIEW 1 Zoom	J2 - J,5 or K,3 - K,5

The *Zoom* options described next can be typed or selected by any method. Unlike most commands, AutoCAD provides a tool (icon button) for each option of the *Zoom* command. If you type *Zoom*, the options can be invoked by typing the first letter of the desired option or pressing Enter for the *Realtime* option.

The Command prompt appears as follows.

Command: *zoom*
Specify corner of window, enter a scale factor (nX or nXP), or
[All/Center/Dynamic/Extents/Previous/Scale/Window] <real time>:

Realtime (RTZOOM)
Realtime is the default option of *Zoom*. If you type *Zoom*, just press Enter to activate the *Realtime* option. Alternately, you can type *Rtzoom* to invoke this option directly.

Figure 10-6

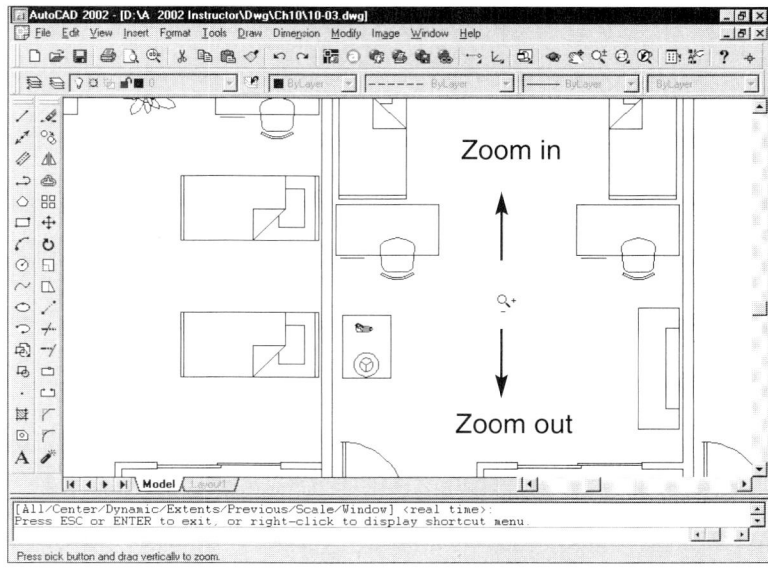

With the *Realtime* option you can interactively zoom in and out with vertical cursor motion. When you activate the *Realtime* option, the cursor changes to a magnifying glass with a plus (+) and a minus (-) symbol. Move the cursor to any location on the drawing area and hold down the PICK (left) mouse button. Move the cursor up to zoom in and down to zoom out (Fig. 10-6). Horizontal movement has no effect. You can zoom in or out repetitively. Pressing Esc or Enter exits *Realtime Zoom* and returns to the Command prompt.

The current drawing window is used to determine the zooming factor. Moving the cursor half the window height (from the center to the top or bottom) zooms in or out to a zoom factor of 100%. Starting at the bottom or top allows zooming in or out (respectively) at a factor of 200%.

If you have zoomed in or out repeatedly, you can reach a zoom limit. In this case the plus (+) or minus (-) symbol disappears when you press the left mouse button. To zoom further with *Rtzoom*, first type *Regen*.

Pressing the right mouse button (or button #2 on digitizing pucks) produces a small cursor menu with other viewing options (Fig. 10-7). This same menu and its options can also be displayed by right-clicking during *Realtime Pan*. The options of the cursor menu are described below.

Figure 10-7—

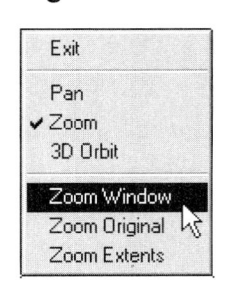

Exit
Select *Exit* to exit *Realtime Pan* or *Zoom* and return to the Command prompt. The Escape or Enter key can be used to accomplish the same action.

Pan
This selection switches to the *Realtime* option of *Pan*. A check mark appears here if you are currently using *Realtime Pan*. See "*Pan*" next in this chapter.

Zoom
A check mark appears here if you are currently using *Realtime Zoom*. If you are using the *Realtime* option of *Pan*, check this option to switch to *Realtime Zoom*.

190 Viewing Commands

Zoom Window
Select this option if you want to display an area to zoom in to by specifying a *Window*. This feature operates differently but accomplishes the same action as the *Window* option of *Zoom*. With this option, window selection is made with one PICK. In other words, PICK for the first corner, hold down the left button and drag, then release to establish the other corner. With the *Window* option of the *Zoom* command, PICK once for each of two corners. See "*Window.*"

Zoom Original
Selecting this option automatically displays the area of the drawing that appeared on the screen immediately before using the *Realtime* option of *Zoom* or *Pan*. Using this option successively has no effect on the display. This feature is different from the *Previous* option of the *Zoom* command, in which ten successive previous views can be displayed.

Zoom Extents
Use this option to display the entire drawing in its largest possible form in the drawing area. This option is identical to the *Extents* option of the *Zoom* command. See "*Extents.*"

Window

To *Zoom* with a *Window* is to draw a rectangular window around the desired viewing area. You PICK a first and a second corner (diagonally) to form the rectangle. The windowed area is magnified to fill the screen (Fig. 10-8). It is suggested that you draw the window with a 4 x 3 (approximate) proportion to match the screen proportion. If you type *Zoom* or *Z*, *Window* is an automatic option so you can begin selecting the first corner of the window after issuing the *Zoom* command without indicating the *Window* option as a separate step.

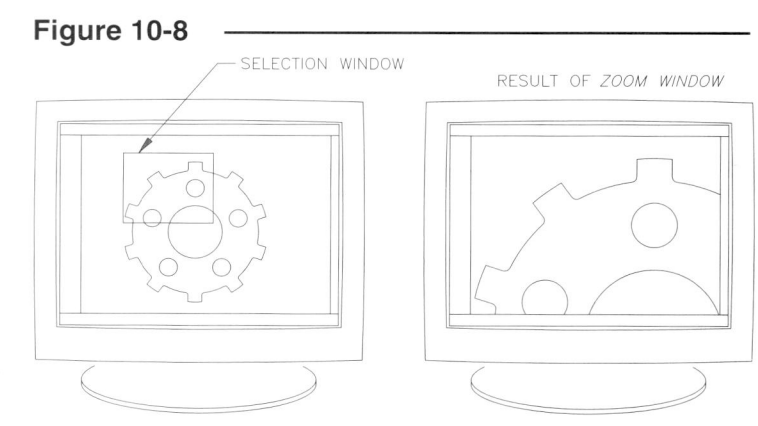

Figure 10-8

All

This option displays all of the objects in the drawing and all of the *Limits*. In Figure 10-9, notice the effects of *Zoom All*, based on the drawing objects and the drawing *Limits*.

Extents

This option results in the largest possible display of all of the objects, disregarding the *Limits*. (*Zoom All* includes the *Limits*.) See Figure 10-9.

Invoking *Zoom Extents* causes AutoCAD to execute two steps: (1) a true *Zoom Extents* is performed which causes the geometry to be maximized so it is "pressed" against the edge of

Figure 10-9

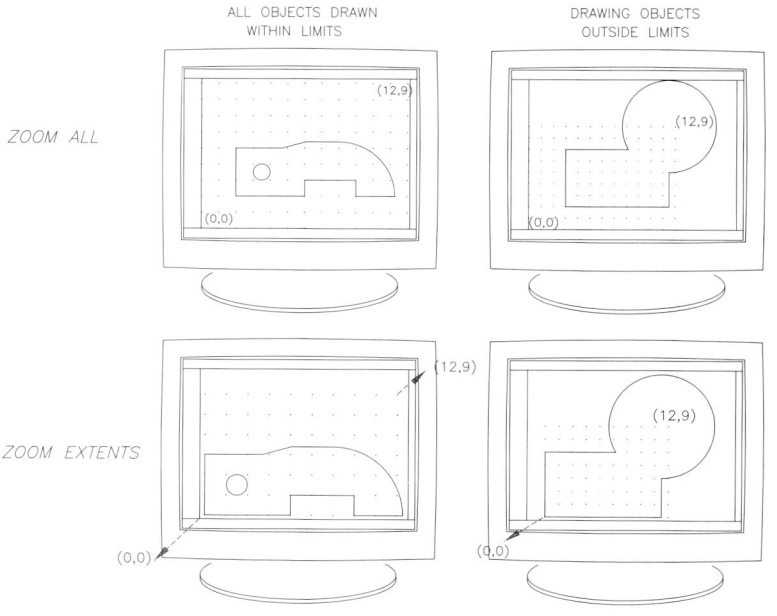

the drawing area window, and (2) a *Zoom Scale .95X* is performed which causes the geometry to be scaled down slightly to fit just within the window (see the *Scale* option).

Scale (X/XP)

This option allows you to enter a scale factor for the desired display. The value that is entered can be relative to the full view (*Limits*) or to the current display (Fig. 10-10). A value of **1**, **2**, or **.5** causes a display that is 1, 2, or .5 times the size of the *Limits*, centered on the current display. A value of **1X**, **2X**, or **.5X** yields a display 1, 2, or .5 times the size of the current display. If you are using paper space and viewing model space, a value of **1XP**, **2XP**, or **.5XP** yields a model space display scaled 1, 2, or .5 times paper space units.

If you type *Zoom* or *Z*, you can enter a scale factor at the *Zoom* command prompt without having to type the letter *S*.

Figure 10-10

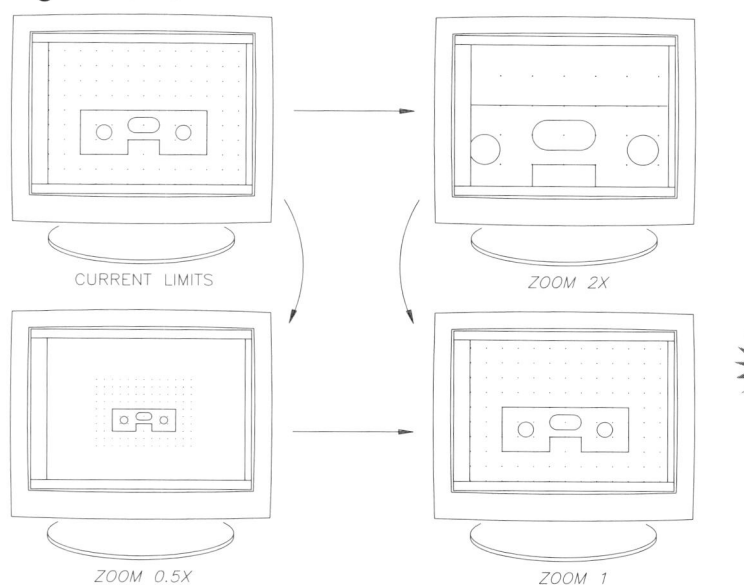

In

Zoom In magnifies the current display by a factor of 2X (2 times the current display). Using this option is the same as entering a *Zoom Scale* of 2X.

Out

Zoom Out makes the current display smaller by a factor of .5X (.5 times the current display). Using this option is the same as entering a *Zoom Scale* of .5X.

Center

First, specify a location as the center of the zoomed area; then specify either a *Magnification* factor (see *Scale X/XP*), a *Height* value for the resulting display, or PICK two points forming a vertical to indicate the height for the resulting display (Fig. 10-11).

Figure 10-11

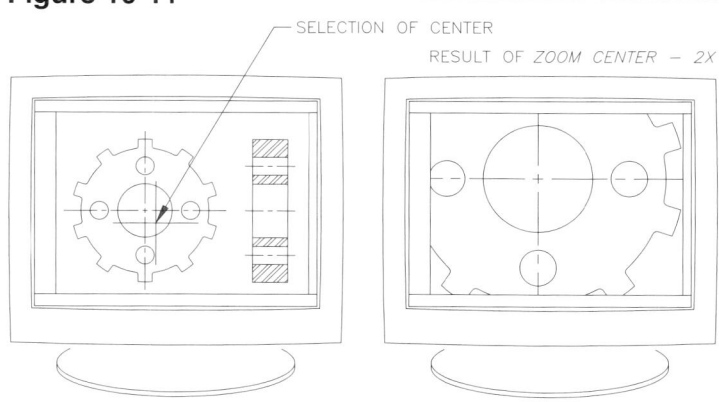

Previous

Selecting this option automatically changes to the previous display. AutoCAD saves the previous ten displays changed by *Zoom*, *Pan*, and *View*. You can successively change back through the previous ten displays with this option.

Dynamic

With this option you can change the display from one windowed area in a drawing to another without using *Zoom All* (to see the entire drawing) as an intermediate step. *Zoom Dynamic* causes the screen to display the drawing *Extents* bounded by a box. The current view or window is bounded by a box in a broken-line pattern. The view box is the box with an "X" in the center which can be moved to the desired location (Fig. 10-12). The desired location is selected by pressing **Enter** (not the PICK button as you might expect).

Figure 10-12

Pressing the left button allows you to resize the view box (displaying an arrow instead of the "X") to the desired size. Move the mouse or puck left and right to increase and decrease the window size. Press the left button again to set the size and make the "X" reappear.

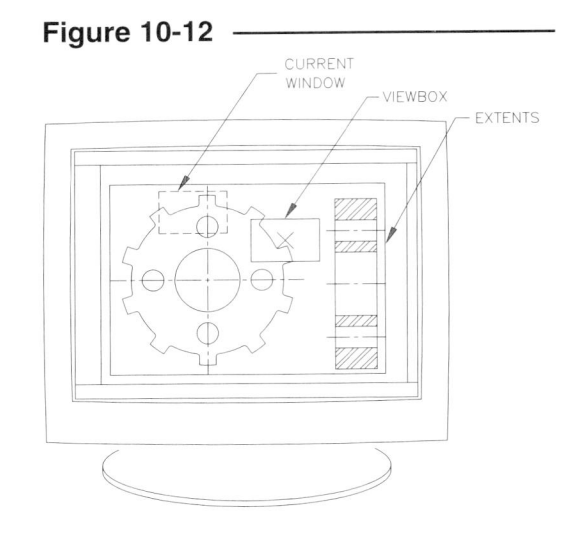

When changing the display of a drawing from one windowed area to another, *Zoom Dynamic* can be a fast method to use because (1) a *Zoom All* is not required as an intermediate step and (2) the view box can be moved <u>while</u> AutoCAD is "redrawing" the display.

Zoom Is Transparent

Zoom is a <u>transparent</u> command, meaning it can be invoked while another command is in operation. You can, for example, begin the *Line* command and at the "Specify next point or [Undo]:" prompt '*Zoom* with a *Window* to better display the area for selecting the endpoint. This transparent feature is automatically entered if *Zoom* is invoked by the screen, pull-down, or tablet menus or toolbar icons, but if typed it must be prefixed by the apostrophe (') symbol, e.g., "Specify next point or [Undo]: '*Zoom*." If *Zoom* has been invoked transparently, the >> symbols appear at the command prompt before the listed options as follows:

```
Command: line
Specify first point: 'zoom
>>Specify corner of window, enter a scale factor (nX or nXP), or
[All/Center/Dynamic/Extents/Previous/Scale/Window] <real time>:
```

PAN

Pull-down Menu	COMMAND (TYPE)	ALIAS (TYPE)	Short-cut	Screen (side) Menu	Tablet Menu
View Pan	PAN or -PAN	P or -P	(Default Menu) Pan	VIEW 1 Pan	N,10-P,11

Using the *Pan* command by typing or selecting from the tool, screen, or digitizing tablet menu produces the *Realtime* version of *Pan*. The *Point* option is available from the pull-down menu or by typing -*Pan*. The other "automatic" *Pan* options (*Left, Right, Up, Down*) can only be selected from the pull-down menu.

The *Pan* command is useful if you want to move (pan) the display area slightly without changing the <u>size</u> of the current view window. With *Pan*, you "drag" the drawing across the screen to display an area outside of the current view in the drawing area.

Realtime (RTPAN)

Realtime is the default option of *Pan*. It is invoked by selecting *Pan* by any method, including typing *Pan* or *Rtpan*. *Realtime Pan* is essentially the same as using the mouse wheel or third button to pan. That is, *Realtime Pan* allows you to interactively pan by "pulling" the drawing across the screen with the cursor motion. After activating the command, the cursor changes to a hand cursor. Move the hand to any location on the drawing, then hold the PICK (left) mouse button down and drag the drawing around on the screen to achieve the desired view (see Figure 10-5). When you release the mouse button, panning is discontinued. You can move the hand to another location and pan again without exiting the command.

You must press Escape or Enter or use *Exit* from the pop-up cursor menu to exit *Realtime Pan* and return to the Command prompt. The following Command prompt appears during *Realtime Pan*.

Command: **pan**
Press Esc or Enter to exit, or right-click to display shortcut menu.

If you press the right mouse button (#2 button), a small cursor menu pops up. This is the same menu that appears during *Realtime Zoom* (see Figure 10-7). This menu provides options for you to *Exit, Zoom* or *Pan* (realtime), *Zoom Window, Zoom Original,* or *Zoom Extents.* See "*Zoom*" earlier in this chapter.

Point

The *Point* option is available only through the pull-down menu or by typing "*-Pan*" (some commands can be typed with a hyphen [-] prefix to invoke the command line version of the command without dialog boxes, etc.). Using this option produces the following Command prompt.

Command: **-pan**
Specify base point or displacement: **PICK** or (**value**)
Specify second point: **PICK** or (**value**)
Command:

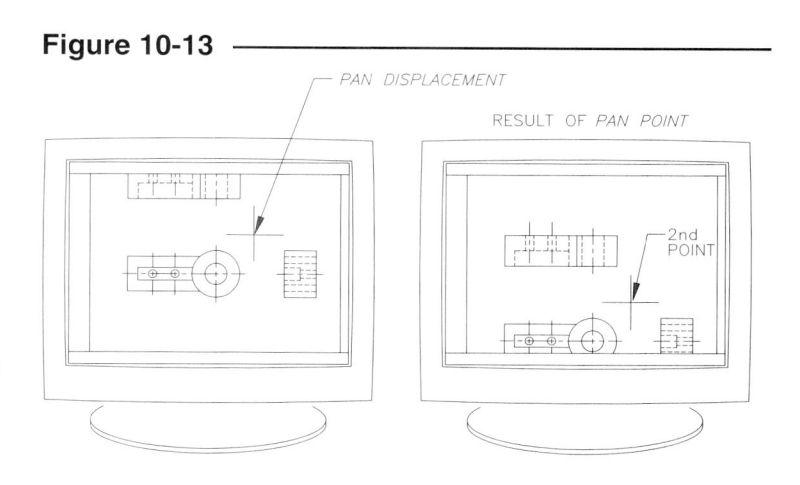

Figure 10-13

The "base point:" can be thought of as a "handle," or point to move from, and the "second point:" as the new location to move to (Fig. 10-13). You can PICK each of these points with the cursor.

You can also enter coordinate values rather than interactively PICKing points with the cursor. The command syntax is as follows:

Command: **-pan**
Specify base point or displacement: **0,-2**
Specify second point: **Enter**
Command:

Entering coordinate values allows you to *Pan* to a location outside of the current display. If you use the interactive method (PICK points), you can *Pan* only within the range of whatever is visible on the screen.

The following *Pan* options are available from the *View* pull-down menu only.

Left
Automatically pans to the left, equal to about 1/8 of the current drawing area width.

Right
Pans to the right—similar to but opposite of *Left*.

Up
Automatically pans up, equal to about 1/8 of the current drawing area height.

Down
Pans down—similar to but opposite of *Up*.

Pan **Is Transparent**
Pan, like *Zoom*, can be used as a transparent command. The transparent feature is automatically entered if *Pan* is invoked by the screen, pull-down, or tablet menus or icons. However, if you type *Pan* during another command operation, it must be prefixed by the apostrophe (') symbol.

```
Command: line
Specify first point: 'pan
>>Press ESC or ENTER to exit, or right-click to display shortcut menu. (Pan, then) Enter
Resuming LINE command.
Specify next point or [Undo]:
```

Scroll Bars

Figure 10-14

You can also use the horizontal and vertical scroll bars directly below and to the right of the graphics area to pan the drawing (Fig. 10-14). These scroll bars pan the drawing in one of three ways: (1) click on the arrows at the ends of the scroll bar, (2) move the thumb wheel, or (3) click inside the scroll bar. The scroll bars can be turned off (removed from the screen) by using the *Display* tab in the *Options* dialog box.

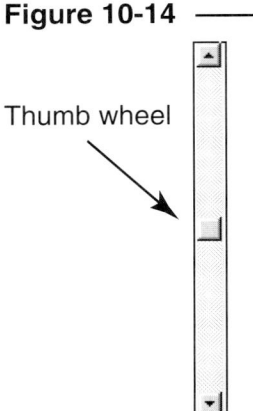

Thumb wheel

VIEW

Pull-down Menu	COMMAND (TYPE)	ALIAS (TYPE)	Short-cut	Screen (side) Menu	Tablet Menu
View Named Views...	VIEW or -VIEW	V or -V	...	VIEW 1 Ddview	M,5

The *View* command provides a dialog box for you to create *New* views of a specified display window and restore them (make *Current*) at a later time. For typical applications you would first *Zoom* in to achieve a desired display, then use *View, New* to save that display under an assigned name. Later in the drawing session, the named *View* can be made *Current* any number of times. This method is preferred to continually *Zooming* in and out to display several of the same areas repeatedly. Making a named *View* the *Current* view requires no regeneration time.

A *View* can be sent to the plotter as a separate display as long as the view is named and saved. Do this by toggling the *View* option in the *Plot Area* section of the *Plot* dialog box. Plotting is discussed in detail in Chapter 14.

Using *View* invokes the *Named Views* tab of the *View* dialog box (Fig. 10-15) and *–View* produces the Command line format. The options of the *View* command are described on the next page.

Figure 10-15

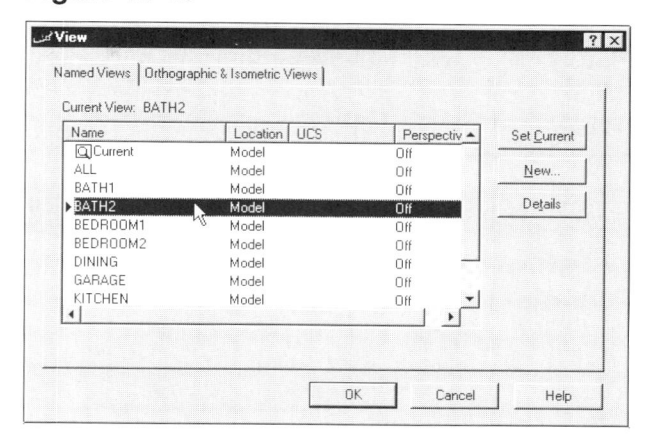

To save a new *View*, first use *Zoom* or *Pan* to achieve the desired display; then invoke the *View* dialog box and select *New*. In the *New View* dialog box that appears (Fig. 10-16), enter the desired name in the *View Name:* edit box. Ensure the *Current Display* button is selected.

Figure 10-16

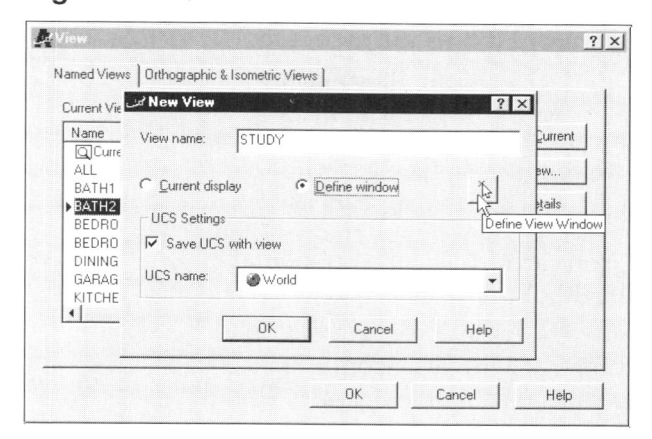

If you do not *Zoom* into the desired display before using *View*, you can select the *Define Window* button, then press the *Define View Window* button on the right (see the pointer in Figure 10-16). This action temporarily removes the dialog box so you can *Zoom* with a window to define the new view area.

To *Restore* a named *View*, double-click on the name from the list in the *Named View* tab, or select the name (one click) and select *Set Current*. This action makes the selected view current in the drawing area.

In the *UCS Settings* section of the *New View* dialog box, you can decide whether or not to save the UCS (User Coordinate System) with the view. This feature is used primarily with 3D drawings. See Chapter 30, User Coordinate Systems, for more information on this feature and the *UCSVIEW* system variable.

The *Orthographic and Isometric Views* tab of the *View* dialog box is also used for 3D models. This tab and related options are discussed in Chapter 29, 3D Display and Viewing.

If you prefer the Command line version of the command, type *–View* to produce the following prompt.

Command: *-view*
Enter an option [?/Orthographic/Delete/Restore/Save/Ucs/Window]:

The options here are generally the same as those available in the dialog box.

?	Displays the list of named views.
Orthographic	Provides orthographic and isometric viewpoints for 3D models.
Delete	Deletes one or more saved views that you enter.
Restore	Displays the named view you request.
Save	Saves the current display as a *View* with a name you assign.
UCS	Allows you to save the UCS with the view.
Window	Allows you to specify a window in the current display and save it as a named *View*.

VIEWRES

Pull-down Menu	COMMAND (TYPE)	ALIAS (TYPE)	Short-cut	Screen (side) MenuMenu	Tablet
Tools Options... Display Display Resolution	VIEWRES	...	...	TOOLS2 Options Display ... Display Resolution	...

Viewres controls the resolution of curved shapes for the screen display only. Its purpose is to speed regeneration time by displaying curved shapes as linear approximations of curves; that is, a curved shape (such as an *Arc*, *Circle*, or *Ellipse*) appears as several short, straight line segments (Fig. 10-17). The drawing database and the plotted drawing, however, always define a true curve. The range of *Viewres* is from 1 to 20,000 with 100 being the default. The higher the value (called "circle zoom percentage"), the more accurate the display of curves and the slower the regeneration time. The lower the *Viewres* value, the faster is the regeneration time but the more "jagged" the curves. A value of 500 to 1,000 is suggested for most applications. The command syntax is shown here.

Figure 10-17

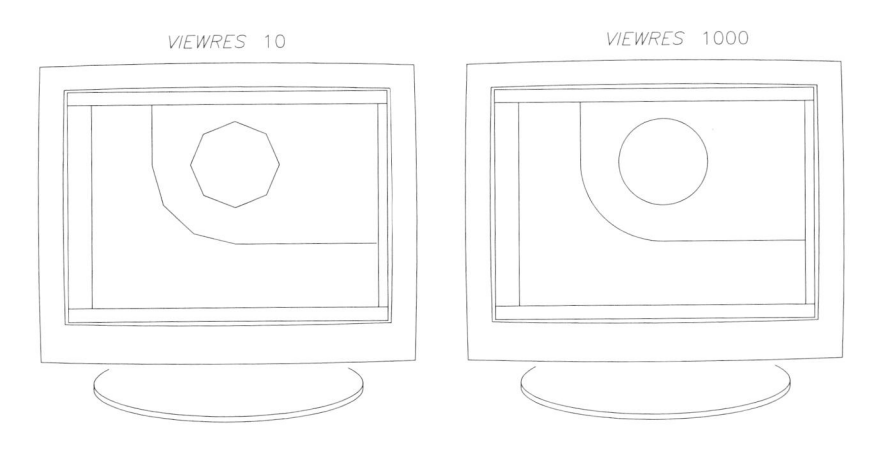

VIEWRES 10   VIEWRES 1000

Command: **viewres**
Do you want fast zooms? [Yes/No] <Y>: **Y**
Enter circle zoom percent (1-20000) <100>: **1000**
Command:

2002 — Fast zooms are no longer a functioning part of this command, so your response to "Do you want fast zooms?" is irrelevant. The prompt is kept only for script compatibility.

UCSICON

Pull-down Menu	COMMAND (TYPE)	ALIAS (TYPE)	Short-cut	Screen (side) Menu	Tablet Menu
View Display > UCS Icon >	UCSICON	...	...	VIEW 2 UCSicon	L,2

The icon that appears in the lower-left corner of the AutoCAD Drawing Editor is the Coordinate System Icon (Fig. 10-18). The icon is sometimes called the "UCS icon" because it automatically orients itself to the new location of a coordinate system that you create, called a "UCS" (User Coordinate System).

Figure 10-18

The *Ucsicon* command controls the appearance and positioning of the Coordinate System Icon. It is important to display the icon when working with 3D drawings so that you can more easily visualize the orientation of the X, Y, and Z coordinate system. However, when you are working with 2D drawings, it is not necessary to display this icon. (By now you are accustomed to normal orientation of the positive X and Y axes—X positive is to the right and Y positive is up.)

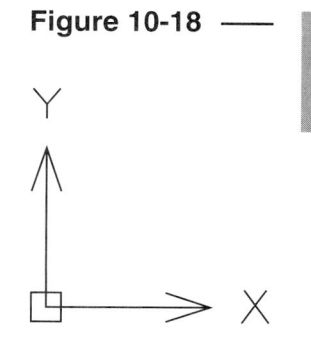

You can use the *Off* option of the *Ucsicon* command to remove the icon from the display. Use *Ucsicon* again with the *On* option to display the icon again. The Command prompt is as follows:

Command: *ucsicon*
Enter an option [ON/OFF/All/Noorigin/ORigin/Properties] <OFF>:

The *Properties* option can be used to change the type and properties of icon that are displayed. Using *Properties* produces the *UCS Icon* dialog box displayed in Figure 10-19. Here you can select from the *3D* (3-pole) icon or the *2D* (flat, or planar) icon. Either icon is suitable for 2D and 3D work. In fact, even though the older 2D-style icon only shows the direction for the X and Y axes, many AutoCAD users prefer it for 3D work because it makes the orientation of the XY plane much more apparent than the newer "3D" style icon.

Figure 10-19

You can also control the visibility of the icon using the *View* pull-down menu. Select *Display,* then *UCS Icon,* then remove the check by the word *On.* Make sure you turn on the Coordinate System Icon when you begin working in 3D.

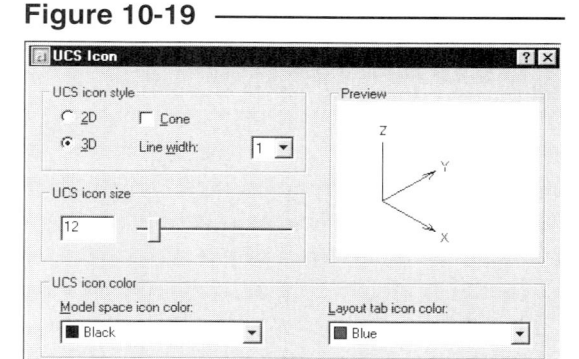

VPORTS

Pull-down Menu	COMMAND (TYPE)	ALIAS (TYPE)	Short-cut	Screen (side) Menu	Tablet Menu
View Viewports>	VPORTS or -VPORTS	...	...	VIEW 1 Vports	M,3 and M,4

A "viewport" is one of several simultaneous views of a drawing on the screen. The *Vports* command invokes the *Viewports* dialog box. With this dialog box you can create model space viewports or paper space viewports, depending on which space is current when you invoke *Vports*.

Model tab (model space) viewports are sometimes called "tiled" viewports because they fit together like tiles with no space between. *Layout* tab (paper space) viewports can be any shape and configuration and are used mainly for setting up several views on a sheet for plotting. Only an introduction to *Model* tab viewports is given in this chapter. *Layout* tab (paper space) viewports are discussed in Chapter 13.

If the *Model* tab is active and you use the *Vports* command, you will create model space viewports. Several viewport configurations are available. *Vports* in this case simply divides the screen into multiple sections, allowing you to view different parts of a drawing. Seeing several views of the same drawing simultaneously can make construction and editing of complex drawings more efficient than repeatedly using other display commands to view detailed areas. Model space *Vports* affect only the screen display. The viewport configuration cannot be plotted. If the *Plot* command is used from Model space, only the current viewport is plotted.

Figure 10-20 displays the AutoCAD drawing editor after the *Vports* command was used to divide the screen into viewports. Tiled viewports always fit together like tiles with no space between. The shape and location of the viewports are not variable as with paper space viewports.

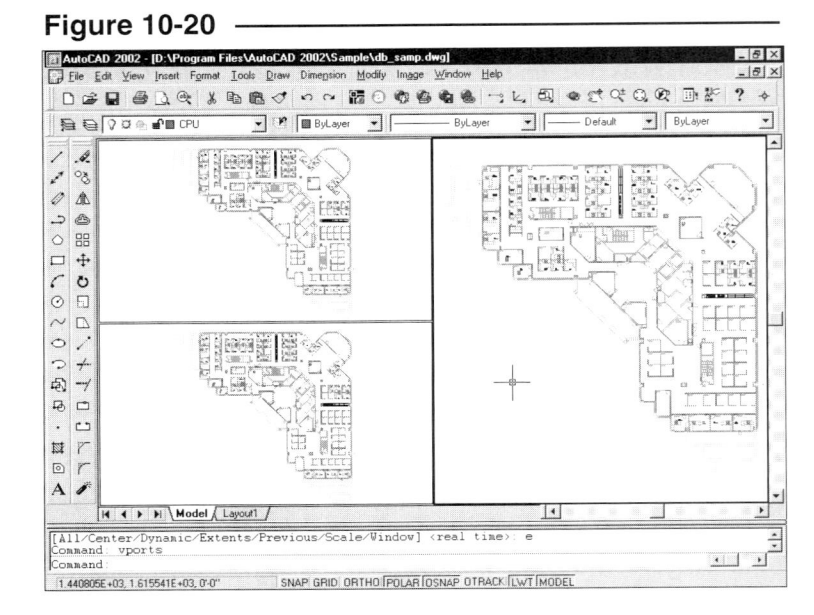

Figure 10-20

The *View* pull-down menu (see Figure 10-1) can be used to display the *Viewports* dialog box (Fig. 10-21). The dialog box allows you to PICK the configuration of the viewports that you want. If you are working on a 2D drawing, make sure you select *2D* in the *Setup* box.

After you select the desired layout, the previous display appears in each of the viewports. For example, if a full view of the office layout is displayed when you use the *Vports* command, the resulting display in each viewport would be the same full view of the office (see Figure 10-20). It is up to you then to use viewing commands (*Zoom*, *Pan*, etc.) in each viewport to specify what areas of the drawing you want to see in each viewport. There is no automatic viewpoint configuration option for *Vports* for 2D drawings.

Figure 10-21

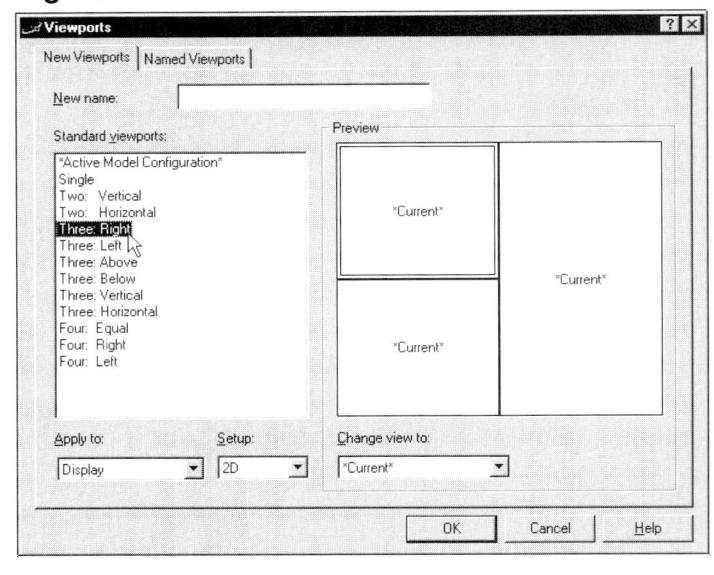

A popular arrangement of viewpoints for construction and editing of 2D drawings is a combination of an overall view and one or two *Zoomed* views (Fig. 10-22). You cannot draw or project from one viewport to another. Keep in mind that there is only one model (drawing) but several views of it on the screen. Notice that the active or current viewport displays the cursor, while moving the pointing device to another viewport displays only the pointer (small arrow) in that viewport.

Figure 10-22

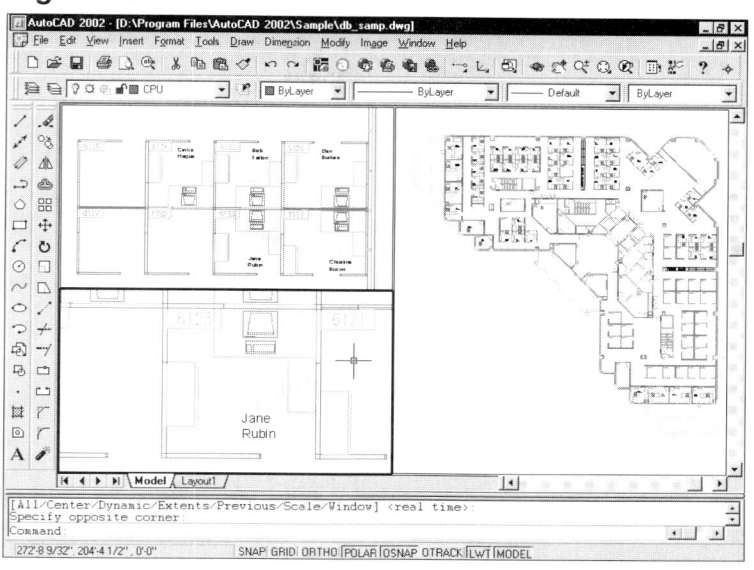

A viewport is made active by PICKing in it. Any display commands (*Zoom, Pan, Redraw,* etc.) and some drawing aids (*SNAP, GRID*) used affect only the current viewport. Draw and modify commands that affect the model are potentially apparent in all viewports (for every display of the affected part of the model). *Redrawall* and *Regenall* can be used to redraw and regenerate all viewports.

You can begin a drawing command in one viewport and finish in another. In other words, you can toggle viewports within a command. For example, you can use the *Line* command to PICK the "first point:" in one viewport, then make another viewport current to PICK the "next point:".

You can save a viewport configuration (including the views you prepare) by entering a name in the *New Name* box. If you save several viewport configurations for a drawing, you can then select from the list in the *Named Viewports* tab.

You can type *–Vports* to use the Command line version. The format is as follows:

Command: *-vports*
Enter an option [Save/Restore/Delete/Join/SIngle/?/2/3/4] <3>:

Save
Allows you to assign a name and save the current viewport configuration. The configuration can be *Restored* at a later time.

Restore
Redisplays a previously *Saved* viewport configuration.

Delete
Deletes a named viewport configuration.

Join
This option allows you to combine (join) two adjacent viewports. The viewports to join must share a common edge the full length of each viewport. For example, if four equal viewports were displayed, two adjacent viewports could be joined to produce a total of three viewports. You must select a *dominant* viewport. The *dominant* viewport determines the display to be used for the new viewport.

SIngle
Changes back to a single screen display using the current viewport's display.

200 Viewing Commands

?
Displays the identification numbers and screen positions of named (saved) and active viewport configurations. The screen positions are relative to the lower-left corner of the screen (0,0) and the upper-right corner (1,1).

2, 3, 4
Use these options to create 2, 3, or 4 viewports. You can choose the configuration. The possibilities are illustrated if you use the *Viewports* dialog box.

CHAPTER EXERCISES

1. *Zoom Extents, Realtime, Window, Previous, Center*

   Open the sample drawing supplied with AutoCAD called **DB_SAMP.DWG**. If your system has the standard installation of AutoCAD, the drawing is located in the **C:\Program Files\AutoCAD 2002\Sample** directory. The drawing shows an office building layout (Fig. 10-23). Use *SaveAs* and rename the drawing to **ZOOM TEST** and locate it in your working folder (directory). (When you view sample drawings, it is a good idea to copy them to another name so you do not accidentally change the original drawings.)

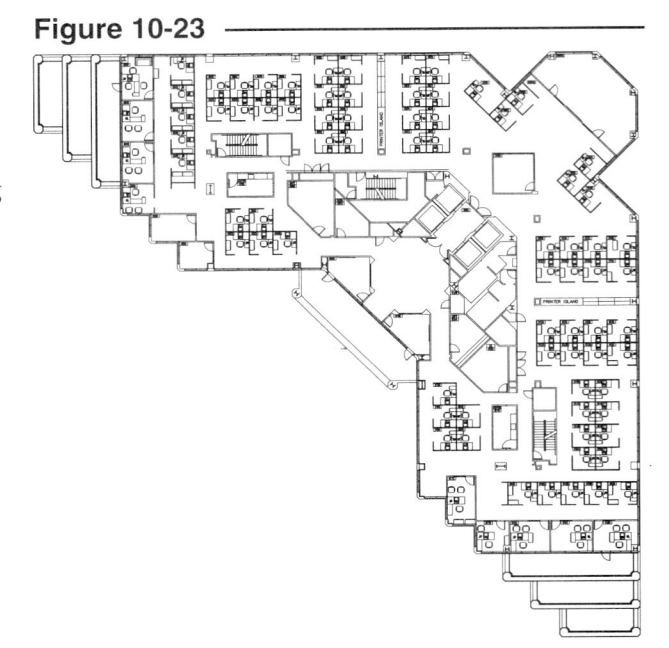

   Figure 10-23

   A. First, use *Zoom Extents* to display the office drawing as large as possible on your screen. Next, use *Zoom Realtime* and zoom in to the center of the layout (the center of your screen). You should be able to see rooms 6002 and 6198 located against the exterior wall (Fig. 10-24). Right-click and use *Zoom Extents* from the cursor pop-up menu to display the entire layout again. Press **Esc**, **Enter**, or right-click and select *Exit* to exit *Realtime*.

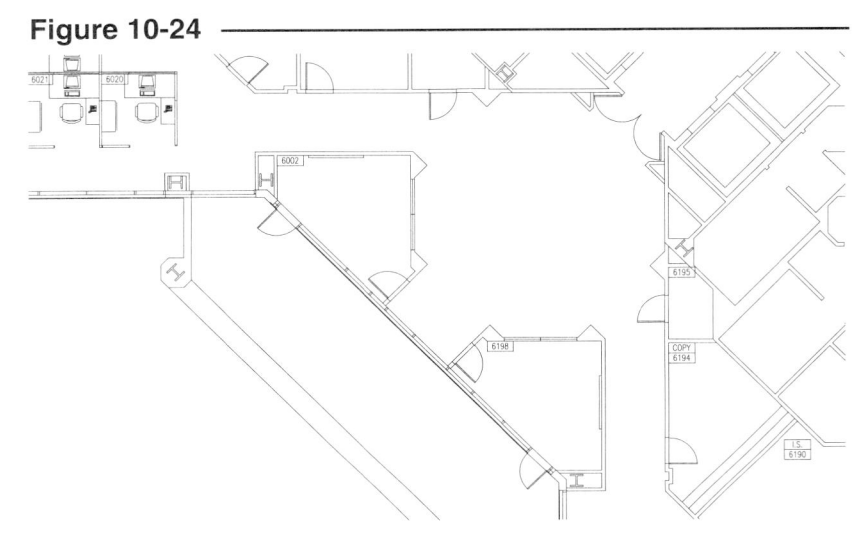

   Figure 10-24

B. Use *Zoom Window* to closely examine room 6050 at the top left of the building. Whose office is it? What items on the desk are magenta in color? What item is cyan (light blue)?

C. Use *Zoom Center*. **PICK** the telephone on the desk next to the computer as the center. Specify a height of **24** (2'). You should see a display showing only the telephone. How many keys are on the telephone's number pad (*Point* objects)?

D. Next use *Zoom Previous* repeatedly until you see the same display of rooms 6002 and 6198 as before (Fig. 10-24). Use *Zoom Previous* repeatedly until you see the office drawing *Extents*.

2. **Pan Realtime**

    A. Using the same drawing as exercise 1 (ZOOM TEST.DWG, originally DB_SAMP.DWG), use *Zoom Extents* to ensure you can view the entire drawing. Next *Zoom* with a *Window* to an area about 1/4 the size of the building.

    B. Invoke the *Real Time* option of *Pan*. *Pan* about the drawing to find a coffee room. Can you find another coffee room? How many coffee rooms are there?

3. **Pan and Zoom with the Mouse Wheel**

    In the following exercise, *Zoom* and *Pan* using the wheel on your mouse (turn the wheel to *Zoom* and press and drag the wheel to *Pan*). If you do not have this capability, use *Realtime Pan* and *Zoom* in concert to examine specific details of the office layout.

    A. Find your new office. It is room 6100. (HINT: It is centrally located in the building.) Your assistant is in the office just to the right. What is your assistant's name and room number?

    B. Although you have a nice office, there are some disadvantages. How far is it from your office door to the nearest coffee room (nearest corner of the coffee room)? HINT: Use the *Dist* command with *Endpoint OSNAP* to select the two indicated points.

    C. *Pan* and/or *Zoom* to find the copy room 6006. How far is it to the copy room (direct distance, door to door)?

    D. Perform a *Zoom Extents*. Use the wheel to zoom in to Kathy Ragerie's office (room 6150) in the lower-right corner of the building. Remember to locate the cursor at the spot in the drawing that you want to zoom in to. If you need to pan, hold down the wheel and "drag" the drawing across the screen until you can view all of room 6150 clearly. How many chairs are in Kathy's office? Finally use *Zoom Extents* to size the drawing to the screen.

4. **Pan Point**

    A. *Zoom* in to your new office (room 6100) so that you can see the entire room and room number. Assume you were listening to music on your computer with the door open and calculated it could be heard about 100' away. Naturally, you are concerned about not bothering the company CEO who has an office in room 6048. Use *Pan Point* to see if the CEO's office is within listening range. (HINT: enter **100'<0** at the "Specify base point or displacement:" prompt.) Should you turn down the system?

    B. *Exit* the **ZOOM TEST** drawing.

202 Viewing Commands

5. *Viewres, Zoom Dynamic*

    A. *Open* the **TABLET2000** drawing from the **Sample** folder (C:\Program Files\AutoCAD 2002\Sample). Use *SaveAs*, name the new drawing **TABLET TEST**, and locate it in your working folder. Use *Zoom* with the *Window* option (or you can try *Realtime Zoom*). Zoom in to the first several command icons located in the middle left section of the tablet menu. Your display should reveal icons for *Zoom* and other viewing commands. *Pan* or *Zoom* in or out until you see the *Zoom* icons clearly.

    B. Notice how the *Zoom* command icons are shown as polygons instead of circles? Use the *Viewres* command and change the value to **1000**. Do the icons now appear as circles? Use *Zoom Realtime* again. Does the new setting change the speed of zooming?

    C. Use *Zoom Dynamic*, then try moving the view box immediately. Can you move the box before the drawing is completely regenerated? Change the view box size so it is approximately equal to the size of one icon. Zoom in to the command in the lower-right corner of the tablet menu. What is the command? What is the command in the lower-left corner? For a drawing of this complexity and file size, which option of *Zoom* is faster and easier for your system—*Realtime* or *Dynamic*?

    D. If you wish, you can *Exit* the **TABLET TEST** drawing.

6. *View*

    A. Make the **ZOOM TEST** drawing that you used in previous exercises the current drawing. *Zoom* in to the office that you will be moving into (room 6100). Make sure you can see the entire office. Use the *View* command and create a *New* view named **6100**. Next, *Zoom* or *Pan* to room 6048. *Save* the display as a *View* named **6048**.

    B. Use the *View* dialog box again and make view **6100** *Current*.

    C. In order for the CEO to access information on your new computer, a cable must be stretched from your office to room 6048. Find out what length of cable is needed to connect the upper left corner of office 6100 to the upper right corner of office 6048. (HINT: Use *Dist* to determine the distance. Type the '–*View* command transparently [prefix with an apostrophe and hyphen] during the *Dist* command and use **Running Endpoint OSNAP** to select the two indicated office corners. The *View* dialog box is not transparent.) What length of cable is needed? Do not *Exit* the drawing.

7. *Zoom All, Extents*

    A. Begin a *New* drawing. Turn on the *SNAP* (**F9**) and *GRID* (**F7**). Draw two *Circles*, each with a **1.5** unit *radius*. The *Circle* centers are at **3,5** and at **5,5**. See Figure 10-25.

    B. Use *Zoom All*. Does the display change? Now use *Zoom Extents*. What happens? Now use *Zoom All* again. Which option always shows all of the *Limits*?

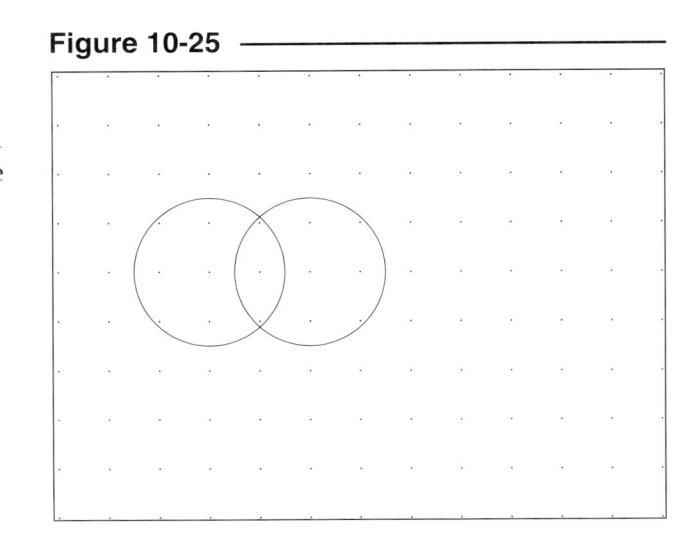

Figure 10-25

C. Draw a *Circle* with the center at **10,10** and with a *radius* of **5**. Now use *Zoom All*. Notice the *GRID* appears only on the area defined by the *Limits*. Can you move the cursor to 0,0? Now use *Zoom Extents*. What happens? Can you move the cursor to 0,0?

D. *Erase* the large *Circle*. Use *Zoom All*. Can you move the cursor to 0,0? Use *Zoom Extents*. Can you find point 0,0?

E. *Exit* the drawing and discard changes.

8. *Vports*

A. *Open* the **ZOOM TEST** drawing again. Perform a *Zoom Extents*. Invoke the *Viewports* dialog box by any method. When the dialog box appears, choose the *Three: Right* option. The resulting viewport configuration should appear as Figure 10-20, shown earlier in the chapter.

B. Using *Zoom* and *Pan* in the individual viewports, produce a display with an overall view on the top and two detailed views as shown in Figure 10-22.

C. Type the *Vports* command. Use the *New* option to save the viewport configuration as **3R**.

D. Click in the bottom-left viewport to make it the current viewport. Use *Vports* again and change the display to a *Single* screen. Now use *Realtime Zoom* and *Pan* to locate and zoom to room 6100.

E. Use the *Viewports* dialog box again. Select the *Named Viewports* tab and select **3R** as the viewport configuration to restore. Ensure the **3R** viewport configuration was saved as expected, including the detail views you prepared.

F. Invoke a *Single* viewport again. Finally, use the *Vports* dialog box to create another viewport configuration of your choosing. *Save* the viewport configuration and assign an appropriate name. *Save* the **ZOOM TEST** drawing.

204 Viewing Commands

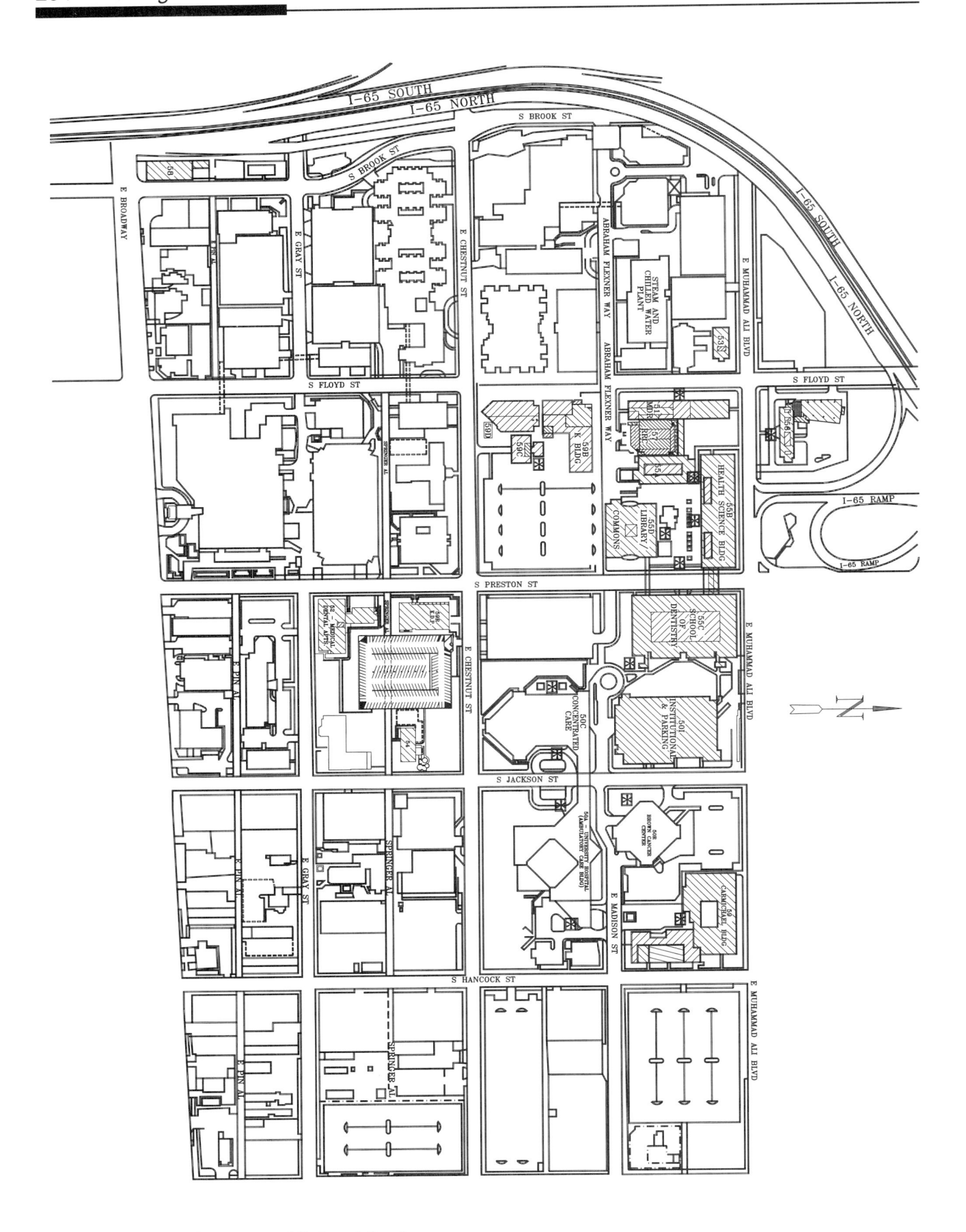

HSCMAP.DWG Courtesy, Michael Anderson

11

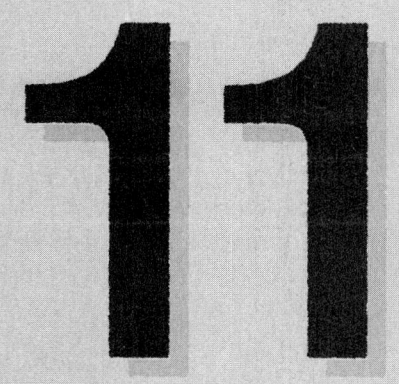

LAYERS AND OBJECT PROPERTIES

Chapter Objectives

After completing this chapter you should:

1. understand the strategy of grouping related geometry with *Layers*;
2. be able to create *Layers*;
3. be able to assign *Color, Linetype,* and *Lineweight* to *Layers*;
4. be able to control a layer's properties and visibility settings (*On, Off, Freeze, Thaw, Lock, Unlock*);
5. be able to assign *Color, Linetype,* and *Lineweight* to objects;
6. be able to set *LTSCALE* to adjust the scale of linetypes globally;
7. understand the concept of object properties;
8. be able to change an object's properties (layer, color, linetype, linetype scale, and lineweight) with the *Object Properties* toolbar, the *Properties* window, and with *Match Properties*.

CONCEPTS

In a CAD drawing, layers are used to group related objects in a drawing. Objects (*Lines*, *Circles*, *Arcs*, etc.) that are created to describe one component, function, or process of a drawing are perceived as related information and, therefore, are typically drawn on one layer. A single CAD drawing is generally composed of several components and, therefore, several layers. Use of layers provides you with a method to control visible features of the components of a drawing. For each layer, you can control its color on the screen, the linetype and lineweight it will be displayed with, and its visibility setting (on or off). You can also control if the layer is plotted or not.

Figure 11-1

Layers in a CAD drawing can be compared to clear overlay sheets on a manual drawing. For example, in a CAD architectural drawing, the floor plan can be drawn on one layer, electrical layout on another, plumbing on a third layer, and HVAC (heating, ventilating, and air conditioning) on a fourth layer (Fig. 11-1). Each layer of a CAD drawing can be assigned a different color, linetype, lineweight, and visibility setting similar to the way clear overlay sheets on a manual drawing can be used. Layers can be temporarily turned *Off* or *On* to simplify drawing and editing, like overlaying or removing the clear sheets. For example, in the architectural CAD drawing, only the floor plan layer can be made visible while creating the electrical layout, but can later be cross-referenced with the HVAC layout by turning its layer on. Layers can also be made "non-plottable." Final plots can be made of specific layers for the subcontractors and one plot of all layers for the general contractor by controlling the layers' *Plot/No Plot* icon before plotting.

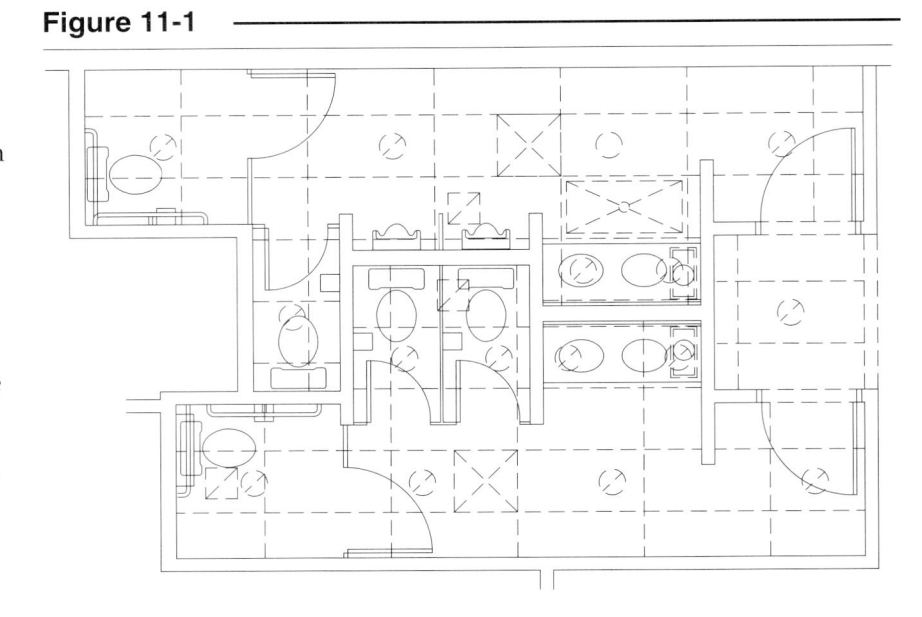

AutoCAD allows you to create a practically unlimited number of layers. You should assign a name to each layer when you create it. The layer names should be descriptive of the information on the layer.

Assigning Colors, Linetypes, and Lineweights

There are two strategies for assigning colors, linetypes, and lineweights in a drawing: assign these properties to layers or assign them to objects.

<u>Assign colors, linetypes, and lineweights to layers</u>
Usually, layers are assigned a color, linetype, and lineweight so that all objects drawn on a single layer have the same color, linetype, and lineweight. Assigning colors, linetypes, and lineweights to layers is called *ByLayer* color, linetype, and lineweight setting. Using the *ByLayer* method makes it visually apparent which objects are related (on the same layer). All objects on the same layer have the same linetype and color.

Assign colors, linetypes, and lineweights to individual objects
Alternately, you can assign colors, linetypes, and lineweights to specific objects, overriding the layer's color, linetype, and lineweight setting. This method is fast and easy for small drawings and works well when layering schemes are not used or for particular applications. However, using this method makes it difficult to see which layers the objects are located on.

Object Properties

Another way to describe the assignment of color, linetype, and lineweight properties is to consider the concept of Object Properties. Each object has properties such as a layer, a color, a linetype, and a lineweight. The color, linetype, and lineweight for each object can be designated as *ByLayer* or as a specific color, linetype, or lineweight. Object properties for the two drawing strategies (schemes) are described as follows.

Object Properties

	1. *ByLayer* Drawing Scheme	2. Object-Specific Drawing Scheme
Layer Assignment	Layer name descriptive of geometry on the layer	Layer name descriptive of geometry on the layer
Color Assignment	*ByLayer*	*Red*, *Green*, *Blue*, *Yellow*, or other specific color setting
Linetype Assignment	*ByLayer*	*Continuous*, *Hidden*, *Center*, or other specific linetype setting
Lineweight Assignment	*ByLayer*	0.05 mm, 0.15 mm, 0.010", or other specific lineweight setting

It is recommended that beginners use only one method for assigning colors, linetypes, and lineweights. After gaining some experience, it may be desirable to combine the two methods only for specific applications. Usually, the *ByLayer* method is learned first, and the object-specific color, linetype, and lineweight assignment method is used only when layers are not needed or when complex applications are needed. (The *ByBlock* color, linetype, and lineweight assignment has special applications for *Blocks* and is discussed in Chapter 20.)

LAYERS AND LAYER PROPERTIES CONTROLS

Layer Control Drop-Down List

The Object Properties toolbar contains a drop-down list for making layer control quick and easy (Fig. 11-2). The window normally displays the current layer's name, visibility setting, and properties. When you pull down the list, all the layers (unless otherwise specified in the *Named Layer Filters* dialog box) and their settings are displayed. Selecting any layer <u>name</u> makes it current. Clicking on any of the visibility/properties icons changes the layers' setting as described (see "Layer"). Several layers can be changed in one "drop." You cannot change a layer's color, linetype, or lineweight, nor can you create new layers using this drop-down list.

Figure 11-2

LAYER

Pull-down Menu	COMMAND (TYPE)	ALIAS (TYPE)	Short-cut	Screen (side) Menu	Tablet Menu
Format Layer	LAYER or -LAYER	LA or -LA	...	FORMAT Layer	U,5

The way to gain complete layer control is through the *Layer Properties Manager* (Fig. 11-3). The *Layer Properties Manager* is invoked by using the icon button (shown above), typing the *Layer* command or *LA* command alias, or selecting *Layer* from the *Format* pull-down or screen menu.

Figure 11-3

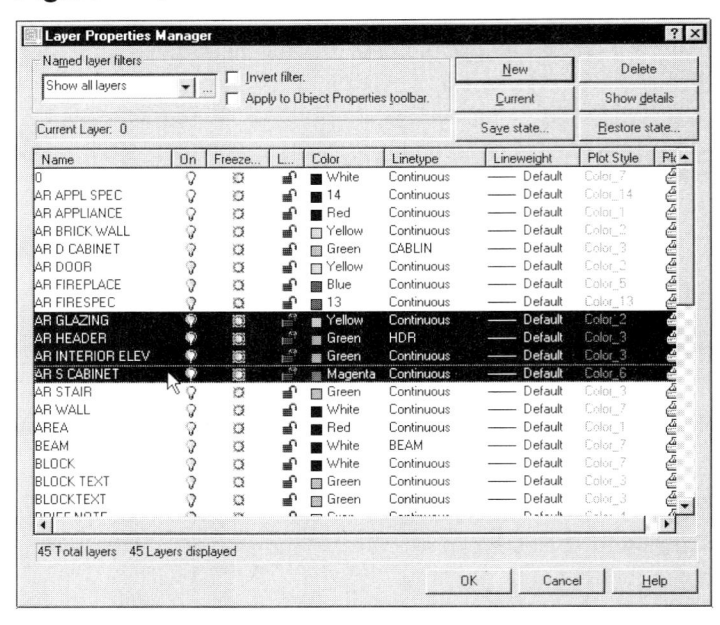

This dialog box allows full control for all layers in a drawing. Layers existing in the drawing appear in the list at the central area (only Layer 0 exists in new drawings and in those created from standard templates such as ACAD.DWT). New layers can be created by selecting the *New* button near the upper right corner of the dialog box.

All properties and visibility settings of layers can be controlled by highlighting the layer name and then selecting one of the icons for the layer such as the light bulb icon (*On, Off*), sun/snowflake icon (*Thaw/Freeze*), padlock icon (*Lock/Unlock*), *Color* tile, *Linetype, Lineweight*, or *Plot/No plot* icon.

Typical Windows dialog box features apply as explained in this paragraph. Multiple layers can be highlighted (see Figure 11-3) by holding down the Ctrl key while PICKing (to highlight one at a time) or holding down the Shift key while PICKing (to select a range of layers between and including two selected names). Right-clicking in the list area displays a cursor menu allowing you to *Select All* or *Clear All* names in the list and other options (see Figure 11-5). You can rename a layer by clicking twice slowly on the name (this is the same as PICKing an already highlighted name). The column widths can be changed by moving the pointer to the "crack" between column headings until double arrows appear (Fig. 11-4). You can also resize the entire dialog box by placing your pointer on the extreme border or corner until double arrows appear, then clicking and dragging.

Figure 11-4

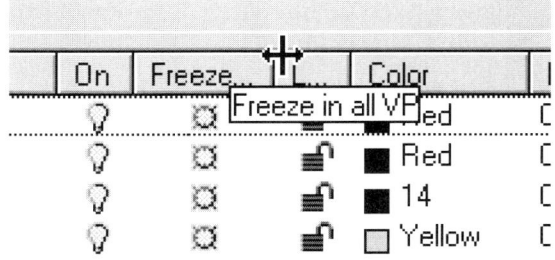

A particularly useful feature is the ability to sort the layers in the list by any one of the headings (*Name, On, Freeze, Linetype*, etc.) by clicking on the heading tile. For example, you can sort the list of names in alphabetical order (or reverse order) by clicking once (or twice) on the *Name* heading above the list of names. Or you may want to sort the *Frozen* and *Thawed* layers or sort layers by *Color* by clicking on the column heading.

The *Show details* tile near the top-right corner of the dialog box displays the details (properties and visibility settings) of the highlighted layer in the list (Fig. 11-5). This area is basically an <u>alternative</u> method of controlling layer properties and visibility settings (instead of using the icons in the list area).

Figure 11-5

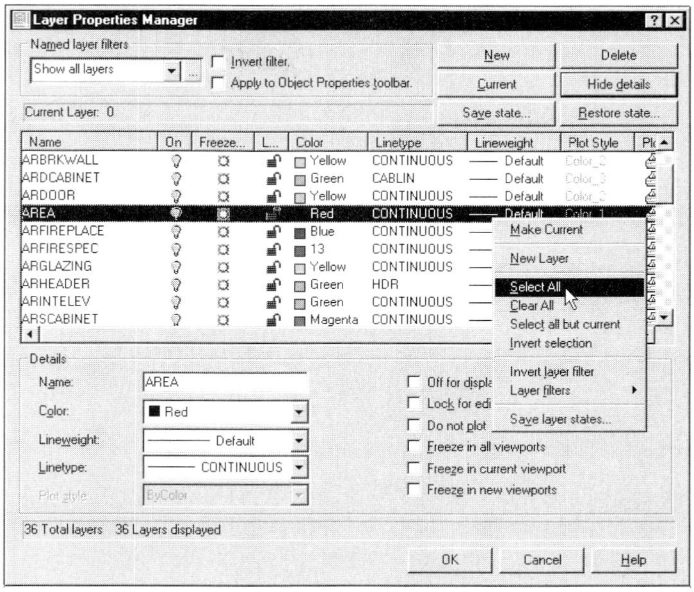

As an alternative to the dialog box, the *Layer* command can be used in command line format by typing *Layer* with a hyphen (-) prefix. The Command line format of *Layer* shows all of the available options.

    Command: *-layer*
    Current layer: "0"
    Enter an option
    [?/Make/Set/New/ON/OFF/Color/Ltype/LWeight/Plot/PStyle/Freeze/Thaw/LOck/Unlock/stAte]:

Detailed explanations of the options for controlling layer properties and visibility settings, whether using the *Layer Properties Manager* or using *-Layer* in Command line format, are listed next.

Current or *Set*

To *Set* a layer as the *Current* layer is to make it the active drawing layer. Any objects created with draw commands are created on the *Current* layer. You can, however, edit objects on any layer, but draw only on the current layer. Therefore, if you want to draw on the FLOORPLAN layer (for example), use the *Set* or *Current* option or double-click on the layer name. If you want to draw with a certain *Color* or *Linetype*, set the layer with the desired *Color, Linetype,* and *Lineweight* as the *Current* layer. Any layer can be made current, but only <u>one layer at a time</u> can be current.

To set the current layer with the *Layer Properties Manager* (see Figures 11-3 and 11-5), select the desired layer from the list and then select the *Current* tile. You can instead double-click on the desired layer name. Alternately, if you are typing, use the *Set* option of the *-Layer* command to make a layer the current layer.

On, Off

If a layer is *On*, it is visible. Objects on visible layers can be edited or plotted. Layers that are *Off* are not visible. Objects on layers that are *Off* will not plot and cannot be edited (unless the *ALL* selection option is used, such as *Erase, All*). It is not advisable to turn the current layer *Off*.

Freeze, Thaw

Freeze and *Thaw* override *On* and *Off*. *Freeze* is a more protected state than *Off*. Like being *Off*, a frozen layer is not visible, nor can its objects be edited or plotted. Objects on a frozen layer cannot be accidentally *Erase*d with the *ALL* option. *Freezing* also prevents the layer from being considered when *Regen*s occur. *Freezing* unused layers speeds up computing time when working with large and complex drawings. *Thawing* reverses the *Freezing* state. Layers can be *Thawed* and also turned *Off*. Frozen layers are not visible even though the light bulb icon is on.

Lock, Unlock

Layers that are *Locked* are protected from being edited but are still visible and can be plotted. *Locking* a layer prevents its objects from being changed even though they are visible. Objects on *Locked* layers cannot be selected with the *ALL* selection option (such as *Erase, All*). Layers can be *Locked* and *Off*.

Color, Linetype, Lineweight, and Other Properties

Layers have properties of *color, linetype,* and *lineweight* such that (generally) an object that is drawn on, or changed to, a specific layer assumes the layer's linetype and color. Using this scheme (*ByLayer*) enhances your ability to see what geometry is related by layer. It is also possible, however, to assign specific color, linetype, and lineweight to objects which will override the layer's color, linetype, and lineweight (see "*Color, Linetype,* and *Lineweight* Commands" and "Changing Object Properties").

Color

Figure 11-6

Selecting one of the small color boxes in the list area of the *Layer Properties Manager* causes the *Select Color* dialog box to pop up (Fig. 11-6). The desired color can then be selected or the name or color number (called the ACI—AutoCAD Color Index) can be typed in the edit box. This action retroactively changes the color assigned to a layer. Since the color setting is assigned to the layer, all objects on the layer that have the *ByLayer* setting change to the new layer color. Objects with specific color assigned (not *ByLayer*) are not affected. Alternately, the *Color* option of the *–Layer* command (hyphen prefix) can be typed to enter the color name or ACI number.

The actual number of colors that are available depends on the type of monitor and graphics controller card that are configured. Although most setups allow up to 16 million colors and more, only 256 colors are shown in the standard ACI pallet in AutoCAD.

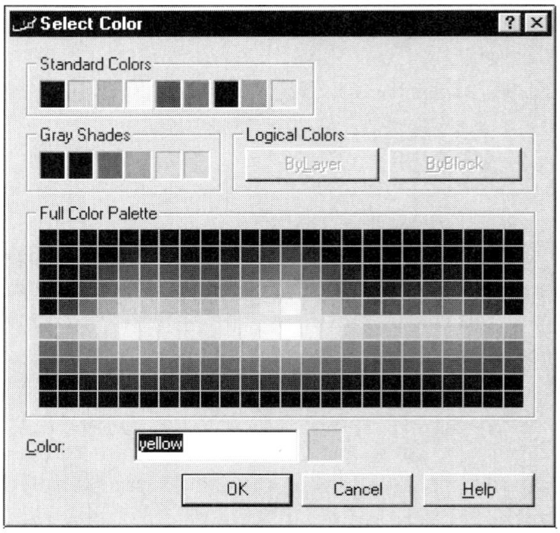

Linetype

Figure 11-7

To set a layer's linetype, select the *Linetype* (word such as *Continuous* or *Hidden*) in the layer list, which in turn invokes the *Select Linetype* dialog box (Fig. 11-7). Select the desired linetype from the list. Alternately, the *Linetype* option of the *–Layer* command (hyphen prefix) can be typed. Similar to changing a layer's color, all objects on the layer with *ByLayer* linetype assignment are retroactively displayed in the selected layer linetype while non-*ByLayer* objects remain unchanged.

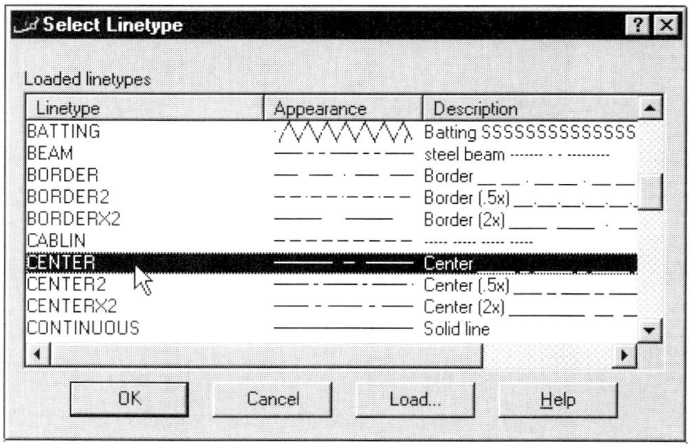

The ACAD.DWT template drawing as supplied by Autodesk has only one linetype available (*Continuous*). Before you can use other linetypes in a drawing, you must load the linetypes by selecting the *Load* tile (see the *Linetype* command) or by using a template drawing that has the desired linetypes already loaded.

Lineweight

The lineweight for a layer can be set by selecting the *Lineweight* (word such as *Default* or *0.20 mm, 0.50 mm, 0.010"* or *0.020"*, etc.) in the layer list. This action produces the *Lineweight* dialog box (Fig. 11-8). Select the desired lineweight from the list. Alternately, the *Lineweight* option of the *–Layer* command (hyphen prefix) can be typed. Like the *Color* and *Linetype* properties, all objects on the layer with *ByLayer* lineweight assignment are retroactively displayed in the lineweight assigned to the layer while non-*ByLayer* objects remain unchanged. You can set the units for lineweights (mm or inches) using the *Lineweight* command.

Figure 11-8

Plot Style

This column of the *Layer Properties Manager* designates the plot style assigned to the layer. If the drawing has a color-dependent Plot Style Table attached, this section is disabled since the plot styles are automatically assigned to colors. If a named Plot Style Table is attached to the drawing you can select this section to assign plot styles from the table to layers.

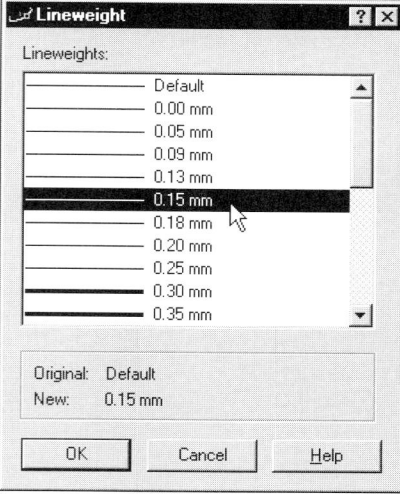

Plot/No Plot

You can prevent a layer from plotting by clicking the printer icon so a red circle appears over the printer symbol. There are actually three methods for preventing a layer from appearing on the plot: *Freeze* it, turn it *Off*, or change the *Plot/ No Plot* icon. The *Plot/ No Plot* icon is generally preferred since the other two options prevent the layer from appearing in the drawing (screen) as well as in the plot.

Current Viewport Freeze, New Viewport Freeze

These options are used and are displayed in the *Layer Properties Manager* only when paper space viewports exist in the drawing. Using these options, you can control what geometry (layers) appears in specific viewports.

New

The *New* option allows you to make new layers. There is only one layer in the AutoCAD template drawing (ACAD.DWT) as it is provided to you "out of the box." That layer is Layer 0. Layer 0 is a part of every AutoCAD drawing because it cannot be deleted. You can, however, change the *Color, Linetype,* and *Lineweight* of Layer 0 from the defaults (*Continuous* linetype, *Default* lineweight, and color #7 *White*). Layer 0 is generally used as a construction layer or for geometry not intended to be included in the final draft of the drawing. Layer 0 has special properties when creating *Blocks* (see Chapter 20).

You should create layers for each group of related objects and assign appropriate layer names for that geometry. You can use up to 256 characters (including spaces) for layer names. There is practically no limit to the number of layers that can be created (although 32,767 has been found to be the actual limit). To create layers in the *Layer Properties Manager* box, select the *New* tile. A new layer named "Layer1" (or other number) then appears in the list with the default color (*White*), linetype (*Continuous*), and lineweight (*Default*). In AutoCAD 2002, if an existing layer name is highlighted when you make a new layer, the highlighted layer's properties (*color, linetype, lineweight,* and *plot style*) are used as a template for the new layer. The new layer name initially appears in the rename mode so you can immediately assign a more appropriate and descriptive name for the layer. You can create many new names quickly by typing (renaming) the first layer name, then typing a comma before other names. A comma forces a new "blank" layer name to appear. Colors, linetypes, and lineweights should be assigned as the next step.

If you want to create layers by typing, the *New* and *Make* options of the *Layer* command can be used. *New* allows creation of one or more new layers. *Make* allows creation of one layer (at a time) and sets it as the current layer.

Delete
The *Delete* tile allows you to delete layers. Only layers with no geometry can be deleted. You cannot delete a layer that has objects on it, nor can you delete Layer 0, the current layer, or layers that are part of externally referenced (*Xref*) drawings. If you attempt to *Delete* such a layer, accidentally or intentionally, a warning appears.

Named Layer Filters
The *Named Layer Filters* drop-down list determines what names appear in the list of layers in the *Layer* tab (Fig. 11-9). The choices are as follows.

Figure 11-9

Show all layers	shows all layers in the drawing
Show all used layers	shows all layers that have objects on them
Show all Xref dependent layers	shows all layers that are part of externally referenced drawings

Displaying an Object's Properties and Visibility Settings

When the Layer Control drop-down list in the Object Properties toolbar is in the normal position (not "dropped down"), it can be used to display an object's layer properties and visibility settings. Do this by selecting an object (with the pickbox, window, or crossing window) when no commands are in use.

Figure 11-10

When an object is selected, the Object Properties toolbar displays the layer, color, linetype, lineweight, and plot style of the selected object (Fig. 11-10). The Layer Control box displays the selected object's layer name and layer settings rather than that of the current layer. The Color Control, Linetype Control, Lineweight Control, and Plot Style Control boxes in the Object Properties toolbar (just to the right) also temporarily reflect the color, linetype, lineweight, and plot style properties of the selected object. Pressing the Escape key causes the list boxes to display the current layer name and settings again, as would normally be displayed. If more than one object is selected, and the objects are on different layers or have different color, linetype, and lineweight properties, the list boxes go blank until the objects become unhighlighted or a command is used.

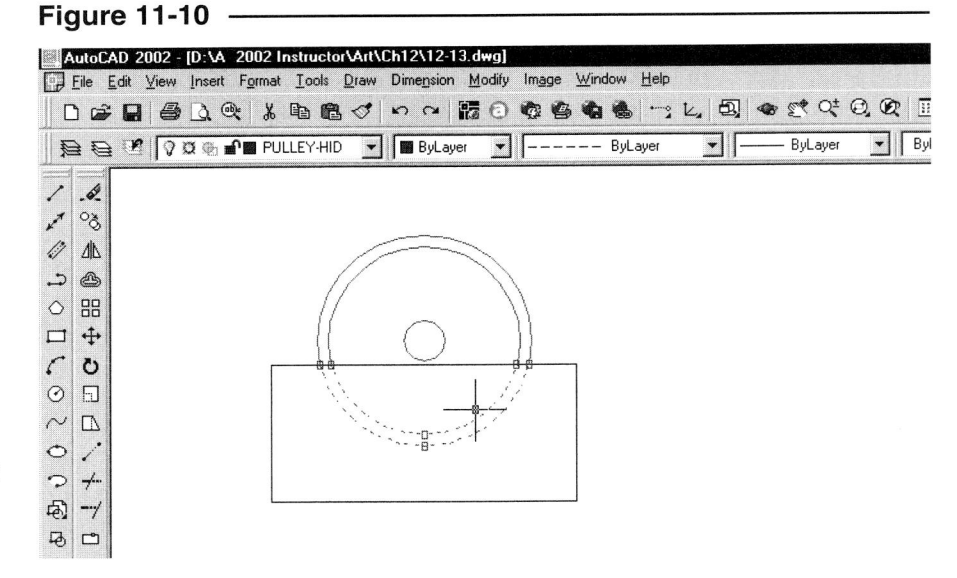

These sections of the Object Properties toolbar have another important new feature that allows you to change a highlighted object's (or set of objects) properties. See "Changing Object Properties" near the end of this chapter.

Make Object's Layer Current

This productive feature can be used to make a desired layer current simply by selecting <u>any object on the layer</u>. This option is generally faster than using the *Layer Properties Manager* or *-Layer* command to set a layer current. The *Make Object's Layer Current* feature is particularly useful when you want to draw objects on the same layer as other objects you see, but are not sure of the layer name. This feature (not a formal command) is only available by selecting the icon on the far left of the Object Properties toolbar (see Figure 11-10). It cannot be invoked by any other method.

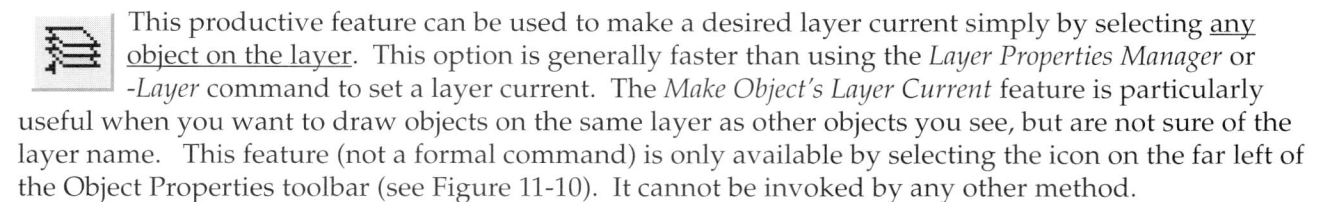

There are two steps in the procedure to make an object's layer current: (1) select the icon and (2) select any object on the desired layer. The selected object's layer becomes current and the new current layer name immediately appears in the Layer Control list box in the Object Properties toolbar.

LAYERP

Pull-down Menu	COMMAND (TYPE)	ALIAS (TYPE)	Short-cut	Screen (side) Menu	Tablet Menu
...	LAYERP	...	...	...	...

Layerp (*Layer Previous*) is an *Undo* command only for layers. In other words, when you use *Layerp*, all the changes you made the last time you used the *Layer Properties Manager*, the *Layer Control* drop-down list, or the *–Layer* command are undone. Using *Layerp* does not affect any other activities that occurred to the drawing since the last layer settings, such as creating or editing geometry or viewing controls like *Pan* or *Zoom*.

Use *Layerp* to return the drawing to the previous layer settings. For example, if you froze several layers and changed some of the geometry in a drawing, but now want to thaw those frozen layers again without affecting the geometry changes, use *Layerp*. Or, if you changed the color and linetype properties of several layers but later decide you prefer the previous property settings, use *Layerp* to undo the changes and restore the original layer settings.

Layerp affects only layer-related activities; however, *Layerp* does not undo the following changes:

   Renamed layers: If you rename a layer and change its properties, *Layerp* restores the
      original properties but not the original layer name.
   Deleted layers: If you delete or purge a layer, using *Layerp* does not restore it.
   New layers: If you create a new layer in a drawing, using *Layerp* does not remove it.

The *LAYERPMODE* system variable setting (*On* or *Off*) enables or disables the *Layerp* command and the related layer tracking function. Since there is a modest performance loss for *Layerp* tracking, you can suspend layer tracking when you don't need it, such as when you run large scripts.

OBJECT-SPECIFIC PROPERTIES CONTROLS

The commands in this section are used to control the *color*, *linetype*, and *lineweight* properties of individual objects. This method of object property assignment is used only in special cases—when you want the objects' *color*, *linetype*, and *lineweight* to override those properties assigned to the layers on which the objects reside. Using this method makes it difficult to see which objects are on which layers. In most cases, the *Bylayer* property is assigned instead to individual objects so the objects assume the properties of their layers.

LINETYPE

Pull-down Menu	COMMAND (TYPE)	ALIAS (TYPE)	Short-cut	Screen (side) Menu	Tablet Menu
Format Linetype...	LINETYPE or -LINETYPE	LT or -LT	...	FORMAT Linetype	U,3

Invoking this command presents the *Linetype Manager* (Fig. 11-11). Even though this looks similar to the dialog box used for assigning linetypes to layers (shown earlier in Figure 11-7), beware!

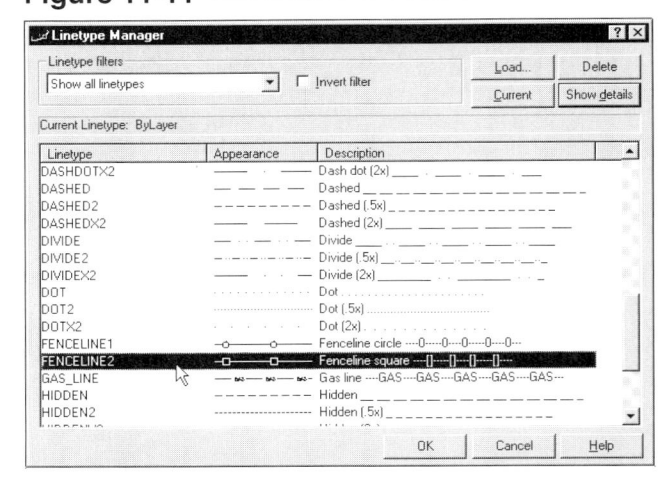

When linetypes are selected and made *Current* using the *Linetype Manager*, they are assigned to objects—not to layers. That is, selecting a linetype by this manner (making it *Current*) causes all objects from that time on to be drawn using that linetype, regardless of the layer that they are on (unless the *ByLayer* type is selected). In contrast, selecting linetypes during the *Layer* command (using the *Layer Properties Manager*) results in assignment of linetypes to layers. Remember that using both of these methods for color, linetype, and lineweight assignment in one drawing can be very confusing until you have some experience using both methods.

Figure 11-11

Linetype Control Drop-Down List

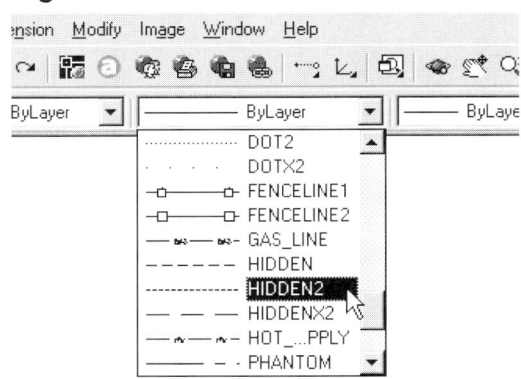

The Object Properties toolbar contains a drop-down list for selecting linetypes (Fig. 11-12). Although this appears to be quick and easy, you can only assign linetypes to objects by this method unless *ByLayer* is selected to use the layers' assigned linetypes. Any linetype you select from this list becomes the current object linetype. If you want to select linetypes for layers, make sure this list displays the *ByLayer* setting, then use the *Layer Properties Manager* to select linetypes for layers.

Figure 11-12

Keep in mind that the Linetype Control drop-down list as well as the Layer Control, Color Control, and Lineweight Control drop-down lists display the current settings for selected objects (objects selected when no commands are in use). When linetypes are assigned to layers rather than objects, the layer drop-down list reports a *ByLayer* setting.

LWEIGHT

Pull-down Menu	COMMAND (TYPE)	ALIAS (TYPE)	Short-cut	Screen (side) Menu	Tablet Menu
Format Lineweight...	LWEIGHT	LW	...	...	...

The *Lweight* command (short for lineweight) produces the *Lineweight Settings* dialog box (Fig. 11-13). This dialog box is the lineweight equivalent of the *Linetype Manager*—that is, this dialog box assigns lineweights to objects, not layers (unless the *ByLayer* or *Default* lineweight is selected). See the previous discussion under "*Linetype*." The *ByBlock* setting is discussed in Chapter 20.

Figure 11-13

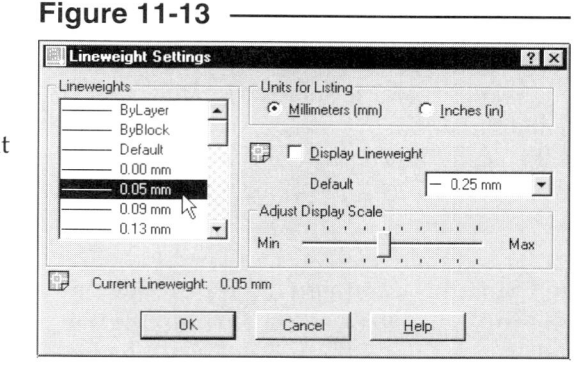

Any lineweight selected in the *Lineweight Settings* dialog box automatically becomes current (without having to select a *Current* button or double-click). The current lineweight is assigned to all subsequently drawn objects. That is, selecting a lineweight by this manner (making it current) causes all objects from that time on to be drawn using that lineweight, regardless of the layer that they are on (unless *ByLayer* or *Default* is selected). In contrast, selecting lineweights during the *Layer* command (using the *Layer Properties Manager*) results in assignment of lineweights to layers.

As a reminder, this type of drawing method (object-specific property assignment) can be difficult for beginning drawings and for complex drawings. If you want to draw objects in the layer's assigned lineweight, select *ByLayer*. If you want all layers to have the same lineweight, select *Default*.

Lineweight Control Drop-Down List

The Object Properties toolbar contains a drop-down list for selecting lineweights (Fig. 11-14). Similar to the function of the Linetype Control drop-down list, you can only assign lineweights to objects by this method unless *ByLayer* is selected to use the layers' assigned lineweights. Any lineweight you select from this list becomes the current object lineweight. If you want to select lineweights for layers, make sure this list displays the *ByLayer* setting, then use the *Layer Properties Manager* to select lineweights for layers.

Figure 11-14

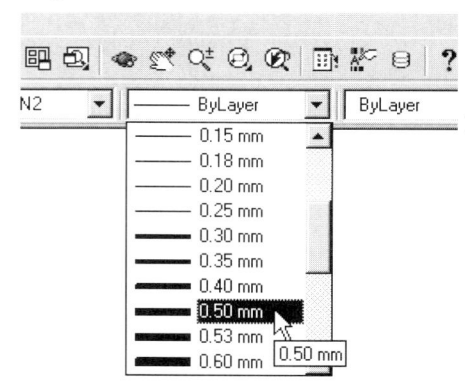

Remember that the Lineweight Control drop-down list as well as the Layer Control, Color Control, and Linetype Control drop-down lists display the current settings for selected objects (objects selected when no commands are in use). When lineweights are assigned to layers rather than objects, the layer drop-down list reports a *ByLayer* setting.

COLOR

Pull-down Menu	COMMAND (TYPE)	ALIAS (TYPE)	Short-cut	Screen (side) Menu	Tablet Menu
Format Color...	COLOR	COL	...	FORMAT Color	U,4

Similar to linetypes and lineweights, colors can be assigned to layers or to objects. Using the *Color* command assigns a color for all newly created objects, regardless of the layer's color designation (unless the *ByLayer* color is selected). This color setting overrides the layer color for any newly created objects so that all new objects are drawn with the specified color no matter what layer they are on. This type of color designation prohibits your ability to see which objects are on which layers by their color; however, for some applications object color setting may be desirable. Use the *Layer Properties Manager* to set colors for layers.

Invoking this command by the menus or by typing *Color* presents the *Select Color* dialog box shown in Figure 11-15. This is essentially the same dialog box used for assigning colors to layers; however, the *ByLayer* and *ByBlock* tiles are accessible. (The buttons are grayed-out when this dialog is invoked from the *Layer Properties Manager* because, in that case, any setting is a *ByLayer* setting.) *ByBlock* color assignment is discussed in Chapter 20.

Figure 11-15

Color Control Drop-Down List

When an object-specific color has been set, the top item in the Object Properties toolbar Color Control drop-down list displays the current color (Fig. 11-16). Beware—using this list to select a color assigns an object-specific color unless *ByLayer* is selected. Any color you select from this list becomes the current object color. Selecting *Other...* from the bottom of the list invokes the *Select Color* dialog box (see Figure 11-15). If you want to assign colors to layers, make sure this list displays the *ByLayer* setting, then use the *Layer Properties Manager* to select colors for layers.

Figure 11-16

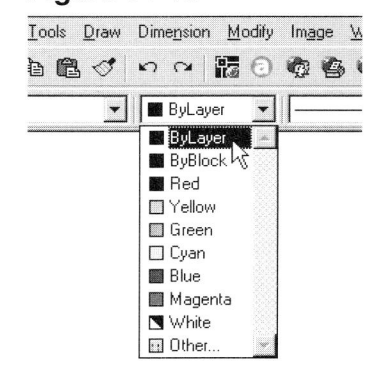

This drop-down list, as well as the others in the Object Properties toolbar, displays the current settings for selected objects (objects selected when no commands are in use). A *ByLayer* setting indicates that properties are assigned to layers rather than objects.

CONTROLLING LINETYPE SCALE

LTSCALE

Pull-down Menu	COMMAND (TYPE)	ALIAS (TYPE)	Short-cut	Screen (side) Menu	Tablet Menu
Format Linetype... Show details>> Global scale factor	LTSCALE	LTS	...	FORMAT Linetype Show details>> Global scale factor	...

Hidden, dashed, dotted, and other linetypes that have spaces are called non-continuous linetypes. When drawing objects that have non-continuous linetypes (either *ByLayer* or object-specific linetype designations), the linetype's dashes or dots are automatically created and spaced. The *LTSCALE* (Linetype Scale) system variable controls the length and spacing of the dashes and/or dots. The value that is specified for *LTSCALE* affects the drawing globally and retroactively. That is, all existing non-continuous lines in the drawing as well as new lines are affected by *LTSCALE*. You can therefore adjust the drawing's linetype scale for all lines at any time with this one command.

Figure 11-17

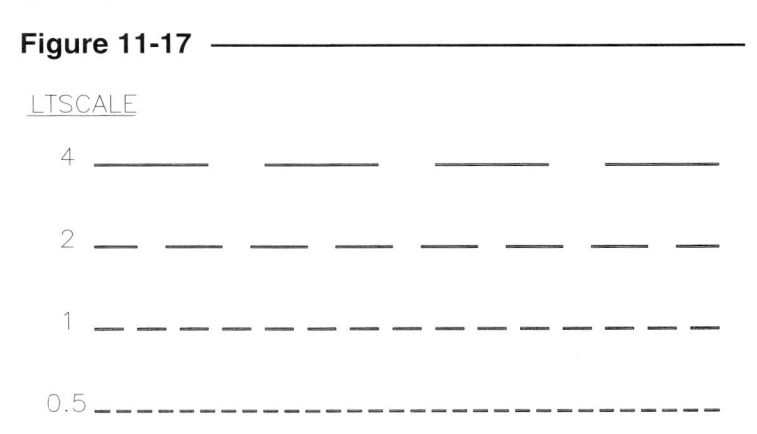

If you choose to make the dashes of non-continuous lines smaller and closer together, reduce *LTSCALE*; if you desire larger dashes, increase *LTSCALE*. The *Hidden* linetype is shown in Figure 11-17 at various *LTSCALE* settings. Any positive value can be specified.

LTSCALE can be set in the *Details* section of the *Linetype Manager* or in Command line format. In the *Linetype Manager*, select the *Details* button to allow access to the *Global scale factor* edit box (Fig. 11-18). Changing the value in this edit box sets the *LTSCALE* variable. Changing the value in the *Current object scale* edit box sets the linetype scale for the current object only (see "*CELTSCALE*"), but does not affect the global linetype scale (*LTSCALE*).

Figure 11-18

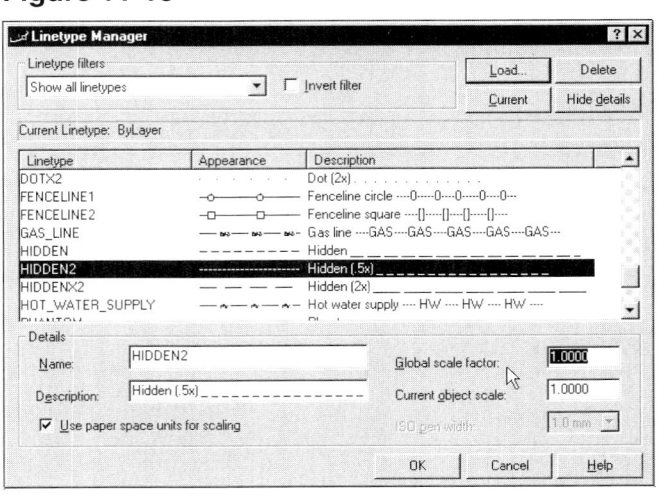

LTSCALE can also be used in the Command line format.

    Command: ltscale
    Enter new linetype scale factor <1.0000>:
    (value) (Enter any positive value.)
    Command:

The *LTSCALE* for the default template drawing (ACAD.DWT) is 1. This value represents an appropriate *LTSCALE* for objects drawn within the default *Limits* of 12 x 9.

> As a general rule, you should change the *LTSCALE* proportionally when *Limits* are changed (more specifically, when *Limits* are changed to other than the intended plot sheet size). For example, if you increase the drawing area defined by *Limits* by a factor of 2 (to 24 x 18) from the default (12 x 9), you might also change *LTSCALE* proportionally to a value of 2. Since *LTSCALE* is retroactive, it can be changed at a later time or repeatedly adjusted to display the desired spacing of linetypes.

If you load the *Hidden* linetype, it displays dashes (when plotted 1:1) of 1/4" with a *LTSCALE* of 1. The ANSI and ISO standards state that hidden lines should be displayed on drawings with dashes of approximately 1/8" or 3mm. To accomplish this, you can use the *Hidden* linetype and change *LTSCALE* to .5. It is recommended, however, that you use the *Hidden2* linetype that has dashes of 1/8" when plotted 1:1. Using this strategy, *LTSCALE* remains at a value of 1 to create the standard linetype sizes for *Limits* of 12 x 9 and can easily be changed in proportion to the *Limits*. (For more information on *LTSCALE*, see Chapter 12 and Chapter 13.)

The ACAD_ISO*n*W100 linetypes are intended to be used with metric drawings. Changing *Limits* to metric sheet sizes automatically displays these linetypes with appropriate linetype spacing. For these linetypes only, *LTSCALE* is changed automatically to the value selected in the *ISO Pen Width* box of the *Linetype* tab (see Figure 11-18). Using both ACAD_ISO*n*W100 linetypes and other linetypes in one drawing is discouraged due to the difficulty managing two sets of linetype scales.

Even though you have some control over the size of spacing for non-continuous lines, you have almost no control over the placement of the dashes for non-continuous lines. For example, the short dashes of center lines cannot always be controlled to intersect at the centers of a series of circles. The spacing can only be adjusted globally (all lines in the drawing) to reach a compromise. You can also adjust individual objects' linetype scale (*CELTSCALE*) to achieve the desired effect. See "*CELTSCALE*" (next) and Chapter 22, Multiview Drawing, for further discussion and suggestions on this subject.

CELTSCALE

Pull-down Menu	COMMAND (TYPE)	ALIAS (TYPE)	Short-cut	Screen (side) Menu	Tablet Menu
Format Linetype... Show details>> Current object scale	CELTSCALE	...	...	*FORMAT Linetype Show details>> Current object scale*	...

CELTSCALE stands for Current Entity Linetype Scale. *CELTSCALE* is actually a system variable for linetype scale stored with each object—an object property. This setting changes the object-specific linetype scale proportional to the *LTSCALE*. The *LTSCALE* value is global and retroactive, whereas *CELTSCALE* sets the linetype scale for all newly created objects and is not retroactive. *CELTSCALE* is object-specific. Using *CELTSCALE* to set an object linetype scale is similar to setting an object color, linetype, and lineweight in that the properties are assigned to the specific object.

For example, if you wanted all non-continuous lines (dashes and spaces) in the drawing to be two times the default size, set *LTSCALE* to 2 (*LTSCALE* is global and retroactive). If you then wanted only, say, a select two or three lines to have smaller spacing, change *CELTSCALE* to .5, draw the new lines, and then change *CELTSCALE* back to 1.

You can also use the *Linetype Manager* to set the *CELTSCALE* (see Figure 11-18). Set the desired *CELTSCALE* by listing the *Details* and changing the value in the *Current object scale* edit box. This setting affects all linetypes for all newly created objects. The linetype scale for individual objects can also be changed retroactively using the method explained in the following paragraph.

Using <u>CELTSCALE is not the recommended method</u> for adjusting individual lines' linetype scale. Setting this variable each time you wanted to create a new entity with a different linetype scale would be too time consuming and confusing. The <u>recommended</u> method for adjusting linetypes in the drawing to different scales is <u>not to use CELTSCALE</u>, but to use the following strategy.

1. Set *LTSCALE* to an appropriate value for the drawing.
2. Create all objects in the drawing with the desired linetypes.
3. Adjust the *LTSCALE* again if necessary to globally alter the linetype scale.
4. Use *Properties* or *Matchprop* to <u>retroactively</u> change the linetype scale (*CELTSCALE*) <u>for selected objects</u> (see *"Properties"* and *"Matchprop"*).

CHANGING OBJECT PROPERTIES RETROACTIVELY

Often it is desirable to change the properties of an object after it has been created. For example, an object's *Layer* property could be changed. This can be thought of as "moving" the object from one layer to another. When an object is changed from one layer to another, it assumes the new layer's color, linetype, and lineweight, provided the object was originally created with color, linetype, and lineweight assigned *ByLayer*, as is generally the case. In other words, if an object was created on the wrong layer (possibly with the wrong linetype or color), it could be "moved" to the desired layer, therefore assuming the new layer's linetype and color.

Another example is to change an individual object's linetype scale. In some cases an individual object's linetype scale requires an adjustment to other than the global linetype scale (*LTSCALE*) setting. One of several methods can be used to adjust the individual object's linetype scale (*CELTSCALE*) to an appropriate value.

Several methods can be used to retroactively change properties of selected objects. The Object Properties toolbar, the *Properties* window, and the *Match Properties* command can be used to <u>retroactively</u> change the properties of individual objects. Properties that can be changed by these three methods are *Layer*, *Linetype*, *Lineweight*, *Color*, (object-specific) *Linetype scale*, *Plot Style*, and other properties.

Although some of the commands discussed in this section have additional capabilities, the discussion is limited to changing the object properties covered in this chapter—specifically layer, color, linetype, and lineweight. For this reason, these commands and features are also discussed in Chapter 16, Modify Commands II, and in other chapters.

Object Properties Toolbar

The five drop-down lists in the Object Properties toolbar (when not "dropped down") generally show the current layer, color, linetype, lineweight, and plot style. However, if an object or set of objects is selected, the information in these lists changes to display the current objects' settings. <u>You can change the selected objects' settings</u> by "dropping down" any of the lists and making another selection.

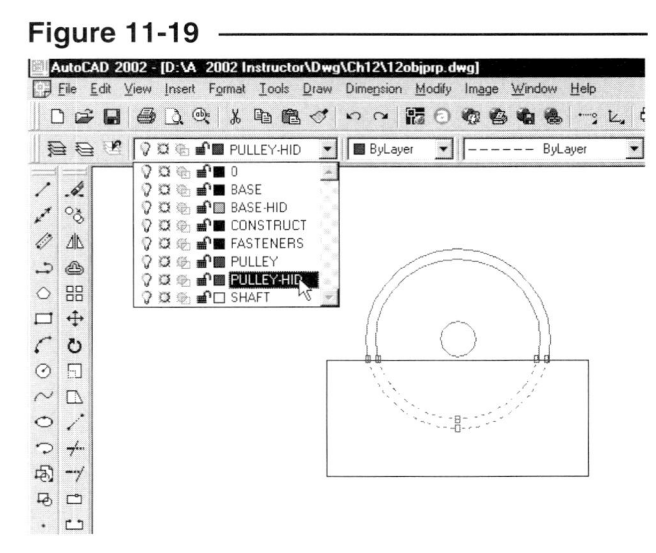

Figure 11-19

First, select (highlight) an object when no commands are in use. Use the pickbox (that appears on the crosshairs), window or crossing window to select the desired object or set of objects. The entries in the five lists (Layer Control, Color Control, Linetype Control, Lineweight Control, and Plot Style Control) then change to display the settings for the selected object or objects. If several objects are selected that have different properties, the boxes display no information (go "blank"). Next, use any of the drop-down lists to make another selection (see Figure 11-19). The highlighted object's properties are changed to those selected in the lists. Press the Escape key to complete the process.

TIP: Remember that in most cases color, linetype, and lineweight settings are assigned *ByLayer*. In this type of drawing scheme, to change the linetype or color properties of an object, you would change the object's layer (see Figure 11-19). If you are using this type of drawing scheme, refrain from using the Color Control, Linetype Control, and Lineweight Control drop-down lists for changing properties.

PROPERTIES

Pull-down Menu	COMMAND (TYPE)	ALIAS (TYPE)	Short-cut	Screen (side) Menu	Tablet Menu
Modify Properties	PROPERTIES	PROPS or CH	(Edit Mode) Properties or Ctrl+1	MODIFY1 Property	Y,12 to Y,13

The *Properties* window (Fig. 11-20) gives you complete access to one or more objects' properties. The contents of the window change based on what type and how many objects are selected.

You can use the window two ways: you can invoke the window, then select (highlight) one or more objects, or you can select objects first and then invoke the window. Once opened, this window remains on the screen until dismissed by clicking the "X" in the upper right corner. You can even toggle the window on and off with Ctrl+1 (to appear and disappear). When multiple objects are selected, a drop-down list in the top of the window appears. Select the object(s) whose properties you want to change.

Figure 11-20

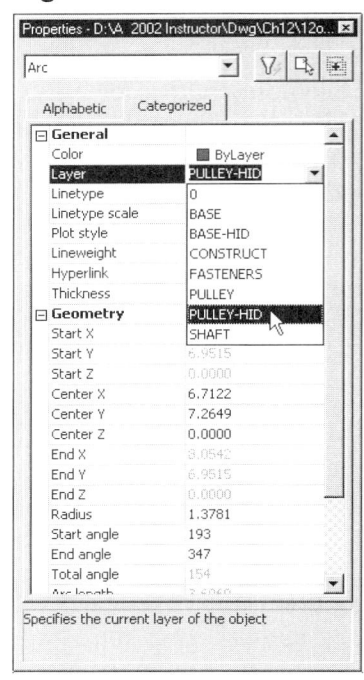

To change an object's layer, linetype, lineweight, or color properties, highlight the desired objects in the drawing and select the properties you want to change from the right side of the window. In the *Categorized* tab, the layer, linetype, lineweight, and color properties are located in the top half of the list.

TIP: If you use the *ByLayer* strategy of linetype, lineweight, and color properties assignment, you can change an object's linetype, lineweight, and color simply by changing its layer (see Figure 11-20). This method is recommended for most applications.

MATCHPROP

Pull-down Menu	COMMAND (TYPE)	ALIAS (TYPE)	Short-cut	Screen (side) Menu	Tablet Menu
Modify Match Properties	MATCHPROP or PAINTER	MA	...	MODIFY1 Matchprp	Y,14 and Y,15

Matchprop is used to "paint" the properties of one object to another. The process is simple. After invoking the command, select the object that has the desired properties (source object), then select the object you want to "paint" the properties to (destination object). The command prompt is as follows.

    Command: matchprop
    Select source object: PICK
    Current active settings: Color Layer Ltype Ltscale Lineweight Thickness PlotStyle Text Dim Hatch
    Select destination object(s) or [Settings]: PICK
    Select destination object(s) or [Settings]: Enter
    Command:

Only one "source object" can be selected, but its properties can be painted to several "destination objects." The "destination object(s)" assume all of the properties of the "source object" (listed as "Current active settings").

Use the *Settings* option to control which of several possible properties and other settings are "painted" to the destination objects. At the "Select destination object(s) or [Settings]:" prompt, type S to display the *Property Settings* dialog box (Fig. 11-21). In the dialog box, designate which of the *Basic Properties* or *Special Properties* are to be painted to the "destination objects." The following *Basic Properties* correspond to properties discussed in this chapter.

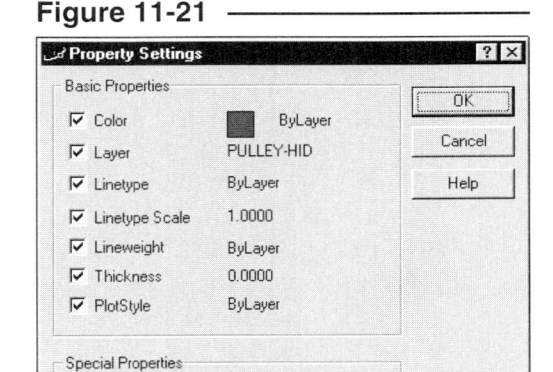

Figure 11-21

Color	paints the object-specific or *ByLayer* color
Layer	moves selected objects to Source Object layer
Linetype	paints the object-specific or *ByLayer* linetype
Lineweight	paints the object-specific or *ByLayer* lineweight
Linetype Scale	changes the individual object's linetype scale (*CELTSCALE*), not global (*LTSCALE*)

The *Special Properties* of the *Property Settings* dialog box are discussed in Chapter 16, Modify Commands II. Also see Chapter 16 for a full explanation of the *Properties* window (*Properties*).

CHAPTER EXERCISES

1. *Layer Properties Manager,* **Layer Control drop-down list,** and *Make Object's Layer Current*

   Open the **WILHOME.DWG** sample drawing located in the C:\Program Files\AutoCAD 2002\Sample\ directory. Activate model space by selecting the *Model* tab. Next, use *Saveas* to save the drawing in your working directory as **WILHOME2**.

   A. Invoke the *Layer* command in Command line format by typing *-LAYER* or *-LA*. Use the *?* option to yield the list of layers in the drawing. Notice that all the layers have the default linetype except one. Which layer has a different linetype?

   B. Use the **Object Properties** toolbar to indicate the layers of selected items. On the right side of the drawing, **PICK** the yellow lines around the floor plan. What layer are the dimensions on? Press the **Esc** key twice to cancel the selection. Next, **PICK** the cyan (light blue) text around the semi-circular room near the top of the floor plan on the right. What layer is the text on? Press the **Esc** key twice to cancel the selection.

   C. Invoke the *Layer Properties Manager* by selecting the icon button or using the *Format* pull-down menu. Turn *Off* the **Dimensions** layer (click the light bulb icon). Select *OK* to return to the drawing and check the new setting. Are the dimensions displayed? Next, use the *Layerp* command to turn the **Dimensions** layer back *On*.

   D. Type *Regen* and count the number of seconds it takes for the drawing to regenerate. Next, use the **Object Properties** toolbar to indicate the name of the layer of the green elevations on the left side of the drawing (**PICK** a green object and note the layer name that appears). Use the **Layer Control** drop-down list to turn *Off* that layer (light bulb icon) and the layer with the cyan text. Now type *Regen* again and count the number of seconds it takes for the regeneration. Is there any change? Finally, *Freeze* both layers (snowflake/sun icon), then type *Regen* again and count the number of seconds it takes for the regeneration. Is there any difference between *Off* and *Freeze* for regenerations?

   E. Use the *Layer Properties Manager* and **PICK** the *Select All* option (right-click for menu). *Freeze* all layers. Which layer will not *Freeze*? Notice the light bulb icon indicates the layers are still *On* although the layers do not appear in the drawing (*Freeze* overrides *On*). Use the same procedure to select all layers and *Thaw* them.

   F. Use the *Layer Properties Manager* again and drop down the *Named Layer Filters* list. Are there any layers in the drawing that are unused (select *Show All Used Layers*, then *Invert Filter*)? Use the *Named Layer Filters* list again and list *All* layers.

   G. Now we want to *Freeze* all the *Yellow* layers. Using the *Layer Properties Manager* again, sort the list by color. **PICK** the first layer name in the list of the four yellow ones. Now hold down the **Shift** key and **PICK** the last name in the list. All four layers should be selected. *Freeze* all four by selecting any one of the sun/snowflake icons. Select *OK* to return to the drawing.

   H. Assuming you wanted to work only with the layers beginning with "AR," you can set the filter to display only those layers. First, activate the *Named Layer Filters* dialog by selecting the small button to the right of the drop-down list of the *Layer Properties Manager.* In the *Named Layer Filters* dialog box, enter **AR*** in the *Layer Names* edit box, then assign the name **AR Layers** in the *Filter Name* edit box and select *Add*. *Close* the dialog box. In the *Layer Properties Manager*, check *Apply to Object Properties toolbar.* Find **AR Layers** in the *Named layer filters* drop-down list and select it. Examine the list.

Now select **OK** in the *Layer Properties Manager* to return to the drawing. Examine the **Layer Control** drop-down list. Does it display only layer names beginning with AR? Return to the *Layer Properties Manager* and select *Show All Layers* in the *Named layer filters* list. Select **OK** and examine the **Layer Control** drop-down list again. It should also show all layer names in the list.

I. Ensure all layers are *Thawed*. Sort the list alphabetically by *Name*. *Freeze* all layers beginning with **A** and **B**. Select **OK** and then use the drop-down list to display the names. Assuming you were working only with the thawed layers, it would be convenient to display <u>only</u> those names in the drop-down list. Use the *Named Layer Filters* dialog box to set a filter named **Thawed Layers**. In the *Layer Properties Manager*, find **Thawed Layers** in the *Named layer filters* drop-down list and select it. Ensure the *Apply to Object Properties toolbar* checkbox is checked. Return to the drawing and examine the drop-down list. Do only names of *Thawed* layers appear?

J. Next, you will "move" objects from one layer to another using the **Object Properties** toolbar. Make layer **0** the current layer by using the drop-down list and selecting the name. Use the *Layer Properties Manager* and right-click to *Select All* names. **PICK** one icon to *Freeze* all layers (except the current layer). Then select *Clear All* and *Thaw* layer **General Note**. Select **OK** and return to the drawing. Use a crossing window to select all of the objects in the drawing (the small blue Grips may appear). The **Object Properties** toolbar should indicate the layer on which the objects reside. Next, list the layers using the drop-down list and **PICK** layer **0**. Wait several seconds. Press **Esc** twice when the drawing regenerates. The objects should change color and be on layer 0. You actually "moved" the objects to layer 0. Finally, type *U* to undo the last action so the notes are on layer Gennote (cyan). Also, *Thaw* all layers.

K. Let's assume you want to draw some additional objects on the text layer and on the dimensions layer. First, *Zoom* in to the upper half of the floor plan on the right. Now select the *Make Object's Layer Current* icon button (on the far left of the Object Properties toolbar). At the "Select object whose layer will become current" prompt, select a text object (cyan color). The **General Note** layer name should appear in the Layer Control box as the current layer. Draw a *Line* and it should be on layer Gennote and cyan in color. Next, use *Make Object's Layer Current* again and select a dimension object. The **Dimensions** layer should appear as the current layer. Draw another *Line*. It should be on layer Dimensions and yellow in color. Keep in mind that this method of setting a layer current works especially well when you do not know the layer names or know what objects are on which layers.

L. *Close* the drawing. Do not save the changes.

2. *Layer Properties Manager* and *Linetype Manager*

   A. Begin a *New* drawing and use *Save* to assign the name **CH11EX2**. Set *Limits* to **11 x 8.5**, then *Zoom All*. Use the *Linetype Manager* to list the loaded linetypes. Are any linetypes already loaded? **PICK** the *Load* button then right-click to *Select All* linetypes. Select **OK**, then close the *Linetype Manager*.

   B. Use the *Layer Properties Manager* to create 3 new layers named **OBJ, HID,** and **CEN**. Assign the following colors, linetypes, and lineweights to the layers by clicking on the *Color, Linetype,* and *Lineweight* columns.

OBJ	*Red*	*Continuous*	*Default*
HID	*Yellow*	*Hidden2*	*Default*
CEN	*Green*	*Center2*	*Default*

224 Layers and Object Properties

C. Make the **OBJ** layer *Current* and PICK the *OK* tile. Verify the current layer by looking at the *Layer Control* drop-down list, then draw the visible object *Lines* and the *Circle* only (not the dimensions) as shown in Figure 11-22.

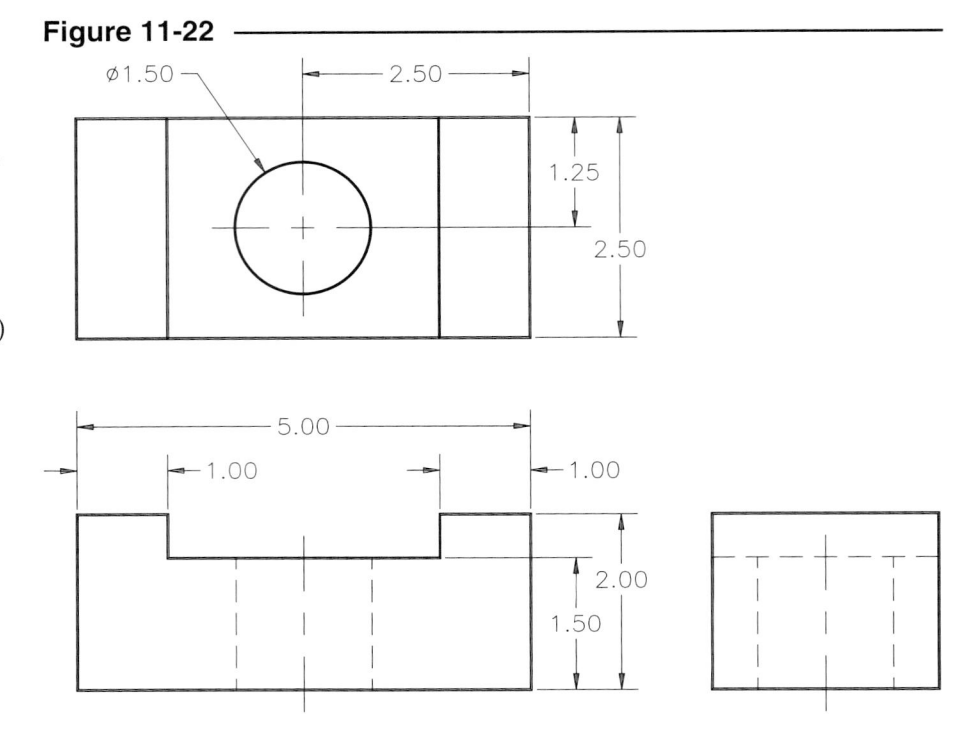

Figure 11-22

D. When you are finished drawing the visible object lines, create the necessary hidden lines by making layer **HID** the *Current* layer, then drawing *Lines*. Notice that you only specify the *Line* endpoints as usual and AutoCAD creates the dashes.

E. Next, create the center lines for the holes by making layer **CEN** the *Current* layer and drawing *Lines*. Make sure the center lines extend <u>slightly beyond</u> the *Circle* and beyond the horizontal *Lines* defining the hole. *Save* the drawing.

F. Now create a *New* layer named **BORDER** and draw a border and title block of your design on that layer. The final drawing should appear as that in Figure 11-23.

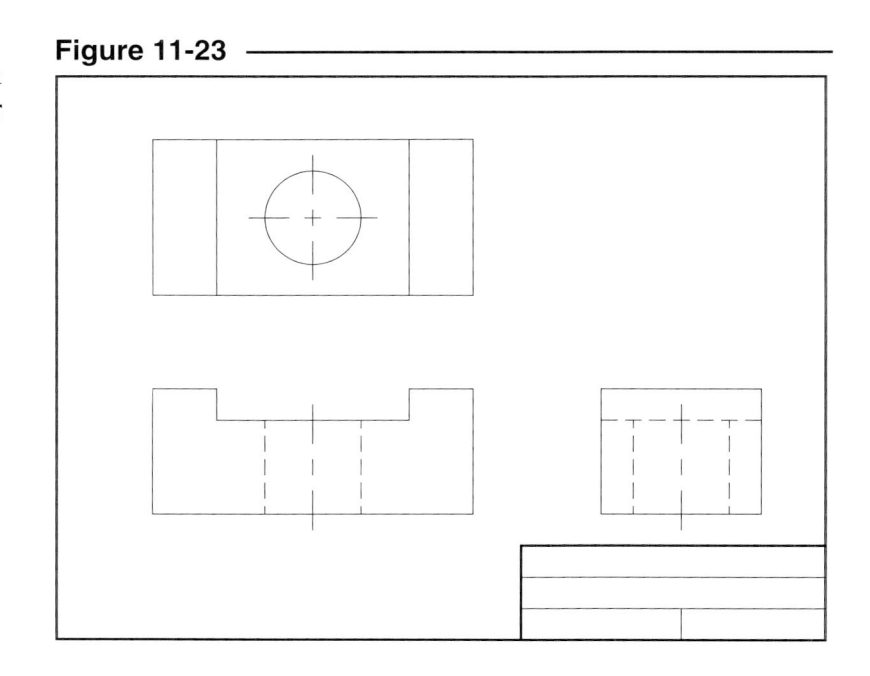

Figure 11-23

G. Open the *Linetype Manager* again and right-click to *Select All*, then select the *Delete* button. All unused linetypes should be removed from the drawing.

H. *Save* the drawing, then *Close* the drawing.

3. *LTSCALE, Properties* **window**, *Matchprop*

   A. Begin a *New* drawing and use *Save* to assign the name **CH11EX3**. Create the same four *New* layers that you made for the previous exercise (**OBJ, HID, CEN**, and **BORDER**) and assign the same linetypes, lineweights, and colors. Set the *Limits* equal to a "C" size sheet, **22 x 17,** then *Zoom All*.

   B. Create the part shown in Figure 11-24. Draw on the appropriate layers to achieve the desired linetypes.

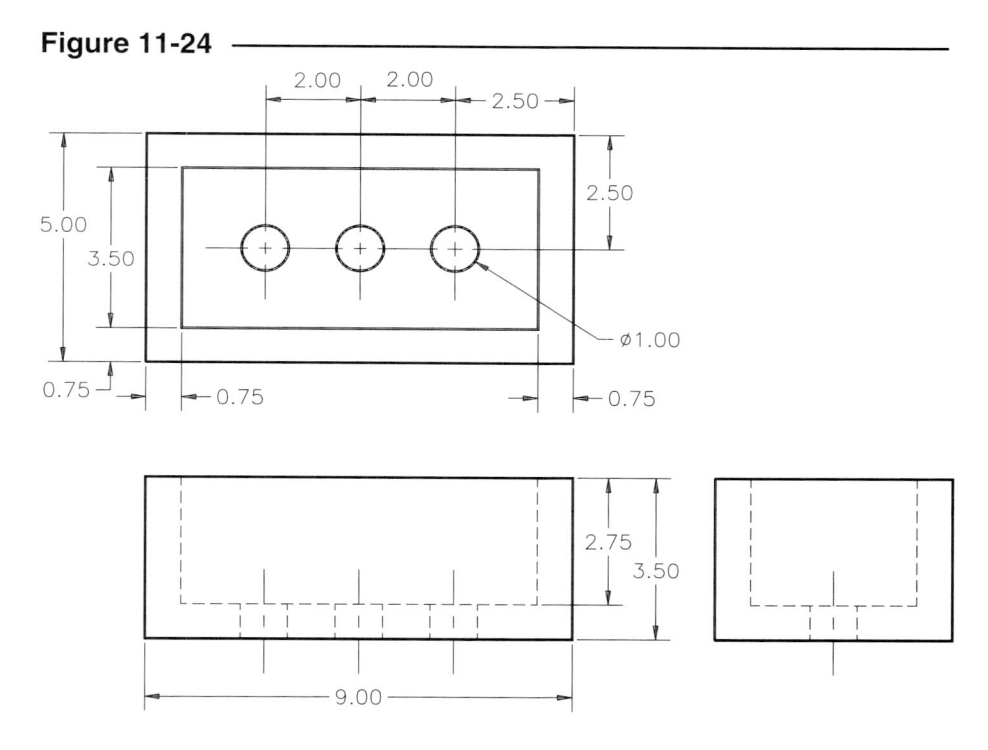

   Figure 11-24

   C. Notice that the *Hidden* and *Center* linetype dashes are very small. Use **LTSCALE** to adjust the scale of the non-continuous lines. Since you changed the *Limits* by a factor of slightly less than 2, try using **2** as the **LTSCALE** factor. Notice that all the lines are affected (globally) and that the new *LTSCALE* is retroactive for existing lines as well as for new lines. If the dashes appear too long or short because of the small hidden line segments, *LTSCALE* can be adjusted. Remember that you cannot control where the multiple short centerline dashes appear for one line segment. Try to reach a compromise between the hidden and center line dashes by adjusting the *LTSCALE*.

D. The six short vertical hidden lines in the front view and two in the side view should be adjusted to a smaller linetype scale. Invoke the *Properties* window by selecting **Properties** from the *Modify* pull-down menu or selecting the **Properties** icon button. Select <u>only the two short lines in the side view</u>. Use the window to retroactively adjust the two individual objects' **Linetype Scale** to **0.6000**.

E. When the new *Linetype Scale* (actually, the *CELTSCALE*) for the two selected lines is adjusted correctly, use *Matchprop* to "paint" the *Linetype Scale* to the several other short lines in the front view. Select **Match Properties** from the *Modify* pull-down menu or select the "paint brush" icon. When prompted for the "source object," select one of the two short vertical lines in the side view. When prompted for the "destination object(s)," select <u>each</u> of the short lines in the front view. This action should "paint" the *Linetype Scale* to the lines in the front view.

F. When you have the desired linetype scales, **Exit** AutoCAD and **Save Changes**. Keep in mind that the drawing size may appear differently on the screen than it will on a print or plot. Both the plot scale and the paper size should be considered when you set *LTSCALE*. This topic will be discussed further in Chapter 12, Advanced Drawing Setup.

12

ADVANCED DRAWING SETUP

Chapter Objectives

After completing this chapter you should:

1. know the steps for setting up a drawing;

2. be able to determine an appropriate *Limits* setting for the drawing;

3. be able to calculate and apply the "drawing scale factor";

4. be able to access existing and create new template drawings;

5. know what setup steps can be considered for creating template drawings.

CONCEPTS

When you begin a drawing, there are several steps that are typically performed in preparation for creating geometry, such as setting *Units, Limits,* and creating *Layers* with *linetypes, lineweights,* and *colors.* Some of these basic concepts were discussed in Chapters 6 and 11. This chapter discusses setting *Limits* for correct plotting as well as other procedures, such as layer creation and variables settings, that help prepare a drawing for geometry creation.

To correctly set up a drawing for printing or plotting to scale, <u>any two</u> of the following three variables must be known. The third can be determined from the other two.

Drawing *Limits*
Print or plot sheet (paper) size
Print or plot scale

The method given in this text for setting up a drawing (Chapters 6 and 12) is intended for plotting from both the *Model* tab (model space) and from *Layout* tabs (paper space). The method applies when you plot from the *Model* tab or plot with <u>one viewport</u> in a *Layout* tab. If multiple viewports are created in one layout, the same general steps would be taken; however, the plot scale might vary for each viewport based on the size and number of viewports. (See Chapters 13 and 14 for more information on layouts, viewports, and plotting layouts.)

> **TIP** Rather than performing the steps for drawing setup each time you begin, you can use "template drawings." Template drawings have many of the setup steps performed but contain no geometry. AutoCAD provides several template drawings and you can create your own template drawings. A typical engineering, architectural, design, or construction office generally produces drawings that are similar in format and can benefit from creation and use of individualized template drawings. The drawing similarities may be subject of the drawing (geometry), plot scale, sheet size, layering schemes, dimensioning styles, and/or text styles. Template drawing creation and use are discussed in this chapter.

STEPS FOR DRAWING SETUP

Assuming that you have in mind the general dimensions and proportions of the drawing you want to create, and the drawing will involve using layers, dimensions, and text, the following steps are suggested for setting up a drawing:

1. Determine and set the *Units* that are to be used.
2. Determine and set the drawing *Limits;* then *Zoom All.*
3. Set an appropriate *Snap Type* (*Polar Snap* or *Grid Snap*), *Snap* spacing, and *Polar* spacing.
4. Set an appropriate *Grid* value.
5. Change the *LTSCALE* value based on the new *Limits.* Set *PSLTSCALE* to 0.
6. Create the desired *Layers* and assign appropriate *linetype, lineweight,* and *color* settings.
7. Create desired *Text Styles* (optional, discussed in Chapter 18).
8. Create desired *Dimension Styles* (optional, discussed in Chapter 27).
9. Activate a *Layout* tab, set it up for the plot or print device and paper size, and create a viewport (if not already existing).
10. Create or insert a title block and border in the layout.

Each of the steps for drawing setup is explained in detail here.

1. **Set *Units***

    This task is accomplished by using the *Units* command or the *Quick Setup* or *Advanced Setup* wizard. Set the linear units and precision desired. Set angular units and precision if needed. (See Chapter 6 for details on the *Units* command and setting *Units* using the wizards.)

2. **Set *Limits***

   Before beginning to create an AutoCAD drawing, determine the size of the drawing area needed for the intended geometry. Using the actual *Units,* appropriate *Limits* should be set in order to draw the object or geometry to the real-world size. *Limits* are set with the *Limits* command by specifying the lower-left and upper-right corners of the drawing area. Always *Zoom All* after changing *Limits*. *Limits* can also be set using the *Quick Setup* or *Advanced Setup* wizard. (See Chapter 6 for details on the *Limits* command and the setup wizards.)

   If you are planning to plot the drawing to scale, *Limits* should be set to a proportion of the sheet size you plan to plot on. For example, if the sheet size is 11" x 8.5", set *Limits* to 11 x 8.5 if you want to plot full size (1"=1"). Setting *Limits* to 22 x 17 (2 times 11 x 8.5) provides 2 times the drawing area and allows plotting at 1/2 size (1/2"=1") on the 11" x 8.5" sheet. Simply stated, set *Limits* to a proportion of the paper size.

   Setting *Limits* to the paper size allows plotting at 1=1 scale. Setting *Limits* to a proportion of the sheet size allows plotting at the reciprocal of that proportion. For example, setting *Limits* to 2 times an 11" x 8.5" sheet allows you to plot 1/2 size on that sheet. Or setting *Limits* to 4 times an 11" x 8.5" sheet allows you to plot 1/4 size on that sheet. (Standard paper sizes are given in Chapter 14.)

   Even if you plan to use a layout tab (paper space) and create one viewport, setting model space *Limits* to a proportion of the paper size is recommended. In this way you can easily calculate the viewport scale (the proportion of paper space units to model space units); therefore, the model space geometry can be plotted to a standard scale.

   If you plan to plot from a layout, *Limits* in paper space should be set to the actual paper size; however, setting the paper space *Limits* values is automatically done for you when you select a *Plot Device* and *Paper Size* from the *Page Setup* or *Plot* dialog box (see Chapter 13, Layouts and Viewports, and Chapter 14, Printing and Plotting).

   To set model space *Limits*, you can use the individual commands for setting *Units* and *Limits* (*Units* command and *Limits* command) or use the *Quick Setup* wizard or *Advanced Setup* wizard.

Drawing Scale Factor

The proportion of the *Limits* to the intended print or plot sheet size is the "drawing scale factor." This factor can be used as a general scale factor for other size-related drawing variables such as *LTSCALE,* dimension *Overall Scale (DIMSCALE),* and *Hatch* pattern scale. The drawing scale factor can also be used to determine the viewport scales.

Most size-related AutoCAD drawing variables are set to 1 by default. This means that variables (such as *LTSCALE*) that control sizing and spacing of objects are set appropriately for creating a drawing plotted full size (1=1).

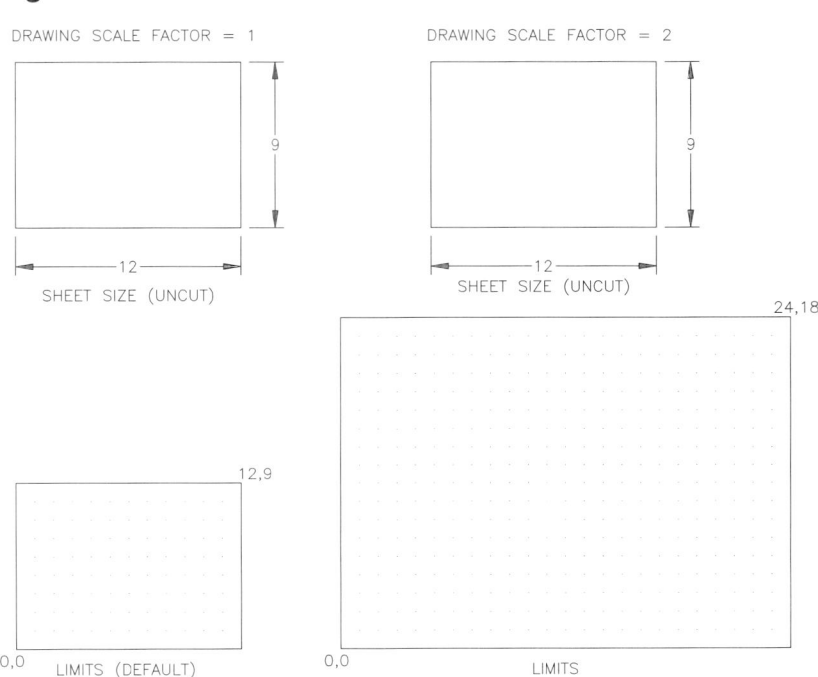

Figure 12-1

Therefore, when you set *Limits* to the sheet size and print or plot at 1=1, the sizing and spacing of linetypes and other variable-controlled objects are correct. When *Limits* are changed to some proportion of the sheet size, the size-related variables should also be changed proportionally. For example, if you intend to plot on a 12 x 9 sheet and the default *Limits* (12 x 9) are changed by a factor of 2 (to 24 x 18), then 2 becomes the drawing scale factor. Then, as a general rule, the values of variables such as *LTSCALE, DIMSCALE (Overall Scale)*, and other scales should be multiplied by a factor of 2 (Fig. 12-1, on the previous page). When you then print the 24 x 18 area onto a 12" x 9" sheet, the plot scale would be 1/2, and all sized features would appear correct.

Limits should be set to the paper size or to a proportion of the paper size used for plotting or printing. In many cases, "cut" paper sizes are used based on the 11" x 8.5" module (as opposed to the 12" x 9" module called "uncut sizes"). Assume you plan to print on an 11" x 8.5" sheet. Setting *Limits* to the sheet size provides plotting or printing at 1=1 scale. Changing the *Limits* to a proportion of the sheet size <u>by some multiplier</u> makes that value the drawing scale factor. The <u>reciprocal</u> of the drawing scale factor is the scale for plotting on that sheet (Fig. 12-2).

Figure 12-2

```
        |←— 11 —→|
        ┌─────────┐   ↑
        │  PLOT   │
        │  SHEET  │  8 1/2
        │         │   ↓
        └─────────┘
      (CUT PAPER SIZE)
```

LIMITS	SCALE FACTOR	PLOT SCALE	
11 x 8.5	1	1=1	(FULL SIZE)
22 x 17	2	1=2	(HALF SIZE)
44 x 34	4	1=4	(1/4 SIZE)
110 x 85	10	1=10	(1/10 SIZE)

Since the drawing scale factor (DSF) is the proportion of *Limits* to the sheet size, you can use this formula:

$$DSF = \frac{Limits}{Sheet\ size}$$

Because plot/print scale is the reciprocal of the drawing scale factor:

$$Plot\ scale = \frac{1}{DSF}$$

The term "drawing scale factor" is not a variable or command that can be found in AutoCAD or its official documentation. This concept has been developed by AutoCAD users to set system variables appropriately for printing and plotting drawings to standard scales.

3. **Set *Snap***

   Use the *Snap* command or *Snap and Grid* tab of the *Drafting Settings* dialog box to set the *Snap* type (*Grid Snap* and/or *Polar Snap*) and appropriate *Snap Spacing* and/or *Polar Spacing* values. The *Snap Spacing* and *Polar Spacing* values are dependent on the <u>interactive</u> drawing accuracy that is desired.

   The accuracy of the drawing and the size of the *Limits* should be considered when setting the *Snap Spacing* and *Polar Spacing* values. On one hand, to achieve accuracy and detail, you want the values to be the smallest dimensional increments that would commonly be used in the drawing. On the other hand, depending on the size of the *Limits*, the *Snap* value should be large enough to make interactive selection (PICKing) fast and easy. As a starting point for determining appropriate *Snap Spacing* and *Polar Spacing* values, the default values can be multiplied by the "drawing scale factor."

   If you are creating a template drawing (.DWT file), it is of no help to set the *Snap Type* (Grid Snap or Polar Snap), *Polar Spacing* value, or *Polar Tracking* angles since these values are stored in the system registry, not in the drawing. The *Grid Snap Spacing* value, however, is stored in the drawing file.

4. **Set *Grid***

   The *Grid* value setting is usually set equal to or proportionally larger than that of the *Grid Snap* value. *Grid* should be set to a proportion that is easily visible and to some value representing a regular increment (such as .5, 1, 2, 5, 10, or 20). Setting the value to a proportion of *Grid Snap* gives visual indication of the *Grid Snap* increment. For example, a *Grid* value of 1X, 2X, or 5X would give visual display of every 1, 2, or 5 *Grid Snap* increments, respectively. If you are not using *Snap* because only extremely small or irregular interval lengths are needed, you may want to turn *Grid* off. (See Chapter 6 for details on the *Grid* command.)

5. **Set the *LTSCALE* and *PSLTSCALE***

   A change in *Limits* (then *Zoom All*) affects the display of non-continuous (hidden, dashed, dotted, etc.) lines. The *LTSCALE* variable controls the spacing of non-continuous lines. The drawing scale factor can be used to determine the *LTSCALE* setting. For example, if the default *Limits* (based on sheet size) have been changed by a factor of 2, the drawing scale factor=2; so set the *LTSCALE* to 2. In other words, if the DSF=2 and plot scale=1/2, you would want to double the linetype dashes to appear the correct size in the plot, so *LTSCALE*=2. If you prefer a *LTSCALE* setting of other than 1 for a drawing in the default *Limits*, multiply that value by the drawing scale factor.

   American National Standards Institute (ANSI) requires that hidden lines be plotted with 1/8" dashes. An AutoCAD drawing plotted full size (1 unit=1") with the default *LTSCALE* value of 1 plots the *Hidden* linetype with 1/4" dashes. Therefore, it is recommended for English drawings that you use the *Hidden2*, *Center2*, and other *2 linetypes with dash lengths .5 times as long so an *LTSCALE* of 1 produces the ANSI standard dash lengths. Multiply your *LTSCALE* value by the drawing scale factor when plotting to scale. If individual lines need to be adjusted for linetype scale, use *Properties* when the drawing is nearly complete. (See Chapter 11 for details on the *LTSCALE* and *CELTSCALE* variables.)

   Assuming you are planning to create only one viewport in a *Layout* tab, set the *PSLTSCALE* value (paper space *LTSCALE*) to 0. AutoCAD's default setting is 1. A setting of 0 forces the linetype scale to appear the same in *Layout* tabs as it does in the *Model* tab. To set the *PSLTSCALE* value, similar to setting the *LTSCALE*, simply type *PSLTSCALE* at the Command prompt. (See Chapter 13 for details on *PSLTSCALE*.)

6. **Create *Layers*; Assign *Linetypes*, *Lineweights*, and *Colors***

   Use the *Layer Properties Manager* command to create the layers that you anticipate needing. Multiple layers can be created with the *New* option. You can type in several new names separated by commas. Assign a descriptive name for each layer, indicating its type of geometry, part name, or function. Include a *linetype* designator in the layer name if appropriate; for example, PART1-H and PART1-V indicate hidden and visible line layers for PART1.

   Once the *Layers* have been created, assign a *Color*, *Linetype*, and *Lineweight* to each layer. *Colors* can be used to give visual relationships to parts of the drawing. Geometry that is intended to be plotted with different pens (pen size or color) or printed with different plotted appearances should be drawn in different screen colors (especially when color-dependent plot styles are used). Use the *Linetype* command to load the desired linetypes. You should also select *Lineweights* for each layer (if desired) using the *Layer Properties Manager*. (See Chapter 11 for details on creating layers and assigning linetypes, lineweights, and colors.)

7. **Create Text Styles**

   AutoCAD has only one text style as part of the standard template drawings (ACAD.DWT and ACADISO.DWT) and one or two text styles for most other template drawings. If you desire other *Text Styles*, they are created using the *Style* command or the *Text Style...* option from the *Format* pull-down menu. (See Chapter 18, Creating and Editing Text.) If you desire engineering standard text, create a *Text Style* using the ROMANS.SHX or .TTF font file.

8. Create Dimension Styles

If you plan to dimension your drawing, Dimension Styles can be created at this point; however, they are generally created during the dimensioning process. Dimension Styles are names given to groups of dimension variable settings. (See Chapter 27 for information on creating Dimension Styles.)

Although you do not have to create Dimension Styles until you are ready to dimension the geometry, it is helpful to create Dimension Styles as part of a template drawing. If you produce similar drawings repeatedly, your dimensioning techniques are probably similar. Much time can be saved by using a template drawing with previously created Dimension Styles. Several of the AutoCAD-supplied template drawings have prepared Dimension Styles.

9. Activate a *Layout* Tab, Set the Plot Device, and Create a Viewport

This step can be done just before making a print or plot, or you can complete this and the next step before creating your drawing geometry. Completing these steps early allows you to see how your printed drawing might appear and how much area is occupied by the titleblock and border.

Activate a *Layout* tab. If your *Options* are set as explained in Chapter 6, Basic Drawing Setup, the layout is automatically set up for your default print/plot device and an appropriate-sized viewport is created. Otherwise, use the *Page Setup* dialog box (*Pagesetup* command) to set the desired print/plot device and paper size. Then use the *Vports* command to create one or more viewports. This topic is discussed briefly in Chapter 6, Basic Drawing Setup, and is discussed in more detail in Chapter 13, Layouts and Viewports.

10. Create a Titleblock and Border

For 2D drawings, it is helpful to insert a titleblock and border early in the drawing process. This action gives a visual drawing boundary as well as reserves the space needed by the titleblock.

Rather than create a new titleblock and border for each new drawing, a common practice is to use the *Insert* command to insert an existing titleblock and border as a *Block*.

You can draw or *Insert* a titleblock and border in the *Model* tab (model space) or into a layout (paper space), depending on how you want to plot the drawing. If you plan to plot from the *Model* tab, multiply the actual size of the titleblock and border by the drawing scale factor to determine their sizes for the drawing. If you want to plot from a layout, draw or insert the titleblock and border at the actual size (1:1) since a layout represents the plot sheet. (See Chapter 20 for information on *Block* creation and *Insertion*.)

USING AND CREATING TEMPLATE DRAWINGS

Instead of going through the steps for setup each time you begin a new drawing, you can create one or more template drawings or use one of the AutoCAD-supplied template drawings. AutoCAD template drawings are saved as a .DWT file format. A template drawing has the initial setup steps (*Units*, *Limits*, *Layers*, *Linetypes*, *Lineweights*, *Colors*, etc.) completed and saved, but no geometry has been created yet. For some templates, *Layout* tabs have been set up with titleblocks, viewports, and plot settings. Template drawings are used as a template or starting point each time you begin a new drawing. AutoCAD actually makes a copy of the template you select to begin the drawing. The creation and use of template drawings can save hours of preparation.

Using Template Drawings

To use a template drawing, select the *Template* option in the *Startup* or *Create New Drawing* dialog box (Fig. 12-3, on the next page) or in the *Today* window. This option allows you to select the desired template (.DWT) drawing from a list including all templates in the TEMPLATE folder (usually in the C:\Program

Files\AutoCAD 2002\Template subdirectory). Any template drawings that you create using the *SaveAs* command (as described in the next section) appear in the list of available templates. Select the *Browse...* option to browse other folders.

Several template drawings are supplied with AutoCAD 2002 (see the selections in Figure 12-3). Generally, the template drawing files include the titleblocks. In some cases, Dimension Styles have been created but other changes have not been made to the template drawings such as layer setups or loading linetypes.

Figure 12-3

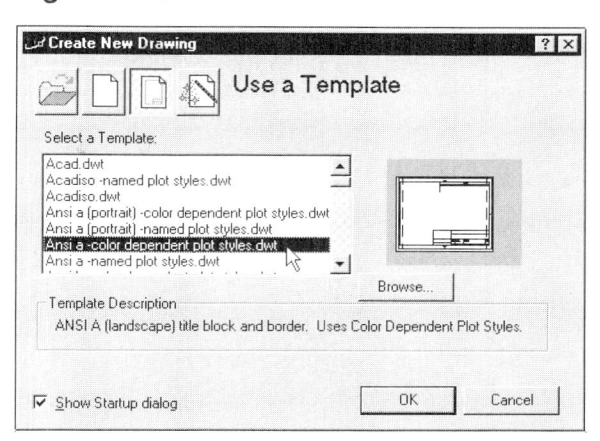

Creating Template Drawings

To make a template drawing, begin a *New* drawing using a default template drawing (ACAD.DWT) or other template drawing (*.DWT), then make the initial drawing setups. Alternately, *Open* an existing drawing that is set up as you need (layers created, linetypes loaded, *Limits* and other settings for plotting or printing to scale, etc.), and then *Erase* all the geometry. Next, use *SaveAs* to save the drawing under a different descriptive name with a .DWT file extension. Do this by selecting the *AutoCAD Drawing Template File (*.dwt)* option in the *Save Drawing As* dialog box (Fig. 12-4).

Figure 12-4

AutoCAD automatically (by default) saves the drawing in the folder where other template (*.DWT) files are found (usually in the C:\Program Files\AutoCAD 2002\Template subdirectory).

After you assign the file name in the *Save Drawing As* dialog box, you can enter a description of the template drawing in the *Template Description* dialog box that appears (Fig. 12-5). The description is saved with the template drawing and appears in the *Startup* and *Create New Drawing* dialog boxes when you highlight the template drawing name (see Figure 12-3).

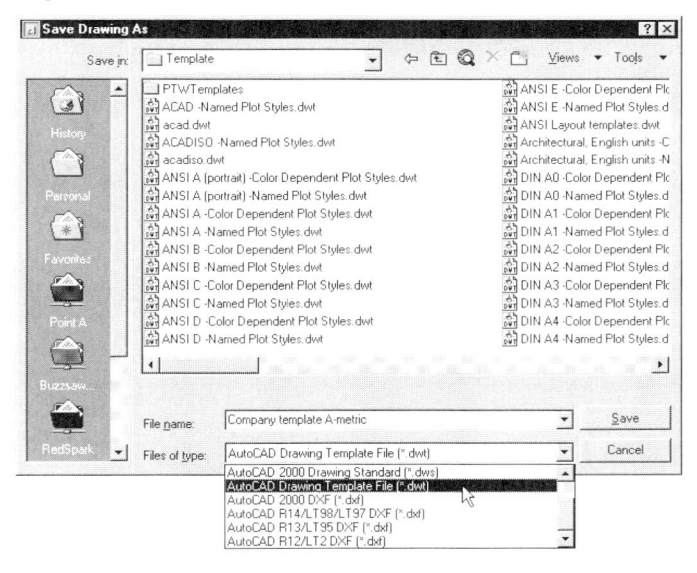

Figure 12-5

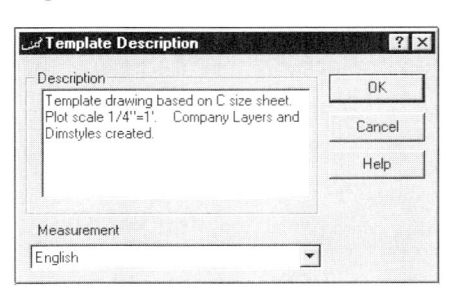

Multiple template drawings can be created, each for a particular situation. For example, you may want to create several templates, each having the setup steps completed but with different *Limits* and with the intention to plot or print each in a different scale or on a different size sheet. Another possibility is to create templates with different layering schemes. There are many possibilities, but the specific settings used depend on your applications (geometry), typical scales, or your plot/print devices.

Typical drawing steps that can be considered for developing template drawings are listed below:

Set *Units*
Set *Limits*
Set *Snap*
Set *Grid*
Set *LTSCALE* and *PSLTSCALE*
Create *Layers* with color and linetypes assigned
Create *Text Styles* (see Chapter 18)
Create Dimension Styles (see Chapter 27)
Create *Layout* tabs, each with titleblock, border and plot settings saved (plot device, paper size, plot scale, etc.). (See Chapter 13, Layouts and Viewports.)

You can create template drawings for different sheet sizes and different plot scales. If you plot from the *Model* tab, a template drawing should be created for each case (scale and sheet size). When you learn to create *Layouts*, you can save one template drawing that contains several layouts, each layout for a different sheet size and/or scale.

Additional Drawing Setup Concepts

The drawing setup described in this chapter is appropriate for learning the basic concepts of setting up a drawing in preparation for creating layouts and viewports and plotting to scale. In Chapter 13, Layouts and Viewports, you will learn the basics of creating layouts and viewports—particularly for drawings that use one large viewport in a layout.

CHAPTER EXERCISES

For the first three exercises, you will set up several drawings that will be used in other chapters for creating geometry. Follow the typical steps for setting up a drawing given in this chapter.

1. **Create a Metric Template Drawing**
   You are to create a template drawing for use with metric units. The drawing will be printed full size on an "A" size sheet (not on an A4 metric sheet), and the dimensions are in millimeters. Start a *New* drawing, select *Start from Scratch*, and select the **English** default settings. Follow these steps.

   A. Set *Units* to *Decimal* and *Precision* to **0.00**.
   B. Set *Limits* to a metric sheet size, **279.4 x 215.9**; then *Zoom All* (scale factor is approximately 25).
   C. Set *Snap Spacing* (for *Grid Snap*) to **1**. (*Polar Tracking* and *Polar Snap* settings do not have to be preset since they are saved in the system registry.)
   D. Set *Grid* to **10**.
   E. Set *LTSCALE* to **25** (drawing scale factor of 25.4). Set *PSLTSCALE* to **0**.
   F. *Load* the *Center2* and *Hidden2 Linetypes*.

G. Create the following *Layers* and assign the *Colors, Linetypes*, and *Lineweights* as shown:

OBJECT	red	continuous	0.40 mm
CONSTR	white	continuous	0.25 mm
CENTER	green	center2	0.25 mm
HIDDEN	yellow	hidden2	0.25 mm
DIM	cyan	continuous	0.25 mm
TITLE	white	continuous	0.25 mm
VORTS	white	continuous	0.25 mm

H. Use the *SaveAs* command. In the dialog box under *Save as type:*, select *AutoCAD Drawing Template File (.dwt)*. Assign the name **A-METRIC**. Enter **Metric A Size Drawing** in the *Template Description* dialog box.

2. ***Quick Setup* wizard**
   A drawing of a mechanical part is to be made and plotted full size (1"=1") on an 11" x 8.5" sheet. Use the *Quick Setup* wizard to assist with the setup.

   A. Begin a *New* drawing. When the *Startup* or *Create New Drawing* dialog box or *Today* window appears, select **Wizard**. Select *Quick Setup*.
   B. In the *Quick Setup* dialog box, select **Decimal** in the **Units** step.
   C. Press *Next* to proceed to the *Area* step. Enter **11** and **8.5** in the two edit boxes so 11 appears as the *Width* (X direction) and 8.5 appears as the *Length* (in the Y direction).
   D. Select the *Finish* button.
   E. Check to ensure the *Limits* are set to 11,8.5 and *Grid* is set to .5.
   F. Change the *Snap* to .250.
   G. Set *LTSCALE* to 1. Set *PSLTSCALE* to 0.
   H. *Load* the *Center2* and *Hidden2 Linetypes*.
   I. Create the following *Layers* and assign the given *Colors, Linetypes,* and *Lineweights*:

OBJECT	red	continuous	0.40 mm
CONSTR	white	continuous	0.25 mm
CENTER	green	center2	0.25 mm
HIDDEN	yellow	hidden2	0.25 mm
DIM	cyan	continuous	0.25 mm
TITLE	white	continuous	0.25 mm
VPORTS	white	continuous	0.25 mm

   J. Save the drawing as **BARGUIDE** for use in another chapter exercise.

3. ***Quick Setup* wizard**
   A floor plan of an apartment has been requested. Dimensions are in feet and inches. The drawing will be plotted at 1/2"=1' scale on an 24" x 18" sheet.

   A. Begin a *New* drawing and select the *Quick Setup* wizard.
   B. Select *Architectural* in the **Units** step.
   C. In the *Area* step, set *Width* and *Length* to **48' x 36'** (576" x 432"), respectively (use the apostrophe symbol for feet). Select *Finish*.
   D. Set the *Snap* value to **1** (inch).
   E. Set the *Grid* value to **12** (inches).
   F. Set the *LTSCALE* to **12**. Set *PSLTSCALE* to **0**.

G. Create the following *Layers* and assign the *Colors* and *Lineweights*:

FLOORPLN	red	0.016"
CONSTR	white	0.010"
TEXT	green	0.010"
TITLE	yellow	0.010"
DIM	cyan	0.010"
VPORTS	white	0.010"

H. *Save* the drawing and assign the name **APARTMENT** to be used later.

4. *Quick Setup* **wizard**
   A drawing is to be created and printed in 1/4"=1" scale on an A size (11" x 8.5") sheet. The *Quick Setup* wizard can be used to automate the setup.

   A. Begin a *New* drawing. Select *Quick Setup* wizard.
   B. In the *Quick Setup* dialog box, select *Decimal* in the *Units* step.
   C. Press *Next* to proceed to the *Area* step. Enter **44** and **34** in the two edit boxes so 44 appears in the *Width* edit box (X direction) and 34 appears in the *Length* edit box (Y direction).
   D. Select the *Finish* button.
   E. Check to ensure the *Limits* are set to 44,34 and *Snap* and *Grid* are set to 1.
   F. Save the drawing as **CH12EX4**.

5. **Create a Template Drawing**
   In this exercise, you will create a "generic" template drawing that can be used at a later time. Creating templates now will save you time later when you begin new drawings.

   A. Create a template drawing for use with decimal dimensions and using standard paper "A" size format. Begin a *New* drawing and select *Start from Scratch*. Select the *English* default settings.
   B. Set *Units* to *Decimal* and *Precision* to **0.00**.
   C. Set *Limits* to **11 x 8.5**.
   D. Set *Snap* to **.25**.
   E. Set *Grid* to **1**.
   F. Set *LTSCALE* to **1**. Set *PSLTSCALE* to **0**.
   G. Create the following *Layers*, assign the *Linetypes* and *Lineweights* as shown, and assign your choice of *Colors*. Create any other layers you think you may need or any assigned by your instructor.

OBJECT	continuous	0.016"
CONSTR	continuous	0.010"
TEXT	continuous	0.010"
TITLE	continuous	0.010"
VPORTS	continuous	0.010"
DIM	continuous	0.010"
HIDDEN	hidden2	0.010"
CENTER	center2	0.010"
DASHED	dashed2	0.010"

   H. Use *SaveAs* and name the template drawing **ASHEET**. Make sure you select *AutoCAD Drawing Template File (*.dwt)* from the *Save as Type:* drop-down list in the *Save Drawing As* dialog box. Also, from the *Save In:* drop-down list on top, select your working directory as the location to save the template.

13

LAYOUTS AND VIEWPORTS

Chapter Objectives

After completing this chapter you should:

1. know the difference between paper space and model space and between tiled viewports and paper space viewports;

2. know that the purpose of a layout is to prepare model space geometry for plotting;

3. know the "Guidelines for Using Layouts and Viewports";

4. be able to create layouts using the *Layout Wizard* and the *Layout* command;

5. know how to set up a layout for plotting using the *Page Setup* dialog box;

6. be able to use the options of *Vports* to create and control viewports in paper space;

7. and know how to scale the model geometry displayed in paper space viewports.

CONCEPTS

You are already somewhat familiar with model space (the *Model* tab) and paper space (*Layout* tabs). Model space (the *Model* tab) is used for construction of geometry, whereas paper space (*Layout* tabs) is used for preparing to print or plot the model geometry. In order to view the model geometry in a layout, one or more viewports must be created. This can be done automatically by AutoCAD when you activate a *Layout* tab for the first time, or you can use the *Vports* command to create viewports in a layout. An introduction to these concepts is given in Chapter 6, Basic Drawing Setup, "Introduction to Layouts and Printing." However, this chapter (Chapter 13) gives a full explanation of paper space, layouts, and creating paper space viewports.

Paper Space and Model Space

The two drawing spaces that AutoCAD provides are model space and paper space. Model space is activated by selecting the *Model* tab and paper space is activated by selecting the *Layout1*, *Layout2*, or other layout tab. When you start AutoCAD and begin a drawing, model space is active by default. Objects that represent the subject of the drawing (model geometry) are normally drawn in model space. Dimensioning is traditionally performed in model space because it is associative—directly associated to the model geometry. The model geometry is usually completed before using paper space. (In AutoCAD 2002, it is possible to create dimensions in paper space that are attached to objects in model space; however, this practice is recommended only for certain applications. See "Dimensioning in Paper Space Layouts" in Chapter 27.)

Paper space represents the paper that you plot or print on. When you enter paper space (by activating a *Layout* tab) for the first time, you may see only a blank "sheet," unless your system is configured to automatically generate a viewport as explained in Chapter 6. Normally the only geometry you would create in paper space is the title block, drawing border, and possibly some other annotation (text). In order to see any model geometry from paper space, viewports must be created (like cutting rectangular holes) so you can "see" into model space. Viewports are created either automatically by AutoCAD or by using the *Vports* command. Any number or size of rectangular or non-rectangular viewports can be created in paper space. Since there is only one model space in a drawing, you see the same model space geometry in each viewport initially. You can, however, control the size and area of the geometry displayed and which layers are *Frozen* and *Thawed* in each viewport. Since paper space represents the actual paper used for plotting, you should plot from paper space at a scale of 1:1.

The *TILEMODE* system variable controls which space is active—paper space (*TILEMODE*=0) or model space (*TILEMODE*=1). *TILEMODE* is automatically set when you activate the *Model* tab or a *Layout* tab. The *TILEMODE* system variable also controls which type of viewports can be created with the *Vports* command—paper space viewports (when *TILEMODE*=0) or model space (tiled) viewports (when *TILEMODE*=1).

Layouts

In AutoCAD Release 14 and older versions, there was only one paper space and one model space. Although there is still only one model space, you can now have multiple paper spaces, currently known as layouts. By default there is a *Layout1* tab and a *Layout2* tab (no *Layout2* tab exists when opening drawings created with Release 14 or earlier releases). You can produce any configuration of layouts by creating new layouts or copying, renaming, deleting, or moving existing layouts.

A shortcut menu is available (right-click while your pointer is on a layout tab) that allows you to *Rename* the selected layout as well as create a *New Layout* from scratch or *From Template*, *Delete*, *Move or Copy* a layout, *Select all Layouts*, or set up for plotting (Fig. 13-1).

Figure 13-1

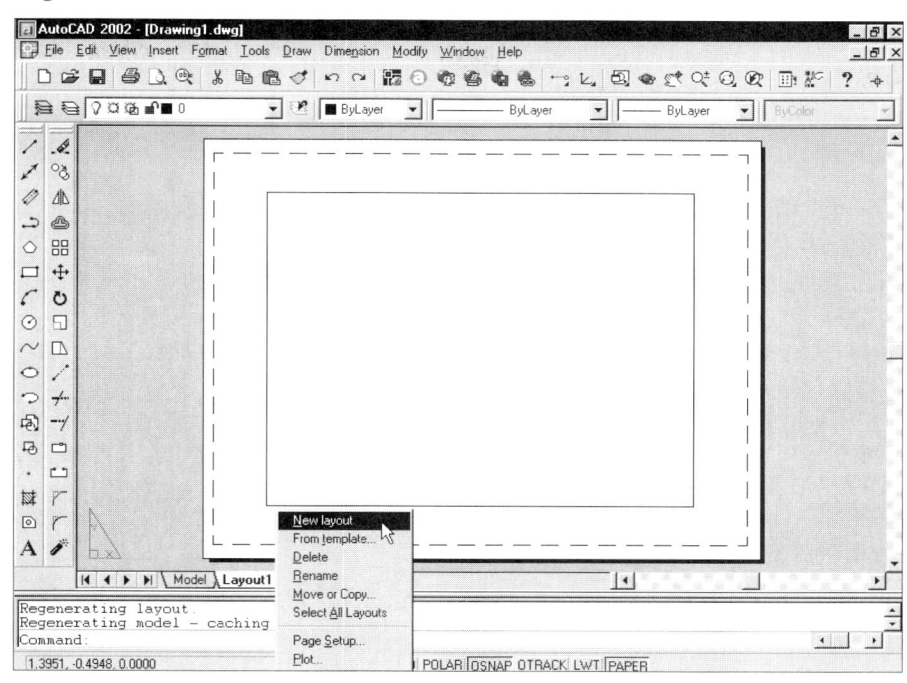

The primary function of a layout is to prepare the model geometry for plotting. A layout simulates the sheet of paper you will plot on and allows you to specify plotting parameters and see the changes in settings that you make, similar to a plot preview. Plot specifications are made using the *Page Setup* dialog box to select such parameters as plot device, paper size and orientation, scale, and Plot Style Tables to attach.

Since you can have multiple layouts in AutoCAD, you can set up multiple plot schemes. For example, assume you have a drawing (in model space) of an office complex. You can set up one layout to plot the office floor plan on a "C" size sheet with an electrostatic plotter in 1/4"=1' scale and set up a second layout to plot the same floor plan on a "A" size sheet with a laser jet printer at 1/8"=1' scale. Since the plotting setup is saved with each layout, you can plot any layout again without additional setup.

Viewports

When you activate a layout tab (enter paper space) for the first time, you may see a "blank sheet." However, if *Create Viewport in New Layouts* is checked in the *Display* tab of the *Options* dialog box (see Figure 13-2, lower-left corner), you may see a viewport that is automatically created. If no viewport exists, you can create one or more with the *Vports* command. A viewport in paper space is a window into model space. Without a viewport, no model geometry is visible in a layout.

Figure 13-2

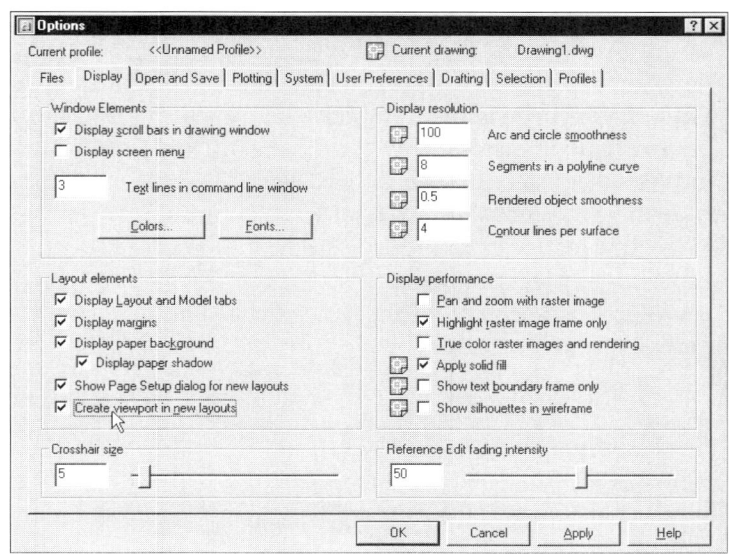

240   *Layouts and Viewports*

There are two types of viewports in AutoCAD, (1) viewports in model space that divide the screen like "tiles" described in Chapter 10 (known as model space viewports or tiled viewports) and (2) viewports in paper space described in this chapter (known as paper space viewports or floating viewports). The term "floating viewports" is used because these viewports can be moved or resized and can take on any shape. Floating viewports are objects that can be *Erased*, *Moved*, *Copied*, stretched with grips, or the viewport's layer can be turned *Off* or *Frozen* so no viewport "border" appears in the plot.

The command generally used to create viewports, *Vports*, can create either tiled viewports or floating viewports, but only one type at a time depending on the active space—*Model* tab or *Layout* tab. When the *Model* tab is active (*TILEMODE*=1), *Vports* creates tiled viewports. When a *Layout* tab is active (*TILEMODE*=0), *Vports* creates paper space viewports.

Figure 13-3

Typically there is only one viewport per layout; however, you can create multiple viewports in one layout. This capability gives you the flexibility to display two or more areas of the same model (model space) in one layout. For example, you may want to show the entire floor plan in one viewport and a detail (enlarged view) in a second viewport, both in the same layout (Fig. 13-3).

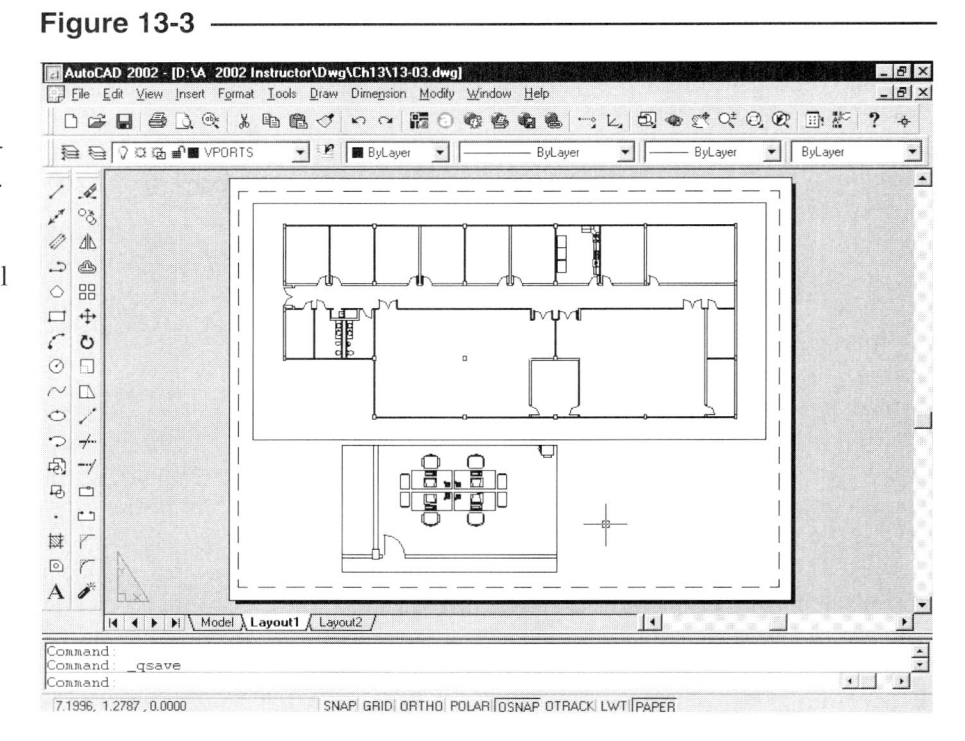

You can control the layer visibility in each viewport. In other words, you can control which layers are visible in which viewports. Do this by using the *Current VP Freeze* button and *New VP Freeze* button in the *Layer Properties Manager* (Fig. 13-4, last two columns). For example, notice in Figure 13-3 that the furniture layer is on in the detail viewport, but not in the overall view of the office floor plan above.

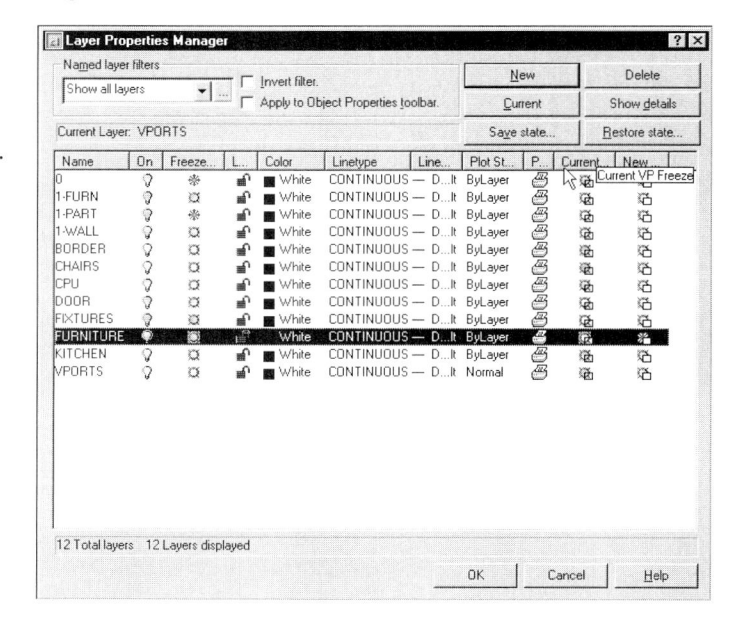

Figure 13-4

You can also scale the display of the geometry that appears in a viewport. You can set the viewport scale by using the *Viewports* toolbar drop-down list (Fig. 13-5), using the *Properties* dialog box, or entering a *Zoom XP* factor. This action sets the proportion (or scale) of paper space units to model space units.

Figure 13-5

Since a layout represents the plotted sheet, you normally plot the layout at 1:1 scale. However, the geometry in each viewport is scaled to achieve the scale for the drawing objects. For example, you could have two or more viewports in one layout, each displaying the geometry at different scales, as is the case for Figure 13-6. Or, you can set up one layout to display model geometry in one scale and set up a similar layout to display the same geometry at a different scale.

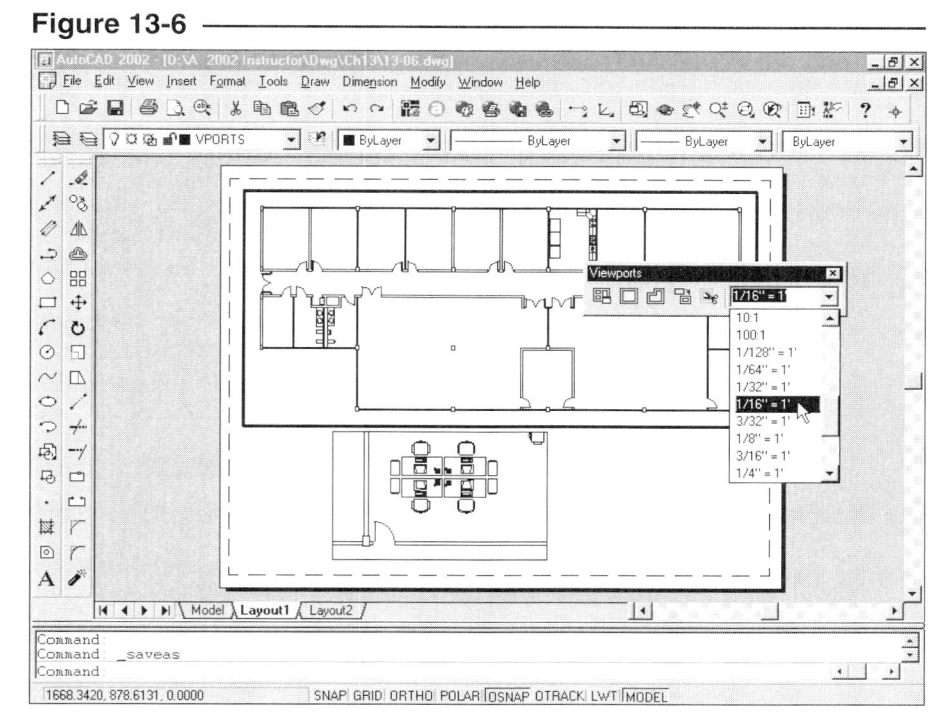

Figure 13-6

An important concept to remember when using layouts is that the *Plot Scale* that you select in the *Page Setup* and *Plot* dialog boxes should almost always be 1:1. This is because the layout is the actual paper size. Therefore, the geometry displayed in the viewports must be scaled to the desired plot scale (the reciprocal of the "drawing scale factor," which can be set by using the *Viewports* toolbar, the *Properties* dialog box for the viewport, or a *Zoom XP* factor.

Layouts and Viewports Example

Consider this brief example to explain the basics of using layouts and viewports. In order to keep this example simple, only one viewport is created in the layout to set up a drawing for plotting. This example assumes that no automatic viewports or page setups are created. Your system may be configured to automatically create viewports when a *Layout* tab is activated (as recommended in Chapter 6 to simplify viewport creation). Disabling automatic setups is accomplished by ensuring *Show Page Setup Dialog for New Layouts* and *Create Viewport in New Layouts* is unchecked in the *Display* tab in the *Options* dialog box.

242 Layouts and Viewports

First, the part geometry is created in the *Model* tab as usual (Fig. 13-7). Associative dimensions are also normally created in the *Model* tab. This step is the same method that you would have normally used to create a drawing.

When the part geometry is complete, enable paper space by double-clicking a *Layout* tab. AutoCAD automatically changes the setting of the *TILEMODE* variable to 0. When you enable paper space for the first time in a drawing, a "blank sheet" appears (assuming no automatic viewports or page setups are created as a result of settings in the *Options* dialog box).

Figure 13-7

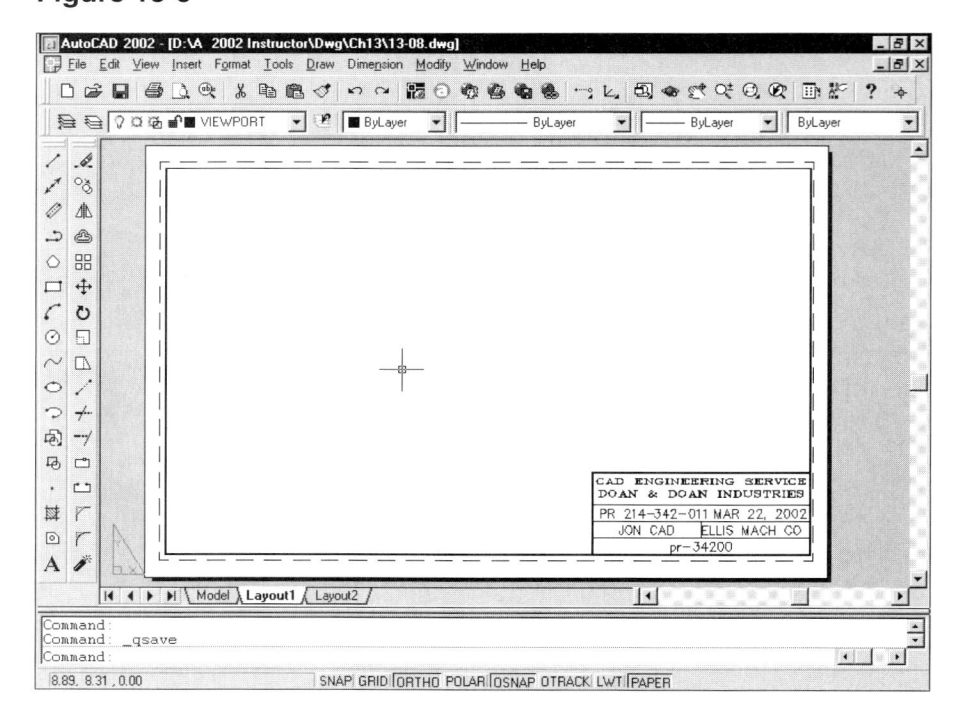

By activating a *Layout* tab, you are automatically switched to paper space. When that occurs, the paper space *Limits* are set automatically based on the selected plot device and sheet size previously selected for plotting. (The default plot device is set in the *Plotting* tab of the *Options* dialog box.) Objects such as a title block and border should be created in the paper space layout (Fig. 13-8). Normally, only objects that are annotations for the drawing (title blocks, tables, border, company logo, etc.) are drawn in the layout.

Figure 13-8

Next, create a viewport in the "paper" with the *Vports* command in order to "look" into model space. All of the model geometry in the drawing initially appears in the viewport. In this example, the *Vports* command is selected from the *Viewports* toolbar. At this point, there is no specific scale relation between paper space units and the size of the geometry in the viewport (Fig. 13-9).

Figure 13-9

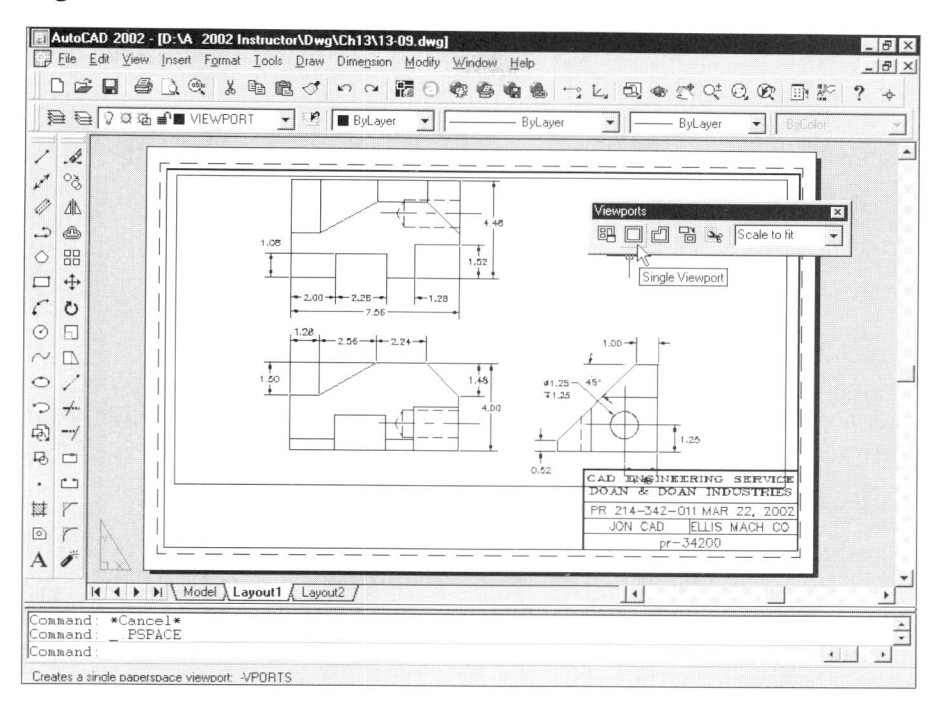

You can control what part of model space geometry you see in a viewport and control the scale of model space to paper space. Two methods are used to control the geometry that is visible in a particular viewport.

1. **Display commands and *Viewport Scale***
    The *Zoom, Pan, View, 3Dorbit* and other display commands allow you to specify what area of model space you want to see in a viewport. In addition, the scale of the model geometry in the viewport can be set by using the *Viewports* toolbar, the *Properties* dialog box for the viewport, or a *Zoom XP* factor.

2. **Viewport-specific layer visibility control**
    You can control what layers are visible in specific viewports. This function is often used for displaying different model space geometry in separate viewports. You can control which layers appear in which viewports by using the *Layer Properties Manager*. (See Chapter 11 for information on viewport-specific layer visibility control.)

Figure 13-10

For example, the *Viewports* toolbar could be used to scale the model geometry in a viewport. In this example, double-click inside the viewport and select *1:2* to scale the model geometry to 1/2 size (model space units equal 1/2 times paper space units).

While in a layout with a viewport created to display model space geometry, you can double-click inside or outside the viewport. This action allows you to switch between model space (inside a viewport) and paper space (outside a viewport) so you can draw or edit in either space. You could instead use the *Mspace* (Model Space) and *Pspace* (Paper Space) commands or single-click the words *MODEL* or *PAPER* on the Status Bar to switch between inside and outside the viewport. The "crosshairs" can move completely across the screen when in paper space, but only within the viewport when in model space (model space inside a viewport).

An object <u>cannot be in both spaces</u>. You can, however, draw in paper space and <u>OSNAP to objects in model space</u>. Commands that are used will affect the objects or display of the current space.

When drawing is completed, activate paper space and plot at 1:1 since the layout size (paper space *Limits*) is set to the <u>actual paper size</u>. Typically, the *Page Setup* dialog box or the *Plot* dialog box is used to prepare the layout for the plot. The plot scale to use for the layout is almost always 1:1 since the layout represents the plotted sheet and the model geometry is already scaled.

Guidelines for Using Layouts and Viewports

Although there are other alternatives, the typical steps used to set up a layout and viewports for plotting are listed here. (These guidelines assume that no automatic viewports or page setups are created, as described in Chapter 6, "Introduction to Layouts and Printing.")

1. Create the part geometry in model space. Associative dimensions are typically created in model space.

2. Single-click on a *Layout* tab. When this is done, AutoCAD automatically sets up the size of the layout (*Limits*) for the default plot device, paper size, and orientation. Use the *Page Setup* dialog box to ensure the correct device, paper size, and orientation are selected. If not, make the desired selections. (If the layout does not automatically reflect the desired settings, check the *Output* tab of the *Options* dialog box to ensure that *Use the Device Paper Size* is checked.)

3. Set up a title block and border as indicated.

    A. Make a layer named BORDER or TITLE and *Set* that layer as current.
    B. Draw, *Insert*, or *Xref* a border and a title block.

4. Make viewports in paper space.

    A. Make a layer named VIEWPORT (or other descriptive name) and set it as the *Current* layer (it can be turned *Off* later if you do not want the viewport objects to appear in the plot).
    B. Use the *Vports* command to make viewports. Each viewport contains a view of model space geometry.

5. Control the display of model space geometry in each viewport. Complete these steps for each viewport.

    A. Double-click inside the viewport or use the *MODEL/PAPER* toggle to "go into" model space (in the viewport). Move the cursor and PICK to activate the desired viewport if several viewports exist.
    B. Use the *Viewport Scale Control* drop-down list, the *Properties* dialog box for the viewport, or *Zoom XP* to scale the display of the paper space units to model units. This action dictates the plot scale for the model space geometry. The viewport scale is the same as the plot scale factor that would otherwise be used (reciprocal of the "drawing scale factor").
    C. Control the layer visibility for each viewport by using the icons in the *Layer Properties Manager*.

6. Plot from paper space at a scale of 1:1.

   A. Use the *MODEL/PAPER* toggle or double-click outside the viewport to switch to paper space.
   B. If desired, turn *Off* the VIEWPORT layer so the viewport borders do not plot.
   C. Invoke the *Plot* dialog box or the *Page Setup* dialog box to set the plot scale and other options. Normally, set the plot scale for the layout to *1:1* since the layout is set to the actual paper size. The paper space geometry will plot full size, and the resulting model space geometry will plot to the scale selected in the *Viewport Scale* drop-down list, *Properties* window, or by the designated *Zoom XP* factor.
   D. Make a *Full Preview* to check that all settings are correct. Make changes if necessary. Finally, make the plot.

CREATING AND SETTING UP LAYOUTS

Listed below are several ways to create layouts.

Shortcut menu	Right-click on a *Layout* tab and select *New Layout, From Template*, or *Move or Copy* (see Figure 13-1).
Layout command	Type or select from the *Insert* pull-down menu. Use the *New*, *Copy*, or *Template* option.
Layoutwizard command	Type or select from the *Insert* pull-down menu.

There are several steps involved in correctly setting up a layout, as listed previously in the "Guidelines for Using Layouts and Viewports." The steps involve specifying a plot device, setting paper size and orientation, setting up a new *Layout* tab, and creating the viewports and scaling the model geometry. One of the best methods for accomplishing all these tasks is to use the *Layout Wizard*.

LAYOUTWIZARD

Pull-down Menu	COMMAND (TYPE)	ALIAS (TYPE)	Short-cut	Screen (side) Menu	Tablet Menu
Insert Layout> Layout Wizard	LAYOUTWIZARD	...	...	...	...

The *Layoutwizard* command produces a wizard to automatically lead you through eight steps to correctly set up a layout tab for plotting. The steps are essentially the same as those listed earlier in the "Guidelines for Using Layouts and Viewports."

Figure 13-11

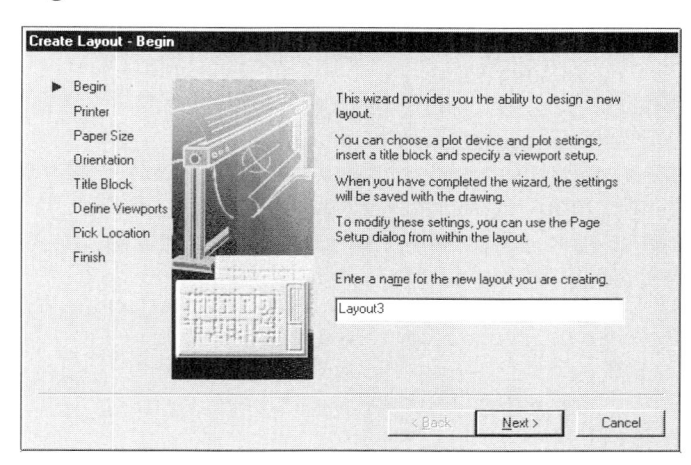

Begin
The first step is intended for you to enter a name for the layout. The default name is whatever layout number is next in the sequence, for example, *Layout3*. Enter any name that has not yet been used in this drawing. You can change the name later if you want using the *Rename* option of the *Layout* command.

Printer

Because the *Limits* in paper space are automatically set based on the size and orientation of the selected paper, choosing a print or plot device is essential at this step. The list includes all previously configured printer and plotter devices. If you intend to use a different device, it may be wise to *Cancel*, use *Plotter Manager* to configure the new device, then begin *Layoutwizard* again.

Figure 13-12

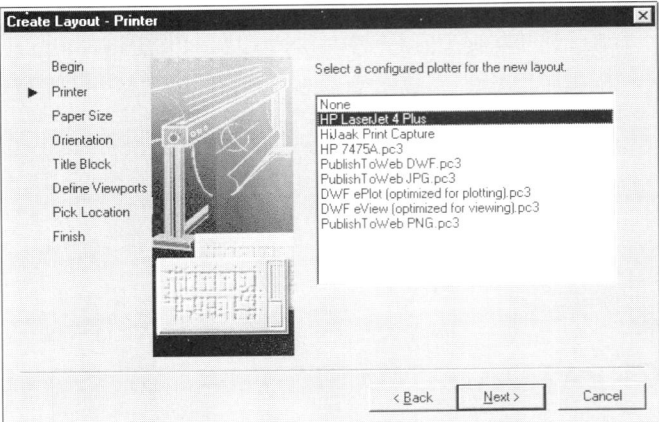

Paper Size

Select the paper size you expect to use for the layout (Fig. 13-13). Make sure you also select the units for the paper (*Drawing Units*) since the layout size (*Limits* in paper space) is automatically set in the selected units.

Figure 13-13

Orientation

Select either *Portrait* or *Landscape* (not shown). *Landscape* plots horizontal lines in the drawing along the long axis of the paper, whereas *Portrait* plots horizontal lines in the drawing along the short axis of the paper.

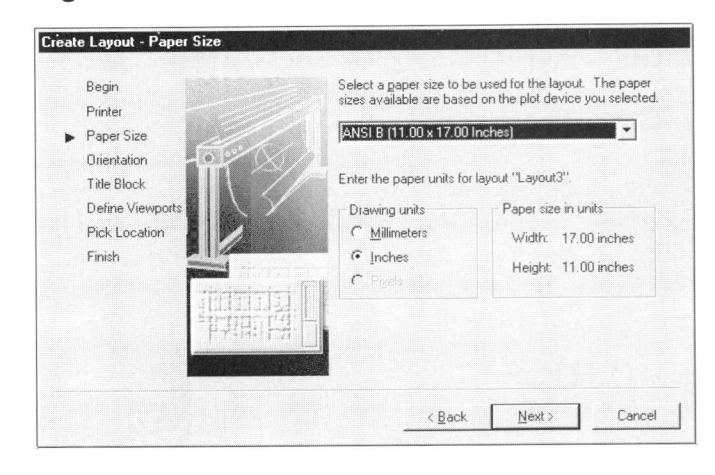

Title Block

In this step, you can select *None* or select from several AutoCAD-supplied ANSI, DIN, ISO and JIS standard title blocks. The selected title block is inserted into the paper space layout. You can also specify whether you want the title block to be inserted as a *Block* or as an *Xref*. Normally inserting the title block as a *Block* object is safer because it becomes a permanent part of the drawing although it occupies a small bit of drawing space (increased file size). (See Chapter 20, Blocks and DesignCenter for more information on *Blocks*.) You can also create your own title block drawings and save them in the AutoCAD 2002\Template folder so they will appear in this list.

Figure 13-14

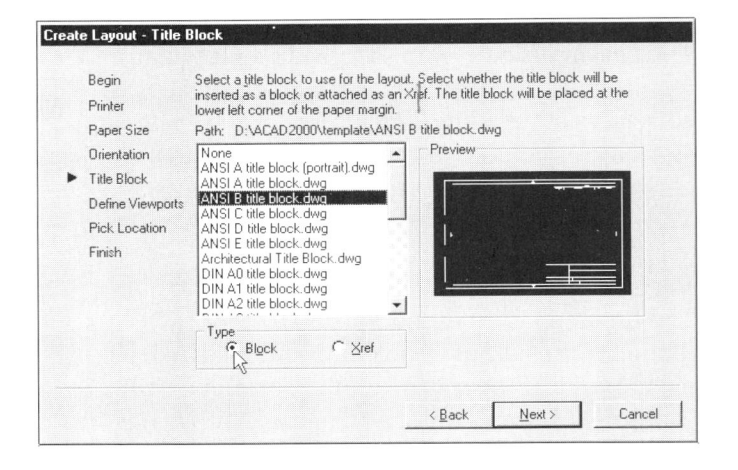

Define Viewports
This step has two parts. First select the type and number of viewports to set up. Use *None* if you want to set up viewports at a later time or want to create a non-rectangular viewport. Normally, use *Single* if you need only one rectangular viewport. Use *Std. 3D Engineering Views* if you have a 3D model. This option sets up four viewports with a top, front, side, and isometric view. The *Array* option creates multiple viewports with the number of *Rows* and *Columns* you enter in the edit boxes below. If you select *Std. 3D Engineering Views* or *Array*, you can then specify the *Spacing between rows* and *Spacing between columns*.

Figure 13-15

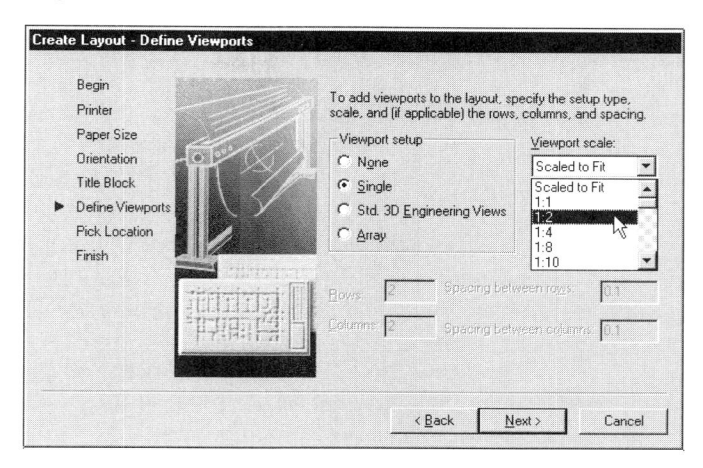

Pick Location
In this step, pick *Select Location* to temporarily exit the wizard and return to the layout to pick diagonal corners for the viewport(s) to fit within. Normally, you do not want the corners of the viewport(s) to overlap the title block or border. You can bypass this step to have AutoCAD automatically draw the viewport border at the extents of the printable area.

Figure 13-16

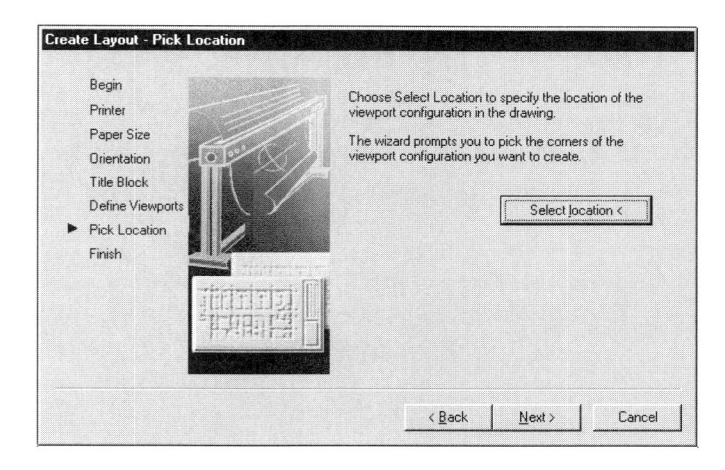

Finish
This is not really a step, rather a confirmation (not shown). In all other steps, you can use *Back* to go backward in the process to change some of the specifications. In this step, *Back* is disabled, so if you get this far, you are finished (with the *Layout Wizard*, anyway).

At this time, the new layout is activated so you can see the setup you specified. If you want to make changes to the plot device, paper size, orientation, or viewport scale, you can do so by using the *Page Setup* dialog box. You can also change the viewport sizes or configuration using *Vports*, *Vpclip*, or grips (explained later in this chapter).

LAYOUT

Pull-down Menu	COMMAND (TYPE)	ALIAS (TYPE)	Short-cut	Screen (side) Menu	Tablet Menu
Insert Layout> New Layout	LAYOUT	LO or -LO	...	...	...

If you prefer to create a layout without using the *Layout Wizard*, use the *Layout* command by typing it or selecting from the *Insert* pull-down menu or *Layout* toolbar. Using the wizard leads you through all the steps required to create a new layout including selecting plot device, paper size and orientation, and creating viewports. Use the *Layout* command to create viewports "manually" (if your system is not set to automatically create viewports). You will then have to use *Vports* to create viewports in the layout.

It is a good idea to ensure the correct plot device is specified before you use *Layout* to create a new layout. See the *New* option for more information. Besides creating a new layout, using a template, and copying an existing layout, you can rename, save, and delete layout.

   Command: `layout`
   Enter layout option [Copy/Delete/New/Template/Rename/SAveas/Set/?] <set>:

Copy
Use this option if you want to copy an existing layout including its plot specifications. If you do not provide the name of a layout to copy, AutoCAD assumes you want to copy the active layout tab. If you do not assign a new name, AutoCAD uses the name of the copied layout (*Floor Plan*, for example) and adds an incremental number in parentheses, such as *Floor Plan (2)*.

   Enter layout to copy <current>:
   Enter layout name for copy <default>:

Delete
Use this option to *Delete* a layout. If you do not enter a name, the most current layout is deleted by default.

   Enter name of layout to delete <current>:

Wildcard characters can be used when specifying layout names to delete. You can select several *Layout* tabs to delete by holding down Shift while you pick. If you select all layouts to delete, all the layouts are deleted, and a single layout tab remains named *Layout1*. The *Model* tab cannot be deleted.

New

This option creates a new layout tab. AutoCAD creates a new layout from scratch based on settings defined by the default plot device and paper size (in the *Options* dialog box). If you choose not to assign a new name, the layout name is automatically generated (*Layout3*, for example).

   Enter new layout name <Layout#>:

Keep in mind that when the new layout is created, its size (*Limits*) and shape are determined by the *Use as Default Output Device* setting specified in the *Plotting* tab of the *Options* dialog. Therefore, it is a good idea to ensure the correct plot device is specified before you use the *New* option. You can, however, change the layout at a later time using the *Page Setup* dialog box, assuming that *Use Plot Device Paper Size* is checked in the *Plotting* tab of the *Options* dialog box (see *Options* in this chapter).

Template

You can create a new template based on an existing layout in a template file (.DWT) or drawing file (.DWG) with this option. The layout, viewports, associated layers, and all the paper space geometry in the layout (from the specified template or drawing file) are inserted into the current drawing. No dimension styles or other objects are imported. The standard file navigation dialog box is displayed for you to select a file to use as a template. Next, the *Insert Layout* dialog box appears for you to select the desired layout from the drawing (Fig. 13-17).

Figure 13-17

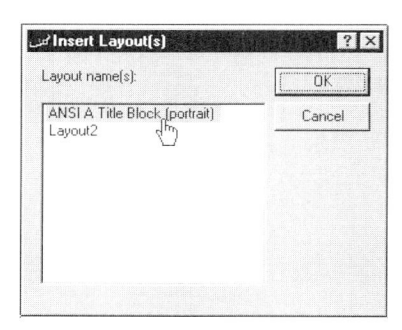

Rename
Use *Rename* to change the name of an existing layout. The last current layout is used as the default for the layout to rename.

   Enter layout to rename <current>:
   Enter new layout name:

Layout names can contain up to 255 characters and are not case sensitive. Only the first 32 characters are displayed in the tab.

Save
This option creates a .DWT file. The file contains all of the (paper space) geometry in the layout and all of the plot settings for the layout, but no model space geometry. All *Block* definitions appearing in the layout, such as a titleblock, are also saved to the .DWT file, but not unused *Block* definitions. All layouts are stored in the template folder as defined in the *Options* dialog box. The last current layout is used as the default for the layout to save.

   Enter layout to save to template <current>:

The standard file selection dialog box is displayed in which you can specify a file name for the .DWT file. When you later create new layout from a template (using the *Template* option of *Layout*), all the saved information (plot settings, layout geometry and *Block* definitions) are imported into the new layout.

Set
This option simply makes a layout current. This option has identical results as picking a layout tab.

? (List Layouts)
Use this option to list all the layouts when you have turned off the layout tabs in the *Display* tab of the *Options* dialog box. Otherwise, layout tabs are visible at the bottom of the drawing area.

Inserting Layouts with AutoCAD DesignCenter

A feature in AutoCAD, called DesignCenter, allows you to insert content from any drawing into another drawing. Content that can be inserted includes drawings, *Blocks*, *Dimstyles*, *Layers*, *Layouts*, *Linetypes*, *Textstyles*, *Xrefs*, raster images, and URLs (web site addresses). If you have created a layout in a drawing and want to insert it into the current drawing, you can use the *Template* option of *Layout* (as previously explained) or use DesignCenter to drag and drop the layout name into the drawing. With DesignCenter, only layouts from .DWG files (not .DWT files) can be imported. See Chapter 20, Blocks and DesignCenter, for more information.

Setting Up the Layout

When you do not use the *Layout Wizard*, the steps involved in setting up the layout must be done individually. In cases when you need flexibility, such as creating non-rectangular viewports or when some factors are not yet known, it is desirable to take each step individually. These steps are listed in "Guidelines for Using Layouts and Viewports."

OPTIONS

Pull-down Menu	COMMAND (TYPE)	ALIAS (TYPE)	Short-cut	Screen (side) Menu	Tablet Menu
Tools Options...	OPTIONS	OP	(Default) Options....	TOOLS2 Options	Y,10

When you create *New* layouts using the *Layout* command, the size and shape of the layout is determined by the selected plot device, paper size, and paper size. Default settings are made in the *Plotting* tab of the *Options* dialog box (Fig. 13-18). Although most settings concerned with plotting a layout can later be changed using the *Page Setup* dialog box, it may be efficient to make the settings in the *Options* dialog box <u>before</u> creating layouts from scratch. Four settings in this dialog box affect layouts.

Figure 13-18

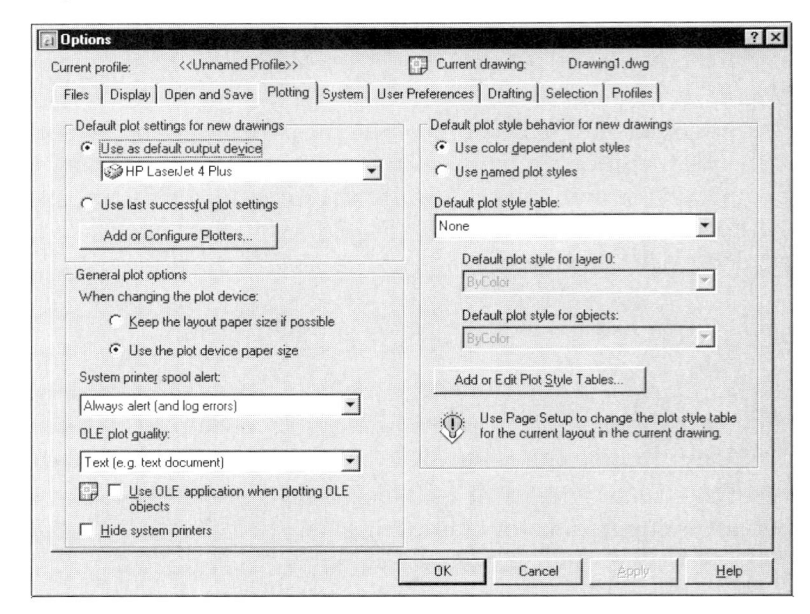

Default Plot Settings for New Drawings
Your choice in this section determines what plot device is used to automatically set the paper size (paper space *Limits*) when new layouts are created. Your setting here applies to <u>new drawings and the current drawing</u>. (Paper size information for each device is stored either in the plotter configuration file [.PC3] or in the default system settings if the output device is a system printer.)

Use as Default Output Device
The selected device is used to determine the paper size (paper space *Limits*) that are set automatically when a new layout is created in the current drawing and for new drawings. This drop-down list contains all configured plot or print devices (any plotter configuration files [PC3] and any system printers that are configured in the system).

Use Last Successful Plot Settings
This button uses the plotting settings (device and paper size) according to those of the last successful plot that was made on the system (a plot must be made from the layout for this option to be valid) and applies them to new layouts that are created. This setting affects new layouts created for the current drawing as well as new drawings. This option overrides the output device appearing in the *Use as Default Output Device* drop-down list.

General Plot Options
These options apply only when plot devices are changed for existing layouts <u>and</u> the layout has a viewport created. Plot devices for layouts can be changed using the *Page Setup* or *Plot* dialog boxes. When you change the plot device for an existing layout, the paper size (*Limits* setting) of the layout is determined by your choice in this section.

Keep the Layout Paper Size if Possible
If this button is pressed, AutoCAD attempts to use the paper size initially specified for the layout if you decide later to change the plot device for the layout. This option is useful if you have two devices that

can use the same size paper as specified in the layout, so you can select either device to use without affecting the layout. However, if the selected output device cannot plot to the previously specified paper size, AutoCAD displays a warning message and uses the paper size specified by the plot device you select in the *Page Setup* or *Plot* dialog box. If no viewports have been created in the layout, you can change the plot device and automatically reset the layout limits without consequence. This button sets the *PAPERUPDATE* system variable to 0 (layout paper size is not updated).

Use the Plot Device Paper Size
As the alternative to the previous button, this one sets the paper size (*Limits*) for an existing layout to the paper size of whichever device is selected for that layout in the *Page Setup* or *Plot* dialog box. This option sets *PAPERUPDATE* to 1.

PAGESETUP

Pull-down Menu	COMMAND (TYPE)	ALIAS (TYPE)	Short-cut	Screen (side) Menu	Tablet Menu
File Page Setup...	PAGESETUP	...	...	...	V,25

After ensuring that the desired settings have been made in the *Plotting* tab of the *Options* dialog box, the next step in setting up a layout (assuming that you are not using the *Layout Wizard*) is to use the *Page Setup* dialog box. Here you select or change the selected plot device, paper size, and orientations for the layout. All settings made in the *Page Setup* dialog box are saved with the layout. In this way, plot settings are already prepared whenever you want to make a plot.

Using the *Page Setup* dialog box (Fig. 13-19) at this time is highly recommended after you create a *New* layout. If you are using a *Template* layout, this step is also suggested to ensure the plot specifications are set as you expect.

Figure 13-19

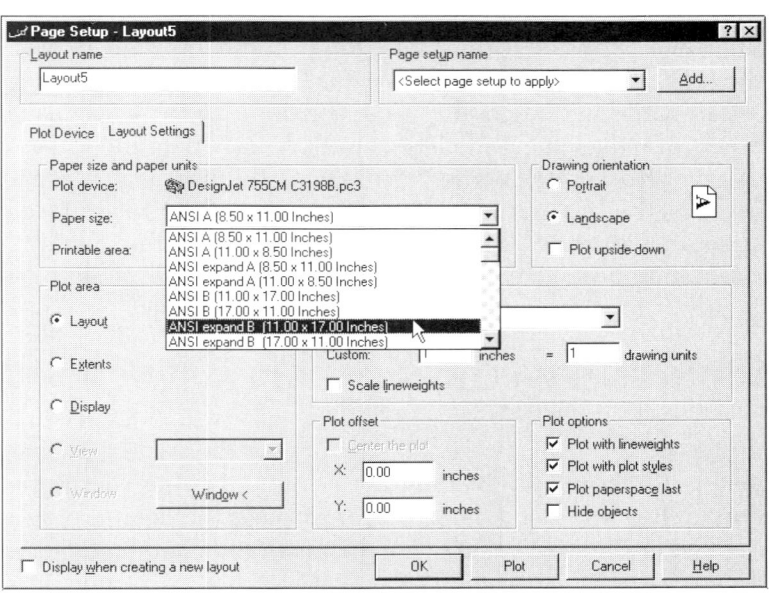

The primary selections to make are as follows:

1. Select a plotter or printer in the *Plot Device* tab.
2. Select *Paper Size* and *Drawing Orientation* in the *Layout Settings* tab.
3. Specify the *Plot area* and *Plot options*.
4. Ensure that *Plot Scale* is set to 1:1.

Remember that the size of the layout (paper space *Limits*) is automatically set based on *Paper Size* and *Drawing Orientation* you select here. Depending on your selection of *Keep the Layout Size if Possible* or *Use the Plot Device Paper Size* in the *Plotting* tab of the *Options* dialog box (see previous section), the layout size and shape may or may not change to the new plot device settings. You can, however, select a new plot device, paper size, or orientation after creating the layout, but before creating viewports.

Since the layout size and shape may change with a new device, viewports that were previously created may no longer fit on the page. In that case, the viewports must be changed to a new size or shape. Alternately, you can *Erase* the viewports and create new ones. Either of these choices is not recommended. Instead, ensure you have the desired plot device and paper settings before creating viewports. (See the "Guidelines for Using Layouts and Viewports.")

Plot Scale is almost always set to *1:1* since the layout represents the sheet you will be plotting on. The model geometry that appears in the layout (in a viewport) will be scaled by setting a viewport scale (see "Scaling Viewport Geometry").

Layout Name
The *Layout Name* at the top of the *Page Setup* dialog box generally displays the name appearing on the active layout tab. You can change the name of the tab here or use the *Rename* option of *Layout* or right-click on the tab.

Page Setup Name
Page setups are automatically saved using the name appearing in the *Layout Name* edit box. If you want to use a setup previously created for an existing layout, select it from the list. For example, if you wanted to copy the setup you made for *Layout1* to the active layout (*Layout2*), select *Layout1* from the drop-down list.

Add
Use the *Add* button to specify a unique name for the layout (a name other than the name appearing on the tab). In this case, the *User Defined Page Setup* dialog box appears allowing you to specify a new name or select another page setup from the drawing. You can also import user-defined page setups from other drawings using the *Import* button. If you define a user-defined page setup, make all the settings before assigning the name. Once a user-defined page setup name is defined, you cannot edit it. This feature is helpful in cases where you want to plot the same drawing with different settings (such as attached Plot Style Tables). Additionally, this feature could be used to import company standard setups as an alternative to including the standards in layout templates.

USING VIEWPORTS IN PAPER SPACE

If you use the *Layout Wizard*, viewports can be created during that process. If you do not use the wizard, viewports should be created for a layout only after specifying the plot device, paper size, and orientation in the *Page Setup* dialog box or through the default plot options set in the *Options* dialog box. You can create viewports with the *Vport* or *-Vport* commands. The Command line version (*-Vports*) produces several options not available in the dialog box, including options to create non-rectangular viewports.

VPORTS

Pull-down Menu	COMMAND (TYPE)	ALIAS (TYPE)	Short-cut	Screen (side) Menu	Tablet Menu
View Viewports>	VPORTS or -VPORTS	...	...	VIEW 1 Vports	M,3 and M,4

The *Vports* command produces the *Viewports* dialog box. Here you can select from several configurations to create rectangular viewports as shown in Figure 13-20 (next page). Making a selection in the *Standard Viewports* list on the left of the dialog box in turn displays the selected *Preview* on the right.

After making the desired selection from the dialog box, AutoCAD issues the following prompt.

    Command: *vports*
    Specify first corner or [Fit] <Fit>:

You can pick two diagonal corners to specify the area for the viewports to fill. Generally, pick just inside the titleblock and border if one has been inserted. If you use the *Fit* option, AutoCAD automatically fills the printable area with the specified number of viewports.

The *Setup: 3D* and *Change View to:* options are intended for use with 3D models.

You may notice that the *Viewports* dialog box is the same dialog box that appears when you use *Vports* while in the *Model* tab. However, when *Vports* dialog is used in the *Model* tab, tiled viewports are created, whereas when the *Vports* is used in a layout (paper space), paper space viewports are created.

Figure 13-20

Paper space viewports differ from tiled viewports. AutoCAD treats paper space viewports created with *Vports* as objects. Like other objects, paper space viewports can be affected by most editing commands. For example, you could use *Vports* to create one viewport, then use *Copy* or *Array* to create other viewports. You could edit the size of viewports with *Stretch* or *Scale*. You can use Grips to change paper space viewports. Additionally, you can use *Move* (Fig. 13-21) to relocate the position of viewports. Delete a viewport using *Erase*. To edit a paper space viewport, you must be in paper space to PICK the viewport objects (borders).

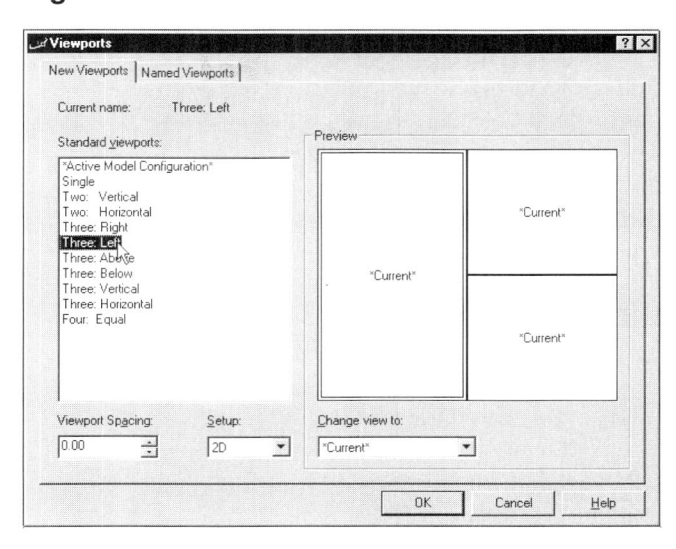

Figure 13-21

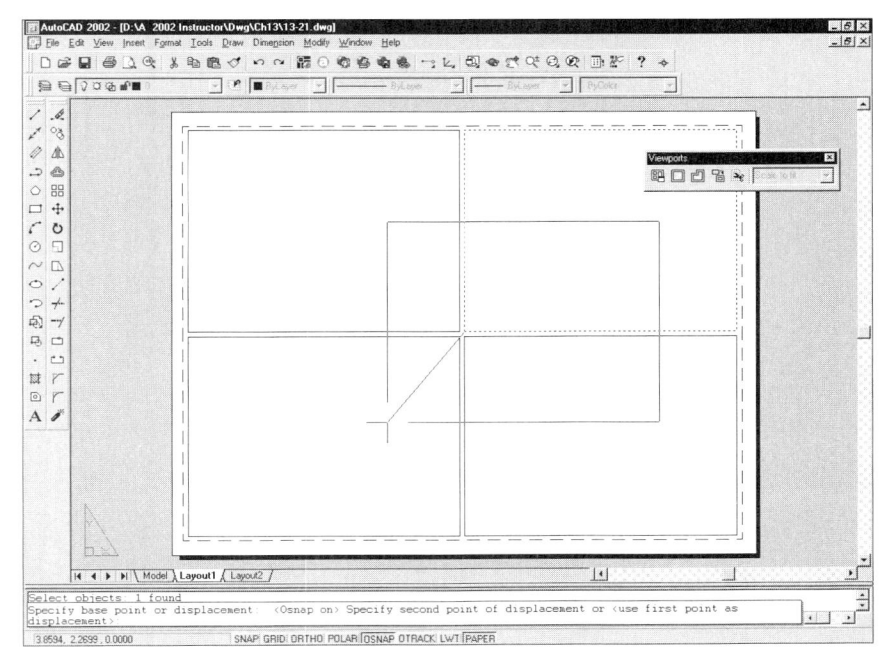

Paper Space viewports can also overlap. This feature makes it possible for geometry appearing in different viewports to occupy the same area on the screen or on a plot.

NOTE: Avoid creating one viewport completely within another's border because visibility and selection problems may result.

254   Layouts and Viewports

Another difference between tiled (*Model* tab) viewports and paper space viewports is that viewports in layouts can have space between them. The option in the bottom left of the *Viewports* dialog box called *Viewport Spacing* allows you to create space between the viewports (if you choose any option other than *Single*) by entering a value in the edit box. Figure 13-22 illustrates a layout with three viewports after entering .5 in the *Viewport Spacing* edit box. Also note several of the –*Vports* command options are available in the right-click shortcut menu that appears when a viewport object is selected.

Figure 13-22

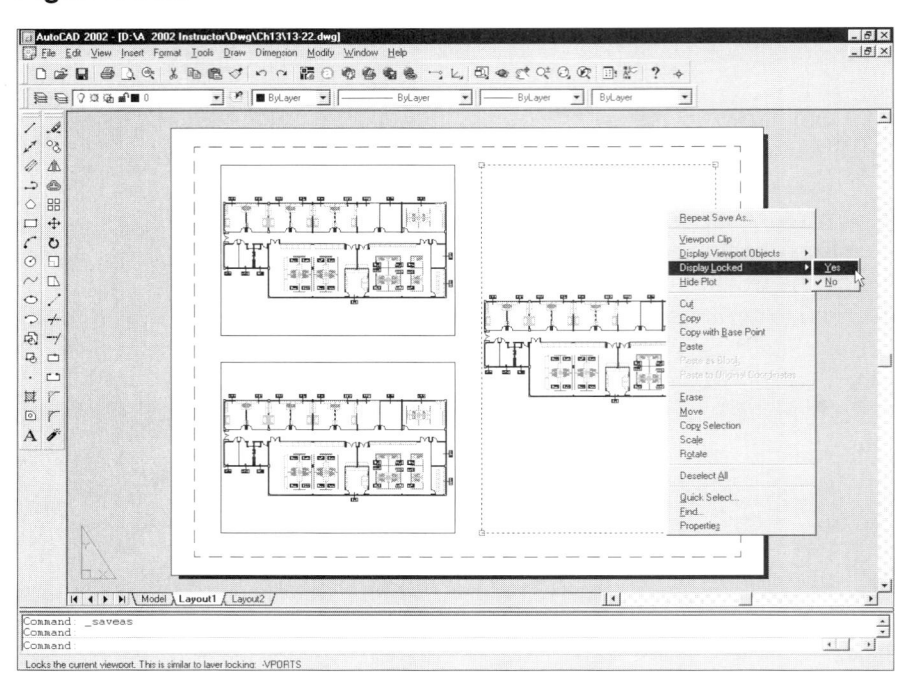

PSPACE

Pull-down Menu	COMMAND (TYPE)	ALIAS (TYPE)	Short-cut	Screen (side) Menu	Tablet Menu
...	PSPACE	PS	...	VIEW1 Pspace	L,5

The *Pspace* command switches from model space (inside a viewport) to paper space (outside a viewport). The cursor is displayed in paper space and can move <u>across the entire screen</u> when in paper space, while the cursor appears only inside the current viewport if the *Mspace* command is used. This command accomplishes the same action as double-clicking outside the viewport in a layout or using the *MODEL/PAPER* toggle on the Status Bar.

If you type this command while in the *Model* tab, the following message appears:

    Command: *pspace*
    PSPACE
    ** Command not allowed in Model Tab **

MSPACE

Pull-down Menu	COMMAND (TYPE)	ALIAS (TYPE)	Short-cut	Screen (side) Menu	Tablet Menu
...	MSPACE	MS	...	VIEW Mspace	L,4

The *Mspace* command switches from paper space to <u>model space inside a viewport</u>, similar to double-clicking inside a viewport. If several viewports exist in the layout when you use the command, you are switched to the last active viewport. The cursor appears only in that viewport. The current viewport displays a heavy border. You can switch to another viewport (make another current) by PICKing in it. To use *Mspace*, there must be at least one viewport in the layout, and it must be *On* (not turned *Off* by the -*Vports* command) or AutoCAD issues a message and cancels the command.

MODEL

Pull-down Menu	COMMAND (TYPE)	ALIAS (TYPE)	Short-cut	Screen (side) Menu	Tablet Menu
...	MODEL	...	...	...	...

The *Model* command accomplishes the same action as selecting the *Model* tab; that is, *Model* makes the *Model* tab active. If you issue this command while you are in the *Model* tab, nothing happens.

Scaling the Display of Viewport Geometry

Paper space (layout) objects such as title block and border should correspond to the paper at a 1:1 scale. A paper space layout is intended to represent the print or plot sheet. *Limits* in a layout are automatically set to the exact paper size, and the finished drawing is plotted from the layout to a scale of 1:1. The model space geometry, however, should be true scale in real-world units, and *Limits* in model space are generally set to accommodate that geometry.

When model space geometry appears in a viewport in a layout, the size of the displayed geometry can be controlled so that it appears and plots in the correct scale. There are three ways to scale the display of model space geometry in a paper space viewport. You can (1) use the *Viewport Scale Control* drop-down list, (2) use *Properties* dialog box, or (3) use the *Zoom* command and enter an XP factor. Once you set the scale for the viewport, use the *Lock* option of –*Vports* or the *Properties* dialog box to lock the scale of the viewport so it is not accidentally changed when you *Zoom* or *Pan* inside the viewport.

When you use layouts, the Plot Scale specified in the *Plot* dialog box and *Page Setup* dialog box dictates the scale for the layout, which should normally be set to 1:1. Therefore, the viewport scale actually determines the plot scale of the geometry in a finished plot. Set the viewport scale to the reciprocal of the "drawing scale factor," as explained in Chapter 12, Advanced Drawing Setup.

Viewport Scale Control Drop-Down List

The *Viewport Scale Control* drop-down list is located in the *Viewports* toolbar (see Figures 13-5 and 13-10). Invoke the *Viewports* toolbar using the *Toolbars...* option at the bottom of the *View* pull-down menu.

To set the viewport scale (scale of the model geometry in the viewport), first make the viewport current by double-clicking in it or typing *Mspace*. When the crosshairs appear in the viewport and the viewport border is highlighted (a wide border appears), select the desired scale from the list.

The *Viewport Scale Control* drop-down list provides only standard scales for decimal, metric, and architectural scales. You can select from the list or enter a value. Values can be entered as a decimal, a fraction (proper or improper), a proportion using a colon symbol (:), or an equation using an equal symbol (=). Feet (') and inch (") unit symbols can also be entered. For example, if you wanted to scale the viewport geometry at 1/2"=1', you could enter any of the following, as well as other, values.

    1/2"=1'
    .5"=1'
    1/2=12
    1/24
    1:24
    1=24
    .0417

Remember that the viewport scale is actually the desired plot scale for the geometry in the viewport, which is the reciprocal of the "drawing scale factor" (see Chapter 12).

Properties

The *Properties* dialog box can also be used to specify the scale for the display of model space geometry in a viewport. Do this by first double-clicking in paper space or issuing the *Pspace* command, then invoking *Properties*. When (or before) the dialog box appears, select the viewport object (border). Ensure the word "Viewport" appears in the top of the box (Fig. 13-23).

Figure 13-23

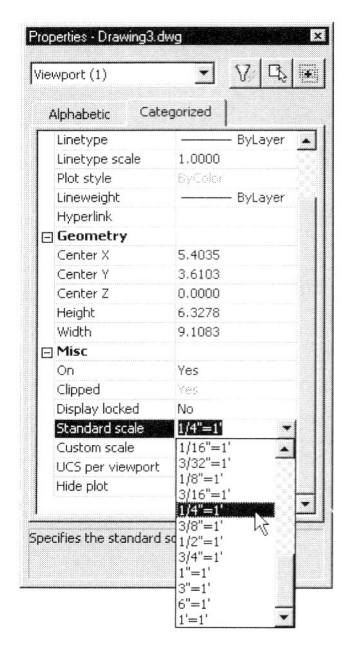

Under *Standard Scale*, select the desired scale from the drop-down list. This method offers only standard scales. Alternately, you can enter any values in the *Custom Scale* box. (See *Properties*, Chapter 16.) Similar to using the *Viewport Scale Control* edit box, you can enter values as a decimal or fraction and use a colon symbol or an equal symbol. Feet and inch unit symbols can also be entered. You can also *Lock* the viewport scale using the *Display Locked* option in this dialog box. (See *Properties*, Chapter 16.)

ZOOM XP **Factors**

To scale the display of model space geometry relative to paper space (in a viewport), you can also use the *XP* option of the *Zoom* command. *XP* means "times paper space." Thus, model space geometry is *Zoomed* to some factor "times paper space."

Since paper space is set to the actual size of the paper, the *Zoom XP* factor that should be used for a viewport is equivalent to the plot scale that would otherwise be used for plotting that geometry in model space. The *Zoom XP* factor is the reciprocal of the "drawing scale factor." *Zoom XP* only while you are "in" the desired model space viewport. Fractions or decimals are accepted.

For example, if the model space geometry would normally be plotted at 1/2"=1" or 1:2, the *Zoom* factor would be .5XP or 1/2XP. If a drawing would normally be plotted at 1/4"=1', the *Zoom* factor would be 1/48XP. Other examples are given in the following list.

1:5	*Zoom* .2XP or 1/5 XP
1:10	*Zoom* .1XP or 1/10XP
1:20	*Zoom* .05XP or 1/20XP
1/2"=1"	*Zoom* 1/2XP
3/8"=1"	*Zoom* 3/8XP
1/4"=1"	*Zoom* 1/4XP
1/8"=1"	*Zoom* 1/8XP
3"=1'	*Zoom* 1/4XP
1"=1'	*Zoom* 1/12XP
3/4'=1'	*Zoom* 1/16XP
1/2"=1'	*Zoom* 1/24XP
3/8"=1'	*Zoom* 1/32XP
1/4"=1'	*Zoom* 1/48XP
1/8"=1'	*Zoom* 1/96XP

Refer to the "Tables of Limits Settings," Chapter 14, for other plot scale factors.

Locking Viewport Geometry

Once the viewport scale has been set to yield the correct plot scale, you should lock the scale of the viewport so it is not accidentally changed when you *Zoom* or *Pan* inside the viewport. If viewport lock is on and you use *Zoom* or *Pan* when the viewport is active, AutoCAD automatically switches to paper space (*Pspace*) for zooming and panning.

You can lock the viewport scale by the following methods.

1. Use the *Lock* option of the *–Vports* command.
2. Use the *Display Locked* setting in the *Properties* dialog box (see Figure 13-23).
3. Use the *Display Locked* option in the right-click shortcut edit menu that appears when a viewport object is selected (see Figure 13-22).

Linetype Scale in Viewports—*PSLTSCALE*

The *LTSCALE* (linetype scale) setting controls how hidden, center, dashed, and other non-continuous lines appear in the *Model* tab. *LTSCALE* can be set to any value. The *PSLTSCALE* (paper space linetype scale) setting determines if those lines appear and plot the same in *Layout* tabs as they do in the *Model* tab. *PSLTSCALE* can be set to 1 or 0 (on or off).

If *PSLTSCALE* is set to 0, the non-continuous line scaling that appears in *Layout* tabs (and in viewports) is the same as that in the *Model* tab. In this way, the linetype spacing always looks the same relative to model space units, no matter whether you view it from the *Model* tab or from a viewport in a *Layout*. For example, if in the *Model* tab a particular center line shows one short dash and two long dashes, it will look the same when viewed in a viewport in a layout—one short dash and two long dashes. When *PSLTSCALE* is 0, linetype scaling is controlled exclusively by the *LTSCALE* setting. A *PSLTSCALE* setting of 0 is recommended for most drawings unless they contain multiple viewports or layouts.

If *PSLTSCALE* is set to 1 (the default setting for all AutoCAD template drawings), the linetype scale for non-continuous lines in viewports is automatically changed relative to the viewport scale; therefore, the scale of those lines in *Layout* tabs can appear differently than they do in the *Model* tab. A *PSLTSCALE* setting of 1 is recommended for drawings that contain multiple viewports or layouts, especially when they are at different scales. In such a situation, the line dashes would appear the same size in different viewports relative to paper space, even though the viewport scales were different. Technically speaking, if *PSLTSCALE* is set to 1, *LTSCALE* controls linetype scaling globally for the drawing, but lines that appear in viewports are automatically scaled to the *LTSCALE* times the viewport scale.

Until you gain more experience with layouts and viewports, and in particular creating multiple viewports, a *PSLTSCALE* setting of 0 is the simplest strategy. This setting also agrees with the strategy used in this text discussed earlier (in Chapter 6 and Chapter 12) for drawing setup. You can change the default *PSLTSCALE* setting of 1 to 0 by typing *PSLTSCALE* at the Command prompt.

CHAPTER EXERCISES

1. *Pagesetup, Vports*

   A. *Open* the **A-METRIC.DWT** template drawing that you created in the Chapter 12 Exercises. Activate a *Layout* tab. If your *Options* are set as explained in Chapter 6, the layout is automatically set up for your default print/plot device. Otherwise, use the *Page Setup* dialog box (*Pagesetup* command) to set the desired print/plot device and to select an **11 x 8.5** paper size.

   B. If a viewport already exists in the layout, *Erase* it. Make **VPORTS** the *Current* layer. Then use the *Vports* command and select a *Single* viewport and accept the default (*Fit*) option to create one viewport at the maximum size for the printable area. Ensure that *PSLTSCALE* is set to **0** by typing it at the Command prompt. Finally, make the *Model* tab active. *Save* and *Close* the drawing.

2. *Pagesetup, Vports*

   A. *Open* the **BARGUIDE** drawing that you set up in the Chapter 12 Exercises. Activate a *Layout* tab. If your *Options* are set as explained in Chapter 6, the layout is automatically set up for your default print/plot device. Otherwise, use *Pagesetup* to set the desired print/plot device and to select an **11 x 8.5** paper size.

   B. If a viewport already exists in the layout, *Erase* it. Make **VPORTS** the *Current* layer. Then use the *Vports* command and select a *Single* viewport and accept the default (*Fit*) option to create one viewport at the maximum size for the printable area. Type *PSLTSCALE* and set the value to **0**. Finally, make the *Model* tab active. *Save* and *Close* the drawing.

3. **Create a Layout and Viewport for the ASHEET Template Drawing**

   A. *Open* the **ASHEET.DWT** template drawing you created in the Chapter 12 Exercises. Activate a *Layout* tab. Unless already set up, use *Pagesetup* to set the desired print/plot device and to select an **11 x 8.5** paper size.

   B. If a viewport already exists in the layout, *Erase* it. Make **VPORTS** the *Current* layer. Then use the *Vports* command and select a *Single* viewport and accept the default (*Fit*) option to create one viewport at the maximum size for the printable area. Ensure that *PSLTSCALE* is set to **0**. Finally, make the *Model* tab active. *Save* and *Close* the ASHEET template drawing.

4. **Create a New BSHEET Template Drawing**
   Complete this exercise if your system is configured with a "B" size printer or plotter.

   A. Using the template drawing in the previous exercise, create a template for a standard engineering "B" size sheet. First, use the *New* command and select the *Template* option from the *Create New Drawing* dialog box or *Today* window. Select the *Browse* button. When the *Select a Template File* dialog box appears, select the **ASHEET.DWT**. When the drawing opens, set the model space *Limits* to **17 x 11** and *Zoom All*.

B. Activate the *Layout 1* tab. *Erase* the existing viewport. Activate the *Page Setup* dialog box. In the *Plot Device* tab, select a print or plot device that can use a B size sheet (17 x 11). In the *Layout Settings* tab, select ANSI B (11 x 17 Inches) in the *Paper Size* drop-down list. Select *OK* to dismiss the *Page Setup* dialog box.

C. Make **VPORTS** the *Current* layer. Then use the *Vports* command and select a *Single* viewport and accept the default (*Fit*) option to create one viewport at the maximum size for the printable area. Finally, make the *Model* tab active. All other settings and layers are okay as they are. Use *Saveas* and save the new drawing as a template (.**DWT**) drawing in your working directory. Assign the name **BSHEET**.

5. **Create Multiple Layouts for Plotting on C and D Size Sheets**
   Complete this exercise if your system is configured with a "C" and "D" sized plotter.

   A. Using the ASHEET.DWT template drawing from an earlier exercise, create a template for standard engineering "C" and "D" size sheets. First, use the *New* command and select the *Template* option from the *Create New Drawing* dialog box or *Today* window. Select the *Browse* button. When the *Select a Template File* dialog box appears, select the **ASHEET.DWT**. When the drawing opens, set the model space *Limits* to **34** x **22** and *Zoom All*.

   B. Activate the *Layout1* tab. *Erase* the existing viewport. Activate the *Page Setup* dialog box. In the *Plot Device* tab, select a plotter that can use a C size sheet (22 x 17). In the *Layout Settings* tab, select ANSI C (22 x 17 Inches) in the *Paper Size* drop-down list. Select *OK* to dismiss the *Page Setup* dialog box.

   C. Make **VPORTS** the *Current* layer. Then use the *Vports* command and select a *Single* viewport and use the *Fit* option to create one viewport at the maximum size for the printable area. Right-click on the *Layout 1* tab and select *Rename* from the shortcut menu. Rename the layout to "**C Sheet**."

   D. Activate the *Layout2* tab. *Erase* the existing viewport if one exists. Activate the *Page Setup* dialog box. In the *Plot Device* tab, select a print or plot device that can use a D size sheet (34 x 22). (This can be the same device you selected in step B, as long as it supports a D size sheet.) In the *Layout Settings* tab, select ANSI D (34 x 22 Inches) in the *Paper Size* drop-down list. Select *OK* to dismiss the *Page Setup* dialog box.

   E. Make **VPORTS** the *Current* layer. Then use the *Vports* command and select *Single* to create one viewport. *Rename* the layout "**D Sheet**."

   F. Finally, make the *Model* tab active. All other settings and layers are okay as they are. Use *Saveas* and save the new drawing as a template (.**DWT**) drawing in your working directory. Assign the name **C-D-SHEET**.

6. *Layout Wizard*
   In this exercise, you will open an existing drawing and use the *Layout Wizard* to set up a layout for plotting.

   A. *Open* the **DB_SAMP** drawing that is located in the AutoCAD 2002\Sample folder. Use the *Saveas* command and assign the name **VP_SAMP** and specify your working folder as the location to save.

B. Create a new layer named **VPORTS** and make it the *Current* layer. Invoke the *Layout Wizard*. In the *Begin* step, enter a name for the new layout, such as **Layout 3**. Select *Next*.

C. In the *Printer* step, the list of available printers in your lab or office appears. The default printer should be highlighted. Choose any device that you know will operate for your lab or office and select *Next*. If you are unsure of which devices are usable, keep the default selection and select *Next*.

D. For *Paper Size*, select the largest paper size supported by your device. Ensure *Inches* is selected unless you are using a standard metric sheet. Select *Next*.

E. Select **Landscape** for the *Orientation*.

F. Select a *Title Block* that matches the paper size you specified. If you are unsure of which to choose, refer to the Layout Templates table in the chapter. Insert the title block as a *Block*. Select *Next*.

G. In the *Define Viewports* step, select a **Single** viewport. Under *Viewport Scale*, select **Scaled to Fit**. Proceed to the next step.

Figure 13-24

H. Bypass the next step in the wizard, *Pick Location*, by selecting *Next*. Also, when the *Finish* step appears, select *Next*. This action causes AutoCAD to create a viewport that fits the extents of the printable area. The resulting layout should look similar to that shown in Figure 13-24. As you can see, the model space geometry extends beyond the title block. Your layout may appear slightly different depending on the sheet size and title block you selected. *Save* the drawing.

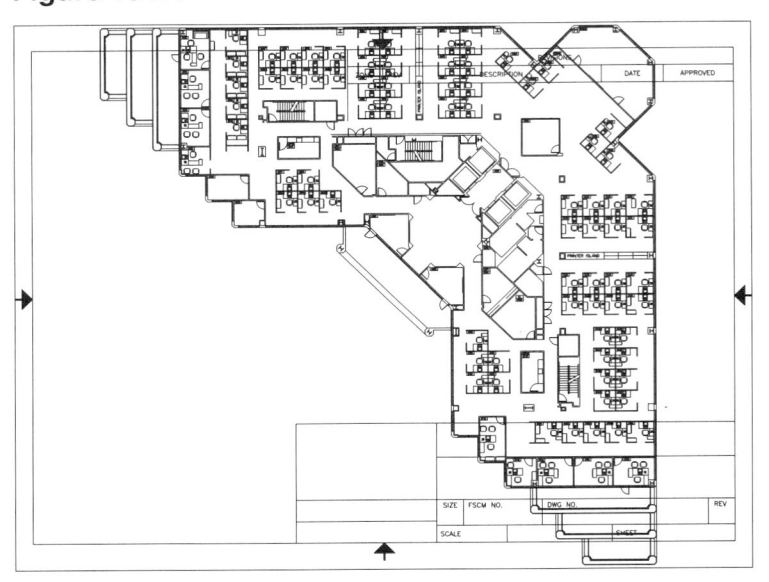

Figure 13-25

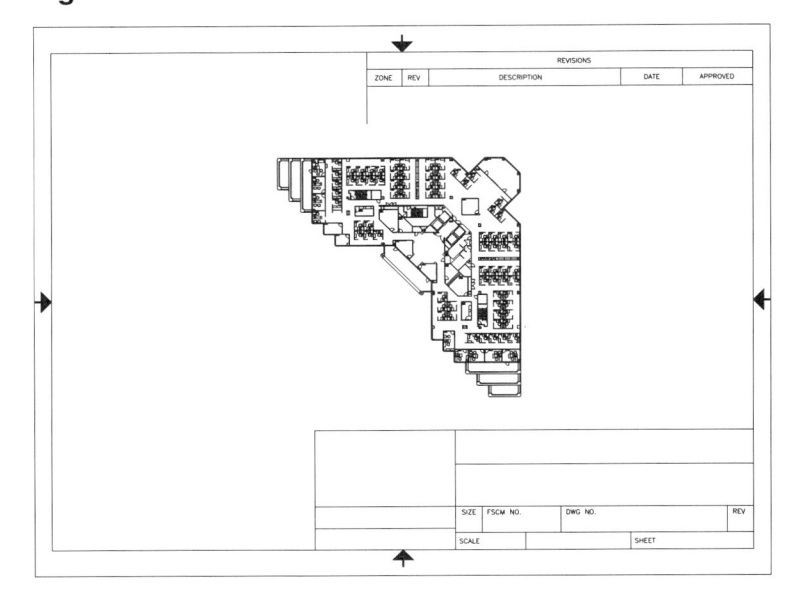

I. Use the *Mspace* command or double-click inside the viewport to activate model space. To specify a standard scale for the geometry in the viewport, invoke the *Viewports* toolbar. In the *Viewport Scale Control* drop-down list, select a scale (such as *1/64"=1'*) until the model geometry appears completely within the title block border (Fig. 13-25).

J. *Freeze* layer **VPORTS** so the viewport border does not appear. *Save* the drawing. With the layout active, access the *Plot* dialog box and ensure the *Plot Scale* is set to **1:1**. Select *Plot*.

7. *Page Setup*, *Vports*, and *Zoom XP*

   In this exercise, you will use an existing metric drawing, access *Layout1*, draw a border and title block, create one viewport, and use *Zoom XP* to scale the model geometry to paper space. Finally, you will *Plot* the drawing to scale on an "A" size sheet.

   A. *Open* the **CBRACKET** drawing you worked on in Chapter 9 Exercises. If you have already created a title block and border, *Erase* them.

   B. Invoke the *Options* dialog box and access the *Display* tab. In the *Layout Elements* section (lower left), ensure that *Show Page Setup Dialog for New Layouts* is checked and *Create Viewport in New Layouts* is not checked.

   C. Select the *Layout1* tab. The *Page Setup* dialog box should appear. In the **Plot Device** tab, select a plot device from the list that will allow you to plot on an A size sheet (11" x 8.5"). In the *Layout Settings* tab, select the correct sheet size and select *Landscape* orientation. Pick *OK* to finish the setup.

   D. Set layer **TITLE** *Current*. Draw a title block and border (in paper space). HINT: Use *Pline* with a *width* of .02, and draw the border with a .5 unit margin within the edge of the *Limits* (border size of 10 x 7.5). Provide spaces in the title block for your school or company name, your name, part name, date, and scale as shown in the following figure. Use *Saveas* and assign the name **CBRACKET-PS**.

   E. Create layer **VPORTS** and make it *Current*. Use the *Vports* command to create a *Single* viewport. Pick diagonal corners for the viewport at **.5,1.5** and **9.5,7**. The CBRACKET drawing should appear in the viewport at no particular scale, similar to Figure 13-26.

   Figure 13-26

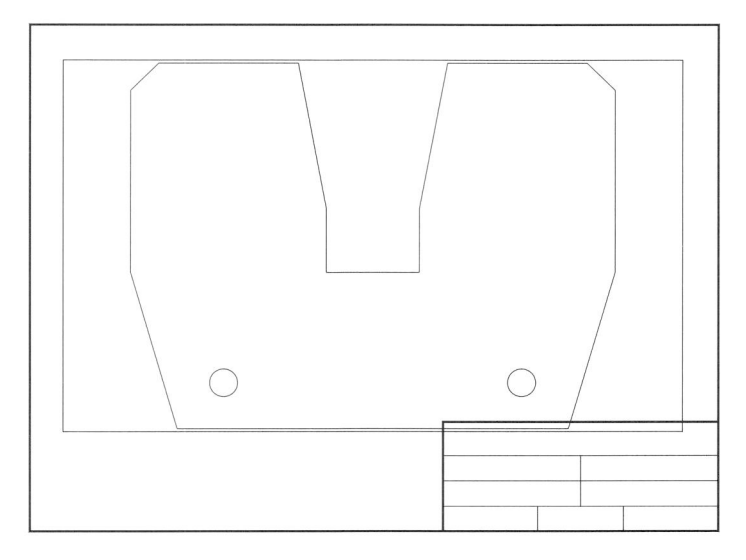

   F. Next, use the *Mspace* command, the **MODEL/PAPER** toggle on the Status Bar, or double-click inside the viewport to bring the cursor into model space. Use *Zoom* with a *.03937XP* factor to scale the model space geometry to paper space. (The conversion from inches to millimeters is 25.4 and from millimeters to inches is .03937. Model space units are millimeters and paper space units are inches; therefore, enter a *Zoom XP* factor of .03937.)

G. Finally, activate paper space by using the *Pspace* command, the *MODEL/PAPER* toggle, or double-clicking outside the viewport. Use the *Layer Properties Manager* to make layer **VPORTS** *non-plottable*. Your completed drawing should look like that in Figure 13-27. *Save* the drawing. *Plot* the drawing from the layout at **1:1** scale.

Figure 13-27

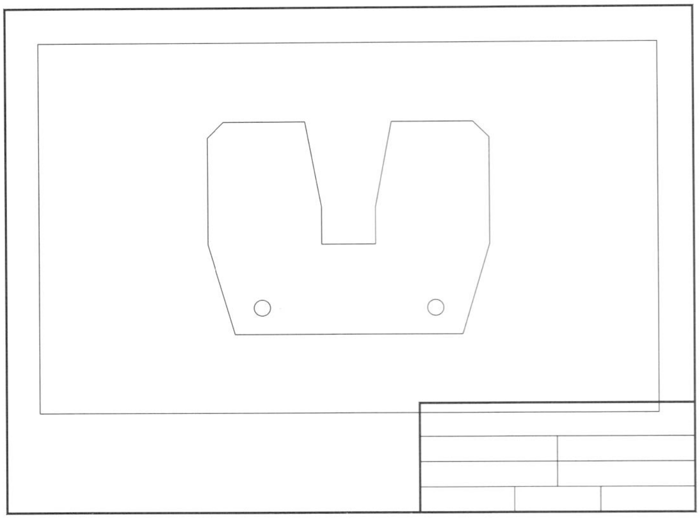

8. **Using a Layout Template**

   This exercise gives you experience using a layout template already set up with a title block and border. You will draw the hammer (see Figure 13-28) and use a layout template to plot on a B size sheet. It is necessary to have a plot or print device configured for your system that can use a "B" size sheet.

   A. Start AutoCAD. In the *Options* dialog box, select the *System* tab. In the *General Options* section, select **Show Traditional Startup Dialog**.

   B. Now access the *Display* tab in the *Options* dialog box. In the *Layout Elements* section (lower left), remove the check for both *Show Page Setup Dialog for New Layouts* and *Create Viewport in New Layouts*. Select *OK*.

   C. Begin a *New* drawing. In the *Create New Drawing* dialog box that appears, select **Use a Wizard** and select the **Advanced Setup Wizard**. In the first step, *Units*, select **Decimal** units and a *Precision* of **0.00**. Accept the defaults for the next three steps. In the *Area* step, enter a **Width** of **17** and a **Length** of **11**.

   D. Right-click on an existing layout tab and use the **From Template** option from the shortcut menu. Select the **ANSI B -Color Dependent Plot Styles** template drawing (.DWT) in the *Select Template From File* dialog box, then select the **ANSI B Title Block** in the *Insert Layout(s)* dialog box. Access the new tab. Note that the template loads a layout with a title block and border in paper space and has one viewport created. Access the **Page Setup** dialog box and select the matching plot device and sheet size for the selected ANSI B title block.

   E. Pick the **Model** tab. Set up appropriate **Snap** and **Grid** (for the *Limits* of 17 x 11) if desired. Also set **LTSCALE** and any running **Osnaps** you want to use. Type **PSLTSCALE** and set the value to **0**.

   F. Create the following layers and assign linetypes and line weights. Assign colors of your choice.

   GEOMETRY   *Continuous*
   CENTER   *Center2*

   Notice the *Title Block* and *Viewport* layers already exist as part of the template drawing.

G. Draw the hammer according to the dimensions given in Figure 13-28. *Save* the drawing as **HAMMER**.

H. When you are finished with the drawing, select the *ANSI B Title Block* tab. Now, scale the hammer to plot to a standard scale. You can use either the *Viewports* toolbar or the *Properties* window to set the viewport scale to **1:1**. Alternately, use *Zoom 1XP* to correctly size model space units to paper space units.

Figure 13-28

I. Your completed drawing should look like that in Figure 13-29. Use the *Layer Properties Manager* to make layer **Viewports** *non-plottable*. *Save* the drawing. Make a *Plot* from the layout at a scale of **1:1**.

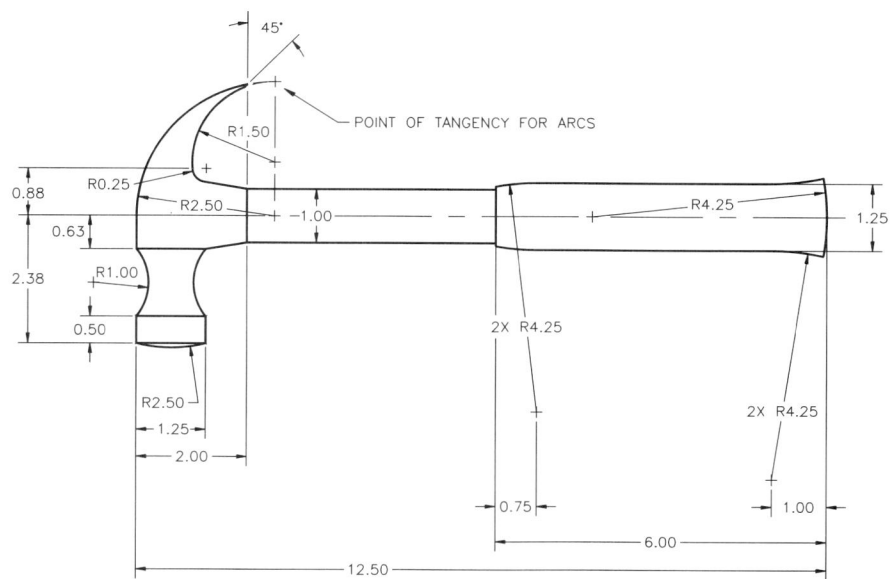

Figure 13-29

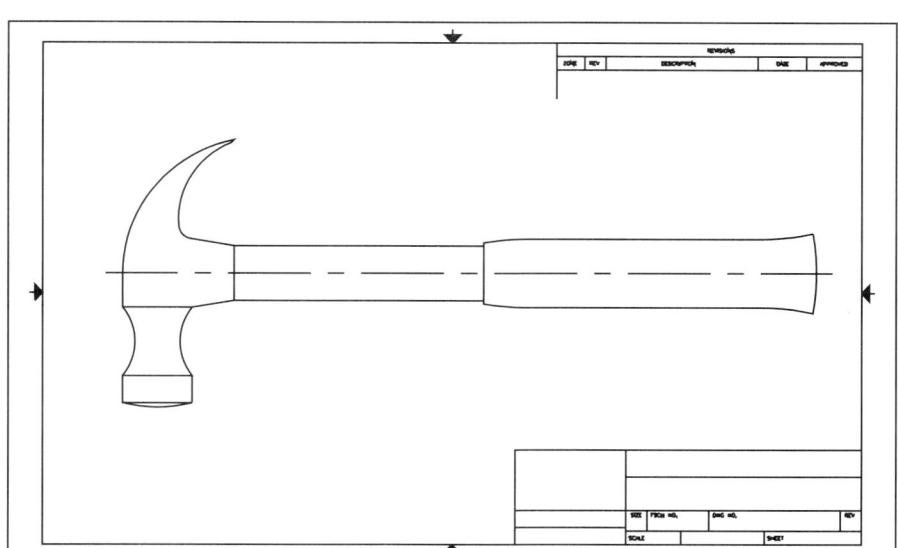

9. **Using a Template Drawing**

This exercise gives you experience using a template drawing that includes a title block and border. You will construct the wedge block in Figure 13-30 and plot on a B size sheet. It is necessary to have a plot or print device configured for your system that can use a "B" size sheet.

A. Complete steps 8.A. and 8.B. from the previous exercise.

B. Begin a *New* drawing. In the *Create New Drawing* dialog box that appears, select **Use a Template**. Select *ANSI B -Color Dependent Plot Styles.Dwt*.

C. Select the *ANSI B Title Block* tab. Then invoke the *Page Setup* dialog box and select a matching plot or print device and sheet size for the ANSI B title block.

D. Pick the *Model* tab. Set up model space with *Limits* of **17 x 11**. Set up appropriate *Snap* and *Grid* if desired. Also set *LTSCALE* and any running *Osnaps* you want to use. Type **PSLTSCALE** and set the value to **0**.

E. Create the following layers and assign linetypes and line weights. Assign colors of your choice.

   GEOMETRY   *Continuous*
   CENTER   *Center2*
   HIDDEN   *Hidden2*

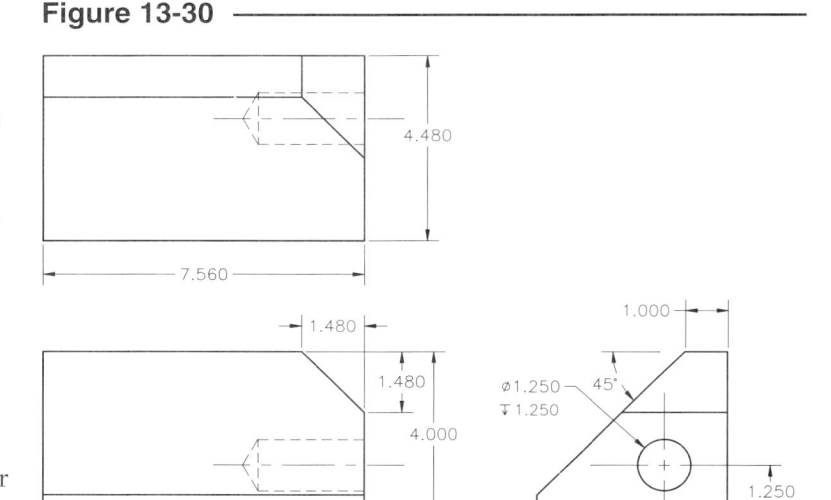

Notice the *Title Block* and *Viewport* layers already exist as part of the template drawing.

F. Draw the wedge block according to the dimensions given (Fig. 13-30). *Save* the drawing as **WEDGEBK**.

Figure 13-30

G. When you are finished with the views, select the *ANSI B Title Block* tab. To scale the model space geometry to paper space, you can use either the *Viewports* toolbar or the *Properties* window to set the viewport scale to **1:2**. You may also need to use *Pan* to center the geometry.

Figure 13-31

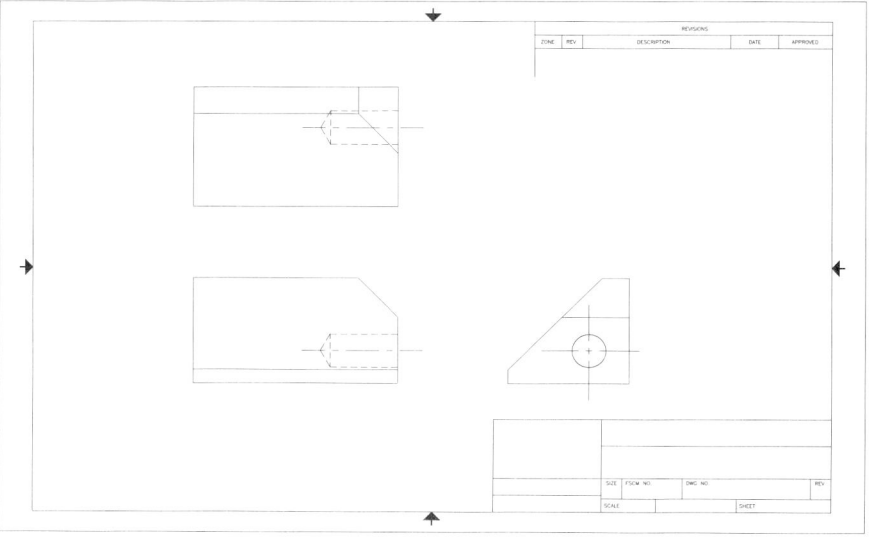

H. Next, use the *Layer Properties Manager* to make layer **Viewports** *non-plottable*. Your completed drawing should look like that in Figure 13-31. *Save* the drawing. From paper space, plot the drawing on a B size sheet and enter a scale of **1:1**

10. **Multiple Layouts**

   Assume a client requests that a copy of the HAMMER drawing be faxed to him immediately; however, your fax machine cannot accommodate the B size sheet that contains the plotted drawing. In this exercise you will create a second layout using a layout template to plot the same model geometry on an A size sheet.

   A. Open the drawing you created in a previous exercise named **HAMMER**. Access the *Display* tab in the *Options* dialog box. In the *Layout Elements* section (lower left), remove the check for both *Show Page Setup Dialog Box for New Layouts* and *Create Viewport in New Layouts*. Select *OK*.

   B. Right-click on an existing layout tab and use the *From Template* option from the shortcut menu. Select the *ANSI A -Color Dependent Plot Styles* template drawing (.DWT) to import. When the layout is imported, click the new *ANSI A Title Block* tab. Note that the template loads a layout with a title block and border in paper space and has one viewport created. Access the *Page Setup* dialog box and select the matching print or plot device and sheet size for the selected ANSI A title block.

   C. Double-click inside the viewport or use *Mspace* or the *MODEL/PAPER* toggle. Now scale the hammer to plot to a standard scale by using *Zoom 3/4XP* or enter .75 in the appropriate edit box in the *Viewports* toolbar or *Properties* window to correctly scale the model space geometry. Use *Pan* in the viewport to center the geometry within the viewport.

   D. Your new layout should look like that in Figure 13-32. Access the *ANSI B Title Block* tab you created earlier. Note that you now have two layouts, each layout has settings saved to plot the same geometry using different plot devices and at different scales.

   E. Access the new layout tab. *Save* the drawing. Make a *Plot* from the layout at a scale of **1:1** and send the fax.

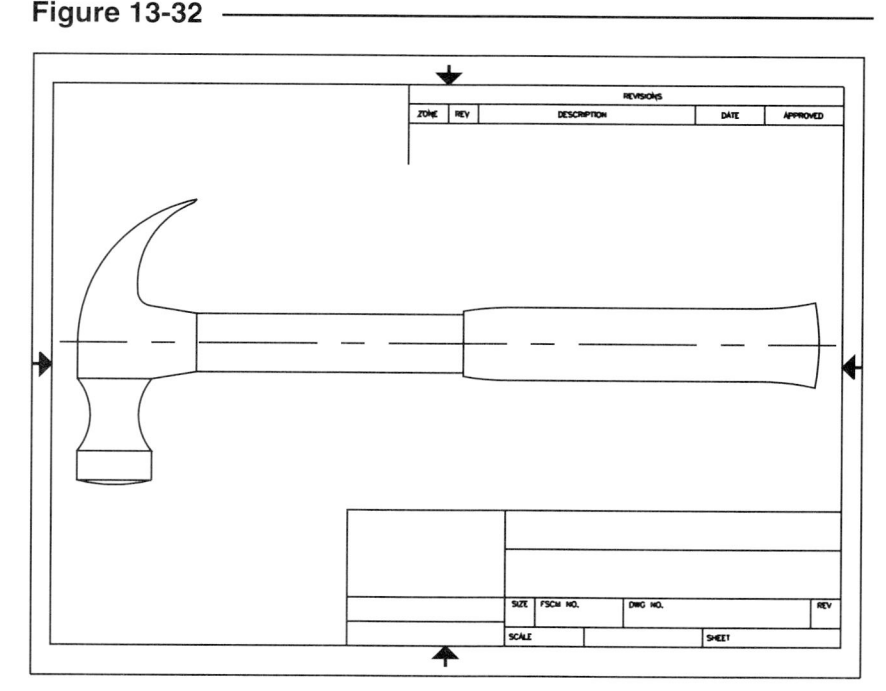

Figure 13-32

266 Layouts and Viewports

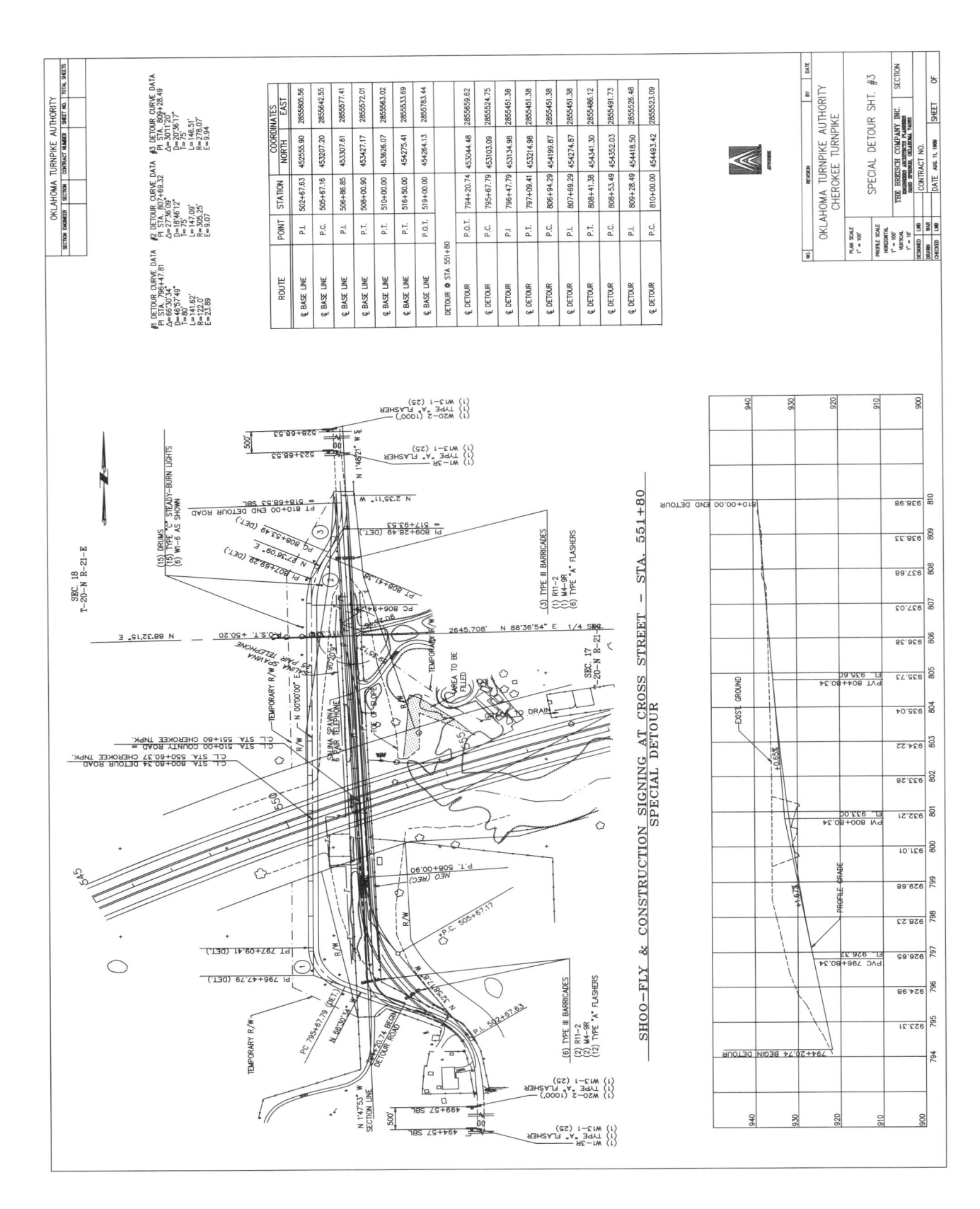

14

PRINTING AND PLOTTING

Chapter Objectives

After completing this chapter you should:

1. know the typical steps for printing or plotting;

2. be able to invoke and use the *Plot* and *Page Setup* dialog boxes;

3. be able to select from available plotting devices and set the paper size and orientation;

4. be able to specify what area of the drawing you want to plot;

5. be able to preview the plot before creating a plotted drawing;

6. be able to specify a scale for plotting a drawing;

7. know how to set up a drawing for plotting to a standard scale on a standard size sheet; and

8. be able to use the Tables of *Limits* Settings to determine *Limits*, scale, and paper size settings.

CONCEPTS

There are several concepts, procedures, and tools in AutoCAD related to printing and plotting. Many of these topics have already been discussed, such as setting up a drawing to draw true size, creating layouts and viewports, specifying the viewport scale, and using *Page Setup* to specify plot or print options. This chapter explains the typical steps to plotting from either the *Model* tab or *Layout* tabs, using both the *Page Setup* and *Plot* dialog boxes, and plotting to scale.

Figure 14-1

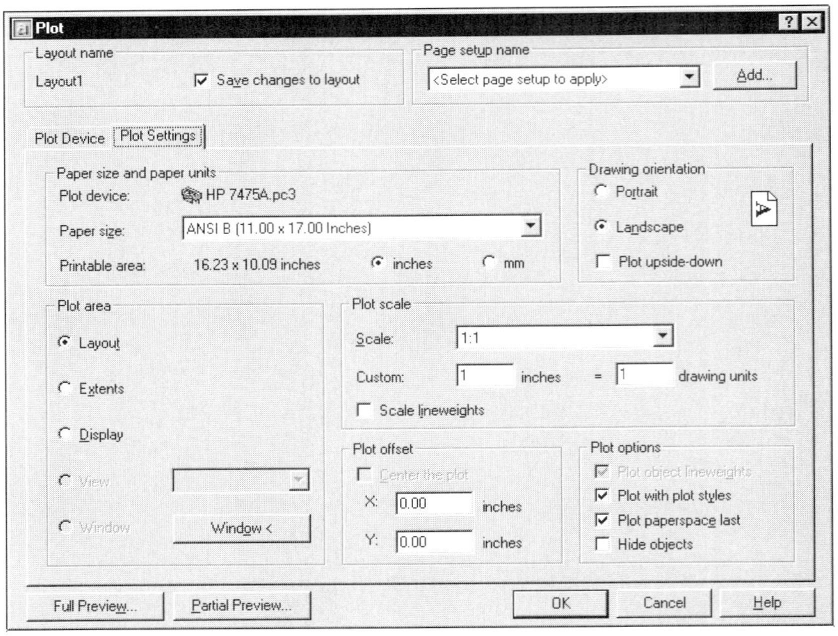

In AutoCAD, the term "plotting" can refer to plotting on a plot device (such as a pen plotter or electrostatic plotter) or printing with a printer (such as a laser jet printer). The *Plot* command is used to create a plot or print by producing the *Plot* dialog box. However, creating the plot or print can be accomplished using either the *Plot* dialog box or the *Page Setup* dialog box.

The *Plot* dialog box and the *Page Setup* dialog box have almost the same features and functions. The basic difference is that the *Page Setup* dialog box is used to specify and save plot and print specifications for each *Layout* or *Model* tab, whereas the *Plot* dialog box does not save the settings unless you plot, but has a few more functions including a plot preview. This chapter explains all features and distinctions of both dialog boxes.

The *Plot* dialog box, similar to the *Page Setup* dialog box, has two tabs, the *Plot Settings* tab (Fig. 14-1) and the *Plot Device* tab (see Figure 14-2). Use the *Plot Settings* tab to specify parameters of how the drawing will appear on the plot, such as orientation on the paper and paper size, what area of the drawing to plot, plot scale, and other options. Use the *Plot Device* tab to specify what type of printer or plotter you want to use, what plot style tables are attached, and other options.

TYPICAL STEPS TO PLOTTING

Assuming your CAD system and plotting devices have been properly configured, the typical basic steps to plotting the *Model* tab or a *Layout* tab are listed below.

1. Use *Save* to ensure the drawing has been saved in its most recent form before plotting (just in case some problem arises while plotting).

2. Make sure the plotter or printer is turned on, has paper (and pens for some devices) loaded, and is ready to accept the plot information from the computer.

3. Invoke the *Plot* dialog box.

4. Use the *Plot Device* tab to select the intended plot device from the drop-down list. The list includes all devices currently configured for your system.

5. In the *Plot Settings* tab, ensure the desired *Drawing Orientation* and *Paper Size* are selected.

6. Determine and select the desired *Plot Area* for the drawing: *Layout, Limits, Extents, Display, Window,* or *View.*

7. Select the desired scale from the *Scale* drop-down list or enter a *Custom* scale. If no standard scale is needed, select *Scaled to Fit.* If you are plotting from a *Layout*, normally set the scale to 1:1.

8. If necessary, specify a *Plot Offset* or *Center the Plot* on the sheet.

9. If necessary, specify the *Plot Options*, such as plotting with lineweights or plot styles.

10. Always preview the plot to ensure the drawing will plot as you expect. Select a *Full Preview* to view the drawing objects as they will plot. If the preview does not display the plot as you intend, make the appropriate changes. Otherwise, needless time and media could be wasted.

11. If all settings are acceptable, selecting the *OK* tile causes the drawing to be sent to the plotter or printer. All settings are saved in the drawing file when you plot, but are not saved if you *Cancel*.

Alternately, if you want to set up all the parameters for plotting and save them, but do not actually want to make a plot yet, you can use the *Page Setup* dialog box to do so. The *Page Setup* dialog box is almost the same as the *Plot* dialog box and can be used to set up and save plotting parameters.

USING THE *PLOT* AND *PAGE SETUP* DIALOG BOXES

PLOT

Pull-down Menu	COMMAND (TYPE)	ALIAS (TYPE)	Short-cut	Screen (side) Menu	Tablet Menu
File Plot...	PLOT, PRINT, or -PLOT	...	Ctrl+P	FILE Plot	W,25

Using *Plot* invokes the *Plot* dialog box. The *Plot* dialog box has two tabs, the *Plot Settings* tab (see Figures 14-1 and 14-3) and the *Plot Device* tab (Figure 14-2).

The *Plot* dialog box allows you to set plotting parameters such as scale, orientation, offset, plot style, paper size, and plotting device. Settings made to plotting options can be saved for the *Model* tab and each *Layout* tab in the drawing so they do not have to be re-entered the next time, but only if you create the plot.

Layout Name
At the top-left corner of the *Plot* dialog box, the *Layout Name* corresponds to the *Model* or *Layout* tab that is currently active (when the *Plot* command is used).

Save Changes to Layout
Check this box if you want settings you make to be saved with the drawing for the selected *Model* tab or *Layout* tab. You can save the settings for each layout. This is a great advantage since you can set up

several tabs, each showing a different aspect of the drawing and specifying different plot options such as scales, plot devices, layer combinations, plot styles, and so on. Using this feature allows you to load the drawing at a later time and reproduce plots with all plot specifications pre-set.

Page Setup Name
Typically page setups (plot specifications made in this dialog box) are automatically saved for each layout tab (including the *Model* tab) using the name of the tab as the *Page Setup Name*. For some situations, you may want to apply a previously saved *Page Setup* to a new layout by selecting a name from the list. This idea is discussed in Chapter 13.

Plot Device Tab

Plotter Configuration
Many devices (printers and plotters) can be configured for use with AutoCAD, and all configurations are saved for your selection. For example, you can have both an "A" size and a "D" size plotter as well as a laser printer; any one or more could be used to plot the current drawing.

When you install AutoCAD, it automatically configures itself to use the Windows system printer. Additional devices for use specifically with AutoCAD can be configured within AutoCAD by using the *Plotter Manager*. The *Plotter Manager* can be accessed several ways.

Figure 14-2

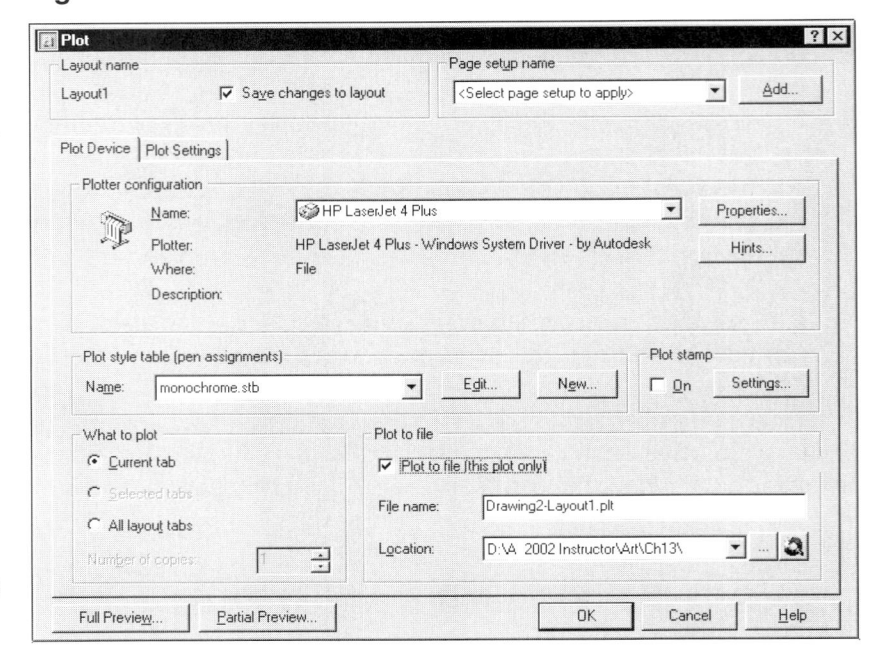

Name:
Select the desired device from the list. The list is composed of all previously configured devices for your system.

Properties...
This button invokes the *Plotter Configuration Editor*. In the editor, you can specify a variety of characteristics for the specific plot device shown in the *Name*: drop-down list. The *Plotter Configuration Editor* is also accessible through the *Plotter Manager*.

Hints...
Use this button to invoke help on the device configuration.

Plot Style Table (pen assignments)
If desired, select a plot style table to use with your plot device. A plot style table can contain several plot styles. Each plot style can assign specific line characteristics (lineweight, screening, dithering, end and joint types, fill patterns, etc.). Normally, a plot style table can be assigned to the *Model* tab, or to *Layout* tabs.

Plot Stamp
In AutoCAD 2002 you have the capability of adding an informational text "stamp" on each plot. For example, the drawing name, date, and time of plot can be added to each drawing that you plot for tracking and verification purposes. To specify the plot stamp, you can select the *Settings* button or use the *Plotstamp* command. Select the *On* button to apply the stamp you specified to the plotted drawing.

What to Plot
In this section you can specify which of the *Layout* or *Model* tabs you want to plot. You can plot the *Current tab*, *Selected tabs* or *All Layout tabs*. Multiple tabs can be selected with your pointer by holding down the Shift key, selecting the tabs, then using *Plot*. Multiple copies can also be generated.

Plot to File
Choosing this option writes the plot to a file instead of making a plot. This action generally creates a .PLT file type. The format of the file (for example, PCL or HP/GL language) depends on the device brand and model that is configured. The plot file can be printed or plotted later without AutoCAD, assuming the correct interpreter for the device is available. Specify the *File Name* and *Location* (path) or use the browse button (...) to specify the name and location for the plot file. You can also access a Web site to send the plot file to by selecting the *Search the Web* button (lower-right corner of the dialog box).

Plot Settings Tab

Paper Size and Paper Units
Depending in the *Plot Device* selected, several paper sizes may be available. Select the desired *Paper Size* from the drop-down list, then select which units you want to use.

Inches or *mm* should be selected to correspond to the units used in the drawing. The selected units affect the *Scale* and *Plot Offset* units. For example, assuming the drawing was created using the actual units, a scale for plotting can be calculated without millimeter to inch conversion or vice versa.

Drawing Orientation
Normally, *Landscape* is selected to match the orientation of the AutoCAD drawing area—that is, horizontal lines in the drawing plot lengthwise on the paper. *Portrait* positions horizontal lines in the drawing along the short axis of the paper.

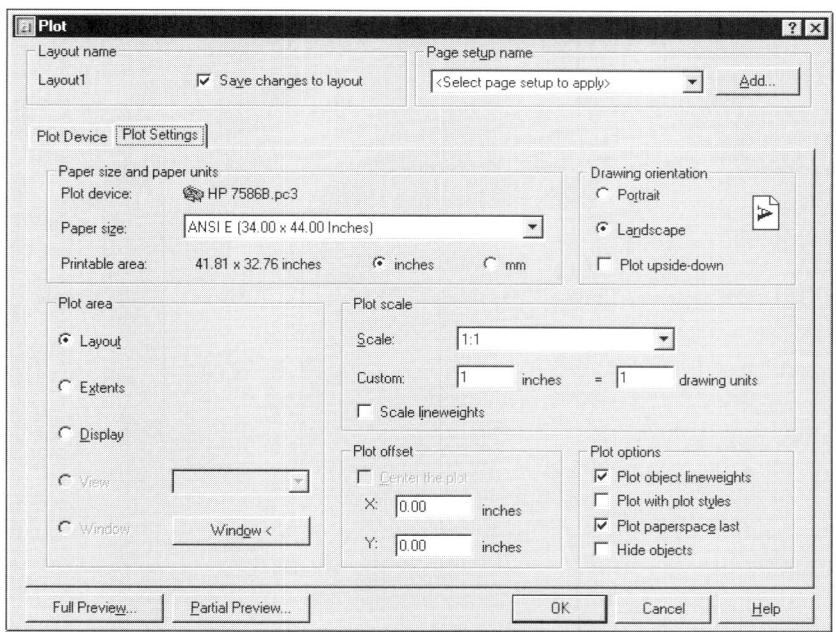

Figure 14-3

Plot Area
Specify what part of the drawing you want to plot by selecting the desired button from the lower-left corner of the dialog box (see Figure 14-3). The choices are listed next.

Limits or *Layout*
This button name changes depending on whether you invoke the *Plot* or *Page Setup* dialog box while the *Model* tab or a *Layout* tab is current. If the *Model* tab is current, this selection plots the area defined by the *Limits* command (unless plotting a 3D object from other than the plan view). If a *Layout* tab is current, the layout is plotted as defined by the selected paper size.

Extents
Plotting *Extents* is similar to *Zoom Extents*. This option plots the entire drawing (all objects), disregarding the *Limits*. Use the *Zoom Extents* command in the drawing to make sure the extents have been updated before using this plot option.

Display
This option plots the current display on the screen. If using viewports, it plots the current viewport.

View
With this option you can plot a view previously saved using the *View* command. Creating *Views* of specific areas of a drawing is very helpful when you plan to make repetitive plots of the same area or plots of multiple areas of one drawing.

Window
Window allows you to plot any portion of the drawing. You must specify the window by picking or supplying coordinates for the lower-left and upper-right corners.

Plot Scale
This is one of the most important areas of the dialog box, assuming you want to plot the drawing to a standard scale. You can select a standard scale from the drop-down list or enter a custom scale.

Scale
In the drop-down list (Fig. 14-4) you have two basic choices: scale the drawing to automatically fit on the paper or select a specific scale for the drawing to be plotted. By selecting *Scaled to Fit*, the drawing is automatically sized to fit on the sheet based on the specified area to plot (*Display, Limits, Extents*, etc.). If you want to plot the drawing to a specific scale, first ensure the correct paper units are selected, then select the desired scale from the list. Available options include ratios such as 1:2, 1:4, and 2:1, which are normally used for civil or mechanical engineering (decimal) drawings, and fractional inch scales such as 1/4"=1' and 1/8"=1' which are normally used for architectural and construction drawings. Fractional inch scales such as 3/8"=1" or 5/8"=1" used for older mechanical drawings (pre-ANSI 1994 standards) must be specified in the *Custom* edit boxes.

Figure 14-4

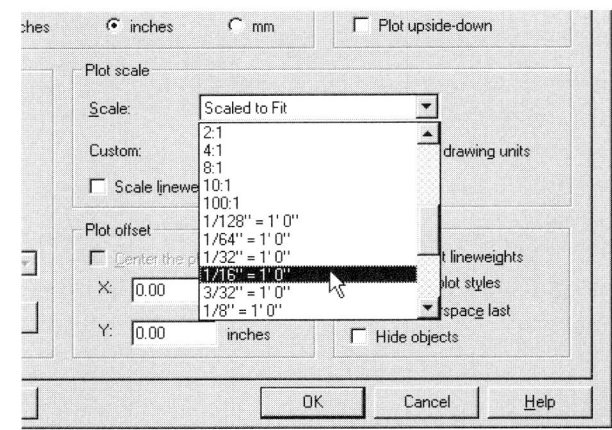

If you are plotting the *Model* tab, select an option directly from the *Scale* drop-down list to specify the desired plot scale for the drawing. If you are plotting from a *Layout* tab, first set the scale for the viewport using the *Viewport* toolbar, *Properties* window, or using *Zoom* with an *XP* value, then select *1:1* from the *Scale* drop-down list in the *Plot* or *Page Setup* dialog box. In either case, the plot scale for the drawing is equal to the reciprocal of the drawing scale factor and can be determined by referencing the Tables of *Limits* Settings. See "Plotting to Scale."

Custom
Enter the desired ratio in the boxes (see Figure 14-3). Decimals or fractions can be entered. For example, to prepare for a plot of 3/8 size (3/8"=1"), the following ratios can be entered to achieve a plot at the desired scale: 3=8, 3/8=1, or .375=1 (inches=drawing units). For guidelines on plotting a drawing to scale, see "Plotting to Scale" in this chapter.

Scale Lineweights
This option scales the lineweights for layouts only. When checked, lineweights are scaled proportionally with the plot scale, so a 1mm lineweight, for example, would plot at .5mm when the drawing is plotted at 1:2 scale. Normally (no check) lineweights are plotted at their absolute value (0.010", 0.016", 0.25 mm, etc.) at any scale the drawing is plotted.

Plot Offset
Plotting and printing devices cannot plot to the edge of the paper, as noted by the difference in the *Paper Size* and *Printable Area* given in the *Paper Size and Paper Units* section of the dialog box. Therefore, the lower-left corner of your drawing will not plot at exactly the lower-left corner of the paper. You can choose to reposition, or offset, the drawing on the paper by entering positive or negative values in the X and Y edit boxes. Home position for plotters is the lower-left corner (landscape orientation) and for printers it is the upper-left corner of the paper (portrait orientation). If plotting the *Model* tab, you can choose to *Center the Plot* on the paper.

Plot Options
These checkboxes control the following options.

Plot Object Lineweights
Lineweights on a plot can be assigned by one of two ways: use lineweights in your drawing and plot them as they appear on the screen, or use plot styles to override lineweights assigned in the drawing. If you have assigned lineweights in your drawing, checking this option plots the lineweights as they appear in the drawing. Checking *Plot With Plot Styles* disables this option.

Plot with Plot Styles
You must have a plot style table attached (see *Plot Device* Tab, *Plot Style Table*) for this option to have effect. If you are using lineweights in the drawing and also have lineweights assigned in the attached plot style table, the plot style table lineweights have precedence for plots.

Plot Paperspace Last
Checking this option plots paper space geometry last if you are plotting a layout. Normally, paper space geometry is plotted before model space geometry.

Hide Objects
This option removes hidden lines (edges obscured from view) if you are plotting a 3D surface or solid model only when plotting the *Model* tab. Hidden line removal for *Layouts* is controlled by the *Hideplot* option of the *–Vports* command or the *Properties* window. In that case, hidden lines are not plotted or displayed in a full preview, but are displayed in the layout (on screen).

Full Preview

Selecting the *Full Preview* button displays the complete drawing as it will plot on the sheet (Fig. 14-5). This function is particularly helpful to ensure the drawing is plotted as you expect. When you invoke a *Full Preview*, the drawing editor displays a simulated sheet of paper with the completed print or plot. This is helpful particularly when plot style tables are attached that create plots that appear differently than the *Model* tab or *Layout* tabs display. The *Zoom* feature is on by default so you can check specific areas of the drawing before making the plot. Right-clicking produces the shortcut, allowing you to use *Pan* and other *Zoom* options. Changing the display during a *Full Preview* does not change the area of the drawing to be plotted or printed.

Figure 14-5

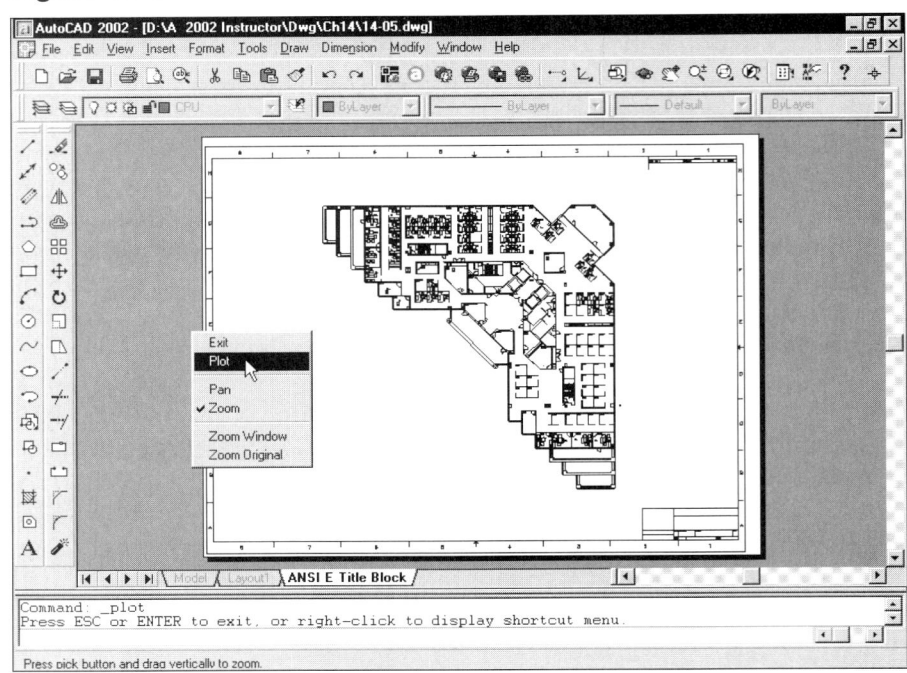

Partial Preview

Selecting the *Partial Preview* tile displays the effective plotting area as shown in Figure 14-6. Use the *Partial Preview* option to get a quick check showing how the drawing will fit on the sheet. The dotted border indicates the Printable Area and the blue filled Effective Area represents the size of the drawing as plotted on the sheet. The Effective Area is based on the selection made in the *Plot Area* section of the *Plot* dialog box (*Display, Extents, Limits,* etc.) and the selected *Plot Scale*. The red triangle designates the home position (location for the drawing's 0,0 point) for the plotting device.

Figure 14-6

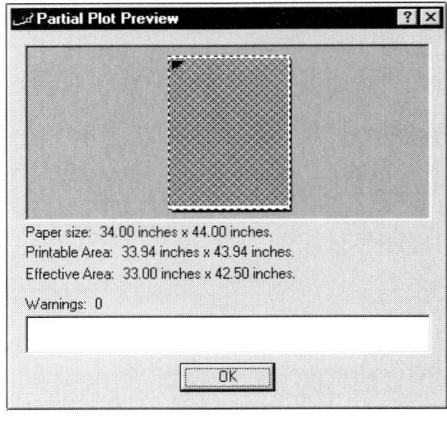

-PLOT

Entering *–Plot* at the Command line invokes the Command line version of *Plot*. This is an alternate method for creating a plot if you want to use the current settings or if you enter a page setup name to define the plot settings.

```
Command: -plot
Detailed plot configuration? [Yes/No] <No>:
Enter a layout name or [?] <ANSI E Title Block>:
Enter a page setup name <>:
Enter an output device name or [?] <HP 7586B.pc3>:
Write the plot to a file [Yes/No] <N>:
Save changes to layout [Yes/No]? <N>
Proceed with plot [Yes/No] <Y>:
Effective plotting area:  33.54 wide by 42.60 high
Plotting viewport 2.
Command:
```

PAGESETUP

Pull-down Menu	COMMAND (TYPE)	ALIAS (TYPE)	Short-cut	Screen (side) Menu	Tablet Menu
File Page Setup...	PAGESETUP	...	...	...	V,25

Produce the *Page Setup* dialog box by any method shown in the command table above or right-click on a *Layout* or *Model* tab and select *Page Setup*. To cause the *Page Setup* dialog box to appear automatically when you create a new layout, use the *Display* tab of the *Options* dialog box. However, you can toggle this feature off by removing the check at the lower-left corner of the dialog box (Fig. 14-7).

The *Page Setup* dialog box is similar and has the same functions as the *Plot* dialog box with a few exceptions. The name of the second tab in this dialog box is *Layout Settings* instead of *Plot Settings*, although options in the tab are almost identical to the *Plot Settings* tab of the *Plot* dialog box.

Figure 14-7

Should You Use the *Plot* Dialog Box or *Page Setup* Dialog Box?

You will notice that the *Plot* dialog box and *Page Setup* dialog box are almost identical (see previous Figure 14-1 and Figure 14-7). Both dialog boxes have essentially the same two tabs and features, and you can create a plot from either dialog box. There are four main differences described below.

1. Settings made in the *Page Setup* dialog box can be saved for the *Layout* or *Model* tab without making a plot, whereas settings made in the *Plot* dialog box cannot be saved unless you plot. You can exit the *Plot* dialog box only by *OK*, which makes a plot, or *Cancel*, which cancels everything. You can exit the *Page Setup* dialog box by *OK*, which automatically saves the settings, *Plot*, which saves settings and makes a plot, or *Cancel*.
2. The *Preview* button only appears in the *Plot* dialog box. Although you cannot preview from within the *Page Setup* dialog, you can exit and use the *Preview* button on the Standard toolbar.
3. Only the *Plot* dialog box allows you to *Plot to File*, plot multiple layouts, and apply a plot stamp.
4. Only with the *Page Setup* dialog box can you *Display Plot Styles* in layout tabs. When plot styles are attached, this option causes the display in the *Layout* tab to look like a plot or a *Full Preview*.

So which dialog box should you use? If you want to make settings for plotting but are not ready to plot, use the *Page Setup* dialog box to make the settings and save the changes (select *OK*), then select *Preview* from the standard toolbar. When you are ready to plot, use the *Plot* dialog box where you can change settings and *Preview* them immediately before you plot. Don't expect to save the changes in the *Plot* dialog box unless you plot.

PREVIEW

Pull-down Menu	COMMAND (TYPE)	ALIAS (TYPE)	Short-cut	Screen (side) Menu	Tablet Menu
File Plot Preview	PREVIEW	PRE	...	...	X,24

The *Preview* command accomplishes the same action as selecting a *Full Preview* in the *Plot* dialog box. The advantage to using this command is being able to see a full print preview directly without having to invoke the *Plot* dialog box first (see Figure 14-5). This command is especially helpful if you prefer to use the *Page Setup* dialog box instead of the *Plot* dialog box to specify plotting options. All functions of this preview (right-click for shortcut menu to *Pan* and *Zoom*, etc.) operate the same as a full preview from the *Plot* dialog box including one important option, *Plot*.

You can print or plot during *Preview* directly from the shortcut menu (Fig. 14-8). Using the *Plot* option creates a print or plot based on your settings in the *Plot* or *Page Setup* dialog box. If you use this feature, ensure you first set the plotting parameters in the *Page Setup* dialog box.

Figure 14-8

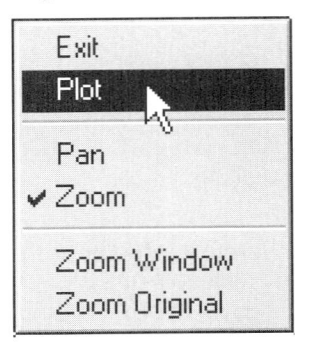

PLOTTING TO SCALE

When you create a manual drawing, a scale is determined before you can begin drawing. The scale is determined by the proportion between the size of the object on the paper and the actual size of the object. You then complete the drawing in that scale so that the actual object is proportionally reduced or enlarged to fit on the paper.

With a CAD drawing you are not restricted to drawing on a sheet of paper, so the geometry can be created full size. Set *Limits* to provide an appropriate amount of drawing space; then the geometry can be drawn using the actual dimensions of the object. The resulting drawing on the CAD system is a virtual full-size replica of the actual object. Not until the CAD drawing is plotted on a fixed size sheet of paper, however, is it scaled to fit on the sheet.

You can specify the plot scale of an AutoCAD drawing in one of two ways. If you are plotting a *Layout* tab, specify the scale of the geometry that appears in the viewport by using the *Viewport* toolbar, the *Properties* window, or the *Zoom XP* factor, and then select 1:1 in the *Plot* or *Page Setup* dialog box. If you plot the *Model* tab, select the desired scale directly in the *Plot* or *Page Setup* dialog box. In either case, the value or ratio that you specify is the proportion of paper space units to model space units—the same as the proportion of the paper size to the *Limits*. This ratio is also equal to the reciprocal of the "drawing scale factor," as explained in Chapter 12. Exceptions to this rule are cases in which multiple viewports are used.

For example, if you want to draw an object 15" long and plot it on an 11" x 8.5" sheet, you can set the model space *Limits* to 22 x 17 (2 x the sheet size). The value of **2** is then the drawing scale factor, and the plot scale is **1:2** (or **1/2**, the reciprocal of the drawing scale factor). If, in another case, the calculated drawing scale factor is **4**, then **1:4** (or **1/4**) would be the plot scale selected to achieve a drawing plotted at 1/4 actual size. In order to calculate *Limits* and drawing scale factors accurately, it is helpful to know the standard paper sizes.

Standard Paper Sizes

Size	Engineering (")	Architectural (")
A	8.5 x 11	9 x 12
B	11 x 17	12 x 18
C	17 x 22	18 x 24
D	22 x 34	24 x 36
E	34 x 44	36 x 48

Size	Metric (mm)
A4	210 x 297
A3	297 x 420
A2	420 x 594
A1	594 x 841
A0	841 x 1189

Calculating the Drawing Scale Factor

Assuming you are planning to plot from the *Model* tab or from a *Layout* tab using one viewport that fills almost all of the printable area, use the following method to determine the "drawing scale factor" and, therefore, the plot or print scale to specify.

1. In order to provide adequate space to create the drawing geometry full size, it is recommended to set the model space *Limits*, then *Zoom All*. *Limits* should be set to the intended sheet size used for plotting times a factor, if necessary, that provides enough area for drawing. This factor, the proportion of the *Limits* to the sheet size, is the "drawing scale factor."

$$DSF = \frac{Limits}{sheet\ size}$$

   You should also use a proportion that will yield a standard drawing scale (1/2"=1", 1/8"=1', 1:50, etc.), instead of a scale that is not a standard (1/3"=1", 3/5"=1', 1:23, etc.). See the "Tables of *Limits* Settings" for common standard drawing scales.

2. The "drawing scale factor" is used as the value (at least a starting point) for changing all size-related variables (*LTSCALE, Overall Scale* for dimensions, text height, *Hatch* pattern scale, etc.). The reciprocal of the drawing scale factor is the plot scale to use for plotting or printing the drawing.

$$Plot\ Scale = \frac{1}{DSF}$$

3. If the drawing is to be created using millimeter dimensions, set *Units* to *Decimal* and use metric values for the sheet size. In this way, the reciprocal of the drawing scale factor is the plot scale, the same as feet and inch drawings. However, multiply the drawing's scale factor by 25.4 (25.4mm = 1") to determine the factor for changing all size-related variables (*LTSCALE, Overall Scale* for dimensions, etc.).

$$DSF(mm) = \frac{Limits}{sheet\ size} \times 25.4$$

If drawing or inserting a border on the sheet, its maximum size cannot exceed the *Printable Area*. The *Printable Area* is given near the upper-left corner of the *Plot* or *Page Setup* dialog box, but the values can be different for different devices. Since plotters or printers do not draw or print all the way to the edge of the paper, the border should not be drawn outside of the *Printable Area*. Generally, approximately 1/4" to 1/2" (6mm to 12mm) offset from each edge of the paper (margin) is required. When plotting a *Layout*, the printable area is denoted by a dashed border around the sheet. If plotting the *Model* tab, you should multiply the margin (1/2 of the difference between the *Paper Size* and *Printable Area*) by the "drawing scale factor" to determine the margin size in model space drawing units.

Guidelines for Plotting the *Model* Tab to Scale

Even though model space *Limits* can be changed and plot scale can be reset at <u>any time</u> in the drawing process, it is helpful to begin the process during the initial drawing setup (also see Chapter 12, Advanced Drawing Setup).

1. Set *Units* (*Decimal, Architectural, Engineering*, etc.) and *Precision* to be used in the drawing.
2. Set model space *Limits*. Set the *Limits* values to a proportion of the desired sheet size that provides enough area for drawing geometry as described in "Calculating the Drawing Scale Factor." The resulting value is the drawing scale factor.
3. Create the drawing geometry in the *Model* tab. Use the DSF as the initial value (or multiplier) for linetype scale, text size, dimension scale, hatch scale, etc.
4. Use the DSF to determine the scale factor to create, *Insert* or *Xref* the title block and border in model space, if one is needed.
5. Access the *Page Setup* or *Plot* dialog box. Select the desired scale in the *Scale* drop-down list or enter a value in the *Custom* edit boxes. The value to enter is 1/DSF. (Values are also given in the "Tables of *Limits* Settings.") Make a plot *Preview*, then make the plot or make the needed adjustments.

Guidelines for Plotting *Layouts* to Scale

Assuming you are using one viewport that fills almost all of the printable area, plot from the *Layout* tab using this method.

1. Set *Units* (*Decimal, Architectural, Engineering*, etc.) and *Precision* to be used in the drawing.
2. Set model space *Limits* and create the geometry in the *Model* tab. Use the DSF to determine values for linetype scale, text size, dimension scale, hatch scale, etc. Complete the drawing geometry in model space.
3. Click a *Layout* tab to begin setting up the layout. You can also use the *Layout Wizard* to create a new layout and complete steps 2 through 5.
4. Use *Page Setup* to select the desired *Plot Device* and *Paper Size*.
5. Create, *Insert*, or *Xref* the title block and border in the layout. If you are using one of AutoCAD's template drawings or another template, the title block and border may already exist.
6. Create a viewport using *Vports*.
7. Click inside the viewport to set the viewport scale. You can set the scale using the *Viewports* toolbar, the *Properties* window, or use *Zoom* and enter an XP factor. Use 1/DSF as the viewport scale.
8. Activate the *Plot* or *Page Setup* dialog box and ensure the scale in the *Scale* drop-down list is 1:1. Make a plot *Preview*, then make the plot or make the needed adjustments.

If you want to print the same drawing using different devices and different scales, you can create multiple layouts. With each layout you can save the specific plot settings. Beware—a drawing printed at different scales may require changing or setting different values for linetype scale, text size, dimension variables, hatch scales, etc. for each layout.

Simplifying the process of plotting to scale and calculating model space *Limits* and the "drawing scale factor" can be accomplished by preparing template drawings. One method is to create a template for each sheet size that is used in the lab or office. In this way, the CAD operator begins the session by selecting the template drawing representing the sheet size and then multiplies those *Limits* by some factor to achieve the desired *Limits* and drawing scale factor. Another method is to create multiple layouts in the template drawing, one for each device and sheet size that you have available. In this way, a template drawing can be selected with the final layouts, title blocks, plot devices, and sheet sizes already specified, then you need only set the model space *Limits* and plot scale(s). (See Chapter 12 for help creating template drawings.)

TABLES OF *LIMITS* SETTINGS

Rather than making calculations of *Limits*, drawing scale factor, and plot scale for each drawing, the Tables of *Limits* Settings on the following pages can be used to make calculating easier. There are five tables, one for each of the following applications:

Mechanical Engineering
Architectural
Metric (using ISO standard metric sheets)
Metric (using U.S. standard engineering sheets)
Civil Engineering

To use the tables correctly, you must know any two of the following three variables. The third variable can be determined from the other two.

Approximate drawing *Limits*
Desired print or plot paper size
Desired print or plot scale

1. If you know the ***Scale*** and **paper size**:
   Assuming you know the scale you want to use, look along the top row to find the desired scale that you eventually want to print or plot. Find the desired paper size by looking down the left column. The intersection of the row and column yields the model space *Limits* settings to use to achieve the desired plot scale.

2. If you know the approximate model space ***Limits*** and **paper size**:
   Calculate how much space (minimum *Limits*) you require to create the drawing actual size. Look down the left column of the appropriate table to find the paper size you want to use for plotting. Look along that row to find the next larger *Limits* settings than your required area. Use these values to set the drawing *Limits*. The scale to use for the plot is located on top of that column.

3. If you know the ***Scale*** and approximate model space ***Limits***:
   Calculate how much space (minimum *Limits*) you need to create the drawing objects actual size. Look along the top row of the appropriate table to find the desired scale that you eventually want to plot or print. Look down that column to find the next larger *Limits* than your required minimum area. Set the model space *Limits* to these values. Look to the left end of that row to give the paper size you need to print or plot the *Limits* in the desired scale.

NOTE: Common scales are given in the Tables of *Limits* Settings. If you need to create a print or plot in a scale that is not listed in the tables, you may be able to use the tables as a guide by finding the nearest table and standard scale to match your needs, then calculating proportional settings.

MECHANICAL TABLE OF *LIMITS* SETTINGS
(X axis x Y axis)

Paper Size (Inches)	Drawing Scale Factor Scale 1″ = 1″ Proportion 1:1	1.33 3/4″ = 1″ 3:4	2 1/2″ = 1″ 1:2	2.67 3/8″ = 1″ 3:8	4 1/4″ = 1″ 1:4	5.33 3/6″ = 1″ 3:16	8 1/8″ = 1″ 1:8
A 11 x 8.5 In.	11.0 x 8.5	14.7 x 11.3	22.0 x 17.0	29.3 x 22.7	44.0 x 34.0	58.7 x 45.3	88.0 x 68.0
B 17 x 11 In.	17.0 x 11.0	22.7 x 14.7	34.0 x 22.0	45.3 x 29.3	68.0 x 44.0	90.7 x 58.7	136.0 x 88.0
C 22 x 17 In.	22.0 x 17.0	29.3 x 22.7	44.0 x 34.0	58.7 x 45.3	88.0 x 68.0	117.0 x 90.7	176.0 x 136.0
D 34 x 22 In.	34.0 x 22.0	45.3 x 29.3	68.0 x 44.0	90.7 x 58.7	136.0 x 88.0	181.3 x 117.3	272.0 x 176.0
E 44 x 34 In.	44.0 x 34.0	58.7 x 45.3	88.0 x 68.0	117.3 x 90.7	176.0 x 136.0	235.7 x 181.3	352.0 x 272.0

ARCHITECTURAL TABLE OF *LIMITS* SETTINGS
(X axis x Y axis)

Paper Size (Inches)	Drawing Scale Factor: 12 Scale: 1" = 1' Proportion: 1:12	16 3/4" = 1' 1:16	24 1/2" = 1' 1:24	32 3/8" = 1' 1:32	48 1/4" = 1' 1:48	64 3/16" = 1' 1:64	96 1/8" = 1' 1:96
A 12 x 9 Ft In.	12 x 9 144 x 108	16 x 12 192 x 144	24 x 18 288 x 216	32 x 24 384 x 288	48 x 36 576 x 432	64 x 48 768 x 576	96 x 72 1152 x 864
B 18 x 12 Ft In.	18 x 12 216 x 144	24 x 16 288 x 192	36 x 24 432 x 288	48 x 32 576 x 384	64 x 48 768 x 576	96 x 64 1152 x 768	128 x 96 1536 x 1152
C 24 x 18 Ft In.	24 x 18 288 x 216	32 x 24 384 x 288	48 x 36 576 x 432	64 x 48 768 x 576	96 x 72 1152 x 864	128 x 96 1536 x 1152	192 x 144 2304 x 1728
D 36 x 24 Ft In.	36 x 24 432 x 288	48 x 32 576 x 384	72 x 48 864 x 576	96 x 64 1152 x 768	144 x 96 1728 x 1152	192 x 128 2304 x 1536	288 x 192 3456 x 2304
E 48 x 36 Ft In.	48 x 36 576 x 432	64 x 48 768 x 576	96 x 72 1152 x 864	128 x 96 1536 x 1152	192 x 144 2304 x 1728	256 x 192 3072 x 2304	384 x 288 4608 x 3456

METRIC TABLE OF *LIMITS* SETTINGS
FOR METRIC SHEET SIZES
(X axis x Y axis)

Paper Size (mm)	Drawing Scale Factor Scale Proportion	25.4 1:1 1:1	50.8 1:2 1:2	127 1:5 1:5	254 1:10 1:10	508 1:20 1:20	1270 1:50 1:50	2540 1:100 1:100
A4 297 x 210 mm m		297 x 210 .297 x .210	594 x 420 .594 x .420	1485 x 1050 1.485 x 1.050	2970 x 2100 2.97 x 2.10	5940 x 4200 5.94 x 4.20	14,850 x 10,500 14.85 x 10.50	29,700 x 21,000 29.70 x 21.00
A3 420 x 297 mm m		420 x 297 .420 x .297	840 x 594 .840 x .594	2100 x 1485 2.100 x 1.485	4200 x 2970 4.20 x 2.97	8400 x 5940 8.40 x 5.94	21,000 x 14,850 21.00 x 14.85	42,000 x 29,700 42.00 x 29.70
A2 594 x 420 mm m		594 x 420 .594 x .420	1188 x 840 1.188 x .840	2970 x 2100 2.97 x 2.10	5940 x 4200 5.94 x 4.20	11,880 x 8400 11.88 x 8.40	29,700 x 21,000 29.70 x 21.00	59,400 x 42,000 59.40 x 42.00
A1 841 x 594 mm m		841 x 594 .841 x .594	1682 x 1188 1.682 x 1.188	4205 x 2970 4.205 x 2.970	8410 x 5940 8.41 x 5.94	16,820 x 11,880 16.82 x 11.88	42,050 x 29,700 42.05 x 29.70	84,100 x 59,400 84.10 x 59.40
A0 1189 x 841 mm m		1189 x 841 1.189 x .841	2378 x 1682 2.378 x 1.682	5945 x 4205 5.945 x 4.205	11,890 x 8410 11.89 x 8.41	23,780 x 16,820 23.78 x 16.82	59,450 x 42,050 59.45 x 42.05	118,900 x 84,100 118.90 x 84.10

METRIC TABLE OF *LIMITS* SETTINGS
FOR ENGINEERING (8.5 x 11 Format) SHEET SIZES
(X axis x Y axis)

Paper Size (mm)	Drawing Scale Factor Scale Proportion	25.4 1:1 1:1	50.8 1:2 1:2	127 1:5 1:5	254 1:10 1:10	508 1:20 1:20	1270 1:50 1:50	2540 1:100 1:100
A 279.4 x 215.9 mm m		279.4 x 215.9 0.2794 x 0.2159	558.8 x 431.8 0.5588 x 0.4318	1397 x 1079.5 1.397 x 1.0795	2794 x 2159 2.794 x 2.159	5588 x 4318 5.588 x 4.318	13,970 x 10,795 13.97 x 10.795	27,940 x 21,590 27.94 x 21.59
B 431.8 x 279.4 mm m		431.8 x 279.4 0.4318 x 0.2794	863.6 x 558.8 0.8636 x 0.5588	2159 x 1397 2.159 x 1.397	4318 x 2794 4.318 x 2.794	8636 x 5588 8.636 x 5.588	21,590 x 13,970 21.59 x 13.97	43,180 x 27,940 43.18 x 27.94
C 558.8 x 431.8 mm m		558.8 x 431.8 0.5588 x 0.4318	1117.6 x 863.6 1.1176 x 0.8636	2794 x 2159 2.794 x 2.159	5588 x 4318 5.588 x 4.318	11,176 x 8636 11.176 x 86.36	27,940 x 21,590 27.94 x 21.59	55,880 x 43,180 55.88 x 43.18
D 863.6 x 558.8 mm m		863.6 x 558.8 0.8636 x 0.5588	1727.2 x 1117.6 1.7272 x 1.1176	4318 x 2794 4.318 x 2.794	8636 x 5588 8.636 x 5.588	17,272 x 11,176 17.272 x 11.176	43,180 x 27,940 43.18 x 27.94	86,360 x 55,880 86.36 x 55.88
E 1117.6 x 863.6 mm m		1117.6 x 863.6 1.1176 x 0.8636	2235.2 x 1727.2 2.2352 x 1.7272	5588 x 4318 5.588 x 4.318	11,176 x 8636 11.176 x 8.636	22,352 x 17,272 22.352 x 17.272	55,880 x 43,180 55.88 x 43.18	111,760 x 86,360 111.76 x 86.36

CIVIL TABLE OF *LIMITS* SETTINGS
(For Engineering Units)
(X axis x Y axis)

Paper Size (Inches)	Drawing Scale Factor 120 Scale Tab 1" = 10' Proportion 1:120	240 1" = 20' 1:240	360 1" = 30' 1:360	480 1" = 40' 1:480	600 1" = 50' 1:600
A. 11 x 8.5 In. Ft	1320 x 1020 110 x 85	2640 x 2040 220 x 170	3960 x 3060 330 x 255	5280 x 4080 440 x 340	6600 x 5100 550 x 425
B. 17 x 11 In. Ft	2040 x 1320 170 x 110	4080 x 2640 340 x 220	6120 x 3960 510 x 330	8160 x 5280 680 x 440	10,200 x 6600 850 x 550
C. 22 x 17 In. Ft	2640 x 2040 220 x 170	5280 x 4080 440 x 340	7920 x 6120 660 x 510	10,560 x 8160 880 x 680	13,200 x 10,200 1100 x 850
D. 34 x 22 In. Ft	4080 x 2640 340 x 220	8160 x 5280 680 x 440	12,240 x 7920 1020 x 660	16,320 x 10,560 1360 x 880	20,400 x 13,200 1700 x 1100
E. 44 x 34 In. Ft	5280 x 4080 440 x 340	10,560 x 8160 880 x 680	15,840 x 12,240 1320 x 1020	21,120 x 16,320 1760 x 1360	26,400 x 20,400 2200 x 1700

Examples for Plotting to Scale

Following are several hypothetical examples of drawings that can be created using the "Guidelines for Plotting to Scale." As you read the examples, try to follow the logic and check the "Tables of *Limits* Settings."

A. A one-view drawing of a mechanical part that is 40" in length is to be drawn requiring an area of approximately 40" x 30", and the drawing is to be plotted on an 8.5" x 11" sheet. The template drawing *Limits* are preset to 0,0 and 11,8.5. (AutoCAD's default *Limits* are set to 0,0 and 12,9, which represents an uncut sheet size. Template *Limits* of 11 x 8.5 are more practical in this case.) The expected plot scale is 1/4"=1". The following steps are used to calculate the new *Limits*.

1. *Units* are set to *decimal*. Each unit represents 1.00 inches.
2. Multiplying the *Limits* of 11 x 8.5 by a factor of **4**, the new *Limits* should be set to **44 x 34**, allowing adequate space for the drawing. The "drawing scale factor" is **4**.
3. All size-related variable default values (**LTSCALE, Overall Scale** for dimensions, etc.) are multiplied by **4**. The plot scale (for *Layout* tabs, entered in the *Viewport* toolbar, etc., or for the *Model* tab, entered in the *Plot* or *Page Setup* dialog box) is **1:4** to achieve a plotted drawing of 1/4"=1".

B. A floorplan of a residence will occupy 60' x 40'. The drawing is to be plotted on a "D" size architectural sheet (24" x 36"). No prepared template drawing exists, so the standard AutoCAD default drawing (12 x 9) *Limits* are used. The expected plot scale is 1/2"=1'.

1. *Units* are set to *Architectural*. Each unit represents 1".
2. The floorplan size is converted to inches (60' x 40' = 720" x 480"). The sheet size of 36" x 24" (or 3' x 2') is multiplied by **24** to arrive at *Limits* of **864" x 576"** (72' x 48'), allowing adequate area for the floor plan. The *Limits* are changed to those values.
3. All default values of size-related variables (**LTSCALE, Overall Scale** for dimensions, etc.) are multiplied by **24**, the drawing scale factor. The plot scale (for *Layout* tabs, entered in the *Viewport* toolbar, etc., or for the *Model* tab, entered in the *Plot* or *Page Setup* dialog box) is **1/2"=1'** (or 1/24, which is the reciprocal of the drawing scale factor).

C. A roadway cloverleaf is to be laid out to fit in an acquired plot of land measuring 1500' x 1000'. The drawing will be plotted on "D" size engineering sheet (34" x 22"). A template drawing with *Limits* set equal to the sheet size is used. The expected plot scale is 1"=50'.

1. *Units* are set to *Engineering* (feet and decimal inches). Each unit represents 1.00".
2. The sheet size of 34" x 22" (or 2.833' x 1.833') is multiplied by **600** to arrive at *Limits* of **20400" x 13200"** (1700' x 1100'), allowing enough drawing area for the site. The *Limits* are changed to **1700' x 1100'**.
3. All default values of size-related variables (**LTSCALE, Overall Scale** for dimensions, etc.) are multiplied by **600**, the drawing scale factor. The plot scale (for *Layout* tabs, entered in the *Viewport* toolbar, etc., or for the *Model* tab, entered in the *Plot* or *Page Setup* dialog box) is **1/600** (*inches = drawing units*) to achieve a drawing of 1"=50' scale.

NOTE: Many civil engineering firms use one unit in AutoCAD to represent one foot. This simplifies the problem of using decimal feet (10 parts/foot rather than 12 parts/foot); however, problems occur if architectural layouts are inserted or otherwise combined with civil work.

CHAPTER EXERCISES

1. **Print GASKETA from the *Model* Tab**

    A. *Open* the **GASKETA** drawing that you created in Chapter 9 Exercises. Activate the *Model* tab. What are the model space *Limits*?

    B. From the *Model* tab, **Plot** the drawing *Extents* on an 11" x 8.5" sheet. Select the *Scale to Fit* option from the *Scale* drop-down list in the *Plot* dialog box.

    C. Next, **Plot** the drawing from the *Model* tab again. Plot the *Limits* on the same size sheet. *Scale to Fit*.

    D. Now, **Plot** the drawing *Limits* as before but plot the drawing at **1:1**. Measure the drawing and compare the accuracy with the dimensions given in the exercise in Chapter 9.

    E. Compare the three plots. What are the differences and why did they occur? (When you finish, there is no need to *Save* the changes.)

2. **Print GASKETA from a *Layout* Tab in Two Scales**

    A. Using the **GASKETA** drawing again, activate the *Layout1* tab. If a viewport exists in the layout, *Erase* it. Activate the *Page Setup* dialog box and configure the layout to use a printer to print on a *Letter* ("A" size) sheet.

    B. Next, make layer **0 *Current***. Use **Vports** to create a *Single* viewport using the *Fit* option. Double-click inside the viewport and **Zoom All** to see all of the model space *Limits*. Next, set the viewport scale to **1:1** using any method (*Zoom XP*, the *Viewport* toolbar, or the *Properties* window). You may need to *Pan* to center the geometry, but do <u>not</u> *Zoom*.

    C. (Optional) Create a title block and border for the drawing in paper space.

    D. Activate the *Plot* dialog box. Ensure the *Scale* edit box is set at **1:1**. Plot the *Layout*. When complete, measure the plot for accuracy by comparing with the dimensions given for the exercise in Chapter 9. *Save* the drawing.

    E. Reset the viewport scale to **2:1** using any method (*Zoom XP*, the *Viewport* toolbar, or the *Properties* window). Activate the *Plot* dialog box and make a second plot at the new scale. *Close* the drawing, but <u>do not save</u> it.

3. **Print the MANFCELL Drawing in Two Standard Scales**

    A. *Open* the **MANFCELL** drawing from Chapter 9. Check to make sure that the *Limits* settings are at **0,0** and **40',30'**.

    B. Activate the *Layout1* tab. If a viewport exists, *Erase* it. Use *Pagesetup* to select a printer to print on a letter ("A") sheet size. Next, create a *Single* viewport with the *Fit* option on layer **0**.

    C. In this step you will make one print of the drawing at the largest possible size using a standard architectural scale. Use any method to set the viewport scale. Use the "Architectural Table of *Limits* Settings" in the chapter for guidance on an appropriate scale to set. (HINT: Look in the "A" size sheet row and find the next larger *Limits* settings that will accommodate your current *Limits*, then use the scale shown above at the top of the column.) *Save* the drawing. Make a print using the *Plot* or *Page Setup* dialog box.

D. Now make a plot of the drawing using the next smaller standard architectural scale. This time determine the appropriate scale to set using either the "Architectural Table of *Limits* Settings" or by selecting it from the *Properties* window or *Viewport Scale Control* drop-down list.

4. **Use *Plotter Manager* to Configure a "D" Size Plot Device**
Even if your system currently has a "D" size plot device configured, this exercise will give you practice with *Plotter Manager*. This exercise does not require that the device actually exist in your lab or office or that it be physically attached to your system.

   A. Invoke the ***Plotter Manager***. In *Plotter Manager*, double-click on ***Add-A-Plotter Wizard***.

   B. Select the ***Next*** button in the *Introduction* page. In the *Begin* page, select ***My Computer***. In the *Plotter Model* page, select any *Manufacturer* and *Model* that can plot on a "D" size sheet, such as **CalComp 68444 Color Electrostatic, Hewlett-Packard Draft-Pro EXL (7576A)**, or **Oce 9800 FBBS R3.x**.

   C. Select ***Next*** in both the *Import PCP or PC2* and *Ports* pages. In the *Plotter Name* step, select the default name but insert the manufacturer's name as a prefix, such as "HP" or "CalComp."

   D. Finally, activate the ***Plot*** or ***Page Setup*** dialog box, select the ***Plot Device*** tab, and ensure your new device is listed in the *Name* drop-down list.

5. **Create Multiple Layouts for Different Plot Devices, Set *PSLTSCALE***
In this exercise you will create two layouts—one for plotting on an engineering "A" size sheet and one on a "C" size sheet. Check the table of "Standard Paper Sizes" in the chapter and find the correct size in inches for both engineering "A" and "C" size sheets. You will also adjust *LTSCALE* and *PSLTSCALE* appropriately for the layouts.

   A. ***Open*** drawing **CH11EX3**. Check to ensure the model space ***Limits*** are set at **22 x 17**. If you plan to print on a "A" size sheet, what is the drawing scale factor? Is the *LTSCALE* set appropriately?

   B. Activate the ***Layout1*** tab. Right-click and use ***Rename*** to rename the tab to **A Sheet**. If a viewport exists in the layout, ***Erase*** it. Activate the ***Page Setup*** dialog box and configure the layout to use a printer to print on a ***Letter*** ("A" size) sheet. Next, use ***Vports*** to create a viewport on layer **0** using the ***Fit*** option. Double-click inside the viewport and set the viewport scale to **1:2**.

   C. Toggle between ***Model*** tab and the ***A Sheet*** tab to examine the hidden and center lines. Do the dashes appear the same in both tabs? If the dashes appear differently in the two tabs, type ***PSLTSCALE*** and ensure the value is set to **0**. (Make sure you always use ***Regenall*** after using ***PSLTSCALE***.)

   D. Use the ***Page Setup*** or ***Plot*** dialog box to make the plot. Measure the plot for accuracy by comparing it with the dimensions given for the exercise in Chapter 11. To refresh your memory, that exercise involved adjusting the *LTSCALE* factor. Does the *LTSCALE* in the drawing yield hidden line dashes of 1/8" (for the long lines) on the plot? If not, make the *LTSCALE* adjustment and plot again. Use *SaveAs* to save and rename the drawing as **CH14EX5**.

   E. Next, activate the ***Layout2*** tab. Right-click and use ***Rename*** to rename the tab to **C Sheet**. If a viewport exists in the layout, ***Erase*** it. Activate the ***Page Setup*** dialog box and configure the layout to use a plotter for a ***ANSI C (22 x 17 Inches)*** sheet. Next, on layer **0**, use ***Vports*** to create a viewport using the ***Fit*** option. Double-click inside the viewport and set the viewport scale. Since the model space limits are set to 22 x 17 (same as the sheet size), set the viewport scale to **1:1**.

F. Do the hidden lines appear the same in all tabs? *PSLTSCALE* set to 0 ensures that the hidden and center lines appear with the same *LTSCALE* in the viewports as in the *Model* tab. If everything looks okay, make the plot. Does the plot show hidden line dashes of 1/8"? *Save* the drawing.

6. **Create an Architectural Template Drawing**
   In this exercise you will create a new template for architectural applications to plot at 1/8"=1' scale on a "D" size sheet.

   A. Begin a *New* drawing using the template **ASHEET.DWT** you worked on in Chapter 13 Exercises. Use *Save* to create a new template **(.DWT)** named **D-8-AR** in your working folder (not where the AutoCAD-supplied templates are stored).

   B. Set *Units* to *Architectural*. Use the "Architectural Table of *Limits* Settings" to determine and set the new model space *Limits* for an architectural "D" size sheet to plot at **1/8"=1'**. Multiply the existing *LTSCALE* (**1.0**) times the drawing scale factor shown in the table. Turn *Snap* to *Polar Snap* and turn *Grid* off.

   C. Activate the *Layout2* tab. Right-click to produce the shortcut menu and select *Rename*. Rename the tab to **D Sheet**. *Erase* the viewport if one exists.

   D. Activate the *Page Setup* dialog box. In the *Plot Device* tab, select a device that can use an architectural D size sheet (36 x 24), such as the device you configured for your system in Exercise 4. In the *Layout Settings* tab, select **ARCH D (24 x 36 Inches)** in the *Paper Size* drop-down list. Select *OK* to dismiss the *Page Setup* dialog box and save the settings.

   E. Make **VPORTS** the current layer. Use the *Vports* command to create a *Single* viewport using the *Fit* option. Activate the *Viewports* toolbar. Double-click inside the viewport and set the viewport scale to **1/8"=1'**.

   F. Make the *Model* tab active and set the **OBJECT** layer current. Finally, *Save* as a template (.DWT) drawing in your working folder.

7. **Create a Civil Engineering Template Drawing**
   Create a new template for civil engineering applications to plot at 1"=20' scale on a "C" size sheet.

   A. Begin a *New* drawing using the template **C-D-SHEET**. Use *Save* to create a new template (.DWT) named **C-CIVIL-20** in your working folder (not where the AutoCAD-supplied templates are stored).

   B. Set *Units* to *Engineering*. Use the "Civil Table of *Limits* Settings" to determine and set the new *Limits* for a "C" size sheet to plot at **1"=20'**. Multiply the existing *LTSCALE* (.5) times the scale factor shown. Turn *Snap* to *Polar* and *Grid* Off.

   C. Activate the *D Sheet* layout tab. Use any method to set the viewport scale to **1:240**. *Save* the drawing as a template (.DWT) drawing in your working folder.

15

DRAW COMMANDS II

Chapter Objectives

After completing this chapter you should:

1. be able to create construction lines using the *Xline* and *Ray* commands;
2. be able to create *Polygons* by the *Circumscribe*, *Inscribe*, and *Edge* methods;
3. be able to create *Rectangles*;
4. be able to use *Donut* to create circles with width;
5. be able to create *Spline* curves passing exactly through the selected points;
6. be able to create *Ellipses* using the *Axis End* method, the *Center* method, and the *Arc* method;
7. be able to use *Divide* to add points at equal parts of an object and *Measure* to add points at specified segment lengths along an object;
8. be able to use the *Sketch* command to create "freehand" sketch lines;
9. be able to create a *Boundary* by PICKing inside a closed area;
10. know that objects forming a closed shape can be combined into one *Region* object.

CONCEPTS

Remember that *Draw* commands create AutoCAD objects. The draw commands addressed in this chapter create more complex objects than those discussed in Chapter 8, Draw Commands I. The draw commands covered previously (*Line, Circle, Arc, Point,* and *Pline*) create simple objects composed of one object. The shapes created by the commands covered in this chapter are more complex. Most of the objects appear to be composed of several simple objects, but each shape is actually treated by AutoCAD as one object. Only the *Divide, Measure* and *Sketch* commands create multiple objects. The following commands are explained in this chapter.

Xline, Ray, Polygon, Rectangle, Donut, Spline, Ellipse, Divide, Measure, Sketch, Boundary, and *Region*

These draw commands can be accessed using any of the command entry methods, including the menus and icon buttons, as illustrated in Chapter 8. Buttons that appear by the Command tables in this chapter are available in the default *Draw* toolbar or can be activated by creating your own custom toolbar.

COMMANDS

XLINE

Pull-down Menu	COMMAND (TYPE)	ALIAS (TYPE)	Short-cut	Screen (side) Menu	Tablet Menu
Draw Construction Line	XLINE	XL	...	DRAW 1 Xline	L,10

When you draft with pencil and paper, light "construction" lines are used to lay out a drawing. These construction lines are not intended to be part of the finished object lines but are helpful for preliminary layout such as locating intersections, center points, and projecting between views.

There are two types of construction lines—*Xline* and *Ray*. An *Xline* is a line with infinite length, therefore having no endpoints. A *Ray* has one "anchored" endpoint and the other end extends to infinity. Even though these lines extend to infinity, they do not affect the drawing *Limits* or *Extents* or change the display or plot area in any way.

An *Xline* has no endpoints (*Endpoint Osnap* cannot be used) but does have a root, which is the theoretical midpoint (*Midpoint Osnap* can be used). If *Trim* or *Break* is used with *Xlines* or *Rays* such that two endpoints are created, the construction lines become *Line* objects. *Xlines* and *Rays* are drawn on the current layer and assume the current linetype and color (object-specific or *ByLayer*).

There are many ways that you can create *Xlines* as shown by the options appearing at the command prompt. All options of *Xline* automatically repeat so you can easily draw multiple lines.

Command: *xline*
Specify a point or [Hor/Ver/Ang/Bisect/Offset]:

Specify a point
The default option only requires that you specify two points to construct the *Xline* (Fig. 15-1). The first point becomes the root and anchors the line for the next point specification. The second point, or "through point," can be PICKed at any location and can pass through any point (*Polar Snap*, *Polar Tracking*, *Osnaps*, and *Objects Snap Tracking* can be used). If horizontal or vertical *Xlines* are needed, *ORTHO* can be used in conjunction with the "Specify a point:" option.

Command: *xline*
Specify a point or [Hor/Ver/Ang/Bisect/Offset]:
Specify through point:

Figure 15-1

Hor
This option creates a horizontal construction line. Type the letter "H" at the first prompt. You only specify one point, the through point or root (Fig. 15-2).

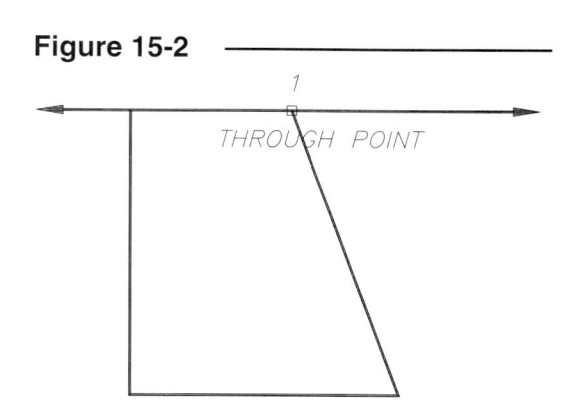

Figure 15-2

Ver
Ver creates a vertical construction line. You only specify one point, the through point (root) (Fig. 15-3).

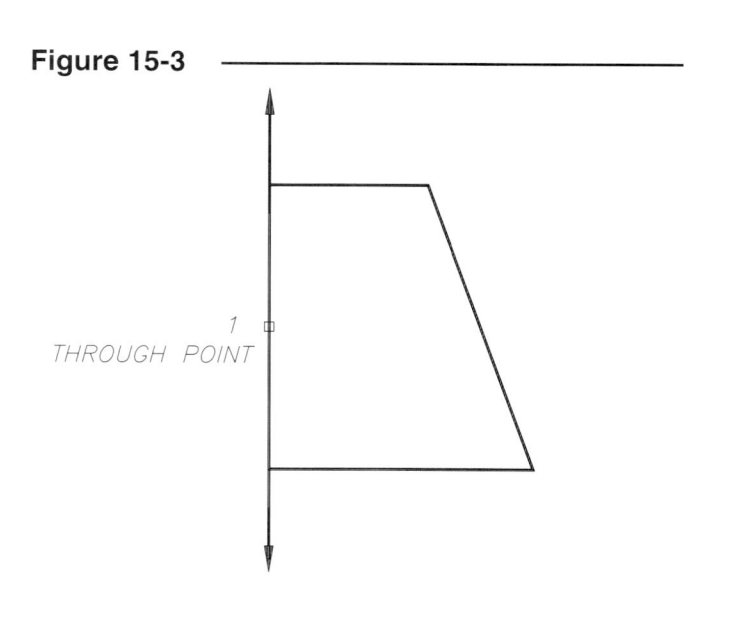

Figure 15-3

Ang
The *Ang* option provides two ways to specify the desired angle. You can (1) *Enter angle* or (2) select a *Reference* line (*Line, Xline, Ray,* or *Pline*) as the starting angle, then specify an angle from the selected line (in a counterclockwise direction) for the *Xline* to be drawn (Fig. 15-4).

    Command: *xline*
    Specify a point or [Hor/Ver/Ang/Bisect/Offset]: *a*
    Enter angle of xline (0) or [Reference]:

Figure 15-4

Bisect
This option draws the *Xline* at an angle between two selected points. First, select the angle vertex, then two points to define the angle (Fig. 15-5).

    Command: *xline*
    Specify a point or [Hor/Ver/Ang/Bisect/Offset]: *b*
    Specify angle vertex point: PICK
    Specify angle start point: PICK
    Specify angle end point: PICK
    Specify angle end point: Enter
    Command:

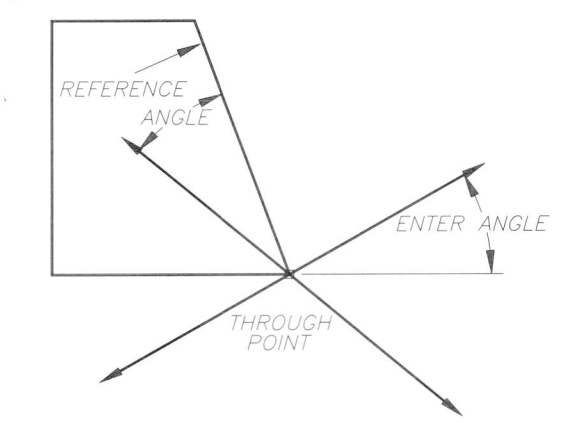

Figure 15-5

Offset
Offset creates an *Xline* parallel to another line. This option operates similarly to the *Offset* command. You can (1) specify a *Distance* from the selected line or (2) PICK a point to create the *Xline Through*. With the *Distance* option, enter the distance value, select a line (*Line, Xline, Ray,* or *Pline*), and specify on which side to create the offset *Xline*.

    Command: *xline*
    Specify a point or [Hor/Ver/Ang/Bisect/Offset]: *o*
    Specify offset distance or [Through] <1.0000>:
    (**value**) or PICK
    Select a line object: PICK
    Specify side to offset: PICK
    Select a line object: Enter
    Command:

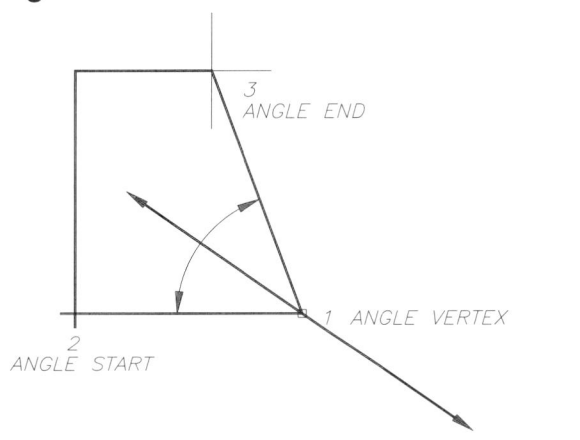

Figure 15-6

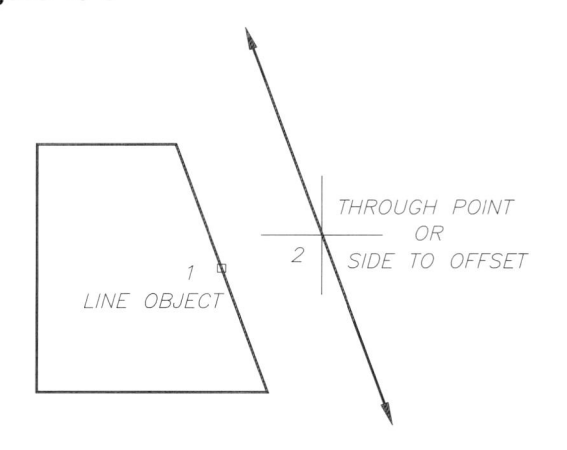

Using the *Through* option, select a line (*Line, Xline, Ray,* or *Pline*); then specify a point for the *Xline* to pass through. In each case, the anchor point of the *Xline* is the "root." (See "*Offset*," Chapter 9.)

```
Command: xline
Specify a point or [Hor/Ver/Ang/Bisect/Offset]: o
Specify offset distance or [Through] <1.0000>: t
Select a line object: PICK
Specify through point: PICK
Select a line object: Enter
Command:
```

RAY

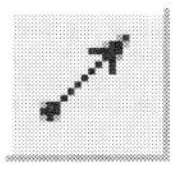

Pull-down Menu	COMMAND (TYPE)	ALIAS (TYPE)	Short-cut	Screen (side) Menu	Tablet Menu
Draw Ray	RAY	...	...	DRAW 1 Ray	K,10

A *Ray* is also a construction line (see *Xline*), but it extends to infinity in only <u>one direction</u> and has one "anchored" endpoint. Like an *Xline*, a *Ray* extends past the drawing area but does not affect the drawing *Limits* or *Extents*. The construction process for a *Ray* is simpler than for an *Xline*, only requiring you to establish a "start point" (endpoint) and a "through point" (Fig. 15-7). Multiple *Rays* can be created in one command.

Figure 15-7

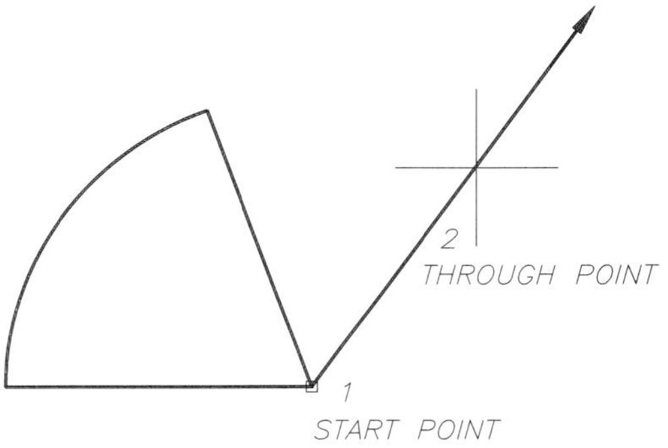

```
Command: ray
Specify start point: PICK or (coordinates)
Specify through point: PICK or (coordinates)
Specify through point: PICK or (coordinates)
Specify through point: Enter
Command:
```

Rays are especially helpful for construction of geometry about a central reference point or for construction of angular features. In each case, the geometry is usually constructed in only one direction from the center or vertex. If horizontal or vertical *Rays* are needed, just toggle on *ORTHO*. To draw a *Ray* at a specific angle, use relative polar coordinates or Polar Tracking. *Endpoint* and other appropriate *Osnaps* can be used with *Rays*. A *Ray* has one *Endpoint* but no *Midpoint*.

POLYGON

Pull-down Menu	COMMAND (TYPE)	ALIAS (TYPE)	Short-cut	Screen (side) Menu	Tablet Menu
Draw Polygon	POLYGON	POL	...	DRAW 1 Polygon	P,10

The *Polygon* command creates a regular polygon (all angles are equal and all sides have equal length). A *Polygon* object appears to be several individual objects but, like a *Pline*, is actually <u>one</u> object. In fact, AutoCAD uses *Pline* to create a *Polygon*. There are two basic options for creating *Polygons*: you can specify an *Edge* (length of one side) or specify the size of an imaginary circle for the *Polygon* to *Inscribe* or *Circumscribe*.

Inscribe/Circumscribe
The command sequence for this default method follows:

    Command: *polygon*
    Enter number of sides <4>: (**value**)
    Specify center of polygon or [Edge]: **PICK** or (**coordinates**)
    Enter an option [Inscribed in circle/Circumscribed about circle] <I>: *I* or *C*
    Specify radius of circle: **PICK** or (**coordinates**)
    Command:

The orientation of the *Polygon* and the imaginary circle are shown in Figure 15-8. Note that the *Inscribed* option allows control of one-half of the distance <u>across the corners</u>, and the *circumscribed* option allows control of one-half of the distance <u>across the flats</u>.

Figure 15-8

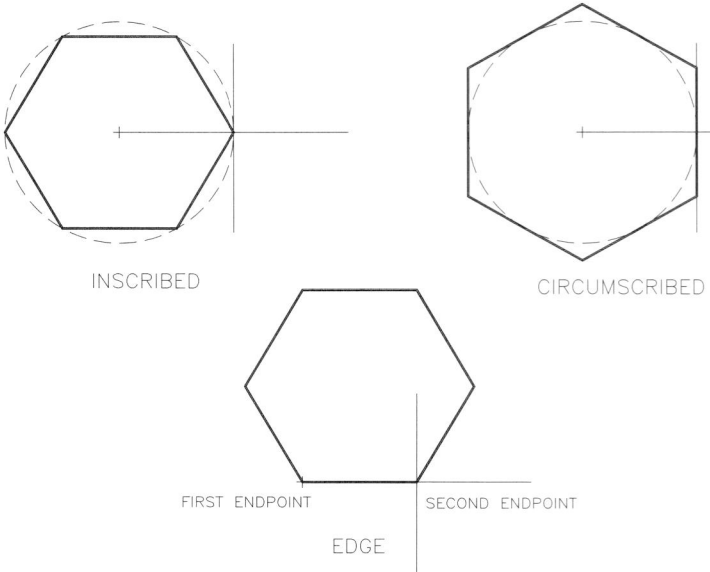

Using ORTHO ON with specification of the *radius of circle* forces the *Polygon* to a 90 degree orientation.

Edge
The *Edge* option only requires you to indicate the number of sides desired and to specify the two endpoints of one edge (Fig. 15-8).

    Command: *polygon*
    Enter number of sides <4>: (**value**)
    Specify center of polygon or [Edge]: *e*
    Specify first endpoint of edge: **PICK** or (**coordinates**)
    Specify second endpoint of edge: **PICK** or (**coordinates**)
    Command:

Because *Polygon*s are created as *Pline*s, *Pedit* can be used to change the line width or edit the shape in some way (see "*Pedit*," Chapter 16). *Polygon*s can also be *Exploded* into individual objects similar to the way other *Pline*s can be broken down into component objects (see "*Explode*," Chapter 16).

RECTANG

Pull-down Menu	COMMAND (TYPE)	ALIAS (TYPE)	Short-cut	Screen (side) Menu	Tablet Menu
Draw Rectangle	RECTANG	REC	...	DRAW 1 Rectangle	Q,10

The *Rectang* command only requires the specification of two diagonal corners for construction of a rectangle, identical to making a selection window (Fig. 15-9, on the next page). The corners can be PICKed or dimensions can be entered. The rectangle can be any proportion, but the sides are always horizontal and vertical. The completed rectangle is <u>one AutoCAD object</u>, not four separate objects.

    Command: *rectang*
    Specify first corner point or [Chamfer/Elevation/Fillet/Thickness/Width]: **PICK** or (**coordinates**)
    Specify other corner point or [Dimensions]: *D* or **PICK** or (**coordinates**)

Several options of the *Rectangle* command affect the shape as if it were a *Pline* (see *Pline* and *Pedit*). For example, a *Rectangle* can have *width*:

Specify first corner point or
[Chamfer/Elevation/Fillet/Thickness/Width]: **w**
Specify line width for rectangles <0.0000>: (**value**)

The *Fillet* and *Chamfer* options allow you to specify values for the fillet radius or chamfer distances:

Specify first corner point or
[Chamfer/Elevation/Fillet/Thickness/Width]: **f**
Specify fillet radius for rectangles <0.0000>: (**value**)

Specify first corner point or
[Chamfer/Elevation/Fillet/Thickness/Width]: **c**
Specify first chamfer distance for rectangles <0.0000>: (**value**)
Specify second chamfer distance for rectangles <0.5000>: (**value**)

The *Dimensions* option prompts you for *length* and *width* dimensions for rectangles. (*Thickness* and *Elevation* are 3D properties.)

Figure 15-9

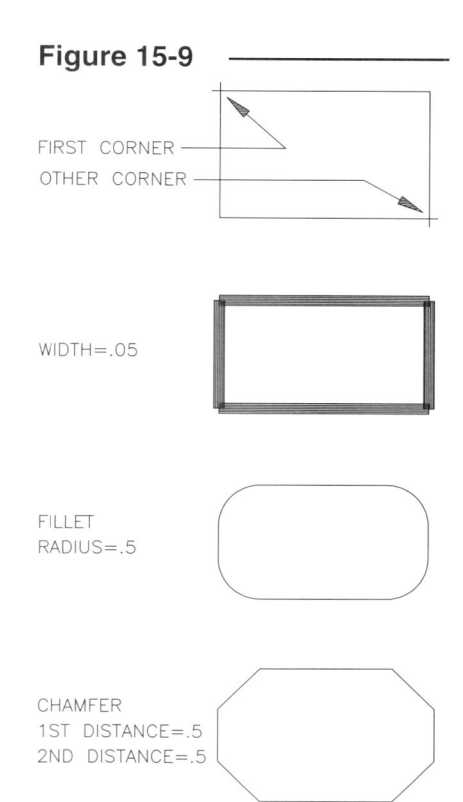

DONUT

Pull-down Menu	COMMAND (TYPE)	ALIAS (TYPE)	Short-cut	Screen (side) Menu	Tablet Menu
Draw Donut	DONUT	DO	...	DRAW 1 Donut	K,9

A *Donut* is a circle with width (Fig. 15-10). Invoking the command allows changing the inside and outside diameters and creating multiple *Donuts*:

Command: **donut**
Specify inside diameter of donut <0.5000>: (**value**)
Specify outside diameter of donut <1.0000>: (**value**)
Specify center of donut or <exit>:
PICK
Specify center of donut or <exit>:
Enter

Figure 15-10

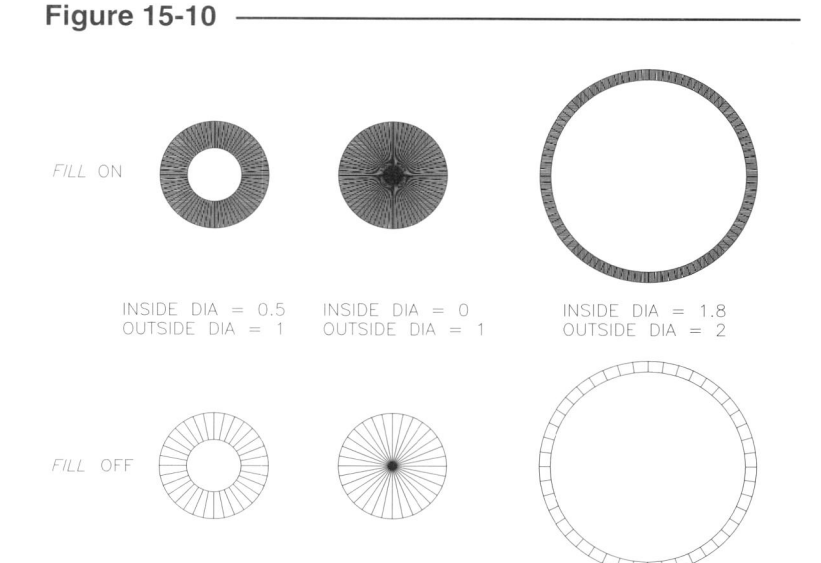

*Donut*s are actually solid filled circular *Pline*s with width. The solid fill for *Donut*s, *Pline*s, and other "solid" objects can be turned off with the *Fill* command or *FILLMODE* system variable.

SPLINE

Pull-down Menu	COMMAND (TYPE)	ALIAS (TYPE)	Short-cut	Screen (side) Menu	Tablet Menu
Draw Spline	SPLINE	SPL	...	DRAW 1 Spline	L,9

The *Spline* command creates a NURBS (non-uniform rational bezier spline) curve. The non-uniform feature allows irregular spacing of selected points to achieve sharp corners, for example. A *Spline* can also be used to create regular (rational) shapes such as arcs, circles, and ellipses. Irregular shapes can be combined with regular curves, all in one spline curve definition.

Spline is the newer and more functional version of a *Spline*-fit *Pline* (see "*Pedit*," Chapter 16). The main difference between a *Spline*-fit *Pline* and a *Spline* is that a *Spline* curve passes through the points selected, while the points selected for construction of a *Spline*-fit *Pline* only have a "pull" on the curve. Therefore, *Splines* are more suited to accurate design because the curve passes exactly through the points used to define the curve (data points) (Fig. 15-11).

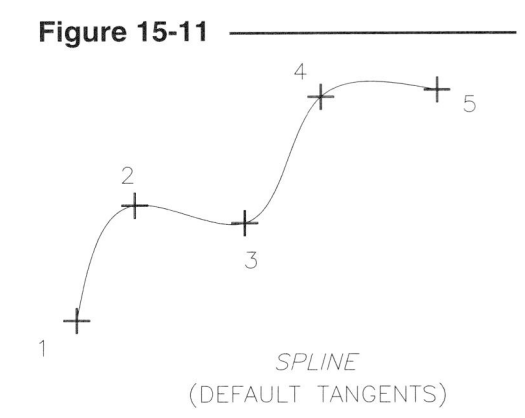

Figure 15-11

The construction process involves specifying points that the curve will pass through and determining tangent directions for the two ends (for non-closed *Splines*) (Fig. 15-12).

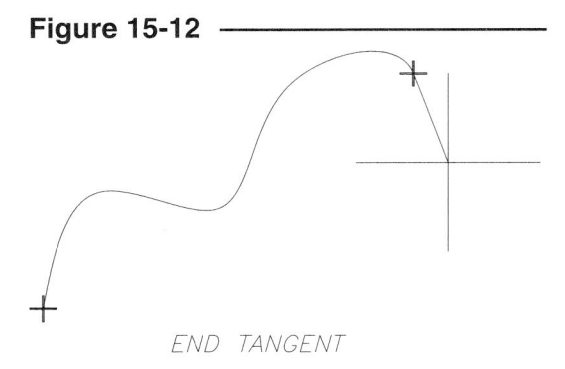

Figure 15-12

The *Close* option allows creation of closed *Splines* (these can be regular curves if the selected points are symmetrically arranged) (Fig. 15-13).

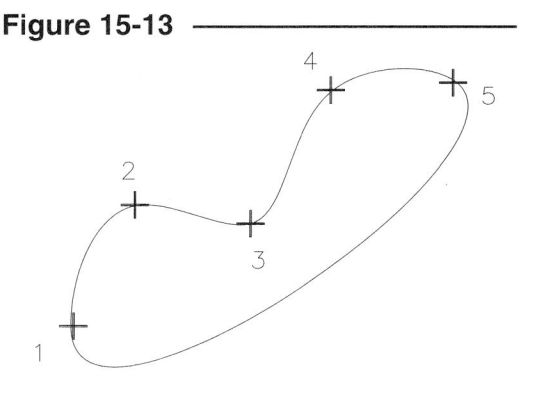

Figure 15-13

Command: *spline*
Specify first point or [Object]: **PICK** or (**coordinates**)
Specify next point: **PICK** or (**coordinates**)
Specify next point or [Close/Fit tolerance] <start tangent>: **PICK** or (**coordinates**)
Specify next point or [Close/Fit tolerance] <start tangent>: **PICK** or (**coordinates**)
Specify next point or [Close/Fit tolerance] <start tangent>: **Enter**
Specify start tangent: **PICK** or **Enter** (Select direction for tangent or Enter for default)
Specify end tangent: **PICK** or **Enter** (Select direction for tangent or Enter for default)
Command:

The *Object* option allows you to convert *Spline*-fit *Plines* into NURBS *Splines*. Only *Spline*-fit *Plines* can be converted (see "*Pedit*," Chapter 16).

Command: *spline*
Specify first point or [Object]: **o**
Select objects to convert to splines ...
Select objects: **PICK**
Select objects: **Enter**
Command:

A *Fit Tolerance* applied to the *Spline* "loosens" the fit of the curve. A tolerance of 0 (default) causes the *Spline* to pass exactly through the data points. Entering a positive value allows the curve to fall away from the points to form a smoother curve (Fig. 15-14).

Figure 15-14

Specify next point or [Close/Fit tolerance] <start tangent>: **f**
Specify fit tolerance <0.0000>: (Enter a positive value)

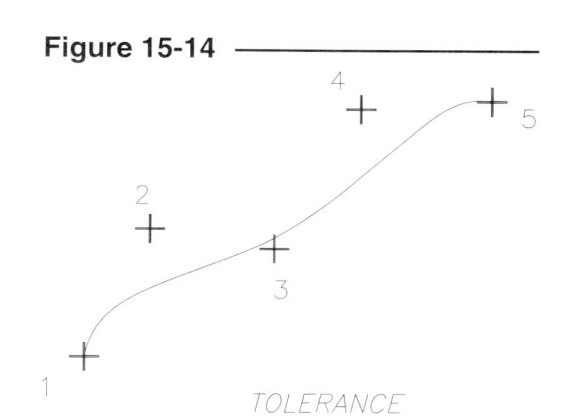

ELLIPSE

Pull-down Menu	COMMAND (TYPE)	ALIAS (TYPE)	Short-cut	Screen (side) Menu	Tablet Menu
Draw Ellipse	ELLIPSE	EL	...	DRAW 1 Ellipse	M,9

An *Ellipse* is one object. AutoCAD *Ellipses* are (by default) NURBS curves (see *Spline*). There are three methods of creating *Ellipses* in AutoCAD: (1) specify one <u>axis</u> and the <u>end</u> of the second, (2) specify the <u>center</u> and the ends of each axis, and (3) create an elliptical <u>arc</u>. Each option also permits supplying a rotation angle rather than the second axis length.

Command: *ellipse*
Specify axis endpoint of ellipse or [Arc/Center]: **PICK** or (**coordinates**)
(This is the first endpoint of either the major or minor axis.)
Specify other endpoint of axis: **PICK** or (**coordinates**)
(Select a point for the other endpoint of the first axis.)
Specify distance to other axis or [Rotation]: **PICK** or (**coordinates**)
(This is the distance measured perpendicularly from the established axis.)
Command:

Axis End

This default option requires PICKing three points as indicated in the command sequence above (Fig. 15-15).

Figure 15-15

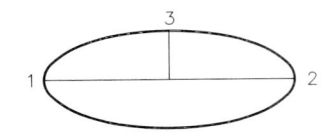

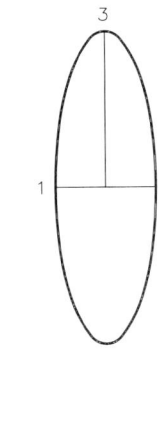

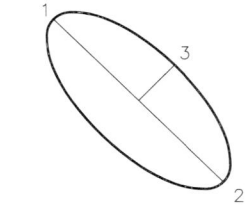

Rotation

If the *Rotation* option is used with the *Axis End* method, the following syntax is used:

    Specify distance to other axis or
    [Rotation]: r
    Specify rotation around major
    axis: PICK or (value)

The specified angle is the number of degrees the shape is rotated from the circular position (Fig. 15-16).

Figure 15-16

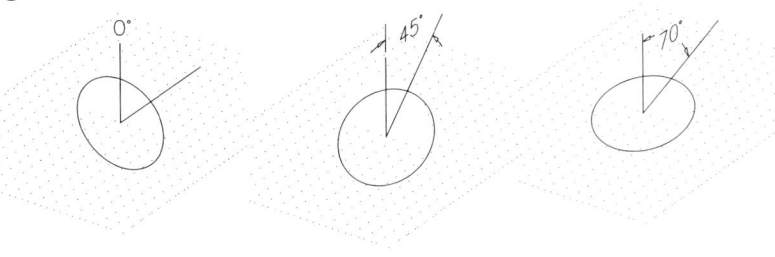

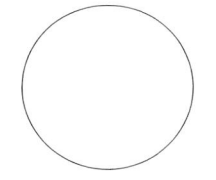

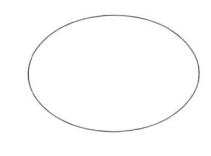

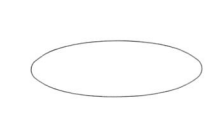

ROTATION = 0   ROTATION = 45   ROTATION = 70

Center

TIP With many practical applications, the center point of the ellipse is known, and therefore the *Center* option should be used (Fig. 15-17):

Figure 15-17

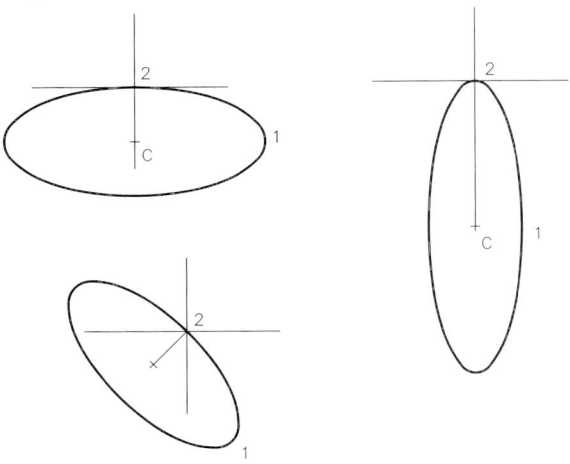

Command: *ellipse*
Specify axis endpoint of ellipse or [Arc/Center]: **c**
Specify center of ellipse: **PICK** or (**coordinates**)
Specify endpoint of axis: **PICK** or (**coordinates**)
(Select a point for the other endpoint of the first axis.)
Specify distance to other axis or [Rotation]: **PICK** or (**coordinates**) (This is the distance measured perpendicularly from the established axis.)
Command:

The *Rotation* option appears and can be invoked after specifying the *Center* and first *Axis endpoint*.

Arc
Use this option to construct an elliptical arc (partial ellipse). The procedure is identical to the *Center* option with the addition of specifying the start- and endpoints for the arc (Fig. 15-18):

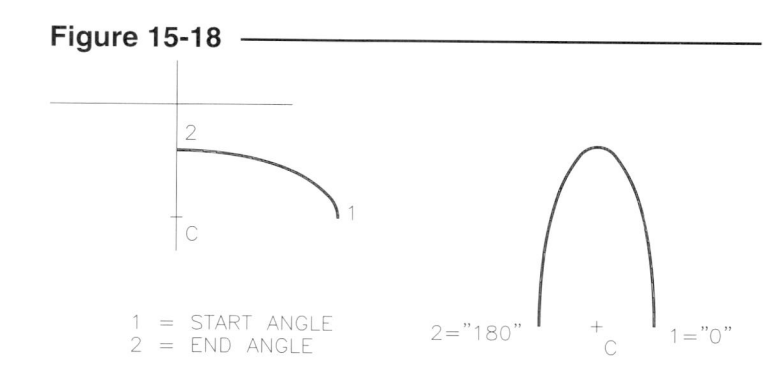

Figure 15-18

Command: *ellipse*
Specify axis endpoint of ellipse or [Arc/Center]: **a**
Specify axis endpoint of elliptical arc or [Center]: **PICK** or (**coordinates**)
Specify other endpoint of axis: **PICK** or (**coordinates**)
Specify distance to other axis or [Rotation]: **PICK** or (**coordinates**)
Specify start angle or [Parameter]: **PICK** or (**angular value**)
Specify end angle or [Parameter/Included angle]: **PICK** or (**angular value**)
Command:

The *Parameter* option allows you to specify the start point and endpoint for the elliptical arc. The parameters are based on the parametric vector equation: p(u)=c+a*cos(u)+b*sin(u), where c is the center of the ellipse and a and b are the major and minor axes.

DIVIDE

Pull-down Menu	COMMAND (TYPE)	ALIAS (TYPE)	Short-cut	Screen (side) Menu	Tablet Menu
Draw Point> Divide	DIVIDE	DIV	...	DRAW 2 Divide	V,13

The *Divide* and *Measure* commands add *Point* objects to existing objects. Both commands are found in the *Draw* pull-down menu under *Point* because they create *Point* objects.

The *Divide* command finds equal intervals along an object such as a *Line*, *Pline*, *Spline*, or *Arc* and adds a *Point* object at each interval. The object being divided is not actually broken into parts—it remains as one object. *Point* objects are automatically added to display the "divisions."

The point objects that are added to the object can be used for subsequent construction by allowing you to *OSNAP* to equally spaced intervals (*Node*s).

The command sequence for the *Divide* command is as follows:

Command: `divide`
Select object to divide: **PICK** (Only one object can be selected.)
Enter the number of segments or [Block]: (**value**)
Command:

Point objects are added to divide the object selected into the desired number of parts. Therefore, there is <u>one less</u> *Point* added than the number of segments specified.

After using the *Divide* command, the *Point* objects may not be visible unless the point style is changed with the *Point Style* dialog box (*Format* pull-down menu) or by changing the *PDMODE* variable by command line format (see Chapter 8). Figure 15-19 shows *Points* displayed using a *PDMODE* of 3.

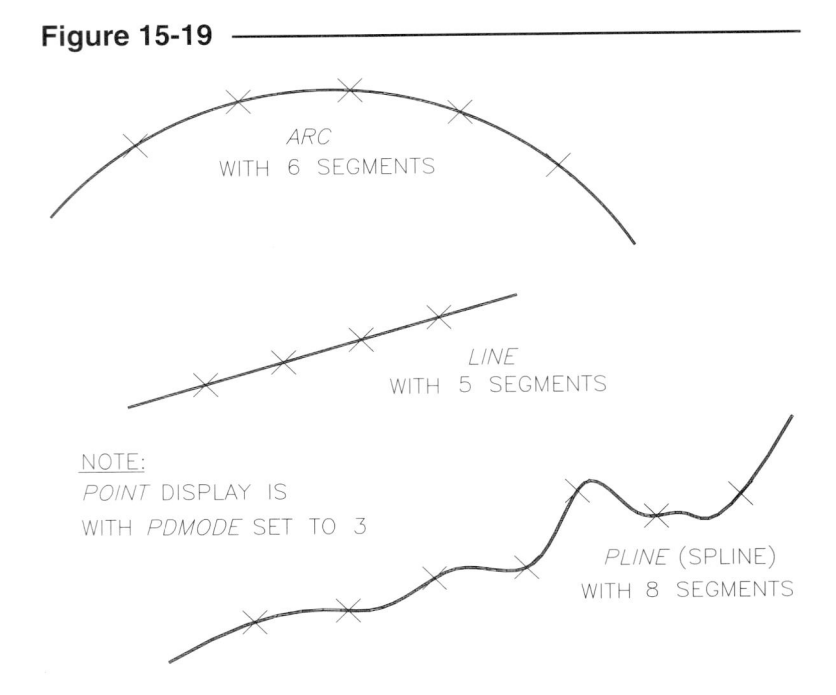

Figure 15-19

You can request that *Block*s be inserted rather than *Point* objects along equal divisions of the selected object. Figure 15-20 displays a generic rectangular-shaped block inserted with *Divide*, both aligned and not aligned with a *Line*, *Arc*, and *Pline*. In order to insert a *Block* using the *Divide* command, the name of an <u>existing</u> *Block* must be given. (See Chapter 20, Blocks and DesignCenter.)

Figure 15-20

MEASURE

Pull-down Menu	COMMAND (TYPE)	ALIAS (TYPE)	Short-cut	Screen (side) Menu	Tablet Menu
Draw Point > Measure	MEASURE	ME	...	DRAW 2 Measure	V,12

The *Measure* command is similar to the *Divide* command in that *Point* objects (or *Blocks*) are inserted along the selected object. The *Measure* command, however, allows you to designate the length of segments rather than the number of segments as with the *Divide* command.

    Command: measure
    Select object to measure: PICK (Only one object can be selected.)
    Specify length of segment or [Block]: (value) (Enter a value for length of one segment.)
    Command:

Point objects are added to the selected object at the designated intervals (lengths). One *Point* is added for each interval beginning at the end nearest the end used for object selection (Fig. 15-21). The intervals are of equal length except possibly the last segment, which is whatever length is remaining.

You can request that *Block*s be inserted rather than *Point* objects at the designated intervals of the selected object. Inserting *Blocks* with *Measure* requires that an existing *Block* be used.

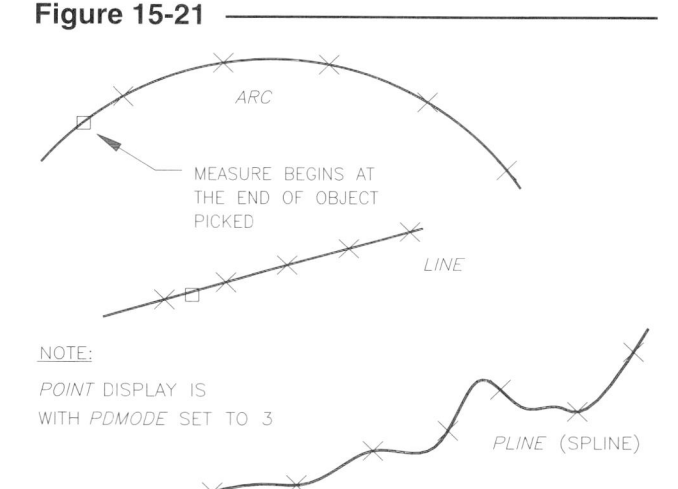

Figure 15-21

SKETCH

Pull-down Menu	COMMAND (TYPE)	ALIAS (TYPE)	Short-cut	Screen (side) Menu	Tablet Menu
...	SKETCH	...	...	...	...

The *Sketch* command is unlike other draw commands. *Sketch* quickly creates many short line segments (individual objects) by following the motions of the cursor. *Sketch* is used to give the appearance of a freehand "sketched" line, as used in Figure 15-22 for the tree and bushes. You do not specify individual endpoints, but rather draw a freehand line by placing the pen down, moving the cursor, and then picking up the pen. This action creates a large number of short *Line* segments. *Sketch* line segments are temporary (displayed in a different color) until *Recorded* or until *eXiting Sketch*.

Figure 15-22

CAUTION: *Sketch* can increase the drawing file size greatly due to the relatively large number of line segment endpoints required to define the *Sketch* line.

```
Command: sketch
Record increment <0.1000>: (value) or Enter
```
(Enter a value to specify the segment increment length or press Enter to accept the default increment.)
Sketch. Pen eXit Quit Record Erase Connect. (`letter`) (Enter "p" or press button #1 to put the pen down and begin drawing or enter another letter for another option. After drawing a sketch line, enter "p" or press button #1 again to pick up the pen.)

It is important to specify an *increment* length for the short line segments that are created. This *increment* controls the "resolution" of the *Sketch* line (Fig. 15-23).

Figure 15-23

Too large of an *increment* makes the straight line segments apparent, while too small of an *increment* unnecessarily increases file size. The default *increment* is 0.1 (appropriate for default *Limits* of 12 x 9) and should be changed proportionally with a change in *Limits*. Generally, multiply the default *increment* of 0.1 times the drawing scale factor.

Another important rule to consider whenever using *Sketch* is to turn SNAP Off and ORTHO Off, unless a "stair-step" effect is desired (see Figure 15-23). *Polar Snap* and *Polar Tracking* do not affect *Sketch* lines.

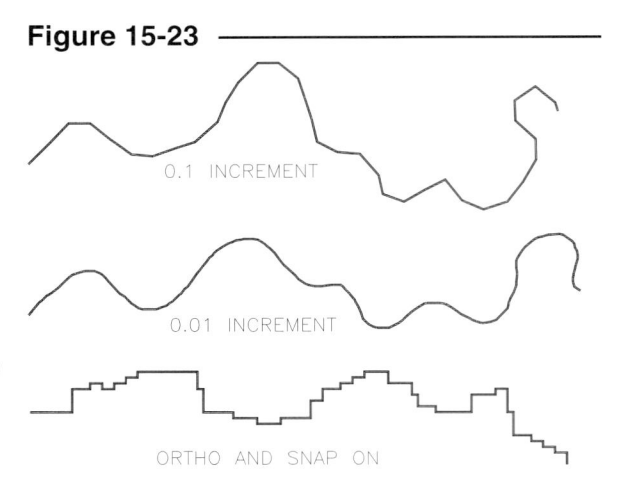

Sketch Options

Options of the *Sketch* command can be activated either by entering the corresponding letter(s) shown in uppercase at the *Sketch* command prompt or by pressing the desired mouse or puck button. For example, putting the pen up or down can be done by entering **P** at the Command line or by pressing button #1. The options of *Sketch* are as follows.

Pen   Lifts and lowers the pen. Position the cursor at the desired location to begin the line. Lower the pen and draw. Raise the pen when finished with the line.

Record   Records all temporary lines sketched so far without changing the pen position. After recording, the lines cannot be *Erased* with the *Erase* option of *Sketch* (although the normal *Erase* command can be used).

eXit   Records all temporary lines entered and returns to the Command: prompt.

Quit   Discards all temporary lines and returns to the Command: prompt.

Erase   Allows selective erasing of temporary lines (before recording). To erase, move backward from last sketch line toward the first. Press "p" to indicate the end of an erased area. This method works easily for relatively straight sections. To erase complex sketch lines, *eXit* and use the normal *Erase* command with window, crossing, or pickbox object selection.

Connect   Allows connection to the end of the last temporary sketch line (before recording). Move the cursor to the last sketch line and the pen is automatically lowered.

(period)   Draws a straight line (using *Line*) from the last sketched line to the cursor. After adding the straight lines, the pen returns to the up position.

Several options are illustrated in Figure 15-24. The *Pen* option lifts the pen up and down. A *period* (.) causes a straight line segment to be drawn from the last segment to the cursor location. *Erase* is accomplished by entering *E* and making a reverse motion.

As an alternative to the *Erase* option of *Sketch*, the *Erase* command can be used to erase all or part of the *Sketch* lines. Using a *Window* or *Crossing Window* is suggested to make selection of all the objects easier.

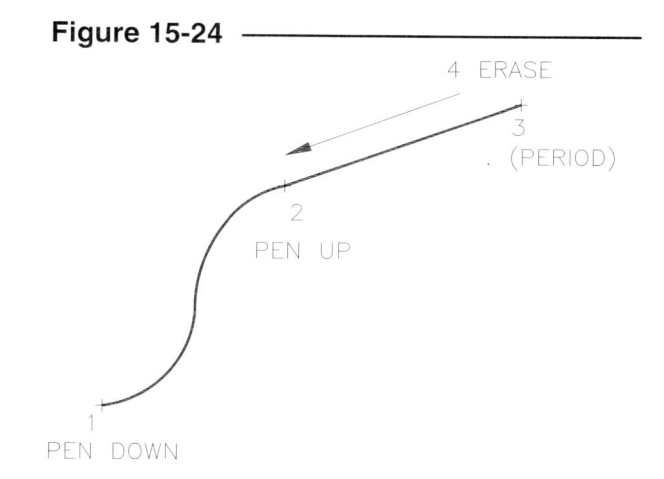

Figure 15-24

The *SKPOLY* Variable

The *SKPOLY* system variable controls whether AutoCAD creates connected *Sketch* line segments as one *Pline* (one object) or as multiple *Line* segments (separate objects). *SKPOLY* affects newly created *Sketch* lines only (Fig. 15-25).

SKPOLY=0   This setting generates *Sketch* segments as individual *Line* objects. This is the default setting.

SKPOLY=1   This setting generates connected *Sketch* segments as one Pline object.

Figure 15-25

Using editing commands with *Sketch* lines can normally be tedious (when *SKPOLY* is set to the default value of 0). In this case, editing *Sketch* lines usually requires *Zooming* in since the line segments are relatively small. However, changing *SKPOLY* to 1 simplifies operations such as using *Erase* or *Trim*. For example, *Sketch* lines are sometimes used to draw break lines (Fig. 15-26) or to represent broken sections of mechanical parts, in which case use of *Trim* is helpful. If you expect to use *Trim* or other editing commands with *Sketch* lines, change *SKPOLY* to 1 before creating the *Sketch* lines.

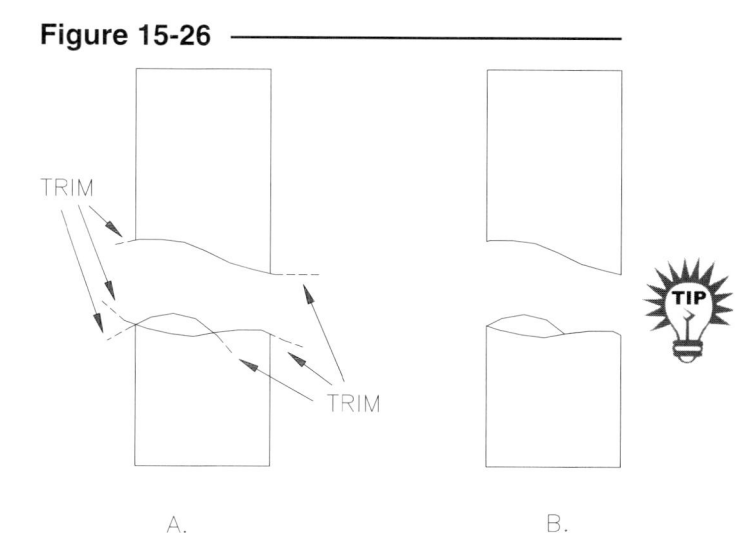

Figure 15-26

BOUNDARY

Pull-down Menu	COMMAND (TYPE)	ALIAS (TYPE)	Short-cut	Screen (side) Menu	Tablet Menu
Draw Boundary...	BOUNDARY or -BOUNDARY	BO or -BO	...	DRAW 2 Boundary	Q,9

Boundary finds and draws a boundary from a group of connected or overlapping shapes forming an enclosed area. The shapes can be any AutoCAD objects and can be in any configuration, as long as they form a totally enclosed area. *Boundary* creates either a *Polyline* or *Region* object forming the boundary shape (see *Region* next). The resulting *Boundary* does not affect the existing geometry in any way. *Boundary* finds and includes internal closed areas (called "islands") such as circles (holes) and includes them as part of the *Boundary*.

Figure 15-27

To create a *Boundary*, select an internal point in any enclosed area (Fig. 15-27). *Boundary* finds the boundary (complete enclosed shape) surrounding the internal point selected. You can use the resulting *Boundary* to construct other geometry or use the generated *Boundary* to determine the *Area* for the shape (such as a room in a floor plan). The same technique of selecting an internal point is also used to determine boundaries for sectioning using *Bhatch* (Chapter 24).

Boundary operates in dialog box mode (Fig. 15-28), or type *-Boundary* for Command line mode. After setting the desired options, select the *Pick Points* tile to select the desired internal point (as shown in Fig. 15-27). The *Boundary Creation* dialog box is a subset of the *Advanced* tab of the *Boundary Hatch* dialog box; therefore, many options are disabled. (See "*Bhatch*" in Chapter 24.) The options are as follows.

Figure 15-28

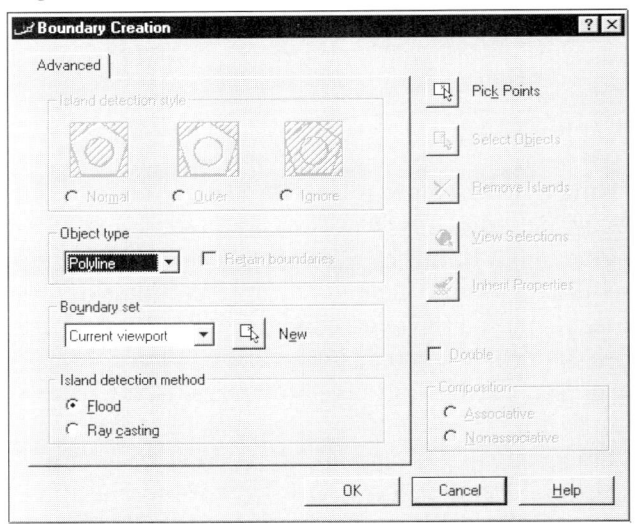

Object Type
Construct either a *Region* or *Polyline* boundary (see *Region* next). If islands are found, two or more *Plines* are formed but only one *Region*.

Boundary Set
The *Current Viewport* option is sufficient for most applications; however, for large drawings you can make a new boundary set by selecting a smaller set of objects to be considered for the possible boundary.

Island Detection
This options tells AutoCAD whether or not to include interior enclosed objects in the Boundary. Flood automatically includes islands as boundary objects. Ray Casting sends a line out from the point you pick to the nearest object and then traces the boundary in a counterclockwise direction; therefore, islands are excluded as boundary objects.

Pick Points
Use this option to PICK an enclosed area in the drawing that is anywhere <u>inside</u> the desired boundary.

REGION

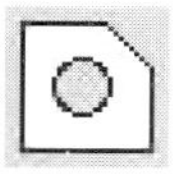

Pull-down Menu	COMMAND (TYPE)	ALIAS (TYPE)	Short-cut	Screen (side) Menu	Tablet Menu
Draw *Region...*	REGION	REG	...	DRAW 2 *Region*	R,9

The *Region* command converts one object or a set of objects forming a closed shape into one object called a *Region*. This is similar to the way in which a set of objects (*Lines, Arcs, Plines,* etc.) forming a closed shape and having matching endpoints can be converted into a closed *Pline*. A *Region*, however, has special properties:

1. Several *Regions* can be combined with Boolean operations known as *Union, Subtract,* and *Intersect* to form a "composite" *Region*. This process can be repeated until the desired shape is achieved.

2. A *Region* is considered a planar surface. The surface is defined by the edges of the *Region* and no edges can exist within the *Region* perimeter. *Regions* can be used with surface modeling.

In order to create a *Region*, a closed shape must exist. The shape can be composed of one or more objects such as a *Line, Arc, Pline, Circle, Ellipse,* or anything composed of a *Pline* (*Polygon, Rectangle, Boundary*). If more than one object is involved, the <u>endpoints must match (having no gaps or overlaps)</u>. Simply invoke the *Region* command and select all objects to be converted to the *Region*.

    Command: *region*
    Select Objects: **PICK**
    Select Objects: **PICK**
    Select Objects: **PICK**
    Select Objects: **Enter**
    1 loop extracted.
    1 Region created.
    Command:

Consider the shape shown in Figure 15-29 composed of four connecting *Lines*. Using *Region* combines the shape into one object, a *Region*. The appearance of the object does not change after the conversion, even though the resulting shape is one object.

Although the *Region* appears to be no different than a closed *Pline*, it is more powerful because several *Regions* can be combined to form complex shapes (composite *Regions*) using the three Boolean operations explained in Chapter 16.

Figure 15-29

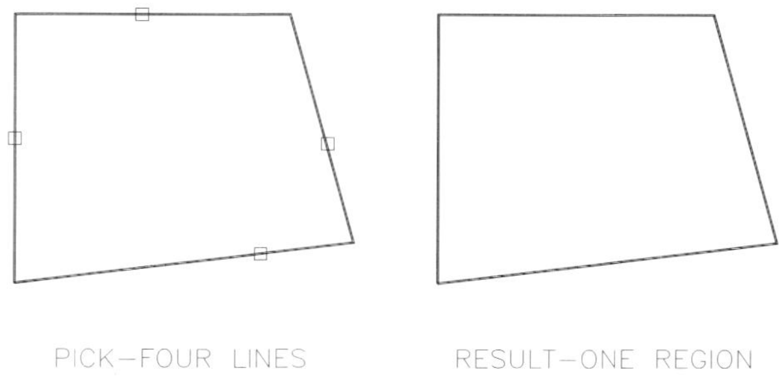

PICK—FOUR LINES   RESULT—ONE REGION

As an example, a set of *Regions* (converted *Circles*) can be combined to form the sprocket with only <u>one</u> *Subtract* operation (Fig. 15-30). Compare the simplicity of this operation to the process of using *Trim* to delete <u>each</u> of the unwanted portions of the small circles.

The Boolean operators, *Union, Subtraction,* and *Intersection,* can be used with *Regions* as well as solids. Any number of these commands can be used with *Regions* to form complex geometry (see Chapter 16, Modify Commands II).

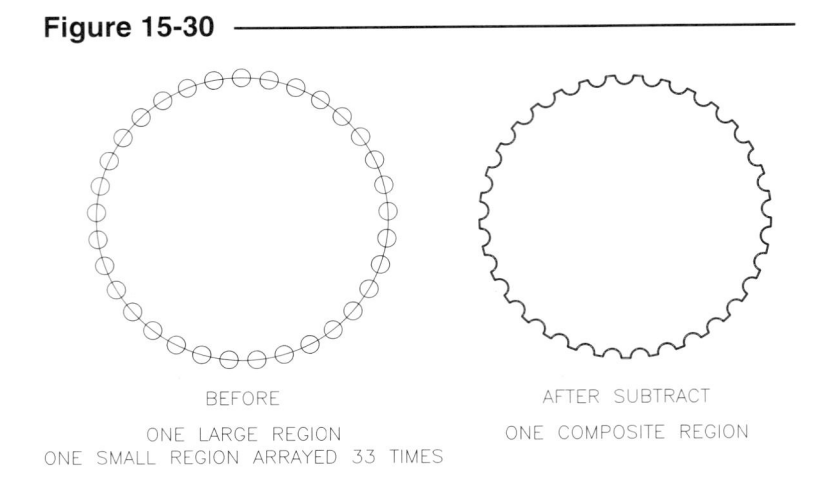

Figure 15-30

BEFORE
ONE LARGE REGION
ONE SMALL REGION ARRAYED 33 TIMES

AFTER SUBTRACT
ONE COMPOSITE REGION

CHAPTER EXERCISES

1. *Polygon*

    Open the **PLINE1** drawing you created in Chapter 8 Exercises. Use *Polygon* to construct the polygon located at the *Center* of the existing arc, as shown in Figure 15-31. When finished, *SaveAs* **POLYGON1**.

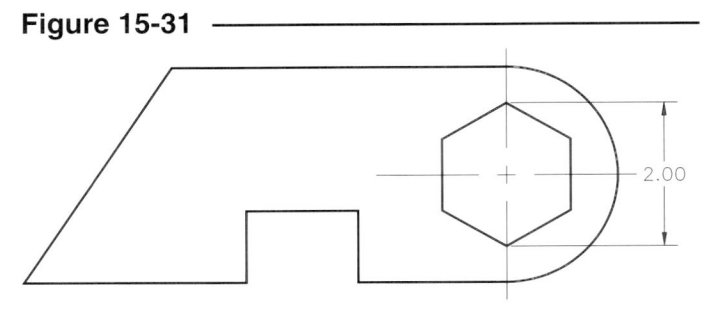

Figure 15-31

2. *Xline, Ellipse, Rectangle*

    Open the **APARTMENT** drawing that you created in Chapter 12. Draw the floor plan of the efficiency apartment on layer FLOORPLAN as shown in Figure 15-32. Use *SaveAs* to assign the name **EFF-APT**. Use *Xline* and *Line* to construct the floor plan with 8" width for the exterior walls and 5" for interior walls. Use the *Ellipse* and *Rectangle* commands to design and construct the kitchen sink, tub, wash basin, and toilet. *Save* the drawing but do not exit AutoCAD. Continue to Exercise 3.

Figure 15-32

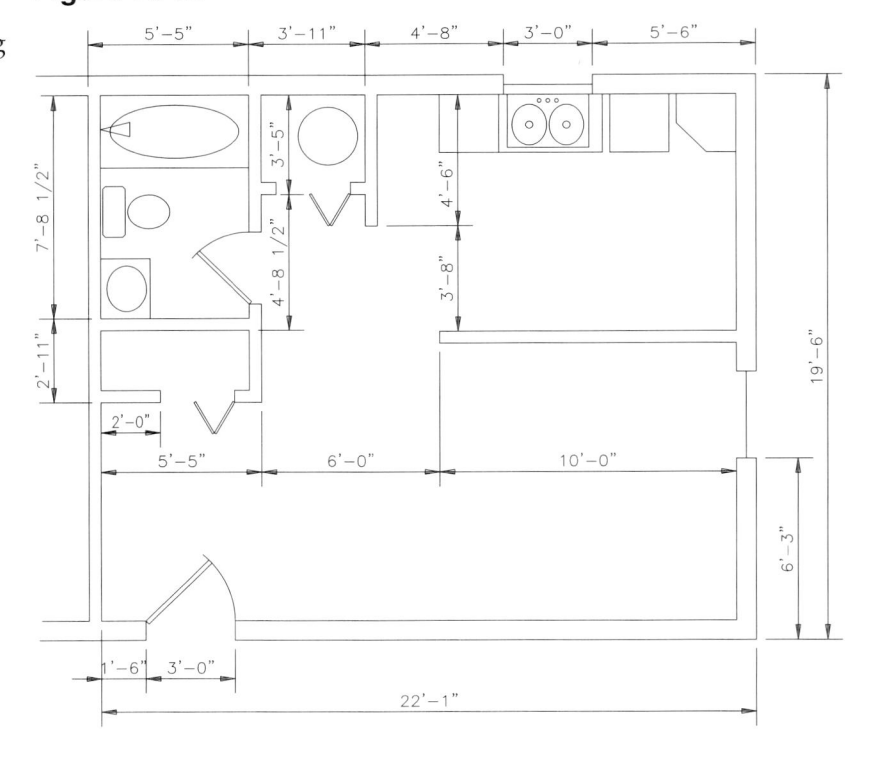

3. *Polygon, Sketch*

    Create a plant for the efficiency apartment as shown in Figure 15-33. Locate the plant near the entry. The plant is in a hexagonal pot (use *Polygon*) measuring 18" across the corners. Use *Sketch* to create 2 leaves as shown in figure A. Create a *Polar Array* to develop the other leaves similar to Figure B. *Save* the **EFF-APT** drawing.

    Figure 15-33

4. *Donut*

    Open **SCHEMATIC1** drawing you created in Chapter 9 Exercises. Use the *Point Style* dialog box (*Format* pull-down menu) to change the style of the *Point* objects to dots instead of small circles. Turn on the *Node Running Osnap* mode. Use the *Donut* command and set the *Inside diameter* and the *Outside diameter* to an appropriate value for your drawing such that the *Inside diameter* is 1/2 the value of the *Outside diameter*. Create a *Donut* at each existing *Node*. *SaveAs* **SCHEMATIC1B**. Your finished drawing should look like that in Figure 15-34.

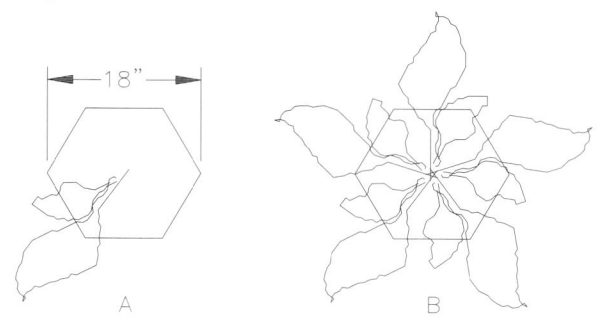

    Figure 15-34

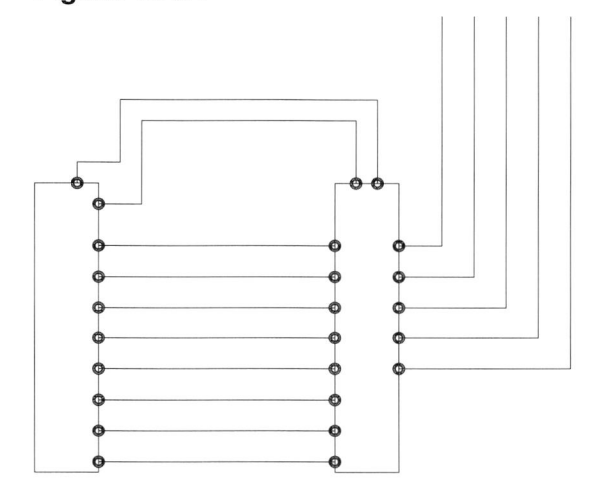

5. *Ellipse, Polygon*

    Begin a *New* drawing using the template drawing that you created in Chapter 12 named **A-METRIC**. Use *SaveAs* and assign the name **WRENCH**. Complete the construction of the wrench shown in Figure 15-35. Center the drawing within the *Limits*. Use *Line*, *Circle*, *Arc*, *Ellipse*, and *Polygon* to create the shape. For the "break lines," create one connected series of jagged *Lines*, then *Copy* for the matching set. Utilize *Trim*, *Rotate*, and other edit commands where necessary. HINT: Draw the wrench head in an orthogonal position; then rotate the entire shape 15 degrees. Notice the head is composed of 1/2 of a *Circle* (R25) and 1/2 of an *Ellipse*. *Save* the drawing when completed. Activate a *Layout* tab, configure a print or plot device using *Pagesetup*, make one *Viewport*, and *Plot* the drawing at a standard scale.

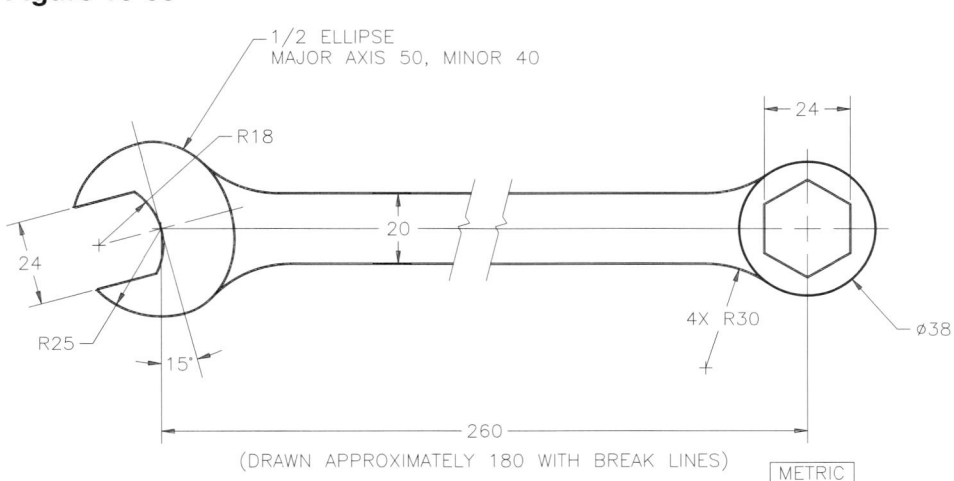

    Figure 15-35

6. *Xline, Spline*

   Create the handle shown in Figure 15-36. Construct multiple *Horizontal* and *Vertical Xlines* at the given distances on a separate layer (shown as dashed lines in Figure 15-36). Construct two *Splines* that make up the handle sides by PICKing the points indicated at the *Xline* intersections. Note the *End Tangent* for one *Spline*. Connect the *Splines* at each end with horizontal *Lines*. *Freeze* the construction (*Xline*) layer. *Save* the drawing as **HANDLE**.

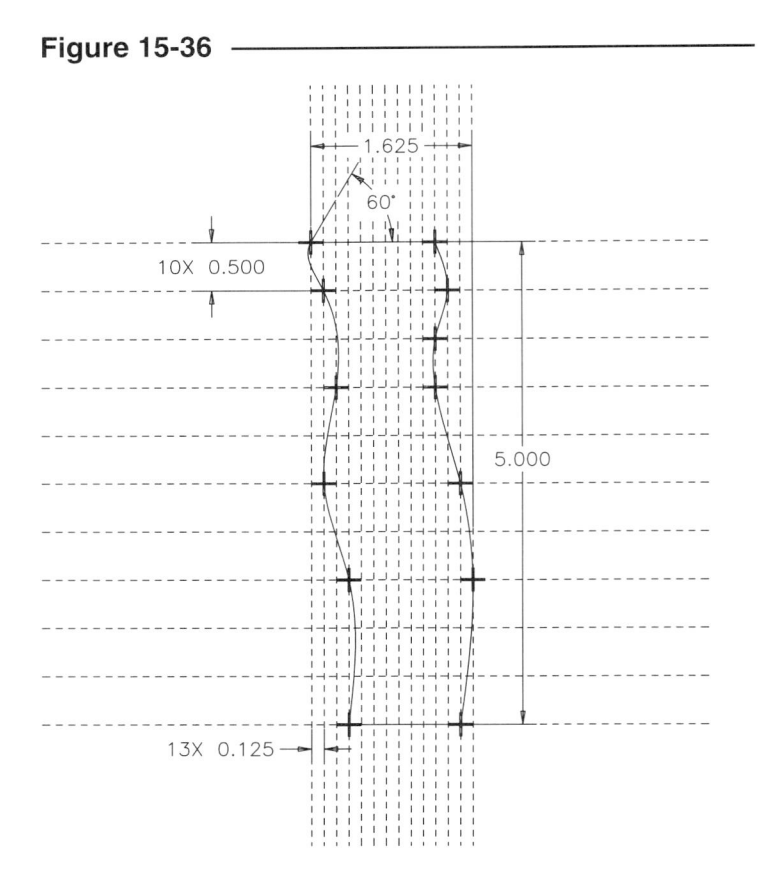

   Figure 15-36

7. *Divide, Measure*

   Use the **ASHEET** template drawing, use *SaveAs*, and assign the name **BILL-MATL**. Create the table in Figure 15-37 to be used as a bill of materials. Draw the bottom *Line* (as dimensioned) and a vertical *Line*. Use *Divide* along the bottom *Line* and *Measure* along the vertical *Line* to locate *Points* as desired. Create *Offsets Through* the *Points* using *Node OSNAP*. (*ORTHO* and *Trim* may be of help.)

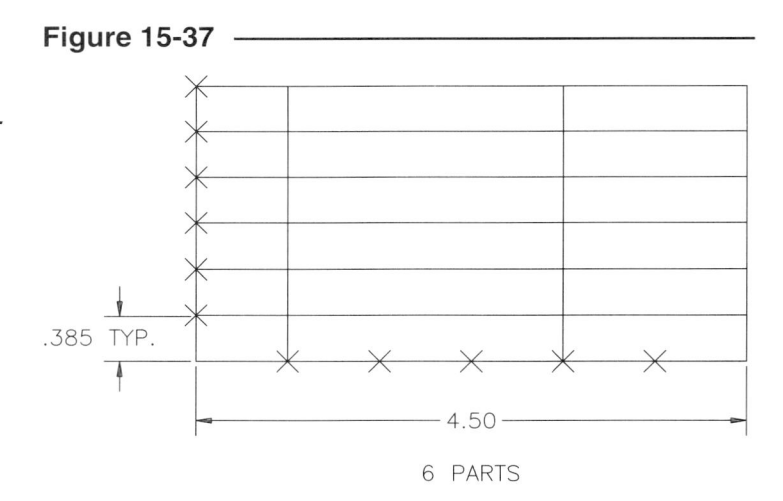

   Figure 15-37

8. *Boundary*

   Open the **GASKETA** drawing that you created in Chapter 9 Exercises. Use the *Boundary* command to create a boundary of the <u>outside shape only</u> (no islands). Use *Move Last* to displace the new shape to the right of the existing gasket. Use *SaveAs* and rename the drawing to **GASKET-AREA**. Keep this drawing to determine the *Area* of the shape after reading Chapter 17.

9. *Region*

    Create a gear, using *Regions*. Begin a *New* drawing, *Start from Scratch* with the *English* defaults settings. Use *Save* and assign the name **GEAR-REGION**.

    A. Set *Limits* at **8 x 6** and *Zoom All*. Create a *Circle* of **1** unit *diameter* with the center at **4,3**. Create a second concentric *Circle* with a *radius* of **1.557**. Create a closed *Pline* (to represent one gear tooth) by entering the following coordinates:

From point:	5.571,2.9191
To point:	@.17<160
To point:	@.0228<94
To point:	@.0228<86
To point:	@.17<20
To point:	c

    Figure 15-38

    The gear at this stage should look like that in Figure 15-38.

    B. Use the *Region* command to convert the three shapes (two *Circles* and one "tooth") to three *Regions*.

    C. *Array* the small *Region* (tooth) in a *Polar* array about the center of the gear. There are **40** items that are rotated as they are copied. This action should create all the teeth of the gear (Fig. 15-39).

    Figure 15-39

    D. *Save* the drawing. The gear will be completed in Chapter 16 Exercises by using the *Subtract* Boolean operation.

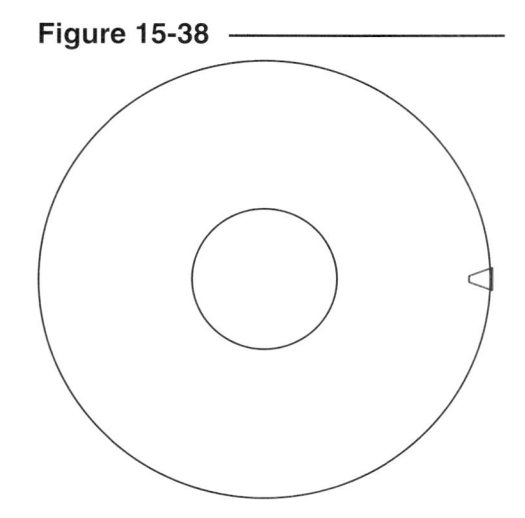

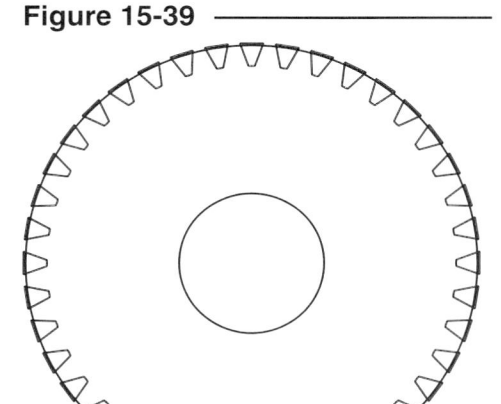

310 Draw Commands II

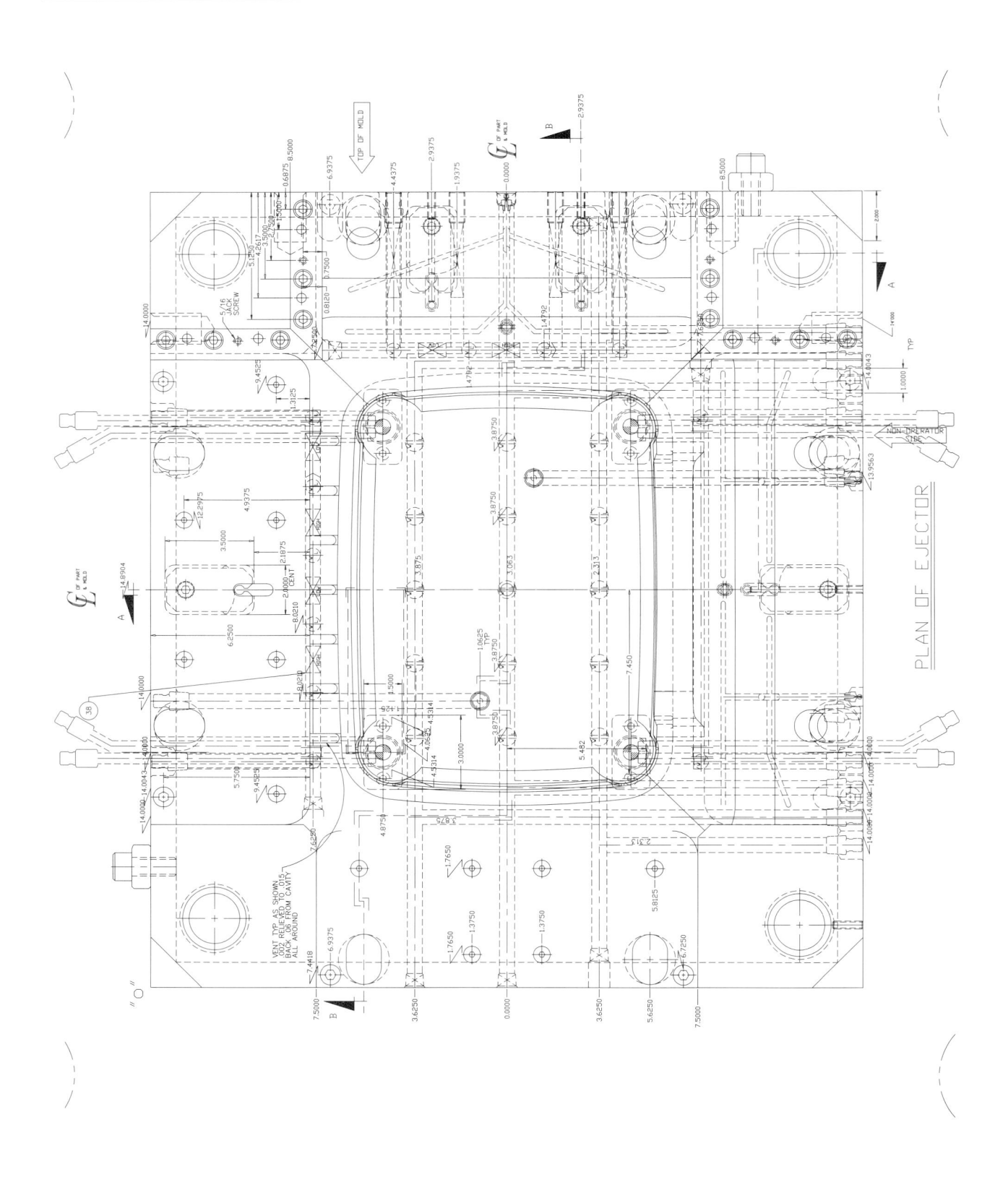

SINGLE CAVITY MOLD.DWG Courtesy, Autodesk, Inc.

16

MODIFY COMMANDS II

Chapter Objectives

After completing this chapter you should:

1. be able to use *Properties* to modify any type of object;
2. know that you can double-click on any object to produce the *Properties* window or a more specific editing tool;
3. be able to use *Matchprop* and the *Object Properties* toolbar to change properties of an object;
4. be able to use *Chprop* to change an object's properties;
5. know how to use *Change* to change points or properties;
6. be able to *Explode* a *Pline* into its component objects;
7. be able to *Align* objects with other objects;
8. be able to use all the *Pedit* options to modify *Plines* and to convert *Lines* and *Arcs* to *Plines*;
9. be able to modify *Splines* with *Splinedit*;
10. know that composite *Regions* can be created with *Union*, *Subtract*, and *Intersect*.

CONCEPTS

This chapter examines commands that are similar to, but generally more advanced and powerful than, those discussed in Chapter 9, Modify Commands I. None of the commands in this chapter create new duplicate objects from existing objects but instead modify the properties of the objects or convert objects from one type to another. Several commands are used to modify specific types of objects such as *Pedit* (modifies *Plines*), *Splinedit* (modifies *Splines*), and *Union, Subtract,* and *Intersect* (modify *Regions*). Only one command in this chapter, *Align,* does not modify object properties but combines *Move* and *Rotate* into one operation. Several of the commands and features discussed in this chapter were mentioned in Chapter 11 (*Properties, MatchProp,* and *Object Properties* toolbar) but are explained completely here.

Only some commands in this chapter that modify general object properties have icon buttons that appear in the AutoCAD Drawing Editor by default. For example, you can access *Properties* and *Matchprop* from the *Object Properties* toolbar and *Explode* by using its icon from the *Modify* toolbar.

Other commands that modify specific objects such as *Pedit* and *Splinedit* appear in the *Modify II* toolbar. The Boolean operators (*Union, Subtract,* and *Intersect*) appear in the *Solids Editing* toolbar. Activate a toolbar by right-clicking on any tool (icon button) and selecting from the list (Fig. 16-1).

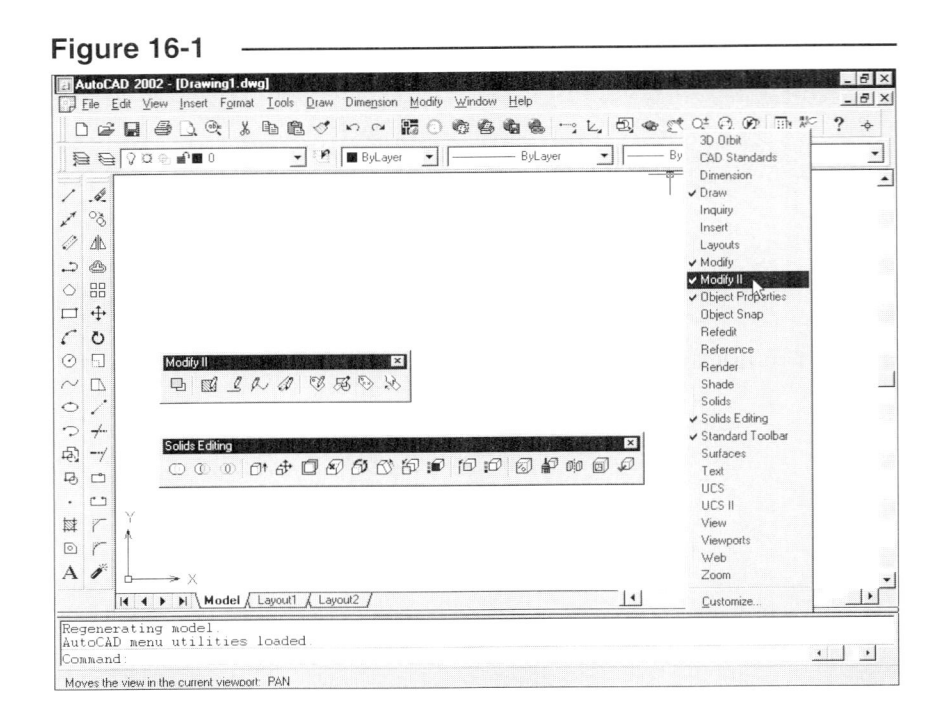

Figure 16-1

COMMANDS

You can use several methods to change properties of an object or of several objects. If you want to change an object's layer, color, or linetype only, the quickest method is by using the *Object Properties* toolbar (see "*Object Properties*" toolbar). If you want to change many properties of one or more objects (including layer, color, linetype, linetype scale, text style, dimension style, or hatch style) to match the properties of another existing object, use *Matchprop*. If you want to change any property (including coordinate data) for any object, use *Properties*.

In AutoCAD 2002 you can double-click on any object (assuming the *Dblclkedit* command is set to *On*) to produce the *Properties* window or a more specific editing tool, such as the *Hatch Edit* dialog box or the *Multiline Text Editor*, depending on the type of object you select.

Considerations for Changing Basic Properties of Objects

The *Properties* window, the *Object Properties* toolbar, and the *Matchprop*, *Chprop*, and *Change* commands can be used to change the basic properties of objects. Here are several basic properties that can be changed and considerations when doing so.

Layer
By changing an object's *Layer*, the selected object is effectively moved to the designated layer. In doing so, if the object's *Color, Linetype,* and *Lineweight* are set to BYLAYER, the object assumes the color, linetype, and lineweight of the layer to which it is moved.

Color
It may be desirable in some cases to assign explicit *Color* to an object or objects independent of the layer on which they are drawn. The properties editing commands allow changing the color of an existing object from one object color to another, or from BYLAYER assignment to an object color. An object drawn with an object-specific color can also be changed to BYLAYER with this option.

Linetype
An object can assume the *Linetype* assigned BYLAYER or can be assigned an object-specific *Linetype*. The *Linetype* option is used to change an object's *Linetype* to that of its layer (BYLAYER) or to any object-specific linetype that has been loaded into the current drawing. An object drawn with an object-specific linetype can also be changed to BYLAYER with this option.

Linetype Scale
When an individual object is selected, the object's individual linetype scale can be changed with this option, but not the global linetype scale (*LTSCALE*). This is the recommended method to alter an individual object's linetype scale. First, draw all the objects in the current global *LTSCALE*. One of the properties editing commands could then be used with this option to retroactively adjust the linetype scale of specific objects. The result would be similar to setting the *CELTSCALE* before drawing the specific objects. Using this method to adjust a specific object's linetype scale does not reset the global *LTSCALE* or *CELTSCALE* variables.

Thickness
An object's *Thickness* can be changed by this option. *Thickness* is a three-dimensional quality (Z dimension) assigned to a two-dimensional object.

Lineweight
An object can assume the *Lineweight* assigned ByLayer or can be assigned an object-specific *Linetype*. The *Lineweight* option is used to change an object's *Lineweight* to that of its layer (ByLayer) or to any object-specific linetype that has been loaded into the current drawing. An object drawn with an object-specific linetype can also be changed to ByLayer with this option.

Plotstyle
Use this option to change an object's *Plotstyle*. Plot styles assigned as ByLayer or to individual objects can change the way the objects appear in plots, such as having certain screen patterns, line end joints, plotted colors, plotted lineweights, and so on.

314   Modify Commands II

PROPERTIES

Pull-down Menu	COMMAND (TYPE)	ALIAS (TYPE)	Short-cut	Screen (side) Menu	Tablet Menu
Modify Properties...	PROPERTIES	PROPS or CH	(Edit Mode) Properties or Ctrl+1	MODIFY1 Property	Y12 to Y,13

The *Properties* window (Fig. 16-2, right side) gives you complete access to one or more objects' properties. You can edit the properties simply by changing the entries in the right column.

Once opened, this window remains on the screen until dismissed by clicking the "X" in the upper-right corner. The *Properties* window can be "docked" on the side of the screen (Fig. 16-3) or can be "floating." You can toggle the window on and off with Ctrl+1 (to appear and disappear).

Figure 16-2

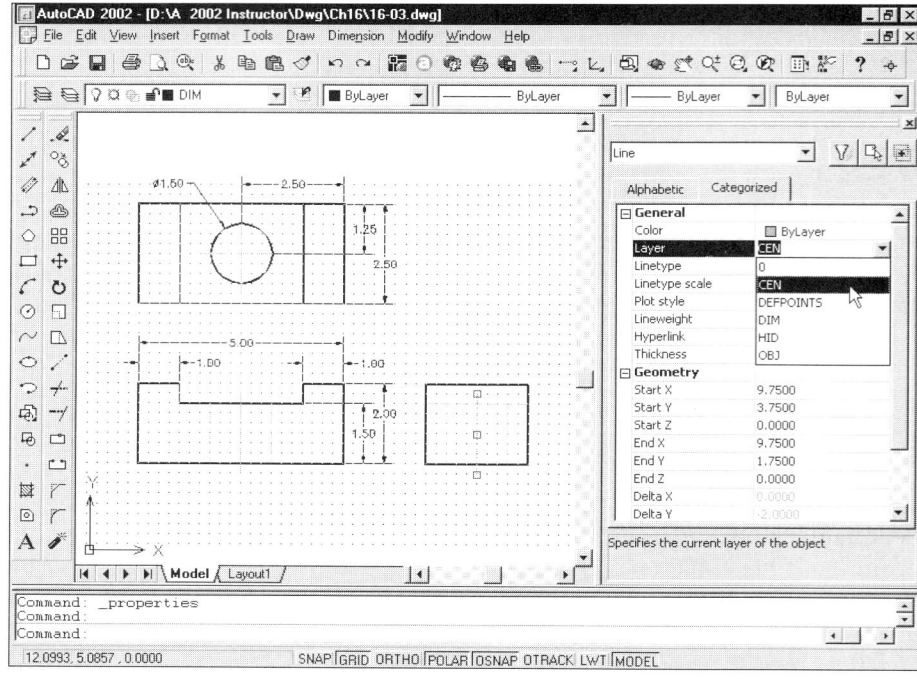

The contents of the window change based on what type and how many objects are selected. The power of this feature is apparent because the contents of the window are specific to the type of object that you select. For example, if you select a *Line*, entries specific to that *Line* appear, allowing changes to any properties that the *Line* possesses (see Figure 16-2). Or, if you select a *Circle* or some *Text*, a window appears specific to the *Circle* or *Text* properties.

You can use the window three ways: you can invoke the window, then select (highlight) one or more objects, you can select objects first and then invoke the window, or you can double-click on an object (for most objects) to produce the *Properties* window (see *Dblclkedit*). When multiple objects are selected, a drop-down list in the top of the window appears. Select the listed object(s) whose properties you want to change.

If you prefer to leave the window on your screen, select the objects you want to change when no commands are in use with the pickbox or *Auto* window (see Figure 16-2, highlighted line). After changing the objects' properties, press Escape to deselect the objects. Pressing Escape does not dismiss the *Properties* window, nor does it undo the changes as long as the cursor is outside the window.

Figure 16-3

If no objects are selected, the *Properties* window displays "No selection" in the drop-down list at the top (Fig. 16-3) and gives the current settings for the drawing. Drawing-wide settings can be changed such as the *LTSCALE* (see Figure 16-3).

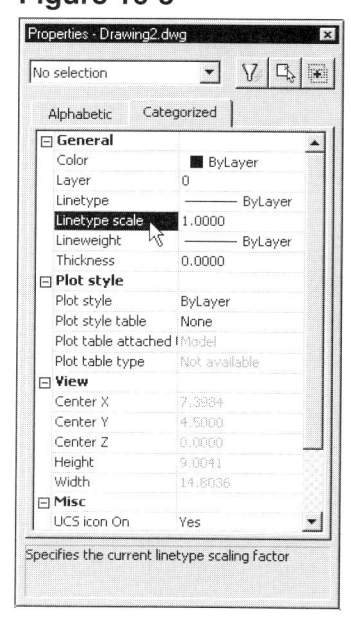

When objects are selected, the dialog box gives access to properties of the selected objects. Figure 16-2 displays the dialog box after selecting a *Line*. Notice the selection (*Line*) in the drop-down box at the top of the window. The properties displayed in the central area of the window are specific to the object or group of objects highlighted in the drawing and selected from the drop-down list. In this case (see Figure 16-2), any aspect of the *Line* can be modified.

If a *Pline* is selected for example, any property of the *Pline* can be changed. For example, the *Width* of the *Pline* can be changed by entering a new value in the *Properties* window (Fig. 16-4).

Figure 16-4

If a dimension is selected, for example, properties specific to a dimension, or to a group of dimensions if selected, can be changed. Note the list of variables that can be changed for a single dimension in Figure 16-5.

Figure 16-5

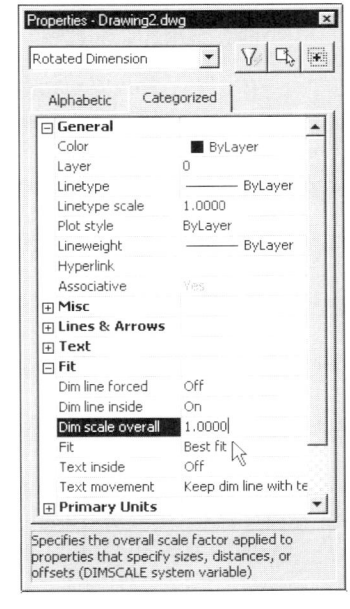

Categorized tab
The *Categorized* tab groups properties specific to the selected object(s). Each group name is indicated in bold, and each group can be contracted or expanded by clicking the minus (-) or plus (+) symbol (see Figure 16-5).

The *General* group lists basic properties for an object such as *Layer*, *Color*, or *Linetype*. To change an object's layer, linetype, lineweight, or color properties, highlight the desired objects in the drawing and select the properties you want to change from the right side of the window. If you use the *ByLayer* strategy of linetype, lineweight, and color properties assignment, you can change an object's linetype, lineweight, and color simply by changing its layer (see Figure 16-5). This method is recommended for most applications and is described in Chapter 11.

The *Geometry* group lists and allows changing any values that control the selected object's geometry. For example, to change the diameter of a *Circle*, highlight the property and change the value in the right side of the window (Fig. 16-6).

Figure 16-6

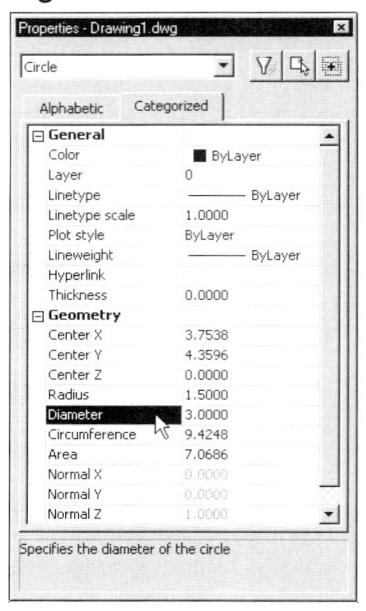

Alphabetical **tab**
The *Alphabetical* tab lists all properties for an object by name sorted alphabetically but does not categorize them into groups. Some of these properties are common for all objects (*Layer*, *Linetype*, *Color*, etc.) and some properties are specific to the selected object. For example, the *Diameter* property is specific to a *Circle* but is not grouped and only sorted alphabetically (Fig. 16-7).

Quick Select

The *Quick Select* button appears near the upper-right corner of the *Properties* window (Fig. 16-7). Selecting this button produces the *Quick Select* dialog box in which a selection set can be constructed based on criteria you choose from the dialog box.

Figure 16-7

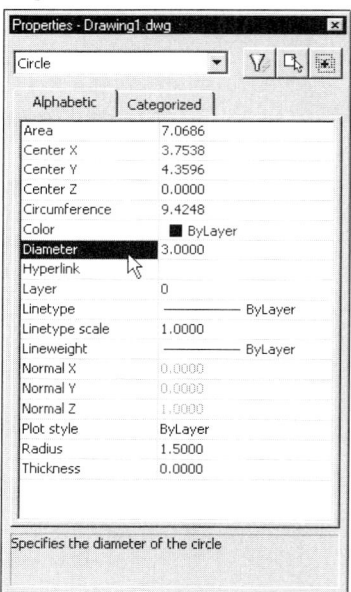

Select Objects

Normally you can select objects by the typical methods any time the *Properties* window is open. When objects are selected, AutoCAD searches its database and presents the properties of each object, one at a time, in the *Properties* window. Instead, you can use the *Select Objects* button to save time posting information on each object individually to the window. Using *Select objects*, the properties of the total set of selected objects are not posted to the *Properties* window until you press Enter.

Toggle Value of Pickadd Sysvar

This button toggles the *PICKADD* system variable on or off (1 or 0). This variable affects object selection at all times, not only for use with the *Properties* window. *PICKADD* determines whether objects you PICK are added to the current selection set (normal setting, *PICKADD*=1) or replace the previous object or set (*PICKADD*=0). Changing the *PICKADD* variable to 1 works well with the *Properties* window because selecting an object replaces the previous set of properties appearing in the window with the new object's properties.

The button position of "+" (plus symbol) indicates a current setting of 1, or on, for *PICKADD* (PICKed objects are added). Confusing as it appears, a button position of "1" indicates a current setting of 0, or off, for *PICKADD* (PICKed objects replace the previous ones).

NOTE: Since the *PICKADD* setting affects object selection anytime, <u>make sure you set this variable back to your desired setting (usually on, or "+") before dismissing the *Properties* window</u>.

DBLCLKEDIT

Pull-down Menu	COMMAND (TYPE)	ALIAS (TYPE)	Short-cut	Screen (side) Menu	Tablet Menu
...	DBLCLKEDIT	...	...	...	...

In AutoCAD 2002 you can double-click on an object to produce the *Properties* window or other similar dialog box to edit the object. The *Dblclkedit* command (double-click edit) controls whether double-clicking an object produces a dialog box.

   Command: ***dblclkedit***
   Enter double-click editing mode [ON/OFF] <ON>:

If double-click editing is turned on, the *Properties* window or other dialog box is displayed when an object is double-clicked. When you double-click most objects, the *Properties* window is displayed.

However, double-clicking some types of objects displays editing tools that are specific to the type of object. For example, double-clicking a line of *Text* produces the *Text Edit* dialog box. The object types (and resulting editing tool) <u>that do not produce the *Properties* window</u> when the object is double-clicked are listed below. These objects and the related editing tools are discussed fully in upcoming chapters of this text.

Attribute	Displays the *Edit Attribute Definition* dialog box (*Ddedit*).
Attribute within a block	Displays the *Enhanced Attribute Editor* (*Eattedit*).
Block	Displays the *Reference Edit* dialog box (*Refedit*).
Hatch	Displays the *Hatch Edit* dialog box (*Hatchedit*).
Leader text	Displays the *Multiline Text Editor* dialog box (*Ddedit*).
Mline	Displays the *Multiline Edit Tools* dialog box (*Mledit*).
Mtext	Displays the *Multiline Text Editor* dialog box (*Ddedit*).
Text	Displays the *Edit Text* dialog box (*Ddedit*).
Xref	Displays the *Reference Edit* dialog box (*Refedit*).

Dblclkedit is a command, not a system variable.

NOTE: The *PICKFIRST* system variable must be on (set to 1) for the *Properties* window or other editing tool to appear when an object is double-clicked.

MATCHPROP

Pull-down Menu	COMMAND (TYPE)	ALIAS (TYPE)	Short-cut	Screen (side) Menu	Tablet Menu
Modify Match Properties	MATCHPROP or PAINTER	MA	...	MODIFY1 Matchprp	Y,14 and Y,15

Matchprop is explained briefly in Chapter 11 but is explained again in this chapter with the full details of the *Special Properties*.

Matchprop is used to "paint" the properties of one object to another. Simply invoke the command, select the object that has the desired properties ("source object"), then select the object you want to "paint" the properties to ("destination object"). The Command prompt is as follows:

Command: *matchprop*
Select source object: **PICK**
Current active settings: Color Layer Ltype LTSCALE Lineweight Thickness PlotStyle Text Dim Hatch
Select destination object(s) or [Settings]: **PICK**
Select destination object(s) or [Settings]: **PICK**
Select destination object(s) or [Settings]: **Enter**
Command:

You can select several destination objects. The destination object(s) assume all of the "Current active settings" of the source object.

The *Property Settings* dialog box can be used to set which of the *Basic Properties* and *Special Properties* are to be painted to the destination objects (Fig. 16-8). At the "Select destination object(s) or [Settings]:" prompt, type S to display the dialog box. You can control the following *Basic Properties*. Only the checked properties are painted to the destination objects.

Figure 16-8

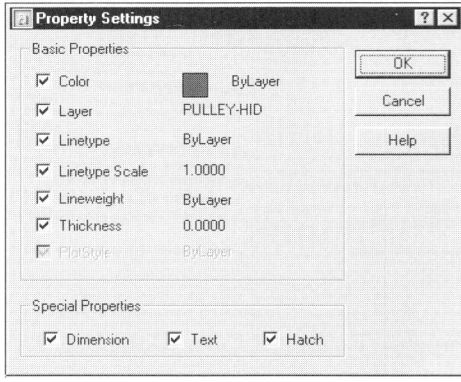

Color	This option paints the object-specific or *ByLayer* color.	
Layer	Move selected objects to the source object layer with this option checked.	
Linetype	Paint the object-specific or *ByLayer* linetype of the source object to the destination object.	
Linetype Scale	This option changes the individual object's linetype scale (*CELTSCALE*) not global linetype scale (*LTSCALE*).	
Thickness	Thickness is a 3-dimensional quality.	
Lineweight	You can paint the object-specific or *ByLayer* lineweight of the source object to the destination object.	
PlotStyle	The plot style of the source object is painted to the destination object when this setting is checked.	

The *Special Properties* section allows you to specify features of dimensions, text, and hatch patterns to match, as explained:

Dimension	This setting paints the *Dimension Style*. A *Dimension Style* defines the appearance of a dimension such as text style, size of arrows and text, tolerances if used, and many other features. (See Chapter 27.)

Text   This setting paints the source object's text *Style*. The text *Style* defines the text font and many other parameters that affect the appearance of the text. (See Chapter 18.)

Hatch Checking this box paints the hatch properties of the source object to the destination object(s). The properties can include the hatch *Pattern, Angle, Scale,* and other characteristics. (See Chapter 24.)

Matchprop is a simple and powerful command to use, especially if you want to convert dimensions, text or hatch patterns to look like other objects in the drawing. This method works only when you have existing objects in the drawing you want to "match." If you want to convert only layer, linetype, and color properties without changing the other properties or if you do not have existing objects to match, you can use the *Object Properties* toolbar.

Object Properties Toolbar

The five drop-down lists in the Object Properties toolbar (when not "dropped down") generally show the current layer, color, linetype, lineweight, and plot style. However, if an object or set of objects is selected, the information in these boxes changes to display the current object's settings. You can change an object's settings by picking an object (when no commands are in use), then dropping down any of the three lists and selecting a different layer, linetype, or color, etc.

Make sure you select (highlight) an object when no commands are in use. Use the pickbox (that appears on the cursor), *Window*, or *Crossing Window* to select the desired object or set of objects. The entries in the five boxes (Layer Control, Color Control, Linetype Control, etc.) then change to display the settings for the selected object or objects. If several objects are selected that have different properties, the boxes display no information (go "blank").

Next, use any of the drop-down lists to make another selection (Fig. 16-9). The highlighted object's properties are changed to those selected in the lists. Press the Escape key to complete the process.

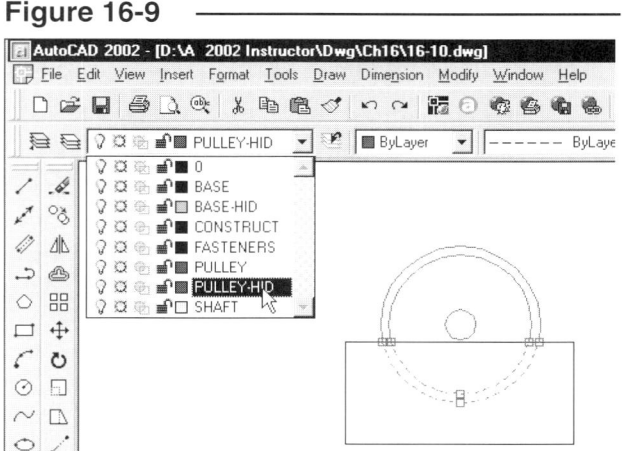

Figure 16-9

Remember that in most cases color, linetype, lineweight, and plot style settings are assigned *ByLayer*. In this type of drawing scheme, to change the linetype or color properties of an object, you would change only the object's layer (see Figure 16-9). If you are using this type of drawing scheme, refrain from using the Color Control and Linetype Control drop-down lists for changing properties.

This method for changing object properties is about as quick and easy as *Matchprop*; however, only the layer, linetype, color, lineweight, or plot style can be changed with the *Object Properties* toolbar. The *Object Properties* toolbar method works well if you do not have other existing objects to match or if you want to change only layer, linetype, color, lineweight, and plot style without matching text, dimension, and hatch styles.

320 Modify Commands II

CHPROP

Pull-down Menu	COMMAND (TYPE)	ALIAS (TYPE)	Short-cut	Screen (side) Menu	Tablet Menu
...	CHPROP	...	...	...	...

Chprop allows you to change basic properties of one or more objects using Command line format. If you want to change properties of several objects, pick several objects or select with a window or crossing window at the "Select objects:" prompt.

    Command: *chprop*
    Select objects: **PICK**
    Select objects: **Enter**
    Enter property to change [Color/LAyer/LType/ltScale/LWeight/Thickness/PLotstyle]:

CHANGE

Pull-down Menu	COMMAND (TYPE)	ALIAS (TYPE)	Short-cut	Screen (side) Menu	Tablet Menu
...	CHANGE	-CH	...	...	...

The *Change* command allows changing three options: *Points*, *Properties*, or *Text*.

Point
This option allows changing the endpoint of an object or endpoints of several objects to one new position:

    Command: *change*
    Select objects: **PICK**
    Select objects: **Enter**
    Specify change point or [Properties]: **PICK** (Select a point to establish as new endpoint of all objects. *OSNAP*s can be used.)

The endpoint(s) of the selected object(s) nearest the new point selected at the "Specify change point or [Properties]:" prompt is changed to the new point (Fig. 16-10).

Figure 16-10

BEFORE   AFTER
ONE OBJECT   CHANGE POINT

SEVERAL OBJECTS   CHANGE POINT

Properties
These options are discussed previously. The *Elevation* property (a 3D property) of an object can also be changed only with *Change*.

    Specify change point or [Properties]: *p*
    Enter property to change
      [Color/Elev/LAyer/LType/ltScale/
        LWeight/Thickness/PLotstyle]:

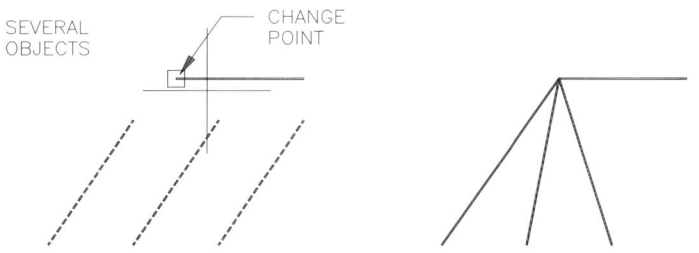

Text

Although the word *"Text"* does <u>not</u> appear as an option, the *Change* command recognizes text created with *Dtext* if selected. <u>*Change* does not change *Mtext*</u>. (See Chapter 18 for information on *Dtext* and *Mtext*.)

You can change the following characteristics of *Dtext* objects:

Text insertion point
Text style
Text height
Text rotation angle
Textual content

To change text, use the following command syntax:

Command: *change*
Select objects: **PICK** (Select one or several lines of text)
Select objects: **Enter**
Specify change point or [Properties]: **Enter**
Specify new text insertion point <no change>: **PICK** or **Enter**
Enter new text style <Standard>: (**text style name**) or **Enter**
Specify new height <0.2000>: (**value**) or **Enter**
Specify new rotation angle <0>: (**value**) or **Enter**
Enter new text <text>: (**new text**) or **Enter** (Enter the complete new line of text)

EXPLODE

Pull-down Menu	COMMAND (TYPE)	ALIAS (TYPE)	Short-cut	Screen (side) Menu	Tablet Menu
Modify Explode	EXPLODE	X	...	MODIFY2 Explode	Y,22

Many graphical shapes can be created in AutoCAD that are made of several elements but are treated as one object, such as *Plines, Polygons, Blocks, Hatch* patterns, and dimensions. The *Explode* command provides you with a means of breaking down or "exploding" the complex shape from one object into its many component segments (Fig. 16-11). Generally, *Explode* is used to allow subsequent editing of one or more of the component objects of a *Pline, Polygon,* or *Block,* etc., which would otherwise be impossible while the complex shape is considered one object.

Figure 16-11

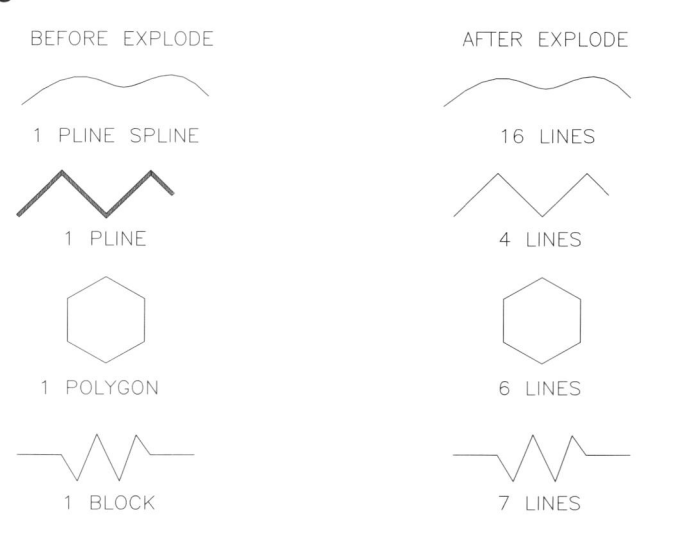

BEFORE EXPLODE
1 PLINE SPLINE
1 PLINE
1 POLYGON
1 BLOCK
EACH SHAPE IS ONE OBJECT

AFTER EXPLODE
16 LINES
4 LINES
6 LINES
7 LINES
EACH SHAPE IS EXPLODED INTO SEVERAL OBJECTS

322 Modify Commands II

The *Explode* command has no options and is simple to use. You only need to select the objects to *Explode*.

> Command: *explode*
> Select Objects: **PICK** (Select one or more *Plines*, *Blocks*, etc.)
> Select Objects: **Enter** (Indicates selection of objects is complete.)

> **TIP:** When *Plines*, *Polygons*, *Blocks*, or hatch patterns are *Exploded*, they are transformed into *Line*, *Arc*, and *Circle* objects. Beware, *Plines* having *width* lose their width information when *Exploded* since *Line*, *Arc*, and *Circle* objects cannot have width. *Exploding* objects can have other consequences such as losing "associativity" of dimensions and hatch objects and increasing file sizes by *Exploding Blocks*.

ALIGN

Pull-down Menu	COMMAND (TYPE)	ALIAS (TYPE)	Short-cut	Screen (side) Menu	Tablet Menu
Modify 3D Operation> Align	ALIGN	AL	...	MODIFY2 Align	X,14

Align provides a means of aligning one shape (a simple object, group of objects, *Pline*, *Boundary*, *Region*, *Block*, or a 3D object) with another shape. *Align* provides a complex motion, usually a combined translation (like *Move*) and rotation (like *Rotate*), in one command.

The alignment is accomplished by connecting source points (on the shape to be moved) to destination points (on the stationary shape). You should use *OSNAP* modes to select the source and destination points to assure accurate alignment. Either a 2D or 3D alignment can be accomplished with this command. The command syntax for alignment in a <u>2D alignment</u> is as follows:

> Command: *align*
> Select objects: **PICK**
> Select objects: **Enter**
> Specify first source point: **PICK** (with *Osnap*)
> Specify first destination point: **PICK** (with *Osnap*)
> Specify second source point: **PICK** (with *Osnap*)
> Specify second destination point: **PICK** (with *Osnap*)
> Specify third source point or <continue>: **Enter**
> Scale objects based on alignment points? [Yes/No] <N>: **Enter** (or *Y* to scale the source object to match destination object)
> Command:

This command performs a translation and a rotation in one motion if needed to align the points as designated (Fig. 16-12).

Figure 16-12

First, the first source point is connected to (actually touches) the first destination point (causing a translation). Next, the vector defined by the first and second source points is aligned with the vector defined by the first and second destination points (causing rotation).

If no third destination point is given (needed only for a 3D alignment), a 2D alignment is assumed and performed on the basis of the two sets of points.

Note that you can scale the source object based on the distance between the source points and the destination points. If you answer "Y" to the "Scale objects based on alignment points?" prompt, the source object is enlarged or reduced so the distance between its alignment points matches that of the destination points.

PEDIT

Pull-down Menu	COMMAND (TYPE)	ALIAS (TYPE)	Short-cut	Screen (side) Menu	Tablet Menu
Modify Object > Polyline	PEDIT	PE	...	MODIFY1 Pedit	Y,17

This command provides numerous options for editing *Polylines* (*Plines*). As an alternative, *Properties* can be used to change many of the *Pline*'s features in dialog box form (see Figure 16-4).

The list of options below emphasizes the great flexibility possible with *Polylines*. The first step after invoking *Pedit* is to select the *Pline* to edit.

    Command: *pedit*
    Select polyline or [Multiple]: **PICK**
    Enter an option [Close/Join/Width/Edit vertex/Fit/Spline/Decurve/Ltype gen/Undo]:

Multiple
The *Multiple* option allows multiple *Plines* to be edited simultaneously. The selected *Plines* can be totally separate objects and do not have to be connected in any way. Once the *Multiple* option is invoked and the objects are selected, any *Pline* option, such as *Close*, *Open*, *Join*, *Width*, *Fit*, *Spline*, *Decurve*, or *Ltype gen*, operates on the selected *Plines*.

    Command: *Pedit*
    Select polyline or [Multiple]: **m**
    Select objects: **PICK**
    Select objects: **PICK**
    Select objects: **Enter**
    Enter an option [Close/Open/Join/Width/Fit/Spline/Decurve/Ltype gen/Undo]:

For example, you can change the width of all *Plines* in a drawing simultaneously using the *Multiple* option, selecting all *Plines*, and using the *Width* option.

Close
Close connects the last segment with the first segment of an existing "open" *Pline*, resulting in a "closed" *Pline* (Fig. 16-13). A closed *Pline* is one continuous object having no specific start or endpoint, as opposed to one closed by PICKing points. A *Closed Pline* reacts differently to the *Spline* option and to some commands such as *Fillet*, *Pline* option (see "*Fillet*," Chapter 9).

Figure 16-13

OPEN   CLOSE

Open
Open removes the closing segment if the *Close* option was used previously (Fig. 16-13).

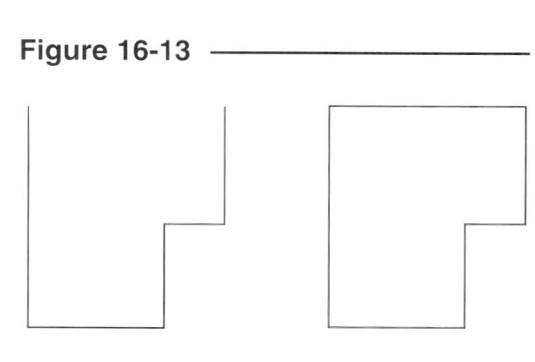

Join

The *Join* option combines two or more objects into one *Pline*. In previous releases of AutoCAD, object endpoints had to match <u>exactly</u> for the objects to be joined into one *Pline*. With the AutoCAD 2002 enhanced version of *Pedit*, the line segments do not have to meet exactly for *Join* to work. You must first use the *Multiple* option, then *Join*.

```
Command: Pedit
Select polyline or [Multiple]: m
Select objects: PICK
Select objects: Enter
Enter an option [Close/Open/Join/Width/Fit/Spline/Decurve/Ltype gen/Undo]: j
Join Type = Extend
Enter fuzz distance or [Jointype] <0.0000>: .5
1 segments added to polyline
Enter an option [Close/Open/Join/Width/Fit/Spline/Decurve/Ltype gen/Undo]:
```

If the ends of the line segments do not touch but are within a distance that you can set, called the *fuzz distance*, the ends can be joined by *Pedit*. *Pedit* handles this automatically by either extending and trimming the line segments or by adding a new line segment based on your setting for *Jointype*.

```
Select objects: PICK
Enter an option [Close/Open/Join/Width/Fit/Spline/Decurve/Ltype gen/Undo]: j
Join Type = Extend
Enter fuzz distance or [Jointype] <2.0000>: j
Enter join type [Extend/Add/Both] <Extend>:
```

Extend
This option causes AutoCAD to join the selected polylines by extending or trimming the segments to the nearest endpoints (see Figure 16-14).

Add
Use this option to add a straight segment between the nearest endpoints (see Figure 16-14).

Both
If you use this option, the selected polylines are joined by extending or trimming if possible. If not, as in the case of near parallel lines when an extension would be outside the fuzz distance, a straight segment is added between the nearest endpoints.

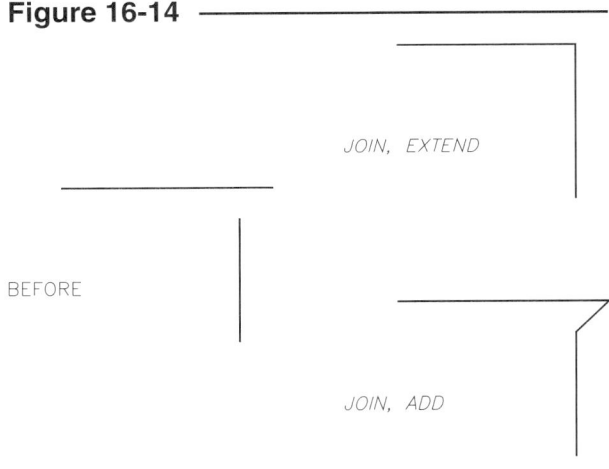

Figure 16-14

If the properties of several objects being joined into a polyline differ, the resulting polyline inherits the properties of the first object you select.

Figure 16-15

Width
Width allows specification of a uniform width for *Pline* segments (Fig. 16-15). Non-uniform width can be specified with the *Edit vertex* option.

Edit vertex
This option is covered in the next section.

Fit
This option converts the *Pline* from straight line segments to arcs. The curve consists of two arcs for each pair of vertices. The resulting curve can be radical if the original *Pline* consists of sharp angles. The resulting curve passes through all vertices (Fig. 16-16).

Figure 16-16

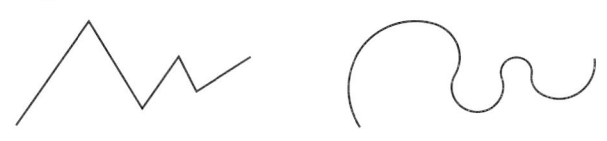

BEFORE FIT   AFTER FIT

Spline
This option converts the *Pline* to a B-spline (Bezier spline) (Fig. 16-17). The *Pline* vertices act as "control points" affecting the shape of the curve. The resulting curve passes through only the end vertices. A *Spline*-fit *Pline* is not the same as a spline curve created with the *Spline* command. This option produces a less versatile version of the newer *Spline* object.

Decurve
Decurve removes the *Spline* or *Fit* curve and returns the *Pline* to its original straight line segments state (Fig. 16-17).

Figure 16-17

BEFORE SPLINE   AFTER SPLINE
AFTER DECURVE   BEFORE DECURVE

When you use the *Spline* option of *Pedit*, the amount of "pull" can be affected by setting the SPLINETYPE system variable to either 5 or 6 before using the *Spline* option. SPLINETYPE applies either a quadratic (5=more pull) or cubic (6=less pull) B-spline function (Fig. 16-18).

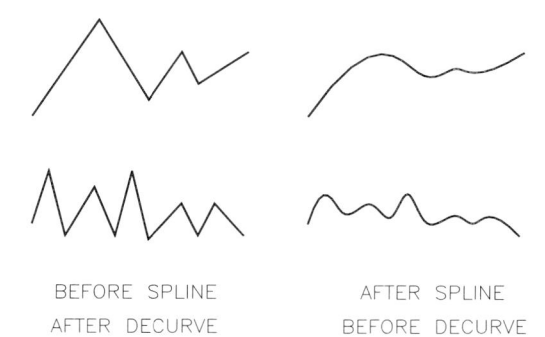

Figure 16-18

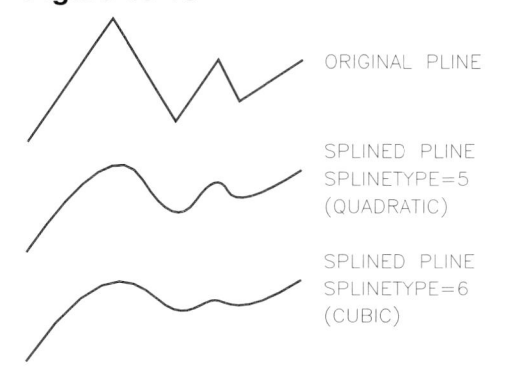

ORIGINAL PLINE

SPLINED PLINE
SPLINETYPE=5
(QUADRATIC)

SPLINED PLINE
SPLINETYPE=6
(CUBIC)

The SPLINESEGS system variable controls the number of line segments created when the *Spline* option is used. The variable should be set before using the option to any value (8=default): the higher the value, the more line segments. The actual number of segments in the resulting curve depends on the original number of *Pline* vertices and the value of the SPLINETYPE variable (Fig. 16-19).

Figure 16-19

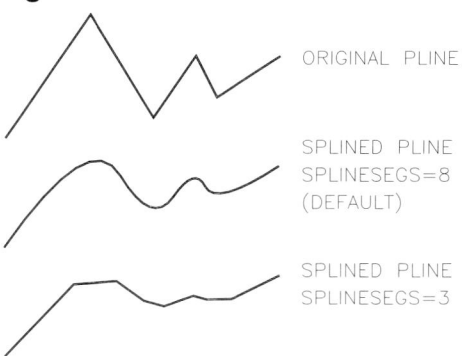

ORIGINAL PLINE

SPLINED PLINE
SPLINESEGS=8
(DEFAULT)

SPLINED PLINE
SPLINESEGS=3

326 *Modify Commands II*

> **TIP:** Changing the *SPLFRAME* variable to 1 causes the *Pline* frame (the original straight segments) to be displayed for *Splined* or *Fit Plines*. *Regen* must be used after changing the variable to display the original *Pline* "frame" (Fig. 16-20).

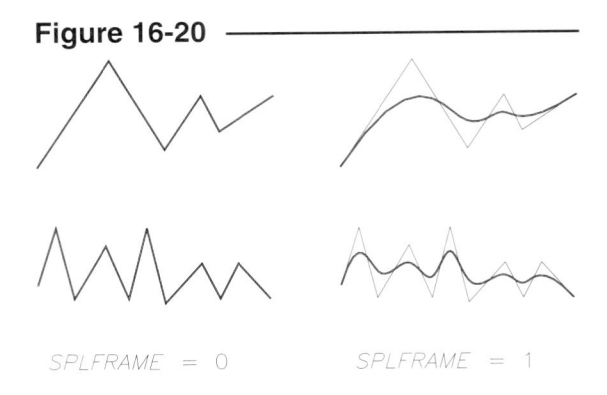

Figure 16-20

Ltype gen
This setting controls the generation of non-continuous linetypes for *Plines*. If *Off*, non-continuous linetype dashes start and stop at each vertex, as if the *Pline* segments were individual *Line* segments. For dashed linetypes, each line segment begins and ends with a full dashed segment (Fig. 16-21). If *On*, linetypes are drawn in a consistent pattern, disregarding vertices. In this case, it is possible for a vertex to have a space rather than a dash. Using the *Ltype gen* option retroactively changes *Plines* that have already been drawn. *Ltype gen* affects objects composed of *Plines* such as *Polygons, Rectangles,* and *Boundaries*.

Figure 16-21

Similarly, the *PLINEGEN* system variable controls how new non-continuous linetypes are drawn for *Plines*. A setting of 1 creates a consistent linetype pattern, disregarding vertices (like *Ltype gen On*). A *PLINEGEN* setting of 0 creates linetypes stopping and starting at each vertex (like *Ltype gen Off*). However, *PLINEGEN* is not retroactive—it affects only new *Plines*.

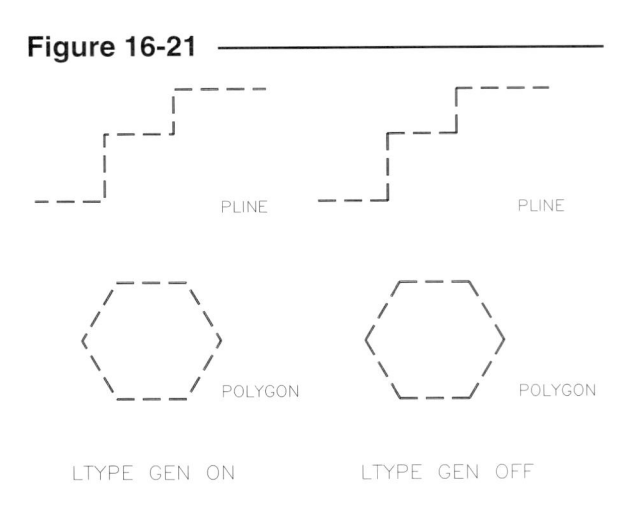

Undo
Undo reverses the most recent *Pedit* operation.

eXit
This option exits the *Pedit* options, keeps the changes, and returns to the Command prompt.

Vertex Editing

Upon selecting the *Edit Vertex* option from the *Pedit* options list, the group of suboptions is displayed on the screen menu and Command line:

   Command: `pedit`
   Select polyline or [Multiple]: `PICK`
   Enter an option [Close/Join/Width/Edit vertex/Fit/Spline/Decurve/Ltype gen/Undo]: `e`
   Enter a vertex editing option
   [Next/Previous/Break/Insert/Move/Regen/Straighten/Tangent/Width/eXit] <N>:

Next
AutoCAD places an **X** marker at the first endpoint of the *Pline*. The *Next* and *Previous* options allow you to sequence the marker to the desired vertex (Fig. 16-22).

Previous
See *Next* above.

Figure 16-22

Break
This selection causes a break between the marked vertex and another vertex you then select using the *Next* or *Previous* option (Fig. 16-23).

　　Enter an option [Next/Previous/Go/eXit] <N>:

Selecting *Go* causes the break. An endpoint vertex cannot be selected.

Figure 16-23

Insert
Insert allows you to insert a new vertex at any location <u>after</u> the vertex that is marked with the **X** (Fig. 16-24). Place the marker before the intended new vertex, use *Insert*, then PICK the new vertex location.

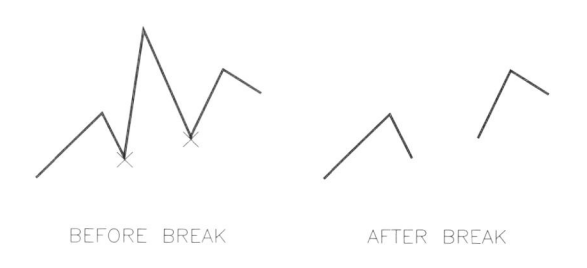

Figure 16-24

Move
You are prompted to indicate a new location to *Move* the marked vertex (Fig. 16-25).

Regen
In older releases of AutoCAD, *Regen* should be used after the *Width* option to display the new changes.

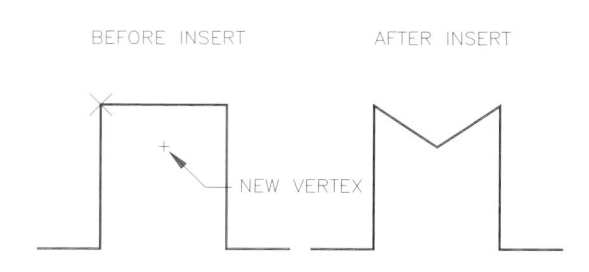

Figure 16-25

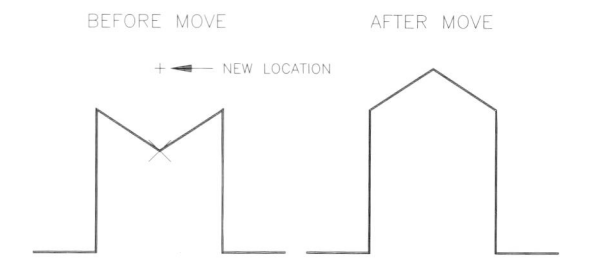

Straighten

You can *Straighten* the *Pline* segments between the current marker and the other marker that you then place by one of these options:

Enter an option [Next/Previous/Go/eXit] <N>:

Selecting *Go* causes the straightening to occur (Fig. 16-26).

Figure 16-26

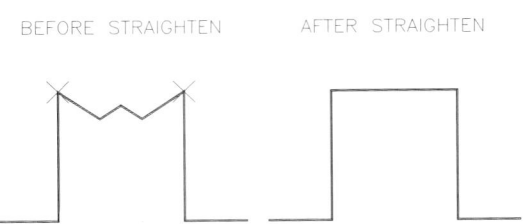

BEFORE STRAIGHTEN   AFTER STRAIGHTEN

Tangent

Tangent allows you to specify the direction of tangency of the current vertex for use with curve *Fitting*.

Width

This option allows changing the *Width* of the *Pline* segment immediately following the marker, thus achieving a specific width for one segment of the *Pline* (Fig. 16-27). *Width* can be specified with different starting and ending values.

Figure 16-27

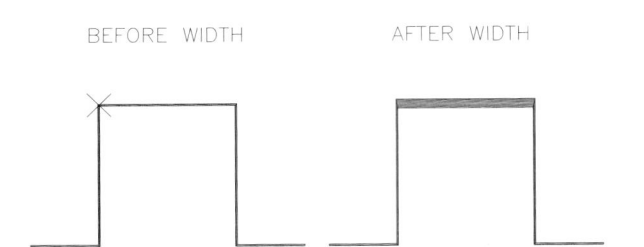

BEFORE WIDTH   AFTER WIDTH

eXit

This option exits from vertex editing, saves changes, and returns to the main *Pedit* prompt.

Grips

Plines can also be edited easily using Grips (see Chapter 21). A Grip appears on each vertex of the *Pline*. Editing *Plines* with Grips is sometimes easier than using *Pedit* because Grips are more direct and less dependent on the command interface.

Converting *Lines* and *Arcs* to *Plines*

A very important and productive feature of *Pedit* is the ability to convert *Lines* and *Arcs* to *Plines* and closed *Pline* shapes. Potential uses of this option are converting a series of connected *Lines* and *Arcs* to a closed *Pline* for subsequent use with *Offset* or for inquiry of the area (*Area* command) or length (*List* command) of a single shape. The only requirement for conversion of *Lines* and *Arcs* to *Plines* is that the selected objects must have <u>exact</u> matching endpoints.

To accomplish the conversion of objects to *Plines*, simply select a *Line* or *Arc* object and request to turn it into one:

```
Command: pedit
Select polyline or [Multiple]: PICK  (Select only one Line or Arc)
Object selected is not a polyline
Do you want to turn it into one? <Y> Enter
Enter an option [Close/Join/Width/Edit vertex/Fit/Spline/Decurve/Ltype gen/Undo]: j  (Use the Join option)
Select objects: PICK
Select objects: Enter
1 segments added to polyline
Enter an option [Close/Join/Width/Edit vertex/Fit/Spline/Decurve/Ltype gen/Undo]: Enter
Command:
```

The resulting conversion is a closed *Polyline* shape.

SPLINEDIT

Pull-down Menu	COMMAND (TYPE)	ALIAS (TYPE)	Short-cut	Screen (side) Menu	Tablet Menu
Modify Object > Splinedit	SPLINEDIT	SPE	...	MODIFY1 Splinedt	Y,18

Splinedit is an extremely powerful command for changing the configuration of existing *Splines*. You can use multiple methods to change *Splines*. All of the *Splinedit* methods fall under two sets of options.

The two groups of options that AutoCAD uses to edit *Splines* are based on two sets of points: data points and control points. Data points are the points that were specified when the *Spline* was created—the points that the *Spline* actually passes through (Fig. 16-28).

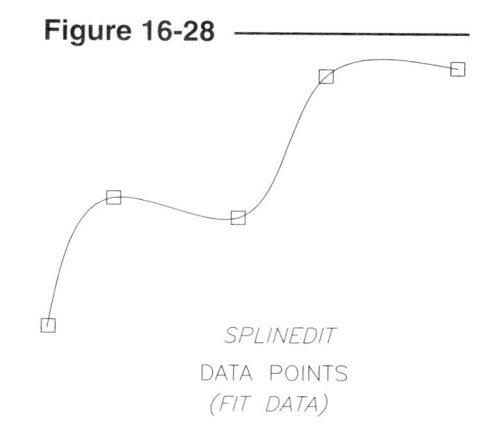

Figure 16-28

SPLINEDIT
DATA POINTS
(FIT DATA)

Control points are other points outside of the path of the *Spline* that only have a "pull" effect on the curve (Fig. 16-29).

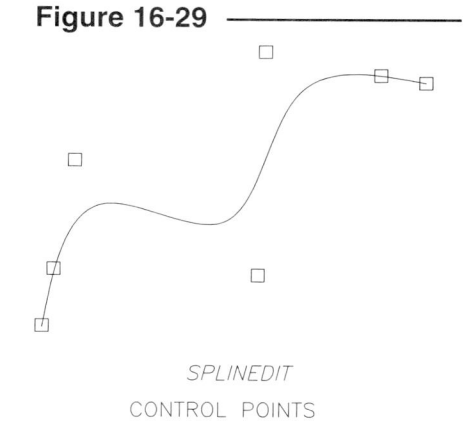

Figure 16-29

SPLINEDIT
CONTROL POINTS

Editing *Spline* Data Points

The command prompt displays several levels of options. The top level of options uses the control points method for editing. Select *Fit Data* to use data points for editing. The *Fit Data* methods are recommended for most applications. Since the curve passes directly through the data points, these options offer direct control of the curve path.

    Command: *splinedit*
    Select spline: **PICK**
    Enter an option [Fit data/Close/Move vertex/Refine/rEverse/Undo]: *f*
    Enter a fit data option [Add/Close/Delete/Move/Purge/Tangents/toLerance/eXit] <eXit>:

Add
You can add points to the *Spline*. The *Spline* curve changes to pass through the new points. First, PICK an existing point on the curve. That point and the next one in sequence (in the order of creation) become highlighted. The new point will change the curve between those two highlighted data points (Fig. 16-30).

> Specify control point: **PICK** (Select an existing point on the curve before the intended new point.)
> Specify new point: **PICK** (PICK a new point location between the two marked points.)

Figure 16-30

SPLINEDIT   FIT DATA   ADD

Close/Open
The *Close* option appears only if the existing curve is open, and the *Open* prompt appears only if the curve is closed. Selecting either option automatically forces the opposite change. *Close* causes the two ends to become tangent, forming one smooth curve (Fig. 16-31). This tangent continuity is characteristic of *Closed Splines* only. *Splines* that have matching endpoints do not have tangent continuity unless the *Close* option of *Spline* or *Splinedit* is used.

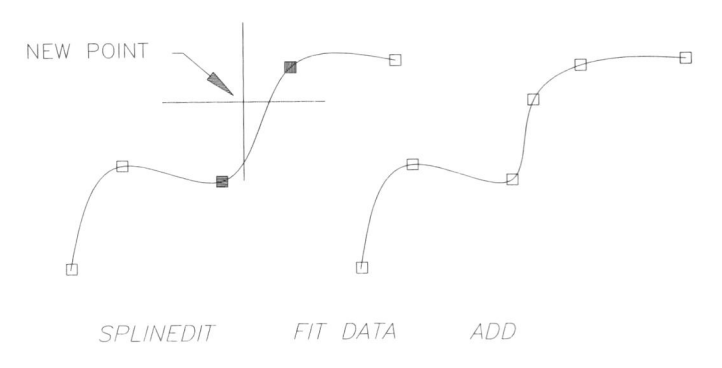

Figure 16-31

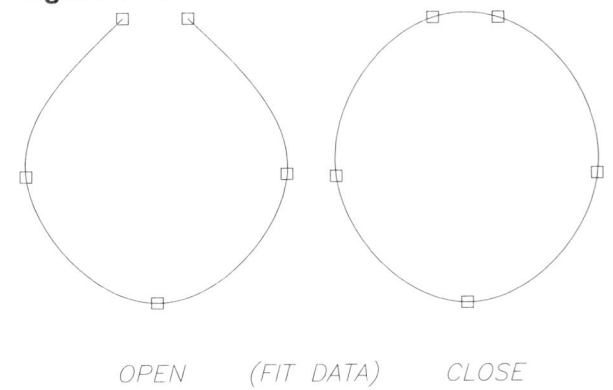

OPEN   (FIT DATA)   CLOSE

Move
You can move any data point to a new location with this option (Fig. 16-32). The beginning endpoint (in the order of creation) becomes highlighted. Type *N* for next or *S* to select the desired data point to move; then PICK the new location.

> Specify new location or [Next/Previous/Select point/eXit] <N>:

Figure 16-32

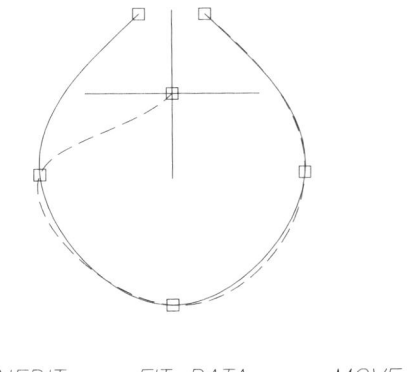

Purge
Purge deletes all data points and renders the *Fit Data* set of options unusable. You are returned to the control point options (top level). To reinstate the points, use *Undo*.

SPLINEDIT   FIT DATA   MOVE

Tangents
You can change the directions for the start and endpoint tangents with this option. This action gives the same control that exists with the "Enter start tangent" and "Enter end tangent" prompts of the *Spline* command used when the curves were created (see Figure 15-12, *Spline, End Tangent*).

toLerance
Use *toLerance* to specify a value, or tolerance, for the curve to "fall" away from the data points. Specifying a tolerance causes the curve to smooth out, or fall, from the data points. The higher the value, the more the curve "loosens." The *toLerance* option of *Splinedit* is identical to the *Fit Tolerance* option available with *Spline* (see Figure 15-14, *Spline, Tolerance*).

Editing *Spline* Control Points

Use the top level of command options (except *Fit Data*) to edit the *Spline's* control points. These options are similar to those used for editing the data points; however, the results are different since the curve does not pass through the control points.

```
Command: splinedit
Select spline: PICK
Enter an option [Fit data/Close/Move vertex/Refine/rEverse/Undo]:
```

Fit Data
Discussed previously.

Close/Open
These options operate similar to the *Fit Data* equivalents; however, the resulting curve falls away from the control points (Fig. 16-33; see also Fig. 16-31, *Fit Data, Close*).

Figure 16-33

FIT DATA CLOSE

CLOSE (CONTROL POINTS)

Move Vertex
Move Vertex allows you to move the location of any control points. This is the control points' equivalent to the *Move* option of *Fit Data* (see Figure 16-32, *Fit Data, Move*). The method of selecting points (*Next/Previous/Select point/eXit/*) is the same as that used for other options.

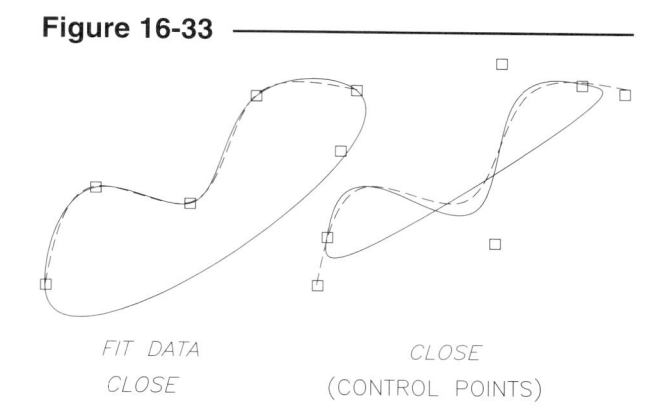

Refine
Selecting the *Refine* option reveals another level of options.

```
Enter a refine option [Add control point/Elevate order/Weight/eXit] <eXit>:
```

Add control points is the control points' equivalent to *Fit Data Add* (see Figure 16-30). At the "Select a point on the Spline" prompt, simply PICK a point near the desired location for the new point to appear. Once the *Refine* option has been used, the *Fit Data* options are no longer available.

Elevate order allows you to increase the number of control points uniformly along the length of the *Spline*. Enter a value from *n* to 26, where *n* is the current number of points + one. Once a *Spline* is elevated, it cannot be reduced.

Weight is an option that you use to assign a value to the amount of "pull" that a specific control point has on the *Spline* curve. The higher the value, the more "pull," and the closer the curve moves toward the control point. The typical method of selecting points (*Next/Previous/Select point/eXit/*) is used.

rEverse
The *rEverse* option reverses the direction of the *Spline*. The first endpoint (when created) then becomes the last endpoint. Reversing the direction may be helpful for selection during the *Move* option.

Grips

Splines can also be edited easily using *Grips* (see Chapter 21). The *Grip* points that appear on the *Spline* are identical to the *Fit Data* points. Editing with *Grips* is a bit more direct and less dependent on the command interface.

Boolean Commands

Region combines one or several objects forming a closed shape into one object, a *Region*. The appearance of the object(s) does not change after the conversion, even though the resulting shape is one object (see "*Region,*" Chapter 15). Although the *Region* appears to be no different than a closed *Pline*, it is more powerful because several *Regions* can be combined to form complex shapes (called "composite *Regions*") using the three Boolean operations explained next. As an example, a set of *Regions* (converted *Circles*) can be combined to form the sprocket with only one *Subtract* operation, as shown in Chapter 15, Figure 15-30.

The Boolean operators, *Union*, *Subtract*, and *Intersect*, can be used with *Regions* as well as solids. Any number of these commands can be used with *Regions* to form complex geometry. To illustrate each of the Boolean commands, consider the shapes shown in Figure 16-34. The *Circle* and the closed *Pline* are first converted to *Regions*; then *Union*, *Subtract*, or *Intersection* can be used.

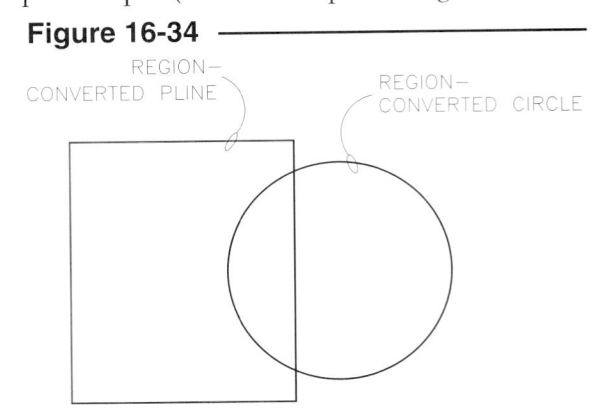

Figure 16-34

UNION

Pull-down Menu	COMMAND (TYPE)	ALIAS (TYPE)	Short-cut	Screen (side) Menu	Tablet Menu
Modify Solids Editing> Union	UNION	UNI	...	MODIFY2 Union	X,15

Union combines two or more *Regions* (or solids) into one *Region* (or solid). The resulting composite *Region* has the encompassing perimeter and area of the original *Regions*. Invoking *Union* causes AutoCAD to prompt you to select objects. You can select only existing *Regions* (or solids).

    Command: *union*
    Select Objects: **PICK** (*Region*)
    Select Objects: **PICK** (*Region*)
    Select Objects: **Enter**
    Command:

The selected *Regions* are combined into one composite *Region* (Fig. 16-35). Any number of Boolean operations can be performed on the *Region*(s).

Several *Regions* can be selected in response to the "Select objects:" prompt. For example, a composite *Region* such as that in Figure 16-36 can be created with one *Union*.

Two or more *Regions* can be *Unioned* even if they do not overlap. They are simply combined into one object although they still appear as two.

Figure 16-35

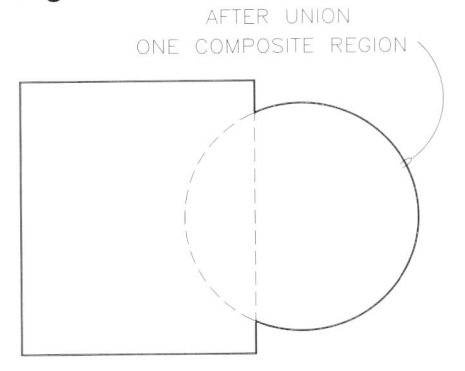

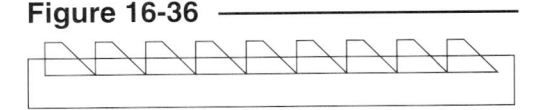

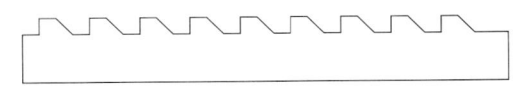

Figure 16-36

SUBTRACT

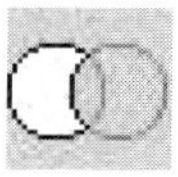

Pull-down Menu	COMMAND (TYPE)	ALIAS (TYPE)	Short-cut	Screen (side) Menu	Tablet Menu
Modify Solids Editing > Subtract	SUBTRACT	SU	...	MODIFY2 Subtract	X,16

Subtract enables you to remove one *Region* (or set of *Regions*) from another. The *Regions* must be created before using *Subtract* (or another Boolean operator). *Subtract* also works with solids (as do the other Boolean operations).

There are two steps to *Subtract*. First, you are prompted to select the *Region* or set of *Regions* to "subtract from" (those that you wish to keep), then to select the *Regions* "to subtract" (those you want to remove). The resulting shape is one composite *Region* comprising the perimeter of the first set minus the second (sometimes called "difference").

Figure 16-37

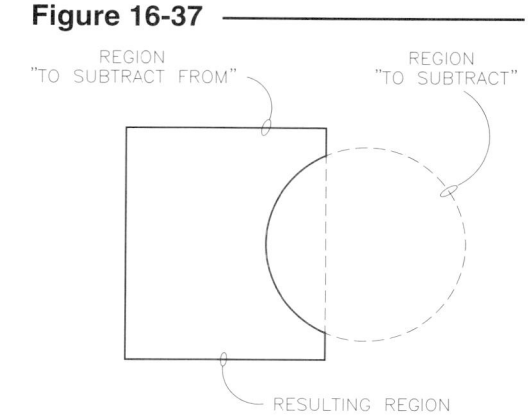

```
Command: subtract
Select solids and regions to subtract from...
Select Objects: PICK
Select Objects: Enter
Select solids and regions to subtract...
Select Objects: PICK
Select Objects: Enter
Command:
```

Consider the two shapes previously shown in Figure 16-34. The resulting *Region* shown in Figure 16-37 is the result of *Subtracting* the circular *Region* from the rectangular one.

Keep in mind that multiple *Regions* can be selected as the set to keep or as the set to remove. For example, the sprocket illustrated previously (Fig. 15-30) was created by subtracting several circular *Regions* in one operation. Another example is the removal of material to create holes or slots in sheet metal.

INTERSECT

Pull-down Menu	COMMAND (TYPE)	ALIAS)TYPE)	Short-cut	Screen (side) Menu	Tablet Menu
Modify Solids Editing> Intersect	INTERSECT	IN	...	MODIFY2 Intrsect	X,17

Intersect is the Boolean operator that finds the common area from two or more *Regions*.

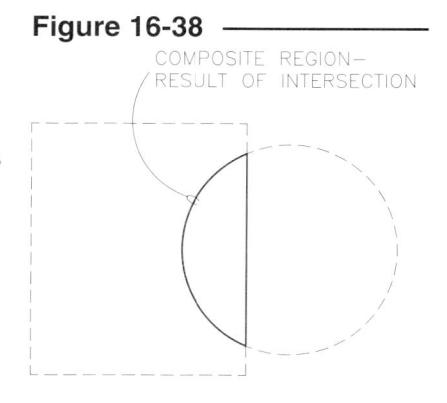

Consider the rectangular and circular *Regions* previously shown (Fig. 16-34). Using the *Intersect* command and selecting both shapes results in a *Region* comprising only that area that is shared by both shapes (Fig. 16-38):

Figure 16-38

```
Command: intersect
Select Objects: PICK
Select Objects: PICK
Select Objects: Enter
Command:
```

If more than two *Regions* are selected, the resulting *Intersection* is composed of only the common area from all shapes (Fig. 16-39). If all of the shapes selected do not overlap, a null *Region* is created (all shapes disappear because no area is common to all).

Intersect is a powerful operation when used with solid modeling. Techniques for saving time using Boolean operations are discussed in Chapter 31, Solid Modeling Construction.

Figure 16-39

CHAPTER EXERCISES

1. *Chprop* or *Properties* **Window**

   Open the **PIVOTARM CH9** drawing that you worked on in Chapter 9 Exercises. *Load* the **Hidden2** and **Center2** *Linetypes*. Make two *New Layers* named **HID** and **CEN** and assign the matching linetypes and yellow and green colors, respectively. Check the *Limits* of the drawing; then calculate and set an appropriate *LTSCALE*. Use *Chprop* or *Properties* window to change the *Lines* representing the holes in the front view to the **HID** layer as shown in Figure 16-40. Use *SaveAs* and name the drawing **PIVOTARM CH16**.

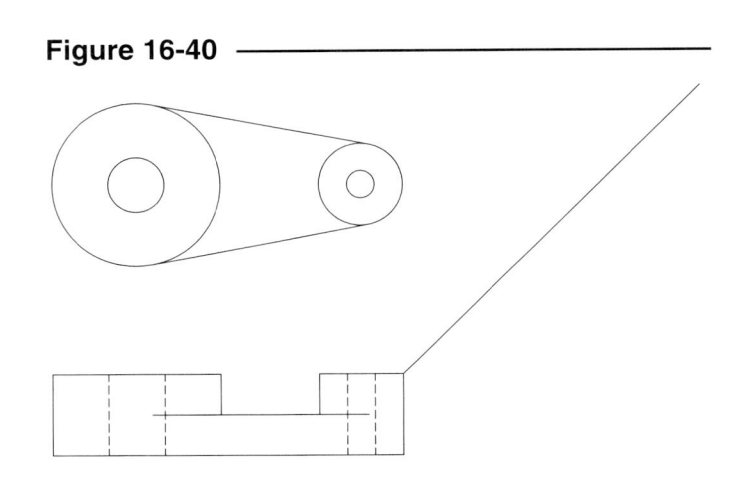

   Figure 16-40

2. *Change*

   Open **CH8EX3**. *Erase* the *Arc* at the top of the object. *Erase* the *Points* with a window. Invoke the *Change* command. When prompted to *Select objects*, **PICK** all of the inclined *Lines* near the top. When prompted to "Specify change point," enter coordinate **6,8**. The object should appear as that in Figure 16-41. Use *SaveAs* and assign the name **CH16EX2**.

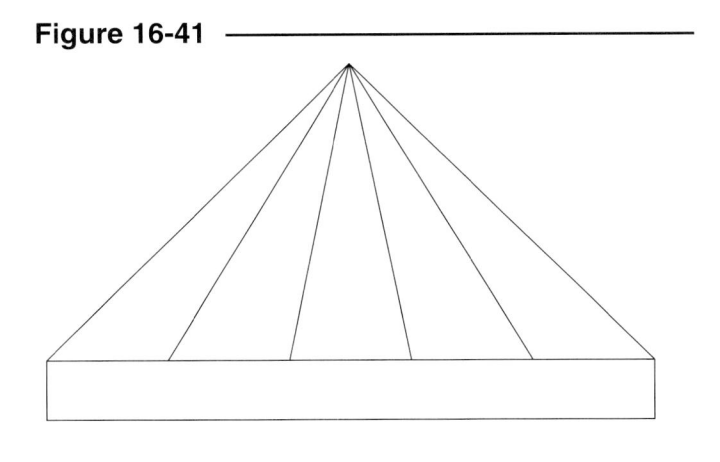

   Figure 16-41

3. *Properties* **Window**

   A design change is required for the bolt holes in **GASKETA** (from Chapter 9 Exercises). *Open* **GASKETA** and invoke the *Properties* window. Change each of the four bolt holes to **.375** diameter (Fig. 16-42). *Save* the drawing.

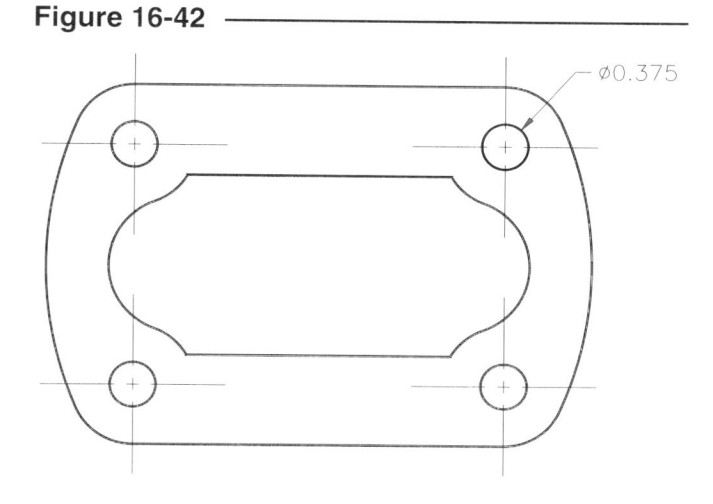

   Figure 16-42

4. *Align*

   Open the **PLATES** drawing you created in Chapter 9. The three plates are to be stamped at one time on a single sheet of stock measuring 15" x 12". Place the three plates together to achieve optimum nesting on the sheet stock.

   A. Use *Align* to move the plate in the center (with 9 holes). Select the *First source* and *destination points* (1S, 1D) and *Second source* and *destination points* (2S, 2D) as shown in Figure 16-43.

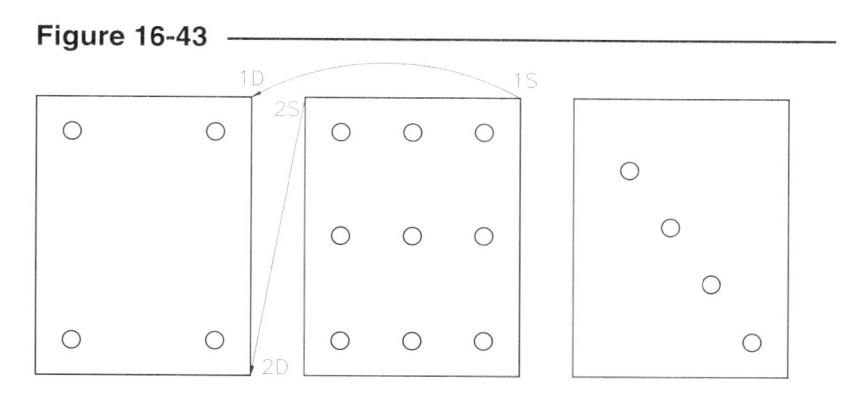

   Figure 16-43

   B. After the first alignment is complete, use *Align* to move the plate on the right (with 4 diagonal holes). The *source* and *destination points* are indicated in Figure 16-44.

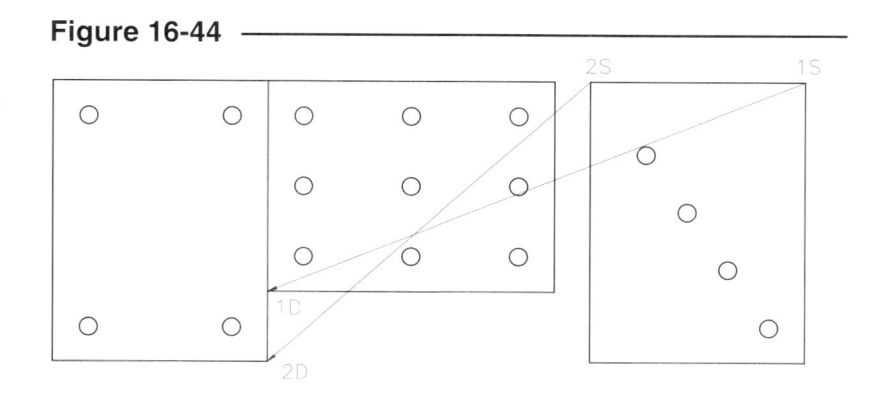

   Figure 16-44

C. Finally, draw the sheet stock outline (15" x 12") using *Line* as shown in Figure 16-45. The plates are ready for production. Use *SaveAs* and assign the name **PLATENEST**.

Figure 16-45

5. *Explode*

   Open the **POLYGON1** drawing that you completed in Chapter 15 Exercises. You can quickly create the five-sided shape shown (continuous lines) in Figure 16-46 by *Exploding* the *Polygon*. First, *Explode* the *Polygon* and *Erase* the two *Lines* (highlighted). Draw the bottom *Line* from the two open *Endpoints*. Do not exit the drawing.

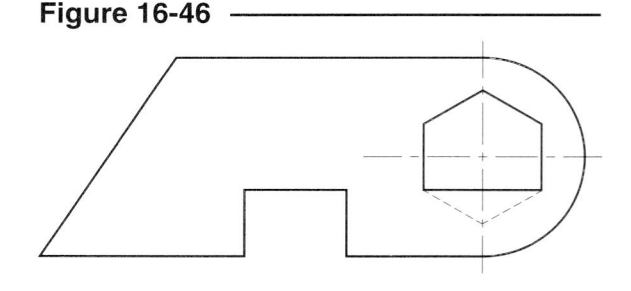

Figure 16-46

6. *Pedit*

   Use *Pedit* with the *Edit vertex* options to alter the shape as shown in Figure 16-47. For the bottom notch, use *Straighten*. For the top notch, use *Insert*. Use *SaveAs* and change the name to **PEDIT1**.

Figure 16-47

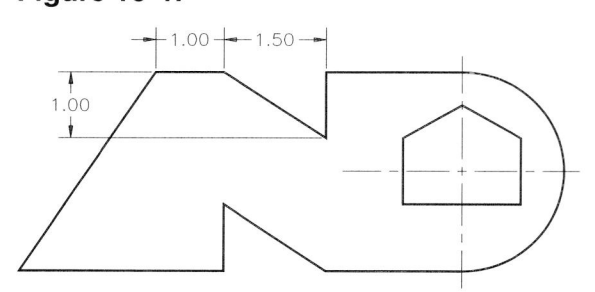

7. *Pline, Pedit*

   A. Create a line graph as shown in Figure 16-48 to illustrate the low temperatures for a week. The temperatures are as follows:

X axis	Y axis
Sunday	20
Monday	14
Tuesday	18
Wednesday	26
Thursday	34
Friday	38
Saturday	27

   Use equal intervals along each axis. Use a *Pline* for the graph line. *Save* the drawing as **TEMP-A**. (You will label the graph at a later time.)

Figure 16-48

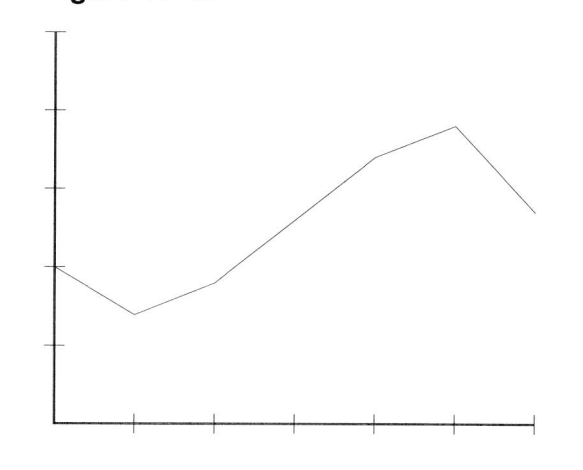

B. Use *Pedit* to change the *Pline* to a *Spline*. Note that the graph line is no longer 100% accurate because it does not pass through the original vertices (see Fig. 16-49). Use the *SPLFRAME* variable to display the original "frame" (*Regen* must be used after).

Use the *Properties* window and try the *Cubic* and *Quadratic* options. Which option causes the vertices to have more pull? Find the most accurate option. Set *SPLFRAME* to 0 and *SaveAs* **TEMP-B**.

C. Use *Pedit* to change the curve from *Pline* to *Fit Curve*. Does the graph line pass through the vertices? *Saveas* **TEMP-C**.

Figure 16-49

D. *Open* drawing **TEMP-A**. *Erase* the *Splined Pline* and construct the graph using a *Spline* instead (Fig. 16-50, see Exercise 7A for data). *SaveAs* **TEMP-D**. Compare the *Spline* with the variations of *Plines*. Which of the four drawings (A, B, C, or D) is smoothest? Which is the most accurate?

Figure 16-50

E. It was learned that there was a mistake in reporting the temperatures for that week. Thursday's low must be changed to 38 degrees and Friday's to 34 degrees. *Open* **TEMP-D** (if not already open) and use *Splinedit* to correct the mistake. Use the *Move* option of *Fit Data* so that the exact data points can be altered as shown in Figure 16-51. *SaveAs* **TEMP-E**.

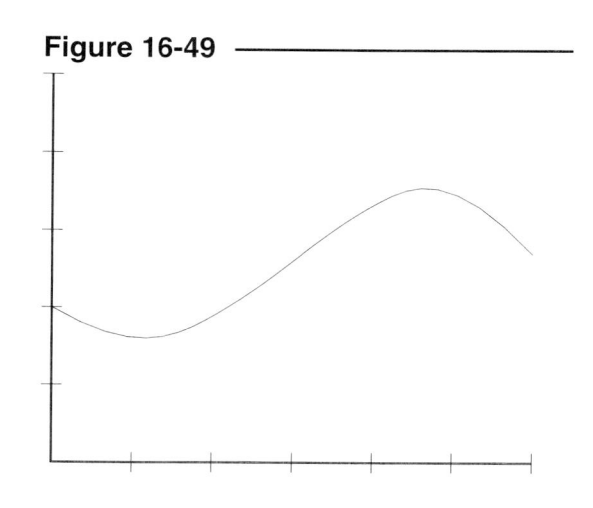

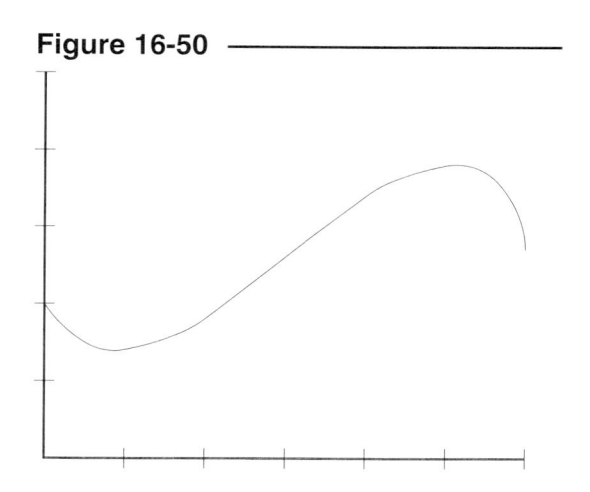

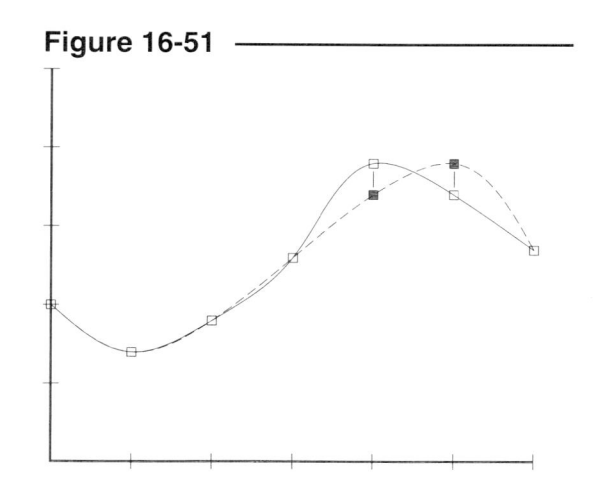

Figure 16-51

338   Modify Commands II

8. **Converting *Lines*, *Circles*, and *Arcs* to *Plines***

    A. Begin a *New* drawing and use the **A-METRIC** template (that you worked on in Chapter 13 Exercises). Use *Save* and assign the name **GASKETC**. Change the *Limits* for plotting on an A sheet at 2:1 (refer to the Metric Table of *Limits* Settings and set *Limits* to 1/2 x *Limits* specified for 1:1 scale). Change the *LTSCALE* to **12**. First, draw only the <u>inside</u> shape using *Lines* and *Circles* (with *Trim*) or *Arcs* (Fig. 16-52). Then convert the *Lines* and *Arcs* to one closed *Pline* using *Pedit*. Finally, locate and draw the 3 bolt holes. *Save* the drawing.

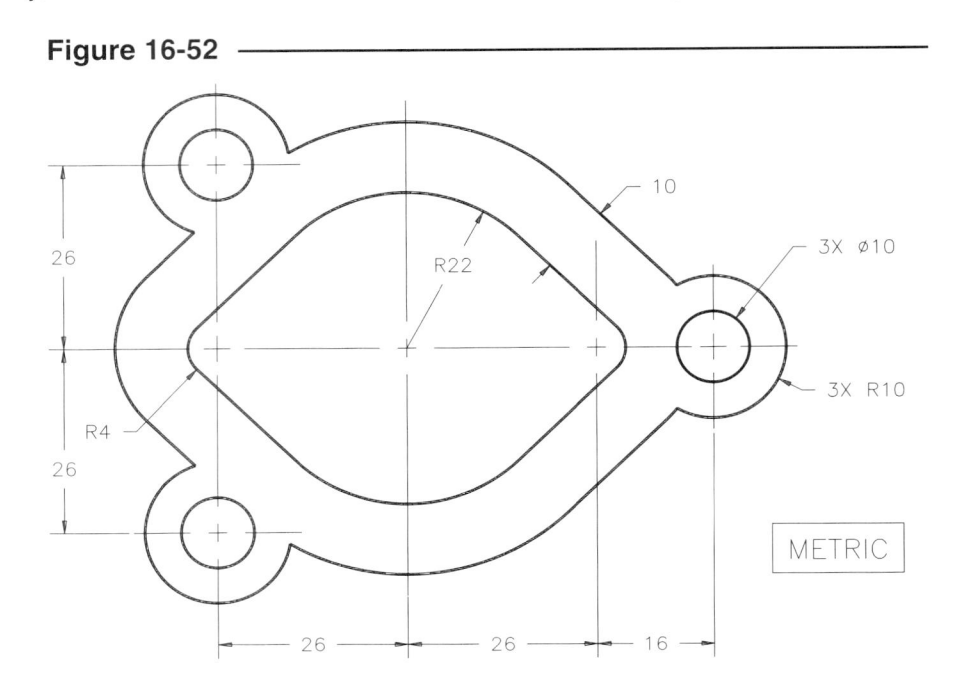

    Figure 16-52

    B. *Offset* the existing inside shape to create the outside shape. Use *Offset* to create concentric circles around the bolt holes. Use *Trim* to complete the gasket. *Save* the drawing and create a plot at 2:1 scale.

9. *Splinedit*

    Figure 16-53 —   Figure 16-54 —

    Open the **HANDLE** drawing that you created in Chapter 15 Exercises. During the testing and analysis process, it was discovered that the shape of the handle should have a more ergonomic design. The finger side (left) should be flatter to accommodate varying sizes of hands, and the thumb side (right) should have more of a protrusion on top to prevent slippage.

    First, add more control points uniformly along the length of the left side with *Spinedit, Refine. Elevate* the *Order* from 4 to **6**, then use *Move Vertex* to align the control points as shown in Figure 16-53.

    On the right side of the handle, *Add* two points under the *Fit Data* option to create the protrusion shown in Figure 16-54. You may have to *Reverse* the direction of the *Spline* to add the new points between the two highlighted ones. *SaveAs* **HANDLE2**.

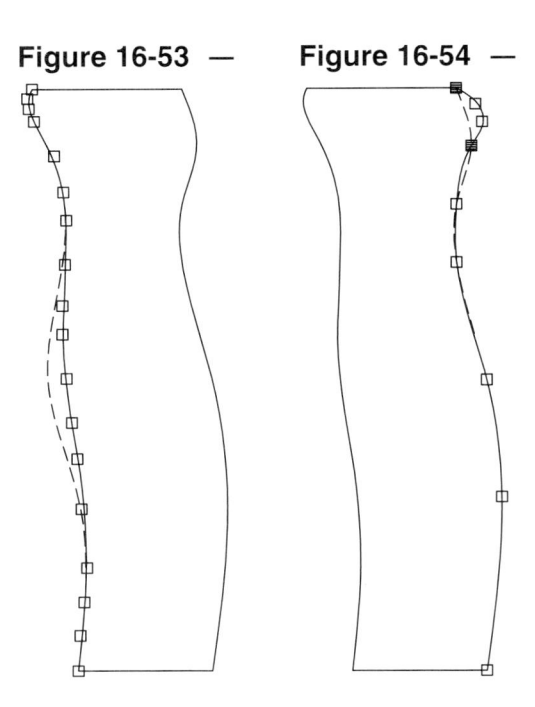

10. *Line, Stretch*

    Draw the floor plan of the storage room shown in Figure 16-55. Use *Line* objects for the walls and *Plines* for the windows and doors. When your drawing is complete according to the given specifications, use *Stretch* to center the large window along the top wall. Save the drawing as **STORE ROOM**.

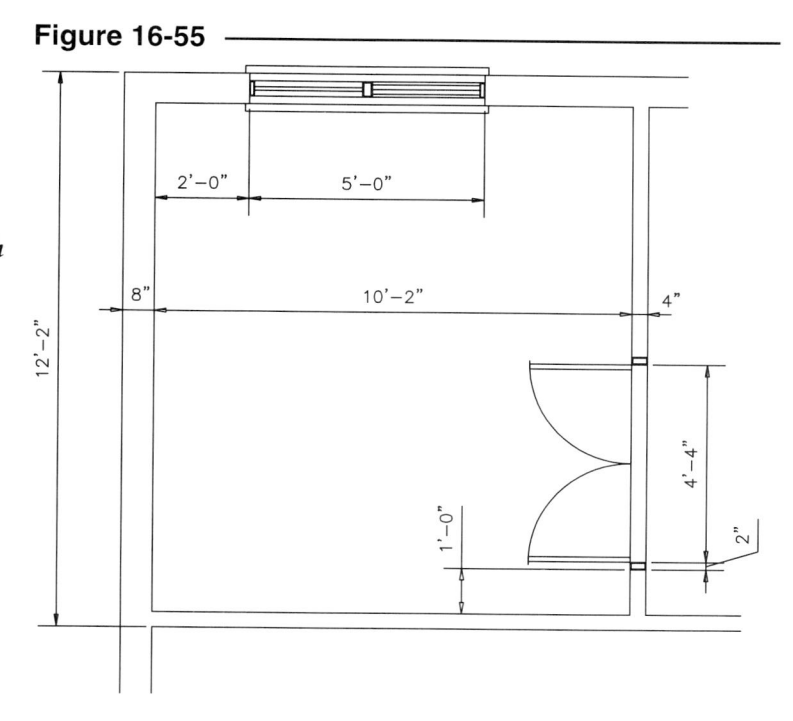

    Figure 16-55

11. **Gear Drawing**

    Complete the drawing of the gear you began in Chapter 15 Exercises called **GEAR-REGION**. If you remember, three shapes were created (two *Circles* and one "tooth") and each was converted to a *Region*. Finally, the "tooth" was *Arrayed* to create the total of 40 teeth.

    Figure 16-56

    A. To complete the gear, *Subtract* the small circular *Region* and all of the teeth from the large circular *Region*. First, use *Subtract*. At the "Select solids and regions to subtract from..." prompt, PICK the large circular *Region*. At the "Select solids and regions to subtract..." prompt, use a window to select every-thing (the large circular *Region* is automatically filtered out). The resulting gear should resemble Figure 16-56. *Save* the drawing as **GEAR-REGION 2**.

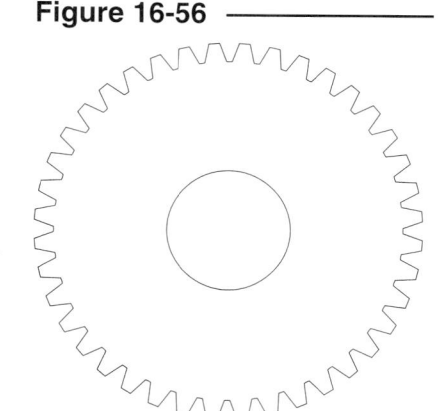

    B. Consider the steps involved if you were to create the gear (as an alternative) by using *Trim* to remove 40 small sections of the large *Circle* and all unwanted parts of the teeth. *Regions* are clearly easier in this case.

12. **Wrench Drawing**

    Create the same wrench you created in Chapter 15 again, only this time use region modeling. Refer to Chapter 15, Exercise 5 for dimensions (exclude the break lines). Begin a *New* drawing and use the **A-METRIC** template. Use *SaveAs* and assign the name **WRENCH-REG**.

    A. Set *Limits* to **372,288** to prepare the drawing for plotting at 3:4 (the drawing scale factor is 33.87). Set the *GRID* to **10**.

340 Modify Commands II

B. Draw a *Circle* and an *Ellipse* as shown on the left in Figure 16-57. The center of each shape is located at **60,150**. *Trim* half of each shape as shown (highlighted). Use *Region* to convert the two remaining halves into a *Region* as shown on the right of Figure 16-57.

Figure 16-57

C. Next, create a *Circle* with the center at **110,150** and a diameter as shown in Figure 16-58. Then draw a closed *Pline* in a rectangular shape as shown. The height of the rectangle must be drawn as specified; however, the width of the rectangle can be drawn <u>approximately</u> as shown on the left. Convert each shape to a *Region*; then use *Intersect* to create the region as shown on the right side of the figure.

Figure 16-58

D. *Move* the rectangular-shaped region **68** units to the left to overlap the first region as shown in Figure 16-59. Use *Subtract* to create the composite region on the right representing the head of the wrench.

Figure 16-59

E. Complete the construction of the wrench in a manner similar to that used in the previous steps. Refer to Chapter 15 Exercises for dimensions of the wrench. Complete the wrench as one *Region*. *Save* the drawing as **WRENCH-REG**.

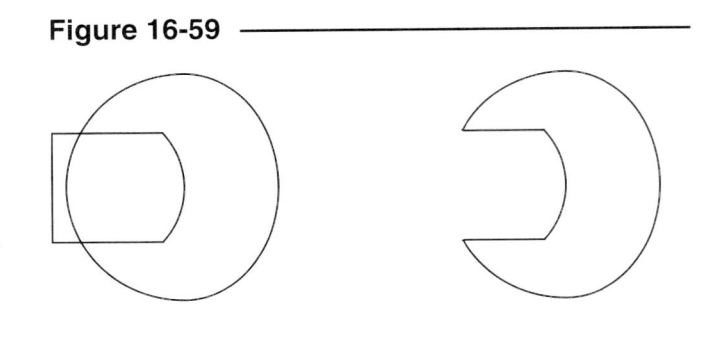

13. **Retaining Wall**

    Open the **RET-WALL** drawing that you created in Chapter 8 Exercises. Mark the wall with 50 unit stations as shown in Figure 16-60.

    HINT: Convert the centerline of the retaining wall to a *Pline*; then use the *Measure* command to place *Point* objects at the 50 unit stations. *Offset* the wall on both sides to provide a construction aid in the creation of the perpendicular tick marks. Use *Node* and *Perpendicular* Osnaps. *Save* the drawing as **RET-WAL3**.

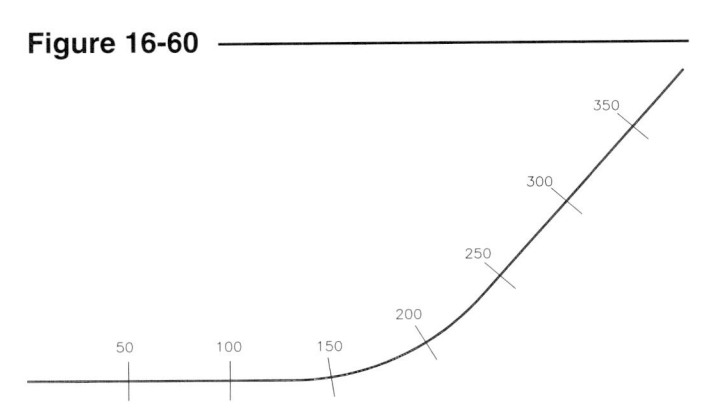

Figure 16-60

17

INQUIRY COMMANDS

Chapter Objectives

After completing this chapter you should:

1. be able to list the *Status* of a drawing;

2. be able to *List* the AutoCAD database information about an object;

3. know how to list the entire database of all objects with *Dblist*;

4. be able to calculate the *Area* of a closed shape with and without "islands";

5. be able to find the distance between two points with *Dist*;

6. be able to report the coordinate value of a selected point using the *ID* command;

7. know how to list the *Time* spent on a drawing or in the current drawing session;

8. be able to use *Setvar* to change system variable settings or list current settings.

CONCEPTS

AutoCAD provides several commands that allow you to find out information about the current drawing status and specific objects in a drawing. These commands as a group are known as "*Inquiry* commands" and are grouped together in the menu systems. The *Inquiry* commands are located in the *Inquiry* toolbar (Fig. 17-1). You can also use the *Tools* pull-down menu to access the *Inquiry* commands (Fig. 17-2).

Figure 17-1

Using *Inquiry* commands, you can find out such information as the amount of time spent in the current drawing, the distance between two points, the area of a closed shape, the database listing of properties for specific objects (coordinates of endpoints, lengths, angles, etc.), and current settings for system variables as well as other information. The *Inquiry* commands are:

Status, List, Dblist, Area, Dist, ID, Time, and *Setvar*

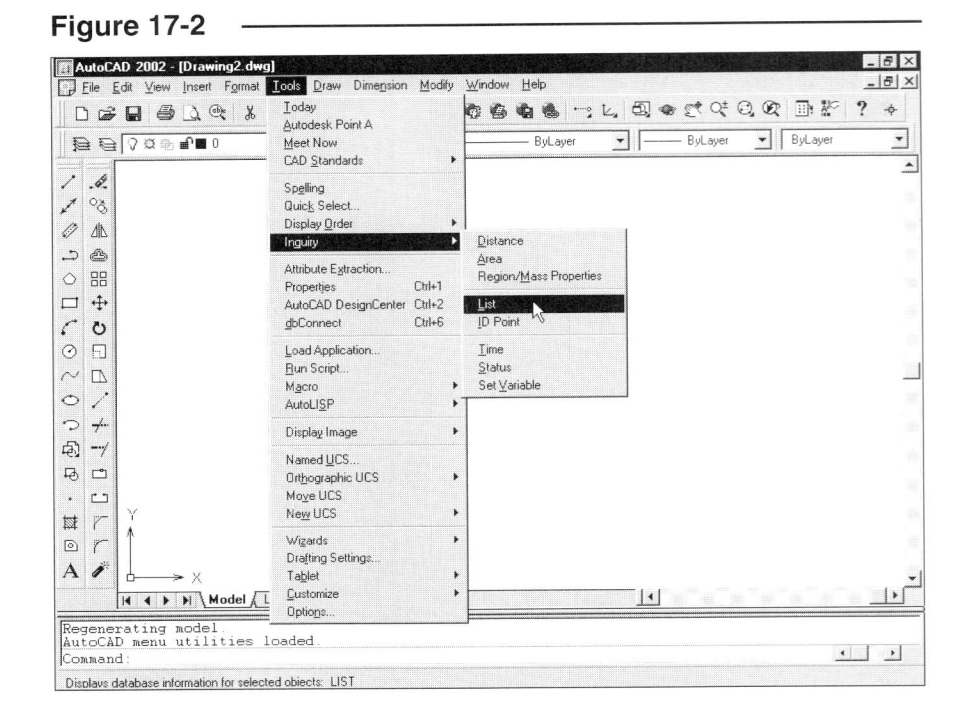

Figure 17-2

COMMANDS

STATUS

Pull-down Menu	COMMAND (TYPE)	ALIAS (TYPE)	Short-cut	Screen (side) Menu	Tablet Menu
Tools Inquiry > Status	STATUS	...	...	TOOLS 1 Status	...

The *Status* command gives many pieces of information related to the current drawing. Typing or **PICK**ing the command from the icon or one of the menus causes a text screen to appear similar to that shown in Figure 17-3, on the next page. The information items are:

  Total number of objects in the current drawing
  Paper space limits: values set by the *Limits* command in Paper Space (*Limits* in Paper Space are set
      by selecting a *Paper size* in the *Page Setup* or *Plot* dialog box.)
  Paper space uses: area used by the objects (drawing extents) in Paper Space
  Model space limits: values set by the *Limits* command in Model Space
  Model space uses: area used by the objects (drawing extents) in Model Space

Display shows: current display or windowed area
Insertion basepoint: point specified by the *Base* command or default (0,0)
Snap resolution: value specified by the *Snap* command
Grid spacing: value specified by the *Grid* command
Current space: Paper Space or Model Space
Current layer: name
Current color: current color assignment
Current linetype: current linetype assignment
Current lineweight: current lineweight assignment
Current plot style: current plot style assignment
Current elevation, thickness: 3D properties—current height above the XY plane and Z dimension
On or off status: *FILL, GRID, ORTHO, QTEXT, SNAP, TABLET*
Object Snap Modes: current *Running OSNAP* modes
Free dwg disk: space on the current drawing hard disk drive
Free temp disk: space on the current temporary files hard disk drive
Free physical memory: amount of free RAM (total RAM)
Free swap file space: amount of free swap file space (total allocated swap file)

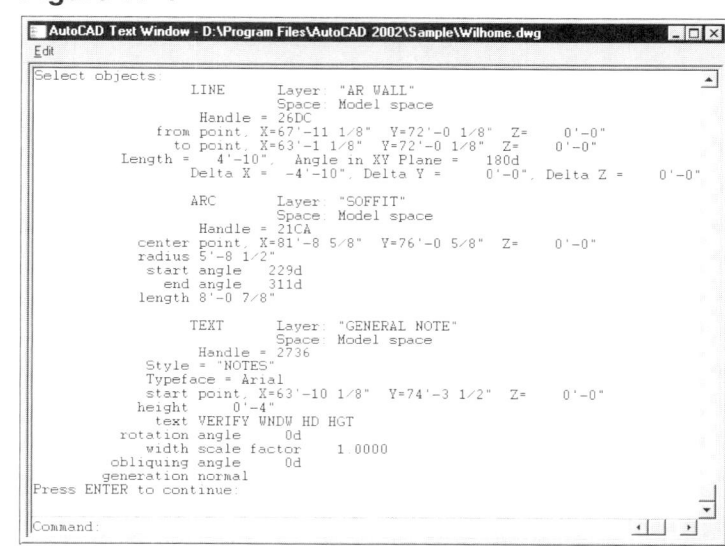

Figure 17-3

LIST

Pull-down Menu	COMMAND (TYPE)	ALIAS (TYPE)	Short-cut	Screen (side) Menu	Tablet Menu
Tools Inquiry > List	LIST	LS or LI	...	TOOLS 1 List	U,8

The *List* command displays the database list of information in text window format for one or more specified objects. The information displayed depends on the type of object selected. Invoking the *List* command causes a prompt for you to select objects. AutoCAD then displays the list for the selected objects (see Figs. 17-4 and 17-5, on the next page). A *List* of a *Line* and an *Arc* is given in Figure 17-4.

For a *Line*, coordinates for the endpoints, line length and angle, current layer, and other information are given.

For an *Arc*, the center coordinate, radius, start and end angles, and length are given.

Figure 17-4

344 Inquiry Commands

The *List* for a *Pline* is shown in Figure 17-5. The location of each vertex is given, as well as the length and perimeter of the entire *Pline*.

Since Release 14, *Plines* are created and listed as *Lwpolylines*, or "lightweight polylines." In previous releases, complete data for each *Pline* vertex (starting width, ending width, color, etc.) were stored along with the coordinate values of the vertex, then repeated for each vertex. In AutoCAD 2002, the data common to all vertices are stored only once, and only the coordinate data are stored for each vertex. Because this data structure saves file space, the new *Plines* are known as lightweight *Plines*.

Figure 17-5

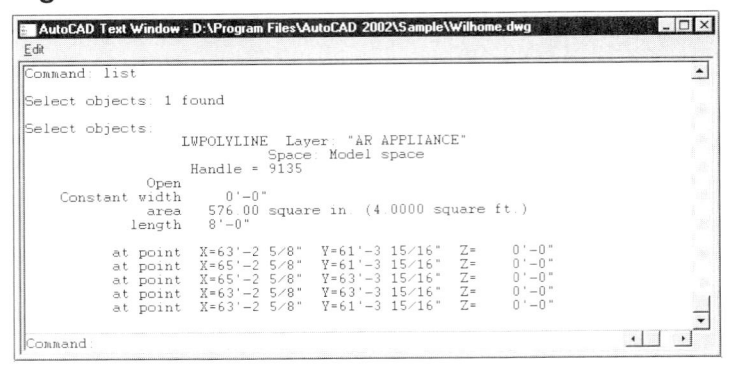

```
AutoCAD Text Window - D:\Program Files\AutoCAD 2002\Sample\Wilhome.dwg
Edit
Command: list
Select objects: 1 found
Select objects:
             LWPOLYLINE  Layer: "AR APPLIANCE"
                         Space: Model space
                Handle = 9135
         Open
Constant width    0'-0"
         area    576.00 square in  (4.0000 square ft.)
       length    8'-0"
   at point   X=63'-2 5/8"   Y=61'-3 15/16"   Z=   0'-0"
   at point   X=65'-2 5/8"   Y=61'-3 15/16"   Z=   0'-0"
   at point   X=65'-2 5/8"   Y=63'-3 15/16"   Z=   0'-0"
   at point   X=63'-2 5/8"   Y=63'-3 15/16"   Z=   0'-0"
   at point   X=63'-2 5/8"   Y=61'-3 15/16"   Z=   0'-0"
Command:
```

DBLIST

Pull-down Menu	COMMAND (TYPE)	ALIAS (TYPE)	Short-cut	Screen (side) Menu	Tablet Menu
...	DBLIST	...	...	...	...

The *Dblist* command is similar to the *List* command in that it displays the database listing of objects; however, *Dblist* gives information for <u>every</u> object in the current drawing! This command is generally used when you desire to send the list to a printer or when only a few objects are in the drawing. If you use this command in a complex drawing, be prepared to page through many screens of information. Press Escape to cancel *Dblist* and return to the Command: prompt. Press F2 to open and close the text window.

AREA

Pull-down Menu	COMMAND (TYPE)	ALIAS (TYPE)	Short-cut	Screen (side) Menu	Tablet Menu
Tools Inquiry > Area	AREA	AA	...	TOOLS 1 Area	T,7

The *Area* command is helpful for many applications. With this command AutoCAD calculates the area and the perimeter of any enclosed shape in a matter of milliseconds. You specify the area (shape) to consider for calculation by PICKing the *Object* (if it is a closed *Pline, Polygon, Circle, Boundary, Region* or other closed object) or by PICKing points (corners of the outline) to define the shape. The options are given below.

Specify first corner point
The command sequence for specifying the area by PICKing points is shown below. This method should be used only for shapes with <u>straight</u> sides. An example of the *Point* method (PICKing points to define the area) is shown in Figure 17-6.

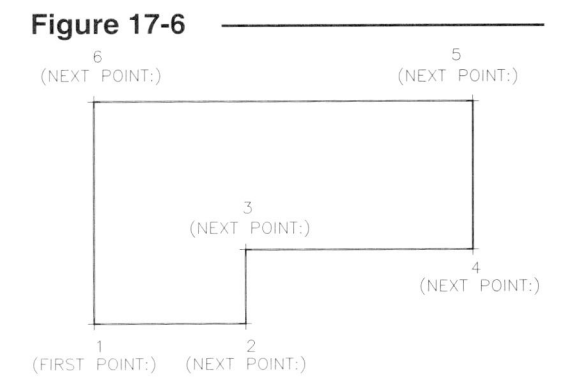

Figure 17-6

```
   6                                    5
(NEXT POINT:)                       (NEXT POINT:)

                        3
                   (NEXT POINT:)
                                        4
                                   (NEXT POINT:)

   1             2
(FIRST POINT:) (NEXT POINT:)
```

Command: `area`
Specify first corner point or [Object/Add/Subtract]: **PICK** (Locate the first corner to define the shape.)
Specify next corner point or press ENTER for total: **PICK** (Locate the second corner on the shape.)
Specify next corner point or press ENTER for total: **PICK** (Locate the next corner.)
Specify next corner point or press ENTER for total: **PICK** (Continue selecting points until all corners have been defined.)
Specify next corner point or press ENTER for total: **Enter**
Area = *nn.nnn* perimeter = *nn.nnn*
Command:

Object

If the shape for which you want to find the area and perimeter is a *Circle, Polygon, Ellipse, Boundary, Region,* or closed *Pline*, the *Object* option of the *Area* command can be used. Select the shape with one PICK (since all of these shapes are considered as one object by AutoCAD).

The ability to find the area of a closed *Pline, Region,* or *Boundary* is extremely helpful. Remember that any closed shape, even if it includes *Arcs* and other curves, can be converted to a closed *Pline* with the *Pedit* command (as long as there are no gaps or overlaps) or can be used with the *Boundary* command. This method provides you with the ability to easily calculate the area of any shape, curved or straight. In short, convert the shape to a closed *Pline, Region,* or *Boundary* and find the *Area* with the *Object* option.

Add, Subtract

Add and *Subtract* provide you with the means to find the area of a closed shape that has islands, or negative spaces. For example, you may be required to find the surface area of a sheet of material that has several punched holes. In this case, the area of the holes is subtracted from the area defined by the perimeter shape. The *Add* and *Subtract* options are used specifically for that purpose. The following command sequence displays the process of calculating an area and subtracting the area occupied by the holes.

Command: `area`
Specify first corner point or [Object/Add/Subtract]: **a** (Use the *Add* option to begin a running total.)
Specify first corner point or [Object/Subtract]: **o** (Use the *Object* option to select the outside shape.)
(ADD mode) Select objects: **PICK** (Select the closed object.)
Area = 13.31, Perimeter = 14.39
Total area = 13.31

(ADD mode) Select objects: **Enter** (Completion of *Add* mode.)
Specify first corner point or [Object/Subtract]: **s** (Switch to *Subtract* mode.)
Specify first corner point or [Object/Add]: **o** (Use *Object* mode.)
(SUBTRACT mode) Select objects: **PICK** (Select the first *Circle* to subtract.)
Area = 0.69, Length = 2.95
Total area = 12.62

(SUBTRACT mode) Select objects: **PICK** (Select the second *Circle* to subtract.)
Area = 0.69, Length = 2.95
Total area = 11.93

(SUBTRACT mode) Select objects: **Enter** (Completion of *Subtract* mode.)
Specify first corner point or [Object/Add]: **Enter** (Completion of *Area* command.)
Command:

Make sure that you press Enter between the *Add* and *Subtract* modes.

346 Inquiry Commands

An example of the last command sequence used to find the area of a shape minus the holes is shown in Figure 17-7. Notice that the object selected in the first step is a closed *Pline* shape, including an *Arc*.

Figure 17-7

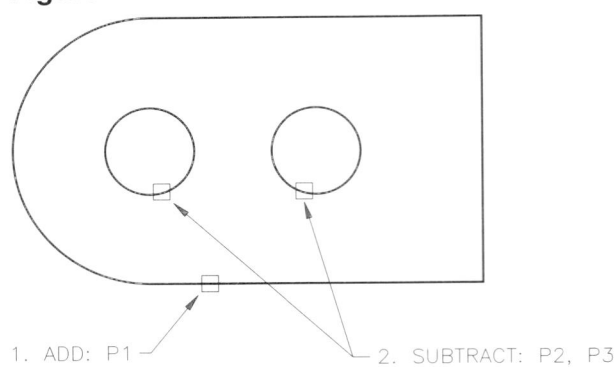

1. ADD: P1
2. SUBTRACT: P2, P3

DIST

Pull-down Menu	COMMAND (TYPE)	ALIAS (TYPE)	Short-cut	Screen (side) Menu	Tablet Menu
Tools Inquiry > Distance	DIST	DI	...	TOOLS 1 Dist	T,8

The *Dist* command reports the distance between any two points you specify. *OSNAPs* can be used to snap to the existing points. This command is helpful in many engineering or architectural applications, such as finding the clearance between two mechanical parts, finding the distance between columns in a building, or finding the size of an opening in a part or doorway. The command is easy to use.

    Command: *dist*
    Specify first point: PICK (Use *Osnaps* if needed.)
    Specify second point: PICK (Use *Osnaps* if needed.)
    Distance = 3.63, Angle in XY Plane = 165, Angle from XY Plane = 0
    Delta X = -3.50, Delta Y = 0.97, Delta Z = 0.00
    Command:

AutoCAD reports the absolute and relative distances as well as the angle of the line between the points.

ID

Pull-down Menu	COMMAND (TYPE)	ALIAS (TYPE)	Short-cut	Screen (side) Menu	Tablet Menu
Tools Inquiry > ID Point	ID	...	...	TOOLS 1 ID	U,9

The *ID* command reports the coordinate value of any point you select with the cursor. If you require the location associated with a specific object, an *OSNAP* mode (*Endpoint, Midpoint, Center,* etc.) can be used.

    Command: *id*
    Specify point: PICK (Use *Osnaps* if needed.)
    X = 7.63   Y = 6.25   Z = 0.00
    Command:

NOTE: *ID* also sets AutoCAD's "last point." The last point can be referenced in commands by using the @ (at) symbol with relative rectangular or relative polar coordinates.

TIME

Pull-down Menu	COMMAND (TYPE)	ALIAS (TYPE)	Short-cut	Screen (side) Menu	Tablet Menu
Tools Inquiry > Time	TIME	...	...	TOOLS 1 Time	...

This command is useful for keeping track of the time spent in the current drawing session or total time spent on a particular drawing. Knowing how much time is spent on a drawing can be useful in an office situation for bidding or billing jobs.

The *Time* command reports the information shown in Figure 17-8. The *Total editing time* is automatically kept, starting from when the drawing was first created until the current time. Plotting and printing time is not included in this total, nor is the time spent in a session when changes are discarded.

Figure 17-8

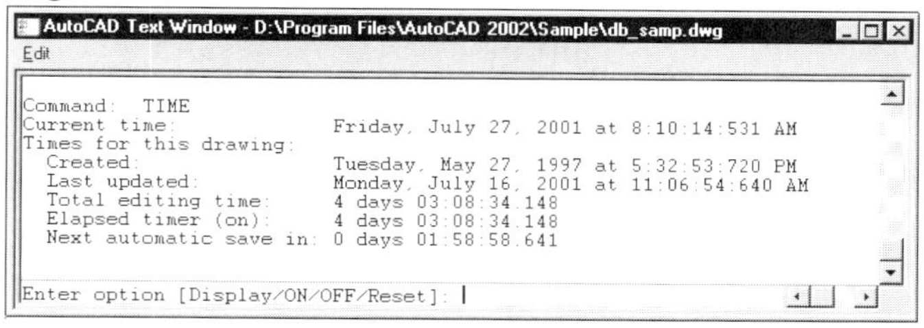

Display
The *Display* option causes *Time* to repeat the display with the updated times.

ON/OFF/Reset
The *Elapsed timer* is a separate compilation of time controlled by the user. The *Elapsed timer* can be turned *ON* or *OFF* or can be *Reset*.

Time also reports when the next automatic save will be made. The time interval of the Automatic Save feature is controlled by the *SAVETIME* system variable. To set the interval between automatic saves, type *SAVETIME* at the Command line and specify a value for time (in minutes). The default interval is 120 minutes.

SETVAR

Pull-down Menu	COMMAND (TYPE)	ALIAS (TYPE)	Short-cut	Screen (side) Menu	Tablet Menu
Tools Inquiry > Set Variable	SETVAR	SET	...	TOOLS 1 Setvar	U,10

The settings (values or on/off status) that you make for many commands, such as *Limits, Grid, Snap, Running Osnaps, Fillet* values, *Pline* width, etc. are saved in system variables. In AutoCAD 2002 there are 364 system variables. The variables store the settings that are used to create and edit the drawing. *Setvar* ("set variable") gives you access to the system variables. (See AutoCAD *Help* for a complete list of the system variables, including an explanation, default setting, and possible settings for each.)

Setvar allows you to perform two functions: (1) change the setting for any system variable and (2) display the current setting for one or all system variables. To change a setting for a system variable using *Setvar*, just use the command and enter the name of the variable. For example, the following syntax lists values for the *GRID* setting and current *Fillet* radius:

348 Inquiry Commands

Command: *setvar*
Enter variable name or [?]: *gridunit*
Enter new value for GRIDUNIT <1.00,1.00>:

Command: *setvar*
Enter variable name or [?]: *filletrad*
Enter new value for FILLETRAD <0.50>:

With recent releases of AutoCAD, *Setvar* is not needed to set system variables. You can enter the variable name directly at the Command prompt without using *Setvar* first.

Command: *filletrad*
Enter new value for FILLETRAD <0.50>:

TIP To list the current settings for all system variables, use *Setvar* with the *?* (question mark) option. The complete list of system variables is given in a text window with the current setting for each variable (Fig. 17-9).

Figure 17-9

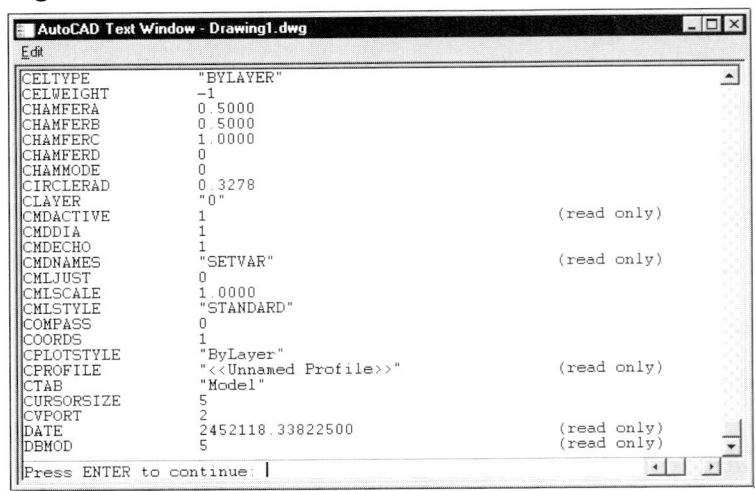

```
AutoCAD Text Window - Drawing1.dwg
Edit
CELTYPE       "BYLAYER"
CELWEIGHT     -1
CHAMFERA      0.5000
CHAMFERB      0.5000
CHAMFERC      1.0000
CHAMFERD      0
CHAMMODE      0
CIRCLERAD     0.3278
CLAYER        "0"
CMDACTIVE     1                           (read only)
CMDDIA        1
CMDECHO       1
CMDNAMES      "SETVAR"                    (read only)
CMLJUST       0
CMLSCALE      1.0000
CMLSTYLE      "STANDARD"
COMPASS       0
COORDS        1
CPLOTSTYLE    "ByLayer"
CPROFILE      "<<Unnamed Profile>>"       (read only)
CTAB          "Model"
CURSORSIZE    5
CVPORT        2
DATE          2452118.33822500            (read only)
DBMOD         5                           (read only)
Press ENTER to continue:
```

You can also use *Help* to list the system variables with a short explanation for each. In the *Help Topics* dialog box, select the *Contents* tab, select *Command Reference*, then *System Variables*.

MASSPROP
The *Massprop* command is used to give mass and volumetric properties for AutoCAD solids. See Chapter 32, Advanced Solids Features, for use of *Massprop*.

CHAPTER EXERCISES

1. *List*

   A. *Open* the **PLATENEST** drawing from Chapter 16 Exercises. Assume that a laser will be used to cut the plates and holes from the stock, and you must program the coordinates. Use the *List* command to give information on the *Line*s and *Circles* for the one plate with four holes in a diagonal orientation. Determine and write down the coordinates for the 4 corners of the plate and the centers of the 4 holes.

   B. *Open* the **EFF-APT** drawing from Chapter 15 Exercises. Use *List* to determine the area of the inside of the tub. If the tub were filled with 10" of water, what would be the volume of water in the tub?

2. *Area*

   A. *Open* the **EFF-APT** drawing. The entry room is to be carpeted at a cost of $12.50 per square yard. Use the *Area* command (with the PICK points option) to determine the cost for carpeting the room, not including the closet.

   B. *Open* the **PLATENEST** drawing from Chapter 16 Exercises. Use *SaveAs* to create a file named **PLATE-AREA**. Using the *Area* command, calculate the wasted material (the two pieces of stock remaining after the 3 plates have been cut or stamped). HINT: Use *Boundary* to create objects from the waste areas for determining the *Area*.

   C. Create a *Boundary* (with islands) of the plate with four holes arranged diagonally. *Move* the new boundary objects **10** units to the right. Using the *Add* and *Subtract* options of *Area*, calculate the surface area for painting 100 pieces, both sides. (Remember to press Enter between the *Add* and *Subtract* operations.) *Save* the drawing.

3. *Dist*

   A. *Open* the **EFF-APT** drawing. Use the *Dist* command to determine the best location for installing a wall-mounted telephone in the apartment. Where should the telephone be located in order to provide the most equal access from all corners of the apartment? What is the farthest distance that you would have to walk to answer the phone?

   B. Using the *Dist* command, determine what length of pipe would be required to connect the kitchen sink drain (use the center of the far sink) to the tub drain (assume the drain is at the far end of the tub). Calculate only the direct distance (under the floor).

4. *ID*

   Open the **PLATENEST** drawing once again. You have now been assigned to program the laser to cut the plate with 4 holes in the corners. Use the *ID* command (with *OSNAP*s) to determine the coordinates for the 4 corners and the hole centers.

5. *Time*

   Using the *Time* command, what is the total amount of editing time you spent with the **PLATENEST** drawing? How much time have you spent in this session? How much time until the next automatic save?

6. *Setvar*

   Using the **PLATENEST** drawing again, use *Setvar* to list system variables. What are the current settings for *Fillet* radius (**FILLETRAD**), the last point used (**LASTPOINT**), *Linetype Scale* (**LTSCALE**), automatic file save interval (**SAVETIME**), and text *Style* (**TEXTSTYLE**).

18

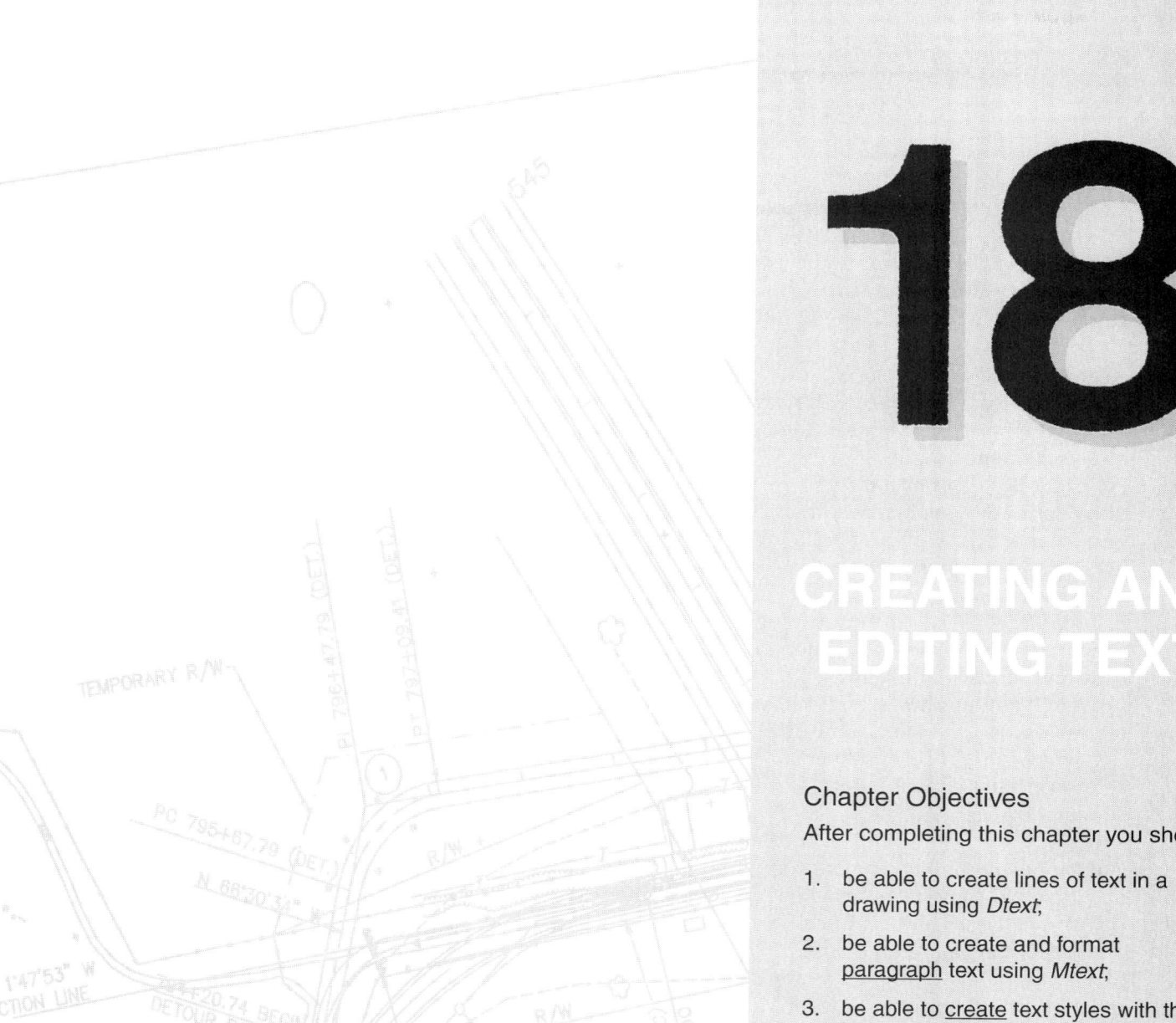

CREATING AND EDITING TEXT

Chapter Objectives

After completing this chapter you should:

1. be able to create lines of text in a drawing using *Dtext*;
2. be able to create and format paragraph text using *Mtext*;
3. be able to create text styles with the *Style* command;
4. know that *Ddedit* can be used to edit the content of existing text and *Properties* can be used to modify any property of existing text;
5. be able to use *Spell* to check spelling and be able to *Find and Replace* text in a drawing;
6. know how to change the justification and scale of existing text using *Scaletext* and *Justifytext*;
7. know how features such as *Qtext* (quick text) and *TEXTFILL* can be used to control the appearance of text in a drawing or print/plot.

CONCEPTS

The *Dtext* and *Mtext* commands provide you with a means of creating text in an AutoCAD drawing. "Text" in CAD drawings usually refers to sentences, words, or notes created from alphabetical or numerical characters that appear in the drawing. The numeric values that are part of specific dimensions are generally not considered "text," since dimensional values are a component of the dimension created automatically with the use of dimensioning commands.

Text in technical drawings is typically in the form of notes concerning information or descriptions of the objects contained in the drawing. For example, an architectural drawing might have written descriptions of rooms or spaces, special instructions for construction, or notes concerning materials or furnishings (Fig. 18-1). An engineering drawing may contain, in addition to the dimensions, manufacturing notes, bill of materials, schedules, or tables (Fig. 18-2). Technical illustrations may contain part numbers or assembly notes. Title blocks also contain text.

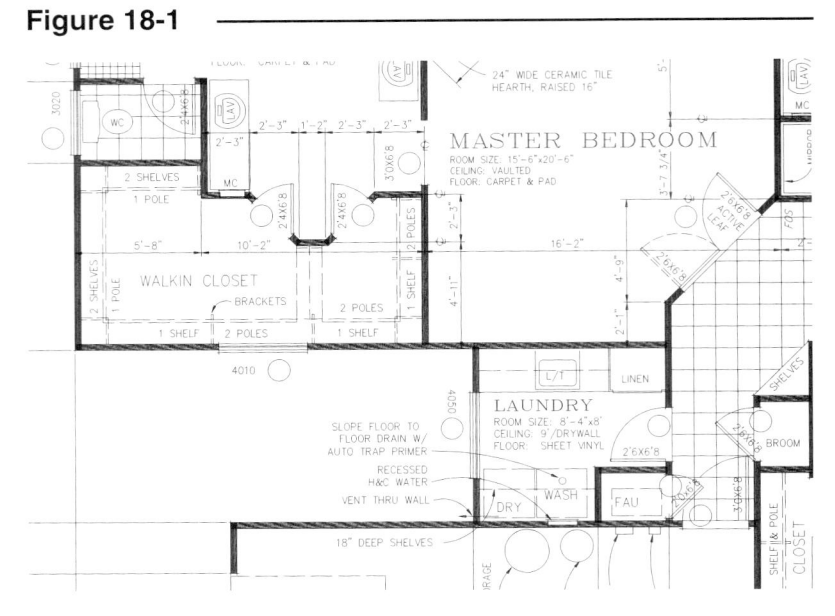

Figure 18-1

Figure 18-2

A line of text or paragraph of text in an AutoCAD drawing is treated as an object, just like a *Line* or a *Circle*. Each text object can be *Erased*, *Moved*, *Rotated*, or otherwise edited as any other graphical object. The letters themselves can be changed individually with special text editing commands. A spell checker is available by using the *Spell* command. Since text is treated as a graphical element, the use of many lines of text in a drawing can slow regeneration time and increase plotting time significantly.

The *Dtext* and *Mtext* commands perform basically the same function; they create text in a drawing. *Mtext* is the newer and more sophisticated method of text entry. With *Mtext* (multiline text) you can create a paragraph of text that "wraps" within a text boundary (rectangle) that you specify. An *Mtext* paragraph is treated as one AutoCAD object. The *Dtext* command is intended to be used for creating single or multiple independent lines of text. The *Text* command, available in AutoCAD Release 14 and previous releases, has been converted to operate identically to the *Dtext* command.

Many options for text justification are available. *Justification* is the method of aligning multiple lines of text. For example, if text is right justified, the right ends of the lines of text are aligned. The form or shape of the individual letters is determined in AutoCAD by the text *Style*. Creating a *Style* begins with selecting a Windows standard TrueType or AutoCAD-supplied font file. Font files supplied with AutoCAD have file extensions of .TTF (TrueType) or .SHX (AutoCAD compiled shape files). The AutoCAD .SHX fonts are located in the C:\Program Files\AutoCAD 2002\Fonts directory (by the default installation). The TrueType fonts are installed with the other Windows fonts in the C:\Windows\Fonts directory. Additional fonts can be purchased or may already be on your computer (supplied with Windows, word processors, or other software).

After a font for the *Style* is selected, other parameters (such as width and obliquing angle) can be specified to customize the *style* to your needs. Only one *style*, called *Standard*, has been created as part of the traditional default template drawing (ACAD.DWT) and uses the TXT.SHX font file. Other template drawings may have two or more created *styles*.

If any other style of text is desired, it can be created with the *Style* command. When a new *style* is created, it becomes the current one used by the *Dtext* or *Mtext* command. If several *styles* have been created in a drawing, a particular one can be recalled or made current by using the *style* option of the *Dtext* or *Mtext* commands.

In summary, the *Style* command allows you to design new styles with your choice of options, such as fonts, width factor, and obliquing angle, whereas the *style* option of *Dtext* and *Mtext* allows you to select from existing styles in the drawing that you previously created.

Commands related to creating or editing text in an AutoCAD drawing include:

Dtext   Places individual lines of text in a drawing and allows you to see each letter as it is typed.
Mtext   Places text in paragraph form (with word wrap) within a text boundary and allows many methods of formatting the appearance of the text.
Style   Creates text styles for use with any of the text creation commands. You can select from font files, specify other parameters to design the appearance of the letters, and assign a name for each style.
Spell   Checks the spelling of existing text in a drawing.
Find   Used to find or replace text strings globally in the drawing. Text created with *Dtext* or *Mtext* can be located with *Find*.
Ddedit   Invokes a dialog box for editing text. If you select *Text* or *Dtext* for editing, you can change the text (characters) only; if you select *Mtext* objects, you can edit individual characters and change the appearance of individual characters or the entire paragraph(s).
Scaletext   Allows you to change the scale of multiple text objects without altering the text insertion points.
Justifytext   Allows you to change the insertion point and justification of existing text without changing the text position.
Qtext   Short for quick-text, temporarily displays a line of text as a box instead of individual characters in order to speed up regeneration time and plotting time.

TEXT CREATION COMMANDS

The commands for creating text are formally named *Dtext* and *Mtext* (these are the commands used for typing). The *Draw* pull-down and screen (side) menus provide access to the two commonly used text commands, *Multiline Text...* (*Mtext*) and *Single-Line Text* (*Dtext*) (Fig. 18-3). Only the *Mtext* command has an icon button (by default) near the bottom of the *Draw* toolbar (Fig. 18-4).

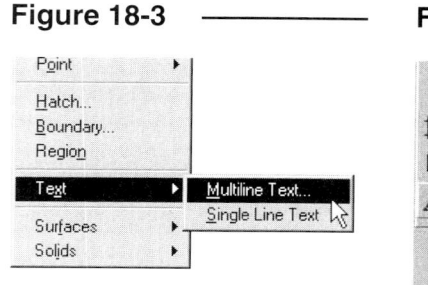

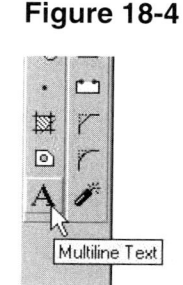

DTEXT

Pull-down Menu	COMMAND (TYPE)	ALIAS (TYPE)	Short-cut	Screen (side) Menu	Tablet Menu
Draw Text > Single Line Text...	DTEXT	DT	...	DRAW 2 Dtext	K,8

Dtext (dynamic text) lets you insert single lines of text into an AutoCAD drawing. *Dtext* displays each character in the drawing as it is typed. You can enter multiple lines of text without exiting the *Dtext* command. The lines of text do not wrap. The options are presented below:

    Command: dtext
    Current text style: "Standard" Text height: 0.2000
    Specify start point of text or [Justify/Style]:

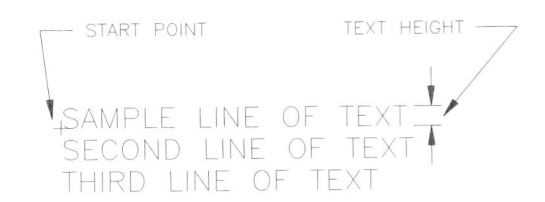

Start Point

The *Start point* for a line of text is the left end of the baseline for the text (Fig. 18-5). *Height* is the distance from the baseline to the top of upper case letters. Additional lines of text are automatically spaced below and left justified. The *rotation angle* is the angle of the baseline (Fig. 18-6).

Figure 18-5

Figure 18-6

The command sequence for this option is:

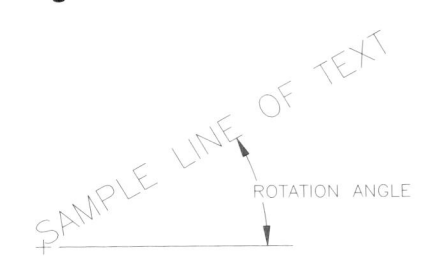

    Command: dtext
    Current text style: "Standard" Text height: 0.2000
    Specify start point of text or [Justify/Style]: PICK or (coordinates)
    Specify height <0.2000>: Enter or (value)
    Specify rotation angle of text <0>: Enter or (value)
    Enter text: (Type the desired line of text and press Enter)
    Enter text: (Type another line of text and press Enter)
    Enter text: Enter
    Command:

NOTE: When the "Enter text:" prompt appears, you can also PICK a new location for the next line of text anywhere in the drawing.

Justify

If you want to use one of the justification methods, invoking this option displays the choices at the prompt:

    Command: dtext
    Current text style: "Standard" Text height: 0.2000
    Specify start point of text or [Justify/Style]: J (Justify option)
    Enter an option [Align/Fit/Center/Middle/Right/TL/TC/TR/ML/MC/MR/BL/BC/BR]:

After specifying a justification option, you can enter the desired text in response to the "Enter text:" prompt. The text is not justified until after you press Enter.

Align
Aligns the line of text between the two points specified (P1, P2). The text height is adjusted automatically (Fig. 18-7).

Fit
Fits (compresses or extends) the line of text between the two points specified (P1, P2). The text height does not change (Fig. 18-7).

Figure 18-7

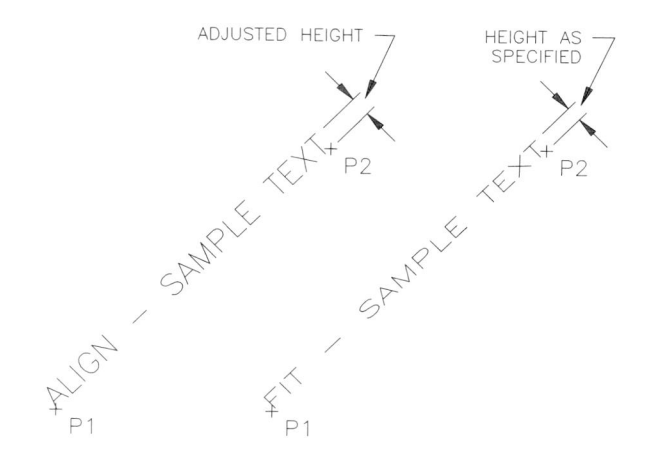

Center
Centers the baseline of the first line of text at the specified point. Additional lines of text are centered below the first (Fig. 18-8).

Middle
Centers the first line of text both vertically and horizontally about the specified point. Additional lines of text are centered below it (Fig. 18-8).

Right
Creates text that is right justified from the specified point (Fig. 18-8).

Figure 18-8

TL
Top Left. Places the text in the drawing so the top line (of the first line of text) is at the point specified and additional lines of text are left justified below the point. The top line is defined by the upper case and tall lower case letters (Fig. 18-9).

Figure 18-9

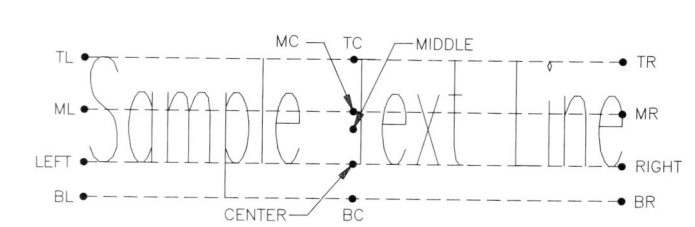

TC
Top Center. Places the text so the top line of text is at the point specified and the line(s) of text are centered below the point (Fig. 18-9).

TR
Top Right. Places the text so the top right corner of the text is at the point specified and additional lines of text are right justified below that point (Fig. 18-9).

ML
Middle Left. Places text so it is left justified and the middle line of the first line of text aligns with the point specified. The middle line is half way between the top line and the baseline, not considering the bottom (extender) line (Fig. 18-9).

MC
Middle Center. Centers the first line of text both vertically and horizontally about the midpoint of the middle line. Additional lines of text are centered below that point (Fig. 18-9).

MR
Middle Right. Justifies the first line of text at the right end of the middle line. Additional lines of text are right justified (Fig. 18-9).

BL
Bottom Left. Attaches the bottom (extender) line of the first line of text to the specified point. The bottom line is determined by the lowest point of lower case extended letters such as y, p, q, j, and g. If only upper-case letters are used, the letters appear to be located above the specified point. Additional lines of text are left justified (Fig. 18-9).

BC
Bottom Center. Centers the first line of text horizontally about the bottom (extender) line (Fig. 18-9).

BR
Bottom Right. Aligns the bottom (extender) line of the first line of text at the specified point. Additional lines of text are right justified (Fig. 18-9).

NOTE: Because there is a separate baseline and bottom (extender) line, the *MC* and *Middle* points do not coincide and the *BL*, *BC*, *BR* and *Left, Center, Right* options differ. Also, because of this feature, when all uppercase letters are used, they ride above the bottom line. This can be helpful for placing text in a table because selecting a horizontal *Line* object for text alignment with *BL*, *BC*, or *BR* options automatically spaces the text visibly above the *Line*.

Style   (option of *Dtext* or *Mtext*)
The *style* option of the *Dtext* or *Mtext* command allows you to select from the existing text styles that have been previously created as part of the current drawing. The style selected from the list becomes the current style and is used when placing text with *Dtext* or *Mtext*.

Since only one text *style, Standard,* is available in the traditional (English) template drawing (ACAD.DWT) and the metric template drawing (ACADISO.DWT), other styles must be created before the *style* option of *Dtext* is of any use. Various text styles are created with the *Style* command (this topic is discussed later).

Use the *Style* option of *Dtext* or *Mtext* to list existing styles for the drawing. An example listing is shown below:

```
Command: dtext
Current text style: "Standard" Text height: 0.2000
Specify start point of text or [Justify/Style]: S  (Style option)
Enter style name or [?] <STANDARD>: ?  (list option)
Enter text style(s) to list <*>: Enter
Text styles:

  Style name: "FUTURALS"   Font files: simplex.shx
    Height: 0.00  Width factor: 1.00  Obliquing angle: 0.000
    Generation: Normal

   Style name: "FUTURALS2"   Font files: simplex.shx
    Height: 0.00  Width factor: 1.00  Obliquing angle: 0.000
    Generation: Normal

   Style name: "ROMAND"    Font files: romand.shx
    Height: 0.00  Width factor: 1.00  Obliquing angle: 0.000
    Generation: Normal
```

Style name: "ROMANS"   Font files: romans
  Height: 1.50 Width factor: 1.00 Obliquing angle: 0.000
  Generation: Normal

Style name: "STANDARD"   Font files: txt
  Height: 0.00 Width factor: 1.00 Obliquing angle: 0.000
  Generation: Normal
Press ENTER to continue:

Current text style: "STANDARD"
Text height: 0.30
Specify start point of text or [Justify/Style]:

MTEXT

Pull-down Menu	COMMAND (TYPE)	ALIAS (TYPE)	Short-cut	Screen (side) Menu	Tablet Menu
Draw Text > Multiline Text...	MTEXT or -MTEXT	T, -T or MT	...	DRAW 2 Mtext	J,8

Multiline Text (*Mtext*) has more editing options than other text commands. You can apply underlining, color, bold, italic, font, and height changes to individual characters or words within a paragraph or multiple paragraphs of text.

Mtext allows you to create paragraph text defined by a text boundary. The <u>text boundary</u> is a reference rectangle that specifies the paragraph width. The *Mtext* object that you create can be a line, one paragraph, or several paragraphs. AutoCAD references *Mtext* (created with one use of the *Mtext* command) as one object, regardless of the amount of text supplied. Like *Dtext*, several justification methods are possible.

  Command: **mtext**
  Current text style: "Standard" Text height: 0.2000
  Specify first corner: **PICK**
  Specify opposite corner or [Height/Justify/Line spacing/Rotation/Style/Width]: **PICK** or (**option**)

You can PICK two corners to invoke the *Multiline Text Editor*, or enter the first letter of one of these options: *Height, Justify, Rotation, Style, Line Spacing*, or *Width*. <u>All of the options</u> can also be accessed within the *Multiline Text Editor*.

Using the default option the *Mtext* command, you supply a "first corner" and "opposite corner" to define the diagonal corners of the text boundary (like a window). Although this boundary confines the text on two or three sides, one or two arrows indicate the direction text flows if it "spills" out of the boundary (Fig. 18-10). (See "Text Flow and *Justification*.")

Figure 18-10

After you PICK the two points defining the text boundary, the *Multiline Text Editor* appears ready for you to enter the text (Fig. 18-11). Enter the desired text. The text wraps based on the width you defined for the text boundary. You can right-click for a menu allowing you to *Cut*, *Copy*, and *Paste* selected text. Select the *OK* button to have the text entered into the drawing within the text boundary.

Figure 18-11

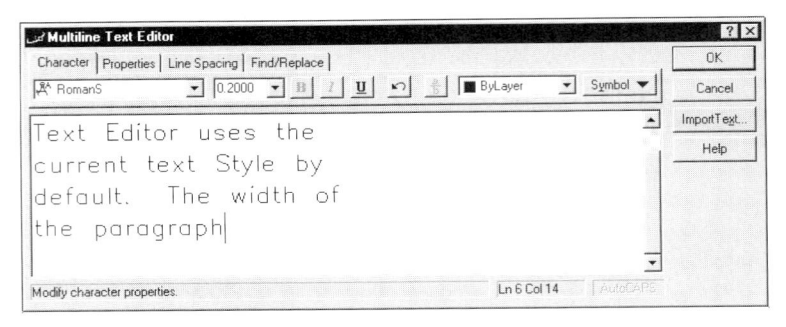

There are four tabs in the *Multiline Text Editor*: *Character* tab, *Properties* tab, *Line Spacing* tab, and *Find/Replace* tab. Using the options in these tabs is interactive—text in the editor immediately reflects the changes made for most options in these tabs. There is also a button to *Import Text*.

Import Text
Use this button to bring an external text file into the editor. See AutoCAD *Help* for more information.

Find/Replace **Tab**
See "Editing Text" later in this chapter.

Properties **Tab**

Use the *Properties* tab to specify the format of the entire paragraph. Although this is the second tab, it is recommended that you format the entire paragraph(s) here before editing individual characters using the *Character* tab (Fig. 18-12). The following options are available.

Figure 18-12

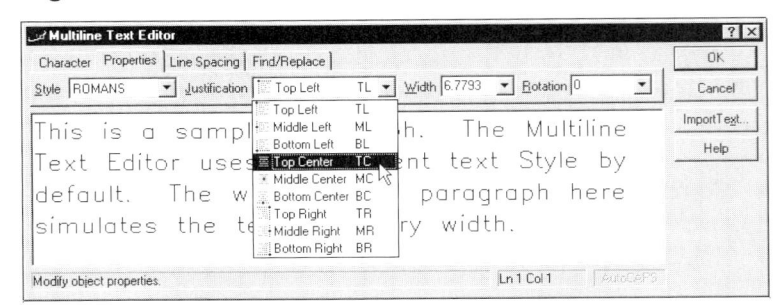

Style
Choose from a drop-down list of existing text styles (see "*Style*" command).

Justification
This property determines how the paragraph is located and direction of flow with respect to the text boundary (see "Text Flow and *Justification*").

Width
Previous paragraph widths used are displayed in this drop-down list. You can enter a new value in the edit box to change the width of the existing text boundary. If a *Width* of 0 is entered or "no wrap" is selected, the lines of text will not wrap within the text boundary.

Rotation
The entire paragraph can be rotated to any angle. Changes made here are not reflected in the text appearing in the editor but only in the drawing itself. (See "Text Flow and *Justification*.")

You can type *Mtprop* at the Command line to directly access the *Properties* tab of the *Multiline Text Editor*.

Character Tab

After formatting the entire paragraph, use the *Character* tab to alter individual characters in the paragraph(s) (Fig. 18-13). Using the options in this tab, first select (highlight) the desired characters or words, then set the desired options. The following options are available.

Figure 18-13

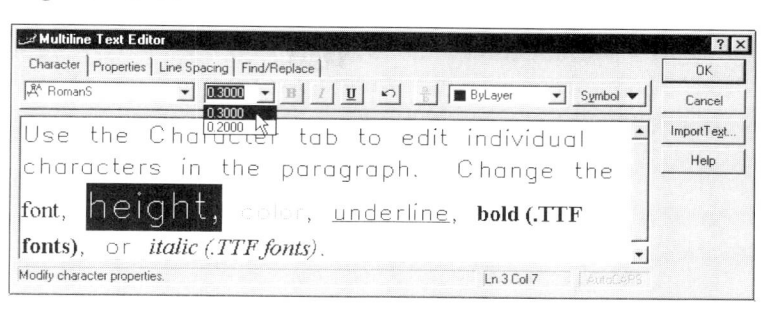

Font
Choose from any font in the drop-down list. Your selection here <u>overrides the text *Style*</u> used for the entire paragraph(s). Even though you can change the font for the entire *Mtext* object (paragraph), it is recommended to set the paragraph to the desired *Style* (in the *Properties* tab), rather than changing all characters to a different font here. See following NOTE.

Height
Select from the list or enter a new value for the height of selected words or letters. Your selection overrides the text *Height* used globally for the paragraph.

Bold, Italic, Underline
Select (highlight) the desired letters or words, then PICK the desired button. Only authentic TrueType fonts (not the AutoCAD-supplied .SHX equivalents) can be bolded or italicized.

Stack/Unstack
If creating a stacked fraction, use a slash (/) between the numerator and denominator. If creating stacked text, place a caret (^) before the bottom text. Highlight the fraction or text, then use this option to stack or unstack the fraction or text. (Also see "*AutoStack Properties.*")

Text Color
Select individual text, then use this drop-down list to select a color for the selected text. This selection overrides the layer color.

Symbol
Common symbols (plus/minus, diameter, degrees) can be inserted. Selecting *Other...* produces a character map to select symbols from (see "*Other Symbols*" below).

Other Symbols
The steps for inserting symbols from the *Unicode Character Map* dialog box (Fig. 18-14) are as follows:

1. Highlight the symbol.
2. Double-click or PICK *Select* so the item appears in the *Characters to copy:* edit box.
3. Select the *Copy* button to copy the item(s) to the Windows Clipboard.
4. *Close* the dialog box.
5. In the *Multiline Text Editor,* move the cursor to the desired location to insert the symbol.
6. Finally, right-click and select *Paste* from the menu.

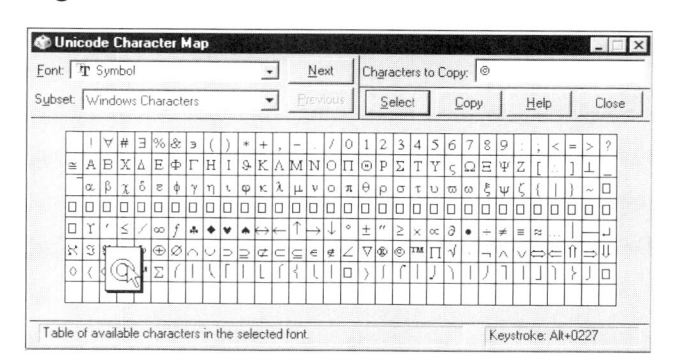

Figure 18-14

NOTE: The font, color, and height options in the *Character* tab override the properties of the entire paragraph. For example, changing the font in the *Character* tab overrides the text style's font used for the paragraph so it is possible to have one font used for the paragraph and others for individual characters within the paragraph. This is analogous to object-specific color and linetype assignment in that you can have a layer containing objects with different linetypes and colors than the linetype and color assigned to the layer. To avoid confusion, it is recommended to create text *Styles* to be used globally for the paragraphs, layer color to determine color for the paragraphs, and text *Height* for global paragraph formatting. Then if needed, use the *Character* tab to change fonts, colors, or height for <u>selected</u> text rather than for the entire paragraph.

Line Spacing Tab

The *Line Spacing* tab sets the spacing between lines for the paragraph (Fig. 18-15). The spacing increment is the vertical distance between the bottom (or baseline) of one line of text and the bottom of the next line of text. Unlike the *Character* tab, this tab is <u>retroactive</u>, meaning that you can enter the text first, then set the line spacing without highlighting the text. The options are described next.

Figure 18-15

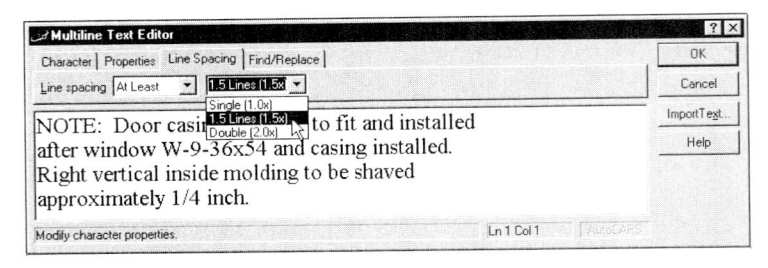

At Least/Exactly
If you have different size characters in the paragraph, select *At Least* (the default setting). This option automatically adds space between lines based on the height of the largest character in the line, so the spacing can vary depending on the size of the characters.

The *Exactly* option forces the line spacing to be the same for all lines of text in the *Mtext* object. Use this option to insert text into a table or to ensure that line spacing is identical in multiple *Mtext* objects. Using *Exactly* can cause text in lines above or below lines with large font characters to overlap the larger characters.

Single (1x), etc.
You can set the spacing increment to multiples of single-line spacing, or to an absolute value. Single spacing is 1.66 times the height of the text characters. Options include *Single (1x)*, *1.5 lines (1.5x)*, and *Double (2x)*. You can select an option from the list or enter any number followed by x to indicate a multiple of single spacing. You can also enter an absolute value, such as 1 for spacing of exactly 1.0 units, regardless of the text height.

Mtext Right-Click Shortcut Menu

A right-click shortcut menu is available in the *Multiline Text Editor* (Fig. 18-16). This menu offers some options not available by any other means. To enable all options in the menu, you must first select (highlight) characters in the paragraph or use the *Select all* option from the menu. The options are described here.

Figure 18-16

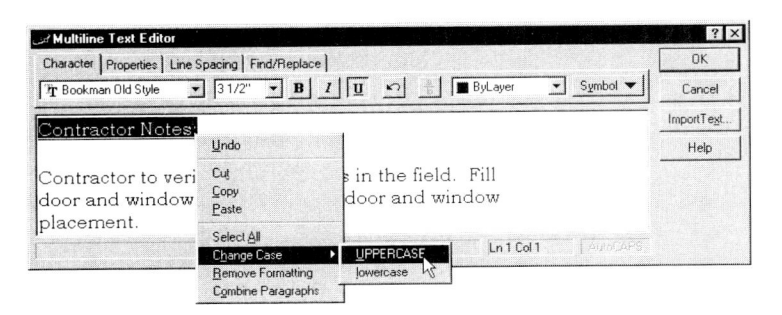

Undo
Use this option to undo the last formatting action performed.

Cut, Copy, Paste
These selections allow you to *Cut* (erase) highlighted text from the paragraph, *Copy* highlighted text, and *Paste* (the *Cut* or *Copied*) text to the current cursor position. Text that you *Cut* or *Copy* is held in the Clipboard for *Pasting*.

Select All
Use this to select all text in the *Multiline Text Editor*.

Change Case
These options allow you to change the case for the selected text to all upper or all lower case.

Remove Formatting
You can remove formatting to highlighted text that was applied in the *Character* tab. *Fonts*, *Height*, *Bold*, *Italic*, and *Underline* can be removed. Stacking, colors, and line spacing cannot be removed using this option.

Text Flow and *Justification*

When you PICK two corners to define the text boundary, you determine the width of the paragraph and the direction that the text flows. You can draw the text boundary in any direction from the "first corner" to the "other corner." Text does not always fit within the boundary but is confined on two or three sides and "spills" out of the boundary (up, down, or both) in the direction of the arrow(s) displayed when you draw the boundary (see Figure 18-10).

Justification is the method of aligning the text with respect to the text boundary. The *Justification* options are *TL*, *TC*, *TR*, *ML*, *MC*, *MR*, *BL*, *BC*, and *BR* (*Top Left*, *Top Centered*, *Top Right*, *Middle Left*, *Bottom Left*, etc.) The text paragraph (*Mtext* object) is effectively "attached" to the boundary based on the *Justification* option selected. *Justification* can be specified by two methods: in command line format before specifying the text boundary or in the *Properties* tab of the *Multiline Text Editor*.

The *Justificaiton* methods are illustrated in Figure 18-17. The illustration shows the relationship among the *Justification* option, the text boundary, the direction of flow, and the resulting text paragraph.

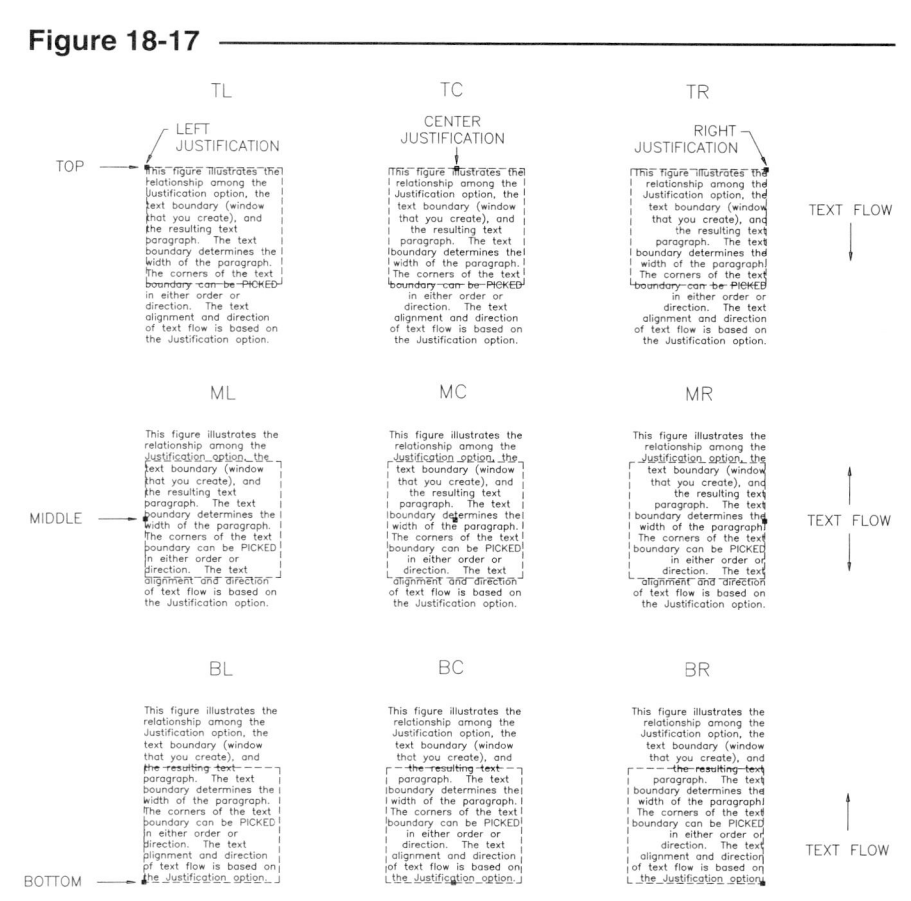

Figure 18-17

362 Creating and Editing Text

If you want to adjust the text boundary after creating the *Mtext* object, you can use *Ddedit* or *Properties* (see "Editing Text") or use Grips (Chapter 21) to stretch the text boundary. If you activate the Grips, four grips appear at the text boundary corners and one grip appears at the defined *Justification* point (Fig. 18-18).

Figure 18-18

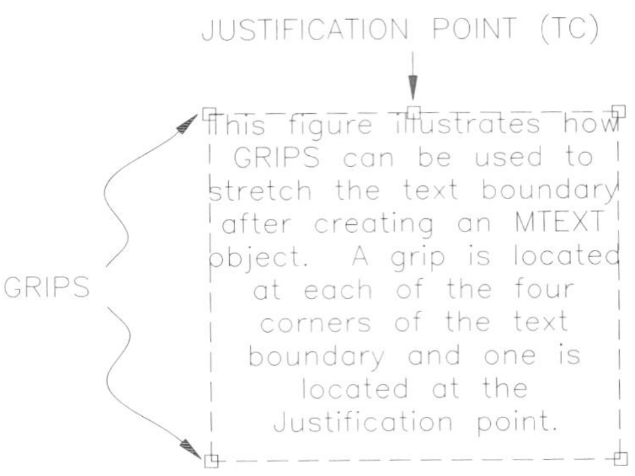

Rotation
The *Rotation* option specifies the rotation angle of the entire *Mtext* object (paragraph) including the text boundary. *Rotation* can be specified by two methods: in Command line format before specifying the text boundary or in the *Properties* tab of the *Multiline Text Editor*. It is recommended to use the Command line method to specify a rotation angle so you can see the rotated text boundary as you specify the corners (Fig. 18-19). Using this method you can enter a value, or the angle can be specified by PICKing two points. If you use the *Rotation* option within the *Properties* tab of the *Multiline Text Editor*, the text boundary is rotated after the fact.

Figure 18-19

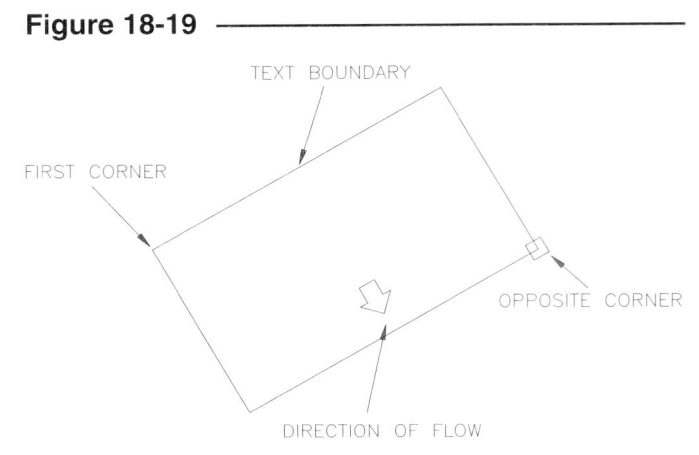

AutoStack Properties Dialog Box

Entering two or more numbers separated by a slash (/), carat (^), or pound (#) in the *Multiline Text Editor* automatically invokes the *AutoStack Properties* dialog box (Fig. 18-20). Stacking is the process of converting numbers to fractions. The options in the dialog box are explained here.

Figure 18-20

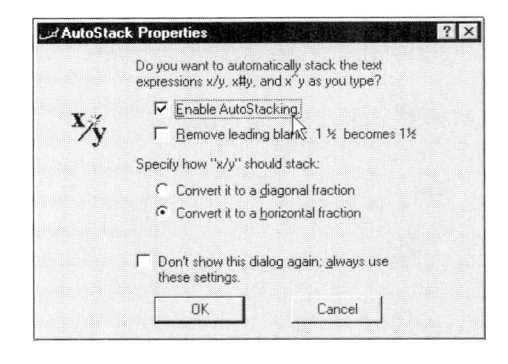

Enable AutoStacking
When this box is checked, fractions are automatically stacked as you type. Anytime you enter numeric characters before and after the carat, slash, or pound character, the numbers are automatically stacked. Each of these characters causes the following style of stacking:

slash (/) causes a vertical stack separated by a horizontal line
pound (#) causes a diagonal stack separated by a diagonal line
carat (^) causes a vertical stack without a horizontal separator line

Remove Leading Blank
This option removes blanks between a whole number and a fraction. For example, you would normally type one and one-half with a space (1 1/2) which would convert to a whole number plus a space before the fraction.

Convert it to a Diagonal Fraction and *Convert it to a Horizontal Fraction*
These options convert the slash character to a diagonal or horizontal fraction when AutoStack is on. Whether AutoStack is on or off, the pound character is always converted to a diagonal fraction, and the carat character is always converted to a tolerance format.

Don't Show This Dialog Again; Always Use These Settings
When this option is cleared, the AutoStack Properties dialog box is automatically displayed if you type two numbers separated by a slash, carat, or pound sign followed by a space or nonnumeric character.

Calculating Text Height for Scaled Drawings

To achieve a specific height in a drawing intended to be plotted to scale, multiply the desired text height for the plot by the drawing scale factor. (See Chapter 14, Printing and Plotting.) For example, if the drawing scale factor is 48 and the desired text height on the plotted drawing is 1/8", enter **6** (1/8 x 48) in response to the "Height:" prompt of the *Dtext* or *Mtext* command.

If you know the plot scale (for example, 1/4"= 1') but not the drawing scale factor (DSF), calculate the reciprocal of the plot scale to determine the DSF, then multiply the intended text height for the plot by the DSF, and enter the value in response to the "Height:" prompt. For example, if the plot scale is 1/4"=1', then the DSF = 48 (reciprocal of 1/48).

If *Limits* have already been set and you do not know the drawing scale factor, use the following steps to calculate a text height to enter in response to the "Height:" prompt to achieve a specific plotted text height.

1. Determine the sheet size to be used for plotting (for example, 36" x 24").
2. Decide on the text height for the finished plot (for example, .125").
3. Check the *Limits* of the current drawing (for example, 144' x 96' or 1728" x 1152").
4. Divide the *Limits* by the sheet size to determine the drawing scale factor (1728"/36" = 48).
5. Multiply the desired text height by the drawing scale factor (.125 x 48 = 6).

TEXT STYLES AND FONTS

STYLE

Pull-down Menu	COMMAND (TYPE)	ALIAS (TYPE)	Short-cut	Screen (side) Menu	Tablet Menu
Format Text Style...	STYLE or -STYLE	ST	...	FORMAT Style:	U,2

Text styles can be created by using the *Style* command. A text *Style* is created by selecting a font file as a foundation and then specifying several other parameters to define the configuration of the letters.

Using the *Style* command invokes the *Text Style* dialog box (Fig. 18-21). All options for creating and modifying text styles are accessible from this device. Selecting the font file is the initial step in creating a style. The font file selected then becomes a foundation for "designing" the new style based on your choices for the other parameters (*Effects*).

Figure 18-21

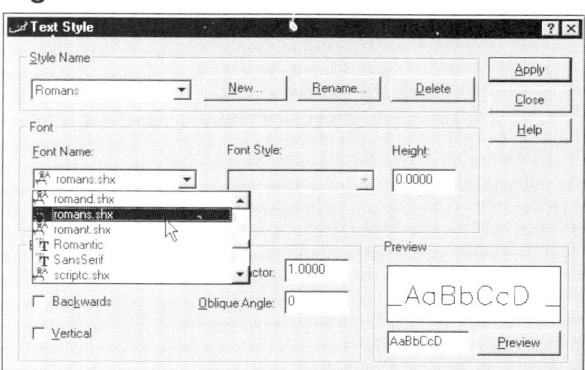

Recommended steps for creating text *Styles* are as follows:

1. Use the *New* button to create a new text *Style*. This button opens the *New Text Style* dialog box (Fig. 18-22). By default, AutoCAD automatically assigns the name Style*n*, where *n* is a number that starts at 1. Enter a descriptive name into the edit box. Style names can be up to 256 characters long and can contain letters, numbers, and the special characters dollar sign ($), underscore (_), and hyphen (-).
2. Select a font to use for the style from the *Font Name* drop-down list (see Figure 18-21). TrueType equivalents of the traditional AutoCAD fonts files (.SHX) and authentic TrueType fonts are available.
3. Select parameters in the *Effects* section of the dialog box (Fig. 18-23, lower left). Changes to the *Upside down*, *Backwards*, *Width Factor*, or *Oblique Angle* are displayed immediately in the *Preview* tile.
4. You can enter specific words or characters in the edit box in the lower-right corner, then press *Preview* to view the characters in the new style.
5. Select *Apply* to save the changes you made in the *Effects* section to the new style.
6. Create other new *Styles* using the same procedure listed in steps 1 through 5.
7. Select *Close*. The new (or last created) text style that is created automatically becomes the current style inserted when *Dtext* or *Mtext* is used.

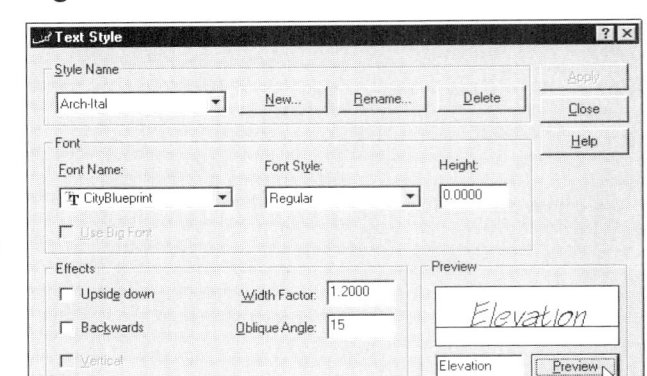

Figure 18-22

Figure 18-23

You can modify existing styles using this dialog box by selecting the existing *Style* from the list, making the changes, then selecting *Apply* to save the changes to the existing *Style*.

Rename
Select an existing *Style* from the drop-down list, then select *Rename*. Enter a new name in the edit box.

Delete
Select an existing *Style* from the drop-down list, then select *Delete*. You cannot delete the current *Style* or *Styles* that have been used for creating text in the drawing.

Alternately, you can enter *-Style* (use the hyphen prefix) at the command prompt to display the Command line format of *Style* as shown below:

```
Command: -style
Enter name of text style or [?] <Standard>: name (Enter new style name.)
New style.
Specify full font name or font filename (TTF or SHX) <txt>: name (Enter desired font file.)
Specify height of text <0.0000>: Enter or (value)
Specify width factor <1.0000>: Enter or (value)
Specify obliquing angle <0>: Enter or (value)
Display text backwards? [Yes/No] <N>: Enter or Y
Display text upside-down? [Yes/No] <N>: Enter or Y
Vertical? <N> Enter or Y
"(new style)" is now the current text style.
Command:
```

The options that appear in both the *Text Style* dialog box and in the Command line format are described in detail here.

Height <0.000>

The height should be 0.000 if you want to be prompted again for height each time the *Dtext* or *Mtext* command is used. In this way, the height is variable for the style each time you create text in the drawing. If you want the height to be constant, enter a value other than 0. Then, *Dtext* or *Mtext* will not prompt you for a height since it has already been specified. A specific height assignment with the *Style* command also overrides the *DIMTXT* setting (see Chapter 27).

Width factor <1.000>

A *width factor* of 1 keeps the characters proportioned normally. A value less than 1 compresses the width of the text (horizontal dimension) proportionally; a value of greater than 1 extends the text proportionally.

Obliquing angle <0>

An angle of 0 keeps the font file as vertical characters. Entering an angle of 15, for example, would slant the text forward from the existing position, or entering a negative angle would cause a back-slant on a vertically oriented font (Fig. 18-24).

Figure 18-24

0° OBLIQUING ANGLE

15° OBLIQUING ANGLE

−15° OBLIQUING ANGLE

Backwards? <Y/N>

Backwards characters can be helpful for special applications, such as those in the printing industry.

Figure 18-25

UPSIDE DOWN TEXT

TEXT ROTATED 180°

Upside-down? <Y/N>

Each letter is created upside-down in the order as typed (Fig. 18-25). This is different than entering a rotation angle of 180 in the *Dtext* command. (Turn this book 180 degrees to read the figure.)

Vertical? <Y/N>

Vertical letters are shown in Figure 18-26. The normal rotation angle for vertical text when using *Dtext* or *Mtext* is 270. Only .SHX fonts can be used for this option. *Vertical* text does not display in the *Preview* image tile.

Figure 18-26

V T
E E
R X
T T
I
C
A
L

TIP Since specification of a font file is an initial step in creating a style, it seems logical that different styles could be created using one font file but changing the other parameters. It is possible, and in many cases desirable, to do so. For example, a common practice is to create styles that reference the same font file but have different obliquing angles (often used for lettering on isometric planes). This can be done by using the *Style* command to assign the same font file for each style, but assign different parameters (obliquing angle, width factor, etc.) and a unique name to each style. The relationship between fonts, styles, and resulting text is shown in Figure 18-27.

Figure 18-27

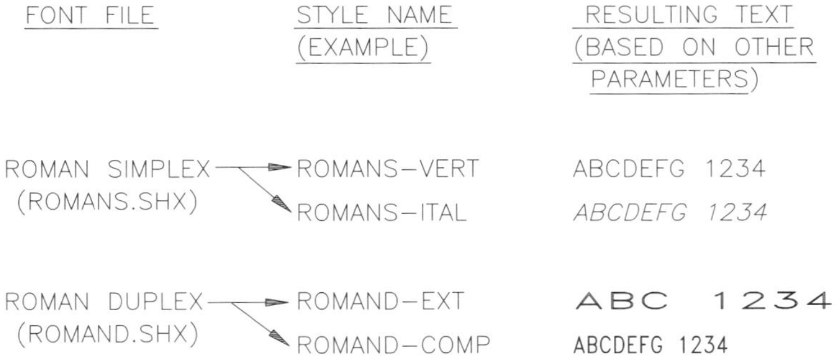

FONT FILE	STYLE NAME (EXAMPLE)	RESULTING TEXT (BASED ON OTHER PARAMETERS)
ROMAN SIMPLEX (ROMANS.SHX)	ROMANS−VERT	ABCDEFG 1234
	ROMANS−ITAL	*ABCDEFG 1234*
ROMAN DUPLEX (ROMAND.SHX)	ROMAND−EXT	A B C 1 2 3 4
	ROMAND−COMP	ABCDEFG 1234

EDITING TEXT

SPELL

Pull-down Menu	COMMAND (TYPE)	ALIAS (TYPE)	Short-cut	Screen (side) Menu	Tablet Menu
Tools Spelling	SPELL	SP	...	TOOLS 1 Spell	T,10

AutoCAD has an internal spell checker that can be used to spell check and correct existing text in a drawing after using *Dtext* or *Mtext*. In AutoCAD 2002, you can also select *Block* attributes and *Blocks* containing text to spell check. The *Check Spelling* dialog box (Fig. 18-28) has options to *Ignore* the current word or *Change* to the suggested word. The *Ignore All* and *Change All* options treat every occurrence of the highlighted word.

AutoCAD matches the words in the drawing to the words in the current dictionary. If the speller indicates that a word is misspelled but it is a proper name or an acronym you use often, it can be added to a custom dictionary. Choose *Add* if you want to leave a word unchanged but add it to the current custom dictionary.

Figure 18-28

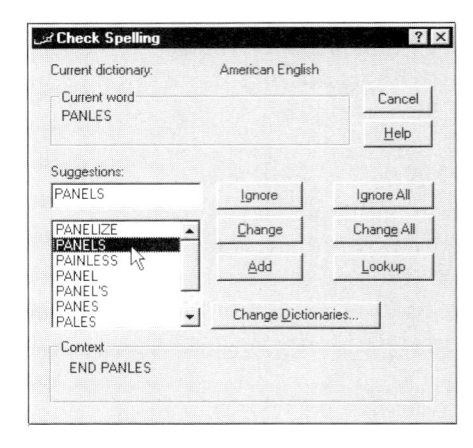

Selecting the *Change Dictionaries...* tile produces the *Change Dictionaries* dialog box (see Figure 18-29). You can select from other *Main Dictionaries* that are provided with your version of AutoCAD. The current main dictionary can also be changed in the *Files* tab of the *Options* dialog box, and the name is stored in the *DCTMAIN* system variable.

The AutoCAD default custom dictionary is SAMPLE.CUS (Fig. 18-29). If you use the *Add* function (in the *Check Spelling* dialog box), the selected word is added to the current custom dictionary. You can also create a custom dictionary "on the fly" by entering any name in the *Custom dictionary* edit box. A file extension of .CUS should be used with the name (although other extensions will work). A custom dictionary name must be specified before you can add words. Words can be added by entering the desired word in the *Custom dictionary words* edit box or by using the *Add* tile in the *Check Spelling* dialog box. Custom dictionaries can be changed during a spell check. The custom dictionary can also be changed in the *Files* tab of the *Options* dialog box, and its name is stored in the *DCTCUST* system variable.

Figure 18-29

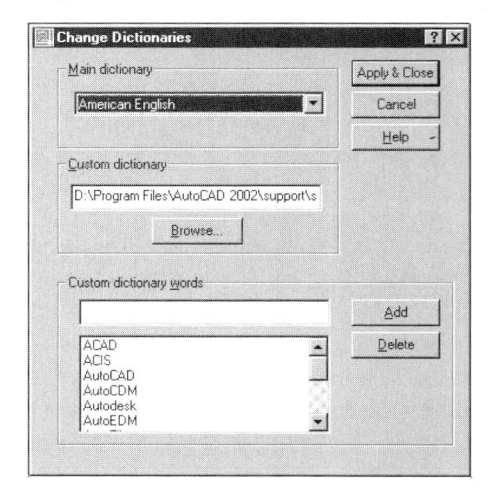

DDEDIT

Pull-down Menu	COMMAND (TYPE)	ALIAS (TYPE)	Short-cut	Screen (side) Menu	Tablet Menu
Modify Object > Text> Edit...	DDEDIT	ED		Modify1 Ddedit	Y,21

Ddedit invokes a dialog box for editing existing text in a drawing. You can edit <u>individual characters</u> or the entire line or paragraph. If the selected text was created by the *Dtext* command, the *Edit Text* dialog appears displaying one line of text (Fig. 18-30). If *Mtext* created the selected text, the *Multiline Text Editor* box appears (or current text editor).

Figure 18-30

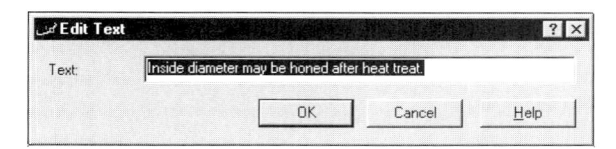

PROPERTIES

Pull-down Menu	COMMAND (TYPE)	ALIAS (TYPE)	Short-cut	Screen (side) Menu	Tablet Menu
Modify Properties...	PROPERTIES	PROPS or CH	(Default Menu) Find or Ctrl + 1	MODIFY1 Modify	Y,14

The *Properties* command invokes the *Properties* window. As described in Chapter 16, the window that appears is <u>specific</u> to the type of object that is PICKed—kind of a "smart" properties window. Remember that you can select objects and then invoke the window, or you can keep the *Properties* window on the screen and select objects in the drawing whose properties you want to change. (See Chapter 16 for more information on the *Properties* window.)

If a line of *Dtext* is selected, the window displays its properties. The *Properties* window allows you to change almost any properties associated with the text, including the *Contents* (text), *Style*, *Justification*, *Height*, *Rotation*, *Width*, and *Obliquing*. Simply locate the property in the left column and change the value in the right column.

If you select an *Mtext* paragraph, the *Properties* window appears allowing you to change the properties as for a *Dtext* object. However, if you select the *Contents* section, a small button on the right appears (Fig. 18-31). Pressing this button invokes the *Multiline Text Editor* with the *Mtext* contents and allows you to edit the text with the full capabilities of the editor.

Figure 18-31

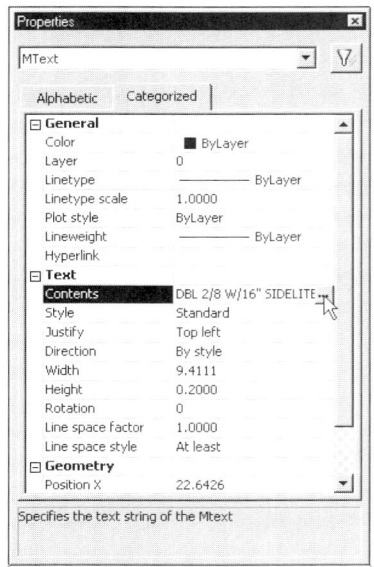

FIND

Pull-down Menu	COMMAND (TYPE)	ALIAS (TYPE)	Short-cut	Screen (side) Menu	Tablet Menu
Edit Find...	FIND	...	(Default menu) Find...	...	X,10

Find is used to find or replace text strings globally in a drawing. It can find and replace any text in a drawing, whether it is *Dtext* or *Mtext*. *Find* produces the *Find and Replace* dialog box (see Figure 18-32).

The *Find* command was introduced in AutoCAD 2000. The older *Find/Replace* tab in the *Multiline Text Editor* (introduced in AutoCAD Release 14) operates only for *Mtext* and only for one *Mtext* object at a time. The new *Find* command can search a drawing globally and operates with all types of text, including dimensions and *Block* attributes.

Figure 18-32

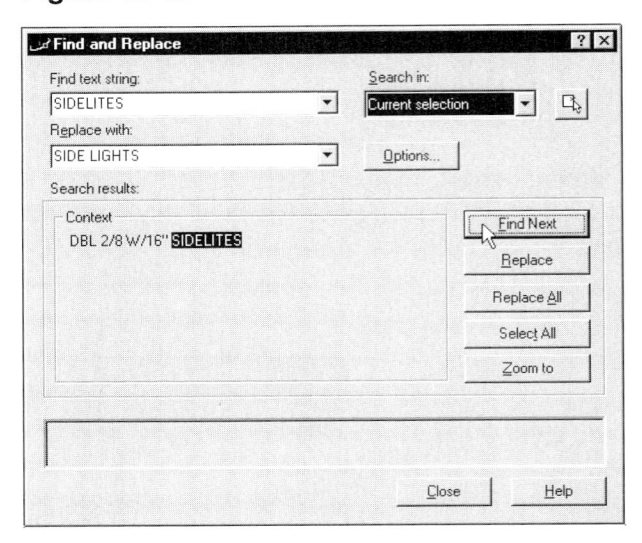

Find Text String, Find/Find Next
To find text only (not replace), enter the desired string in the *Find Text String* edit box and press *Find* (Fig. 18-32). When matching text is found, it appears in the context of the sentence or paragraph in the *Search Results* area. When one instance of the text string is located, the *Find* button changes to *Find Next*.

Replace With, Replace
You can find any text string in a drawing, or you can search for text and replace it with other text. If you want to search for a text string and replace it with another, enter the desired text strings in the *Find Text String* and *Replace With* edit boxes, then press the *Find/Find Next* button. You can verify and replace each instance of the text string found by alternately using *Find Next* and *Replace*, or you can select *Replace All* to globally replace all without verification. The status area confirms the replacements and indicates the number of replacements that were made.

Select Objects
The small button in the upper-right corner of the dialog box allows you to select objects with a pickbox or window to determine your selection set for the search. Press Enter when objects have been selected to return to the dialog box. Once a selection set has been specified, you can search the *Entire drawing* or limit the search to the *Current selection* by choosing these options in the *Search in*: drop-down list.

Select All
This options finds and selects (highlights) all objects in the current selection set containing instances of the text that you enter in *Find Text String*. This option is available only when you use *Select Objects* and set *Search In* to *Current Selection*. When you choose *Select All*, the dialog box closes and AutoCAD displays a message on the Command line indicating the number of objects that it found and selected. Note that *Select All* does not replace text; AutoCAD ignores any text in *Replace With*.

Zoom To
This feature is extremely helpful for finding text in a large drawing. Enter the desired text string to search for in the *Find Text String* edit box, select *Find/Find Next* button, then use *Zoom to*. AutoCAD automatically locates the desired text string and zooms in to display the text (Fig. 18-33). Although AutoCAD searches model space and all layouts defined for the drawing, you can zoom only to text in the current *Model* or *Layout* tab.

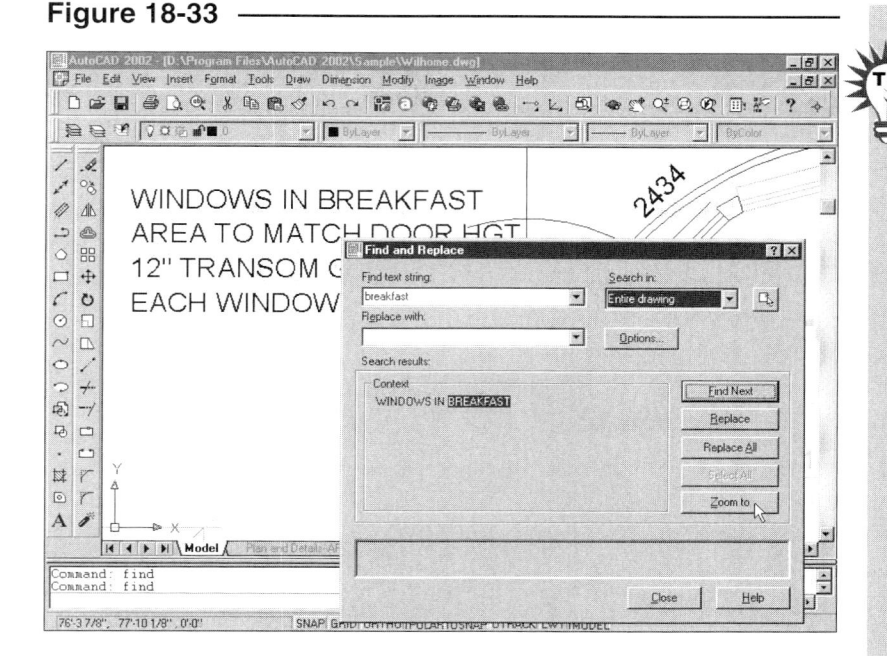

Figure 18-33

Options
Selecting the *Options* button in the *Find and Replace* dialog box produces the *Find and Replace Options* dialog box (Fig. 18-34). Note that the *Find* feature can locate and replace *Block Attribute Values*, *Dimension Annotation Text*, *Text (Mtext, Dtext, Text)*, *Hyperlink Description*, and *Hyperlinks*. Removing any check disables the search for that text type.

Figure 18-34

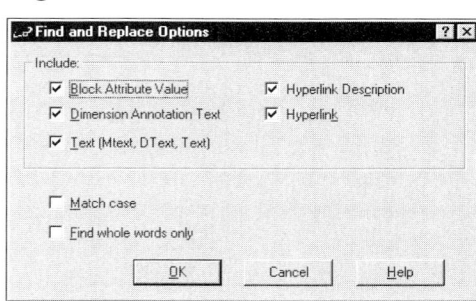

If it is important that the search exactly match the case (upper and lowercase letters) you entered in the *Find Text String* and *Replace With* edit boxes, check the *Match Case* box. If you want to search for complete words only, check the *Find whole words only* box. For example, without a check in either of these boxes, entering "door" in the *Find Text String* edit box would yield a find of "Doors" in the *Search Results* area. With a check in either *Match Case* or *Find whole words only*, a search for "door" would not find "Doors."

SCALETEXT

Pull-down Menu	COMMAND (TYPE)	ALIAS (TYPE)	Short-cut	Screen (side) Menu	Tablet Menu
Modify Object > Text > Scale	SCALETEXT	...	...	...	...

You cannot effectively scale multiple lines of text (*Dtext* objects) or multiple paragraphs of text (*Mtext* objects) using the normal *Scale* command since all text objects become scaled with relation to only one base point; therefore, individual lines of text lose their original insertion point.

In AutoCAD 2002 a new text editing command, *Scaletext*, allows you to scale multiple text objects using this one command, and each line or paragraph is scaled relative to its individual justification point.

370   Creating and Editing Text

For example, consider the three *Dtext* objects shown in Figure 18-35. Assume each line of text was created using the *Start point* option (left-justified) and with a height of .20. If you wanted to change the height of all three text objects to .25, the following sequence could be used.

Figure 18-35

SCALETEXT scales each text object individually. Each line is scaled relative to the justification point. These sample lines are created by DTEXT.

```
Command: scaletext
Select objects: PICK
Select objects: PICK
Select objects: PICK
Select objects: Enter
Enter a base point option for scaling
[Existing/Left/Center/Middle/Right/TL/TC/TR/ML/MC/MR/BL/BC/BR] <Existing>: Enter
Specify new height or [Match object/Scale factor] <0.2000>: .25
Command:
```

JUSTIFYTEXT

Pull-down Menu	COMMAND (TYPE)	ALIAS (TYPE)	Short-cut	Screen (side) Menu	Tablet Menu
Modify Object > Text > Justify	JUSTIFYTEXT	...	...	...	...

The *Justifytext* command is new in AutoCAD 2002. *Justifytext* changes the justification point of selected text objects without changing the text locations. Technically, this command relocates the insertion point (sometimes called attachment point) for the text object (*Mtext* objects, *Dtext* objects, leader text objects, and block attributes) and then justifies the text to the new insertion point.

Figure 18-36

These are two Dtext objects.
The lines are left−justified.

This is a paragraph of text created with the Mtext command. Justification method (and therefore, insertion point) is top left.

Consider the two *Dtext* objects and the single *Mtext* object shown in Figure 18-36. In each case, the default justification methods were used when creating the text, resulting in left-justified text and insertion points as shown by the small "blip" (the "blip" is not created as part of the text—it is shown here only for illustration).

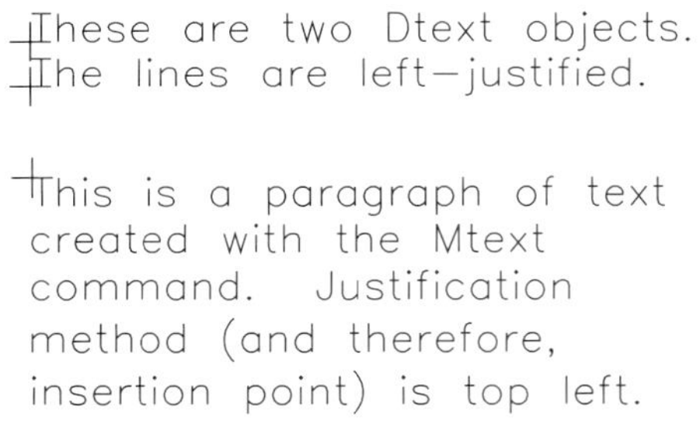

```
Command: justifytext
Select objects: PICK
Select objects: PICK
Select objects: PICK
Select objects: Enter
Enter a justification option
[Left/Align/Fit/Center/Middle/Right/TL/TC/TR/ML/MC/MR/BL/BC/BR] <TL>: br
Command:
```

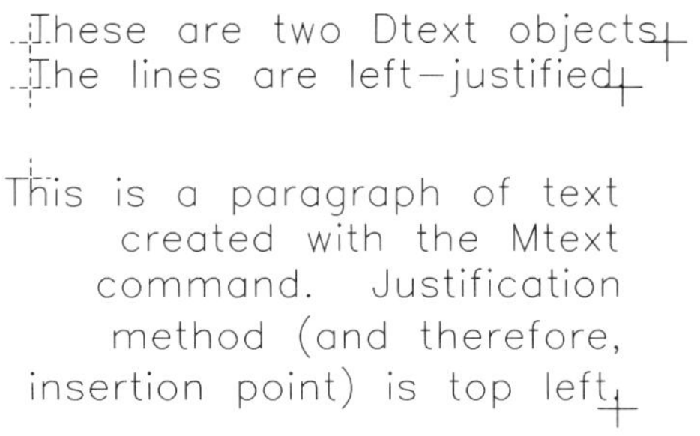

The resulting changes are shown in Figure 18-37. Compare these changes to the original text objects in Figure 18-36. Notice that the *Justifytext* command keeps the text in the same location, although the insertion points have been changed. Each text object has been justified in relation to its own new insertion point.

Figure 18-37

These are two Dtext objects.
The lines are left−justified.

This is a paragraph of text created with the Mtext command. Justification method (and therefore, insertion point) is top left.

QTEXT

Pull-down Menu	COMMAND (TYPE)	ALIAS (TYPE)	Short-cut	Screen (side) Menu	Tablet Menu
...	QTEXT	...	...	...	...

Qtext (quick text) allows you to display a line of text <u>as a box</u> in order to speed up drawing and plotting times. Because text objects are treated as graphical elements, a drawing with much text can be relatively slower to regenerate and take considerably more time plotting than the same drawing with little or no text.

When *Qtext* is turned *ON* and the drawing is regenerated, each text line is displayed as a rectangular box (Fig. 18-38). Each box displayed represents one line of text and is approximately equal in size to the associated line of text.

Figure 18-38

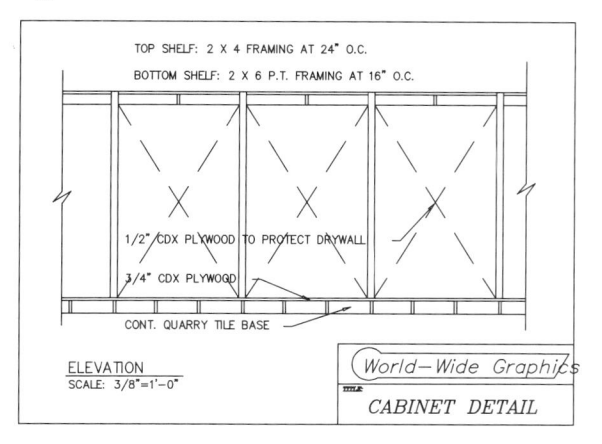

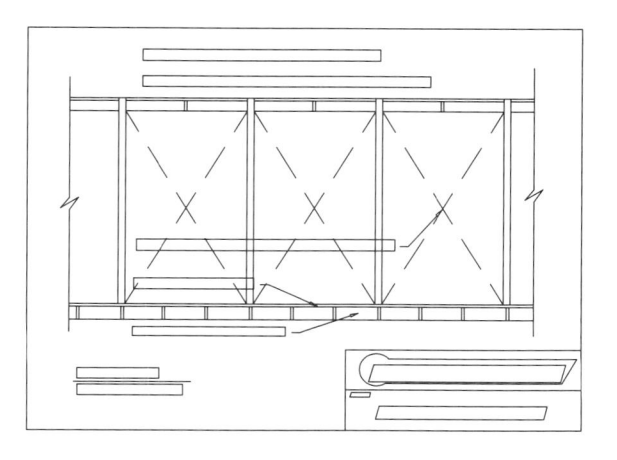

QTEXT OFF   QTEXT ON

For drawings with considerable amounts of text, *Qtext ON* noticeably reduces regeneration time. For check plots (plots made during the drawing or design process used for checking progress), the drawing can be plotted with *Qtext ON*, requiring considerably less plotting time. *Qtext* is then turned *OFF* and the drawing must be *Regenerated* to make the final plot.

When *Qtext* is turned *ON*, the text remains in a readable state until a *Regen* is invoked or caused. When *Qtext* is turned *OFF*, the drawing must be regenerated to read the text again.

TEXTFILL

Pull-down Menu	COMMAND (TYPE)	ALIAS (TYPE)	Short-cut	Screen (side) Menu	Tablet Menu
...	TEXTFILL	...	...	...	...

The *TEXTFILL* variable controls the display of TrueType fonts for <u>printing and plotting only</u>. *TEXTFILL* does not control the display of fonts for the screen—fonts always appear filled in the Drawing Editor. If *TEXTFILL* is set to 1 (on), these fonts print and plot with solid-filled characters (Fig. 18-39). If *TEXTFILL* is set to 0 (off), the fonts print and plot as outlined text (Fig. 18-40). The variable controls text display globally and retroactively.

Figure 18-39

TEXTFILL ON

Figure 18-40

TEXTFILL OFF

CHAPTER EXERCISES

1. *Dtext*

   Open the **EFF-APT-PS** drawing. Make *Layer* **TEXT** *Current* and use *Dtext* to label the three rooms: **KITCHEN, LIVING ROOM,** and **BATH**. Use the *Standard style* and the *Start point* justification option. When prompted for the *Height:*, enter a value to yield letters of 3/16" on a 1/4"=1' plot (3/16 x the drawing scale factor = text height). *Save* the drawing.

2. *Style, Dtext, Ddedit*

   Open the drawing of the temperature graph you created as **TEMP-D**. Use *Style* to create two styles named **ROMANS** and **ROMANC** based on the *romans.shx* and *romanc.shx* font files (accept all defaults). Use *Dtext* with the *Center* justification option to label the days of the week and the temperatures (and degree symbols) with the **ROMANS** style as shown in Figure 18-41. Use a *Height* of **1.6**. Label the axes as shown using the **ROMANC** style. Use *Ddedit* for editing any mistakes. Use *SaveAs* and name the drawing **TEMP-GRPH**.

   Figure 18-41

3. *Style, Dtext*

   Open the **BILLMATL** drawing created in the Chapter 15 Exercises. Use *Style* to create a new style using the *romans.shx* font. Use whatever justification methods you need to align the text information (not the titles) as shown in Figure 18-42. Next, type the *Style* command to create a new style that you name as **ROMANS-ITAL**. Use the *romans.shx* font file and specify a **15** degree *obliquing angle*. Use this style for the **NO., PART NAME,** and **MATERIAL**. *SaveAs* **BILLMAT2**.

   Figure 18-42

NO.	PART NAME	MATERIAL
1	Base	Cast Iron
2	Centering Screw	N/A
3	Slide Bracket	Mild Steel
4	Swivel Plate	Mild Steel
5	Top Plate	Cast Iron

4. *Edit Text*

   Open the **EFF-APT-PS** drawing. Create a new style named **ARCH1** using the *CityBlueprint* (.TTF) font file. Next, invoke the *Properties* command. Use this dialog box to modify the text style of each of the existing room names to the new style as shown in Figure 18-43. *SaveAs* **EFF-APT2**.

   Figure 18-43

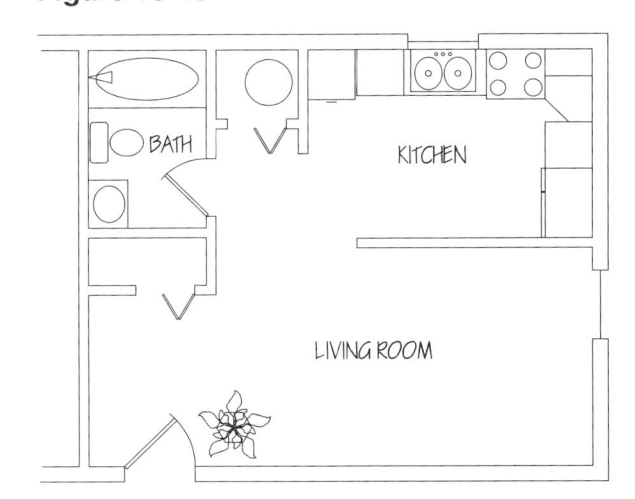

5. *Dtext, Mtext*

   Open the **CBRACKET** drawing from Chapter 9 Exercises. Using *romans.shx* font, use *Dtext* to place the part name and METRIC annotation (Fig. 18-44). Use a *Height* of **5** and **4**, respectively, and the *Center Justification* option. For the notes, use *Mtext* to create the boundary as shown. Use the default *Justify* method (*TL*) and a *Height* of **3**. Use *Ddedit* or *Properties* if necessary. *SaveAs* **CBRACTXT**.

   Figure 18-44

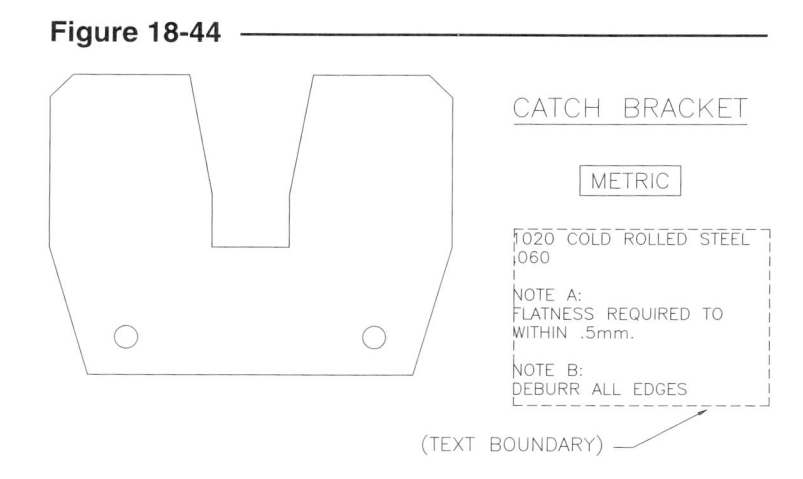

6. *Style*

   Create two new styles for each of your template drawings: **ASHEET**, **BSHEET**, and **C-D-SHEET**. Use the *romans.shx* style with the default options for engineering applications or *CityBlueprint* (.TTF) for architectural applications. Next, design a style of your choosing to use for larger text as in title blocks or large notes.

7. *Create a Title Block*

   A. Begin a *New* drawing and assign the name **TBLOCK**. Create the title block as shown in Figure 18-45 or design your own, allowing space for eight text entries. The dimensions are set for an A size sheet. Draw on *Layer* **0**. Use a *Pline* with .02 *width* for the boundary and *Lines* for the interior divisions. (No *Lines* are needed on the right side and bottom because the title block will fit against the border lines.)

   Figure 18-45

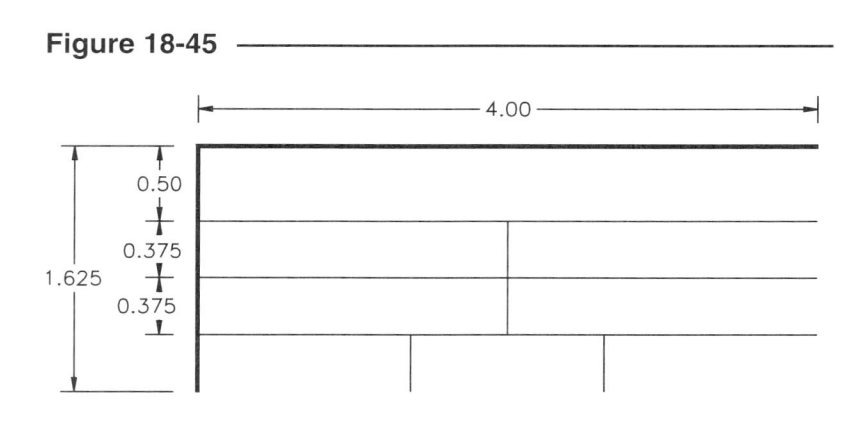

   B. Create two text *Styles* using *romans.shx* and *romanc.shx* font files. Insert text similar to that shown in Figure 18-46. Examples of the fields to create are:

   Company or School Name
   Part Name or Project Title
   Scale
   Designer Name
   Checker or Instructor Name
   Completion Date
   Check or Grade Date
   Project or Part Number

   Figure 18-46

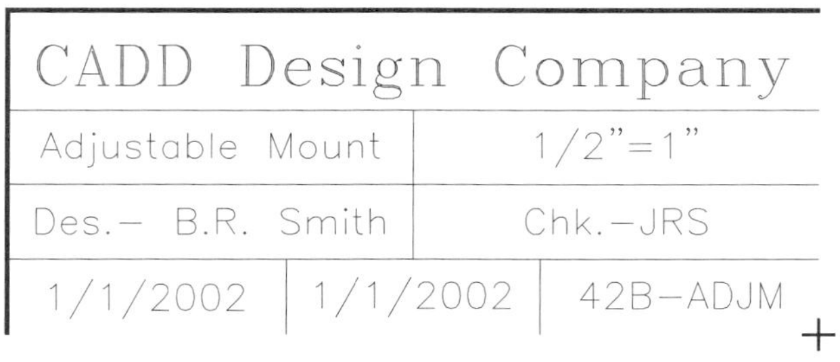

   Choose text items relevant to your school or office. *Save* the drawing.

8. *Mtext*

   Open the **STORE ROOM** drawing that you created in Chapter 16 Exercises. Use *Mtext* to create text paragraphs giving specifications as shown in Figure 18-47. Format the text below as shown in the figure. Use *CityBlueprint* (.TTF) as the base font file and specify a base *Height* of **3.5**". Use a *Color* of your choice and *CountryBlueprint* (.TTF) font to emphasize the first line of the Room paragraph. All paragraphs use the *TC Justify* methods except the Contractor Notes paragraph, which is *TL*. Save the drawing as **STORE ROOM2**.

   Room: STORAGE ROOM
       11'-2" x 10'-2"
       Cedar Lined - 2 Walls

   Doors: 2 - 2268 DOORS
       Fire Type A
       Andermax

   Windows: 2 - 2640 CASEMENT WINDOWS
       Triple Pane Argon Filled
       Andermax

   Notes: Contractor Notes:

       Contractor to verify all dimensions in field. Fill door and window roughouts after door and window placement.

Figure 18-47

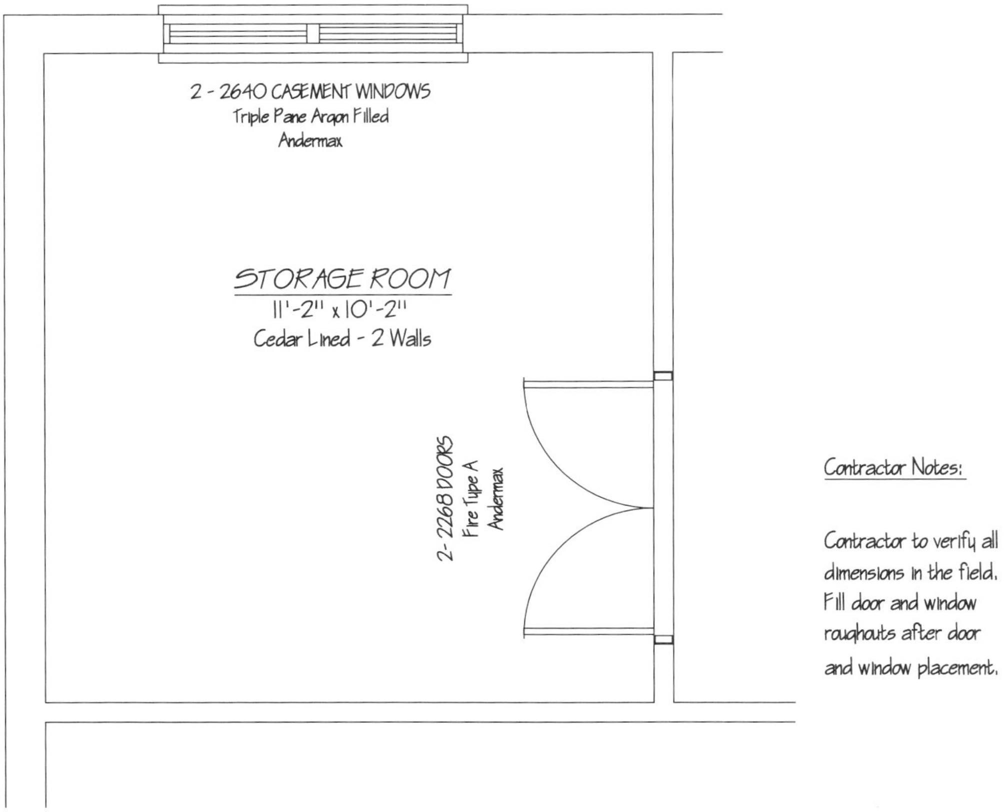

19

INTERNET TOOLS

Chapter Objectives

After completing this chapter you should:

1. be able to use the *Today* window to access *My Drawings*, the *Bulletin Board*, and *Autodesk PointA*;

2. know how to use the *Autodesk PointA* Web site to increase awareness and productivity;

3. be able to use the *Browser* command to launch your Web browser from within AutoCAD;

4. be able to use *Hyperlink* to link your drawings to other drawings, text documents, or spreadsheets on your computer, your network, or the Internet;

5. be able to use *Publishtoweb* to create Web pages displaying AutoCAD drawings;

6. know how to attach drawings to your email messages using *Etransmit*.

CONCEPTS

In recent years Autodesk has created features to allow AutoCAD users to fully utilize the latest Internet-related technological advances to enhance their productivity. During the summer of 2000 Autodesk introduced AutoCAD 2000i, an Internet-enabled release, providing multiple tools to collaborate and share your designs with colleagues and clients using Internet technologies.

AutoCAD 2002 includes all of the Internet-related features of AutoCAD 2000i and previous releases. These features allow you to be more connected to Web technology by enabling capabilities such as teleconferencing and sharing AutoCAD drawings; publishing Web pages including drawings; emailing drawings; linking other remotely located drawings and documents to your drawing; creating drawings that are viewable in standard Web browsers; and connecting to the Autodesk-provided Web sites to increase awareness and productivity.

The commands explained in this chapter are listed below.

Today The *Today* window gives you access to *My Drawings*, the *Bulletin Board*, and the *PointA* Web site.

PointA Autodesk's *PointA* Web site provides an enormous amount of resources for your awareness and productivity.

Browser Launch your Web browser from within AutoCAD with the *Browser* command so you can have instant access to the Internet while in your drawing environment.

Hyperlink Use the *Hyperlink* command to create pointers in your AutoCAD drawings that provide jumps to other files you want to associate to the current drawing. The other files can be located on your local computer or network or can link to Web pages on the Internet. In this way your AutoCAD drawings can "contain" additional graphical, numerical, or textual information.

Publishtoweb With this powerful tool, you can generate Web pages complete with imbedded AutoCAD drawing images in .JPG, .PNG, and .DWF format. This tool is automatic so no experience with .HTML is required.

Etransmit *Etransmit* opens your email system and allows you to include an AutoCAD drawing (and any associated files) as an email attachment.

Additional exciting Internet-related features that are included in AutoCAD 2002 are described briefly below. These capabilities are beyond the scope of an introductory AutoCAD course and are therefore not included in this text. See AutoCAD 2002 *Help* for more information.

MeetNow This unique feature allows you to connect with other AutoCAD users and collaborate over the Internet. You can view another user's AutoCAD drawing screen and even take control of the remote AutoCAD session. You can also communicate using a chat room, sound, and video capabilities.

ePlot/eView You can create a Drawing Web Format (.DWF) file of the current drawing using an *ePlot* or *eView* device from the *Plot* dialog box. A .DWF file is a compact, vector-based file that is viewable without AutoCAD and easily transportable over the Internet. You can view .DWF files using Volo View, Volo View Express, or Microsoft Internet Explorer.

Chapter Nineteen 377

TODAY

Pull-down Menu	COMMAND (TYPE)	ALIAS (TYPE)	Short-cut	Screen (side) Menu	Tablet Menu
Tools Today	TODAY	...	...	...	...

One of the most visible new features of AutoCAD 2002 is the *Today* window (Fig. 19-1). When you open AutoCAD, this window appears by default instead of the traditional *Startup* dialog box. You can also produce the *Today* window any time in a drawing session by using the *Today* command. The three sections of the *Today* window are described briefly here and explained in detail in the following pages.

Figure 19-1

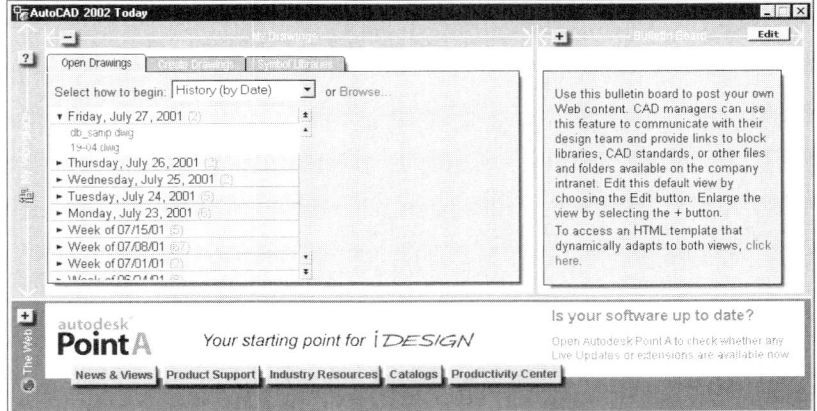

My Drawings
This tool allows you to *Open* existing drawings and create *New* drawings (see Chapter 2). It also provides access to symbol libraries through DesignCenter (see Chapter 20).

Bulletin Board
The *Bulletin Board* displays HTML documents and .GIF or .JPG images (as local files or from Web pages) that can be used as reminders, instructions, news, calendars, libraries, links, or other information.

Autodesk Point A
This section connects to the Autodesk PointA Web site by automatically launching your default Web browser. Autodesk PointA is an invaluable resource of information for any AutoCAD user.

The *Today* window operates as a separate application in Windows. Therefore, it runs in the background during your AutoCAD session. Use the task bar or Alt+Tab key sequence to open the window again. Closing the window closes the application completely.

If you prefer, you can suppress the *Today* window that appears when you open AutoCAD and specify the traditional *Startup* dialog box or no dialog box at all using the *System* tab of the *Options* dialog box (Fig. 19-2). If you suppress this feature for startup, it is still available during a drawing session using the *Today* command.

Figure 19-2

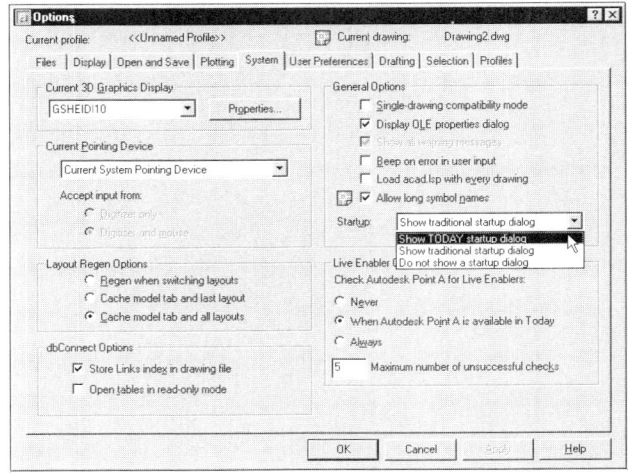

My Drawings, Open Drawings **Tab**
This section provides an automated *Open* command that allows you to select drawings to open using any of several sort options. The preview area (see Figure 19-3, on the right) displays a thumbnail sketch of the file highlighted by the pointer. Sort options are *Most Recently Used*, *History (by Date)* (shown in Figure 19-3), *History (by Filename)* (alphabetical listing), and *History (by Location)* (alphabetical by drive letter and path). Alternately, you can select *Browse…*, which produces the *Select File* dialog box.

Figure 19-3

My Drawings, Create Drawings **Tab**
This tab offers the *Template*, *Start from Scratch*, and *Wizards* methods of creating new drawings (Fig. 19-4). These functions operate similarly to the *Startup* dialog box (see Chapter 2).

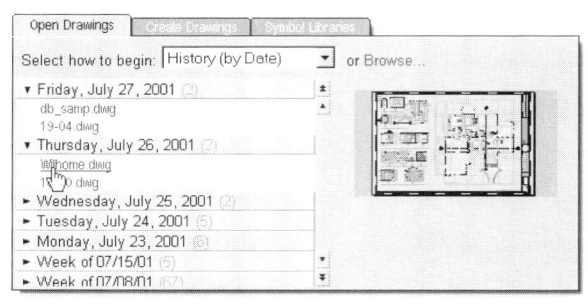

Figure 19-4

My Drawings, Symbol Libraries **Tab**
The *Symbols Libraries* tab (Fig. 19-5) provides a link to AutoCAD-provided and user-provided symbol libraries. Keep in mind this feature is useful anytime during an AutoCAD session and can be produced using the *Today* command.

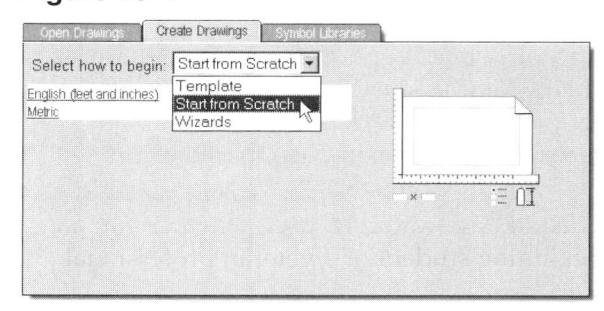

Figure 19-5

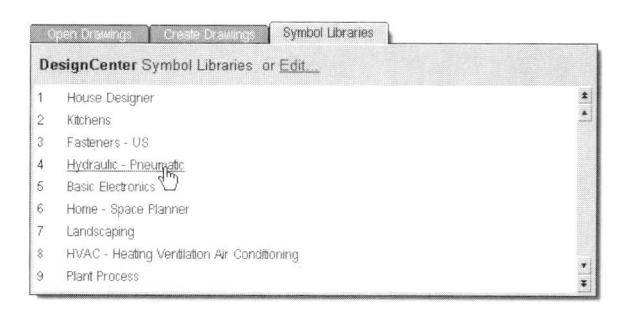

Selecting any of the listed libraries launches DesignCenter and displays the selected symbols ready to drag and drop into your AutoCAD session (Fig. 19-6). The symbols are inserted into the current drawing as *Blocks*. (For information on Blocks and DesignCenter, see Chapter 20.)

Figure 19-6

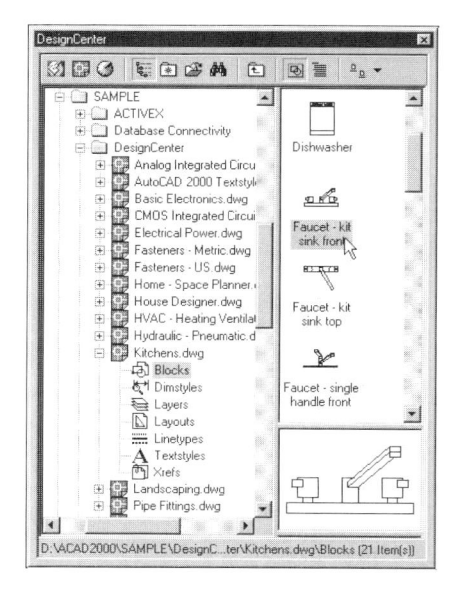

Bulletin Board

The *Bulletin Board* is intended to serve as a source of communication at your school or office (Fig. 19-7). Generally, your institution's teacher or CAD manager controls the information displayed in the *Bulletin Board*. The *Bulletin Board* can display any specified .HTM file, .GIF file, .JPG file, or an entire Web page. In other words, you can create an HTML document as you would a Web page and display it in the *Bulletin Board* section of the *Today* window. Therefore, the *Bulletin Board* can be used to display news, a calendar, daily instructions, reminders, or links to Web sites. Smaller documents typically display better in the limited space provided.

Figure 19-7

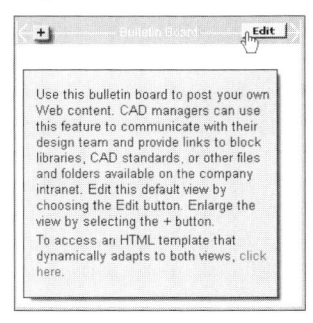

Autodesk PointA

Figure 19-8

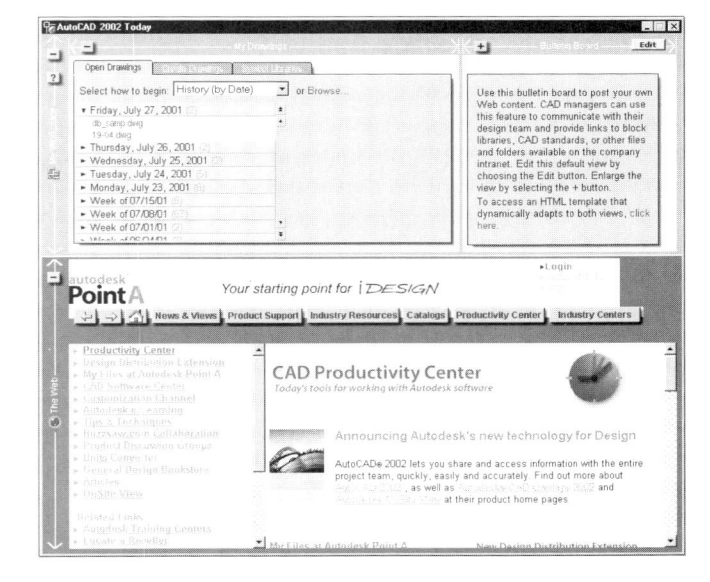

Autodesk PointA (Fig. 19-8) is a Web site operated by Autodesk that provides AutoCAD users with an incredible source of AutoCAD and industry-specific related services and information. Exploring this Web site reveals many benefits for any AutoCAD user, whether you are a beginning student or a veteran professional.

Assuming you have a live Internet connection, click on this section of the *Today* window to connect to the Autodesk PointA site. Once you connect, this section of the *Today* window automatically links to PointA whenever the *Today* window is opened.

Autodesk Point A

Pull-down Menu	COMMAND (TYPE)	ALIAS (TYPE)	Short-cut	Screen (side) Menu	Tablet Menu
Tools Autodesk Point A	...	...	...	...	...

Selecting the *PointA* button or the *Autodesk Point A* option from the *Tools* menu launches your default browser and displays the Autodesk PointA Web site (there is no **PointA** command you can enter at the Command prompt). Either action is essentially the same as accessing PointA from the *Today* window with two exceptions. Otherwise, all sections and options on the Web site are identical.

Using the *PointA* button or the menu differs from the *Today* window in two ways. First, this method launches your Web browser as a separate window so you have more viewing area available on the screen without the other options of the *Today* window (*My Drawings* and *Bulletin Board*) to occupy your viewing area. Second, access to PointA by this method requires you to join and log in but provides a customizable interface for the display of sections and services of your choice.

BROWSER

Pull-down Menu	COMMAND (TYPE)	ALIAS (TYPE)	Short-cut	Screen (side) Menu	Tablet Menu
...	BROWSER	...	...	...	...

AutoCAD 2002 includes this command specifically to launch your Web browser from within AutoCAD. An icon button on the *Web* toolbar (Fig. 19-9) and in the file navigation dialog boxes is available to invoke the *Browser* command.

Figure 19-9

The information about your browser (location and name of the executable file) is stored in the system registry. Therefore, when you use *Browser*, AutoCAD automatically locates and launches your current (default) Web browser from within AutoCAD. This action makes it possible to view any Web site or select any viewable files located on your computer. If you have Volo View, you can view .DWF files after you have created them, view other .DWF files on the Internet, and drag-and-drop related .DWG files from the Internet directly into your AutoCAD session.

USING HYPERLINKS

The *Hyperlink* command allows you to attach a link to any AutoCAD graphical object. The link is actually a file name or Web address (URL) associated with the object. Therefore, *Hyperlink* can be used to link to files on the local computer or network (without Internet access) or to connect to files over the Internet, depending on what information is entered as the link.

Once a hyperlink is created, you can right-click on the object to display a shortcut menu allowing you to *Open* the link. The link automatically opens the related program (word processor, spreadsheet, or CAD program) and displays the file, or it opens your Web browser and displays the page or image. When the cursor passes over an AutoCAD object that has a hyperlink attached, the *Hyperlink* symbol and the related link information are displayed (Fig. 19-10).

Figure 19-10

A *Hyperlink* can be used to create links within two types of AutoCAD files: (1) to create links in the current AutoCAD drawing (.DWG) and (2) to create links for a .DWF image that is created from the current drawing but is to be viewed from a Web browser. To create links to be used in .DWF images, use the *Hyperlink* command as you would normally to attach links to objects in the current drawing, then plot the drawing using the *DWF ePlot* or *DWF eView* device to create the related .DWF file.

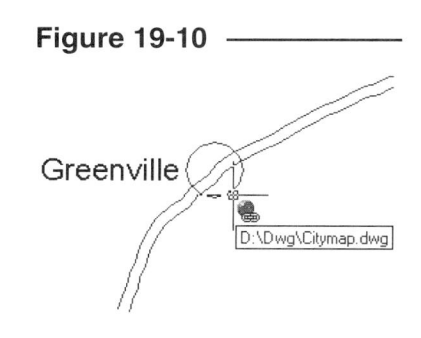

Keep in mind the power and potential usefulness of hyperlinks in AutoCAD drawings. Any hyperlink can call another file. For example, a drawing of a mechanical assembly could contain hyperlinks for each component of the assembly such that passing the cursor over each component and selecting *Open* from the right-click menu could open the related component drawing in AutoCAD. Or, as another example, opening another hyperlink for the entire assembly could in turn open a bill of materials in a word processing or spreadsheet program. The same capability hyperlinks add to an AutoCAD drawing can be contained in a .DWF image of the drawing viewed in a Web browser if those hyperlinks call other images or HTML files normally viewable in a browser.

HYPERLINK

Pull-down Menu	COMMAND (TYPE)	ALIAS (TYPE)	Short-cut	Screen (side) Menu	Tablet Menu
Insert Hyperlink...	HYPERLINK	...	Ctrl+K	...	...

When the *Hyperlink* command is invoked, you must first select the object you want to attach the link to. If you select graphical objects that do not already contain hyperlinks, AutoCAD displays the *Insert Hyperlink* dialog box (Fig. 19-11). If the graphical objects already contain hyperlinks, the *Edit Hyperlink* dialog box is displayed, which has the same options as the *Insert Hyperlink* dialog box.

Figure 19-11

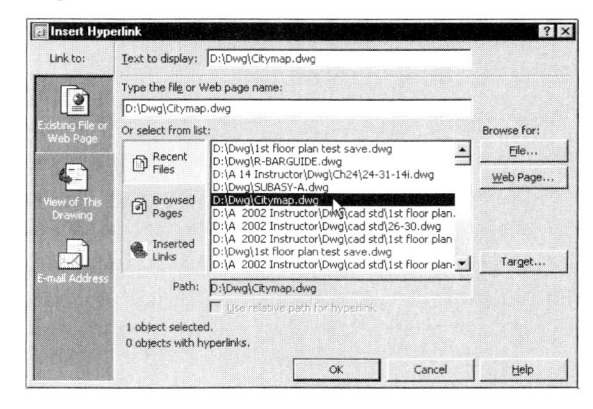

```
Command: hyperlink
Select objects: PICK
Select objects: Enter
```
(The *Insert Hyperlink* or *Edit Hyperlink* dialog box appears.)
Command:

When the *Insert Hyperlink* dialog box appears, first select (from the buttons on the left side of the dialog box) what kind of link you want to attach to the selected object—an *Existing File or Web Page*, a view of the current drawing (by selecting *View of this Drawing*), or an *Email Address*.

Existing File or Web Page
Selecting the *Existing File or Web Page* button allows you to type the information or select from *Recent Files*, *Browsed Pages*, or *Inserted Links*. In each case, files, pages, and URLs you recently used are displayed. There are also buttons to select another *File...* or *Web Page...* that is not displayed in the list. Selecting the *Target* button opens the *Select Place in Document* dialog box where you can select a named *View* in a drawing to link to (not the current drawing).

The text you enter in the *Text to display:* edit box is the text that appears in the drawing when the cursor passes over the object containing the hyperlink.

The *Use Relative Path for Hyperlink* checkbox toggles the use of a relative path for the current drawing. If this option is selected, the full path to the linked file is not stored in the drawing with the hyperlink.

View of This Drawing
If you want a specific view or layout of the current drawing to be associated to the selected object as a hyperlink, you can use this button to select from a list of *Views* that are contained in the current drawing (Fig. 19-12). In other words, this feature allows you to select an object in the current drawing and attach a hyperlink that causes a jump to a different layout or saved view of the drawing. AutoCAD displays that portion of the drawing (view or layout) when the hyperlink is opened.

Figure 19-12

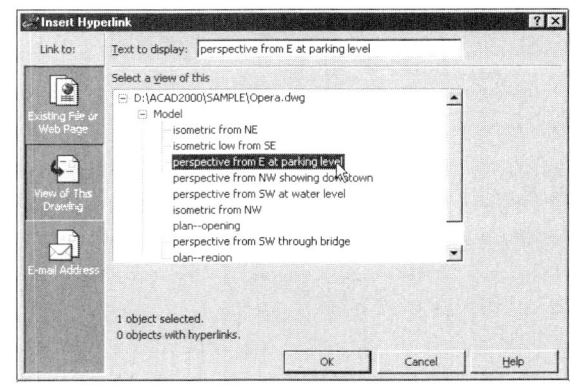

Email Address

This feature allows you to attach a hyperlink to an object so that when this hyperlink is opened, it causes the computer's default email system to open ready to type a new message. The email address contained in the link (appearing in the *E-mail address:* edit box of the *Insert Hyperlink* dialog box, Figure 19-13) is automatically entered in the "Send to:" box of your new email message and the subject text in the link (contained in the *Subject:* edit box) is automatically entered in the "Subject:" box of your email message. You simply type the message and select *Send* in your email program. Use the *Text to display:* edit box to supply the text that appears in the drawing when the cursor passes over the object containing the hyperlink.

Figure 19-13

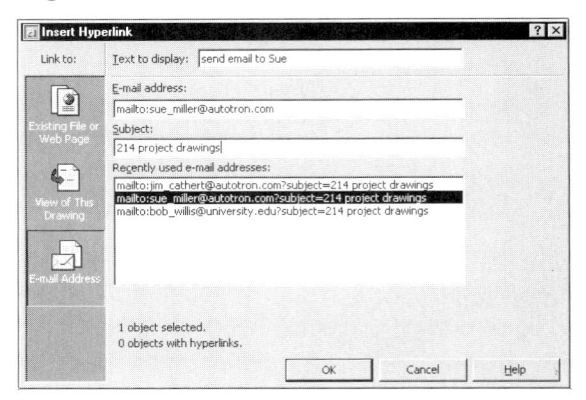

The *Remove Link* button appears only in the *Edit Hyperlink* dialog box. You can use this option to remove a previously attached hyperlink from the selected object.

SELECTURL

Pull-down Menu	COMMAND (TYPE)	ALIAS (TYPE)	Short-cut	Screen (side) Menu	Tablet Menu
...	SELECTURL	...	...	...	...

Use *Selecturl* to "select," or highlight, all objects in the drawing that have hyperlinks (URLs) attached. The highlighting action essentially enables grips on the objects. Press Escape to cancel the grips.

PUBLISHTOWEB

Pull-down Menu	COMMAND (TYPE)	ALIAS (TYPE)	Short-cut	Screen (side) Menu	Tablet Menu
Files Publish to Web	PUBLISHTOWEB	PTW	...	...	...

The Publish To Web feature in AutoCAD 2002 is a surprisingly simple and quick method of generating a Web page (HTML document) complete with embedded AutoCAD drawing images (.JPG, .PNG, or .DWF files). You need no previous experience creating Web pages or HTML code—AutoCAD does it all for you automatically! All you need to begin are the AutoCAD drawings you want to make images of and, if you are ready to post the page to the Web, a location on a server to store the files.

An example Web page created by PublishToWeb is shown in Figure 19-14 on the next page. The general structure of the page is determined by a template you select. You supply the page title and subtitle as well as the AutoCAD drawings to use. Each image and its accompanying description are hyperlinks to display a full-screen version of the image.

The *PublishToWeb* command produces a wizard that leads you through the process. The sequence is almost self-explanatory; however, here is some additional help you will need.

NOTE: You can publish only saved drawings. You can publish an open drawing if it has been previously saved. Because of the confusing AutoCAD alert that may appear, it is suggested that you *Save* the current drawing (even if it is a *New* drawing) before you use *PublishToWeb*.

Figure 19-14

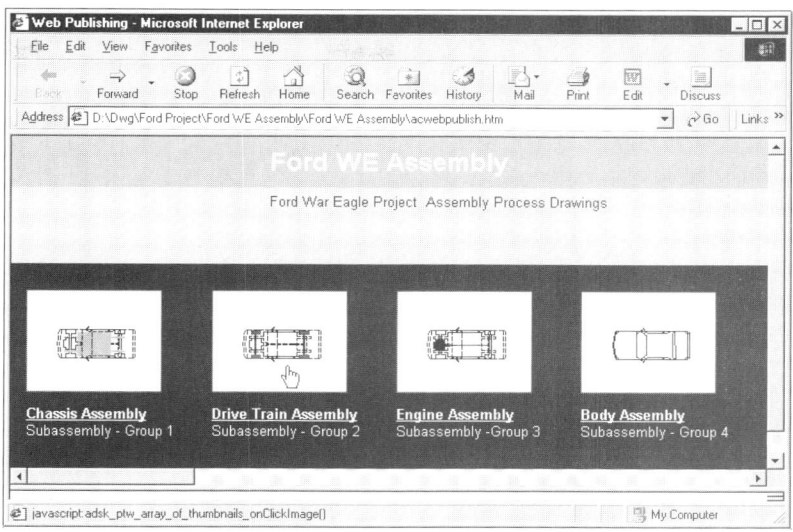

Begin
The first step in the *Publish to Web* wizard (not shown) is simply a choice of *Create a New Web Page* or *Edit Existing Web Page*. If this is your first run, select *Create Web Page*, then *Next*. If a confusing AutoCAD alert appears, select *Cancel* and see the NOTE above.

Create Web Page
In this step (Fig. 19-15) you supply the *name of your Web* page (in the first box), the *parent directory* where a folder will be created to store the files (use the browse button, right center), and a *description* (in the bottom box). The *name* appears on top of the final Web page (see previous Figure 19-14). The *name* you enter also determines the name of the folder (under the parent directory you select) that AutoCAD creates to store the files. The *description* appears on the Web page under the title (see previous Figure 19-14).

Figure 19-15

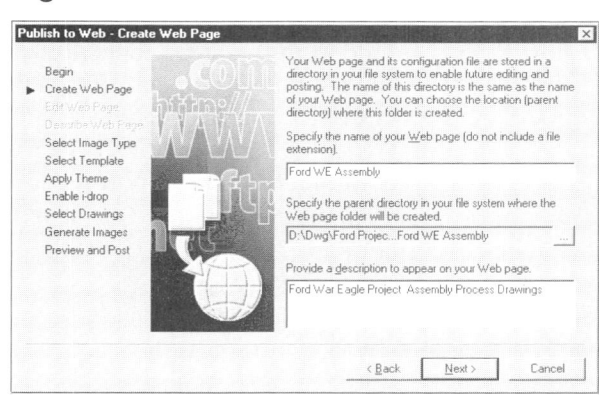

Select Image Type
Here you can select from a *DWF*, *JPEG*, or *PNG* format for the images that will be created (Fig. 19-16). A brief explanation for each image type is given <u>after</u> you make the selection. For *JPEG* and *PNG* images, you can select from *Small*, *Medium*, *Large*, or *Extra Large* image sizes.

Figure 19-16

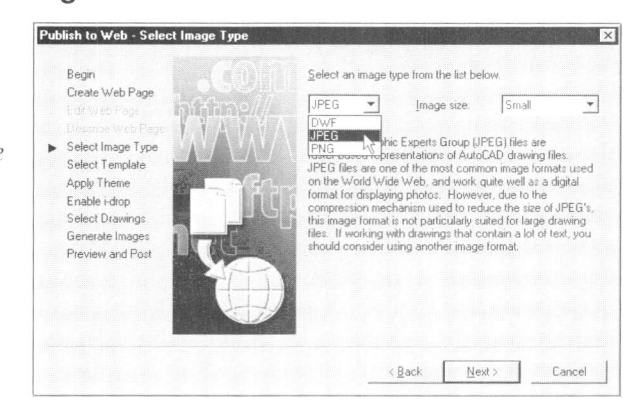

Select Template

Four basic templates are available to choose from (Fig. 19-17). The image tile on the right gives a simplified display of what your resulting Web page will look like. The two *Array* options display an array of images on the page so the viewer can click on any image to see a larger view of the drawing. The two *List* options allow the viewer to select an entry from the list to view a large version of the drawing in an image frame. Summary information (if selected) is the text that appears for each drawing in its *Summary* tab of the *Drawing Properties* dialog box. You can assign a description for a drawing using the *DWGPROPS* command.

Figure 19-17

Apply Theme

AutoCAD provides several color schemes for you to choose for your Web page (Fig. 19-18). The colors are applied to different sections of the Web page, such as the title, description, and summary.

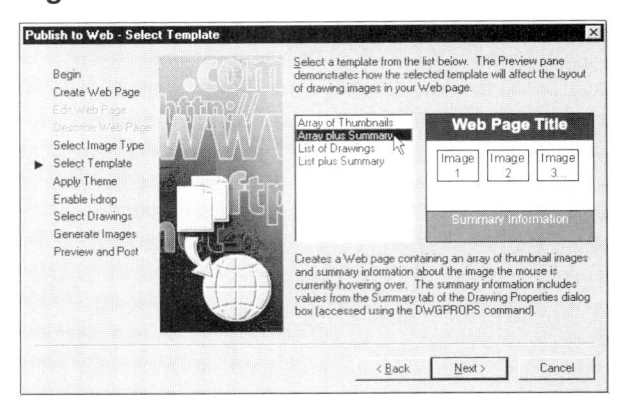

Figure 19-18

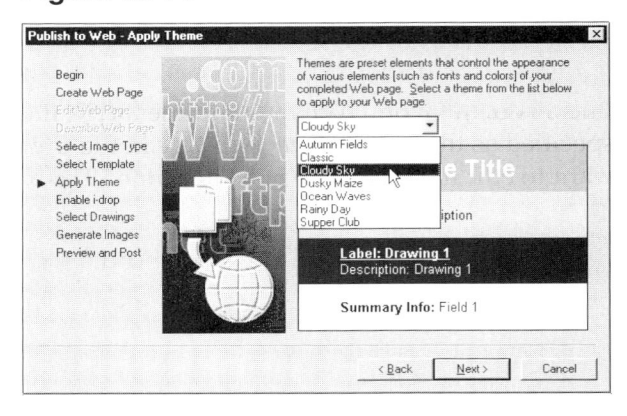

Enable i-Drop

The *Enable i-Drop* page (not shown) simply allows you to check whether you want to use this feature or not. If you enable i-Drop, copies of your drawing files are automatically included in the folder on the computer or server that contains your Web page. This allows viewers of the page to drag and drop the drawing files over the Internet directly into their AutoCAD session. This option is recommended only if you want to provide the page viewers access to your drawings. - This option removes the paths (stored drive and directory location information) from each attached Xref or other file contained in the transmittal set (useful when *Preserve Directory Structure* is not checked).

Select Drawings

In this step (Fig. 19-19), you select the images that you want to use for the page and enter the text you want to appear under each image. First, select an AutoCAD drawing you want to use for an image (use the browse button). Next, select the *Model* (tab) or a *Layout* in the drawing you want to display. The text you enter in the *Label* box appears immediately under the image on the resulting Web page (see Figure 19-14, underlined text) and is also used as the label displayed in the *Image list*, so you must enter a descriptive label if you use multiple images on the page. The *Description* will appear on the Web page under the *Label* text. Select the *Add->* button to add all those entries and selections that appear on the left side of the wizard to the *Image list*. To make changes, highlight any item from the *Image list* on the right, make the changes on the left, then select *Update->*. The *Remove* button removes any highlighted selections from the *Image list*. The images appear on the page in the same order as the *Image list*. Use *Move Up* or *Move Down* to change the order of a selected item in the list. Select *Next* to proceed.

Figure 19-19

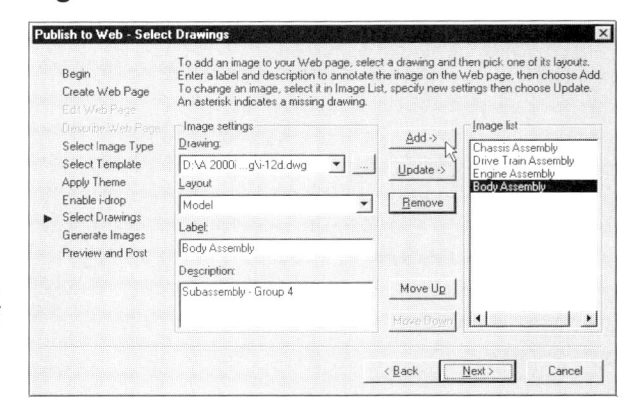

Generate Images

Selecting *Next* in this dialog box (not shown) causes AutoCAD to create the .PNG, .JPG, or .DWF images and generate the HTML document to display the images. No other input is required since the default option (*Regenerate images for drawings that have changed*) causes all new and all changed images to generate. Select *Regenerate all images* if you edit an existing Web page and want a duplicate set of images to be created in a second location.

Preview and Post

This last step (not shown) allows you to *Preview* the newly created Web page or to *Post Now*. The *Preview* option launches your Web browser with the page exactly as it will appear and operate when an outside viewer sees it. *Close* the browser to return to the *Publish to Web* wizard.

You can select *Back* at any step in the *Publish to Web* wizard, allowing you to change any aspect of the Web page that you specified in earlier steps. Use this feature to make corrections or changes before you post the Web page.

If you select *Post Now*, you are prompted for a location to store the files. Select the location on your Web server as specified by your system administrator.

The .PNG, .JPG, or .DWF files and the .HTM file that are generated by AutoCAD reside in a folder you specified as the *name of your Web page* in the *Create Web Page* step. A .PTW file is used by AutoCAD if you want to edit the Web page at a later time. This folder is located in the parent directory you also specified in that step. You can view the page and images at any time using your browser by selecting *Open* from the *File* menu, then using the *Browse* button to locate the .HTM file. If you select *Post Now* in the last step of the wizard, AutoCAD copies the files (not the .PTW file) to the location that you specify. The original files are not deleted.

ETRANSMIT

Pull-down Menu	COMMAND (TYPE)	ALIAS (TYPE)	Short-cut	Screen (side) Menu	Tablet Menu
File eTransmit	ETRANSMIT	...	...	...	...

eTransmit is a new feature that greatly simplifies the process of sending AutoCAD drawings by email. The process is straightforward: *Open* the drawing you want to send, use *eTransmit*, enter any notes and select your preferences in the *Create Transmittal* dialog box. AutoCAD then creates a compressed .EXE or .ZIP file of the drawing package and opens your email software with the transmittal file attached. Just write your email note and send it!

eTransmit operates on the current drawing. First *Open* the drawing you want to transmit. If the drawing contains Xrefs, fonts, plot style tables, or compiled shapes, you can specify each item you want to include in the transmittal. The *eTransmit* command produces the *Create Transmittal* dialog box (Fig. 19-20).

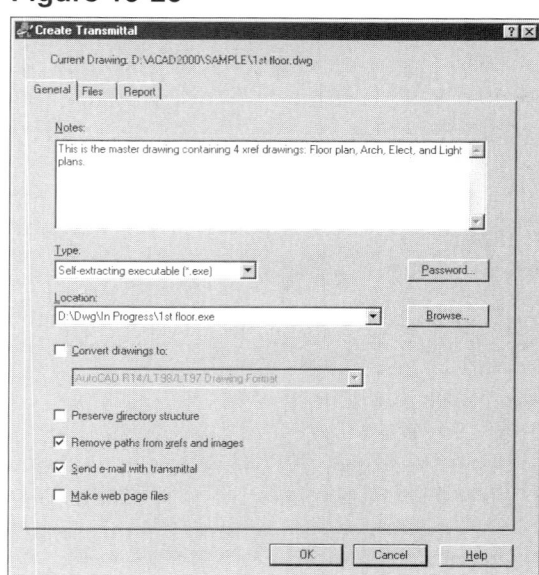

Figure 19-20

General Tab

Notes:
Enter any notes you want to include with the transmittal report (see *Report* tab). Alternately, you can specify a template of default notes to be included with all your transmittal sets by creating an ASCII text file called ETRANSMIT.TXT. Then, specify the location of the ETRANSMIT.TXT file in the *Support File Search Path* option on the *Files* tab of the *Options* dialog box.

Type
Here you can select from *Folder (set of files)*, *Self-extracting executable (*.exe)*, or *Zip (*.zip)*. The *Folder* option creates a new folder or uses an existing folder and creates the uncompressed transmittal files. The *Self-extracting executable (*.exe)* option creates one self-executable file that the recipient can double-click to decompress and restore the original files. The *Zip (*.zip)* option creates a transmittal set of files as one compressed .ZIP file.

Password
Here you can set a password that is required to decompress the files. Make sure you notify the recipient of the password that is required to decompress the files.

Location
Use this section to specify the location on your computer or network where the transmittal set is to be created. You can use the *Browse...* button to specify a new location.

Convert Drawings To
If you want to convert the drawing(s) to an earlier version of AutoCAD or AutoCAD LT, check this box and select from the drop-down list.

Preserve Directory Structure
Checking this box causes AutoCAD to preserve the directory structure of all files in the transmittal set during decompression and installation on another system. This feature may be helpful to the recipient if the set contains Xrefs, fonts, or shapes, etc. If this option is cleared, all files are installed to the target directory when the transmittal set is installed.

Remove Paths from Xrefs and Images
This option removes the paths (stored drive and directory location information) from each attached Xref or other file contained in the transmittal set (useful when *Preserve Directory Structure* is not checked).

Figure 19-21

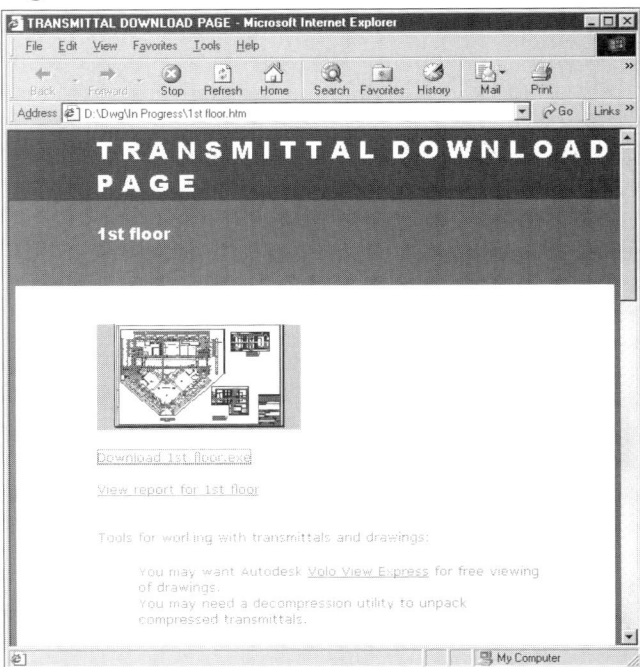

Send E-mail with Transmittal
Check this box to launch your default email program when the transmittal set is created. When the program opens, AutoCAD automatically attaches the transmittal set to the email message and enters the drawing set title in the "Subject:" line of your note.

Make Web Page Files
This feature is especially powerful. Checking this box causes AutoCAD to generate a Web page (.HTM and .BMP files) that includes a bitmap image and a link to download the transmittal set (Fig. 19-21).

Files **Tab**
The *Files* tab lists the files to be included in the transmittal set. When you first access the tab, all files associated with the current drawing (such as related Xrefs, plot styles, and fonts) are listed. You can remove files from the transmittal set or use the *Add File* button to add additional files to the set.

List View
The left-most button above the list toggles the display of files to list view. In this format, files are listed in alphabetical order by default but can be sorted by *Filename, Type, Size,* or *Date* in normal or reverse order by clicking on the column heading.

Tree View
This arrangement displays a hierarchical listing—especially helpful for determining which Xrefs, fonts, and plot styles are related to which drawing files (see Figure 19-22).

Add File
Use this button to open the standard file selection dialog box where you can select additional files to include in the transmittal set.

Report **Tab**
This tab displays a report that is automatically generated and included with the transmittal set. The generated report is a .TXT file that has the same name as the current drawing. You can add additional notes in the report by entering the information in the *Notes* section of the *General* tab. The information automatically generated by AutoCAD explains what steps must be taken by the recipient for the transmittal set to work properly. For example, if AutoCAD detects .SHX fonts in one of the transmittal drawings, the report instructs the recipient where to copy these files so AutoCAD can detect the files when the transmittal drawing is opened.

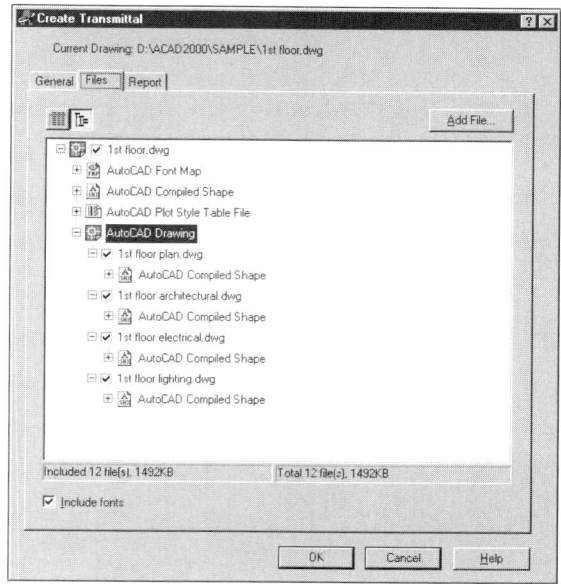

Figure 19-22

CHAPTER EXERCISES

1. ***Publish to Web***

    A. Start AutoCAD if not already running. *Save* the current drawing. (It is recommended to always save the current drawing before using *Publishtoweb*.)

    B. Invoke *Publishtoweb*. Follow the necessary steps in the wizard to create a new Web page named **Project Drawings** and locate it in your working directory. Write an appropriate *description* when prompted. Specify **JPEG** image type and **Small** image size. For the template, specify **Array plus Summary**. Select a *Theme* of your choice. In the *Select Drawings* step, specify four of your favorite drawings that you completed in the previous Chapter Exercises. Proceed to generate the images and *Preview* them. If the Web page needs improving, go *Back* and make the changes, or use *Publishtoweb* again and select **Edit an Existing Web Page** to generate new images. Finally, *Post* the Web page to the **Project Drawings** folder.

C. Start Microsoft Internet Explorer, locate and open the **ACWEBPUBLISH.HTM** file to view your Web page. If you like, you can improve the page as you see fit by changing the drawings in AutoCAD and regenerating the Web page again with *Publishtoweb*.

2. *Etransmit*

   A. Start AutoCAD if not already running. *Open* a drawing that you want to transmit to a friend. If you make any changes, *Save* the current drawing.

   B. Invoke *eTransmit*. The *Create Transmittal* dialog box should appear. In the *Notes* section of the *General* tab, enter "Here is the drawing that I would like for you to review," or similar message. Also specify a *Self-extracting executable (*exe)* as the *Type* of file to create. Use the *Browse* button to specify your working directory as the *Location*. Clear the check boxes for the *Preserve directory structure* and *Make web page files* options. Check *Remove paths from Xrefs and images*. In the *Files* tab, make sure all associated files are checked. Do not yet select OK.

   C. If you have an email system available on the system, select *Send e-mail with transmittal* in the *General* tab. If not, clear the check for this option. Select **OK**.

   D. Depending on your selection in step C, your email system may open with the .EXE file as an attachment. If you chose not have AutoCAD open your email program, copy the .EXE file to disk, and attach it to an email message at a later time. In either case, send an email to a friend and ask him or her to extract the file and view the drawing in AutoCAD. Request a confirmation and a reply email reporting on the success of the transmittal.

3. **Insert** *Hyperlinks*

   A. In AutoCAD, *Open* the **GASKETA** drawing. Activate the *Model* tab. Use *SaveAs* to save the drawing as **GASKETA2**. *Close* the drawing.

   B. *Open* the **T-PLATE** drawing. Activate the *Model* tab. Use *SaveAs* to save the drawing as **T-PLATE2**. *Close* the drawing.

   C. *Open* the **GASKETA2** drawing. Use *Dtext* to place text at the bottom of the drawing. Enter "Click here to view the T-PLATE2 drawing." The text should appear similar to that shown in Figure 19-23. Create a *Hyperlink* for the entire line of text. Associate the **T-PLATE2.DWG** as the file to link. *Save* and *Close* the drawing.

   Figure 19-23

   D. *Open* the **T-PLATE2** drawing. Use *Dtext* to place text at the bottom of the drawing stating "Click here to view the GASKETA2 drawing." Create a *Hyperlink* for the text that links to the **GASKETA2.DWG**. *Save* and *Close* the drawing.

   Click here to view the T-PLATE2 drawing

   E. *Open* either drawing. Select the text, right-click, and select *Hyperlink*, then the *Open* option. You should be able to do the same in both drawings to "toggle" between the two drawings.

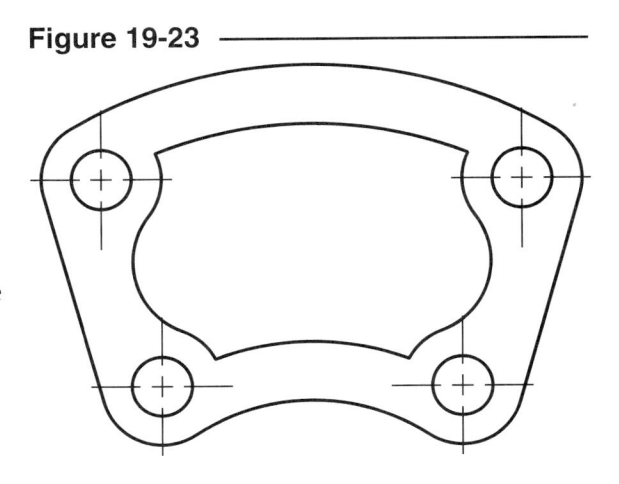

20

BLOCKS and DesignCenter

Chapter Objectives

After completing this chapter you should:

1. understand the concept of creating and inserting symbols in AutoCAD drawings;

2. be able to use the *Block* command to transform a group of objects into one object that is stored in the current drawing's block definition table;

3. be able to use the *Insert* and *Minsert* commands to bring *Blocks* into drawings;

4. know that *color*, *linetype* and *lineweight* of *Blocks* are based on conditions when the *Block* is made;

5. be able to convert *Blocks* to individual objects with *Explode*;

6. be able to use *Wblock* to prepare .DWG files for insertion into other drawings;

7. be able to redefine and globally change previously inserted *Blocks*;

8. be able to use DesignCenter™ to drag and drop *Blocks* from other drawings into the current drawing.

CONCEPTS

A *Block* is a group of objects that are combined into one object with the *Block* command. The typical application for *Blocks* is in the use of symbols. Many drawings contain symbols, such as doors and windows for architectural drawings, capacitors and resistors for electrical schematics, or pumps and valves for piping and instrumentation drawings. In AutoCAD, symbols are created first by constructing the desired geometry with objects like *Line*, *Arc*, and *Circle*, then transforming the set of objects comprising the symbol into a *Block*. A description of the objects comprising the *Block* is then stored in the drawing's "block definition table." The *Blocks* can then each be *Inserted* into a drawing many times and treated as a single object. Text can be attached to *Blocks* (called *Attributes*) and the text can be modified for each *Block* when inserted.

Figure 20-1 compares a shape composed of a set of objects and the same shape after it has been made into a *Block* and *Inserted* back into the drawing. Notice that the original set of objects is selected (highlighted) individually for editing, whereas, the *Block* is only one object.

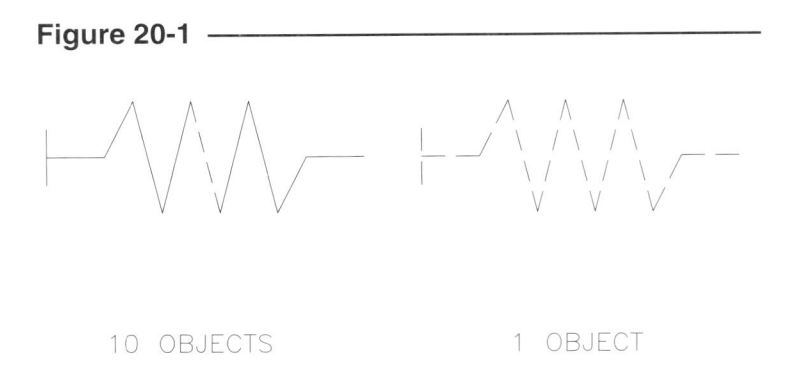

Figure 20-1

Since an inserted *Block* is one object, it uses less file space than a set of objects that is copied with *Copy*. The *Copy* command creates a duplicate set of objects, so that if the original symbol were created with 10 objects, 3 copies would yield a total of 40 objects. If instead the original set of 10 were made into a *Block* and then *Inserted* 3 times, the total objects would be 13 (the original 10 + 3).

Upon *Inserting* a *Block*, its scale can be changed and rotational orientation specified without having to use the *Scale* or *Rotate* commands (Fig. 20-2). If a design change is desired in the *Blocks* that have already been *Inserted*, the original *Block* can be redefined and the previously inserted *Blocks* are automatically updated. *Blocks* can be made to have explicit *Linetype*, *Lineweight* and *Color*, regardless of the layer they are inserted onto, or they can be made to assume the *Color*, *Linetype*, and *Lineweight* of the layer onto which they are *Inserted*.

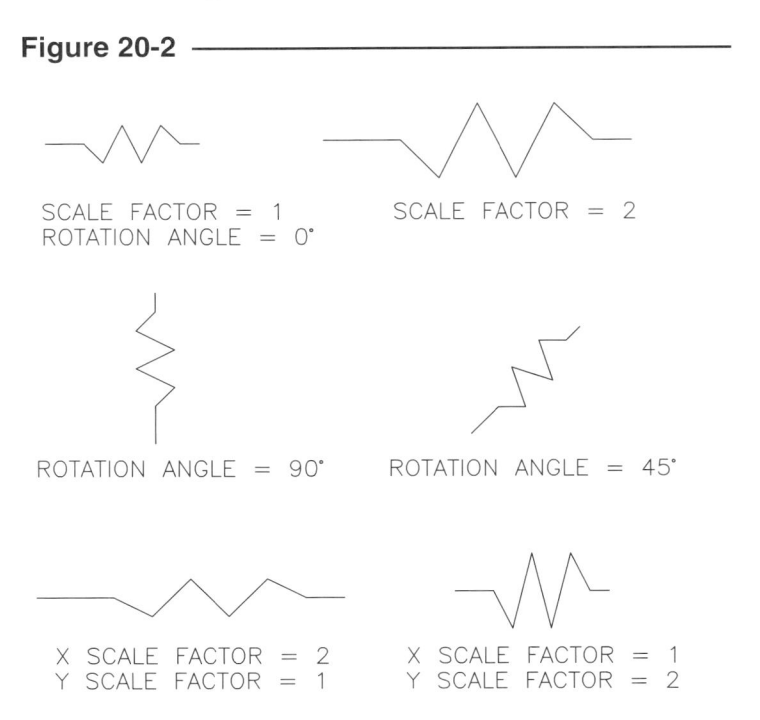

Figure 20-2

Blocks can be nested; that is, one *Block* can reference another *Block*. Practically, this means that the definition of *Block* "C" can contain *Block* "A" so that when *Block* "C" is inserted, *Block* "A" is also inserted as part of *Block* "C" (Fig. 20-3).

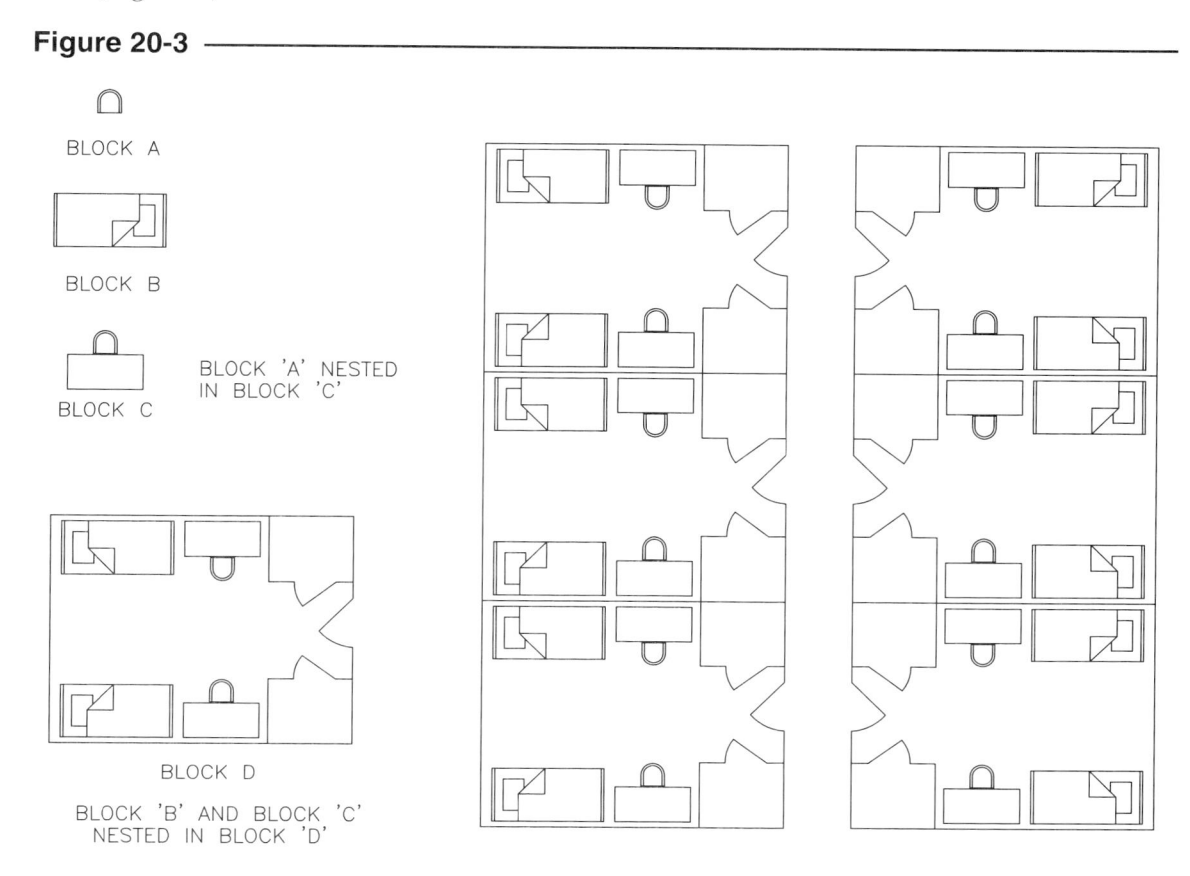

Figure 20-3

Blocks created within the current drawing can be copied to disk as complete and separate drawing files (.DWG file) by using the *Wblock* command (Write Block). This action allows you to *Insert* the *Blocks* into other drawings. Specifically, when you use the *Insert* command, AutoCAD first searches for the supplied *Block* name in the current drawing's block definition table. If the designated *Block* is not located there, AutoCAD searches the directories for a .DWG file with the designated name.

Commands related to using *Blocks* are:

Block	Creates a *Block* from individual objects
Insert	Inserts a *Block* into a drawing
Minsert	Permits a multiple insert in a rectangular pattern
Explode	Breaks a *Block* into its original set of multiple objects
Wblock	Writes an existing *Block* or a set of objects to a file on disk
Base	Allows specification of an insertion base point
Purge	Deletes uninserted *Blocks* from the block definition table
Rename	Allows renaming *Blocks*
Adcenter	Invokes DesignCenter, which allows you to drag and drop *Blocks*, *Dimension Styles*, *Layers*, *Layouts*, *Linetypes*, *Text Styles*, and *Xrefs* into the current drawing (separate window)

DesignCenter provides an easy method for inserting *Blocks* into a drawing. The unique characteristic of DesignCenter is that you can drag and drop any named objects (including *Blocks*) from other drawings into the current drawing. DesignCenter is fully explained near the end of this chapter.

COMMANDS

BLOCK

Pull-down Menu	COMMAND (TYPE)	ALIAS (TYPE)	Short-cut	Screen (side) Menu	Tablet Menu
Draw Block > Make...	BLOCK, or -BLOCK	B or -B	...	DRAW2 Bmake	N,9

Selecting the icon button, using the pull-down or screen menu, or typing *Block* or *Bmake*, produces the *Block Definition* dialog box shown in Figure 20-4. This dialog box provides the same functions as using the *-Block* command (a hyphen prefix produces the Command line equivalent).

Figure 20-4

To make a *Block*, first create the *Lines*, *Circles*, *Arcs*, or other objects comprising the shapes to be combined into the *Block*. Next, use the *Block* command to transform the objects into one object—a *Block*.

In the *Block Definition* dialog box, enter the desired *Block* name in the *Name* edit box. Then use the *Select Objects* button (top center) to return to the drawing temporarily to select the objects you wish to comprise the *Block*. After selection of objects, the dialog box reappears. Use the *Specify Insertion Base Point* button (in the *Base Point* section of the dialog box) if you want to use a point other than the default 0,0,0 as the "insertion point" when the *Block* is later inserted. Usually select a point in the corner or center of the set of objects as the base point. When you select *OK*, the new *Block* is defined and stored in the drawing's block definition table awaiting future insertions.

If *Delete* is selected in the *Objects* section of the dialog box, the original set of "template" objects comprising the *Block* disappear even though the definition of the *Block* remains in the table. Checking *Retain* forces AutoCAD to retain the original objects (similar to using *Oops* after the *Block* command), or selecting *Convert to Block* keeps the original set of objects visible in the drawing but transforms them into a *Block*.

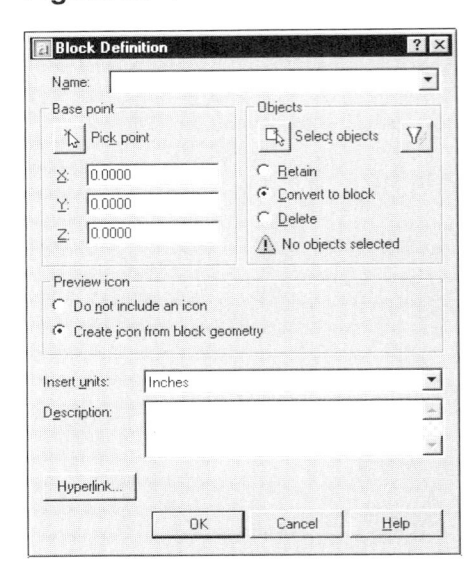

Figure 20-5

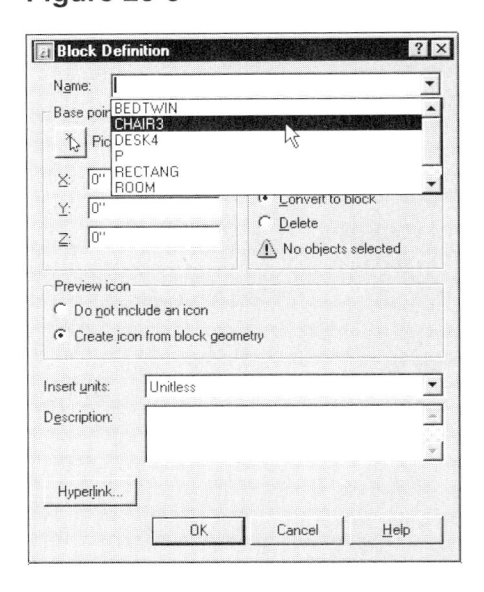

The bottom half of the dialog box is used to specify how *Blocks* are described when DesignCenter is used to drag and drop the *Blocks* into a drawing instead of using the *Insert* command. In DesignCenter, you can preview the *Blocks* and read the description of the *Blocks*. Generally the *Create Icon from Block Geometry* option is used to create the preview thumbnail sketch unless the *Block* geometry is very complex, in which case another icon file could be selected. The *Block Units* options are described later in this chapter. Use the *Hyperlink* button to produce the *Insert Hyperlink* dialog box for attaching a hyperlink to a *Block*. The *Names* drop-down list (Fig. 20-5) is used to select existing *Blocks* if you want to redefine a *Block* (see "Redefining Blocks" later in this chapter).

If you prefer to type, use -*Block* to produce the Command line equivalent of the *Block Definition* dialog box. The command syntax is as follows:

> Command: *-Block*
> Block name (or ?): (**name**) (Enter a descriptive name for the *Block* up to 255 characters.)
> Insertion base point: **PICK** or (**coordinates**) (Select a point to be used later for insertion.)
> Select objects: **PICK**
> Select objects: **PICK** (Continue selecting all desired objects.)
> Select objects: **Enter**

The *Block* then <u>disappears</u> as it is stored in the current drawing's "block definition table." The *Oops* command can be used to restore the original set of "template" objects (they reappear), but the definition of the *Block* remains in the table. Using the *?* option of the *Block* command lists the *Blocks* stored in the block definition table.

Block Color, Linetype, and Lineweight **Settings**

The <u>color</u>, <u>linetype</u>, and <u>lineweight</u> of the *Block* are determined by one of the following settings when the *Block* is created:

1. When a *Block* is inserted, it is drawn on its original layer with its original *color, linetype* and *lineweight* (when the objects were <u>created</u>), regardless of the layer or *color, linetype* and *lineweight* settings that are current when the *Block* is inserted (unless conditions 2 or 3 exist).

2. If a *Block* is created on Layer 0 (Layer 0 is current when the original objects comprising the *Block* are created), then the *Block* assumes the *color, linetype* and *lineweight* of any layer that is current when it is inserted (Fig. 20-6).

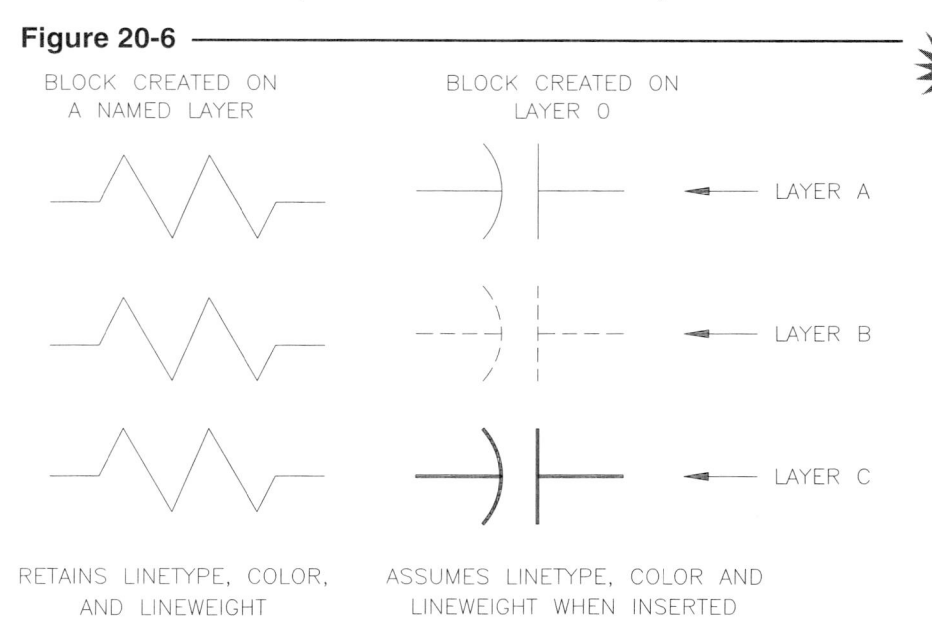

Figure 20-6

BLOCK CREATED ON A NAMED LAYER

BLOCK CREATED ON LAYER 0

◀— LAYER A

◀— LAYER B

◀— LAYER C

RETAINS LINETYPE, COLOR, AND LINEWEIGHT

ASSUMES LINETYPE, COLOR AND LINEWEIGHT WHEN INSERTED

3. If the *Block* is created with the special BYBLOCK *color, linetype* and *lineweight* setting, the *Block* is inserted with the *color, linetype* and *lineweight* settings that are <u>current during insertion</u>, whether the BYLAYER or explicit object *color, linetype* and *lineweight* settings are current.

394 Blocks and DesignCenter

INSERT

Pull-down Menu	COMMAND (TYPE)	ALIAS (TYPE)	Short-cut	Screen (side) Menu	Tablet Menu
Insert Block...	INSERT or -INSERT	I or -I	...	INSERT Ddinsert	T,5

Once the *Block* has been created, it is inserted back into the drawing at the desired location(s) with the *Insert* command. The *Insert* command produces the *Insert* dialog box (Fig. 20-7) which allows you to select which *Block* to insert and to specify the *Insertion Point*, *Scale*, and *Rotation*, either interactively (*On-screen*) or by specifying values.

First, select the *Block* you want to insert. All *Blocks* located in the drawing's block definition table are listed in the *Name* drop-down list. Next, determine the parameters for *Insertion Point*, *Scale*, and *Rotation*. You can enter values in the edit boxes if you have specific parameters in mind or check *Specify On-screen* to interactively supply the parameters. For example, with the settings shown in Figure 20-7, AutoCAD would allow you to preview the *Block* as you dragged it about the screen and picked the *Insertion Point*. You would not be prompted for a *Scale* or *Rotation* angle since they are specified in the dialog box as 1.0000 an 0 degrees, respectively. Entering any other values in the *Scale* or *Rotation* edit boxes causes AutoCAD to preview the *Block* at the specified scale and rotation angle as you drag it about the drawing to pick the insertion point. Remember that *Osnaps* can be used when specifying the parameters interactively. Check *Uniform Scale* to ensure the X, Y, and Z values are scaled proportionally. *Explode* can also be toggled, which would insert the *Block* as multiple objects (see "*Explode*").

Figure 20-7

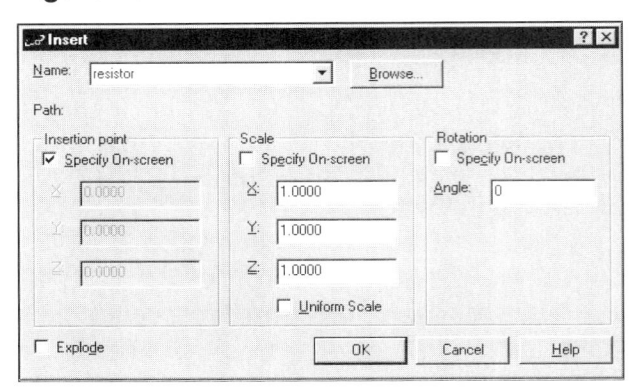

Inserting Other Drawings as *Blocks*
Selecting the *Browse* tile in the *Insert* dialog box produces the *Select Drawing* dialog box (Fig. 20-8). Here you can select any drawing (.DWG file) for insertion. When one drawing is *Inserted* into another, the entire drawing comes into the current drawing as a *Block*, or as one object. If you want to edit individual objects in the inserted drawing, you must *Explode* the object.

Figure 20-8

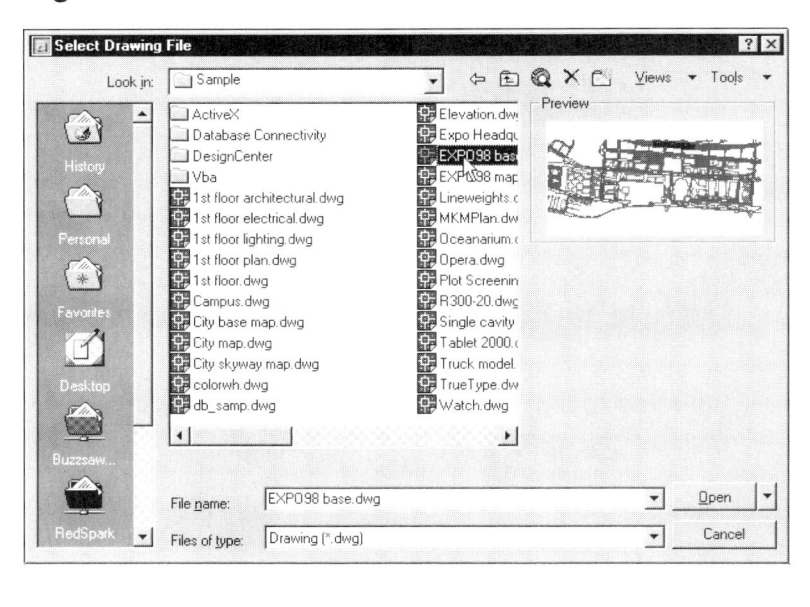

If you prefer the Command line equivalent, type *–Insert*.

    Command: -insert
    Enter block name or [?]: name
    (Type the name of an existing block or .DWG file to insert.)
    Specify insertion point or
    [Scale/X/Y/Z/Rotate/PScale/PX/PY/PZ/PRotate]: PICK or option
    Enter X scale factor, specify opposite corner, or [Corner/XYZ] <1>: value, PICK or option
    Enter Y scale factor <use X scale factor>: value or PICK
    Specify rotation angle <0>: value or PICK

Sometimes it is desirable to see the *Block* in the intended scale factor or rotation angle before you choose the insertion point. Presets ("PScale/PC/PY/PZ/PRotate") allow you to specify a rotation angle or scale factor <u>before</u> you dynamically drag the *Block* to pick the insertion point. (Normally, you would have to select the insertion point before the prompts for scale factor and rotation angle appear.)

MINSERT

Pull-down Menu	COMMAND (TYPE)	ALIAS (TYPE)	Short-cut	Screen (side) Menu	Tablet Menu
...	MINSERT	...	...	...	...

This command allows a <u>multiple insert</u> in a rectangular pattern (Fig. 20-9). *Minsert* is actually a combination of the *Insert* and the *Array Rectangular* commands. The *Blocks* inserted with *Minsert* are associated (the group is treated as one object) and cannot be edited independently (unless *Exploded*).

Examining the command syntax yields the similarity to a *Rectangular Array*.

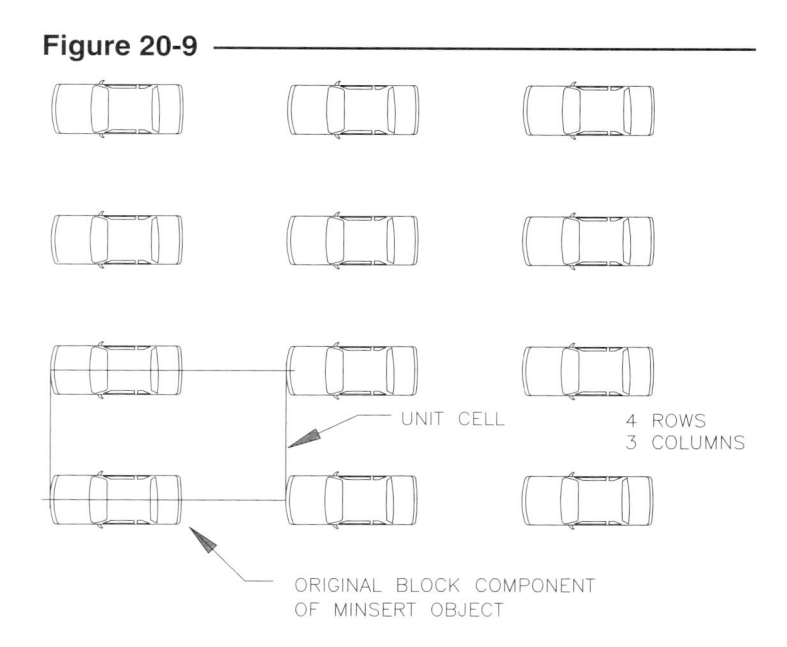

Figure 20-9

Command: **Minsert**
Enter block name [or ?] <current>: **name**
Specify insertion point or [Scale/X/Y/Z/Rotate/PScale/PX/PY/PZ/PRotate]: (**value**), **PICK** or option
Enter X scale factor, specify opposite corner, or [Corner/XYZ] <1>: (**value**) or **Enter**
Enter Y scale factor <use X scale factor>: (**value**) or **Enter**
Specify rotation angle <0>: (**value**) or **Enter**
Enter number of rows (—-) <1>: (**value**)
Enter number of columns (|||) <1>: (**value**)
Enter distance between rows or specify unit cell (—-): (**value**) or **PICK** (Value specifies Y distance from *Block* corner to *Block* corner; PICK allows drawing a unit cell rectangle.)
Distance between columns: (**value**) or **PICK** (Specifies X distance between *Block* corners.)
Command:

EXPLODE

Pull-down Menu	COMMAND (TYPE)	ALIAS (TYPE)	Short-cut	Screen (side) Menu	Tablet Menu
Modify Explode	Explode	X	...	MODIFY2 Explode	Y,22

Explode breaks a <u>previously</u> inserted *Block* back into its original set of objects (Fig. 20-10, on the next page), which allows you to edit individual objects comprising the shape. *Blocks* that have been *Minsert*ed cannot be *Explode*d. There are no options for this command.

Command: *explode*
Select objects: **PICK**
Select objects: **Enter**
Command:

Figure 20-10

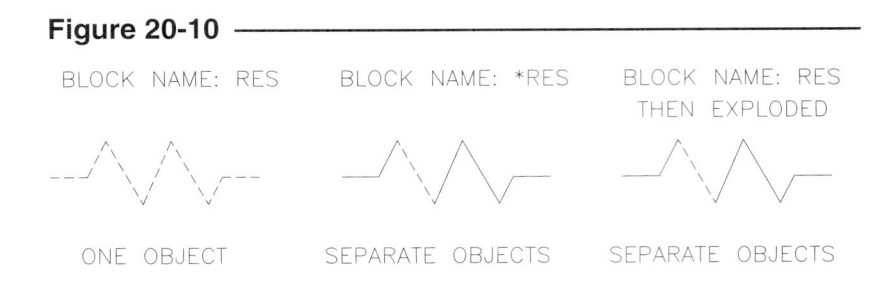

BLOCK NAME: RES   BLOCK NAME: *RES   BLOCK NAME: RES THEN EXPLODED

ONE OBJECT   SEPARATE OBJECTS   SEPARATE OBJECTS

XPLODE

Pull-down Menu	COMMAND (TYPE)	ALIAS (TYPE)	Short-cut	Screen (side) Menu	Tablet Menu
...	XPLODE	XP	...	...	...

When you *Insert* a *Block* into a drawing, all layers, linetypes, and colors contained in the *Block* are also inserted into the parent drawing if they do not already exist. If you use *Explode* to break down the *Block* into its component entities, those entities retain their native properties of layer, linetype, and color. For example, you may insert a *Block* that also inserts its layer named FLOORPLAN. If that *Block* is *Exploded*, its objects remain on layer FLOORPLAN. You may instead want to *Explode* the *Block* and have the objects reside on a layer existing in the current drawing, AR-WALL, for example.

The *Xplode* command is an expanded version of *Explode*. The *Xplode* command allows you to specify new properties for the objects that are exploded. When you use *Xplode*, you can also specify the new *Layer*, *Linetype*, *Color*, *Lineweight*, or choose to *Explode* the *Block* normally.

Command: **xplode**
Select objects to XPlode.
Select objects: **PICK**
Select objects: **Enter**
Enter an option [All/Color/LAyer/LType/Inherit from parent block/Explode] <Explode>:

Color
Use this option to specify a color for the exploded objects. Entering *ByLayer* causes the component objects to inherit the color of the exploded object's layer. Entering *ByBlock* causes the component objects to inherit the object-specific color of the exploded object.

Enter an option [All/Color/LAyer/LType/Inherit from parent block/Explode] <Explode>: **c**
Enter new color for exploded objects.
[Red/Yellow/Green/Cyan/Blue/Magenta/White/BYLayer/BYBlock] <BYLAYER>:

Layer
With this option you can specify the layer of the component objects after you explode them. The default option is to inherit the current layer rather than the layer of the exploded object.

Linetype
You can enter the name of any linetype that is loaded in the drawing. The exploded objects assume the specified linetype. Entering *ByLayer* causes the component objects to inherit the linetype of the exploded object's layer. Entering *ByBlock* causes the component objects to inherit the object-specific linetype of the exploded object.

Inherit from parent block
This option sets the color, linetype, lineweight, and layer of the exploded objects to that of the *Block* if the *Block* was created using *ByBlock* color, linetype, and lineweight and the objects were drawn on layer 0.

WBLOCK

Pull-down Menu	COMMAND (TYPE)	ALIAS (TYPE)	Short-cut	Screen (side) Menu	Tablet Menu
File Export...	WBLOCK	W	...	FILE Export	W,24

The *Wblock* command writes a *Block* out to disk as a separate and complete drawing (.DWG) file. The *Block* used for writing to disk can exist in the current drawing's *Block* definition table or can be created by the *Wblock* command. Remember that the *Insert* command inserts *Blocks* (from the current drawing's block definition table) or finds and accepts .DWG files and treats them as *Blocks* upon insertion.

There are two ways to create a *Wblock* from the current drawing, (1) using an existing *Block* and (2) using a set of objects not previously defined in the current drawing as a *Block*. If you are using an existing *Block*, a copy of the *Block* is essentially transformed by the *Wblock* command to create a complete AutoCAD drawing (.DWG) file. The original block definition remains in the current drawing's block definition table. In this way, *Blocks* that were originally intended for insertion into the current drawing can be inserted into other drawings.

Figure 20-11 illustrates the relationship among a *Block*, the current drawing, and a *WBlock*. In the figure, SCHEM1.DWG contains several *Blocks*. The RES block is written out to a .DWG file using *Wblock* and named RESISTOR. RESISTOR is then *Inserted* into the SCHEM2 drawing.

Figure 20-11

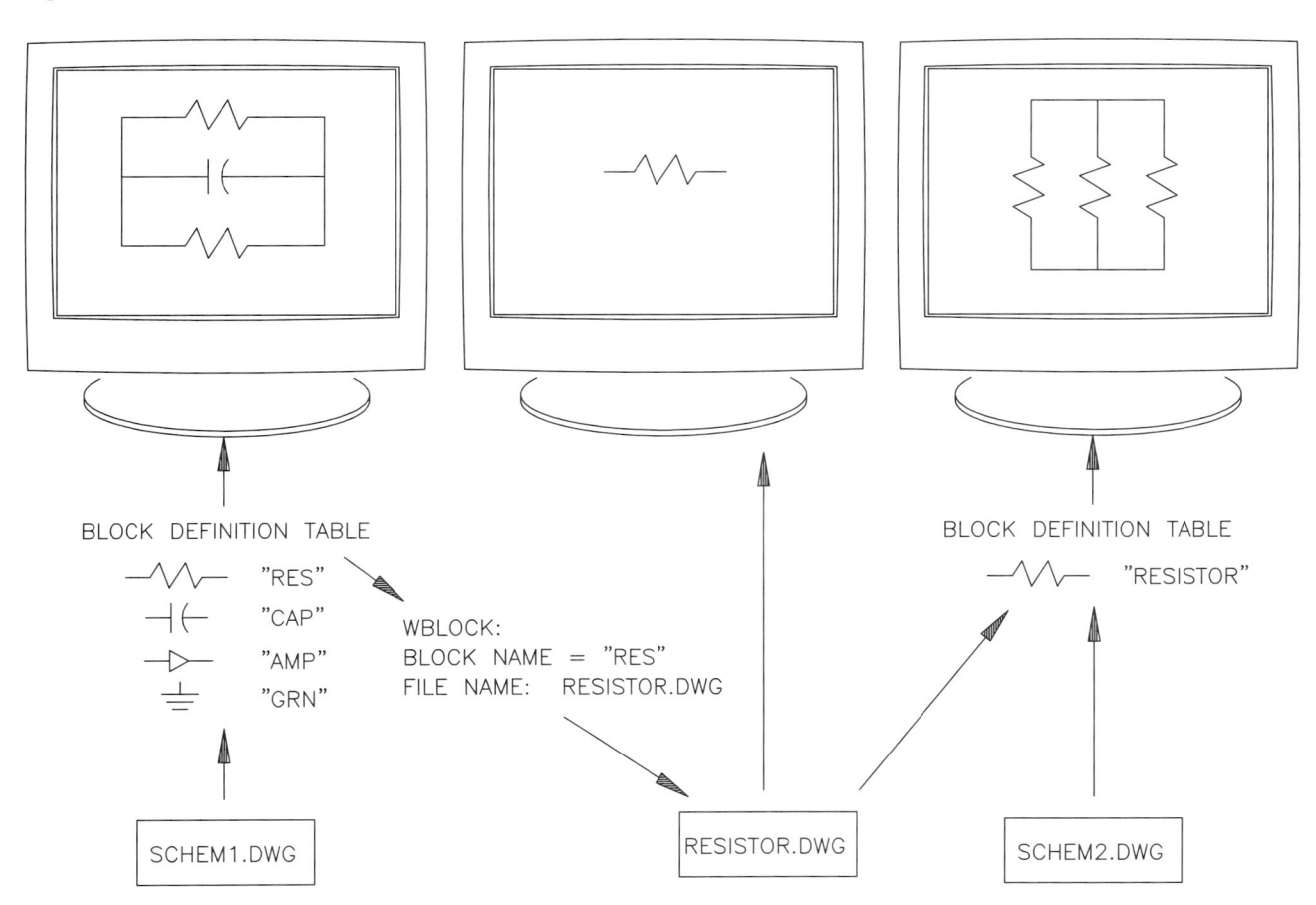

> If you want to transform a set of objects into a *Block* to be used in other drawings but not in the current one, you can use *Wblock* to transform (a copy of) the objects in the current drawing into a separate .DWG file. This action does not create a *Block* in the current drawing. As an alternative, if you want to create symbols specifically to be inserted into other drawings, each symbol could be created initially as a separate .DWG file.

The *Wblock* command produces the *Write Block* dialog box (Fig. 20-12). You should notice similarities to the *Block Definition* dialog box. Under *Source*, select *Block* if you want to write out an existing *Block* and select the *Block* name from the list, or select *Objects* if you want to transform a set of objects (not a previously defined *Block*) into a separate .DWG file.

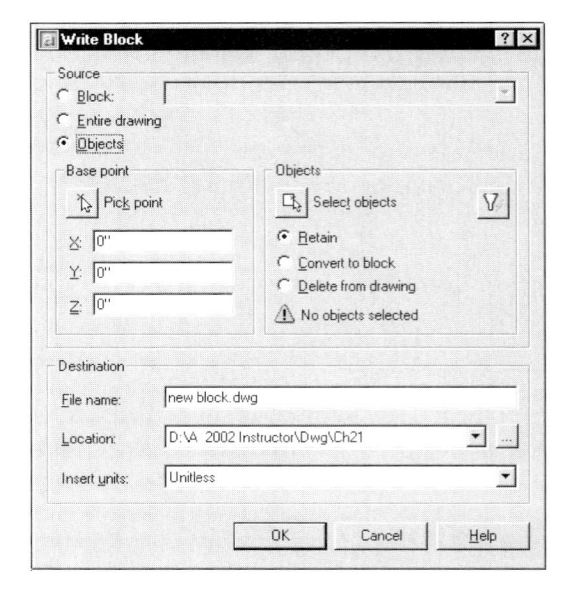

Figure 20-12

The *Base Point* section allows you to specify a base point to use upon insertion of the *Block*. Enter coordinate values or use the *Specify Insert Base Point* button to pick a location in the drawing. The *Objects* section allows you to specify how you want to treat selected objects if you create a new .DWG file from objects (not from an existing *Block*). You can *Retain* the objects in their current state, *Convert to Block*, or *Delete from Drawing*.

The *Destination* section defines the desired *File Name*, *Location*, and *Insert Units*. Your choice for *Insert Units* is applicable only when you drag and drop a *Block*, as with DesignCenter. See "DesignCenter" later in this chapter for an explanation of this subject.

If you prefer the Command line equivalent to the *Write Block* dialog box, type *-Wblock* and follow the prompt sequence shown here to create *Wblocks* (.DWG files) <u>from existing *Blocks*</u>,

    Command: *-wblock*
    (At this point, the *Create Drawing File* dialog box appears, prompting you to supply a name for the .DWG file to be created. Typically, a new descriptive name would be typed in the edit box rather than selecting from the existing names.)
    Enter name of existing block or [= (block=output file)/* (whole drawing)] <define new drawing>:
    (Enter the name of the desired existing *Block*. If the file name given in the previous step is the same as the existing *Block* name, an "=" symbol can be entered, or enter an asterisk to write out the entire drawing.)
    Command:

A copy of the existing *Block* is then created in the selected directory as a *Wblock* (.DWG file).

When *Wblocks* are *Inserted*, the *Color*, *Linetype*, and *Lineweight* settings of the *Wblock* are determined by the settings current when the original objects comprising the *Wblock* were created. The three possible settings are the same as those for *Blocks* (see "Block," Color, Linetype, and Lineweight Settings).

> When a *Wblock* is *Inserted*, its parent (original) layer is also inserted into the current drawing. *Freezing* <u>either</u> the parent layer or the layer that was current during the insertion causes the *Wblock* to be frozen.

Redefining *Blocks*

If you want to change the configuration of a *Block*, even after it has been inserted, it can be accomplished by redefining the *Block*. In doing so, all of the previous *Block* insertions are automatically and globally updated (Fig. 20-13). AutoCAD stores two fundamental pieces of information for each *Block* insertion—the insertion point and the *Block* name. The actual block definition is stored in the block definition table. Redefining the *Block* involves changing that definition.

Figure 20-13

BLOCK "RES" REDEFINITION OF BLOCK "RES"

BEFORE REDEFINING BLOCK "RES" AFTER REDEFINING BLOCK "RES"

To redefine a *Block*, use the *Block* command. First, draw the new geometry or change the original "template" set of objects. (The change cannot be made using an inserted *Block* unless it is *Exploded* because a *Block* cannot reference itself.) Next, use the *Block* command and select the new or changed geometry. The old *Block* is redefined with the new geometry as long as the original Block name is used.

The *Refedit* command can also be used to redefine a *Block*. *Refedit* "opens" the *Block* for editing, then you make the necessary changes, and finally use *Refclose* to "close" the editing session and save the *Block*. This process has the same result as redefining the *Block*. See "*Refedit*" in AutoCAD 2002 *Help*.

BASE

Pull-down Menu	COMMAND (TYPE)	ALIAS (TYPE)	Short-cut	Screen (side) Menu	Tablet Menu
Draw Block > Base	BASE	...	...	DRAW2 Base	...

The *Base* command allows you to specify an "insertion base point" (see the *Block* command) in the current drawing for subsequent insertions. If the *Insert* command is used to bring a .DWG file into another drawing, the insertion base point of the .DWG is 0,0 by default. The *Base* command permits you to specify another location as the insertion base point. The *Base* command is used in the symbol drawing, that is, used in the drawing to be inserted. For example, while creating separate symbol drawings (.DWGs) for subsequent insertion into other drawings, the *Base* command is used to specify an appropriate point on the symbol geometry for the *Insert* command to use as a "handle" other than point 0,0 (Fig. 20-14).

Figure 20-14

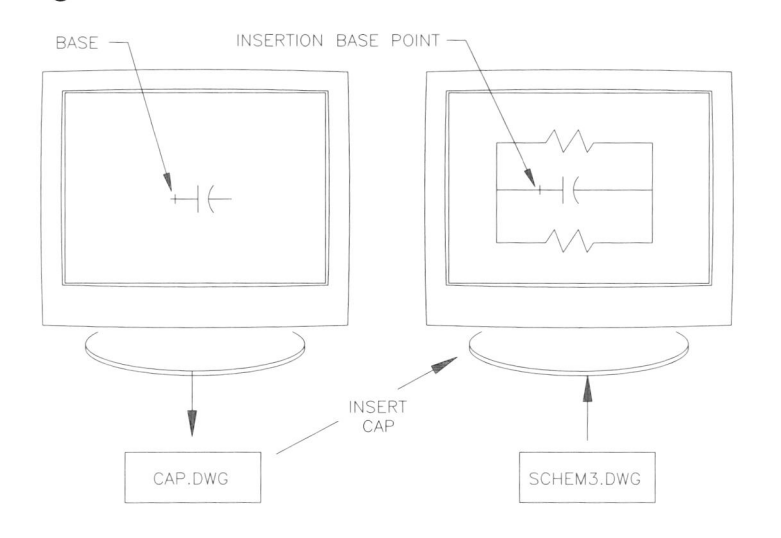

BASE INSERTION BASE POINT

CAP.DWG INSERT CAP SCHEM3.DWG

PURGE

Pull-down Menu	COMMAND (TYPE)	ALIAS (TYPE)	Short-cut	Screen (side) Menu	Tablet Menu
File Drawing Utilities > Purge...	PURGE	PU	...	FILE Purge	X,25

Purge allows you to selectively delete a *Block* that is not referenced in the drawing. In other words, if the drawing has a *Block* defined but not appearing in the drawing, it can be deleted with *Purge*. In fact, *Purge* can selectively delete any named object that is not referenced in the drawing. Examples of unreferenced named objects are:

Blocks that have been defined but not *Inserted*;
Layers that exist without objects residing on the layers;
Dimstyles that have been defined, yet no dimensions are created in the style (see Chapter 27);
Linetypes that were loaded but not used;
Shapes that were loaded but not used;
Text Styles that have been defined, yet no text has been created in the *Style*;
Mlstyles (multiline styles) that have been defined, yet no *Mlines* have been drawn in the style.

Purge is especially useful for getting rid of unused *Blocks* because unused *Blocks* can occupy a huge amount of file space compared to other named objects (*Blocks* are the only geometry-based named objects). Although some other named objects can be deleted by other methods (such as selecting *Delete* in the *Text Style* dialog box), *Purge* is the only method for deleting *Blocks*.

The *Purge* command produces the *Purge* dialog box (Fig. 20-15). Here you can view all named objects in the drawing, but you can purge only those items that are not used in the current drawing.

Figure 20-15

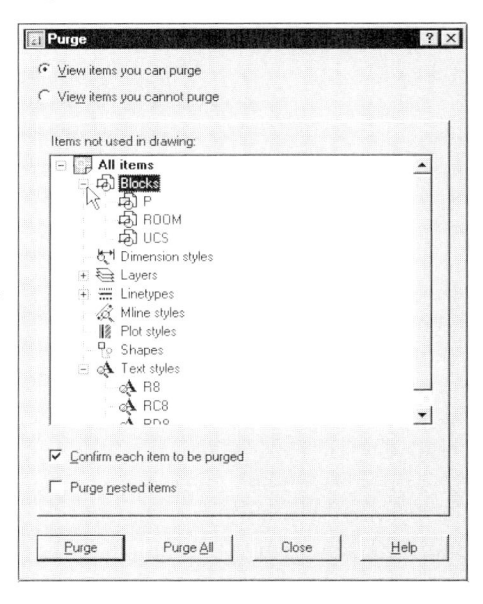

View items you can purge
Generally this option is the default choice since you can remove unused named objects from your drawing only if this option is checked.

View items you cannot purge
This option is useful if you want to view all the named objects contained in the drawing. You can also get an explanation appearing at the bottom of the dialog box as to why a selected item cannot be removed.

Purge
Use the *Purge* button to remove only the items you select from the list. First, select the items, then select *Purge*. You can select more than one item using the Ctrl or Shift keys while selecting multiple items or a range, respectively.

Purge All
Use this button to remove all unused named objects from the drawing. You do not have to first select items from the list. To be safe, you should also select *Confirm each item to be purged* when using this option.

Confirm each item to be purged
If the *Confirm each item to be purged* box is checked, AutoCAD asks for confirmation before removing each object. If this box is not checked and you *Purge All*, all unused named objects in the drawing are removed at once without notification.

RENAME

Pull-down Menu	COMMAND (TYPE)	ALIAS (TYPE)	Short-cut	Screen (side) Menu	Tablet Menu
Format Rename...	RENAME or -RENAME	REN or -REN	...	FORMAT Rename	V,1

This utility command allows you to rename a *Block* or <u>any named object</u> that is part of the current drawing. *Rename* allows you to rename the named objects listed here:

Blocks, Dimension Styles, Layers, Linetypes, Plot Styles, Text Styles, User Coordinate Systems, Views, and *Viewport* configurations.

You can type *Rename* or select *Rename...* from the *Format* pull-down menu to access the dialog box shown in Figure 20-16.

Figure 20-16

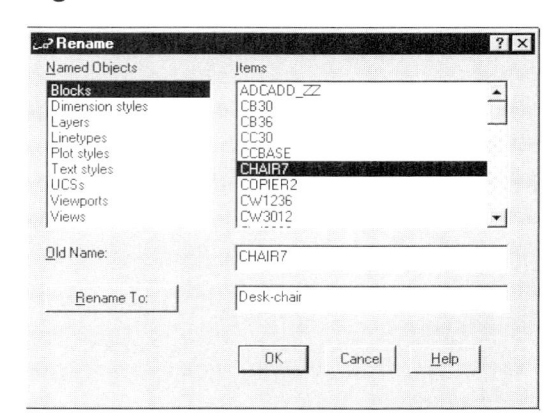

You can select from the *Named Objects* list to display the related objects existing in the drawing. Then select or type the old name so it appears in the *Old Name:* edit box. Specify the new name in the *Rename To:* edit box. <u>You must then PICK the *Rename To:* tile</u> and the new names will appear in the list. Then select the *OK* tile to confirm.

DesignCenter

The DesignCenter window (Fig. 20-17) allows you to navigate, find, and preview a variety of content, including *Blocks*, located anywhere accessible to your workstation, then allows you to open or insert the content using drag-and-drop. "Content" that can be viewed and managed includes other drawings, *Blocks, Dimstyles, Layers, Layouts, Linetypes, Textstyles, Xrefs*, raster images, and URLs (Web site addresses). In addition, if you have multiple drawings open, you can streamline your drawing process by copying and pasting content, such as layer definitions, between drawings.

Figure 20-17

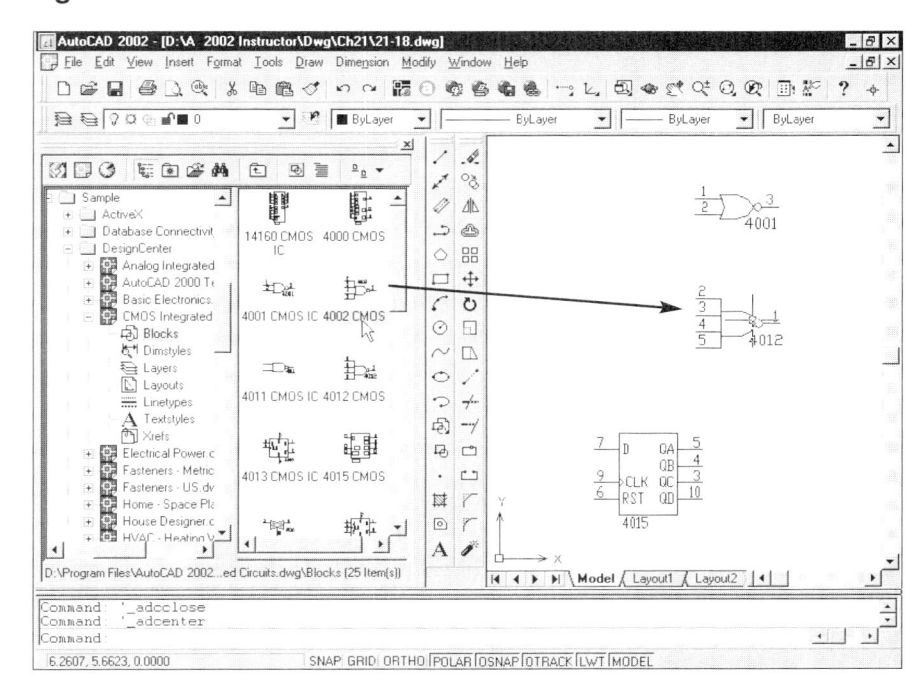

402 *Blocks and DesignCenter*

ADCENTER

Pull-down Menu	COMMAND (TYPE)	ALIAS (TYPE)	Short-cut	Screen (side) Menu	Tablet Menu
Tools AutoCAD DesignCenter...	ADCENTER	ADC	Ctrl + 2	...	...

Accessing *Adcenter* by any method produces the DesignCenter window (see Figure 20-18). By default, DesignCenter is docked at the left side of the drawing area (see Figure 20-17). There are two sections to the window. The left side is called the Tree View and displays a Windows Explorer-type hierarchical directory (folder) structure of the local system (Fig. 20-18). The right side is called the Palette and displays lists, icons, or thumbnail sketches of the content selected in the Tree View. The content shown in the Palette can be *Blocks*, *Dimstyles*, *Layers*, or a variety of content.

Generally, the Palette is used to drag and drop the icons or thumbnails into the current drawing as shown in Figure 20-18. However, you can also streamline a variety of tasks such as those listed here:

- Browse sources of drawing content including open drawings, other drawings, raster images, content within the drawings (*Blocks*, *Dimstyles*, *Layers*, and so on), content on network drives, or content on a Web page.

- Insert, attach, or copy and paste the content (drawings, images, *Blocks*, *Layers*, etc.) into the current drawing.

- Create shortcuts to drawings, folders, and Internet locations that you access frequently.

- Use a special search engine to find drawing content on your computer or network drives. You can specify criteria for the search based on key words, names of *Blocks*, *Dimstyles*, *Layers*, etc., or the date a drawing was last saved. Once you have found the content, you can load it into DesignCenter or drag it into the current drawing.

- Open drawings by dragging a drawing (.DWG) file from the Palette into the drawing area.

Figure 20-18

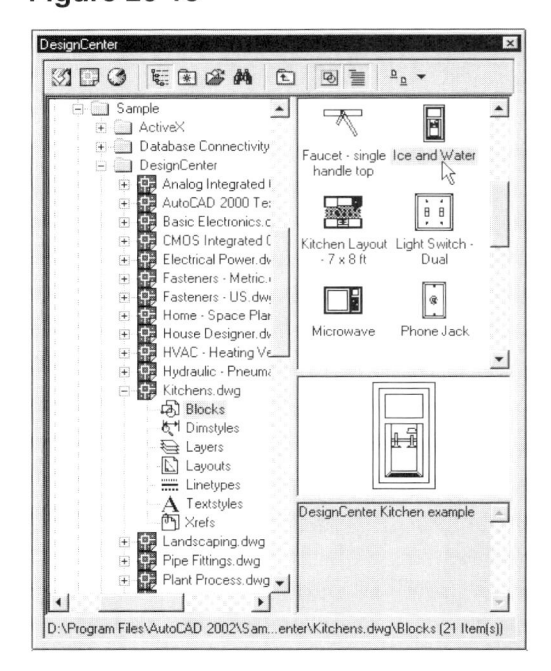

These features of DesignCenter are explained in the following descriptions of the DesignCenter toolbar icons.

Moving and Resizing the DesignCenter Window
You can resize DesignCenter by moving the bar between the Palette and the Tree View or by moving the lower-right corner with your pointing device and dragging the window to the required size. Undock DesignCenter by clicking the top bar, dragging it away from the left side, and dropping it to make it a floating window. You can move the floating window anywhere on the screen and change its width and height.

Tree View Options

In addition to selecting the icons from the toolbar, you can right-click the Palette background to produce the shortcut menu and choose the desired option (Fig. 20-19).

Desktop

Desktop is the default display for the Tree View side of DesignCenter. This choice displays a hierarchical structure of the desktop (local workstation). Because this arrangement is similar to Windows Explorer, you can navigate and locate content anywhere accessible to your system, including network drives. Figure 20-19 displays a typical hierarchical structure on the Tree View side. For example, you may want to import layer definitions (including color and linetype information) from a drawing into the current drawing by dragging and dropping.

Figure 20-19

Open Drawings

This option changes the Tree View to display all open drawings (Fig. 20-20). This feature is helpful when you have several drawings open and want to locate content from one drawing and import it into the current drawing. As shown in Figure 20-20, you may want to locate *Block* definitions from one drawing and *Insert* them into another drawing. Ensure you make the desired "target" drawing current in the drawing area before you drag and drop content from the Palette.

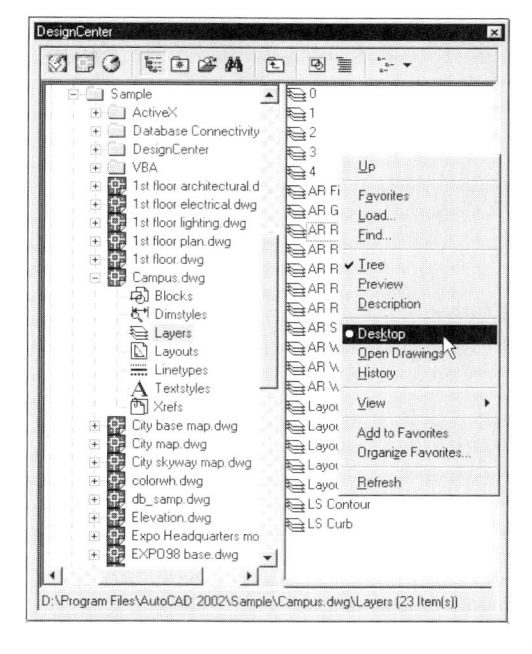

Figure 20-20

History

This option displays a history (chronological list) of the last 20 file locations accessed through DesignCenter (Fig. 20-21). The purpose of this feature is simply to locate the file and load it into the Palette. Load the file into the Palette by double-clicking on it.

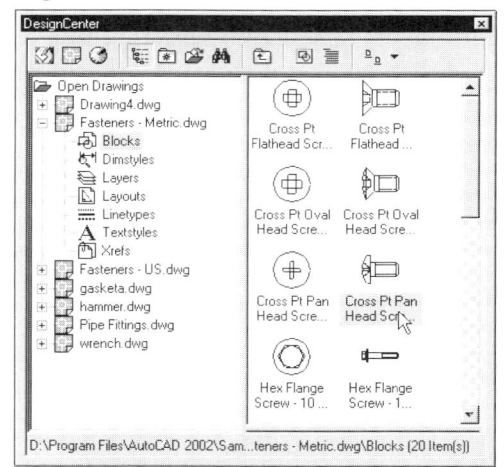

Figure 20-21

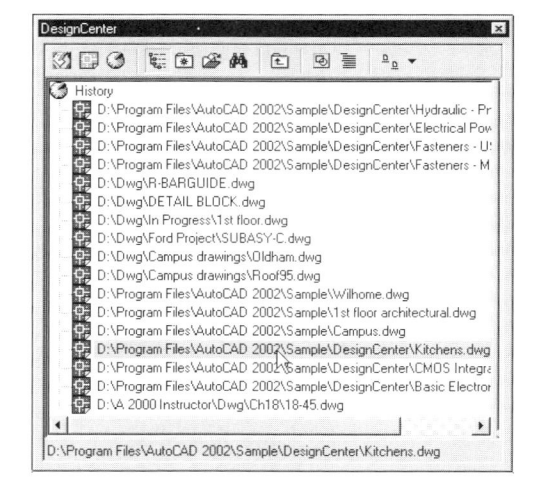

Tree View **Toggle**

Tree View is helpful for navigating your system for content. Once the desired folder or drawing is found and highlighted in Tree View, you may want to toggle Tree View off so only the Palette is displayed with the desired content. The desired content may be drawings, *Blocks*, images, or a variety of other content. For example, consider previous Figure 20-20 which displays Tree View with *Blocks* for FASTENERS-METRIC.DWG selected. Figure 20-22 illustrates the same selection in DesignCenter with Tree View toggled off and only the Palette displayed. The resulting configuration displays only the *Blocks* contained in the selected drawing. Keep in mind that you can also change the Views of the Palette to display *Large Icons*, *Small Icons*, a *List*, or *Details* (see "Views").

Figure 20-22

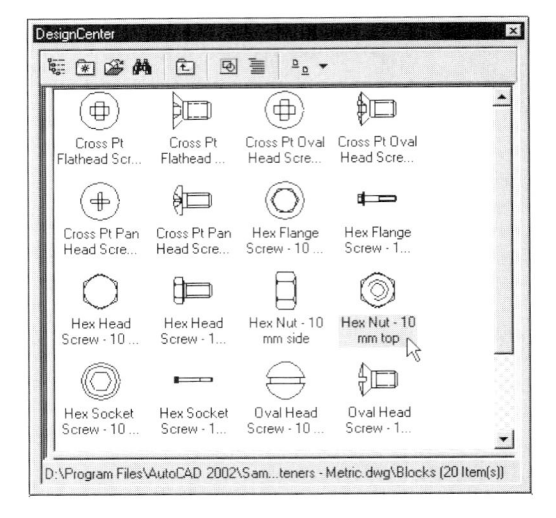

Favorites

This button displays the contents of the AutoCAD Favorites folder in the Palette (Fig. 20-23). The Tree View section displays the highlighted folder in the Desktop view. You can add folders and files to Favorites by highlighting an item in Tree View or the Palette, right-clicking on it, and selecting *Add to Favorites* from the shortcut menu.

Figure 20-23

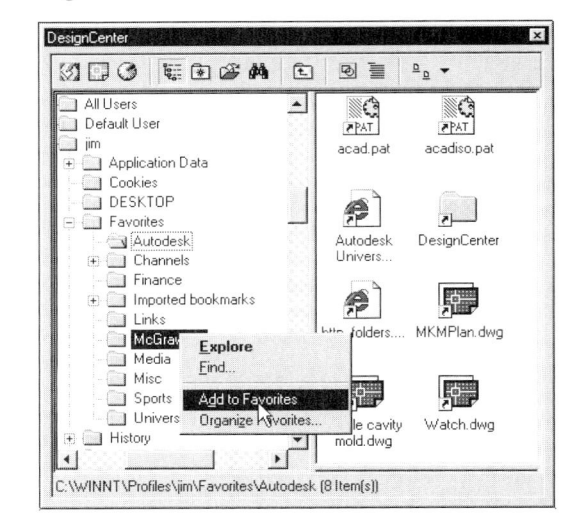

Load

Displays the *Load DesignCenter Palette* dialog box (not shown), in which you can load the Palette with content from anywhere accessible from your system. The *Load DesignCenter Palette* dialog box is identical to the *Select File* dialog box (see "*Open*," Chapter 2, Working with Files). After selecting a file, DesignCenter automatically finds the file in Tree View and loads its content (*Blocks*, *Layers*, *Dimstyles*, etc.) into the Palette. Note that the *Find File* and *Locate* buttons can be used to locate files to load. You can also load files into the Palette using Windows Explorer (see "Loading the Palette with Windows Explorer" at the end of this section).

Find

This button invokes the *Find* dialog box (Fig. 20-24), in which you can specify search criteria to locate drawing files, *Blocks*, *Layers*, *Dimstyles*, and other content within drawings. <u>Once the desired content is found in the list at the bottom of the dialog box, double-click on it to load it into the Palette.</u>

This feature is extremely powerful and easy to use. The list of possible items to search for is displayed in the *Look For* drop-down list and includes the following choices:

Blocks
Dimstyles
Drawings
Drawings and Blocks
Layers
Layouts
Linetypes
Text Styles
Xrefs

Figure 20-24

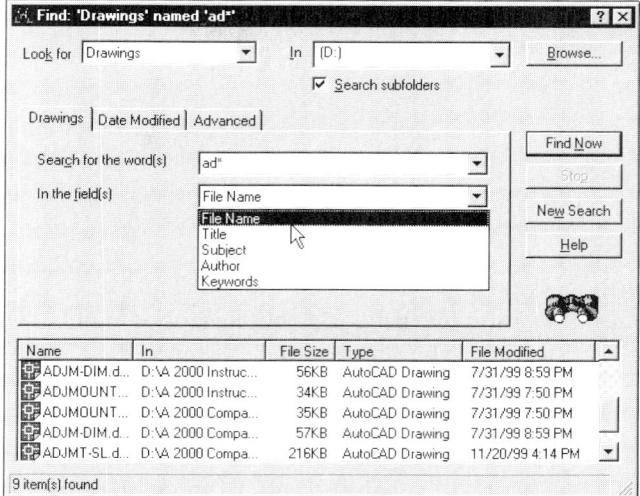

If *Drawings* is selected in the list, three tabs (shown in Figure 20-24) appear to allow you to refine the search criteria.

Drawings
Enter a text string in the *Search for words* edit box. Wildcards can be used. From the *In the field(s)* drop-down list, select what field you want your text string to apply to:

File Name
Title
Subject
Author
Keywords

Title, *Subject*, *Author*, and *Keywords* are searched for only in the *Drawing Properties* (description), so if you are searching for pre-AutoCAD 2000 drawings or have not specified drawing properties, these fields are not useful (see AutoCAD 2002 *Help* for information on *Drawing Properties*).

Date Modified
You can search by *Date Modified* by specifying modification between two dates or during the previous X number of days or X number of months.

Advanced
The *Advanced* tab allows you to search for text strings in the *Drawing and Block description*, *Block Name*, *Attribute Tag*, or *Attribute Value*.

Up

Use the *Up* tool to display the next higher level in the Tree View hierarchy.

Preview

This button causes a preview image (thumbnail sketch) of the selected item to appear at the bottom of the Palette. If there is no preview image saved with the selected item, the *Preview* area is empty. You can resize the preview image by dragging the bar between the Palette and the preview pane.

To make a preview image appear, you must select an item (drawing, *Block*, or image file) from the Palette. Selecting items in the Tree View does not cause a preview to appear. Figure 20-25 displays a drawing preview. Figure 20-18 displays a *Block* preview.

Figure 20-25

Description

This button displays a text description of the selected item at the bottom of the Palette (Fig. 20-26). The description is displayed for *Blocks* (from the *Description* field when the *Block* is created) and for drawings.

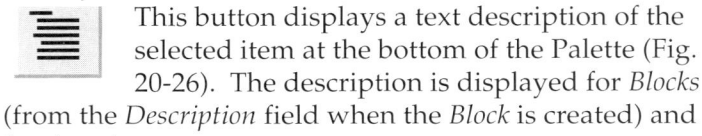

The description area can be resized. If you display both *Preview* and *Description* panes, the *Description* pane is displayed below the *Preview* pane, separated from it by a bar (see Figure 20-18).

You cannot edit the text description in DesignCenter; however, you can copy it to the Clipboard. Select the text you want to copy, right-click inside the pane, and then select *Copy to Clipboard* from the shortcut menu.

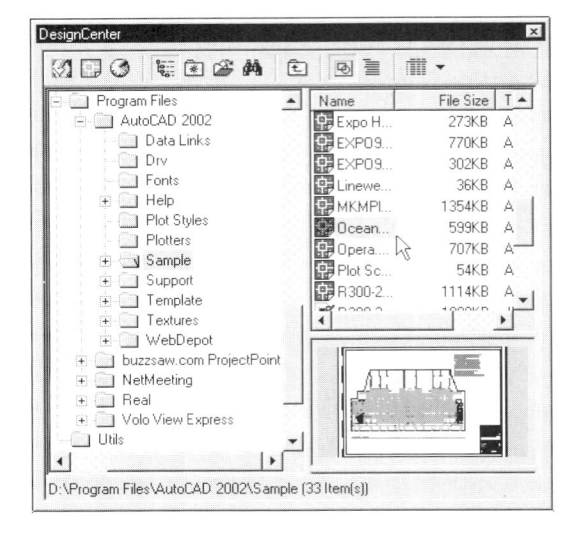

Figure 20-26

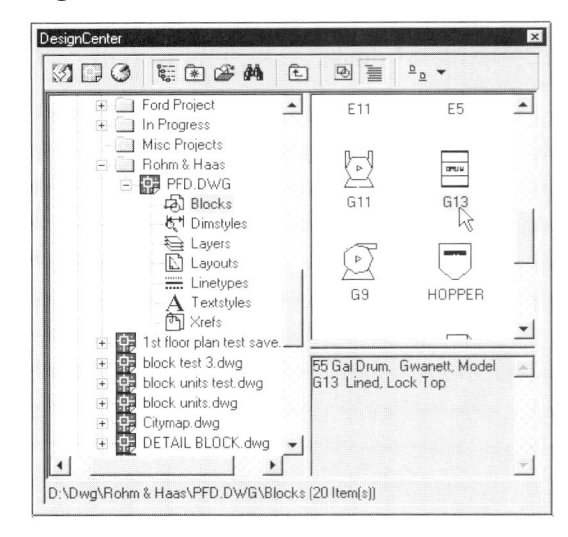

Views

Four possible *Views* of the Palette area are possible. Make the choice by selecting the down arrow to display the options (Fig. 20-27) or repeatedly click on the *View* icon as it cycles through the *Views*.

Large Icons
Small Icons
List
Details

Figure 20-27

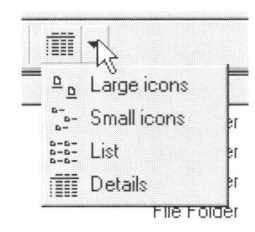

It is helpful to reset the configuration of the Palette depending on the job you are performing. For example, if the *Preview* pane is displayed, you can set the Palette to display only a *List* or *Details* (see Figure 20-25). Or if you prefer to "see" all the drawing files or *Blocks* in the Palette, it would be better to set the View to *Large Icons* and toggle *Preview* off (see Figure 20-26).

Generally, items are sorted alphabetically by name in the Palette. If you change the *View* to *Details*, you can sort files (not *Blocks*) by name, size, type, and other properties, depending on the type of content displayed in the Palette (see Figure 20-25). Click on the column header for the column to sort by (one click for ascending order, two clicks for descending order).

Loading the Palette with Windows Explorer
You can use Windows Explorer to load content into the Palette. For example, if you are browsing a network drive in Windows Explorer and locate a drawing file, you can drag the selected file directly into the Palette. The selected file must be dropped in to the Palette area, not the Tree View, Preview, or description areas. If you drop the file into any area other than the Palette, the drawing is opened into the AutoCAD session.

To Open a Drawing from DesignCenter

In addition to using DesignCenter to view and import content contained within drawing files, it is possible to use DesignCenter to *Open* drawings. While holding down the Ctrl key, drag the icon of the drawing file you want to open from the Palette and drop it in the drawing area. Remember, you must drag it from the Palette, not from the Tree View area.

Usually, you would want to drop the selected drawing into a blank (*New*) drawing. Unless the selected drawing is a titleblock and border, drop it into model space, not into a layout (paper space). Make sure the background (drawing area) is visible. You may need to resize the windows displaying any currently open drawings.

A drawing file that is dropped into AutoCAD is actually *Inserted* as a *Block*. The typical Command line prompts appear for insertion point, scale factors, and rotation angle. Generally, enter 0,0 as the insertion point and accept the defaults for X and Y scale factors and rotation angle. If you want to edit the geometry, you will have to *Explode* the drawing.

To *Insert* a *Block* Using DesignCenter

One of the primary functions of DesignCenter is to insert *Block* definitions into a drawing. When you insert a *Block* into a drawing, the block definition is copied into the drawing database. Any instance of that *Block* that you *Insert* into the drawing from that time on references the original *Block* definition.

You cannot add *Blocks* to a drawing while another command is active. If you try to drop a *Block* into AutoCAD while a command is active at the Command line, the icon changes to a slash circle indicating the action is invalid.

There are two methods for inserting *Blocks* into a drawing using DesignCenter: (1) using drag-and-drop with Autoscaling and (2) using the *Insert* dialog box with explicit insertion point, scale, and rotation value entry.

Block **Insertion with Drag-and-Drop**
When you drag-and-drop a *Block* from DesignCenter into a drawing, you are not prompted for an insertion point, X and Y scale factors, or a rotation angle. Although you specify an insertion point interactively when you "drop" the *Block* icon, AutoCAD uses Autoscaling to determine the scaling parameters. Autoscaling is a process of comparing the specified units of the *Block* definition (when the *Block* was created) and the *Drawing Units for Block Inserts* set in the *Drawing Units* dialog box of the target drawing. The value options are *Unitless, Inches, Feet, Miles, Millimeters, Centimeters, Kilometers*, and many other choices.

The *Drawing Units for Block Inserts* set in the *Drawing Units* dialog box (Fig. 20-28) should be set in the target drawing and controls how the *Block* units are scaled when the *Block* is dropped (into the target drawing). It is helpful that the setting here can be changed immediately before dropping a *Block* into the drawing. (Also see "*Units*" in Chapter 6.)

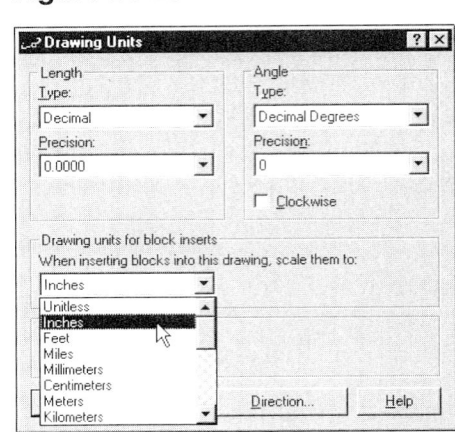

Figure 20-28

CHAPTER EXERCISES

1. *Block, Insert*
   In the next several exercises, you will create an office floor plan, then create pieces of furniture as *Blocks* and *Insert* them into the office. All of the block-related commands are used.

   A. Start a *New* drawing. Select **Start from Scratch** and use the **English** defaults. Use *Save* and assign the name **OFF-ATT**. Set up the drawing as follows:

1. *Units*	Architectural	1/2" Precision		
2. *Limits*	48' x 36'	(1/4"=1' scale on an A size sheet), drawing scale factor = 48		
3. *Snap, Grid*	3			
4. *Grid*	12			
5. *Layers*	FLOORPLAN	continuous	.014	colors of your choice
	FURNITURE	continuous	.060	
	ELEC-HDWR	continuous	.060	
	ELEC-LINES	hidden 2	.060	
	DIM-FLOOR	continuous	.060	
	DIM-ELEC	continuous	.060	
	TEXT	continuous	.060	
	TITLE	continuous	.060	
6. *Text Style*	CityBlueprint	CityBlueprint (TrueType font)		
7. *Ltscale*	48			

B. Create the floor plan shown in Figure 20-29. Center the geometry in the *Limits*. Draw on layer **FLOORPLAN**. Use any method you want for construction (e.g., *Line, Pline, Xline, Mline, Offset*).

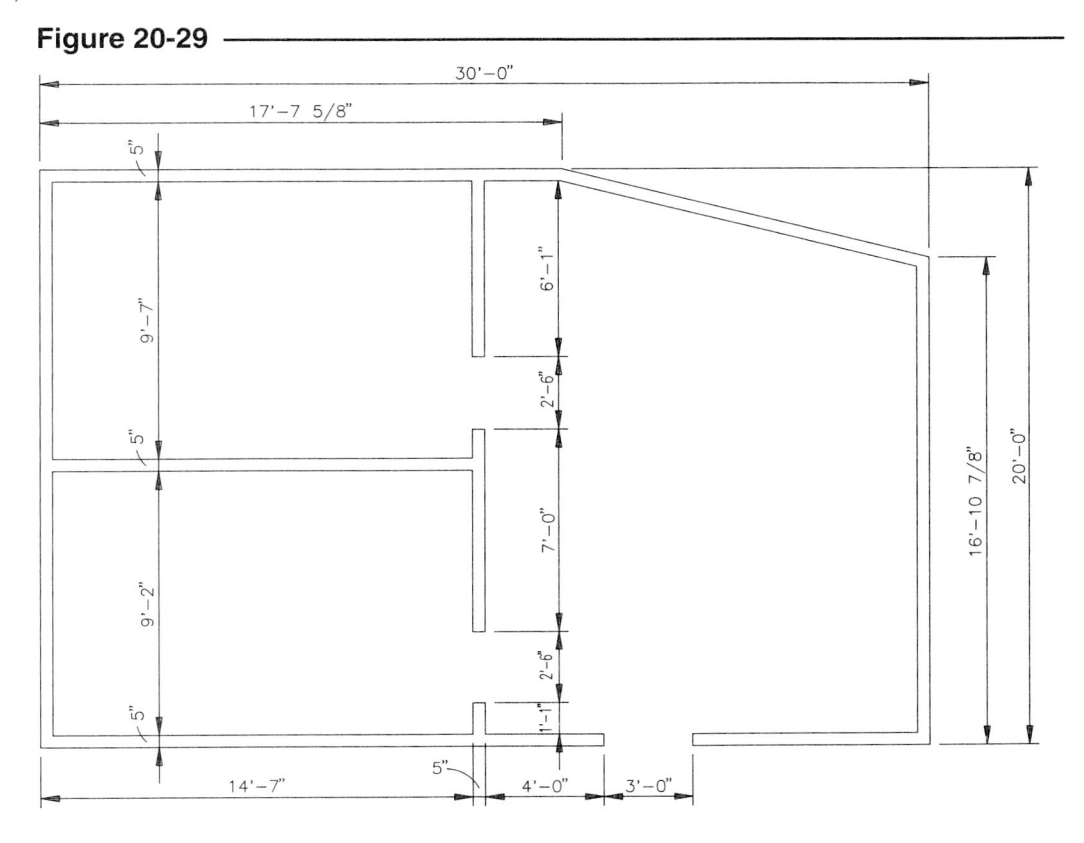

Figure 20-29

C. Create the furniture shown in Figure 20-30. Draw on layer **FURNITURE**. Locate the pieces anywhere for now.

Now make each piece a *Block*. Use the *name* as indicated and the *insertion base point* as shown by the "blip." Next, use the *Block* command again but to list the block definition table. Use *SaveAs* and rename the drawing **OFFICE**.

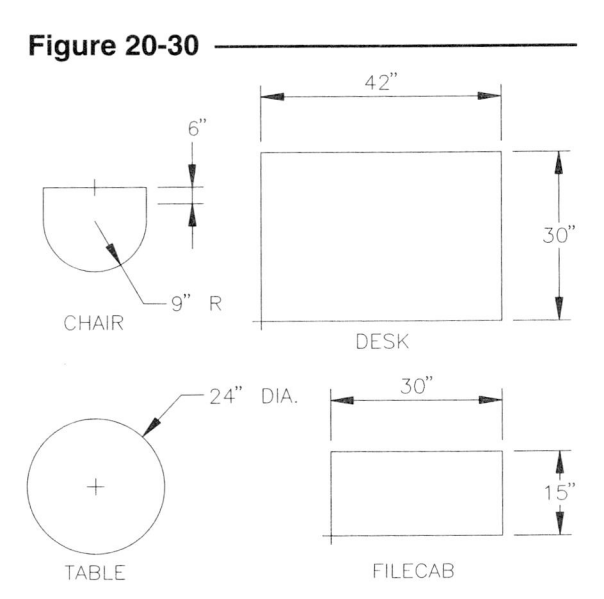

Figure 20-30

410 *Blocks and DesignCenter*

D. Use *Insert* to insert the furniture into the drawing, as shown in Figure 20-31. You may use your own arrangement for the furniture, but *Insert* the same number of each piece as shown. *Save* the drawing.

Figure 20-31

2. **Creating a .DWG file for** *Insertion, Base*

Figure 20-32

Begin a *New* drawing. Assign the name **CONFTABL**. Create the table as shown in Figure 20-32 on *Layer* **0**. Since this drawing is intended for insertion into the **OFFICE** drawing, use the *Base* command to assign an insertion base point at the lower-left corner of the table.

When you are finished, *Save* the drawing.

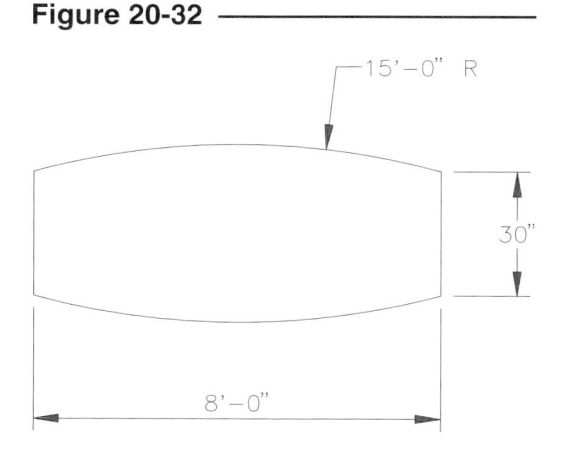

3. *Insert, Explode, Divide*

Figure 20-33

A. *Open* the **OFFICE** drawing. Ensure that layer **FURNITURE** is current. Use **DesignCenter** to bring the **CONFTABL** drawing in as a *Block* in the placement shown in Figure 20-33.

Notice that the CONFTABL assumes the linetype and color of the current layer, since it was created on layer **0**.

B. *Explode* the CONFTABL. The *Exploded* CONFTABL returns to *Layer* **0**, so use the *Properties* window to change it back to *Layer* **FURNITURE**. Then use the *Divide* command (with the *Block* option) to insert the **CHAIR** block as shown in Figure 20-33. Also *Insert* a **CHAIR** at each end of the table. *Save* the drawing.

4. *Wblock, BYBLOCK setting*

    Figure 20-34

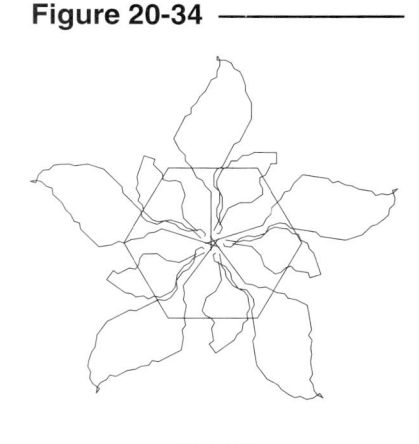

    A. *Open* the **EFF-APT** drawing you worked on in Chapter 15 Exercises. Use the *Wblock* command to transform the plant into a .DWG file (Fig. 20-34). Use the name **PLANT** and specify the *Insertion base point* at the center. Do not save the EFF-APT drawing. *Close* **EFF-APT**.

    B. *Open* the **OFFICE** drawing and use **DesignCenter** to *Insert* the **PLANT** into one of the three rooms. The plant probably appears in a different color than the current layer. Why? Check the *Layer* listing to see if any new layers came in with the PLANT block. *Erase* the PLANT block.

    PLANT

    C. *Open* the **PLANT** drawing. Change the *Color, Linetype*, and *Lineweight* setting of the plant objects to *BYBLOCK*. *Save* the drawing.

    D. *Open* the **OFFICE** drawing again and *Insert* the **PLANT** onto the **FURNITURE** layer. It should appear now in the current layer's *color*, *linetype*, and *lineweight*. *Insert* a **PLANT** into each of the 3 rooms. *Save* the drawing.

5. **Redefining a** *Block*

    Figure 20-35

    After a successful meeting, the client accepts the proposed office design with one small change. The client requests a slightly larger chair than that specified. *Explode* one of the **CHAIR** blocks. Use the *Scale* command to increase the size slightly or otherwise redesign the chair in some way. Use the *Block* command to redefine the **CHAIR** block. All previous insertions of the CHAIR should reflect the design change. *Save* the drawing. Your design should look similar to that shown in Figure 20-35. *Plot* to a standard scale based on your plotting capabilities.

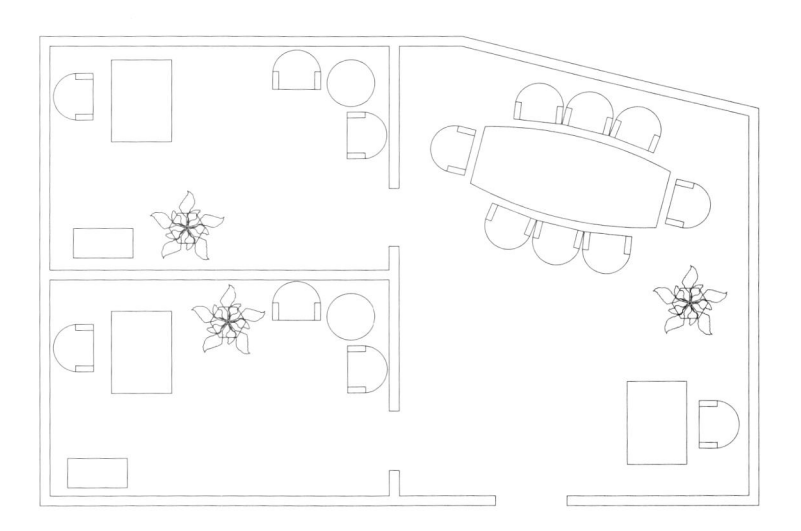

6. *Rename*

    Open the **OFFICE** drawing. Enter *Rename* (or select from the *Format* pull-down menu) to produce the *Rename* dialog box. Rename the *Blocks* as follows:

New Name	Old Name
DSK	DESK
CHR	CHAIR
TBL	TABLE
FLC	FILECAB

412 Blocks and DesignCenter

Next, use the *Rename* dialog box again to rename the *Layers* as indicated below:

New Name	Old Name
FLOORPLN-DIM	**DIM-FLOOR**
ELEC-DIM	**DIM-ELEC**

Use the *Rename* dialog box again to change the *Text Style* name as follows:

New Name	Old Name
ARCH-FONT	**CITYBLUEPRINT**

Use *SaveAs* to save and rename the drawing to **OFFICE-REN**.

7. *Purge*

   Using the Windows Explorer, check the file size of **OFFICE-REN.DWG**. Now use *Purge* to remove any unreferenced named objects. *Exit AutoCAD* and *Save Changes*. Check the file size again. Is the file size slightly smaller?

8. Create the process flow diagram shown in Figure 20-36. Create symbols (*Blocks*) for each of the valves and gates. Use the names indicated (for the *Blocks*) and include the text in your drawing. *Save* the drawing as **PFD**.

Figure 20-36

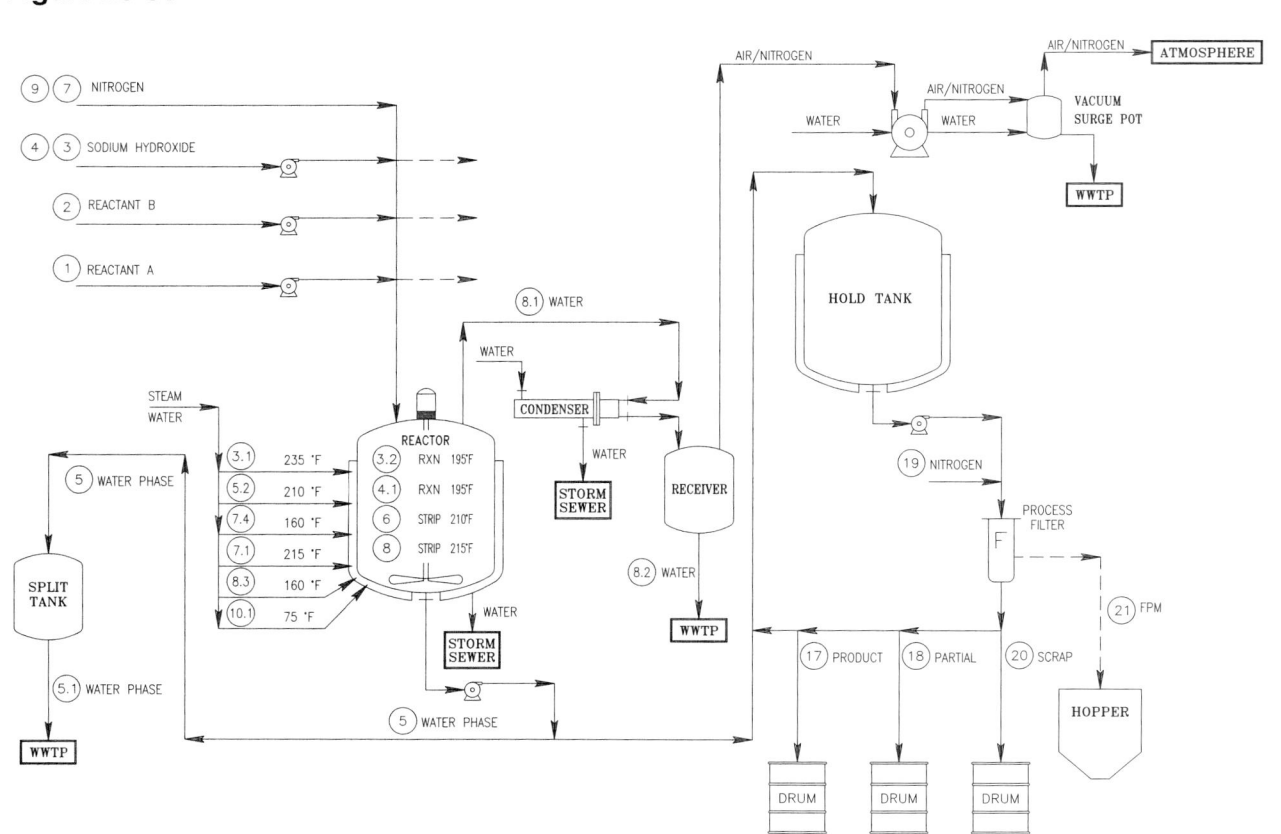

21

GRIP EDITING

Chapter Objectives

After completing this chapter you should:

1. be able to use the *GRIPS* variable to enable or disable object grips;

2. be able to activate the grips on any object;

3. be able to make an object's grips warm, hot, or cold;

4. be able to use each of the grip editing options, namely, STRETCH, MOVE, ROTATE, SCALE, AND MIRROR;

5. be able to use the Copy and Base suboptions;

6. be able to use the auxiliary grid that is automatically created when the Copy suboption is used.

CONCEPTS

Grips provide an alternative method of editing AutoCAD objects. The object grips are available for use by setting the *GRIPS* variable to **1**. Object grips are small squares appearing on selected objects at endpoints, midpoints, or centers, etc. The object grips are activated (made visible) by **PICK**ing objects with the cursor pickbox only <u>when no commands are in use</u> (at the open Command prompt). Grips are like small, magnetic *OSNAPs* (*Endpoint*, *Midpoint, Center, Quadrant*, etc.) that can be used for snapping one object to another, for example. If the cursor is moved within the small square, it is automatically "snapped" to the grip. Grips can replace the use of *OSNAP* for many applications. The grip option allows you to STRETCH, MOVE, ROTATE, SCALE, MIRROR, or COPY objects without invoking the normal editing commands or *OSNAP*s.

As an example, the endpoint of a *Line* could be "snapped" to the endpoint of an *Arc* (shown in Fig. 21-1) by the following steps:

Figure 21-1

1. Activate the grips by selecting both objects. Selection is done when no commands are in use (during the open Command prompt).
2. Select the grip at the endpoint of the *Line* (1). The grip turns **hot** (red).
3. The ** STRETCH ** option appears in place of the Command prompt.
4. STRETCH the *Line* to the endpoint grip on the *Arc* (2). **PICK** when the cursor "snaps" to the grip.
5. The *Line* and the *Arc* should then have connecting endpoints. The Command prompt reappears. Press Escape to cancel (deactivate) the grips.

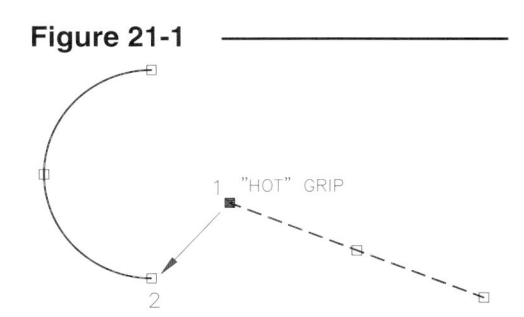

GRIPS FEATURES

GRIPS and *DDGRIPS*

Pull-down Menu	COMMAND (TYPE)	ALIAS (TYPE)	Short-cut	Screen (side) Menu	Tablet Menu
Tools Options... Selection	GRIPS or DDGRIPS	GR	...	TOOLS 2 Options... Grips	...

Grips are enabled or disabled by changing the setting of the system variable, *GRIPS*. A setting of 1 enables or turns ON *GRIPS* and a setting of **0** disables or turns OFF *GRIPS*. This variable can be typed at the Command prompt, or Grips can be invoked from the *Selection* tab of the *Options* dialog box, or by typing *Ddgrips* (Fig. 21-2). Using the dialog box, toggle *Enable Grips* to turn *GRIPS ON*. The default setting in AutoCAD for the *GRIPS* variable is **1** (*ON*).

Figure 21-2

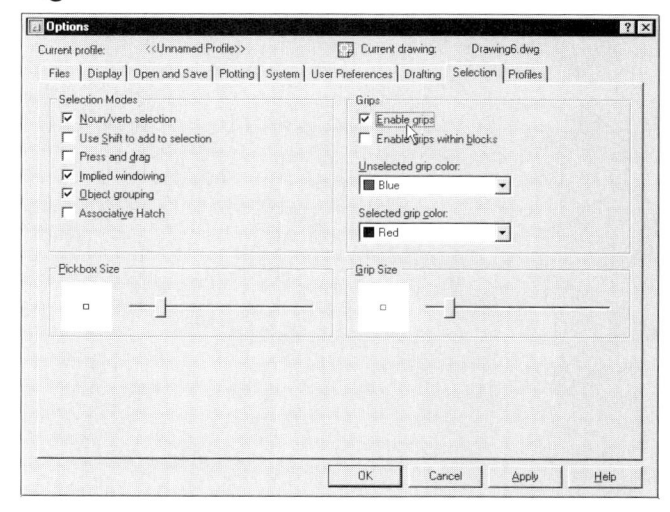

The *GRIPS* variable is saved in the system registry rather than in the current drawing as with most other system variables. Variables saved in the system registry are effective for any drawing session on that particular computer, no matter which drawing is current. The reasoning is that grip-related variables (and selection set-related variables) are a matter of personal preference and therefore should remain constant for a particular CAD station.

When *GRIPS* have been enabled, a small pickbox (3 pixels is the default size) appears at the center of the cursor crosshairs. (The pickbox also appears if the *PICKFIRST* system variable is set to **1**.) This pickbox operates in the same manner as the pickbox appearing during the "Select objects:" prompt. Only the pickbox, *AUTO window*, or *AUTO crossing window* methods can be used for selecting objects to activate the grips. (These three options are the only options available for Noun/Verb object selection as well.)

Activating Grips on Objects

Figure 21-3

The grips on objects are activated by selecting desired objects with the cursor pickbox, window, or crossing window. This action is done when no commands are in use (at the open Command prompt). When an object has been selected, two things happen: the grips appear and the object is highlighted. The grips are the small blue (default color) boxes appearing at the endpoints, midpoint, center, quadrants, vertices, insertion point, or other locations depending on the object type (Fig. 21-3). Highlighting indicates that the object is included in the selection set.

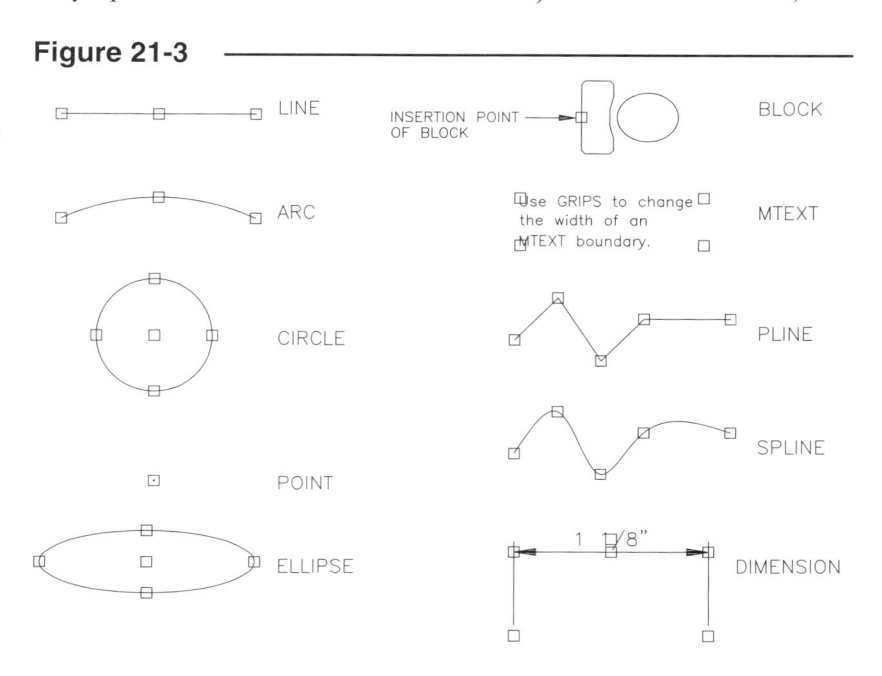

Warm, Hot, and Cold Grips

Figure 21-4

Grips can have three states: **warm, hot,** or **cold** (Fig. 21-4).

A grip is always **warm** first. When an object is selected, it is **warm**—its grips are displayed in blue (default color) and the object is highlighted. The grips can then be made **hot** or **cold**. **Cold** grips are also blue, but the object becomes unhighlighted.

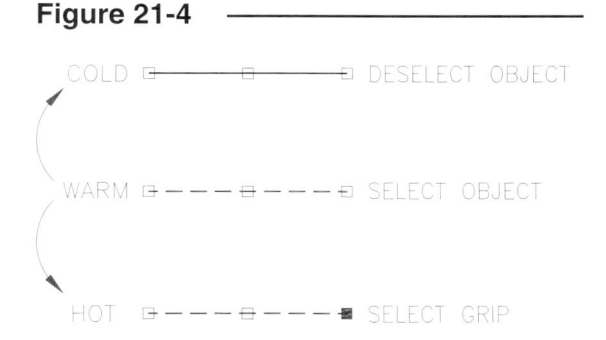

Cold grips are created by deselecting a highlighted object that has **warm** grips. In other words, removing the **warm** grip object from the selection set with SHIFT+#1, or using the *Remove* mode, makes its grips **cold**. A **cold** grip can be used as a point to snap to. A **cold** grip's object is not in the selection set and therefore is not affected by the MOVE, ROTATE, or other editing action. Pressing Escape deactivates all grips.

A **hot** grip is red (by default) and its object is almost always highlighted. Any grip can be changed to **hot** by selecting the grip itself. A hot grip is the default base point used for the editing action such as MOVE, ROTATE, SCALE, or MIRROR, or is the stretch point for STRETCH. When a **hot** grip exists, a new series of prompts appear in place of the Command prompt that displays the various grip editing options. The grip editing options are also available from a right-click cursor menu (Fig. 21-5). A grip must be changed to **hot** before the editing options appear. A grip can be transformed directly from **cold** to **hot**. Two or more grips can be made **hot** simultaneously by pressing Shift while selecting each grip.

Figure 21-5

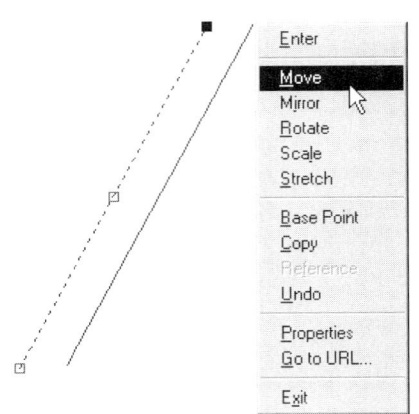

NOTE: Beware that there are two right-click (shortcut) menus available when grips are activated. The Grip menu (see previous Figure 21-5) is available only when a grip is hot (red). If you make any grip hot, then right-click, the Grip menu appears. The Grip menu contains grip editing options. However, if grips are warm and you right-click, the Edit Mode menu appears (Fig. 21-6). This menu appears any time one or more objects are selected, no commands are active, and you right-click (see Chapter 1, Edit Mode Menu). Although the commands displayed in the two menus appear to be the same, they are not. The Edit Mode menu contains the full commands. For example, *Move* in Figure 21-5 is the MOVE grip option, whereas *Move* in Figure 21-6 is the *Move* command.

Figure 21-6

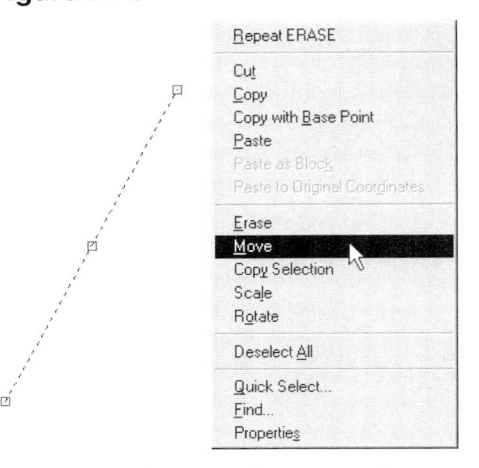

If you have made a *Grip* **hot** and want to deactivate it, possibly to make another *Grip* **hot** instead, press Escape once. This returns the object to a **warm** state. In effect, this is an undo only for the **hot** Grip. Pressing Escape again cancels all *Grips*.

The three states of grips can be summarized as follows:

Cold	blue grips unhighlighted object	The object is not part of the selection set, but cold grips can be used to "snap" to or used as an alternate base point.
Warm	blue grips highlighted object	The object is included in selection set and is affected by the editing action.
Hot	red grips highlighted object	The base point or control point for the editing action depending on which option is used. The object is included in selection set.

Pressing Escape demotes hot grips or cancels all grips:

Grip State	Press Escape once	Press Escape twice
only **cold**	grips are deactivated	
warm, cold	grips are deactivated	
hot, **warm**, cold	**hot** demoted to **warm**	grips are deactivated

Grip Editing Options

When a **hot grip** has been activated, the grip editing options are available. The Command prompt is replaced by the STRETCH, MOVE, ROTATE, SCALE, or MIRROR grip editing options. You can sequentially cycle through the options by pressing the Space bar or Enter key. The editing options are displayed in Figures 21-7 through 21-11.

Alternately, you can right-click when a grip is **hot** to activate the grip menu (see Figure 21-5). This menu has the same options available in Command line format with the addition of *Reference, Properties…*, and *Go to URL….* The options are described in the following figures.

** STRETCH **
Specify stretch point or
[Basepoint/Copy/Undo/eXit]:

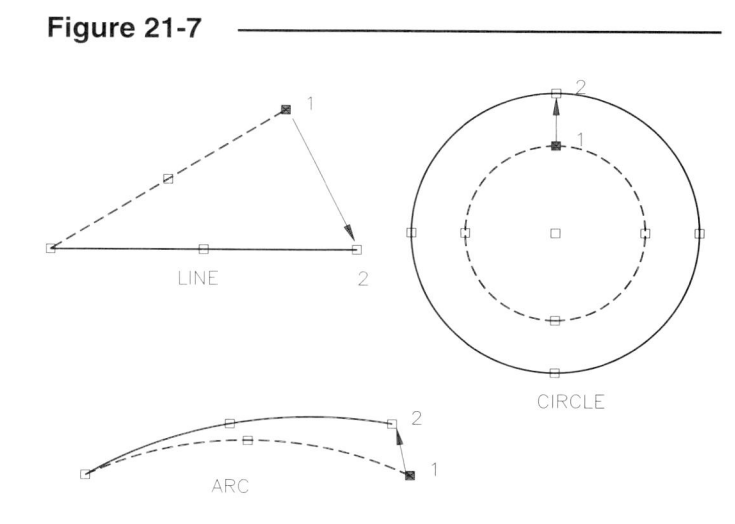

Figure 21-7

** MOVE **
Specify move point or
[Base point/Copy/Undo/eXit]:

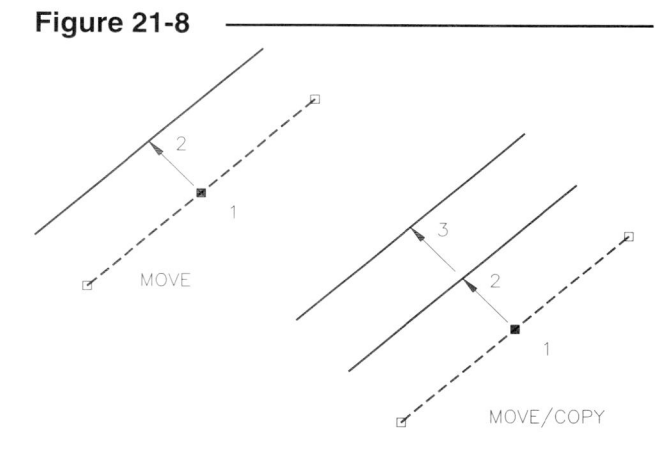

Figure 21-8

** ROTATE **
Specify rotation angle or
[Base point/Copy/Undo/Reference/eXit]:

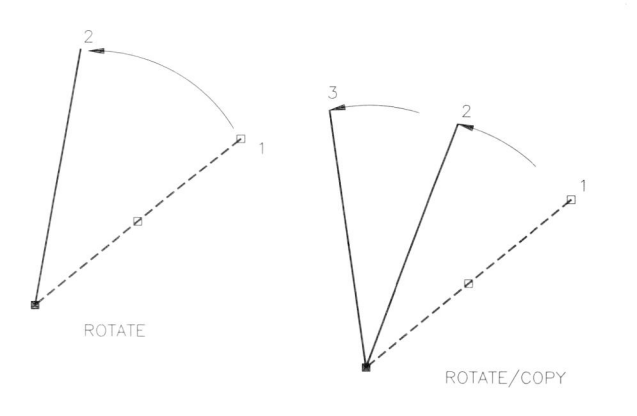

Figure 21-9

418 Grip Editing

** SCALE **
Specify scale factor or
[Base point/Copy/Undo/Reference/eXit]:

Figure 21-10

SCALE   SCALE/COPY

** MIRROR **
Specify second point or
[Base point/Copy/Undo/eXit]:

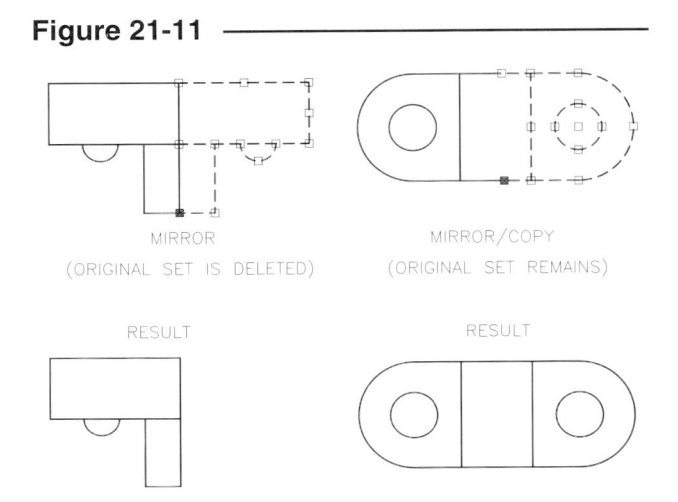

Figure 21-11

MIRROR (ORIGINAL SET IS DELETED)   MIRROR/COPY (ORIGINAL SET REMAINS)

RESULT   RESULT

The *Grip* options are easy to understand. Each option operates like the full AutoCAD command by the same name. Generally, the editing option used (except for STRETCH) affects all highlighted objects. The **hot** grip is the base point for each operation. The suboptions, Base and Copy, are explained next.

NOTE: The STRETCH option differs from other options in that STRETCH affects only the object that is attached to the **hot** grip, rather than affecting all highlighted (**warm**) objects.

Base
The Base suboption appears with all of the main grip editing options (STRETCH, MOVE, etc.). Base allows using any other grip as the base point instead of the **hot** grip. Type the letter *B* or select from the right-click cursor menu to invoke this suboption.

Copy
Copy is a suboption of every main choice. Activating this suboption by typing the letter *C* or selecting from the right-click cursor menu invokes a Multiple copy mode, such that whatever set of objects is STRETCHed, MOVEd, ROTATEd, etc., becomes the first of an unlimited number of copies (see the previous five figures). The Multiple mode remains active until exiting back to the Command prompt.

Undo
The Undo option, invoked by typing the letter *U* or selecting from the right-click cursor menu will undo the last Copy or the last Base point selection. Undo functions only after a Base or Copy operation.

Reference

This option operates similarly to the reference option of the *Scale* and *Rotate* commands. Use *Reference* to enter or PICK a new reference length (SCALE) or angle (ROTATE). (See "*Scale* and *Rotate*," Chapter 9.) With grips, *Reference* is only enabled when the SCALE or ROTATE options are active.

Properties... (Right-Click Menu Only)

Selecting this option from the right-click grip menu (see Figure 21-5) activates the *Properties* window. All highlighted objects become subjects of the window. Any property of the selected objects can be changed with this window. (See "*Properties*," Chapter 16.)

Go to URL... (Right-Click Menu Only)

If a URL (Internet address) is associated to the hot grip object, this option opens your Web browser and connects to the address. URLs can be attached to objects using the *Hyperlink* command (see "*Hyperlink*," Chapter 19, Internet Tools).

Auxiliary Grid

An auxiliary grid is <u>automatically</u> established on creating the first Copy (Fig. 21-12). The grid is activated by pressing Shift while placing the subsequent copies. The subsequent copies are then "snapped" to the grid in the same manner that *SNAP* functions. The spacing of this auxiliary grid is determined by the location of the first Copy, that is, the X and Y intervals between the base point and the second point.

Figure 21-12

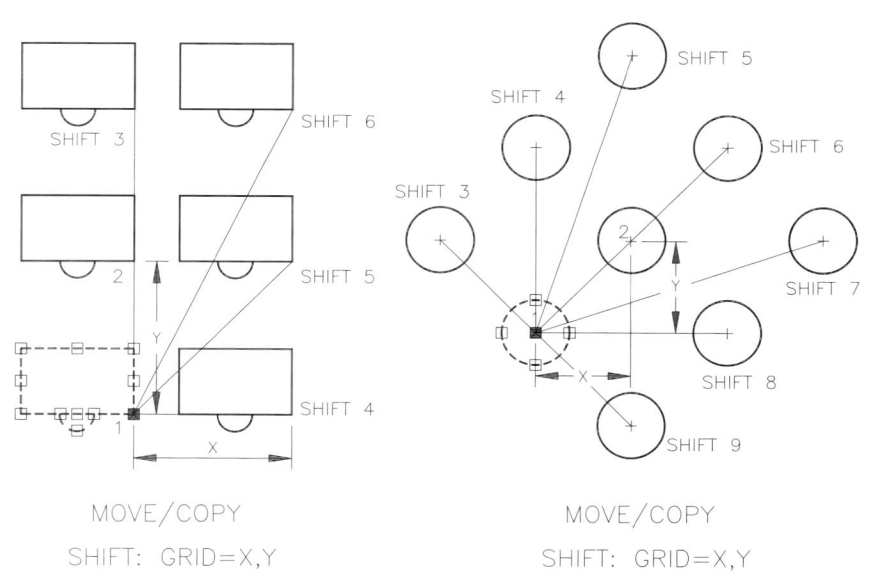

MOVE/COPY
SHIFT: GRID=X,Y

MOVE/COPY
SHIFT: GRID=X,Y

For example, a "polar array" can be simulated with grips by using ROTATE with the Copy suboption (Fig. 21-13).

The "array" can be constructed by making one Copy, then using the auxiliary grid to achieve equal angular spacing. The steps for creating a "polar array" are as follows:

1. Select the object(s) to array.

2. Select a grip on the set of objects to use as the center of the array. Cycle to the ROTATE option by pressing Enter or selecting from the right-click cursor menu. Next, invoke the Copy suboption.

3. Make the first copy at any desired location.

Figure 21-13

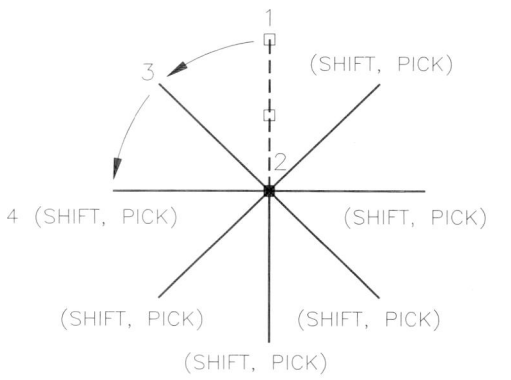

ROTATE/COPY
SHIFT — ANGLES EQUAL

4. After making the first copy, activate the auxiliary angular grid by holding down Shift while making the other copies.

5. Cancel the grips or select another command from the menus.

Editing Dimensions

One of the most effective applications of grips is as a dimension editor. Because grips exist at the dimension's extension line origins, arrowhead endpoints, and dimensional value, a dimension can be changed in many ways and still retain its associativity (Fig. 21-14). See Chapter 26, Dimensioning, for further information about dimensions, associativity, and editing dimensions with *Grips*.

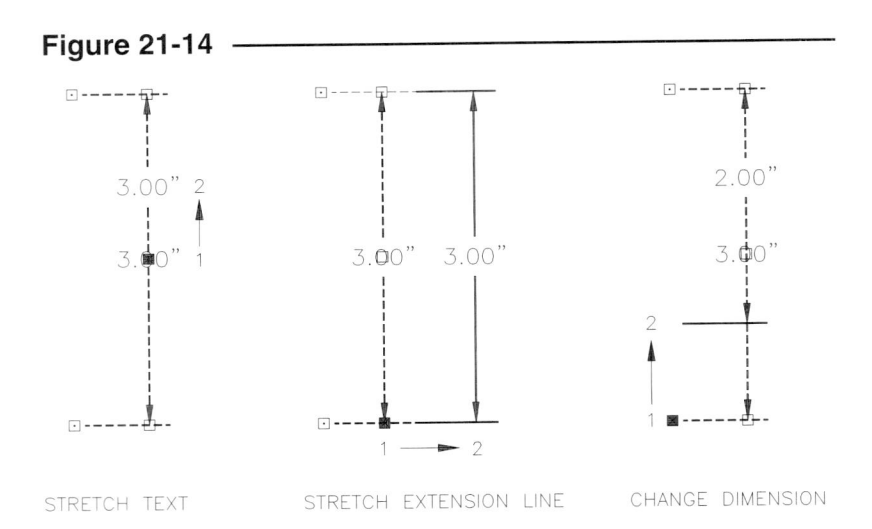

Figure 21-14

STRETCH TEXT   STRETCH EXTENSION LINE   CHANGE DIMENSION

GUIDELINES FOR USING GRIPS

Although there are many ways to use grips based on the existing objects and the desired application, a general set of guidelines for using grips is given here:

1. Create the objects to edit.

2. Select **warm** and **cold** grips first. This is usually accomplished by selecting all grips first (**warm** state), then deselecting objects to establish the desired **cold** grips.

3. Select the desired **hot** grip(s). The *Grip* options should appear in place of the Command line.

4. Press Space or Enter to cycle to the desired editing option (STRETCH, MOVE, ROTATE, SCALE, MIRROR) or select from the right-click cursor menu.

5. Select the desired suboptions, if any. If the *Copy* suboption is needed or the base point needs to be re-specified, do so at this time. *Base* or *Copy* can be selected in either order.

6. Make the desired STRETCH, MOVE, ROTATE, SCALE, or MIRROR.

7. Cancel the grips by pressing Escape or selecting a command from a menu.

CHAPTER EXERCISES

1. *Open* the **TEMP-D** drawing from Chapter 16 Exercises. Use the grips on the existing *Spline* to generate a new graph displaying the temperatures for the following week. **PICK** the *Spline* to activate **warm** grips. Use the **STRETCH** option to stretch the first data point grip to the value indicated below for Sunday's temperature. **Cancel** the grips; then repeat the steps for each data point on the graph. *Save* the drawing as **TEMP-F**.

   Figure 21-15

X axis	Y axis
Sunday	29
Monday	32
Tuesday	40
Wednesday	33
Thursday	44
Friday	49
Saturday	45

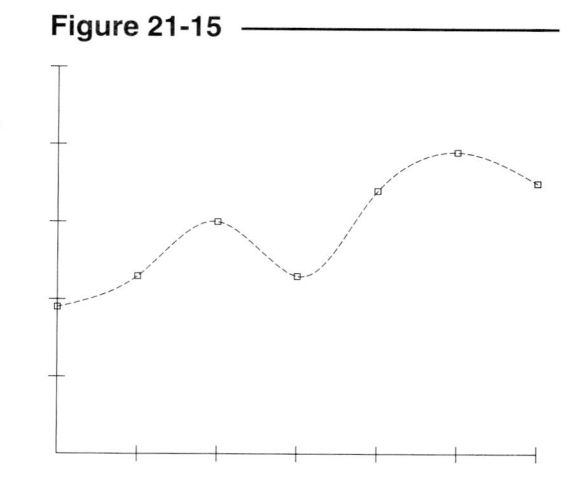

2. *Open* **CH16EX2** drawing. Activate **grips** on the *Line* to the far right. Make the top grip **hot**. Use the **STRETCH** option to stretch the top grip to the right to create a vertical *Line*. **Cancel** the grips.

   Next, activate the **grips** for all the *Lines* (including the vertical one on the right); then make the vertical *Line* grips **cold** as shown in Figure 21-16. **STRETCH** the top of all inclined *Lines* to the top of the vertical *Line* by making the common top grips **hot**, then stretching to the cold **grip** of the vertical *Line*. *Save* the drawing as **CH21EX2**.

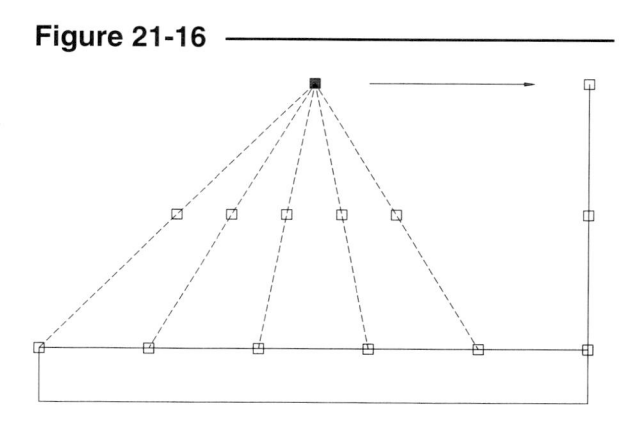

   Figure 21-16

3. This exercise involves all the options of grip editing to create a Space Plate. Begin a *New* drawing or use the **ASHEET** *template* and use *SaveAs* to assign the name **SPACEPLT**.

   Figure 21-17

   A. Set the *Snap* value to **.125**. Set **Polar Snap** to **.125** and turn on **Polar Tracking**. Draw the geometry shown in Figure 21-17 using the *Line* and *Circle* commands.

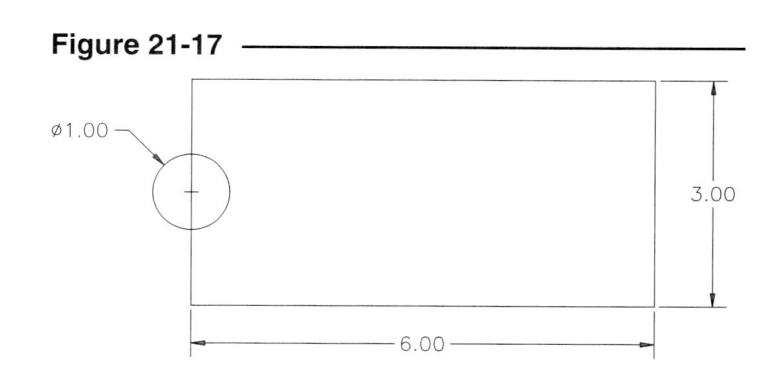

422   Grip Editing

B. Activate the **grips** on the *Circle*. Make the center grip **hot**. Cycle to the **MOVE** option. Enter *C* for the **Copy** option. You should then get the prompt for **MOVE (multiple)**. Make two copies as shown in Figure 21-18. The new *Circles* should be spaced evenly, with the one at the far right in the center of the rectangle. If the spacing is off, use **grips** with the **STRETCH** option to make the correction.

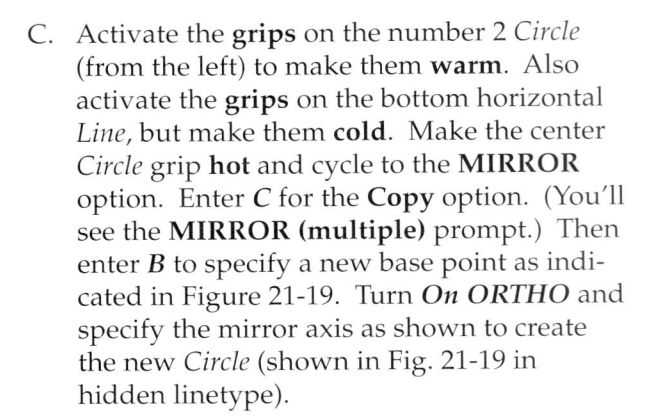

Figure 21-18

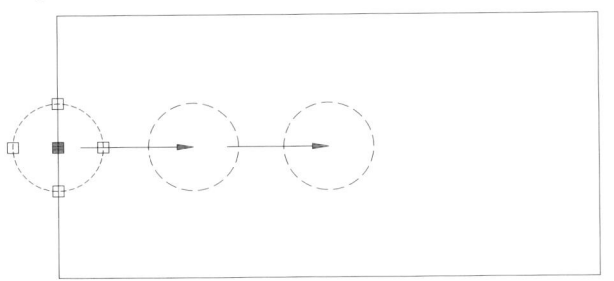

C. Activate the **grips** on the number 2 *Circle* (from the left) to make them **warm**. Also activate the **grips** on the bottom horizontal *Line*, but make them **cold**. Make the center *Circle* grip **hot** and cycle to the **MIRROR** option. Enter *C* for the **Copy** option. (You'll see the **MIRROR (multiple)** prompt.) Then enter *B* to specify a new base point as indicated in Figure 21-19. Turn *On* **ORTHO** and specify the mirror axis as shown to create the new *Circle* (shown in Fig. 21-19 in hidden linetype).

Figure 21-19

D. Use *Trim* to trim away the outer half of the *Circle* and the interior portion of the vertical *Line* on the left side of the Space Plate. Activate the **grips** on the two vertical *Lines* and the new *Arc*. Make the common grip **hot** (on the *Arc* and *Line* as shown in Fig. 21-20) and **STRETCH** it downward .5 units. (Note how you can affect multiple objects by selecting a common grip.) **Stretch** the upper end of the *Arc* upward .5 units.

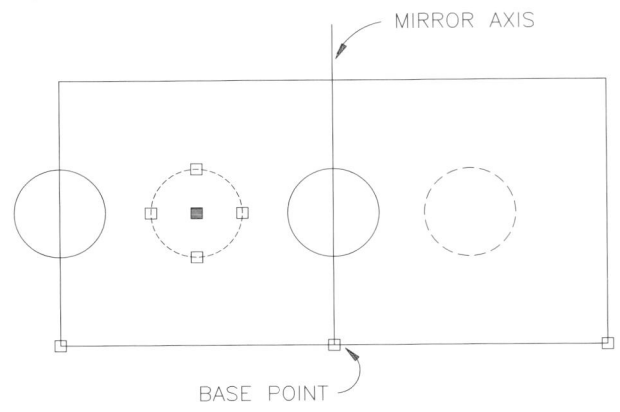

Figure 21-20

E. *Erase* the *Line* on the right side of the Space Plate. Use the same method that you used in step C to **MIRROR** the *Lines* and *Arc* to the right side of the plate (as shown in Fig. 21-21).

(REMINDER: Make sure that you make the grips on the bottom *Line* **cold**. After you select the **hot** grip, use the **Copy** option and the **Base** point option. Use *ORTHO*.)

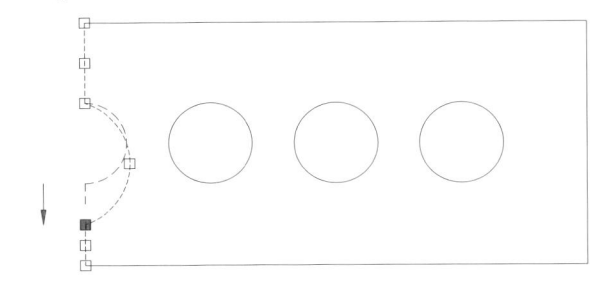

Figure 21-21

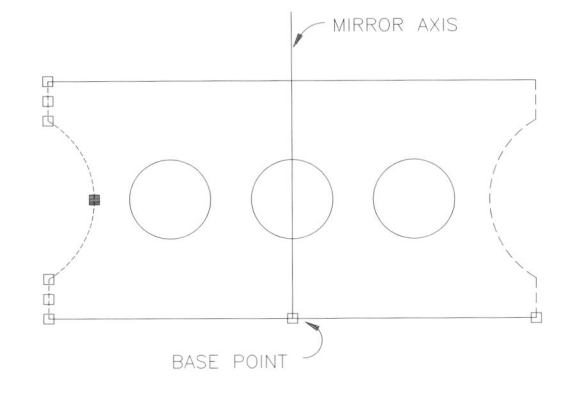

F. In this step, you will **STRETCH** the top edge upward one unit and the bottom edge downward one unit by selecting multiple **hot** grips.

Select the desired horizontal and attached vertical *Lines*. Hold down Shift while selecting each of the endpoint grips, as shown in Figure 21-22. Although they appear **hot**, you must select one of the two again to activate the **STRETCH** option.

Figure 21-22

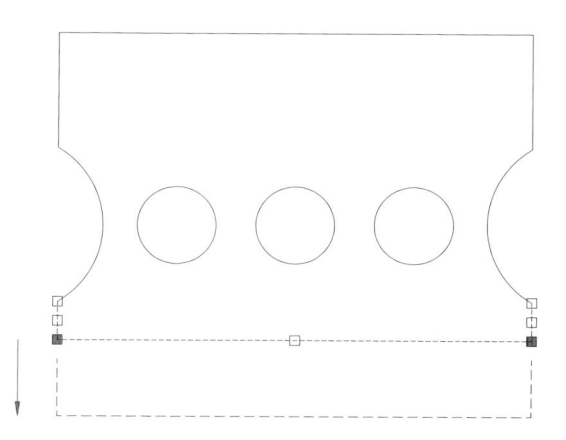

G. In this step, two more *Circles* are created by the **ROTATE** option (see Figure 21-23). Select the three existing *Circles* to make the grips **warm**. Deselect the center *Circle* so that it will not be copied. PICK the center grip to make it **hot**. Cycle to the **ROTATE** option. Enter *C* for the **Copy** option. Make sure *ORTHO* is *On* and create the new *Circles*.

Figure 21-23

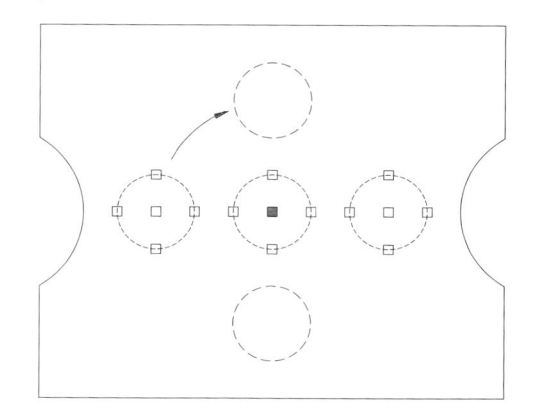

H. Select the center *Circle* and make the center grip **hot** (Fig. 21-24). Cycle to the **SCALE** option. The scale factor is **1.5**. Since the *Circle* is a 1 unit diameter, it can be interactively scaled (watch the *Coords* display), or you can enter the value. The drawing is complete. *Save* the drawing as **SPACEPLT**.

Figure 21-24

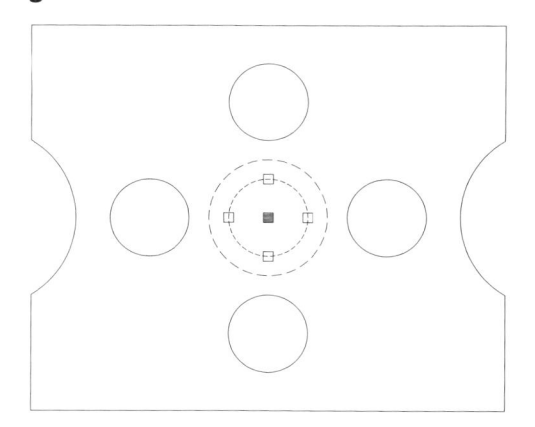

4. *Open* the **STORE ROOM2** drawing from Chapter 18 Exercises. Use the grips to **STRETCH, MOVE,** or **ROTATE** each text paragraph to achieve the results shown in Figure 21-25. *SaveAs* **STORE ROOM3**.

Figure 21-25

5. *Open* the **T-PLATE** drawing that you created in Chapter 9. An order has arrived for a modified version of the part. The new plate requires two new holes along the top, a 1" increase in the height, and a .5" increase from the vertical center to the hole on the left (Fig. 21-26). Use grips to **STRETCH** and **Copy** the necessary components of the existing part. Use *SaveAs* to rename the part to **TPLATEB**.

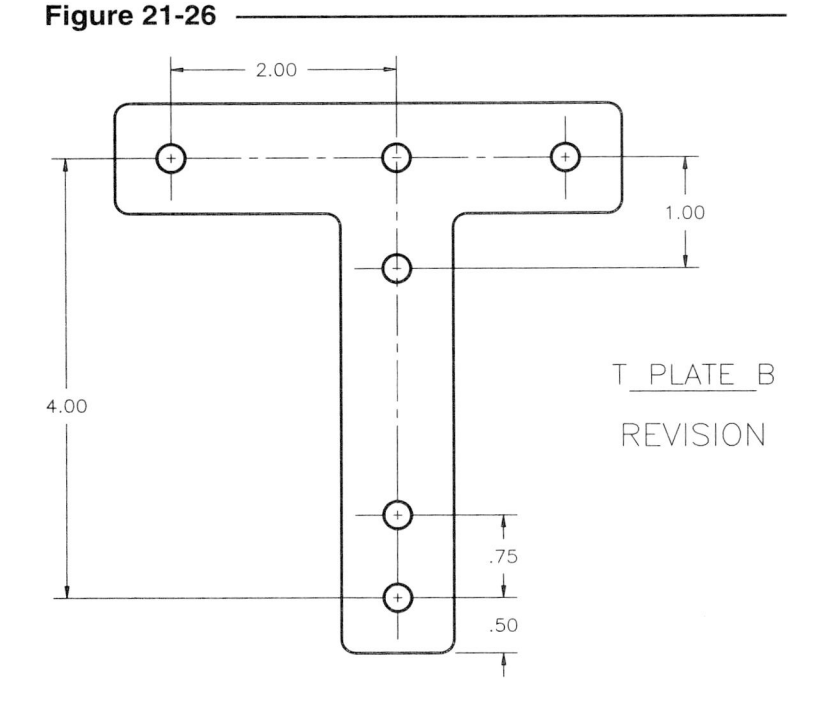

Figure 21-26

22

MULTIVIEW DRAWING

Chapter Objectives

After completing this chapter you should:

1. be able to draw projection lines using *ORTHO* and *SNAP*, and *Polar Tracking*;

2. be able to use *Xline* and *Ray* to create construction lines;

3. know how to use *Offset* for construction of views;

4. be able to use *Object Snap Tracking* for alignment of lines and views;

5. be able to use construction layers for managing construction lines and notes;

6. be able to use linetypes, lineweights, and layers to draw and manage ANSI standards;

7. know how to create fillets, rounds, and runouts;

8. know the typical guidelines for creating a three-view multiview drawing.

CONCEPTS

Multiview drawings are used to represent 3D objects on 2D media. The standards and conventions related to multiview drawings have been developed over years of using and optimizing a system of representing real objects on paper. Now that our technology has developed to a point that we can create 3D models, some of the methods we use to generate multiview drawings have changed, but the standards and conventions have been retained so that we can continue to have a universally understood method of communication.

This chapter illustrates methods of creating 2D multiview drawings with AutoCAD (without a 3D model) while complying with industry standards. Many techniques can be used to construct multiview drawings with AutoCAD because of its versatility. The methods shown in this chapter are the more common methods because they are derived from traditional manual techniques. Other methods are possible.

PROJECTION AND ALIGNMENT OF VIEWS

Projection theory and the conventions of multiview drawing dictate that the views be aligned with each other and oriented in a particular relationship. AutoCAD has particular features, such as *SNAP*, *ORTHO*, construction lines (*Xline, Ray*), Object Snap, Polar Tracking and Object Snap Tracking, that can be used effectively for facilitating projection and alignment of views.

Using *ORTHO* and *OSNAP* to Draw Projection Lines

ORTHO (F8) can be used effectively in concert with *OSNAP* to draw projection lines during construction of multiview drawings. For example, drawing a Line interactively with *ORTHO ON* forces the *Line* to be drawn in either a horizontal or vertical direction.

Figure 22-1 simulates this feature while drawing projection *Lines* from the top view over to a 45 degree miter line (intended for transfer of dimensions to the side view). The "first point:" of the *Line* originated from the *Endpoint* of the *Line* on the top view.

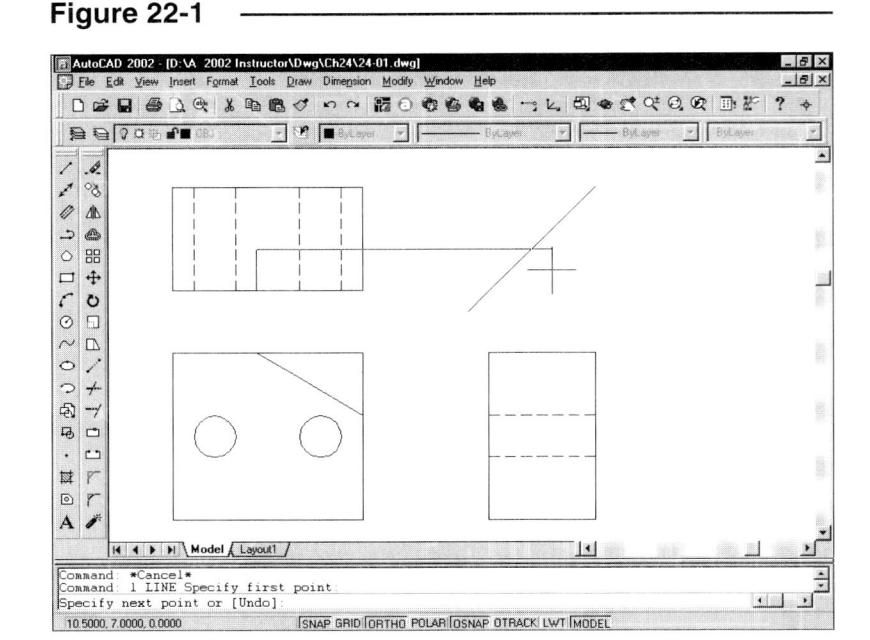

Figure 22-1

Figure 22-2 illustrates the next step. The vertical projection *Line* is drawn from the *Intersection* of the 45 degree line and the last projection line. *ORTHO* forces the *Line* to the correct vertical alignment with the side view.

Remember that any draw or edit command that requires PICKing is a candidate for *ORTHO* and/or *OSNAP*.

NOTE: *OSNAP* overrides *ORTHO*. If *ORTHO* is *ON* and you are using an *OSNAP* mode to PICK the "next point:" of a *Line*, the *OSNAP* mode has priority; and, therefore, the construction may not result in an orthogonal *Line*.

Figure 22-2

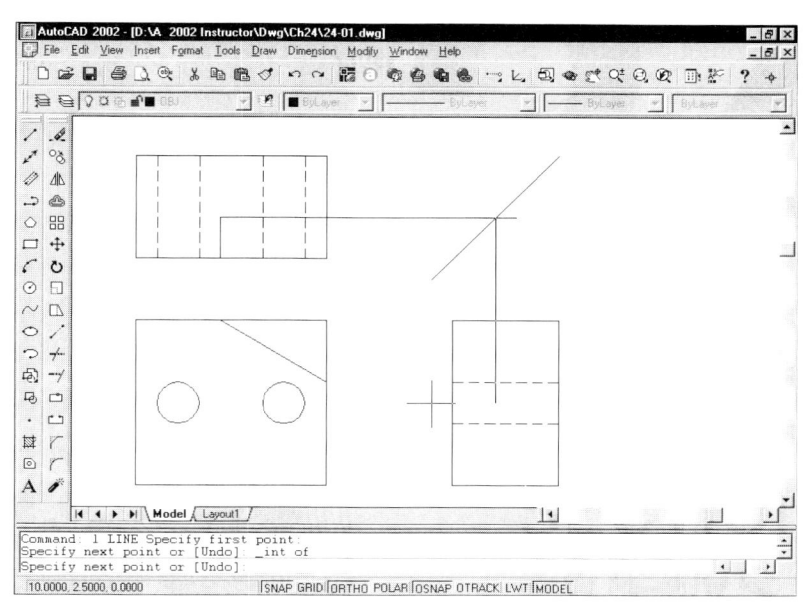

Using *Polar Tracking* to Draw Projection Lines

Similar to using *ORTHO* and *SNAP*, you can use *Polar Tracking* and *SNAP* options to draw projection lines at 90-degree increments. Using the previous example, *Polar Tracking* is used to project dimensions between the top and side views through the 45-degree miter line.

To use *Polar Tracking*, set the desired *Polar Angle Settings* in the *Polar Tracking* tab of the *Drafting Settings* dialog box. Ensure POLAR is appears recessed on the Status Bar. You can also set a *Polar Snap* increment or a *Grid Snap* increment to use with *Polar Tracking* (see Chapter 3 for more information on these settings).

For example, you could use the *Endpoint OSNAP* option to snap to the *Line* in the top view, then *Polar Tracking* forces the line to a previously set polar increment (0 degrees in this case).

Note that you can use *OSNAPs* in conjunction with *Polar Tracking*, as shown in Figure 22-3, such that the current horizontal line *OSNAPs* to its *Intersection* with the 45-degree miter line. (*OSNAPS* cannot be used with *ORTHO*, since *ORTHO* overrides any *OSNAPs*.)

Figure 22-3

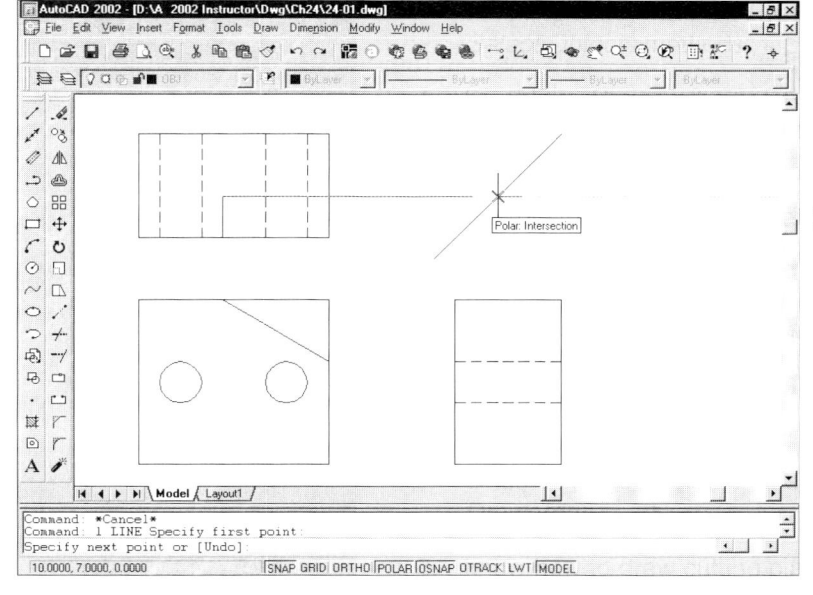

428 Multiview Drawing

In the following step, use *Intersection OSNAP* option to snap to the intersection of the previous *Line* and the 45-degree miter line. Again *Polar Tracking* forces the *Line* to a vertical position (270 degrees).

Figure 22-4

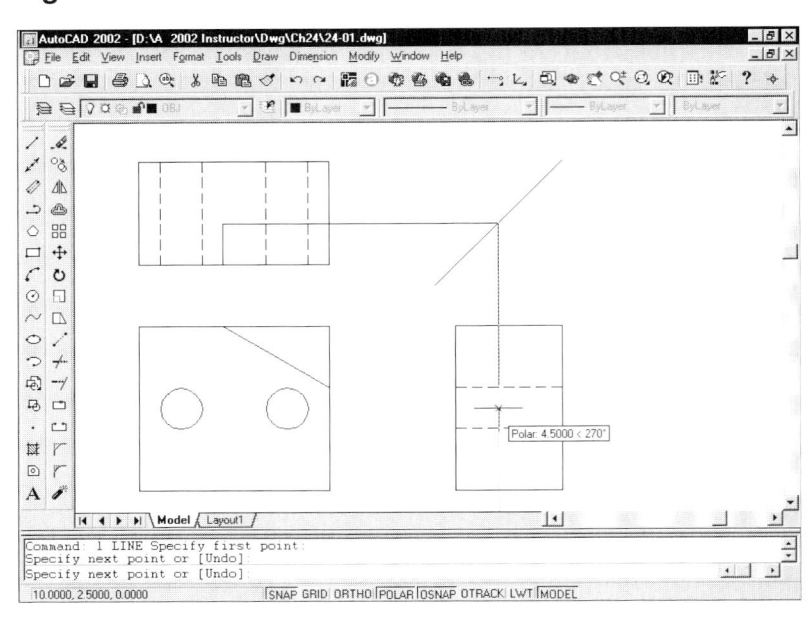

Using *Object Snap Tracking* to Draw Projection Lines Aligned with *OSNAP* Points

This feature, although the most complex, provides the greatest amount of assistance in constructing multiview drawings. The advantage is the availability of the features described in the previous method (*Polar Tracking* and *OSNAP* for alignment of vertical and horizontal *Lines*) in addition to the creation of *Lines* and other objects that align with *OSNAP* points (*Endpoint, Intersection, Midpoint,* etc.) of other objects.

To use *Object Snap Tracking*, first set the desired running *OSNAP* options (*Endpoint, Intersection,* etc.). Next, toggle on *OTRACK* at the Status Bar. Use *Polar Tracking* in conjunction with *Object Snap Tracking* by setting polar angle increments (see previous discussion) and toggling on *POLAR* on the Status Bar.

For example, to begin the *Line* in Figure 22-5, first "acquire" the *Endpoint* of the *Line* shown in the front view. Note that the "first point" of the new *Line* (in the top view) aligns vertically with the acquired *Endpoint*. At this point in time, the short vertical *Line* for the top view could be drawn from the intersection shown. *Object Snap Tracking* prevents having to draw the vertical projection line from the front up to the top view.

Figure 22-5

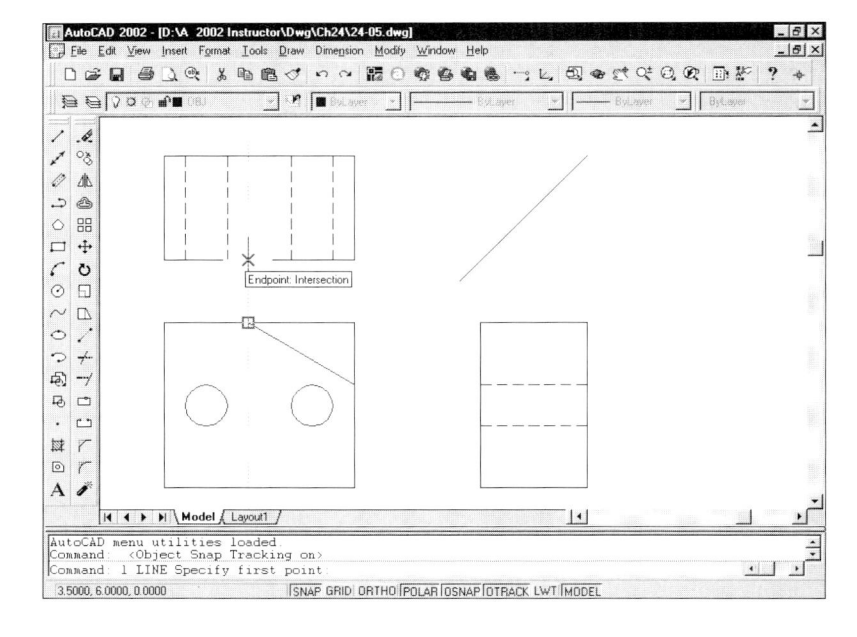

After constructing the two needed *Lines* for the top view, a tracking vector aligns with the acquired *Endpoint* of the indicated line in the top view and the *Intersection* of the 45-degree miter line. Note that a horizontal projection line is not needed since *Object Snap Tracking* ensures the "first point" aligns with the appropriate point from the top view.

Figure 22-6

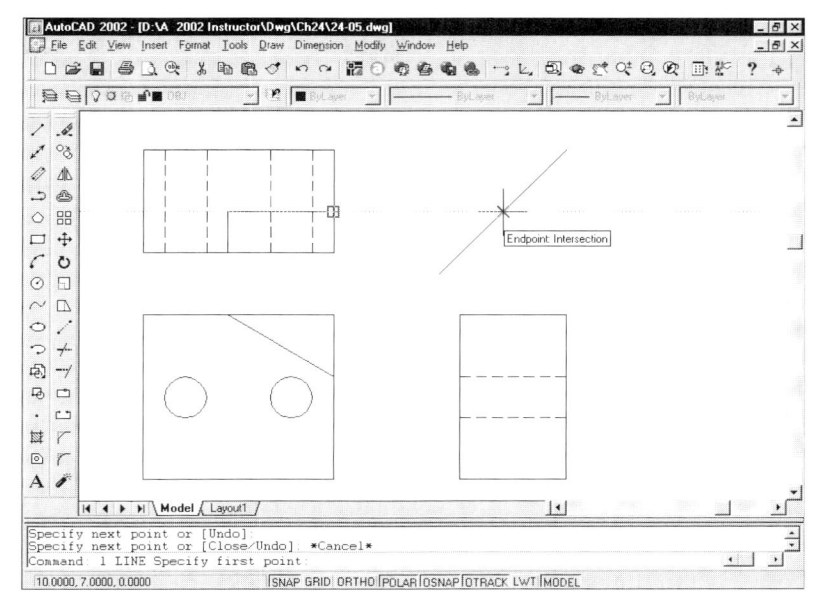

The next step involves drawing the "next point" of the vertical projection line from the 45-degree miter line to the appropriate point in the side view (see horizontal tracking vector) that also aligns with the related *Endpoint* in the front view. Since this *Line* is constructed to the correct endpoint, Trimming is unnecessary here, but it would be needed in the previous two methods.

Figure 22-7

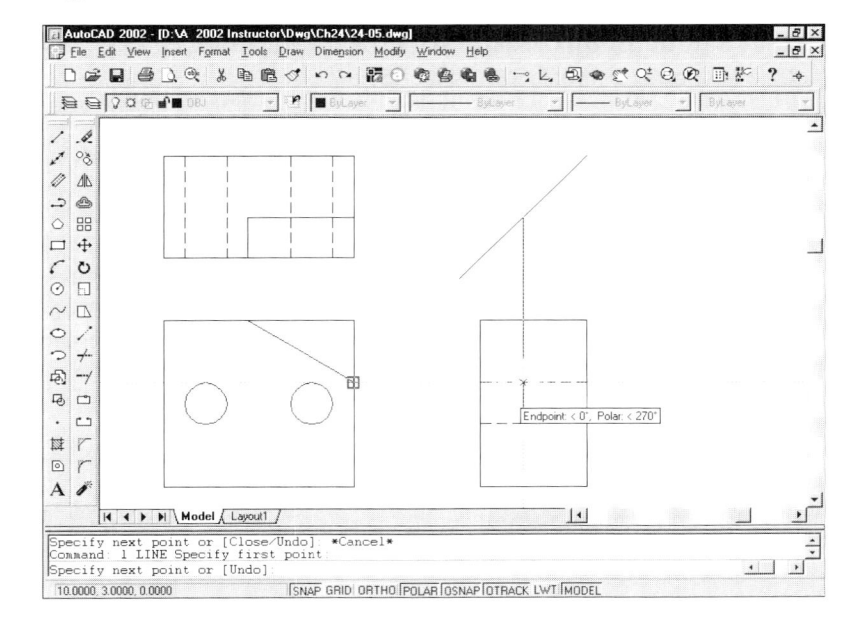

These AutoCAD features (*Object Snap Tracking* in conjunction with *Polar Tracking*) are probably the most helpful features for construction of multiview drawing since the introduction of AutoCAD in 1982.

430 Multiview Drawing

Using *Xline* and *Ray* for Construction Lines

Another strategy for constructing multiview drawings is to make use of the AutoCAD construction line commands *Xline* and *Ray*.

Xlines can be created to "box in" the views and ensure proper alignment. *Ray* is suited for creating the 45 degree miter line for projection between the top and side views. The advantage to using this method is that horizontal and vertical *Xlines* can be created quickly.

These lines can be *Trimmed* to become part of the finished view, or other lines could be drawn "on top of" the construction lines to create the final lines of the views. In either case, construction lines should be drawn on a separate layer so that the layer can be frozen before plotting. If you intend to *Trim* the construction lines so that they become part of the final geometry, draw them originally on the view layers.

Figure 22-8

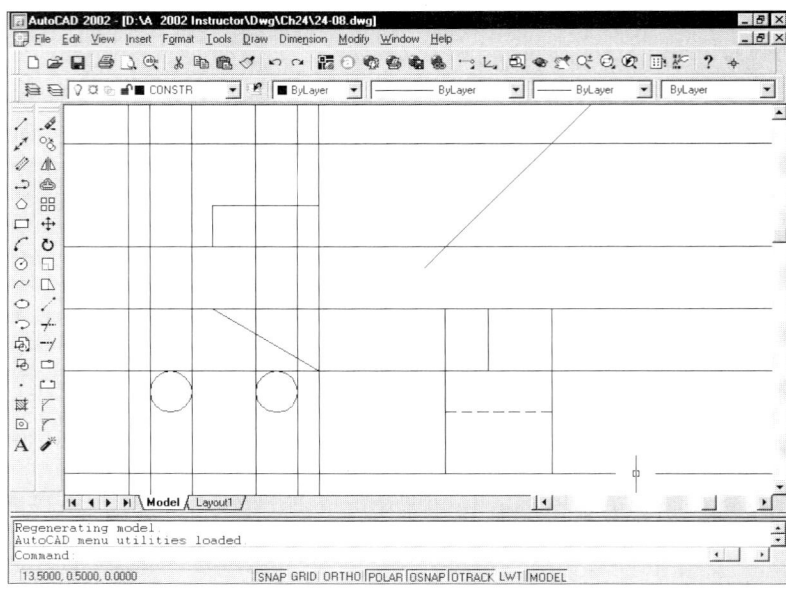

Using *Offset* for Construction of Views

An alternative to using the traditional miter line method for construction of a view by projection, the *Offset* command can be used to transfer distances from one view and to construct another. The *Distance* option of *Offset* provides this alternative.

For example, assume that the side view was completed and you need to construct a top view (Fig. 22-9). First, create a horizontal line as the inner edge of the top view (shown highlighted) by *Offset* or other method. To create the outer edge of the top view (shown in phantom linetype), use *Offset* and PICK points (1) and (2) to specify the *distance*. Select the existing line (3) as the "*Object to Offset:*", then PICK the "*side to offset*" at the current cursor position.

Figure 22-9

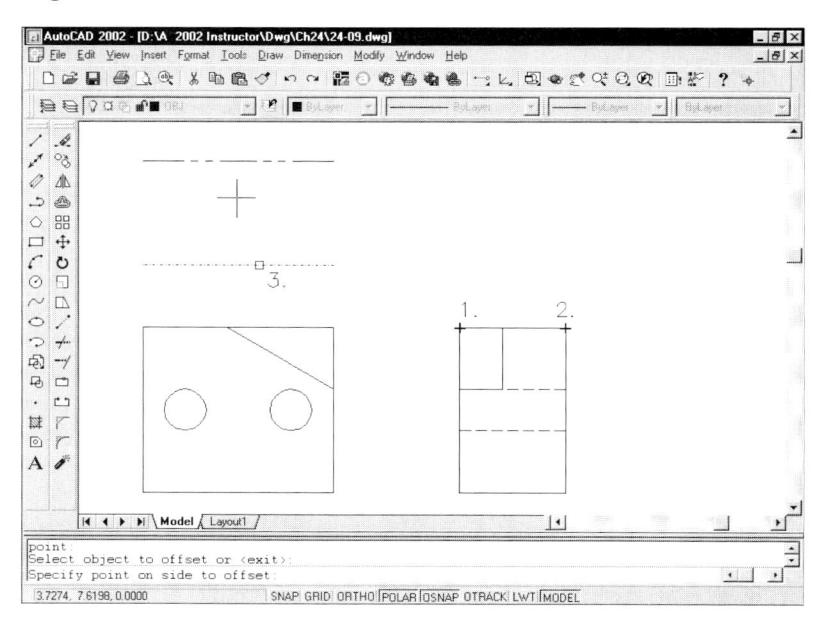

Realignment of Views Using *Polar Snap* and *Polar Tracking*

Another application of *Polar Snap* and *Polar Tracking* is the use of *Move* to change the location of an entire view while retaining its orthogonal alignment.

For example, assume that the views of a multiview (Fig. 22-10) are complete and ready for dimensioning; however, there is not enough room between the front and side views. You can invoke the *Move* command, select the entire view with a window or other option, and "slide" the entire view outward. *Polar Tracking* ensures proper orthogonal alignment. *Polar Snap* forces the movement to a regular increment so the coordinate points of the geometry retain a relationship to the original points (for example, moving exactly 1 unit).

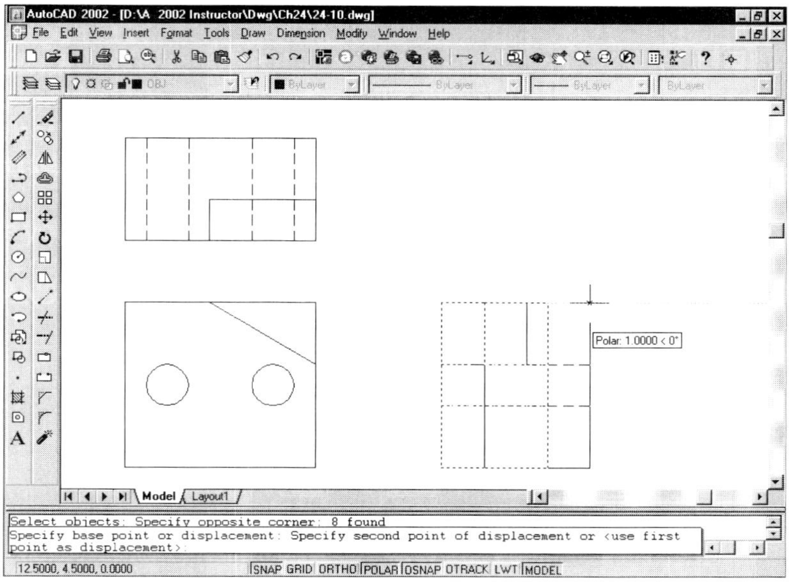

An alternative to *Mov*ing the view interactively is use of coordinate specification (absolute, relative rectangular, or relative polar or direct distance entry). (See Chapter 9 for an example of moving a view with direct distance entry.)

Figure 22-10

USING CONSTRUCTION LAYERS

The use of layers for isolating construction lines can make drawing and editing faster and easier. The creation of multiview drawings can involve construction lines, reference points, or notes that are not intended for the final plot. Rather than *Erasing* these construction lines, points, or notes before plotting, they can be created on a separate layer and turned *Off* or made *Frozen* before running the final plot. If design changes are required, as they often are, the construction layers can be turned *On*, rather than having to recreate the construction.

There are two strategies for creating construction objects on separate layers:

1. Use Layer 0 for construction lines, reference points, and notes. This method can be used for fast, simple drawings.

2. Create a new layer for construction lines, reference points, and notes. Use this method for more complex drawings or drawings involving use of *Blocks* on Layer 0.

432   Multiview Drawing

For example, consider the drawing during construction in Figure 22-11. A separate layer has been created for the construction lines, notes, and reference points.

Figure 22-11

In Figure 22-12, the same drawing is shown ready for making the final plot. Notice that the construction Layer has been *Frozen*.

If you are plotting the *Limits*, and the construction layer has objects <u>outside</u> the *Limits*, the construction layer should be *Frozen*, rather than being turned *Off*, unless only *Xlines* and *Rays* exist on the layer.

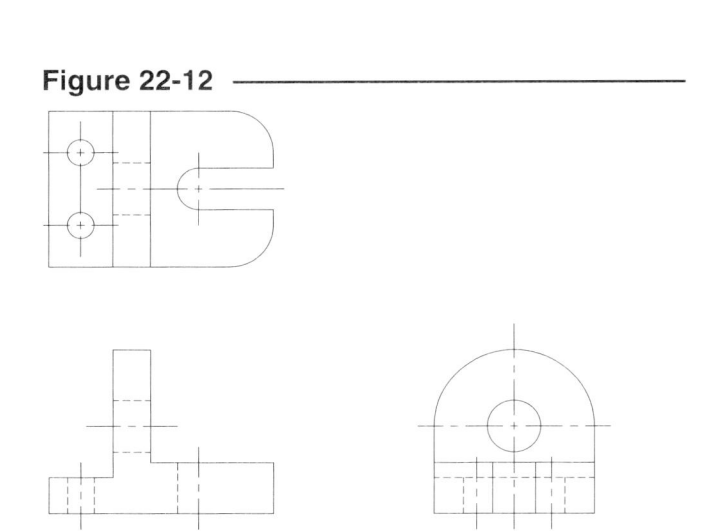

Figure 22-12

USING LINETYPES

Different types of lines are used to represent different features of a multiview drawing. Linetypes in AutoCAD are accessed by the *Linetype* command or through the *Layer Properties Manager*. *Linetypes* can be changed retroactively by the *Properties* window. *Linetypes* can be assigned to individual objects specifically or to layers (*ByLayer*). See Chapter 11 for a full discussion on this topic.

AutoCAD complies with the ANSI and ISO standards for linetypes. The principal AutoCAD linetypes used in multiview drawings and the associated names are shown in Figure 22-13.

Many other linetypes are provided in AutoCAD. Refer to Chapter 11 for the full list and illustration of the linetypes.

Figure 22-13

CONTINUOUS	————————————	DARK, WIDE
HIDDEN	— — — — — — — —	MEDIUM
CENTER	—— — —— — ——	MEDIUM
PHANTOM	——— — — ———	VARIES
DASHED	— — — — —	VARIES

Other standard lines are created by AutoCAD automatically. For example, dimension lines can be automatically drawn when using dimensioning commands (Chapter 26), and section lines can be automatically drawn when using the *Hatch* command (Chapter 24).

Objects in AutoCAD can have lineweight. This is accomplished by using the *Lineweight* command or by assigning *Lineweight* in the *Layer Properties Manager* (see Chapter 11). Additionally, lineweights can be assigned by using plot styles or assigning plot device lineweights or pen thickness.

Drawing Hidden and Center Lines

Figure 22-14

Figure 22-14 illustrates a typical application of AutoCAD *Hidden* and *Center* linetypes. Notice that the horizontal center line in the front view does not automatically locate the short dashes correctly, and the hidden lines in the right side view incorrectly intersect the center vertical line.

Although AutoCAD supplies ANSI standard linetypes, the application of those linetypes does not always follow ANSI standards. For example, you do not have control over the placement of the individual dashes of center lines and hidden lines. (You have control of only the endpoints of the lines and the *Ltscale*.) Therefore, the short dashes of center lines may not cross exactly at the circle centers, or the dashes of hidden lines may not always intersect as desired.

You do, however, have control of the endpoints of the lines. Draw lines with the *Center* linetype such that the endpoints are symmetric about the circle or group of circles. This action assures that the short dash occurs at the center of the circle (if an odd number of dashes are generated). Figure 22-15 illustrates correct and incorrect technique.

Figure 22-15

CORRECT SYMMETRY   INCORRECT SYMMETRY

You can also control the relative size of non-continuous linetypes with the *Ltscale* variable. In some cases, the variable can be adjusted to achieve the desired results.

For example, Figure 22-16 demonstrates the use of *Ltscale* to adjust the center line dashes to the correct spacing. Remember that *Ltscale* adjusts linetypes globally (all linetypes across the drawing).

Figure 22-16

When the *Ltscale* has been optimally adjusted for the drawing globally, use the *Properties* window to adjust the linetype scale of individual objects. In this way, the drawing lines can originally be created to the global linetype scale without regard to the *Celtscale*. The finished drawing linetype scale can be adjusted with *Ltscale* globally; then only those objects that need further adjusting can be fine-tuned retroactively with *Properties*.

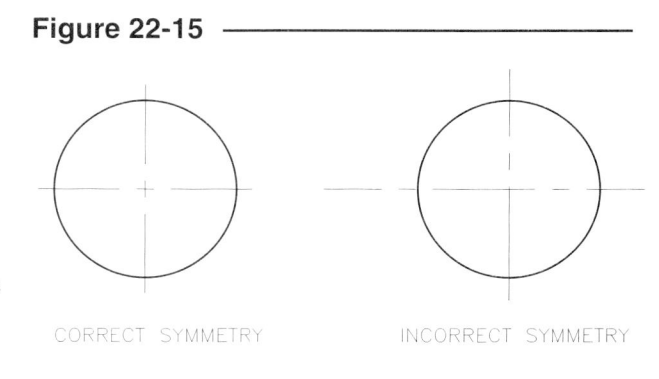

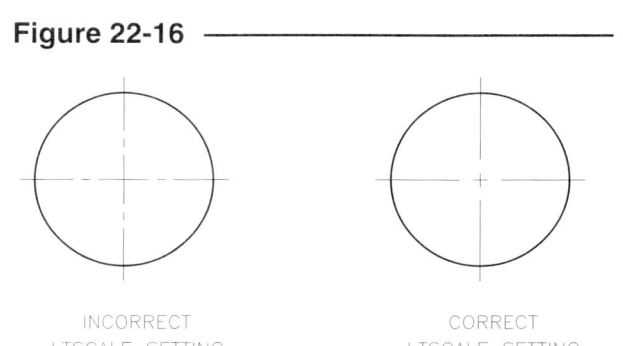

INCORRECT LTSCALE SETTING   CORRECT LTSCALE SETTING

TIP: The *Center* command (a dimensioning command) can be used to draw center lines automatically with correct symmetry and spacing (see Chapter 26, Dimensioning).

ANSI Standards require that multiview drawings are created with object lines having a dark lineweight, and hidden, center, dimension, and other reference lines created in a medium lineweight (see Figure 22-13).

You can assign *Lineweight* to objects using the *Lineweight* command or assign *Lineweight* to layers in the *Layer Properties Manager*. As described previously in Chapter 11, *Lineweight* can be assigned to individual objects or to layers (*ByLayer*). Use the *Lineweight Settings* dialog box to assign *Lineweight* property to objects. Use the *Layer Properties Manager* to assign *Lineweight* to layers. Generally, the *ByLayer Lineweight* assignment is preferred, similar to the *ByLayer* method of assigning *Color* and *Linetype*.

Additionally, lineweights can be assigned by using plot styles or assigning plot device lineweights or pen thickness.

Managing Linetypes, Lineweights, and Colors

There are two strategies for assigning linetypes, lineweights, and colors: *ByLayer* and object-specific assignment. In either case, thoughtful layer utilization for linetypes will make your drawings more flexible and efficient.

BYLAYER Linetypes, Lineweights, and Colors

The <u>*ByLayer* linetype, lineweight, and color settings are recommended when you want the most control over linetype visibility and plotting</u>. This is accomplished by creating layers with the *Layer Properties Manager* and assigning linetype, lineweight, and color for each layer. After you assign *Linetypes* to specific layers, you simply set the layer (with the desired linetype) as the *Current* layer and draw on that layer in order to draw objects in a specific linetype.

A template drawing for creating typical multiview drawings could be set up with the layer and linetype assignments similar to that shown in Figure 22-17. Notice the layer names, associated colors, and assigned linetypes and lineweights.

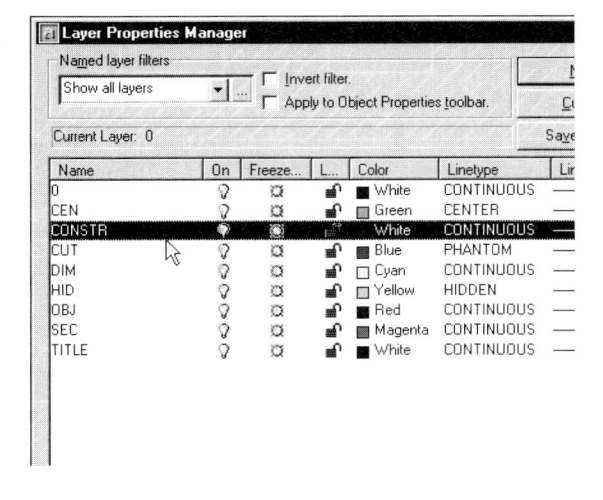

Figure 22-17

TIP: Using this strategy (*ByLayer Linetype*, *Lineweight*, and *Color* assignment) gives you flexibility. You can control the linetype visibility by controlling the layer visibility (show only object layers or hidden line layers). You can also <u>retroactively</u> change the linetype, lineweight, and color of an existing object by changing the object's *Layer* property with the *Properties* window. Objects changed to a different layer assume the *color*, *linetype*, and *lineweight* of the new layer.

Another strategy for multiview drawings involving several parts, such as an assembly, is to create layers for each linetype specific to each part, as shown in Figure 22-18. With this strategy, each part has the complete set of linetypes, but only one color per part in order to distinguish the part from others in the display.

Related layer groups can be selected within the dialog box using the *Named Layer Filters* dialog box. If the *-Layer* command is used instead, wildcards can be typed for layer selection. For example, entering "?????-HID" selects all of the layers with hidden lines, or entering "MOUNT*" would select all of the layers associated with the "MOUNT" part.

Object *Linetypes, Lineweights,* **and** *Colors*
Although this method can be complex, object-specific linetype, lineweight, and color assignment can also be managed easier by skillful utilization of layers. One method is to create one layer for each part or each group of related geometry (Fig. 22-19). The color, lineweight, and linetype settings should be left to the defaults. Then object-specific linetype settings can be assigned using the *Linetype*, *Lineweight*, and *Color* commands.

For assemblies, all lines related to one part would be drawn on one layer. Remember that you can draw everything with one linetype and color setting, then use *Properties* to retroactively set the desired color and linetype for each set of objects. Visibility of parts can be controlled by *Freezing* or *Thawing* part layers. You cannot isolate and control visibility of linetypes or colors by this method.

Figure 22-18

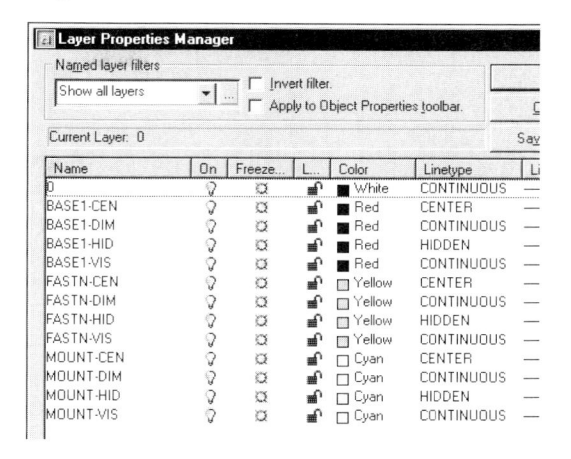

Figure 22-19

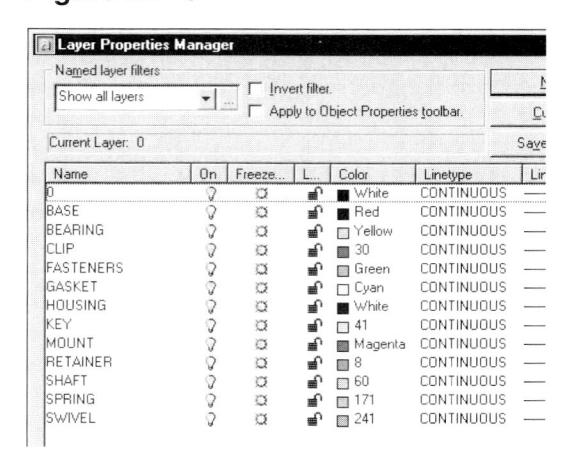

CREATING FILLETS, ROUNDS, AND RUNOUTS

Many mechanical metal or plastic parts manufactured from a molding process have slightly rounded corners. The otherwise sharp corners are rounded because of limitations in the molding process or for safety. A convex corner is called a round and a concave corner is called a fillet. These fillets and rounds are created easily in AutoCAD by using the *Fillet* command.

The example in Figure 22-20 shows a multiview drawing of a part with sharp corners before the fillets and rounds are drawn.

Figure 22-20

436 Multiview Drawing

The corners are rounded using the *Fillet* command. First, use *Fillet* to specify the *Radius*. Once the *Radius* is specified, just select the desired lines to *Fillet* near the end to round.

If the *Fillet* is in the middle portion of a *Line* instead of the end, *Extend* can be used to reconnect the part of the *Line* automatically trimmed by *Fillet*, or *Fillet* can be used in the *Notrim* mode.

Figure 22-21

A <u>runout is a visual representation</u> of a complex fillet or round. For example, when two filleted edges intersect at less than a 90 degree angle, a runout should be drawn as shown in the top view of the multiview drawing (Fig. 22-22).

The finish marks (V-shaped symbols) indicate machined surfaces. Finished surfaces have sharp corners.

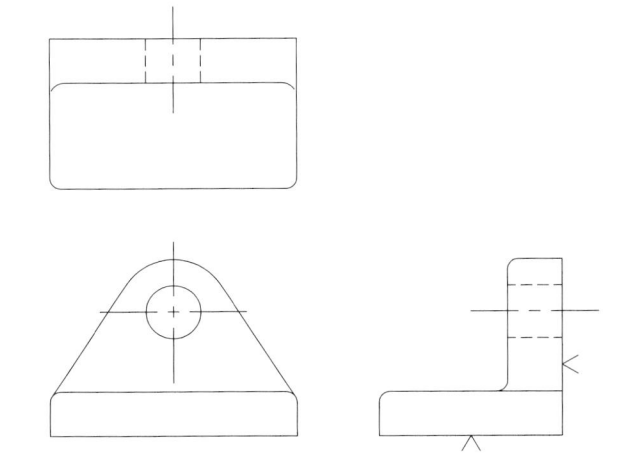

Figure 22-22

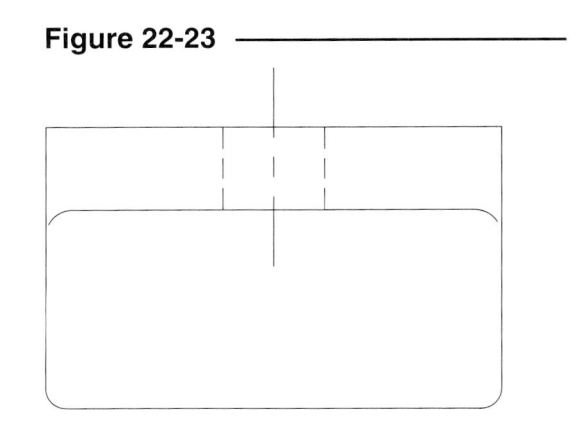

A close-up of the runouts is shown in Figure 22-23. No AutoCAD command is provided for this specific function. The *3point* option of the *Arc* command can be used to create the runouts, although other options can be used. Alternately, the *Circle TTR* option can be used with *Trim* to achieve the desired effect. As a general rule, use the same radius or slightly larger than that given for the fillets and rounds, but draw it less than 90 degrees.

Figure 22-23

GUIDELINES FOR CREATING A TYPICAL THREE-VIEW DRAWING

Following are some guidelines for creating the three-view drawing in Figure 22-24. This object is used only as an example. The steps or particular construction procedure may vary, depending on the specific object drawn. Dimensions are shown in the figure so you can create the multiview drawing as an exercise.

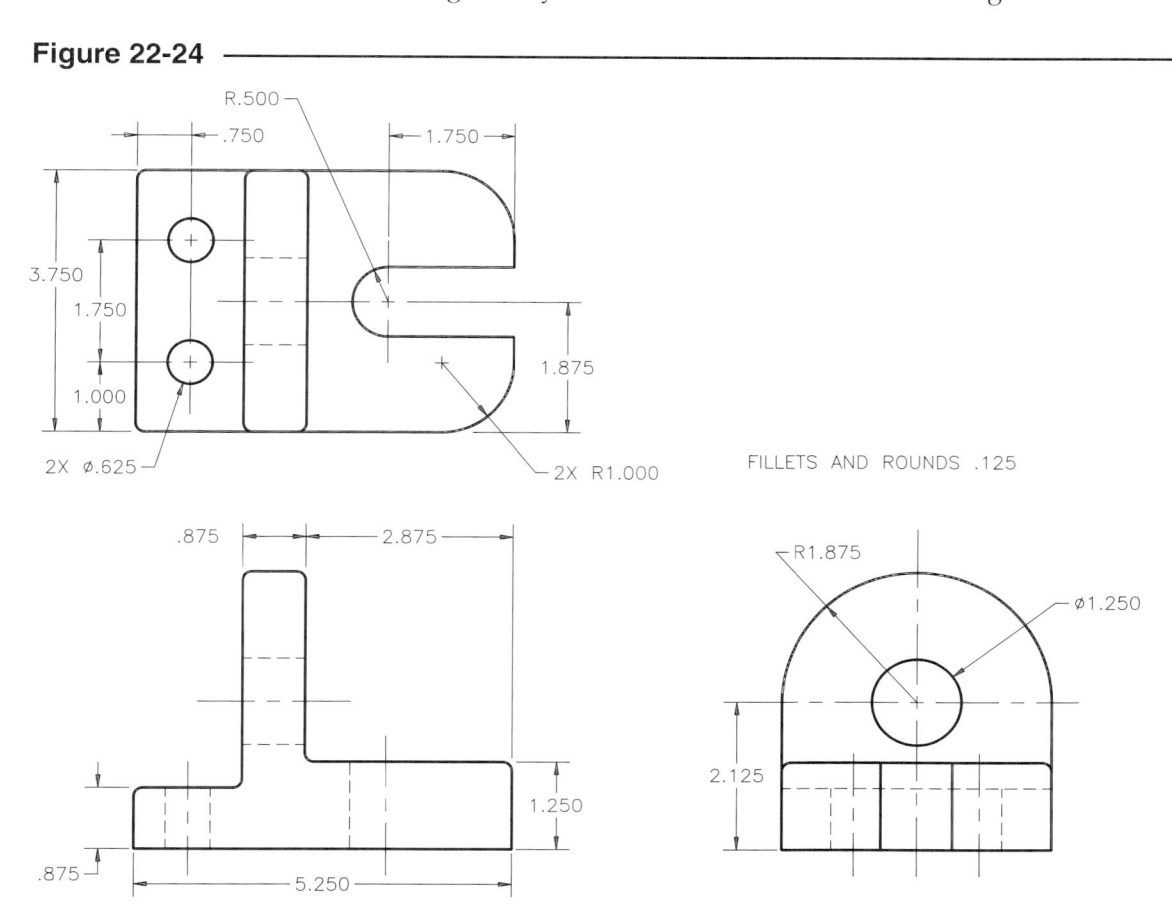

Figure 22-24

1. Drawing Setup

   Units are set to *Decimal* with 3 places of *Precision*. *Limits* of 22 x 17 are set to allow enough drawing space for both views. The finished drawing can be plotted on a B size sheet at a scale of 1"=1" or on an A size sheet at a scale of 1/2"=1". *Snap* is set to an increment of .125. *Grid* is set to an increment of .5. Although it is recommend that you begin drawing the views using *Grid Snap*, a *Polar Snap* increment of .125 is set, and *Polar Tracking* angles are set. Turn POLAR on. Set the desired running *OSNAPs*, such as *Endpoint*, *Midpoint*, *Intersection*, *Quadrant*, and *Center*. Turn on OTRACK. *Ltscale* is not changed from the default of 1. Layers are created (OBJ, HID, CEN, DIM, BORDER, and CONSTR) with appropriate *Linetypes*, *Lineweights*, and *Colors* assigned.

438   Multiview Drawing

2. An outline of each view is "blocked in" by drawing the appropriate *Lines* and *Circles* on the OBJ layer similar to that shown in Figure 22-25. Ensure that *SNAP* is *ON*. *ORTHO* or *Polar Tracking* should be turned *ON* when appropriate. Use the cursor to ensure that the views align horizontally and vertically. Note that the top edge of the front view was determined by projecting from the *Circle* in the right side view.

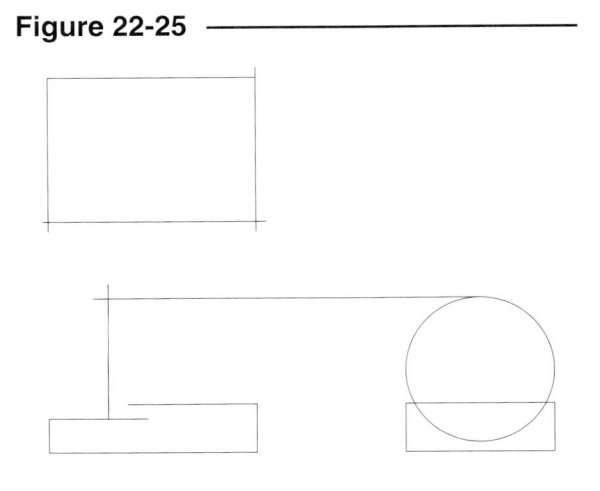

Figure 22-25

Another method for construction of a multiview drawing is shown in Figure 22-26. This method uses the *Xline* and *Ray* commands to create construction and projection lines. Here all the views are blocked in and some of the object lines have been formed. The construction lines should be kept on a separate layer, except in the case where *Xlines* and *Rays* can be trimmed and converted to the object lines. (The following illustrations do not display this method because of the difficulty in seeing which are object and construction lines.)

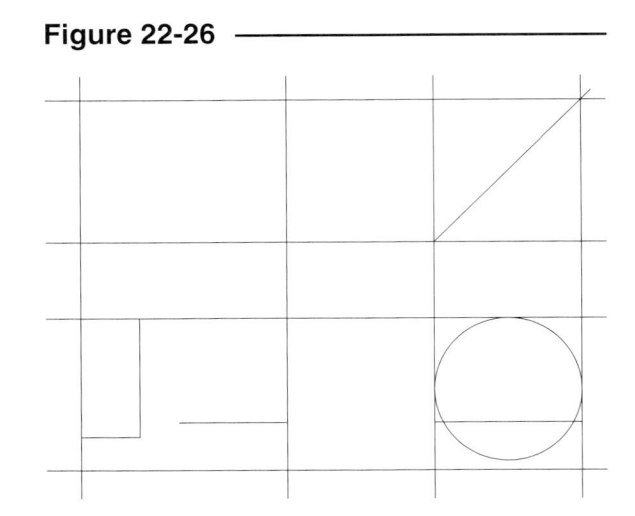

Figure 22-26

3. This drawing requires some projection between the top and side views (Fig. 22-27). The CONSTR layer is set as *Current*. Two *Lines* are drawn from the inside edges of the two views (using *OSNAP* and *ORTHO* for alignment). A 45 degree miter line is constructed for the projection lines to "make the turn." A *Ray* is suited for this purpose.

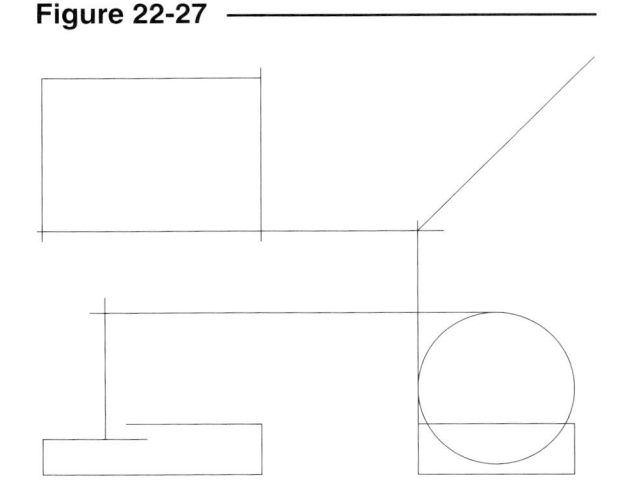

Figure 22-27

4. Details are added to the front and top views (Fig. 22-28). The projection line from the side view to the front view (previous figure) is *Trimmed*. A *Circle* representing the hole is drawn in the side view and projected up and over to the top view and to the front view. The object lines are drawn on layer OBJ and some projection lines are drawn on layer CONSTR. The horizontal projection lines from the 45 degree miter line are drawn on layer HID awaiting *Trimming*. Alternately, those two projection lines could be drawn on layer CONSTR and changed to the appropriate layer with *Properties* after *Trimming*. Keep in mind that *Object Snap Tracking* can be used here to ensure proper alignment with object features and to prevent having to actually draw some construction lines.

Figure 22-28

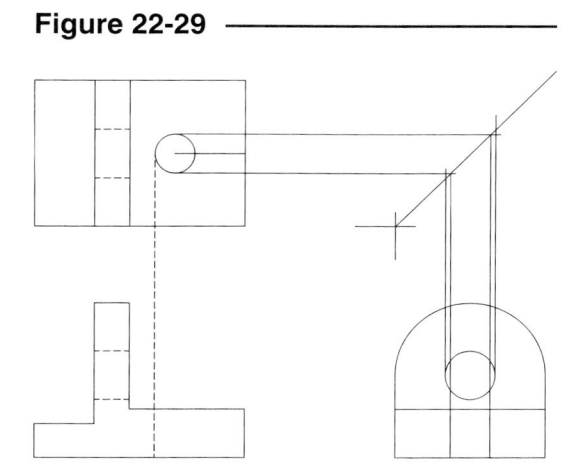

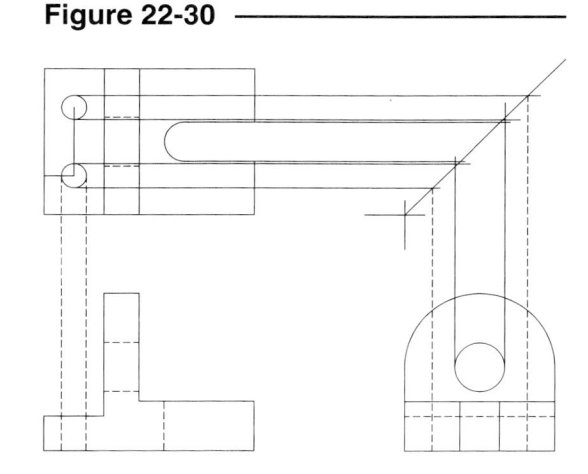

5. The hidden lines used for projection to the top view and front view (previous figure) are *Trimmed*. The slot is created in the top view with a *Circle* and projected to the side and front views. It is usually faster and easier to draw round object features in their circular view first, then project to the other views. Make sure you use the correct layers (OBJ, CONSTR, HID) for the appropriate features. If you do not, *Properties* can be used retroactively.

Figure 22-29

6. The lines shown in the previous figure as projection lines or construction lines for the slot are *Trimmed* or *Erased*. The holes in the top view are drawn on layer OBJ and projected to the other views. The projection lines and hidden lines are drawn on their respective layers.

Figure 22-30

440 Multiview Drawing

7. *Trim* the appropriate hidden lines. *Freeze* layer CONSTR. On layer OBJ, use *Fillet* to create the rounded corners in the top view. Draw the correct center lines for the holes on layer CEN. The value for *Ltscale* should be adjusted to achieve the optimum center line spacing. The *Properties* window can be used to adjust individual object linetype scale.

Figure 22-31 ─────────

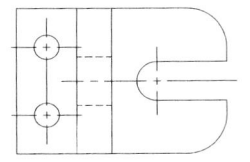

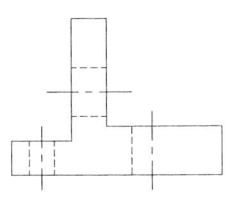

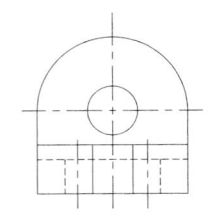

8. Fillets and rounds are added using the *Fillet* command. The runouts are created by drawing a *3point Arc* and *Trimming* or *Extending* the *Line* ends as necessary. Use *Zoom* for this detail work.

Figure 22-32 ─────────

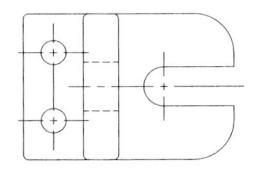

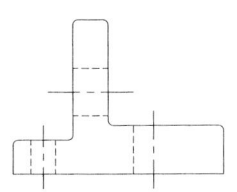

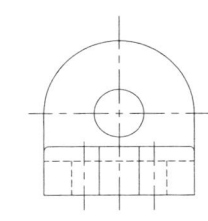

9. Activate a *Layout* tab, configure a print or plot device using *Pagesetup*, make one *Viewport*, and set the *Viewport scale* to a standard scale. Add a border and a title block using *Pline*. Include the part name, company, draftsperson, scale, date, and drawing file name in the title block. The drawing is ready for dimensioning and manufacturing notes.

Figure 22-33 ─────────

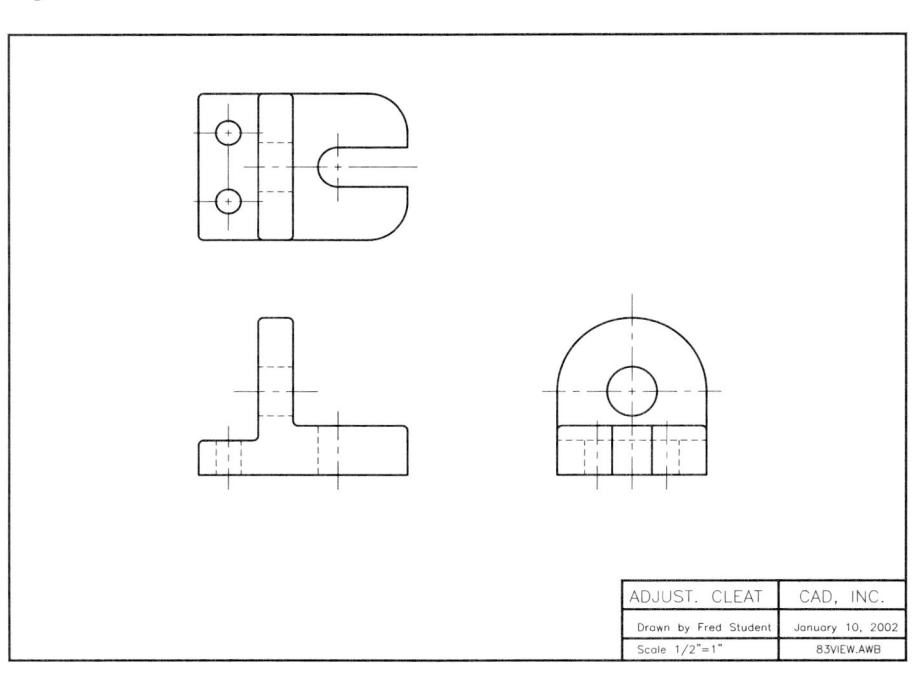

CHAPTER EXERCISES

1. *Open* the **PIVOTARM CH16** drawing.

   A. Create the right side view. Use *OSNAP* and *ORTHO* or *Polar Tracking* to create *Lines* or *Rays* to the miter line and down to the right side view as shown in Figure 22-34. *Offset* may be used effectively for this purpose instead. Use *Extend*, *Offset*, or *Ray* to create the projection lines from the front view to the right side view. Use *Object Snap Tracking* when appropriate.

   Figure 22-34

   B. *Trim* or *Erase* the unwanted projection lines, as shown in Figure 22-35. Draw a *Line* or *Ray* from the *Endpoint* of the diagonal *Line* in the top view down to the front to supply the boundary edge for *Trimming* the horizontal *Line* in the front view as shown.

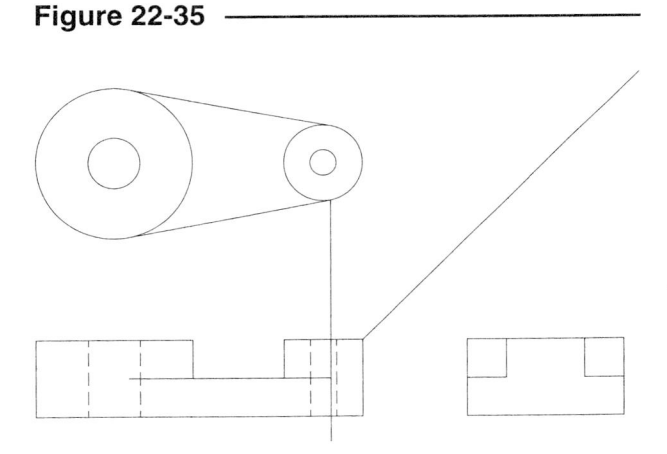

   Figure 22-35

   C. Next, create the hidden lines for the holes by the same fashion as before. Use previously created *Layers* to achieve the desired *Linetypes*. Complete the side view by adding the horizontal hidden *Line* in the center of the view.

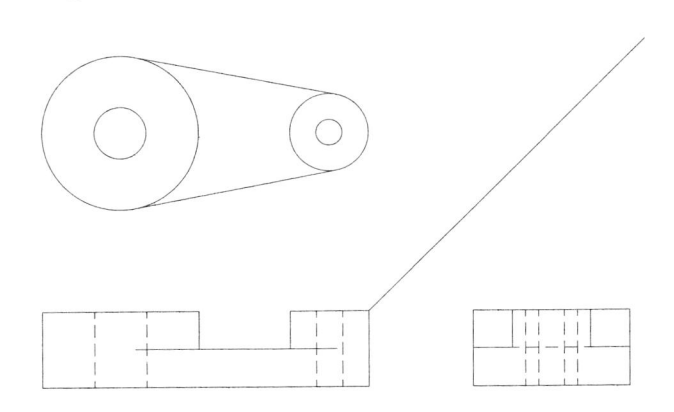

   Figure 22-36

442 Multiview Drawing

D. Another hole for a set screw must be added to the small end of the Pivot Arm. Construct a *Circle* of **4**mm diameter with its center located **8**mm from the top edge in the side view as shown in Figure 22-37. Project the set screw hole to the other views.

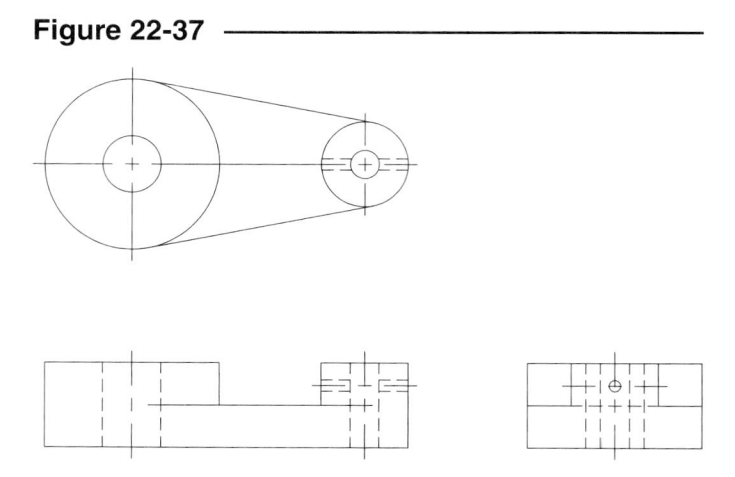

Figure 22-37

E. Make new layers **CONSTR**, **OBJ**, and **TITLE** and change objects to the appropriate layers with *Properties*. *Freeze* layer **CONSTR**. Add centerlines on the **CEN** layer as shown in Figure 22-37. Change the *Ltscale* to **18**. Activate a *Layout* tab, configure a print or plot device using *Pagesetup*, make one *Viewport*, and set the *Viewport scale* to a standard scale. To complete the PIVOTARM drawing, draw a *Pline* border (*width* **.02** x scale factor) in the layout and *Insert* the **TBLOCK** drawing that you created in Chapter 18 Exercises. *SaveAs* **PIVOTARM CH22**.

For exercises 2 through 5, construct and plot the multiview drawings as instructed. Use an appropriate *template* drawing for each exercise unless instructed otherwise. Use conventional practices for *layers*, *linetypes*, and *linetypes*. Draw a *Pline* border with the correct *width* and *Insert* your **TBLOCK** drawing.

2. Make a two-view multiview drawing of the Clip. *Plot* the drawing full size (**1=1**). Use the **ASHEET** template drawing to achieve the desired plot scale. *Save* the drawing as **CLIP**.

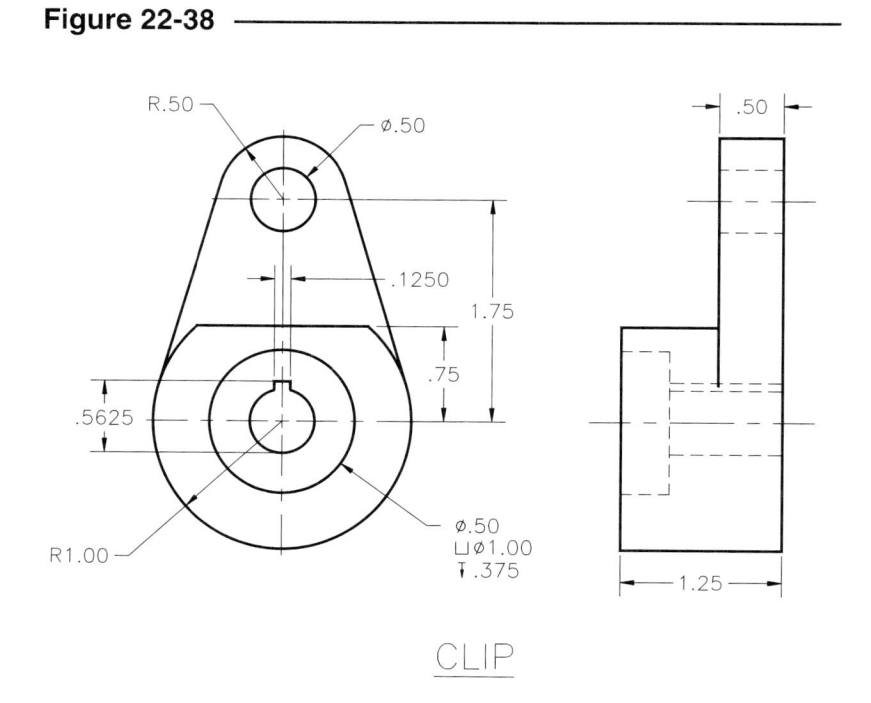

Figure 22-38

3. Make a three-view multiview drawing of the Bar Guide and *Plot* it **1=1**. Use the **BARGUIDE** drawing you set up in Chapter 12 Exercises. Note that a partial *Ellipse* will appear in one of the views.

Figure 22-39

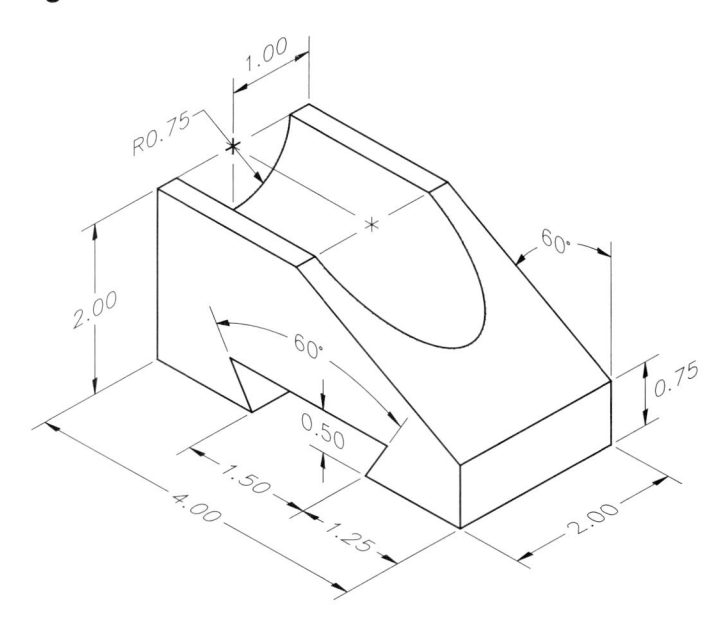

4. Construct a multiview drawing of the Saddle. Three views are needed. The channel along the bottom of the part intersects with the saddle on top to create a slotted hole visible in the top view. *Plot* the drawing at **1=1**. *Save* as **SADDLE**.

Figure 22-40

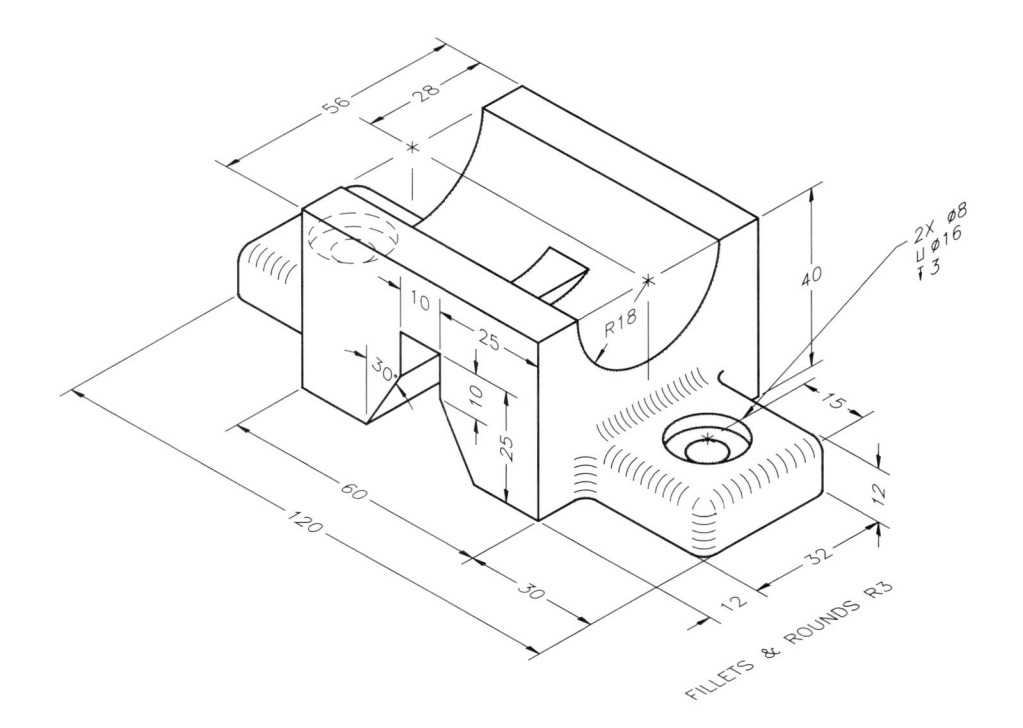

5. Draw a multiview of the Adjustable Mount shown in Figure 22-41. Determine an appropriate template drawing to use and scale for plotting based on your plotter capabilities. *Plot* the drawing to an accepted scale. *Save* the drawing as **ADJMOUNT**.

Figure 22-41

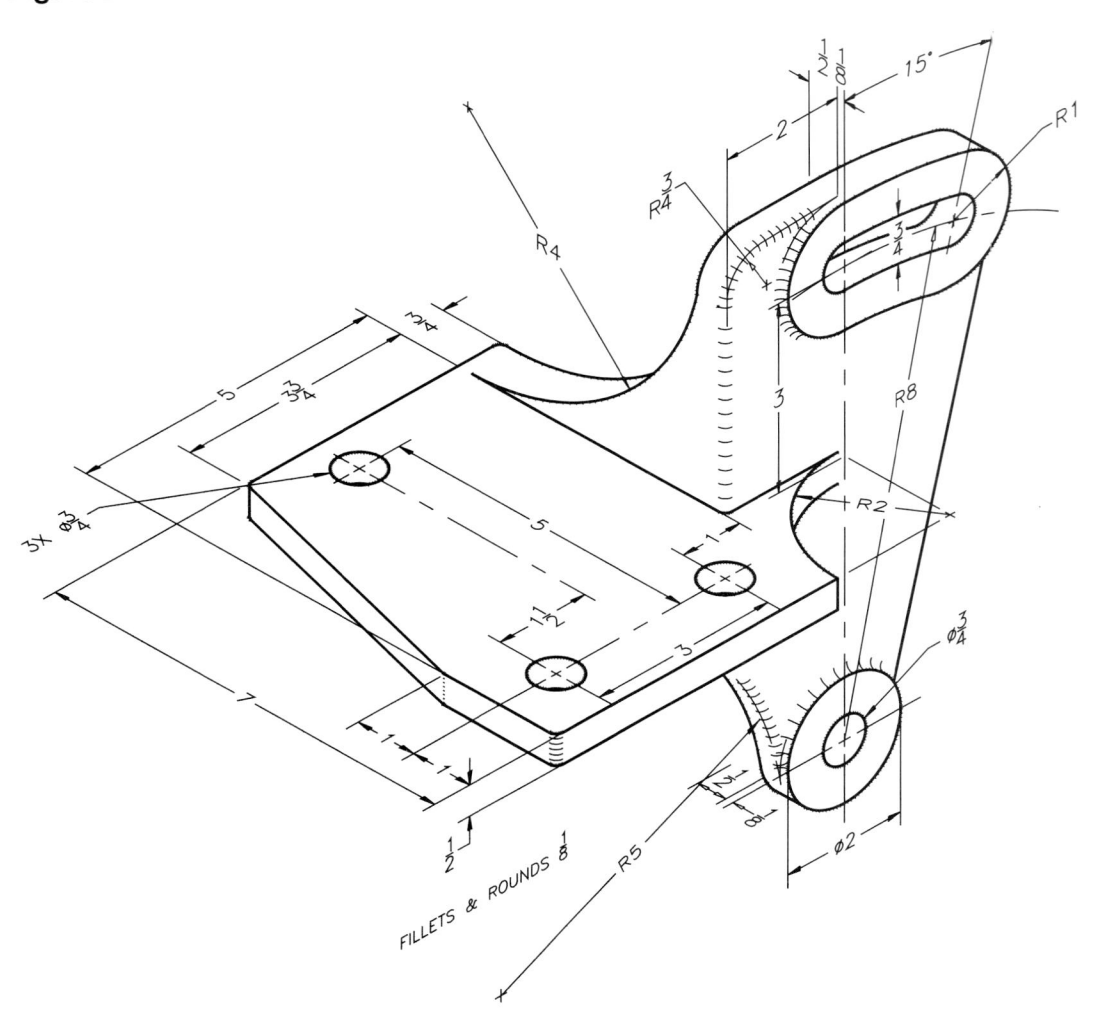

23

PICTORIAL DRAWINGS

Chapter Objectives

After completing this chapter you should be able to:

1. activate the *Isometric Style* of *Snap* for creating isometric drawings;

2. draw on the three isometric planes by toggling *Isoplane* using Ctrl+E;

3. create isometric ellipses with the *Isocircle* option of *Ellipse*;

4. construct an isometric drawing in AutoCAD;

5. create Oblique Cavalier and Cabinet drawings in AutoCAD.

CONCEPTS

Isometric drawings and oblique drawings are pictorial drawings. Pictorial drawings show three principal faces of the object in one view. A pictorial drawing is a drawing of a 3D object as if you were positioned to see (typically) some of the front, some of the top, and some of the side of the object. All three dimensions of the object (width, height, and depth) are visible in a pictorial drawing.

Multiview drawings differ from pictorial drawings because a multiview only shows two dimensions in each view, so two or more views are needed to see all three dimensions of the object. A pictorial drawing shows all dimensions in the one view. Pictorial drawings depict the object similar to the way you are accustomed to viewing objects in everyday life, that is, seeing all three dimensions. Figure 23-1 and Figure 23-2 show the same object in multiview and in pictorial representation, respectively. Notice that multiview drawings use hidden lines to indicate features that are normally obstructed from view, whereas <u>hidden lines are normally omitted</u> in isometric drawings (unless certain hidden features must be indicated for a particular function or purpose).

Figure 23-1

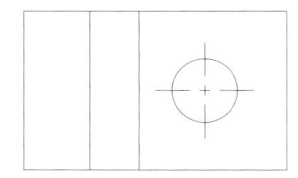

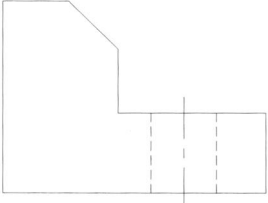

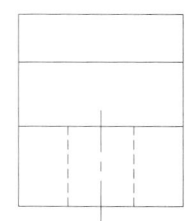

Figure 23-2

Types of Pictorial Drawings

Pictorial drawings are classified as follows:

1. Axonometric drawings
   a. Isometric drawings
   b. Dimetric drawings
   c. Trimetric drawings
2. Oblique drawings

Axonometric drawings are characterized by how the angle of the edges or axes (axon-) are measured (-metric) with respect to each other.

Isometric drawings are drawn so that each of the axes have equal angular measurement. ("Isometric" means equal measurement.) The isometric axes are always drawn at 120 degree increments (Fig. 23-3). All rectilinear lines on the object (representing horizontal and vertical edges—not inclined or oblique) are drawn on the isometric axes.

A 3D object seen "in isometric" is thought of as being oriented so that each of three perpendicular faces (such as the top, front, and side) are seen equally. In other words, the angles formed between the line of sight and each of the principal faces are equal.

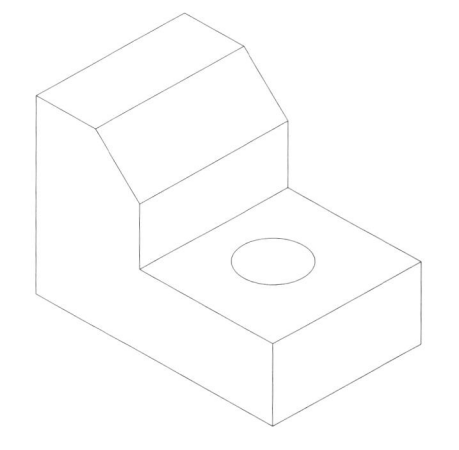

Figure 23-3

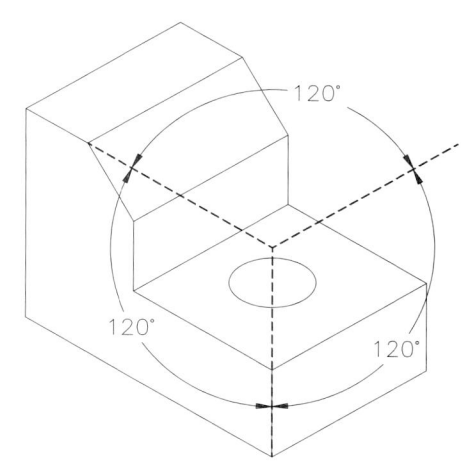

Dimetric drawings are constructed so that the angle between any two of the three axes is equal. There are many possibilities for dimetric axes. A common orientation for dimetric drawings is shown in Figure 23-4. For 3D objects seen from a dimetric viewpoint, the angles formed between the line of sight and each of two principal faces are equal.

Figure 23-4

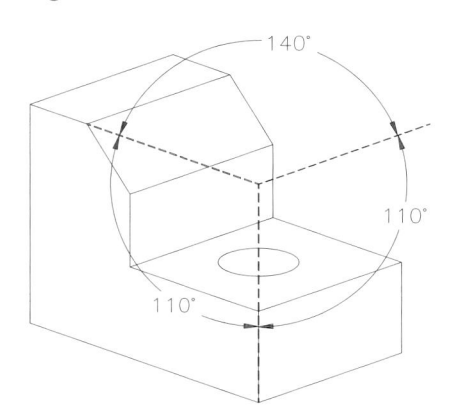

Trimetric drawings have three unequal angles between the axes. Numerous possibilities exist. A common orientation for trimetric drawings is shown in Figure 23-5.

Figure 23-5

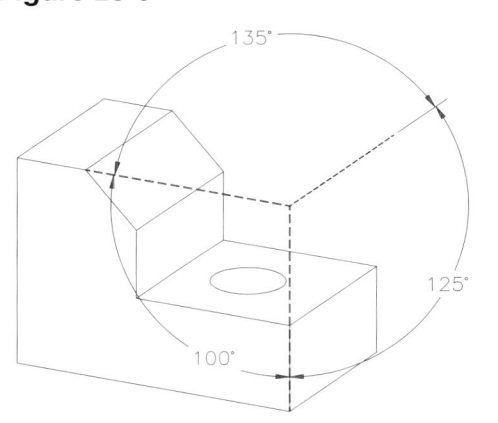

Oblique drawings are characterized by a vertical axis and horizontal axis for the two dimensions of the front face and a third (receding) axis of either 30, 45, or 60 degrees (Fig. 23-6). Oblique drawings depict the true size and shape of the front face, but add the depth to what would otherwise be a typical 2D view. This technique simplifies construction of drawings for objects that have contours in the profile view (front face) but relatively few features along the depth. Viewing a 3D object from an oblique viewpoint is not possible.

Figure 23-6

This chapter will explain the construction of isometric and oblique drawings in AutoCAD.

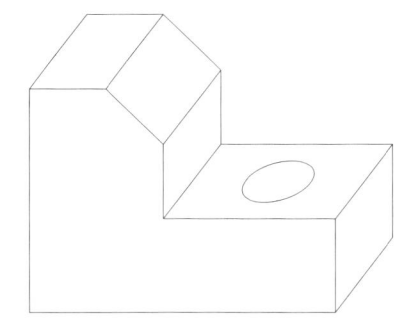

Pictorial Drawings Are 2D Drawings

Isometric, dimetric, trimetric, and oblique drawings are 2D drawings, whether created with AutoCAD or otherwise. Pictorial drawing was invented before the existence of CAD and therefore was intended to simulate a 3D object on a 2D plane (the plane of the paper). If AutoCAD is used to create the pictorial, the geometry lies on a 2D plane—the XY plane. All coordinates defining objects have X and Y values with a Z value of 0. When the *Isometric* style of *Snap* is activated, an isometrically structured *SNAP* and *GRID* appear on the XY plane.

448 Pictorial Drawings

Figure 23-7 illustrates the 2D nature of an isometric drawing created in AutoCAD. Isometric lines are created on the XY plane. The *Isometric SNAP* and *GRID* are also on the 2D plane. (The *Vpoint* command was used to give other than a *Plan* view of the drawing in this figure.)

Although pictorial drawings are based on the theory of projecting 3D objects onto 2D planes, it is physically possible to achieve an axonometric (isometric, dimetric, or trimetric) viewpoint of a 3D object using a 3D CAD system. In AutoCAD, the *Vpoint* and *3Dorbit* commands are used to specify the observer's position in 3D space with respect to a 3D model. Chapter 29 discusses the specific commands and values needed to attain axonometric viewpoints of a 3D model.

Figure 23-7

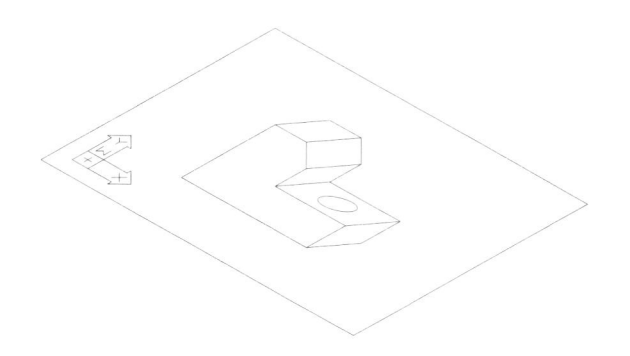

ISOMETRIC DRAWING IN AutoCAD

AutoCAD provides the capability to construct isometric drawings. An isometric *SNAP* and *GRID* are available, as well as a utility for creation of isometrically correct ellipses. Isometric lines are created with the *Line* command. There are no special options of *Line* for isometric drawing, but isometric *SNAP* and *GRID* can be used to force *Lines* to an isometric orientation. Begin creating an isometric drawing in AutoCAD by activating the *Isometric Style* option of the *Snap* command. This action can be done using any of the options listed in the following Command table.

SNAP

Pull-down Menu	COMMAND (TYPE)	ALIAS (TYPE)	Short-cut	Screen (side) Menu	Tablet Menu
Tools Drafting Settings... Snap and Grid	SNAP	SN	F9 or Ctrl+B	TOOLS 2 Grid	W,10

Command: *snap*
Specify snap spacing or
[ON/OFF/Aspect/Rotate/Style/Type] <0.5000>: *s*
Enter snap grid style [Standard/Isometric] <S>: *i*
Specify vertical spacing <0.5000>: **Enter**
Command:

Alternately, toggling the indicated checkbox in the lower-right corner of the *Drafting Settings* dialog box activates the *Isometric Snap* and *Grid Snap* (Fig. 23-8).

Figure 23-8

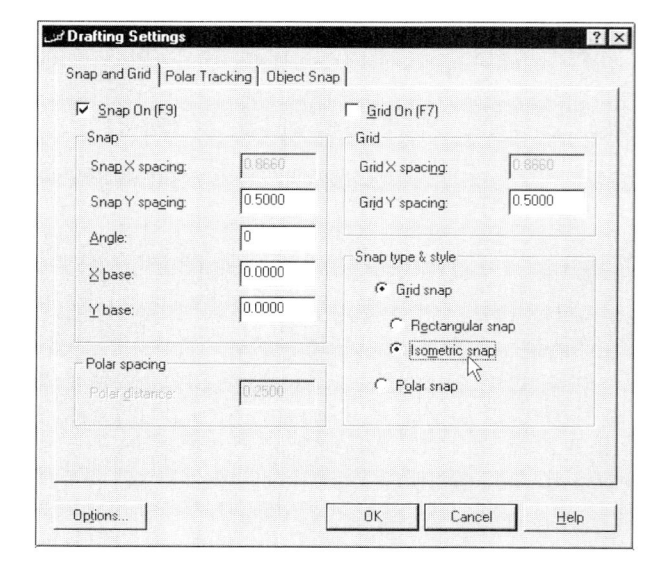

Figure 23-9 illustrates the effect of setting the *Isometric SNAP* and *GRID*. Notice the new position of the cursor.

Figure 23-9

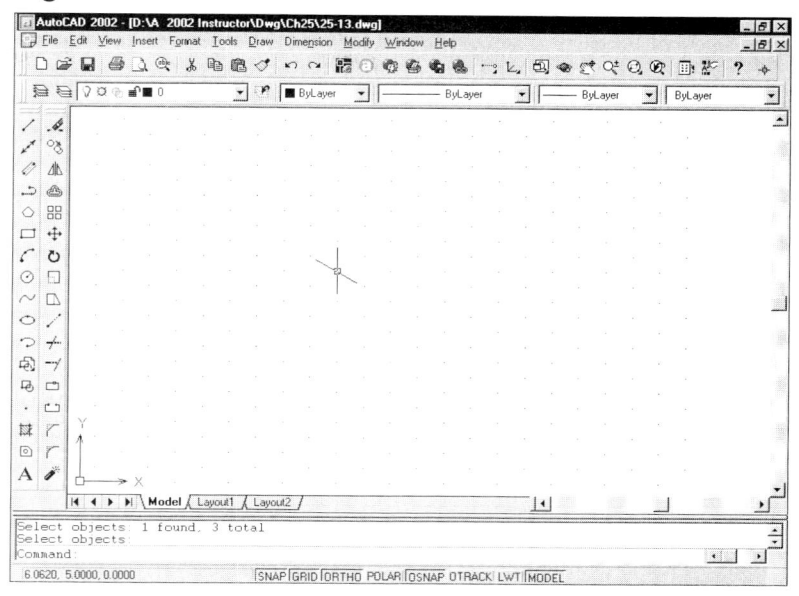

Using Ctrl+E (pressing the Ctrl key and the letter "E" simultaneously) toggles the cursor to one of three possible *Isoplanes* (AutoCAD's term for the three faces of the isometric pictorial). If *ORTHO* is *ON*, only isometric lines are drawn; that is, you can only draw *Lines* aligned with the isometric axes. *Lines* can be drawn on only two axes for each isoplane. Ctrl+E allows drawing on the two axes aligned with another face of the object. *ORTHO* is *OFF* in order to draw inclined or oblique lines (not on the isometric axis). The functions of *GRID* (F7) and *SNAP* (F9) remain unchanged.

With *SNAP ON*, toggle *Coords* (F6) several times and examine the readout as you move the cursor. The absolute coordinate format is of no particular assistance while drawing in isometric because of the configuration of the *GRID*. The relative polar format, however, is very helpful. Use relative polar format for *Coords* while drawing in isometric (Fig. 23-10).

Figure 23-10

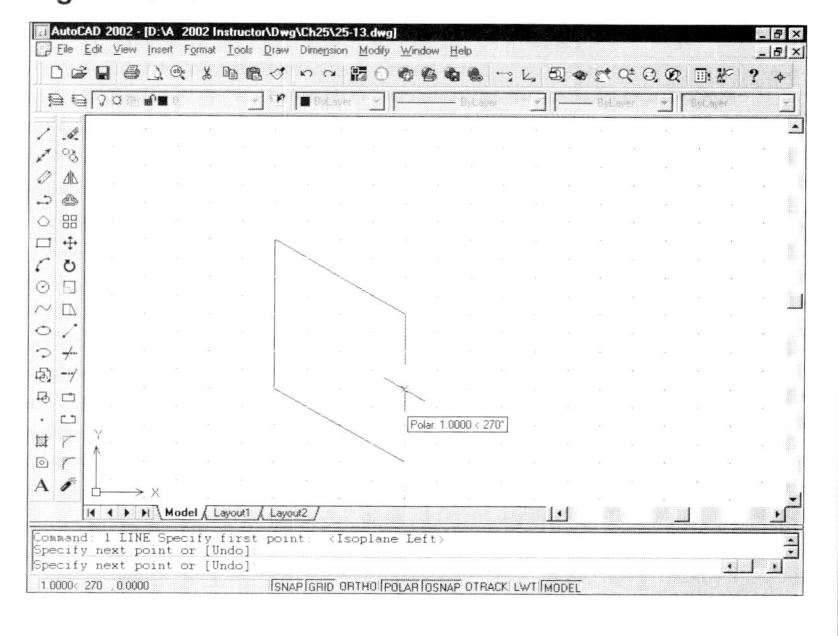

Alternately, use *Polar Tracking* instead of *ORTHO*. Set the *Polar Angle Settings* to 30 degrees. The advantage of using *Polar Tracking* is that the current line length is given on the polar tracking tip (see Figure 23-10). The disadvantage is that it is possible to draw non-isoplane lines accidentally. Only one setting at a time can be used in AutoCAD—*Polar Tracking* or *ORTHO*. (The remainder of the figures illustrate the use of *ORTHO*.)

The effects of changing the *Isoplane* are shown in the following figures. Press Ctrl+E to change *Isoplane*.

With *ORTHO ON*, drawing a *Line* is limited to the two axes of the current *Isoplane*. Only one side of a cube, for example, can be drawn on the current *Isoplane*. Watch *Coords* (in a polar format) to give the length of the current *Line* as you draw (lower-left corner of the screen).

Toggling Ctrl+E switches the cursor and the effect of *ORTHO* to another *Isoplane*. One other side of a cube can be constructed on this *Isoplane* (Fig. 23-11).

Figure 23-11

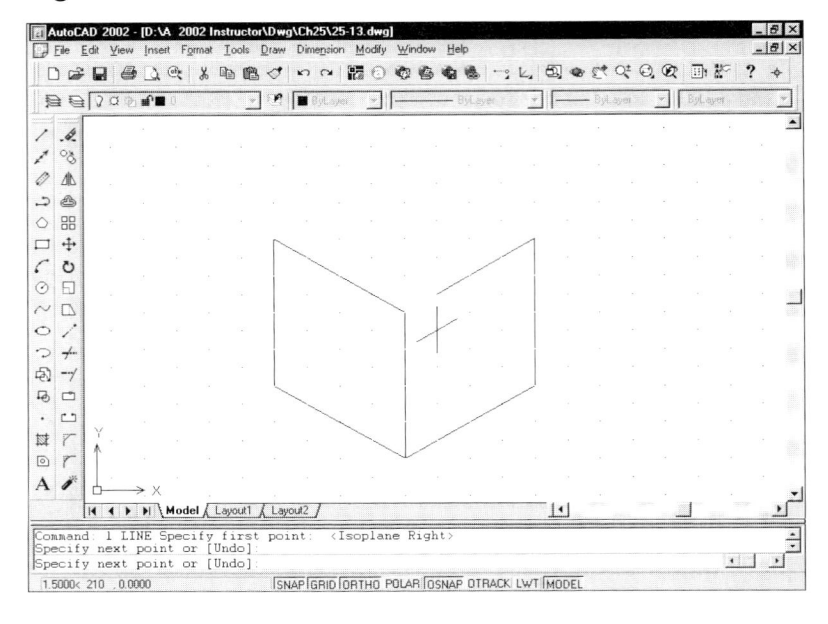

Direct Distance Entry can be of great help when drawing isometric lines. Use Ctrl+E and *ORTHO* to force the *Line* to the correct orientation, then enter the desired distance value at the Command line.

Isometric Ellipses

Isometric ellipses are easily drawn in AutoCAD by using the *Isocircle* option of the *Ellipse* command. This option appears only when the isometric *SNAP* is *ON*.

ELLIPSE

Pull-down Menu	COMMAND (TYPE)	ALIAS (TYPE)	Short-cut	Screen (side) Menu	Tablet Menu
Draw Ellipse	ELLIPSE	EL	...	DRAW 1 Ellipse	M,9

Although the *Isocircle* option does not appear in the pull-down or digitizing tablet menus, it can be invoked as an option of the *Ellipse* command. The *Isocircle* option of *Ellipse* appears only when the *Snap Type* is set to *Isometric*. You must type "I" to use the *Isocircle* option. The command syntax is as follows:

Command: `ellipse`
Specify axis endpoint of ellipse or [Arc/Center/Isocircle]: `i`
Specify center of isocircle: **PICK** or (`coordinates`)
Specify radius of isocircle or [Diameter]: **PICK** or (`coordinates`)
Command:

After selecting the center point of the *Isocircle*, the isometrically correct ellipse appears on the screen on the current *Isoplane*. Use Ctrl+E to toggle the ellipse to the correct orientation. When defining the radius interactively, use *ORTHO* to force the rubberband line to an isometric axis (Fig. 23-12, next page).

Since isometric angles are equal, all isometric ellipses have the same proportion (major to minor axis). The only differences in isometric ellipses are the size and the orientation (*Isoplane*).

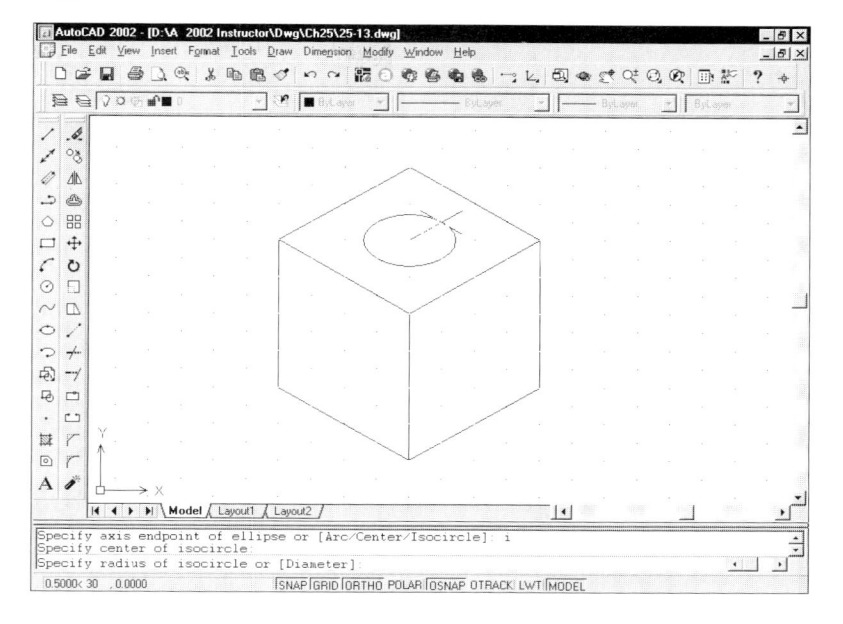

Figure 23-12

Figure 23-13 shows three ellipses correctly oriented on their respective faces. Use Ctrl+E to toggle the correct *Isoplane* orientation: *Isoplane Top*, *Isoplane Left*, or *Isoplane Right*.

Figure 23-13

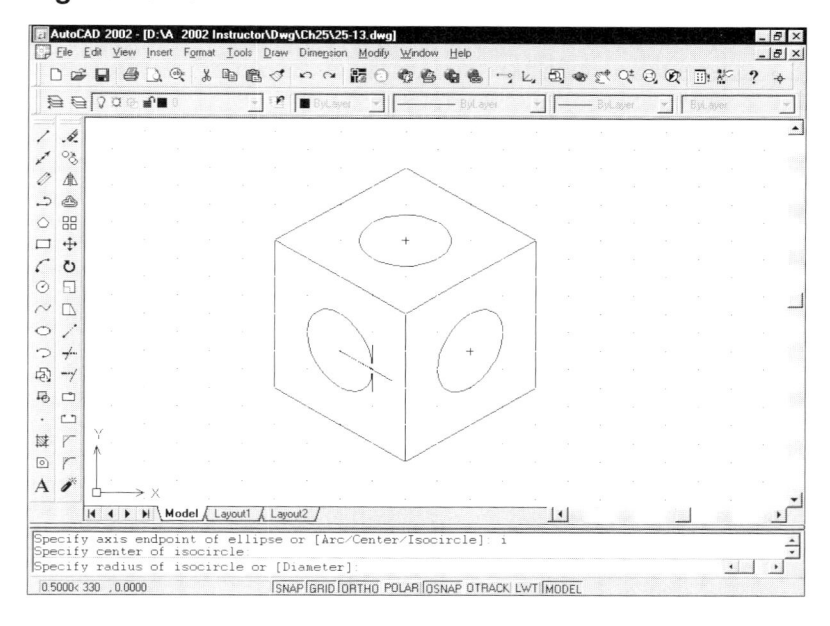

When defining the radius or diameter of an ellipse, it should always be measured in an isometric direction. In other words, an isometric ellipse is always measured on the two isometric axes (or center lines) parallel with the plane of the ellipse.

If you define the radius or diameter interactively, use *ORTHO ON*. If you enter a value, AutoCAD automatically applies the value to the correct isometric axes.

Creating an Isometric Drawing

In this exercise, the object in Figure 23-14 is drawn in isometric.

The initial steps to create an isometric drawing begin with the typical setup (see Chapter 6, Drawing Setup):

1. Set the desired *Units*.

2. Set appropriate *Limits*.

3. Set the *Isometric Style* of *Snap* and specify an appropriate value for spacing.

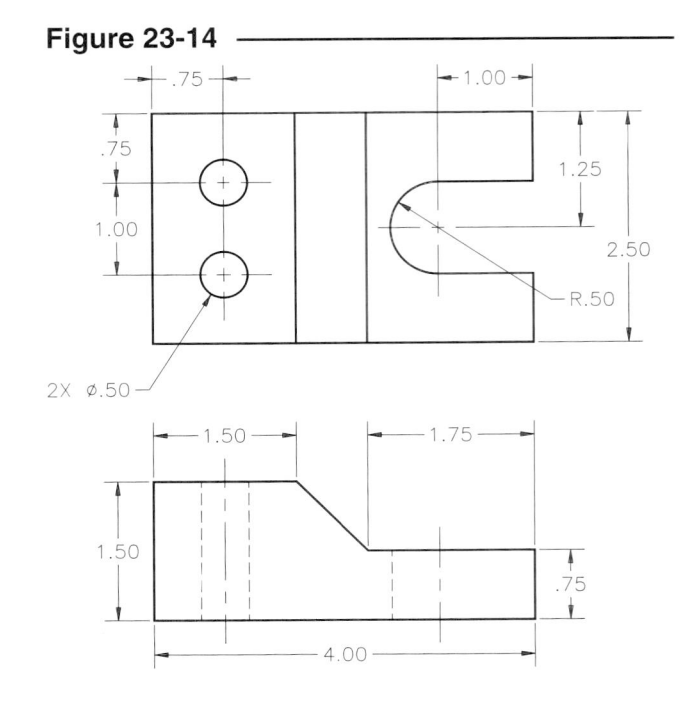

Figure 23-14

4. The next step involves creating an isometric framework of the desired object. In other words, draw an isometric box equal to the overall dimensions of the object. Using the dimensions given in Figure 23-14, create the encompassing isometric box with the *Line* command (Fig. 23-15).

    Use *ORTHO* to force isometric *Lines*. Watch the *Coords* display (in a relative polar format) to give the current lengths as you draw or use direct distance entry.

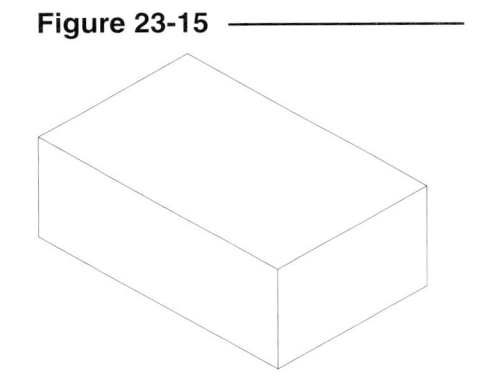

Figure 23-15

5. Add the lines defining the lower surface. Define the needed edge of the upper isometric surface as shown.

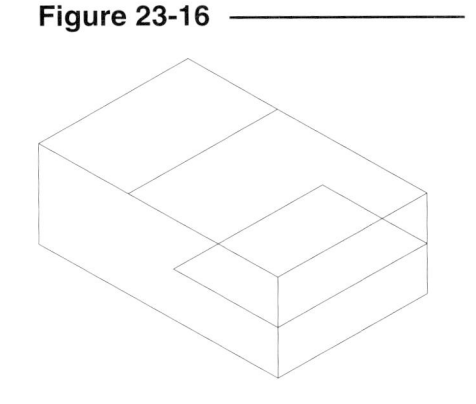

Figure 23-16

6. The inclined edges of the inclined surface can be drawn (with *Line*) only when *ORTHO* is *OFF*. Inclined lines in isometric cannot be drawn by transferring the lengths of the lines, but only by defining the ends of the inclined lines on isometric lines, then connecting the endpoints. Next, *Trim* or *Erase* the necessary *Lines*.

Figure 23-17

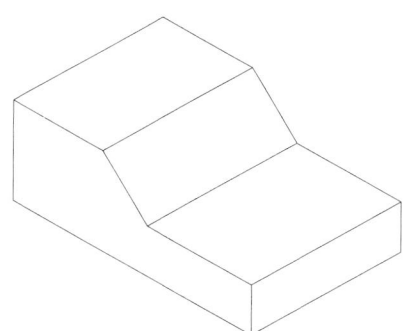

7. Draw the slot by constructing an *Ellipse* with the *Isocircle* option. Draw the two *Lines* connecting the circle to the right edge. *Trim* the unwanted part of the *Ellipse* (highlighted) using the *Lines* as cutting edges.

Figure 23-18

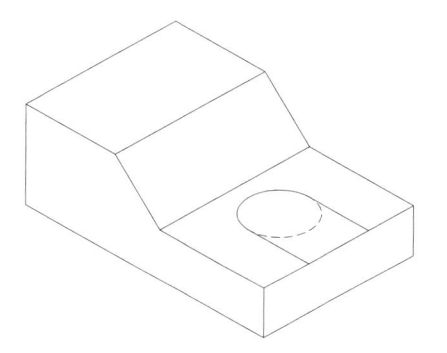

8. *Copy* the far *Line* and the *Ellipse* down to the bottom surface. Add two vertical *Lines* at the end of the slot.

Figure 23-19

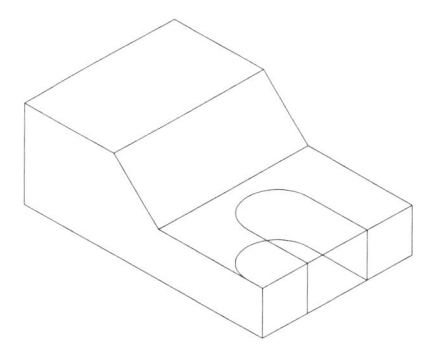

9. Use *Trim* to remove the part of the *Ellipse* that would normally be hidden from view. *Trim* the *Lines* along the right edge at the opening of the slot.

Figure 23-20

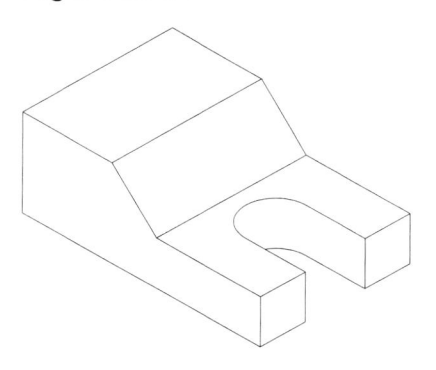

454 Pictorial Drawings

10. Add the two holes on the top with *Ellipse, Isocircle* option. Use ORTHO ON when defining the radius. *Copy* can also be used to create the second *Ellipse* from the first.

Figure 23-21

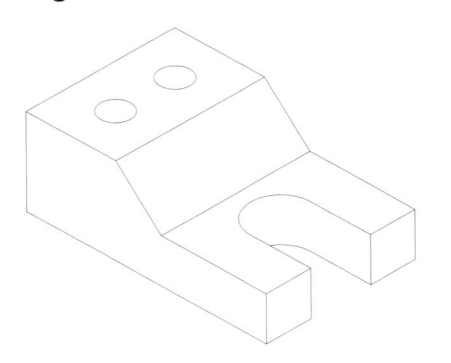

OBLIQUE DRAWING IN AutoCAD

Figure 23-22

Oblique drawings are characterized by having two axes at a 90 degree orientation. Typically, you should locate the <u>front face</u> of the object along these two axes. Since the object's characteristic shape is seen in the front view, an oblique drawing allows you to create all shapes parallel to the front face true size and shape as you would in a multi-view drawing. Circles on or parallel to the front face can be drawn as circles. The third axis, the receding axis, can be drawn at a choice of angles, 30, 45, or 60 degrees, depending on whether you want to show more of the top or the side of the object.

Figure 23-22 illustrates the axes orientation of an oblique drawing, including the choice of angles for the receding axis.

Another option allowed with oblique drawings is the measurement used for the receding axis. Using the full depth of the object along the receding axis is called <u>Cavalier</u> oblique drawing. This method depicts the object (a cube with a hole in this case) as having an elongated depth (Fig. 23-23).

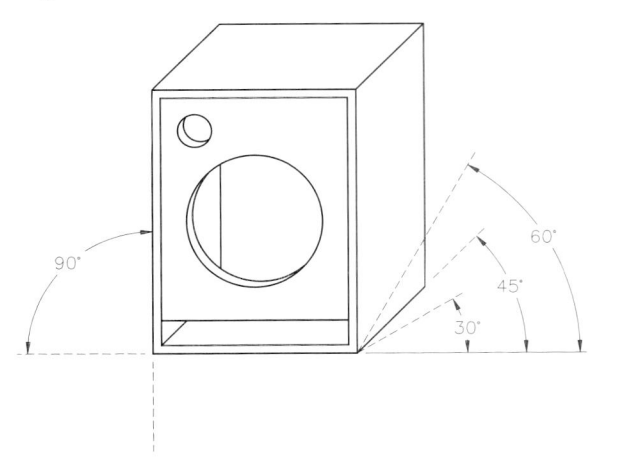

Figure 23-23

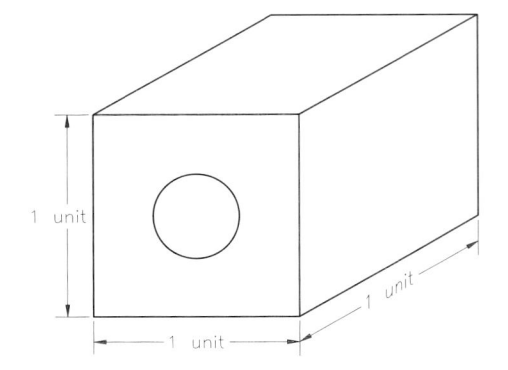

Using 1/2 or 3/4 of the true depth along the receding axis gives a more realistic pictorial representation of the object. This is called a <u>Cabinet</u> oblique (Fig. 23-24).

Figure 23-24

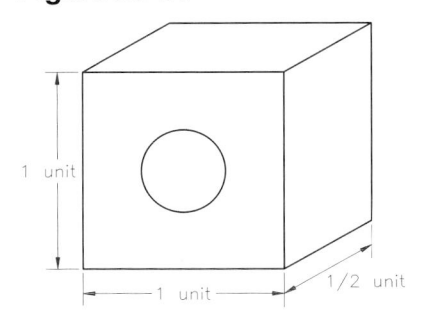

No functions or commands in AutoCAD are intended specifically for oblique drawing. However, *Polar Snap* and *Polar Tracking* can simplify the process of drawing lines on the front face of the object and along the receding axis. The steps for creating a typical oblique drawing are given next.

Chapter Twenty-Three 455

The object in Figure 23-25 is used for the example. From the dimensions given in the multiview, create a cabinet oblique with the receding axis at 45 degrees.

Figure 23-25

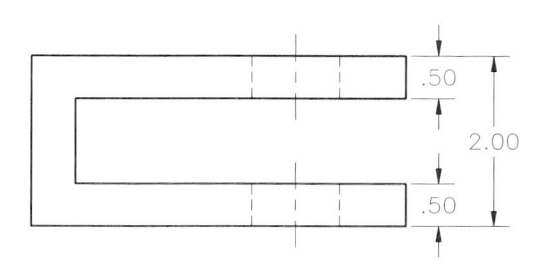

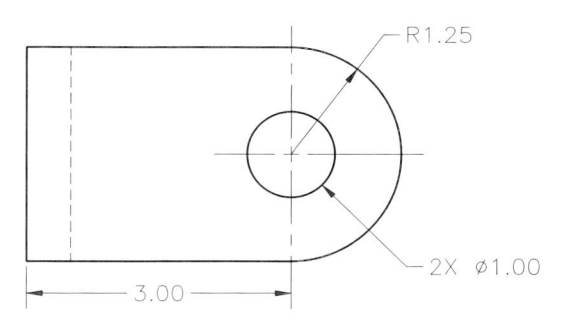

1. Create the characteristic shape of the front face of the object as shown in the front view.

Figure 23-26

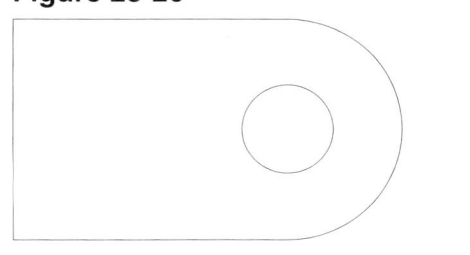

2. Use *Copy* with the *Multiple* option to copy the front face back on the receding axis. *Polar Snap* and *Polar Tracking* can be used to specify the "second point of displacement" as shown in Figure 23-27. Notice that a distance of .25 along the receding axis is used (1/2 of the actual depth) for this cabinet oblique.

Figure 23-27

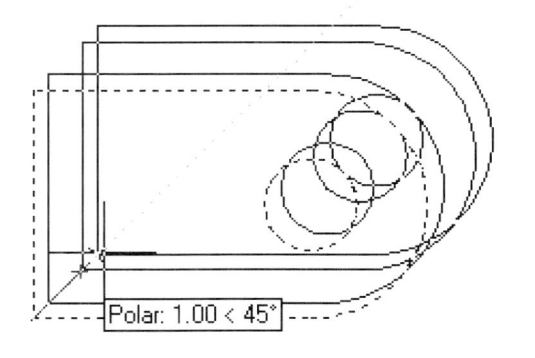

3. Draw the *Line* representing the edge on the upper-left of the object along the receding axis. Use *Endpoint OSNAP* to connect the *Lines*. Make a *Copy* of the *Line* or draw another *Line* .5 units to the right. Drop a vertical *Line* from the *Intersection* as shown.

Figure 23-28

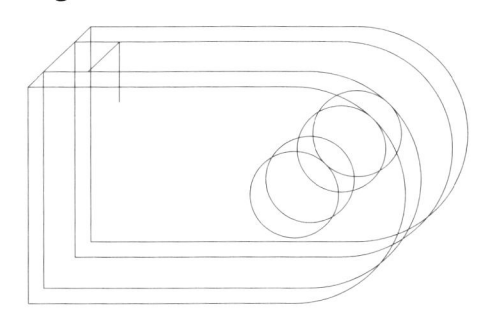

456 Pictorial Drawings

4. Use *Trim* and *Erase* to remove the unwanted parts of the *Lines* and *Circles* (those edges that are normally obscured).

Figure 23-29

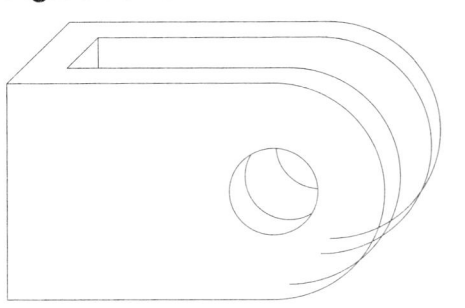

5. *Zoom* with a *window* to the lower-right corner of the drawing. Draw a *Line Tangent* to the edges of the arcs to define the limiting elements along the receding axis. *Trim* the unwanted segments of the arcs.

Figure 23-30

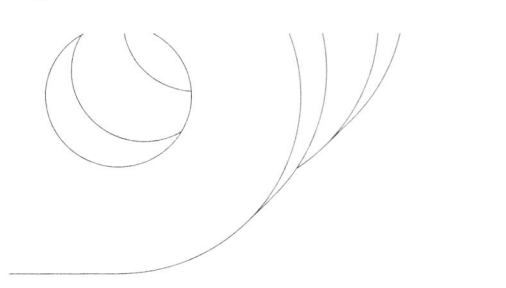

The resulting cabinet oblique drawing should appear like that in Figure 23-31.

Figure 23-31

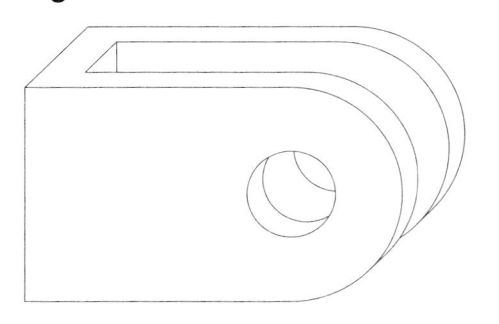

CHAPTER EXERCISES

Isometric Drawing

For exercises 1, 2, and 3 create isometric drawings as instructed. To begin, use an appropriate template drawing and draw a *Pline* border and insert the **TBLOCK** drawing.

Figure 23-32

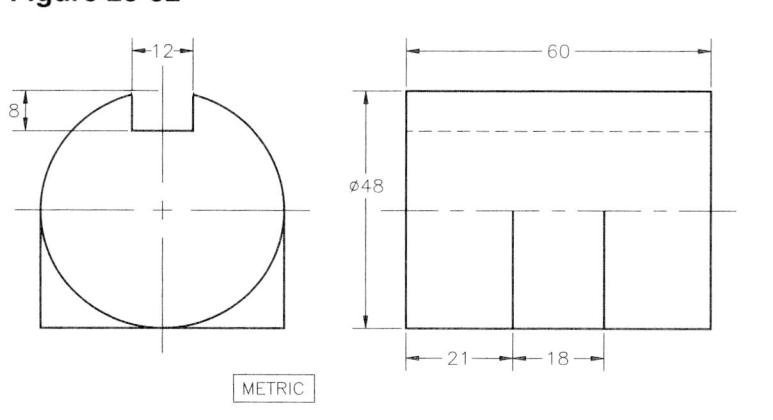

1. Create an isometric drawing of the cylinder shown in Figure 23-32. *Save* the drawing as **CYLINDER** and *Plot* so the drawing is *Scaled to Fit* on an A size sheet.

METRIC

2. Make an isometric drawing of the Corner Brace shown in Figure 23-33. *Save* the drawing as **CRNBRACE**. *Plot* at **1=1** scale on an A size sheet.

Figure 23-33

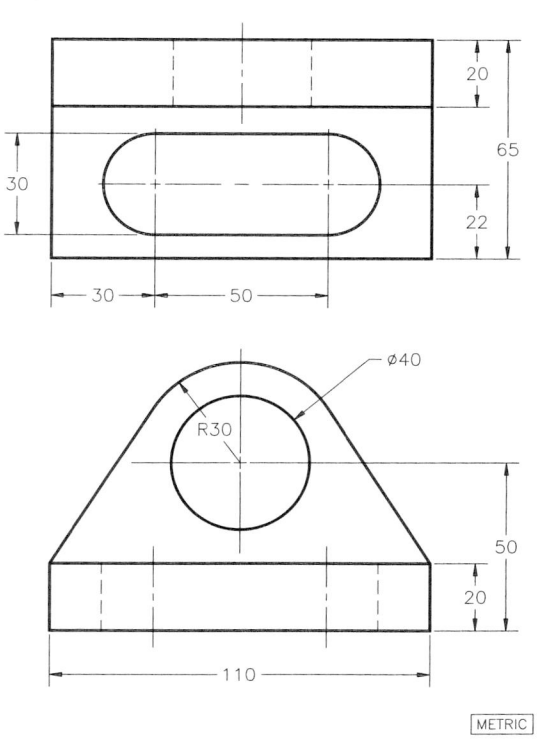

3. Draw the Support Bracket (Fig. 23-34) in isometric. The drawing can be *Plotted* at **1=1** scale on an A size sheet. *Save* the drawing and assign the name **SBRACKET**.

Figure 23-34

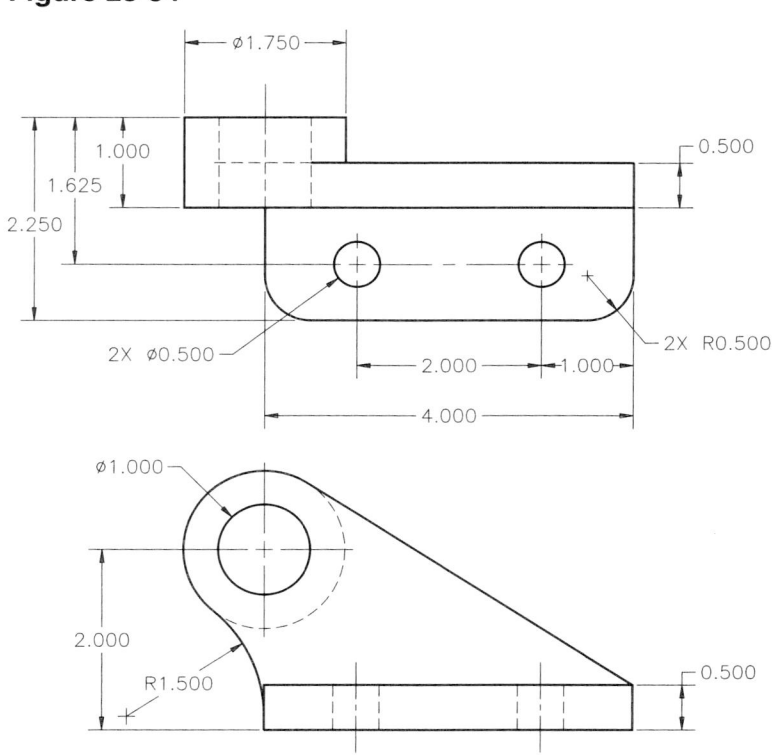

Oblique Drawing

For exercises 4 and 5, create oblique drawings as instructed. To begin, use an appropriate template drawing and draw a *Pline* border and *Insert* the **TBLOCK** drawing.

4. Make an oblique cabinet projection of the Bearing shown in Figure 23-35. Construct all dimensions on the receding axis 1/2 of the actual length. Select the optimum angle for the receding axis to be able to view the 15 x 15 slot. *Plot* at **1=1** scale on an A size sheet. *Save* the drawing as **BEARING**.

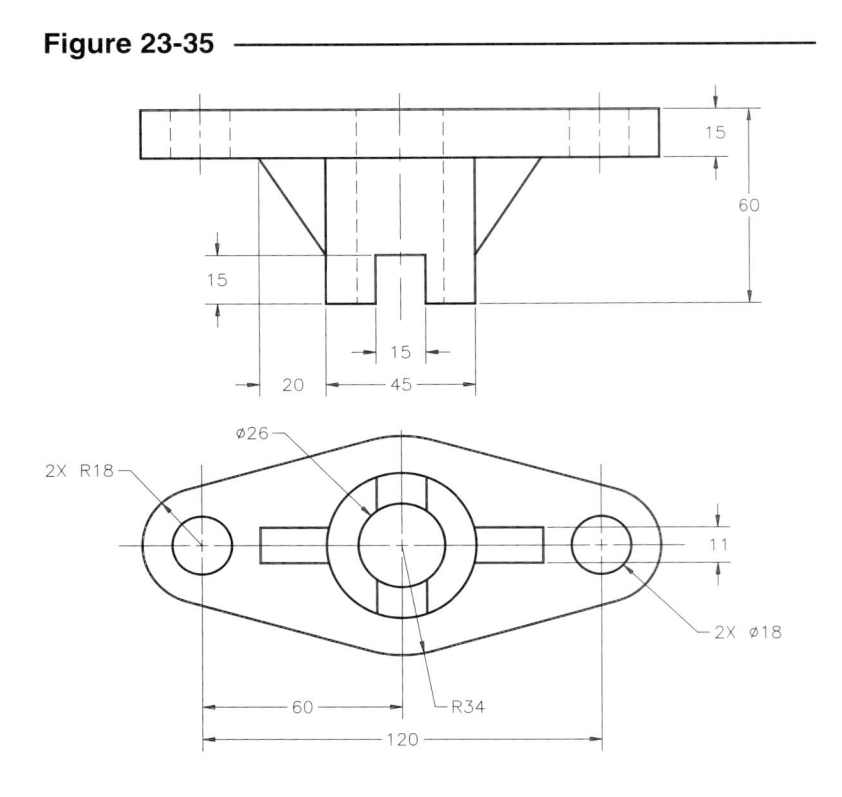

Figure 23-35

5. Construct a cavalier oblique drawing of the Pulley showing the circular view true size and shape. The illustration in Figure 23-36 gives only the side view. All vertical dimensions in the figure are diameters. *Save* the drawing as **PULLEY** and make a *Plot* on an A size sheet at **1=1**.

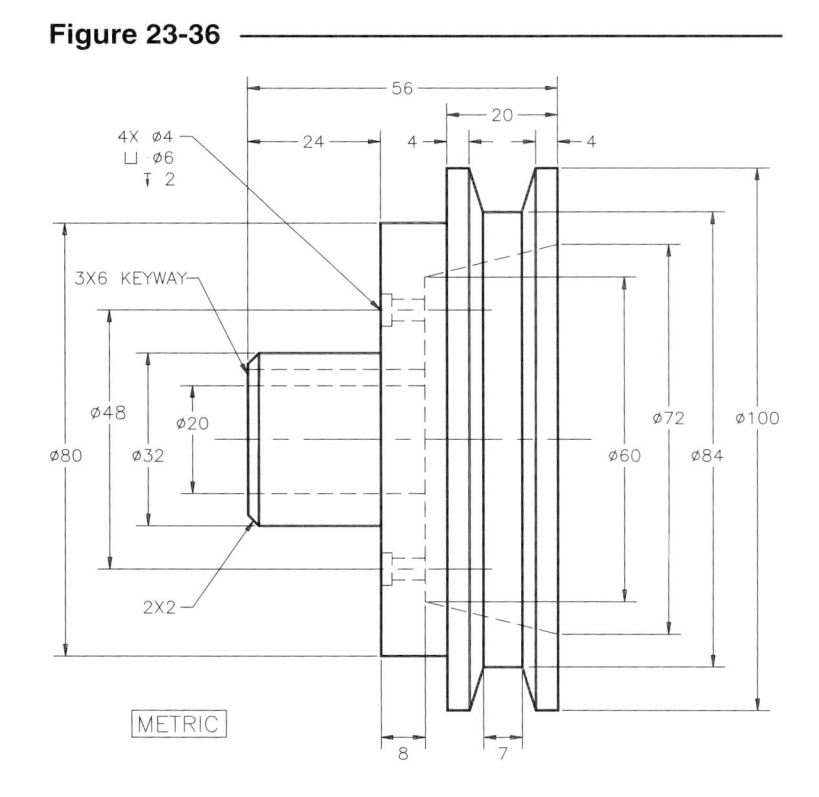

Figure 23-36

24

SECTION VIEWS

Chapter Objectives

After completing this chapter you should:

1. be able to use the *Bhatch* command to select associative hatch patterns;

2. be able to specify a *Scale* and *Angle* for hatch lines;

3. know how to define a boundary for hatching using the *Pick Points* and *Select Objects* methods;

4. be able to *Preview* the hatch and make necessary adjustments, then apply the hatch pattern;

5. be able to use *Hatch* to create non-associative hatch lines and discard the boundary;

6. know how to use *Hatchedit* to modify parameters of existing hatch patterns in the drawing;

7. be able to edit hatched areas using Grips, selection options, and *Draworder*;

8. know how to draw cutting plane lines for section views.

CONCEPTS

A section view is a view of the interior of an object after it has been imaginarily cut open to reveal the object's inner details. A section view is only one of two or more views of a multiview drawing describing the object. For example, a multiview drawing of a machine part may contain three views, one of which is a section view.

Hatch lines (also known as section lines) are drawn in the section view to indicate the solid material that has been cut through. Each combination of section lines is called a hatch pattern, and each pattern is used to represent a specific material. In full and half section views, hidden lines are omitted since the inside of the object is visible.

ANSI (American National Standards Institute) and ISO (International Organization of Standardization) have published standard configurations for section lines, and AutoCAD supports those standards. However, ANSI no longer specifies section line pattern standards.

A cutting plane line is drawn in an adjacent view to the section view to indicate the plane that imaginarily cuts through the object. Arrows on each end of the cutting plane line indicate the line of sight for the section view. ANSI dictates that a thick dashed or phantom line be used as the standard cutting plane line.

This chapter discusses the AutoCAD methods used to draw hatch lines for section views and related cutting plane lines. The *Bhatch* (boundary hatch) command allows you to select an enclosed area and select the hatch pattern and the parameters for the appearance of the hatch pattern; then AutoCAD automatically draws the hatch (section) lines. Existing hatch lines in the drawing can be modified using *Hatchedit*. Cutting plane lines are created in AutoCAD by using a dashed linetype. Cutting plane lines should be assigned a heavy *Lineweight* or should be drawn with a *Pline* with *Width*.

DEFINING HATCH PATTERNS AND HATCH BOUNDARIES

A hatch pattern is composed of many lines that have a particular linetype, spacing, and angle. Many standard hatch patterns are provided by AutoCAD for your selection. Rather than having to draw each section line individually, you are required only to specify the area to be hatched and AutoCAD fills the designated area with the selected hatch pattern. An AutoCAD hatch pattern is inserted as one object. For example, you can *Erase* the inserted hatch pattern by selecting only one line in the pattern, and the entire pattern in the area is *Erased*.

Figure 24-1

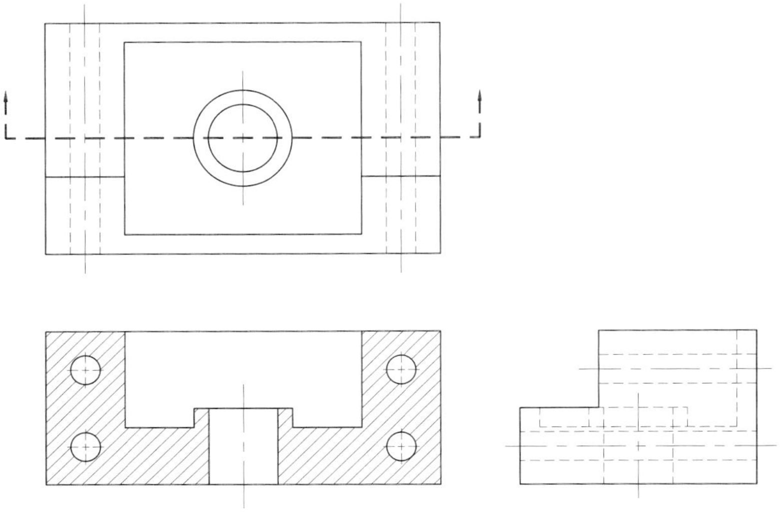

In a typical section view (Fig. 24-1), the hatch pattern completely fills the area representing the material that has been cut through. With the *Bhatch* command you can define the boundary of an area to be hatched simply by pointing inside of an enclosed area.

Both the *Hatch* and the *Bhatch* commands fill a specified area with a selected hatch pattern. *Hatch* requires that you select <u>each object</u> defining the boundary, whereas the *Bhatch* command finds the boundary automatically when you point inside it. Additionally, *Hatch* operates in command line format, whereas *Bhatch* operates in dialog box fashion. For these reasons, *Bhatch* is superior to *Hatch* and is recommended in most cases for drawing section views.

Hatch patterns created with *Bhatch* are <u>associative</u>. Associative hatch patterns are associated to the boundary geometry such that when the shape of the boundary changes (by *Stretch, Scale, Rotate, Move, Properties, Grips,* etc.), the hatch pattern automatically reforms itself to conform to the new shape (Fig. 24-2). For example, if a design change required a larger diameter for a hole, *Properties* could be used to change the diameter of the hole, and the surrounding section lines would automatically adapt to the new diameter.

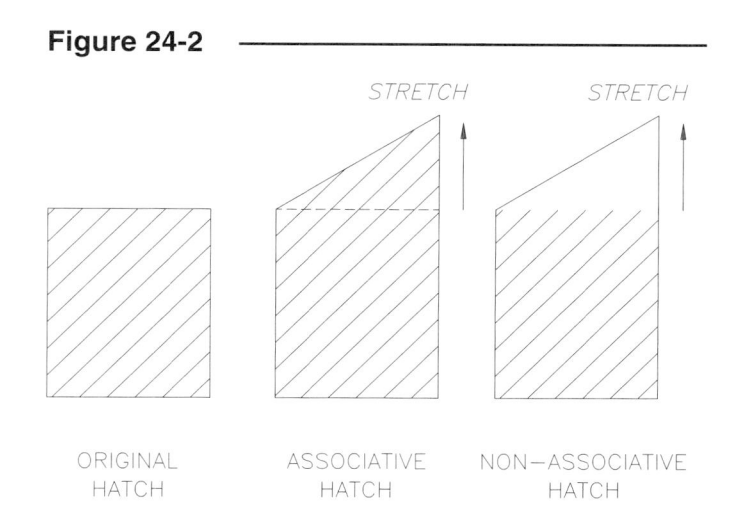

Figure 24-2

Once the hatch patterns have been drawn, any feature of the existing hatch pattern (created with *Bhatch*) can be changed retroactively using *Hatchedit*. The *Hatchedit* dialog box gives access to the same options that were used to create the hatch (in the *Boundary Hatch* dialog box). Changing the scale, angle, or pattern of any existing section view in the drawing is a simple process.

Steps for Creating a Section View Using the *Bhatch* Command

1. Create the view that contains the area to be hatched using typical draw commands such as *Line, Arc, Circle,* or *Pline*. If you intend to have text or dimensions inside the area to be hatched, add them before hatching.

2. Invoke the *Bhatch* command. The *Boundary Hatch* dialog box appears (see Figure 24-3).

3. Specify the *Type* to use. Select the desired pattern from the *Pattern* drop-down list or select the *Swatch* tile to allow you to select from the *Hatch Pattern Palette* image tiles.

4. Specify the *Scale* and *Angle* in the dialog box.

5. Define the area to be hatched by PICKing an internal point (*Pick Points* button) or by individually selecting the objects (*Select objects* button).

6. *Preview* the hatch to make sure everything is as expected. Adjust hatching parameters as necessary and *Preview* again.

7. Apply the hatch by selecting *OK*. The hatch pattern is automatically drawn and becomes an associated object in the drawing.

8. If other areas are to be hatched, additional internal points or objects can be selected to define the new area for hatching. The parameters used previously appear again in the *Boundary Hatch* dialog box by default. You can also *Inherit Properties* from a previously applied hatch.

9. For mechanical drawings, draw a cutting plane line in a view adjacent to the section view. A *Lineweight* is assigned or a *Pline* with a *Dashed* or *Phantom* linetype is used. Arrows at the ends of the cutting plane line indicate the line of sight for the section view.

10. If any aspect of the hatch lines needs to be edited at a later time, *Hatchedit* can be used to change those properties. If the hatch boundary is changed by *Stretch, Rotate, Scale, Move, Properties*, etc., the hatched area will conform to the new boundary.

BHATCH

Pull-down Menu	COMMAND (TYPE)	ALIAS (TYPE)	Short-cut	Screen (side) Menu	Tablet Menu
Draw Hatch...	BHATCH or -BHATCH	BH or H	...	DRAW 2 Bhatch	P,9

Bhatch allows you to create hatch lines for a section view (or for other purposes) by simply PICKing inside a closed boundary. A closed boundary refers to an area completely enclosed by objects. *Bhatch* locates the closed boundary automatically by creating a temporary *Pline* that follows the outline of the hatch area, fills the area with hatch lines, and then deletes the boundary (default option) after hatching is completed. *Bhatch* ignores all objects or parts of objects that are not part of the boundary.

Any method of invoking *Bhatch* yields the *Boundary Hatch* dialog box (Fig. 24-3). Typically, the first step in the *Boundary Hatch* dialog box is the selection of a hatch pattern.

Figure 24-3

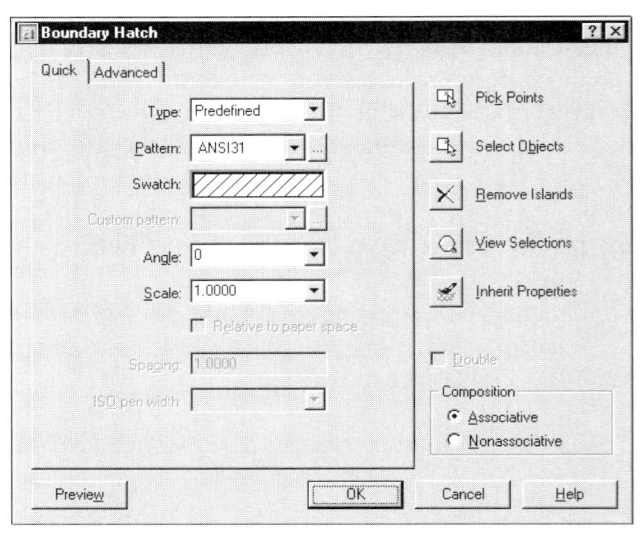

After you select the *Pattern* and other parameters for the hatch in the *Boundary Hatch* dialog box, you select *Pick Points* or *Select Objects* to return to the drawing and indicate the area to fill with the pattern. While in the drawing, you can right-click to produce a shortcut menu, providing access to other options without having first to return to the dialog box (Fig. 24-4).

Figure 24-4

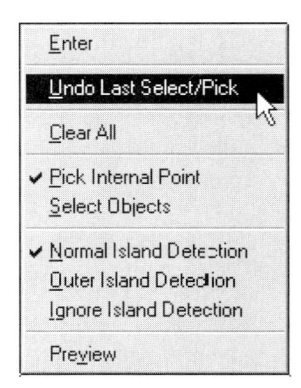

Boundary Hatch **Dialog Box**—*Quick* **Tab**

Type
This option allows you to specify the type of the hatch pattern: *Predefined*, *User-defined*, or *Custom*. Use *Predefined* for standard hatch pattern styles that AutoCAD provides (in the ACAD.PAT file).

Predefined
There are two ways to select from *Predefined* patterns: (1) select the *Swatch...* tile to produce the *Hatch Pattern Palette* dialog box displaying hatch pattern names and image tiles (Fig. 24-5), or (2) select the *Pattern:* drop-down list to PICK the pattern name.

The *Hatch Pattern Palette* dialog box allows you to select a predefined pattern by its image tile or by its name. There are four tabs of image tiles: *ANSI, ISO, Other Predefined*, and *Custom*.

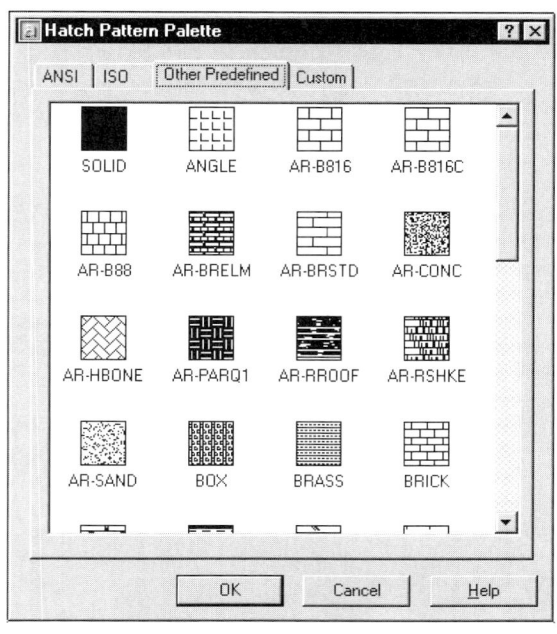

Figure 24-5

User-defined
To define a simple hatch pattern "on the fly," select the *User-defined* tile. This causes the *Pattern* and *Scale* options to be disabled and the *Angle, Spacing,* and *Double* options to be enabled. Creating a *User-defined* pattern is easy. Specify the *Angle* of the lines, the *Spacing* between lines, and optionally create *Double* (perpendicular) lines. All *User-defined* patterns have continuous lines.

Custom
Custom patterns are previously created user-defined patterns stored in other than the ACAD.PAT file. Custom patterns can contain continuous, dashed, and dotted line combinations. See the AutoCAD Customization Guide for information on creating and saving custom hatch patterns.

Pattern...
Selecting the *Pattern* drop-down list (see Figure 24-3) displays the name of each predefined pattern. Making a selection dictates the current pattern and causes the pattern to display in the *Swatch* window.

Swatch:
Click in the *Swatch* tile to produce the *Hatch Pattern Palette* dialog box (see Figure 24-5). You can select from *ANSI, ISO, Other Predefined*, and *Custom* (if available) hatch patterns.

464   Section Views

Figure 24-6 displays each of the AutoCAD hatch patterns defined in the ACAD.PAT file. Note that the patterns are not shown to scale.

Figure 24-6

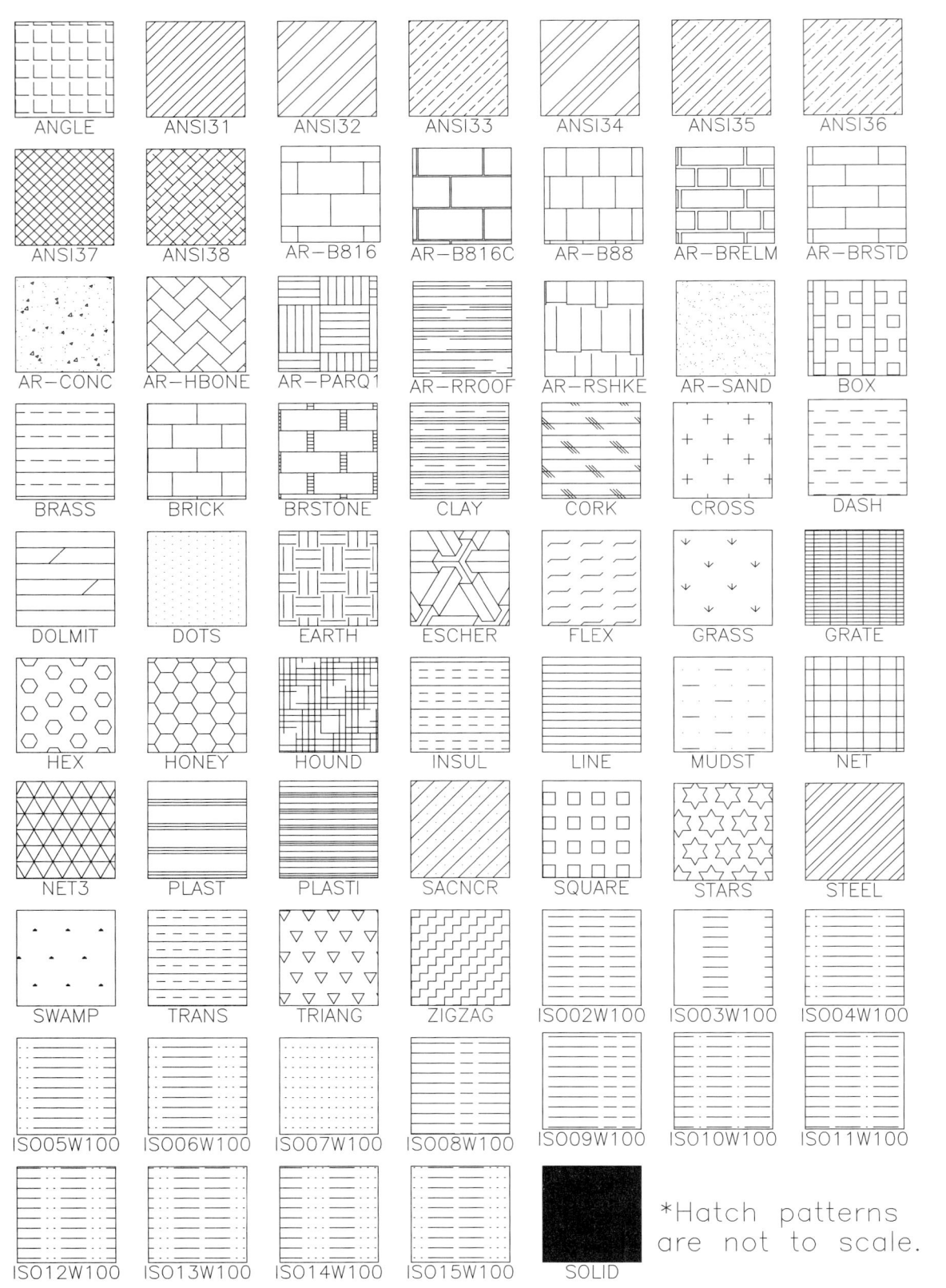

ANGLE, ANSI31, ANSI32, ANSI33, ANSI34, ANSI35, ANSI36
ANSI37, ANSI38, AR-B816, AR-B816C, AR-B88, AR-BRELM, AR-BRSTD
AR-CONC, AR-HBONE, AR-PARQ1, AR-RROOF, AR-RSHKE, AR-SAND, BOX
BRASS, BRICK, BRSTONE, CLAY, CORK, CROSS, DASH
DOLMIT, DOTS, EARTH, ESCHER, FLEX, GRASS, GRATE
HEX, HONEY, HOUND, INSUL, LINE, MUDST, NET
NET3, PLAST, PLASTI, SACNCR, SQUARE, STARS, STEEL
SWAMP, TRANS, TRIANG, ZIGZAG, ISO02W100, ISO03W100, ISO04W100
ISO05W100, ISO06W100, ISO07W100, ISO08W100, ISO09W100, ISO10W100, ISO11W100
ISO12W100, ISO13W100, ISO14W100, ISO15W100, SOLID

*Hatch patterns are not to scale.

Hatch patterns are created using the current linetype. Therefore, the <u>Continuous linetype should be set current</u> when hatching to ensure the selected area is filled with the pattern as it appears in the image tile. After selecting a pattern, specify the desired *Scale* and *Angle* of the pattern or ISO pen width.

Scale
The value entered in this edit box is a scale factor that is applied to the existing selected pattern. Normally, this scale factor should be changed proportionally with changes in the drawing *Limits*. Like many other scale factors (*LTSCALE, DIMSCALE*), AutoCAD defaults are set to a value of 1, which is appropriate for the default *Limits* of 12 x 9. If you have calculated the drawing scale factor, enter that value in the *Scale* edit box (see Chapter 12 for information on the "Drawing Scale Factor"). The *Scale* value is stored in the *HPSCALE* system variable.

ISO hatch patterns are intended for use with metric drawings; therefore, the scale (spacing between hatch lines) is much greater than for inch drawings. Because *Limits* values for metric sheet sizes are greater than for inch-based drawings (25.4 times greater than comparable inch drawings), the ISO hatch pattern scales are automatically compensated. If you want to use an ISO pattern with inch-based drawings, calculate a hatch pattern scale based on the drawing scale factor and multiply by .039 (1/25.4).

Angle
The *Angle* specification determines the angle (slant) of the hatch pattern. The default angle of 0 represents whatever angle is displayed in the pattern's image tile. Any value entered deflects the existing pattern (as it appears in the swatch) by the specified value (in degrees). The value entered in this box is held in the *HPANG* system variable.

Relative to Paper Space
This option changes the hatch pattern scale relative to paper space units. Checking this box automatically changes the *HPSCALE* variable based on the proportion of the paper space units to model space units (viewport scale or *Zoom XP* factor). Using this option, you can easily display hatch patterns at a scale that is appropriate for your layout. You can use this option only when you have a layout with a viewport and invoke *Bhatch* from the layout.

Spacing
This option is enabled if *User-defined* pattern is specified. Enter a value for the distance between lines.

ISO Pen Width
You must select an ISO hatch pattern for this tile to be enabled. Selecting an *ISO Pen Width* from the drop-down list automatically sets the scale and enters the value in the *Scale* edit box. See "*Scale*."

Double
Only for a *User-defined* pattern, check this box to have a second set of lines drawn at 90 degrees to the original set.

Selecting the Hatch Area

Once the hatch pattern and options have been selected, you must indicate to AutoCAD what area(s) should be hatched. Either the *Pick Points* method, *Select Objects* method, or a combination of both can be used to accomplish this.

Pick Points
This tile should be selected if you want AutoCAD to automatically determine the boundaries for hatching. You only need to select a point <u>inside</u> the area you want to hatch. The point selected must be inside a <u>completely closed shape</u>. When the *Pick Points* tile is selected, AutoCAD gives the following prompts:

```
Select internal point: PICK
Selecting everything...
Selecting everything visible...
Analyzing the selected data...
Analyzing the internal islands...
Select internal point: PICK another area or Enter
```

When an internal point is PICKed, AutoCAD traces and highlights the boundary (Fig. 24-7). The interior area is then analyzed for islands to be included in the hatch boundary (if you select the *Flood* island detection method). Multiple boundaries can be designated by selecting multiple internal points. Type *U* to undo the last one, if necessary.

Figure 24-7

The location of the point selected is usually not critical. However, the point must be PICKed inside the expected boundary. If there are any gaps in the area, a complete boundary cannot be formed and a boundary error message appears (Fig. 24-8).

Figure 24-8

Select Objects
Alternately, you can designate the boundary with the *Select Objects* method. Using the *Select Objects* method, you specify the boundary objects rather than let AutoCAD locate a boundary. With the *Select Objects* method, no temporary *Pline* boundary is created as with the *Pick Points* method. Therefore, the selected objects must form a closed shape with no gaps or overlaps. If gaps exist (Fig. 24-9) or if the objects extend past the desired hatch area (Fig. 24-10), AutoCAD cannot interpret the intended hatch area correctly, and problems will occur.

Figure 24-9

The *Select Objects* option can be used after the *Pick Points* method to select specific objects for *Bhatch* to consider before drawing the hatch pattern lines. For example, if you had created text objects or dimensions within the boundary found by the *Pick Points* method, you may then have to use the *Select Objects* option to select the text and dimensions (if *Ray Casting* is your choice of island detection method). Using this procedure, the hatch lines are automatically "trimmed" around the text and dimensions.

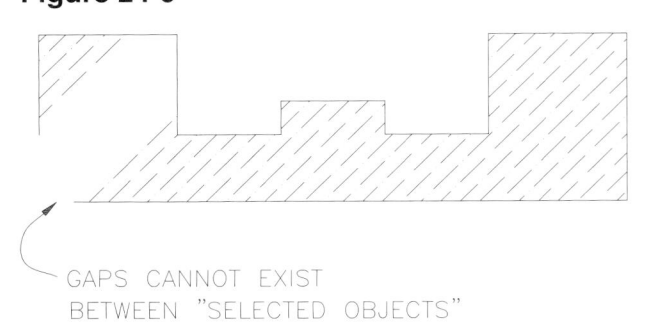

Figure 24-10

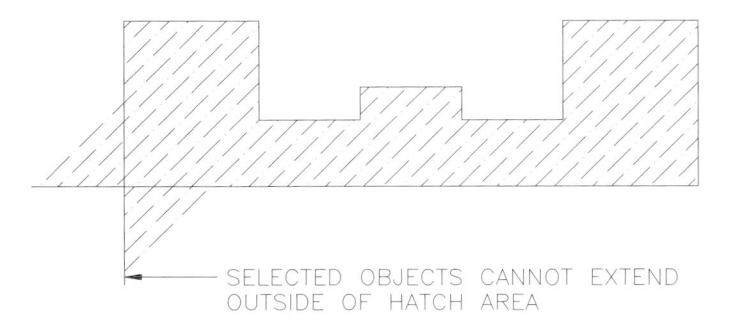

Remove Islands
PICKing this tile allows you to select specific islands (internal objects) to remove from those AutoCAD has found within the outer boundary. If hatch lines have been drawn, be careful to *Zoom* in close enough to select the desired boundary and not the hatch pattern.

View Selections
Clicking the *View Selections* tile causes AutoCAD to highlight all selected boundaries. This can be used as a check to ensure the desired areas are selected.

Inherit Properties
This option allows you to select a hatch pattern from one existing in the drawing. This option operates similarly to *Matchprop* because you PICK the existing hatch pattern from the drawing and copy it to another area by selecting an internal point.

    Select associative hatch object: **PICK**
    Inherited Properties: Name <ANSI32>, Scale <1.0000>, Angle <0>
    Select internal point:

Associative / Nonassociative
This checkbox toggles (on or off) the associative property for newly applied hatch patterns. Associative hatch patterns automatically update by conforming to the new boundary shape when the boundary is changed (see Figure 24-2). A non-associative hatch pattern is static even when the boundary changes.

Preview
You should always use the *Preview* option after specifying the hatch parameters and selecting boundaries, but before you apply the hatch. This option allows you to temporarily look at the hatch pattern in your drawing with the current settings applied and allows you to adjust the settings, if necessary, before using *OK*. After viewing the drawing, press Enter to redisplay the *Boundary Hatch* dialog box, allowing you to make adjustments.

Advanced Tab

The *Advanced* tab produces the options shown in Figure 24-11 and described below.

Figure 24-11

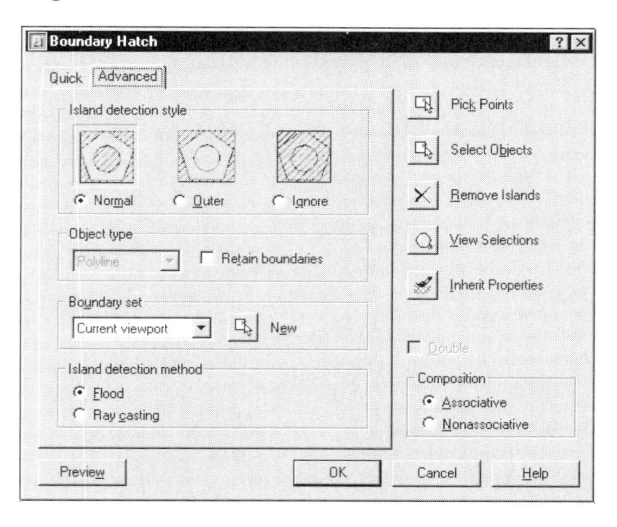

Island Detection Style
This section allows you to specify how the hatch pattern is drawn when the area inside the defined boundary contains text or closed areas (islands). Select the text from the list or PICK the icons displayed in the window. These options are applicable only when interior objects (islands) have been included in the selection set (considered for hatching). Otherwise, if only the outer shape is included in the selection set, the results are identical to the *Ignore* option. (See also "Island Detection Method.")

Normal
This should be used for most applications of *Bhatch*. Text or closed shapes within the outer border are considered in such cases. Hatching will begin at the outer boundary and move inward alternating between applying and not applying the pattern as interior shapes or text are encountered (Fig. 24-12).

Outer
This option causes AutoCAD to hatch only the outer closed shape. Hatching is turned off for all interior closed shapes (Fig. 24-12).

Figure 24-12

Ignore
Ignore draws the hatch pattern from the outer boundary inward ignoring any interior shapes. The resulting hatch pattern is drawn through the interior shapes (Fig. 24-12).

Object Type
Bhatch creates *Polyline* or *Region* boundaries. This option is enabled only when the *Retain Boundaries* box is checked (see Figure 24-11, on the previous page).

Retain Boundaries
When AutoCAD uses the *Pick Points* method to locate a boundary for hatching, a temporary *Pline* or *Region* is created for hatching, then discarded after the hatching process. Checking this box forces AutoCAD to keep the boundary. When the box is checked, *Bhatch* creates two objects—the hatch pattern and the boundary object. (You can specify whether you want to create a *Pline* or a *Region* boundary in the *Object Type* drop-down list.) Using this option and erasing the hatch pattern accomplishes the same results as using the *Boundary* command.

After using *Bhatch*, the *Pline* or *Region* can be used for other purposes. To test this function, complete a *Bhatch* with the *Retain Boundaries* box checked; then use *Move* (make sure you select the boundary) to reveal the new boundary object (Fig. 24-13).

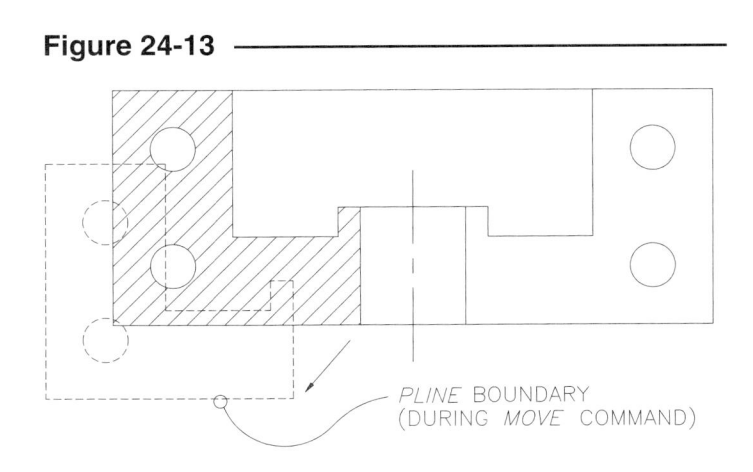

Figure 24-13

Boundary Set
By default, AutoCAD examines all objects in the viewport when determining the boundaries by the *Pick Points* method. (*Current Viewport* is selected by default when you begin the *Bhatch* command.) For complex drawings, examining all objects can take some time. In that case, you may want to specify a smaller boundary set for AutoCAD to consider. Clicking the *New* tile clears the dialog boxes and permits you to select objects or select a window to define the new set.

Island Detection Method
This section specifies whether to include objects within the outermost boundary as boundary objects. These internal objects are known as islands.

Flood
Selecting the *Flood* method automatically includes islands as boundary objects and places the hatch lines around them, so the *Normal* or *Outer* island detection style can be applied.

Ray Casting
When you pick an internal point, a ray is cast from the point you specify to the nearest object. It then traces the boundary in a counterclockwise direction, thus excluding islands as boundary objects. If you want to include islands using this method, you must select them using *Select Objects*.

HATCH

Pull-down Menu	COMMAND (TYPE)	ALIAS (TYPE)	Short-cut	Screen (side) Menu	Tablet Menu
...	HATCH	-H	...	...	...

Hatch must be typed at the keyboard since it is not available from the menus. *Hatch* operates in a Command line format with many of the options available in the *Bhatch* dialog box. However, *Hatch* creates a non-associative hatch pattern and does not use the *Pick points* method ("select internal point:") to create a boundary. Generally, *Bhatch* would be used instead of *Hatch* except for special cases.

You can use *Hatch* to find a boundary from existing objects. The objects must form a complete closed shape, just as with *Bhatch*. The shape can be composed of one closed object such as a *Pline, Polygon, Spline,* or *Circle* or composed of several objects forming a closed area such as *Lines* and *Arcs*. For example, the shape in Figure 24-14 can be hatched correctly with *Hatch*, whether it is composed of one object (*Pline, Region,* etc.) or several objects (*Lines, Arc,* etc.).

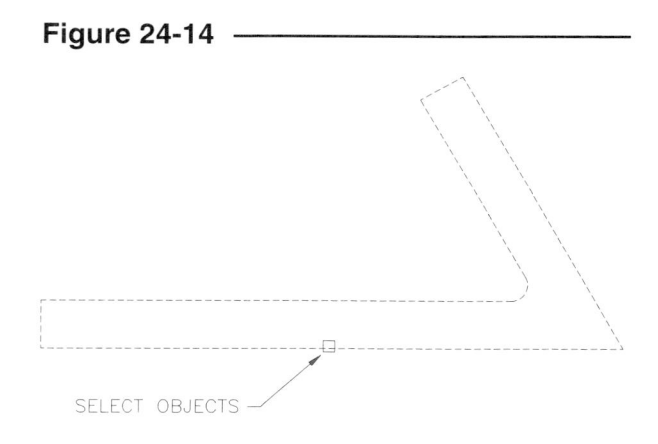

Figure 24-14

If you select several objects to define the boundary, the objects must comprise only the boundary shape and not extend past the desired boundary. If objects extend past the desired boundary (Fig. 24-15) or if there are gaps, the resulting hatch will be incorrect as shown in Figures 24-9 and 24-10.

Figure 24-15

Command: **hatch**
Enter a pattern name or [?/Solid/User defined] <ANSI31>: (**pattern name** or **style**)
Specify a scale for the pattern <1.0000>: (**value**) or **Enter**
Specify an angle for the pattern <0>: (**value**) or **Enter**
Select objects to define hatch boundary or <direct hatch>,
Select objects: **PICK**
Select objects: **Enter**
Command:

Direct Hatch
With the *Direct Hatch* method of *Hatch*, you specify points to define the boundary, not objects. The significance of this method is that you can create a hatch pattern without using existing objects as the boundary. In addition, you can select whether you want to retain or discard the boundary after the pattern is applied (Fig. 24-16).

Figure 24-16

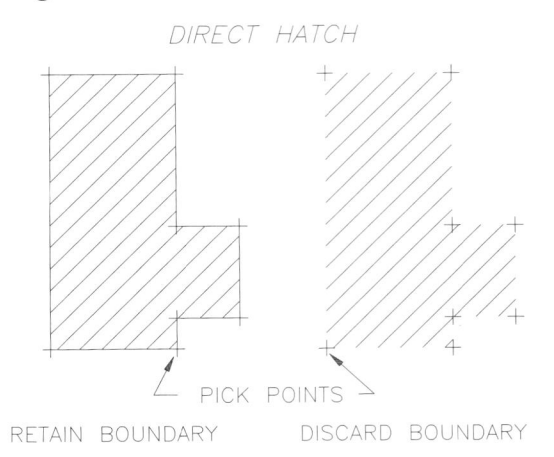

```
Command: hatch
Enter a pattern name or [?/Solid/User defined] <ANSI31>: (pattern name or style)
Specify a scale for the pattern <1.0000>: (value) or Enter
Specify an angle for the pattern <0>: (value) or Enter
Select objects to define hatch boundary or <direct hatch>,
Select objects: Enter
Retain polyline boundary? [Yes/No] <N>: (option)
Specify start point: PICK
Specify next point or [Arc/Close/Length/Undo]: PICK or (option)
Specify next point or [Arc/Close/Length/Undo]: PICK or (option)
Specify next point or [Arc/Close/Length/Undo]: PICK or (option)
Specify next point or [Arc/Close/Length/Undo]: c
Specify start point for new boundary or <apply hatch>: Enter
Command:
```

TIP For special cases when you do not want a hatch area boundary to appear in the drawing (Fig. 24-17), you can use the *Direct Hatch* method to specify a hatch area and discard the boundary *Pline*.

Figure 24-17

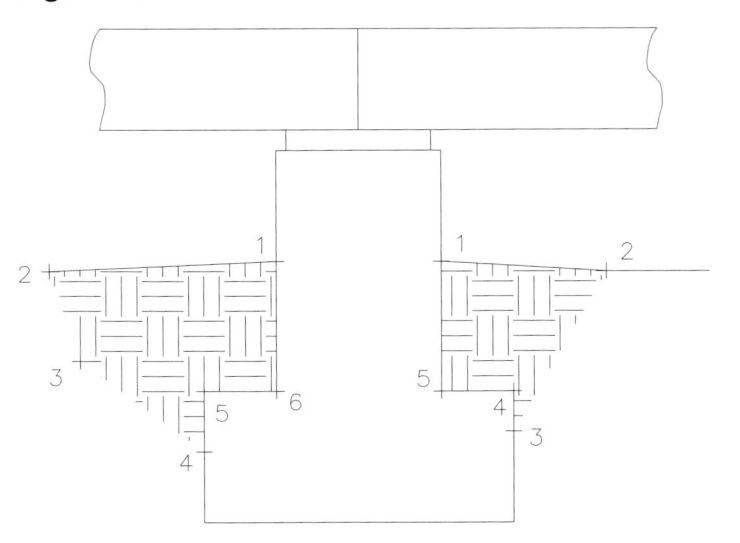

Drag and Drop Hatch Patterns

You can use DesignCenter to drag and drop hatch patterns directly into a closed area of a drawing you want to hatch. You do not access this feature from the *Boundary Hatch* dialog box, rather through DesignCenter.

First, invoke DesignCenter. In the Tree View list, locate any drawing that contains the hatch pattern you want, or locate the ACAD.PAT file in the AutoCAD 2002\Support folder. When you highlight the drawing name in the list, hatch patterns contained within the drawing are displayed in the Palette. Simply drag the desired pattern from the Palette into an open drawing and drop it into the intended closed area (Fig. 24-18).

Figure 24-18

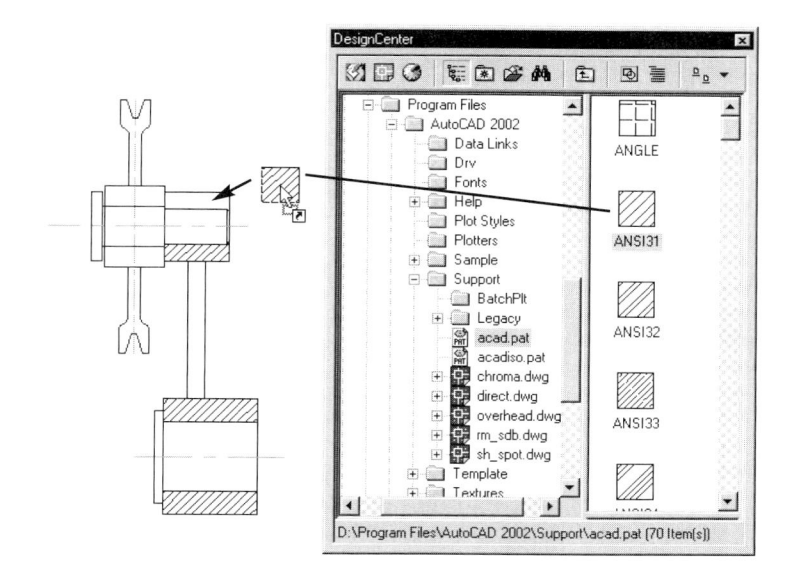

When you drag and drop a hatch pattern from DesignCenter, the hatch pattern takes on the parameters currently specified in the *Boundary Hatch* dialog box. You can quickly access this dialog box by right-clicking on any pattern in the Palette (Fig. 24-19). Set the desired parameters in the *Boundary Hatch* dialog box, close the dialog box, then drag and drop.

Figure 24-19

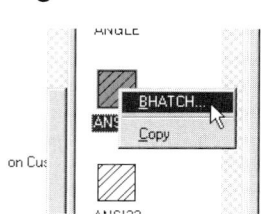

EDITING HATCH PATTERNS AND BOUNDARIES

HATCHEDIT

Pull-down Menu	COMMAND (TYPE)	ALIAS (TYPE)	Short-cut	Screen (side) Menu	Tablet Menu
Modify Object > Hatch...	HATCHEDIT or -HATCHEDIT	HE	...	MODIFY1 Hatchedt	Y,16

Hatchedit allows you to modify an existing associative hatch pattern in the drawing. This feature of AutoCAD makes the hatching process more flexible because you can hatch several areas to quickly create a "rough" drawing, then retroactively fine-tune the hatching parameters when the drawing nears completion with *Hatchedit*.

You can produce the *Hatchedit* dialog box (Fig. 24-20) by any method shown in the Command table as well as double-clicking on any hatched area in the drawing (assuming *DBLCLKEDIT* is set to *On* and *PICKSTYLE* is set to 1). The dialog box provides options for changing the *Pattern, Scale, Angle,* and *Type* properties of the existing hatch. You can also change the hatch to *Associative* or *Nonassociative*.

Figure 24-20

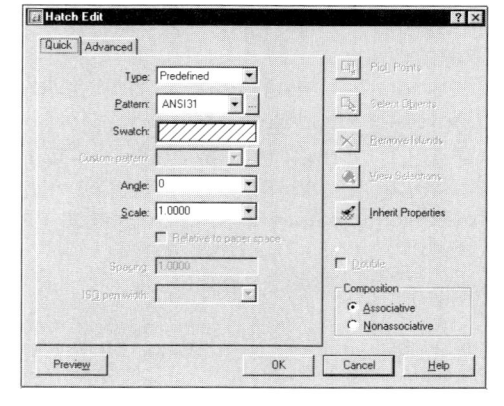

Apparent in Figure 24-20, the *Hatch Edit* dialog box is essentially the same as the *Boundary Hatch* dialog box with some of the options disabled. For an explanation of the options that are available with *Hatchedit*, see *Boundary Hatch* earlier in this chapter.

Using Grips with Hatch Patterns

Figure 24-21

Grips can be used effectively to edit hatch pattern boundaries of associative hatches. Do this by selecting the boundary with the pickbox or automatic window/crossing window (as you would normally activate an object's grips). Associative hatches <u>retain the association</u> with the boundary after the boundary has been changed using grips (see Figure 24-2, earlier in this chapter).

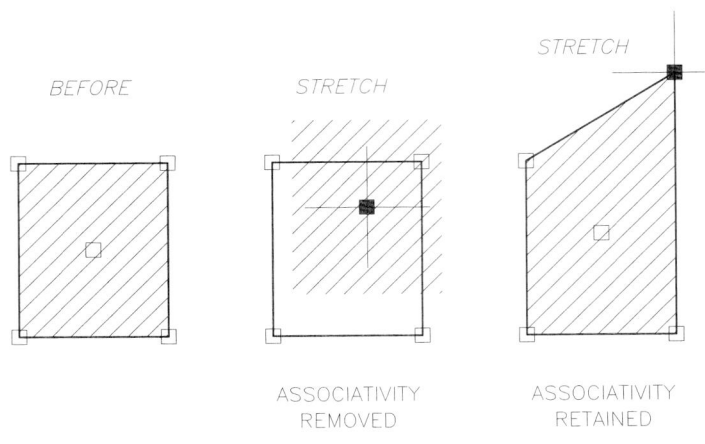

Beware, when selecting the hatch boundary, ensure you <u>edit the boundary and not the hatch object</u>. If you edit the hatch object only, the associative feature is lost. This can happen if you select the <u>hatch object</u> grip (Fig. 24-21).

If you activate grips on <u>both</u> the hatch object and the boundary and make a boundary grip hot, the boundary and related hatch retain associativity (see Figure 24-21).

DRAWORDER

Pull-down Menu	COMMAND (TYPE)	ALIAS (TYPE)	Short-cut	Screen (side) Menu	Tablet Menu
Tools Display Order >	DRAWORDER	DR	...	TOOLS 1 Drawordr	T,9

It is possible to use solid hatch patterns and raster images in combination with filled text, other hatch patterns and solid images, etc. With these solid areas and images, some control of which objects are in "front" and "back or "above" and "under" must be provided. This control is provided by the *Draworder* command.

Figure 24-22

For example, if a company logo were created using filled *Dtext*, a *Sand* hatch pattern, and a *Solid* hatched circle, the object that was created last would appear "in front" (Fig. 24-22, top logo). The *Draworder* command is used to control which objects appear in front and back or above and under. This capability is necessary for creating prints and plots in black and white and in color.

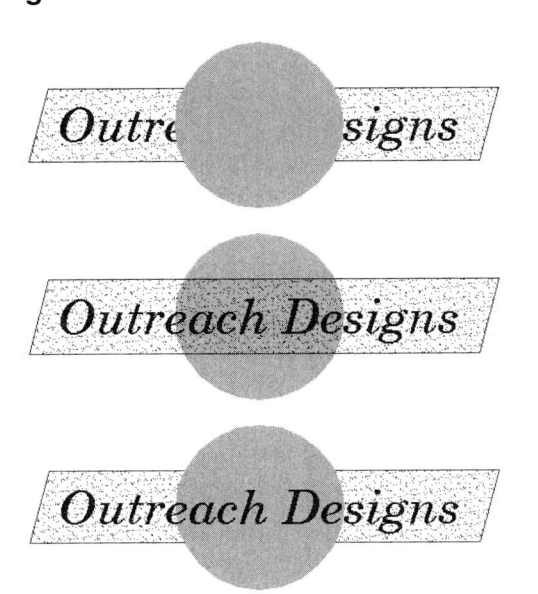

The *Draworder* command is simple to use. Simply select objects and indicate *Front*, *Back*, *Above*, or *Under*.

    Command: *draworder*
    Select objects: **PICK**
    Select objects: **Enter**
    Enter object ordering option [Above object/Under object/Front/Back] <Back>: (**option**)
    Regenerating drawing.
    Command:

In Figure 24-22, three (of four) possibilities are shown for the three objects—*Dtext*, *Solid* hatch and boundary, and *Sand* hatch and boundary.

DRAWING CUTTING PLANE LINES

Most section views (full, half, and offset sections) require a cutting plane line to indicate the plane on which the object is cut. The cutting plane line is drawn in a view adjacent to the section view because the line indicates the plane of the cut from its edge view. (In the section view, the cutting plane is perpendicular to the line of sight, therefore, not visible as a line.)

Standards provide two optional line types for cutting plane lines. In AutoCAD, the two linetypes are *Dashed* and *Phantom,* as shown in Figure 24-23. Arrows at the ends of the cutting plane line indicate the line-of-sight for the section view.

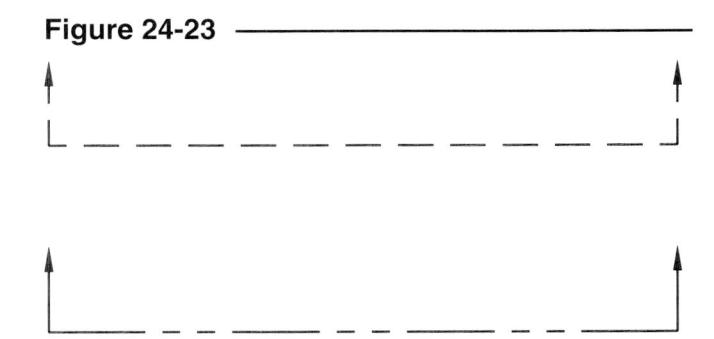

Figure 24-23

Cutting plane lines should be drawn or plotted in a heavy lineweight. Two possible methods can be used to accomplish this: use a *Pline* with *Width* or assign a *Lineweight* to the line or layer. If you prefer drawing the cutting plane line using a *Pline*, use a *Width* of .02 or .03 times the drawing scale factor. Assigning a *Lineweight* to the line or layer is a simpler method, but the *Lineweight* appears only in a plot preview. A *Lineweight* of .8mm or .031" is appropriate for cutting plane lines.

For the example section view, a cutting plane line is created in the top view to indicate the plane on which the object is cut and the line of sight for the section view (Fig. 24-24). (The cutting plane could be drawn in the side view, but the top would be much clearer.) First, a *New* layer named CUT is created and the *Dashed linetype* is assigned to the layer. The layer is *Set* as the *Current* layer. Next, construct a *Pline* with the desired *Width* to represent the cutting plane line.

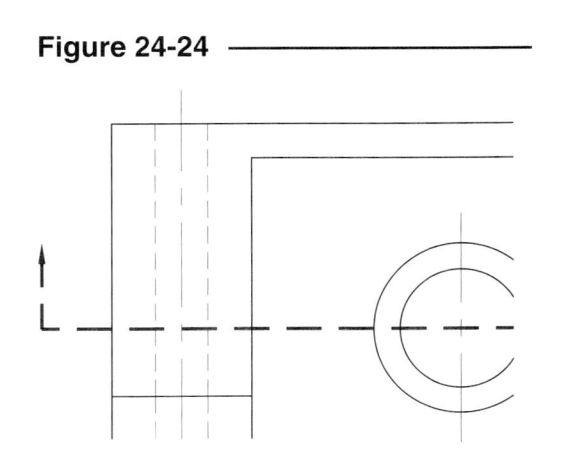

Figure 24-24

The resulting cutting plane line appears as shown in Figure 24-25 but without the arrow heads. The horizontal center line for the hole (top view) was *Erased* before creating the cutting plane line.

Figure 24-25

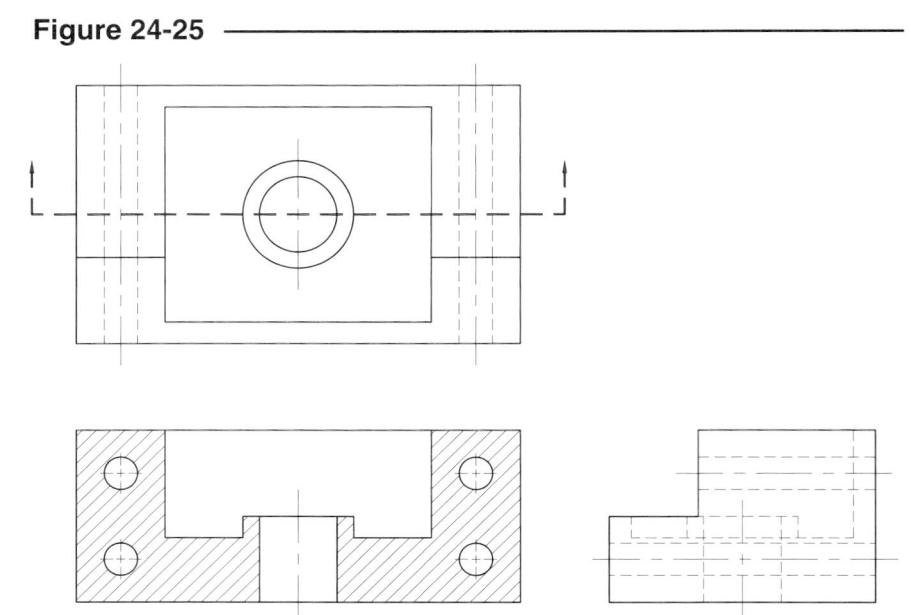

The last step is to add arrowheads to the ends of the cutting plane line. Arrowheads can be drawn by three methods: (1) use the *Solid* command to create a three-sided solid area; (2) use a tapered *Pline* with beginning width of 0 and ending width of .08, for example; or (3) use the *Leader* command to create the arrowhead (see Chapter 26, Dimensioning, for details on *Leader*). The arrowhead you create can be *Scaled*, *Rotated*, and *Copied* if needed to achieve the desired size, orientation, and locations. The resulting multiview drawing with section view is complete and ready for dimensioning, drawing a border, and inserting a title block (Fig. 24-25).

CHAPTER EXERCISES

For the following exercises, create the section views as instructed. Use an appropriate template drawing for each unless instructed otherwise. Include a border and title block in the layout for each drawing.

1. Make a multiview drawing of the Bearing shown in Figure 24-26. Convert the front view to a full section view. Add the necessary cutting plane line in the top view. *Save* the drawing as **BEAR-SEC**. Make a *Plot* at full size.

2. *Open* the **SADDLE** drawing that you created in Chapter 22. Convert the front view to a full section. *Save* the drawing as **SADL-SEC**. *Plot* the drawing at **1=1** scale.

Figure 24-26

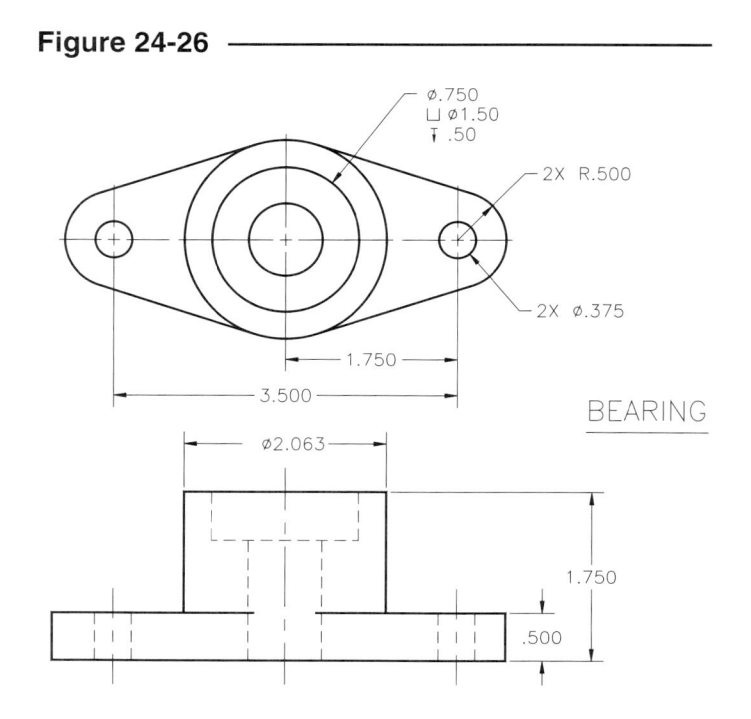

3. Create a multiview drawing of the Clip shown in Figure 24-27. Include a side view as a full section view. You can use the **CLIP** drawing you created in Chapter 22 and convert the side view to a section view. Add the necessary cutting plane line in the front view. *Plot* the finished drawing at **1=1** scale and *SaveAs* **CLIP-SEC**.

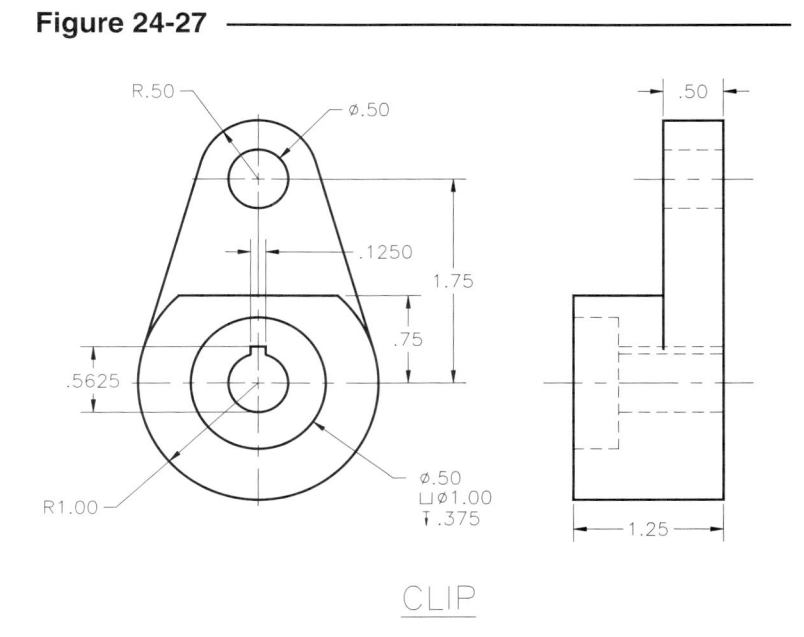

Figure 24-27

4. Create a multiview drawing, including two full sections of the Stop Block as shown in Figure 24-28. Section B–B' should replace the side view shown in the figure. *Save* the drawing as **SPBK-SEC**. *Plot* the drawing at **1=2** scale.

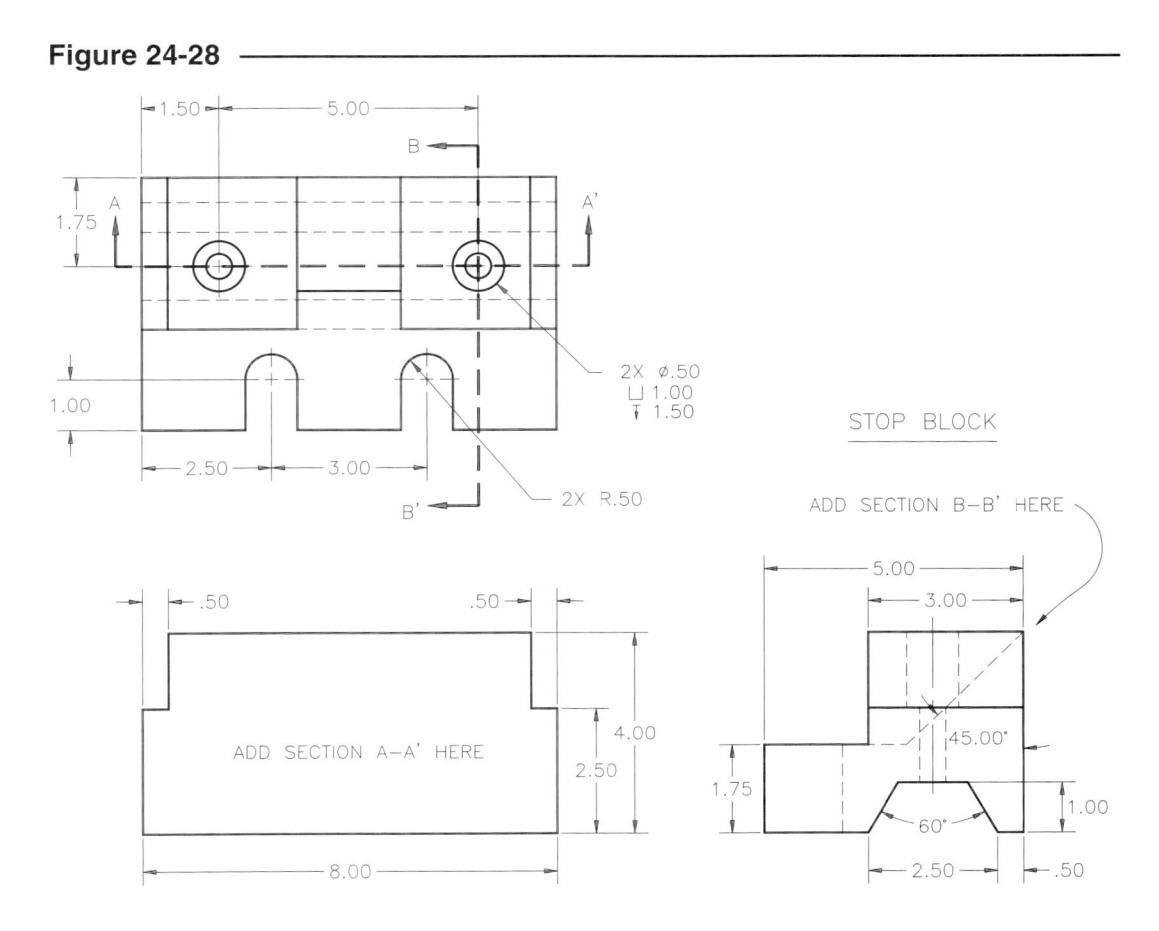

Figure 24-28

476 Section Views

5. Make a multiview drawing, including a half section of the Pulley. All vertical dimensions are diameters. Two views (including the half section) are sufficient to describe the part. Add the necessary cutting plane line. *Save* the drawing as **PULSEC** and make a *Plot* at **1:1** scale.

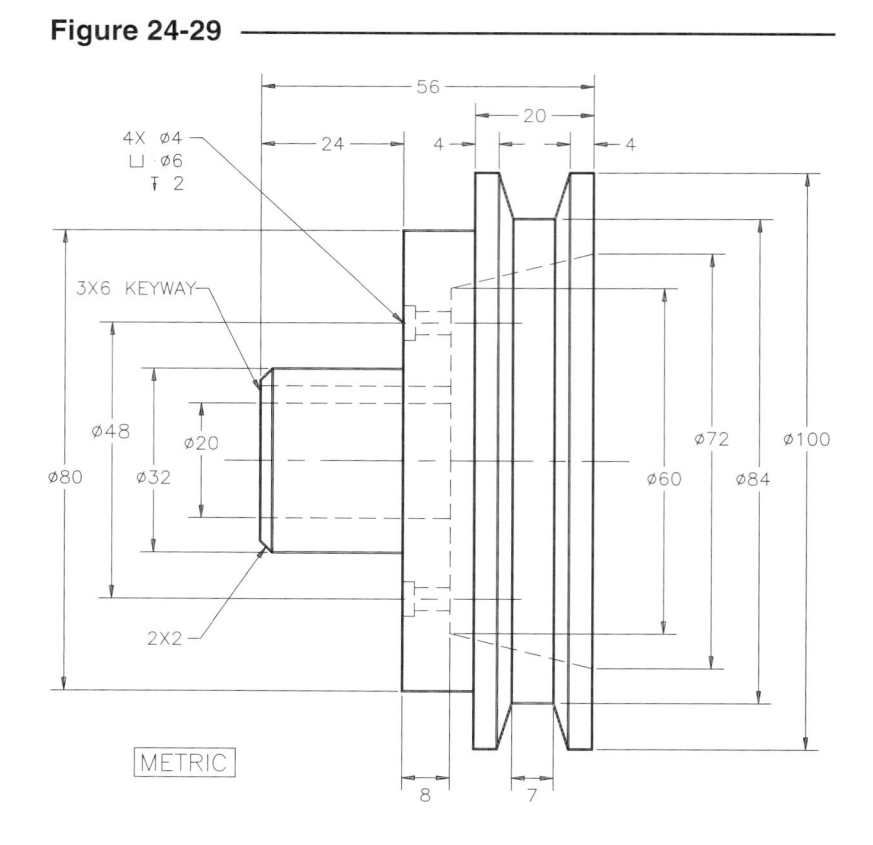

Figure 24-29

6. Draw the Grade Beam foundation detail in Figure 24-30. Do not include the dimensions in your drawing. Use the **Bhatch** command to hatch the concrete slab with **AR-CONC** hatch pattern. Use *Sketch* to draw the grade line and *Hatch* with the **EARTH** hatch pattern. *Save* the drawing as **GRADBEAM**.

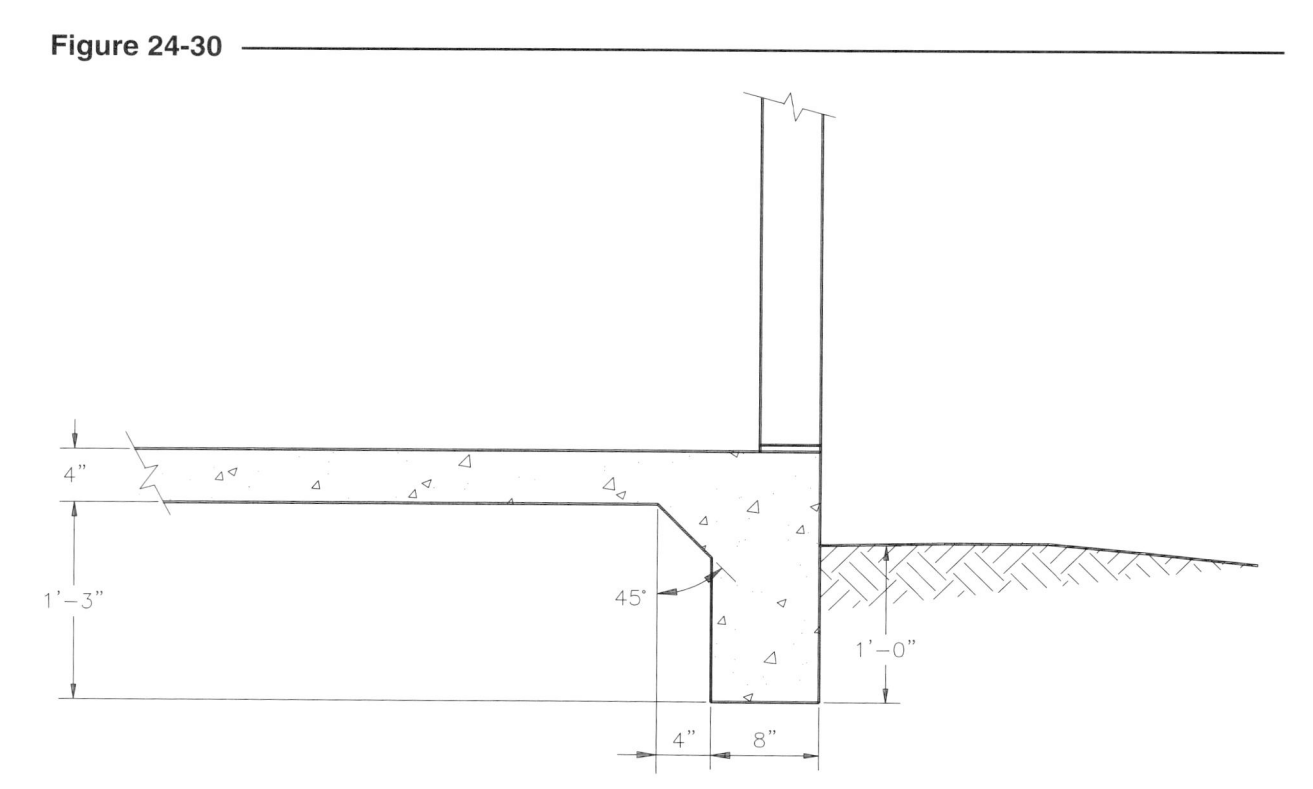

Figure 24-30

25

AUXILIARY VIEWS

Chapter Objectives

After completing this chapter you should:

1. be able to use the *Rotate* option of *Snap* to change the angle of the *SNAP*, *GRID*, and *ORTHO*;

2. be able to set the *Increment angle* and *Additional angles* for creating auxiliary views using *Polar Tracking*;

3. know how to use the *Offset* command to create parallel line copies;

4. be able to use *Xline* and *Ray* to create construction lines for auxiliary views.

478   Auxiliary Views

CONCEPTS

AutoCAD provides no features explicitly for the creation of auxiliary views in a 2D drawing. No new commands are discussed in this chapter. However, four particular features that have been discussed earlier can assist you in construction of auxiliary views. Those features are the *SNAP* rotation, *Polar Tracking*, the *Offset* command, and the *Xline* and *Ray* commands.

An auxiliary view is a supplementary view among a series of multiviews. The auxiliary view is drawn in addition to the typical views that are mutually perpendicular (top, front, side). An auxiliary view is one that is normal (the line-of-sight is perpendicular) to an inclined surface of the object. Therefore, the auxiliary view is constructed by projecting in a 90 degree direction from the edge view of an inclined surface in order to show the true size and shape of the inclined surface. The edge view of the inclined surface could be at any angle (depending on the object), so lines are typically drawn parallel and perpendicular relative to that edge view. Hence, the *SNAP* rotation feature, *Polar Tracking*, the *Offset* command, and the *Xline* and *Ray* commands can provide assistance in this task.

Figure 25-1

An example mechanical part used for the application of these AutoCAD features related to auxiliary view construction is shown in Figure 25-1. As you can see, there is an inclined surface that contains two drilled holes. To describe this object adequately, an auxiliary view should be created to show the true size and shape of the inclined surface.

This chapter explains the construction of a partial auxiliary view for the example object in Figure 25-1.

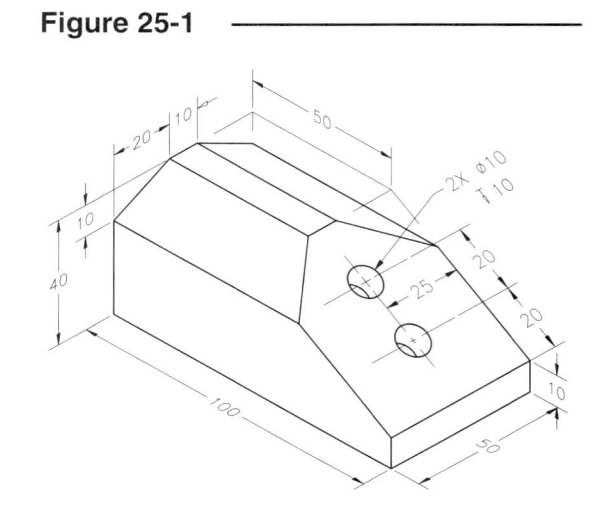

CONSTRUCTING AN AUXILIARY VIEW

Setting Up the Principal Views

Figure 25-2

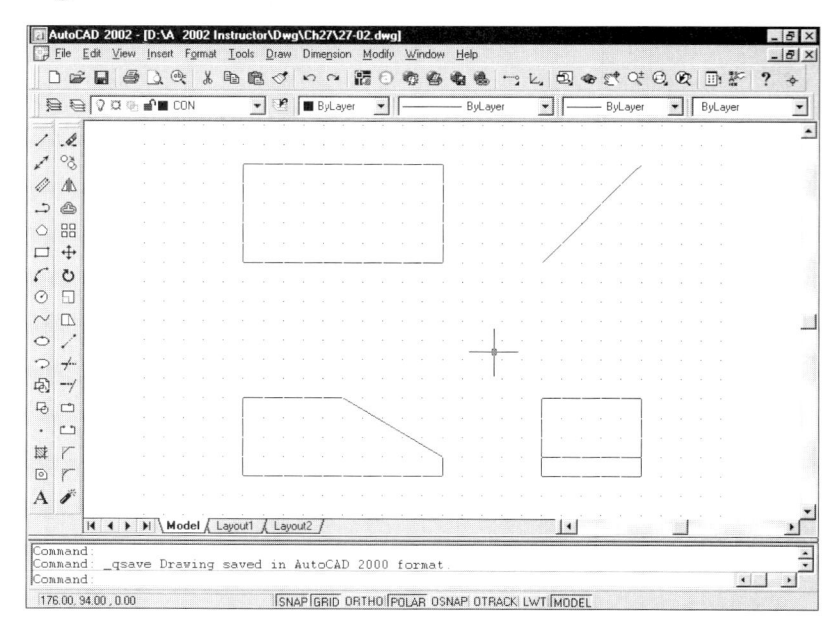

To begin this drawing, the typical steps are followed for drawing setup (Chapter 12). Because the dimensions are in millimeters, *Limits* should be set accordingly. For example, to provide enough space to draw the views full size and to plot full size on an A sheet, *Limits* of 279 x 216 are specified.

In preparation for the auxiliary view, the principal views are "blocked in," as shown in Figure 25-2. The purpose of this step is to ensure that the desired views fit and are optimally spaced within the allocated *Limits*. If there is too little or too much room, adjustments can be made to the *Limits*.

Notice that space has been allotted between the views for a partial auxiliary view to be projected from the front view.

Before additional construction on the principal views is undertaken, initial steps in the construction of the partial auxiliary view should be performed. The projection of the auxiliary view requires drawing lines perpendicular and parallel to the inclined surface. One or more of the three alternatives (explained next) can be used.

Using *Snap Rotate* and *ORTHO*

One possibility to construct an auxiliary view is to use the *Snap* command with the *Rotate* option. This action permits you to rotate the *SNAP* to any angle about a specified base point. The *GRID* automatically follows the *SNAP*. Turning *ORTHO ON* forces *Lines* to be drawn orthogonally with respect to the rotated *SNAP* and *GRID*.

Figure 25-3 displays the *SNAP* and *GRID* after rotation. The command syntax is given below.

Figure 25-3

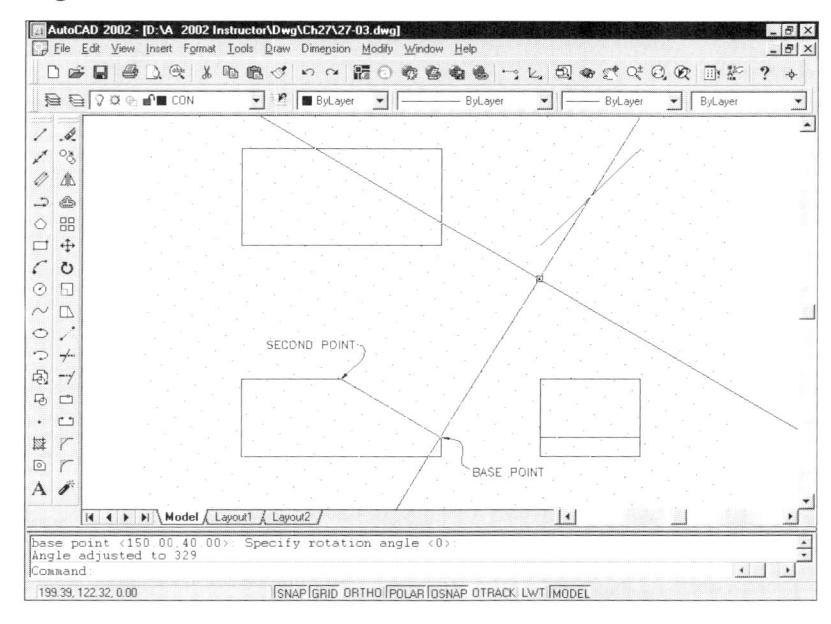

In this figure, the cursor size is changed from the default 5% of screen size to 100% of screen size to help illustrate the orientation of the *SNAP*, *GRID*, and cursor when *SNAP* is *Rotated*. You can change the cursor size in the *Display* tab of the *Options* dialog box.

```
Command: snap
Specify snap spacing or [ON/OFF/Aspect/Rotate/Style/Type] <0.5000>: r
Specify base point <0.0000,0.0000>: PICK or (coordinates) (PICK starts a rubberband line.)
Specify rotation angle <0>: PICK or (value) (PICK to specify second point to define the angle. See Figure 25-3.)
Command:
```

PICK (or specify coordinates for) the endpoint of the *Line* representing the inclined surface as the "base point." At the "rotation angle:" prompt, a value can be entered or another point (the other end of the inclined *Line*) can be PICKed. Use *OSNAP* when PICKing the *Endpoints*. If you want to enter a value but don't know what angle to rotate to, use *List* to display the angle of the inclined *Line*. The *GRID*, *SNAP*, and crosshairs should align with the inclined plane as shown in Figure 25-3.

(An option for simplifying construction of the auxiliary view is to create a new *UCS* [User Coordinate System] with the origin at the new base point. Use the *3Point* option and turn on *ORTHO* to select the three points. See Chapter 30, User Coordinate Systems, for more information on this procedure.)

After rotating the *SNAP* and *GRID*, the partial auxiliary view can be "blocked in," as displayed in Figure 25-4. Begin by projecting *Lines* up from and perpendicular to the inclined surface. (Make sure *ORTHO* is *ON*.) Next, two *Lines* representing the depth of the view should be constructed parallel to the inclined surface and perpendicular to the previous two projection lines. The depth dimension of the object in the auxiliary view is equal to the depth dimension in the top or right view. *Trim* as necessary.

Figure 25-4

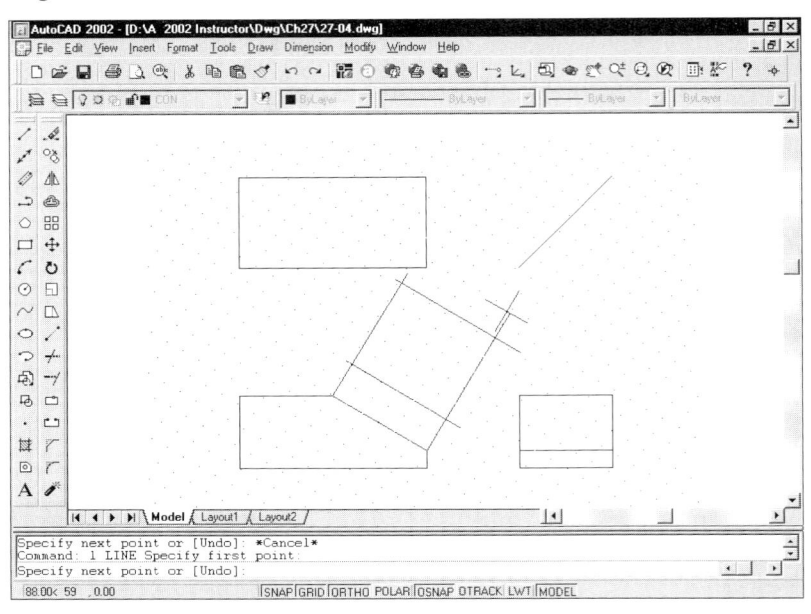

Locate the centers of the holes in the auxiliary view and construct two *Circles*. It is generally preferred to construct circular shapes in the view in which they appear as circles, then project to the other views. That is particularly true for this type of auxiliary since the other views contain ellipses. The centers can be located by projection from the front view or by *Offsetting Lines* from the view outline.

Next, project lines from the *Circles* and their centers back to the inclined surface. Use of a hidden line layer can be helpful here. While the *SNAP* and *GRID* are rotated, construct the *Lines* representing the bottom of the holes in the front view. (Alternately, *Offset* could be used to copy the inclined edge down to the hole bottoms; then *Trim* the unwanted portions of the *Lines*.)

Figure 25-5

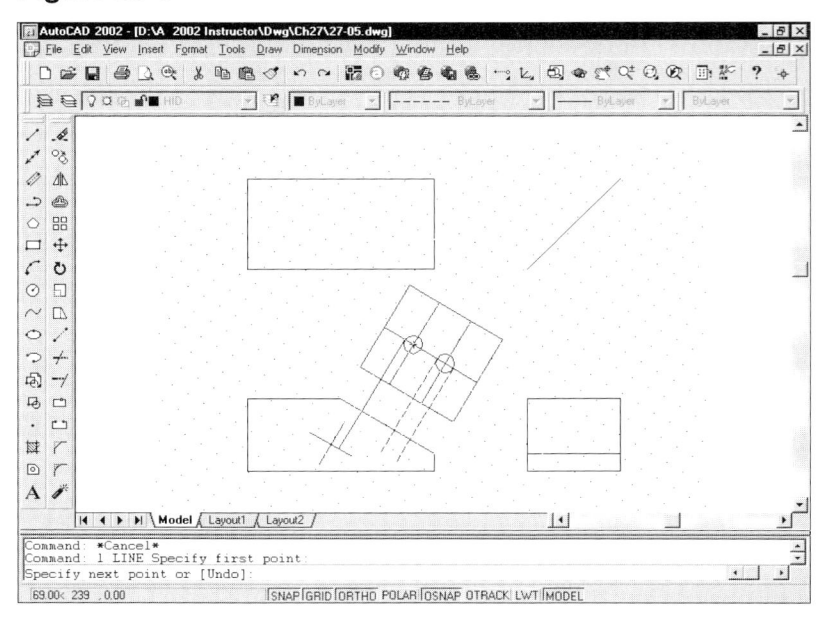

Rotating *SNAP* Back to the Original Position

Before details can be added to the other views, the *SNAP* and *GRID* should be rotated back to the original position. It is very important to rotate back using the <u>same base point</u>. Fortunately, AutoCAD remembers the original base point so you can accept the default for the prompt.

Next, enter a value of **0** when rotating back to the original position. (When using the *Snap Rotate* option, the value entered for the angle of rotation is absolute, not relative to the current position. For example, if the *Snap* was rotated to 45 degrees, rotate back to 0 degrees, not -45.)

Command: *snap*
Specify snap spacing or [ON/OFF/Aspect/Rotate/Style/Type] <2.00>: *r*
Specify base point <150.00,40.00>: **Enter** (AutoCAD remembers the previous base point.)
Specify rotation angle <329>: 0
Command:

Construction of multiview drawings with auxiliaries typically involves repeated rotation of the *SNAP* and *GRID* to the angle of the inclined surface and back again as needed.

With the *SNAP* and *GRID* in the original position, details can be added to the other views as shown in Figure 25-6. Since the two circles appear as ellipses in the top and right side views, project lines from the circles' centers and limiting elements on the inclined surface. Locate centers for the two *Ellipses* to be drawn in the top and right side views.

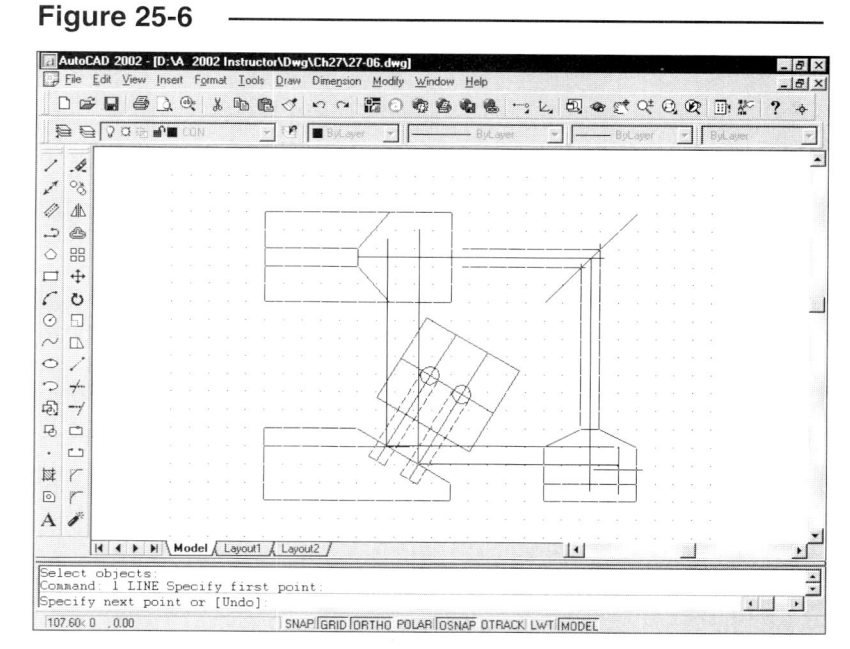

Figure 25-6

Use the *Ellipse* command to construct the ellipses in the top and right side views. Using the *Center* option of *Ellipse*, specify the center by PICKing with the *Intersection OSNAP* mode. *OSNAP* to the appropriate construction line *Intersection* for the first axis endpoint. For the second axis endpoint (as shown), use the actual circle diameter, since that dimension is not foreshortened.

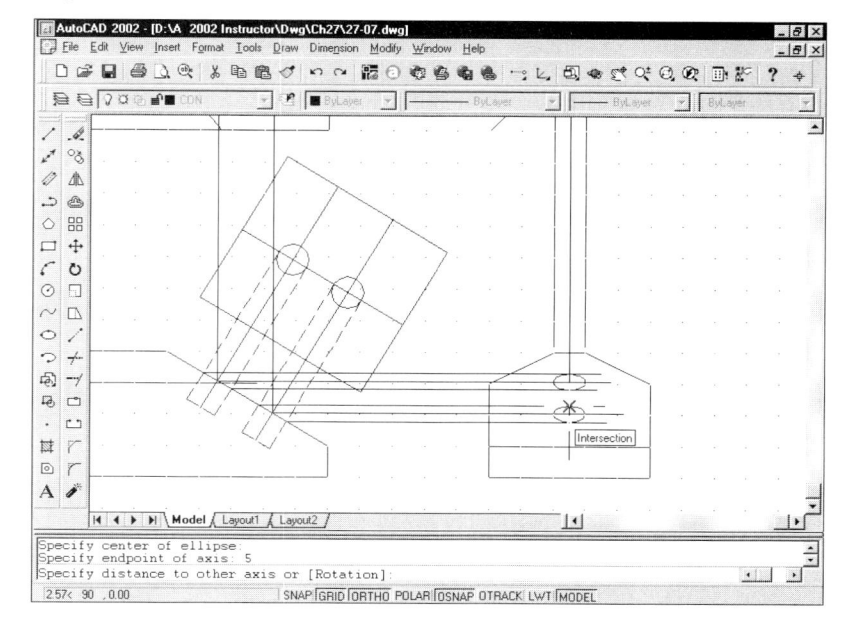

Figure 25-7

The remaining steps for completing the drawing involve finalizing the perimeter shape of the partial auxiliary view and *Copying* the *Ellipses* to the bottom of the hole positions. The *SNAP* and *GRID* should be rotated back to project the new edges found in the front view (Fig. 25-8).

At this point, the multiview drawing with auxiliary view is ready for centerlines, dimensioning, setting up a layout, and construction or insertion of a border and title block.

Figure 25-8

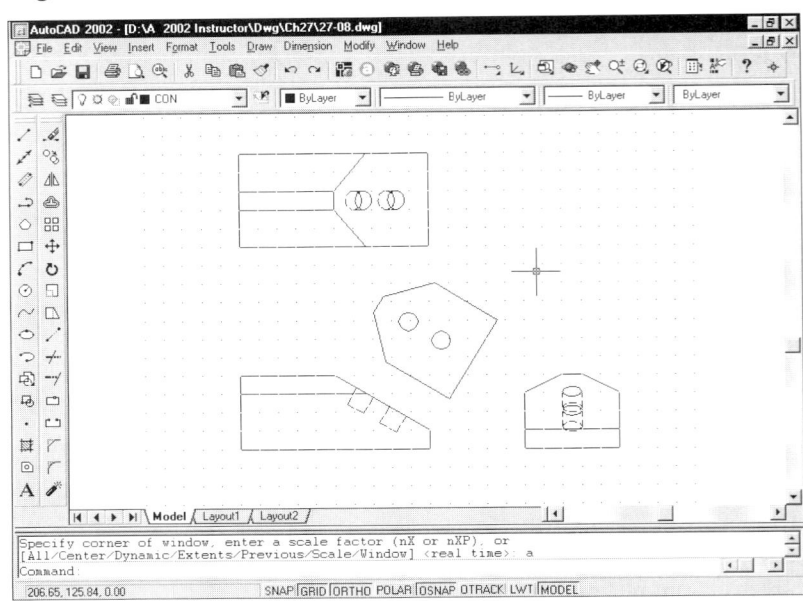

Using *Polar Tracking*

Polar Tracking can be used to create auxiliary views by facilitating construction of *Lines* at specific angles. To use *Polar Tracking* for auxiliary view construction, first use the *List* command to determine the angle of the inclined surface you want to project from. (Keep in mind that, by default, AutoCAD reports angles in whole numbers [no decimals or fractions], so use the *Units* command to increase the *Precision* of *Angular* units before using *List*.)

Once the desired angles are determined, specify the *Polar Angle Settings* in the *Drafting Settings* dialog box. Access the dialog box by right-clicking on the word *POLAR* at the Status Bar, typing *Dsettings*, or selecting *Drafting Settings* from the *Tools* pull-down menu. In the *Polar Tracking* tab of the dialog box, select the desired angles if appropriate from the *Increment angle* drop-down list. If the inclined plane is not at a regular angle offered from the drop-down list, specify the desired angle in the *Additional angles* edit box by selecting *New* and inputting the angles. Enter four angles in 90-degree increments (Fig. 25-9). Once the angles are set, ensure *POLAR* is on.

Figure 25-9

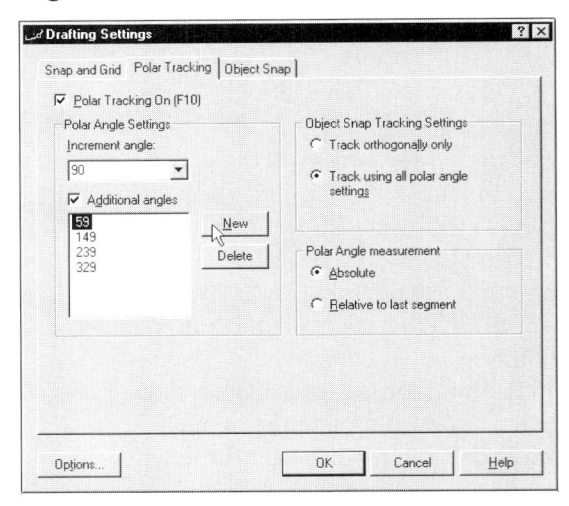

You may also want to set a *Polar Increment* in the *Snap and Grid* tab of the dialog box. This action makes the line lengths snap to regular intervals. (See Chapter 3 for more information on *Polar Snap* and *Polar Tracking*.)

Using *OSNAP*, begin constructing *Lines* perpendicular to the inclined plane. *Polar Tracking* should facilitate the line construction at the appropriate angles (Fig. 25-10).

Figure 25-10

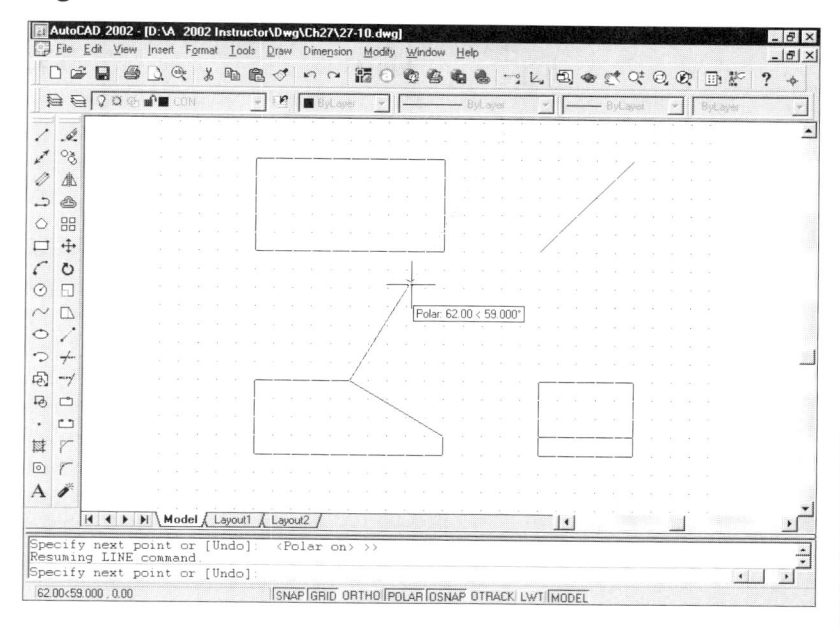

Perpendicular *Line* construction is also assisted by *Polar Tracking* (Fig. 25-11). Continue with this process, constructing necessary lines for the auxiliary view. The remainder of the auxiliary drawing, as described on previous pages (see Figures 25-5 through 25-8), could be constructed using *Polar Tracking*. The advantage of this method is that drawing horizontal and vertical *Lines* is also possible while *Polar Tracking* is on.

Object Snap Tracking can also be employed to construct objects aligned (at the specified polar angles) with object snap locations (*Endpoint*, *Midpoint*, *Intersection*, *Extension*, etc.). See Chapter 7 for more information on *Object Snap Tracking*.

Figure 25-11

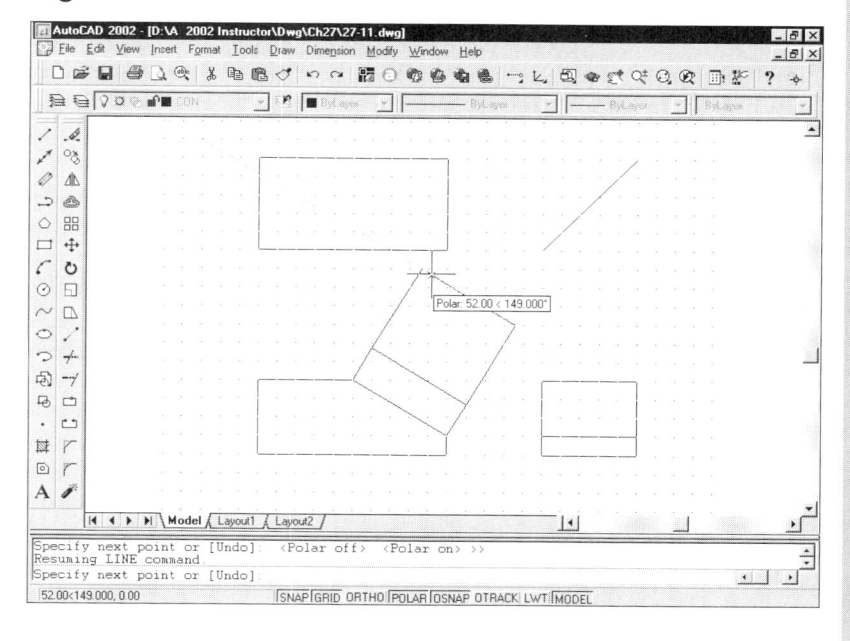

Using the *Offset* Command

Another possibility, and an alternative to the *SNAP* rotation, is to use *Offset* to make parallel *Lines*. This command can be particularly useful for construction of the "blocked in" partial auxiliary view because it is not necessary to rotate the *SNAP* and *GRID*.

OFFSET

Pull-down Menu	COMMAND (TYPE)	ALIAS (TYPE)	Short-cut	Screen (side) Menu	Tablet Menu
Modify Offset	OFFSET	O	...	MODIFY1 Offset	V,17

Invoke the *Offset* command and specify a distance. The first distance is arbitrary. Specify an appropriate value between the front view inclined plane and the nearest edge of the auxiliary view (20 for the example). *Offset* the new *Line* at a distance of 50 (for the example) or PICK two points (equal to the depth of the view).

Note that the *Offset* lines have lengths equal to the original and therefore require no additional editing (Fig. 25-12).

Figure 25-12

Next, two *Lines* would be drawn between *Endpoints* of the existing offset lines to complete the rectangle. *Offset* could be used again to construct additional lines to facilitate the construction of the two circles in the partial auxiliary view (Fig. 25-13).

From this point forward, the construction process would be similar to the example given previously (Figs. 25-5 through 25-8). Even though *Offset* does not require that the *SNAP* be *Rotated*, the complete construction of the auxiliary view could be simplified by using the rotated *SNAP* and *GRID* in conjunction with *Offset*.

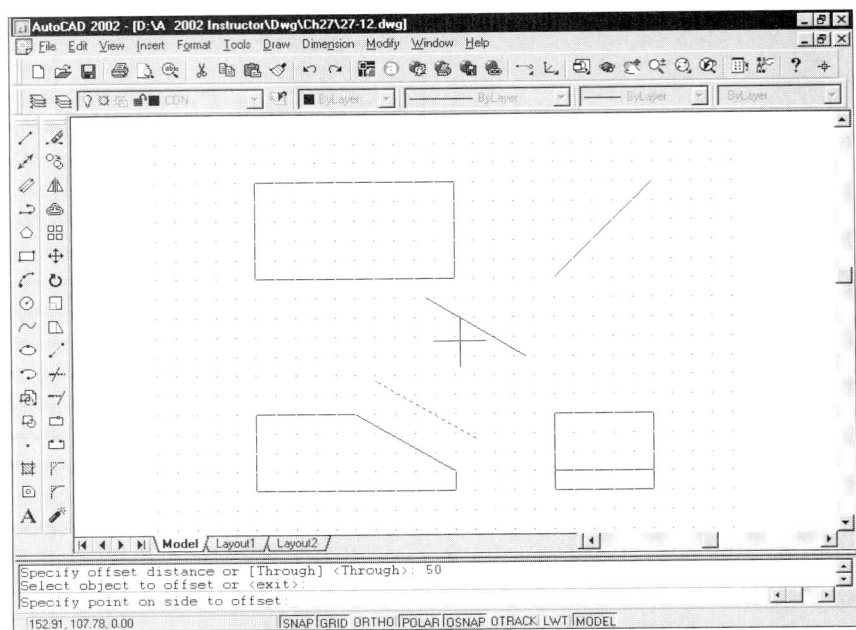

Figure 25-13

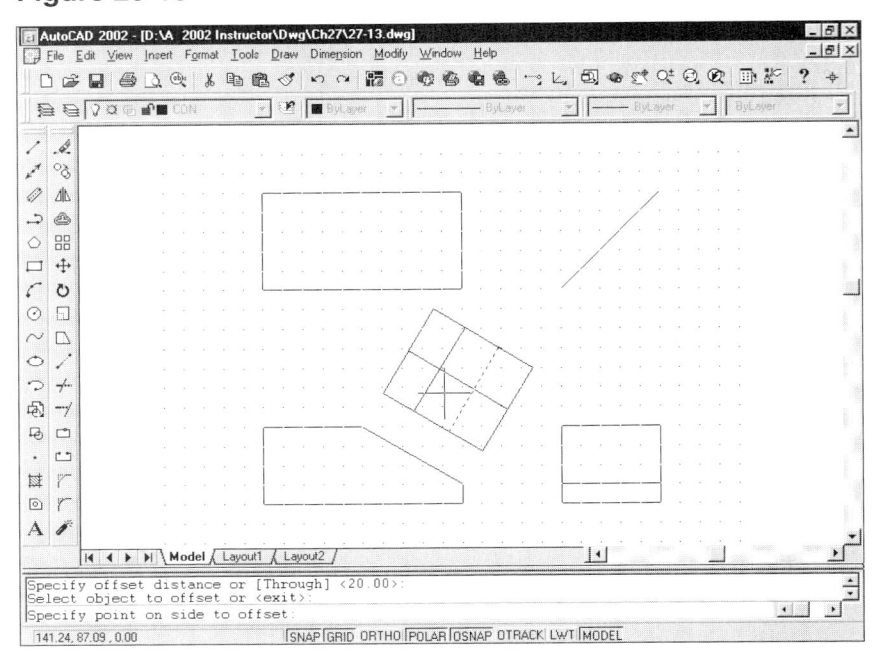

Using the *Xline* and *Ray* Commands

As a fourth alternative for construction of auxiliary views, the *Xline* and *Ray* commands could be used to create construction lines.

XLINE

Pull-down Menu	COMMAND (TYPE)	ALIAS (TYPE)	Short-cut	Screen (side) Menu	Tablet Menu
Draw Construction Line	XLINE	XL	...	DRAW 1 Xline	L,10

RAY

Pull-down Menu	COMMAND (TYPE)	ALIAS (TYPE)	Short-cut	Screen (side) Menu	Tablet Menu
Draw Ray	RAY	...	...	DRAW 1 Ray	K,10

The *Xline* command offers several options shown below:

Command: *xline*
Specify a point or [Hor/Ver/Ang/Bisect/Offset]:

The *Ang* option can be used to create a construction line at a specified angle. In this case, the angle specified would be that of the inclined plane or perpendicular to the inclined plane. The *Offset* option works well for drawing construction lines parallel to the inclined plane, especially in the case where the angle of the plane is not known.

Figure 25-14 illustrates the use of *Xline Offset* to create construction lines for the partial auxiliary view. The *Offset* option operates similarly to the *Offset* command described previously. Remember that an *Xline* extends to infinity but can be *Trimmed*, in which case it is converted to *Ray* (*Trim* once) or to a *Line* (*Trim* twice). See Chapter 15 for more information on the *Xline* command.

Figure 25-14

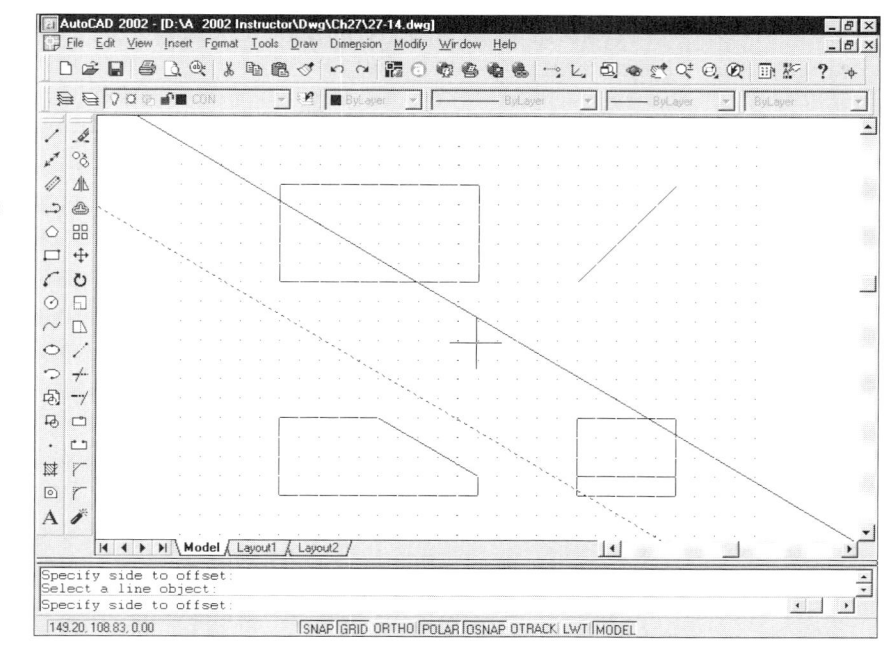

486 Auxiliary Views

The *Ray* command also creates construction lines; however, the *Ray* has one anchored point and the other end extends to infinity.

    Command: ray
    Specify start point: PICK or (coordinates)
    Specify through point: PICK or (coordinates)

Rays are helpful for auxiliary view construction when you want to create projection lines perpendicular to the inclined plane. In Figure 25-15, two *Rays* are constructed from the *Endpoints* of the inclined plane and *Perpendicular* to the existing *Xlines*. Using *Xlines* and *Rays* in conjunction is an excellent method for "blocking in" the view.

Figure 25-15

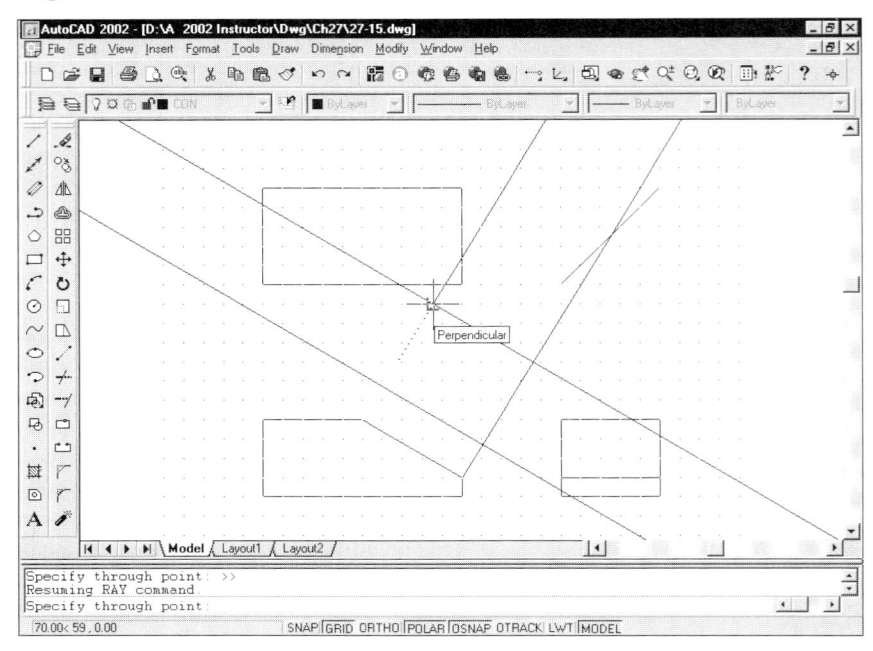

There are two strategies for creating drawings using *Xlines* and *Rays*. First, these construction lines can be created on a separate layer and set up as a framework for the object lines. The object lines would then be drawn on top of the construction lines using *Osnaps*, but would be drawn on the object layer. The construction layer would be *Frozen* for plotting. The other strategy is to create the construction lines on the object layer. Through a series of *Trims* and other modifications, the *Xlines* and *Rays* are transformed to the finished object lines.

Now that you are aware of several methods for constructing auxiliary views, use any one method or a combination of methods for your drawings. No matter which methods are used, the final lines that are needed for the finished auxiliary view should be the same. It is up to you to use the methods that are the most appropriate for the particular application or are the easiest and quickest for you personally.

Constructing Full Auxiliary Views

The construction of a full auxiliary view begins with the partial view. After initial construction of the partial view, begin the construction of the full auxiliary by projecting the other edges and features of the object (other than the inclined plane) to the existing auxiliary view.

The procedure for constructing full auxiliary views in AutoCAD is essentially the same as that for partial auxiliary views. Use of the *Offset, Xline,* and *Ray* commands, *SNAP* and *GRID* rotation, and *Polar Tracking* should be used as illustrated for the partial auxiliary view example. Because a full auxiliary view is projected at the same angle as a partial, the same rotation angle and basepoint would be used for the *SNAP,* or the same *Increment angle* or *Additional angles* should be used for *Polar Tracking*.

CHAPTER EXERCISES

For the following exercises, create the multiview drawing, including the partial or full auxiliary view as indicated. Use the appropriate template drawing based on the given dimensions and indicated plot scale.

1. Make a multiview drawing with a partial auxiliary view of the example used in this chapter. Refer to Figure 25-1 for dimensions. *Save* the drawing as **CH25EX1**. *Plot* on an A size sheet at **1=1** scale.

2. Recreate the views given in Figure 25-16 and add a partial auxiliary view. *Save* the drawing as **CH25EX2** and *Plot* on an A size sheet at **2=1** scale.

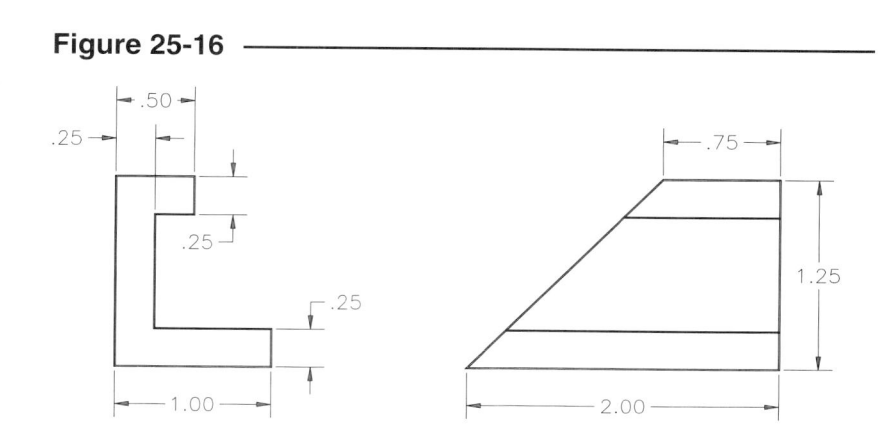

Figure 25-16

3. Recreate the views shown in Figure 25-17 and add a partial auxiliary view. *Save* the drawing as **CH25EX3**. Make a *Plot* on an A size sheet at **1=1** scale.

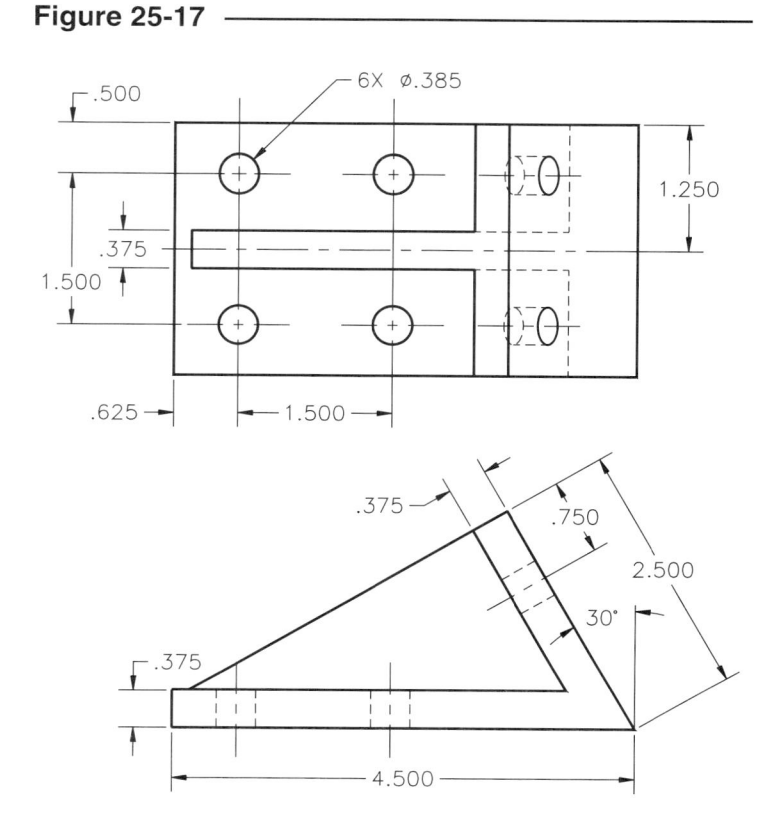

Figure 25-17

4. Make a multiview drawing of the given views in Figure 25-18. Add a full auxiliary view. *Save* the drawing as **CH25EX4**. *Plot* the drawing at an appropriate scale on an A size sheet.

Figure 25-18

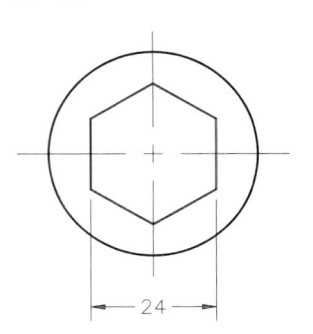

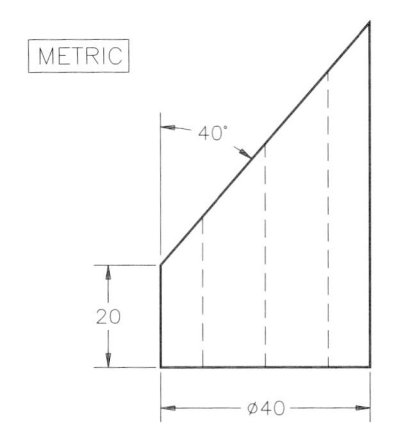

5. Draw the front, top, and a partial auxiliary view of the holder. Make a *Plot* full size. Save the drawing as **HOLDER**.

Figure 25-19

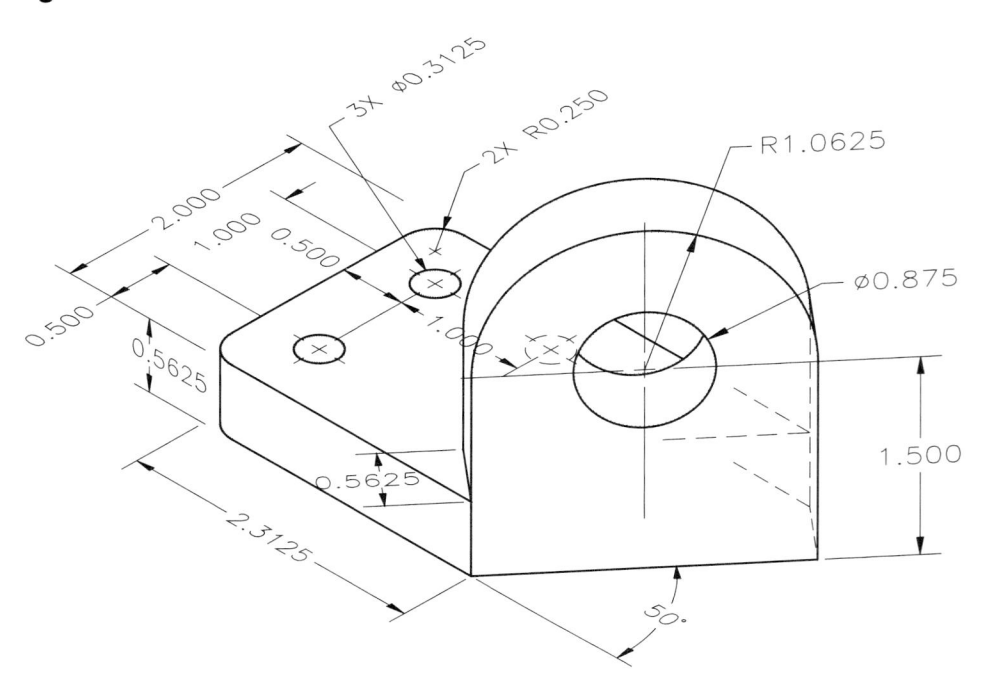

6. Draw three principal views and a full auxiliary view of the V-block shown in Figure 25-20. *Save* the drawing as **VBLOCK**. *Plot* to an accepted scale.

Figure 25-20

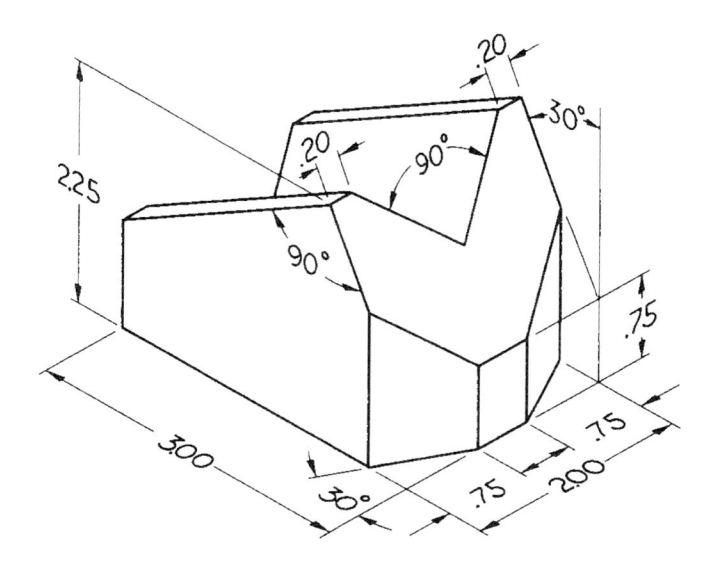

7. Draw two principal views and a partial auxiliary of the angle brace. *Save* as **ANGLBRAC**. Make a plot to an accepted scale on an A or B size sheet. Convert the fractions to the current ANSI standard, decimal inches.

Figure 25-21

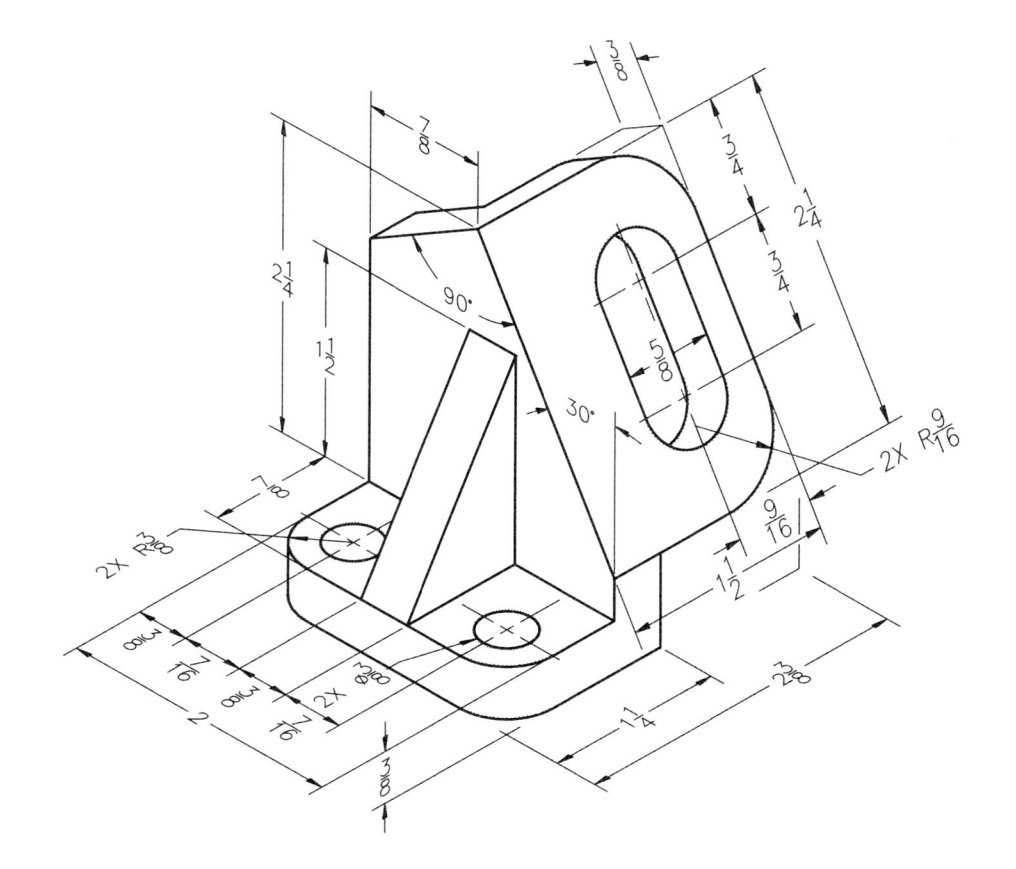

490 Auxiliary Views

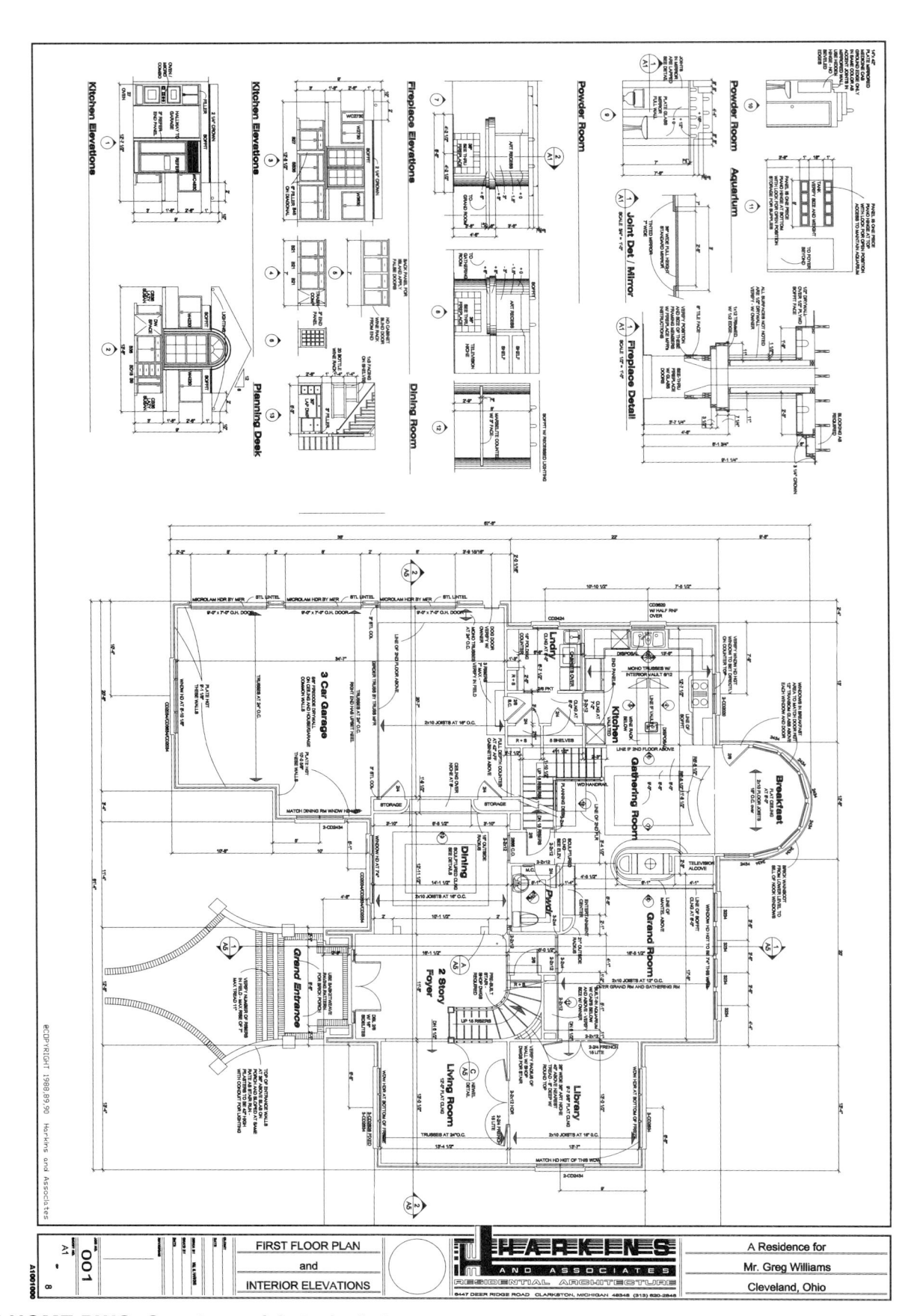

26

DIMENSIONING

Chapter Objectives

After completing this chapter you should:

1. be able to create linear dimensions with *Dimlinear*;

2. be able to append *Dimcontinue* and *Dimbaseline* dimensions to existing dimensions;

3. be able to create *Angular*, *Diameter*, and *Radius* dimensions;

4. know how to affix notes to drawings with *Leaders*;

5. know that *Dimordinate* can be used to specify Xdatum and Ydatum dimensions;

6. be able to use the new *Qdim* command to quickly create a variety of nonassociative dimensions;

7. know the possible methods for editing associative dimensions and dimensioning text.

CONCEPTS

As you know, drawings created with CAD systems should be constructed with the same dimensions and units as the real-world objects they represent. In this way, the features of the object that you apply dimensions to (lengths, diameters, angles, etc.) are automatically measured by AutoCAD and the correct values are displayed in the dimension text. So if the object is drawn accurately, the dimension values will be created correctly and automatically. Generally, dimensions should be created on a separate layer named DIMENSIONS, or similar, and dimensions should be created in model space (see Chapter 27 for information on dimensioning in paper space).

The main components of a dimension are:

1. Dimension line
2. Extension lines
3. Dimension text (usually a numeric value)
4. Arrowheads or tick marks

Figure 26-1

AutoCAD dimensioning is <u>semi-automatic</u>. When you invoke a command to create a linear dimension, AutoCAD only requires that you PICK an object or specify the extension line origins (where you want the extension lines to begin) and PICK the location of the dimension line (distance from the object). AutoCAD then measures the feature and draws the extension lines, dimension line, arrowheads, and dimension text.

Figure 26-2

TIP For linear dimensioning commands, there are <u>two ways</u> to specify placement for a dimension in AutoCAD: you can PICK the <u>object</u> to be dimensioned or you can PICK the two <u>extension line origins</u>. The simplest method is to select the object because it requires only one PICK (Fig. 26-2):

    Command: *dimlinear*
    Specify first extension line origin or <select object>: **Enter**
    Select object to dimension: **PICK**

Figure 26-3

The other method is to PICK the extension line origins (Fig. 26-3). *Osnaps* should be used to PICK the object (endpoints in this case) so that the dimension is <u>associated</u> with the object.

    Command: *dimlinear*
    Specify first extension line origin or <select object>: **PICK**
    Specify second extension line origin: **PICK**

Once the dimension is attached to the object, you specify how far you want the dimension to be placed from the object (called the "dimension line location").

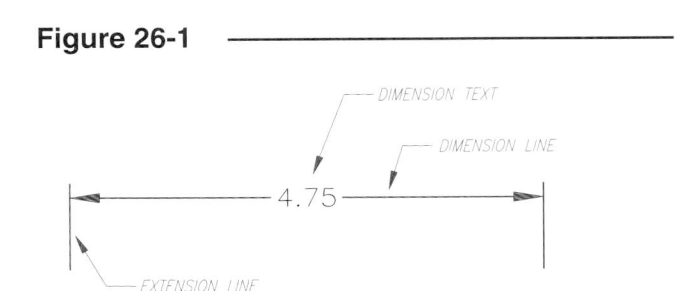

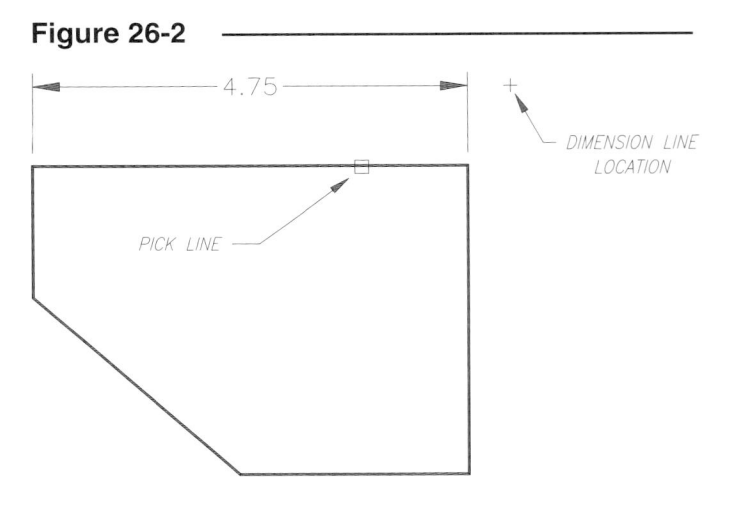

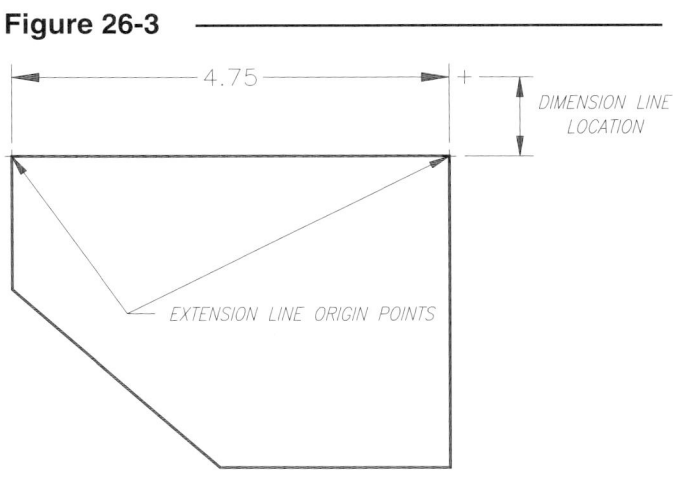

Dimensioning in AutoCAD is associative (by default). Because the extension line origins are "associated" with the geometry, the dimension text automatically updates if the geometry is *Stretched*, *Rotated*, *Scaled*, or likewise edited using grips.

Because dimensioning is semi-automatic, dimensioning variables are used to control the way dimensions are created. Dimensioning variables can be used to control features such as text or arrow size, direction of the leader arrow for radial or diametrical dimensions, format of the text, and many other possible options. Groups of variable settings can be named and saved as Dimension Styles. Dimensioning variables and dimension styles are discussed in Chapter 27.

Dimensioning commands can be invoked by the typical methods. The *Dimension* pull-down menu contains the dimension creation and editing commands (Fig. 26-4). A *Dimension* toolbar (Fig. 26-5) can also be activated by using the *Toolbars* list.

Figure 26-4

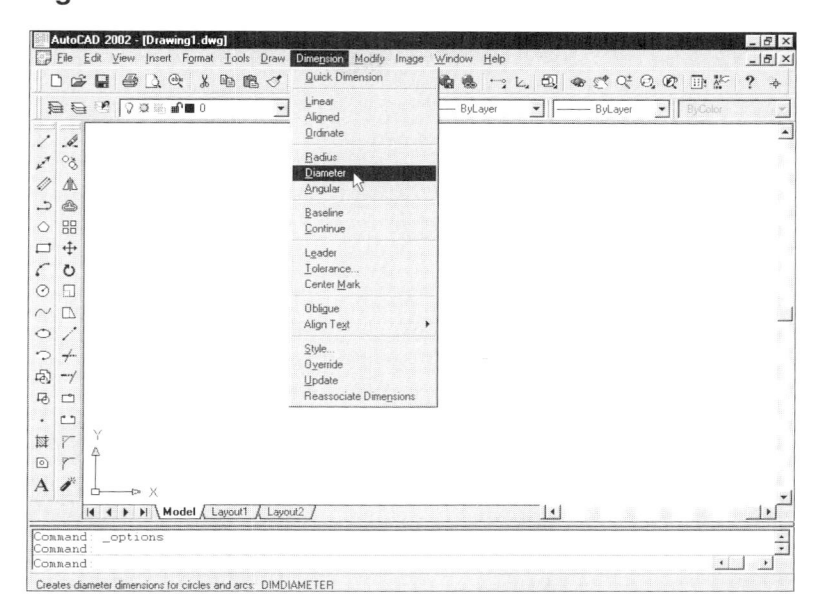

Figure 26-5

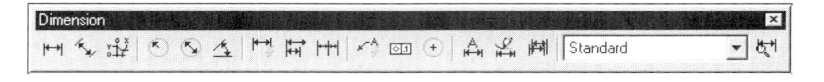

DIMENSION DRAWING COMMANDS

DIMLINEAR

Pull-down Menu	COMMAND (TYPE)	ALIAS (TYPE)	Short-cut	Screen (side) Menu	Tablet Menu
Dimension Linear	DIMLINEAR	DIMLIN or DLI	...	DIMNSION Linear	W,5

Dimlinear creates a horizontal, vertical, or rotated dimension. If the object selected is a horizontal line (or the extension line origins are horizontally oriented), the resulting dimension is a horizontal dimension. This situation is displayed in the previous illustrations (see Figures 26-2 and 26-3), or if the selected object or extension line origins are vertically oriented, the resulting dimension is vertical (Fig. 26-6).

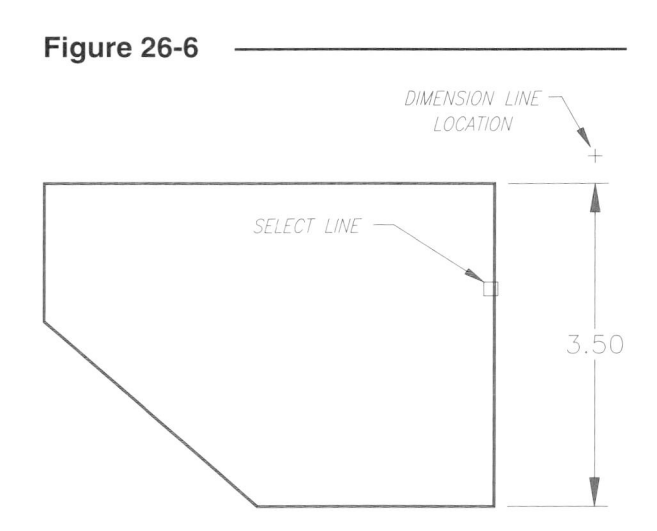

Figure 26-6

When you dimension an inclined object (or if the selected extension line origins are diagonally oriented), a vertical or horizontal dimension can be made, depending on where you drag the dimension line in relation to the object. If the dimension line location is more to the side, a vertical dimension is created (Fig. 26-7), or if you drag farther up or down, a horizontal dimension results (Fig. 26-8).

If you select the extension line origins, it is very important to PICK the object's endpoints if the dimensions are to be truly associative (associated with the geometry). *OSNAP* should be used to find the object's *Endpoint*, *Intersection*, etc.

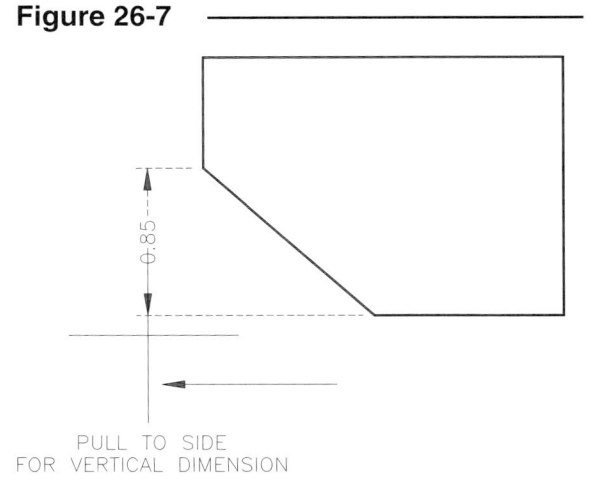

Command: *dimlinear*
Specify first extension line origin or <select object>: **PICK** (use *Osnaps*)
Specify second extension line origin: **PICK** (use *Osnaps*)
Specify dimension line location or [Mtext/Text/Angle/Horizontal/Vertical/Rotated]: **PICK** (where you want the dimension line to be placed)
Dimension text = *n.nnnn*
Command:

When you pick the location for the dimension line, AutoCAD automatically measures the object and inserts the correct numerical value. The other options are explained next.

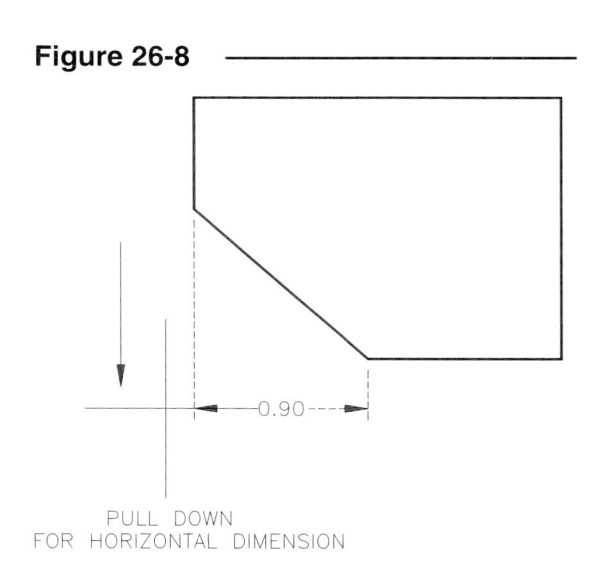

Figure 26-7

Figure 26-8

Rotated
If you want the dimension line to be drawn at an angle instead of vertical or horizontal, use this option. Selecting an inclined line, as in the previous two illustrations, would normally create a horizontal or vertical dimension. The *Rotated* option allows you to enter an angular value for the dimension line to be drawn. For example, selecting the diagonal line and specifying the appropriate angle would create the dimension shown in Figure 26-9. This object, however, could be more easily dimensioned with the *Dimaligned* command (see "*Dimaligned*").

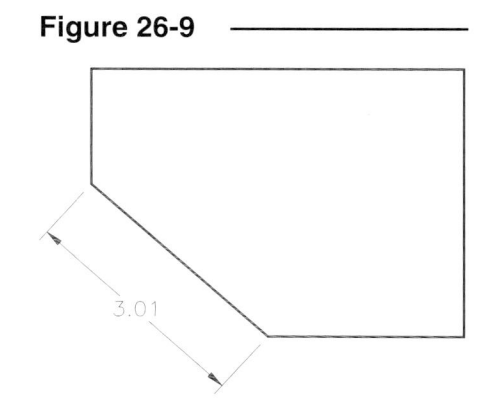

Figure 26-9

A *Rotated* dimension should be used when the geometry has "steps" or any time the desired dimension line angle is different than the dimensioned feature (when you need extension lines of different lengths). Figure 26-10 illustrates the result of using a *Rotated* dimension to give the correct dimension line angle and extension line origins for the given object. In this case, the extension line origins were explicitly PICKed. The feature of *Rotated* that makes it unique is that you specify the angle that the dimension line will be drawn.

Figure 26-10

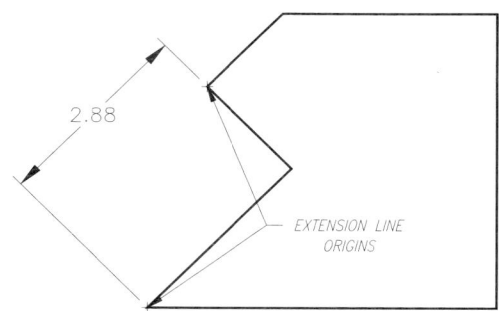

Text
Using the *Text* option allows you to enter any value or characters in place of the AutoCAD-measured text. The measured value is given as a reference at the command prompt.

    Command: *dimlinear*
    Specify first extension line origin or <select object>: **PICK**
    Specify second extension line origin: **PICK**
    Specify dimension line location or
    [Mtext/Text/Angle/Horizontal/Vertical/Rotated]: *T*
    Enter dimension text <*n.nnnn*>: (**text**) or (**value**)

Figure 26-11

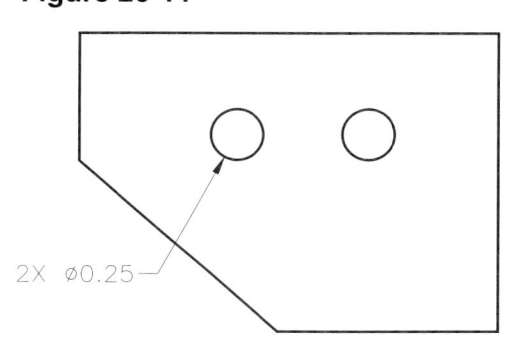

Entering a value or text at the prompt (above) causes AutoCAD to display that value or text instead of the AutoCAD-measured value. If you want to keep the AutoCAD-measured value but add a prefix or suffix, use the less-than, greater-than symbols (< >) to represent the actual (AutoCAD) value:

    Enter dimension text <0.25>: **2X <>**
    Specify dimension line location or [Mtext/Text/Angle/Horizontal/Vertical/Rotated]: **PICK**
    Dimension text = 0.25
    Command:

The "2X <>" response produces the dimension text shown in Figure 26-11.

NOTE: Changing the AutoCAD-measured value should be discouraged. If the geometry is drawn accurately, the dimensional value is correct. If you specify other dimension text, the text value is not updated in the event of *Stretching, Rotating,* or otherwise editing the associative dimension.

Mtext
This option allows you to change the existing or insert additional text to the AutoCAD-supplied numerical value. The text is entered by the *Multiline Text Editor.* The less-than and greater-than symbols (< >) represent the AutoCAD-supplied dimensional value. Place text or numbers inside the symbols if you want to override the correct measurement (not advised), or place text outside the symbols if you want to add annotation to the numerical value. For example, entering "2X " before the symbols (Fig. 26-12) creates a dimensional value as shown in Figure 26-11.

Figure 26-12

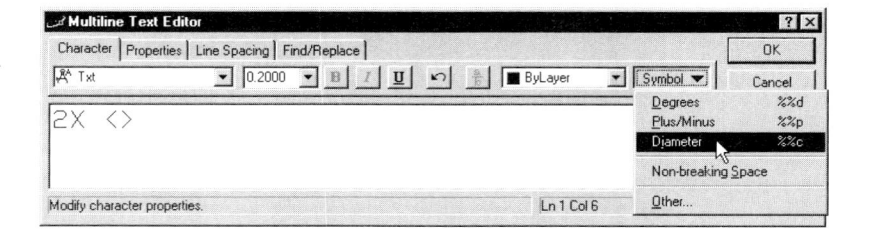

Keep in mind—the power of using the *Multiline Text Editor* is the availability of all the text creation and editing features in the dialog box. For example, you can specify fonts, text height, bold, italic, underline,

496 Dimensioning

stacked text or fractions, and layer from the *Character* tab. In addition, you can use the *Symbol* drop-down box to insert a degree, plus/minus, or diameter symbol with the text value. The *Properties* tab gives access to text style, justification, and rotation angle options.

Angle
This creates text drawn at the angle you specify. Use this for special cases when the text must be drawn to a specific angle other than horizontal. (It is also possible to make the text automatically align with the angle of the dimension line using the *Dimension Style Manager*. See Chapter 27.)

Horizontal
Use the *Horizontal* option when you want to force a horizontal dimension for an inclined line and the desired placement of the dimension line would otherwise cause a vertical dimension.

Vertical
This option forces a *Vertical* dimension for any case.

DIMALIGNED

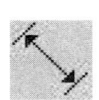

Pull-down Menu	COMMAND (TYPE)	ALIAS (TYPE)	Short-cut	Screen (side) Menu	Tablet Menu
Dimension Aligned	DIMALIGNED	DIMALI or DAL	...	DIMNSION Aligned	W,4

An *Aligned* dimension is aligned with (at the same angle as) the selected object or the extension line origins. For example, when *Aligned* is used to dimension the angled object shown in Figure 26-13, the resulting dimension aligns with the *Line*. This holds true for either option—PICKing the object or the extension line origins. If a *Circle* is PICKed, the dimension line is aligned with the selected point on the *Circle* and its center. The command syntax for the *Dimaligned* command accepting the defaults is:

Figure 26-13

```
Command: dimaligned
Specify first extension line origin or <select object>: PICK
Specify second extension line origin: PICK
Specify dimension line location or [Mtext/Text/Angle]: PICK
Dimension text = n.nnnn
Command:
```

The three options (*Mtext/Text/Angle*) operate similar to those for *Dimlinear*.

Mtext
The *Mtext* option calls the *Multiline Text Editor*. You can alter the AutoCAD-supplied text value or modify other visible features of the dimension text such as fonts, text height, bold, italic, underline, stacked text or fractions, layer, symbols, text style, justification, and rotation angle (see "*Dimlinear*," *Mtext*).

Text
You can change the AutoCAD-supplied numerical value or add other annotation to the value in command line format (see "*Dimlinear*," *Text*).

Angle
Enter a value for the angle that the text will be drawn.

> **TIP** The typical application for *Dimaligned* is for dimensioning an angled but straight feature of an object, as shown in Figure 26-13. *Dimaligned* should not be used to dimension an object feature that contains "steps," as shown in Figure 26-10. *Dimaligned* always draws extension lines of equal length.

DIMBASELINE

Pull-down Menu	COMMAND (TYPE)	ALIAS (TYPE)	Short-cut	Screen (side) Menu	Tablet Menu
Dimension Baseline	DIMBASELINE	DIMBASE or DBA	...	DIMNSION Baseline	...

Dimbaseline allows you to create a dimension that uses an extension line origin from a previously created dimension. Successive *Dimbaseline* dimensions can be used to create the style of dimensioning shown in Figure 26-14.

A baseline dimension must be connected to an existing dimension. If *Dimbaseline* is invoked immediately after another dimensioning command, you are required only to specify the second extension line origin since AutoCAD knows to use the previous dimension's first extension line origin:

Command: *dimbaseline*
Specify a second extension line origin or [Undo/Select] <Select>: **PICK**
Dimension text = *n.nnnn*

Figure 26-14

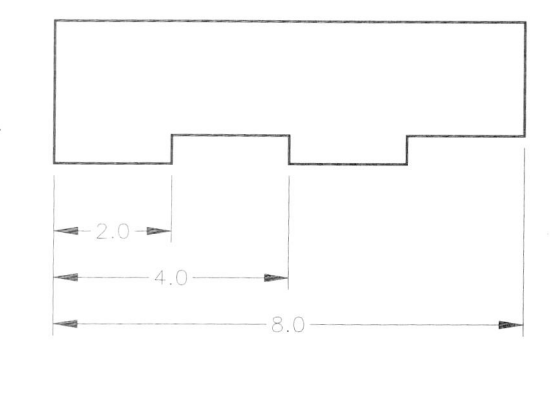

The previous dimension's first extension line is used also for the baseline dimension (Fig. 26-15). Therefore, you specify only the second extension line origin. Note that you are not required to specify the dimension line location. AutoCAD spaces the new dimension line automatically, based on the setting of the dimension line increment variable (Chapter 27).

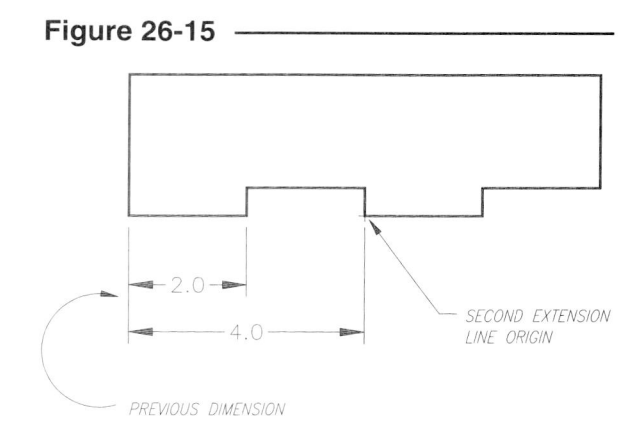

Figure 26-15

If you wish to create a *Dimbaseline* dimension using a dimension other than the one just created, use the "Select" option (Fig. 26-16):

Command: *dimbaseline*
Specify a second extension line origin or [Undo/Select] <Select>: **S** or **Enter**
Select base dimension: **PICK**
Specify a second extension line origin or [Undo/Select] <Select>: **Enter**
Dimension text = *n.nnnn*

Figure 26-16

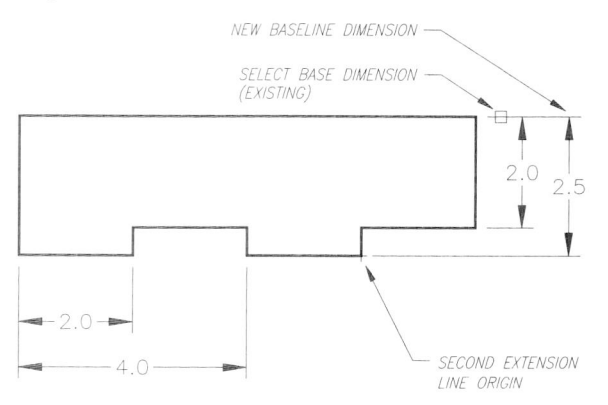

498 Dimensioning

The extension line selected as the base dimension becomes the first extension line for the new *Dimbaseline* dimension.

The *Undo* option can be used to undo the last baseline dimension created in the current command sequence.

Dimbaseline can be used with rotated, aligned, angular, and ordinate dimensions.

DIMCONTINUE

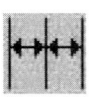

Pull-down Menu	COMMAND (TYPE)	ALIAS (TYPE)	Short-cut	Screen (side) Menu	Tablet Menu
Dimension Continue	DIMCONTINUE	DIMCONT or DCO	...	DIMNSION Continue	...

Dimcontinue dimensions continue in a line from a previously created dimension. *Dimcontinue* dimension lines are attached to, and drawn the same distance from, the object as an existing dimension.

Dimcontinue is similar to *Dimbaseline* except that an existing dimension's <u>second</u> extension line is used to begin the new dimension. In other words, the new dimension is connected to the <u>second</u> extension line, rather than to the <u>first</u>, as with a *Dimbaseline* dimension (Fig. 26-17).

The command syntax is as follows:

    Command: **dimcontinue**
    Specify a second extension line origin or [Undo/Select]
    <Select>: **PICK**
    Dimension text = *n.nnnn*

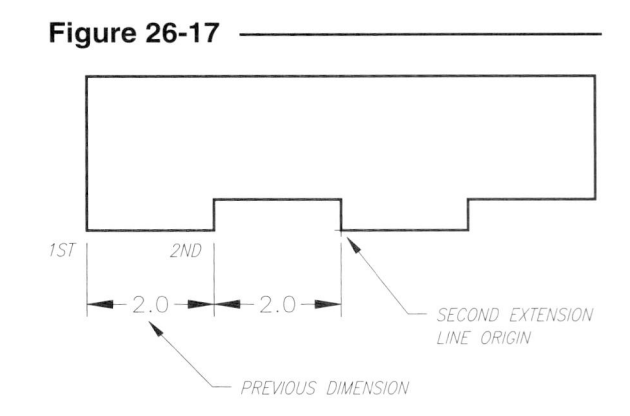

Figure 26-17

Assuming a dimension was just drawn, *Dimcontinue* could be used to place the next dimension, as shown in Figure 26-17.

Figure 26-18

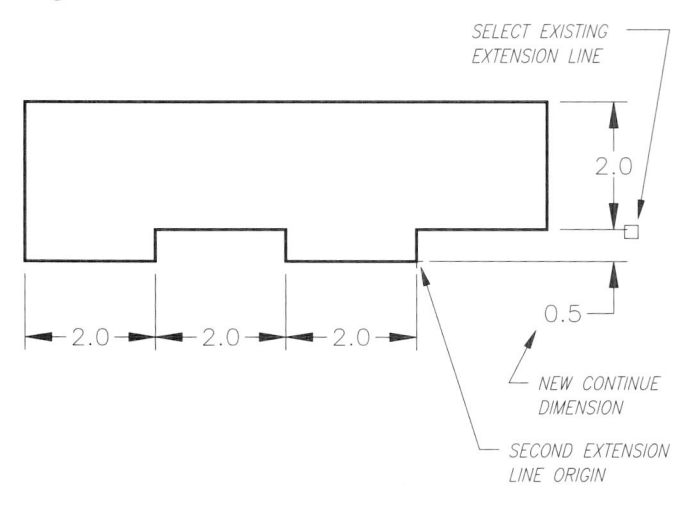

If you want to create a continued dimension and attach it to an extension line <u>other</u> than the previous dimension's second extension line, you can use the *Select* option to pick an extension line of any other dimension. Then, use *Dimcontinue* to create a continued dimension from the selected extension line (Fig. 26-18).

The *Undo* option can be used to undo the last continued dimension created in the current command sequence. *Dimcontinue* can be used with rotated, aligned, angular, and ordinate dimensions.

DIMDIAMETER

Pull-down Menu	COMMAND (TYPE)	ALIAS (TYPE)	Short-cut	Screen (side) Menu	Tablet Menu
Dimension Diameter	DIMDIAMETER	DIMDIA or DDI	...	DIMNSION Diameter	X,4

The *Dimdiameter* command creates a diametrical dimension by selecting any *Circle*. Diametrical dimensions should be used for full 360 degree *Circles* and can be used for *Arcs* of more than 180 degrees.

```
Command: dimdiameter
Select arc or circle: PICK
Dimension text = n.nnnn
Specify dimension line location or [Mtext/Text/Angle]: PICK
Command:
```

Figure 26-19

You can PICK the circle at any location. AutoCAD allows you to adjust the position of the dimension line to any angle or length (Fig. 26-19). Dimension lines for diametrical or radial dimensions should be drawn to a regular angle, such as 30, 45, or 60 degrees, never vertical or horizontal.

A typical diametrical dimension appears as the example in Figure 26-19. According to ANSI standards, a diameter dimension line and arrow should point inward (toward the center) for holes and small circles where the dimension line and text do not fit within the circle. Use the default variable settings for *Dimdiameter* dimensions such as this.

For dimensioning large circles, ANSI standards suggest an alternate method for diameter dimensions where sufficient room exists for text and arrows inside the circle (Fig. 26-20). To create this style of dimensioning, set the variables to *Text* and *Place text manually when dimensioning* in the *Fit* tab of the *Dimension Style Manager* (see Chapter 27).

Figure 26-20

Notice that the *Diameter* command creates center marks at the *Circle's* center. Center marks can also be drawn by the *Dimcenter* command (discussed later). AutoCAD uses the center and the point selected on the *Circle* to maintain its associativity.

Mtext/Text

The *Mtext* or *Text* options can be used to modify or add annotation to the default value. The *Mtext* option summons the *Multiline Text Editor* and the *Text* option uses Command line format. Both options operate similar to the other dimensioning commands (see *Dimlinear*, *Mtext*, and *Text*).

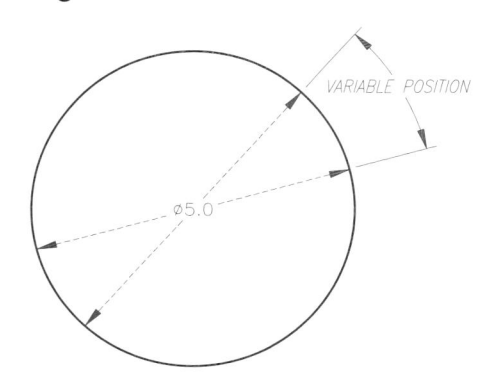

Figure 26-21

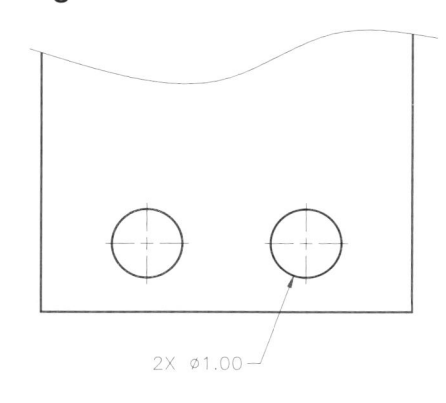

Notice that with diameter dimensions AutoCAD automatically creates the Ø (phi) symbol before the dimensional value. This is the latest ANSI standard for representing diameters. If you prefer to use a prefix before or suffix after the dimension value, it can be accomplished with the *Mtext* or *Text* option. Remember that the < > symbols represent the AutoCAD-measured value so the additional text should be inserted on <u>either side</u> of the symbols. Inserting a prefix by this method <u>does not override</u> the Ø (phi) symbol (Fig. 26-21).

500 Dimensioning

A prefix or suffix can alternately be added to the measured value by using the *Dimension Style Manager* and entering text or values in the *Prefix* or *Suffix* edit boxes. Using the dialog box method, however, overrides the Ø symbol (see Chapter 27).

Angle
With this option, you can specify an angle (other than the default) for the text to be drawn by entering a value.

DIMRADIUS

Pull-down Menu	COMMAND (TYPE)	ALIAS (TYPE)	Short-cut	Screen (side) Menu	Tablet Menu
Dimension Radius	DIMRADIUS	DIMRAD or DRA	...	DIMNSION Radius	X,5

Dimradius is used to create a dimension for an arc of anything less than half of a circle. ANSI standards dictate that a *Radius* dimension line should point outward (from the arc's center), unless there is insufficient room, in which case the line can be drawn on the outside pointing inward, as with a leader. The text can be located inside an arc (if sufficient room exists) or is forced outside of small *Arcs* on a leader.

```
Command: dimradius
Select arc or circle: PICK
Dimension text = n.nnnn
Specify dimension line location or [Mtext/Text/Angle]: PICK
Command:
```

Figure 26-22

Assuming the defaults, a *Dimradius* dimension can appear on either side of an arc, as shown in Figure 26-22. Placement of the dimension line is variable. Dimension lines for arcs and circles should be positioned at a regular angle such as 30, 45, or 60 degrees, never vertical or horizontal.

When the radius dimension is dragged outside of the arc, a center mark is automatically created (Fig. 26-22). When the radius dimension is dragged inside the arc, no center mark is created. (Center marks can be created using the *Center* command discussed next.)

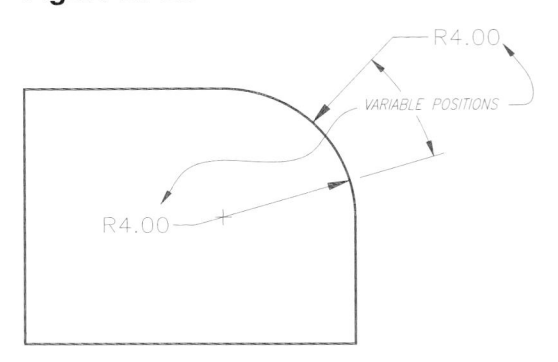

Figure 26-23

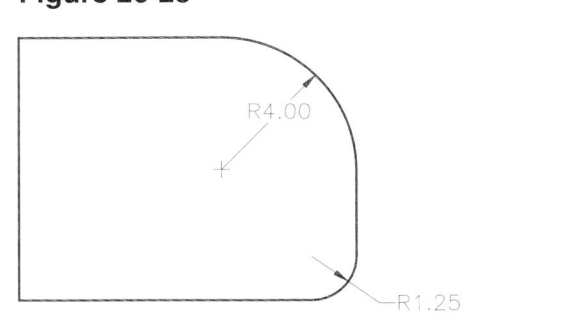

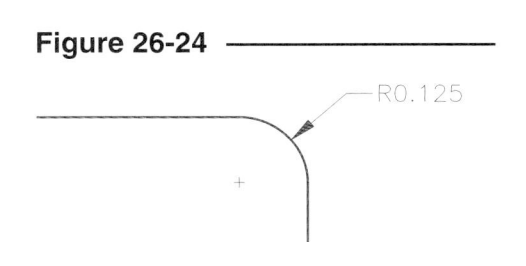

According to ANSI standards, the dimension line and arrow should point outward from the center for radial dimensions (Fig. 26-23). The text can be placed inside or outside of the *Arc*, depending on how much room exists. To create a *Dimradius* dimension to comply with this standard, dimension variables must be changed from the defaults. To achieve *Dimradius* dimensions as shown in Figure 26-23, set the variables to *Text* and *Place text manually when dimensioning* in the *Fit* tab of the *Dimension Style Manager* (see Chapter 27).

Figure 26-24

For very small radii, such as that shown in Figure 26-24, there is insufficient room for the text and arrow to fit inside the *Arc*. In this case, AutoCAD automatically forces the text outside with the leader pointing inward toward the center. No changes have to be made to the default settings for this to occur.

Mtext/Text
These options can be used to modify or add annotation to the default value. AutoCAD automatically inserts the letter "R" before the numerical text whenever a *Dimradius* dimension is created. This is the correct notation for radius dimensions. The *Mtext* option calls the *Multiline Text Editor*, and the *Text* option uses the Command line format for entering text. Remember that AutoCAD uses the < > symbols to represent the AutoCAD-supplied value. Entering text inside the < > symbols overrides the measured value. Entering text before the symbols adds a prefix without overriding the "R" designation. Alternately, text can be added by using the *Prefix* and *Suffix* options of the *Dimension Style Manager* series; however, a *Prefix* entered in the edit box replaces the letter "R." (See Chapter 27.)

Angle
With this option, you can specify an angle (other than the default) for the text to be drawn by entering a value.

DIMCENTER

Pull-down Menu	COMMAND (TYPE)	ALIAS (TYPE)	Short-cut	Screen (side) Menu	Tablet Menu
Dimension Center Mark	DIMCENTER	DCE	...	DIMNSION Center	X,2

The *Dimcenter* command draws a center mark on any selected *Arc* or *Circle*. As shown earlier, the *Dimdiameter* command and the *Dimradius* command sometimes create the center marks automatically.

The command requires you only to select the desired *Circle* or *Arc* to acquire the center marks:

    Command: **dimcenter**
    Select arc or circle: **PICK**
    Command:

Figure 26-25

No matter if the center mark is created by the *Center* command or by the *Diameter* or *Radius* commands, the center mark can be either a small cross or complete center lines extending past the *Circle* or *Arc* (Fig. 26-25). The type of center mark drawn is controlled by the *Mark* or *Line* setting in the *Lines and Arrows* tab of the *Dimension Style Manager*. It is suggested that a new *Dimension Style* be created with these settings just for drawing center marks. (See Chapter 27.)

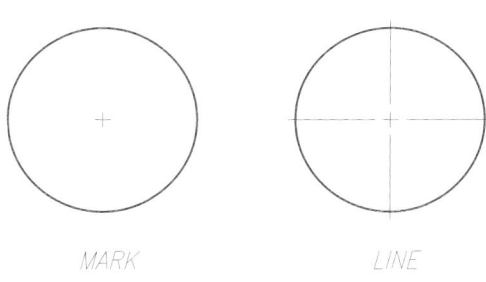

Figure 26-26

When you are dimensioning, short center marks should be used for *Arcs* of less than 180 degrees, and full center lines should be drawn for *Circles* and for *Arcs* of 180 degrees or more (Fig. 26-26).

NOTE: Since the center marks created with the *Dimcenter* command are not associative, they may be *Trimmed*, *Erased*, or otherwise edited, as shown in Figure 26-27. The center marks created with the *Dimradius* or *Dimdiameter* commands are associative and cannot be edited.

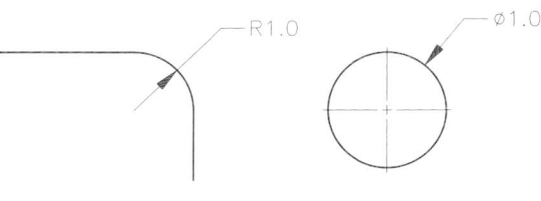

Figure 26-27

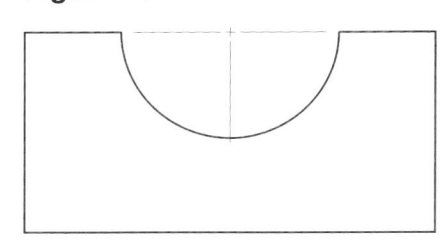

The lines comprising the center marks created with *Dimcenter* can be *Erased* or otherwise edited. In the case of a 180 degree *Arc*, two center mark lines can be shortened using *Break*, and one line can be *Erased* to achieve center lines as shown in Figure 26-27.

DIMANGULAR

Pull-down Menu	COMMAND (TYPE)	ALIAS (TYPE)	Short-cut	Screen (side) Menu	Tablet Menu
Dimension Angular	DIMANGULAR	DIMANG or DAN	...	DIMNSION Angular	X,3

The *Dimangular* command provides many possible methods of creating an angular dimension.

A typical angular dimension is created between two *Lines* that form an angle (of other than 90 degrees). The dimension line for an angular dimension is radiused with its center at the vertex of the angle (Figure 26-28). A *Dimangular* dimension automatically adds the degree symbol (°) to the dimension text. The dimension text format is controlled by the current settings for *Units* in the *Dimension Style* dialog box.

Figure 26-28

AutoCAD automates the process of creating this type of dimension by offering options within the command syntax. The default options create a dimension, as shown here:

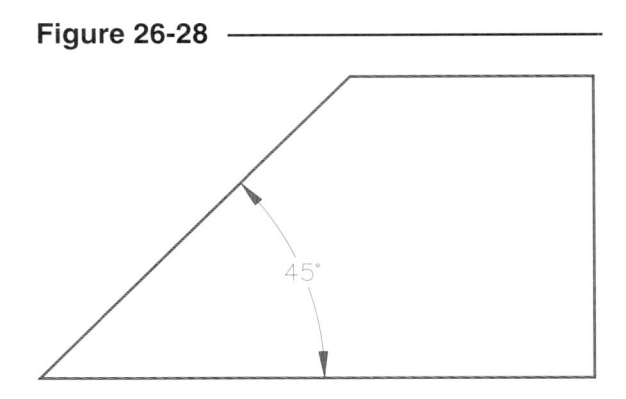

   Command: `dimangular`
   Select arc, circle, line, or <specify vertex>: **PICK**
   Select second line: **PICK**
   Specify dimension arc line location or [Mtext/Text/Angle]: **PICK**
   Dimension text = *nn*
   Command:

Dimangular dimensioning offers some very useful and easy-to-use options for placing the desired dimension line and text location.

At the "Specify dimension arc line location or [Mtext/Text/Angle]:" prompt, you can move the cursor around the vertex to dynamically display possible placements available for the dimension. The dimension can be placed in any of four positions as well as any distance from the vertex (Fig. 26-29). Extension lines are automatically created as needed.

Figure 26-29

The *Dimangular* command offers other options, including dimensioning angles for *Arcs*, *Circles*, or allowing selection of any three points.

If you select an *Arc* in response to the "Select arc, circle, line, or <specify vertex>:" prompt, AutoCAD uses the *Arc's* center as the vertex and the *Arc's* endpoints to generate the extension lines. You can select either angle of the *Arc* to dimension (Fig. 26-30).

Figure 26-30

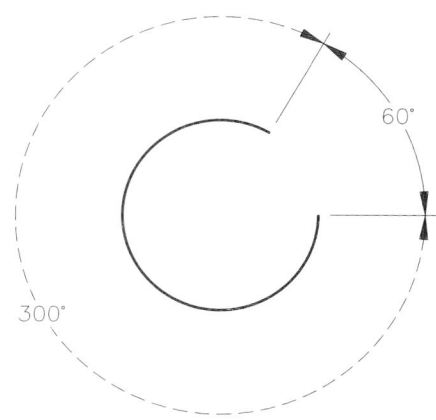

If you select a *Circle*, AutoCAD uses the PICK point as the first extension line origin. The second extension line origin does not have to be on the *Circle*, as shown in Figure 26-31.

Figure 26-31

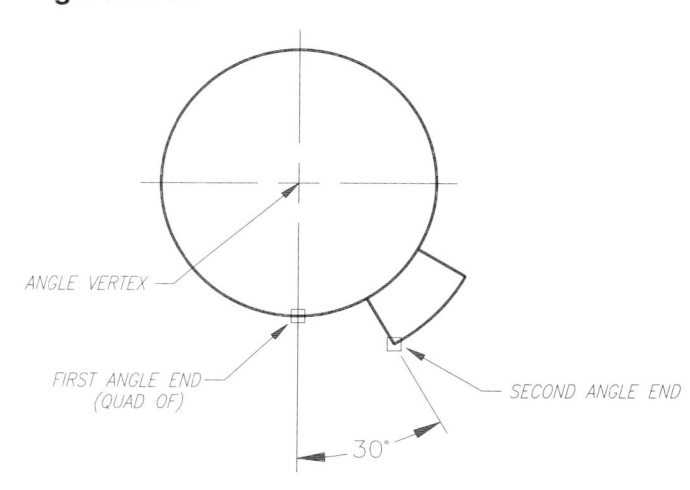

If you press Enter in response to the "Select arc, circle, line, or <specify vertex>:" prompt, AutoCAD responds with the following:

Specify angle vertex: **PICK**
Specify first angle endpoint: **PICK**
Specify second angle endpoint: **PICK**

This option allows you to apply an *Angular* dimension to a variety of shapes.

LEADER

Pull-down Menu	COMMAND (TYPE)	ALIAS (TYPE)	Short-cut	Screen (side) Menu	Tablet Menu
...	LEADER	LEAD	...	...	...

The *Leader* command (not *Qleader*) allows you to create an associative leader similar to that created with the *Diameter* command. The *Leader* command is intended to give dimensional notes such as the manufacturing or construction specifications shown here:

Figure 26-32

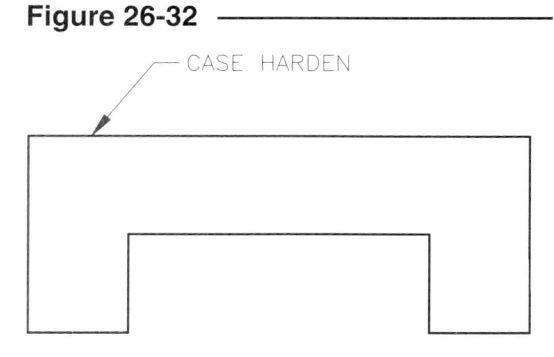

Command: *leader*
Specify leader start point: **PICK**
Specify next point: **PICK**
Specify next point or [Annotation/Format/Undo] <Annotation>: **Enter**
Enter first line of annotation text or <options>: **CASE HARDEN**
Enter next line of annotation text: **Enter**
Command:

At the "start point:" prompt, select the desired location for the arrow. You should use an *OSNAP* option (such as *Nearest*, in this case) to ensure the arrow touches the desired object.

A short horizontal line segment called the "hook line" is automatically added to the last line segment drawn if the leader line is 15 degrees or more from horizontal. Note that command syntax for the *Leader* in Figure 26-32 indicates only one line segment was PICKed. A *Leader* can have as many segments as you desire (Fig. 26-33).

If you do not enter text at the "Annotation:" prompt, another series of options are available:

Command: *leader*
Specify leader start point: PICK
Specify next point: PICK
Specify next point or [Annotation/Format/Undo] <Annotation>: **Enter**
Enter first line of annotation text or <options>: **Enter**
Enter an annotation option [Tolerance/Copy/Block/None/Mtext] <Mtext>:

Figure 26-33

Format
This option produces another list of choices:

Enter leader format option [Spline/STraight/Arrow/None] <Exit>:

Spline/STraight
You can draw either a *Spline* or straight version of the leader line with these options. The resulting *Spline* leader line has the characteristics of a normal *Spline*. An example of a *Splined* leader is shown in Figure 26-34.

Figure 26-34

Arrow/None
This option draws the leader line with or without an arrowhead at the start point.

Annotation
This option prompts for text to insert at the end of the *Leader* line.

Mtext
The *Multiline Text Editor* appears with this option. Text can be entered into paragraph form using the *Mtext* text boundary mechanism (see "*Mtext*," Chapter 18).

Tolerance
This option produces a feature control frame using the *Geometric Tolerances* dialog boxes.

Copy
You can copy existing *Text, Dtext,* or *Mtext* objects from the drawing to be placed at the end of the *Leader* line. The copied object is associated with the *Leader* line.

Block
An existing *Block* of your selection can be placed at the end of the *Leader* line. The same prompts as the *Insert* command are used.

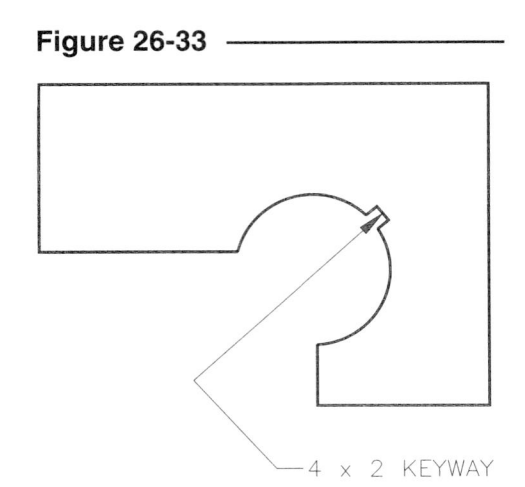

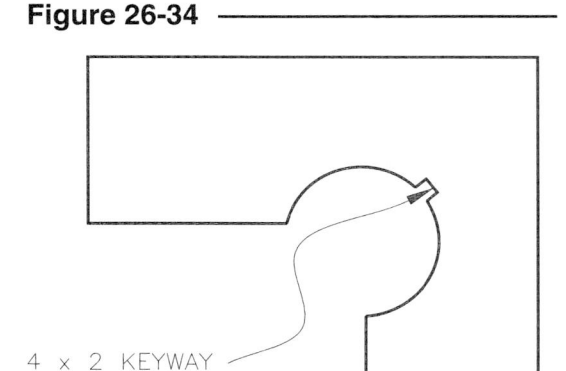

None
Using this option draws the *Leader* line with no annotation.

Undo
This option undoes the last vertex point of the *Leader* line.

Since a *Leader* is associative, it is affected by the current dimension style settings. You can control the *Leader's* arrowhead type, scale, color, etc., with the related dimensioning variables (see Chapter 27).

QLEADER

Pull-down Menu	COMMAND (TYPE)	ALIAS (TYPE)	Short-cut	Screen (side) Menu	Tablet Menu
Dimension Leader	QLEADER	LE	...	DIMNSION Leader	W,2

Qleader (quick leader) creates a leader with or without text similar in appearance to a leader created with *Leader*. However, *Qleader* allows you to specify preset parameters for the configuration of the text and the leader. These presets prevent you from having to specify the same parameters repeatedly when you create several leaders in a similar fashion for the current drawing. For example, you can specify what type of text object (*Mtext*, *Block*, *Tolerance*, etc.), how many line segments for the leader, what angle to draw the leader lines, what type of arrowheads, and other parameters for the leaders you create.

    Command: **qleader**
    Specify first leader point, or [Settings]: **PICK** (arrow location)
    Specify next point: **PICK** (end of leader)
    Specify next point: **Enter**
    Specify text width <0.0000>: **Enter**
    Enter first line of annotation text <Mtext>: Enter text or press **Enter**
    Enter next line of annotation text: Enter text or press **Enter**
    Command:

The command sequence above is used to create a one-segment leader with one string of text using the default *Qleader* settings. If you enter a value in response to the "Specify text width <0.0000>:" prompt, the value determines the width of the *Mtext* paragraph. As an alternative, you can press Enter to produce the *Mtext* dialog box.

With *Qleader*, you specify the preset configurations by pressing Enter or entering the letter "s" at the first prompt: "Specify first leader point, or [Settings]<Settings>:." This action produces the *Leader Settings* dialog box. Settings you make in this dialog box control the appearance of subsequent leaders you create in the drawing until a setting is changed. The settings are saved in the current drawing and are not registered in your system. There are three tabs in the dialog box. The *Annotation* tab is described next.

Annotation Tab

This tab allows you to control presets for the appearance of the leader text as further explained below (Fig. 26-35).

Annotation Type

This section specifies the type of text objects created (*Mtext*, *Copy an Object*, *Tolerance*, *Block Reference*, or *None*) and the related prompts when you use *Qleader*.

The *Mtext* option creates an *Mtext* object. Depending on your settings in the *Mtext Options* section, you can specify the width of each paragraph you create by responding to the "Specify text width <0.0000>:" prompt, use the *Mtext* dialog box, or press Enter to input several individual lines of text.

Figure 26-35

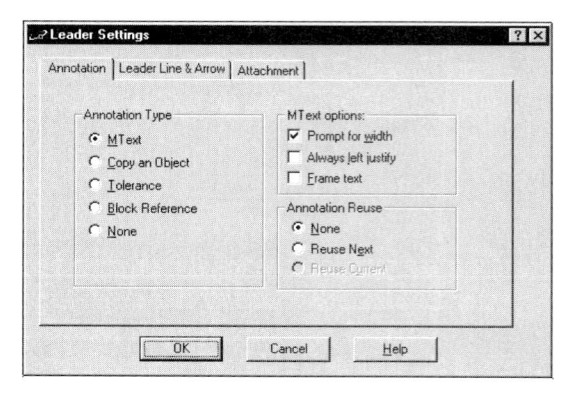

Specifying *Copy an Object* causes *Qleader* to prompt you to select an existing text object to copy. The text object is then automatically attached to the leader end.

Use the *Tolerance* button to cause *Qleader* to produce the *Geometric Tolerance* dialog box. Here you specify text and symbol content for feature control frames that are attached to the leader end (see AutoCAD 2002 *Help* for details on geometric tolerancing and this dialog box).

Use the *Block Reference* option if you want to attach a *Block* to the leader end. *Qleader* prompts you to specify an existing *Block* reference defined in the current drawing by producing the "Enter block name or [?]:" prompt. Enter a question mark (?) to list existing *Blocks* in the current drawing.

To create leaders without text attached, select *None* as the *Annotation Type*.

Mtext Options

This area is enabled only when *Mtext* is the specified *Annotation Type*. You can select none, one or two options in the section.

When *Prompt for width* is checked, this setting produces the "Specify text width <0.0000>:" prompt at the command line. Enter a value for the paragraph width or press Enter to produce the *Mtext* dialog box. When this setting is not checked, the width prompt does not appear, but you can still produce the *Mtext* dialog box by pressing Enter at the "Enter first line of annotation text:" prompt.

When a leader is drawn from right to left so it is attached to the right side of the text paragraph (Fig. 26-36), AutoCAD normally right-justifies the text. Checking the *Always Left Justify* option always draws the text left justified.

Figure 26-36

Always Left Justify is checked

Always Left Justify is not checked

The *Frame Text* option creates a frame around the text (Fig. 26-37).

Figure 26-37

Frame Text

Annotation Reuse
When you want to add the same specification note or manufacturing note at several locations in a drawing, the *Annotation Reuse* option can save considerable time. To use the same annotation repeatedly, check *Reuse Next*. After setting this option, create one leader with the desired text. Subsequent leaders automatically appear with the same text. The *Reuse Current* button becomes active as a reminder.

Leader Line & Arrow Tab

This tab controls the appearance of the leader line and allows you to specify an arrowhead type to use (Fig. 26-38).

Figure 26-38

Leader Line
You can select from a *Straight* leader (*Line* segments) or a *Spline* (smooth curved) leader. Set the *Number of Points* to 2 (minimum setting) if you want one line segment (two endpoints) or check *No Limit* if you are creating *Spline* leaders, since you need several points to control the shape of a curved leader.

Arrowhead
In the *Arrowhead* section, use the drop-down list to select from many different possibilities for the leaders.

Angle Constraints
This section allows you to specify (force) an angle for the *Qleader* segments to be drawn. For example, selecting 45 causes the *Qleader* to always be drawn at 45 degrees.

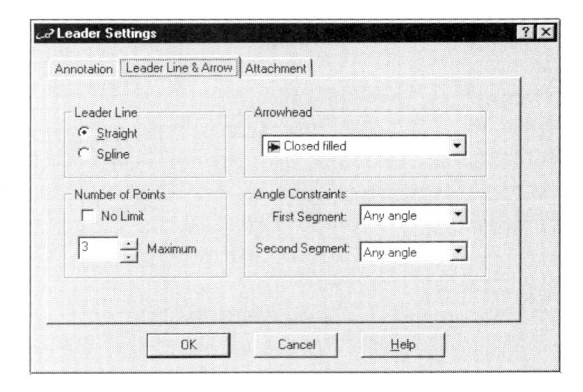

Figure 26-39

Attachment Tab

The *Attachment* tab (Fig. 26-39) is used for setting the attachment point when *Mtext* is the selected annotation type. This tab is disabled unless *Mtext* is specified in the *Annotation* tab.

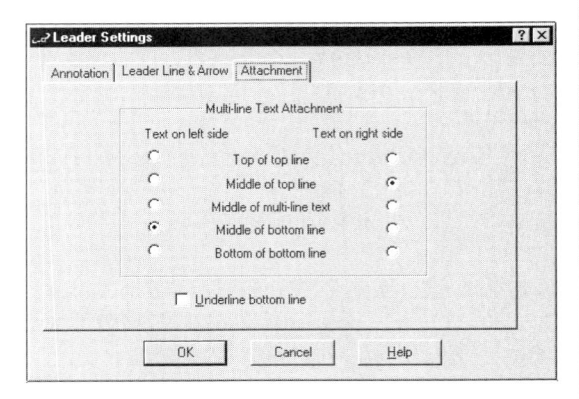

Figure 26-40

Text on the right side
(Middle of top line)

Text on the left side
(Middle of bottom line)

The *Multiline Text Attachment* options can be specified for cases when the text is located on the right or left side of the leader (text is located on the right when a leader is drawn from left to right and vice versa). The default position for text attachment (Fig. 26-40) is *Middle of top line* for text on the right side and *Middle of bottom line* for text on the left side.

DIMORDINATE

Pull-down Menu	COMMAND (TYPE)	ALIAS (TYPE)	Short-cut	Screen (side) Menu	Tablet Menu
Dimension Ordinate	DIMORDINATE	DIMORD or DOR	...	DIMNSION Ordinate	W,3

Ordinate dimensioning is a specialized method of dimensioning used in the manufacturing of flat components such as those in the sheet metal industry. Because the thickness (depth) of the parts is uniform, only the width and height dimensions are specified as Xdatum and Ydatum dimensions.

Dimordinate dimensions give an Xdatum or a Ydatum distance between object "features" and a reference point on the geometry treated as the origin, usually the lower-left corner of the part. This method of dimensioning is relatively simple to create and easy to understand. Each dimension is composed only of one leader line and the aligned numerical value.

To create ordinate dimensions in AutoCAD, the *UCS* command should be used first to establish a new 0,0 point. *UCS*, which stands for user coordinate system, allows you to establish a new coordinate system with the origin and the orientation of the axes anywhere in 3D space (see Chapters 28 and 30 for complete details). In this case, we need to change only the location of the origin and leave the orientation of the axes as is. Type *UCS* and use the *Origin* option to PICK a new origin as shown (Fig. 26-42).

When you create a *Dimordinate* dimension, AutoCAD only requires you to (1) PICK the object "Feature" and then (2) specify the "Leader endpoint:". The dimension text is automatically aligned with the leader line.

Figure 26-41

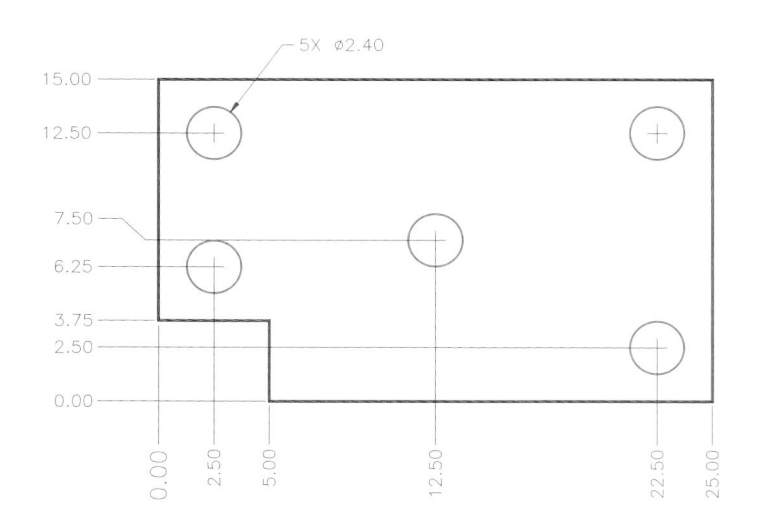

Figure 26-42

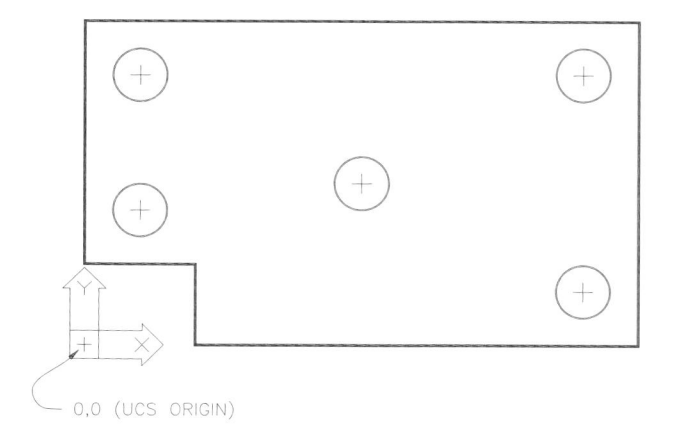

It is not necessary in most cases to indicate whether you are creating an Xdatum or a Ydatum. Using the default option of *Dimordinate*, AutoCAD makes the determination based on the direction of the leader you specify (step 2). If the leader is perpendicular (or almost perpendicular) to the X axis, an Xdatum is created. If the leader is (almost) perpendicular to the Y axis, a Ydatum is created.

The command syntax for a *Dimordinate* dimension is this:

Command: **dimordinate**
Specify feature location: **PICK**
Specify leader endpoint or [Xdatum/Ydatum/Mtext/Text/Angle]: **PICK**
Dimension text = *n.nnnn*
Command:

Figure 26-43

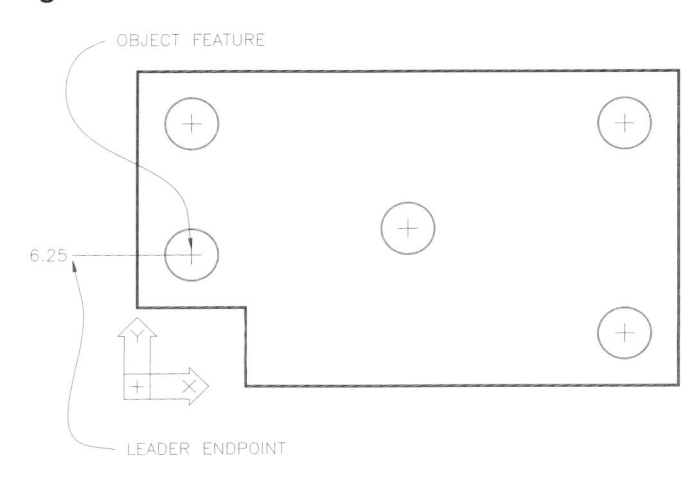

A *Dimordinate* dimension is created in Figure 26-43 by PICKing the object feature and the leader endpoint. That's all there is to it. The dimension is a Ydatum; yet AutoCAD automatically makes that determination, since the leader is perpendicular to the Y axis. It is a good practice to turn *ORTHO* or *POLAR ON* in order to ensure the leader lines are drawn horizontally or vertically.

QDIM

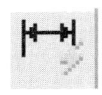

Pull-down Menu	COMMAND (TYPE)	ALIAS (TYPE)	Short-cut	Screen (side) Menu	Tablet Menu
Dimension QDIM	QDIM	...	...	...	W,1

Qdim (Quick Dimension) is a powerful dimensioning feature that creates <u>nonassociative</u> dimensions. *Qdim* simplifies the task of dimensioning by creating <u>multiple dimensions with one command</u>. *Qdim* can create *Continuous, Baseline, Radius, Diameter,* and *Ordinate* dimensions.

Command: **qdim**
Select geometry to dimension: **PICK**
Select geometry to dimension: **Enter**
Specify dimension line position, or [Continuous/Staggered/Baseline/Ordinate/Radius/Diameter/ datumPoint/ Edit] <Continuous>: **PICK**
Command:

When *Qdim* prompts to "Select geometry to dimension," you can specify the geometry using any selection method such as a window, crossing window, or pickbox. The only other step is to select where (how far from the geometry) you want the dimensions to appear. The resulting dimensions differ based on the *Qdim* options used and the current *Dimension Style* settings (see Chapter 27 for information on Dimension Styles).

Although *Qdim* is a relatively quick and easy method for creating dimensions, in AutoCAD 2002 *Qdim* creates <u>nonassociative</u> dimensions (see "Associative Dimensions," next). For this reason, *Qdim* is not a recommended method for dimensioning. The other dimensioning commands explained previously are recommended for creating associative dimensions.

EDITING DIMENSIONS

Associative Dimensions

AutoCAD creates associative dimensions by default. Associative dimensions contain "definition points" that are associated to the related geometry. For example, when you use the *Dimlinear* command and then *Osnap* to two points on the drawing geometry in response to the "Specify first extension line origin" and "Specify second extension line origin," two definition points are created at the ends of the extension lines and are attached, or "associated," to the geometry. You may have noticed these small dots at the ends of the extension lines. The definition points are generally located where you PICK, such as at the extension line origins for most dimensioning commands, or the arrowhead tips for the leader commands and the arrowhead tips and centers for diameter and radius commands. All of the dimensioning commands are associative by default except *Qdim* dimensions.

When you create the first associative dimension, AutoCAD automatically creates a new layer called DEFPOINTS. The layer is set to not plot and that property cannot be changed. AutoCAD manages this layer automatically, so do not attempt to alter this layer.

There are two advantages to associative dimensions. First, if the associated geometry is modified by typical editing methods, the associated definition points automatically change; therefore, the dimension line components (dimension lines and extension lines) automatically change and the numerical value automatically updates to reflect the geometry's new length, angle, radius, or diameter. Second, existing associative dimensions in a drawing automatically update when their dimension style is changed (see Chapter 27 for information on dimension styles).

Examples of typical editing methods that can affect associative dimensions are: *Chamfer*, *Extend*, *Fillet*, *Mirror*, *Move*, *Rotate*, *Scale*, *Stretch*, *Trim* (linear dimensions only), *Array* (if rotated in a *Polar* array), and grip editing options.

In AutoCAD 2002, dimensions can have three possible associativity settings controlled by the *DIMASSOC* system variable.

Associative dimensions   (*DIMASSOC* = 2) Associative dimensions automatically adjust their locations, orientations, and measurement values when the geometric objects associated with them are modified. Associative dimensions in a drawing will update when the dimension style is modified. These fully associative dimensions are new in AutoCAD 2002.

Nonassociative dimensions   (*DIMASSOC* = 1) Nonassociative dimensions do not change automatically when the geometric objects they measure are modified. These dimensions have definition points, but the definition points must be included (selected) if you want the dimensions to change when the geometry they measure is modified. The existing dimensions in a drawing will update when the dimension style is modified. In AutoCAD 2000 and previous releases, these dimensions were called associative.

Exploded dimensions   (*DIMASSOC* = 0) This type of dimension is actually a group of separate objects (lines, arrows, text) rather than a single dimension object. These dimensions do not have definition points and do not change when the geometry is modified. Exploded dimensions do not update when the dimension style is changed. These dimensions can also be created using *Explode* on existing associative or nonassociative dimensions.

Characteristics of the three types of dimensions (associative, nonassociative, and exploded) are explained in the three illustrations here.

An associative dimension in AutoCAD 2002 is fully associative; that is, if the geometry is changed (in this case by the *Scale* command) the related dimensions components (dimension lines, extension lines, and measured value) automatically update. The shape in Figure 26-44 is a *Pline*, so the *Scale* command affects the entire object. Note that only the single *Pline* object is selected to scale, not the dimensions.

Figure 26-44

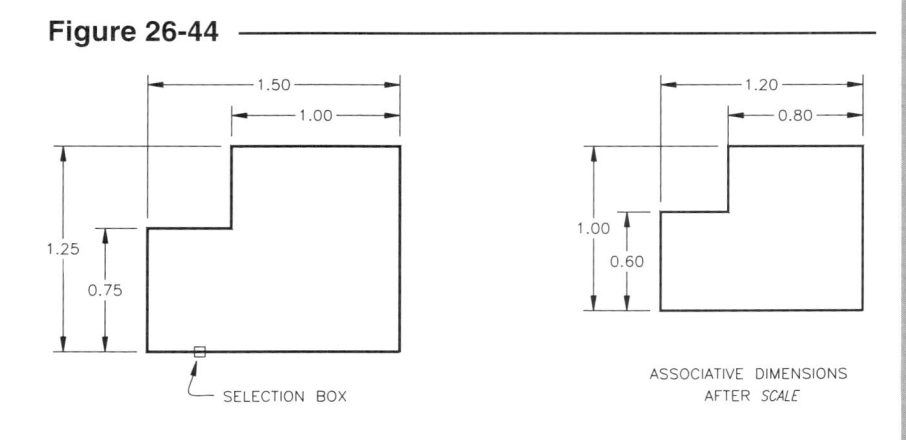

Figure 26-45 displays a similar action—using *Scale* to change an object—but with nonassociative dimensions. Nonassociative dimensions do not automatically change when the related objects are modified unless the dimension definition points are included in the selection set. Note in this case the *Pline* and only two vertical dimensions (1.25 and 0.75) are selected and are therefore scaled, while the horizontal dimensions (1.50 and 1.00) are not selected and therefore are not scaled.

Figure 26-45

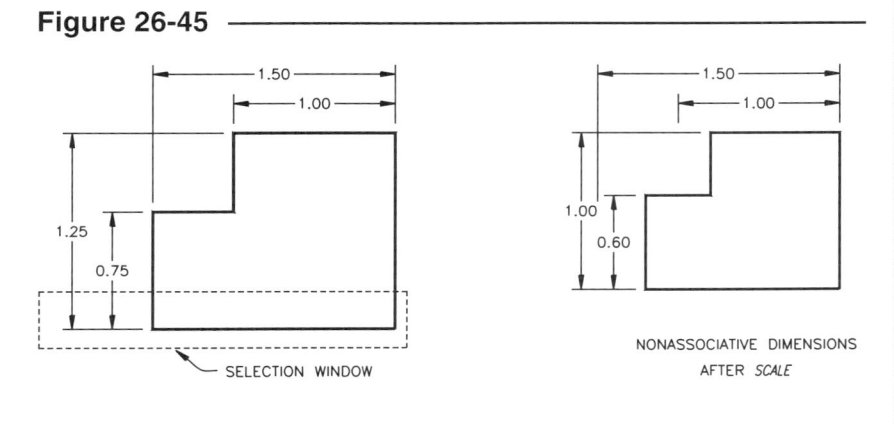

The characteristics of exploded dimensions are explained in Figure 26-46. Exploded dimensions are composed of several individual and unrelated components (arrows, text, and lines). When *Scale* is used, only the selected individual dimension components are affected, so the resulting dimensions are unusable—essentially, the dimensions fall apart. In addition, exploded dimensions do not update if their original dimension style is modified.

Figure 26-46

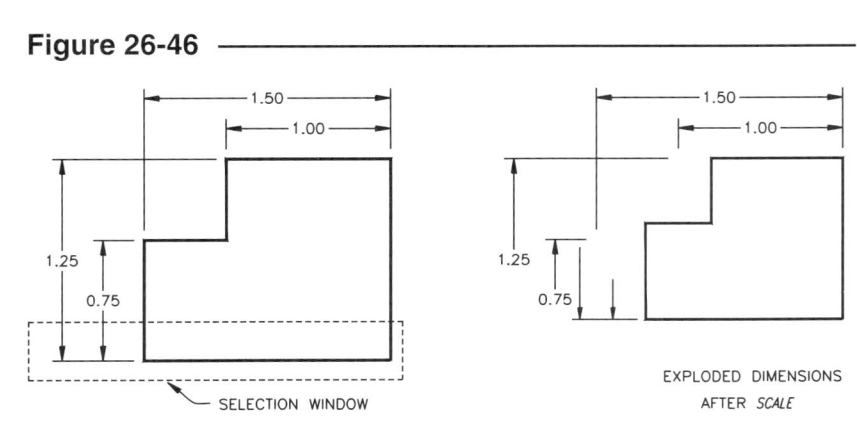

If you want to determine whether a dimension is associative or nonassociative, you can use the *Properties* window or the *List* command. You can also use the *Quick Select* dialog box to filter the selection of associative or nonassociative dimensions. A dimension is considered associative even if only one end of the dimension is associated with a geometric object.

512 Dimensioning

Remember that the associativity status for <u>newly created</u> dimensions (associative, nonassociative, or exploded) is controlled by the *DIMASSOC* system variable (2, 1, 0, respectively). You can set the *DIMASSOC* system variable at the Command line or use the *Options* dialog box. In the *User Preferences* tab, a check in the *Associate new dimensions with objects* box sets *DIMASSOC* to 2, while no check in this box sets *DIMASSOC* to 1.

DIMREASSOCIATE

Pull-down Menu	COMMAND (TYPE)	ALIAS (TYPE)	Short-cut	Screen (side) Menu	Tablet Menu
Dimension Reassociate Dimension	DIMREASSOCIATE	...	...	...	...

AutoCAD 2002 provides the *Dimreassociate* command. With *Dimreassociate*, you can convert nonassociative dimensions to associative dimensions. In addition, you can change the feature of the drawing geometry (line endpoint, etc.) that an existing definition point is associated with. This command is especially helpful in several situations, such as:

- if you are updating a drawing completed in AutoCAD 2000 or a previous release and want to convert the dimensions to fully associative AutoCAD 2002 dimensions;

- if you created nonassociative dimensions in AutoCAD 2002 but want to convert them to associative dimensions;

- if you altered the position of an associative dimension's definition points accidentally with grips or other method and want to reattach them;

- if you want to change the attachment point for an existing associative dimension;

- if you used *Qdim* to create dimensions (*Qdim* creates only nonassociative dimensions) and want to make them associative.

The *Dimreassociate* command prompts for the geometric features (line endpoints, etc.) that you want the dimension to be associated with. Depending on the type of dimension (linear, radial, angular, diametrical, etc.), you are prompted for the appropriate geometric feature as the attachment point. For example, if a linear dimension is selected to reassociate, AutoCAD prompts you to specify (select attachment points for) the extension line origins.

```
Command: dimreassociate
Select dimensions to reassociate ...
Select objects: PICK  (linear dimension selected)
Select objects: Enter
Specify first extension line origin or [Select object] <next>:
PICK
Specify second extension line origin <next>: PICK
```

Figure 26-47

When you select the "dimensions to reassociate," the associative and nonassociative elements of the selected dimension(s) are displayed one at a time. A marker is displayed for each extension line origin (or other definition point) for each dimension selected (Fig. 26-47).

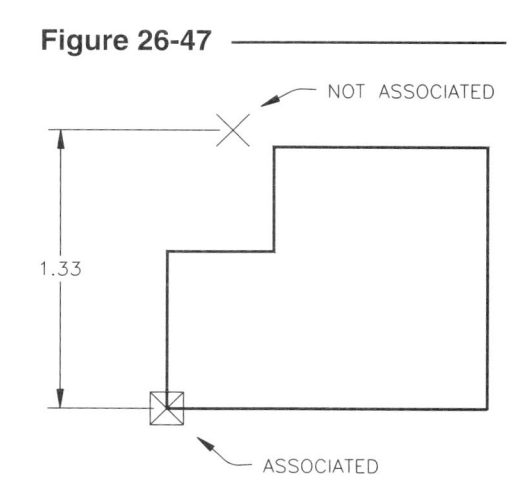

If the definition point is not associated, an "X" marker appears, and if the definition point is associated, an "X" within a box appears. To reassociate a definition point, simply pick the new object feature you want the dimension to be attached to. If an extension line is already associated and you do not want to make a change, press Enter when its definition point marker is displayed. Figure 26-47 displays a partly associated linear dimension with its extension line origin markers during the *Dimreassociate* command.

DIMDISASSOCIATE

Pull-down Menu	COMMAND (TYPE)	ALIAS (TYPE)	Short-cut	Screen (side) Menu	Tablet Menu
...	DIMDISASSOCIATE	...	...	...	...

The *Dimdisassociate* command automatically converts selected associative dimensions to nonassociative. The command prompts for you to select only the dimensions you want to disassociate, then AutoCAD makes the conversion and reports the number of dimensions disassociated.

    Command: *dimdisassociate*
    Select dimensions to disassociate ...
    Select objects: PICK
    Select objects: Enter
    nn disassociated.
    Command:

You can use any selection method including *Qselect* to "Select objects." *Dimdisassociate* filters the selection set to include only associative dimensions that are in the current space (model space or paper space layout) and not on locked layers.

DIMREGEN

Pull-down Menu	COMMAND (TYPE)	ALIAS (TYPE)	Short-cut	Screen (side) Menu	Tablet Menu
...	DIMREGEN	...	...	...	...

Dimregen updates the locations of all associative dimensions in the current drawing. *Dimregen* does not alter the associativity features of dimensions; it affects only the display of associative dimensions. *Dimregen* may be needed to regenerate the display of associative dimensions in three cases:

after panning or zooming with a wheel mouse within a viewport in a paper space layout when dimensions created in paper space are associated with drawing objects in model space (see Chapter 27, "Dimensioning in Paper Space Layouts");

after opening an AutoCAD 2002 drawing that has been modified with a previous version of AutoCAD and the dimensioned objects have been modified;

after opening a drawing containing *Xrefs* if the external reference geometry is dimensioned in the current drawing and the *Xrefs* have been modified.

Grip Editing Dimensions

Grips can be used effectively for editing dimensions. Any of the grip options (STRETCH, MOVE, ROTATE, SCALE, and MIRROR) is applicable. Depending on the type of dimension (linear, radial, angular, etc.), grips appear at several locations on the dimension when you activate the grips by selecting the dimension at the Command: prompt. Associative dimensions offer the most powerful editing possibilities, although nonassociative dimension components can also be edited with grips. There are many ways in which grips can be used to alter the measured value and configuration of dimensions.

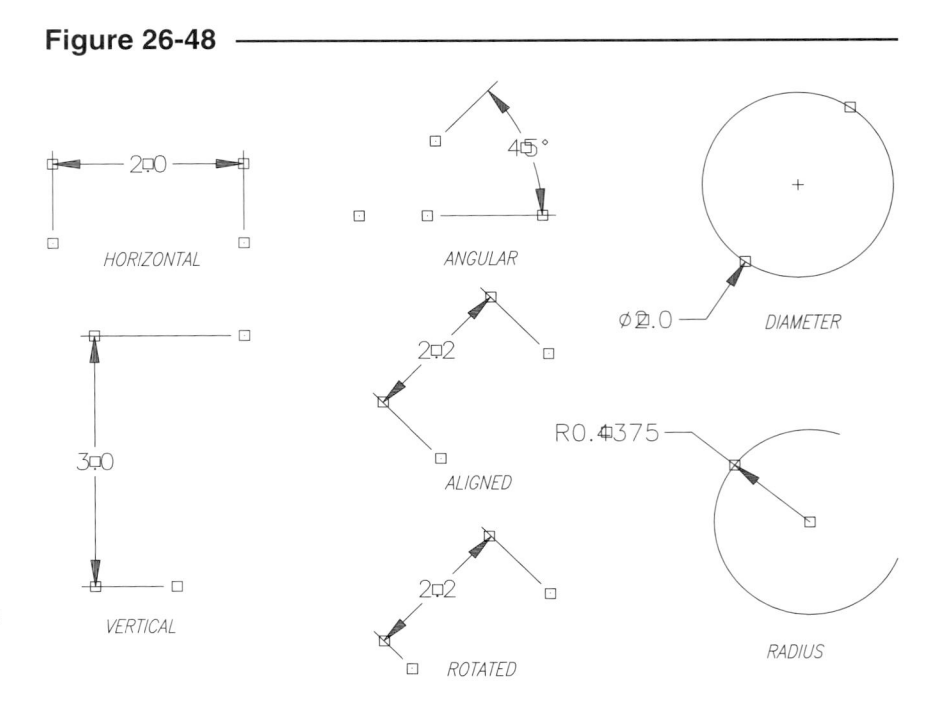

Figure 26-48

Figure 26-48 shows the grips for each type of associative dimension. Linear dimensions (horizontal, vertical, and rotated) and aligned and angular dimensions have grips at each extension line origin, the dimension line position, and a grip at the text. A diameter dimension has grips defining two points on the diameter as well as one defining the leader length. The radius dimension has center and radius grips as well as a leader grip.

For example (Fig. 26-49), a horizontal dimension value can be increased by stretching an extension line origin grip in a horizontal direction. A vertical direction movement changes the length of the extension line. The dimension line placement is changed by stretching its grips. The dimension text can be stretched to any position by manipulating its grip.

Figure 26-49

An angular dimension can be increased by stretching the extension line origin grip. The numerical value automatically updates.

Stretching a rotated dimension's extension line origin allows changing the length of the dimension as well as the length of the extension line. An aligned dimension's extension line origin grip also allows you to change the aligned angle of the dimension.

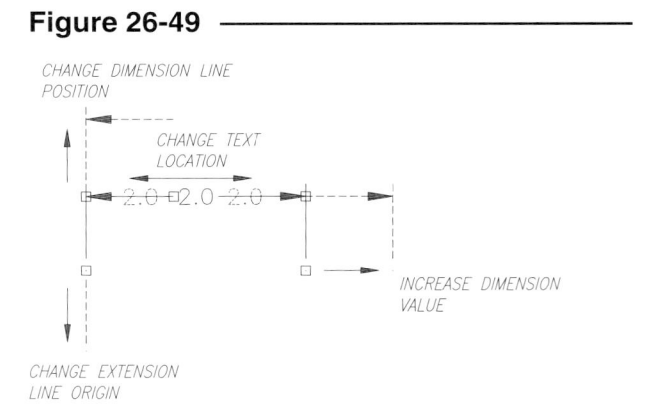

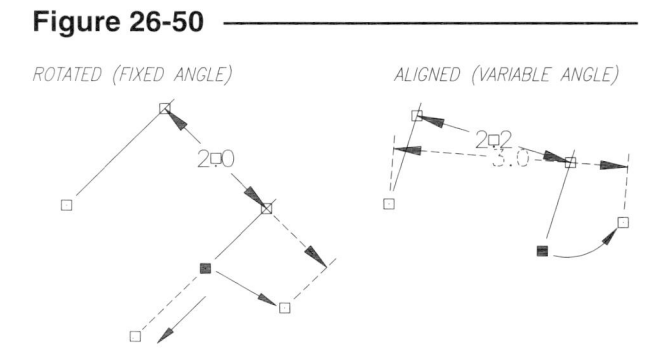

Figure 26-50

Rotating the center grip of a radius dimension (with the ROTATE option) allows you to reposition the location of the dimension around the *Arc*. Note that the text remains in its original horizontal orientation (Fig. 26-51).

Figure 26-51

You can move the dimension text independent of the dimension line with grips for dimensions that have a *DIMTMOVE* variable setting of 2. Use an existing dimension, and change the *DIMTMOVE* setting to 2 (or change the *Fit* setting in the *Dimension Style Manager* to *Over the dimension line without a leader*). The dimension text can be moved to any location without losing its associativity. See "*Fit (DIMTMOVE),*" Chapter 27.

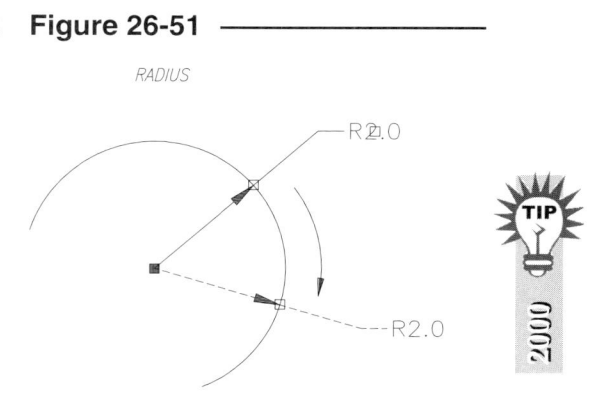

Many other possibilities exist for editing dimensions using grips. Experiment on your own to discover some of the possibilities that are not shown here.

Exploding Associative Dimensions

Associative dimensions are treated as one object. If *Erase* is used with an associative dimension, the entire dimension (dimension line, extension lines, arrows, and text) is selected and *Erased*.

Explode can be used to break an associative dimension into its component parts. The individual components can then be edited. For example, after *Exploding*, an extension line can be *Erased* or text can be *Moved*.

There are two main drawbacks to *Exploding* associative dimensions. First, the associative property is lost. Editing commands or *Grips* cannot be used to change the entire dimension but affect only the component objects. More important, Dimension Styles cannot be used with unassociative dimensions.

Dimension Editing Commands

Several commands are provided to facilitate easy editing of existing dimensions in a drawing. Most of these commands are intended to allow variations in the appearance of dimension text. These editing commands operate with associative and nonassociative dimensions, but not with exploded dimensions.

DIMTEDIT

Pull-down Menu	COMMAND (TYPE)	ALIAS (TYPE)	Short-cut	Screen (side) Menu	Tablet Menu
Dimension Align Text >	DIMTEDIT	DIMTED	...	DIMNSION Dimtedit	Y,2

Dimtedit (text edit) allows you to change the position or orientation of the text for a single associative dimension. To move the position of text, this command syntax is used:

Figure 26-52

    Command: *Dimtedit*
    Select dimension: **PICK**
    Specify new location for dimension text or
    [Left/Right/Center/Home/Angle]: **PICK**

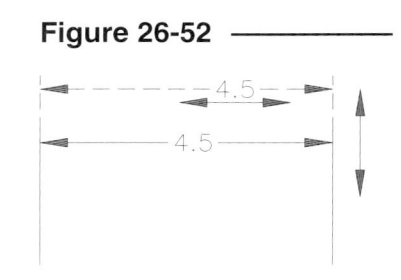

At the "Specify new location for dimension text" prompt, drag the text to the desired location. The selected text and dimension line can be changed to any position, while the text and dimension line retain their associativity.

516 Dimensioning

Angle

The *Angle* option works with any *Horizontal, Vertical, Aligned, Rotated, Radius*, or *Diameter* dimensions. You are prompted for the new text angle (Fig. 26-53):

Figure 26-53

```
Command: dimtedit
Select dimension: PICK
Specify new location for dimension text or [Left/Right/Center/Home/Angle]: a
Specify angle for dimension text: 45
Command:
```

Home

The text can be restored to its original (default) rotation angle with the *Home* option. The text retains its right/left position.

Figure 26-54

Right/Left

The *Right* and *Left* options automatically justify the dimension text at the extreme right and left ends, respectively, of the dimension line. The arrow and a short section of the dimension line, however, remain between the text and closest extension line (Fig. 26-54).

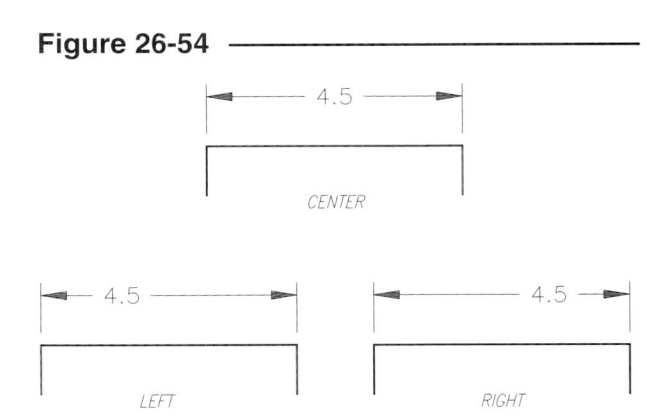

Center

The *Center* option brings the text to the center of the dimension line and sets the rotation angle to 0 (if previously assigned; Fig. 26-54).

DIMEDIT

Pull-down Menu	COMMAND (TYPE)	ALIAS (TYPE)	Short-cut	Screen (side) Menu	Tablet Menu
Dimension Oblique	DIMEDIT	DIMED or DED	...	DIMNSION Dimedit	Y,1

The *Dimedit* command allows you to change the angle of the extension lines to an obliquing angle and provides several ways to edit dimension text. Two of the text editing options (*Home, Rotate*) duplicate those of the *Dimtedit* command. Another feature (*New*) allows you to change the text value and annotation.

```
Command: dimedit
Enter type of dimension editing [Home/New/Rotate/Oblique] <Home>:
```

Home

This option moves the dimension text back to its original angular orientation angle without changing the left/right position. *Home* is a duplicate of the *Dimtedit* option of the same name.

New

You can change the text value and annotation of existing text. The same mechanism appears when you create the dimension and use the *Mtext* option—that is, the *Multiline Text Editor* appears. Remember that AutoCAD draws the dimension value in place of the < > characters. Add a prefix before these characters or a suffix after them. Entering text to replace the < > characters (erase them) causes your new text to appear instead of the AutoCAD-supplied value. Placing text or numbers inside the symbols overrides the correct measurement and creates static, nonassociative text.

The original AutoCAD-measured value <u>can be restored</u>, however, using the *New* option. Simply invoke *Dimedit* and *New* but do not enter any value in the *Multiline Text Editor*. After selecting the desired dimension, the AutoCAD-measured value is restored.

Rotate
AutoCAD prompts for an angle to rotate the text. Enter an absolute angle (relative to angle 0). This option is identical to the *Angle* option of *Dimtedit*.

Oblique
This option is unique to *Dimedit*. Entering an angle at the prompt affects the extension lines. Enter an absolute angle:

Enter obliquing angle (press ENTER for none):

Normally, the extension lines are perpendicular to the dimension lines. In some cases, it is desirable to set an obliquing angle for the extension lines, such as when dimensions are crowded and hard to read (Fig. 26-55) or for dimensioning isometric drawings.

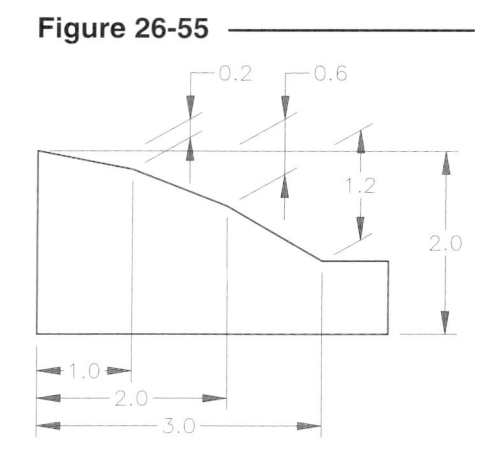

Figure 26-55

PROPERTIES

Pull-down Menu	COMMAND (TYPE)	ALIAS (TYPE)	Short-cut	Screen (side) Menu	Tablet Menu
Modify Properties...	PROPERTIES	MO or PROPS	(Edit Mode) Properties	MODIFY1 Modify	Y,14

The *Properties* command (discussed in Chapter 16) can be used effectively for a wide range of editing purposes. Double-clicking on any dimension produces the *Properties* window. The window gives a list of all properties of the dimension including dimension variable settings.

In the *Categorized* tab, the dimension variable settings for the selected dimension are listed in the *Lines & Arrows*, *Text*, *Fit*, *Primary Units*, *Alternate Units*, and *Tolerances* sections. Each section is expandable, so you can change any property (dimension variable setting) in each section for the dimension (Fig. 26-56). Each entry in the left column represents a dimension variable; changing the entries in the right column changes the variable's setting.

Each of the six categories (*Lines & Arrows*, *Text*, *Fit*, *Primary Units*, *Alternate Units*, and *Tolerances*) corresponds to the tabs in the *Dimension Style Manager*. Typically, you create dimension styles in the *Dimension Style Manager* by selecting settings for the <u>dimensioning variables</u>, then draw the dimensions using one of the styles. Each style has different settings for dimension variables, so the resulting dimensions for each style appear differently.

Figure 26-56

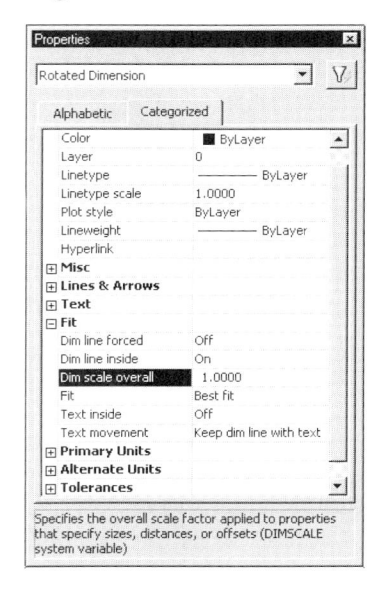

TIP 2000: Using the *Properties* window, you can change the dimension variable settings for any one or more dimensions retroactively. Making a change to a dimension variable with this method does not change the previously created dimension style, but it does create an override for that particular dimension. For more information on dimension variables, dimension styles, and dimension style overrides, see Chapter 27, Dimension Styles and Variables.

Customizing Dimensioning Text

As discussed earlier, you can specify dimensioning text other than what AutoCAD measures and supplies as the default text, using the *Text* option of the individual dimension creation commands. In addition, the text can be modified at a later time using the *Properties* command or the *New* option of *Dimedit*.

The less-than and greater-than symbols (< >) represent the AutoCAD-supplied dimensional value. Placing text or numbers inside the symbols overrides the correct measurement (not advised) and creates static text. Text placed inside the < > symbols is not associative and is not updated in the event of *Stretching*, *Rotating*, or otherwise editing the associative dimension. The original text can only be retrieved using the *New* option of *Dimedit*.

Text placed outside of the < > symbols, however, acts only as a prefix/suffix to the measured value. In the case of a diameter or radius dimension, AutoCAD automatically inserts a diameter symbol (Ø) or radius designator (R) before the dimension value. Inserting a prefix with the *Multiline Text Editor* does not override the AutoCAD symbols; however, entering a *Prefix* or *Suffix* using the *Dimension Style Manager* overrides the symbols. If you use the *Properties* window to change the text using the *Text override* edit box, enter a text string and include the < and > symbols (such as "2X <>") to retain the AutoCAD-measured value.

DIMENSIONING VARIABLES INTRODUCTION

TIP: Although you can change many settings that control the appearance of dimensions, there is one dimensioning variable that is the most universal, that is, the *Overall Scale* (*DIMSCALE*) of the dimensions. The *Overall Scale* controls all of the size-related features of dimension objects, such as the arrowhead size, the text size, the gap between extension lines and the object, and so on. Even though each of those features can be changed individually, changing the *Overall Scale* controls all of the size-related features of dimensions together.

To change the *Overall Scale*, open the *Dimension Style Manager* (by selecting it from the *Dimension* pull-down menu, *Dimension* toolbar, or by typing D). Select the *Modify* button, access the *Fit* tab, and enter the desired value in the *Use overall scale of* edit box (Fig. 26-57). This action changes the overall size for all dimensions created using that particular dimension style.

For example, assume you created a drawing and applied the first few dimensions, but noticed that the dimension text was so small that it was barely readable. Instead of changing the dimension variable that controls the size of just the dimension text, change the *Overall Scale*.

Figure 26-57

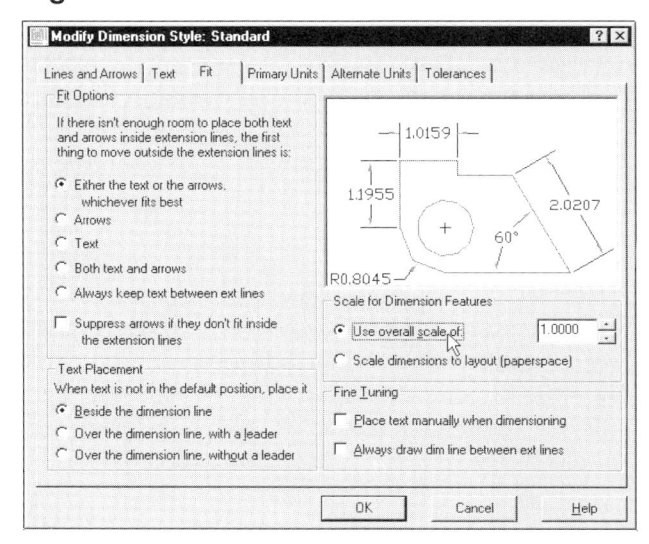

This practice ensures that the text, arrowheads, gaps, extension, and all size-related features of the dimension are changed together and are therefore correctly proportioned. Using the *Dimension Style Manager* to change this variable also ensures that all the dimensions created in that dimension style will have the same size.

Chapter 27 explains all the dimensioning variables, including *Overall Scale,* as well as dimension styles. While working on the exercises in this chapter, you will be able to create dimensions without having to change any dimension variables or dimension styles, although you may want to experiment with the *Overall Scale*. Knowledge and experience with both the dimensioning commands and dimensioning variables is essential for AutoCAD users in a real-world situation. Make sure you read Chapter 27 before attempting to dimension drawings other than those in these exercises.

CHAPTER EXERCISES

Only four exercises are offered in this chapter to give you a start with the dimensioning commands. Many other dimensioning exercises are given at the end of Chapter 27, Dimension Styles and Variables. Since most dimensioning practices in AutoCAD require the use of dimensioning variables and dimension styles, that information should be discussed before you can begin dimensioning effectively.

The units that AutoCAD uses for the dimensioning values are based on the *Unit format* and the *Precision* settings in the *Primary Units* tab of the *Dimension Style Manager*. You can dimension these drawings using the default settings, or if you are adventurous, you can change the *Units* and *Precision* settings for each drawing to match the dimensions shown in the figures. (Type *Ddim*, select the *Primary Units* tab.)

Other dimensioning features may appear different than those in the figures because of the variables set in the AutoCAD default STANDARD dimension style. For example, the default settings for diameter and radius dimensions may draw the text and dimension lines differently than you desire. After reading Chapter 27, those features in your exercises can be changed retroactively by changing the dimension style.

1. *Open* the **PLATES** drawing that you created in Chapter 9 Exercises. *Erase* the plate on the right. Use *Move* to move the remaining two plates apart, allowing 5 units between. Create a *New* layer called **DIM** and make it *Current*. Dimension the two plates, as shown in Figure 26-58. *Save* the drawing as **PLATES-D**.

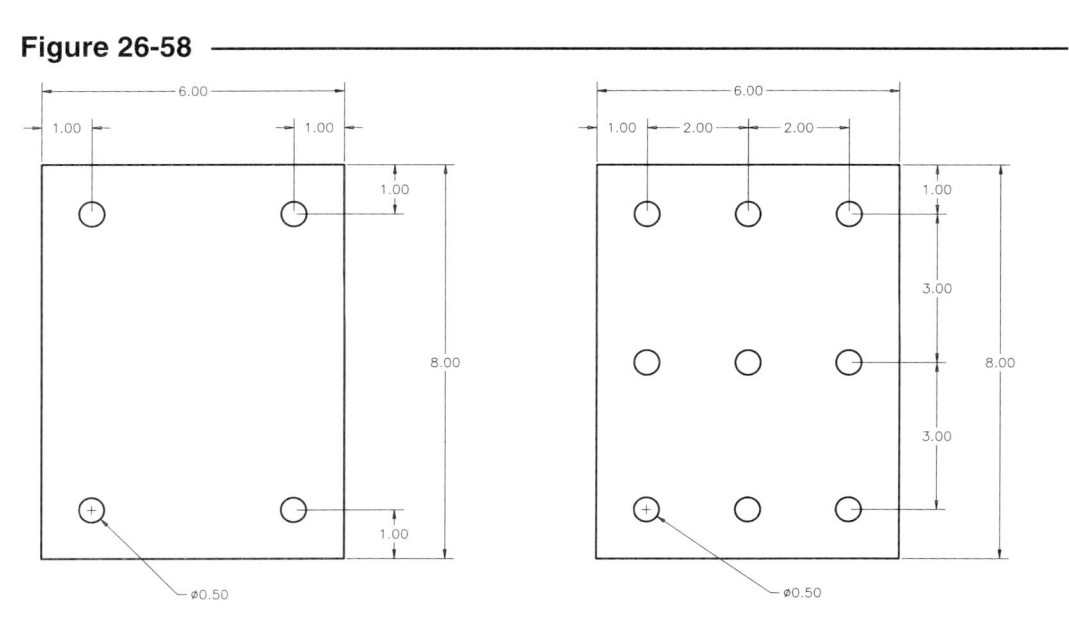

Figure 26-58

520 Dimensioning

2. *Open* the **PEDIT1** drawing. Create a *New* layer called **DIM** and make it *Current*. Dimension the part as shown in Figure 26-59. *Save* the drawing as **PEDIT1-DIM**. Draw a *Pline* border of **.02** *width* and *Insert* the **TBLOCK** drawing. *Plot* on an A size sheet using *Scale to Fit*.

Figure 26-59

3. *Open* the **GASKETA** drawing that you created in Chapter 9 Exercises. Create a *New* layer called **DIM** and make it *Current*. Dimension the part as shown in Figure 26-60. *Save* the drawing as **GASKETD**.

   Draw a *Pline* border with **.02** *width* and *Insert* the **TBLOCK** drawing with an **8/11** scale factor. *Plot* the drawing *Scaled to Fit* on an A size sheet.

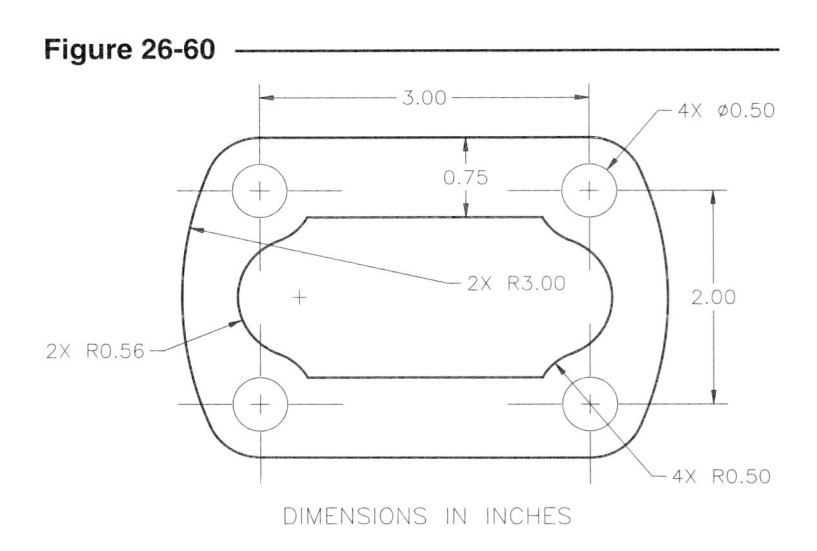

Figure 26-60

4. *Open* the **BARGUIDE** multiview drawing that you created in Chapter 22 Exercises. Create the dimensions on the **DIM** layer. Keep in mind that you have more possibilities for placement of dimensions than are shown in Figure 26-61.

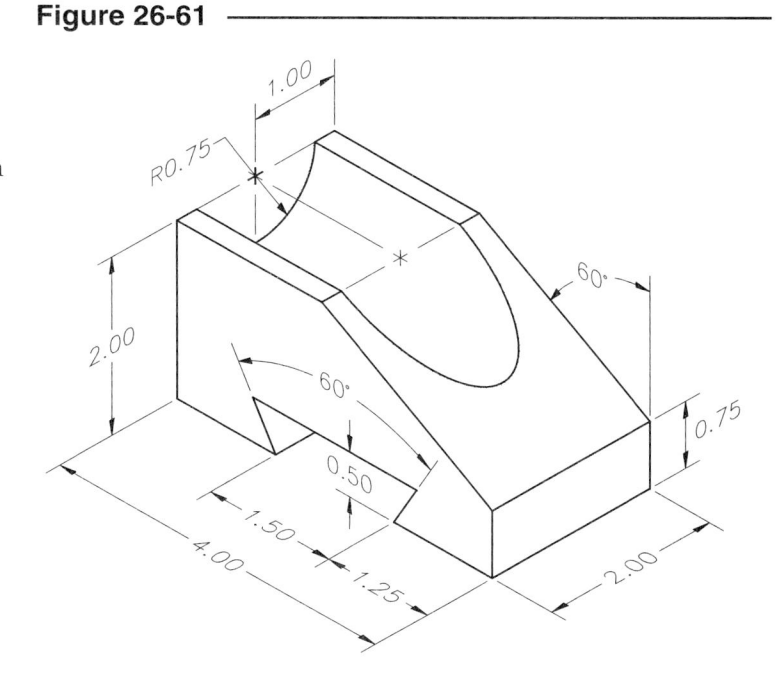

Figure 26-61

27

DIMENSION STYLES AND VARIABLES

Chapter Objectives

After completing this chapter you should:

1. be able to control dimension variable settings using the *Dimension Style Manager*;
2. be able to save and restore dimension styles with the *Dimension Style Manager*;
3. know how to create dimension style families and specify variables for each child;
4. know how to create and apply dimension style overrides;
5. be able to modify dimensions using *Update*, *Dimoverride*, and *Matchprop*;
6. know the guidelines for dimensioning;
7. be able to create associative dimensions in paper space.

CONCEPTS

Dimension Variables

Since a large part of AutoCAD's dimensioning capabilities are automatic, some method must be provided for you to control the way dimensions are drawn. A set of 70 dimension variables allows you to affect the way dimensions are drawn by controlling sizes, distances, appearance of extension and dimension lines, and dimension text formats.

An example of a dimension variable is *DIMSCALE* (if typed) or *Overall Scale* (if selected from a dialog box, Fig. 27-1). This variable controls the overall size of the dimension features (such as text, arrowheads, and gaps). Changing the value from 1.00 to 1.50, for example, makes all of the size-related features of the drawn dimension 1.5 times as large as the default size of 1.00 (Fig. 27-2). Other examples of features controlled by dimension variables are arrowhead type, orientation of the text, text style, units and precision, suppression of extension lines, fit of text and arrows (inside or outside of extension lines), and direction of leaders for radii and diameters (pointing in or out). The dimension variable changes that you make affect the appearance of the dimensions that you create.

Figure 27-1

Figure 27-2

There are two basic ways to control dimensioning variables:

1. Use the *Dimension Style Manager* (Fig. 27-3).
2. Type the dimension variable name in Command line format.

The dialog boxes employ "user-friendly" terminology and selection, while the Command line format uses the formal dimension variable names.

Changes to dimension variables are usually made before you create the affected dimensions. Dimension variable changes are not always retroactive, in contrast to *LTSCALE*, for example, which can be continually modified to adjust the spacing of existing non-continuous lines. Changes to dimensioning variables affect only newly created dimensions unless those changes are *Saved* to an existing *Dimension Style* that was in effect when previous dimensions were created. Generally, dimensioning variables should be set before creating the desired dimensions although it is possible to modify dimensions and *Dimension Styles* retroactively.

Dimension Styles

Associative and nonassociative dimensions are part of a dimension style. The default (and the only supplied) dimension style is named STANDARD (Fig. 27-3). Logically, this dimension style has all of the default dimension variable settings for creating a dimension with the typical size and appearance. Similar to layers, you can create, name, and specify settings for any number of dimension styles. Each dimension style contains the dimension variable settings that you select. A dimension style is a group of dimension variable settings that has been saved under a name you assign. When you set a dimension *Current*, AutoCAD remembers and resets that particular combination of dimension variable settings.

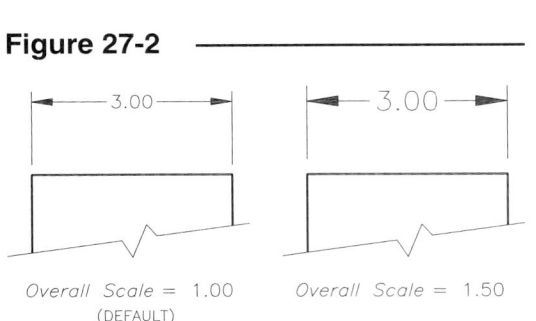

Figure 27-3

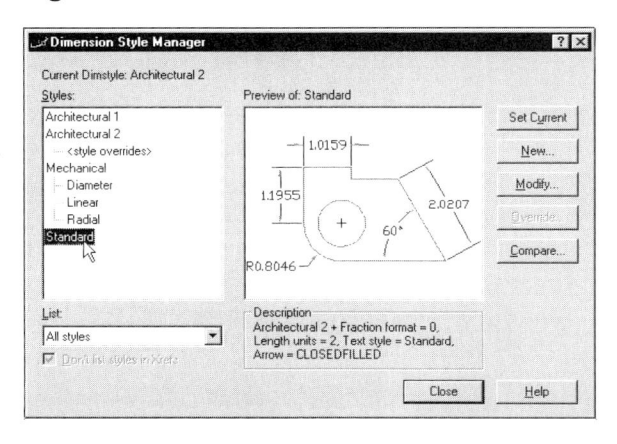

Imagine dimensioning a complex drawing <u>without</u> having dimension styles (Fig. 27-4). In order to create a dimension using limit dimensioning (as shown in the diameter dimension), for example, you would change the desired variable settings, then "draw" the dimension. To draw another dimension without extension lines (as shown in the interior slot), you would have to reset the previous variables and make changes to other variables in order to place the special dimensions as you prefer. This same process would be repeated each time you want to create a new type of dimension. If you needed to add another dimension with the limits, you would have to reset the same variables as before.

Figure 27-4

To simplify this process, you can create and save a dimension style for each particular "style" of dimension. Each time you want to draw a particular style of dimension, select the style name from the list and begin drawing. A dimension style could contain all the default settings plus only one or two dimension variable changes or a large number of variable changes.

To create a new dimension style, select the *New* button in the *Dimension Style Manager* (see Figure 27-3), then use the six tabs that appear in the dialog box to specify the appearance of the dimensions. Selecting options in these tabs (see Figure 27-14) actually sets values for related dimension variables that are saved in the drawing. When a dimension style is *Set Current*, the dimensions you create with that style appear with the dimension variable settings you specified. Creating dimensions with the current *Dimension Style* is similar to creating text with the current *Text Style*—that is, the objects created take on the current settings assigned to the style.

Another advantage of using dimension styles is that <u>existing</u> dimensions in the drawing can be globally modified by making a change and saving it to the dimension style(s). To do this, select *Modify* in the *Dimension Style Manager* or use the *Save* option of –*Dimstyle* after changing a dimension variable in Command line format. This action saves the changes for newly created dimensions as well as <u>automatically updating existing dimensions in the style</u>.

Dimension Style Families

In the previous chapter, you learned about the various <u>types</u> of dimensions, such as linear, angular, radial, diameter, ordinate, and leader. You may want the appearance of each of these types of dimensions to vary, for example, all radius dimensions to appear with the dimension line arrow pointing out and all diameter dimensions to appear with the dimension line arrow inside pointing in. It is logical to assume that a new dimension style would have to be created for each variation. However, a new dimension style name is not necessary for each type of dimension. AutoCAD provides <u>dimension style families</u> for the purpose of providing variations within each named dimension style.

The dimension style <u>names</u> that you create are assigned to a dimension style <u>family</u>. Each dimension style family can have six <u>children</u>. The children are <u>linear, angular, diameter, radial, ordinate, and leader</u>. Although the children take on the dimension variable settings assigned to the family, you can assign special settings for one or more children. For example, you can make a radius dimension appear slightly different from a linear dimension in that family. Do this by selecting the *New* button in the *Dimension Style Manager*, then selecting the type of dimension (*linear*, *radius*, *diameter*, etc.) you want to change from the *Create New Dimension Style* dialog box that appears (Fig. 27-5).

Figure 27-5

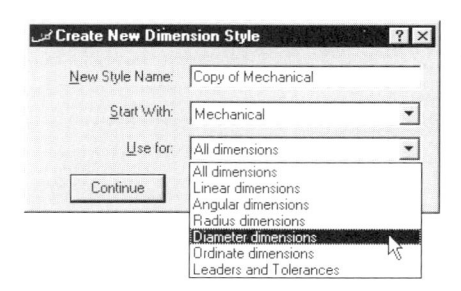

Select *Continue* to specify settings for the child. When a dimension is drawn, AutoCAD knows what type of dimension is created and applies the selected settings to that child.

Dimension styles that have children (special variable settings set for the *linear, radial, diameter,* etc. dimensions) appear in the list of styles in the *Dimension Style Manager* as sub-styles branching from the parent name. For example, notice the "Mechanical" style in Figure 27-3 has special settings for its *Diameter, Linear,* and *Radial* children.

Figure 27-6

For example, you may want to create the "Mechanical" dimension style to draw the arrows and dimension lines inside the arc but drag the dimension text outside of the arc for *Radial* dimensions (as shown in Fig. 27-6). To do this, create a new child for the "Mechanical" family as previously described, and set "Mechanical" as the current style. When a *Radius* dimension is drawn, the dimension line, text, and arrow should appear as shown in Figure 27-6.

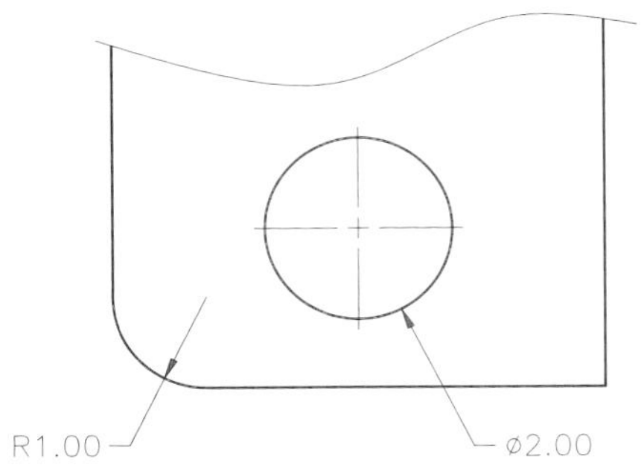

In summary, a dimension style family is simply a set of dimension variables related by name. Variations within the family are allowed, in that each child (type of dimension) can possess its own subset of variables. Therefore, each child inherits all the variables of the family in addition to any others that may be assigned individually.

AutoCAD refers to these children as $0, $2, $3, $4, $6, and $7 as suffixes appended to the family name such as "Standard$0." AutoCAD automatically applies the appropriate suffix code to the dimension style name when a child is created. Although the codes do not appear in the *Dimension Style Manager*, you can display the code by using the *List* command and selecting an existing child dimension.

Child Type	Suffix Code
linear	$0
angular	$2
diameter	$3
radial	$4
ordinate	$6
leader	$7

Dimension Style Overrides

If you want to create one or two dimensions with a special appearance, but do not want to permanently save the new settings to the dimension style, you can create a dimension style override, then create the new dimensions. A dimension style override is a temporary dimension variable setting for the current dimension style. Existing dimensions (in the drawing) are not affected by an override. When another style is made current, the override settings are lost by default.

A dimension style override is created by selecting the style you want to change from the list in the *Dimension Style Manager*, then pressing *Set Current*, then select the *Override* button. Proceed to specify the variable settings from the six tabs that appear. When the override is made, a new branch under the family name appears in the *Dimension Style Manager* named "<style overrides>" (see Figure 27-3 under "Architectural 2"). Finally, create the new dimensions. To clear the overrides, simply make another style current.

Alternately, you can create an override by typing any dimension variable name at the Command line and making a change to the value. For example, type *DIMSCALE* to change the overall scale for new dimensions created in the style. Existing dimensions in the style are not affected. The override is applied only to the newly created dimensions in the current style and clears when another style is made current.

DIMENSION STYLES

DIMSTYLE and *DDIM*

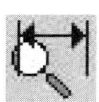

Pull-down Menu	COMMAND (TYPE)	ALIAS (TYPE)	Short-cut	Screen (side) Menu	Tablet Menu
Dimension Style...	DIMSTYLE or DDIM	DIMSTY, DST, or D	...	DIMNSION Dimstyle	Y,3

The *Dimstyle* or *Ddim* command produces the *Dimension Style Manager* (Fig. 27-7). This is the primary interface for creating new dimension styles and making existing dimension styles current. This dialog box also gives access to six tabs that allow you to change dimension variables. Features of the *Dimension Style Manager* that appear on the opening dialog box (shown in Figure 27-7) are described in this section. The tabs that allow you to change dimension variables are discussed in detail later in this chapter in the section "Dimension Variables."

Figure 27-7

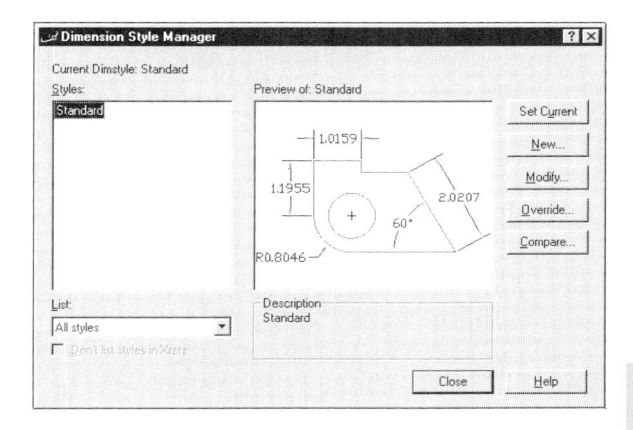

The *Dimension Style Manager* makes the process of creating and using dimension styles easier and more visual than using Command line format. Using the *Dimension Style Manager*, you can *Set Current*, create *New* dimension styles, *Modify* existing dimension styles, and create an *Override*. You also have the ability to view a list of existing styles including children and overrides; see a preview of the selected style, child, or override; examine a description of a dimension style as it relates to any other style; and make an in-depth comparison between styles. You can also control the entries in the list to include or not include Xreferenced dimension styles. The options are explained below.

Styles
This list displays all existing dimension styles in the drawing, including *Xrefs*, depending on your selection in the *List* section below. The list includes the parent dimension styles, the children that are shown on a branch below the parent style, and any overrides also branching from family styles.

The buttons on the right side of the *Dimension Style Manager* affect the highlighted style in the list. For example, to modify an existing style, select the style name from the list and PICK the *Modify* button. If you want to create a *New* style based on an existing style, select the style name you want to use as the "template," then select *New*.

Select (highlight) any dimension style from the *Styles* list to display the appearance of dimensions for that style in the *Preview* tile. Selecting a style from the list also makes the *Description* area list the dimension variables settings compared to the current dimension style.

526 Dimension Styles and Variables

TIP You can right-click to use a shortcut menu for the selected dimension style (Fig. 27-8). The shortcut menu allows you to *Set Current*, *Rename*, or *Delete* the selected dimension style. Only dimension styles that are unreferenced (no dimensions have been created using the style) can be deleted.

Figure 27-8

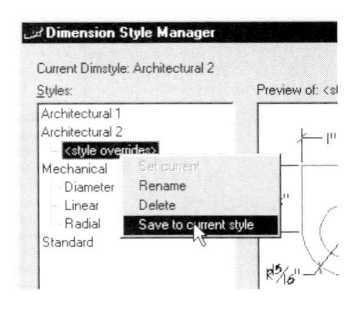

If you have created a style override (by pressing the *Override* button, then setting variables), you can save the overrides to the dimension style using the right-click shortcut menu. Normally, overrides are automatically discarded when another dimension style is made current. This shortcut menu is the only method available in the *Dimension Style Manager* to save overrides. (You can also use the *–Dimstyle* command to save overrides.)

List
Select *All Styles* to display the list of all the previously created or imported dimension styles in the drawing. Selecting *Styles In Use* displays only styles that have been used to create dimensions in the drawing—styles that are saved in the drawing but are not referenced (no dimensions created in the style) do not appear in the list.

Don't List Styles in Xrefs
If the current drawing references another drawing (has an *Xref* attached), the *Xref's* dimension styles can be listed or not listed based on your selection here.

Preview
The *Preview* tile displays the appearance of the highlighted dimension style. When a dimension style name is selected, the *Preview* tile displays an example of all dimension types (linear, radial, etc.) and how the current variable settings affect the appearance of those dimensions. If a child is selected, the *Preview* tile displays only that dimension type and its related appearance. For example, Figure 27-9 displays the appearance of the selected child, *Mechanical: Radial*.

Figure 27-9

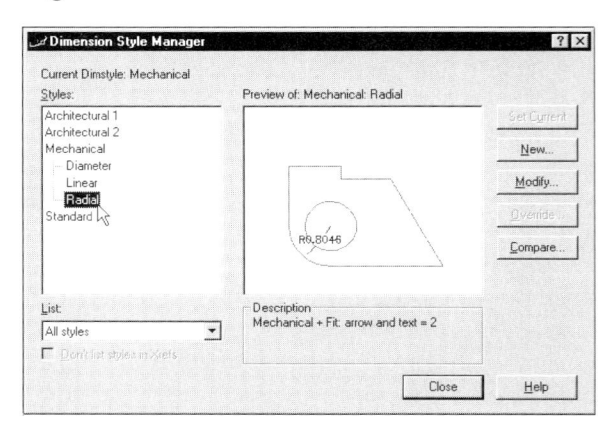

Description
The description area is a valuable tool for determining the variable settings for a dimension style. When a dimension style from the list is selected, the *Description* area lists the differences in variable settings from the current style. For example, assume the "Architectural 1" dimension style was created using "Standard" as the template. To display the dimension variable settings that have been changed (the differences between Standard and Architectural 1), make Standard the current style, then select Architectural 1 and examine the description area, as shown in Figure 27-10. You can use the same procedure to display the variable settings assigned to a child (the differences between the family style and the child), as shown in the *Description* area in Figure 27-9.

Figure 27-10

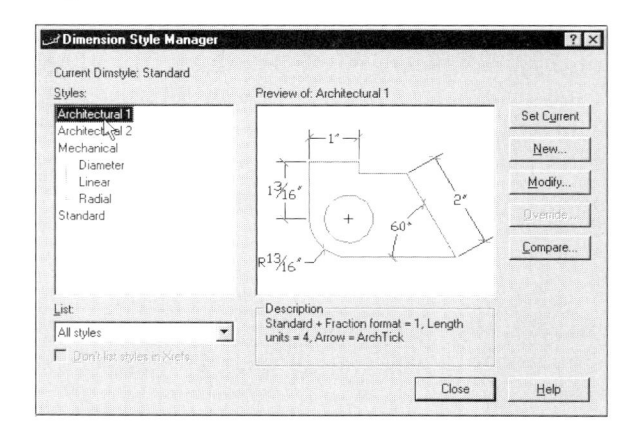

Set Current
Select the desired existing style from the list on the left, then select *Set Current*. The current style is listed at the top of the *Dimension Style Manager*. When you *Close* the dialog box, the current style is used for drawing new dimensions. If you want to set an *Override* to an existing style, you must first make it the current style.

New
Use the *New* button to create a new dimension style. The *New* button invokes the *Create New Dimension Style* dialog box (Fig. 27-11). The options are discussed next.

Figure 27-11

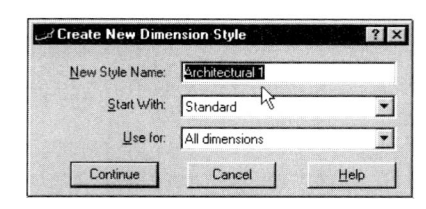

New Style Name
Enter the desired new name in the *New Style Name* edit box. Initially the name that appears is "Copy of (current style)."

Start With
After assigning a new name, select any existing dimension style to use as a "template" from the *Start With* drop-down list. Initially (until dimension variable changes are made for the new style), the new dimension style is actually a copy of the *Start With* style (has identical variable settings). By default, the current style appears in the *Start With* edit box.

Use For
If you want to create a dimension style family, select *All Dimensions*. In this way, all dimension types (*linear*, *radial*, *diameter*, etc.) take on the new variable settings. If you want to specify special dimension variable settings for a child, select the desired dimension type from the drop-down list (see Figure 27-5).

Continue
When the new name and other options have been selected, press the *Continue* button to invoke the *New Dimension Style* dialog box. Here you select from the six tabs to specify settings for any dimension variable to apply to the new style. See the "Dimension Variables" section for information on setting dimension variables.

NOTE: Generally you should not change the Standard dimension style. Rather than change the Standard style, create *New* dimension styles using Standard as the base style. In this way, it is easy to restore and compare to the default settings by making the Standard style current. If you make changes to the Standard style, it can be difficult to restore the original settings.

Modify
Use this button to change dimension variable settings for an existing dimension style. First, select the desired style name from the list, then press *Modify* to produce the *Modify Dimension Style* dialog box. Using *Modify* to change dimension variables automatically saves the changes to the style and updates existing dimensions in the style, as opposed to using *Override*, which does not change the style or update existing dimensions. See the "Dimension Variables" section for information on setting dimension variables.

Override
A dimension style override is a temporary variable setting that affects only new dimensions created with the override. An override does not affect existing dimensions previously created with the style. You can set an override only for the current style, and the overrides are automatically cleared when another dimension style is made current, unless you use the right-click shortcut menu and select *Save to current style* (see Figure 27-8).

528 *Dimension Styles and Variables*

To create a dimension style override, you must first select an existing style from the list and make it the current style. Next, select *Override* to produce the *Override Current Style* dialog box. See the "Dimension Variables" section for information on setting dimension variables to act as overrides.

Compare

This feature is very useful for examining variable settings for any dimension style. Press the *Compare* button to produce the *Compare Dimension Styles* dialog box (Fig. 27-12). This dialog box lists only dimension variable settings; changes to variable settings cannot be made from this interface.

Figure 27-12

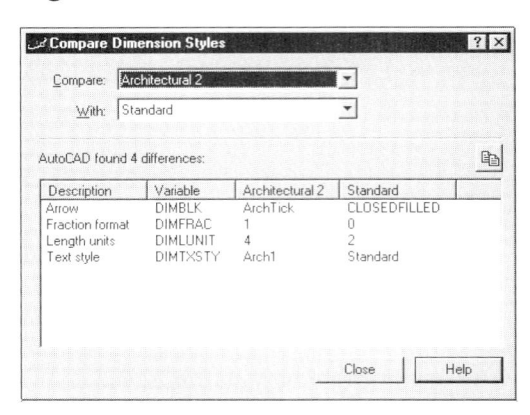

The *Compare Dimension Styles* dialog box lists the differences between the style in the *Compare* box and the style in the *With* box. For example, assume you created a new style named "Architectural 2" from the "Standard" style, but wanted to know what variable changes you had made to the new style. Select Architectural 2 in the *Compare* box and Standard in the *With* box. The variable changes made to Architectural 2 are displayed in the central area of the box. Also note that the settings for each variable are listed for each of the two styles.

A useful feature of this dialog box is that the formal variable names are listed in the *Variable* column. Many experienced users of AutoCAD prefer to use the formal variable names since they do not change from release to release as the dialog box descriptions do.

You can also use the *Compare Dimension Styles* dialog box to give the entire list of variables and related settings for one dimension style. Do this by selecting the desired dimension style name from the *Compare* drop-down list and select *<none>* from the *With* list (Fig. 27-13).

Figure 27-13

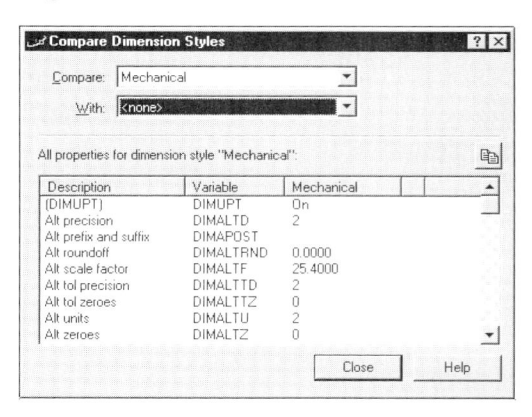

-DIMSTYLE

Pull-down Menu	COMMAND (TYPE)	ALIAS (TYPE)	Short-cut	Screen (side) Menu	Tablet Menu
Dimension Update	-DIMSTYLE	...	...	DIMNSION Dimstyle	Y,3

The *-Dimstyle* command is the Command line format equivalent to the *Dimension Style Manager*. Operations that are performed by the dialog box can also be accomplished with the *-Dimstyle* command. The *Apply* option offers one other important feature that is not available in the dialog box:

Command: *-dimstyle*
Current dimension style: Standard
Enter a dimension style option [Save/Restore/STatus/Variables/Apply/?] <Restore>:

Save
Use this option to create a new dimension style. The new style assumes all of the current dimension variable settings comprised of those from the current style plus any other variables changed by Command line format (overrides). The new dimension style becomes the current style.

   Enter a dimension style option [Save/Restore/STatus/Variables/Apply/?] <Restore>: *s*
   Enter name for new dimension style or [?]:

Restore
This option prompts for an existing dimension style name to be restored. The restored style becomes the current style. You can use the "select dimension" option to PICK a dimension object that references the style you want to restore.

STatus
Status gives the current settings for all dimension variables. The displayed list comprises the settings of the current dimension style and any overrides. All dimension variables that are saved in dimension styles are listed:

   Enter a dimension style option [Save/Restore/STatus/Variables/Apply/?] <Restore>: *st*

   DIMASO   Off   Create dimension objects
   DIMSTYLE   Standard   Current dimension style (read-only)
   DIMADEC   0   Angular decimal places
   DIMALT   Off   Alternate units selected
   DIMALTD   2   Alternate unit decimal places
   etc.

Variables
You can use this option to list the dimension variable settings for any existing dimension style. You cannot modify the settings.

   Enter a dimension style option [Save/Restore/STatus/Variables/Apply/?] <Restore>: *v*
   Enter a dimension style name, [?] or <select dimension>:

Enter the name of any style or use "select dimension" to PICK a dimension object that references the desired style. A list of all variables and settings appears.

~stylename
This variation can be entered in response to the *Variables* option. Entering a dimension style name preceded by the tilde symbol (~) displays the differences between the current style and the (~) named style.

For example, assume that you wanted to display the differences between STANDARD and the current dimension style (Mechanical, for example). Use *Variables* with the ~ symbol. (The ~ symbol is used as a wildcard to mean "all but.") This option is very useful for keeping track of your dimension styles and their variable settings.

   Command: *-dimstyle*
   Current dimension style: Mechanical
   Enter a dimension style option [Save/Restore/STatus/Variables/Apply/?] <Restore>: *v*
   Enter a dimension style name, [?] or <select dimension>: *~standard*
   Differences between STANDARD and current settings:

              STANDARD   Current Setting
   DIMSCALE   1.0000   0.7500
   DIMUPT   Off   On
   DIMTXSTY   Standard   Mech1

Apply

Use *Apply* to update dimension objects that you PICK with the current variable settings. The current variable settings can contain those of the current dimension style plus any overrides. The selected object loses its reference to its original style and is "adopted" by the applied dimension style family.

> Enter a dimension style option [Save/Restore/STatus/Variables/Apply/?] <Restore>: *a*
> Select objects:

This is a useful tool for changing an existing dimension object from one style to another. Simply make the desired dimension style current; then use *Apply* and select the dimension objects to change to that style.

Using the *Apply* option of *-Dimstyle* is identical to using the *Update* command. The *Update* command can be invoked by using the *Dimension Update* button, using *Update* from the *Dimension* pull-down menu, or typing *Dim*: (Enter), then *Update*.

DIMENSION VARIABLES

Now that you understand how to create and use dimension styles and dimension style families, let's explore the dimension variables. Two methods can be used to set dimension variables: the *Dimension Styles Manager* and Command line format. First, we will examine the dialog box (accessible through the *Dimension Styles Manager*) that allows you to set dimension variables. Dimension variables can alternately be set by typing the formal name of the variable (such as *DIMSCALE*) and changing the desired value (discussed after this section on dialog boxes). The dialog box offers a more "user-friendly" terminology, edit boxes, checkboxes, drop-down lists, and a preview image tile that automatically reflects the variables changes you make.

Changing Dimension Variables Using the Dialog Box Method

In this section, the dialog box that contains the six tabs for changing variables is explained (see Figure 27-14). The six tabs indicate different groups of dimension variables. The tab names are:

Lines and Arrows
Text
Fit
Primary Units
Alternate Units
Tolerances

Access to this dialog box from the *Dimension Style Manager* is accomplished by selecting *New, Modify,* or *Override*. Practically, there is only one dialog box that allows you to change dimension variables; however, you might say there are three dialog boxes since the title of the box changes based on your selection of *New, Modify,* or *Override*. Depending on your selection, you can invoke the *Create Dimension Style, Modify Dimension Style,* or *Override Current Style* dialog box. The options in the boxes are identical and only the titles are different; therefore, we will examine only one dialog box.

In the following pages, the heading for the paragraphs below include the dialog box option and the related formal dimension variable name in parentheses. The following figures typically display the AutoCAD default setting for a particular variable and an example of changing the setting to another value. These AutoCAD default settings (in the "Standard" dimension style) are for the ACAD.DWT template drawing and for *Start from Scratch, English Default Settings*. If you use the *Metric Default Settings* or other template drawings, dimension variable settings may be different based on dimension styles that exist in the drawing other than Standard.

Lines and Arrows Tab

The options in this tab change the appearance of dimension lines, extension lines, and arrowheads (Fig. 27-14). The preview image automatically changes based on the settings you select in the *Dimension Lines*, *Extension Lines*, *Arrowheads*, and *Center Marks for Circles* sections.

Figure 27-14

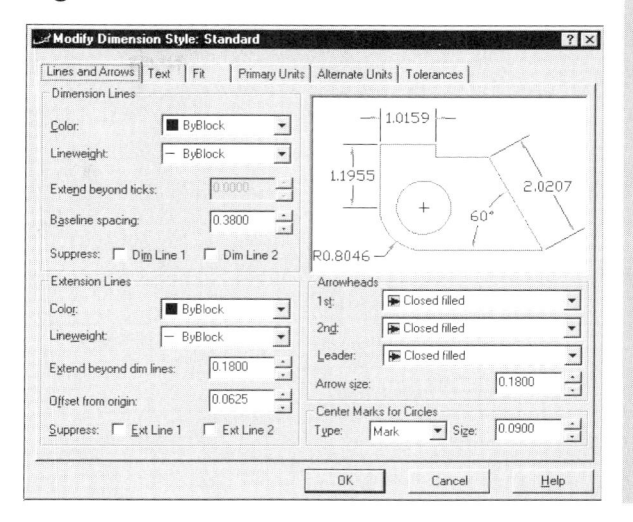

Dimension Lines Section (Lines and Arrows Tab)

Color (DIMCLRD)
The *Color* drop-down list allows you to choose the color for the dimension line (Fig. 27-15). Assigning a specific color to the dimension lines, extension lines, and dimension text gives you more control when color-dependent plot styles are used because you can print or plot these features with different line widths, colors, etc. This feature corresponds to the *DIMCLRD* variable (dim color dimension line).

Figure 27-15

Lineweight (DIMLWD)
Use this option to assign lineweight to dimension lines (Fig. 27-16). Select any lineweight from the drop-down list or enter values in the *DIMLWD* variable (dim lineweight dimension line). Values entered in the variable can be -1 (*ByLayer*), -2 (*ByBlock*), 25 (*Default*), or any integer representing 100th of mm, such as 9 for .09mm.

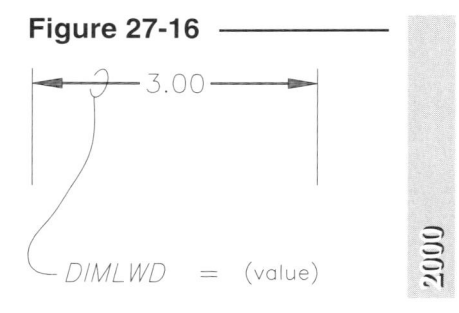

Figure 27-16

Extend beyond ticks (DIMDLE)
The *Extend beyond ticks* edit box is disabled unless the *Oblique, Integral,* or *Architectural Tick* arrowhead type is selected (in the *Arrowheads* section). The *Extend* value controls the length of dimension line that extends past the dimension line (Fig. 27-17). The value is stored in the *DIMDLE* variable (dimension line extension). Generally, this value does not require changing since it is automatically multiplied by the *Overall Scale* (*DIMSCALE*).

Figure 27-17

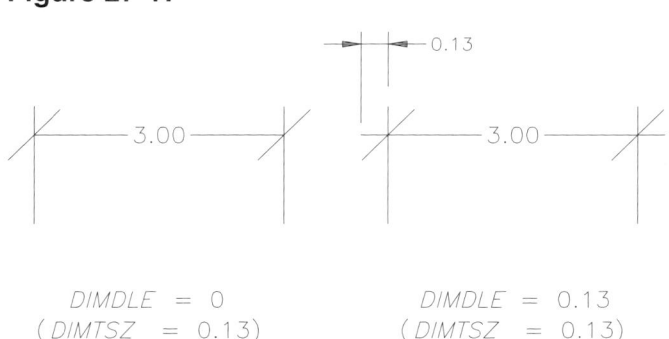

532 Dimension Styles and Variables

Baseline spacing (DIMDLI)

The *Baseline spacing* edit box reflects the value that AutoCAD uses in baseline dimensioning to "stack" the dimension line above or below the previous one (Fig. 27-18). This value is held in the *DIMDLI* variable (dimension line increment). This value rarely requires input since it is affected by *Overall Scale* (*DIMSCALE*).

Figure 27-18

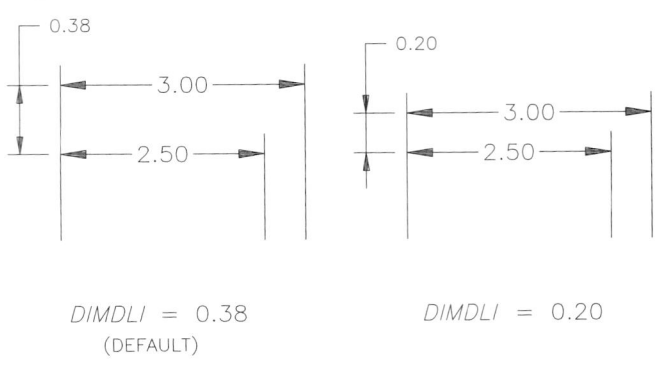

Suppress Dim Line 1, Dim Line 2 (DIMSD1, DIMSD2)

This area allows you to suppress (not draw) the *1st* or *2nd* dimension line or both (Fig. 27-19). The first dimension line would be on the "First extension line origin" side or nearest the end of object PICKed in response to "Select object to dimension." These toggles change the *DIMSD1* and *DIMSD2* dimension variables (suppress dimension line 1, 2).

Figure 27-19

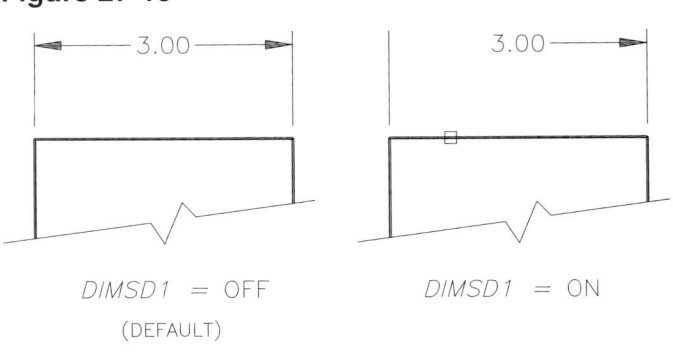

Extension Line Section (Lines and Arrows Tab)

Color (DIMCLRE)

The *Color* drop-down list allows you to choose the color for the extension lines (Fig. 27-20). You can also activate the standard *Select Color* dialog box. The color assignment for extension lines is stored in the *DIMCLRE* variable (dim color extension line).

Figure 27-20

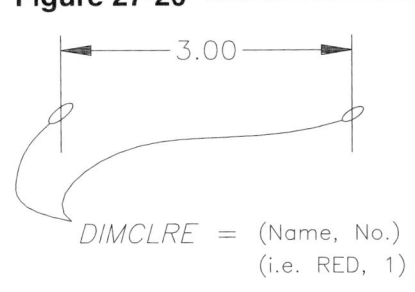

Lineweight (DIMLWE)

This option is similar to that for dimension lines, only the lineweight is assigned to extension lines (Fig. 27-21). Select any lineweight from the drop-down list or enter values in the *DIMLWE* (dim lineweight extension line) variable (*ByLayer* = -1, *ByBlock* = -2, *Default* = 25, or any integer representing 100th of mm).

Figure 27-21

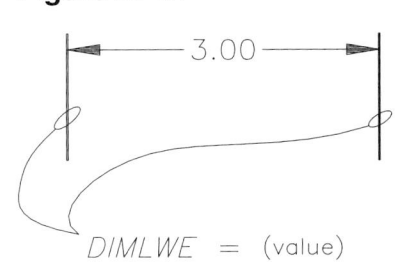

Extend beyond dim lines **(DIMEXE)**
The *Extend beyond dim lines* edit box reflects the value that AutoCAD uses to set the distance for the extension line to extend beyond the dimension line (Fig. 27-22). This value is held in the *DIMEXE* variable (extension line extension). Generally, this value does not require changing since it is automatically multiplied by the *Overall Scale* (*DIMSCALE*).

Figure 27-22

Offset from origin **(DIMEXO)**
The *Offset from origin* value specifies the distance between the origin points and the extension lines (Fig. 27-23). This offset distance allows you to PICK the object corners, yet the extension lines maintain the required gap from the object. This value rarely requires input since it is affected by *Overall Scale* (*DIMSCALE*). The value is stored in the *DIMEXO* variable (extension line offset).

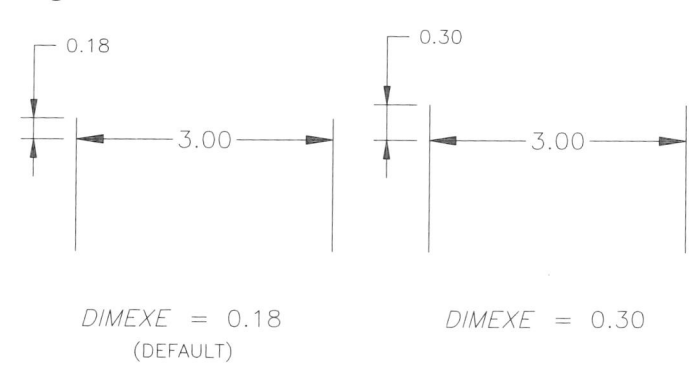

Figure 27-23

Suppress Ext Line 1, Ext Line 2
(DIMSE1, DIMSE2)
This area is similar to the *Dimension Line* area that controls the creation of dimension lines but is applied to extension lines. This area allows you to suppress the *1st* or *2nd* extension line or both (Fig. 27-24). These options correspond to the *DIMSE1* and *DIMSE2* dimension variables (suppress extension line 1, 2).

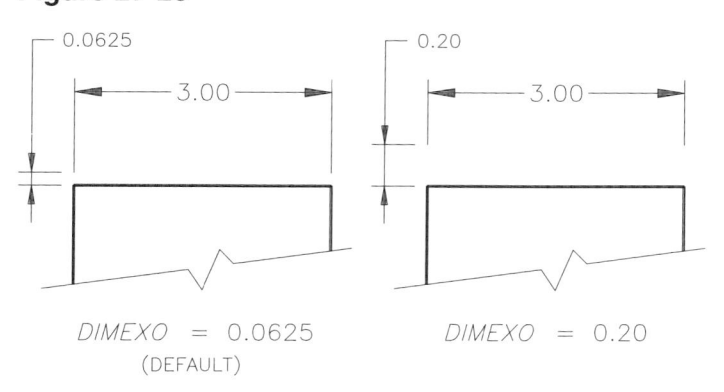

Figure 27-24

<u>*Arrowheads* **Section (*Lines and Arrows* Tab)**</u>

1st, 2nd **(DIMBLK, DIMSAH, DIMBLK1, DIMBLK2)**
This area contains two drop-down lists of various arrowhead types, including dots and ticks. Each list corresponds to the *1st* or *2nd* arrowhead created in the drawn dimension (Fig. 27-25). The image tiles display each arrowhead type selected. Click in the first image tile to change both arrowheads, or click in each to change them individually. The variables affected are *DIMBLK*, *DIMSAH*, *DIMBLK1*, and *DIMBLK2*.

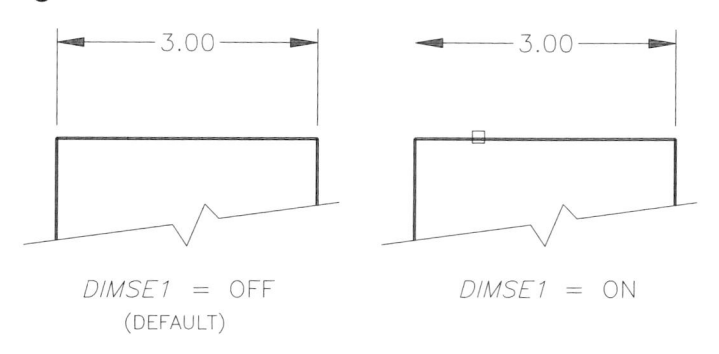

Figure 27-25

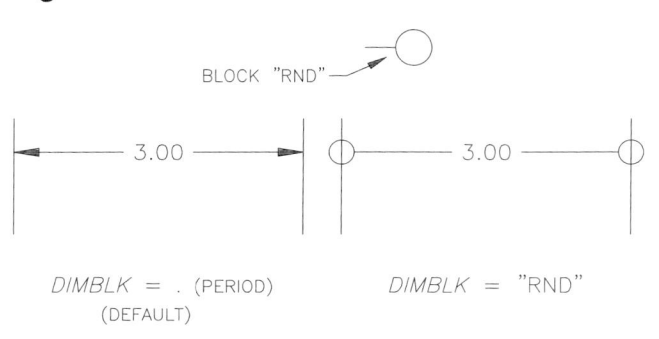

DIMBLK specifies the *Block* to use for arrowheads if both are the same (Fig. 27-25). When *DIMSAH* is on (separate arrow heads), separate arrow heads are allowed for each end defined by *DIMBLK1* and *DIMBLK2*. You can enter any existing *Block* name in the *DIMBLK* variables.

Leader (DIMLDRBLK)

This drop-down list specifies the arrow type for leaders (Fig. 27-26). This setting does not affect the arrowhead types for dimension lines. Select any option from the drop-down list or enter a value in the *DIMLDRBLK* variable (dim leader block). (For a list of arrowhead entries, see AutoCAD 2002 *Help*.)

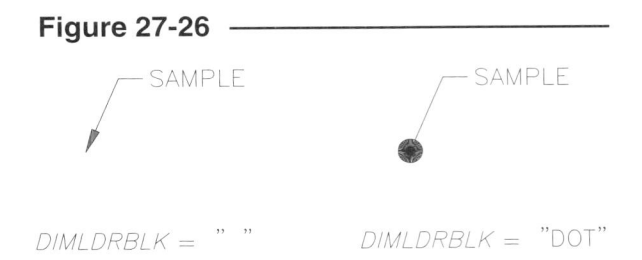

Figure 27-26

DIMLDRBLK = " "   DIMLDRBLK = "DOT"

Arrow size (DIMASZ)

The size of the arrow can be specified in the *Arrow size* edit box. The *DIMASZ* variable (dim arrow size) holds the value (Fig. 27-27). Remember that this value is multiplied by *Overall Scale* (*DIMSCALE*).

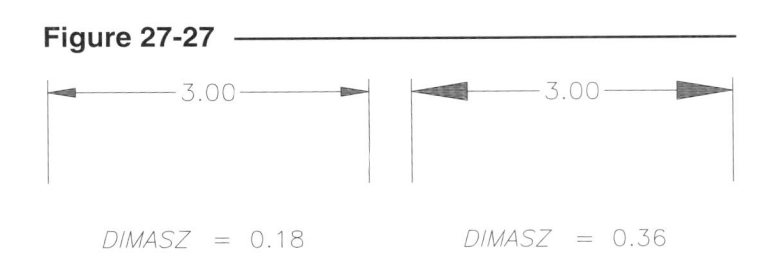

Figure 27-27

DIMASZ = 0.18   DIMASZ = 0.36

Center Marks for Circles Section (Lines and Arrows Tab)

Type (DIMCEN)

The list here determines how center marks are drawn when the dimension commands *Dimcenter*, *Dimdiameter*, or *Dimradius* are used. The image tile displays the *Mark*, *Line*, or *None* feature specified. This area actually controls the value of <u>one</u> dimension variable, *DIMCEN* by using a 0, positive, or negative value (Fig. 27-28). The *None* option enters a *DIMCEN* value of 0.

Figure 27-28

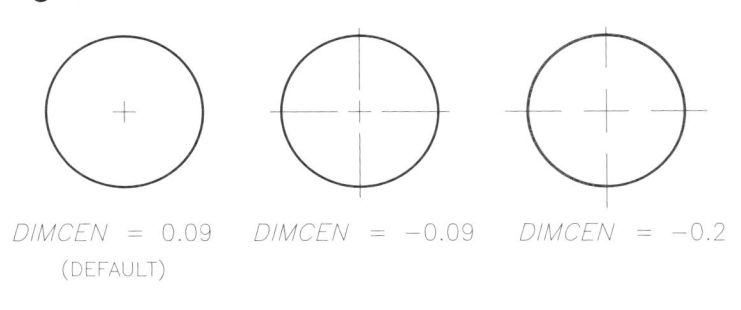

DIMCEN = 0.09   DIMCEN = -0.09   DIMCEN = -0.2
(DEFAULT)

Size (DIMCEN)

The *Size* edit box controls the size of the short dashes and extensions past the arc or circle. The value is stored in the *DIMCEN* variable (Fig. 27-28). Only positive values can be entered in this edit box.

Text Tab

The options in this tab change the appearance, placement, and alignment of the dimension text (Fig. 27-29). The preview image automatically reflects changes you make in the *Text Appearance*, *Text Placement*, and *Text Alignment* sections.

Figure 27-29

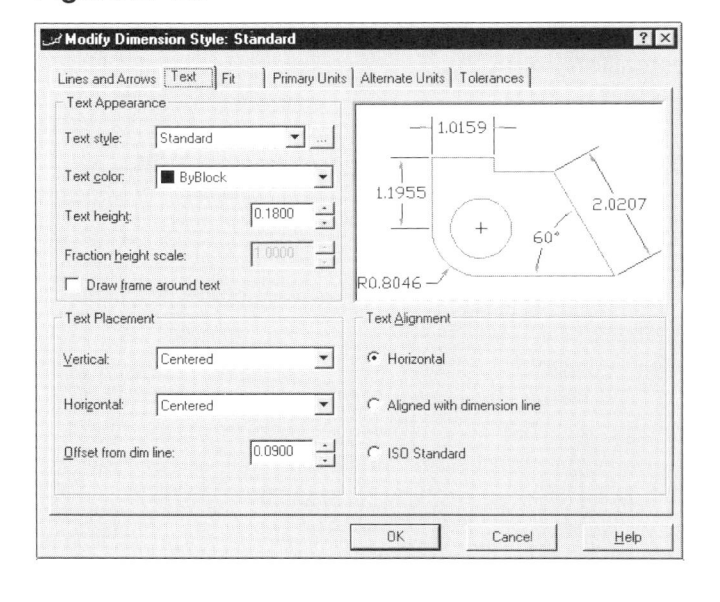

Text Appearance Section (Text Tab)

Text style (DIMTXSTY)
This feature of AutoCAD allows you to have different text styles for different dimension styles. The text styles are chosen from a drop-down list of existing styles in the drawing. You can also create a text style "on the fly" by picking the button just to the right of the *Text style* drop-down list, which produces the standard *Text Style* dialog box. The text style used for dimensions remains constant (as defined by the dimension style) and does not change when other text styles in the drawing are made current (as defined by *Style, Dtext,* or *Mtext*). The *DIMTXSTY* variable (dim text style) holds the text style name for the dimension style.

Text color (DIMCLRT)
Select this drop-down list to select a color or to activate the standard *Select Color* dialog box. The color choice is assigned to the dimension text only (Fig. 27-30). This is useful for controlling the text appearance for printing or plotting when color-dependent plot styles are used. The setting is stored in the *DIMCLRT* variable (dim color text).

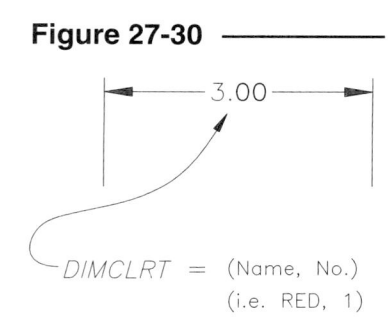

Figure 27-30

Text height (DIMTXT)
This value specifies the primary text height; however, this value is multiplied by the *Overall Scale* (*DIMSCALE*) to determine the actual drawn text height. Change *Text height* only if you want to increase or decrease the text height in relation to the other dimension components (Fig. 27-31). Normally, change the *Overall Scale* (*DIMSCALE*) to change all size-related features (arrows, gaps, text, etc.) proportionally. The text height value is stored in *DIMTXT*.

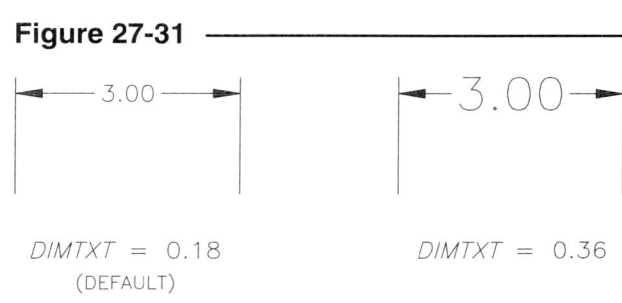

Figure 27-31

Fraction height scale (DIMTFAC)
When fractions or tolerances are used, the height of the fractional or tolerance text can be set to a proportion of the primary text height. For example, 1.0000 creates fractions and tolerances the same height as the primary text; .5000 represents fractions or tolerances at one-half the primary text height (Fig. 27-32). The value is stored in the *DIMTFAC* variable (dim tolerance factor).

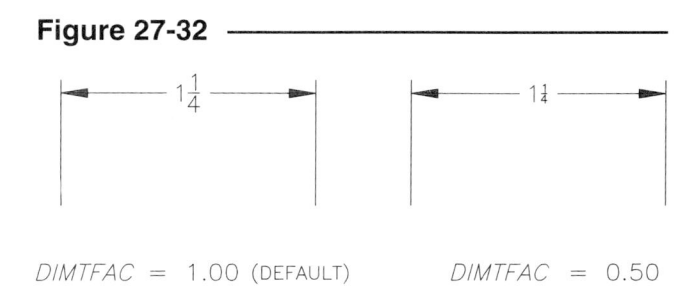

Figure 27-32

Draw frame around text (DIMGAP)
The "frame" or "gap" is actually an invisible box around the text that determines the offset from text to the dimension line. This option sets the *DIMGAP* variable to a negative value which makes the box visible (Fig. 27-33). This practice is standard for displaying a basic dimension. See *"Tolerance"* (geometric dimensioning and tolerancing) in Chapter 26. See also *Offset from dimension line* in the "*Text Placement*" section.

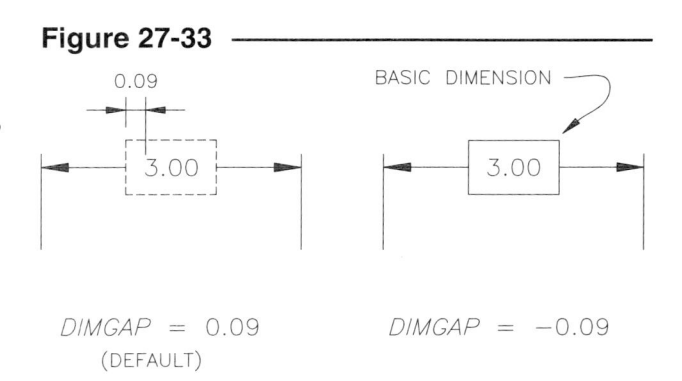

Figure 27-33

536 Dimension Styles and Variables

Text Placement Section (*Text* Tab)

Vertical (DIMTAD)

The *Vertical* option determines the vertical location of the text with respect to the dimension line. There are four possible settings that affect the *DIMTAD* variable (dim text above dimension line). *Centered*, the default option (*DIMTAD* = 0), centers the dimension text between the extension lines. The *Above* option places the dimension text above the dimension line except when the dimension line is not horizontal (*DIMTAD* = 1). *Outside* places the dimension text on the side of the dimension line farthest away from the extension line origin points—away from the dimensioned object (*DIMTAD* = 2). The *JIS* option places the dimension text to conform to Japanese Industrial Standards (*DIMTAD* = 3).

Horizontal (DIMJUST)

The *Horizontal* section determines the horizontal location of the text with respect to the dimension line (dimension justification). The default option (*DIMJUST* = 0) centers the text between the extension lines. The other four choices (*DIMJUST* = 1-3) place the text at either end of the dimension line in parallel and perpendicular positions (Fig. 27-34). You can use the *Horizontal* and *Vertical* settings together to achieve additional text positions.

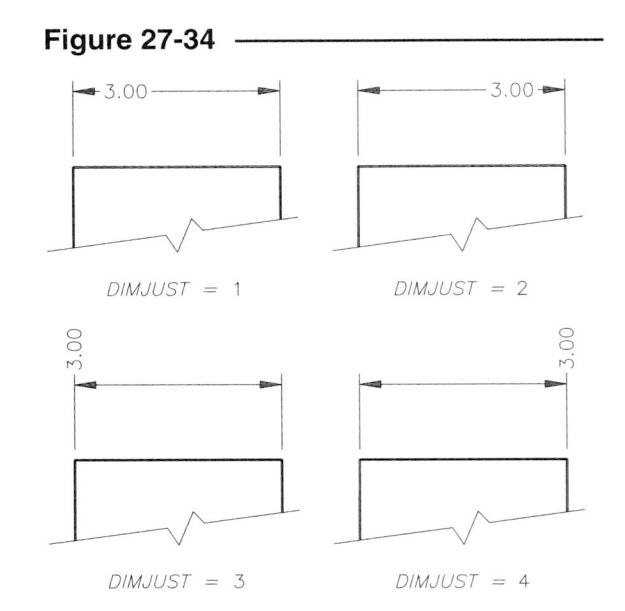

Figure 27-34

Offset from dim line (DIMGAP)

This value sets the distance between the dimension text and its dimension line. The offset is actually determined by an invisible box around the text (Fig. 27-35). Increasing or decreasing the value changes the size of the invisible box. The *Offset from dim line* value is stored in the *DIMGAP* variable. (Also see "*Draw frame around text*.")

Figure 27-35

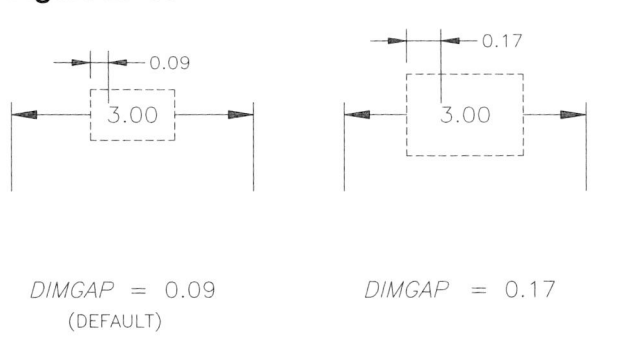

Text Alignment Section, (*Text* Tab)

The *Text Alignment* settings can be used in conjunction with the *Text Placement* settings to achieve a wide variety of dimension text placement options.

Horizontal (DIMTIH, DIMTOH)

This radio button turns on the *DIMTIH* (dim text inside horizontal) and *DIMTOH* (dim text outside horizontal) variables. The text remains horizontal even for vertical or angled dimension lines (see

Figure 27-36

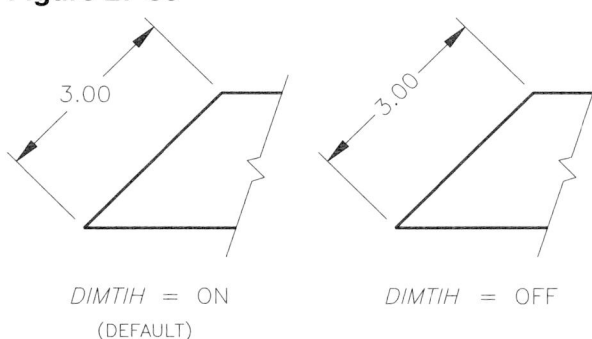

Figures 27-36 and 27-37). This is the correct setting for mechanical drawings (other than ordinate dimensions) according to the ASME Y14.5M-1994 standard, section 1.7.5, Reading Direction.

Aligned with dimension line **(DIMTIH, DIMTOH)**
Pressing this radio button turns off the *DIMTIH* and *DIMTOH* variables so the text aligns with the angle of the dimension line (see Figures 27-36 and 27-37).

ISO Standard **(DIMTIH, DIMTOH)**
This option forces the text inside the dimension line to align with the angle of the dimension line (*DIMTIH* = off), but text outside the dimension line is horizontal (*DIMTOH* = on).

Figure 27-37

DIMTOH = ON (DEFAULT)   DIMTOH = OFF

Fit Tab

The *Fit* tab allows you to determine the *Overall Scale* for dimensioning components, how the text, arrows, and dimension lines fit between extension lines, and how the text appears when it is moved (Fig. 27-38).

Scale for Dimension Features Section (*Fit* Tab)

Although this is not the first section in the dialog box, it is presented first because of its importance.

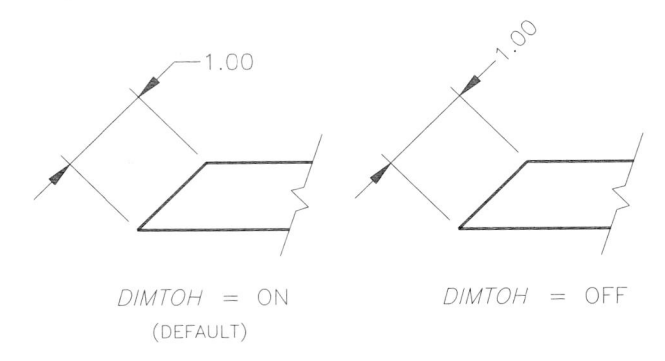

Figure 27-38

Overall Scale **(DIMSCALE)**
The *Overall Scale* value globally affects the scale of all size-related features of dimension objects, such as arrowheads, text height, extension line gaps (from the object), extensions (past dimension lines), etc. All other size-related (variable) values appearing in the dialog box series are multiplied by the *Overall Scale*. Notice how all dimensioning features (text, arrows, gaps, offsets) are all increased proportionally with the *Overall Scale* value (Fig. 27-39). Therefore, to keep all features proportional, change this one setting rather than each of the others individually.

Although this area is located on the right side of the box, it is probably the most important option in the entire series of tabs. Because the *Overall Scale* should be set as a family-wide variable, setting this value is typically the first step in creating a dimension style (Fig. 27-40).

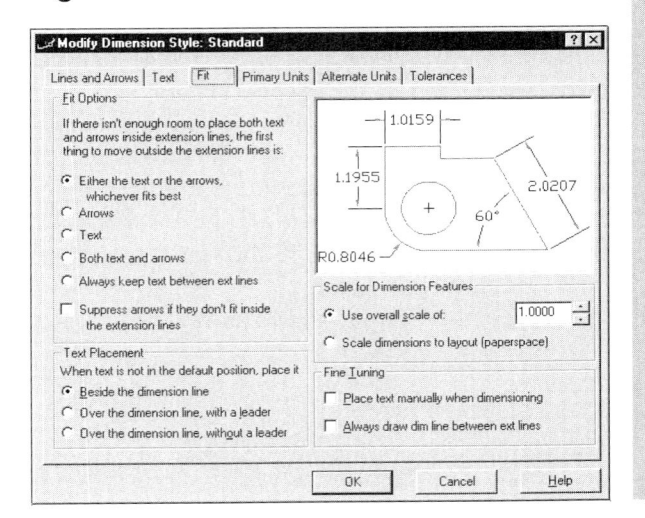

Figure 27-39

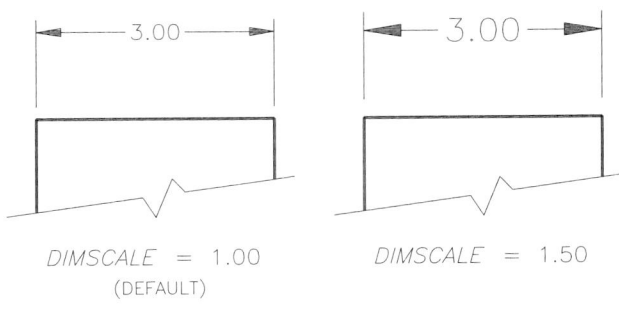

DIMSCALE = 1.00 (DEFAULT)   DIMSCALE = 1.50

Figure 27-40

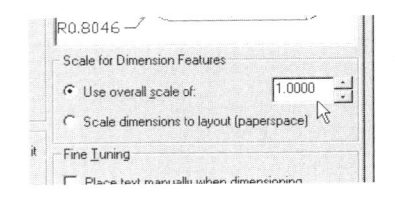

Changes in this variable should be based on the *Limits* and plot scale. You can use the drawing scale factor to determine this value. (See "Drawing Scale Factor," Chapter 12.) The *Overall Scale* value is stored in the *DIMSCALE* variable.

Scale dimensions to layout (paper space) (DIMSCALE)

Checking this box forces dimension components to appear in the same size (*DIMSCALE*) for all viewports in a layout (Paper Space viewports created with *Vports*). Toggling this on sets *DIMSCALE* to 0.

Fit Options Section (Fit Tab)

This section determines which dimension components are <u>forced outside the extension lines only if there is insufficient room</u> for text, arrows, and dimension lines. In most cases where space permits all components to fit inside, the *Fit* settings have no effect on placement. *Linear*, *Aligned*, *Angular*, *Baseline*, *Continue*, *Radius*, and *Diameter* dimensions apply. See "*Radius* and *Diameter* Variable Settings" at the end of this section for information on recommended settings for *Radius* and *Diameter* dimensions.

Either the text or the arrows (DIMATFIT)

This is the default setting for the STANDARD dimension style. AutoCAD makes the determination of whether the text or the arrows are forced outside based on the size of arrows and the length of the text string (Fig. 27-41). This option often behaves similarly to *Arrows*, except that if the text cannot fit, the text is placed outside the extension lines and the arrows are placed inside. However, if the arrows cannot fit either, both text and arrows are placed outside the extension lines. The *DIMATFIT* (dimension arrows/text fit) setting is 3.

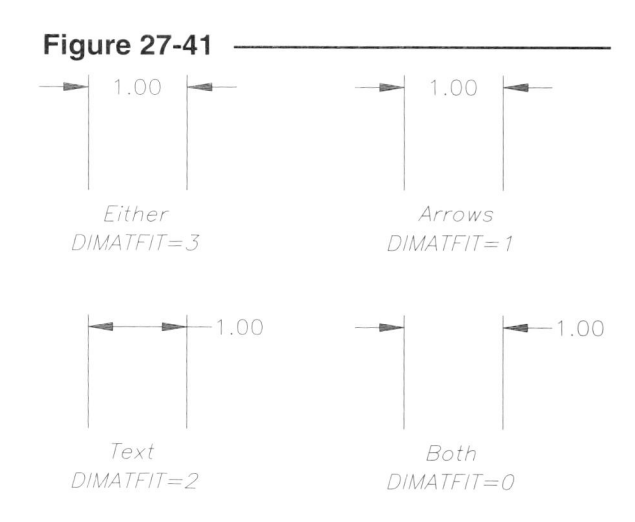

Figure 27-41

Arrows (DIMATFIT)

The *Arrows* option forces the arrows on the outside of the extension lines and keeps the text inside. If the text absolutely cannot fit, it is also placed outside the extension lines (see Figure 27-41). This option sets *DIMATFIT* to 1.

Text (DIMATFIT)

The *Text* option places the text on the outside and keeps the arrows on the inside unless the arrows cannot fit, in which case they are placed on the outside as well (see Figure 27-41). For this option, *DIMATFIT* = 2.

Both text and arrows (DIMATFIT)

The *Both text and arrows* option keeps the text and arrows together always. If space does not permit <u>both</u> features to fit between the extension lines, it places the text and arrows outside the extension lines (see Figure 27-41). You can set *DIMATFIT* to 0 to achieve this placement.

Always keep text between ext lines (DIMTIX)

If you want the text to be forced between the extension line no matter how much room there is, use this option (Fig. 27-42). Pressing this radio button turns *DIMTIX* on (text inside extensions).

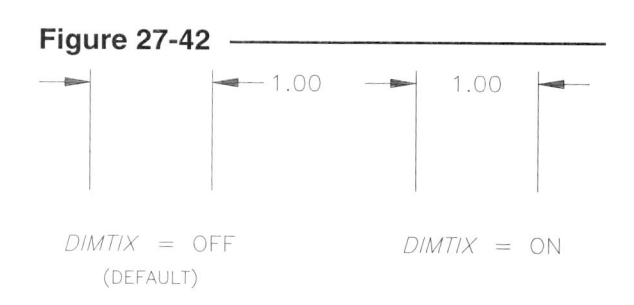

Figure 27-42

Suppress arrows if they don't fit (**DIMSOXD**)
When dimension components are forced outside the extension lines and there are many small dimensions aligned in a row (such as with *Continue* dimensions), the text, arrows, or dimension lines may overlap. In this case, you can prevent the arrows and the dimension lines from being drawn entirely with this option. This option suppresses the arrows and dimension lines <u>only</u> when they are forced outside (Fig. 27-43). The setting is stored as *DIMSOXD* = on (suppress dimension lines outside extensions).

Figure 27-43

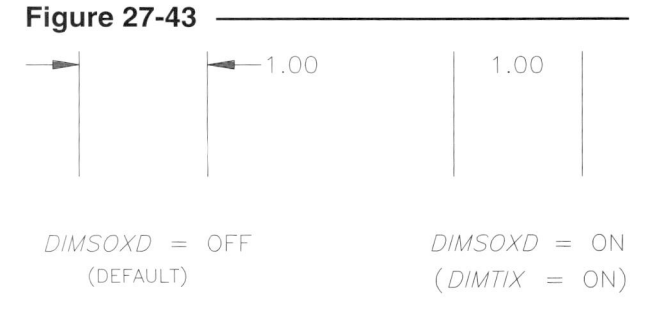

Text Placement **Section (*Fit* Tab)**

This section of the dialog box sets dimension text movement rules. When text is moved either by being automatically forced from between the dimension lines based on the *DIMATFIT* setting (the *Fit Options* above this section) or when you actually move the text with grips or by *Dimtedit*, these rules apply.

Beside the dimension line (**DIMTMOVE**)
This is the normal placement of the text—aligned with and beside the dimension line (Fig. 27-44). The text always moves when the dimension line is moved and vice versa. *DIMTMOVE* (dimension text move) = 0.

Figure 27-44

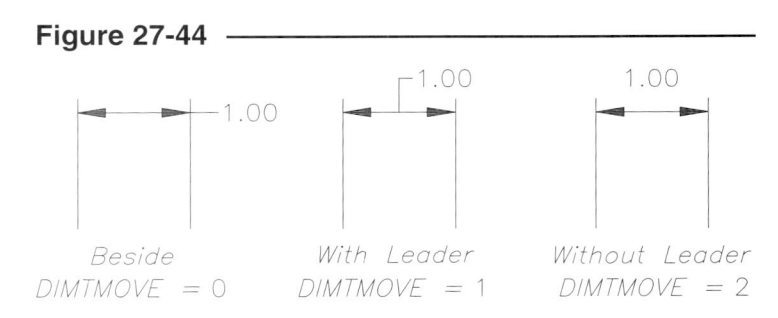

Over the dimension line, with a leader (**DIMTMOVE**)
This option creates a leader between the text and the center of the dimension line whenever the text cannot fit between the extension lines or is moved using grips (see Figure 27-44). *DIMTMOVE* = 1.

Over the dimension line, without a leader (**DIMTMOVE**)
Use this setting to have the text appear above the dimension line, similar to *DIMTMOVE* = 1, but without a leader. This occurs only when there is insufficient room for the text between the extension lines or when you move the text with grips. *DIMTMOVE* = 2.

There is an important benefit to this setting (*DIMTMOVE* = 2). When there is sufficient room for the text and arrows between extension lines, this setting has no effect on the placement of the text. When there is insufficient room, the text moves above without a leader. In either case, if you prefer to move the text to another location with grips or using *Dimtedit*, the text moves as if were "detached" from the dimension line. The text can be moved independently to any location and the dimension retains its associatively. Figure 27-45 illustrates the use of grips to edit the dimension text.

Figure 27-45

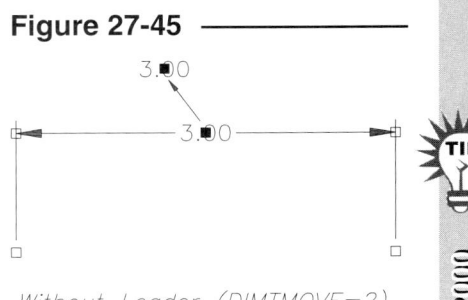

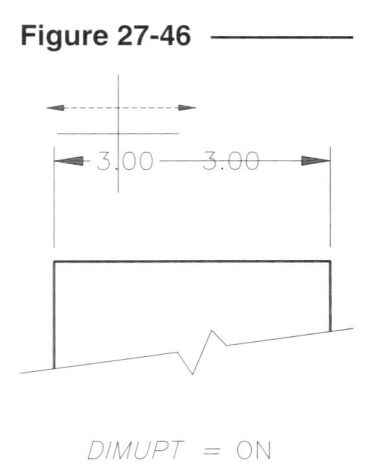

Fine Tuning Section (*Fit* Tab)

Place text manually when dimensioning (**DIMUPT**)
When you press this radio button, you can create dimensions and move the text independently in relation to the dimension and extension lines as you place the dimension line in response to the "specify dimension line location" prompt (Fig. 27-46). Using this option is similar to using *Dimtedit* after placing the dimension. *DIMUPT* (dimension user-positioned text) is on when this box is checked.

Figure 27-46

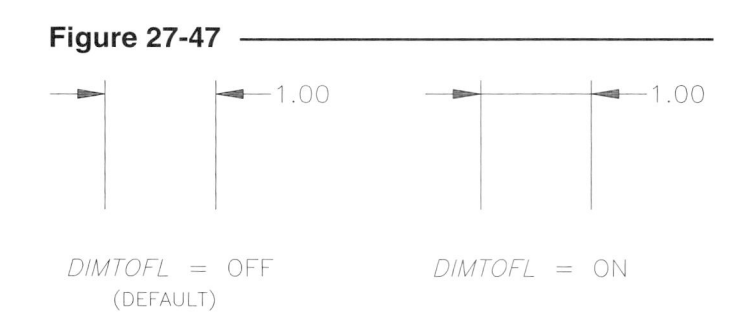

Always draw dim line between ext lines (**DIMTOFL**)
Occasionally, you may want the dimension line to be drawn inside the extension lines even when the text and arrows are forced outside. You can force a line inside with this option (Fig. 27-47). A check in this box turns *DIMTOFL* on (text outside, force line inside).

Figure 27-47

Radius and *Diameter* Variable Settings

For creating mechanical drawing dimensions according to ANSI standards, the default settings in AutoCAD are correct for creating *Diameter* dimensions but not for *Radius* dimensions. The following variable settings are recommended for creating *Radius* and *Diameter* dimensions for mechanical applications.

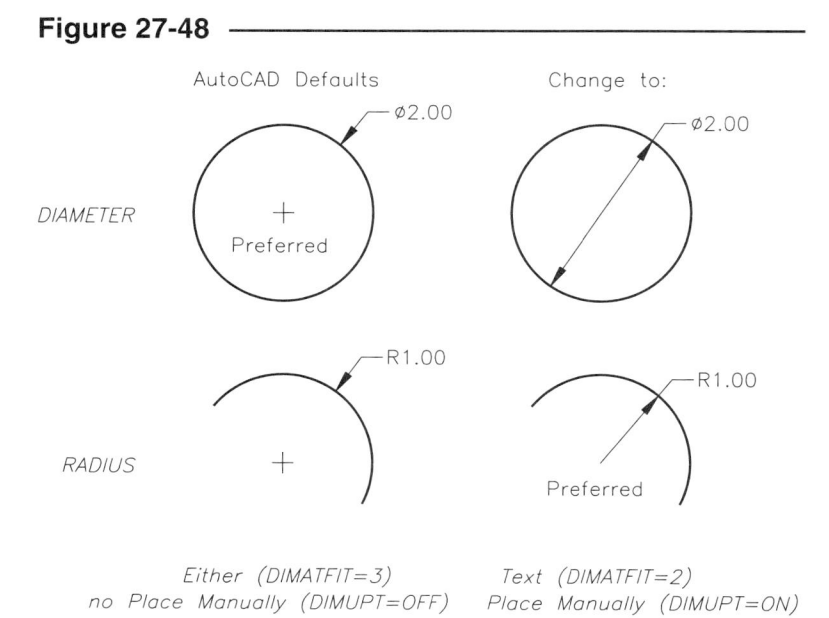

For *Diameter* dimensions, the default settings produce ANSI-compliant dimensions that suit most applications—that is, text and arrows are on the outside of the circle or arc pointing inward toward the center. For situations where large circles are dimensioned, *Fit Options* (*DIMATFIT*) and *Place text manually* (*DIMUPT*) can be changed to force the dimension inside the circle (Fig. 27-48).

Figure 27-48

For *Radius* dimensions the default settings produce incorrect dimensioning practices. Normally (when space permits) you want the dimension line and arrow to be inside the arc, while the text can be inside or outside. To produce ANSI-compliant *Radius* dimensions, set *Fit Options* (*DIMATFIT*) and *Place text manually* (*DIMUPT*) as shown in Figure 27-48. Radius dimensions can be outside the arc in cases where there is insufficient room inside.

Primary Units Tab

The *Primary Units* tab controls the format of the AutoCAD-measured numerical value that appears with a dimension (Fig. 27-49). You can vary the numerical value in several ways such as specifying the units format, precision of decimal or fraction, prefix and/or suffix, zero suppression, and so on. These units are called primary units because you can also cause AutoCAD to draw additional or secondary units called *Alternate Units* (for inch <u>and</u> metric notation, for example).

Figure 27-49

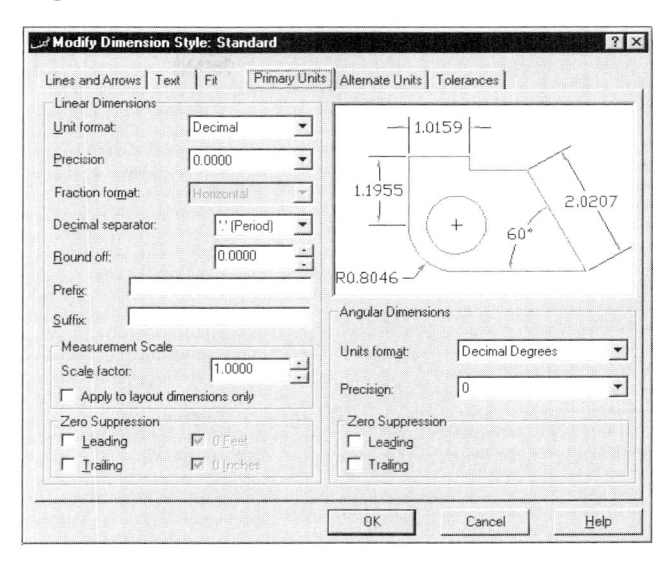

Linear Dimensions Section (*Primary Units* Tab)

This section controls the format of all dimension types except *Angular* dimensions.

Unit format (DIMLUNIT)

The *Units format* section drop-down list specifies the type of units used for dimensioning. These are the same unit types available with the *Units* dialog box (*Decimal, Scientific, Engineering, Architectural,* and *Fractional*) with the addition of *Windows Desktop*. The *Windows Desktop* option displays AutoCAD units based on the settings made for units display in Windows Control Panel (settings for decimal separator and number grouping symbols). Remember that your selection affects the <u>units drawn in dimension objects, not the global drawing units</u>. The choice for *Units format* is stored in the *DIMLUNIT* variable (dimension linear unit). This drop-down list is disabled for an *Angular* family member.

Precision (DIMDEC)

The *Precision* drop-down list in the *Dimension* section specifies the number of places for decimal dimensions or denominator for fractional dimensions. This setting does not alter the drawing units precision. This value is stored in the *DIMDEC* variable (dim decimal).

Fraction format (DIMFRAC)

Use this drop-down list to set the fractional format. The choices are displayed in Figure 27-50. This option is enabled only when DIMLUNIT (*Unit format*) is set to 4 (*Architectural*) or 5 (*Fractional*).

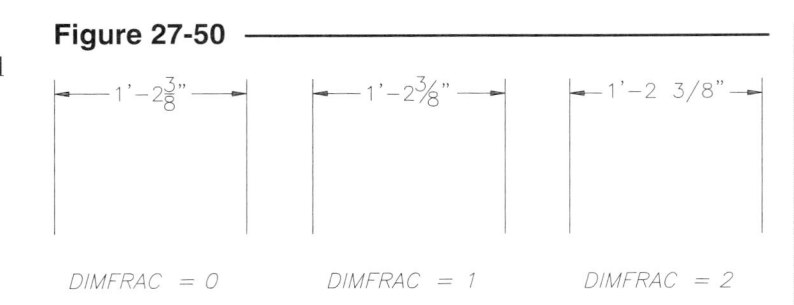

Figure 27-50

Decimal separator (DIMDSEP)

When you are creating dimensions whose unit format is decimal, you can specify a single-character decimal separator. Normally a decimal (period) is used; however, you can also use a comma (,) or a space. The character is stored in the *DIMDSEP* (dimension decimal separator) variable.

Round off (DIMRND)

Use this drop-down list to specify a precision for dimension values to be rounded. Normally, AutoCAD values are kept to 14 significant places but are rounded to the place dictated by the dimension *Precision* (*DIMDEC*). Use this feature to round up or down appropriately to the nearest specified decimal or fractional increment (Fig. 27-51).

Figure 27-51

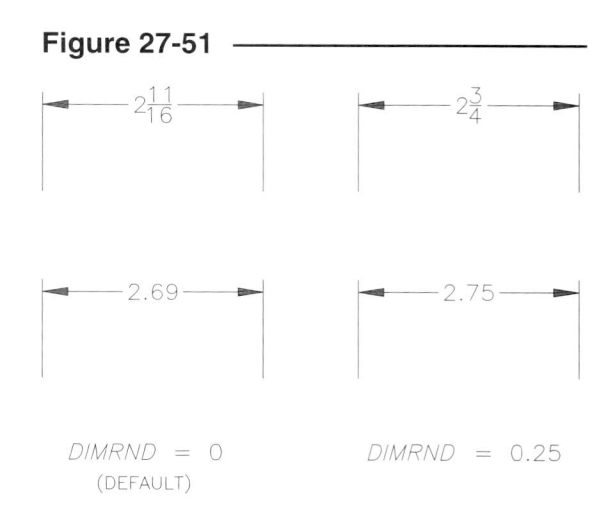

Prefix/Suffix (DIMPOST)

The *Prefix* and *Suffix* edit boxes hold any text that you want to add to the AutoCAD-supplied dimensional value. A text string entered in the *Prefix* edit box appears before the AutoCAD-measured numerical value and a text string entered in the *Suffix* edit box appears after the AutoCAD-measured numerical value. For example, entering the string " mm" or a " TYP." in the *Suffix* edit box would produce text as shown in Figure 27-52. (In such a case, don't forget the space between the numerical value and the suffix.) The string is stored in the *DIMPOST* variable.

Figure 27-52

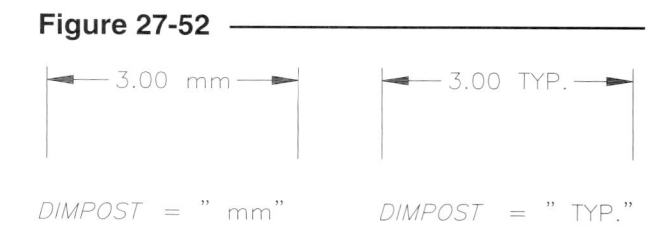

If you use the *Prefix* box to enter letters or values, any AutoCAD-supplied symbols (for radius and diameter dimensions) are overridden (not drawn). For example, if you want to specify that a specific hole appears twice, you should indicate by designating a "2X" before the diameter dimension. However, doing so by this method overrides the phi (Ø) symbol that AutoCAD inserts before the value. Instead, use the *Mtext/Dtext* options within the dimensioning command or use *Dimedit* or *Ddedit* to add a prefix to an existing dimension having an AutoCAD-supplied symbol. Remember that AutoCAD uses the < > brackets in the *Mtext Editor* to represent the AutoCAD-measured value, so place a prefix in front of the brackets. Do not overwrite the brackets unless you want to lose the AutoCAD-measured value.

Measurement Scale Section (*Primary Units* Tab)

Scale factor (DIMLFAC)

Any value placed in the *Scale factor* edit box is a <u>multiplier</u> for the AutoCAD-measured numerical value. The default is 1. Entering a 2 would cause AutoCAD to draw a value two times the actual measured value (Fig. 27-53). This feature might be used when a drawing is created in some scale other than the actual size, such as an enlarged detail view in the same drawing as the full view. In AutoCAD 2002, changing this setting is <u>unnecessary</u> when you create associative dimensions in paper space attached to objects in model space (see "Dimensioning in Paper Space").

Figure 27-53

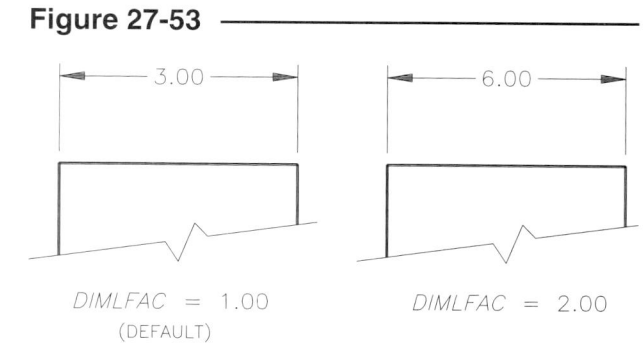

However, if you are using versions of AutoCAD previous to 2002, or if you are using AutoCAD 2002 nonassociative dimensions in paper space, change this setting to create dimensions that display other than the actual measured value. This setting is stored in the *DIMLFAC* variable (dimension length factor).

Apply to layout dimension only (DIMLFAC)
Use this checkbox to apply the *Scale factor* to dimensions placed in a layout for AutoCAD 2002 <u>nonassociative</u> dimensions or for drawings in versions earlier than AutoCAD 2002. For example, assume you have a detail view displayed at 2:1 in a viewport and want to place nonassociative dimensions in paper space, set the *Scale factor* to .5 and check *Apply to layout dimension only* so the measured values adjust for the viewport scale. A check in this box sets the *DIMLFAC* variable to the negative of the *Scale factor* value. Associative dimensions in AutoCAD 2002 drawn in paper space automatically adjust for the viewport scale, so changing this setting is unnecessary (see "Dimensioning in Paper Space").

<u>*Zero Suppression* Section (*Primary Units* Tab)</u>

Figure 27-54

The *Zero Suppression* section controls how zeros are drawn in a dimension when they occur. A check in one of these boxes means that zeros are <u>not drawn for that case</u>. This sets the value for *DIMZIN* (dimension zero indicator).

Leading/Trailing/0 Feet/0 Inches (DIMZIN)
Leading and *Trailing* are enabled for *Scientific*, *Decimal*, *Engineering* and *Fractional* units. The *0 Feet* and *0 Inches* checkboxes are enabled for *Architectural* and *Engineering* units.

For example, assume primary *Units format* was set to *Architectural* and *Zero Suppression* was checked for *0 Inches* only. Therefore, when a measurement displays feet and no inches, the 0 inch value is suppressed (Fig. 27-54, top dimension). On the other hand, since *0 Feet* is not checked, a measurement of less than 1 foot would report 0 feet (Fig. 27-54, bottom dimension).

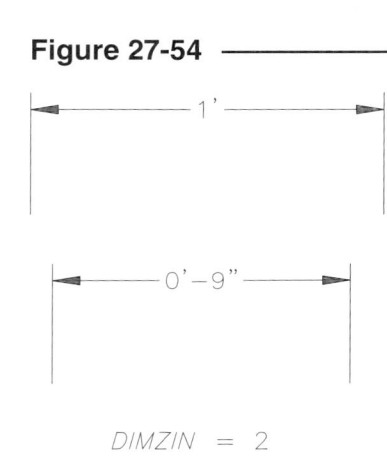

<u>*Angular Dimensions* Section (*Primary Units* Tab)</u>

Units format (DIMAUNIT)
The *Units format* drop-down list sets the unit type for angular dimensions, including *Decimal degrees*, *Degrees Minutes Seconds*, *Gradians*, and *Radians*. This list is enabled only for parent dimension style and *Angular* family members. The variable used for the angular units is *DIMAUNIT* (dimension angular units).

Precision (DIMADEC)
This option sets the number of places of precision (decimal places) for angular dimension text. The selection is stored in the *DIMADEC* variable (dimension angular decimals). This option is enabled only for the parent dimension style and *Angular* family member.

Zero Suppression (DIMAZIN)
Use these two checkboxes to set the desired display for angular dimensions when there are zeros appearing before or after the decimal. A check in one of these boxes means that zeros are <u>not drawn for that case</u>.

Alternate Units Tab

This tab allows you to display alternate, or secondary, dimensioning values along with the primary AutoCAD-measured value when a dimension is created (Fig. 27-55). Typically, alternate units consist of millimeter values given in addition to the decimal inch values. This practice is often called "dual dimensioning." The options in this tab control the display and format of alternate units. Notice that most of these options are equivalent to those found in the *Primary Units* tab.

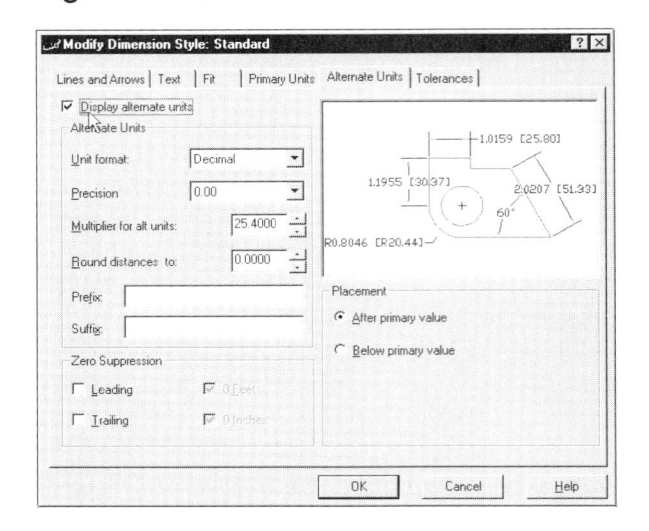

Figure 27-55

Display alternate units (DIMALT)
By default, AutoCAD displays only one value for a dimension—the primary unit. If you want to have AutoCAD measure and create an additional value for each dimension, check this box (Fig. 27-56). All options in this tab are disabled unless *Display alternate units* is checked. The presence of alternate units is controlled by setting the *DIMALT* variable on (dimension alternate).

Figure 27-56

DIMALT = OFF (DEFAULT)   DIMALT = ON

Alternate Unit Section (*Alternate Units* Tab)

Unit format (DIMALTU)
This drop-down list sets the format for alternate units. The setting is stored in the *DIMALTU* variable (dimension alternate units). This is the alternate units equivalent to *Units format* for *Primary Units*.

Precision (DIMALTD)
Use this drop-down list to set the decimal precision for alternate units. Note that the alternate units precision is controlled independently of the primary units precision (Fig. 27-57). The value is stored in the *DIMALTD* variable (dimension alternate decimals).

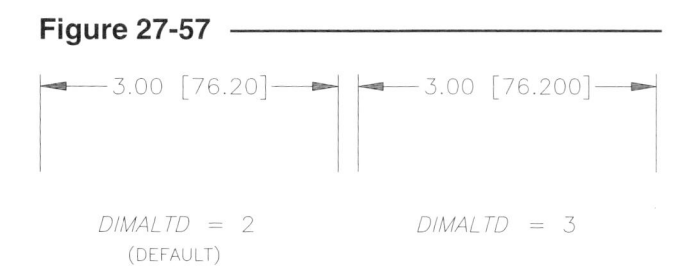

Figure 27-57

DIMALTD = 2 (DEFAULT)   DIMALTD = 3

Multiplier for alt units (DIMALTF)
This is the alternate units equivalent for the primary units scale factor (*DIMLFAC*). In other words, the AutoCAD-measured primary units value is multiplied by this factor to determine the displayed value for alternate units (Fig. 27-58). You can also enter the desired multiplier in the *DIMALTF* (dimension alternate factor) variable.

Figure 27-58

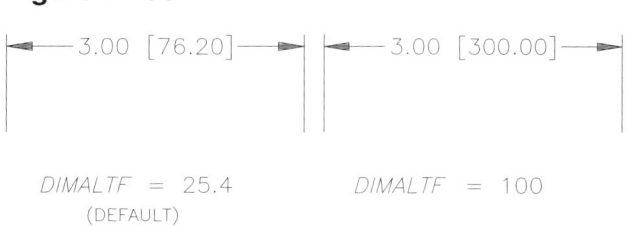

DIMALTF = 25.4 (DEFAULT)   DIMALTF = 100

Round distances to (DIMALTRND)

If you do not want the alternate units value to be displayed as the actual measurement, but to be rounded to nearest regular increment, enter the desired increment in this edit box. For example, you may want alternate units to be displayed with two places of precision (to the right of the decimal) but to round to the nearest millimeter. Alternately, set the increment using the DIMALTRND variable (alternate rounding). This is the alternate units equivalent to the *Round off* option in the *Primary Units* tab (see Figure 27-51).

Prefix/Suffix (DIMAPOST)

This section allows you to include a prefix and/or suffix with the alternate units measured value. For example, you may want to display " mm" after the alternate units value (Fig. 27-59). The prefix and/or suffix is stored in the DIMAPOST variable. (See also *Prefix/Suffix* in the *Primary Units* tab for more information.)

Figure 27-59

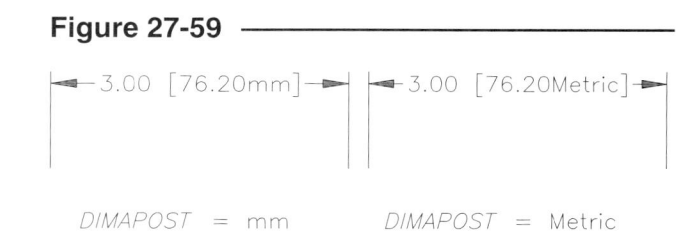

Zero Suppression Section (Alternate Units Tab)

The *Zero Suppression* section controls how zeros are drawn in alternate units values when they occur. A check in one of these boxes means that zeros are <u>not drawn for that case</u>. Your selection sets the value for DIMALTZ. See the description for *Zero Suppression*, *Primary Units* Tab for more information and illustration.

Placement Section (Alternate Units Tab)

This section only has two options; both are related to the DIMPOST variable. Normally, the alternate units values are placed *After primary value*. You can instead toggle *Below primary value* to yield a display as shown in Figure 27-60, right dimension. Selecting *Below primary value* sets the DIMPOST variable to "\x."

Figure 27-60

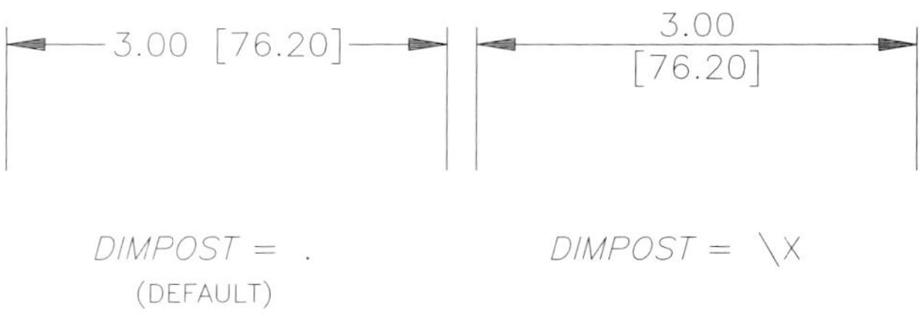

Tolerances Tab

The *Tolerances* tab allows you to create several formats of tolerance dimensions such as limits, two forms of plus/minus dimensions, and basic dimensions (Fig. 27-61). Most options in this tab are disabled until you select a *Method*.

Figure 27-61

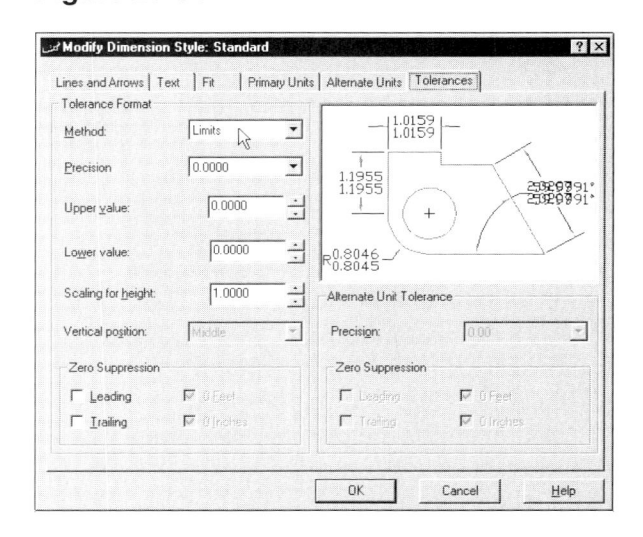

Tolerance Format Section (Tolerances Tab)

Method (DIMTOL, DIMLIM, DIMGAP)
The *Method* option displays a drop-down list with five types: *None, Symmetrical, Deviation, Limits,* and *Basic*. The four possibilities (other than *None*) are illustrated in Figure 27-62. The *Symmetrical* and *Deviation* methods create plus/minus dimensions and turn on the *DIMTOL* variable (dimension tolerance). The *Limits* method creates limit dimensions and turns the *DIMLIM* variable on (dimension limits). The *Basic* method creates a basic dimension by drawing a box around the dimensional value, which is accomplished by changing the *DIMGAP* to a negative value.

Figure 27-62

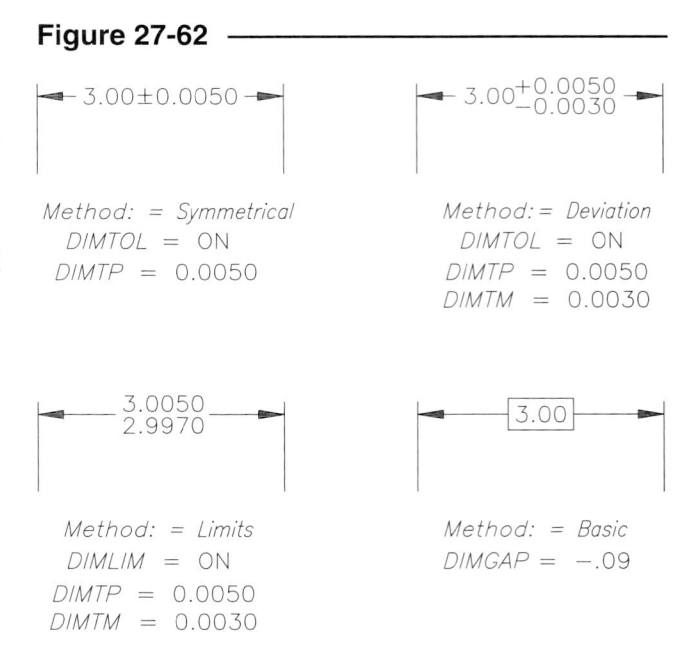

3.00 ± 0.0050
Method: = Symmetrical
DIMTOL = ON
DIMTP = 0.0050

$3.00^{+0.0050}_{-0.0030}$
Method: = Deviation
DIMTOL = ON
DIMTP = 0.0050
DIMTM = 0.0030

3.0050
2.9970
Method: = Limits
DIMLIM = ON
DIMTP = 0.0050
DIMTM = 0.0030

3.00
Method: = Basic
DIMGAP = −.09

Precision (DIMTDEC)
Use the *Precision* drop-down list to set the precision (number of places to the right of the decimal place) for values when drawing *Symmetrical, Deviation,* and *Limits* dimensions. The precision is stored in the *DIMTDEC* variable (dimension tolerance decimals).

Upper value/Lower value (DIMTP, DIMTM)
An *Upper Value* and *Lower Value* can be entered in the edit boxes for the *Deviation* and *Limits* method types. In these cases, the *Upper Value* (*DIMTP*—dimension tolerance plus) is added to the measured dimension and the *Lower Value* (*DIMTP*—dimension tolerance minus) is subtracted (Fig. 27-62). An *Upper Value* only is needed for *Symmetrical* and is applied as both the plus and minus value.

Scaling for height (DIMTFAC)
The height of the tolerance text is controlled with the *Scaling for height* edit box value. The entered value is a <u>proportion</u> of the primary dimension value height. For example, a value of .50 would draw the tolerance text at 50% of the primary text height (Fig. 27-63). The setting affects the *Symmetrical, Deviation,* and *Limits* tolerance methods. The value is stored in the *DIMTFAC* (dimension tolerance factor) variable. Note that this is the same variable that controls the *Fraction height scale* for architectural and fractional values.

Figure 27-63

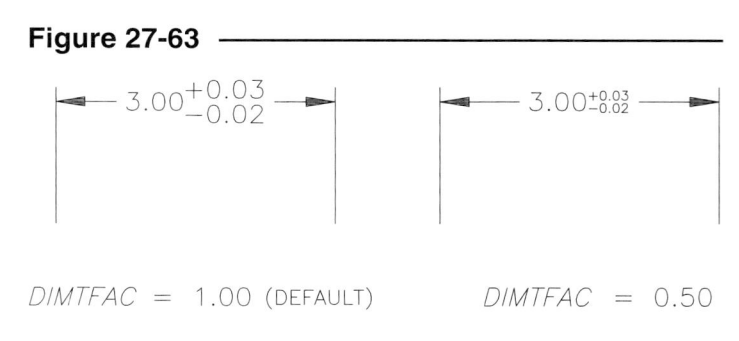

$3.00^{+0.03}_{-0.02}$
DIMTFAC = 1.00 (DEFAULT)

$3.00^{+0.03}_{-0.02}$
DIMTFAC = 0.50

Vertical position (DIMTOLJ)
For *Deviation, Symmetrical,* and *Limits* tolerance methods, you can control the placement of the tolerance values in relation to the primary units values (Fig. 27-64). The choices are *Top, Middle,* and *Bottom* and set the *DIMTOLJ* variable (dimension tolerance justification) to 2, 1, and 0, respectively. The *Top* option aligns the primary text and the top tolerance value, and the *Bottom* option aligns the primary text and the bottom tolerance value.

Figure 27-64

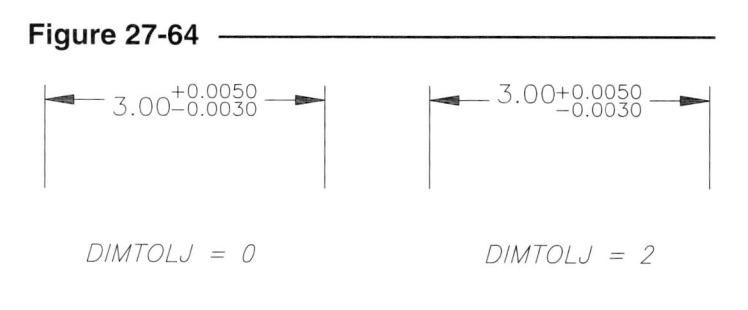

$3.00^{+0.0050}_{-0.0030}$
DIMTOLJ = 0

$3.00^{+0.0050}_{-0.0030}$
DIMTOLJ = 2

Zero Suppression **(DIMTZIN)** (*Tolerance Format* **Section**)
The *Zero Suppression* section controls how zeros are drawn in *Symmetrical*, *Deviation*, and *Limits* tolerance dimensions when they occur. A check in one of these boxes means that zeros are not drawn for that case. The *Zero Suppression* section here operates identically to the *Zero Suppression* section of the *Primary Units* tab, except that it is applied to tolerance dimensions (see *Zero Suppression*, *Primary Units* Tab for more information). The affected variable is *DIMTZIN* (dimension tolerance zero indicator).

<u>*Alternate Unit Tolerance* **Section** (*Tolerances* **Tab**)</u>

When you have specified that alternate units are to be drawn (in the *Alternate Units* tab) <u>and</u> you have turned on some form of tolerance *Method* (*Symmetrical*, *Limits*, etc.), the alternate units will automatically display as a tolerance along with the primary units. In other words, when alternate units are on, both primary and alternate units are drawn the same—with or without tolerances.

Precision **(DIMALTTD)**
This drop-down list sets the decimal precision for the alternate units tolerances (when alternate units and tolerances are on). The *DIMALTTD* variable (dimension alternate tolerance decimal) holds the setting.

Zero Suppression **(DIMALTTZ)**
When alternate units and tolerances are on, use these check boxes to determine how leading and trailing zeros are treated. The setting is stored in the *DIMALTTZ* variable (dimension alternate tolerance zeros). See *Zero Suppression* in the *Primary Units* Tab section for more information on zero suppression.

Changing Dimension Variables Using the Command Line Format

DIM...
(VARIABLE
NAME)

Pull-down Menu	COMMAND (TYPE)	ALIAS (TYPE)	Short-cut	Screen (side) Menu	Tablet Menu
Tools Inquiry > Set Variable	(VARIABLE NAME)	...	...	TOOLS 1 Setvar	U,10

As an alternative to setting dimension variables through the *Dimension Style Manager*, you can type the dimensioning variable name at the Command: prompt. There is a noticeable difference in the two methods—the dialog boxes use <u>different nomenclature</u> than the formal dimensioning variable names used in Command line format; that is, the dialog boxes use descriptive terms that, if selected, make the appropriate change to the dimensioning variable. The formal dimensioning variable names accessed by Command line format, however, all begin with the letters *DIM* and are accessible only by typing.

Another important but subtle difference in the two methods is the act of saving dimension variable settings to a dimension style. Remember that all drawn dimensions are part of a dimension style, whether it is STANDARD or some user-created style. When you change dimension variables by the Command line format, the changes become <u>overrides</u> until you use the *Save* option of the *-Dimstyle* command or the *Dimension Style Manager*. When a variable change is made, it becomes an <u>override that is applied to the current dimension style</u> and affects only the newly drawn dimensions. Variable changes must be *Saved* to become a permanent part of the style and to retroactively affect all dimensions created with that style.

In order to access and change a variable's setting by name, simply type the variable name at the Command: prompt. For dimension variables that require distances, you can enter the distance (in any format accepted by the current *Units* settings) or you can designate by (PICKing) two points.

For example, to change the value of the *DIMSCALE* to .5, this command syntax is used:

```
Command: dimscale
Enter new value for dimscale <1.0000>: .5
```

For a complete list of dimension variables, see AutoCAD 2002 *Help*, *Command Reference*, *System Variables*.

Associative Dimensions

In AutoCAD 2002, dimensions have full associativity. The *DIMASSOC* variable controls the associative feature. The *DIMASSOC* variable setting cannot be saved in a dimension style; therefore, one dimension style can contain associative and nonassociative dimensions, but not exploded dimensions (exploded dimensions can be created from a dimension style, but cannot be updated since they do not reference the style). The *Dimension Style Manager* does not provide access to the *DIMASSOC* variable. *DIMASSOC* must be changed by Command line format or by using the *Options* dialog box (see "Associative Dimensions" in Chapter 26).

MODIFYING EXISTING DIMENSIONS

Even with the best planning, a full understanding of dimension variables, and the correct use of *Dimension Styles*, it is probable that changes will have to be made to existing dimensions in the drawing due to design changes, new plot scales, or new industry/company standards. There are several ways that changes can be made to existing dimensions while retaining associativity and membership to a dimension style. The possible methods are discussed in this section.

Modifying a Dimension Style

Existing dimensions in a drawing can be modified by making one or more variable changes to the dimension style family or child, then *Saving* those changes to the dimension style. This process can be accomplished using either the *Dimension Style Manager* or Command line format. When you use the *Modify* option in the *Dimension Style Manager* or the *Save* option of the *-Dimstyle* command, the existing dimensions in the drawing automatically update to display the new variable settings.

Creating Dimension Style Overrides and Using *Update*

You can modify existing dimensions in a drawing without making permanent changes to the dimension style by creating a dimension style override, then using *Update* to apply the new setting to an existing dimension. This method is preferred if you wish to modify one or two dimensions without modifying all dimensions referencing (created in) the style.

To do this, use either the Command line format or *Dimension Style Manager* to set the new variable. In the *Dimension Style Manager*, select *Override* and make the desired variable settings in the *Override Current Style* dialog box. This creates an override to the current style. In Command line format, simply enter the formal dimension variable name and make the change to create an override to the current style, then use *Update* to apply the current style settings plus the override settings to existing dimensions that you PICK. The overrides remain in effect for the current style unless the variables are reset to the original values or until the overrides are cleared. You can clear overrides for a dimension style by making another dimension style current in the *Dimension Style Manager*.

UPDATE

Pull-down Menu	COMMAND (TYPE)	ALIAS (TYPE)	Short-cut	Screen (side) Menu	Tablet Menu
Dimension Update	DIM UPDATE	DIM UP	...	DIMNSION Update	Y,3

Update can be used to update existing dimensions in the drawing to the current settings. The current settings are determined by the current dimension style and any dimension variable overrides that are in effect (see previous explanation, "Creating Dimension Style Overrides and Using *Update*"). This is an excellent method of modifying one or more existing dimensions without making permanent changes to the dimension style that the selected dimensions reference. *Update* has the same effect as using *-Dimstyle, Apply*.

Update is actually a Release 12 command. In Release 12, dimensioning commands could only be entered at the Dim: prompt. For example, to create a linear dimension, you had to type *Dim* and press Enter, then type *Linear*. In Release 13, all dimensioning commands were upgraded to top-level commands with the *Dim-* prefix added, so you can type *Dimlinear* at the command prompt, for example. *Update*, although very useful, was never upgraded. In AutoCAD 2002, the command is given prominence by making available an *Update* button and an *Update* option in the *Dimension* pull-down menu. However, if you prefer to type, you must first type *Dim*, press Enter, and then enter *Update*. The command syntax is as follows:

    Command: Dim
    Dim: Update
    Select objects: PICK (select a dimension object to update)
    Select objects: Enter
    Dim: press Esc or type Exit
    Command:

For example, if you wanted to change the *DIMSCALE* of several existing dimensions to a value of 2, change the variable by typing *DIMSCALE*. Next use *Update* and select the desired dimension. That dimension is updated to the new setting. The command syntax is the following:

    Command: Dimscale
    New value for DIMSCALE <1.0000>: 2
    Command: Dim
    Dim: Update
    Select objects: PICK (select a dimension object to update)
    Select objects: PICK (select a dimension object to update)
    Select objects: Enter
    Dim: Exit
    Command:

Beware, *Update* creates an override to the current dimension style. You should reset the variable to its original value unless you want to keep the override for creating other new dimensions. You can clear overrides for a dimension style by making another dimension style current in the *Dimension Style Manager*.

DIMOVERRIDE

Pull-down Menu	COMMAND (TYPE)	ALIAS (TYPE)	Short-cut	Screen (side) Menu	Tablet Menu
Dimension Override	DIMOVERRIDE	DIMOVER or DOV	...	...	Y,4

Dimoverrride grants you a great deal of control to edit existing dimensions. The abilities enabled by *Dimoverride* are similar to the effect of using the *Properties* window.

Dimoverride enables you to make variable changes to dimension objects that exist in your drawing without creating an override to the dimension style that the dimension references (was created under). For example, using *Dimoverride*, you can make a variable change and select existing dimension objects to apply the change. The existing dimension does not lose its reference to the parent dimension style nor is the dimension style changed in any way. In effect, you can override the dimension styles for selected dimension objects. There are two steps: set the desired variable and select dimension objects to alter.

>Command: *dimoverride*
>Enter dimension variable name to override or [Clear overrides]: (**variable name**)
>Enter new value for dimension variable <current value>: (**value**)
>Enter dimension variable name to override: **Enter**
>Select objects: **PICK**
>Select objects: **PICK** or **Enter**
>Command:

The *Dimoverride* feature differs from creating dimension style overrides in two ways: (1) *Dimoverride* applies the changes to the selected dimension objects only, so the overrides are not appended to the parent dimension styles, only the objects; and (2) *Dimoverride* can be used once to change dimension objects referencing multiple dimension styles, whereas to make such changes to dimensions by creating dimension style overrides requires changing all the dimension styles, one at a time. The effect of using *Dimoverride* is essentially the same as using the *Properties* window.

Dimoverride is useful as a "backdoor" approach to dimensioning. Once dimensions have been created, you may want to make a few modifications, but you do not want the changes to affect dimension styles (resulting in an update to all existing dimensions that reference the dimension styles). *Dimoverride* offers that capability. You can even make one variable change to affect all dimensions globally without having to change multiple dimension styles. For example, you may be required to make a test plot of the drawing in a different scale than originally intended, necessitating a new global *Dimscale*. Use *Dimoverride* to make the change for the plot:

>Command: *dimoverride*
>Enter dimension variable name to override or [Clear overrides]: *dimscale*
>Enter new value for dimension variable <1.0000>: **.5**
>Enter dimension variable name to override: **Enter**
>Select objects: (window entire drawing) Other corner: 128 found
>Select objects: **Enter**
>Command:

This action results in having all the existing dimensions reflect the new *DIMSCALE*. No other dimension variables or any dimension styles are affected. Only the selected objects contain the overrides. *Dimoverride* does not append changes (overrides) to the original dimension styles, so no action must be taken to clear the overrides from the styles. After making the plot, *Dimoverride* can be used with the *Clear* option to change the dimensions (by object selection) back to their original appearance. The *Clear* option is used to clear overrides from dimension objects, not from dimension styles.

Clear

The *Clear* option removes the overrides from the selected dimension objects. It does not remove overrides from the current dimension style:

>Command: *dimoverride*
>Enter dimension variable name to override or [Clear overrides]: *c*
>Select objects: **PICK**
>Select objects: **Enter**
>Command:

The dimension then displays the variable settings as specified by the dimension style it references without any overrides (as if the dimension were originally created without the overrides). Using *Clear* does not remove any overrides that are appended to the dimension style so that if another dimension is drawn, the dimension style overrides apply.

MATCHPROP

Pull-down Menu	COMMAND (TYPE)	ALIAS (TYPE)	Short-cut	Screen (side) Menu	Tablet Menu
Modify Match Properties	MATCHPROP	MA	...	MODIFY1 Matchprp	Y,14 and Y,15

Matchprop can be used to "convert" an existing dimension to the style (including overrides) of another dimension in the drawing. For example, if you have two linear dimensions, one has *Oblique* arrows and *Romans* text font (*Dimension Style* = "Oblique") and one is a typical linear dimension (*Dimension Style* = "Standard") as in Fig. 27-65, "before." You can convert the typical dimension to the "Oblique" style by using *Matchprop*, selecting the "Oblique" dimension as the "source object" (to match), then selecting the typical dimension as the "destination object" (to convert). The typical dimension then references the "Oblique" dimension style and changes appearance accordingly (Fig. 27-65, "after"). Note that *Matchprop* does not alter the dimension text value.

Figure 27-65

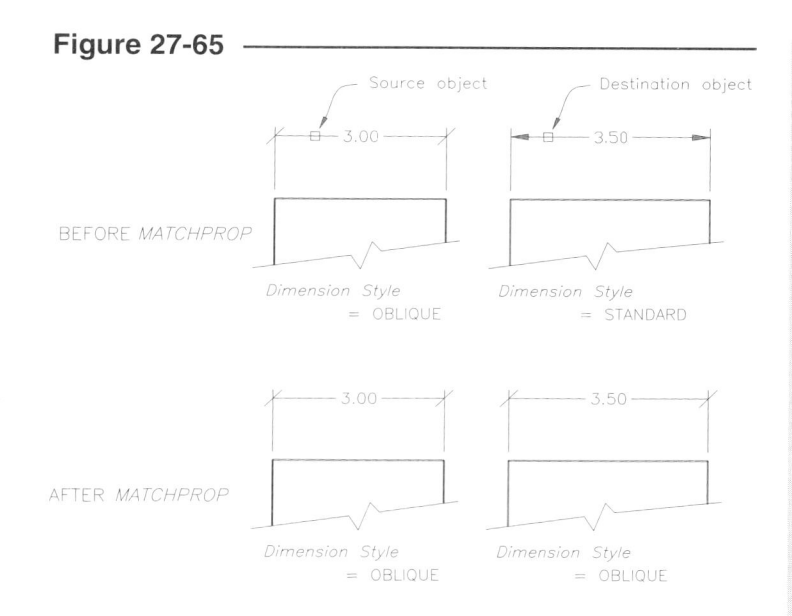

Using the *Settings* option of the *Matchprop* command, you can display the *Property Settings* dialog box (Fig. 27-66). The *Dimension* box under *Special Properties* must be checked to "convert" existing dimensions as illustrated above.

Figure 27-66

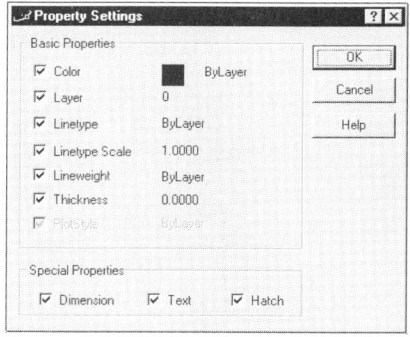

Using *Matchprop* is a fast and easy method for modifying dimensions from one style to another. However, this method is applicable only if you have existing dimensions in the drawing with the desired appearance that you want others to match.

PROPERTIES

Pull-down Menu	COMMAND (TYPE)	ALIAS (TYPE)	Short-cut	Screen (side) Menu	Tablet Menu
Modify Properties...	PROPERTIES	PROPS or CH	(Edit Mode) Properties or Crtl+1	MODIFY1 Property	Y,12 to Y,13

Remember that *Properties* can be used to edit existing dimensions. Using *Properties* and selecting one dimension displays the *Properties* window with all of the selected dimension's properties and dimension variables.

Here you can modify any aspect of one or more dimensions, including text, and access is given to the *Lines and Arrows, Text, Fit, Primary Units, Alternate Units,* and *Tolerances* categories. Any changes to dimension variables through *Properties* result in <u>overrides to the dimension object only</u> but do not affect the dimension style. *Properties* has essentially the same result as *Dimoverride* (see "*Dimoverride*"). *Properties* can also be used to modify the dimension text value. *Properties* of dimensions are also discussed in Chapter 26.

DIMTEDIT, DIMEDIT
Dimension text of existing dimensions can be modified using either *Dimtedit* or *Dimedit*. These commands allow you to change the text position with respect to the dimension line and the angle of the text. *Dimtedit* can be used with the *New* option to restate the original AutoCAD-measured text value if needed. See Chapter 26 for a full explanation of these commands.

Grips
Grips can be used to effectively alter dimension length, position, text location, and more. Keep in mind the possibilities of moving the dimension text with grips for dimensions with *DIMTMOVE* set to 2. See Chapter 26 for a full discussion of the possibilities of Grip editing dimensions.

GUIDELINES FOR DIMENSIONING IN AutoCAD

Listed in this section are some guidelines to use for dimensioning a drawing using dimensioning variables and dimension styles. Although there are other strategies for dimensioning, two strategies are offered here as a framework so you can develop an organized approach to dimensioning.

In almost every case, dimensioning is one of the last steps in creating a drawing, since the geometry must exist in order to dimension it. You may need to review the steps for drawing setup, including the concept of drawing scale factor (Chapter 12).

Strategy 1. Dimensioning a Single Drawing
This method assumes that the fundamental steps have been taken to set up the drawing and create the geometry. Assume this has been accomplished:

>   Drawing setup is completed: *Units, Limits, Snap, Grid, Ltscale, Layers,* border, and titleblock.
>   The drawing geometry (objects comprising the subject of the drawing) has been created.

Now you are ready to dimension the drawing subject (of the multiview drawing, pictorial drawing, floor plan, or whatever type of drawing).

1. Create a *Layer* (named DIM, or similar) for dimensioning if one has not already been created. Set *Continuous* linetype and appropriate color. Make it the *current* layer.

2. Set the *Overall Scale (DIMSCALE)* based on drawing *Limits* and expected plotting size.

   <u>For plotted dimension text of 3/16":</u>

   Multiply *Overall Scale (DIMSCALE)* times the drawing scale factor. The default *Overall Scale* is set to 1, which creates dimensioning text of approximately 3/16" (default *Text Height:* or *DIMTXT* =.18) when plotted full size. All other size-related dimensioning variables' defaults are set appropriately.

For plotted dimension text of 1/8":

Multiply *Overall Scale* times the drawing scale factor, times .7. Since the *Overall Scale* times the scale factor produces dimensioning text of .18, then .18 x .7 = .126 or approximately 1/8". (See "Optional Method for Fixed Dimension Text Height.") (*Overall Scale* [DIMSCALE] is one variable that usually remains constant throughout the drawing and therefore should generally have the same value for every dimension style created.)

3. Make the other dimension variable changes you expect you will need to create most dimensions in the drawing with the appearance you desire. When you have the basic dimension variables set as you want, *Save* this new dimension style family as TEMPLATE or COMPANY_STD (or other descriptive name) style. If you need to make special settings for types of dimensions (*Linear, Diameter,* etc.), create "children" at this stage and save the changes to the style. This style is the fundamental style to use for creating most dimensions and should be used as a template for creating other dimension styles. If you need to reset or list the original (default) settings, the STANDARD style can be restored.

4. Create all the relatively simple dimensions first. These are dimensions that are easy and fast and require no other dimension variable changes. Begin with linear dimensions; then progress to the other types of dimensions.

5. Create the special dimensions next. These are dimensions that require variable changes. Create appropriate dimension styles by changing the necessary variables, then save each set of variables (relating to a particular style of dimension) to an appropriate dimension style name. Specify dimension variables for the classification of dimension (children) in each dimension style when appropriate. The dimension styles can be created "on the fly" or as a group before dimensioning. Use TEMPLATE or COMPANY_STD as your base dimension style when appropriate.

6. When all of the dimensions are in place, make the final adjustments. Several methods can be used:

    A. If modifications need to be made familywide, change the appropriate variables and save the changes to the dimension style. This action automatically updates existing dimensions that reference that style.

    B. To modify the appearance of selected dimensions, you can create dimension style overrides, then use *Update* or *-Dimstyle, Apply* to update the selected dimensions to the new settings. Keep in mind that the dimension style overrides are still in effect and are applied to new dimensions. Clear the overrides by setting the original style current.

    C. Alternately, use *Properties* to change variable settings for selected dimensions. These changes are made as overrides to the selected object only and do not affect the dimension style. This action has the same result as using *Dimoverride* but for only one dimension.

    D. To change one dimension to adopt the appearance of another, use *Matchprop*. Select the dimension to match first, then the dimension(s) to convert. *-Dimstyle, Apply* can be used for this same purpose.

    E. Use *Dimoverride* to make changes to selected dimensions or to all dimensions globally by windowing the entire drawing. *Dimoverride* has the advantage of allowing you to assign the variables to change and selecting the objects to change all in one command. What is more, the changes are applied as overrides only to the selected dimensions but do not alter the original dimension styles in any way.

F. If you want to change only the dimension text value, use *Properties* or *Dimedit, New*. *Dimedit, New* can also be used to reapply the AutoCAD-measured value if text was previously changed. The location of the text can be changed with *Dimtedit* or Grips.

G. Grips can be used effectively to change the location of the dimension text or to move the dimension line closer or farther from the object. Other adjustments are possible, such as rotating a *Radius* dimension text around the arc.

Strategy 2. Creating Dimension Styles as Part of Template Drawings

1. Begin a *New* drawing or *Open* an existing *Template*. Assign a descriptive name.

2. Create a DIM *Layer* for dimensioning with *continuous* linetype and appropriate color and lineweight (if one has not already been created).

3. Set the *Overall Scale* accounting for the drawing *Limits* and expected plotting size. Use the guidelines given in Strategy 1, step 2. Make any other dimension variable changes needed for general dimensioning or required for industry or company standards.

4. Next, *Save* a dimension style named TEMPLATE or COMPANY_STD. This should be used as a template when you create most new dimension styles. The *Overall Scale* is already set appropriately for new dimension styles in the drawing.

5. Create the appropriate dimension styles for expected drawing geometry. Use TEMPLATE or COMPANY_STD dimension style as a base style when appropriate.

6. *Save* and *Exit* the newly created template drawing.

7. Use this template in the future for creating new drawings. Restore the desired dimension styles to create the appropriate dimensions.

Using a template drawing with prepared dimension styles is a preferred alternative to repeatedly creating the same dimension styles for each new drawing.

Optional Method for Fixed Dimension Text Height in Template Drawings

To summarize Strategy 1, step 2., the default *Overall Scale* (DIMSCALE =1) times the default *Text Height* (DIMTXT =.18) produces dimensioning text of approximately 3/16" when plotted to 1=1. To create 1/8" text, multiply *Overall Scale* times .7 (.18 x .7 = .126). As an alternative to this method, try the following.

For 1/8" dimensions, for example, multiply the initial values of the size-related variables by .7; namely

Text Height	(DIMTXT)	.18 x .7 = .126
Arrow Size	(DIMASZ)	.18 x .7 = .126
Extension Line Extension	(DIMEXE)	.18 x .7 = .126
Dimension Line Spacing	(DIMDLI)	.38 x .7 = .266
Text Gap	(DIMGAP)	.09 x .7 = .063

Save these settings in your template drawing(s). When you are ready to dimension, simply multiply *Overall Scale* (1) times the drawing scale factor.

DIMENSIONING IN PAPER SPACE LAYOUTS

Generally, dimensions are created in model space and are attached to model space geometry. In this way, you can make one or more layouts, each with one or more viewports that display the model space geometry, and the dimensions are visible by default in each of the layouts and viewports. Assuming the dimensions are created on a dimensioning layer or layers, you can control the display of the dimensions in each viewport using viewport-specific layer visibility controls in the *Layer Manager*. This strategy is used for almost all AutoCAD drawings previous to AutoCAD 2002, and will most likely be continued for most drawings in the future except for certain cases.

In AutoCAD 2002, it is possible to create dimensions in a paper space layout <u>associated</u> with (attached to) model space geometry inside a viewport. These new dimensions are fully associative and display the actual measurement value of the drawing objects <u>in model space units</u>.

In previous versions of AutoCAD, it was possible to create dimensions in paper space; however, those dimensions displayed the measurement value in paper space units by default. For example, consider the AutoCAD 2000 drawing of a plate shown with two viewports in Figure 27-67. All but one of the dimensions are in model space. The left viewport displays the entire plate including the model space dimensions at 1:1 scale, while the right viewport displays the plate at 2:1 scale. Notice that the dimension features (text, arrows, gaps, etc.) in the right viewport are shown twice as large since the view is scaled 2:1. Below the right viewport a dimension has been created in paper space ("1"). For that dimension, the dimension features (text, arrows, gaps, etc.) appear in the correct size, but the measured value is in paper space units—which is incorrect since the detail view is 2:1 (the measured value should be "2").

In contrast, a similar drawing is shown in Figure 27-68 created in AutoCAD 2002, but this drawing has some dimensions in model space (left viewport) and some dimensions in paper space (right viewport). Note that the dimensions for the small cutout have been omitted in model space (left viewport). All of the dimensions for the detail (right viewport) are created in paper space. Some of the dimensions for the detail are actually outside of the viewport border, but the two short horizontal dimensions (.5 and .25) appear to be in model space even though they were actually created in paper space.

Figure 27-67

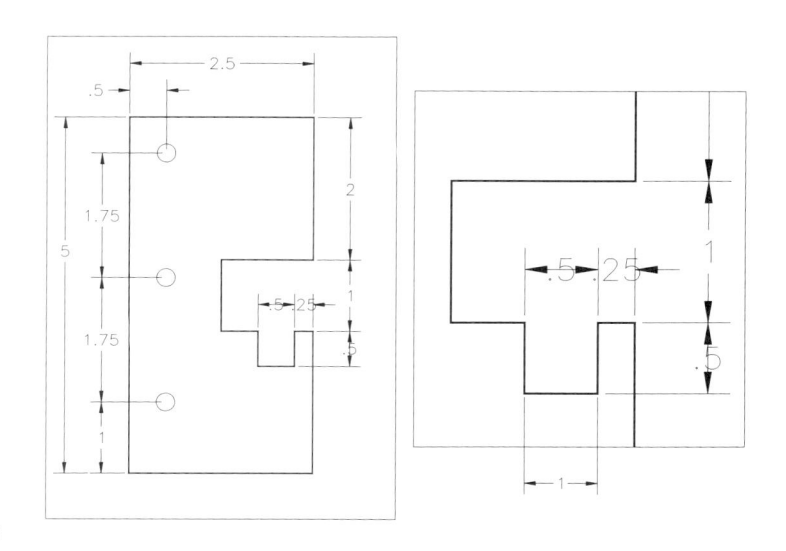

Figure 27-68

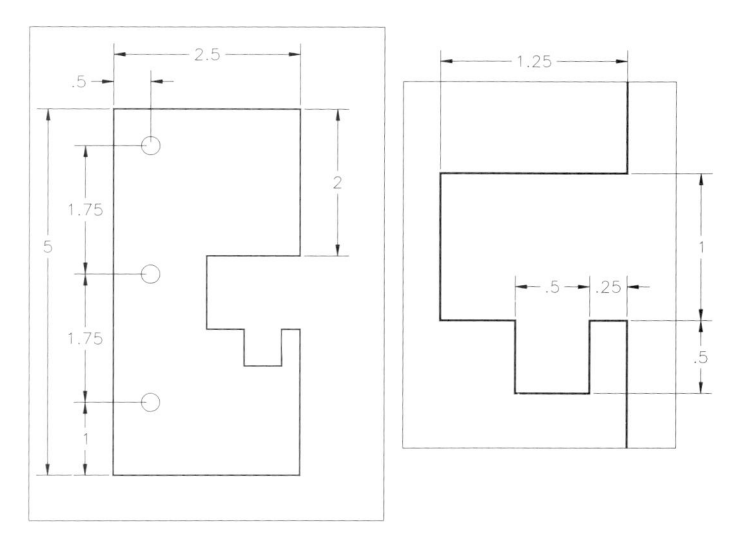

Note that in AutoCAD 2002, all dimensions associated to model space geometry, whether created in model space or paper space, display the correct value of the model feature they are measuring.

Figure 27-69 displays the completed drawing after adding some annotation and freezing the viewport border layer. Note that you cannot detect which dimensions are in model space and which are in paper space.

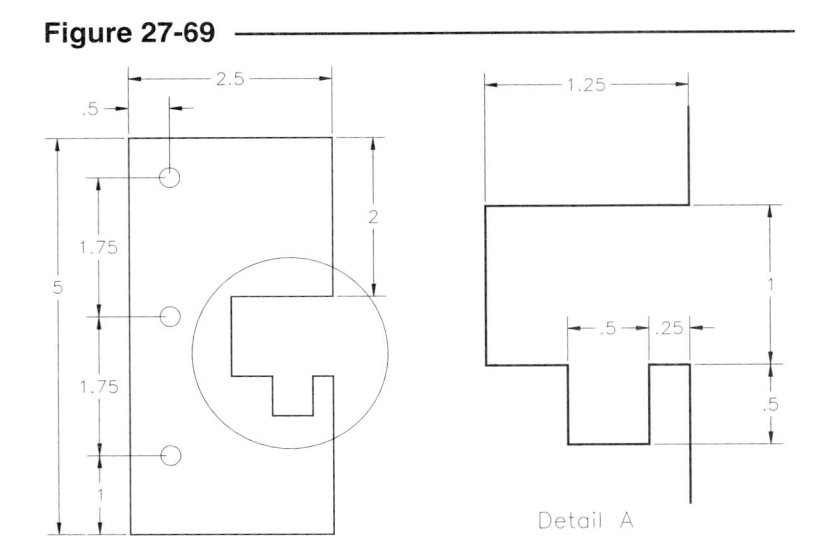

Figure 27-69

When to Dimension in Paper Space

Although it is still recommended to create dimensions in model space for most situations, there are some cases where creating dimensions in paper space could be used. Generally, whenever you want specific dimensions to appear for only one of several viewports, it may be useful to create those dimensions in paper space. Here are two specific examples.

> If you have created a 3D solid model, then generated 2D views (a multiview layout) from the model with each view in a separate paper space viewport, consider creating the dimensions for each view in paper space.

> If you want one or more detail views (small, enlarged sections of a larger drawing) each in a separate viewport or layout and at different scales than the full drawing display, consider creating dimensions for the detail views in paper space.

For the situations listed above, the advantages of creating dimensions in paper space attached to objects in model space (inside a viewport) are these:

> Setting *DIMSCALE* or *Overall Dimension Scale* (the size of the dimension text, arrowheads, etc.) for paper space dimensions is simplified since you are concerned only with paper space units and not with the viewport scale (the scale of the geometry that appears in the viewports). This is especially important when you have several detail views of a drawing and each detail view is at a different scale than the full drawing view. Paper space dimensions appear in one size, whereas model space dimensions appear in different sizes when the drawing is scaled differently in each viewport. This problem occurs in Figure 27-67 but is remedied in Figure 27-68.

> Creating dimensions in paper space ensures that those dimensions appear only for that viewport but not for other displays of the model geometry appearing in other viewports and layouts. Therefore, you do not have to use different layers for different sets of dimensions and the *Layer Manager* to set viewport-specific visibility (which dimension layers you want to appear in which viewports). This problem and the remedy are also shown in Figures 27-67 and 27-68, respectively.

Dimensioning a 3D model requires that dimensions for each side or view of the model appear on different planes (created by using construction planes called User Coordinate Systems or UCSs in AutoCAD). For a multiview-type setup of a 3D model, dimensioning in paper space prevents having to use different UCSs as well as different layers for the dimensions that should appear in each view (on each plane of the 3D model). Also, you do not have to control viewport-specific layer visibility to ensure that dimensions for the top view do not appear in the front view, and so on.

Although it appears that these advantages might outweigh the practice of dimensioning in model space, most drawings require that the dimensions appear in multiple viewports or layouts; therefore, dimensioning in model space is the more common practice. Dimensioning in paper space eliminates the possibility of showing the same dimensions in multiple viewports or layouts. In addition, remember that creating associative dimensions in paper space is available only since AutoCAD 2002, so the practice of dimensioning in paper space is not recommended if you collaborate with clients using previous releases.

CHAPTER EXERCISES

For each of the following exercises, use the existing drawings, as instructed. Create dimensions on the DIM (or other appropriate) layer. Follow the "Guidelines for Dimensioning in AutoCAD" given in the chapter, including setting an appropriate *Overall Scale* based on the drawing scale factor. Use dimension variables and create and use dimension styles when needed.

1. **Dimension One View**

    Open the **HAMMER** drawing you created in Chapter 13 Exercises and use the *Saveas* command to rename it **HAMMER-DIM**. Add all necessary dimensions as shown in Chapter 13 Exercises, Figure 13-28. Create a *New* dimension style and set the variables to generate dimensions as they appear in the figure (set *Precision* to **0.00**, *Text Style* to **Romans** font, and *Overall Scale* to **1** or **1.3**). Create the dimensions in model space. Your completed drawing should look like that in Figure 27-70.

    Figure 27-70

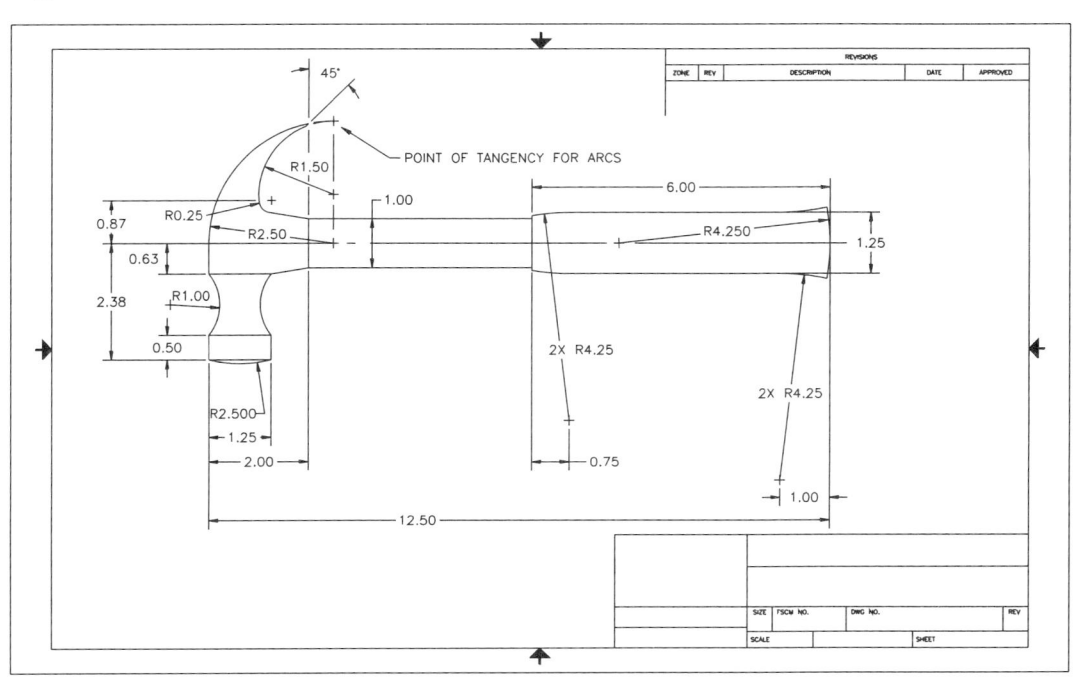

558 Dimension Styles and Variables

2. **Dimensioning a Multiview**

   Figure 27-71

   Open the **SADDLE** drawing that you created in Chapter 22 Exercises. Set the appropriate dimensional *Units* and *Precision*. Add the dimensions as shown. Because the illustration in Figure 27-71 is in isometric, placement of the dimensions can be improved for your multiview. Use optimum placement for the dimensions. *Save* the drawing as **SADDL-DM** and make a *Plot* to scale.

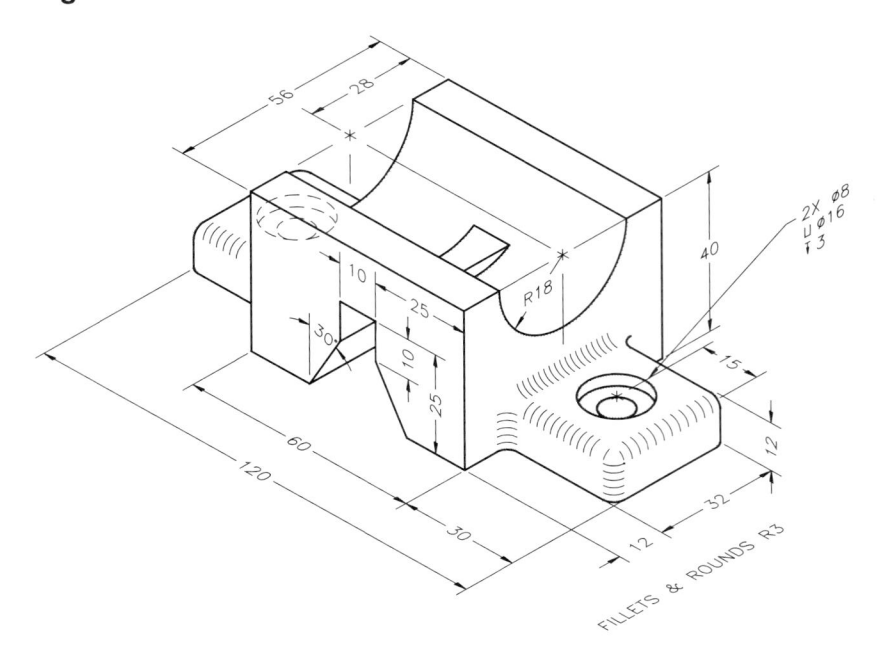

3. **Architectural Dimensioning**

   Open the **OFFICE** drawing that you completed in Chapter 20. Dimension the floor plan as shown in Figure 20-29. Add *Text* to name the rooms. *Save* the drawing as **OFF-DIM** and make a *Plot* to an accepted scale and sheet size based on your plotter capabilities.

4. **Dimensioning an Auxiliary**

   Figure 27-72

   Open the **ANGLBRAC** drawing that you created in Chapter 25. Dimension as shown in Figure 27-72, but convert the dimensions to *Decimal* with **Precision** of **.000**. Use the "Guidelines for Dimensioning." Dimension the slot width as a *Limit* dimension—**.6248/.6255**. *Save* the drawing as **ANGL-DIM** and *Plot* to an accepted scale.

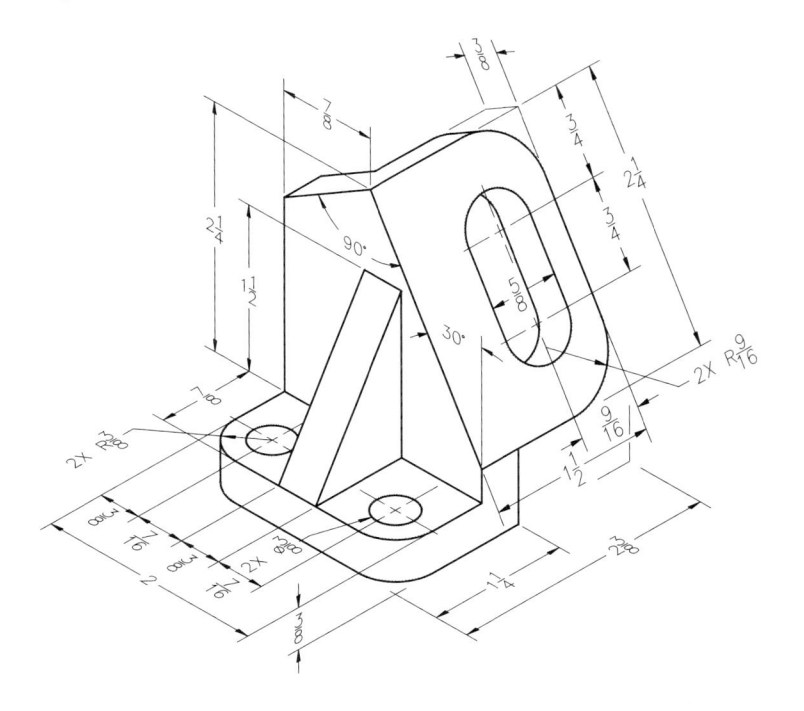

5. **Dimensioning a Multiview**

   Open the **ADJMOUNT** drawing that you completed in Chapter 22. Add the dimensions shown in Figure 27-73 but <u>convert the dimensions to *Decimal*</u> with **Precision** of .000 and check **Trailing** under **Zero Suppression**. Set appropriate dimensional *Units* and *Precision*. Calculate and set an appropriate *Overall Scale*. Use the "Guidelines for Dimensioning" given in this chapter. Save the drawing as **ADJM-DIM** and make a *Plot* to an accepted scale.

   Figure 27-73

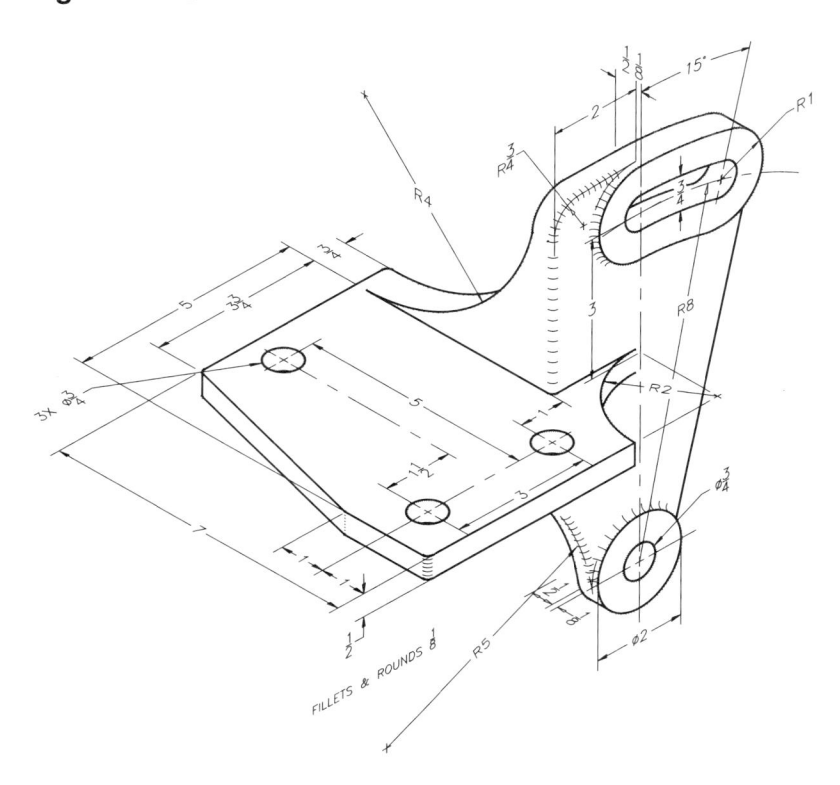

6. **Dimensioning in Model Space and Paper Space**

   A. *Open* the **EFF-APT2** drawing you last worked on in Chapter 18 Exercises. Draw a bed of your own design in the main room.

   B. A small bedroom and bathroom are to be added to the apartment, but a 10′ maximum interior span is allowed. One possible design is shown in Figure 27-74. Other designs you may prefer are possible (consider placing the plumbing walls back-to-back and adding a closet). Draw the new walls for the addition, but design the bedroom and bathroom door locations to your personal specifications. Use *Copy*, *Mirror*, and *Trim* where appropriate to complete the floor plan.

   C. Make a *New* dimension style and set variables to generate dimensions as they appear in Chapter 15 Exercises, Figure 15-32. Examine the dimensions closely to ensure your dimension style includes all the dimensioning features. Create the dimensions in model space on a layer named **DIM**. Add the necessary dimensions for the new bedroom. Use *Dtext* to add the "BEDROOM" label to the new room.

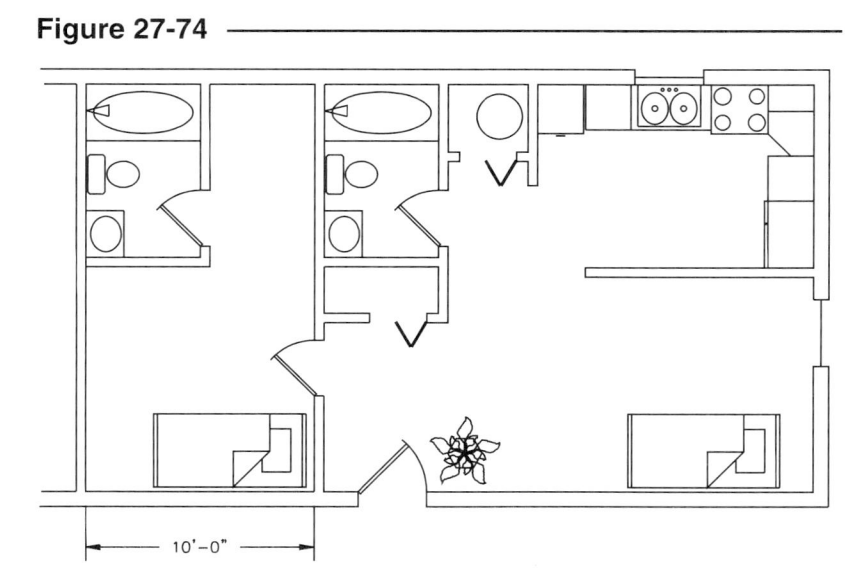

   Figure 27-74

560 Dimension Styles and Variables

D. Create a *New Layout*. Use *Pagesetup* for the layout and select an appropriate *Plot Device* that can use a "**B**" size sheet, such as an HP 7475 plotter. Set the layout to a "B" size sheet. Next, use DesignCenter to locate and insert the **ANSI-B title block** from the Template folder. Make a new layer named **VPORTS** and create one viewport as shown in Figure 27-75. Set the viewport scale to display the apartment floor plan at **1/4"=1'** scale.

Figure 27-75

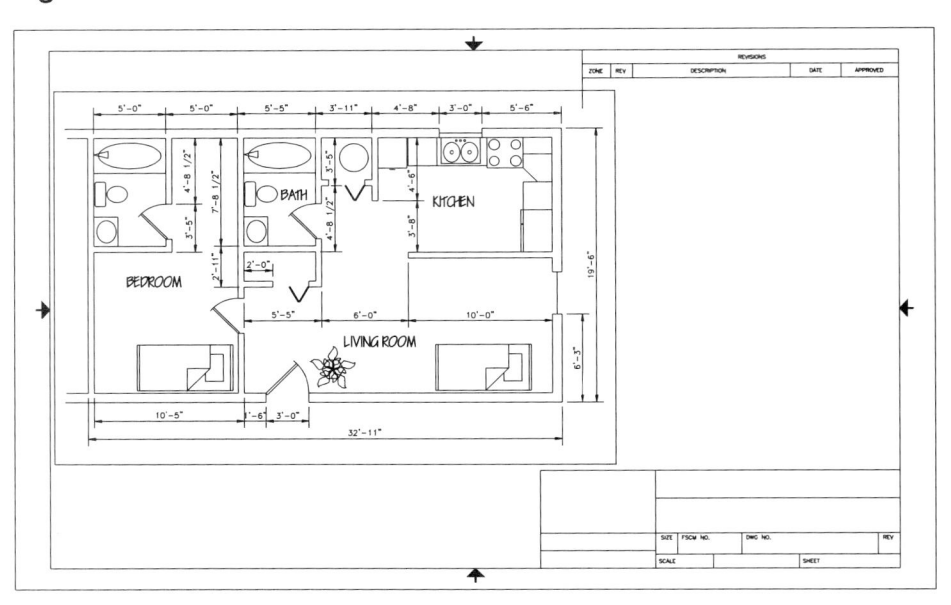

E. Create a smaller viewport on the right to display only one of the bathrooms at 1/2"=1' scale. Use the *Layer Manager* to turn off the display of the **DIM** layer for the new viewport. Create a new layer named **DIM-PS**. Create a *New* dimension style by copying the dimension style you created for model space, but change the *Overall Scale* to **1**. Create the dimensions for the plumbing wall (distance between centers of the fixtures) in paper space as shown in Figure 27-76. Label each viewport giving the scale. *Freeze* the **VPORTS** layer to achieve a drawing like that shown in Figure 27-76. Save the drawing as **APT-DIM**.

Figure 27-76

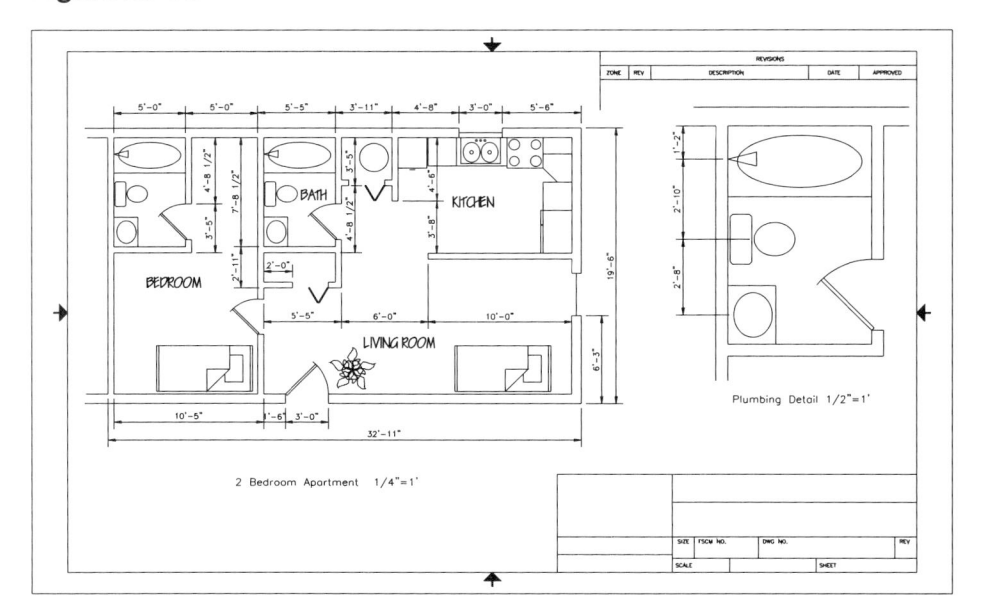

28

3D MODELING BASICS

Chapter Objectives

After completing this chapter you should:

1. know the characteristics of wireframe, surface, and solid models;

2. know the six formats for 3D coordinate entry;

3. understand the orientation of the World Coordinate System (WCS);

4. be able to use the right-hand rule for orientation of the X, Y, and Z axes and for determining positive and negative rotation about an axis;

5. be able to control the appearance and positioning of the Coordinate System icon with the *Ucsicon* command.

CONCEPTS

Three basic types of 3D (three-dimensional) models created by CAD systems are used to represent actual objects. They are:

1. Wireframe models
2. Surface models
3. Solid models

These three types of 3D models range from a simple description to a very complete description of an actual object. The different types of models require different construction techniques, although many concepts of 3D modeling are the same for creating any type of model on any type of CAD system.

Wireframe Models

"Wireframe" is a good descriptor of this type of modeling. A wireframe model of a cube is like a model constructed of 12 coat-hanger wires. Each wire represents an edge of the actual object. The surfaces of the object are not defined; only the boundaries of surfaces are represented by edges. No wires exist where edges do not exist. The model is see-through since it has no surfaces to obscure the back edges. A wireframe model has complete dimensional information but contains no volume. Examples of wireframe models are shown in Figures 28-1 and 28-2.

Wireframe models are relatively easy and quick to construct; however, they are not very useful for visualization purposes because of their "transparency." For example, does Figure 28-1 display the cube as if you are looking toward a top-front edge or looking toward a bottom-back edge? Wireframe models tend to have an optical illusion effect, allowing you to visualize the object from two opposite directions unless another visual clue such as perspective is given.

With AutoCAD a wireframe model is constructed by creating 2D objects in 3D space. The *Line, Circle, Arc,* and other 2D *Draw* and *Modify* commands are used to create the "wires," but 3D coordinates must be specified. The cube in Figure 28-1 was created with 12 *Line* segments. AutoCAD provides all the necessary tools to easily construct, edit, and view wireframe models.

Figure 28-1

Figure 28-2

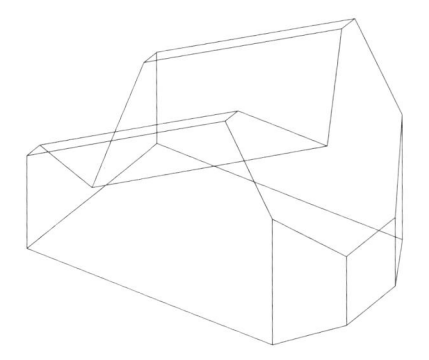

A wireframe model offers many advantages over a 2D engineering drawing. Wireframe models are useful in industry for providing computerized replicas of actual objects. A wireframe model is dimensionally complete and accurate for all three dimensions. Visualization of a wireframe is generally better than a 2D drawing because the model can be viewed from any position or a perspective can be easily attained. The 3D database can be used to test and analyze the object three dimensionally. The sheet metal industry, for example, uses wireframe models to calculate flat patterns complete with bending allowances. A wireframe model can also be used as a foundation for construction of a surface model. Because a wireframe describes edges but not surfaces, wireframe modeling is appropriate to describe objects with planar or single-curved surfaces, but not compound curved surfaces.

Surface Models

Surface models provide a better description of an object than a wireframe, principally because the surfaces as well as the edges are defined. A surface model of a cube is like a cardboard box—all the surfaces and edges are defined, but there is nothing inside. Therefore, a surface model has volume but no mass. A surface model provides an excellent visual representation of an actual 3D object because the front surfaces obscure the back surfaces and edges from view. Figure 28-3 shows a surface model of a cube, and Figure 28-4 displays a surface model of a somewhat more complex shape. Notice that a surface model leaves no question as to which side of the object you are viewing.

Figure 28-3

Surface models require a relatively tedious construction process. Each surface must be constructed individually. Each surface must be created in, or moved to, the correct orientation with respect to the other surfaces of the object. In AutoCAD, a surface is constructed by defining its edges. Often, wireframe models are used as a framework to build and attach surfaces. The complexity of the construction process of a surface is related to the number and shapes of its edges. AutoCAD is not a complete surface modeler. The tools provided allow construction of simple planar and single curved surfaces, but there are few capabilities for construction of double-curved or other complex surfaces. No NURBS (Non-Uniform Rational B-Splines) surfacing capabilities exist in AutoCAD, although these capabilities are available in other Autodesk products such as Mechanical Desktop® and Inventor®.

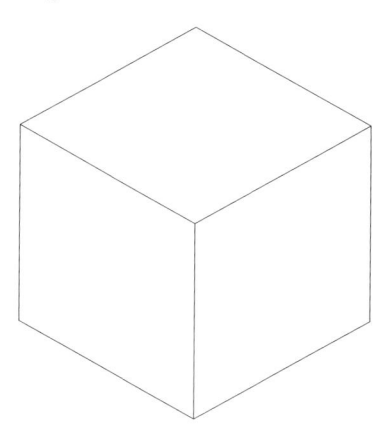

Figure 28-4

Most CAD systems, including AutoCAD, can display surface and solid models in wireframe, hidden, and shaded representation. Figure 28-4 displays an object in "hidden" representation. Figure 28-5 shows the same object in "wireframe" representation. Surface and solid modeling systems can generally display objects in any of the three modes (wireframe, shaded, and hidden) at any time. Although hidden and shaded modes enhance your ability to visualize the object, wireframe representation is often used during the construction process so all edges can be seen and selected, if needed. (See Chapter 29, 3D Display and Viewing, for more information.)

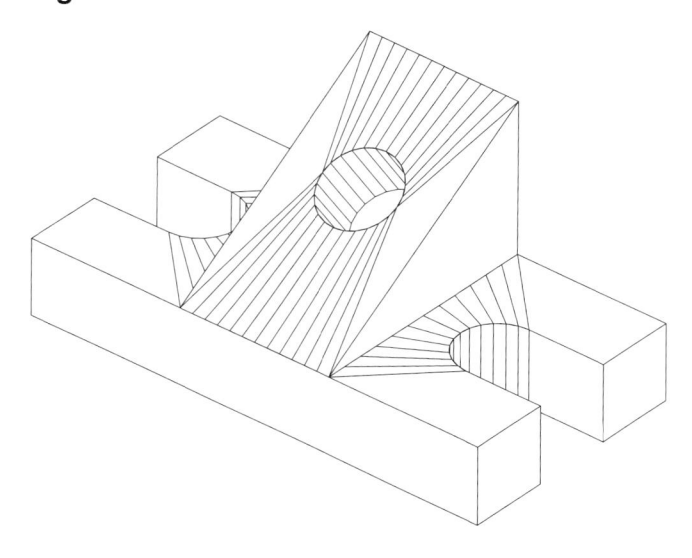

Figure 28-5

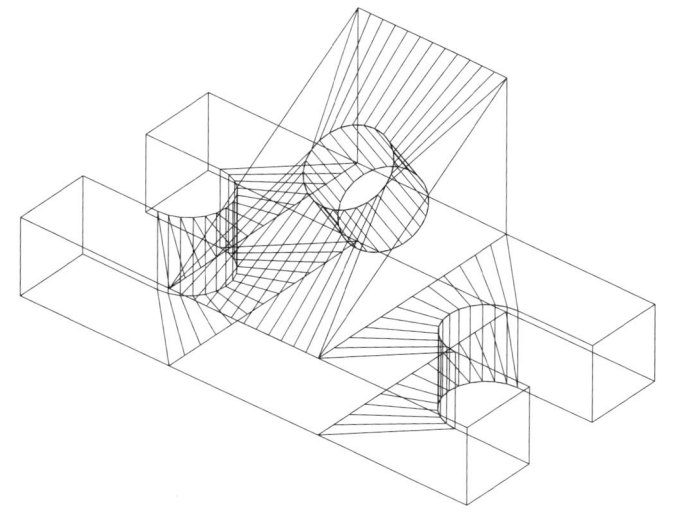

Solid Models

Solid modeling is the most complete and descriptive type of 3D modeling. A solid model is a complete computerized replica of the actual object. A solid model contains the complete surface and edge definition, as well as description of the interior features of the object. If a solid model is cut in half (sectioned), the interior features become visible. Since a solid model is "solid," it can be assigned material characteristics and is considered to have mass. Because solid models have volume and mass, most solid modeling systems include capabilities to automatically calculate volumetric and mass properties.

Figure 28-6

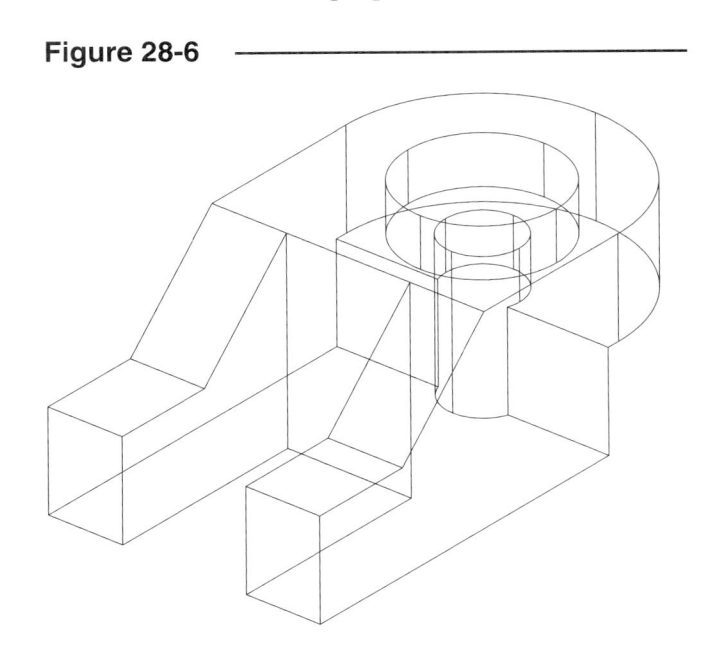

Solid model construction techniques are generally much simpler (and much more fun) than those of surface models. AutoCAD's solid modeler, called ACIS, is a hybrid modeler. That is, ACIS is a combination CSG (Constructive Solid Geometry) and B-Rep (Boundary Representation) modeler. CSG is characterized by its simple and straightforward construction techniques of combining primitive shapes (boxes, cylinders, wedges, etc.) utilizing Boolean operations (*Union*, *Subtract*, and *Intersect*, etc.). Boundary Representation modeling defines a model in terms of its edges and surfaces (boundaries) and determines the solid model based on which side of the surfaces the model lies. The user interface and construction techniques used in ACIS (primitive shapes combined by Boolean operations) are CSG-based, whereas the B-Rep capabilities are invoked automatically to display models in mesh representation and are transparent to the user. Figure 28-6 displays a solid model constructed of simple primitive shapes combined by Boolean operations.

Figure 28-7

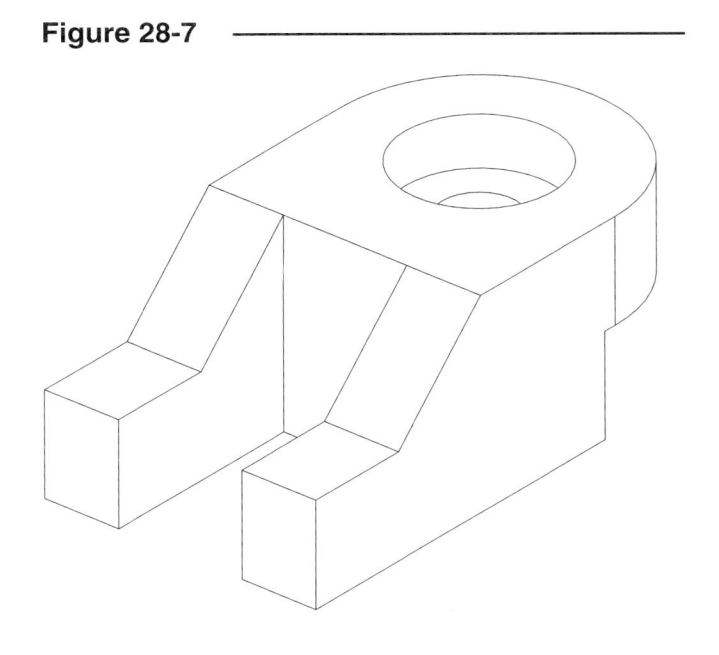

The CSG modeling techniques offer you the advantage of complete and relatively simple editing. CSG construction typically begins by specifying dimensions for simple primitive shapes such as boxes or cylinders, then combines the primitives using Boolean operations to create a "composite" solid. Other primitives and/or composite solids can be combined by the same process. Several repetitions of this process can be continued until the desired solid model is finally achieved. CSG construction techniques are discussed in detail in Chapter 31.

Solid models, like surface models, are capable of wireframe, shaded, or hidden display. Generally, wireframe display is used during construction (see Figure 28-6), and hidden or shaded representation is used to display the finished model (Fig. 28-7).

3D COORDINATE ENTRY

Figure 28-8

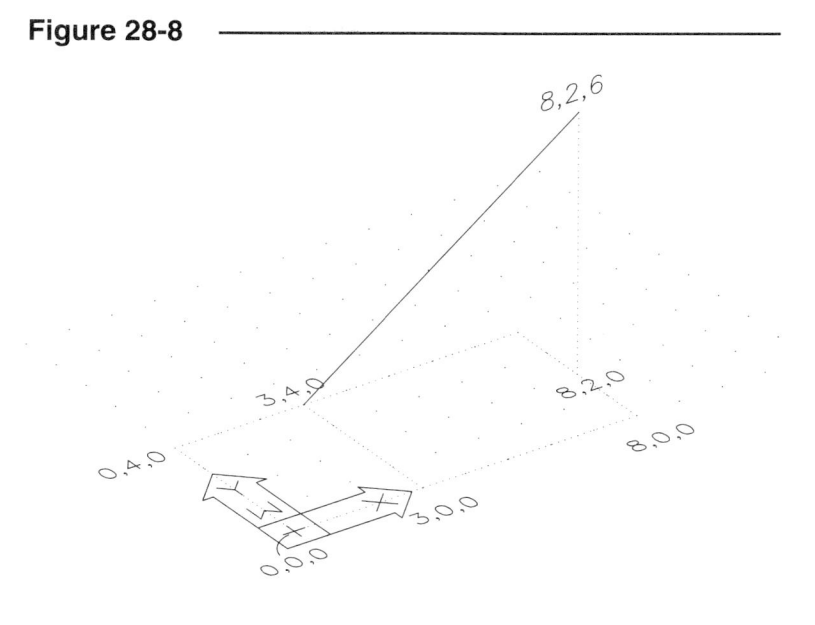

When creating a model in three-dimensional drawing space, the concept of the X and Y coordinate system, which is used for two-dimensional drawing, must be expanded to include the third dimension, Z, which is measured from the origin in a direction perpendicular to the plane defined by X and Y. Remember that two-dimensional CAD systems use X and Y coordinate values to define and store the location of drawing elements such as *Lines* and *Circles*. Likewise, a three-dimensional CAD system keeps a database of X, Y, and Z coordinate values to define locations and sizes of two- and three-dimensional elements. For example, a *Line* is a two-dimensional object, yet the location of its endpoints in three-dimensional space must be specified and stored in the database using X, Y, and Z coordinates (Fig. 28-8). The X, Y, and Z coordinates are always defined in that order, delineated by commas. The AutoCAD Coordinate Display (*Coords*) displays X, Y, and Z values.

The icon that appears in the lower-left corner of the AutoCAD Drawing Editor is the Coordinate System icon (sometimes called the UCS icon) (Fig. 28-9). This UCS icon displays the directions for the X, Y, and the Z axes and can be made to locate itself at the origin, 0,0. The X coordinate values increase going to the right along the X axis, and the Y values increase going upward along the Y axis. This is the UCS default icon that appears when AutoCAD is in the *2D wireframe* option of *Shademode*.

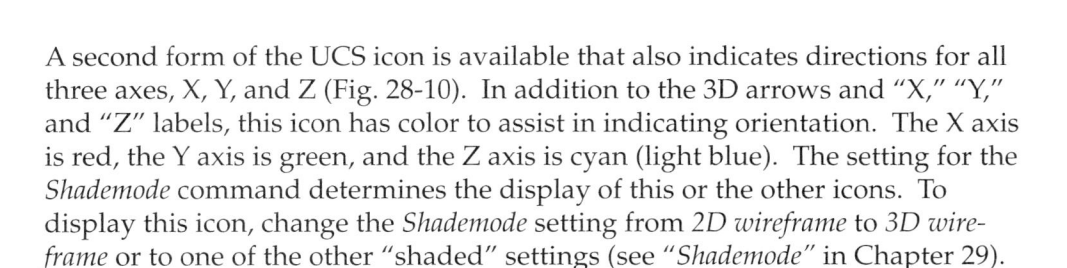

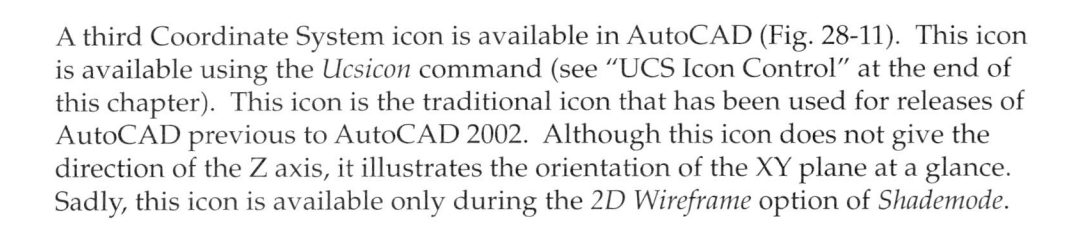

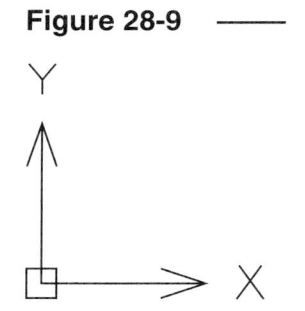

Figure 28-9

A second form of the UCS icon is available that also indicates directions for all three axes, X, Y, and Z (Fig. 28-10). In addition to the 3D arrows and "X," "Y," and "Z" labels, this icon has color to assist in indicating orientation. The X axis is red, the Y axis is green, and the Z axis is cyan (light blue). The setting for the *Shademode* command determines the display of this or the other icons. To display this icon, change the *Shademode* setting from *2D wireframe* to *3D wireframe* or to one of the other "shaded" settings (see "*Shademode*" in Chapter 29).

Figure 28-10

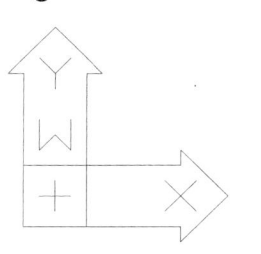

A third Coordinate System icon is available in AutoCAD (Fig. 28-11). This icon is available using the *Ucsicon* command (see "UCS Icon Control" at the end of this chapter). This icon is the traditional icon that has been used for releases of AutoCAD previous to AutoCAD 2002. Although this icon does not give the direction of the Z axis, it illustrates the orientation of the XY plane at a glance. Sadly, this icon is available only during the *2D Wireframe* option of *Shademode*.

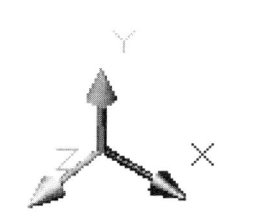

Figure 28-11

You can use any one of the three icons when you construct 3D models. For most of the figures in this text, the traditional "2D" UCS icon is displayed. It is the opinion of this author that, although the Z axis is not shown, the "2D" icon gives the user immediate orientation of the XY plane and, therefore, orientation in 3D space. For the exact reason that the newer, 3-pole icons display all three axes, it is more confusing to distinguish which pole is which and, therefore, takes longer to distinguish 3D orientation—the labels must be read or the colors memorized. Displaying only two of the three axes (as with the "2D" icon) leaves no doubt as to the X and Y directions and the XY plane orientation.

3D Coordinate Entry Formats

Because construction in three dimensions requires the definition of X, Y, and Z values, the methods of coordinate entry used for 2D construction must be expanded to include the Z value. The five methods of command entry used for 2D construction are valid for 3D coordinates with the addition of a Z value specification. Relative polar coordinate specification (@dist<angle) is expanded to form two other coordinate entry methods available explicitly for 3D coordinate entry. The six methods of coordinate entry for 3D construction follow:

1. **Interactive coordinates** PICK Use the cursor to select points on the screen. *OSNAP*, Object Snap Tracking, or point filters must be used to select a point in 3D space; otherwise, points selected are on the XY plane.

2. **Absolute coordinates** X,Y,Z Enter explicit X, Y, and Z values relative to point 0,0.

3. **Relative rectangular coordinates** @X,Y,Z Enter explicit X, Y, and Z values relative to the last point.

4. **Cylindrical coordinates (relative)** @dist<angle,Z Enter a distance value, an angle in the XY plane value, and a Z value, all relative to the last point.

5. **Spherical coordinates (relative)** @dist<angle<angle Enter a distance value, an angle in the XY plane value, and an angle from the XY plane value, all relative to the last point.

6. **Direct distance entry** dist,direction Enter (type) a value, and move the cursor in the desired direction. To draw in 3D effectively, *ORTHO* must be *On*.

Cylindrical and spherical coordinates can be given without the @ symbol, in which case the location specified is relative to point 0,0,0 (the origin). This method is useful if you are creating geometry centered around the origin. Otherwise, the @ symbol is used to establish points in space relative to the last point.

Examples of each of the six 3D coordinate entry methods are illustrated in the following section. In the illustrations, the orientation of the observer has been changed from the default plan view in order to enable the visibility of the three dimensions.

Interactive Coordinate Specification
Figure 28-12 illustrates using the interactive method to PICK a location in 3D space. *OSNAP* must be used in order to PICK in 3D space. Any point PICKed with the input device without *OSNAP* will result in a location on the XY plane. In this example, the *Endpoint OSNAP* mode is used to establish the "Specify next point:" of a second *Line* by snapping to the end of an existing vertical *Line* at 8,2,6:

Figure 28-12

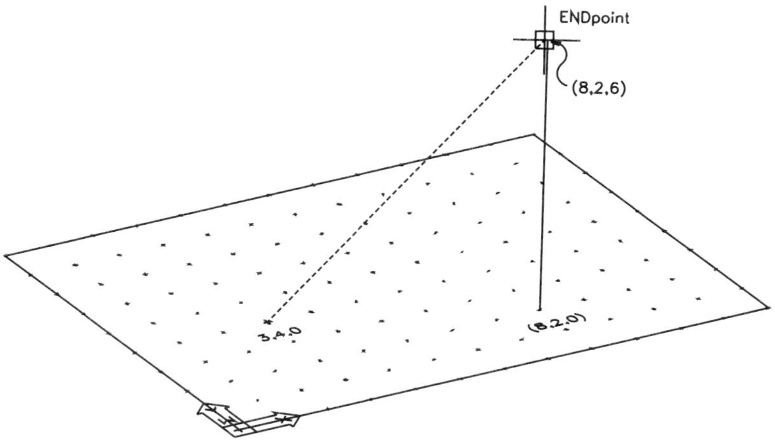

Command: *line*
Specify first point: 3,4,0
Specify next point or [Undo]: *endpoint* of PICK

Absolute Coordinates
Figure 28-13 illustrates the absolute coordinate entry to draw the *Line*. The endpoints of the *Line* are given as explicit X,Y,Z coordinates:

Figure 28-13

Command: *line*
Specify first point: 3,4,0
Specify next point or [Undo]: 8,2,6

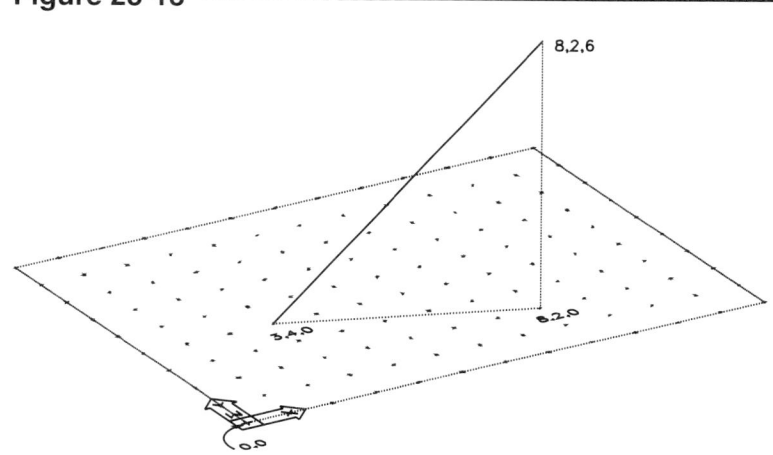

Relative Rectangular Coordinates
Relative rectangular coordinate entry is displayed in Figure 28-14. The "Specify first point:" of the *Line* is given in absolute coordinates, and the "Specify next point:" end of the *Line* is given as X,Y,Z values relative to the last point:

Figure 28-14

Command: *line*
Specify first point: 3,4,0
Specify next point or [Undo]: @5,-2,6

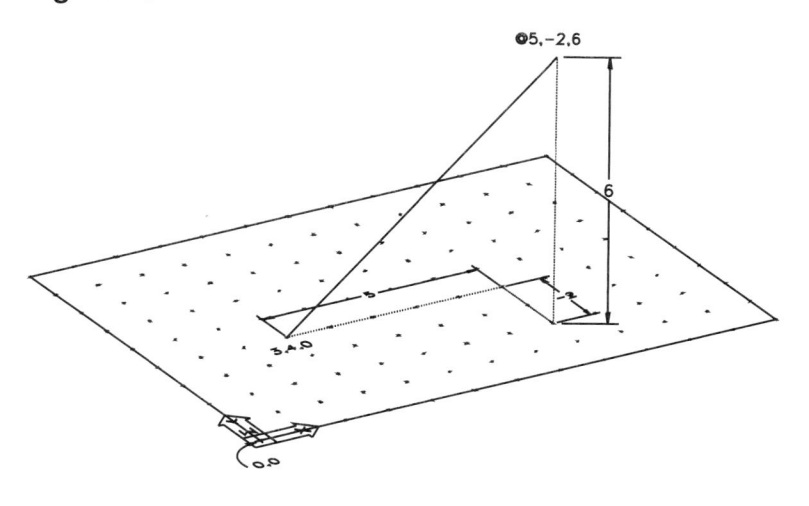

Cylindrical Coordinates (Relative)

Cylindrical and spherical coordinates are an extension of polar coordinates with a provision for the third dimension. Relative cylindrical coordinates give the distance in the XY plane, angle in the XY plane, and Z dimension and can be relative to the last point by prefixing the @ symbol. The *Line* in Figure 28-15 is drawn with absolute and relative cylindrical coordinates. (The *Line* established is approximately the same *Line* as in the previous figures.)

    Command: *line*
    Specify first point: 3,4,0
    Specify next point or [Undo]:
    @5<-22,6

Figure 28-15

Spherical Coordinates (Relative)

Spherical coordinates are also an extension of polar coordinates with a provision for specifying the third dimension in angular format. Spherical coordinates specify a distance, an angle in the XY plane, and an angle from the XY plane and can be relative to the last point by prefixing the @ symbol. The distance specified is a 3D distance, not a distance in the XY plane. Figure 28-16 illustrates the creation of approximately the same line as in the previous figures using absolute and relative spherical coordinates.

    Command: *line*
    Specify first point: 3,4,0
    Specify next point or [Undo]:
    @8<-22<48

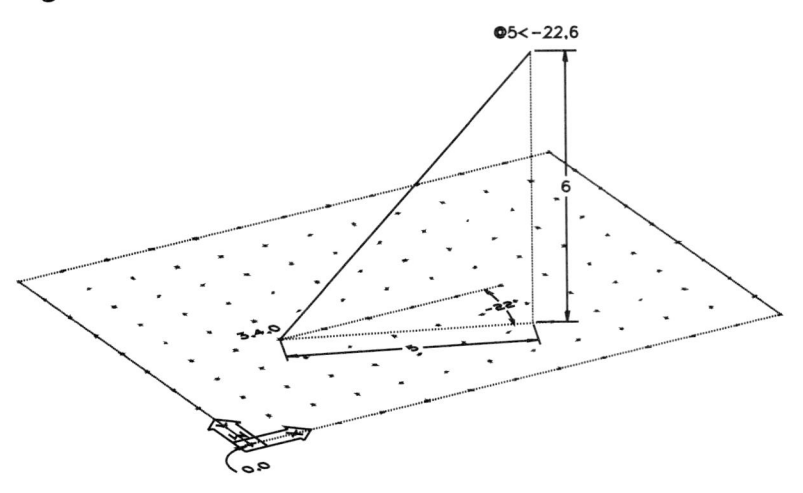

Figure 28-16

Direct Distance Entry

Direct distance entry operates as it does in 2D, except ORTHO must be On in order to draw effectively by this method in 3D space. When ORTHO is *On*, objects drawn in 3D space are limited to a plane parallel with the current XY plane. For example, to draw a square in 3D space with sides of 3 units (Fig. 28-17), begin by using *Line* and specifying the "Specify first point:" at the

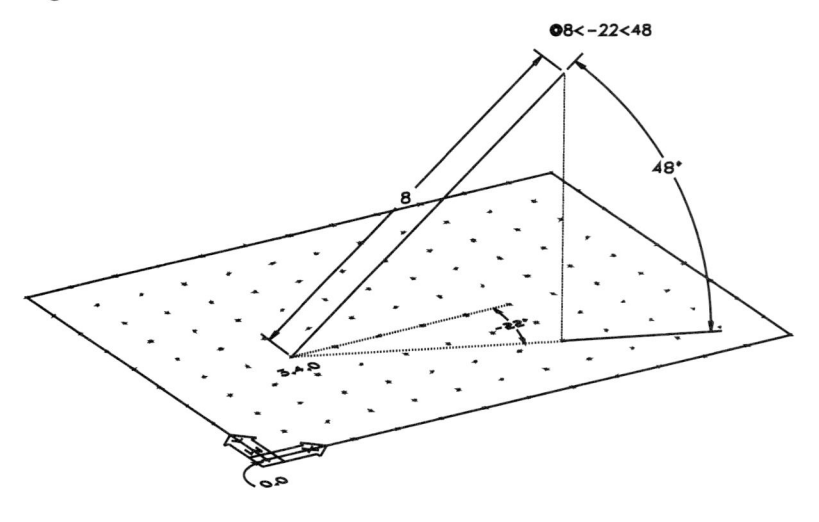

Figure 28-17

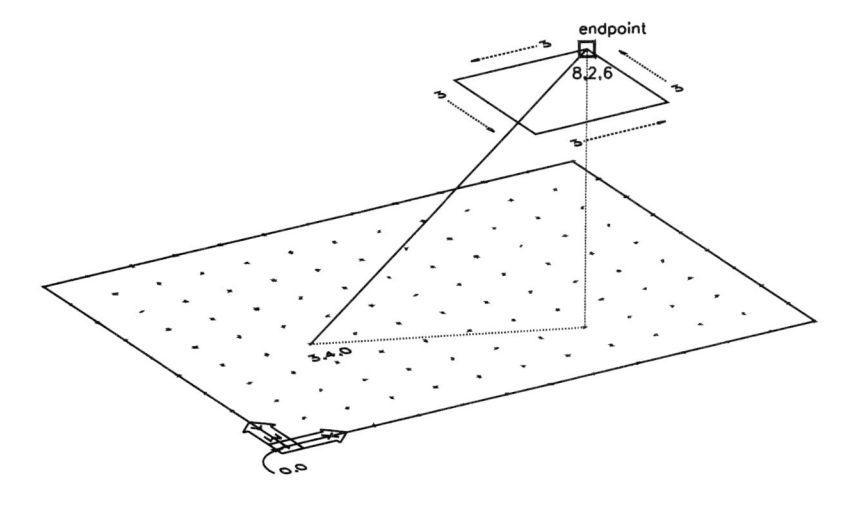

Endpoint of the existing diagonal *Line* shown (point 8,2,6). To draw the first 3 unit segment, turn on ORTHO, move the cursor in the desired direction, and enter "3." The ORTHO feature "locks" the *Line* to a plane parallel to the current XY plane. Next, move the cursor in the desired (90 degree) direction and enter "3," and so on. The command syntax follows:

Command: *line*
Specify first point: *endpoint* of PICK
Specify next point or [Undo]: <Ortho on> 3
Specify next point or [Undo]: 3
Specify next point or [Undo]: 3
Specify next point or [Undo]: 3
Specify next point or [Undo]: **Enter**
Command:

Point Filters

Point filters are used to filter X and/or Y and/or Z coordinate values from a location PICKed with the pointing device. Point filtering makes it possible to build an X,Y,Z coordinate specification from a combination of point(s) selected on the screen and point(s) entered at the keyboard. A .XY (read "point XY") filter would extract, or filter, the X and Y coordinate value from the location PICKed and then prompt you to enter a Z value. Valid point filters are listed below.

.X   Filters (finds) the X component of the location PICKed with the pointing device.
.Y   Filters the Y component of the location PICKed.
.Z   Filters the Z component of the location PICKed.
.XY   Filters the X and Y components of the location PICKed.
.XZ   Filters the X and Z components of the location PICKed.
.YZ   Filters the Y and Z components of the location PICKed.

The .XY filter is the most commonly used point filter for 3D construction and editing. Because 3D construction often begins on the XY plane of the current coordinate system, elements in Z space are easily constructed by selecting existing points on the XY plane using .XY filters and then entering the Z component of the desired 3D coordinate specification by keyboard. For example, in order to draw a line in Z space two units above an existing line on the XY plane, the .XY filter can be used in combination with *Endpoint* OSNAP to supply the XY component for the new line. See Figure 28-18 for an illustration of the following command sequence:

Figure 28-18

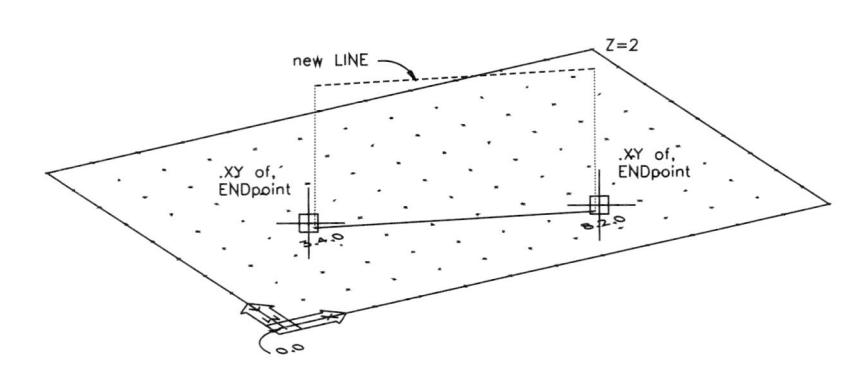

Command: *line*
Specify first point: **.XY** of
endpoint of PICK (Select one *Endpoint* of the existing line on the XY plane.)
(need Z) **2**
Specify next point or [Undo]: **.XY** of
endpoint of PICK (Select the other *Endpoint* of the existing line.)
(need Z) **2**

Typing **.XY** and pressing **Enter** cause AutoCAD to respond with "of" similar to the way "of" appears after typing an *OSNAP* mode. When a location is specified using an .XY filter, AutoCAD responds with "(need Z)" and, likewise, when other filters are used, AutoCAD prompts for the missing component(s).

Object Snap Tracking

Figure 28-19

Object Snap Tracking functions in 3D similar to the way that *OSNAP* and Direct Distance entry operate—that is, when a point is "acquired" by one of the *OSNAP* modes, tracking occurs in the plane of the acquired point. To be more specific, when you acquire a point in 3D space (not on the XY plane), tracking vectors appear originating from the acquired object in a plane parallel to the current XY plane.

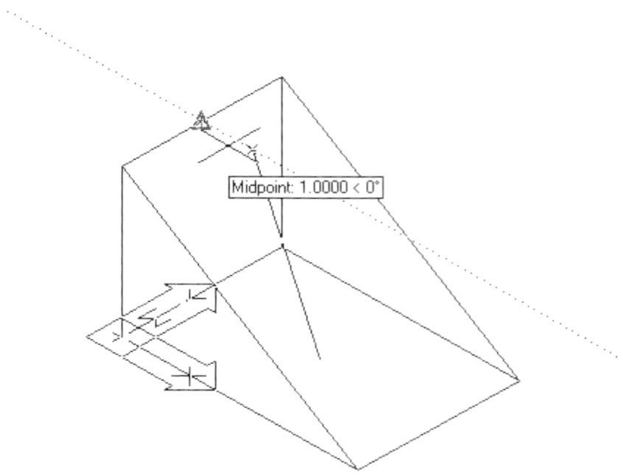

For example, assume a wedge was created with its base on the XY plane. You can track on a plane in 3D space parallel to the current XY plane that passes through the acquired point (Fig. 28-19). In other words, begin the "first point" of a line at a point on the XY plane (at the center of the base, in this case). For the "next point," acquire a point in 3D space with Object Snap Tracking. Note that tracking vectors are generated from that acquired point into space on an imaginary plane that is parallel with the current XY plane (and parallel with the X axis in this case).

COORDINATE SYSTEMS

In AutoCAD two kinds of coordinate systems can exist, the World Coordinate System (WCS) and one or more User Coordinate Systems (UCS). The World Coordinate System always exists in any drawing and cannot be deleted. The user can also create and save multiple User Coordinate Systems to make construction of a particular 3D geometry easier. Only one coordinate system can be active at any one time in any one view or viewport, either the WCS or one of the user-created UCSs. You can have more than one UCS active at any one time if you have several viewports active (model space or layout viewports).

The World Coordinate System (WCS) and WCS Icon

The World Coordinate System (WCS) is the default coordinate system in AutoCAD for defining the position of drawing objects in 2D or 3D space. The WCS is always available and cannot be erased or removed but is deactivated temporarily when utilizing another coordinate system created by the user (UCS). The icon that appears (by default) at the lower-left corner of the Drawing Editor (Fig. 28-9) indicates the orientation of the WCS. The icon, whose appearance (*ON, OFF*) is controlled by the *Ucsicon* command, appears for the WCS and for any UCS.

Remember that three possible UCS icons are available and determined by both your choice in the *Ucsicon* command *Properties* dialog box and your current *Shademode* setting. The "3D" (3-pole) colored icon (see Figure 28-10) that appears with all *Shademode* options except *2D Wireframe* does not indicate if the World Coordinate System is current. Both of the other icons (see Figures 28-9 and 28-11) do indicate if the WCS is active, but these two icons are displayed only during the *2D wireframe* option of *Shademode*. The "3D" version of the icon (see Figure 28-9) displays a small square on the XY plane if the WCS is current, but this square disappears when another UCS is current. The "2D" version of the icon (see Figure 28-11) displays the letter "W" when the WCS is current. This feature is another reason why the older "2D" icon is better for determining 3D orientation. For this reason and other reasons (see discussion at Figure 28-11), the traditional "2D" icon is displayed in most of the figures in this text.

Figure 28-20

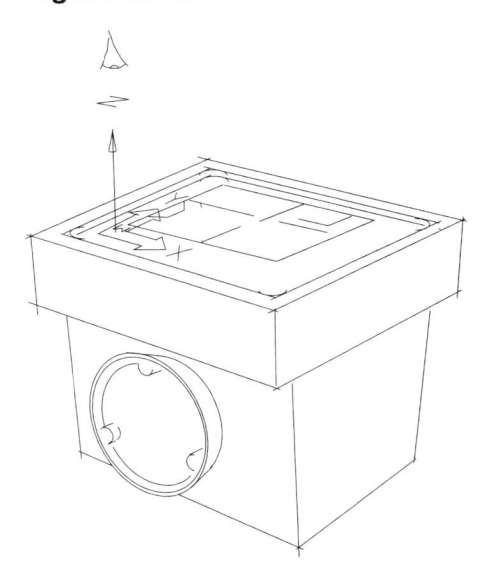

The orientation of the WCS with respect to Earth may be different among CAD systems. In AutoCAD the WCS has an architectural orientation such that the XY plane is a horizontal plane with respect to Earth, making Z the height dimension. A 2D drawing (X and Y coordinates only) is thought of as being viewed from above, sometimes called a plan view. Therefore, in a 3D AutoCAD drawing, X is the width dimension, Y is the depth dimension, and Z is height. This default orientation is like viewing a floor plan—from above (Fig. 28-20).

Some CAD systems that have a mechanical engineering orientation align their World Coordinate Systems such that the XY plane is a vertical plane intended for drawing a front view. In other words, some mechanical engineering CAD systems define X as the width dimension, Y as height, and Z as depth.

User Coordinate Systems (UCS) and Icons

There are no User Coordinate Systems that exist as part of the AutoCAD default template drawing (ACAD.DWT) as it comes "out of the box." UCSs are created to suit the 3D model when and where they are needed.

Figure 28-21

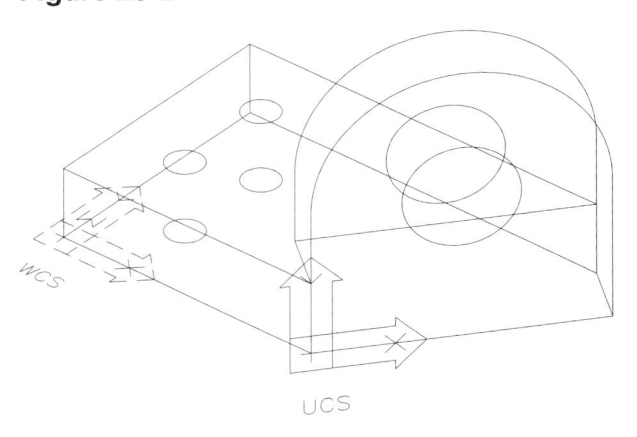

Creating geometry is relatively simple when having to deal only with X and Y coordinates, such as in creating a 2D drawing, or when creating simple 3D geometry with uniform Z dimensions. However, 3D models containing complex shapes on planes not parallel with the XY plane are good candidates for UCSs (Fig. 28-21).

A User Coordinate System is thought of as a construction plane created to simplify creation of geometry on a specific plane or surface of the object. The user creates the UCS, aligning its XY plane with a surface of the object, such as along an inclined plane, with the UCS origin typically at a corner or center of the surface.

Figure 28-22

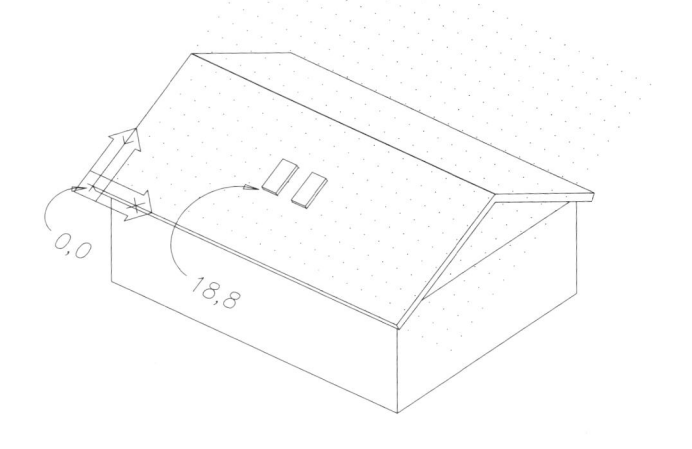

The user can then create geometry aligned with that plane by defining only X and Y coordinate values of the current UCS (Fig. 28-22). The *SNAP*, *Polar Tracking*, and *GRID* automatically align with the current coordinate system, providing *SNAP* points and enhancing visualization of the construction plane. Practically speaking, it is easier in some cases to specify only X and Y coordinates with respect to a specific plane on the object rather than calculating X, Y, and Z values with respect to the World Coordinate System.

User Coordinate Systems can be created by any of several options of the *UCS* command. Once a UCS has been created, it becomes the current coordinate system. Only one coordinate system can be active in one view or viewport; therefore, it is suggested that UCSs be saved (by using the *Save* option of *UCS*) for possible future geometry creation or editing.

When a UCS is created, the icon at the lower-left corner of the screen can be made to automatically align itself with the UCS along with *SNAP* and *GRID*. The letter "W," however, appears only on the "2D" icon and only when the WCS is active (when the WCS is the current coordinate system). When creating 3D geometry, it is recommended that the *ORigin* option of the *Ucsicon* command be used to place the icon always at the origin of the current UCS, rather than in the lower-left corner of the screen. Since the origin of the UCS is typically specified as a corner or center of the construction plane, aligning the Coordinate System icon with the current origin aids your visualization of the UCS orientation. The Coordinate Display (*Coords*) in the Status Bar always displays the X, Y and Z values of the current coordinate system, whether it is WCS or UCS.

THE RIGHT-HAND RULE

AutoCAD complies with the right-hand rule for defining the orientation of the X, Y, and Z axes. The right-hand rule states that if your right hand is held partially open, the thumb, first, and middle fingers define positive X, Y, and Z directions, respectively, and positive rotation about any axis is like screwing in a light bulb.

Figure 28-23

More precisely, if the thumb and first two fingers are held out to be mutually perpendicular, the thumb points in the positive X direction, the first finger points in the positive Y direction, and the middle finger points in the positive Z direction (Fig. 28-23).

In this position, looking toward your hand from the tip of your middle finger is like the default AutoCAD viewing orientation—positive X is to the right, positive Y is up, and positive Z is toward you.

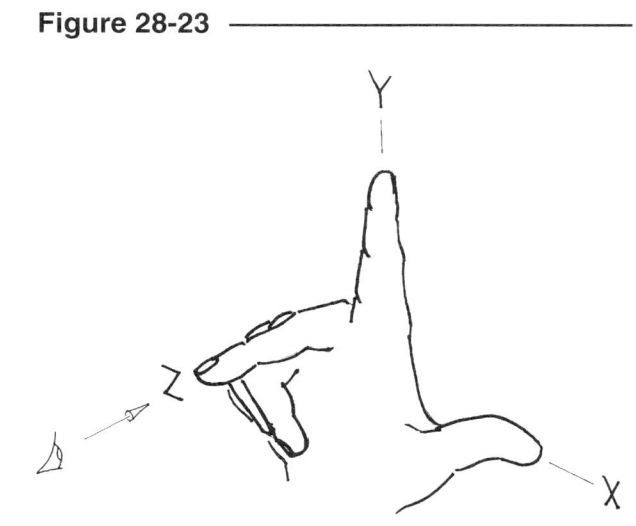

Figure 28-24

Positive rotation about any axis is counterclockwise looking down the axis toward the origin. For example, when you view your right hand, positive rotation about the X axis is as if you look down your thumb toward the hand (origin) and twist your hand counterclockwise (Fig. 28-24).

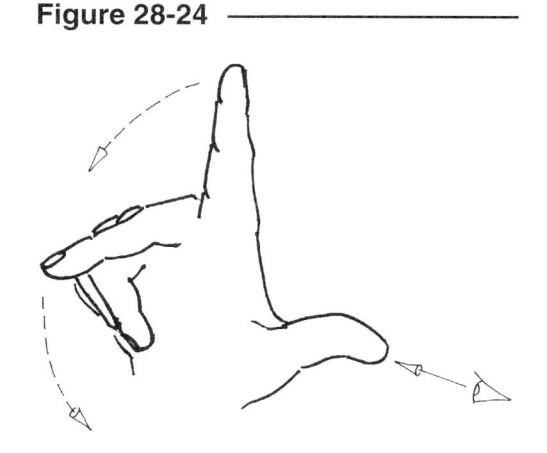

Figure 28-25

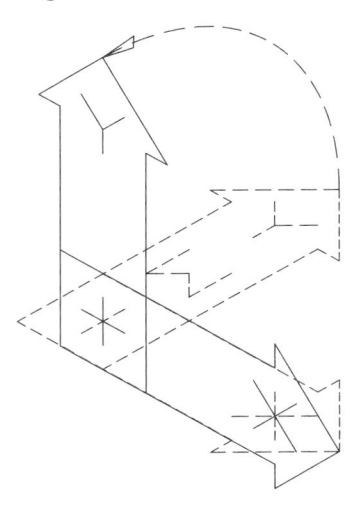

Figure 28-25 shows the 2D UCS icon in a +90 degree rotation about the X axis (like the previous figure). This orientation is typical for setting up a <u>front</u> view UCS for drawing on a plane parallel with the front surface of an object.

Figure 28-26

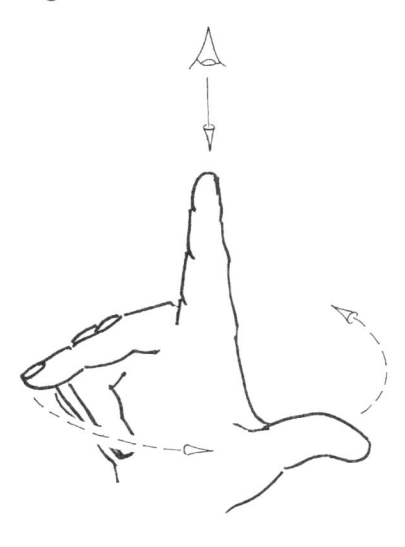

Positive rotation about the Y axis would be as if looking down your first finger toward the hand (origin) and twisting counterclockwise (Fig. 28-26).

Figure 28-27

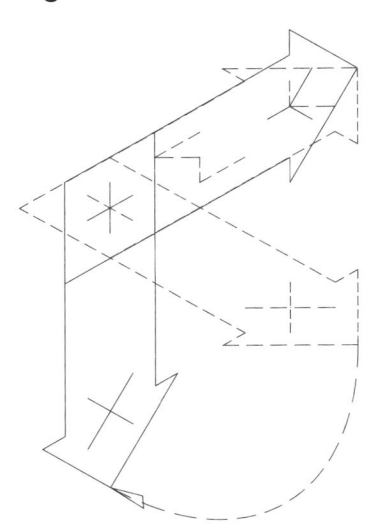

Figure 28-27 illustrates the 2D UCS icon with a +90 degree rotation about the Y axis (like the previous figure). To avoid confusion in relating this figure to Figure 28-26, consider that the icon is oriented on the XY plane as a horizontal plane in its original (highlighted) position, whereas Figure 28-26 shows the hand in an upright position. (Compare Figures 28-20 and 28-10.)

574 3D Modeling Basics

Figure 28-28

Positive rotation about the Z axis would be as if looking toward your hand from the end of your middle finger and twisting counterclockwise (Fig. 28-28).

Figure 28-29

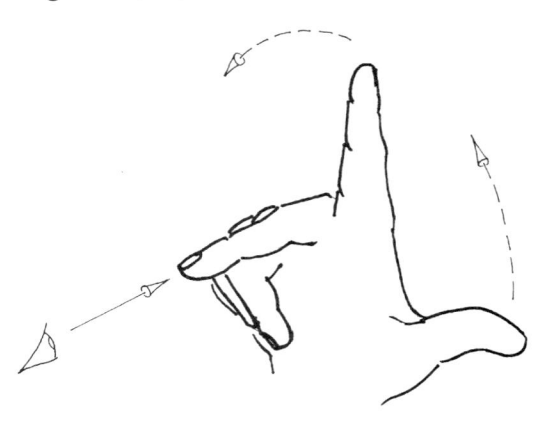

Figure 28-29 shows the 2D UCS icon with a +90 degree rotation about the Z axis. Again notice the orientation of the icon is horizontal, whereas the hand (Fig. 28-28) is upright.

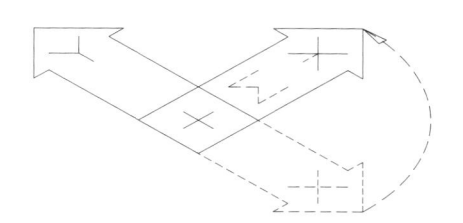

UCS ICON CONTROL

UCSICON

Pull-down Menu	COMMAND (TYPE)	ALIAS (TYPE)	Short-cut	Screen (side) Menu	Tablet Menu
View Display > UCS Icon >	UCSICON	...	...	VIEW 2 UCSicon	L,2

The *Ucsicon* command controls the appearance and positioning of the Coordinate System icon. In order to aid your visualization of the current UCS or the WCS, it is <u>highly</u> recommended that the Coordinate System icon be turned *ON* and positioned at the *ORigin*.

Figure 28-30

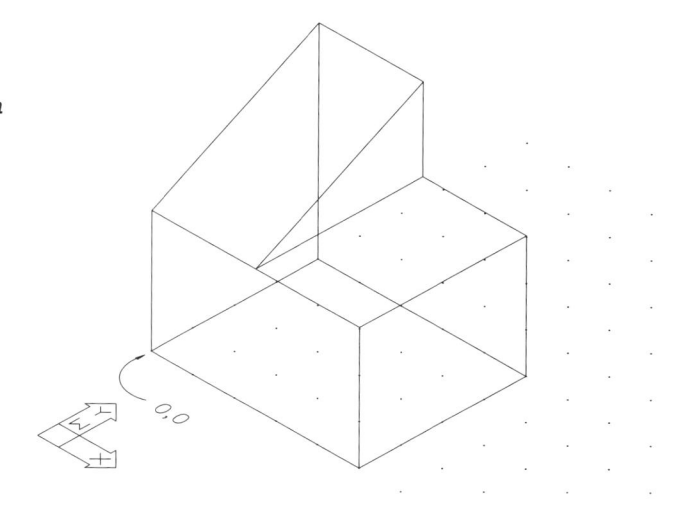

```
Command: ucsicon
Enter an option
[ON/OFF/All/Noorigin/ORigin/Properties] <ON>: on
Command:
```

This option causes the Coordinate System icon to appear (Fig. 28-30).

The *Origin* setting causes the icon to move to the origin of the current coordinate system (Fig. 28-31).

Figure 28-31

The icon does not always appear at the origin after using some viewing commands like *3Dorbit*. This is because *3Dorbit* may force the geometry against the border of the graphics screen (or viewport) area, which prevents the icon from aligning with the origin. In order to cause the icon to appear at the origin, try using *Zoom* with a **.9X** magnification factor. This action usually brings the geometry slightly in from the border and allows the icon to align itself with the origin.

Options of the *Ucsicon* command are:

ON Turns the Coordinate System icon on.

OFF Turns the Coordinate System icon off.

All Causes the *Ucsicon* settings to be effective for all viewports.

Noorigin Causes the icon to appear always in the lower-left corner of the screen, not at the origin.

ORigin Forces the placement and orientation of the icon to align with the origin of the current coordinate system.

Properties

The *Properties* option produces the *UCS Icon* dialog box (Fig. 28-32). Here you can select the *UCS icon style* in the top section. Select from *2D* or *3D*. If *3D* is selected, you also can use the *Cone* option to produce a 3D effect at the X and Y pole ends and increase the *Line width* to make the icon more prominent.

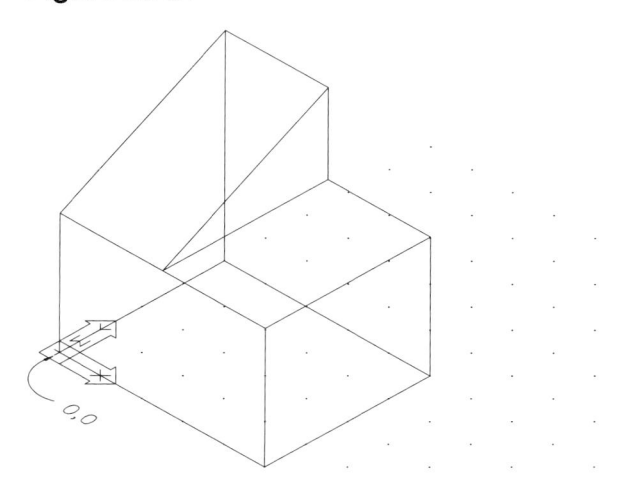

Figure 28-32

This text displays most figures using the *2D UCS icon style* because 1) it displays whether the WCS or other UCS is current, which is not communicated by the 3-pole axis that appears in AutoCAD when any *Shademode* other than *2D Wireframe* is selected, and 2) in the opinion of this author, the "2D" icon readily displays the orientation of the XY plane; therefore, orientation in 3D space is demonstrated with less ambiguity.

Other options in the *UCS Icon* dialog box include an adjustment for the *UCS icon size* (in pixels) and the *UCS icon color* for model space and paper space. (The default option for *Model space icon color* is either *Black* or *White*—the opposite of whatever you set for the *Model* tab background color in the *Options* dialog box.)

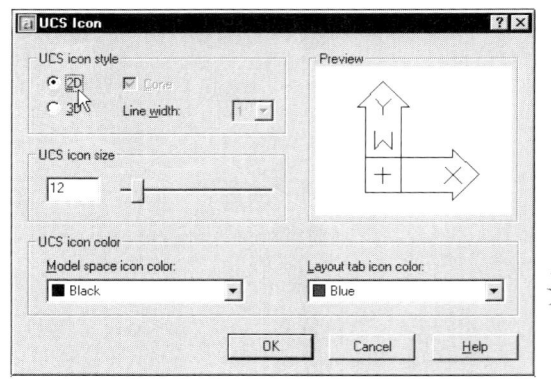

Figure 28-33

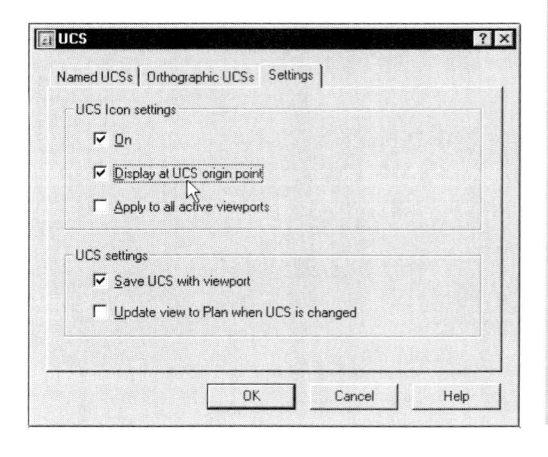

The *UCS Manager* can be invoked (type *Ucsman*) to control the appearance of the UCS icon (Fig. 28-33). Note the same options are available in the *UCS Manager* to turn the icon on or off, display it at the origin or not, and to apply the settings to all viewports. You can also control the appearance of the *Ucsicon* by selecting *Display* from the view pull-down menu.

576 3D Modeling Basics

Now that you know the basics of 3D modeling, you will develop your skills in viewing and displaying 3D models (Chapter 29, 3D Display and Viewing). It is imperative that you are able to view objects from different viewpoints in 3D space before you learn to construct them.

CHAPTER EXERCISES

1. What are the three types of 3D models?

2. What characterizes each of the three types of 3D models?

3. What kind of modeling techniques does AutoCAD's solid modeling system use?

4. What are the five formats for 3D coordinate specification?

5. Examine the 3D geometry shown in Figure 28-34. Specify the designated coordinates in the specified formats below.

    Figure 28-34

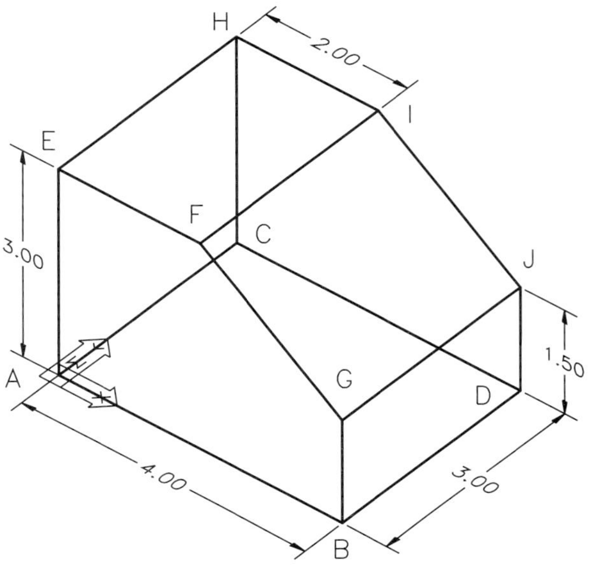

    A. Give the coordinate of corner D in absolute format.

    B. Give the coordinate of corner F in absolute format.

    C. Give the coordinate of corner H in absolute format.

    D. Give the coordinate of corner J in absolute format.

    E. What are the coordinates of the line that define edge F-I?

    F. What are the coordinates of the line that define edge J-I?

    G. Assume point E is the "last point." Give the coordinates of point I in relative rectangular format.

    H. Point I is now the last point. Give the coordinates of point J in relative rectangular format.

    I. Point J is now the last point. Give the coordinates of point A in relative rectangular format.

    J. What are the coordinates of corner G in cylindrical format (from the origin)?

    K. If E is the last point, what are the coordinates of corner C in relative cylindrical format?

29

3D DISPLAY AND VIEWING

Chapter Objectives

After completing this chapter you should:

1. be able to use *Shademode* to generate a wireframe, hidden, and four shaded displays of an object;
2. be able to generate a front, top, side, and several isometric viewpoints of a 3D model with the 3D views;
3. be able to use the *3Dorbit* command to dynamically change the viewpoint of the 3D model;
4. be able to generate perspective views, adjust distance, zoom, pan, and put an object in continuous motion with the 3D orbit commands;
5. be able to use the *Vpoint* command with the *Vector*, *Rotate*, and *Tripod* options to view a 3D model from various viewpoints;
6. be able to create 3D configurations (top, front, side, and isometric views) of viewports in model space with the *Vports* command.

AutoCAD'S 3D VIEWING AND DISPLAY CAPABILITIES

Display Commands

In most CAD systems, <u>surface and solid models are shown in a wireframe display</u> during the construction and editing process. A "hidden" display may hinder your ability to select needed edges of the object during the construction process. Once the model is completed and the desired viewpoint has been attained, these commands can change the appearance of a 3D surface or solid model from the default wireframe representation by displaying the surfaces:

Shademode The *Shademode* command allows you to display surface or solid objects in several shaded or unshaded representations, including wireframe, hidden, and four shaded alternatives. The shading options fill the surfaces with the object color, calculate light reflection, and apply gradient shading.

Render *Render* allows you to create and place lights in 3D space, adjust the light intensity, assign materials (color and reflective qualities) to the surfaces, and create shadows and reflections. This is the most sophisticated of the visualization capabilities offered in AutoCAD.

Viewing Commands

These commands allow you to change the direction from which you view a 3D model or otherwise affect the view of the 3D model:

View The *View* command has several options. You can select from preset orthographic views, including isometric viewpoints. *View* allows you to save and restore 3D viewpoints. *View* also includes an option for saving the current UCS with the view.

3Dorbit *3Dorbit* controls interactive viewing of objects in 3D. You can dynamically view the object from any point, even while the object remains shaded.

3Dcorbit Like *3Dorbit*, the *3Dcorbit* command enables you to set the 3D objects into continuous motion while shaded, hidden, or in any other *Shademode*.

3Dpan Like real-time *Pan*, *3Dpan* enables you to drag the 3D objects about the view interactively while in any *Shademode*.

3Dzoom Like real-time *Zoom*, *3Dzoom* keeps the 3D objects in their current *Shademode*.

3Dclip *3Dclip* allows you to establish front and/or back clipping planes in the *Adjust Clipping Planes* window.

3Ddistance Use this command to make objects appear closer or farther away. If the objects are in perspective mode, you can control the amount of perspective.

Vpoint *Vpoint* allows you to change your viewpoint of a 3D model. The object remains stationary while the viewpoint of the observer changes. Three options are provided: *Vector, Tripod,* and *Rotate*.

Plan The *Plan* command automatically gives the observer a plan (top) view of the object. The plan view can be with respect to the WCS (World Coordinate System) or an existing UCS (User Coordinate System).

Zoom Although *Zoom* operates in 3D just as in 2D, the *Zoom Previous* option restores the previous 3D view.

Vports You can use *Vports* effectively during construction of 3D objects to divide the screen into several sections, each displaying a different standard 3D view.

3D DISPLAY COMMANDS

Commands that are used for changing the appearance of a surface or solid model in AutoCAD are *Shademode*, *Hide*, and *Render*. By default, surface and solid models are shown in wireframe representation in order to speed computing time and enhance visibility during construction. As you know, wireframe representation can somewhat hinder your visualization of a model because it presents the model as transparent. To display surfaces and solid models as opaque and to remove the normally obscured edges, *Shademode*, *Hide*, and *Render* can be used.

Wireframe models are not affected by these commands since they do not contain surfaces. Wireframe models can be displayed only in wireframe representation.

SHADEMODE

Pull-down Menu	COMMAND (TYPE)	ALIAS (TYPE)	Short-cut	Screen (side) Menu	Tablet Menu
View *Shade>*	SHADEMODE	SHA	...	...	N,2

Shademode applies a shaded appearance to surface and solid models and allows you to select the mode, or type, of shading to apply. Shaded surfaces on an object increase the realistic appearance and can greatly enhance your visibility and understanding of the model, especially for complex geometry.

When *Shademode* is used, the model <u>retains</u> the specified shade mode until you change it. For example, if you use *Shademode* to generate a *Hidden* display, you can edit the geometry, add geometry, select a new viewpoint, or use *3Dorbit* and the display mode is retained. In releases previous to AutoCAD 2000, the *Shade* command was used instead of *Shademode*; however, once the model was shaded, the display was temporary and remained only until the next drawing regeneration.

Shademode operates in the current viewport or full-screen display, but is allowed only in model space. If you create a new layout, the layout displays the model in the *Shademode* that was current when the layout was created.

You cannot print or plot a 3D model in one of the four <u>shaded</u> options of *Shademode*—only in the wireframe or hidden representations. You can, however, capture a shaded image with the *Mslide* command and regenerate the same display image with *Vslide* (see AutoCAD 2002 *Help*).

Shademode allows you to specify one of seven options and then displays the objects in the current viewport using the option you choose.

```
Command: shademode
Current mode: Hidden
Enter option [2D wireframe/3D wireframe/Hidden/Flat/Gouraud/fLat+edges/gOuraud+edges] <Hidden>:
(option)
Regenerating model.
Command:
```

The options are explained and illustrated next.

2D Wireframe

This option displays only lines and curves. The 3D geometry is defined only by the edges representing the surface boundaries. The UCS icon is displayed as a "wireframe" icon. Also, linetypes and lineweights are visible. If the *Compass* is on (*COMPASS* system variable is set to 1), it does not appear in the *2D Wireframe* view. Raster and OLE objects are visible in the display. Generally used for 2D geometry, this option can be used for 3D geometry when lineweights and linetypes are important (Fig. 29-1).

Figure 29-1

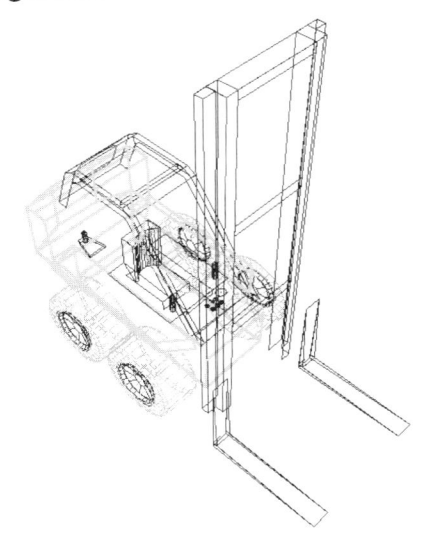

3D Wireframe

Generally used to display 3D geometry when a wireframe representation is needed, this mode displays the objects using lines and curves to represent the surface edges (Fig. 29-2). Since all edges are displayed, use this option for general 3D construction and editing. The shaded 3D UCS icon appears in this mode. Linetypes and lineweights are not displayed and raster and OLE objects are not visible. The *Compass* is visible if turned on. The lines are displayed in the material colors if you have applied *Materials*, otherwise they appear in the defined line or layer color.

Figure 29-2

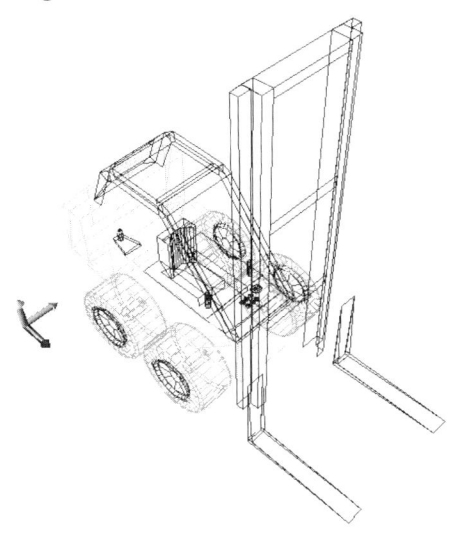

Hidden

Use this mode to display 3D objects similar to the *3D Wireframe* representation but with hidden lines suppressed. In other words, edges that would normally be obscured from view by opaque surfaces become hidden. This option is similar to using the *Hide* command (Fig. 29-3). This option enhances visualization but is not recommended for construction and editing since some edges are not visible.

Figure 29-3

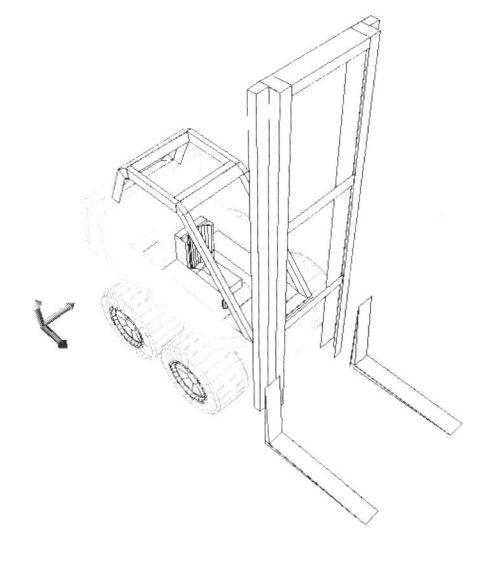

Flat Shaded

This and the other three color shaded options (described next) fill the surfaces with the defined color (by object, by layer, or *Material*). The surfaces are shaded as if a light source were placed at the camera (observer) position. When 3D objects contain curved geometry, this option displays curves as being faceted. For example, a cylinder is displayed as having many small flat surfaces (notice the forklift wheels in Figure 29-4). Therefore, curved objects appear flatter and less smooth than *Gouraud Shaded*. If you have applied *Materials* to objects, they are displayed accordingly with the *Flat Shaded* option.

Figure 29-4

Gouraud Shaded

With this option the surfaces are also filled and shaded like *Flat Shaded*, except here the curved surfaces are smoothed to display true curved geometry (notice the forklift wheels in Figure 29-5). This feature gives the objects a smooth, realistic appearance. Objects are shaded in the layer or object color unless *Materials* have applied to the objects.

Figure 29-5

Flat Shaded, Edges On

Select this option to achieve a combination of the *Flat Shaded* and *3D Wireframe* options. The resulting display has the effect of flat-shading with the edges highlighted (Fig. 29-6).

Figure 29-6

582 3D Display and Viewing

Gouraud Shaded, Edges On

This choice combines the *Gouraud Shaded* and *3D Wireframe* options (Fig. 29-7). The objects are *Gouraud Shaded* with the surface edges showing through. This option is good for cases where visualization is important but minor editing may be necessary and is made easier since edges are visible.

The four shaded options of *Shademode* display lights, materials, textures, and transparency applied with AutoCAD's rendering capabilities; however, some limitations occur such as the inability to display shadows, reflection, refraction, and other features.

Figure 29-7

HIDE

Pull-down Menu	COMMAND (TYPE)	ALIAS (TYPE)	Short-cut	Screen (side) Menu	Tablet Menu
View Hide	HIDE	HI	...	VIEW 2 Hide	M,2

The *Hide* command generates a display of a 3D model similar to the *Hidden* option of *Shademode*. *Hide* suppresses the hidden lines (edges that would normally be obscured from view by opaque surfaces) in the current viewport or full screen display (see Figure 29-3).

There are, however, two differences between *Hide* and the *Hidden* option of *Shademode*. First, *Shademode* is persistent, meaning that once the *Hidden* mode is activated, the model retains its "hidden" display until the *Shademode* is changed, whereas the *Hide* command is a temporary display that remains active only until the view is regenerated. Second, when *Hide* is used on curved surfaces, it can display a meshed or faceted appearance, depending on the setting of the *DISPSILH* variable (see "Display Variables" in Chapter 32 and Figures 32-4 and 32-5).

You cannot plot a model in *Hide* mode. Instead, use the *Hide Objects* option in the *Plot* dialog box or use the *Hideplot* option of *Vports* for paper space viewports to generate a plot with hidden lines removed. To remove the mesh (facet) lines for a plot with hidden lines removed, change the *DISPSILH* variable to 1 before creating the plot (see "*DISPSILH*," Chapter 32).

3D VIEWING COMMANDS

When using the viewing commands *View, 3Dorbit, Vpoint, Plan,* etc., it is important to imagine that the observer moves about the object rather than imagining that the object rotates. The object and the coordinate system (WCS and icon) always remain stationary and always keep the same orientation with respect to Earth. Since the observer moves and not the geometry, the objects' coordinate values retain their integrity, whereas if the geometry rotated within the coordinate system, all coordinate values of the objects would change as the object rotated. The viewing commands change only the viewpoint of the observer.

Although UCSs have not been discussed in detail to this point, the viewing command you use can affect the UCS if you generate an orthographic viewpoint such as the top, front, side, etc. view. The *View* command (discussed next) can automatically change the UCS to align with the view if the *UCSORTHO* system variable is set to 1 (default setting) and you use the *Top*, *Front*, *Right* side, etc. option. All other viewing commands such as the *3Dorbit* commands, *Vpoint*, and *Plan* do not affect the UCS. It is recommended that you set *UCSORTHO* to **0** while you practice with the 3D viewing commands. The *UCSORTHO* system variable is discussed in detail in Chapter 30.

VIEW

Pull-down Menu	COMMAND (TYPE)	ALIAS (TYPE)	Short-cut	Screen (side) Menu	Tablet Menu
View *Named Views...*	*VIEW* or *-VIEW*	*V* or *-V*	...	*VIEW 1* *Ddview*	*M,5*

The *View* command provides several functions. You can assign a name, save, and restore viewed areas of a drawing, and you can control whether you want to save the current UCS with the view. (See *View* in Chapter 10 for the *Save* and *Restore* functions of *View*. See Chapter 30 for the UCS controls included in the *View* command.) This chapter explains how to use *View* to generate preset viewing directions for 3D objects.

The view command can be invoked by several methods. You can also use *View* in command line version by typing *–View*. Invoking *View* (not *–View*) by any method produces the *View* dialog box.

However, if you only want to produce a 3D view, you can bypass the dialog box and generate the desired viewpoint using the icon buttons from the Standard toolbar (Fig. 29-8) or using the *3D Views* option from the *Views* pull-down menu (Fig. 29-9).

Figure 29-8

Keep in mind that if you use the *Top*, *Bottom*, *Front*, *Back*, *Right,* or *Left* option, and the *UCSORTHO* system variable is set to 1 (default setting), the UCS automatically changes to align with the view.

Figure 29-9

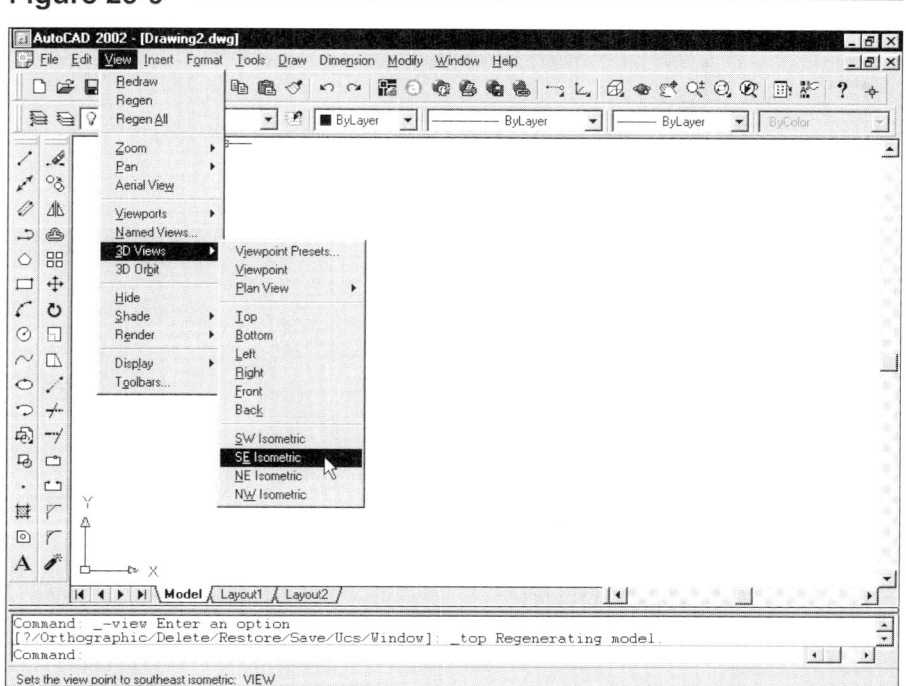

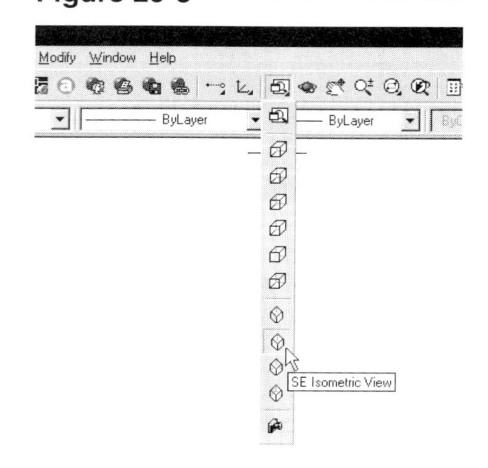

Each of the 3D View options is described next. The FORK-LIFT drawing is displayed for several viewpoints (the hidden shademode was used in these figures for clarity). It is important to remember that for each view imagine you are viewing the object from the indicated position in space. The object does not rotate.

584   3D Display and Viewing

Top

The object is viewed from the top (Fig. 29-10). Selecting this option shows the XY plane from above (the default orientation when you begin a *New* drawing). Notice the position of the WCS icon. This orientation should be used periodically during construction of a 3D model to check for proper alignment of parts.

Figure 29-10

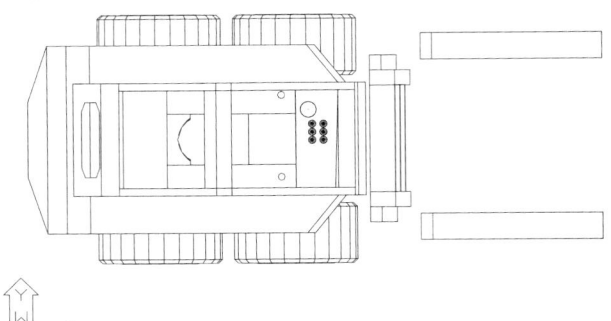

Bottom

This option displays the object as if you are looking up at it from the bottom. The Coordinate System icon appears backward when viewed from the bottom.

Left

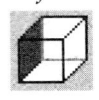

This is looking at the object from the left side.

Figure 29-11

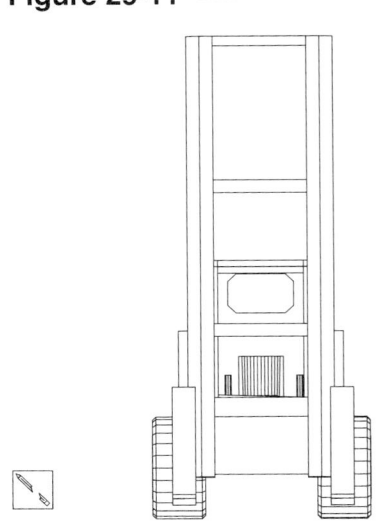

Right

Imagine looking at the object from the right side (Fig. 29-11). This view is used often as one of the primary views for mechanical part drawing. If *UCSORTHO* is set to 0, the "2D" Coordinate System icon does not appear in this view, but a "broken pencil" is displayed instead (meaning that it is not a good idea to draw on the XY plane from this view).

Front

Selecting this option displays the object from the front view. This is a common view that can be used often during the construction process. The front view is usually the "profile" view for mechanical parts (Fig. 29-12).

Figure 29-12

Back

Back displays the object as if the observer is behind the object. Remember that the object does not rotate; the observer moves.

SW Isometric

Isometric views are used more than the "orthographic" views (front, top, right, etc.) for constructing a 3D model. Isometric viewpoints give more information about the model because you can see three dimensions instead of only two dimensions (as in the previously discussed views).

SE Isometric

The southeast isometric is generally the first choice for displaying 3D geometry. If the object is constructed with its base on the XY plane so that X is length, Y is depth, and Z equals height, this orientation shows the front, top, and right sides of the object. Try to use this viewpoint as your principal mode of viewing during construction. Note the orientation of the WCS icon (Fig. 29-13).

NE Isometric

The northeast isometric shows the right side, top, and back (if the object is oriented in the manner described earlier).

Figure 29-13

NW Isometric

This viewpoint allows the observer to look at the left side, top, and back of the 3D object (Fig. 29-14).

NOTE: 3D View displays the object (or orients the observer) with respect to the World Coordinate System by default. For example, the *Top* option always shows the plan view of the WCS XY plane. Even if another coordinate system (UCS) is active, AutoCAD temporarily switches back to the WCS to attain the selected viewpoint.

You can also generate a 3D view (as previously illustrated) by invoking the *View* command instead of using the quicker *3D View* pull-down menu options or icons from the Standard toolbar. Invoking the *View* command produces the *View* dialog box. The *Orthographic & Isometric Views* tab gives access to the 3D views (Fig. 29-15).

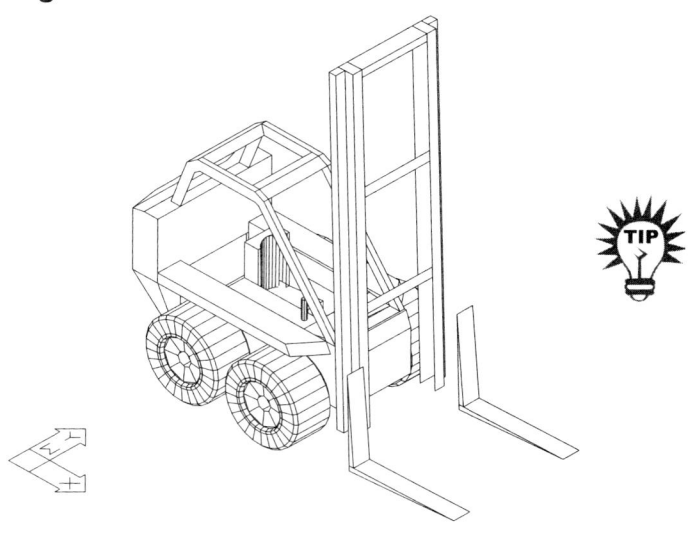

Figure 29-14

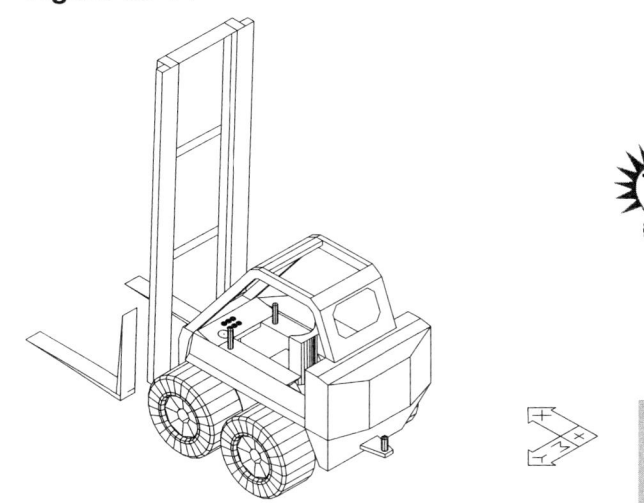

Figure 29-15

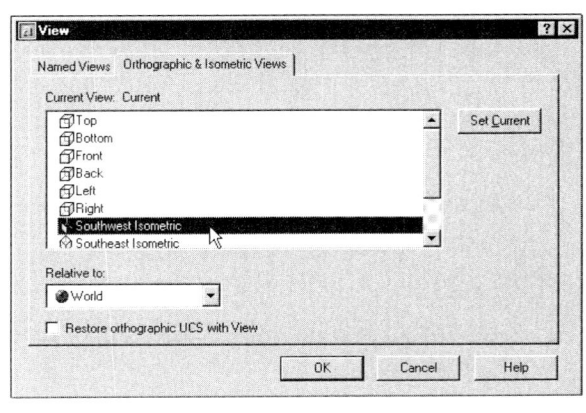

To generate a 3D view using the dialog box, several steps are required. First, select the desired view from the list. Next, make the desired view current by selecting the *Set Current* button. Alternately, you can double-click on the desired view to make it current. Finally, select the *OK* button to produce the view.

NOTE: It is strongly suggested that you ensure the *Relative to:* option is set to *World*. This causes the 3D view that is generated to be relative to the World Coordinate System. Using another option can make the resulting view difficult to predict (*UCSBASE* system variable). See Chapter 30 for more information on this option. Also see Chapter 30 for information concerning the *Restore orthographic UCS with View* option (*UCSORTHO* system variable).

3D Orbit Commands

AutoCAD 2000 introduced a set of new interactive 3D viewing commands, all centered around the *3Dorbit* command. With these commands you can manipulate the view of 3D models by dragging your cursor. When one of the new 3D view commands is activated by typing or selecting from the *3D Orbit* toolbar (Fig. 29-16) all commands in the 3D Orbit group are available through a right-click shortcut menu (Fig. 29-17). The new 3D commands all issue the same simple command prompt (see "*3Dorbit*") because they are interactive, not Command line driven. *3Dorbit* and the related viewing commands are listed and briefly explained at the beginning of this chapter and described in detail on this and the following pages.

Figure 29-16

Figure 29-17

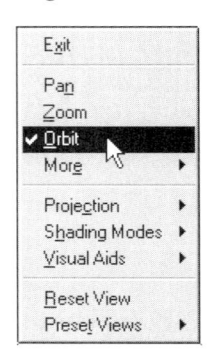

The new *3Dorbit* and related commands essentially replace the function of the older *Dview* (Dynamic View) command and its options. Although the older *Dview* command is still available in AutoCAD, the new 3D commands are more interactive and user-friendly. Because these new commands more directly serve the same function as the older technology, a full description of *Dview* is not discussed in this text.

3DORBIT

Pull-down Menu	COMMAND (TYPE)	ALIAS (TYPE)	Short-cut	Screen (side) Menu	Tablet Menu
View 3D Orbit	3DORBIT	3do	...	VIEW 3dorbit	R,5

3Dorbit controls viewing of 3D objects interactively. By holding down the left button on your pointing device and dragging your cursor, you can control the view of 3D objects by several methods depending on where you place your cursor with respect to the "arc ball."

You can control the view of the entire 3D model or of selected 3D objects. If you want to see only a few objects during the interactive manipulation, select those objects before you invoke the command, otherwise the entire 3D model will be seen. Only a simple command prompt is given since the options are interactive. The same command prompt is issued for all the new 3D viewing commands.

    Command: *3dorbit*
    Press ESC or ENTER to exit, or right-click to display shortcut-menu.
    Regenerating model.
    Command:

The shortcut menu (see previous Figure 29-17) gives access to this and the related commands described on the following pages.

When you invoke *3Dorbit*, the current viewport displays the selected objects and the 3D Orbit "arc ball" (Fig. 29-18). The arc ball is a circle with each quadrant denoted by a smaller circle. Moving the cursor over different areas of the arc ball changes the direction in which the view rotates. The center of the arc ball is the target. (Remember that you are changing the view. When moving the cursor about the arc ball, it appears that the 3D objects rotate; however, in theory the target remains stationary while the observer, or camera, moves around the objects.) Also notice when *3Dorbit* is invoked, the *Ucsicon* (if on) appears as a shaded 3D icon with the X axis as a red arrow, Y axis as green, and Z axis as cyan.

Figure 29-18

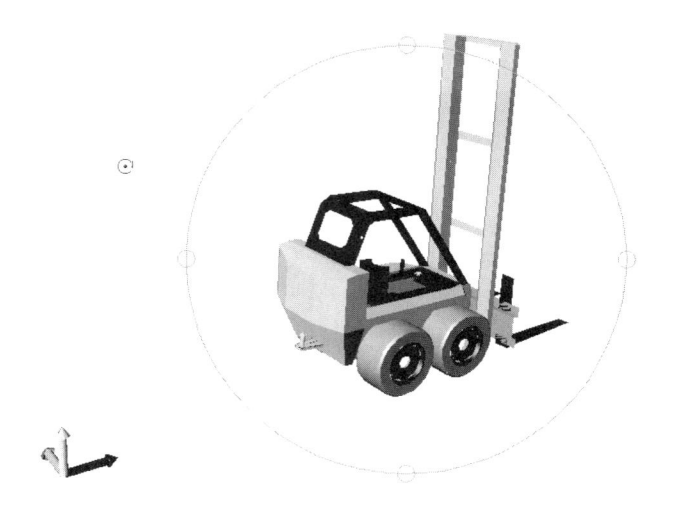

There are four possible methods of rotating your view depending on where you place the cursor and the direction you move it. The cursor icon changes to indicate one of these four methods. The four positions are as follows.

Inside the Arc Ball

This icon is displayed when you move the cursor inside the arc ball. Imagine a sphere around the model. Click and drag the pointing device (hold down the left button) to rotate the sphere (and 3D model) around the target point (center of the sphere) in any direction.

Outside the Arc Ball

Move the cursor outside the arc ball, then and click and drag to "roll" the display. This action rotates the view around the center of the arc ball. Imagine rotating the view about a Z axis protruding perpendicular from the screen.

Left and Right Quadrants

This icon is displayed when you move the cursor over either of the small circles on the left or right quadrants of the arc ball, then click and drag to the left or right. This allows you to rotate the view about a vertical axis that passes through the center of the arc ball or about an imaginary Y axis on the screen.

Top and Bottom Quadrants

To rotate the view about a horizontal axis passing through the center of the arc ball, place the cursor over either of the small circles on the top or bottom quadrants of the arc ball, then click and drag up or down. This is similar to rotation about an imaginary X screen axis.

Remember that the center of the arc ball is the target and remains stationary. The arc ball always appears in the center of your screen or current viewport. To center your objects in the arc ball, use the *Pan* (*3Dpan*) option or *Orbit uses AutoTarget* option from the shortcut menu.

3Dorbit **Shortcut Menu Options**

During the *3Dorbit* or related command, you can use the shortcut menu to access other 3D viewing commands and options (Fig. 29-19). Using commands from the shortcut menu does not exit the *3Dorbit* session (unless, of course, you select *Exit*).

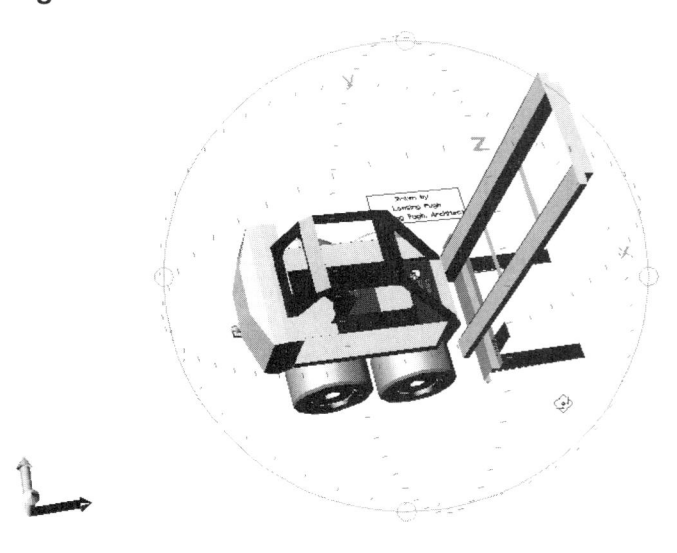

Figure 29-19

More, Orbit Maintains Z

Use this option to keep the Z axis in its current orientation when dragging horizontally inside the arc ball. For example, if the Z axis is vertical before using *3Dorbit*, this option keeps the Z axis in a vertical orientation while using *3Dorbit*. This setting is recommended.

More, Orbit Uses AutoTarget

This option forces the center of the orbit (target) in the center of the objects, rather than on the center of the viewport. This setting is particularly helpful when the objects are not located in the center of the viewport.

Visual Aids, Compass

Selecting the *Compass* toggle from the shortcut menu produces a sphere within the arc ball composed of three circles in 3D space. The three circles represent rotation about each of the three axes, X, Y, and Z (Fig. 29-20).

Figure 29-20

Visual Aids, Grid

Select *Visual Aids*, then *Grid* from the shortcut menu to produce a grid on the XY plane of the current UCS (see Figures 29-22 and 29-23). The number of major grid lines is controlled by the *Grid Spacing* value specified in the *Grid* command or in the *GRIDUNIT* system variable. The number of grid lines between the major lines defaults to 10 but changes as you zoom out, and the grid appears smaller. The *Grid* command cannot be activated during a *3Dorbit* session.

Visual Aids, UCS Icon

Use this option to toggle the UCS icon on and off. The UCS icon can enhance your orientation of the view because it changes as the view changes, even during perspective projection. When *3Dorbit* is invoked, the UCS icon (if on) appears as a shaded 3D icon with the X axis as a red arrow, Y axis as green, and Z axis as cyan. Alternately, use the *Ucsicon* command to control the icon visibility.

Reset View

You can use *Reset View* from the shortcut menu to undo all changes to the view during the *3Dorbit* session and reset to the previous view (when *3Dorbit* was started) and still remain in the *3Dorbit* session.

Exit

If you have achieved the view you want, you must press Enter, Escape, or select *Exit* from the shortcut menu to save the view and use other (non-*3Dorbit*) commands.

Other 3D viewing commands in the shortcut menu are discussed on the following pages.

3DPAN

Pull-down Menu	COMMAND (TYPE)	ALIAS (TYPE)	Short-cut	Screen (side) Menu	Tablet Menu
...	3DPAN	...	...	...	...

The cursor changes to a hand cursor when this command is active. Similar to the real-time *Pan* command, you can click and drag the view of the objects to any location in your drawing area. This tool is particularly useful with *3Dorbit* since the center of the arc ball is the target unless you use *Orbit uses AutoTarget*. Therefore, change the target of your 3D model by dragging it to the center of the arc ball. Keep in mind the arc ball temporarily disappears during *Pan*. Use *3Dpan* with *Perspective* toggled on to get a true 3D effect.

3DZOOM

Pull-down Menu	COMMAND (TYPE)	ALIAS (TYPE)	Short-cut	Screen (side) Menu	Tablet Menu
...	3DZOOM	...	...	...	...

3Dzoom operates similarly to the real-time version of the *Zoom* command. The cursor changes to a magnifying glass. Move the cursor up to zoom in (enlarge image) and down to zoom out (reduce image). If you were previously using *3Dorbit*, the arc ball disappears until you select *Orbit* from the shortcut menu.

Zoom Window
From the shortcut menu (Fig. 29-21) you can also select *More, Zoom Window* to zoom into a smaller area by defining a window, similar to the normal *Zoom* command with the *Window* option. With the *3Dorbit Zoom Window*, you must click and drag to define two diagonal corners of the window rather than click at each corner.

Figure 29-21

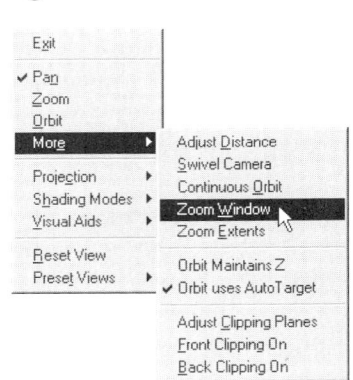

Zoom Extents
Select *More, Zoom Extents* from the shortcut menu to perform a typical *Zoom Extents*, similar to the normal *Zoom* command with the *Extents* option. The view is centered and sized so all selected objects are displayed in the 3D view.

3DCORBIT

Pull-down Menu	COMMAND (TYPE)	ALIAS (TYPE)	Short-cut	Screen (side) Menu	Tablet Menu
...	3DCORBIT	...	...	...	...

The *3Dcorbit* command allows you to set the selected 3D objects into continuous motion. The cursor changes to a sphere with two arrows orbiting it. To set the 3D objects in motion, click in the drawing area and hold down the left button, drag the cursor in any direction, then release the button while the cursor is in motion. The objects continue to spin in the direction of your cursor motion. The speed of the cursor movement determines the speed of the spin. Change the direction and speed of the object movement by clicking and dragging the cursor again. You can stop the motion at any position and discontinue *3Dcorbit* by pressing Escape or Enter. Alternately, right-click and select *Exit* or other option.

3DDISTANCE

Pull-down Menu	COMMAND (TYPE)	ALIAS (TYPE)	Short-cut	Screen (side) Menu	Tablet Menu
...	3DDISTANCE	...	...	...	...

The *3Ddistance* command adjusts the amount of perspective applied to the 3D view. In other 3D views, such as those generated using *Vpoint* or the 3D views obtained through the menus, you see a parallel projection, not a perspective projection. Perspective projection takes into account the distance the observer is from the object, whereas a parallel projection does not.

A parallel projection displays 3D objects unrealistically because parallel edges of an object remain parallel in the display (Fig. 29-22). A good example of this theory is to activate the grid (from the shortcut menu select *Visual Aids*, then *Grid*), then use *3Dorbit* to generate a view displaying the parallel grid lines.

Figure 29-22

A perspective projection simulates the way we actually see 3D objects because it takes into account the distance from our eyes (camera) to the objects (target). Therefore, since objects that are farther away appear smaller, parallel edges on a 3D object converge toward a point as they increase in distance. Figure 29-23 shows the same objects as the previous figure but in perspective projection.

Figure 29-23

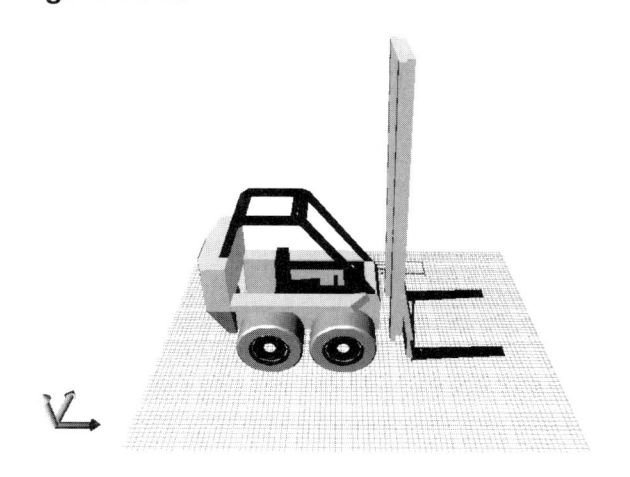

The *3Ddistance* command adjusts the distance between the camera and the target. In other words, it increases or decreases the amount of perspective. Before using *3Ddistance*, you can achieve a reasonable amount of perspective by using the shortcut menu to select *Projection*, then *Perspective*. This action sets the selected 3D objects in perspective mode. Then use *3Ddistance* to increase or decrease the distance, or amount of perspective.

When you invoke *3Ddistance* by typing or selecting from the *3D Orbit* toolbar or shortcut menu, the icon changes to a double-sided arrow shown in perspective. Hold the left mouse button down and move the cursor upward to achieve more perspective (the view becomes closer), or move the cursor down to achieve less perspective (the view becomes farther away).

A very important fact to keep in mind as you use *3Ddistance* is that <u>you should not be concerned with the size of the view</u>, only the distance the view is from the objects(s) and, therefore, the amount of perspective. <u>Use *3Dzoom* to adjust the size of the view after using *3Ddistance*</u>. It is easy to confuse the action of *3Ddistance* with that of *3Dzoom*—the two are related but achieve different results.

3DCLIP

Pull-down Menu	COMMAND (TYPE)	ALIAS (TYPE)	Short-cut	Screen (side) Menu	Tablet Menu
...	3DCLIP	...	...	...	...

Clipping planes are invisible planes that can pass through the view. Anything in front of the front clipping plane or in back of the back clipping plane becomes invisible when the planes are toggled on. The clipping planes are always parallel with the screen and perpendicular to the viewing direction (the screen Z axis).

Using *3Dclip* produces the *Adjust Clip Planes* window (Fig. 29-24). The window displays the current view from above so the clip planes can be seen on edge (appear as horizontal lines). The white (or black) line on the bottom of the window represents the front clipping plane and the green line on the top (if visible) represents the back clipping plane. As you move the clip planes through the 3D objects, the results are displayed in the drawing area. To see the results of adjusting clipping planes, the clipping planes must be toggled on. The controls for the *Adjust Clip Planes* window are available if you select the icon buttons or right-click in the window to produce the shortcut menu.

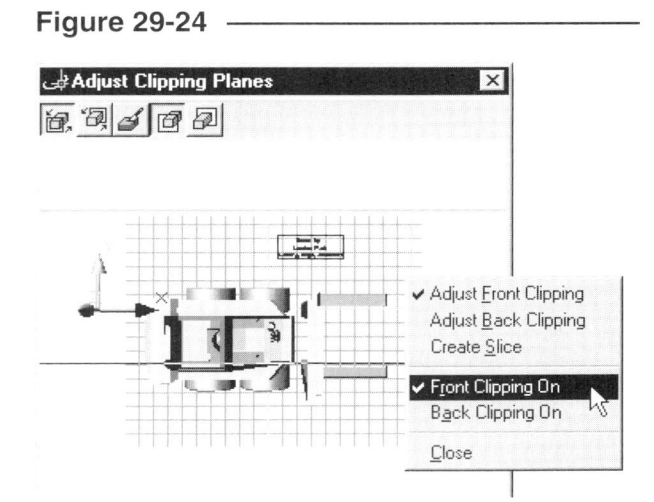

Figure 29-24

Front Clipping On

You must toggle the front clipping plane visibility on (depress the icon button or check the shortcut menu) if you want to see the results of adjusting the front clipping plane in the drawing area. Toggle this option on before using *Adjust Front Clipping*. Once the desired clipping is set, you can toggle this on or off for the display in the drawing area.

Back Clipping On

Toggle this option on before using *Adjust Back Clipping*. Like *Front Clipping On*, you must toggle the back clipping plane on to see the results of adjusting the back clipping plane in the drawing area. Once the desired clipping is set, you can toggle this on or off for the display in the drawing area.

Adjust Front Clipping

Move the horizontal black line near the bottom of the window upward to adjust the front clipping plane into the 3D objects. Any objects or parts of objects in front of the front clipping plane disappear. Notice how most of the fork and lift sections of the forklift are removed by the front clipping plane in Figure 29-25.

Figure 29-25

Adjust Back Clipping

Move the horizontal green line near the top of the window to adjust the back clipping plane. Generally, move the plane downward to pass through the 3D objects. Any objects or parts of objects in back of the back clipping plane disappear as shown in Figure 29-26.

Figure 29-26

Slice

Make this selection if you want to move both front and back clipping planes together. First, adjust either clipping plane to determine the thickness of the slice, then toggle on *Slice*. You can achieve a view to display only a thin section of your 3D model as shown in Figure 29-27.

Figure 29-27

When the *Adjust Clip Planes* window is dismissed, the clipping planes remain active (control the display) <u>during and after</u> the *3Dorbit* session. At any time, control the visibility of clipping planes using the *3Dorbit* shortcut menu by selecting *More*, then the *Front Clipping On* toggle or *Back Clipping On* toggle.

VPOINT

Pull-down Menu	COMMAND (TYPE)	ALIAS (TYPE)	Short-cut	Screen (side) Menu	Tablet Menu
View 3D Views > VPOINT	VPOINT	-VP	...	VIEW 1 Vpoint	N,4

Vpoint, like *3Dorbit*, allows you to achieve a specific view (or viewpoint) of a 3D object; however, *Vpoint* is an older command. Nevertheless, *Vpoint* can be used to specify an exact viewing angle or direction, whereas *3Dorbit* is strictly interactive. *Vpoint* offers fewer options than *3Dorbit* and displays the objects in parallel projection only.

Rotate

The name of this option is somewhat misleading because the object is not rotated. The *Rotate* option prompts for two angles in the WCS (by default) which specify a vector indicating the direction of viewing. The two angles are (1) the angle in the XY plane and (2) the angle from the XY plane. The observer is positioned along the vector looking toward the origin.

Command: ***vpoint***
Current view direction: VIEWDIR=0.0000,0.0000,1.0000
Specify a view point or [Rotate] <display compass and tripod>: **r**
Enter angle in XY plane from X axis <270>: **315**
Enter angle from XY plane <90>: **35.3**
Command:

The first angle is the angle in the XY plane at which the observer is positioned looking toward the origin. This angle is just like specifying an angle in 2D. The second angle is the angle that the observer is positioned up or down from the XY plane. The two angles are given with respect to the WCS (Fig. 29-28).

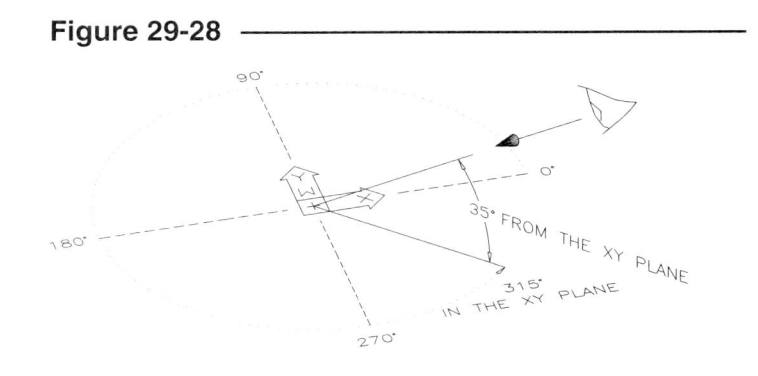

Figure 29-28

Angles of **315** and **35** specified in response to the *Rotate* option display an almost perfect isometric viewing angle. An isometric drawing often displays some of the top, front, and right sides of the object. For some regularly proportioned objects, perfect isometric viewing angles can cause visualization difficulties, while a slightly different angle can display the object more clearly. Figure 29-29 and Figure 29-30 display a cube from an almost perfect isometric viewing angle (**315** and **35**) and from a slightly different viewing angle (**310** and **40**), respectively.

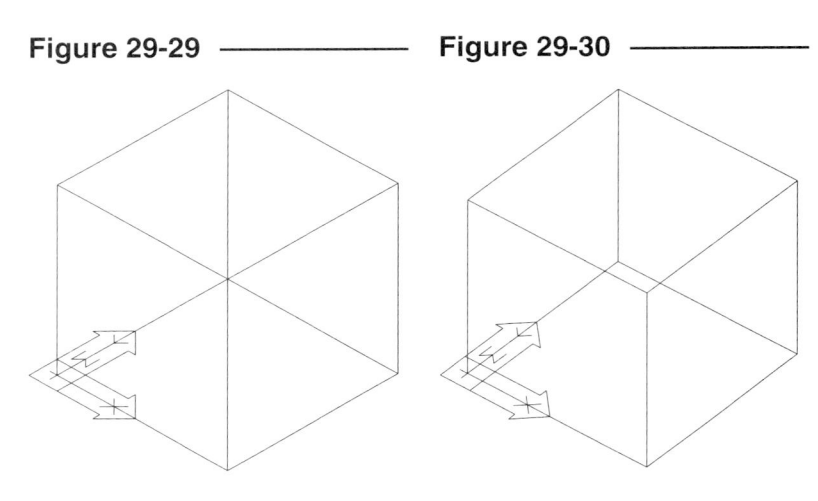

Figure 29-29   **Figure 29-30**

The *Vpoint Rotate* option can be used to display 3D objects from a front, top, or side view. *Rotate* angles for common views are:

Top	270, 90
Front	270, 0
Right side	0, 0
Southeast isometric	315, 35.27
Southwest isometric	225, 35.27

Vector
Another option of the *Vpoint* command is to enter X,Y,Z coordinate values. The coordinate values indicate the position in 3D space at which the observer is located. The coordinate values do not specify an absolute position but rather specify a vector passing through the coordinate position and the origin. In other words, the observer is located at any point along the vector looking toward the origin. Values of 1,-1,1 would generate the same display as 2,-2,2.

The command syntax for specifying a perfect isometric viewing angle is as follows:

Command: *vpoint*
Current view direction: VIEWDIR=0.0000,0.0000,1.0000
Specify a view point or [Rotate] <display compass and tripod>: **1,-1,1**
Command:

Figure 29-31 illustrates positioning the observer in space, using coordinates of 1,-1,1. Using *Vpoint* coordinates of 1,-1,1 generates an isometric display similar to *Rotate* angles of 315, 35, or *SE Isometric*.

Figure 29-31

Other typical views of an object can be easily achieved by entering coordinates at the *Vpoint* command. Consider the following coordinate values and the resulting displays:

Coordinates	Display
0,0,1	Top
0,-1,0	Front
1,0,0	Right side
1,-1,1	Southeast isometric
-1,-1,1	Southwest isometric

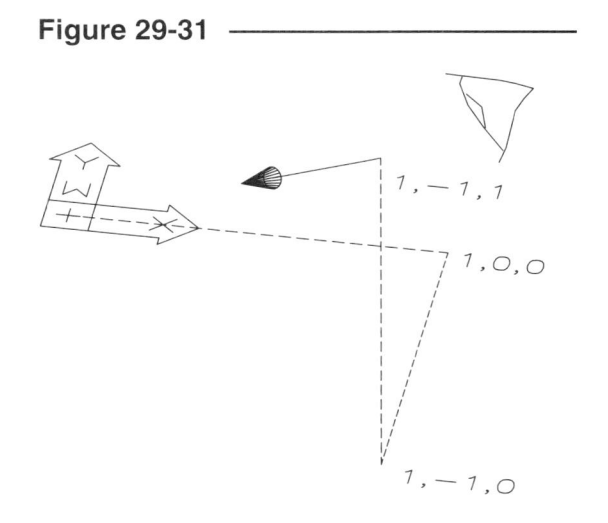

Tripod

When the *Tripod* method is invoked, the current drawing temporarily disappears and a three-pole axes system appears at the center of the screen. The axes are dynamically rotated by moving the cursor (a small cross) in a small "globe" at the upper-right of the screen (Fig. 29-32).

Figure 29-32

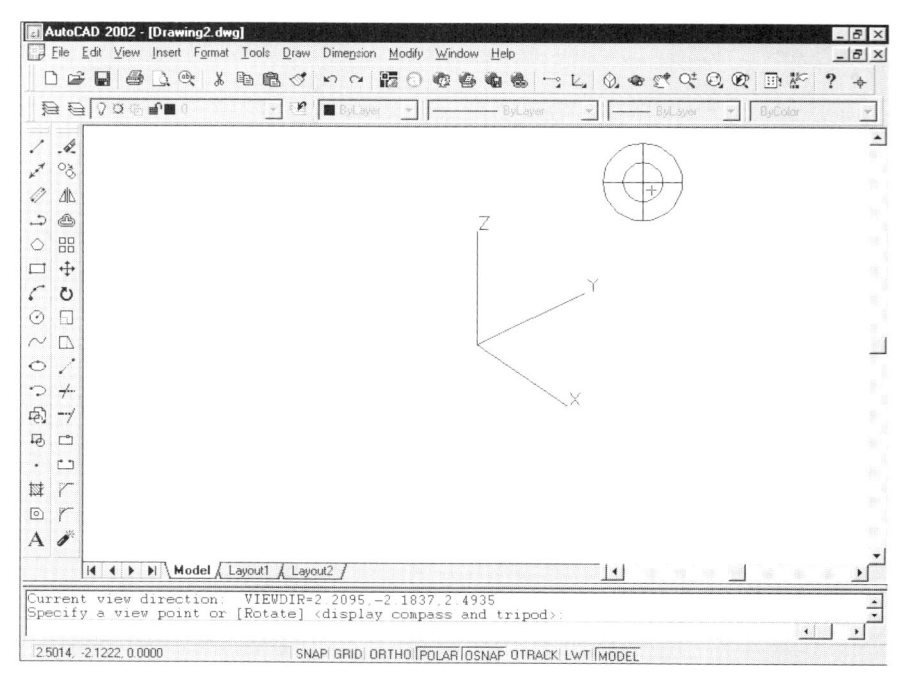

The three-pole axes indicate the orientation of the X, Y, and Z axes for the new *Vpoint*. The center of the "globe" represents the North Pole, so moving the cursor to that location generates a plan, or top, view. The small circle of the globe represents the Equator, so moving the cursor to any location on the Equator generates an elevation view (front, side, back, etc.). The outside circle represents the South Pole, so locating the cursor there shows a bottom view. When you PICK, the axes disappear and the current drawing is displayed from the new viewpoint.

The command format for using the *Tripod* method is as follows:

Command: **vpoint**
Current view direction: VIEWDIR=0.0000,0.0000,1.0000
Specify a view point or [Rotate] <display compass and tripod>: **Enter** (tripod appears)
PICK (Select desired viewing direction.)
Regenerating model
Command:

Using the *Tripod* method of *Vpoint* is quick and easy. Because no exact *Vpoint* can be given, it is difficult to achieve the exact *Vpoint* twice. Therefore, if you are working with a complex drawing that requires a specific viewpoint of a 3D model, use another option of *Vpoint* or save the desired *Vpoint* as a named *View*.

DDVPOINT

Pull-down Menu	COMMAND (TYPE)	ALIAS (TYPE)	Short-cut	Screen (side) Menu	Tablet Menu
View 3D Views > Viewpoint Presets...	DDVPOINT	VP	...	VIEW 1 Ddvpoint	N,5

The *Ddvpoint* command produces the *Viewpoint Presets* dialog box (Fig. 29-33). This tool is an interface for attaining viewpoints that could otherwise be attained with the *3D Viewpoint* or the *Vpoint* command.

Figure 29-33

Ddvpoint serves the same function as the *Rotate* option of *Vpoint*. You can specify angles *From: X Axis* and *XY Plane*. Angular values can be entered in the edit boxes, or you can PICK anywhere in the image tiles to specify the angles. PICKing in the enclosed boxes results in a regular angle (e.g., 45, 90, or 10, 30) while PICKing near the sundial-like "hands" results in irregular angles (see pointer in Figure 29-33). The *Set to Plan View* tile produces a plan view.

The *Relative to UCS* radio button calculates the viewing angles with respect to the current UCS rather than the WCS. Normally, the viewing angles should be absolute to WCS, but certain situations may require this alternative. Viewing angles relative to the current UCS can produce some surprising viewpoints if you are not completely secure with the model and observer orientation in 3D space.

NOTE: When you use this tool, ensure that the *Absolute to WCS* radio button is checked unless you are sure the specified angles should be applied relative to the current UCS.

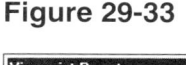

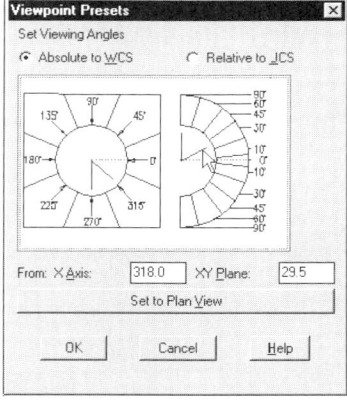

PLAN

Pull-down Menu	COMMAND (TYPE)	ALIAS (TYPE)	Short-cut	Screen (side) Menu	Tablet Menu
View 3D Views > Plan View >	PLAN	...	...	VIEW 1 Plan	N,3

This command is useful for quickly displaying a plan view of any UCS:

Command: *plan*
Enter an option [Current ucs/Ucs/World] <Current>: (**option**) or **Enter**
Regenerating model.
Command:

Responding by pressing Enter causes AutoCAD to display the plan (top) view of the current UCS. Typing *W* causes the display to show the plan view of the World Coordinate System. The *World* option does <u>not</u> cause the WCS to become the active coordinate system but only displays its plan view. Invoking the *UCS* option displays the following prompt:

Enter name of UCS or [?]:

The *?* displays a list of existing UCSs. Entering the name of an existing UCS displays a plan view of that UCS.

ZOOM

The *Zoom* command can be used effectively with 3D models just as with 2D drawings. However, the *Previous* option of *Zoom* is particularly applicable to 3D work.

Previous

This option of *Zoom* restores the previous display. When you are using 3D Views, *Vpoint*, *3Dorbit*, and *Plan* for viewing a 3D model, *Zoom Previous* will display the previous viewpoint.

VPORTS

Pull-down Menu	COMMAND (TYPE)	ALIAS (TYPE)	Short-cut	Screen (side) Menu	Tablet Menu
View Viewports >	VPORTS or -VPORTS	...	...	VIEW 1 Vports	M,3 and M,4

The *Vports* command creates viewports. Either model space viewports or paper space (layout) viewports are created, depending on which space is current when you activate *Vports*. See Chapters 10 and 12 for basic information about *Vports* and using viewports for 2D drawings.

This section discusses the use of viewports to enhance the construction of 3D geometry. Options of *Vports* are available with AutoCAD specifically for this application.

When viewports are created in the *Model* tab (model space), the screen is divided into sections like tiles, unlike paper space viewports. Tiled viewports (viewports in model space) <u>affect only the screen display</u>. The viewport configuration <u>cannot be plotted</u>. If the *Plot* command is used, only the current viewport is plotted.

To use viewports for construction of 3D geometry, use the *Vports* command to produce the *Viewports* dialog box (Fig. 29-34). Make sure you invoke *Vports* in the *Model* tab (model space). To generate a typical orthographic view arrangement in your drawing, first select three or four viewports from the list in the *Standard viewports* list, then select the *3D* option in the *Setup:* drop-down list (see Figure 29-34, bottom center). This action produces a typical top, front, southeast isometric, and side view, depending on whether you selected three or four viewports.

Figure 29-34

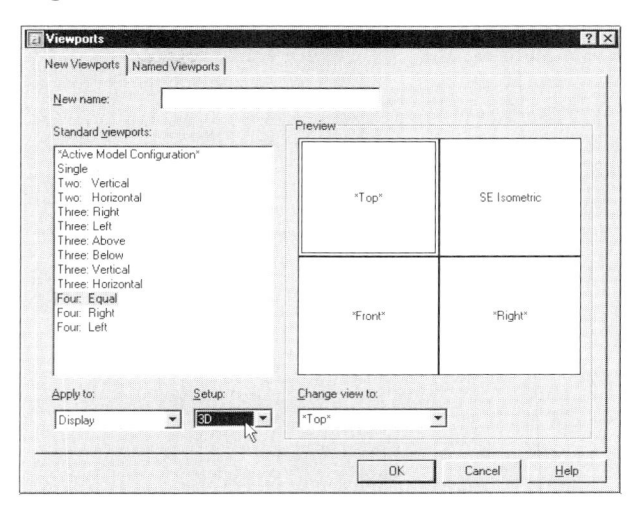

Assuming the options shown in Figure 29-34 were used with the FORKLIFT drawing, the configuration shown in Figure 29-35 would be generated automatically. Note, however, that the objects are not sized proportionally in the viewports. Since a *Zoom Extents* is automatically performed, each view is sized in relation to its viewport, not in relation to the other views.

Figure 29-35

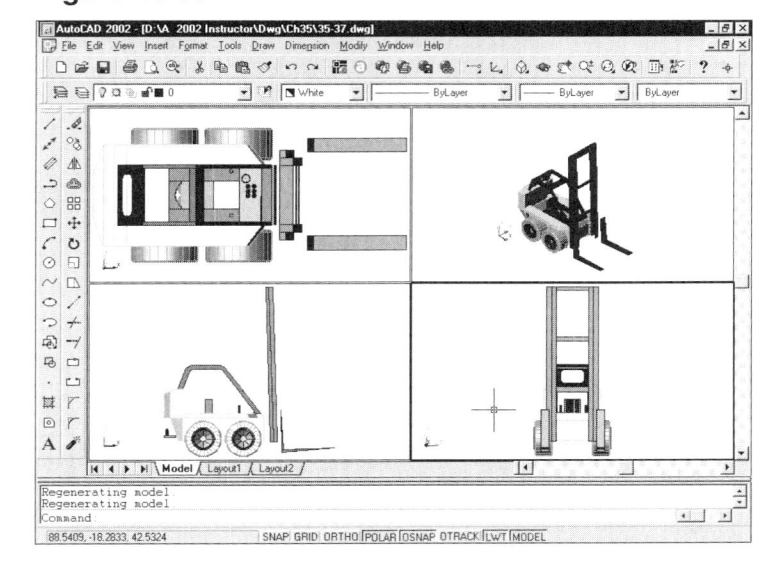

Next, if you desire all views to display the object proportionally, use *Zoom* with a constant scale factor (such as 1, not 1x) in each viewport. This action would generate a display similar to that shown in Figure 29-36.

This *3D Setup* option is available only from the *Viewports* dialog box. If you use one of the other options from the *View* pull-down menu (*3 Viewports*, *4 Viewports*, etc.) or type the command line version (*-Vports*), you cannot automatically generate a typical orthographic view setup.

Figure 29-36

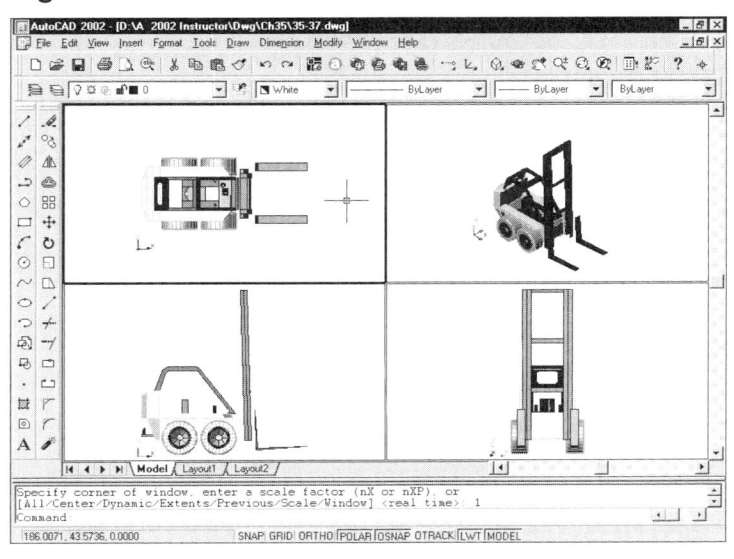

598 **3D Display and Viewing**

> **NOTE:** The setting of *UCSORTHO* has an effect on the resulting UCSs when the *3D Setup* option of the *Viewport* dialog box is used. If *UCSORTHO* is set to 0, no new UCSs are created when *Vports* is used. However, if *UCSORTHO* is set to 1, a new UCS is generated for each new orthographic view created so that the XY plane of each new UCS is parallel with the view. In other words, each orthographic view (not the isometric) now shows a UCS with the XY plane parallel to the screen. See Chapter 30 for more information on the *UCSORTHO* system variable.

CHAPTER EXERCISES

1. In this exercise you will use a sample drawing and practice with the 3D display commands *Shademode* and *Hide*.

   A. *Open* the **TRUCK MODEL** drawing located in the AutoCAD 2002/Sample folder. Select the *Model* tab if not already current.

   B. Use the *Hide* command to generate a hidden display. Your display should look like that in Figure 29-37.

   Figure 29-37

   C. Now use *Shademode* and change the display to *Hidden*. Is there any difference between this *Shademode* setting and the results of the *Hide* command?

   D. Change the *Shademode* setting to *2D wireframe*. Now use the *3D wireframe* option. What is the difference between these two options for the Truck Model?

   E. Change the *Shademode* setting to *Flat*. Now use the *Flat+Edges* option.

   F. Finally, change the setting to *Gouraud*. Now make the edges visible using the *Gouraud+Edges* option.

   G. *Exit* the Truck Model drawing and do not save the changes.

2. In this exercise, you use an AutoCAD sample drawing to practice with the *View* command. *Open* the **OPERA** drawing from the Sample directory. (Assuming AutoCAD is installed using the default location, find the sample drawings in C:/Program Files/AutoCAD 2002/Sample.) Select the *Model* tab. *Freeze* layers **E-AGUAS**, **E-PREDI**, **E-TERRA**, **E-VIDRO**, and **O-PAREDE**. Use *Shademode* to generate a *Gouraud* display.

   A. Use any method to generate a *Top* view. Examine the view and notice the orientation of the coordinate system icon.

B. Generate a *SE Isometric* viewpoint to orient yourself. Examine the icon and find north (Y axis). Consider how you (the observer) are positioned as if viewing <u>from</u> the southeast.

C. Produce a *Front* view. This is a view looking north. Notice the icon. The Y axis should be pointing away from you.

D. Generate a *Right* view. Produce a *Top* view again.

E. Next, view the OPERA from a *SE Isometric* viewpoint. Your view should appear like that in Figure 29-38. Notice the orientation of the WCS icon on your screen.

Figure 29-38

F. Finally, generate a *NW Isometric* and *NE Isometric*. In each case, examine the WCS icon to orient your viewing direction.

G. Experiment and practice more with *3D Viewpoint* if you desire. <u>Do not *Save*</u> the drawing.

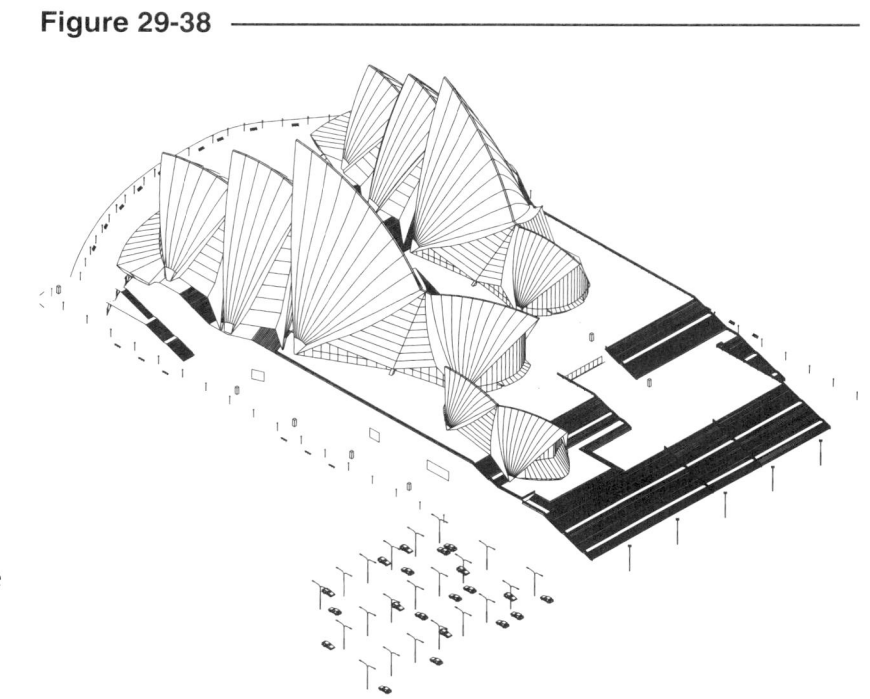

3. In this exercise you will practice using the **3D views**, *Ucsicon*, *Vpoint*, and *Zoom*. *Open* the **TRUCK MODEL** again that you used in Exercise 1.

   A. First, type "*UCS*," then use the *World* option. This action makes the World Coordinate System the current coordinate system. Use the *Ucsicon* command and ensure the icon is at the *Origin*. Notice the orientation of the truck with respect to the WCS. The X axis points outward from the right side of the truck, Y positive is to the rear of the truck, and Z positive is up.

   B. Generate a *Top* view. Note the position of the X, Y, and Z axes. You should be looking down on the XY plane and Z points up toward you.

   C. Generate a *SE Isometric* view. Now use *Zoom* with the *Previous* option two times to give you the original view.

   D. Generate a *Front* view. Notice that AutoCAD performs a *Zoom Extents* so all of the drawing appears on the screen. *Freeze* layer **BASE**. Now generate a *Front* view again.

   E. Use *Vpoint* with the *Tripod* option to view the truck from the southeast direction and slightly from above. HINT: Pick the lower right (southeast) quadrant inside the small circle.

F. Use *Vpoint* with the *Tripod* option again to view the truck from the southwest and slightly above. HINT: Pick the lower left (southwest) quadrant inside the small circle.

G. Close the drawing but <u>do not save</u> the changes.

4. Now you will get some practice with the *3Dorbit* commands. *Open* the **WATCH** drawing from the AutoCAD 2002/Sample folder. Activate the *Model* tab.

   A. First generate a *SE Isometric* view to get a good look at the watch. Notice that the components are disassembled, similar to an exploded view assembly drawing.

   B. Now invoke *3Dorbit*. Note that the *Grid* is on. Right-click to produce the shortcut menu, then under *Visual Aids* toggle off the *Grid*. In the shortcut menu, select *Shading Modes*, then *Gouraud Shaded*.

   C. Now place your cursor inside the "arc ball" and drag your pointing device to change your viewpoint. Next, try "rolling" the display by using your cursor outside the arc ball.

   D. Next, try placing your cursor on the small circle at the right or left quadrant of the arc ball. Hold down the left mouse button and drag left and right to spin the watch about an imaginary vertical axis. Try the same technique from the top and bottom quadrant to spin the watch about an imaginary horizontal axis.

   E. Finally generate your choice of views that shows the watch so all components are visible. Change the *Projection* to *Perspective*. Under the *More* option in the shortcut menu, select *Continuous Orbit*. Hold down the left mouse button, drag the mouse, and release the button to put the watch in a continuous spin.

   F. *Close* the drawing and do not save changes.

5. Use the R300-20 drawing to practice with the *3Dorbit* commands. *Open* the **R300-20** drawing from the **AutoCAD 2002/Sample** folder. Activate the *Model* tab.

   A. Carefully <u>select only the assembled model</u> (not the exploded components), then activate *3Dorbit*. Only the selected components should appear in the arc ball. Right-click to produce the shortcut menu and turn off the *Grid*. Do not exit *3Dorbit*.

   B. Use *Pan* to move the assembly to the center of the arc ball. Next, use *Zoom* to enlarge the display. Use *Pan* again if necessary. Then turn on the *Gouraud Shaded* mode. Do not exit *3Dorbit*.

   C. Now, select *Orbit* from the shortcut menu and change your viewpoint. When you rotate the arc ball, does the assembly stay inside the arc ball? If not, select *More* and *Orbit uses AutoTarget* from the shortcut menu.

   D. Adjust the orbit view to place the assembly in the arc ball. Turn on *Perspective*. Now use *Adjust Distance* to increase the amount of perspective. You will have to *Zoom* then to reduce the size of the assembly. Finally, select *Reset* view to return to the original view.

   E. *Close* the drawing, <u>do not save</u> changes, then *Open* the same drawing again. This time select all the components of the exploded assembly (not the assembled model).

F. Start *3Dorbit*, right-click, and toggle off the *Grid*. Turn on *Gouraud Shaded*.

G. Select *Adjust Clipping* planes from the shortcut menu. The *Adjust Clipping Planes* window should appear. Select the *Adjust Front Clipping Plane* button (you can right-click inside the window to select from a shortcut menu). Notice how adjusting the clipping plane inside the window changes the display of the model in the drawing area to dynamically display the cutting plane as you pass it through the model. Notice also that when you close the window and *Exit* 3Dorbit, the clipping plane remains on in the drawing area and only the portion of the model you selected for *3Dorbit* remains visible.

H. *Close* the drawing and do not save the changes.

6. In this exercise you will create four model space viewports with the *3D* option. *Open* the **WATCH** drawing again. Activate the *Model* tab. Check the setting of the **UCSORTHO** variable by typing it at the command line and examining the setting. A setting of **0** means no new UCSs will be created when new views are generated.

   A. Invoke the **Vports** command (make sure you are in model space). In the *New Viewports* tab of the *Viewports* dialog box, select **Four: equal**. At the bottom of the dialog box in the *Setup* drop-down list, select **3D**. Select the *OK* button.

   B. Your display should show a top, front, right side, and southeast isometric view. Notice that the watch is not sized proportionally with respect to the other views. Activate the front viewport (bottom left). Type *Zoom* at the command line and enter a value of **1.5**. Do the same for the other two orthographic views. Your display should look like that shown in Figure 29-39.

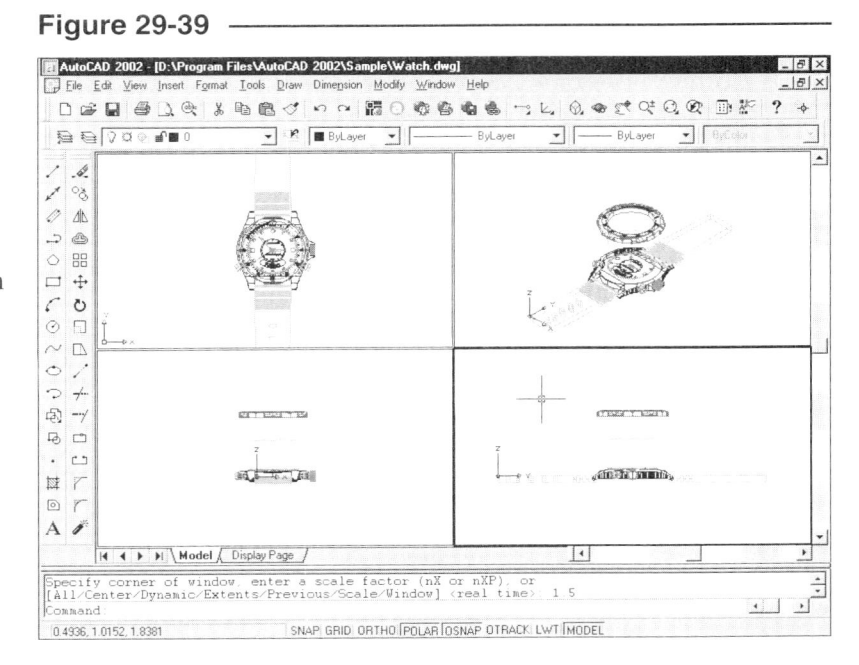

   Figure 29-39

   C. Activate the isometric viewport. Enter *Shademode* at the command line and change the setting to *Gouraud*.

   D. This is a good arrangement to use when you are constructing 3D models because you can see several views of the object on your screen at once. Remember this idea when you begin working with the exercises in the next several chapters. *Close* the drawing but do not save the changes.

7. In this exercise, you will create a simple solid model that you can use for practicing creation of User Coordinate Systems in Chapter 30. You will also get an introduction to some of the solid modeling construction techniques discussed in Chapter 31.

   A. Begin a *New* drawing and name it SOLID1. Turn *On* the *Ucsicon* and force it to appear at the *ORigin*.

   Figure 29-40

   B. Type *Box* to create a solid box. When the prompts appear, use **0,0,0** as the "Corner of box." Next, use the *Length* option and give dimensions for the Length, Width, and Height as **5, 4,** and **3**. A rectangle should appear. Change your viewpoint to *SE Isometric* to view the box. *Zoom* with a magnification factor of *.6X*.

   C. Type the *Wedge* command. When the prompts appear, use **0,0,0** as the "Corner of wedge." Again use the *Length* option and give the dimensions of **4, 2.5,** and **2** for *Length, Width,* and *Height*. Your solid model should appear as that in Figure 29-40.

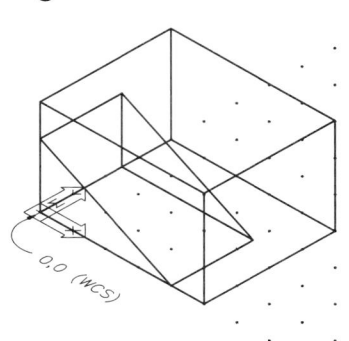

   Figure 29-41

   D. Use *Rotate* and select only the wedge. Use **0,0** as the "Base point" and enter **-90** as the "Rotation angle." The model should appear as Figure 29-41.

   E. Now use *Move* and select the wedge for moving. Use **0,0** as the "Base point" and enter **0,4,3** as the "Second point of displacement."

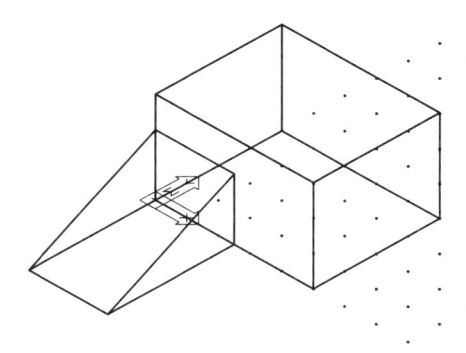

   Figure 29-42

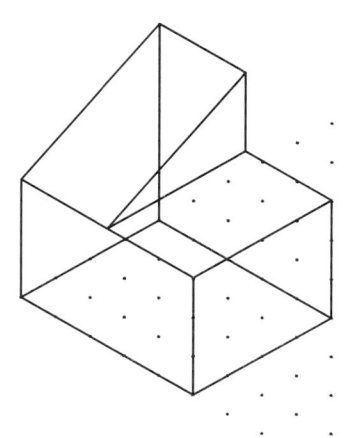

   F. Finally, type *Union*. When prompted to "Select objects:", PICK both the wedge and the box. The finished solid model should look like Figure 29-42. *Save* the drawing.

30

USER COORDINATE SYSTEMS

Chapter Objectives

After completing this chapter you should:

1. know how to create, *Save*, *Restore*, and *Delete* User Coordinate Systems using the *UCS* command;

2. know how to use the *UCS Manager* to create, *Save*, *Restore*, and *Delete* User Coordinate Systems;

3. be able to create UCSs by the *Origin*, *Zaxis*, *3point*, *Object*, *Face*, *View*, *X*, *Y*, and *Z* methods;

4. be able to create *Orthographic* UCSs using the *UCS* command and UCS Manager;

5. know the function of the *UCSVIEW*, *UCSORTHO*, *UCSFOLLOW*, and *UCSVP* system variables.

CONCEPTS

No User Coordinate Systems exist as part of any of the AutoCAD default template drawings as they come "out of the box." UCSs are created to simplify construction of the 3D model when and where they are needed.

When you create a multiview or other 2D drawing, creating geometry is relatively simple since you only have to deal with X and Y coordinates. However, when you create 3D models, you usually have to consider the Z coordinates, which makes the construction process more complex. Constructing some geometries in 3D can be very difficult, especially when the objects have shapes on planes not parallel with, or perpendicular to, the XY plane or not aligned with the WCS (World Coordinate System).

Figure 30-1

UCSs are created when needed to simplify the construction process of a 3D object. For example, imagine specifying the coordinates for the centers of the cylindrical shapes in Figure 30-1, using only world coordinates. Instead, if a UCS were created on the face of the object containing the cylinders (the inclined plane), the construction process would be much simpler. To draw on the new UCS, only the X and Y coordinates (of the UCS) would be needed to specify the centers, since anything drawn on that plane has a Z value of 0. The Coordinate Display (*Coords*) in the Status Bar always displays the X, Y, and Z values of the current coordinate system, whether it is the WCS or a UCS.

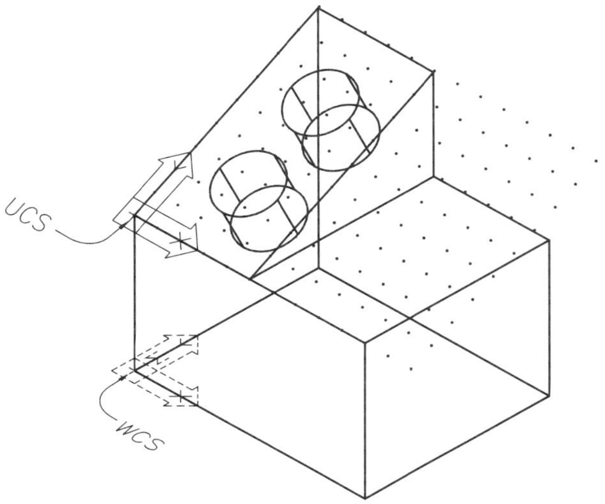

Generally, you would create a UCS aligning its XY plane with a surface of the object, such as along an inclined plane, with the UCS origin typically at a corner or center of the surface. Any coordinates that you specify (in any format) are assumed to be user coordinates (coordinates that lie in or align with the current UCS). Even when you PICK points interactively, they fall on the XY plane of the current UCS. The *SNAP, Polar Snap,* and *GRID* automatically align with the current coordinate system XY plane, enhancing your usefulness of the construction plane.

UCS COMMANDS

User Coordinate Systems are created by using any of several options of the *Ucs* command or UCS Manager (*Ucsman* command). You can create multiple UCSs in one drawing. The *Save* option of the *Ucs* command allows you to assign a name and save the UCSs that you create so you can *Restore* them at a later time.

In releases of AutoCAD previous to AutoCAD 2000, only one UCS could be active in a drawing at one time. If you chose to use viewports (*Vports* command), only one UCS was active and visible in all viewports at a time. This process was relatively simple to understand and manage, but had limitations.

In AutoCAD 2000, several new features were introduced that make UCSs much more flexible, but also more difficult to manage. These features included saving UCSs with *Views*, automatically changing UCSs to align with views when they are created or restored, having multiple UCSs visible when using *Model* tab viewports, and locking UCSs to specific viewports. This chapter will explain the features and assist you in using UCSs wisely.

When creating 3D models, it is recommended that you use the *Ucsicon* command with the *On* option to make the icon visible and the *Origin* option to force the icon to the origin of the current coordinate system rather than its default placement in the lower-left corner of the screen. Most of the figures in this text use the "2D" UCS icon. This icon displays the orientation of the XY plane at a glance and makes the figures more understandable, whereas the 3-pole icons take more effort to distinguish the XY plane orientation.

UCS

Pull-down Menu	COMMAND (TYPE)	ALIAS (TYPE)	Short-cut	Screen (side) Menu	Tablet Menu
Tools Move UCS, New UCS> or Orthographic UCS>	UCS	...	...	TOOLS 2 UCS	W,7

The *UCS* command allows you to create, save, restore, and delete User Coordinate Systems. There are many options available for creating UCSs. The *UCS* command always operates only in Command line mode.

Command: *ucs*
Current ucs name: *WORLD*
Enter an option
[New/Move/orthoGraphic/Prev/Restore/Save/Del/Apply/?/World] <World>:

Although the *UCS* command operates in Command line mode, each *UCS* option has an icon which can be accessed from the Standard toolbar (Fig. 30-2). Choosing one of these icons activates the particular option of the *UCS* command.

Figure 30-2

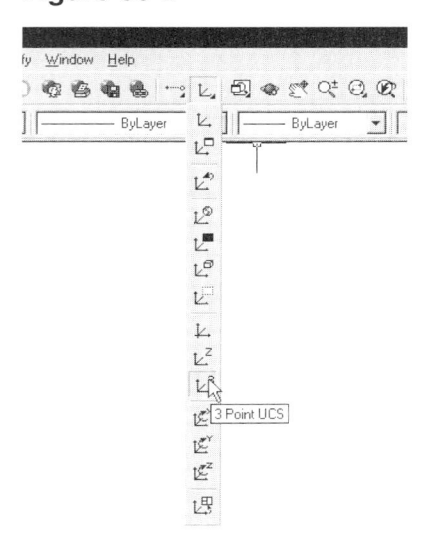

You can also use the *Tools* pull-down menu to select an option of the *UCS* command. There are two main options on the *Tools* pull-down menu: *Orthographic UCS* and *New UCS*. Suboptions are on cascading menus (Fig. 30-3).

In addition to these methods of activating the *UCS* options, a *UCS* toolbar is available (not shown). The *UCS* toolbar has the same icons appearing on the flyout from the Standard toolbar (see Figure 30-2).

Figure 30-3

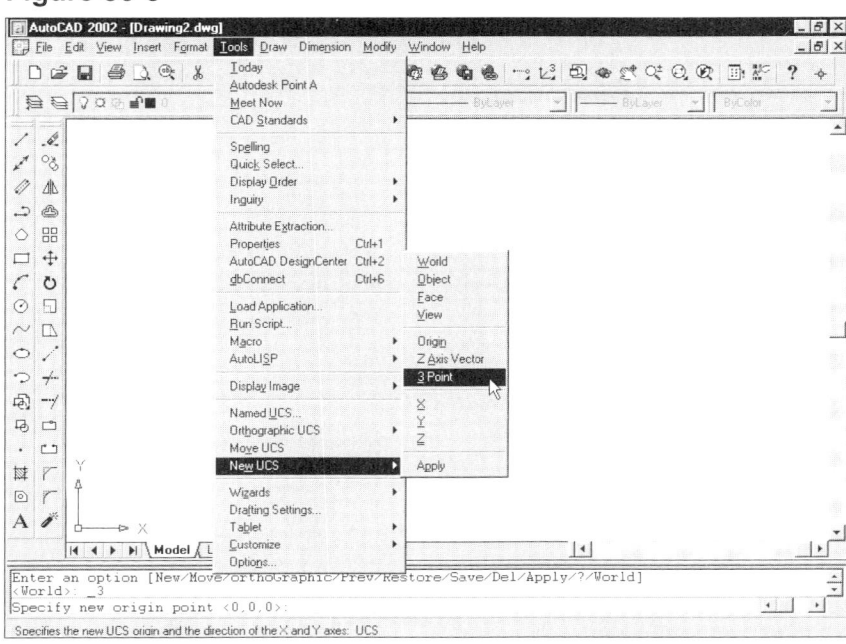

606 User Coordinate Systems

When you create UCSs, AutoCAD prompts for points. These points can be entered as coordinate values at the keyboard or you can PICK points on existing 3D objects. *OSNAP* modes should be used to PICK points in 3D space (not on the current XY plane). In addition, understanding the right-hand rule is imperative when creating some UCSs (see Chapter 28, 3D Modeling Basics).

The *UCS* options are discussed next.

Move, Origin

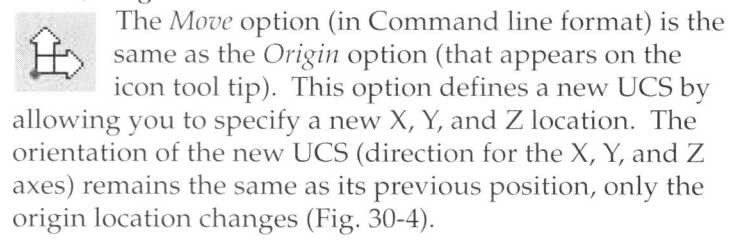

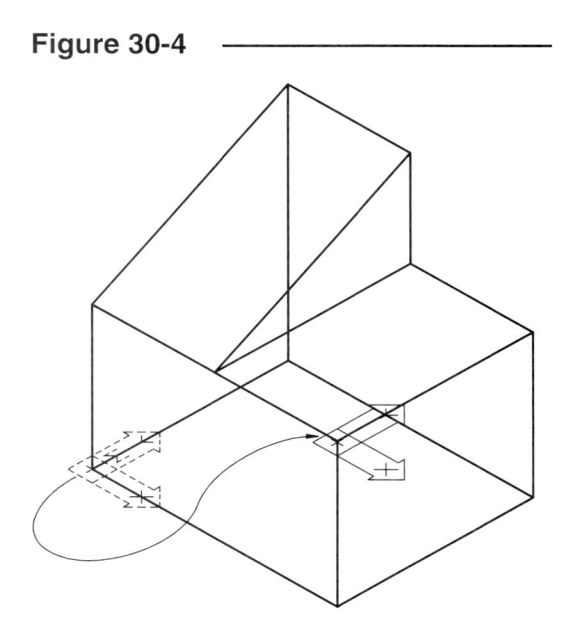

Figure 30-4

The *Move* option (in Command line format) is the same as the *Origin* option (that appears on the icon tool tip). This option defines a new UCS by allowing you to specify a new X, Y, and Z location. The orientation of the new UCS (direction for the X, Y, and Z axes) remains the same as its previous position, only the origin location changes (Fig. 30-4).

```
Command: ucs
Current ucs name:  *WORLD*
Enter an option [New/Move/orthoGraphic/Prev/Restore/Save/Del/Apply/?/World] <World>: m
Specify new origin point or [Zdepth]<0,0,0>: PICK or (value) or Z
Command:
```

Coordinates can be specified in any format or can be PICKed (use *OSNAPs* in 3D space).

New

The *New* option does not actually create a new UCS. It merely gives access to the other options for creating new UCSs. This option is necessary since there are too many *UCS* options to fit on one Command line, so *New* displays a second level of options.

```
Command: ucs
Current ucs name:  *WORLD*
Enter an option [New/Move/orthoGraphic/Prev/Restore/Save/Del/Apply/?/World] <World>: n
Specify origin of new UCS or [ZAxis/3point/OBject/Face/View/X/Y/Z] <0,0,0>:
```

The options displayed on the Command line above are explained next.

ZAxis

You can define a new UCS by specifying an origin and a direction for the Z axis. Only the two points are needed. The X or Y axis generally remains parallel with the current UCS XY plane, depending on how the Z axis is tilted. AutoCAD prompts:

Specify new origin point <0,0,0>: **PICK** or (**coordinates**)
Specify point on positive portion of Z-axis <*n.nnnn, n.nnnn, n.nnnn*>: **PICK** or (**coordinates**)

Figure 30-5

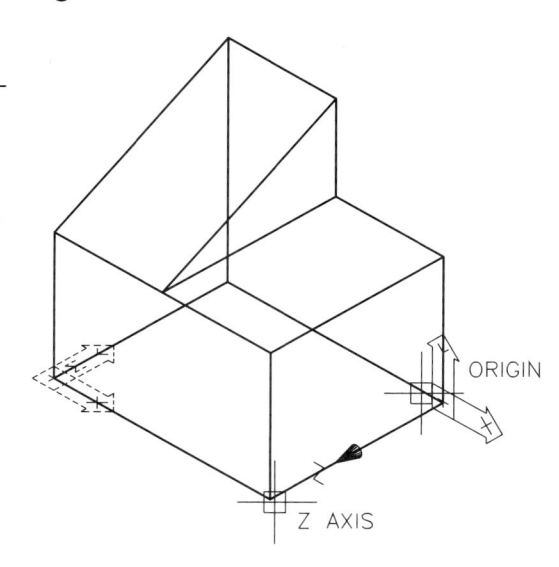

3point

This new UCS is defined by (1) the origin, (2) a point on the X axis (positive direction), and (3) a point on the Y axis (positive direction) or XY plane (positive Y). This is the most universal of all the UCS options (works for most cases). It helps if you have geometry established that can be PICKed with *OSNAP* to establish the points. The prompts are:

Specify origin of new UCS or
[ZAxis/3point/OBject/Face/View/X/Y/Z] <0,0,0>: **3**
Specify new origin point <0,0,0>: **PICK** or (**coordinates**)
Specify point on positive portion of X-axis <*n.nnnn, n.nnnn, n.nnnn*>: **PICK** or (**coordinates**)
Specify point on positive-Y portion of the UCS XY plane <*n.nnnn, n.nnnn, n.nnnn*>: **PICK** or (**coordinates**)
Command:

Figure 30-6

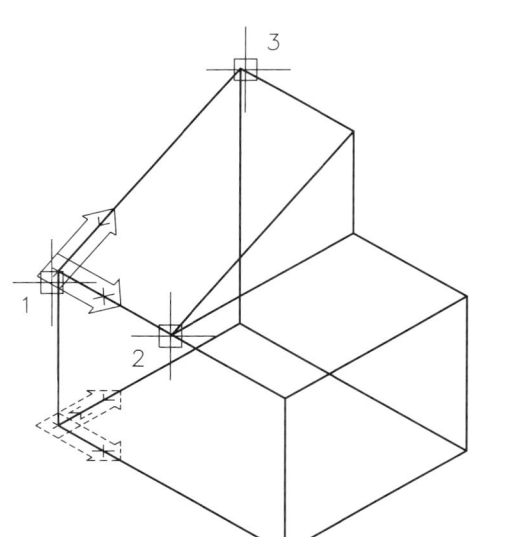

Object

This option creates a new UCS aligned with the selected object. You need to PICK only one point to designate the object. The orientation of the new UCS is based on the type of object selected and the XY plane that was current when the object was created. This option is intended for use primarily with wireframe or surface model objects.

Face

The *Face* option operates with solids only. A *Face* is a planar or curved surface on a solid object, as opposed to an edge. When selecting a *Face*, you can place the cursor directly on the desired face when you PICK, or select an edge of the desired face.

Specify origin of new UCS or [ZAxis/3point/OBject/Face/View/X/Y/Z] <0,0,0>: **f**
Select face of solid object: **PICK**
(Select desired face of 3D solid to attach new UCS. Edges or faces can be selected.)
Enter an option [Next/Xflip/Yflip] <accept>:

Next
Since you cannot actually PICK a surface, or if you PICK an edge, AutoCAD highlights a surface near where you picked or adjacent to a selected edge (Fig. 30-7). Since there are always two surfaces joined by one edge, the *Next* option causes AutoCAD to highlight the other of the two possible faces. Press Enter to accept the highlighted face.

Xflip
Use this option to flip the UCS icon 180 degrees about the X axis. This action causes the Y axis to point in the opposite direction (see Figure 30-7).

Yflip
This option causes the UCS icon to flip 180 degrees about the Y axis. The X axis then points in the opposite direction.

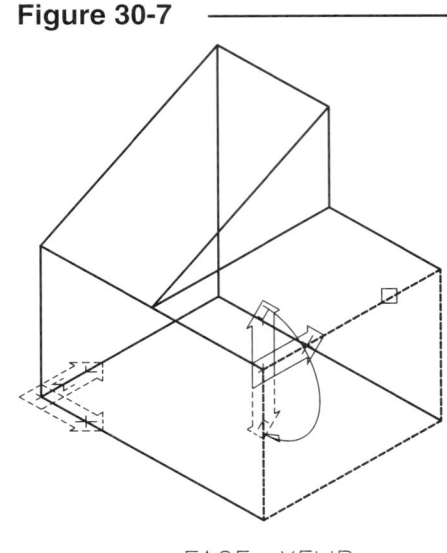

Figure 30-7

FACE, XFLIP

View
This *UCS* option creates a UCS parallel with the screen (perpendicular to the viewing angle). The UCS origin remains unchanged. This option is handy if you wish to use the current viewpoint and include a border, title, or other annotation. There are no options or prompts.

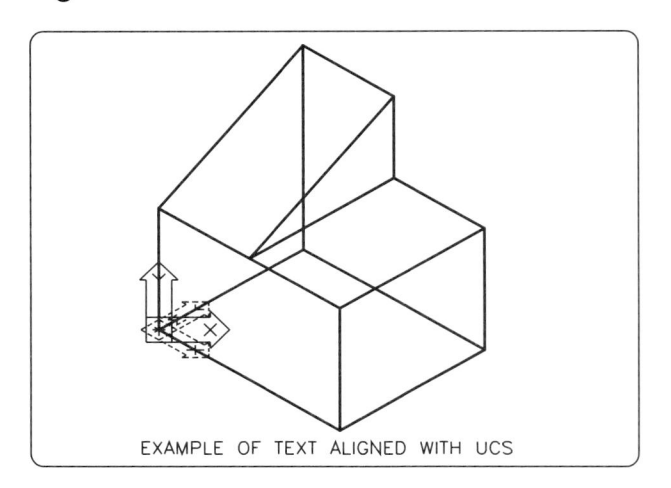

Figure 30-8

EXAMPLE OF TEXT ALIGNED WITH UCS

X Y Z
Each of these options rotates the UCS about the indicated axis according to the right-hand rule. The Command prompt is:

Specify rotation angle about *n* axis <90>:

The angle can be entered by PICKing two points or by entering a value. This option can be repeated or combined with other options to achieve the desired location of the UCS. It is imperative that the right-hand rule is followed when rotating the UCS about an axis (see Chapter 28).

Figure 30-9

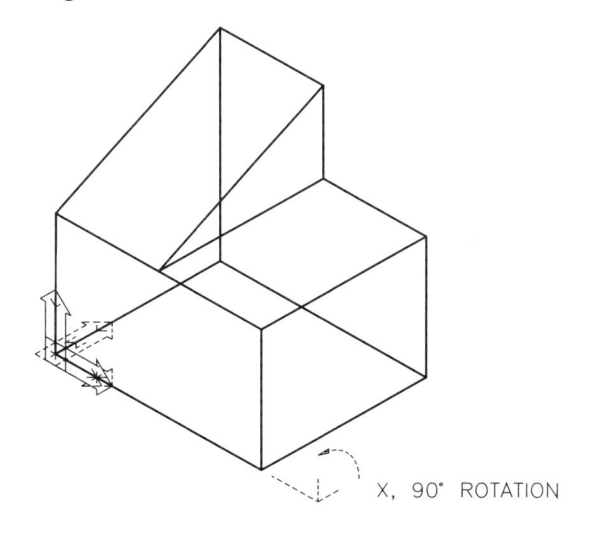

X, 90° ROTATION

Notice that in the prompt shown above for the X, Y, or Z, axis rotate options, the default rotation angle is 90 degrees ("<90>" as shown above). If you press Enter at the prompt, the axis rotates by the displayed amount (90 degrees, in this case). You can set the *UCSAXISANG* system variable to another value which will be displayed in brackets when you next use the X, Y, or Z, axis rotate options. Valid values are: 5, 10, 15, 18, 22.5, 30, 45, 90, 180.

Orthographic
The *Orthographic* option is used to create new UCSs in an orthographic orientation so that the XY plane of the new UCS is parallel to the *Top*, *Front*, *Right*, *Bottom*, *Back*, or *Left* faces. For example, if a *Right* UCS were created, its XY plane would be visible when viewing from the right side view of the object; or a *Plan* view of the *Right* UCS would yield a normal right side view.

When an *Orthographic* option is used, the UCS does not move to the designated side of the object (*Right*, *Top*, *Front*, etc.). Instead, the UCS only rotates to that orientation. The UCS generally rotates in relation to the WCS (when *UCSBASE* system variable is set to the default, *World*). For example, if the current coordinate system were the WCS, selecting the *Right* option would rotate the UCS to the correct orientation keeping the same origin, but would not move to the rightmost face of the object (Fig. 30-10).

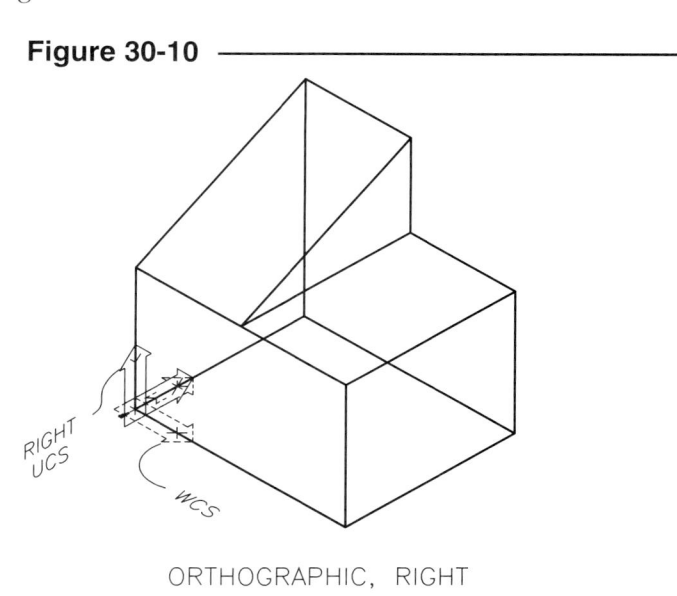

Figure 30-10

ORTHOGRAPHIC, RIGHT

Previous
Use this option to restore the previous UCS. AutoCAD remembers the ten previous UCSs used. *Previous* can be used repeatedly to step back through the UCSs.

World
Using this option makes the WCS (World Coordinate System) the current coordinate system.

Save
Invoking this option prompts for a name for saving the current UCS. Up to 256 characters can be used in the name. Entering a name causes AutoCAD to save the current UCS. The *?* option and the *Dducs* command list all previously saved UCSs.

Restore
Any *Save*d UCS can be restored with this option. The *Restore*d UCS becomes the current UCS. The *?* option and the UCS Manager list the previously saved UCSs.

Delete
You can remove a *Save*d UCS with this option. Entering the name of an existing UCS causes AutoCAD to delete it.

Apply

The *Apply* option allows you to apply the UCS in the current viewport to all or selected viewports.

Since AutoCAD 2000, each view or viewport can have a different UCS. With one view on the screen, only one UCS can be active. If, however, you are using viewports in model space (as described in Chapter 29), you can have a different UCS in each viewport. The *Apply* option allows you to use the UCS in the current viewport and make it the active UCS for any other, or all, viewports.

```
Command: ucs
Current ucs name:  *WORLD*
Enter an option [New/Move/orthoGraphic/Prev/Restore/Save/Del/Apply/?/World] <World>: a
Pick viewport to apply current UCS or [All]<current>: PICK (PICK inside the viewport you want to
change to the current viewport's UCS.)
Command:
```

First, make whichever viewport current that contains the UCS you want to apply. Next, use the *UCS* command and the *Apply* option. Then select the viewport you want to accept the new UCS (from the previous current viewport).

?

This option lists the named UCSs (UCSs that you *Saved*). Also listed are the origin coordinates and directions for the axes.

NOTE: It is important to remember that changing a UCS or creating a new UCS does not change the display (unless the *UCSFOLLOW* system variable is set to 1). Only one coordinate system can be active in a view. If you are using viewports, each viewport can have its own UCS if desired.

DDUCSP

Pull-down Menu	COMMAND (TYPE)	ALIAS (TYPE)	Short-cut	Screen (side) Menu	Tablet Menu
Tools Orthographic UCS...	DDUCSP	UCP	...	TOOLS 2 Ucsp	W,9

The *Dducsp* command invokes the UCS Manager. The UCS Manager was introduced in AutoCAD 2000. It can be used to create new UCSs or to restore existing UCSs. It combines several functions offered in previous releases and offers access to some new features. The UCS Manager can also be used to set specific UCS-related system variables, described later.

The *Dducsp* command activates specifically the *Orthographic UCSs* tab of the UCS Manager (Fig. 30-11). Here you can create new UCSs, similar to the *Orthographic* option of the *Ucs* command. *Orthographic UCSs* are not automatically saved as named UCSs but are created as you select them; that is, you can create, save, and restore additional but separate UCSs with the names Front, Top, Right, etc.

Figure 30-11

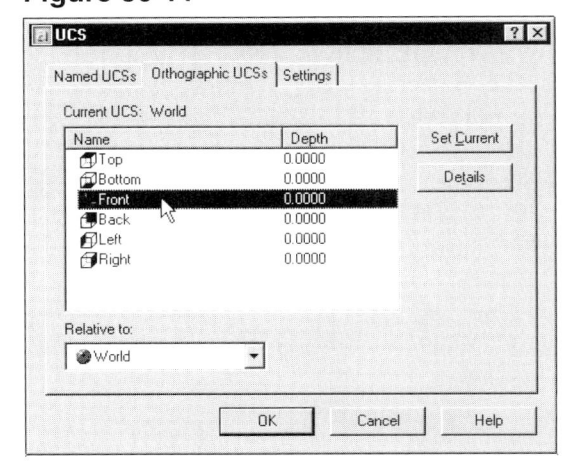

To create one of these *Orthographic UCSs*, you can double-click the desired UCS name, or highlight the desired UCS name (one click) and select the *Current* button. Press *OK* to return to the drawing to see the new UCS. Keep in mind that although the new UCSs are normal to the front, top, right, etc. orthographic views, they are not necessarily attached to the front, top, or right faces of the objects (see Figure 30-10 and previous explanation under *UCS, Orthographic*). The new UCSs typically have the same origin as the WCS, unless a new *Depth* is specified or a named UCS is selected in the *Relative to:* drop-down list.

Depth
The *Orthographic UCS* tab has one feature not available with the *Ucs* command. Notice in the *Orthographic UCS* tab that each UCS can have a specified depth. *Depth* is a Z dimension or distance perpendicular to the XY plane to locate the new UCS origin. Select the *Depth* value of any UCS to produce the *Orthographic UCS Depth* dialog box (Fig. 30-12).

Figure 30-12

Details
Select the *Details* button to see the coordinates for the origin of the highlighted UCS. The direction vector coordinates for the X, Y, and Z axes are also listed. This has the same function as the *?* option of the *UCS* command.

Relative to:
The *Relative to:* drop-down list allows you to select any named UCS as the base UCS. When the *Front*, *Top*, *Right*, or other UCS option is selected, it is created in a orthographic orientation with respect to the origin and orientation of the selected named UCS. Typically, you want to ensure World is selected as the base UCS unless you have some other specific application. The setting in the *Relative to:* list is stored in the *UCSBASE* system variable.

DDUCS

Pull-down Menu	COMMAND (TYPE)	ALIAS (TYPE)	Short-cut	Screen (side) Menu	Tablet Menu
Tools Named UCS...	DDUCS	UC	...	TOOLS 2 Ucsman	W,8

DDUCS produces the *Named UCS* tab of the UCS Manager (Fig. 30-13). The list contains all named UCSs. Use this tab to make named UCSs *Current*, just as you would use the *Restore* option of the *Ucs* command. (You must first create named UCSs using the *Save* option of the *Ucs* command.) In this tab you can double-click the desired UCS name or highlight (single-click) the desired name and then select *Set Current*. Either way, you must then select *OK* to restore the desired UCS. The *Details* button lists the origin coordinates and X, Y, and Z direction vectors of the highlighted UCS.

Figure 30-13

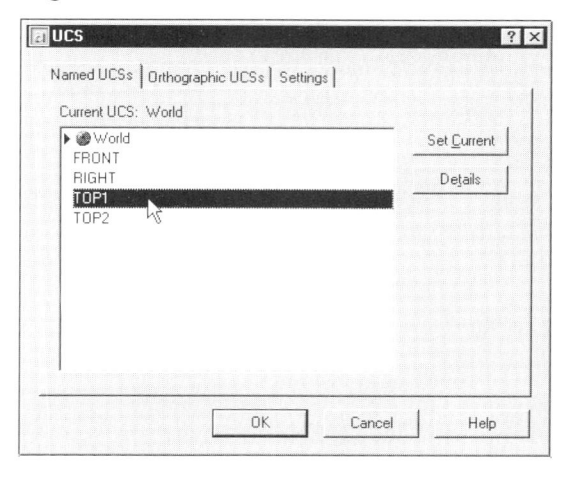

UCSMAN

Pull-down Menu	COMMAND (TYPE)	ALIAS (TYPE)	Short-cut	Screen (side) Menu	Tablet Menu
Tools Named UCS...	UCSMAN	...	...	TOOLS2 Ucsman	W,8

The *Ucsman* command invokes the UCS Manager. The UCS Manager has three tabs: *Named UCSs, Orthographic UCSs* and *Settings*. The UCS Manager can be accessed by several methods shown in the table above. However, the *Named UCSs* and *Orthographic UCSs* tabs of the UCS Manager are also accessed by the *Dducsp* and *Dducs* commands and icon buttons, respectively (see *Dducsp* and *Dducs* commands earlier this chapter). The *Settings* tab is explained below.

Settings **Tab**

The *Settings* tab can be accessed directly only by invoking the UCS Manager. However, you can use the *Dducsp* or *Dducs* commands or buttons explained earlier to produce the other tabs, then change tabs. The *Settings* tab (Fig. 30-14) allows you to control the UCS icon and two system variables.

Figure 30-14

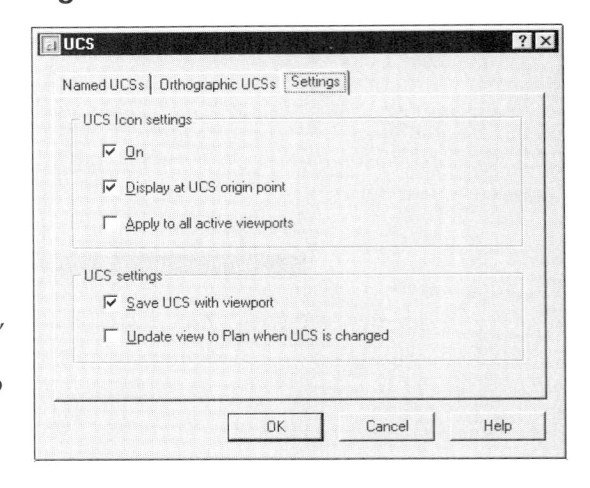

UCS Icon settings

Settings in this section are identical to controls of the *Ucsicon* command (see the *Ucsicon* command in Chapter 28, 3D Modeling Basics). The *On* checkbox toggles the UCS icon on or off. *Display at UCS origin point* moves the icon to the coordinate system origin (otherwise it is located in the lower left corner of the screen) and has the identical functions as the *Origin* and *Noorigin* options of the *Ucsicon* command. *Apply to all viewports* is the same as the *All* option of the *Ucsicon* command; that is, the settings made in this section apply to the icons in all active viewports when multiple viewports are used.

UCS Settings

These two settings control two UCS-related system variables that affect only the <u>currently active viewport</u> when multiple viewports are used. *Save UCS with viewport* controls the *UCSVP* system variable and affects the <u>UCS of the current viewport</u>. When you are using multiple viewports, you have the option of locking the orientation of a UCS with a specific viewport so that when a new UCS is created or an existing UCS is restored, the locked UCS remains unchanged (that is, the UCS of the viewport that is active when the box is checked remains unchanged). *Update view to Plan when UCS is changed* controls the *UCSFOLLOW* system variable which affects the <u>View of the active viewport</u>. The viewport that is active when this box is checked will automatically change to a plan view of a new UCS whenever a new UCS is created or an existing UCS restored. When both boxes are checked, *Update view to Plan when UCS is changed* overrides *Save UCS with viewport*. (See "UCS System Variables" next in this chapter.)

UCS SYSTEM VARIABLES

In releases previous to AutoCAD 2000, UCS control was relatively simple and straight-forward. Because of the added UCS flexibility since AutoCAD 2000, it is easy for first-time users to become overwhelmed. This section is provided to help avoid confusion and to offer guidelines for UCS use. Not all UCS system variables are explained here, but only those that are typically accessed by a user.

UCSBASE

Pull-down Menu	COMMAND (TYPE)	ALIAS (TYPE)	Short-cut	Screen (side) Menu	Tablet Menu
Tools Orthographic UCSs Presets... Relative to	UCSBASE	...	...	TOOLS2 Ucsp Relative to	...

UCSBASE stores the name of the UCS that defines the origin and orientation when *Orthographic* UCSs are created using the *UCS, Orthographic* option or the *Orthographic UCSs* tab of the UCS Manager. Normally the World Coordinate System is used as the base UCS, so when a new orthographic UCS is created it is simply rotated about the WCS origin to be normal to the selected orthographic view (top, front, right, etc.). *UCSBASE* can also be set in the *Relative to:* drop-down list of the *Orthographic UCSs* tab of the UCS Manager (see Figure 30-11).

For beginning users, it is suggested that this variable be left at its initial value, *World*. This variable is saved in the current drawing.

UCSVIEW

Pull-down Menu	COMMAND (TYPE)	ALIAS (TYPE)	Short-cut	Screen (side) Menu	Tablet Menu
View Named Views Named Views New Save UCS with view	UCSVIEW	...	...	VIEW1 Ddview Named Views New Save UCS with view	...

UCSVIEW determines whether the current UCS is saved with a named view. For example, suppose you create a new 3D viewpoint, such as a southeast isometric, to enhance your visualization while you construct one area of a model. You also create a UCS on the desired construction plane. With *UCSVIEW* set to 1, you then *Save* that view and assign a name such as RIGHT. Whenever the RIGHT view is later restored, the UCS is also restored, no matter what UCS is current before restoring the view. Without this control, the current UCS would remain active when the RIGHT view is restored. *UCSVIEW* has two possible settings.

    0   The current UCS is not saved with a named view.
    1   The current UCS is saved whenever a named view is saved (initial setting).

The *UCSVIEW* variable can be accessed in the *New View* dialog box (Fig. 30-15). This dialog box is produced when you create a *New* view from the *Named Views* tab of the *View* dialog box (*View* command). *UCSVIEW* is saved in the current drawing.

UCSVIEW is easily remembered because it saves the *UCS* with the *VIEW*. Remember that the view must be named to save its UCS.

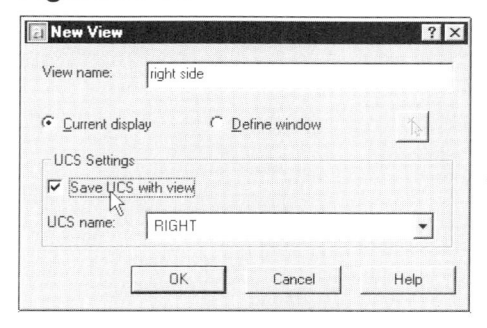

Figure 30-15

UCSORTHO

Pull-down Menu	COMMAND (TYPE)	ALIAS (TYPE)	Short-cut	Screen (side) Menu	Tablet Menu
View Named Views... Restore orthographic UCS with View	UCSORTHO	...		VIEW1 Ddview Restore orthographic UCS with View	...

This variable determines whether an orthographic UCS is automatically created when an orthographic view is created or restored. For example, if *UCSORTHO* is set to 1 and you select or create a front view, a UCS is automatically set up to be parallel with the view. In this way, the XY plane of the UCS is always parallel with the screen when an orthographic view is restored. The settings are as follows.

0   Specifies that the UCS setting remains unchanged when an orthographic view is restored.
1   Specifies that the related orthographic UCS is automatically created or restored when an orthographic view is created or restored (initial setting).

In order for *UCSORTHO* to operate as expected, you must select a preset orthographic view from the *View* pull-down menu, *Orthographic* option of the *View* command, or an orthographic view icon button. *UCSORTHO* does not operate with views created using *Vpoint*, *Plan*, the *3Dorbit* commands, or for any isometric preset views. *UCSORTHO* operates for views as well as viewports. *UCSORTHO* is saved in the current drawing.

A good exercise you can use to understand *UCSORTHO* is to use *Vports* to create four tiled viewports in model space and select the *3D Setup*. With *UCSORTHO* set to 1, the arrangement shown in Figure 30-16 would result. Here, each orthographic view has its own orthographically aligned UCS.

Figure 30-16

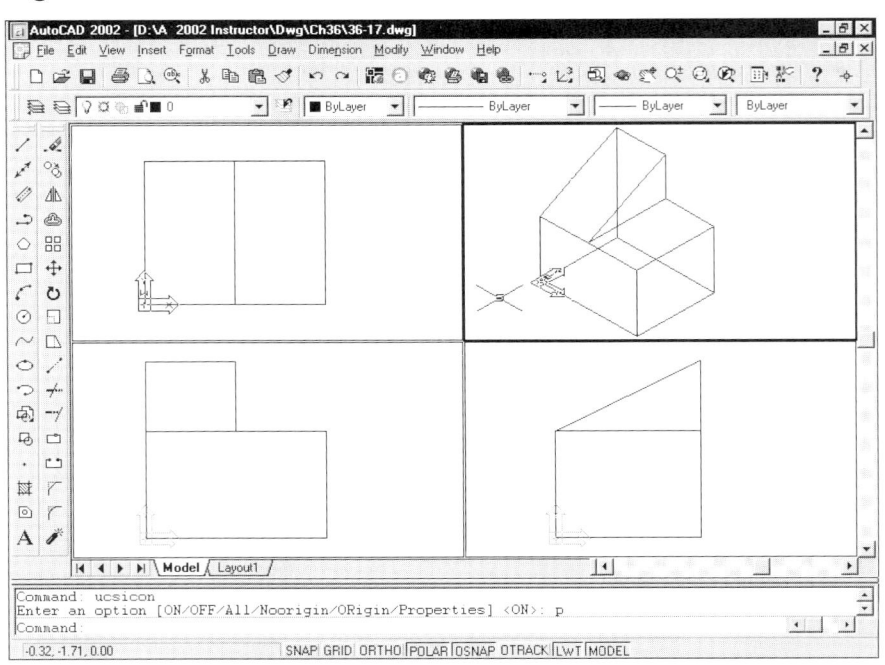

Chapter Thirty   615

If *UCSORTHO* is set to 0 and the *Vports* command is used to create a *3D Setup*, the arrangement shown in Figure 30-17 would be created. Notice that the WCS is the only coordinate system in all viewports.

Figure 30-17

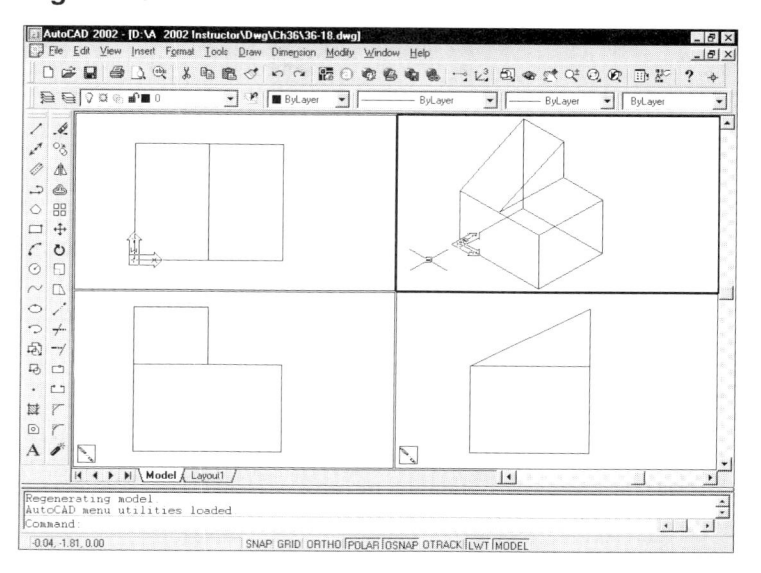

The *UCSORTHO* variable can be set at the Command line as well as in the *View* dialog box (Fig. 30-18).

A point to help remember *UCSORTHO* is when the view changes to an orthographic view, the UCS changes.

Figure 30-18

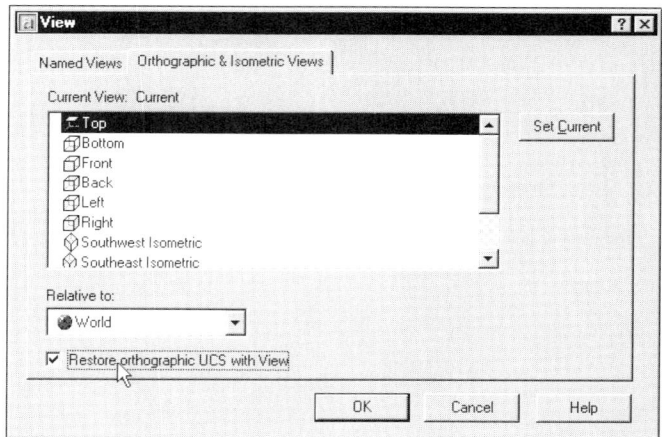

UCSFOLLOW

Pull-down Menu	COMMAND (TYPE)	ALIAS (TYPE)	Short-cut	Screen (side) Menu	Tablet Menu
Tools *Named UCS...* *Settings* *Update view to Plan when UCS is changed*	UCSFOLLOW	...	...	TOOLS2 UCS Follow:	...

The *UCSFOLLOW* system variable, if set to 1 (*On*), causes the plan view of a UCS to be displayed automatically when a UCS is made current. *UCSFOLLOW* can be set separately for each viewport.

616 User Coordinate Systems

For example, consider the case shown in Figure 30-19. Two tiled viewports (*Vports*) are being used, and the *UCSFOLLOW* variable is turned *On* for the left viewport only. Notice that a UCS created on the inclined surface is current. The left viewport shows the plan view of this UCS automatically, while the right viewport shows the model in the same orientation no matter what coordinate system is current. If the WCS were made active, the right viewport would keep the same viewpoint, while the left would automatically show a plan view of the new UCS when the change was made.

Figure 30-19

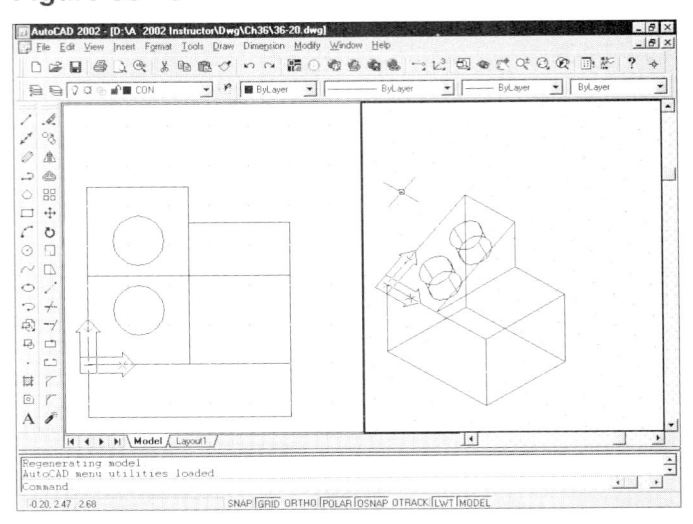

NOTE: When *UCSFOLLOW* is set to 1, the display does not change until a new UCS is created or restored.

The *UCSFOLLOW* settings are as follows.

 0   Changing the UCS does not affect the view (initial setting).
 1   Any UCS change causes a change to plan view of the new UCS in the viewport current when *UCSFOLLOW* is set.

The setting of *UCSFOLLOW* affects only model space. *UCSFOLLOW* is saved in the current drawing and is viewport specific.

TIP A point to help remember *UCSFOLLOW* is when the UCS changes, the view changes.

UCSVP

Pull-down Menu	COMMAND (TYPE)	ALIAS (TYPE)	Short-cut	Screen (side) Menu	Tablet Menu

Tools
Named Views...
Settings
UCS Settings
Save UCS
  with viewport
	UCSVP	...	...	TOOLS2
				Ucsman
				Settings
				UCS Settings
				Save UCS
				  with viewport
					...

UCSVP operates with model space and paper space viewports and is viewport specific (can be set for each viewport). It determines whether the UCS in the active viewport remains locked or changes when another UCS is made current in another viewport. *UCSVP* is similar to *UCSVIEW* where the UCS is saved with the *View*, only with *UCSVP* the UCS is saved with the viewport. The possible *UCSVP* settings are shown below.

 0   The UCS setting is not locked to the viewport so the UCS reflected becomes that of any other viewport that is made active.
 1   The viewport and UCS are locked. Therefore, the UCS is saved in that viewport and does not change when another viewport displaying another UCS becomes active (initial setting).

For example, suppose *UCSVP*, *UCSORTHO*, and *UCSVIEW* are all set to 0 (off) so there is no relationship established between views, viewports, and UCSs. In that case, only one UCS in the drawing could be active and would therefore appear in all viewports, similar to the arrangement shown in Figure 30-20. If a new UCS were created, it would appear in all viewports.

Figure 30-20

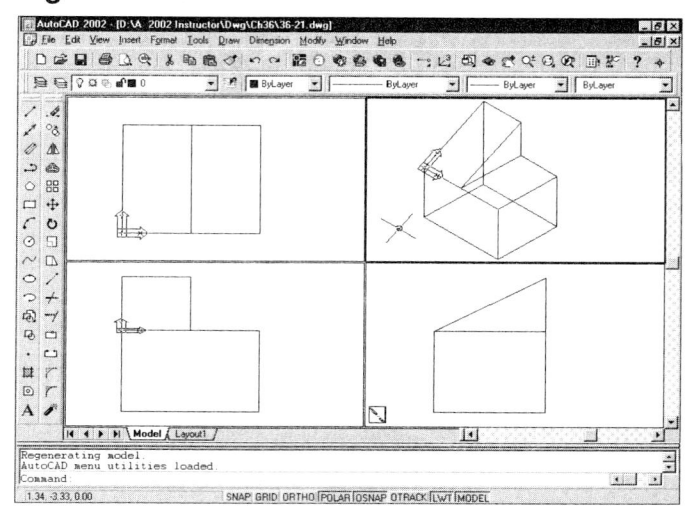

Assume *UCSVP* is then set to 1 in the upper-right (isometric) viewport. Suppose then another viewport were made active and the WCS restored. Now the WCS appears in all viewports <u>except</u> the isometric viewport where the UCS is locked (*UCSVP*=1) as shown in Figure 30-21.

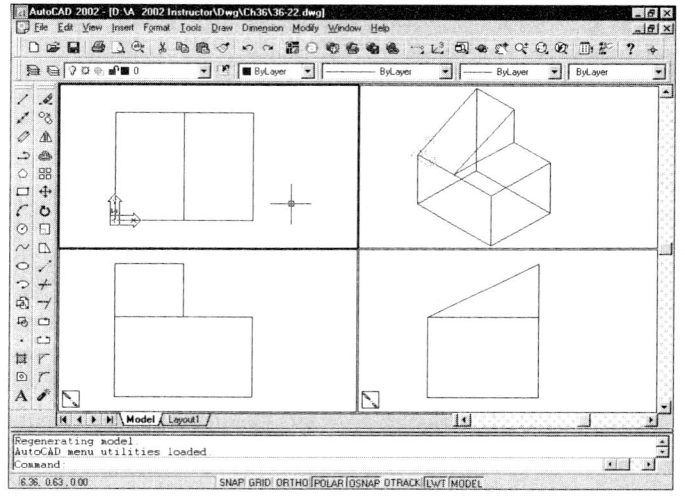

Remember that *UCSVP* saves a UCS to a <u>viewport</u>, whereas *UCSVIEW* saves a UCS to a named *View*.

Figure 30-21

CHAPTER EXERCISES

1. Using the model in Figure 30-22, assume a UCS was created by rotating about the X axis 90 degrees from the existing orientation (World Coordinate System).

    A. What are the absolute coordinates of corner F?

    B. What are the absolute coordinates of corner H?

    C. What are the absolute coordinates of corner J?

Figure 30-22

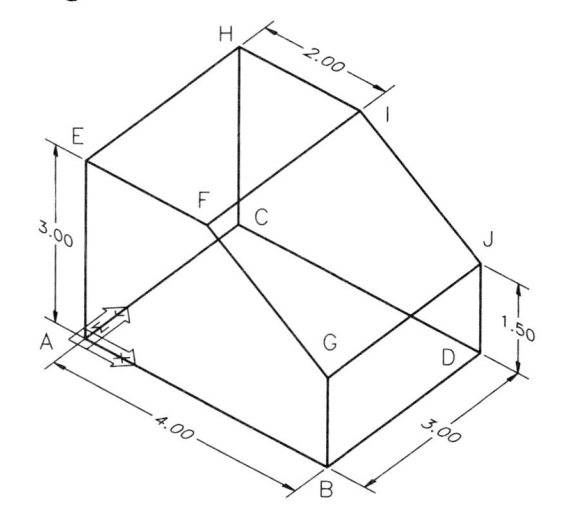

2. Using the model in Figure 30-22, assume a UCS was created by rotating about the X axis 90 degrees from the existing orientation (WCS) and then rotating 90 degrees about the (new) Y.

   A. What are the absolute coordinates of corner F?

   B. What are the absolute coordinates of corner H?

   C. What are the absolute coordinates of corner J?

3. *Open* the **SOLID1** drawing that you created in Chapter 29, Exercise 7. View the object from a *SE Isometric* view. Make sure that the *UCSicon* is *On* and set to the *ORigin*.

   Figure 30-23

   A. Create a *UCS* with a vertical XY plane on the front surface of the model as shown in Figure 30-23. *Save* the UCS as **FRONT**.

   B. Change the coordinate system back to the World. Now, create a *UCS* with the XY plane on the top horizontal surface and with the orientation as shown in Figure 30-23 as TOP1. *Save* the UCS as **TOP1**.

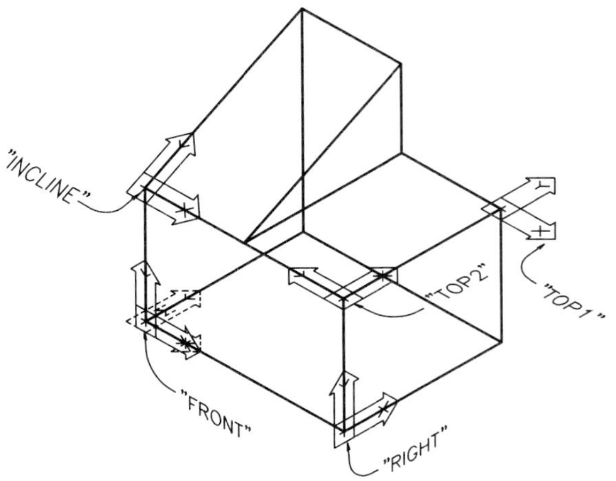

   C. Change the coordinate system back to the World. Next, create the *UCS* shown in the figure as TOP2. *Save* the UCS under the name **TOP2**.

   D. Activate the *WCS* again. Create and *Save* the **RIGHT** UCS.

   E. Use *SaveAs* and change the drawing name to **SOLIDUCS**.

4. In this exercise, you will create two viewport configurations, each with a different setting for *UCSORTHO*. In the first configuration, only one UCS exists, and in the second configuration, a new UCS is created for each viewport.

   A. *Open* the **SOLID1** drawing again. Use *SaveAs* and change the name to **VPUCS1**. Set the *UCSORTHO* variable to **0** so no new UCSs are created when you set up viewports. Ensure you are in model space and use the *Vports* command to create *Four Equal* viewports. Also in the *Viewports* dialog box, select the *3D Setup*. The resulting drawing should look like that in Figure 30-17. Notice that only one UCS appears in all viewports. Next, use any viewport and create a new UCS in the same orientation as the "Front" UCS shown in Figure 30-23. What happens in the other viewports? *Close* **VPUCS1** and *Save* the changes.

   B. *Open* the **SOLID1** drawing again. Use *SaveAs* and change the name to **VPUCS2**. This time, set the *UCSORTHO* variable to **1** so when you set up viewports new orthographically oriented UCSs are created. Set up the same viewport configuration as in the previous step. Your new viewport configuration and the resulting new orthographic UCSs should look like that in Figure 30-16. In the isometric view create a new UCS on the inclined plane (like the "Incline" UCS in Figure 30-23). What happens to the UCSs in the other viewports? *Close* **VPUCS2** and *Save* the changes.

31

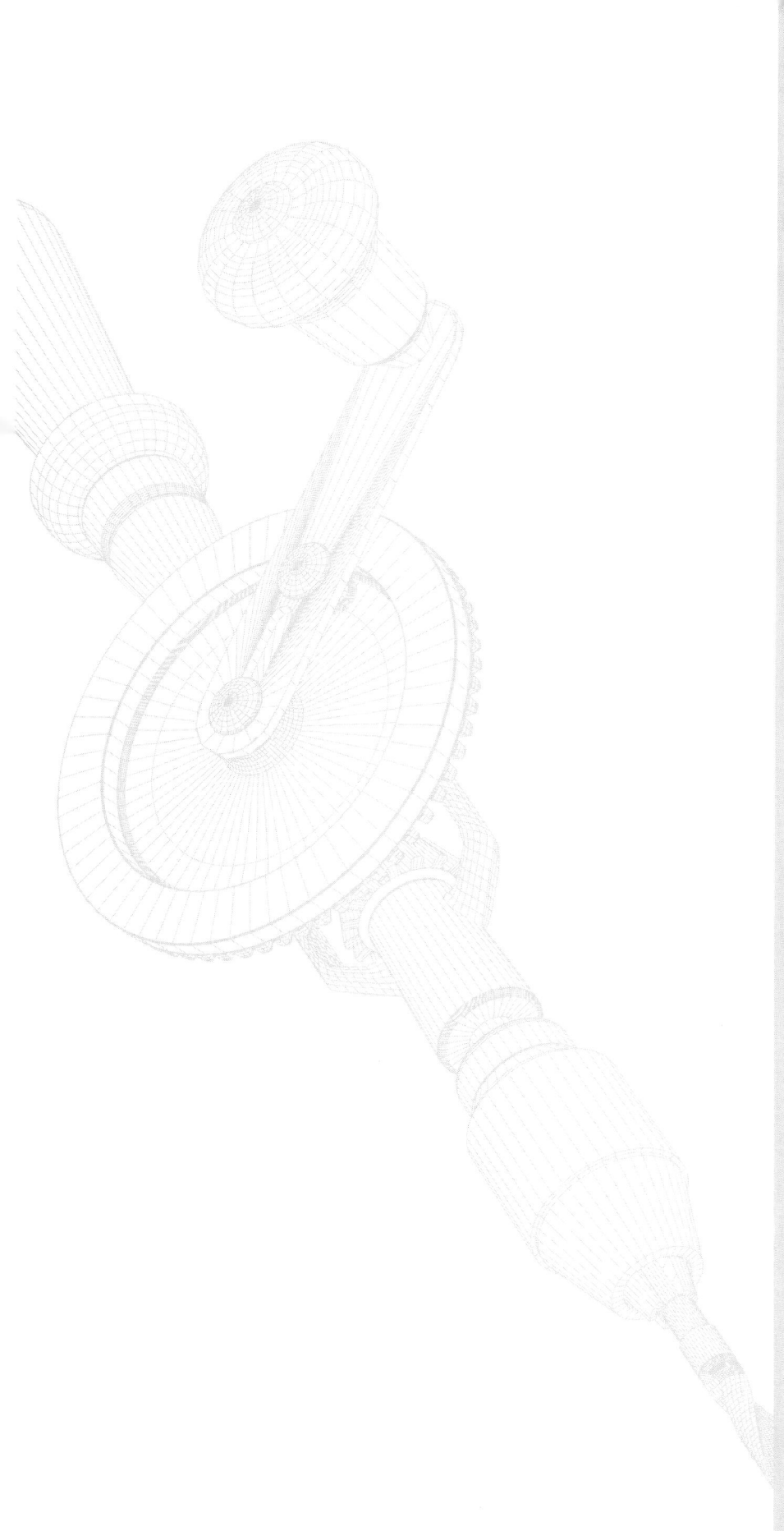

SOLID MODELING CONSTRUCTION

Chapter Objectives

After completing this chapter you should:

1. be able to create solid model primitives using the following commands: *Box*, *Wedge*, *Cone*, *Cylinder*, *Torus*, and *Sphere*;

2. be able to create swept solids from existing 2D shapes using the *Extrude* and *Revolve* commands;

3. be able to move solids in 3D space using *Move*, *Align*, and *Rotate3D*;

4. be able to create new solids in specific relation to other solids using *Mirror3D* and *3DArray*;

5. be able to combine multiple primitives into one composite solid using *Union*, *Subtract*, and *Intersect*;

6. know how to create beveled edges and rounded corners using *Chamfer* and *Fillet*;

7. be able to change composite solids with the vast array of editing tools in *Solidedit*.

CONCEPTS

The ACIS solid modeler is included in AutoCAD. With the ACIS modeler, you can create complex 3D parts and assemblies using Boolean operations to combine simple shapes, called "primitives." This modeling technique is often referred to as CSG or Constructive Solid Geometry modeling. ACIS also provides a method for analyzing and sectioning the geometry of the models (Chapter 32).

The techniques used with ACIS for construction of many solid models follow four general steps:

1. Construct simple 3D primitive solids, or create 2D shapes and convert them to 3D solids by extruding or revolving.

2. Create the primitives in location relative to the associated primitives or move the primitives into the desired location relative to the associated primitives.

3. Use Boolean operations (such as *Union, Subtract,* or *Intersect*) to combine the primitives to form a composite solid.

4. Make necessary design changes to features of a composite solid using the variety of editing tools in *Solidedit*.

A solid model is an informationally complete representation of the shape of a physical object. Solid modeling differs from wireframe or surface modeling in two fundamental ways: (1) the information is more complete in a solid model and (2) the method of construction of the model itself is relatively easy to construct and edit. Solid model construction is accomplished by creating regular geometric 3D shapes such as boxes, cones, wedges, cylinders, etc. (primitive solids) and combining them using union, subtract, and intersect (Boolean operations) to form composite solids.

CONSTRUCTIVE SOLID GEOMETRY TECHNIQUES

AutoCAD uses Constructive Solid Geometry (CSG) techniques for construction of solid models. CSG is characterized by solid primitives combined by Boolean operations to form composite solids. The CSG technique is a relatively fast and intuitive way of modeling that imitates the manufacturing process.

Primitives

Solid primitives are the basic building blocks that make up more complex solid models. The ACIS primitives commands are:

BOX	Creates a solid box or cube
CONE	Creates a solid cone with a circular or elliptical base
CYLINDER	Creates a solid cylinder with a circular or elliptical base
EXTRUDE	Creates a solid by extruding (adding a Z dimension to) a closed 2D object (*Pline, Circle, Region*)
REVOLVE	Creates a solid by revolving a shape about an axis
SHPERE	Creates a solid sphere
TORUS	Creates a solid torus
WEDGE	Creates a solid wedge

Primitives can be created by entering the command name or by selecting from the menus or icons.

Boolean Operations

Primitives are combined to create complex solids by using Boolean operations. The ACIS Boolean operators are listed below. An illustration and detailed description are given for each of the commands.

UNION   Unions (joins) selected solids.

SUBTRACT   Subtracts one set of solids from another.

INTERSECT   Creates a solid of intersection (common volume) from the selected solids.

Primitives are created at the desired location or are moved into the desired location before using a Boolean operator. In other words, two or more primitives can occupy the same space (or share some common space), yet are separate solids. When a Boolean operation is performed, the solids are combined or altered in some way to create one solid. AutoCAD takes care of deleting or adding the necessary geometry and displays the new composite solid complete with the correct configuration and lines of intersection.

Figure 31-1

Consider the two solids shown in Figure 31-1. When solids are created, they can occupy the same physical space. A Boolean operation is used to combine the solids into a composite solid and it interprets the resulting utilization of space.

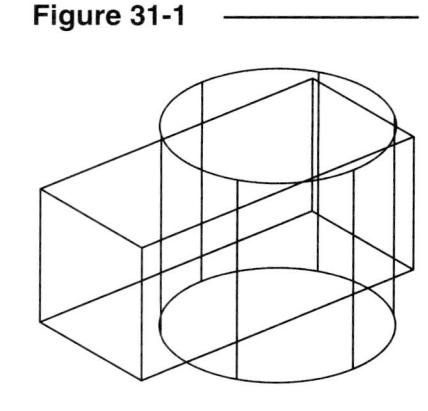

UNION

Union creates a union of the two solids into one composite solid (Fig. 31-2). The lines of intersection between the two shapes are calculated and displayed by AutoCAD. (*Hidden* shademode has been used for this figure.)

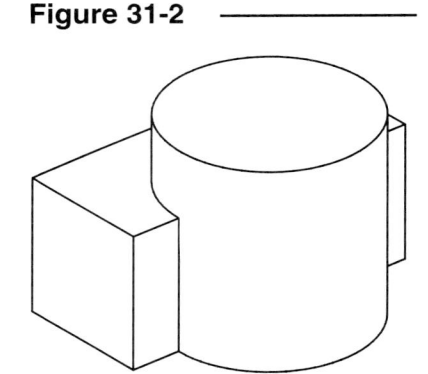

Figure 31-2

SUBTRACT

Subtract removes one or more solids from another solid. ACIS calculates the resulting composite solid. The term "difference" is sometimes used rather than "subtract." In Figure 31-3, the cylinder has been subtracted from the box.

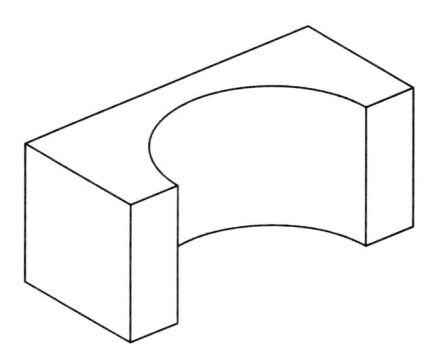

Figure 31-3

INTERSECT

Intersect calculates the intersection between two or more solids. When *Intersect* is used with *Regions* (2D surfaces), it determines the shared area. Used with solids, as in Figure 31-4, *Intersect* creates a solid composed of the shared volume of the cylinder and the box. In other words, the result of *Intersect* is a solid that has only the volume which is part of both (or all) of the selected solids.

Figure 31-4

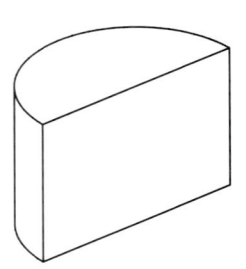

SOLID PRIMITIVES COMMANDS

This section explains the commands that allow you to create primitives used for construction of composite solid models. The commands allow you to specify the dimensions and the orientation of the solids. Once primitives are created, they are combined with other solids using Boolean operations to form composite solids.

NOTE: If you use a pointing device to PICK points, use *OSNAP* when possible to PICK points in 3D space. If you do not use *OSNAP*, the selected points are located on the current XY construction plane, so the true points may not be obvious. It is recommended that you use *OSNAP* or enter values.

Figure 31-5

The solid modeling commands are located in groups. The commands for creating solid primitives are located in a group in the *Draw* menus under *Solids* (Fig. 31-5). Commands for moving solids and the Boolean commands are found in the *Modify* pull-down menu. You can bring the *Solids* toolbar and the *Solids Editing* toolbar (which contains the Boolean commands) to the screen by selecting *Toolbars...* from the *View* pull-down menu (Fig. 31-6).

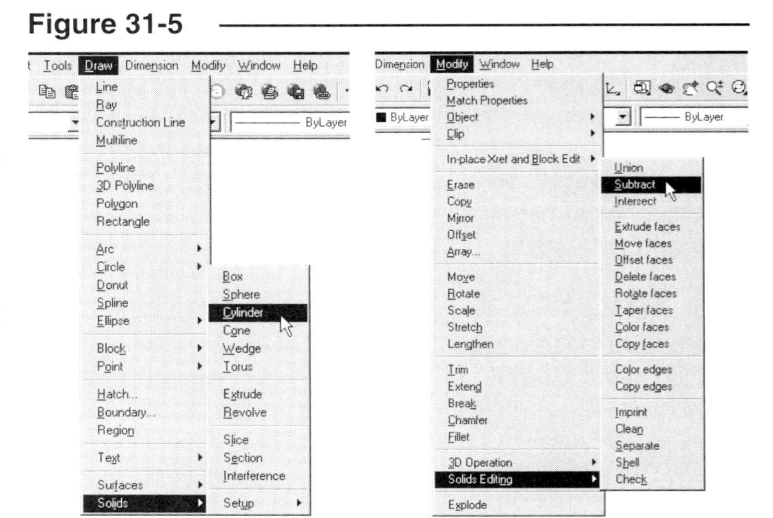

Figure 31-6

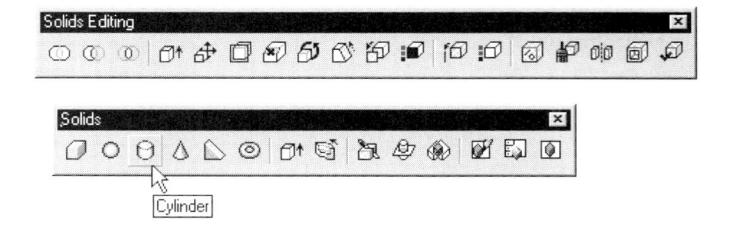

NOTE: When you begin to construct a solid model, create the primitives at location 0,0,0 (or some other known point) rather than PICKing points anywhere in space. It is very helpful to know where the primitives are located in 3D space so they can be moved or rotated in the correct orientation with respect to other primitives when they are assembled to composite solids.

BOX

Pull-down Menu	COMMAND (TYPE)	ALIAS (TYPE)	Short-cut	Screen (side) Menu	Tablet Menu
Draw Solids > Box	BOX	...	...	DRAW 2 SOLIDS Box	J,7

Box creates a solid box primitive to your dimensional specifications. You can specify dimensions of the box by PICKing or by entering values. The box can be defined by (1) giving the corners of the base, then height, (2) by locating the center and height, or (3) by giving each of the three dimensions. The base of the box is oriented parallel to the current XY plane:

Command: *box*
Specify corner of box or [CEnter] <0,0,0>: **PICK** or (**coordinates**) or (**CE**) or **Enter**

Options for the *Box* command are listed as follows.

Corner of box
Pressing **Enter** begins the corner of a box at 0,0,0 of the current coordinate system. In this case, the box can be moved into its desired location later. As an alternative, a coordinate position can be entered or **PICK**ed as the starting corner of the box. AutoCAD responds with:

Specify corner or [Cube/Length]:

The other corner can be PICKed or specified by coordinates. The *Cube* option requires only one dimension to define the cube.

Length
The *Length* option prompts you for the three dimensions of the box in the order of X, Y, and Z. AutoCAD prompts for *Length*, *Width*, and *Height*. These terms are vague since they do not define a specific 3D orientation (in fact, most dictionaries define "length" as the longer of two dimensions). Therefore, substitute X, Y, and Z for AutoCAD's prompts of *Length*, *Width*, and *Height*.

Figure 31-7 shows a box created at 0,0,0 with X, Y, and Z dimensions of 5, 4, and 3.

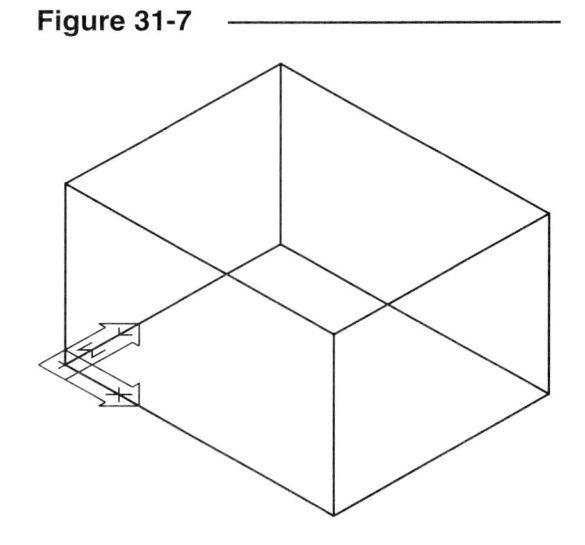

Figure 31-7

Center
With this option, you first locate the center of the box, then specify the three dimensions of the box. AutoCAD prompts:

Specify center of box <0,0,0>: **PICK** or (**coordinates**)

Specify the center location. AutoCAD responds with:

Specify corner or [Cube/Length]:

The resulting solid box is centered about the specified point (0,0,0 for the example; Fig. 31-8).

Figure 31-8

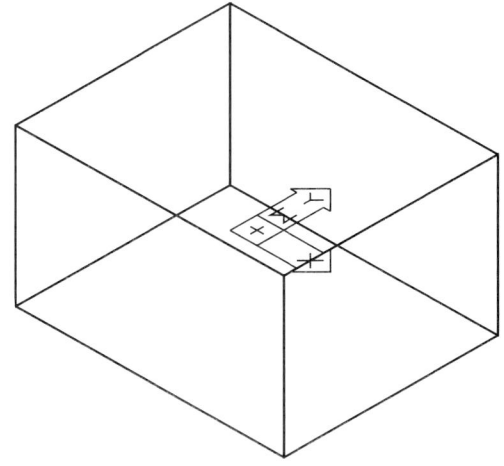

624 Solid Modeling Construction

Figure 31-9 illustrates a *Box* created with the *Center* option using *OSNAP* to snap to the *Center* of the top of the cylinder. Note that the center of the box is the <u>volumetric</u> center, not the center of the base.

Figure 31-9

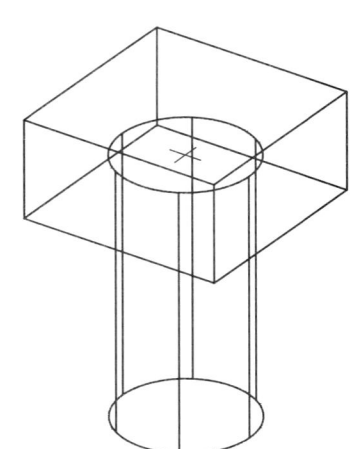

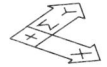

CONE

Pull-down Menu	COMMAND (TYPE)	ALIAS (TYPE)	Short-cut	Screen (side) Menu	Tablet Menu
Draw Solids > Cone	CONE	...	...	DRAW 2 SOLIDS Cone	M,7

Cone creates a right circular or elliptical solid cone ("right" means the axis forms a right angle with the base). You can specify the center location, radius (or diameter), and height. By default, the orientation of the cylinder is determined by the current UCS so that the base lies on the XY plane and height is perpendicular (in a Z direction). Alternately, the orientation can be defined by using the *Apex* option.

Center Point
Using the defaults (center at 0,0,0, PICK a radius, enter a value for height), the cone is generated in the orientation shown in Figure 31-10. The cone here may differ in detail (number of contour lines), depending on the current setting of the *ISOLINES* variable. The default prompts are shown as follows:

Figure 31-10

   Command: **cone**
   Current wire frame density: ISOLINES=4
   Specify center point for base of cone or [Elliptical] <0,0,0>: **PICK** or (**coordinates**)
   Specify radius for base of cone or [Diameter]: **PICK** or (**coordinates**)
   Specify height of cone or [Apex]: **PICK** or (**value**)
   Command:

Invoking the *Apex* option (after *Center point* of the base and *Radius* or *Diameter* have been specified) displays the following prompt:

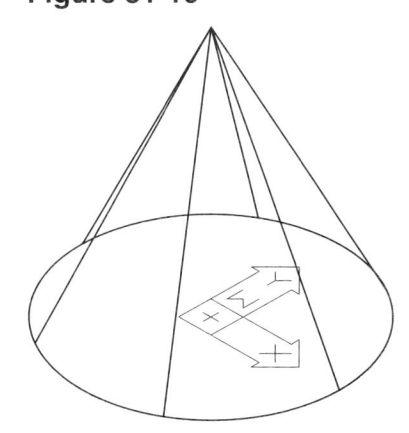

Figure 31-11

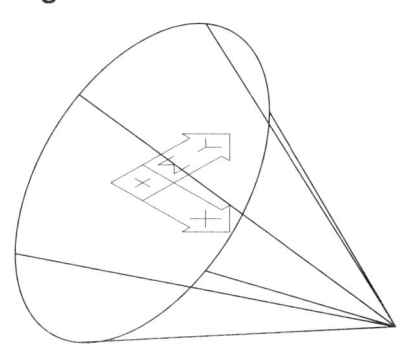

   Specify apex point: **PICK** or (**coordinates**)

Locating a point for the apex defines the height and orientation of the cone (Fig. 31-11). The axis of the cone is aligned with the line between the specified center point and the *Apex* point, and the height is equal to the distance between the two points.

The solid model in Figure 31-12 was created with a *Cone* and a *Cylinder*. The cylinder was created first, then the cone was created using the *Apex* option of *Cone*. The orientation of the *Cone* was generated by PICKing the *Center* of one end of the cylinder for the "Center point" of the base of the cone and the *Center* of the cylinder's other end for the *Apex*. *Union* created the composite model.

Figure 31-12

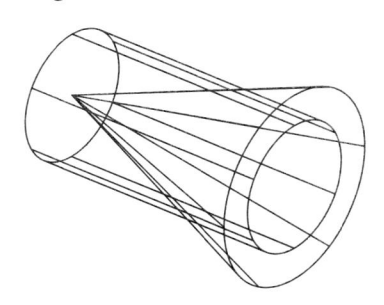

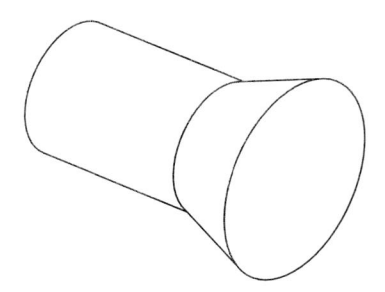

Elliptical

This option draws a cone with an elliptical base (Fig. 31-13). You specify two axis endpoints to define the elliptical base. An elliptical cone can be created using the *Center, Apex,* or *Height* options:

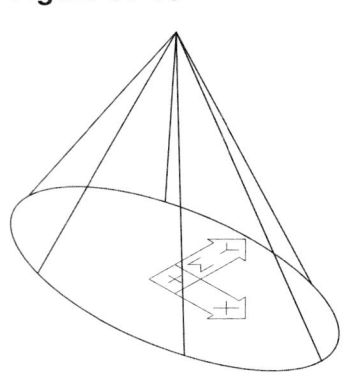

    Specify center point for base of cone or [Elliptical] <0,0,0>: *e*
    Specify axis endpoint of ellipse for base of cone or [Center]: *c*
    Specify center point of ellipse for base of cone <0,0,0>: **PICK** or (**coordinates**)
    Specify axis endpoint of ellipse for base of cone: **PICK** or (**coordinates**)
    Specify length of other axis for base of cone: **PICK** or (**coordinates**)
    Specify height of cone or [Apex]:

Figure 31-13

CYLINDER

Pull-down Menu	COMMAND (TYPE)	ALIAS (TYPE)	Short-cut	Screen (side) Menu	Tablet Menu
Draw Solids > Cylinder	CYLINDER	...	...	DRAW 2 SOLIDS Cylinder	L,7

Cylinder creates a cylinder with an elliptical or circular base with a center location, diameter, and height you specify. Default orientation of the cylinder is determined by the current UCS, such that the circular plane is coplanar with the XY plane and height is in a Z direction. However, the orientation can be defined otherwise by the *Center of other end* option.

Center Point

The default options create a cylinder in the orientation shown in Figure 31-14:

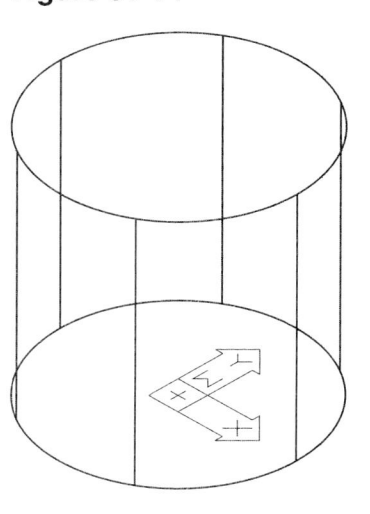

    Command: *cylinder*
    Current wire frame density: ISOLINES=4
    Specify center point for base of cylinder or [Elliptical] <0,0,0>: **PICK** or (**coordinates**)
    Specify radius for base of cylinder or [Diameter]: **PICK** or (**coordinates**)
    Specify height of cylinder or [Center of other end]: **PICK** or (**coordinates**)

Figure 31-14

Elliptical

This option draws a cylinder with an elliptical base (Fig. 31-15). Specify two axis endpoints to define the elliptical base. An elliptical cylinder can be created using the *Center, Center of other end,* or *Height* options.

Figure 31-15

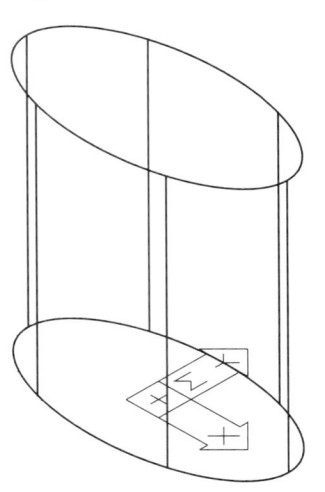

> Specify center point for base of cylinder or [Elliptical] <0,0,0>: **e**
> Specify axis endpoint of ellipse for base of cylinder or [Center]: **c**
> Specify center point of ellipse for base of cylinder <0,0,0>: **PICK** or **(coordinates)**
> Specify axis endpoint of ellipse for base of cylinder: **PICK** or **(coordinates)**
> Specify length of other axis for base of cylinder: **PICK** or **(coordinates)**
> Specify height of cylinder or [Center of other end]: **PICK** or **(value)**

The *Center of other end* option of *Cylinder* is similar to *Cone Apex* option in that the height and orientation are defined by the *Center point* and the *Center of other end.* In Figure 31-16, a hole is created in the *Box* by using *Center of other end* option and *OSNAP*ing to the diagonal lines' *Midpoints*, then *Subtract*ing the *Cylinder* from the *Box*.

Figure 31-16

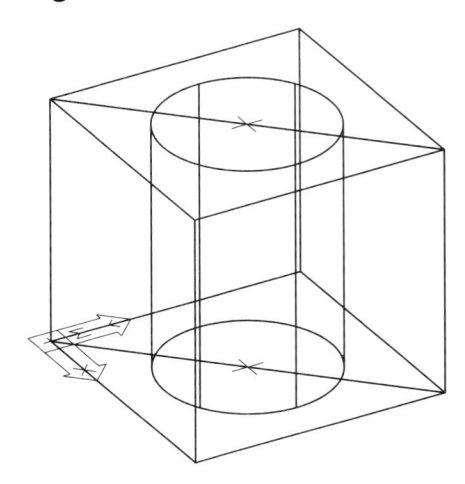

WEDGE

Pull-down Menu	COMMAND (TYPE)	ALIAS (TYPE)	Short-cut	Screen (side) Menu	Tablet Menu
Draw Solids > Wedge	WEDGE	WE	...	DRAW 2 SOLIDS Wedge	N,7

Corner of Wedge

Wedge creates a wedge solid primitive. The base of the wedge is always parallel with the current UCS XY plane, and the slope of the wedge is along the X axis.

Figure 31-17

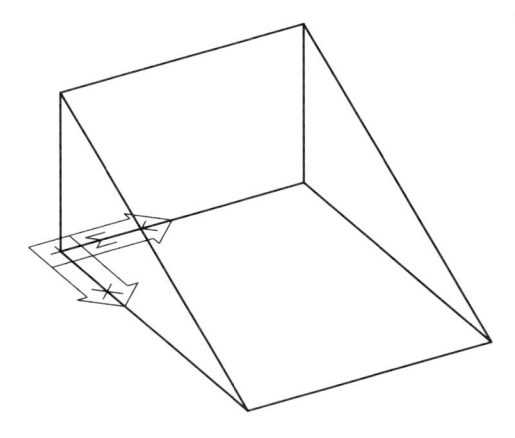

> Command: **wedge**
> Specify first corner of wedge or [CEnter] <0,0,0>: **PICK** or **(coordinates)**
> Specify corner or [Cube/Length]: **PICK** or **(coordinates)**
> Specify height: **PICK** or **(value)**

Accepting all the defaults, a *Wedge* can be created as shown in Figure 31-17, with the slope along the X axis.

Invoking the **Length** option prompts you for the *Length, Width*, and *Height*. More precisely, AutoCAD means width (X dimension), depth (Y dimension), and height (Z dimension).

Center
The point you specify as the center is actually in the center of an imaginary box, half of which is occupied by the wedge. Therefore, the center point is actually at the center of the sloping side of the wedge (Fig. 31-18).

Figure 31-18

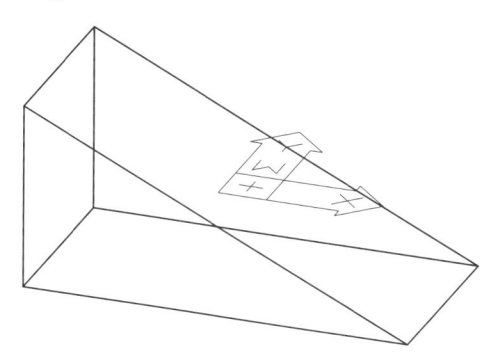

SPHERE

Pull-down Menu	COMMAND (TYPE)	ALIAS (TYPE)	Short-cut	Screen (side) Menu	Tablet Menu
Draw Solids > Sphere	SPHERE	...	...	DRAW 2 SOLIDS Sphere	K,7

Sphere allows you to create a solid sphere by defining its center point and radius or diameter:

   Command: *sphere*
   Current wire frame density: ISOLINES=4
   Specify center of sphere <0,0,0>: **PICK** or (**coordinates**)
   Specify radius of sphere or [Diameter]: **PICK** or (**coordinates**)
   Command: **PICK** or (**value**)

Creating a *Sphere* with the default options would yield a sphere similar to that in Figure 31-19.

Figure 31-19

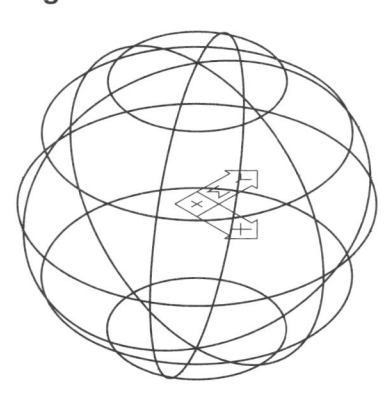

TORUS

Pull-down Menu	COMMAND (TYPE)	ALIAS (TYPE)	Short-cut	Screen (side) Menu	Tablet Menu
Draw Solids > Torus	TORUS	TOR	...	DRAW 2 SOLIDS Tours	O,7

Torus creates a torus (donut shaped) solid primitive using the dimensions you specify. Two dimensions are needed: (1) the radius or diameter of the tube and (2) the radius or diameter from the axis of the torus to the center of the tube. AutoCAD prompts:

   Command: *torus*
   Current wire frame density: ISOLINES=4
   Specify center of torus <0,0,0>: **PICK** or (**coordinates**)
   Specify radius of torus or [Diameter]: **PICK** or (**value**)
   Specify radius of tube or [Diameter]: **PICK** or (**value**)
   Command:

Figure 31-20 shows a *Torus* created using the default orientation with the axis of the tube aligned with the Z axis of the UCS and the center of the torus at 0,0,0. (*Hide* was used for this display.)

Figure 31-20

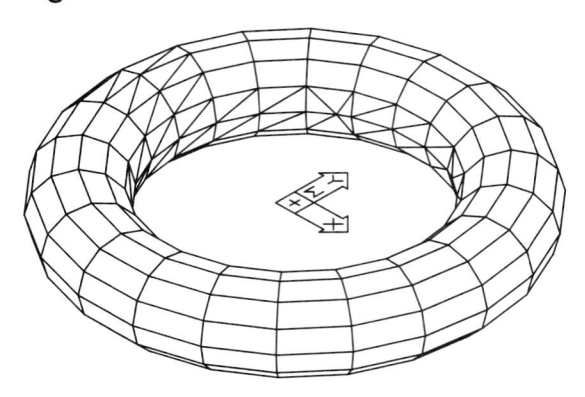

A self-intersecting torus is allowed with the *Torus* command. A self-intersecting torus is created by specifying a torus radius less than the tube radius. Figure 31-21 illustrates a torus with a torus radius of 3 and a tube radius of 4.

Figure 31-21

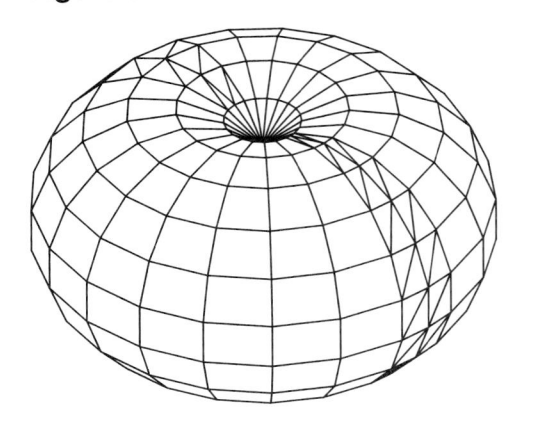

EXTRUDE

Pull-down Menu	COMMAND (TYPE)	ALIAS (TYPE)	Short-cut	Screen (side) Menu	Tablet Menu
Draw Solids > Extrude	EXTRUDE	EXT	...	DRAW 2 SOLIDS Extrude	P,7

Extrude (like *Revolve*) is a "sweeping operation." Sweeping operations use existing 2D objects to create a solid.

Extrude means to add a Z (height) dimension to an otherwise 2D shape. This command extrudes an existing closed 2D shape such as a *Circle, Polygon, Ellipse, Pline, Spline,* or *Region*. Only closed 2D shapes can be extruded. The closed 2D shape cannot be self-intersecting (crossing over itself).

Two methods determine the direction of extruding: perpendicular to the shape and along a *Path*. With the default method, the selected 2D shape is extruded perpendicular to the plane of the shape regardless of the current UCS orientation. Using the *Path* method, you can extrude the existing closed 2D shape along any existing path determined by a *Line, Arc, Spline,* or *Pline*.

The versatility of this command lies in the fact that any closed shape that can be created by (or converted to) a *Pline, Spline, Region*, etc., no matter how complex, can be transformed into a solid and can be extruded perpendicularly or along a *Path*. *Extrude* can simplify the creation of many solids that may otherwise take much more time and effort using typical primitives and Boolean operations. Create the closed 2D shape first, then invoke *Extrude*:

```
Command: extrude
Current wire frame density: ISOLINES=4
Select objects: PICK
Select objects: Enter
Specify height of extrusion or [Path]: PICK or (value)
Specify angle of taper for extrusion <0>: ENTER or (value)
Command:
```

Figure 31-22 shows a *Pline* before and after using *Extrude*.

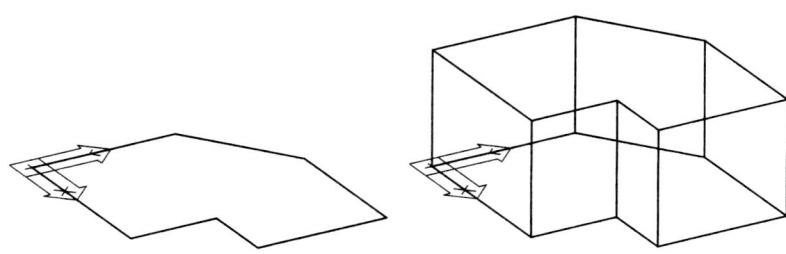

Figure 31-22

A taper angle can be specified for the extrusion. The resulting solid has sides extruded inward at the specified angle (Fig. 31-23). This is helpful for developing parts for molds that require a slight draft angle to facilitate easy removal of the part from the mold. Entering a negative taper angle results in the sides of the extrusion sloping outward.

Figure 31-23

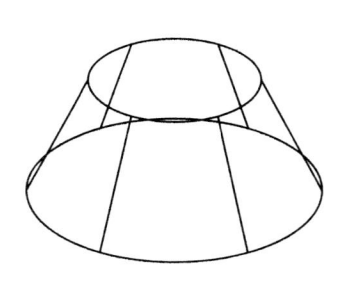

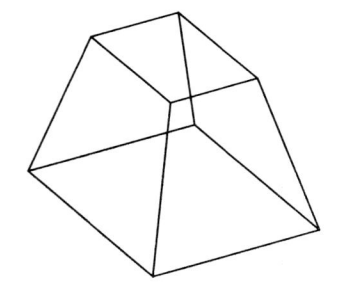

Many complex shapes based on closed 2D *Splines* or *Plines* can be transformed to solids using *Extrude* (Fig. 31-24).

Figure 31-24

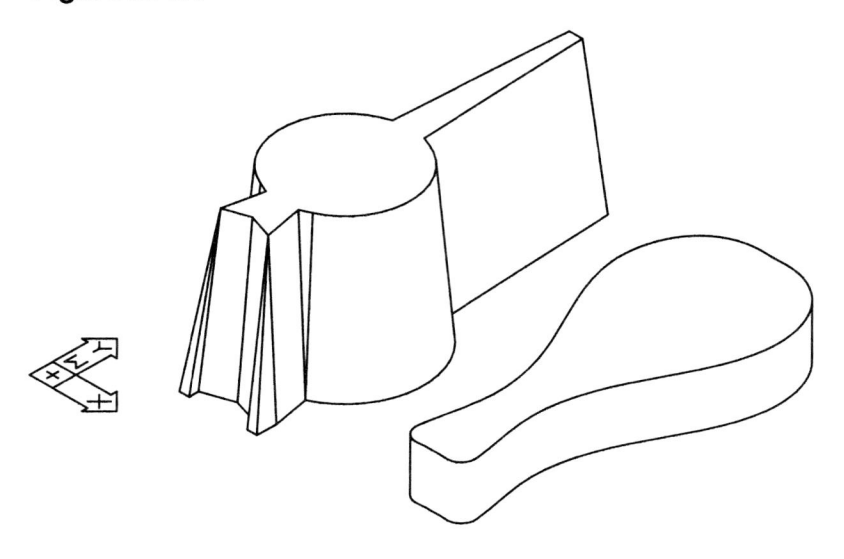

630 Solid Modeling Construction

The *Path* option allows you to sweep the 2D shape along an existing line or curve called a "path." The *Path* can be composed of a *Line, Arc, Ellipse, Pline*, or *Spline* (or can be different shapes converted to a *Pline*). This path must lie in a plane. Figure 31-25 shows a closed *Pline* extruded along a *Pline* path and a curved *Spline* path:

```
Command: extrude
Current wire frame density: ISOLINES=4
Select objects: PICK
Select objects: Enter
Specify height of extrusion or [Path]: path
Select extrusion path: PICK
```

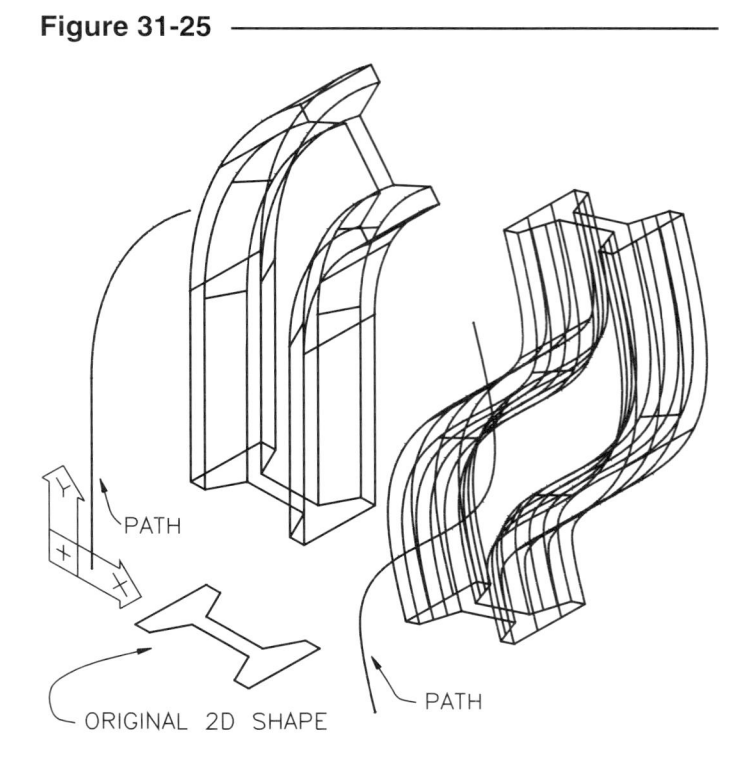

Figure 31-25

The path cannot lie in the same plane as the 2D shape to be extruded, since the plane of the 2D shape is always extruded perpendicular along the path. If one of the endpoints of the path is not located on the plane of the 2D shape, AutoCAD will automatically move the path to the center of the profile temporarily. Notice how the original 2D shape was extruded perpendicularly to the path (Fig. 31-26).

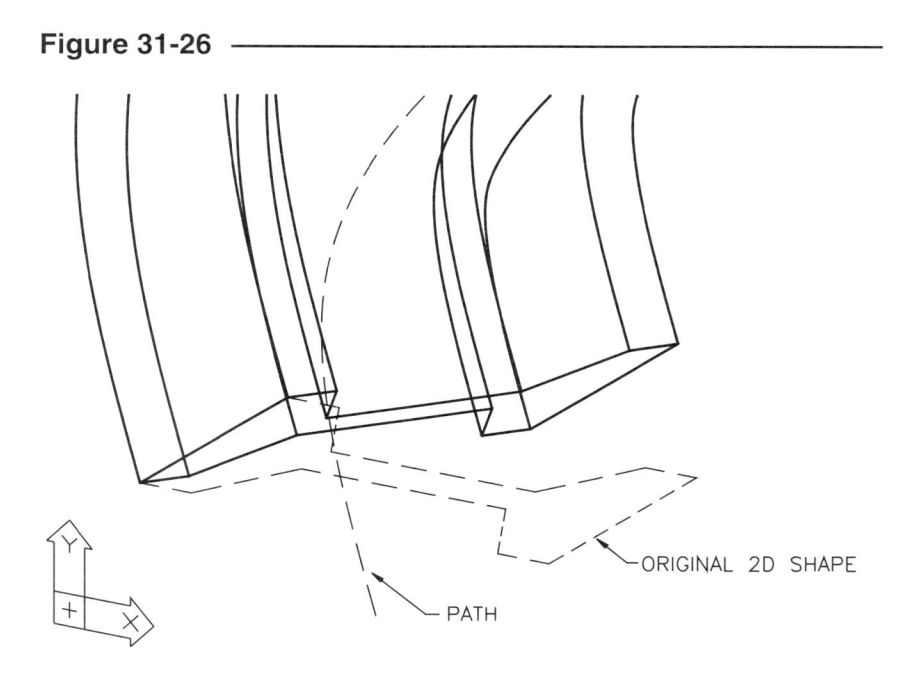

Figure 31-26

REVOLVE

Pull-down Menu	COMMAND (TYPE)	ALIAS (TYPE)	Short-cut	Screen (side) Menu	Tablet Menu
Draw Solids > Revolve	REVOLVE	REV	...	DRAW 2 SOLIDS Revolve	Q,7

Revolve creates a swept solid. *Revolve* creates a solid by revolving a 2D shape about a selected axis. The 2D shape to revolve can be a *Pline, Polygon, Circle, Ellipse, Spline*, or a *Region* object. Only one object at a time can be revolved. *Splines* or *Plines* selected for revolving must be closed. The command syntax for *Revolve* (accepting the defaults) is:

```
Command: revolve
Current wire frame density: ISOLINES=4
Select objects: PICK
Select objects: Enter
Specify start point for axis of revolution or define axis by [Object/X (axis)/Y (axis)]: PICK
Specify endpoint of axis: PICK
Specify angle of revolution <360>: Enter or (value)
```

Figure 31-27 illustrates a possibility for an existing *Pline* shape and the resulting *revolved* shape generated through a full circle.

Figure 31-27

Because *Revolve* acts on an existing object, the 2D shape intended for revolution should be created in or moved to the desired orientation. There are multiple options for selecting an axis of revolution.

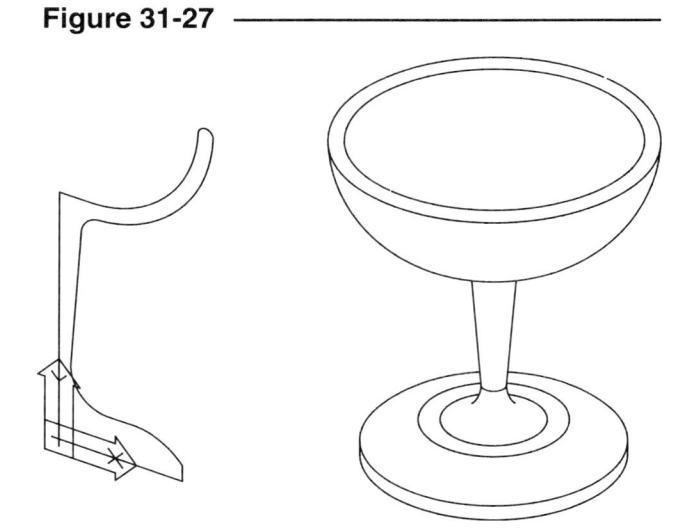

Object
A *Line* or single segment *Pline* can be selected for an axis. The positive axis direction is from the closest endpoint PICKed to the farthest.

X or Y
Uses the positive X or Y axis of the current UCS as the positive axis direction.

Start point of axis
Defines two points in the drawing to use as an axis (length is irrelevant). Select any two points in 3D space. The two points do not have to be coplanar with the 2D shape.

If the *Object* or *Start point* options are used, the selected object or the two indicated points do not have to be coplanar with the 2D shape. The axis of revolution used is always on the plane of the 2D shape to revolve and is aligned with the direction of the object or selected points.

632 Solid Modeling Construction

Figure 31-28 demonstrates a possible use of the *Object* option of *Revolve*. In this case, a *Pline* square is revolved about a *Line* object. Note that the *Line* used for the axis is <u>not</u> on the same plane as the *Pline*. *Revolve* uses an axis <u>on the plane</u> of the revolved shape aligned with the direction of the selected axis. In this case, the endpoints of the *Line* are 0,-1,-1 and 0,1,1 so the shape is actually revolved about the Y axis.

Figure 31-28

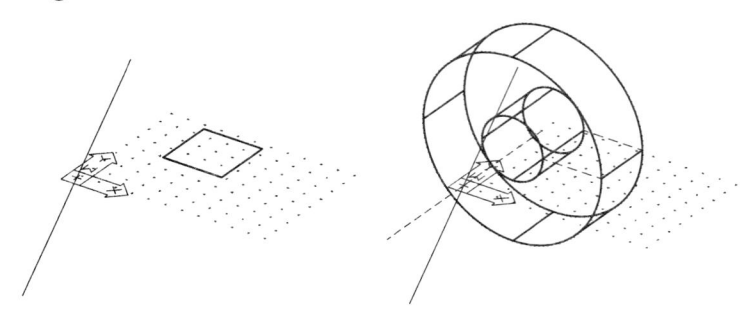

After defining the axis of revolution, *Revolve* requests the number of degrees for the object to be revolved. Any angle can be entered.

Figure 31-29

Another possibility for revolving a 2D shape is shown in Figure 31-29. The shape is generated through 270 degrees.

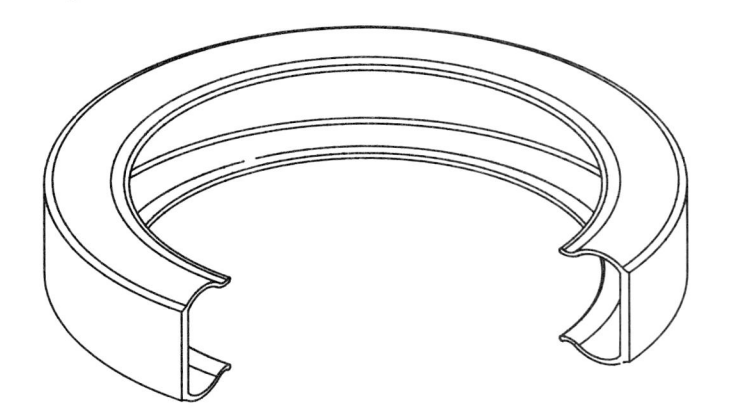

COMMANDS FOR MOVING SOLIDS

When you create the desired primitives, Boolean operations are used to construct the composite solids. However, the primitives must be in the correct position and orientation with respect to each other before Boolean operations can be performed. You can either create the primitives in the desired position during construction (by using UCSs) or move the primitives into position after their creation. Several methods that allow you to move solid primitives are explained in this section.

> When constructing 3D geometry, it is critical that the objects are located in space at some <u>known</u> position. Do <u>not</u> create primitives at any convenient place in the drawing; know the position. The location is important when you begin the process of moving primitives to assemble 3D composite solids. Of course, *OSNAPs* can be used, but sometimes it is necessary to use coordinate values in absolute, rectangular, or polar format. A good practice is to <u>create primitives at the final location</u> if possible (use UCSs when needed) or <u>create the primitives at 0,0,0</u>, then <u>move</u> them preceding the Boolean operations.

MOVE

Pull-down Menu	COMMAND (TYPE)	ALIAS (TYPE)	Short-cut	Screen (side) Menu	Tablet Menu
Modify Move	MOVE	M	...	MODIFY2 Move	V,19

The *Move* command that you use for moving 2D objects in 2D drawings can also be used to move 3D primitives. Generally, *Move* is used to change the position of an object in one plane (translation), which is typical of 2D drawings. *Move* can also be used to move an ACIS primitive in 3D space <u>if</u> *OSNAPs* or 3D coordinates are used.

Move operates in 3D just as you used it in 2D. Previously, you used *Move* only for repositioning objects in the XY plane, so it was only necessary to PICK or use X and Y coordinates. Using *Move* in 3D space requires entering X, Y, and Z values or using *OSNAPs*.

For example, to create the composite solid used in the figures in Chapter 30, a *Wedge* primitive was *Moved* into position on top of the *Box*. The *Wedge* was created at 0,0,0, then rotated. Figure 31-30 illustrates the movement using absolute coordinates described in the syntax below:

    Command: **move**
    Select objects: **PICK**
    Select objects: **Enter**
    Specify base point or displacement: **0,0**
    Specify second point of displacement or <use first point as displacement>: **0,4,3**
    Command:

Alternately, you can use *OSNAPs* to select geometry in 3D space. Figure 31-31 illustrates the same *Move* operation using *Endpoint OSNAPs* instead of entering coordinates.

Figure 31-30

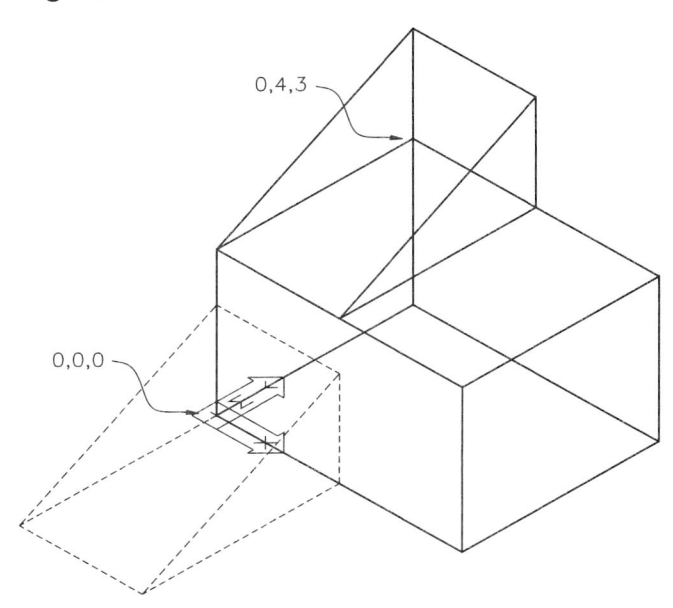

Figure 31-31

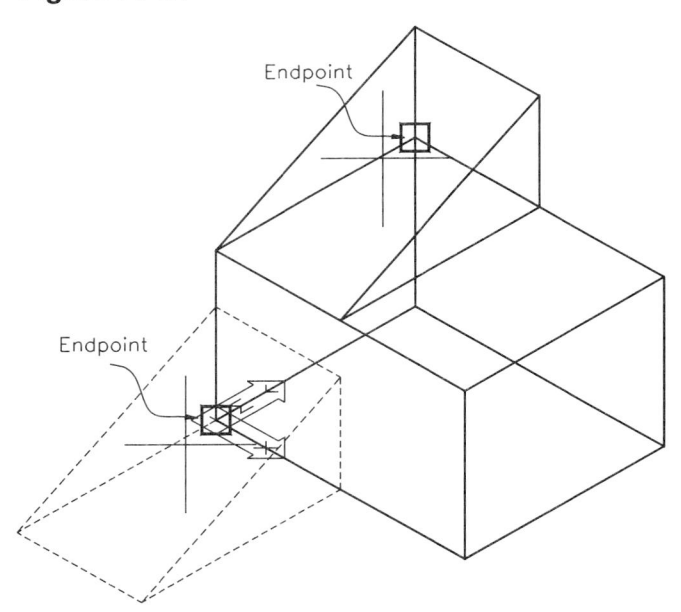

ALIGN

Pull-down Menu	COMMAND (TYPE)	ALIAS (TYPE)	Short-cut	Screen (side) Menu	Tablet Menu
Modify *3D Operation > Align*	ALIGN	AL	...	MODIFY2 *Align*	X,14

Align is an application that is loaded automatically when typing or selecting this command from the menus. *Align* is discussed in Chapter 16 but only in terms of 2D alignment.

634 Solid Modeling Construction

> *Align* is, however, a very powerful 3D command because it automatically performs 3D translation and rotation if needed. *Align* is more intuitive and in many situations is easier to use than *Move*. All you have to do is select the points on two 3D objects that you want to align (connect).

Align provides a means of aligning one shape (an object, a group of objects, a block, a region, or a 3D solid) with another shape. The alignment is accomplished by connecting source points (on the shape to be moved) to destination points (on the stationary shape). You can use *OSNAP* modes to select the source and destination points, assuring accurate alignment. Either a 2D or 3D alignment can be accomplished with this command. The command syntax for 3D alignment is as follows:

    Command: *align*
    Select objects: **PICK**
    Select objects: **Enter**
    Specify first source point: **PICK** (use *OSNAP*)
    Specify first destination point: **PICK** (use *OSNAP*)
    Specify second source point: **PICK** (use *OSNAP*)
    Specify second destination point: **PICK** (use *OSNAP*)
    Specify third source point or <continue>: **PICK** (use *OSNAP*)
    Specify third destination point: **PICK** (use *OSNAP*)
    Command:

Figure 31-32

After the source and destination points have been designated, lines connecting those points temporarily remain until *Align* performs the action (Fig. 31-32).

Align performs a translation (like *Move*) and two rotations (like *Rotate*), each in separate planes to align the points as designated. The motion automatically performed by *Align* is actually done in three steps.

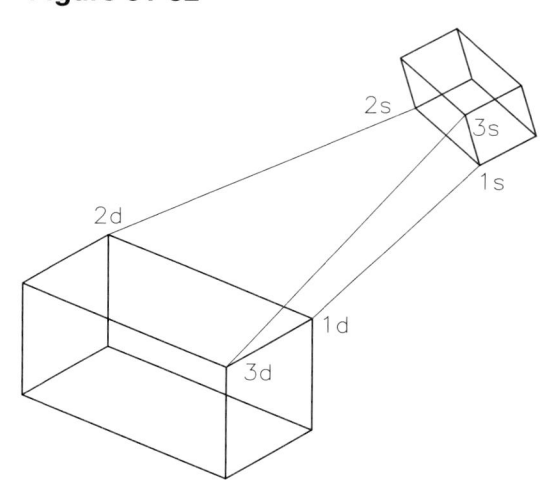

Figure 31-33

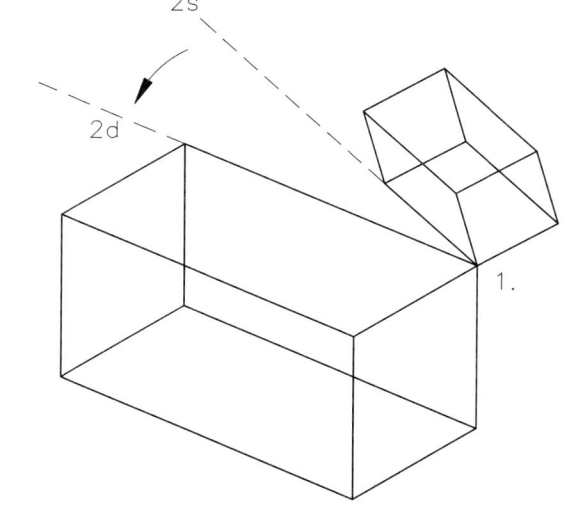

Initially, the first source point is connected to the first destination point (translation). These two points always physically touch.

Next, the vector defined by the first and second source points is aligned with the vector defined by the first and second destination points. The length of the segments between the first and second points on each object is of no consequence because AutoCAD only considers the vector direction. This second motion is a rotation along one axis.

Finally, the third set of points are aligned similarly. This third motion is a rotation along the other axis, completing the alignment.

Figure 31-34

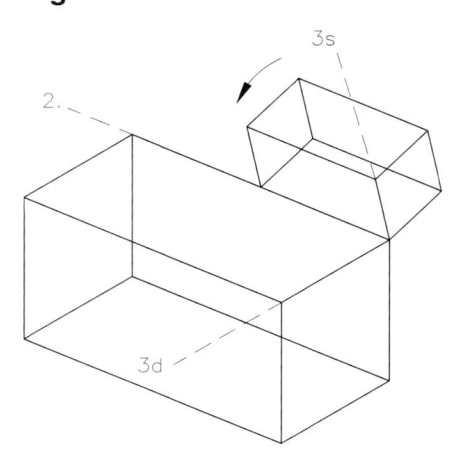

In some cases, such as when cylindrical objects are aligned, only two sets of points have to be specified. For example, if aligning a shaft with a hole (Fig. 31-35), the first set of points (source and destination) specify the attachment of the base of the shaft with the bottom of the hole (use *Center OSNAPs*). The second set of points specify the alignment of the axes of the two cylindrical shapes. A third set of points is not required because the radial alignment between the two objects is not important. When only two sets of points are specified, AutoCAD asks if you want to "continue" based on only two sets of alignment points. The command syntax is as follows:

Figure 31-35

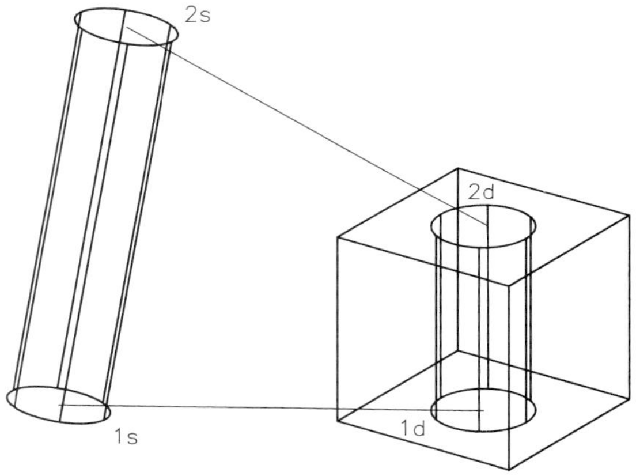

Command: *align*
Select objects: **PICK**
Select objects: **Enter**
Specify first source point: **PICK** (use *OSNAP*)
Specify first destination point: **PICK** (use *OSNAP*)
Specify second source point: **PICK** (use *OSNAP*)
Specify second destination point: **PICK** (use *OSNAP*)
Specify third source point or <continue>: *c*
Scale objects based on alignment points? [Yes/No] <N>: *y* or *n*
Command:

You can also <u>scale the source object</u> to fit between the two selected destination points. If you choose to "Scale objects based on alignment points," the source object is scaled (all dimensions scaled proportionally) and the destination object retains its size.

ROTATE3D

Pull-down Menu	COMMAND (TYPE)	ALIAS (TYPE)	Short-cut	Screen (side) Menu	Tablet Menu
Modify *3D Operation >* *Rotate 3D*	*ROTATE3D*	...	...	*MODIFY2* *Rotate3D*	W,22

Rotate3D is very useful for any type of 3D modeling, particularly with CSG, where primitives must be moved, rotated, or otherwise aligned with other primitives before Boolean operations can be performed.

Rotate3D allows you to rotate a 3D object about any axis in 3D space. Many alternatives are available for defining the desired rotational axis. Following is the command sequence for rotating a 3D object using the default (*2points*) option:

 Command: *rotate3d*
 Current positive angle: ANGDIR=counterclockwise ANGBASE=0
 Select objects: **PICK**
 Select objects: **Enter**
 Specify first point on axis or define axis by [Object/Last/View/Xaxis/Yaxis/Zaxis/2points]: **PICK** or (**option**) (Select the first point to define the rotational axis.)
 Specify second point on axis: **PICK**
 Specify rotation angle or [Reference]: (**value**) or **r**
 Command:

The options are explained next.

2points
The power of *Rotate3D* (over *Rotate*) is that any points or objects in 3D space can be used to define the axis for rotation. When using the default (*2points* option), remember you can use *OSNAP* to select points on existing 3D objects. Figure 31-36 illustrates the *2points* option used to select two points with *OSNAP* on the solid object selected for rotating.

Figure 31-36

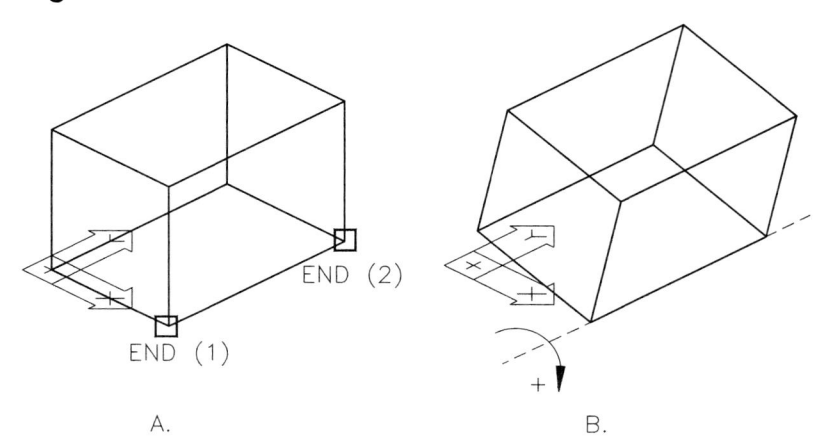

Object
This option allows you to rotate about a selected 2D object. You can select a *Line*, *Circle*, *Arc*, or 2D *Pline* segment. The rotational axis is aligned with the selected *Line* or *Pline* segment. Positive rotation is determined by the right-hand rule and the "arbitrary axis algorithm." When selecting *Arc* or *Circle* objects, the rotational axis is perpendicular to the plane of the *Arc* or *Circle* passing through the center. You cannot select the edge of a solid object with this option.

Last
This option allows you to rotate about the axis used for the last rotation.

Figure 31-37

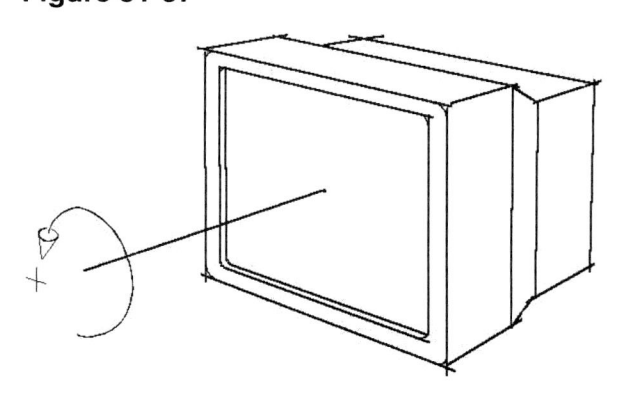

View
The *View* option allows you to pick a point on the screen and rotates the selected object(s) about an axis perpendicular to the screen and passing through the selected point (Fig. 31-37).

Xaxis
With this option, you can rotate the selected objects about the X axis of the current UCS or any axis parallel to the X axis of the current UCS. You are prompted to pick a point on the X axis. The point selected defines an axis for rotation parallel to the current X axis passing through the selected point. You can use *OSNAP* to select points on existing 3D objects (Fig. 31-38). The current X axis can be used if the point you select is on the X axis.

Figure 31-38

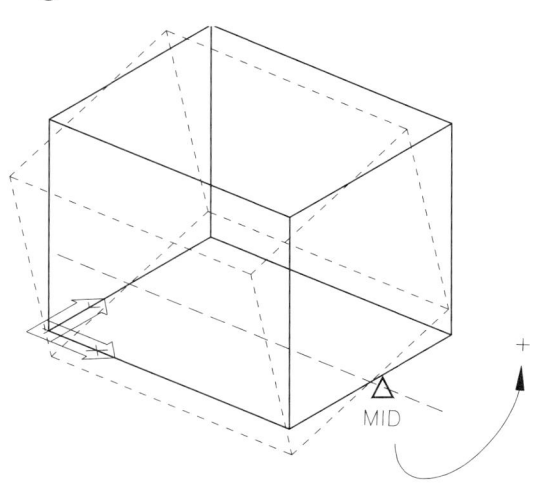

Yaxis
This option allows you to use the Y axis of the current UCS or any axis parallel to the Y axis of the current UCS as the axis of rotation. The point you select defines a rotational axis parallel to the current Y axis passing through the point. *OSNAP* can be used to snap to existing geometry (Fig. 31-39).

Figure 31-39

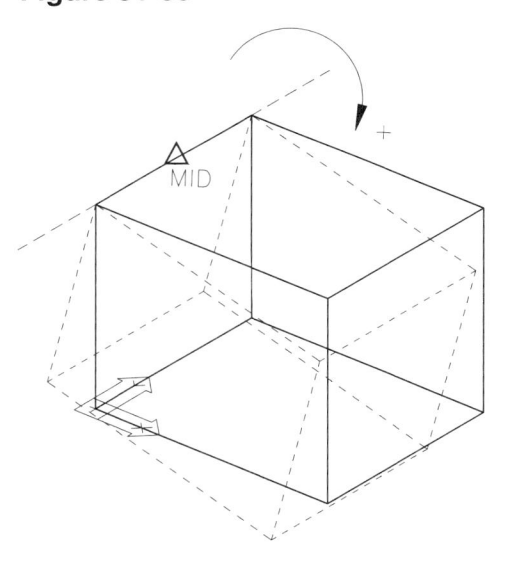

Zaxis
With this option, you can use the Z axis of the current UCS or any axis parallel to the Z axis of the current UCS as the axis of rotation. The point you select defines a rotational axis parallel to the current Z axis passing through the point. Figure 31-40 indicates the use of the *Midpoint OSNAP* to establish a vertical (parallel to Z) rotational axis.

Figure 31-40

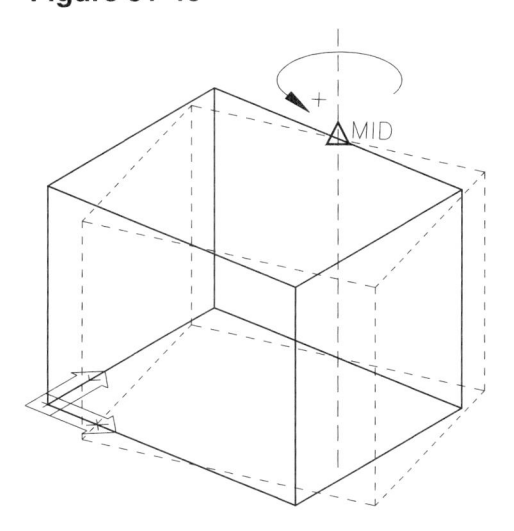

638 Solid Modeling Construction

Reference
After you have specified the axis for rotation, you must specify the rotation angle. You are presented with the following prompt:

   Specify rotation angle or [Reference]: **r**
   Specify the reference angle <0>: **PICK** or (**value**) (PICK two points; *OSNAPs* can be used. You can also enter a value.)
   Specify the new angle: **PICK** or (**value**)
   Command:

TIP The angle you specify for the reference (relative) is used instead of angle 0 (absolute) for the starting position. You can enter either a value or PICK two points to specify the angle. You then specify a new angle. The *Reference* angle you select is rotated to the absolute angle position you specify as the "new angle."

Figure 31-41 illustrates how *Endpoint OSNAPs* are used to select a *Reference* angle. The "new angle" is specified as **90**. AutoCAD rotates the reference angle to the 90 degree position.

Figure 31-41

MIRROR3D

Pull-down Menu	COMMAND (TYPE)	ALIAS (TYPE)	Short-cut	Screen (side) Menu	Tablet Menu
Modify *3D Operation >* *Mirror 3D*	MIRROR3D	...	...	MODIFY2 *Mirror3D*	W,21

Mirror3D operates similar to the 2D version of the command *Mirror* in that mirrored replicas of selected objects are created. With *Mirror* (2D) the selected objects are mirrored about an axis. The axis is defined by a vector lying in the XY plane. With *Mirror3D*, selected objects are mirrored about a plane. *Mirror3D* provides multiple options for specifying the plane to mirror about:

   Command: **mirror3d**
   Select objects: **PICK**
   Select objects: **Enter**
   Specify first point of mirror plane (3 points) or [Object/Last/Zaxis/View/XY/YZ/ZX/3points] <3points>: **PICK** or (**option**)

The options are listed and explained next. A phantom icon is shown in the following figures only to aid your visualization of the mirroring plane.

3points

The *3points* option mirrors selected objects about the plane you specify by selecting three points to define the plane. You can PICK points (with or without *OSNAP*) or give coordinates. *Midpoint OSNAP* is used to define the 3 points in Figure 31-42 A to achieve the result in B.

Figure 31-42

Object

Using this option establishes a mirroring plane with the plane of a 2D object. Selecting an *Arc* or *Circle* automatically mirrors selected objects using the plane in which the *Arc* or *Circle* lies. The plane defined by a *Pline* segment is the XY plane of the *Pline* when the *Pline* segment was created. Using a *Line* object or edge of an ACIS solid is not allowed because neither defines a plane. Figure 31-43 shows a box mirrored about the plane defined by the *Circle* object. Using *Subtract* produces the result shown in B.

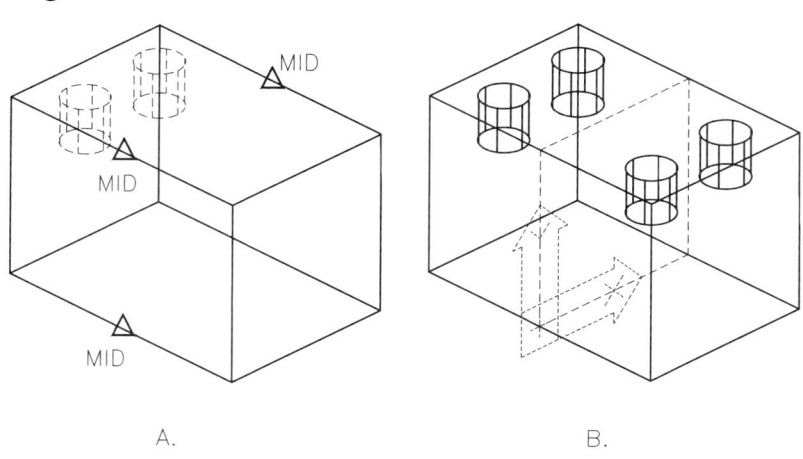

Figure 31-43

Last

Selecting this option uses the plane that was last used for mirroring.

Zaxis

With this option, the mirror plane is the XY plane perpendicular to a Z vector you specify. The first point you specify on the Z axis establishes the location of the XY plane origin (a point through which the plane passes). The second point establishes the Z axis and the orientation of the XY plane (perpendicular to the Z axis). Figure 31-44 illustrates this concept. Note that this option requires only two PICK points.

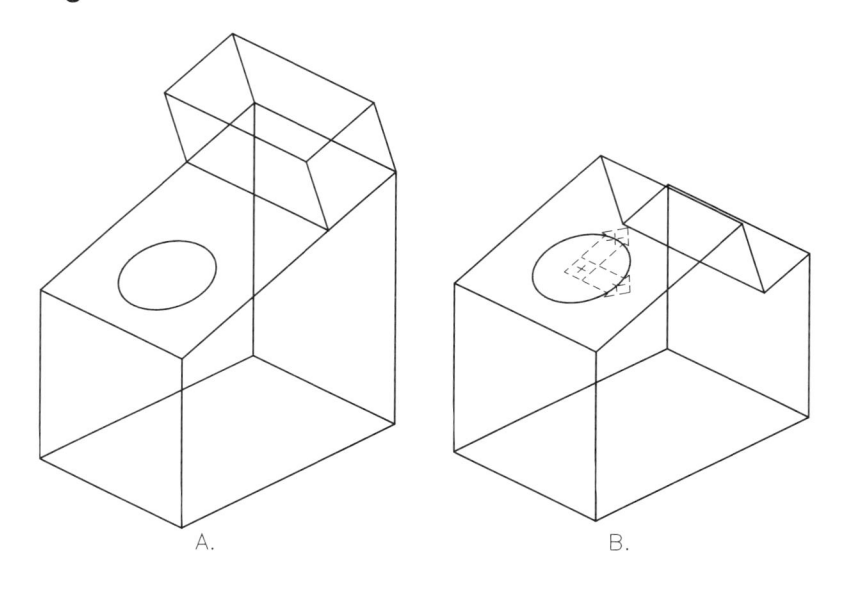

Figure 31-44

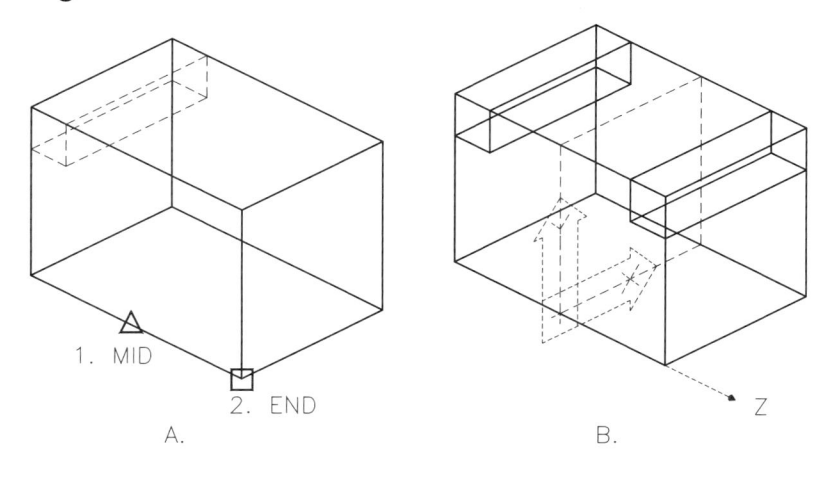

640 Solid Modeling Construction

View

The *View* option of *Rotate3D* uses a mirroring plane parallel with the screen and perpendicular to your line of sight based on your current viewpoint. You are required to select a point on the plane. Accepting the default (0,0,0) establishes the mirroring plane passing through the current origin. Any other point can be selected. You must change your viewpoint to "see" the mirrored objects (Fig. 31-45).

Figure 31-45

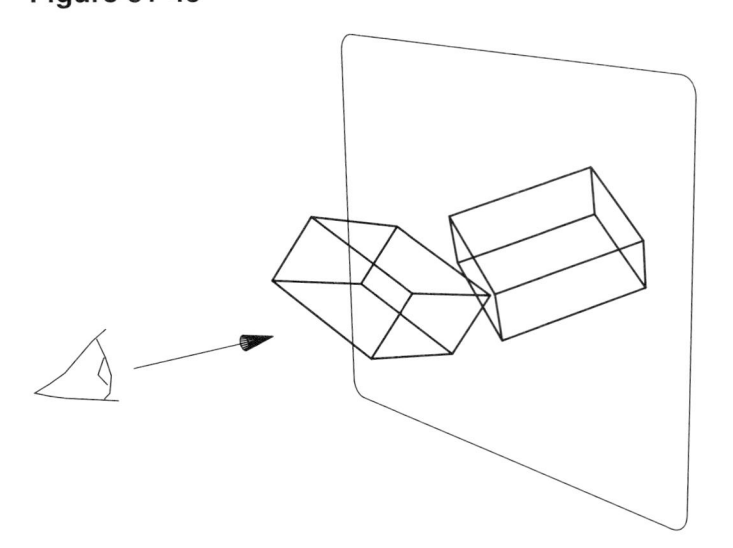

XY

This option situates a mirroring plane parallel with the current XY plane. You can specify a point through which the mirroring plane passes. Figure 31-46 represents a plane established by selecting the *Center* of an existing solid.

Figure 31-46

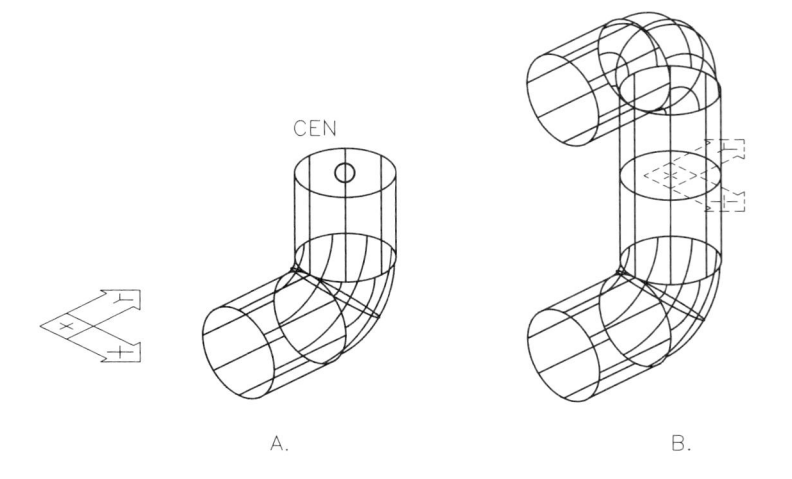

A.   B.

YZ

Using the *YZ* option constructs a plane to mirror about that is parallel with the current YZ plane. Any point can be selected through which the plane will pass (Fig. 31-47).

Figure 31-47

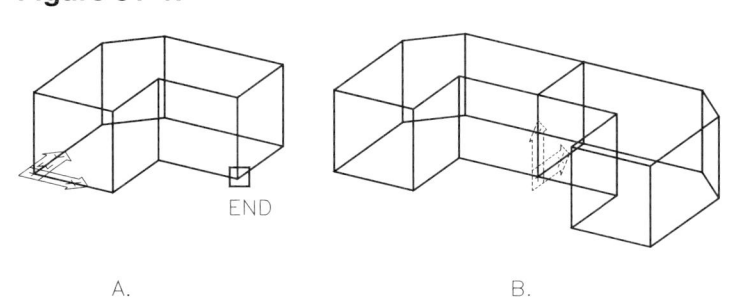

A.   B.

ZX

This option uses a plane parallel with the current ZX plane for mirroring. Figure 31-48 shows a point selected on the *Midpoint* of an existing edge to mirror two holes.

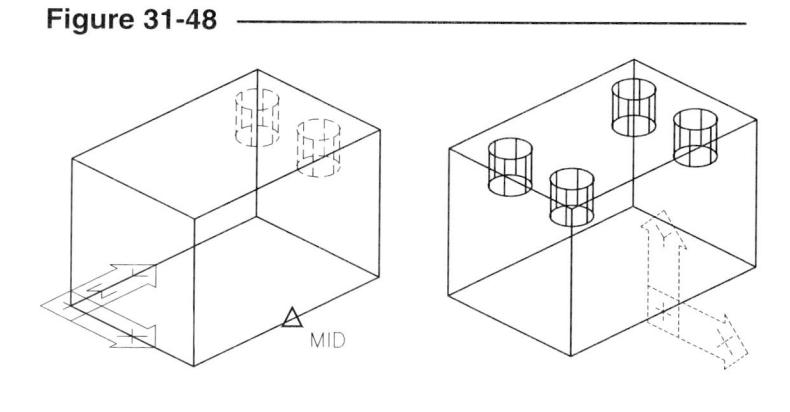

Figure 31-48

A.   B.

3DARRAY

Pull-down Menu	COMMAND (TYPE)	ALIAS (TYPE)	Short-cut	Screen (side) Menu	Tablet Menu
Modify 3D Operation > 3D Array	3DARRAY	3A	...	MODIFY2 3Darray	W,20

Rectangular

With this option of *3Darray*, you create a 3D array specifying three dimensions—the number of and distance between rows (along the Y axis), the number/distance of columns (along the X axis), and the number/distance of levels (along the Z axis). Technically, the result is an array in a prism configuration rather than a rectangle.

```
Command: 3darray
Select objects: PICK
Select objects: Enter
Enter the type of array [Rectangular/Polar] <R>: r
Enter the number of rows (—-) <1>: (value)
Enter the number of columns (|||) <1>: (value)
Enter the number of levels (...) <1>: (value)
Specify the distance between rows (—-): PICK or (value)
Specify the distance between columns (|||): PICK or (value)
Specify the distance between levels (...): PICK or (value)
Command:
```

The selection set can be one or more objects. The entire set is treated as one object for arraying. All values entered must be positive.

642 Solid Modeling Construction

Figures 31-49 and 31-50 illustrate creating a *Rectangular 3Darray* of a cylinder with 3 rows, 4 columns, and 2 levels.

Figure 31-49

ORIGINAL OBJECT

The cylinders are *Subtracted* from the extrusion to form the finished part.

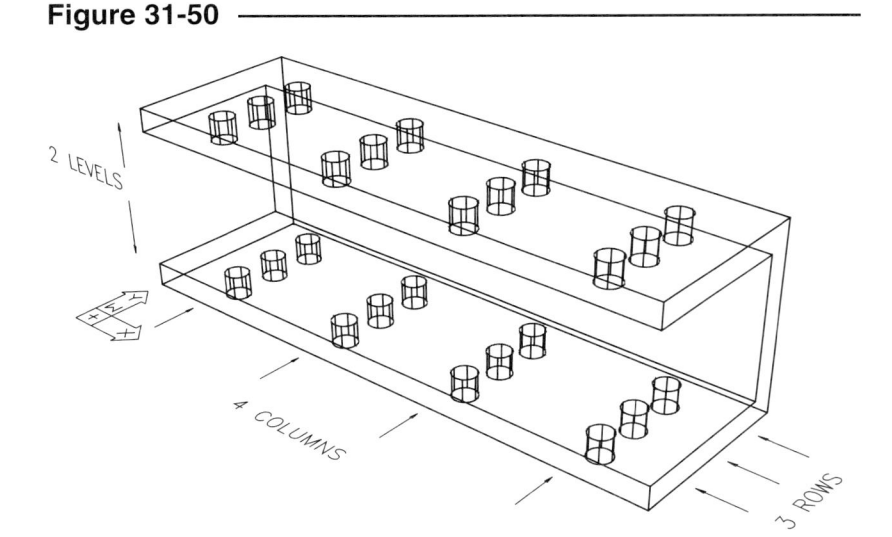

Figure 31-50

2 LEVELS

4 COLUMNS

3 ROWS

Polar
Similar to a *Polar Array* (2D), this option creates an array of selected objects in a circular fashion. The only difference in the 3D version is that an array is created about an axis of rotation (3D) rather than a point (2D). Specification of an axis of rotation requires two points in 3D space:

    Command: *3darray*
    Select objects: **PICK**
    Select objects: **Enter**
    Enter the type of array [Rectangular/Polar] <R>: *p*
    Enter the number of items in the array: (**value**)
    Specify the angle to fill (+=ccw, -=cw) <360>: **PICK** or (**value**)
    Rotate arrayed objects? [Yes/No] <Y>: *y* or *n*
    Specify center point of array: **PICK** or (**coordinates**)
    Specify second point on axis of rotation: **PICK** or (**coordinates**)
    Command:

In Figures 31-51 and 31-52, a *3Darray* is created to form a series of holes from a cylinder. The axis of rotation is the center axis of the large cylinder specified by PICKing the *Center* of the top and bottom circles.

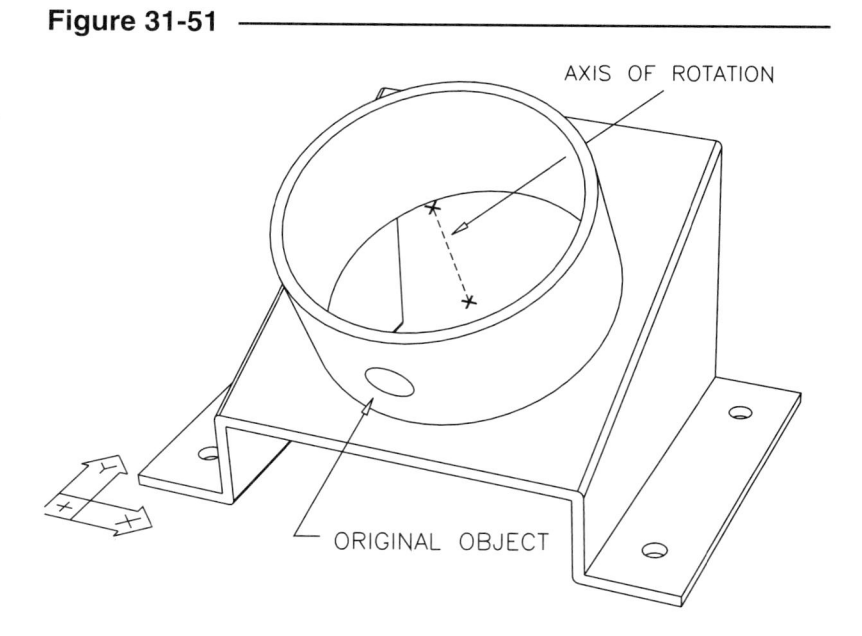

Figure 31-51

After the eight items are arrayed, the small cylinders are subtracted from the large cylinder to create the holes.

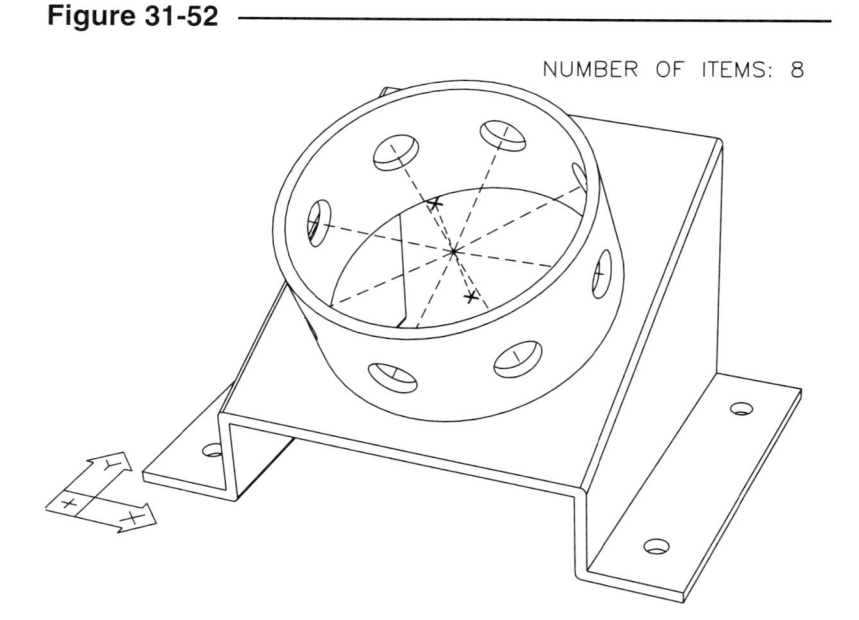

Figure 31-52

BOOLEAN OPERATION COMMANDS

Once the individual 3D primitives have been created and moved into place, you are ready to put together the parts. The primitives can be "assembled" or combined by Boolean operations to create composite solids. The Boolean operations found in AutoCAD are listed in this section: *Union*, *Subtract*, and *Intersect*.

UNION

Pull-down Menu	COMMAND (TYPE)	ALIAS (TYPE)	Short-cut	Screen (side) Menu	Tablet Menu
Modify Solids Editing > Union	UNION	UNI	...	MODIFY2 Union	X,15

Union joins selected primitives or composite solids to form one composite solid. Usually, the selected solids occupy portions of the same space, yet are separate solids. *Union* creates one solid composed of the total encompassing volume of the selected solids. (You can union solids even if the solids do not overlap.) All lines of intersections (surface boundaries) are calculated and displayed by AutoCAD. Multiple solid objects can be unioned with one *Union* command:

    Command: *union*
    Select objects: **PICK** (Select two or more solids.)
    Select objects: **Enter** (Indicate completion of the selection process.)
    Command:

Two solid boxes are combined into one composite solid with *Union* (Fig. 31-53). The original two solids (A) share the same physical space. The resulting union (B) consists of the total contained volume. The new lines of intersection are automatically calculated and displayed. *Hidden* shademode was used to enhance visualization in B.

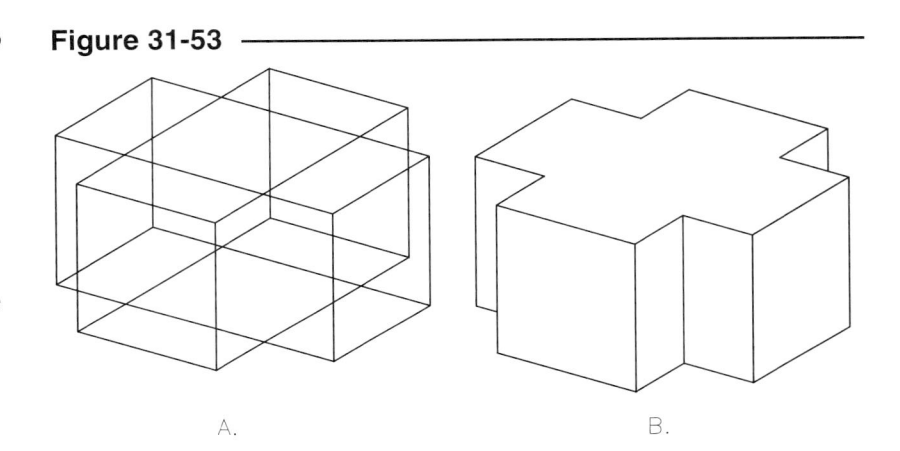

Figure 31-53

A.   B.

Because the volume occupied by any one of the primitives is included in the resulting composite solid, any redundant volumes are immaterial. The two primitives in Figure 31-54 A yield the same enclosed volume as the composite solid B. (*Hidden* shademode has been used for this figure.)

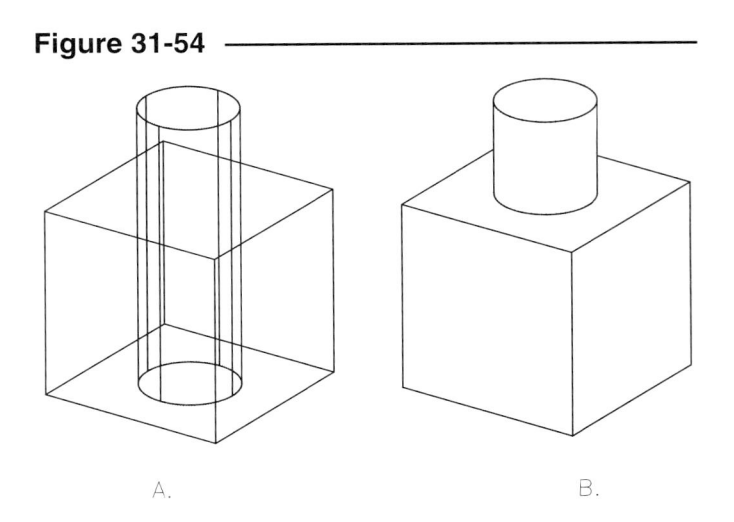

Figure 31-54

A.   B.

Multiple objects can be selected in response to the *Union* "Select objects:" prompt. It is not necessary, nor is it efficient, to use several successive Boolean operations if one or two can accomplish the same result.

Two primitives that have coincident faces (touching sides) can be joined with *Union*. Several "blocks" can be put together to form a composite solid.

Figure 31-55 illustrates how several primitives having coincident faces (A) can be combined into a composite solid (B). Only one *Union* is required to yield the composite solid.

Figure 31-55

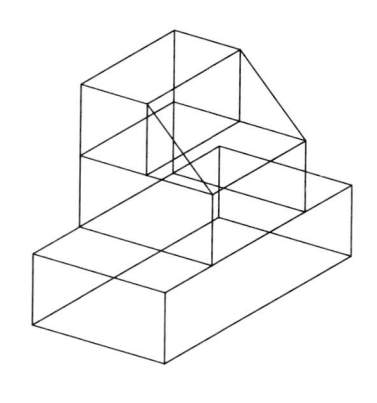

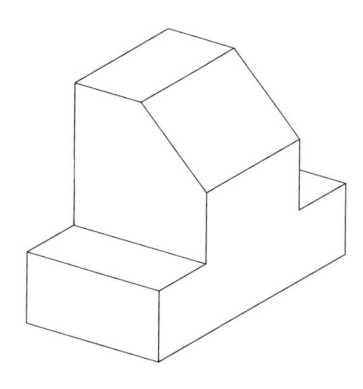

A.   B.

SUBTRACT

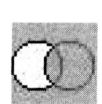

Pull-down Menu	COMMAND (TYPE)	ALIAS (TYPE)	Short-cut	Screen (side) Menu	Tablet Menu
Modify Solids Editing > Subtract	SUBTRACT	SU	...	MODIFY2 Subtract	X,16

Subtract takes the difference of one set of solids from another. *Subtract* operates with *Regions* as well as solids. When using solids, *Subtract* subtracts the volume of one set of solids from another set of solids. Either set can contain only one or several solids. *Subtract* requires that you first select the set of solids that will remain (the "source objects"), then select the set you want to subtract from the first:

    Command: `subtract`
    Select solids and regions to subtract from...
    Select objects: **PICK**
    Select objects: **Enter**
    Select solids and regions to subtract...
    Select objects: **PICK**
    Select objects: **Enter**
    Command:

The entire volume of the solid or set of solids that is subtracted is completely removed, leaving the remaining volume of the source set.

Figure 31-56

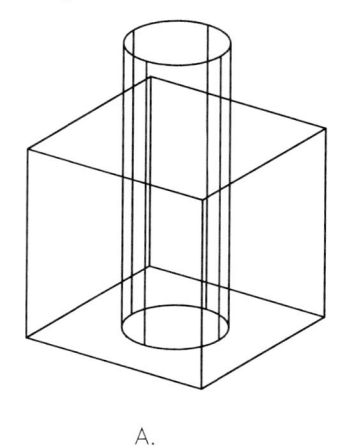

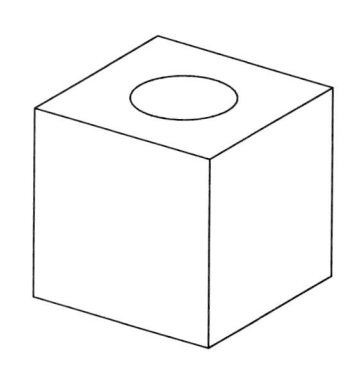

To create a box with a hole, a cylinder is located in the same 3D space as the box (see Figure 31-56). *Subtract* is used to subtract the entire volume of the cylinder from the box. Note that the cylinder can have any height, as long as it is at least equal in height to the box.

Because you can select more than one object for the objects "to subtract from" and the objects "to subtract," many possible construction techniques are possible.

A.   B.

646 Solid Modeling Construction

> **TIP:** If you select multiple solids in response to the select objects "to subtract from" prompt, they are <u>automatically</u> unioned. This is known as an <u>*n*-way Boolean</u> operation. Using *Subtract* in this manner is very efficient and fast.

Figure 31-57 illustrates an *n*-way Boolean. The two boxes (A) are selected in response to the objects "to subtract from" prompt. The cylinder is selected as the objects "to subtract...". *Subtract* joins the source objects (identical to a *Union*) and subtracts the cylinder. The resulting composite solid is shown in B. (*Hidden* shademode has been used for this figure.)

Figure 31-57

A.   B.

INTERSECT

Pull-down Menu	COMMAND (TYPE)	ALIAS (TYPE)	Short-cut	Screen (side) Menu	Tablet Menu
Modify Solids Editing > Intersect	INTERSECT	IN	...	MODIFY2 Intrsect	X,17

Intersect creates composite solids by calculating the intersection of two or more solids. The intersection is the common volume <u>shared</u> by the selected objects. Only the 3D space that is <u>part of all</u> of the selected objects is included in the resulting composite solid. *Intersect* requires only that you select the solids from which the intersection is to be calculated.

    Command: **intersect**
    Select objects: **PICK** (Select all desired solids.)
    Select objects: **Enter** (Indicates completion of the selection process.)
    Command:

An example of *Intersect* is shown in Figure 31-58. The cylinder and the box share common 3D space (A). The result of the *Intersect* is a composite solid that represents that common space (B). (*Hidden* shademode has been used for this figure.)

Figure 31-58

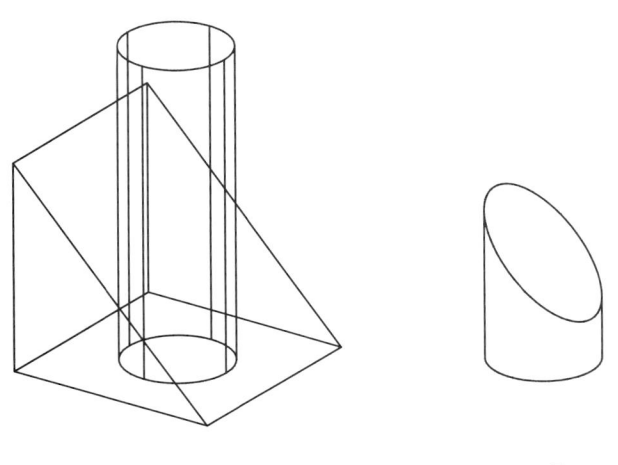

A.   B.

Intersect can be very effective when used in conjunction with *Extrude*. A technique known as reverse drafting can be used to create composite solids that may otherwise require several primitives and several Boolean operations. Consider the composite solid shown in Figure 31-55 A. Using *Union*, the composite shape requires four primitives.

A more efficient technique than unioning several box primitives is to create two *Pline* shapes on vertical planes (Fig. 31-59). Each *Pline* shape represents the outline of the desired shape from its respective view: in this case, the front and side views. The *Pline* shapes are intended to be extruded to occupy the same space. It is apparent from this illustration why this technique is called reverse drafting.

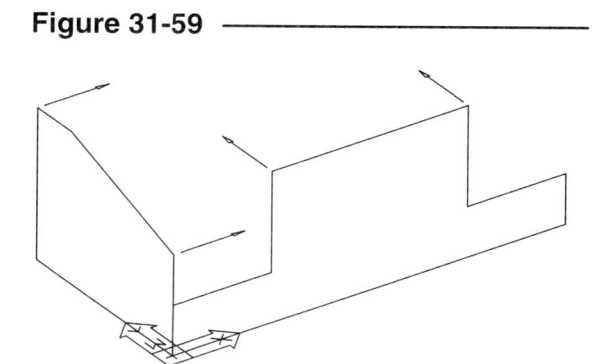

Figure 31-59

The two *Pline* "views" are extruded with *Extrude* to comprise the total volume of the desired solid (Fig. 31-60 A). Finally, *Intersect* is used to calculate the common volume and create the composite solid (B).

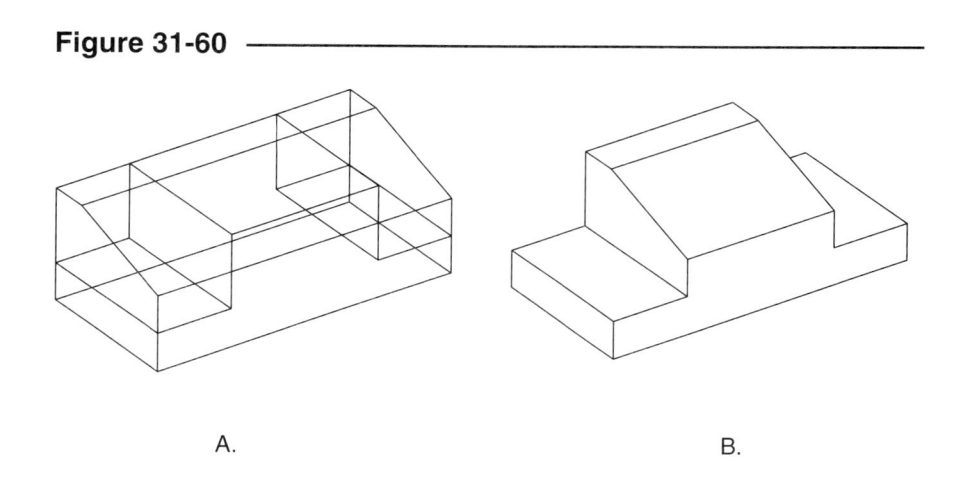

Figure 31-60

A.   B.

CHAMFER

Pull-down Menu	COMMAND (TYPE)	ALIAS (TYPE)	Short-cut	Screen (side) Menu	Tablet Menu
Modify Chamfer	CHAMFER	CHA	...	MODIFY2 Chamfer	W,18

Chamfering is a machining operation that bevels a sharp corner. *Chamfer* chamfers selected edges of an AutoCAD solid as well as 2D objects. Technically, *Chamfer* (used with a solid) is a Boolean operation because it creates a wedge primitive and then adds to or subtracts from the selected solid.

648 Solid Modeling Construction

When you select a solid, *Chamfer* recognizes the object as a solid and switches to the solid version of prompts and options. Therefore, all of the 2D options are not available for use with a solid, only the "distances" method. When using *Chamfer*, you must both select the "base surface" and indicate which edge(s) on that surface you wish to chamfer:

Command: *chamfer*
(TRIM mode) Current chamfer Dist1 = 0.5000, Dist2 = 0.5000
Select first line or [Polyline/Distance/Angle/Trim/Method]: **PICK** (Select solid at desired edge)
Base surface selection...
Enter surface selection option [Next/OK (current)] <OK>: **N** or **Enter**
Specify base surface chamfer distance <0.5000>: (**value**)
Specify other surface chamfer distance <0.5000>: (**value**)
Select an edge or [Loop]: **PICK** (Select edges to be chamfered)
Select an edge or [Loop]: **Enter**
Command:

When AutoCAD prompts to select the "base surface," only an edge can be selected since the solids are displayed in wireframe. When you select an edge, AutoCAD highlights one of the two surfaces connected to the selected edge. Therefore, you must use the "Next/<OK>:" option to indicate which of the two surfaces you want to chamfer (Fig. 31-61 A). The two distances are applied to the object, as shown in Figure 31-61 B.

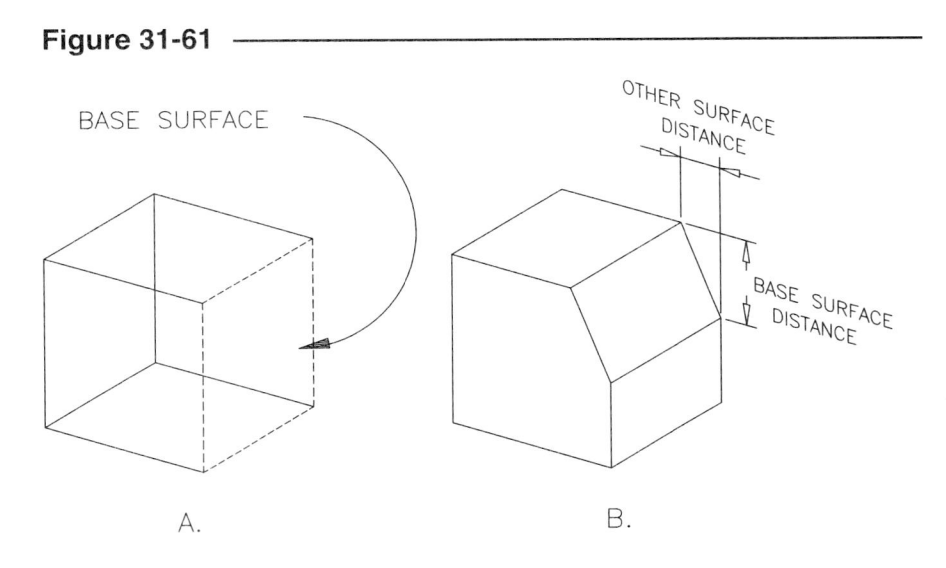

Figure 31-61

A.   B.

You can chamfer multiple edges of the selected "base surface" simply by PICKing them at the "Select an edge or [Loop]:" prompt (Fig. 31-62). If the base surface is adjacent to cylindrical edges, the bevel follows the curved shape.

Figure 31-62

A.   B.

Loop
The *Loop* option chamfers the entire perimeter of the base surface. Simply PICK any edge on the base surface.

Select an edge or [Loop]: **l**
Select an edge loop or [Edge]: **PICK** (Select edges to form loop)
Select an edge or [Loop]: **Enter**
Command:

Edge
The *Edge* option switches back to the "Select edge" method.

Figure 31-63

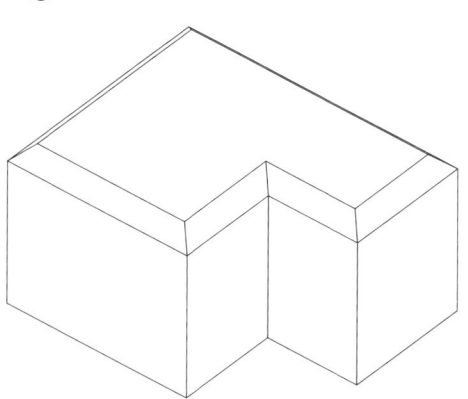

FILLET

Pull-down Menu	COMMAND (TYPE)	ALIAS (TYPE)	Short-cut	Screen (side) Menu	Tablet Menu
Modify Fillet	FILLET	F	...	MODIFY 2 Fillet	W,19

Fillet creates fillets (concave corners) or rounds (convex corners) on selected solids, just as with 2D objects. Technically, *Fillet* creates a rounded primitive and automatically performs the Boolean needed to add or subtract it from the selected solids.

When using *Fillet* with a solid, the command switches to a special group of prompts and options for 3D filleting, and the 2D options become invalid. After selecting the solid, you must specify the desired radius and then select the edges to fillet. When selecting edges to fillet, the edges must be PICKed individually. Figure 31-64 depicts concave and convex fillets created with *Fillet*. The selected edges are highlighted.

Figure 31-64

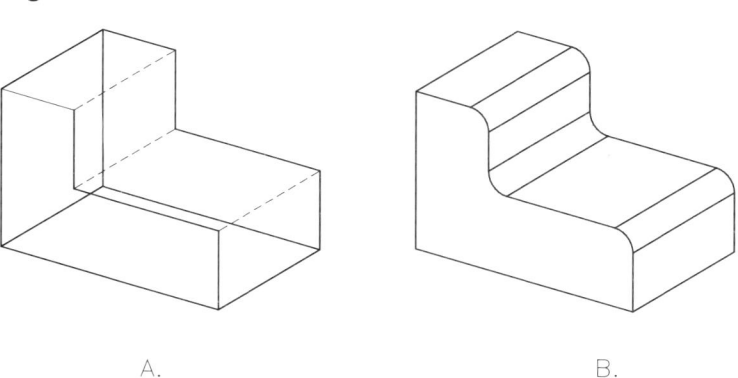

A.   B.

Command: *fillet*
Current settings: Mode = TRIM, Radius = 0.5000
Select first object or [Polyline/Radius/Trim]: **PICK** (Select desired edge to fillet)
Enter fillet radius <0.5000>: (**value**)
Select an edge or [Chain/Radius]: **Enter** or **PICK** (Select additional edges to fillet)
Command:

Curved surfaces can be treated with *Fillet*, as shown in Figure 31-65. If you want to fillet intersecting concave or convex edges, *Fillet* handles your request, provided you specify all edges in <u>one</u> use of the command. Figure 31-65 shows the selected edges (highlighted) and the resulting solid. Make sure you select <u>all</u> edges together (in one *Fillet* command).

Figure 31-65

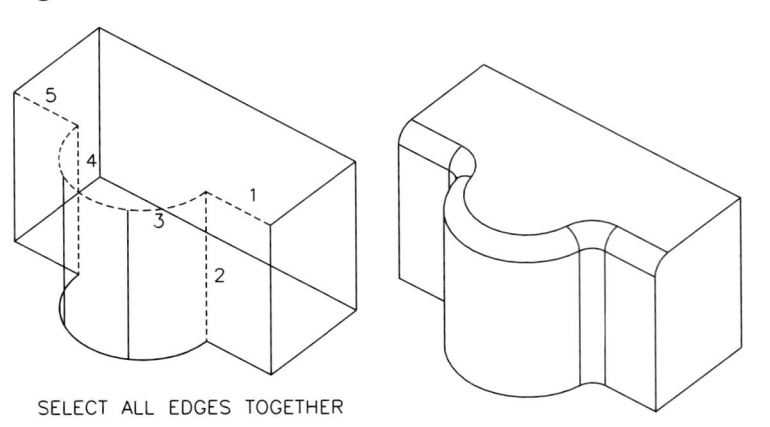

SELECT ALL EDGES TOGETHER

Chain
The *Chain* option allows you to fillet a series of connecting edges. Select the edges to form the chain (Fig. 31-66). If the chain is obvious (only one direct path), you can PICK only the ending edges, and AutoCAD will find the most direct path (series of connected edges):

  Select an edge or [Chain/Radius]: *c*
  Select an edge chain or
  [Edge/Radius]: **PICK**

Figure 31-66

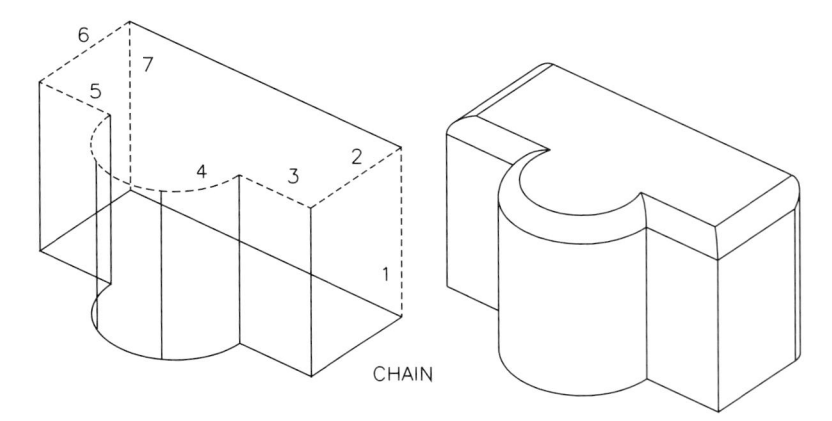

CHAIN

Edge
This option cycles back to the "<Select edge>:" prompt.

Radius
This method returns to the "Enter radius:" prompt.

DESIGN EFFICIENCY

Now that you know the complete sequence for creating composite solid models, you can work toward improving design efficiency. The typical construction sequence is (1) create primitives, (2) ensure the primitives are in place by using UCSs or any of several move and rotate options, and (3) combine the primitives into a composite solid using Boolean operations. The typical step-by-step, "building-block" strategy, however, may not lead to the most efficient design. In order to minimize computation and construction time, you should <u>minimize the number of Boolean operations</u> and, if possible, the <u>number of primitives</u> you use.

For any composite solid, usually several strategies could be used to construct the geometry. You should plan your designs ahead of time, striving to minimize primitives and Boolean operations.

For example, consider the procedure shown in Figure 31-55. As discussed, it is more efficient to accomplish all unions with one *Union*, rather than each union as a separate step. Even better, create a closed *Pline* shape of the profile; then use *Extrude*. Figure 31-57 is another example of design efficiency based on using an *n*-way Boolean. Multiple solids can be unioned automatically by selecting them at the select objects "to subtract from" prompt of *Subtract*. Also consider the strategy of reverse drafting, as shown in Figure 31-59. Using *Extrude* in concert with *Intersect* can minimize design complexity and time.

In order to create efficient designs and minimize Boolean operations and primitives, keep these strategies in mind:

- Execute as many subtractions, unions, or intersections as possible within one *Subtract*, *Union*, or *Intersect* command.

- Use *n*-way Booleans with *Subtract*. Combine solids (union) automatically by selecting multiple objects "to subtract from," and then select "objects to subtract."

- Make use of *Plines* or regions for complex profile geometry; then *Extrude* the profile shape. This is almost always more efficient for complex curved profile creation than using multiple Boolean operations.

- Make use of reverse drafting by extruding the "view" profiles (*Plines* or *Regions*) with *Extrude*, then finding the common volume with *Intersect*.

SOLIDS EDITING

In AutoCAD Releases 13 and 14, editing of 3D solids was not feasible. Once Boolean operators were used to combine primitives, the original 3D primitives could not be edited. For example, if you wanted to change the diameter of a hole, you could only create a new *Cylinder* and *Union* it to "plug" the hole, then make another *Cylinder* with the desired new diameter and *Subtract* it.

AutoCAD 2000 introduced the *Solidedit* command which has options for extruding, moving, rotating, offsetting, tapering, copying, coloring, separating, shelling, cleaning, checking, and deleting faces and edges of 3D solids. The unique feature of this command is that you can change individual primitives and partial internal or external geometry of composite solids. This single command contains all the options which are available on the *Solids Editing* menu from the *Modify* pull-down (Fig. 31-67) and on the *Solids Editing* toolbar (see Figure 31-5).

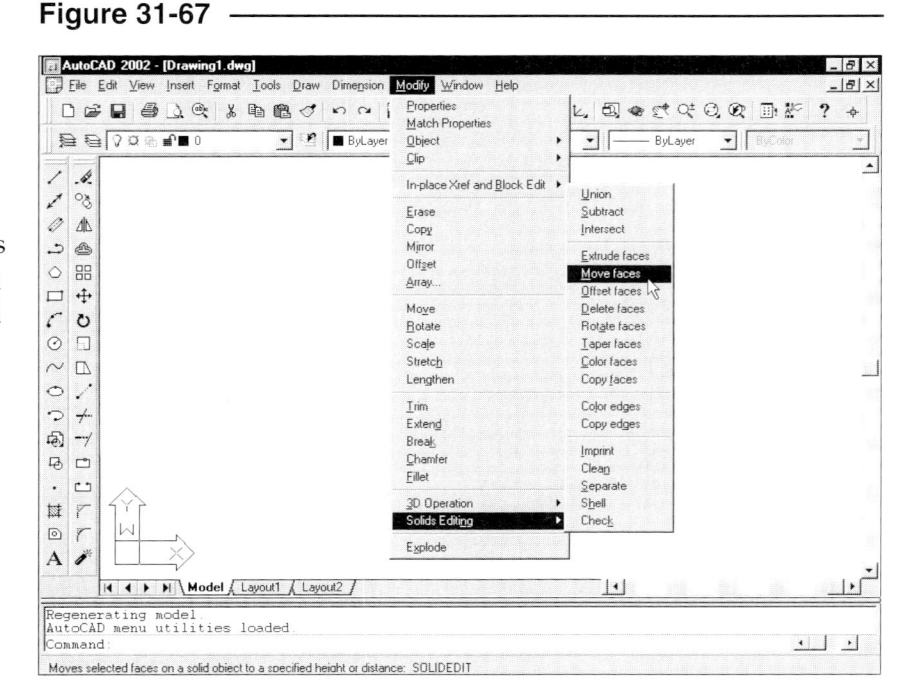

Figure 31-67

SOLIDEDIT

Pull-down Menu	COMMAND (TYPE)	ALIAS (TYPE)	Short-cut	Screen (side) Menu	Tablet Menu
Modify Solids Editing >	SOLIDEDIT	...	...	...	...

The *Solidedit* command has several "levels" of options. The first level prompts you to specify the type of geometry you want to edit—*Face, Edge,* or *Body*. The second level of options depends on your first level response as shown below.

Command: *solidedit*
Solids editing automatic checking: SOLIDCHECK=1
Enter a solids editing option [Face/Edge/Body/Undo/eXit] <eXit>: *f*
Enter a face editing option
[Extrude/Move/Rotate/Offset/Taper/Delete/Copy/coLor/Undo/eXit] <eXit>:

Solidedit operates on three types of geometry: *Edges, Faces* and *Bodies*. An *Edge* is defined as the common edge between two surfaces which has the appearance of a line, arc, circle, or spline (Fig. 31-68 A). *Faces* are planar or curved surfaces of a 3D object (Fig. 31-68 B). A *Body* is defined as an existing 3D solid or a non-solid shape created with a *Solidedit* option.

Figure 31-68

A.   B.

Object selection is an integral part of using *Solidedit*. For example, once you have specified the type of geometry and the particular editing option, the *Solidedit* command prompts to select *Edges, Faces,* or a *Body*.

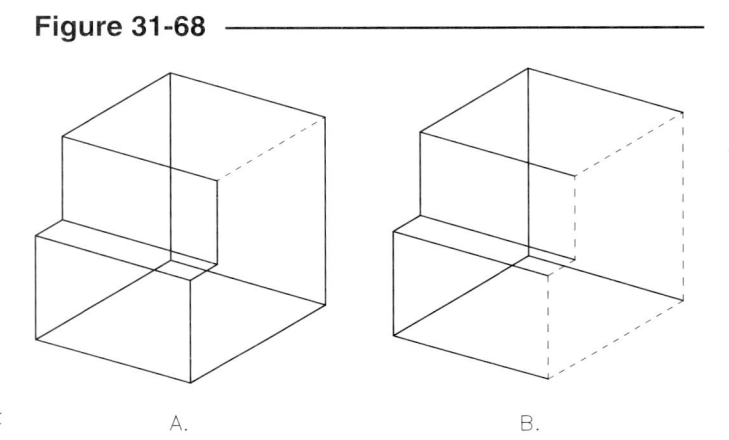

Select faces or [Undo/Remove]: Select a face or enter an option
Select faces or [Undo/Remove/ALL]: Select a face or enter an option

When you are prompted to select *Faces* or *Edges*, you will most likely go through a series of adding and removing geometry until the desired set of lines is highlighted. In addition to the pickbox, you can use the following selection options.

Crossing/Fence/CPolygon/Undo/Remove/ALL

To use a selection method other than the pickbox, you must type the desired choice at the "Select faces or [Undo/Remove]:" prompt since no automatic window or crossing window is available. As an alternative to *Remove*, you can hold down the Shift key and select objects.

Ensure that you highlight exactly the intended geometry before you proceed with editing. If an incorrect set is selected, you can get unexpected results or an error message can appear such as that below.

Modeling Operation Error:
  No solution for an edge.

Chapter Thirty-One   653

For example, many of the *Face* options allow selection of one or multiple faces. In Figure 31-69 both selection sets are valid faces or face combinations, but each will yield different results. Figure 31-69 A indicates selection of one face and B indicates selection of several faces defining an entire primitive. Add and remove faces or edges until you achieve the desired geometry.

Figure 31-69

A.   B.

Face Options

The *Face* options edit existing surfaces and create new surfaces. A *Face* is a planar or curved surface. *Faces* that are part of existing 3D solids can be altered to change the configuration of the 3D composite solid. Individual *Faces* can be edited and multiple *Faces* comprising a primitive can be edited. New surfaces can be created from existing *Faces*, but entirely new independent solids cannot be created using these editing tools.

Extrude

The *Extrude* option of *Solidedit* allows you to extrude any *Face* of a 3D object in a similar manner to using the *Extrude* command to create a 3D solid from a *Pline*. This capability is extremely helpful if you need to make a surface on a 3D solid taller, shorter, or longer. You can select one or more faces to extrude at one time.

```
Command: solidedit
Solids editing automatic checking: SOLIDCHECK=1
Enter a solids editing option [Face/Edge/Body/Undo/eXit] <eXit>: f
Enter a face editing option
[Extrude/Move/Rotate/Offset/Taper/Delete/Copy/coLor/Undo/eXit] <eXit>: e
Select faces or [Undo/Remove]: PICK
Select faces or [Undo/Remove/ALL]: PICK  or remove
Select faces or [Undo/Remove/ALL]: Enter
Specify height of extrusion or [Path]: (value)
Specify angle of taper for extrusion <0>: Enter or (value)
Solid validation started.
Solid validation completed.
```

For example, Figure 31-70 illustrates the extrusion of a face to make the 3D object taller, where A indicates the selected (highlighted) face and B shows the result. This extrusion has a 0 degree taper angle.

Figure 31-70

A.   B.

Specifying a positive angle tapers the face inward (Fig. 31-71 A) and specifying a negative angle tapers the face outward (B) *Height* is always perpendicular to the selected *Face*, not necessarily vertical.

Figure 31-71

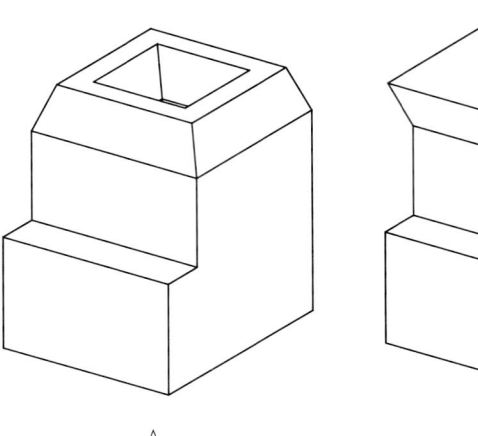

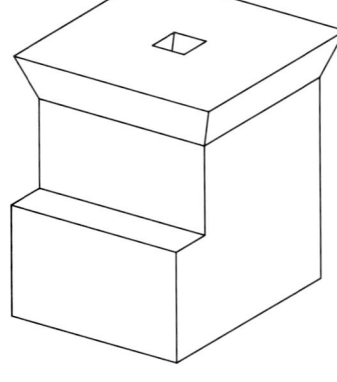

A.   B.

An internal face can be selected for extrusion. In Figure 31-72, a single face is selected (A) and the *Height* value specified is greater than the distance to the outer face of the solid, creating an open side on the object (B). *Extrude* is the only option that allows an internal face to "pass through" an external face. With all other options, the outermost "bounding box" of the solid cannot be extended or opened by editing an internal primitive or face.

Figure 31-72

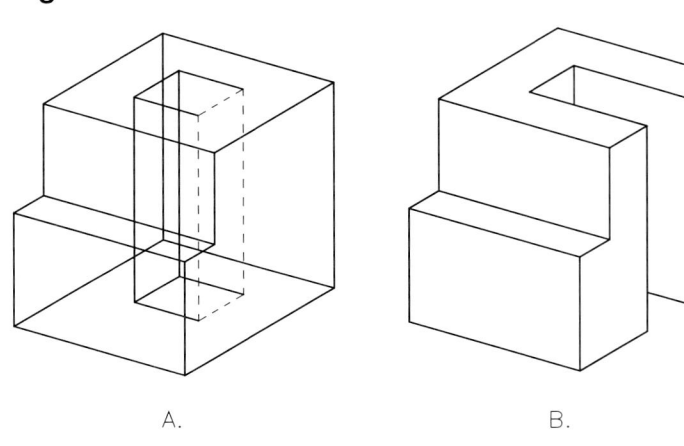

A.   B.

Faces can be extruded along a *Path*. The *Path* object can be a *Line*, *Circle*, *Arc*, *Ellipse*, *Pline*, or *Spline*. For example, Figure 31-73 illustrates extrusion of a face along a *Line* path (A) to yield the result shown on the right (B).

Figure 31-73

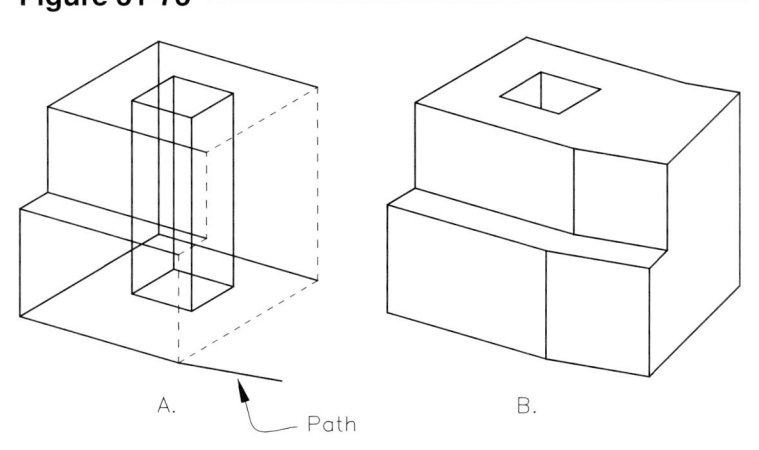

A. Path   B.

Move

With the *Move* option, you can move a single face of a 3D solid or move an entire primitive within a 3D solid. This capability is particularly helpful during geometry construction or when there is a design change and it is required to alter the location of a hole or other feature within the confines of the composite 3D model.

```
[Extrude/Move/Rotate/Offset/Taper/Delete/Copy/coLor/Undo/eXit] <eXit>: m
Select faces or [Undo/Remove]: PICK
Select faces or [Undo/Remove/ALL]: PICK or remove
Select faces or [Undo/Remove/ALL]: Enter
Specify a base point or displacement: PICK
Specify a second point of displacement: PICK
```

For example, Figure 31-74 illustrates using the *Move* option to relocate a hole primitive (four faces) within a composite solid. In this case you must ensure all faces, and only the faces, comprising the primitive are selected.

Figure 31-74

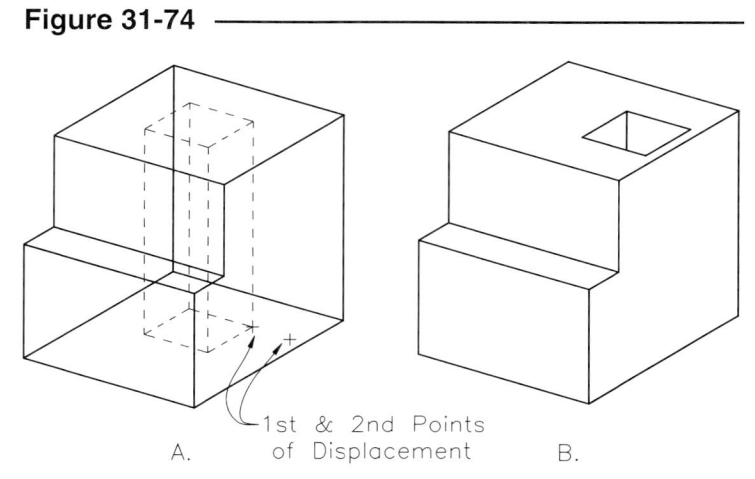

With the *Move* option the internal features <u>cannot</u> be moved to extend into or past the "bounding box" of the composite solid, as they can with *Extrude* (see Figure 31-72). If this condition is attempted, the following message appears.

Modeling Operation Error:
Improper edge/edge intersection.

Another possibility for *Move* is to move only one or selected faces to alter the configuration of an internal feature of a composite solid. An example is shown in Figure 31-75, where only one face is selected to move. Compare the results in Figures 31-74 B and 31-75 B.

Figure 31-75

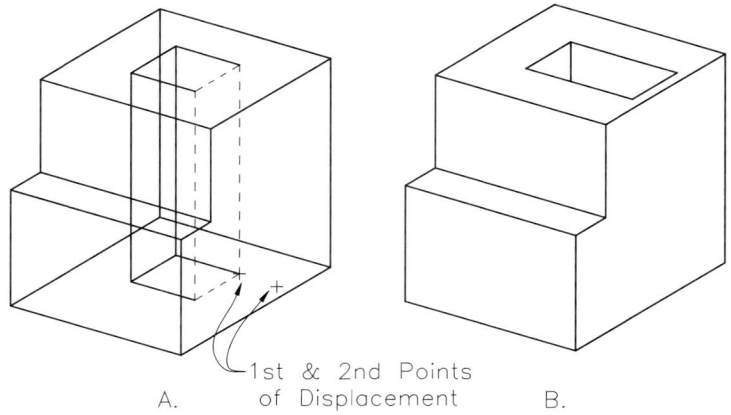

Keep in mind that *Move* can be used to move a singular external face, achieving the same result as *Extrude* (see previous Figure 31-70).

Offset

The 2D *Offset* command makes a parallel copy of a *Line*, *Arc*, *Circle*, *Pline*, etc. The *Offset* option of *Solidedit* makes a 3D offset. A simple application is making a hole larger or smaller.

```
[Extrude/Move/Rotate/Offset/Taper/Delete/Copy/coLor/Undo/eXit] <eXit>: o
Select faces or [Undo/Remove]: PICK
Select faces or [Undo/Remove/ALL]: PICK or remove
Select faces or [Undo/Remove/ALL]: Enter
Specify the offset distance: PICK or (value)
```

656 *Solid Modeling Construction*

You can offset through a point or specify an offset distance. Specify a positive value to increase the size of the solid or a negative value to decrease the size of the solid.

Figure 31-76 demonstrates how an internal feature, such as a hole, can be *Offset* to effectively change the volume of the hole. Since positive values increase the size of the solid, a negative value was used here. In other words, holes inside a solid become smaller when the solid is *Offset* larger.

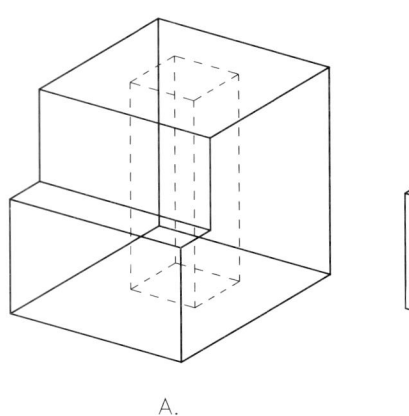

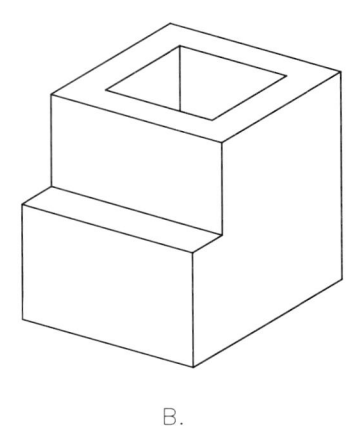

Figure 31-76

A.   B.

Delete

This option of *Solidedit* deletes faces from composite solids. Although you cannot delete one planar face from a solid, you can delete one curved face, such as a cylinder, that comprises an entire primitive. You can also delete multiple faces that comprise a primitive or entire feature.

[Extrude/Move/Rotate/Offset/Taper/Delete/Copy/coLor/Undo/eXit] <eXit>: **d**
Select faces or [Undo/Remove]: **PICK**
Select faces or [Undo/Remove/ALL]: **PICK** or remove
Select faces or [Undo/Remove/ALL]: **Enter**
Solid validation started.
Solid validation completed.

As an example, Figure 31-77 A displays two selected faces to be deleted from a composite solid. The result is displayed in B. In this case, <u>both faces must be highlighted</u> for the deletion to operate.

Figure 31-77

For the same composite solid, if all <u>four</u> internal faces comprising the hole are selected, as shown in previous Figure 31-76 A, the hole could be deleted.

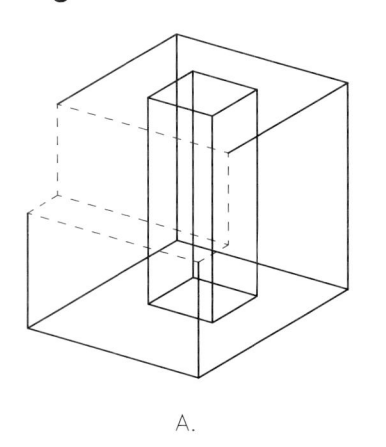

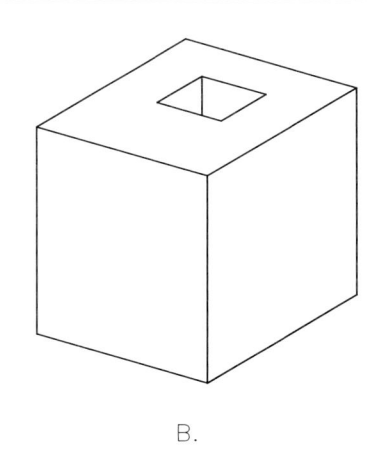

A.   B.

Rotate

Rotate is helpful during geometry construction and for design changes when components within a composite solid must be rotated in some way.

[Extrude/Move/Rotate/Offset/Taper/Delete/Copy/coLor/Undo/eXit] <eXit>: **r**
Select faces or [Undo/Remove]: **PICK**
Select faces or [Undo/Remove/ALL]: **PICK** or remove
Select faces or [Undo/Remove/ALL]: **Enter**
Specify an axis point or [Axis by object/View/Xaxis/Yaxis/Zaxis] <2points>: **PICK**
Specify the second point on the rotation axis: **PICK**
Specify a rotation angle or [Reference]: **PICK** or (**value**)

Several methods of rotation are possible based on the selected axis. These options are essentially the same as those available with the *Rotate3D* command (see "*Rotate3D*" discussed previously).

For example, Figure 31-78 illustrates the rotation of a primitive within a composite solid. Here, the primitive is rotated about two points defining one edge of the hole.

Figure 31-78

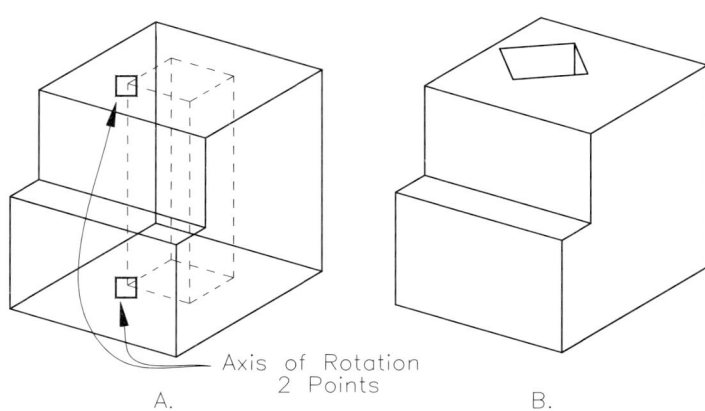

One face of a composite solid can be selected for rotation within a composite solid (Fig. 31-79 A). The selected face is rotated about two points defining one edge of the face.

Figure 31-79

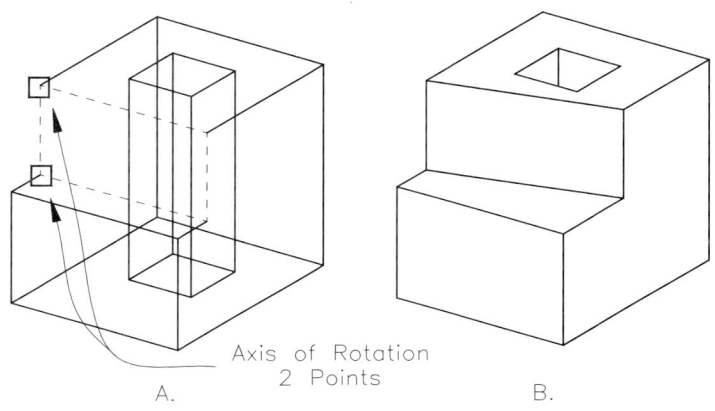

Taper

Taper angles a face. You can use *Taper* to change the angle of planar or curved faces.

    [Extrude/Move/Rotate/Offset/Taper/Delete/Copy/coLor/Undo/eXit] <eXit>: t
    Select faces or [Undo/Remove]: PICK
    Select faces or [Undo/Remove/ALL]: PICK or remove
    Select faces or [Undo/Remove/ALL]: Enter
    Specify the base point: PICK
    Specify another point along the axis of tapering: PICK
    Specify the taper angle: Enter or (value)

The rotation of the taper angle is determined by the order of selection of the base point and second point. Although AutoCAD prompts for the "axis of tapering," the two points actually determine a reference line from which the taper angle is applied. The taper starts at the first base point and tapers away from the second base point. The rotational axis passes through the first base point and is <u>perpendicular</u> to the line between the base points. <u>The line between the base points does not specify the rotational axis, as implied</u>.

Figure 31-80

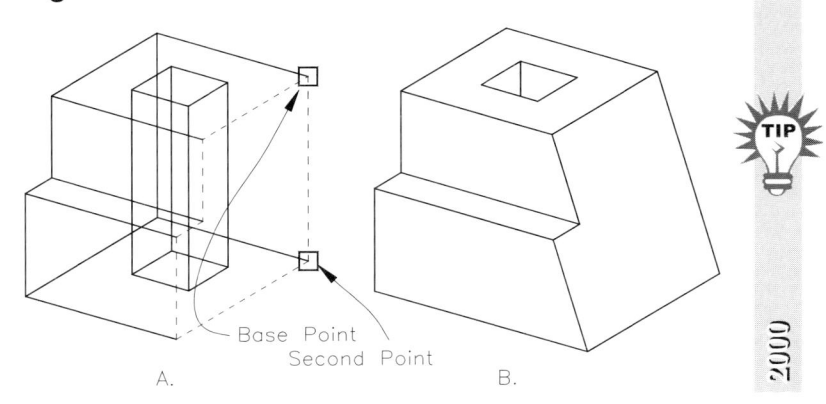

658 Solid Modeling Construction

To explain this idea, Figure 31-80 A (previous page) illustrates the selection of a planar face, the base point, and the second point. An angle of –20 degrees (negative value) is entered to achieve the results shown in B. Positive angles taper the face inward (toward the solid) and negative values taper the face outward (away from the solid).

For some cases, either *Taper* or *Rotate* could be used to achieve the same results. For example, both Figure 31-80 B and previous Figure 31-79 B could be attained using *Taper* or *Rotate* (although different points must be selected depending on the option used).

Curved faces can also be selected for applying a *Taper*. Figure 31-81 A illustrates a cylindrical hole primitive. The entire cylinder primitive is selected as one *Face* object. Note the selection of the base point and second point. *Taper* converts the cylindrical hole into a conical hole (B).

Figure 31-81

Copy

As you would expect, the *Copy* option copies *Faces*. However, the resulting objects are *Regions* or *Bodies*.

```
[Extrude/Move/Rotate/Offset/Taper/Delete/Copy/coLor/Undo/eXit] <eXit>: c
Select faces or [Undo/Remove]: PICK
Select faces or [Undo/Remove/ALL]: PICK  or remove
Select faces or [Undo/Remove/ALL]: Enter
Specify the base point or displacement: PICK or (value)
Specify a second point of displacement: PICK
```

Although any *Face* can be selected (planar or curved), Figure 31-82 shows how *Copy* can be used to create a *Region* from a planar face. Typically, the *Copy* option would be used to create a new *Region* from existing 3D solid geometry. The *Extrude* command could then be used on the *Region* to form a new 3D solid from the original face.

If a curved *Face* is selected to *Copy*, the resulting object is a *Body*. This type of body is a curved surface, not a solid.

Figure 31-82

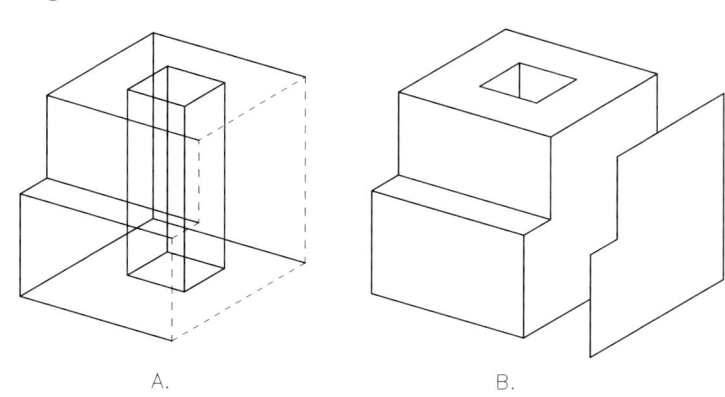

Using the same composite model as in previous figures, assume you were to *Copy* only the five vertical faces indicated in Figure 31-83 A. After selecting the second point of displacement (B), notice only those five vertical surfaces result. Not included are the faces defining the hole or any horizontal faces. Therefore, you can select *All* faces comprising a composite solid to *Copy*, and the resulting model is a surface model composed of *Regions*.

Figure 31-83

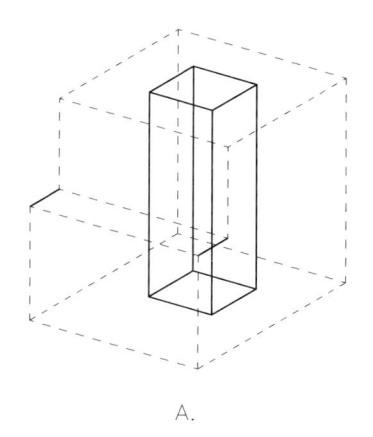

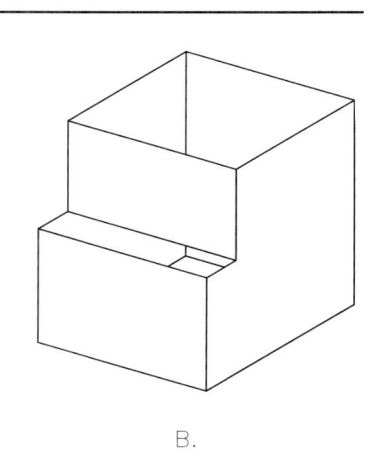

A.   B.

Color

The *Color Face* option of *Solidedit* simply changes the color of selected faces.

[Extrude/Move/Rotate/Offset/Taper/Delete/Copy/coLor/Undo/eXit] <eXit>: **l**
Select faces or [Undo/Remove]: **PICK**
Select faces or [Undo/Remove/ALL]: **PICK** or remove
Select faces or [Undo/Remove/ALL]: **Enter**
Enter new color <BYLAYER>: **(color)**

The resulting display of the 3D solid depends on the *Shademode* setting. Only the object edges appear in the designated color when *2D wireframe*, *3D wireframe*, or *Hidden* settings are used. For the *Flat* and *Gouraud* options of *Shademode*, all surfaces are displayed in shaded intensities of the designated color. For *Flat+edges* or *Gouraud+edges*, the surfaces are shaded in the designated colors and edges are displayed in a lighter intensity of the colors (unless the *Color Edge* option is used).

Edge Options

Edges are lines or curves that define the boundary between *Faces*. Copying an *Edge* would result in a 2D object. The selection process for *Edges* is critical, but not as involved as selecting *Faces*. You can use multiple selection methods (the same options as with *Edges*), and selection may involve a process of adding and removing objects. As with *Face* selection, make sure you have the exact desired set highlighted before continuing with the procedure. The command syntax is as follows.

Command: **solidedit**
Solids editing automatic checking: SOLIDCHECK=1
Enter a solids editing option [Face/Edge/Body/Undo/eXit] <eXit>: **edge**
Enter an edge editing option [Copy/coLor/Undo/eXit] <eXit>:

Copy

The *Copy* option of *Edge* creates wireframe elements, not surfaces or solids. Copied edges become 2D objects such as a *Line*, *Arc*, *Circle*, *Ellipse*, or *Spline*. These elements could be used to create other 3D models such as wireframes, surfaces, or solids.

Enter an edge editing option [Copy/coLor/Undo/eXit] <eXit>: **c**
Select edges or [Undo/Remove]: **PICK**
Select edges or [Undo/Remove]: **PICK** or remove
Select edges or [Undo/Remove]: **Enter**
Specify a base point or displacement: **PICK**
Specify a second point of displacement: **PICK**

660 Solid Modeling Construction

One or multiple *Edges* can be selected. After indicating the second point of displacement, the selected edges are extracted from the solid and copied to the new location (Fig. 31-84).

Keep in mind that the *Copied Edges* are wireframe elements. These new elements can be used for construction of other 3D geometry, such as with the *Body*, *Imprint* option (see "*Body* Options" later in this section).

Figure 31-84

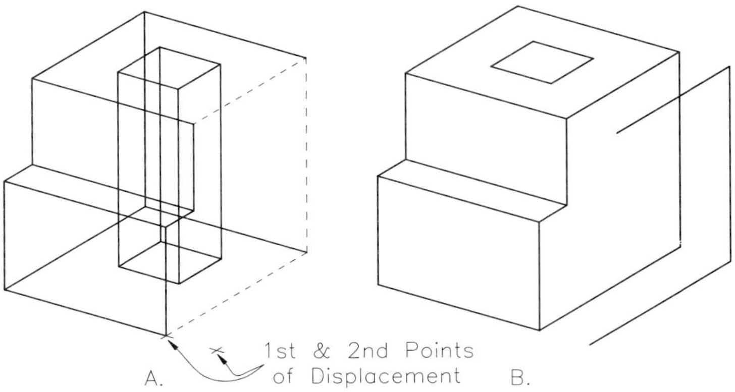

Color

The *Color* option for *Edges* operates similarly to the *Color* option for *Faces* except that only the *Edges* are affected. This option is useful in cases where you need the edges to be more or less visible to bring attention to certain features or components of a solid, or in cases when you want the surfaces of a solid to be shaded in one color and the edges to appear in another.

```
Enter an edge editing option [Copy/coLor/Undo/eXit] <eXit>: l
Select edges or [Undo/Remove]: PICK
Select edges or [Undo/Remove]: PICK or remove
Select edges or [Undo/Remove]: Enter
Enter new color <BYLAYER>: (color)
```

The selected edges appear in the new color. The *Shademode* setting also affects the appearance of the surfaces and edges. When *Shademode* is set to any option that causes the edges to appear (*2D wireframe*, *3D wireframe*, *Hidden*, *Flat+edges*, or *Gouraud+edges*), the edges appear in the selected color. In the *Flat* and *Gouraud* modes, only the surfaces appear without edges. You can use the *Flat+edges* and *Gouraud+edges* modes to display the edges in one color and the surfaces in another.

Body Options

A *Body* is typically a 3D solid. Any primitive or composite solid can be selected as a *Body*. However, some *Solidedit* options can create a *Body* that is a non-solid. For example, selecting a curved *Face* and using *Copy* creates a curved surface that AutoCAD lists as a *Body*. Several options are offered here to alter the configuration of a *Body* or to create or edit 2D geometry used to interact with a *Body*.

Imprint

Imprint is used to attach a 2D object on an existing face of a solid (*Body*). The new *Imprint* can then be used with *Extrusion* to create a new 3D solid.

Objects that can be used to make the *Imprint* can be an *Arc*, *Circle*, *Line*, *Pline*, *Ellipse*, *Spline*, *Region*, or *Solid*. The selected object to *Imprint* must touch or intersect the solid in some way, such as when a *Circle* lies partially on a *Face*, or when two solid volumes overlap. When *Imprint* is used, one or more 2D components becomes attached, or imprinted, to an existing 3D solid face. The resulting *Imprint* has little usefulness of itself, but is an intermediate step to creating a new 3D solid shape on the existing solid.

```
[Imprint/sePrate solids/Shell/cLean/Check/Undo/eXit] <eXit>: i
Select a 3D solid: PICK
Select an object to imprint: PICK
Delete the source object <N>: y
Select an object to imprint: Enter
```

For example, Figure 31-85 A displays a 3D solid with a *Circle* that is on the same plane as the vertical right face of the solid. The *Imprint* option is used, the solid is selected, and the *Circle* is selected as the "object to imprint." Answering "Y" to "Delete the source object," the resulting *Imprint* is shown in B.

Figure 31-85

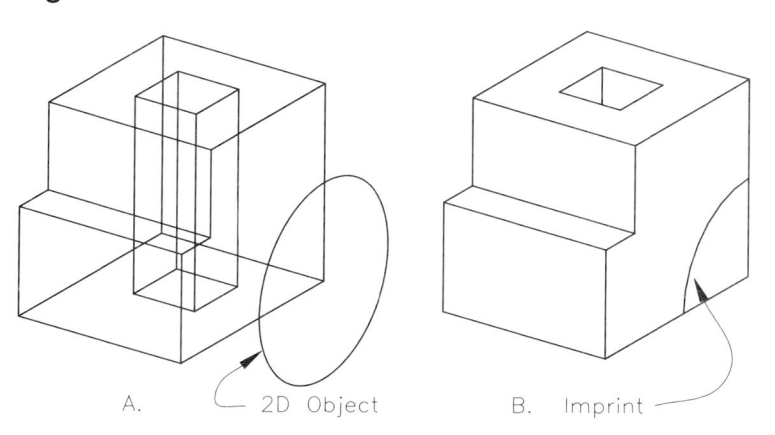

A. 2D Object B. Imprint

The *Imprint* creates a separate *Face* on the object—in this case, a total of two coplanar *Faces* are on the same vertical surface of the object. The new *Face* can be treated independently for use with other *Solidedit* options. For example, *Extrude* could be used to create a new solid feature on the composite solid (Fig. 31-86).

Figure 31-86

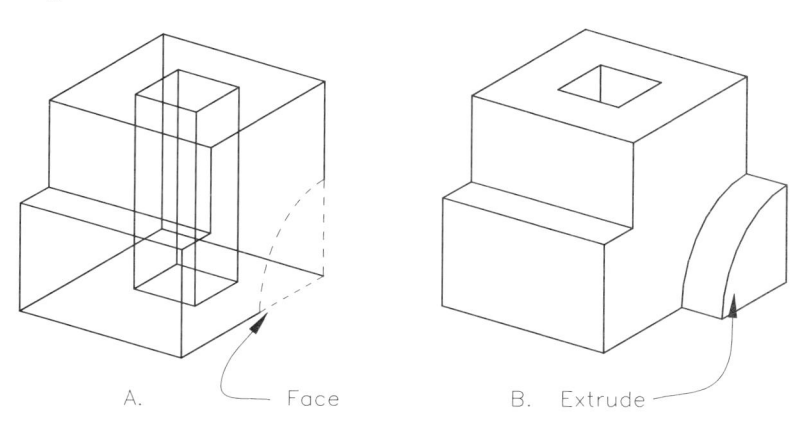

A. Face B. Extrude

Clean

The *Clean* option deletes any 2D geometry on the solid. For example, you may use *Imprint* to "attach" 2D geometry to a *Body* (solid) for the intention of using *Extrude*. *Imprinted* objects become permanently attached to the *Body* and cannot be removed by *Erase*. If you need to remove an *Imprint*, use the *Clean* option to do so. One use of *Clean* removes all *Imprints*.

    [Imprint/seParate solids/Shell/cLean/Check/Undo/eXit] <eXit>: **l**
    Select a 3D solid: **PICK**

Separate

It is possible in AutoCAD to *Union* two solids that do not occupy the same physical space and have two distinct volumes. In that case, the two shapes appear to be separate, but are treated by AutoCAD as one object (if you select one, both become highlighted).

Occasionally you may intentionally or inadvertently use *Union* to combine two or more separate (not physically touching) objects. *Shell* and *Offset*, when used with solids containing holes, can also create two or more volumes from one solid. *Separate* can then be used to disconnect these into discrete independent solids. *Separate* <u>cannot</u> be used to disconnect or "break down" *Unioned*, *Subtracted*, or *Intersected* solids that form one volume.

    [Imprint/seParate solids/Shell/cLean/Check/Undo/eXit] <eXit>: **p**
    Select a 3D solid: **PICK**

Shell

Shell converts a solid into a thin-walled "shell." You first select a solid to shell, then specify faces to remove, and enter an offset distance. An example would be to convert a solid cube into a hollow box—the thickness of the walls equal the "shell offset distance." If no faces are removed from the selection set, the box would have no openings. Faces that are removed become open sides of the box.

```
[Imprint/seParate solids/Shell/cLean/Check/Undo/eXit] <eXit>: s
Select a 3D solid: PICK
Remove faces or [Undo/Add/ALL]: PICK
Remove faces or [Undo/Add/ALL]: PICK
Remove faces or [Undo/Add/ALL]: Enter
Enter the shell offset distance: (value)
Solid validation started.
Solid validation completed.
```

A positive value entered at the "shell offset distance" prompt creates the wall thickness outside of the original solid, whereas a negative value creates the wall thickness inside the existing boundary.

An example is shown in Figure 31-87. The left object (A) indicates the solid with all the selected faces highlighted and the faces removed from the selection set not highlighted. A negative offset distance is used to yield the shape shown on the right (B).

Figure 31-87

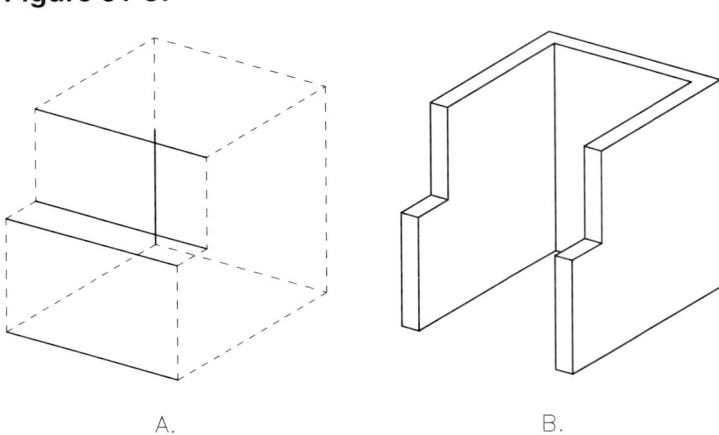

A.   B.

Don't expect to achieve the desired results the first time. As with most *Solidedit* options, the process of adding and removing faces is an inexact procedure since any edge you select can highlight either one of two faces.

Check

Use this option to validate the 3D solid object as a valid ACIS solid, independent of the *SOLIDCHECK* system variable setting.

```
[Imprint/seParate solids/Shell/cLean/Check/Undo/eXit] <eXit>: c
Select a 3D solid: PICK
This object is a valid ACIS solid.
```

The *SOLIDCHECK* variable turns the automatic solid validation on and off for the current AutoCAD session. By default, *SOLIDCHECK* is set to 1, or on, to validate the solid. The *Solidedit* command displays the current status of the variable as shown in the command syntax below.

```
Command: solidedit
Solids editing automatic checking:  SOLIDCHECK=0
Enter a solids editing option [Face/Edge/Body/Undo/eXit] <eXit>
```

CHAPTER EXERCISES

1. What are the typical three steps for creating composite solids?

2. Consider the two solids in Figure 31-88. They are two extruded hexagons that overlap (occupy the same 3D space).

   Figure 31-88

   A. Sketch the resulting composite solid if you performed a *Union* on the two solids.

   B. Sketch the resulting composite solid if you performed an *Intersect* on the two solids.

   C. Sketch the resulting composite solid if you performed a *Subtract* on the two solids.

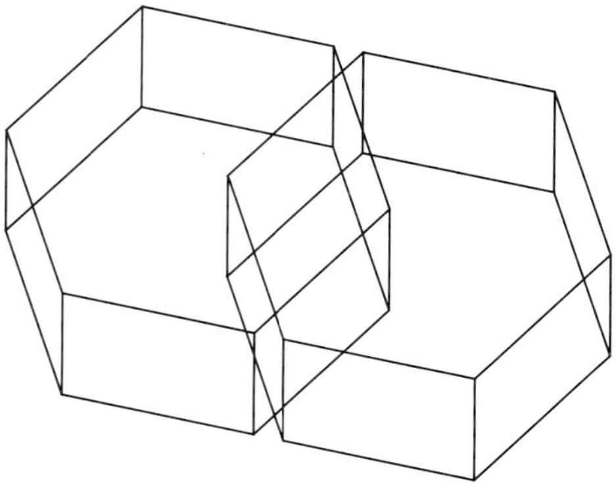

For the following exercises, use a *Template* drawing or begin a *New* drawing. Turn *On* the *Ucsicon* and set it to the *Origin*. Set the *Vpoint* with the *Rotate* option to angles of **310, 30**.

3. Begin a drawing and assign the name **CH31EX3**.

   Figure 31-89

   A. Create a *box* with the lower-left corner at **0,0,0**. The *Lengths* are **5, 4,** and **3**.

   B. Create a second *box* at a new UCS as shown in Figure 31-89 (use the *ORigin* option). The *box* dimensions are **2 x 4 x 2**.

   C. Create a *Cylinder*. Use the same UCS as in the previous step. The *cylinder Center* is at **3.5,2** (of the *UCS*), the *Diameter* is **1.5**, and the *Height* is **-2**.

   D. *Save* the drawing.

   E. Perform a *Union* to combine the two boxes. Next, use *Subtract* to subtract the cylinder to create a hole. The resulting composite solid should look like that in Figure 31-90. *Save* the drawing.

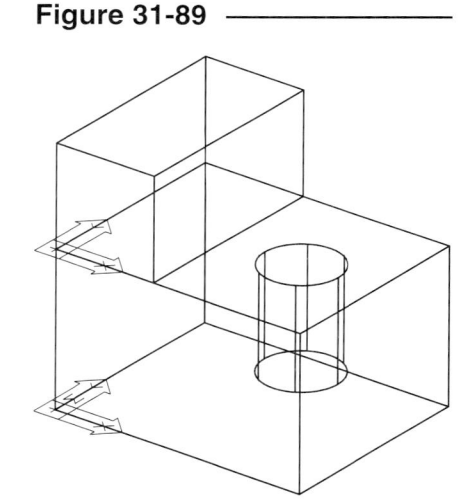

   Figure 31-90

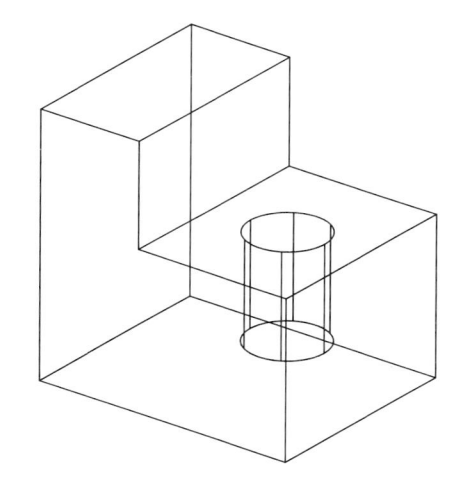

664 Solid Modeling Construction

4. Begin a drawing and assign the name **CH31EX4**.

   A. Create a *Wedge* at point **0,0,0** with the *Lengths* of **5, 4, 3**.

   B. Create a *3point UCS* option with an orientation indicated in Figure 31-91. Create a *Cone* with the *Center* at **2,3** (of the *UCS*) and a *Diameter* of **2** and a *Height* of **-4**.

   Figure 31-91

   C. *Subtract* the cone from the wedge. The resulting composite solid should resemble Figure 31-92. *Save* the drawing.

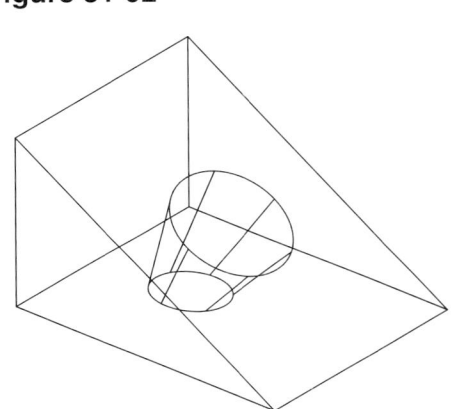

   Figure 31-92

5. Begin a drawing and assign the name **CH31EX5**. Display a *Plan* view.

   A. Create 2 *Circles* as shown in Figure 31-93, with dimensions and locations as specified. Use *Pline* to construct the rectangular shape. Combine the 3 shapes into a *Region* by using the *Region* and *Union* commands or converting the outside shape into a *Pline* using *Trim* and *Pedit*.

   Figure 31-93

   B. Change the display to an isometric-type *Vpoint*. *Extrude* the *Region* or *Pline* with a *Height* of **3** (no *taper angle*).

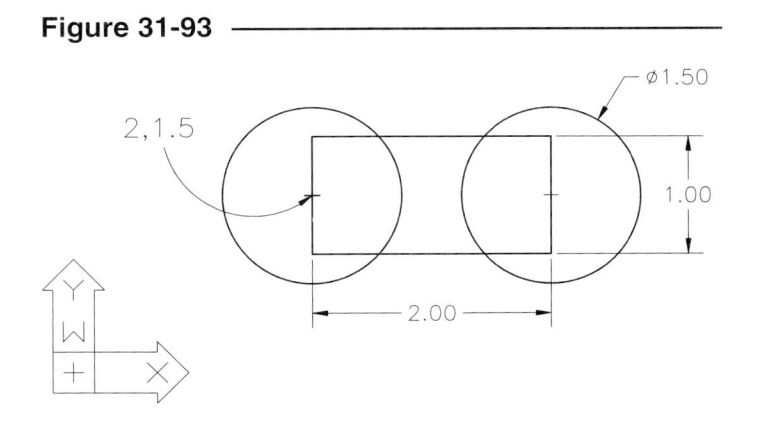

C. Create a *Box* with the lower-left corner at **0,0**. The *Lengths* of the box are **6, 3, 3**.

D. *Subtract* the extruded shape from the box. Your composite solid should look like that in Figure 31-94. *Save* the drawing.

Figure 31-94

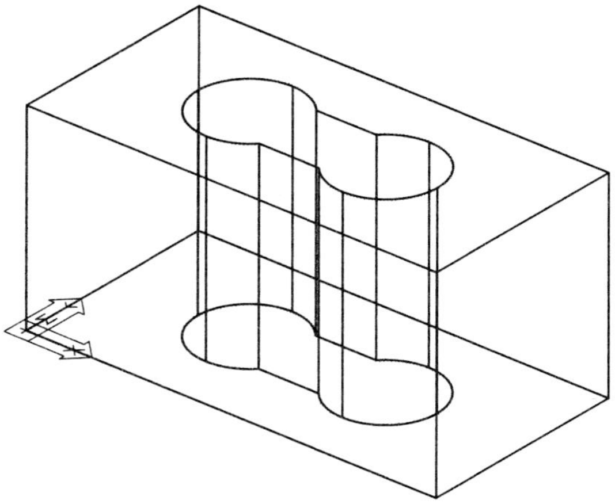

6. Begin a drawing and assign the name **CH31EX6**. Display a *Plan* view.

   A. Create a closed *Pline* shape symmetrical about the X axis with the locational and dimensional specifications given in Figure 31-95.

   B. Change to an isometric-type *Vpoint*. Use *Revolve* to generate a complete circular shape from the closed *Pline*. Revolve about the **Y** axis.

   C. Create a *Torus* with the *Center* at **0,0**. The *Radius of torus* is **3** and the *Radius of tube* is **.5**. The two shapes should intersect.

Figure 31-95

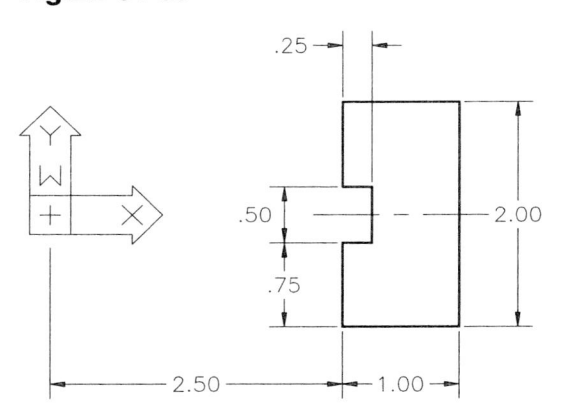

   D. Use *Hide* to generate a display like Figure 31-96.

   E. Create a *Cylinder* with the *Center* at **0,0,0**, a *Radius* of **3**, and a *Height* of **8**.

   F. Use *Rotate3D* to rotate the revolved *Pline* shape **90** degrees about the X axis (the *Pline* shape that was previously converted to a solid—not the torus). Next, move the shape up (positive Z) **6** units with *Move*.

   G. Move the torus up **4** units with *Move*.

Figure 31-96

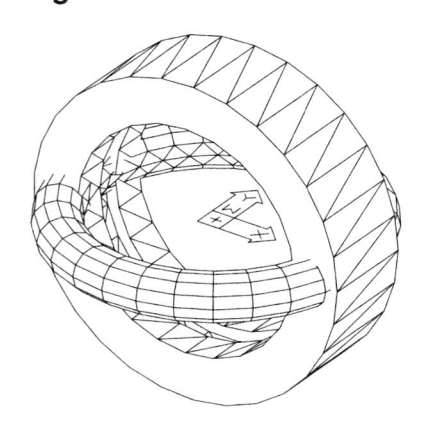

666   Solid Modeling Construction

    H. The solid primitives should appear as those in Figure 31-97. (*Hide* has been used for the figure.)

Figure 31-97

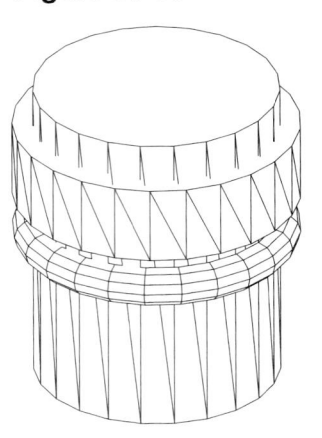

    I. Use *Subtract* to subtract both revolved shapes from the cylinder. Use *Hide*. The solid should resemble that in Figure 31-98. (*Hidden* shademode has been used for this figure.) *Save* the drawing.

Figure 31-98

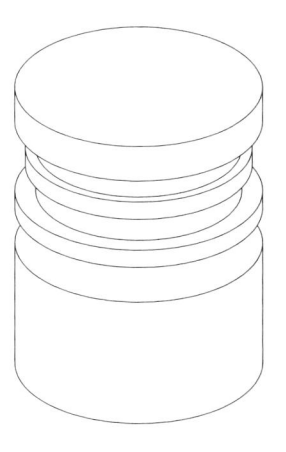

7. Begin a *New* drawing or use a *Template*. Assign the name **FAUCET**.

    A. Draw 3 closed *Pline* shapes, as shown in Figure 31-99. Assume symmetry about the longitudinal axis. Use the WCS and create 2 new *UCS*s for the geometry. Use *3point Arcs* for the "front" profile.

Figure 31-99

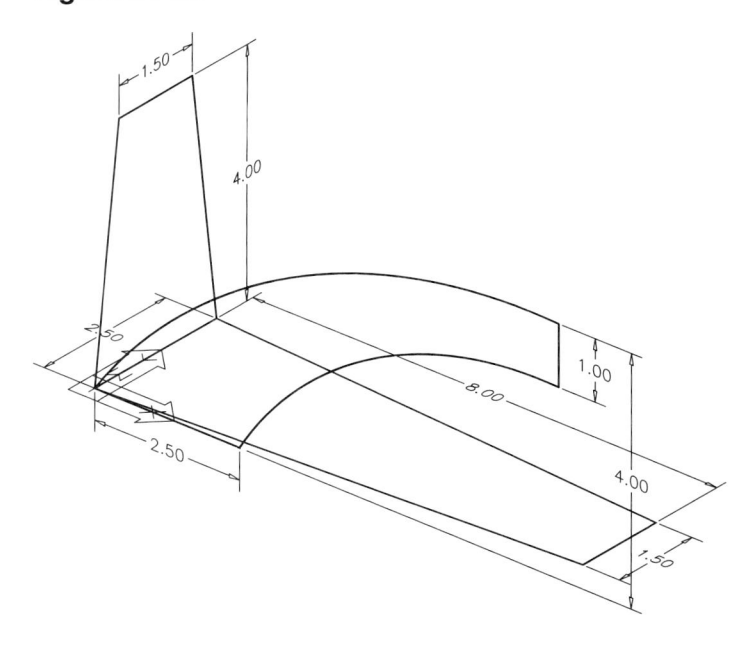

B. *Extrude* each of the 3 profiles into the same space. Make sure you specify the correct positive or negative *Height* value.

C. Finally, use *Intersect* to create the composite solid of the Faucet. *Save* the drawing.

D. (Optional) Create a nozzle extending down from the small end. Then create a channel for the water to flow (through the inside) and subtract it from the faucet.

Figure 31-100

8. Construct a solid model of the bar guide in Figure 31-101. Strive for the most efficient design. It is possible to construct this object with one *Extrude* and one *Subtract*. Save the model as **BGUID-SL**.

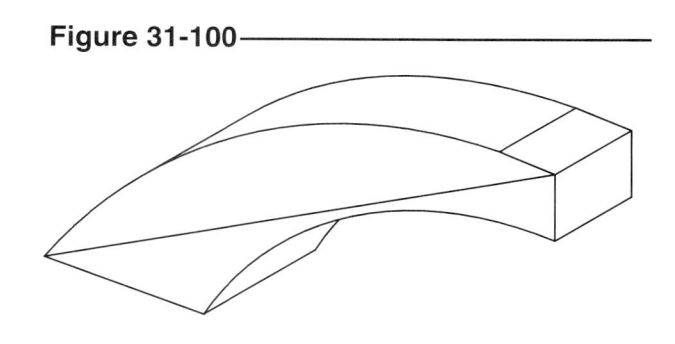

Figure 31-101

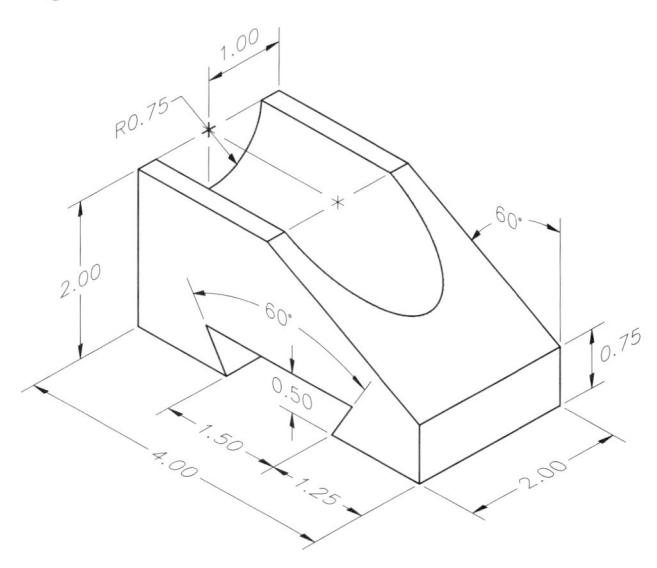

9. Make a solid model of the V-block shown in Figure 31-102. Several strategies could be used for construction of this object. Strive for the most efficient design. Plan your approach by sketching a few possibilities. Save the model as **VBLOK-SL**.

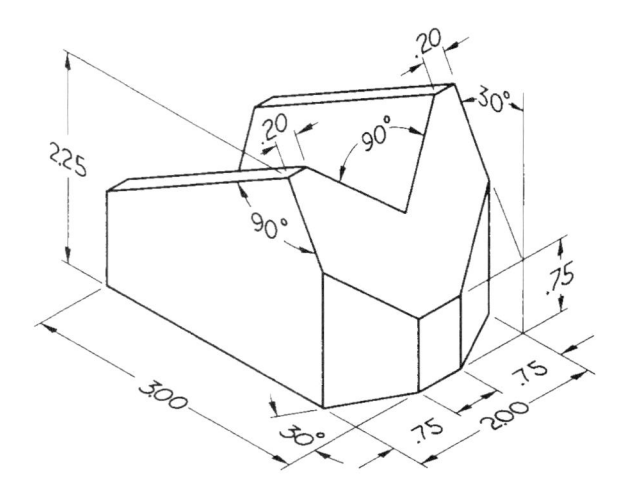

Figure 31-102

668 Solid Modeling Construction

10. Construct a composite solid model of the support bracket using efficient techniques. Save the model as **SUPBK-SL**.

Figure 31-103

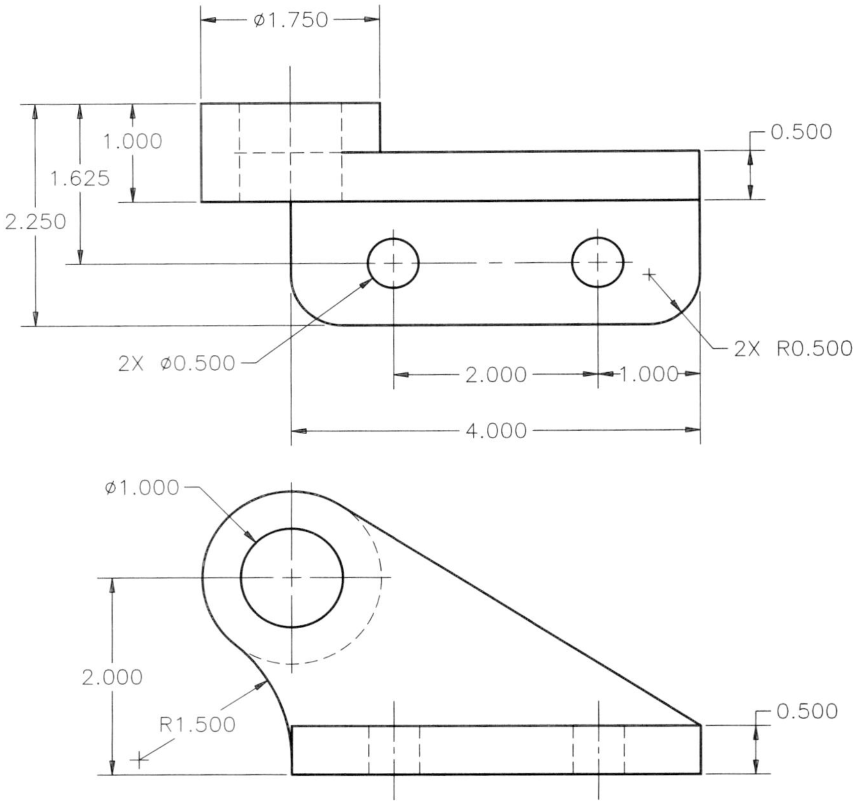

11. Construct the swivel shown in Figure 31-104. The center arm requires *Extruding* the 1.00 x 0.50 rectangular shape along an arc path through 45 degrees. Save the drawing as **SWIVEL**.

Figure 31-104

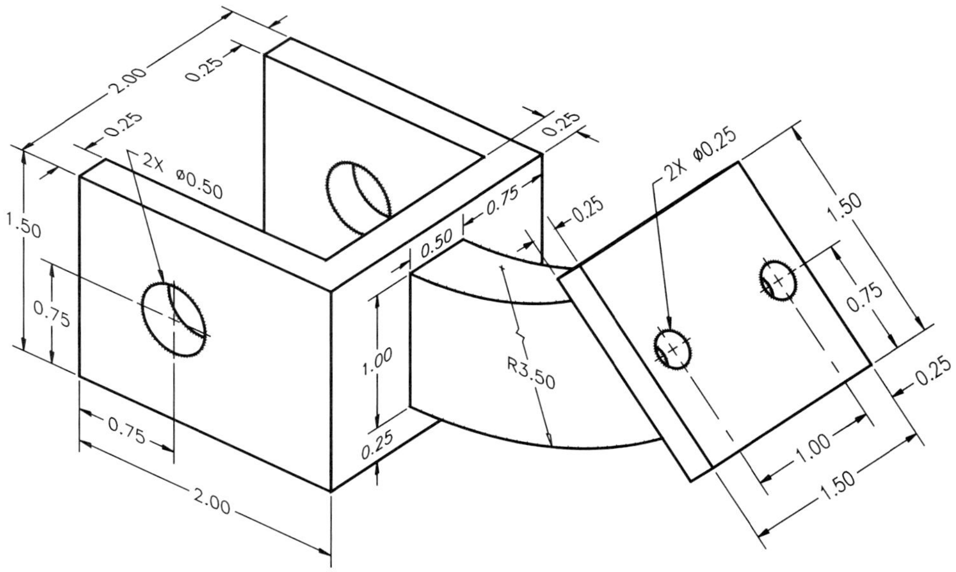

12. Construct a solid model of the angle brace shown in Figure 31-105. Use efficient design techniques. Save the drawing as **AGLBR-SL**.

Figure 31-105

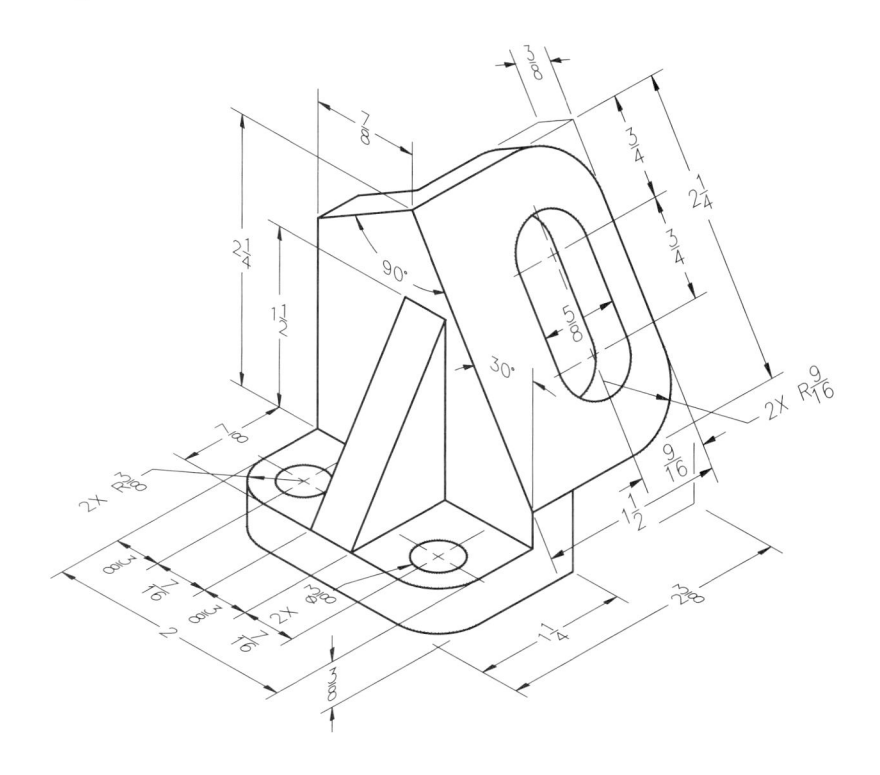

13. Construct a solid model of the saddle shown in Figure 31-106. An efficient design can be utilized by creating *Pline* profiles of the top "view" and the front "view," as shown in Figure 31-107. Use *Extrude* and *Intersect* to produce a composite solid. Additional Boolean operations are required to complete the part. The finished model should look similar to Figure 31-108. Save the drawing as **SADL-SL**.

Figure 31-106

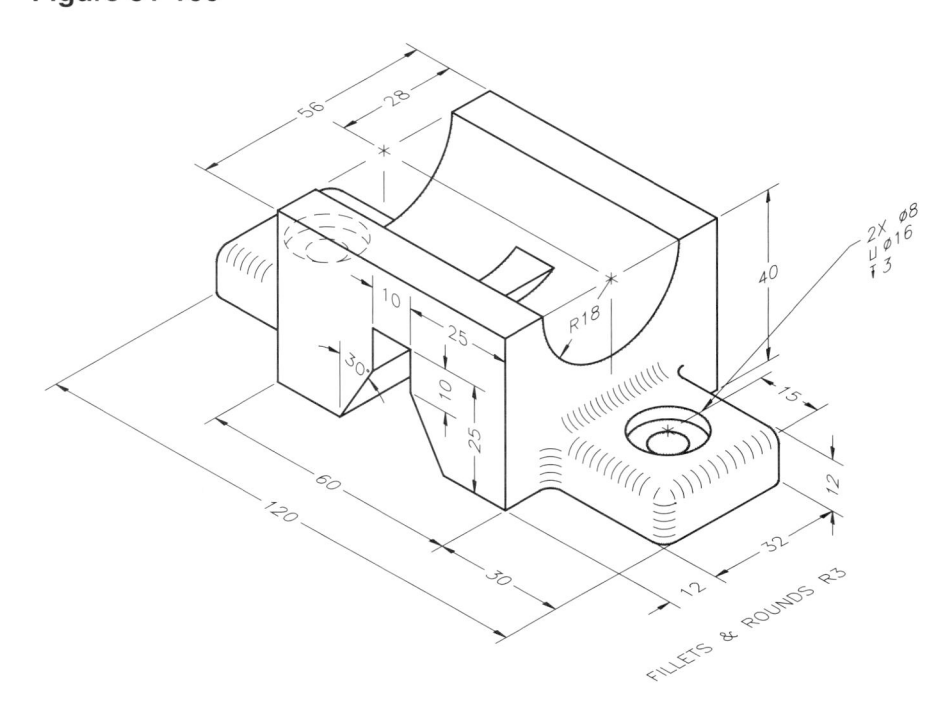

670   Solid Modeling Construction

Figure 31-107

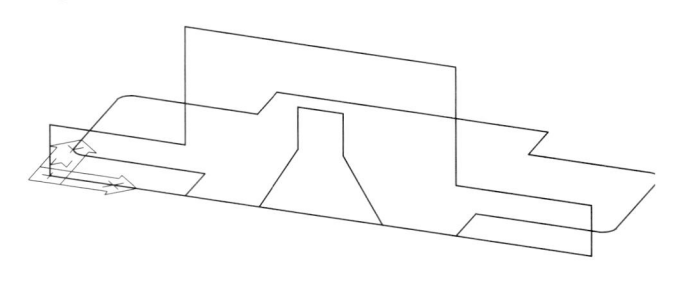

Figure 31-108

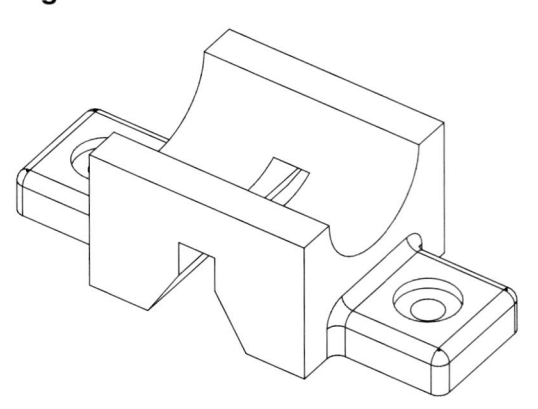

14. Construct a solid model of a bicycle handle bar. Create a center line (Fig. 31-109) as a *Path* to extrude a *Circle* through. Three mutually perpendicular coordinate systems are required: the **WORLD**, the **SIDE**, and the **FRONT**. The center line path consists of three separate *Plines*. First, create the 370 length *Pline* with 60 radii arcs on each end on the WCS. Then create the drop portion of the bars using the SIDE UCS. Create three *Circles* using the FRONT UCS, and extrude one along each *Path*.

Figure 31-109

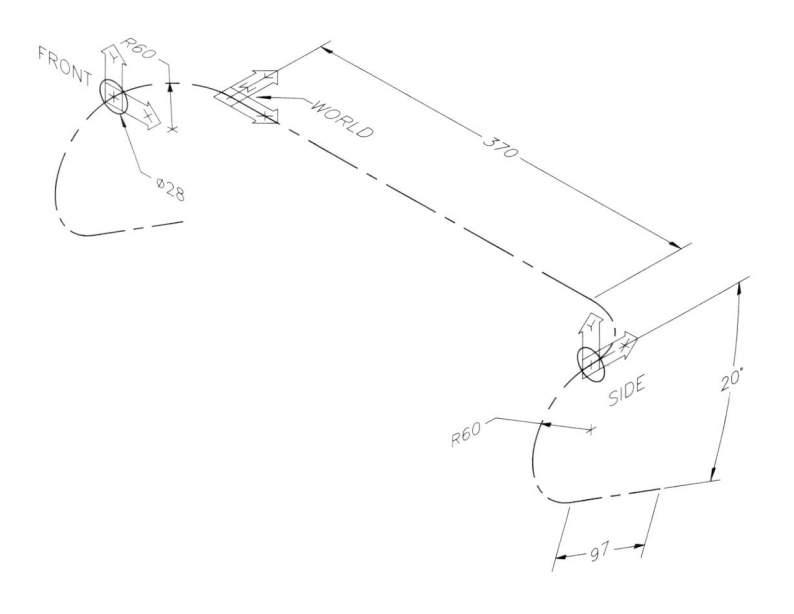

Plot the bar and *Hide Lines* as shown in Figure 31-110. *Save* the drawing as **DROPBAR**.

Figure 31-110

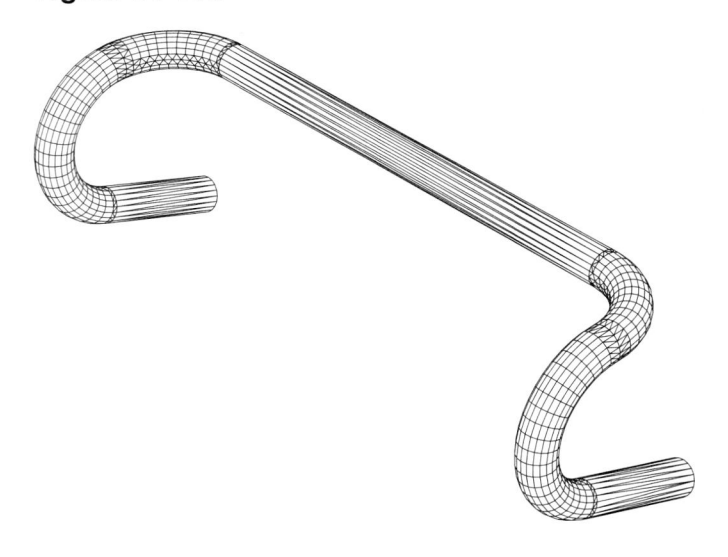

15. Create a solid model of the pulley. All vertical dimensions are diameters. Orientation of primitives is critical in the construction of this model. Try creating the circular shapes on the XY plane (circular axis aligns with Z axis of the WCS). After the construction, use *Rotate3D* to align the circular axis of the composite solid with the Y axis of the WCS. *Save* the drawing as **PULLY-SL**.

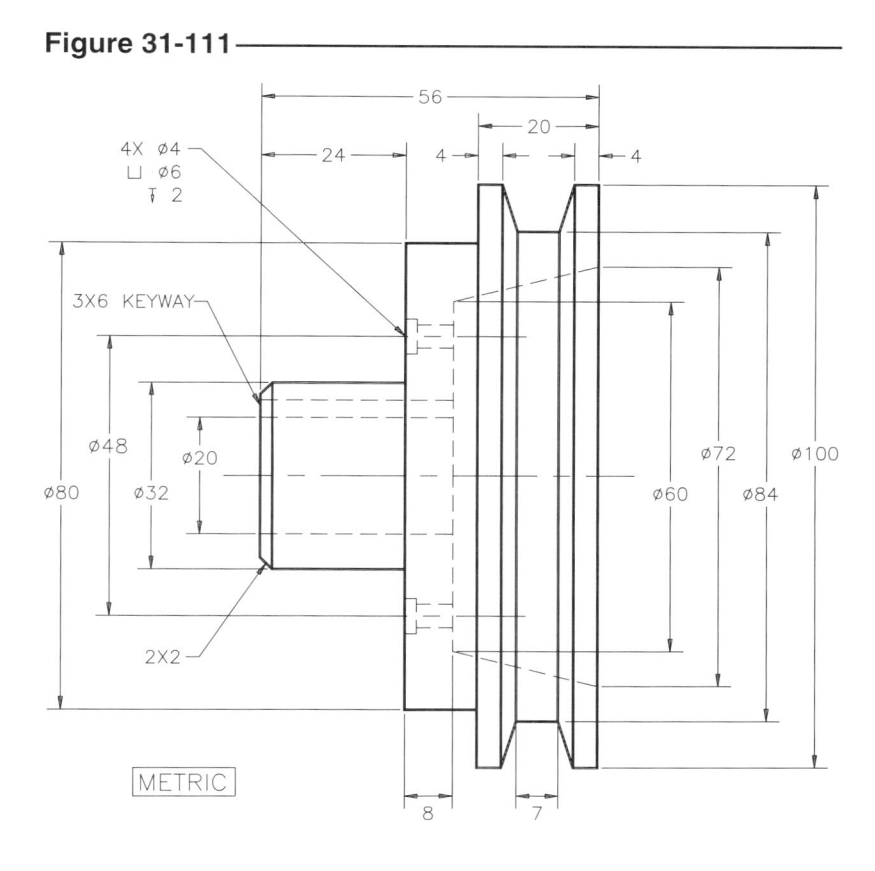

Figure 31-111

16. Create a composite solid model of the adjustable mount (Fig. 31-112). Use *Move* and other methods to move and align the primitives. Use of efficient design techniques is extremely important with a model of this complexity. Assign the name **ADJMT-SL**.

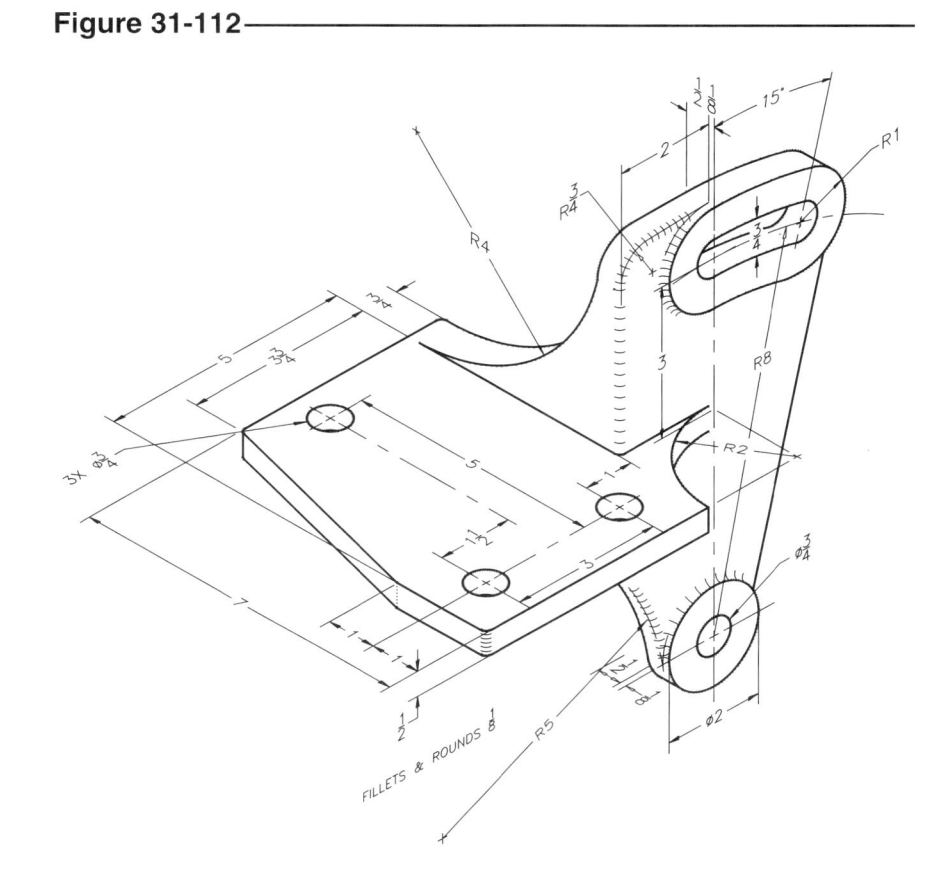

Figure 31-112

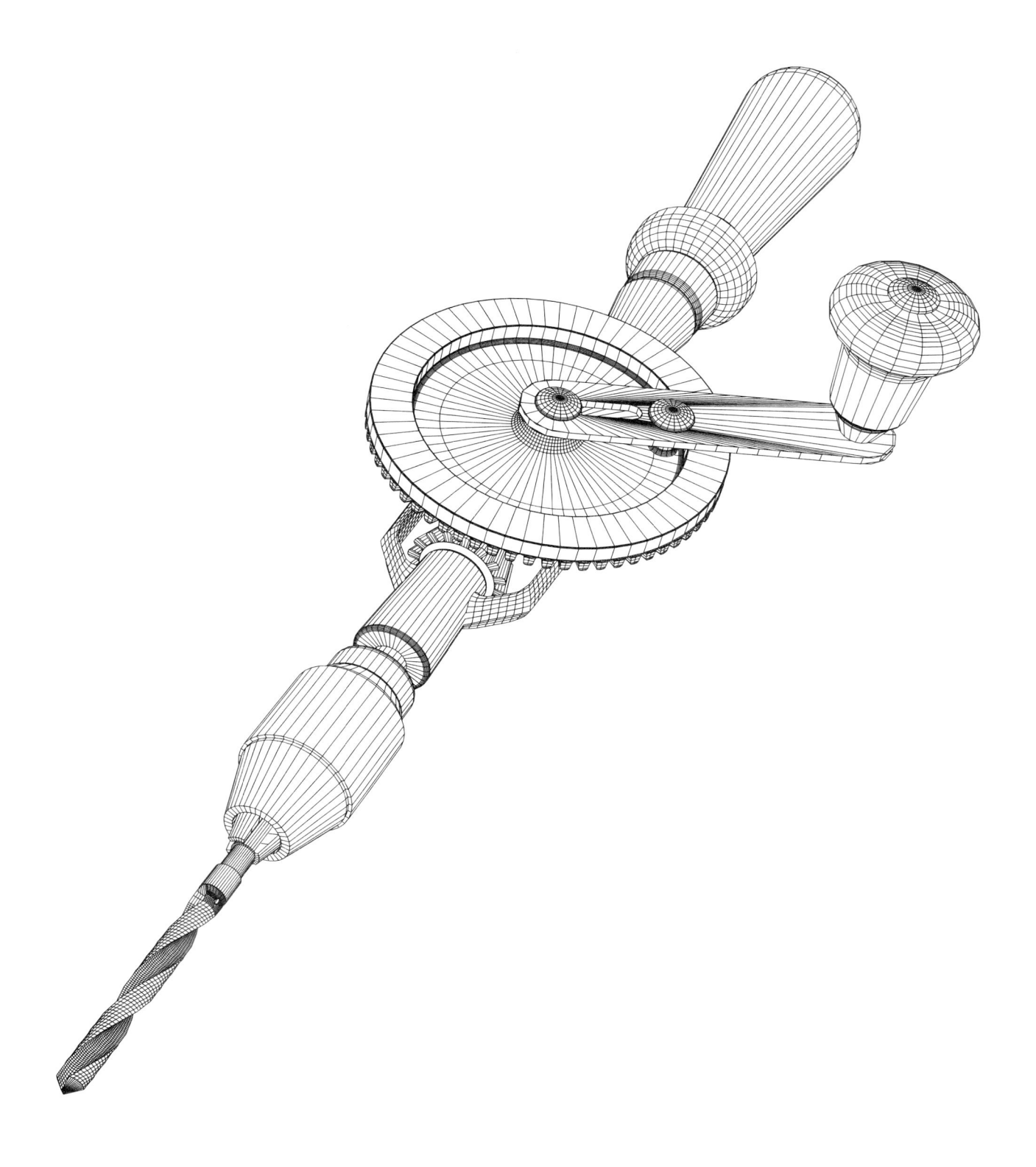

672 *Solid Modeling Construction*

DRILL.DWG, Courtesy of Autodesk, Inc.

32

ADVANCED SOLIDS FEATURES

Chapter Objectives

After completing this chapter you should be able to:

1. use the *ISOLINES*, *DISPSILH*, *FACETRES*, and *FACETRATIO* variables to control the display of tessellation lines, silhouette lines, and mesh density for solid models;

2. calculate mass properties of a solid model using *Massprop*;

3. determine if *Interference* exists between two or more solids and create a solid equal in volume to the interference;

4. create a 2D section view for a solid model using *Section*;

5. use *Slice* to cut a solid model at any desired cutting plane and retain one or both halves.

CONCEPTS

Several topics related to solid modeling capabilities for AutoCAD ACIS models are discussed in this chapter. The topics are categorized in the following sections:

> Solid Modeling Display Variables
> Analyzing Solid Models
> Creating Sections from Solids

SOLID MODELING DISPLAY VARIABLES

AutoCAD solid models are displayed in wireframe representation by default. Wireframe representation requires less computation time and less complex file structure, so your drawing time can be spent more efficiently. When you use *Hide, Shademode,* or *Render,* the solid models are automatically meshed before they are displayed with hidden lines removed or as a shaded image. This meshed version of the model is apparent when you use *Hide* on cylindrical or curved surfaces.

Four variables control the display of solids for wireframe, hidden, and meshed representation. The *ISOLINES* variable controls the number of tessellation lines that are used to visually define curved surfaces for wireframes. The *DISPSILH* variable can be toggled on or off to display silhouette lines for wireframe displays. *FACETRES* is the variable that controls the density of the mesh apparent with *Hide*. *FACETRATIO* creates a two-dimensional mesh for cylinders and cones. The *ISOLINES, DISPSILH,* and *FACETRES* display variables are saved in the drawing file. *FACETRATIO*, however, is not saved.

ISOLINES

Pull-down Menu	COMMAND (TYPE)	ALIAS (TYPE)	Short-cut	Screen (side) Menu	Tablet Menu
Tools Options... Display Coutour lines per surface	ISOLINES	...	...	TOOLS2 Options Display Contour lines per surface	...

This variable sets the number of tessellation lines that appear on a curved surface when shown in wireframe representation. The default setting for *ISOLINES* is 4 (Fig. 32-1). A solid of extrusion shows fewer tessellation lines (the current *ISOLINES* setting less 4) to speed regeneration time.

A higher setting gives better visualization of the curved surfaces but takes more computing time (Fig. 32-2). After changing the *ISOLINES* setting, *Regen* the drawing to see the new display.

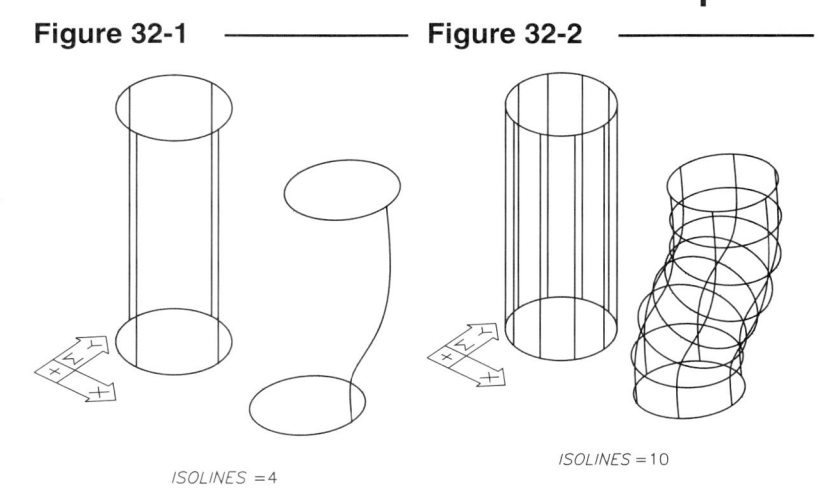

Figure 32-1

ISOLINES = 4

Figure 32-2

ISOLINES = 10

NOTE: In an isometric view attained by *3D Viewpoint*, the 4 lines (like Fig. 32-1) appear to overlap—they align when viewed from any perfect isometric angle.

DISPSILH

Pull-down Menu	COMMAND (TYPE)	ALIAS (TYPE)	Short-cut	Screen (side) Menu	Tablet Menu
Tools Options... Display Show silhouettes in wireframe	DISPSILH	...	...	TOOLS2 Options Display Show silhouettes in wireframe	...

This variable can be turned on to display the limiting element contour lines, or silhouette, of curved shapes for a wireframe display (Fig. 32-3). The default setting is 0 (off). Since the silhouette lines are <u>viewpoint dependent</u>, some computing time is taken to generate the display. You should <u>not</u> leave *DISPSILH* on 1 during construction.

Figure 32-3

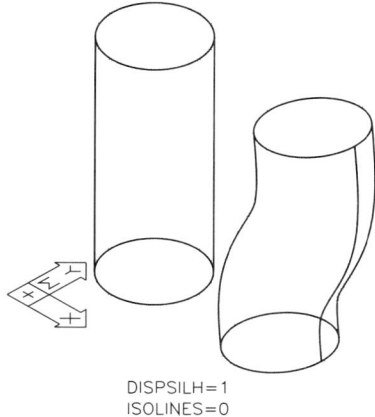

DISPSILH=1
ISOLINES=0

DISPSILH has a special function when used with *Hide* and when creating plots with hidden lines removed. When *DISPSILH* has the default setting of 0 and a *Hide* is performed, the solids appear opaque but display the mesh lines (Fig. 32-4). If *DISPSILH* is set to 1 before *Hide* is performed, the solids appear opaque but do <u>not</u> display the mesh lines (Fig. 32-5).

Figure 32-4

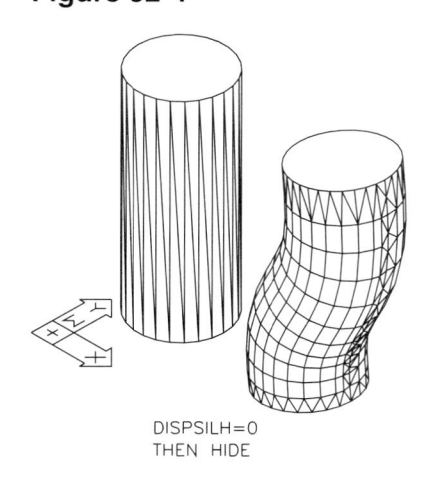

DISPSILH=0
THEN HIDE

Figure 32-5

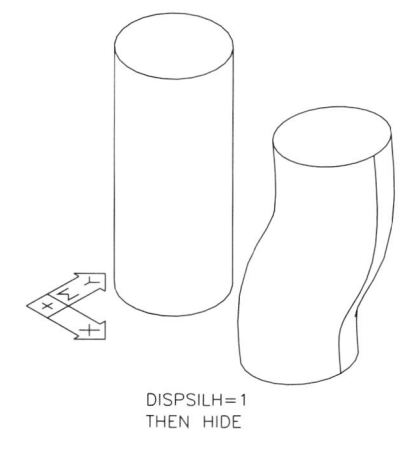

DISPSILH=1
THEN HIDE

FACETRES

Pull-down Menu	COMMAND (TYPE)	ALIAS (TYPE)	Short-cut	Screen (side) Menu	Tablet Menu
Tools Options... Display Rendered object smoothness	FACETRES	...	...	TOOLS2 Options Display Rendered object smoothness	...

FACETRES controls the mesh density of curved surfaces on solid objects. The mesh is most apparent when an object with curved surfaces is displayed using *Hide* and *Shademode* (*Flat*, *Flat+edges*, and *Gouraud+edges* options). The default setting is .5, as shown in Figure 32-6 on the next page.

676 Advanced Solids Features

Decreasing the value produces a coarser mesh (Fig. 32-7), while increasing the value produces a finer mesh. The higher the value, the more computation time involved to generate the display or plot. The density of the mesh is actually a factor of both the *FACETRES* setting and the *VIEWRES* setting. Increasing *VIEWRES* also makes the mesh more dense. *FACETRES* can be set to any value between .01 and 10.

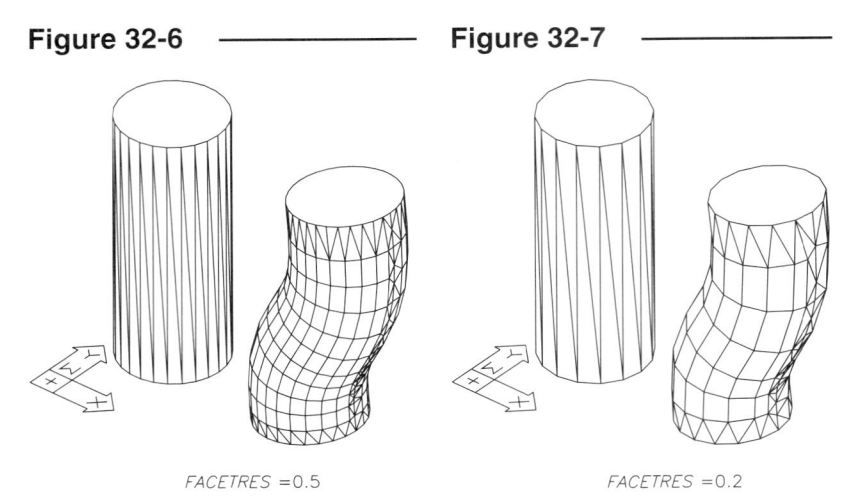

Figure 32-6

FACETRES =0.5

Figure 32-7

FACETRES =0.2

FACETRATIO

Pull-down Menu	COMMAND (TYPE)	ALIAS (TYPE)	Short-cut	Screen (side) Menu	Tablet Menu
...	*FACETRATIO*	...	...	...	...

FACETRATIO is a variable introduced in AutoCAD 2000 that affects only cylindrical and conical solids. When *FACETRATIO* is on (set to 1), an *n* by *m* mesh (two-dimensional mesh) is created. When *FACETRATIO* is off (set to 0), a 1 by *n* mesh (one-dimensional mesh) is created. Compare the two cylinders in Figures 32-7 and 32-8. The cylinder in Figure 32-7 displays a mesh generated in only one dimension—around the circumference—whereas the cylinder in Figure 32-8 displays a cylinder with a 2-dimensional mesh.

Figure 32-8

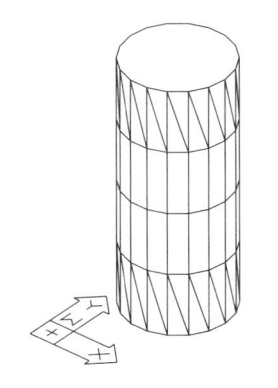

FACETRATIO = 1

The *Options* dialog box can be used to change the settings for the *ISOLINES, DISPSILH,* and *FACETRES* variables (Fig. 32-9). The right side of the *Display* tab has two edit boxes and one checkbox that allow changing these variables as follows.

Figure 32-9

Rendered object smoothness   FACETRES
Contour lines per surface   ISOLINES
Show silhouettes in wireframe   DISPSILH

ANALYZING SOLID MODELS

Two commands in AutoCAD allow you to inquire about and analyze the solid geometry. *Massprop* calculates a variety of properties for the selected ACIS solid model. AutoCAD does the calculation and lists the information in screen or text window format. The data can be saved to a file for future exportation to a report document or analysis package. The *Interfere* command finds the interference of two or more solids and highlights the overlapping features so you can make necessary alterations.

MASSPROP

Pull-down Menu	COMMAND (TYPE)	ALIAS (TYPE)	Short-cut	Screen (side) Menu	Tablet Menu
Tools Inquiry > Mass Properties	MASSPROP	...	...	TOOLS 1 Massprop	U,7

Since solid models define a complete description of the geometry, they are ideal for mass properties analysis. The *Massprop* command automatically computes a variety of mass properties.

Mass properties are useful for a variety of applications. The data generated by the *Massprop* command can be saved to an .MPR file for future exportation in order to develop bills of material, stress analysis, kinematics studies, and dynamics analysis.

Applying the *Massprop* command to a solid model produces a text screen displaying the following list of calculations (Fig. 32-10):

Figure 32-10

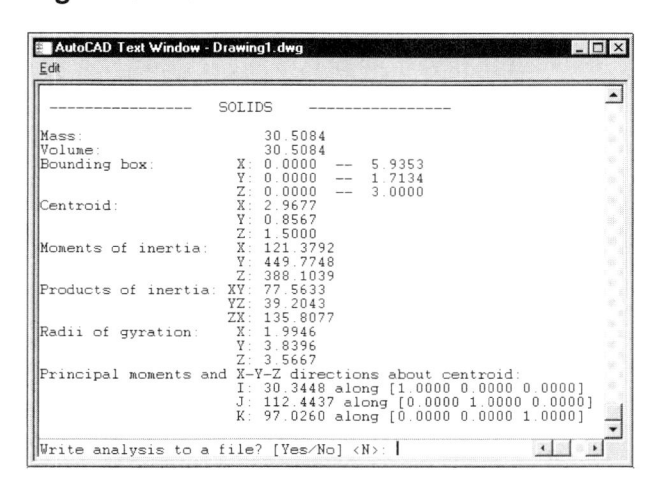

```
----------------- SOLIDS -----------------
Mass:                    30.5084
Volume:                  30.5084
Bounding box:        X:  0.0000  --  5.9353
                     Y:  0.0000  --  1.7134
                     Z:  0.0000  --  3.0000
Centroid:            X:  2.9677
                     Y:  0.8567
                     Z:  1.5000
Moments of inertia:  X:  121.3792
                     Y:  449.7748
                     Z:  388.1039
Products of inertia: XY: 77.5633
                     YZ: 39.2043
                     ZX: 135.8077
Radii of gyration:   X:  1.9946
                     Y:  3.8396
                     Z:  3.5667
Principal moments and X-Y-Z directions about centroid:
                     I:  30.3448 along [1.0000 0.0000 0.0000]
                     J:  112.4437 along [0.0000 1.0000 0.0000]
                     K:  97.0260 along [0.0000 0.0000 1.0000]
Write analysis to a file? [Yes/No] <N>:
```

Mass	Mass is the quantity of matter that a solid contains. Mass is determined by density of the material and volume of the solid. Mass is not dependent on gravity and, therefore, different from but proportional to weight. Mass is also considered a measure of a solid's resistance to linear acceleration (overcoming inertia).
Volume	This value specifies the amount of space occupied by the solid.
Bounding Box	These lengths specify the extreme width, depth, and height of the selected solid.
Centroid	The centroid is the geometrical center of the solid. Assuming the solid is composed of material that is homogeneous (uniform density), the centroid is also considered the center of mass and center of gravity. Therefore, the solid can be balanced when supported only at this point.
Moments of Inertia	Moments convey how the mass is distributed around the X, Y, and Z axes of the current coordinate system. These values are a measure of a solid's resistance to <u>angular</u> acceleration (mass is a measure of a solid's resistance to <u>linear</u> acceleration). Moments of inertia are helpful for stress computations.

Products of Inertia	These values specify the solid's resistance to <u>angular</u> acceleration with respect to two axes at a time (XY, YZ, or ZX). Products of inertia are also useful for stress analysis.
Radii of Gyration	If the object were a concentrated solid mass without holes or other features, the radii of gyration represent these theoretical dimensions (radius about each axis) such that the same moments of inertia would be computed.
Principal Moments and X, Y, Z Directions	In structural mechanics, it is sometimes important to determine the orientation of the axes about which the moments of inertia are at a maximum. When the moments of inertia about centroidal axes reach a maximum, the products of inertia become zero. These particular axes are called the principal axes, and the corresponding moments of inertia with respect to these axes are the principal moments (about the centroid).

TIP: Notice the "Mass:" is reported as having the same value as "Volume." This is because AutoCAD solids cannot have material characteristics assigned to them, so AutoCAD assumes a density value of 1. To calculate the mass of a selected solid in AutoCAD, use a reference guide (such as a machinist's handbook) to find the material density and multiply the value times the reported volume (mass = volume x density).

INTERFERE

Pull-down Menu	COMMAND (TYPE)	ALIAS (TYPE)	Short-cut	Screen (side) Menu	Tablet Menu
Draw Solids > Interference	INTERFERE	INF	...	DRAW 2 SOLIDS *Interfer*	...

In AutoCAD, unlike real life, it is possible to create two solids that occupy the same physical space. *Interfere* checks solids to determine whether or not they interfere (occupy the same space). If there is interference, *Interfere* reports the overlap and allows you to create a new solid from the interfering volume, if you desire. Normally, you specify two sets of solids for AutoCAD to check against each other:

Command: *interfere*
Select first set of solids:
Select objects: **PICK**
Select objects: **Enter**
1 solid selected.
Select second set of solids:
Select objects: **PICK**
Select objects: **Enter**
1 solid selected.
Comparing 1 solid against 1 solid.
Interfering solids (first set): 1
        (second set): 1
Interfering pairs: 1
Create interference solids? [Yes/No] <N>: **y**
Command:

TIP: If you answer "yes" to the last prompt, a new solid is created equal to the exact size and volume of the interference. The original solids are not changed in any way. If no interference is found, AutoCAD reports "Solids do not interfere."

For example, consider the two solids shown in Figure 32-11. (The parts are displayed in wireframe representation.) The two shapes fit together as an assembly. The locating pin on the part on the right should fit in the hole in the left part.

Figure 32-11

Sliding the parts together until the two vertical faces meet produces the assembly shown in Figure 32-12. There appears to be some inconsistency in the assembly of the hole and the pin. Either the pin extends beyond the hole (interference) or the hole is deeper than necessary (no interference). Using *Interfere*, you can find an overlap and create a solid is created by answering "yes" to "Create interference solids?" Use *Move* with the *Last* selection option to view and analyze the solid of interference.

Figure 32-12

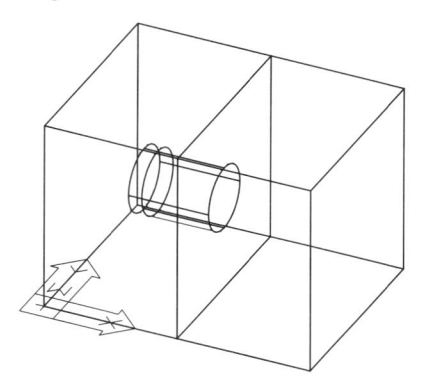

You can compare more than two solids against each other with *Interfere*. This is accomplished by selecting all desired solids at the first prompt and none at the second:

    Select the first set of solids: PICK
    Select the second set of solids: Enter

AutoCAD then compares all solids in the first set against each other. If more than one interference is found, AutoCAD highlights intersecting solids, one pair at a time.

CREATING SECTIONS FROM SOLIDS

Two AutoCAD commands are intended to create sections from solid models. *Section* is a drafting feature that creates a 2D "section view." The cross-section is determined by specifying a cutting plane. A cross-section view is automatically created based on the solid geometry that intersects the cutting plane. The original solid is not affected by the action of *Section*. *Slice* actually cuts the solid at the specified cutting plane. *Slice* therefore creates two solids from the original one and offers the possibility to retain both halves or only one. Many options are available for placement of the cutting plane.

SECTION

Pull-down Menu	COMMAND (TYPE)	ALIAS (TYPE)	Short-cut	Screen (side) Menu	Tablet Menu
Draw Solids > Section	SECTION	SEC	...	DRAW 2 SOLIDS Section	...

Section creates a 2D cross-section of a solid or set of solids. The cross-section created by *Solsect* is considered a traditional 2D section view. The cross-section is defined by a cutting plane, and the resulting section is determined by any solid material that passes through the cutting plane.

The cutting plane can be specified by a variety of methods. The options for establishing the cutting plane are listed in the command prompt:

    Command: *section*
    Select objects: **PICK**
    Select objects: **Enter**
    Specify first point on Section plane by [Object/Zaxis/View/XY/YZ/ZX/3points] <3points>:

For example, assume a cross-section is desired for the geometry shown in Figure 32-13. To create the section, you must define the cutting plane.

Figure 32-13

For this case, the *ZX* option is used and the requested point is defined to establish the position of the plane, as shown in Figure 32-14. The cross-section will be created on this plane.

Once the cutting plane is established, a cross-section is automatically created by *Section*. The resulting geometry is a *Region* created on the <u>current</u> layer.

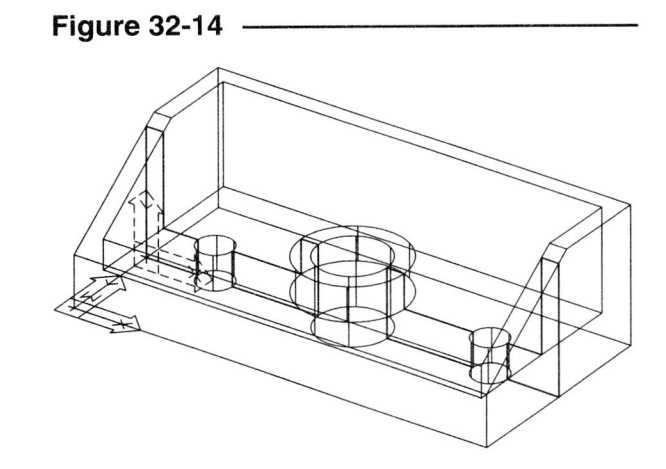

Figure 32-14

The resulting *Region* can be moved if needed as shown in Figure 32-15. If you want to use the *Region* to create a 2D section complete with hatch lines, several steps are required. First, the *Region* must be *Exploded* into individual *Regions* (4, in this case). You may need to create a *UCS* on the plane of the shapes before using *Bhatch*. Other *Lines* may be needed to make a complete section view.

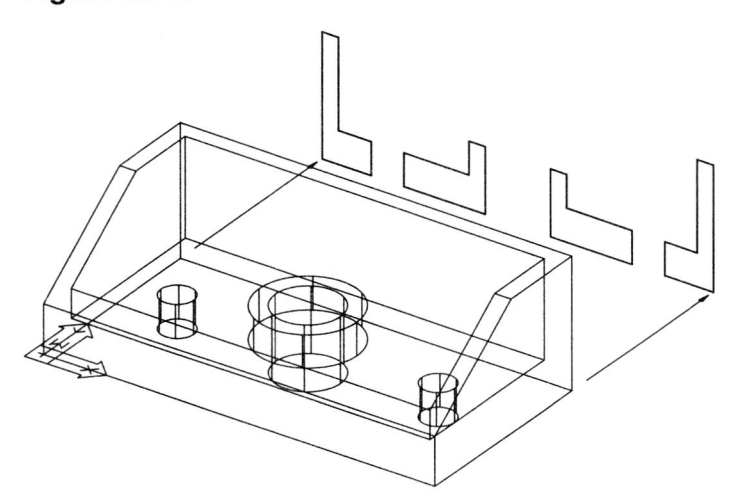

Figure 32-15

SLICE

Pull-down Menu	COMMAND (TYPE)	ALIAS (TYPE)	Short-cut	Screen (side) Menu	Tablet Menu
Draw Solids > Slice	SLICE	SL	...	DRAW 2 SOLIDS Slice	...

Slice creates a true solid section. *Slice* cuts an ACIS solid or set of solids on a specified cutting plane. The original solid is converted to two solids. You have the option to keep both halves or only the half that you specify. Examine the following command syntax:

Command: *slice*
Select objects: **PICK**
Select objects: **Enter**
Specify first point on slicing plane by [Object/Zaxis/View/XY/YZ/ZX/3points] <3points>: (**option**)
(option prompts)
Specify a point on desired side of the plane or [keep Both sides]: **PICK** or **Enter**
Command:

Entering *B* at the "keep Both sides" prompt retains the solids on both sides of the cutting plane. Otherwise, you can pick a point on either side of the plane to specify which half to keep.

For example, using the solid model shown previously in Figure 32-13, you can use *Slice* to create a new sectioned solid shown here. The *ZX* method is used to define the cutting plane midway through the solid. Next, the new solid to retain was specified by PICKing a point on that geometry. The resulting sectioned solid is shown in Figure 32-16. Note that the solid on the near side of the cutting plane was not retained.

Figure 32-16

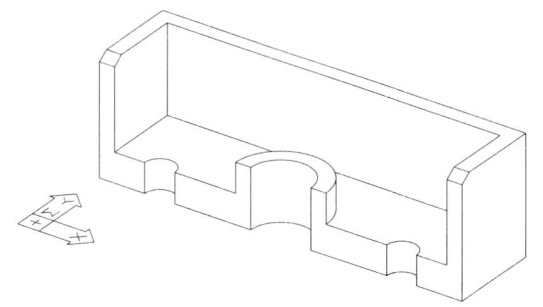

CHAPTER EXERCISES

1. *Open* the **SADL-SL** drawing that you created in Chapter 31 Exercises. Calculate *Mass Properties* for the saddle. Write the report out to a file named **SADL-SL.MPR**. Use a text editor or the DOS TYPE command to examine the file.

2. *Open* the **SADL-SL** drawing again. Use *Slice* to cut the model in half longitudinally. Use an appropriate method to establish the "slicing plane" in order to achieve the resulting model, as shown in Figure 32-17. Use *SaveAs* and assign the name **SADL-CUT**.

Figure 32-17

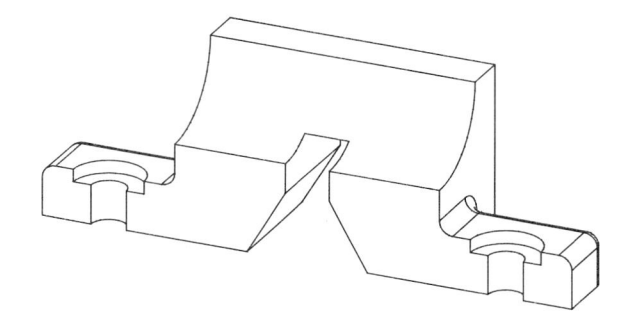

3. *Open* the **PULLY-SL** drawing that you created in Chapter 31 Exercises.

   Make a *New Layer* named **SECTION** and set it *Current*. Then use the *Section* command to create a full section "view" of the pulley. Establish a vertical cutting plane through the center of the model. Remove the section view object (*Region*) with the *Move* command, translating **100** units in the **X** direction. The model and the new section view should appear as in Figure 32-18 (*Hide* was performed on the pulley to enhance visualization). Complete the view by establishing a *UCS* at the section view, then adding the *Bhatch,* as shown in Figure 32-18. Finally, create the necessary *Lines* to complete the view. *SaveAs* **PULLY-SC**.

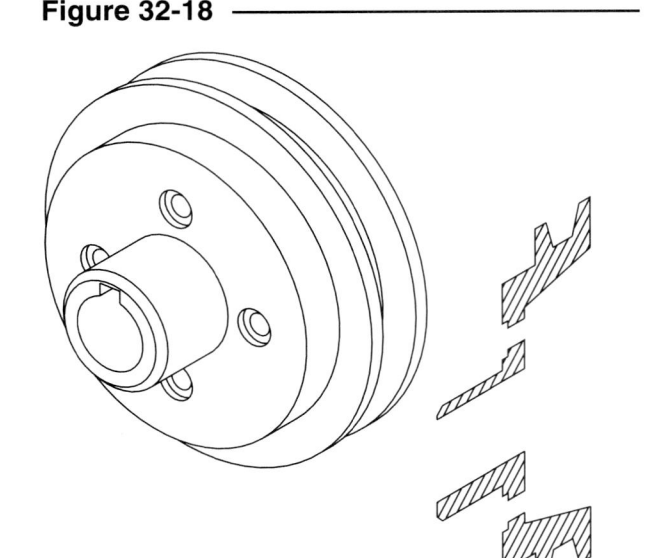

   Figure 32-18

4. *Open* the **SADL-SL** drawing that you created in the Chapter 31 Exercises. Change the *ISOLINES* setting to display **10** tessellation lines. Use the **Hide** command to create a meshed hidden display. Change the *FACETRES* setting to display a coarser mesh and use *Hide* again. Make a plot of the model with the coarse mesh and with hidden lines (check the *Hide Objects* box in the *Plot* dialog box).

   Next, change the *FACETRES* setting to display a fine mesh. Use *Hide* to reveal the change. Make a plot of the model with *Hide Objects* checked to display the fine mesh. Then set the *DISPSILH* variable to **1** and make another plot with lines hidden. What is the difference in the last two plots? *Save* the **SADL-SL** file with the new settings.

A & B

APPENDICES

Contents

Appendix A. AutoCAD 2002 Command Alias List Sorted by Command

Appendix B. Buttons and Special Keys

APPENDIX A

AutoCAD 2002 Command Alias List Sorted by Command

Command	Alias	Command	Alias
3DARRAY	3A	DIMDIAMETER	DDI
3DFACE	3F	DIMDIAMETER	DIMDIA
3DORBIT	3DO	DIMDISASSOCIATE	DDA
3DORBIT	ORBIT	DIMEDIT	DED
3DPOLY	3P	DIMEDIT	DIMED
ADCENTER	ADC	DIMLINEAR	DIMLIN
ALIGN	AL	DIMLINEAR	DLI
APPLOAD	AP	DIMORDINATE	DIMORD
ARC	A	DIMORDINATE	DOR
AREA	AA	DIMOVERRIDE	DIMOVER
ARRAY	AR	DIMOVERRIDE	DOV
-ARRAY	-AR	DIMRADIUS	DIMRAD
ATTDEF	ATT	DIMRADIUS	DRA
-ATTDEF	-ATT	DIMREASSOCIATE	DRE
ATTEDIT	ATE	DIMSTYLE	D
-ATTEDIT	-ATE	DIMSTYLE	DIMSTY
-ATTEDIT	ATTE	DIMSTYLE	DST
BHATCH	BH	DIMTEDIT	DIMTED
BHATCH	H	DIST	DI
BLOCK	B	DIVIDE	DIV
-BLOCK	-B	DONUT	DO
BOUNDARY	BO	DRAWORDER	DR
-BOUNDARY	-BO	DSETTINGS	DS
BREAK	BR	DSETTINGS	SE
CHAMFER	CHA	DSVIEWER	AV
CHANGE	-CH	DTEXT	DT
CIRCLE	C	DVIEW	DV
COLOR	COL	ELLIPSE	EL
COLOR	COLOUR	ERASE	E
COPY	CO	EXPLODE	X
COPY	CP	EXPORT	EXP
DBCONNECT	DBC	EXTEND	EX
DDEDIT	ED	EXTRUDE	EXT
DDGRIPS	GR	FILLET	F
DDRMODES	RM	FILTER	FI
DDUCS	UC	GROUP	G
DDUCSP	UCP	-GROUP	-G
DDVPOINT	VP	HATCH	-H
DIMALIGNED	DAL	HATCHEDIT	HE
DIMALIGNED	DIMALI	HIDE	HI
DIMANGULAR	DAN	IMAGE	IM
DIMANGULAR	DIMANG	-IMAGE	-IM
DIMBASELINE	DBA	IMAGEADJUST	IAD
DIMBASELINE	DIMBASE	IMAGEATTACH	IAT
DIMCENTER	DCE	IMAGECLIP	ICL
DIMCONTINUE	DCO	IMPORT	IMP
DIMCONTINUE	DIMCONT	INSERT	I

Command	Alias	Command	Alias
-INSERT	-I	QUIT	EXIT
INSERTOBJ	IO	RECTANGLE	REC
INTERFERE	INF	REDRAW	R
INTERSECT	IN	REDRAWALL	RA
LAYER	LA	REGEN	RE
-LAYER	-LA	REGENALL	REA
-LAYOUT	LO	REGION	REG
LEADER	LEAD	RENAME	REN
LENGTHEN	LEN	-RENAME	-REN
LINE	L	RENDER	RR
LINETYPE	LT	REVOLVE	REV
-LINETYPE	-LT	ROTATE	RO
LINETYPE	LTYPE	RPREF	RPR
-LINETYPE	-LTYPE	SCALE	SC
LIST	LI	SCRIPT	SCR
LIST	LS	SECTION	SEC
LTSCALE	LTS	SETVAR	SET
LWEIGHT	LINEWEIGHT	SHADE	SHA
LWEIGHT	LW	SLICE	SL
MATCHPROP	MA	SNAP	SN
MEASURE	ME	SOLID	SO
MIRROR	MI	SPELL	SP
MLINE	ML	SPLINE	SPL
MOVE	M	SPLINEDIT	SPE
MSPACE	MS	STRETCH	S
MTEXT	MT	STYLE	ST
MTEXT	T	SUBTRACT	SU
-MTEXT	-T	TABLET	TA
MVIEW	MV	THICKNESS	TH
OFFSET	O	TILEMODE	TI
OPTIONS	OP	TILEMODE	TM
OPTIONS	PR	TOLERANCE	TOL
OSNAP	OS	TOOLBAR	TO
-OSNAP	-OS	TORUS	TOR
PAN	P	TRIM	TR
-PAN	-P	UNION	UNI
-PARTIALOPEN	PARTIALOPEN	UNITS	UN
PASTESPEC	PA	-UNITS	-UN
PEDIT	PE	VIEW	V
PLINE	PL	-VIEW	-V
PLOT	PRINT	VPOINT	-VP
POINT	PO	WBLOCK	W
POLYGON	POL	-WBLOCK	-W
PREVIEW	PRE	WEDGE	WE
PROPERTIES	CH	XATTACH	XA
PROPERTIES	MO	XBIND	XB
PROPERTIES	PROPS	-XBIND	-XB
PROPERTIESCLOSE	PRCLOSE	XCLIP	XC
PSPACE	PS	XLINE	XL
PUBLISHTOWEB	PTW	XREF	XR
PURGE	PU	-XREF	-XR
QLEADER	LE	ZOOM	Z

APPENDIX B

BUTTONS AND SPECIAL KEYS

Mouse and Digitizing Puck Buttons

Depending on the type of mouse or digitizing puck used for cursor control, a different number of buttons are available. The *User Preferences* tab of the *Options* dialog box can be used to control the appearance of shortcut menus and to customize the actions of a right-click. In any case, the buttons have the following default settings.

#1 (left mouse)	**PICK**	Used to select commands or point to locations on screen.
#2 (right mouse)	**Shortcut menu or Enter**	Generally, activates a shortcut menu (see Chapter 1, "Shortcut Menus" and Chapter 2, "Windows Right-Click Shortcut Menus"). Otherwise, performs the same action as the Enter key on the keyboard.
press (center wheel)	*Pan*	Activates the realtime *Pan* command.
turn (center wheel)	*Zoom*	Activates the realtime *Zoom* command.

Function (F) Keys

Function keys in AutoCAD offer a quick method of turning on or off (toggling) drawing aids.

F1	*Help*	Opens a help window providing written explanations on commands and variables (see Chapter 5).
F2	*Flipscreen*	Activates a text window showing the previous command line activity (see Chapter 1).
F3	*Osnap Toggle*	If Running Osnaps are set, toggling this key temporarily turns the Running Osnaps off so that a point can be picked without using Osnaps. If no Running Osnaps are set, F3 produces the *Object Snap* tab of the *Drafting Settings* dialog box (see Chapter 7).
F4	*Tablet*	Turns the TABMODE variable on or off. If TABMODE is on, the digitizing tablet can be used to digitize an existing paper drawing into AutoCAD.
F5	*Isoplane*	When using an *Isometric* style SNAP and GRID setting, toggles the crosshairs (with ORTHO on) to draw on one of three isometric planes (see Chapter 23).
F6	*Coords*	Toggles the coordinate display between cursor tracking mode and off. If used transparently (during a command in operation), displays a polar coordinate format (see Chapter 1, "Drawing Aids").
F7	*GRID*	Turns the *GRID* on or off (see Chapter 1, "Drawing Aids").
F8	*ORTHO*	Turns *ORTHO* on or off (see Chapter 1, "Drawing Aids").
F9	*SNAP*	Turns *SNAP* (Grid Snap or Polar Snap) on or off (see Chapter 1, "Drawing Aids" and Chapter 3, "Polar Tracking and Polar Snap").
F10	*POLAR*	Turns Polar Tracking on or off (see Chapter 1, "Drawing Aids" and Chapter 3, "Polar Tracking and Polar Snap").
F11	*OTRACK*	Turns Object Snap Tracking on or off (see Chapter 7).

Control Key Sequences (Accelerator Keys)

Accelerator keys (holding down the Ctrl key and pressing another key simultaneously) invoke regular AutoCAD commands or produce special functions. Several have the same duties as F3 through F11.

Drawing Aids

Ctrl+F (F3)	*Osnap Toggle*		If Running Osnaps are set, pressing Ctrl+F temporarily turns off the Running Osnaps so that a point can be picked without using Osnaps. If there are no Running Object Snaps set, Ctrl+F produces the *Osnap Settings* dialog box. This dialog box is used to turn on and off Running Object Snaps (discussed in Chapter 7).
Ctrl+T (F4)	*Tablet*		Turns the *TABMODE* variable on or off. If *TABMODE* is on, the digitizing tablet can be used to digitize an existing paper drawing into AutoCAD.
Ctrl+E (F5)	*Isoplane*		When using an *Isometric* style *SNAP* and *GRID* setting, toggles the crosshairs (with *ORTHO* on) to draw on one of three isometric planes.
Ctrl+D (F6)	*Coords*		Toggles the Coordinate Display between cursor tracking mode and off. If used during a command operation, can be toggled to a polar coordinate format.
Ctrl+G (F7)	*GRID*		Turns the *GRID* on or off.
Ctrl+L (F8)	*ORTHO*		Turns *ORTHO* on or off.
Crtl+B (F9)	*SNAP*		Turns *SNAP* on or off.
Ctrl+U (F10)	*POLAR*		Turns *Polar Tracking* on or off.
Ctrl+W (F11)	*OTRACK*		Turns *Object Snap Tracking* on or off.

Windows Copy, Cut, Paste

Ctrl+C	*Copyclip*	Copies the highlighted objects to the Windows clipboard.
Ctrl+X	*Cutclip*	Cuts the highlighted objects from the drawing and copies them to the Windows clipboard.
Ctrl+V	*Pasteclip*	Pastes the clipboard contents into the current AutoCAD drawing.

File Operations

Ctrl+O	*Open*	Invokes the *Open* command to open an existing drawing.
Ctrl+N	*New*	Invokes the *New* command to start a new drawing.
Ctrl+S	*Qsave*	Performs a quick save or produces the *Saveas* dialog box if the file is not yet named.

Other Control Key Sequences

Ctrl+Z	*Undo*	Undoes the last command.
Ctrl+Y	*Redo*	Invokes the *Redo* command.
Ctrl+P	*Plot*	Produces the *Plot* dialog box for creating and controlling prints and plots.
Ctrl+A	*Group*	Toggles selectable *Groups* on or off.
Ctrl+J	*Enter*	Executes the last command.
Ctrl+K	*Hyperlink*	Activates the *Hyperlink* command.
Ctrl+H	*PICKSTYLE*	Toggles *PICKSTYLE* On (1) and Off (0)
Ctrl+1	*Properties*	Toggles the *Properties* window
Ctrl+2	*DesignCenter*	Toggles DesignCenter

Special Key Functions

Esc
: The Escape key cancels a command, menu, or dialog box or interrupts processing of plotting or hatching.

Spacebar
: In AutoCAD, the space bar performs the same action as the Enter key or #2 button. Only when you are entering text into a drawing does the space bar create a space.

Enter
: If Enter or Spacebar is pressed when no command is in use (the open Command: prompt is visible), the last command used is invoked again.

INDEX

690 Index

.BAK, 34
.DWF, 383, 385
.DWG, 26, 31, 32, 33, 248, 397
.DWS, 31, 32, 33
.DWT, 31, 32, 33, 232, 233, 248
.DXF, 3, 31, 32, 33
.EXE, 385
.HTM, 385
.JPG, 383, 385
.PC3, 250
.PLT, 271
.PNG, 383, 385
.PTW, 385
.SHX, 353
.TTF, 353
.ZIP, 385
@ (last point), 346
@ (*Lastpoint*), 566
2D Drawings, Pictorial Drawings, 447
3D Array, 641
3D Coordinate Entry, 565
3D Coordinate Entry Formats, 566
3D DISPLAY AND VIEWING, 577
3D Display Commands, 579
3D MODELING BASICS, 561
3D Orbit, 586
3D Orbit Commands, 586
3D Viewing Commands, 582
3D Views, Viewpoint, 592
3darray, 641
3darray, Polar, 642
3darray, Rectangular, 641
3Dclip, 578, 591
3Dclip, Back Clipping, 591, 592
3Dclip, Front Clipping, 591
3Dclip, Slice, 592
3Dcorbit, 578, 589
3Ddistance, 578, 590
3Dorbit, 578, 586
3Dpan, 578, 589
3Dzoom, 578, 589

-A-
Absolute coordinates, 39
Absolute Coordinates, 3D, 566, 567
ACAD.DWT, 30, 88, 89, 210, 233, 356
ACAD.MNU, 12
ACAD.PAT, 463, 464
ACAD.PGP, 12
ACAD_ISO*n*W100 linetypes, 218
ACADISO.DWT, 30, 88, 89, 356
Accelerator Keys, 15
ACIS, 564
Acquire an object, 113
Acquisition Object Snap Modes, 113
Active Assistance, 79
Active Assistance Settings, 79
Adcenter, 391, 402
Adcenter, Description, 406
Adcenter, Desktop, 403
Adcenter, Favorites, 404
Adcenter, Find, 405
Adcenter, History, 403
Adcenter, Load, 404
Adcenter, Open Drawings, 403
Adcenter, Preview, 406
Adcenter, Tree View, 403, 404
Add, object selection, 67
ADVANCED DRAWING SETUP, 227
Advanced Setup Wizard, 90
ADVANCED SOLIDS FEATURES, 673
Align, 322
Align, 3D, 633
Alignment vector, 113
All, object selection, 66
Angle, Advanced Setup Wizard, 91
Angles in AutoCAD, 3
ANSI, 460, 500

Aperture, 109
Arc, 38, 138
Arc, 3Points, 139
Arc, Center, Start, Angle, 141
Arc, Center, Start, End, 141
Arc, Center, Start, Length, 142
Arc, Continue, 142
Arc, Start, Center, Angle, 139
Arc, Start, Center, End, 139
Arc, Start, Center, Length, 140
Arc, Start, End, Angle, 140
Arc, Start, End, Direction, 141
Arc, Start, End, Radius, 140
Arc ball, *3Dorbit*, 587
Arcs or *Circles*, 142
Area, 344
Area, Add, Subtract, 345
Area, Advanced Setup Wizard, 91
Area, Object, 345
Area, Quick Setup Wizard, 90
Area, Specify first corner point, 344
Array, 168
Array, Polar, 170
Array, Rectangular, 169
Array dialog box, 169
Arrowheads, tick marks, 492
Assist, 79
Assist menu, 63
Associative *Bhatch*, 461
Associative Dimensions, 510, 548
Associative Dimensions, Exploding, 515
Associative Hatch, 472
AUto, 64
AUto (crossing window), 64
AUto (window), 64
AutoCAD DesignCenter, 401
AutoCAD DesignCenter, Inserting Layouts with, 249
AutoCAD Drawing Editor, 5
AutoCAD Drawing Files, 26
AutoCAD File Commands, 29
AutoCAD Objects, 38
AutoCAD Text Window, 17
Autodesk PointA, 379
AutoSnap, 109
AutoSnap Marker, 109
AutoStack Properties dialog box, 362
Auxiliary View, Constructing, 478
Auxiliary View, Setting Up Principal Views, 478
AUXILIARY VIEWS, 477
Auxiliary Views, Full, Constructing, 486
Axonometric drawings, 446

-B-
Backup Files, 34
Base, 391, 399
BASIC DRAWING SETUP, 85
Beginning an AutoCAD Drawing, 26
Bhatch, 462
Bhatch, Angle, 465
Bhatch, Associative, 461
Bhatch, Associative/Nonassociative, 467
Bhatch, Boundary Set, 468
Bhatch, Custom, 463
Bhatch, Double, 465
Bhatch, Inherit Properties, 467
Bhatch, Island Detection Method, 469
Bhatch, Island Detection Style, 467
Bhatch, ISO Pen Width, 465
Bhatch, Pattern, 463
Bhatch, Pre-defined, 463
Bhatch, Preview, 467
Bhatch, Relative to Paper Space, 465
Bhatch, Remove Islands, 467
Bhatch, Retain Boundaries, 468
Bhatch, Scale, 465
Bhatch, Select Objects, 466

Bhatch, Spacing, 465
Bhatch, Type, 463
Bhatch, User-defined, 463
Block, 390, 391, 392
Block Color, Linetype, and *Lineweight* Settings, 393
Block Definition dialog box, 392
Block Insertion with Drag-and-Drop, 407
Blocks, Redefining, 399
BLOCKS and DesignCenter, 389
Boolean Commands, *Regions*, 332
Boolean Operation Commands, 643
Boolean Operations, 620
Boundary, 304
Boundary, Boundary Set, 304
Boundary, Island Detection, 304
Boundary, Object Type, 304
Boundary Hatch dialog box, 462
Boundary Hatch dialog box, *Advanced* tab, 467
Boundary Hatch dialog box, *Quick* tab, 463
Bounding Box, 677
Box, 620, 622
Box, Center, 623
Box, Corner of box, 623
Box, Length, 623
Box, object selection, 66
Break, 162
Break, at Point, 164
Break, Select, 2 Points, 163
Break, Select, Second, 163
Break, Select Point, 164
Browse the Web, 380
Browser, 376, 380
Bulletin Board, 379
Button, 9
Byblock, 393
Bylayer, 393
Bylayer, Color, 216
Bylayer, Linetype, 214
ByLayer drawing scheme, 207

-C-
Cabinet oblique, 454
CAD, 3
CAD Accuracy, 108
CAD Database, 3, 108
CAD/CAM, 108
Calculating Text Height for Scaled Drawings, 363
Cartesian coordinates, 2
Cascade, 18
Cavalier oblique, 454
Celtscale, 217, 218
Celtscale, changing, 313
Centroid, 677
Chamfer, 174
Chamfer, 3D, 647
Chamfer, 3D, Edge, 649
Chamfer, 3D, Loop, 649
Chamfer, Angle, 175
Chamfer, Distance, 174
Chamfer, Method, 174
Chamfer, Polyline, 175
Chamfer, Trim/Notrim, 175
Change, 320
Change, Point, 320
Change, Properties, 320
Change, Text, 321
Check Spelling dialog box, 366
Checkbox, 9
Chprop, 320
Circle, 38, 49, 136
Circle, 2 Points, 137
Circle, 3 Points, 138
Circle, Center, Diameter, 137
Circle, Center, Radius, 137
Circle, Tangent, Tangent, Radius, 138

Close, 33
Closeall, 34
CNC (Computer Numerical Control), 108
Color, 216
Color, Bylayer, 216
Color, Linetype and *Lineweight*, 210
Color Control drop-down list, 216
Color dialog box, 210
Color property, changing, 313
Colors and Linetypes, Assigning, 206
Command Entry, 12
Command Entry Methods Practice, 19
Command Line, 6, 23
Command Reference, Help, 77
Command Tables, 12
Command-Mode, Menu, 13
Commands, Methods for Entering, 12
Compare Dimension Styles dialog box, 528
Computer Numerical Control (CNC), 108
Computer-Aided Design, 3
Cone, 620, 624
Cone, Apex, 624
Cone, Center Point, 624
Cone, Elliptical, 625
Configuring a Default Output Device, 102
Construction Layers, Using, 431
Construction Line, 290
Constructive Solid Geometry (CSG), 564, 620
Constructive Solid Geometry Techniques, 620
Control Key Sequences, 15
Coordinate Display, 6, 10
Coordinate Entry, 135
Coordinate Entry Methods, 39
Coordinate System Icon, 197, 565
Coordinate Systems, 2, 570
Coords, 10, 15
Copy, 71, 165
Copy, Multiple, 166
Create New Dimension Style dialog box, 523, 527
Create New Drawing dialog box, 29, 87
Create Transmittal dialog box, 385
CREATING AND EDITING TEXT, 351
Crosshairs, 6
Crossing Polygon, 66
Crossing Window, 65
CSG (Constructive Solid Geometry), 564
Ctrl+E, 450, 451
Current Entity Linetype Scale, 218
Cursor, 6
Cursor menu, 116
Cursor Tracking, 10
Cursor tracking, 10
Custom Dictionary, 366
Customizing the AutoCAD Screen, 21
Cutting Plane Lines, Drawing, 473
Cylinder, 620, 625
Cylinder, Center Point, 625
Cylinder, Elliptical, 626
Cylindrical Coordinates (Relative), 566, 568

-D-
Dblclkedit, 317, 471
Dblist, 344
Dctcust, 366
Dctmain, 366
Ddedit, 353, 367
Ddgrips, 414
Ddim, 525
Ddptype, 143
Dducs, 611
Dducsp, 610

Ddvpoint, 595
Default Menu, 13
DesignCenter, 401
DesignCenter, *Description*, 406
DesignCenter, *Desktop*, 403
DesignCenter, *Favorites*, 404
DesignCenter, *Find*, 405
DesignCenter, Hatch Patterns, 471
DesignCenter, *History*, 403
DesignCenter, Inserting Layouts with, 249
DesignCenter, *Load*, 404
DesignCenter, *Open Drawings*, 403
DesignCenter, *Preview*, 406
DesignCenter, *Today*, 378
DesignCenter, *Tree View*, 403, 404
DesignCenter Blocks, Drawing Units, 93
DesignCenter *Insert* a *Block*, 407
DesignCenter Open a Drawing, 407
Dialog Box Functions, 27
Dialog Boxes, 9
Dialog-Mode Menu, 14
Dictionary, 366
Digitizing Tablet Menu, 11
Dimadec, 543
Dimaligned, 496
Dimaligned, Angle, 496
Dimaligned, Mtext, 496
Dimaligned, Text, 496
Dimalt, 544
Dimaltd, 544
Dimaltf, 544
Dimaltrnd, 545
Dimalttd, 547
Dimalttz, 547
Dimaltu, 544
Dimangular, 502
Dimapost, 545
Dimaso, 512
Dimassoc, 510, 512, 548
Dimasz, 534, 554
Dimatfit, 538, 540
Dimaunit, 543
Dimazin, 543
Dimbaseline, 497
Dimblk, 533
Dimblk1, 533
Dimblk2, 533
Dimcen, 534
Dimcenter, 501
Dimclrd, 531
Dimclre, 532
Dimclrt, 535
Dimcontinue, 498
Dimdec, 541
Dimdiameter, 499
Dimdiameter, Angle, 500
Dimdiameter, Mtext/Text, 499
Dimdisassociate, 513
Dimdle, 531
Dimdli, 532, 554
Dimdsep, 541
Dimedit, 516
Dimedit, Home, 516
Dimedit, New, 516
Dimedit, Oblique, 517
Dimedit, Rotate, 517
Dimension, *Align Text*, 515
Dimension, *Aligned*, 496
Dimension, *Angular*, 502
Dimension, *Baseline*, 497
Dimension, *Center Mark*, 501
Dimension, *Continue*, 498
Dimension, *Diameter*, 499
Dimension, *Leader (Qleader)*, 505
Dimension, *Linear*, 493
Dimension, *Oblique*, 516
Dimension, *Ordinate*, 508

Dimension, *Overall Scale*, 522
Dimension, *Override*, 549
Dimension, *Qdim*, 509
Dimension, *Radius*, 500
Dimension, *Update*, 528, 549
Dimension Drawing Commands, 493
Dimension Editing Commands, 515
Dimension line, 492
Dimension *Overall Scale*, 518
Dimension pull-down menu, 493
Dimension Style, 525
Dimension Style, Modifying, 548
Dimension Style Children, 523
Dimension Style Families, 523
Dimension Style Manager, 525, 531
Dimension Style Manager, Alternate Units tab, 544
Dimension Style Manager, Compare, 528
Dimension Style Manager, Current, 527
Dimension Style Manager, Description, 526
Dimension Style Manager, Fit tab, 518, 537
Dimension Style Manager, Lines and Arrows tab, 531
Dimension Style Manager, List, 526
Dimension Style Manager, Modify, 527
Dimension Style Manager, New, 527
Dimension Style Manager, Override, 527
Dimension Style Manager, Preview, 526
Dimension Style Manager, Primary Units tab, 541
Dimension Style Manager, Styles, 525
Dimension Style Manager, Text tab, 534
Dimension Style Manager, Tolerance tab, 545
Dimension Style Overrides, 524, 548
Dimension Styles, 232, 522
Dimension Styles, Template Drawings, 554
DIMENSION STYLES AND VARIABLES, 521
Dimension text, 492
Dimension Text Height, Fixed, 554
Dimension toolbar, 493
Dimension Variables, 522
Dimension Variables, Changing, 530, 547
Dimension Variables Introduction, 518
DIMENSIONING, 491
Dimensioning, Guidelines, 552
Dimensioning a Single Drawing, 552
Dimensioning in Layouts, 555, 556
Dimensioning in Paper Space, 555, 556
Dimensioning Text, Customizing, 518
Dimensions, Associative, 493
Dimensions, Editing, 510
Dimensions, Grip Editing, 420, 514
Dimensions, Modifying Existing, 548
Dimetric drawings, 446, 447
Dimexe, 533, 554
Dimexo, 533
Dimfrac, 541
Dimgap, 535, 536, 546, 554
Dimjust, 536
Dimldrblk, 534
Dimlfac, 542, 543
Dimlim, 546
Dimlinear, 493
Dimlinear, Angle, 496
Dimlinear, Horizontal, 496
Dimlinear, Mtext, 495
Dimlinear, Rotated, 494
Dimlinear, Text, 495
Dimlinear, Vertical, 496
Dimlunit, 541
Dimlwd, 531
Dimlwe, 532
Dimordinate, 508
Dimoverride, 549
Dimpost, 542
Dimradius, 500

692 Index

Dimradius, Angle, 501
Dimradius, Mtext/Text, 501
Dimreassociate, 512
Dimregen, 513
Dimrnd, 542
Dimsah, 533
Dimscale, 230, 518
Dimscale, 522
Dimscale, 88, 89, 537, 538
Dimsd1, 532
Dimsd2, 532
Dimse1, 533
Dimse2, 533
Dimsodx, 539
Dimstyle, 525
-Dimstyle, 528
-Dimstyle, Apply, 530
-Dimstyle, Restore, 529
-Dimstyle, Save, 529
-Dimstyle, Status, 529
-Dimstyle, Variables, 529
Dimtad, 536
Dimtdec, 546
Dimtedit, 515
Dimtedit, Angle, 516
Dimtedit, Center, 516
Dimtedit, Dimedit, Dimensioning, 552
Dimtedit, Home, 516
Dimtedit, Right/Left, 516
Dimtfac, 535, 546
Dimtih, 536, 537
Dimtix, 538
Dimtm, 546
Dimtmove, 539
Dimtofl, 540
Dimtoh, 536, 537
Dimtol, 546
Dimtolj, 546
Dimtp, 546
Dimtxsty, 535
Dimtxt, 535, 554
Dimtzin, 547
Dimupt, 540
Dimzin, 543
Direct Distance Entry, 39, 53
Direct Distance Entry, 3D, 566, 568
Direct Distance Entry and Polar Tracking, 57
Display Commands, 578
Display Contour lines per surface, 674
Display Order, 472
Display Resolution, 196
Dispsilh, 582, 675, 676
Dist, 346
Distance, 346
Divide, 299
Donut, 295
Double-click Edit, 317
Drafting Settings, 98, 448
Drafting Settings dialog box, 117, 448
Drafting Settings dialog box, Polar Tracking, 121
Drafting Settings dialog box, *Polar Tracking* tab, 52, 482
Draw and Modify toolbars, 7
Draw Command Access, 134
DRAW COMMAND CONCEPTS, 37
Draw Commands, 142
Draw Commands, Locating, 38
DRAW COMMANDS I, 135
DRAW COMMANDS II, 289
Draw pull-down menu, 39
Draw toolbar, 39
Draw True Size, 3
Drawing Aids, 15
Drawing *Circles* Using the Five Command Entry Methods, 49

Drawing Editor, 5
Drawing Files, 26
Drawing Files, Naming, 26
Drawing Lines Using Polar Tracking and Polar Snap, 54
Drawing *Lines* Using the Five Coordinate Entry Methods, 40
Drawing Scale Factor, 229
Drawing Scale Factor, Calculating, 277
Drawing Setup, Steps, 86, 228
Drawing Setup Options, 87
Draworder, 472
Drop-down list, 9
Dsettings, 52, 98
DSF, 277
Dtext, 353, 354
Dtext, Align, 355
Dtext, Center, 355
Dtext, Fit, 355
Dtext, Justify, 354
Dtext, Middle, 355
Dtext, MR, BL, BC, BR, 356
Dtext, Right, 355
Dtext, Start Point, 354
Dtext, Style, 356
Dtext, TL, TC, TR, ML, MC, 355
-E-
Edit box, 9
Edit Hyperlink Dialog box, 381
Edit Text dialog box, 367
Editing Text, 366
Edit-Mode Menu, 13
Ellipse, 297
Ellipse, Arc, 299
Ellipse, Axis End, 298
Ellipse, Center, 298
Ellipse, Isometric, 450
Ellipse, Parameter, 299
Ellipse, Rotation, 298
Entity, 38
ePlot, 376
Erase, 69, 153
Etransmit, 376, 385
Etransmit, Files tab, 386
Etransmit, General tab, 386
Etransmit, Report tab, 386
ETRANSMIT.TXT, 386
eView, 376
Exit, 5, 34
Explode, 321, 391, 395
Exploded Dimensions, 510
Extend, 161
Extend, Edge, 162
Extend, Projection, 162
Extend, Shift-Select, 162
Extension line, 492
Extrude, 620, 628
Extrude, Path, 630
-F-
F1, 14
F2, 14
F3, 14
F4, 14
F5, 15
F6, 15
F7, 15
F8, 15
F9, 15
F10, 15
F11, 15
Facetratio, 676
Facetres, 675, 676
Fence, 66
File Commands, Accessing, 26
File Management, 35
File Navigation Dialog Box Functions, 27
Fillet, 171

Fillet, 3D, 649
Fillet, 3D, *Chain*, 650
Fillet, 3D, *Edge*, 650
Fillet, 3D, *Radius*, 650
Fillet, Polyline, 173
Fillet, Radius, 172
Fillet, Trim/Notrim, 173
Fillets, Rounds and Runouts, Creating, 435
Fillmode, 295
Find, 353, 368
Find, Find Text String, 368
Find, Options, 369
Find, Replace With, Replace, 368
Find, Zoom to, 369
Find and Replace dialog box, 368
Find and Replace Options dialog box, 369
Find/Replace, Mtext, 369
Flipscreen, 14
Floating Viewports, 240
Flyouts, 7
Function Keys, 14
-G-
GETTING STARTED, 1
Graphics Area, 6
Grid, 15, 16, 97, 231
Grid, Aspect, 98
Grid, On/Off, 97
Grid Snap, 15, 40, 95, 97
Grid Snap and Polar Tracking, 54
Grid, Spacing, 97
Gridunit, 88, 89
GRIP EDITING, 413
Grip Editing Dimensions, 514
Grip Editing Options, 417
Grips, 414
Grips, Auxiliary Grid, 419
Grips, *Base*, 418
Grips, *Copy*, 418
Grips, Dimensioning, 552
Grips, Editing Dimensions, 420
Grips, *Go to URL*, 419
Grips, Guidelines for Using, 420
Grips, *Mirror*, 418
Grips, *Move*, 417
Grips, *Properties*, 419
Grips, *Reference*, 419
Grips, *Rotate*, 417
Grips, *Scale*, 418
Grips, *Stretch*, 417
Grips, Using with Hatch Patterns, 472
Grips, Warm, Hot and Cold, 415
Grips Features, 414
Grips Shortcut Menu, 416
-H-
Hatch, 469
Hatch, Direct Hatch, 470
Hatch Pattern Palette dialog box, 463
Hatch Patterns, AutoCAD, 464
Hatch Patterns, DesignCenter, 471
Hatch Patterns, Drag and Drop, 471
Hatch Patterns, Using Grips, 472
Hatch Patterns and Boundaries, Editing, 471
Hatch Patterns and Hatch Boundaries, 460
Hatchedit, 471
Help, 14, 76
Help, Ask Me, 79
Help, Contents, 77
Help, Favorites, 78
Help, Index, 78
Help, Search, 78
Help message, 7
HELPFUL COMMANDS, 75
Hidden and Center Lines, Drawing, 433
Hide, 582

Index 693

Horizontal Lines, Drawing, 41
HP/GL Language, 271
Hpang, 465
Hpscale, 88, 89, 465
HTML, 382
Hyperlink, 376, 380, 381
Hyperlink, Email Address, 382
Hyperlink, Existing File or Web Page, 381
Hyperlink, View of This Drawing, 381
-I-
ID, 346
i-Drop, 384
Image tile, 9
Inclined Lines, Drawing, 47
INQUIRY COMMANDS, 341
Insert, 391, 394
Insert a Block, DesignCenter, 407
Insert dialog box, 394
Insert Layout dialog box, 248
Insert Presets, 395
Inserting Drawings as Blocks, 394
Interactive Coordinates, 39
Interactive Coordinates, 3D, 566, 567
Interfere, 678
Interference, 678
INTERNET TOOLS, 375
Intersect, 622, 646
Intersect, Regions, 333
Isavebak, 34
ISO, 460
Isolines, 674, 675
Isometric Drawing, Creating, 452
Isometric Drawing in AutoCAD, 448
Isometric Drawings, 446
Isometric Ellipses, 450
Isoplane, 15, 450, 451
-J-
Justifytext, 353, 370
-L-
Last, object selection, 66
Lastpoint, 346
Layer, 208
Layer, Color, 210
Layer, Current, 209
Layer, Current Viewport Freeze, New Viewport Freeze, 211
Layer, Delete, 212
Layer, Freeze, Thaw, 209
Layer, Linetype, 210
Layer, Lineweight, 211
Layer, Lock, Unlock, 210
Layer, Make Object Current, 213
Layer, New, 211
Layer, On, Off, 209
Layer, Plot/Noplot, 211
Layer, Plotstyle, 211
Layer, Set, 209
Layer Control Drop-Down List, 207
Layer Filters, 212
Layer Previous, 213
Layer Properties Manager, 9, 208
Layer Properties Manager, Active VP Freeze, New VP Freeze, 240
Layerp, 213
Layerpmode, 213
LAYERS AND OBJECT PROPERTIES, 205
Layout, 99, 247
Layout, Copy, 248
Layout, Delete, 248
Layout, New, 248
Layout, Rename, 249
Layout, Save, 249
Layout, Set, 249
Layout, Template, 248
Layout Options and Plot Options, Setting, 102

Layout Tab, Activate, 232
Layout Tabs, 19, 99
Layout Wizard, 245
Layout Wizard, Begin, 245
Layout Wizard, Define Viewports, 247
Layout Wizard, Orientation, 246
Layout Wizard, Paper Size, 246
Layout Wizard, Pick Location, 247
Layout Wizard, Printer, 246
Layout Wizard, Title Block, 246
Layouts, 238
Layouts, Creating, 245
Layouts, Setting up, 249
Layouts and Printing, Introduction, 98
LAYOUTS AND VIEWPORTS, 237
Layouts and Viewports, Guidelines for Using, 244
Layouts and Viewports Example, 241
Layoutwizard, 245
Leader, 503
Leader, Annotation, 504
Leader, Arrow/None, 504
Leader, Block, 504
Leader, Copy, 504
Leader, Format, 504
Leader, Mtext, 504
Leader, Spline/Straight, 504
Leader, Tolerance, 504
Leader Settings dialog box, 506
Leader Settings dialog box, *Attachment* tab, 507
Leader Settings dialog box, *Leader Line and Arrow* tab, 507
Lengthen, 158
Lengthen, Delta, 158
Lengthen, Dynamic, 159
Lengthen, Percent, 159
Lengthen, Total, 159
Limits, 6, 93, 229, 276
Limits, On/Off, 95
Limits Settings Table, Architectural, 281
Limits Settings Table, Civil, 284
Limits Settings Table, Mechanical, 280
Limits Settings Table, Metric, for Engineering Sheet Sizes, 283
Limits Settings Table, Metric, for Metric Sheet Sizes, 282
Limits Settings Tables, 279
Limmax, 88, 89
Line, 38, 41, 135
Linetype, 214
Linetype, Bylayer, 214
Linetype Control Drop-Down List, 214
Linetype dialog box, 210
Linetype Manager, 214
Linetype property, changing, 313
Linetype Scale property, changing, 313
Linetypes, Lineweights, and Colors, Managing, 434
Linetypes, Lineweights, and Colors, Multiview Drawings, 434
Linetypes, Using, 432
Lineweight, 215
Lineweight Control Drop-Down List, 215
Lineweight dialog box, 211
Lineweight property, changing, 313
Lineweight Settings dialog box, 215
List, 343
List box, 9
Locking Viewport Geometry, 257
Ltscale, 88, 89, 95, 217, 218, 229, 231, 257
Ltscale, changing, 313
Lunits, 88, 89
Lweight, 215
-M-
Make Object's Layer Current, 213

Mass, 677
Mass Properties, 677
Massprop, 677
Match Properties, 221, 318
Match Properties, Dimensioning, 551
Matchprop, 221, 318
Matchprop, Dimensioning, 551
Maximize, 34
Mbuttonpan, 116, 188
Measure, 301
Measureinit, 6, 30, 87, 89
MeetNow, 376
Minimize, 34
Minsert, 391, 395
Mirror, 166
Mirror 3D, 638
Mirror3d, 3points, 639
Mirror3d, 638
Mirror3d, Last, 639
Mirror3d, Object, 639
Mirror3d, View, 640
Mirror3d, XY, 640
Mirror3d, YZ, 640
Mirror3d, Zaxis, 639
Mirror3d, ZX, 641
Model, 255
Model Space, 18
Model Space and Paper Space, 238
Model tab, 18, 99
Model/Paper toggle, 254
Modify Command Concepts, 62
MODIFY COMMANDS I, 151
MODIFY COMMANDS II, 311
Modify II toolbar, 312
Modify Properties, 517
Modify pull-down menu, 152
Modify toolbar, 152
Moments of Inertia, 677
Mouse and Digitizing Puck Buttons, 14
Mouse Wheel, *Zoom* and *Pan*, 187
Move, 70, 153
Move, 3D, 632
Mspace, 244, 254
Mtext, 353, 357
Mtext, At Least/Exactly, 360
Mtext, Bold, Italic, Underline, 359
Mtext, Character tab, 358
Mtext, dimensions, 495
Mtext, Find/Replace tab, 358
Mtext, Font, 359
Mtext, Height, 359
Mtext, Import Text, 358
Mtext, Justification, 358
Mtext, Line Spacing tab, 360
Mtext, Properties tab, 358
Mtext, Rotation, 358, 362
Mtext, Stack/Unstack, 359
Mtext, Style, 358
Mtext, Symbol, 359
Mtext, Text Color, 359
Mtext, Width, 358
Mtext Shortcut Menu, 360
Multiline Text, 357
Multiline Text Editor, 357
Multiline Text Editor, dimensions, 495
Multiple, object selection, 67
MULTIVIEW DRAWING, 425
-N-
Named Layer Filters, 212
Named Views, 194
New, 29
New Layout, 248
New UCS, 605
New View dialog box, 613
Nonassociative Dimensions, 510
Non-Uniform Rational Bezier Spline (NURBS), 296

694 Index

Noun/Verb Command Syntax Practice, 72
Noun/Verb Syntax, 68
NURBS, 296, 297
-O-
Object, 38
Object Cycling, 67
Object *Linetypes, Lineweights* and *Colors*, 435
Object Properties, 207
Object Properties, Changing Retroactively, 219
Object Properties, Displaying, 212
Object Properties Toolbar, 219
Object Properties Toolbar, 7, 319
Object Properties window, 220
Object Snap, *Apparent intersection*, 112
Object Snap, *Endpoint*, 110
Object Snap, *Extension*, 114
Object Snap, *From*, 112
Object Snap, *Insert*, 110
Object Snap, *Intersection*, 110
Object Snap, *Midpoint*, 111
Object Snap, *Nearest*, 111
Object Snap, *Node*, 111
Object Snap, *None*, 118
Object Snap, *Parallel*, 113
Object Snap, *Perpendicular*, 111
Object Snap, *Quadrant*, 112
Object Snap, *Tangent*, 112
Object Snap, *Temporary Tracking*, 114
OBJECT SNAP AND OBJECT SNAP TRACKING, 107
Object Snap *Center*, 109
Object Snap Cycling, 118
Object Snap Modes, 108, 109
Object Snap Modes, Acquisition, 113
Object Snap toggle, 118
Object Snap toolbar, 115
Object Snap Tracking, 118, 120
Object Snap Tracking, 3D, 570
Object Snap Tracking, Using to Draw Projection Lines, 428
Object Snap Tracking Settings, 121
Object Snap Tracking with Polar Tracking, 120
Object-specific drawing scheme, 207
Object-Specific Properties Controls, 214
Oblique Drawing in AutoCAD, 454
Oblique drawings, 446, 447
Offset, 167, 483, 484
Offset, Auxiliary Views, 483
Offset, Distance, 168
Offset, Through, 168
Offset, Using for Construction of Views, 430
Oops, 80
Open, 31
Open a Drawing from DesignCenter, 407
Options, Default Plot Settings for New Drawings, 250
Options, Display tab, 103
Options, Display tab, 8, 23, 239, 676
Options, General Plot Options, 250
Options, Plotting tab, 102, 250
Options, System tab, 377
Options, Selections tab, 414
Ortho, 15, 17
Ortho, Auxiliary Views, 479
Ortho, Isometric drawings, 449
Ortho and *Osnap*, Using for Projection Lines, 426
Orthographic & Isometric Views, 585
Orthographic UCS, 605
Orthographic UCS, *Depth*, 611
Osnap, 108

Osnap Applications, 122
Osnap Practice, 122
Osnap Running Mode, 117
Osnap Single Point Selection, 114
Osnap Toggle, 14
Osnap Tracking, 15
Otrack, 118
-P-
Page Setup, 275
Page Setup, Add, 252
Page Setup, Layout Name, 252
Page Setup, Page Setup Name, 252
Page Setup dialog box, 251, 275
Page Setup for Layouts, 251
Pagesetup, 251, 275
Painter, 221, 318
Palette, *Adcenter*, 402
Pan, 192
Pan, Down, 194
Pan, Left, 194
Pan, Mouse Wheel or Third Button, 188
Pan, Point, 193
Pan, Real Time, 193
Pan, Right, 194
Pan, Transparent, 194
Pan, Up, 194
Pan and *Zoom*, Mouse Wheel, 187
Paper Sizes, Standard, 277
Paper Space, 19
Paper Space and Model Space, 238
Paper Space Dimensioning, 555, 556
Paper/Model toggle, 254
Paperupdate, 251
PCL Language, 271
Pdmode, 143
Pdsize, 143
Pedit, 323
Pedit, Close, 323
Pedit, Decurve, 325
Pedit, Fit, 325
Pedit, Join, 324
Pedit, Ltype gen, 326
Pedit, Multiple, 323
Pedit, Open, 323
Pedit, Spline, 325
Pedit, Width, 324
Perspective, 590
Pickadd, 317
Pickbox, 64
Pickfirst, 68, 317, 415
Pickstyle, 471, 472
Pictorial Drawings, 2D Drawings, 447
PICTORIAL DRAWINGS, 445
Pictorial Drawings, Types, 446
Plan, 578, 596
Plan View, 596
Pline, 144
Pline, Arc, 145
Pline, Close, 144
Pline, Halfwidth, 144
Pline, Length, 144
Pline, Undo, 144
Pline, Vertex Editing, 326
Pline, Width, 144
Pline Arc Segments, 145
Plinegen, 326
Plines, Converting *Lines* and *Arcs*, 328
Plot, 101, 269
Plot, Display, 272
Plot, Drawing Orientation, 271
Plot, Extents, 272
Plot, Hide Objects, 273
Plot, Layout Name, 269
Plot, Limits or *Layout*, 272
Plot, Page Setup Name, 270
Plot, Paper Size and Paper Units, 271
Plot, Plot Area, 271

Plot, Plot Stamp, 271
Plot, Plot Style Table (pen assignments), 270
Plot, Plot to File, 271
Plot, Plotter Configuration, 270
Plot, Preview, Full, 274
Plot, Preview, Partial, 274
Plot, View, 272
Plot, What to Plot, 271
Plot, Window, 272
Plot Device, Set, 232
Plot dialog box, 101, 268
Plot dialog box, *Plot Device* tab, 270
Plot dialog box, *Plot Settings* tab, 271
Plot dialog box or *Page Setup* dialog box, 275
Plot Offset, 273
Plot Options and Layout Options, Setting, 102
Plot Paperspace Last, 273
Plot Preview, 276
Plot Scale, 272, 230
Plot Scale, Custom, 273
Plot Scale, Scale, 272
Plot Scale, Scale Lineweights, 273
Plot to Scale, 5
Plot with Lineweights, 273
Plot with Plot Styles, 273
Plotstyle property, changing, 313
Plotting, 268
Plotting, Typical Steps, 268
Plotting *Layouts* to Scale, 278
Plotting *Model* Tab to Scale, 278
Plotting to Scale, 276
Plotting to Scale, Examples, 285
Plotting to Scale Guidelines, 278
Point, 38, 143
Point Filters, 3D, 569
Point Style, 143
Point Style dialog box, 143
PointA, 376
Polar, 15, 16
Polar Snap, 15, 40, 51, 53, 95
Polar Snap and Polar Tracking, 56
Polar Snap and *Polar Tracking*, Realignment of Views, 431
Polar Tracking, 51, 52
Polar Tracking, Auxiliary Views, 482
Polar Tracking, Isometric Drawings, 449
Polar Tracking, Using to Draw Projection Lines, 427
Polar Tracking and Direct Distance Entry, 57
Polar Tracking and Grid Snap, 54
Polar Tracking and Polar Snap, 56
Polar Tracking Override, 53
Polar Tracking with Object Snap Tracking, 120
Polarang, 97
Polardist, 97
Polygon, 293
Polygon, Edge, 294
Polygon, Inscribe/Circumscribe, 294
Polyline, 144
Preview, 276
Previous, object selection, 66
Primitives, Solids, 620
Principal Moments, 678
Print, 269
Printable Area, 273
Printing, 268
Printing and Layouts, Introduction, 98
Printing and Plotting, 101
PRINTING AND PLOTTING, 267
Printing the Drawing, 103
Products of Inertia, 678
Projection and Alignment of Views, 426

Properties, 220, 314, 367, 517
Properties, Alphabetical tab, 316
Properties, Categorized tab, 315
Properties, Dimensioning, 551
Properties, Pickadd toggle, 317
Properties, Quick Select, 316
Properties, Select Objects, 316
Properties of Objects, Changing, 313
Properties window, 220, 256, 314, 367
Property Settings dialog box, 221, 318
Psltscale, 231, 257
Pspace, 244, 254
Publish to Web, 382
Publishtoweb, 376, 382
Publishtoweb, Apply Theme, 384
Publishtoweb, Begin, 383
Publishtoweb, Create Web Page, 383
Publishtoweb, Enable i-Drop, 384
Publishtoweb, Generate Images, 385
Publishtoweb, Preview and Post, 385
Publishtoweb, Select Drawings, 384
Publishtoweb, Select Image Type, 383
Publishtoweb, Select Template, 384
Pull-down Menus, 8
Purge, 391, 400
Purge, Confirm each item, 400
Purge, Purge, 400
Purge, Purge All, 400
Purge, View Items, 400
Purge dialog box, 400
-Q-
Qdim, 509
Qleader, 505
Qleader, Annotation tab, 506
Qleader, Annotation type, 506
Qleader, Mtext Options, 506
Qleader, Annotation Reuse, 507
Qsave, 32
Qtext, 353, 371
Quick Dimension, 509
Quick Setup Wizard, 89
-R-
Radii of Gyration, 678
Radio button, 9
Radius and *Diameter* Variable Settings, 540
Ray, 293, 485
Reassociate Dimension, 512
Rectang, 294
Rectangle, 294
Redo, 81
Regen, 82
Region, 305
Relative polar coordinates, 39
Relative polar display, 10
Relative Rectangular Coordinates, 39
Relative Rectangular Coordinates, 3D, 566, 567
Remove, object selection, 67
Rename, 391, 401
Rename dialog box, 401
Revolve, 620, 631
Revolve, Object, 631
Revolve, Start point of axis, 631
Revolve, X or Y, 631
Right-Hand Rule, 572
Rotate, 155
Rotate, Reference, 155
Rotate 3D, 635
Rotate3d, 2points, 636
Rotate3d, Last, 636
Rotate3d, Object, 636
Rotate3d, Reference, 638
Rotate3d, View, 636
Rotate3d, Xaxis, 637
Rotate3d, Yaxis, 637
Rotate3d, Zaxis, 637

Rtpan, 193
Rtzoom, 189
Running Object Snap, 117
Running Object Snap toggle, 118
-S-
Save, 32
Save As, 32
Save Drawing As dialog box, 34, 233
Save UCS with View, 613
Saveas, 32
Savetime, 347
Saving an AutoCAD Drawing, 26
Scale, 156
Scale, Reference, 156
Scale, Text, 369
Scaletext, 353, 369
Scaling Viewport Geometry, 255
Screen Menu, 8
Scroll Bars, 194
Section, 679
Section View, Creating, 461
SECTION VIEWS, 459
Sections, Creating from Solids, 679
Select, 68
Select Color dialog box, 210, 216
Select File dialog box, 31
Select Linetype dialog box, 210
Selection Set Options, 63
SELECTION SETS, 61
Selection Sets, 62
Selection Sets Practice, 68
Selecturl, 382
Set Variable, 347
Setup Commands, 91
Setvar, 347
Shademode, 2D Wireframe, 580
Shademode, 3D Wireframe, 580
Shademode, 578, 579
Shademode, Flat Shaded, 581
Shademode, Flat Shaded, Edges On, 581
Shademode, Gouraud Shaded, 581
Shademode, Gouraud Shaded, Edges On, 582
Shademode, Hidden, 580
Shape files, compiled, 353
Shortcut Menu, *3Dorbit*, 588
Shortcut Menu, Command-Mode, 13
Shortcut Menu, Default, 13
Shortcut Menu, Dialog-Mode, 14
Shortcut Menu, Edit-Mode, 13
Shortcut Menu, Grips, 416
Shortcut Menu, *Mtext*, 360
Shortcut Menus, 13
Shpere, 620
Side Menu, 8
Single, object selection, 67
Single Line Text, 354
Sketch, 301
Sketch Options, 302
Skpoly, 303
Slice, 681
Snap, 15, 95, 230
Snap, Aspect, 96
Snap, Isometric, 448
Snap, On/Off, 96
Snap, Rotate, 96
Snap, Style, 96
Snap, Type, 96
Snap, Value, 96
Snap Marker, 109
Snap Rotate, Auxiliary Views, 479
Snap Tips, 109
Snaptype, 97
Snapunit, 88, 89, 97
SOLID MODELING CONSTRUCTION, 619
Solid Models, 564

Solid Models, Analyzing, 677
Solid Primitives Commands, 622
Solidcheck, 662
Solidedit, 652
Solidedit, Body, Check, 662
Solidedit, Body, Clean, 661
Solidedit, Body, Imprint, 660
Solidedit, Body, Separate, 661
Solidedit, Body, Shell, 662
Solidedit, Body Options, 660
Solidedit, Edge, Color, 660
Solidedit, Edge, Copy, 659
Solidedit, Edge Options, 659
Solidedit, Face, Color, 659
Solidedit, Face, Copy, 658
Solidedit, Face, Delete, 656
Solidedit, Face, Extrude, 653
Solidedit, Face, Move, 655
Solidedit, Face, Offset, 655
Solidedit, Face, Rotate, 656
Solidedit, Face, Taper, 657
Solidedit, Face Options, 653
Solids, Commands for Moving 632
Solids, Design Efficiency, 650
Solids Editing, 651
Special Key Functions, 15
Spell, 353, 366
Spelling, 366
Sphere, 627
Spherical Coordinates (Relative), 566, 568
Splframe, 326
Spline, 296
Spline, Add, 330
Spline, Close, 296
Spline, Close/Open, 330, 331
Spline, Fit Data, 331
Spline, Fit Tolerance, 297
Spline, Grips, 331
Spline, Move, 330
Spline, Move Vertex, 331
Spline, Object, 297
Spline, Purge, 330
Spline, Refine, 331
Spline, Reverse, 331
Spline, Tangents, 330
Spline, Tolerance, 330
Spline Control Points, Editing, 331
Spline Data Points, Editing, 330
Splinedit, 329
Splinesegs, 325
Splinetype, 325
Standard Dimension style, 522
Standard toolbar, 7
Start from Scratch, 30, 87
Start from Scratch English Settings, Table of, 88
Start from Scratch Metric Settings, Table of, 89
Starting AutoCAD, 5
Startup dialog box, 6, 29, 87
Startup Options, 86
Status, 342
Status Bar, 9, 10
Stereo Lithography Apparatus, 108
Stretch, 157
Style, 353, 363
Style, Backwards, 365
Style, Delete, 364
Style, Height, 365
Style, Obliquing angle, 365
Style, Rename, 364
Style, Upside-down, 365
Style, Vertical, 365
Style, Width factor, 365
Style overrides, dimensions, 524
Subtract, 621, 645
Subtract, Regions, 333

Surface Models, 563
Symbols, *Mtext*, 359
Syswindows, 18
-T-
Tablet, 14
Target box, 109
Template, 88, 89
Template, Use a, 30
Template Drawings, 232
Template Drawings, Creating, 232, 233
Template Drawings, Dimensioning, 554
Template Drawings, Using, 232
Text, Justify, 370
Text, Scale, 369
Text Creation Commands, 353
Text Flow and *Justification*, 361
Text Height, Calculating, 363
Text Justification, *TL, TC, TR, ML, MC, MR, BL, BC, BR*, 355, 356
Text Style, 363
Text Style dialog box, 364
Text Styles, 231
Text Styles and Fonts, 363
Text Symbols, 359
Text Window, 17
Textfill, 371
Textsize, 88, 89
Thickness property, changing, 313
Three-View Drawing, Guidelines for Creating, 437
Tile, 18
Tiled Viewports, 240
Tilemode, 238, 244
Time, 347
Time, Display, 347
Time, On/Off/Reset, 347
Title Block, 232
Today, 376, 377
Today, Autodesk PointA, 377
Today, Bulletin Board, 377
Today, Create Drawings tab, 378
Today, My Drawings, 377
Today, Open Drawings tab, 378
Today, Symbol Libraries tab, 378
Today window, 5, 29, 86, 87
Tool Tip, 7
Toolbar, docked, 22
Toolbar, floating, 22
Toolbars, 7, 21
Torus, 620, 627
Trim, 160
Trim, Edge, 160
Trim, Projection, 161
Trim, Shift-Select, 162
Trimetric drawings, 446, 447
Trimmode, 173
TrueType, 353
TXT.SHX, 353
-U-
U, 80
UCS, 571, 605
UCS, 3point, 607
UCS, Apply, 610
UCS, Face, 607
UCS, Move, 606
UCS, New, 606
UCS, Object, 607
UCS, Origin, 606
UCS, Orthographic, 609
UCS, Previous, 609
UCS, Restore, 609
UCS, Save, 609
UCS, View, 608
UCS, World, 609
UCS, X, Y, Z, 608
UCS, Zaxis, 607

UCS Commands, 604
UCS dialog box, 575
UCS Icon, 197, 565, 574
UCS Icon, All, 575
UCS Icon, Noorigin, 575
UCS Icon, On/Off, 575
UCS Icon, Origin, 575
UCS Icon, Properties, 575
UCS Icon Control, 574
UCS Icon dialog box, 197, 575
UCS Icons, 571
UCS Manager, 575, 610, 612
UCS Manager, Named UCSs tab, 611
UCS Manager, Orthographic UCSs tab, 610
UCS Manager, Settings tab, 612
UCS System Variables, 612
Ucsaxisang, 609
Ucsbase, 609, 613
Ucsfollow, 610, 615
Ucsicon, 197, 574
Ucsman, 612
Ucsortho, 583, 585, 598, 614
Ucsview, 195, 613
Ucsvp, 616
Undo, 80
Undo, Auto, 81
Undo, Back, 81
Undo, Begin, 81
Undo, Control, 81
Undo, End, 81
Undo, Mark, 81
Union, 621, 644
Union, Regions, 332
Unitmode, 93
Units, 91, 228
Units, Advanced Setup Wizard, 91
Units, Angles, 92
Units, DesignCenter Blocks, 93
Units, Precision, 92
Units, Quick Setup Wizard, 89
Units Values, Keyboard Input, 93
Update, 548, 549
User Coordinate System, 197
USER COORDINATE SYSTEMS, 603
User Coordinate Systems (UCS), 571
-V-
Verb/Noun Syntax, 68
Vertical Lines, Drawing, 45
View, 194, 578, 583
View, Back, 584
View, Bottom, 584
View, Delete, 196
View, Front, 584
View, Left, 584
View, NE Isometric, 585
View, NW Isometric, 585
View, Orthographic, 196
View, Restore, 196
View, Right, 584
View, Save, 196
View, SE Isometric, 585
View, SW Isometric, 584
View, Top, 584
View, Ucs, 196
View, Window, 196
View dialog box, 195, 585
VIEWING COMMANDS, 185
Viewing Commands, 578
Viewpoint Presets dialog box, 595
Viewport, 99
Viewport, Create, 232
Viewport Geometry, Locking, 257
Viewport Geometry, Scaling, 255
Viewport Scale, 241
Viewport Scale Control Drop-down list, 255
Viewport toolbar, 241
Viewports, 197, 239, 252, 596

Viewports, Creating Automatic, 103
Viewports, Floating, 240
Viewports, Tiled, 240
Viewports and Layouts, Guidelines for Using, 244
Viewports dialog box, 198, 253
Viewports in Paper Space, Using, 252
Viewres, 196
Volume, 677
Vpoint, 578, 592
Vpoint, Rotate, 592
Vpoint, Tripod, 594
Vpoint, Vector, 593
Vports, 2,3,4, 197, 200, 252, 578, 596
Vports, Delete, 199
Vports, Join, 199
Vports, Restore, 199
Vports, Save, 199
Vports, Single, 199
-W-
Wblock, 391, 397
WCS, 570
WCS Icon, 570
Wedge, 620, 626
Wedge, Center, 627
Wedge, Corner of Wedge, 626
Wheel Mouse, *Zoom* and *Pan*, 187
Why Set Up Layouts Before Drawing?, 101
Window, 65
Window Polygon, 65
Windows Right-Click Shortcut Menus, 28
Wireframe Models, 562
Wizard, 89
Wizard, Use a, 30
WORKING WITH FILES, 25
World Coordinate System (WCS), 570
Write Block dialog box, 398
-X-
X,Y,Z coordinates, 2
Xline, 290, 485
Xline, Ang, 292
Xline, Bisect, 292
Xline, Hor, 291
Xline, Offset, 292
Xline, Specify a point, 291
Xline, Ver, 291
Xline and *Ray*, Using for Construction Lines, 430
Xline and *Ray* Commands, Auxiliary Views, 485
Xplode, 396
Xplode, All, 397
Xplode, Color, 396
Xplode, Inherit from Parent Block, 396
Xplode, Layer, 396
Xplode, Linetype, 396
-Z-
Zoom, 188, 578
Zoom, 3D, 596
Zoom, All, 190
Zoom, Center, 191
Zoom, Dynamic, 192
Zoom, Extents, 190
Zoom, In, 191
Zoom, Mouse Wheel, 187
Zoom, Out, 191
Zoom, Previous, 191
Zoom, Realtime, 189
Zoom, Scale (X/XP), 191
Zoom, Transparent, 192
Zoom, Window, 190
Zoom and *Pan*, Mouse Wheel, 187
Zoom XP Factors, 256